DISCRETE-TIME CONTROL SYSTEMS

Katsuhiko Ogata

University of Minnesota

Prentice-Hall International, Inc.

ISBN 0-13-216227-X

A Division of Simon & Schuster
Englewood Cliffs, NJ 07632

Printed in the United States of America

10 9 8 7 6

ISBN 0-13-216227-X 025

Prentice-Hall International (UK) Limited, *London*
Prentice-Hall of Australia Pty. Limited, *Sydney*
Prentice-Hall Canada Inc., *Toronto*
Prentice-Hall Hispanoamericana, S.A., *Mexico*
Prentice-Hall of India Private Limited, *New Delhi*
Prentice-Hall of Japan, Inc., *Tokyo*
Prentice-Hall of Southeast Asia Pte. Ltd., *Singapore*
Editora Prentice-Hall do Brasil, Ltda., *Rio de Janeiro*
Prentice-Hall, *Englewood Cliffs, New Jersey*

Contents

Chapter 5

State Space Analysis 479

Chapter 6

Analysis and Design in State Space 625

Chapter 7

Optimal Control Systems 813

Preface

Many industrial control systems include digital computers as an integral part of their operation. Recent trends toward digital control of dynamic systems, rather than analog control, are mainly due to the recent revolutionary advances in digital computers and to advantages found in working with digital signals rather than continuous-time signals. Also, the availability of low-cost microprocessors and microcomputers established a new trend for even small-scale control systems to include digital computers to obtain optimal performance.

The main purpose of this book is to present a comprehensive treatment of the analysis and design of discrete-time control systems. In particular, this book provides clear and easy-to-understand explanations for concepts involved in the study of discrete-time control systems.

This book can be used as a text book for courses on discrete-time control systems, digital control systems, or discrete-time systems theory. It is written at the level of the senior engineering (electrical, mechanical, aerospace, or chemical) student or beginning graduate student. All the material may be covered in two quarters. In a semester course, the instructor will have some flexibility in choosing the subjects to be covered. This book can also serve as a reference book or self-study book for practicing engineers who wish to study discrete-time control theory.

The prerequisite on the part of the reader is that he or she has had a course on introductory control systems. This book provides all other background materials necessary to study discrete-time control systems.

Since the book is written from the engineer's point of view, the basic concepts involved are emphasized and highly mathematical arguments are carefully avoided in the presentation. All the material has been organized toward a gradual development of discrete-time control theory.

The outline of the book is as follows: Chapter 1 gives an introduction to discrete-time control systems. Chapter 2 presents the z transform theory. Chapter 3 deals with background materials for the z-domain analysis of discrete-time control systems. Chapter 4 discusses design of discrete-time control systems via transform methods. Chapter 5 presents basic state space analyses of discrete-time control systems, including Liapunov stability analysis. Chapter 6 treats controllability and observability, pole placement techniques, design of state observers, and servo systems. The final chapter, Chapter 7, discusses quadratic optimal control problems and other types of optimal control problems. This chapter also discusses system identification by the least-squares method, and Kalman filters. Since a reasonable background of vector-matrix analysis is needed for the state space analysis, a summary of vector-matrix analysis is provided in an Appendix.

At the end of each chapter, except Chapter 1, and at the end of the Appendix, numerous solved problems are provided so that the reader will have a clear understanding of the subject matter presented in the book. In addition to the solved problems, many problems (without solutions) are provided in this book to test the reader's degree of comprehension of the subject matter. Most of the materials presented in this book (including solved and unsolved problems) have been class-tested in senior and first-year graduate level courses on control systems at the University of Minnesota.

I would like to express my appreciation to all those who were helpful in the completion of this writing project. Appreciation is due to anonymous reviewers who provided numerous comments and valuable suggestions at the midstage of this writing project. Sincere appreciation is due to Professor E. I. Jury who made a critical review of the entire manuscript and provided me pages of constructive comments which resulted in the improvement of the presentation of the z transform method, as well as many other subjects throughout the book. Sincere appreciation is also due to Professor Alan Kraus of Naval Post Graduate School who made a careful review of the entire manuscript. His invaluable comments helped to polish the final version of the manuscript.

Appreciation is also due to my former students who solved many problems and made numerous constructive comments about the materials in the book. Needless to say, they raised a variety of questions during lecture hours. Reflecting this, the book provides detailed explanations to answer a number of possible questions that many students may ask in classes.

Finally, I would like to thank Bernard M. Goodwin, Executive Editor; John J. McCanna, Prentice-Hall Representative in Minnesota; and Tim Bozik, Editor, for their enthusiasm in this writing project.

Katsuhiko Ogata

1

Introduction to Discrete-Time Control Systems

1-1 INTRODUCTION

In recent years there has been a rapid increase in the use of digital controllers in control systems. In fact, many industrial control systems include digital computers as an integral part of their operation. Digital controls are used for achieving optimal performance—for example, in the form of maximum productivity, maximum profit, minimum cost, or minimum energy use. The recent evolution of microprocessors and microcomputers which can be used for various control functions has established a new trend toward the inclusion of digital computers even in small-scale control systems to obtain optimal performance.

Most recently, the application of computer control has made possible "intelligent" motion in industrial robots, the optimization of fuel economy in automobiles, and refinements in the operation of household appliances and machines such as microwave ovens and sewing machines, among others. Decision-making capability and flexibility in the control program are major advantages of digital control systems.

The current trend toward digital rather than analog control of dynamic systems is mainly due to the availability of low-cost digital computers and the advantages found in working with digital signals rather than continuous-time signals.

In control engineering, digital computers have been used for two different purposes. First, they have been used for the analysis and synthesis of complex control systems, including digital simulation and digital computation of complex control dynamics. Second, they have been included in control systems as controllers. In this book our emphasis is on digital controllers, rather than digital simulation or digital computation of complex control dynamics.

Types of signals. A continuous-time signal is a signal defined over a continuous range of time. The amplitude may assume a continuous range of values or may assume only a finite number of distinct values. The process of representing a variable by a set of distinct values is called *quantization*, and the resulting distinct values are called *quantized* values. The quantized variable changes only by a set of distinct steps.

An analog signal is a signal defined over a continuous range of time whose amplitude can assume a continuous range of values. Figure 1–1(a) shows a continuous-time analog signal, and Fig. 1–1(b) shows a continuous-time quantized signal (quantized in amplitude only).

Notice that the analog signal is a special case of the continuous-time signal. In practice, however, we frequently use the terminology "continuous-time" in lieu of "analog." Thus in the literature, including this book, the terms "continuous-time signal" and "analog signal" are frequently interchanged, although strictly speaking they are not quite synonymous.

A discrete-time signal is a signal defined only at discrete instants of time (that is, one in which the independent variable t is quantized). In a discrete-time signal, if the amplitude can assume a continuous range of values, then the signal is called a *sampled-data signal*. A sampled-data signal can be generated by sampling an analog signal at discrete instants of time. It is an amplitude-modulated pulse signal. Figure 1–1(c) shows a sampled-data signal.

A digital signal is a discrete-time signal with quantized amplitude. Such a signal can be represented by a sequence of numbers—for example, in the form of binary numbers. (In practice, many digital signals are obtained by sampling analog signals and then quantizing them; it is the quantization that allows these analog signals to be read as finite binary words.) Figure 1–1(d) depicts a digital signal. Clearly, it is a signal quantized both in amplitude and in time. The use of the digital controller requires quantization of signals both in amplitude and in time.

The term "discrete-time signal" is broader than the term "digital signal" or the term "sampled-data signal." In fact, a discrete-time signal can refer either to a digital signal or to a sampled-data signal. In practical usage, the terms "discrete-time" and "digital" are often interchanged. However, the term "discrete-time" is frequently used in theoretical study, while the term "digital" is used in connection with hardware or software realizations.

In control engineering, the controlled object is a plant or process. It may be a physical plant or process or a nonphysical process such as an economic process. Most plants and processes involve continuous-time signals, and therefore if digital controllers are involved in the control systems, signal conversions (analog to digital and digital to analog) become necessary. There are standard techniques available for such signal conversions; we shall discuss them in Sec. 1–4.

Loosely speaking, terminologies such as discrete-time control systems, sampled-data control systems, and digital control systems imply the same type or very similar types of control systems. Precisely speaking, there are, of course, differences in these systems. For example, in a sampled-data control system both continuous-time and

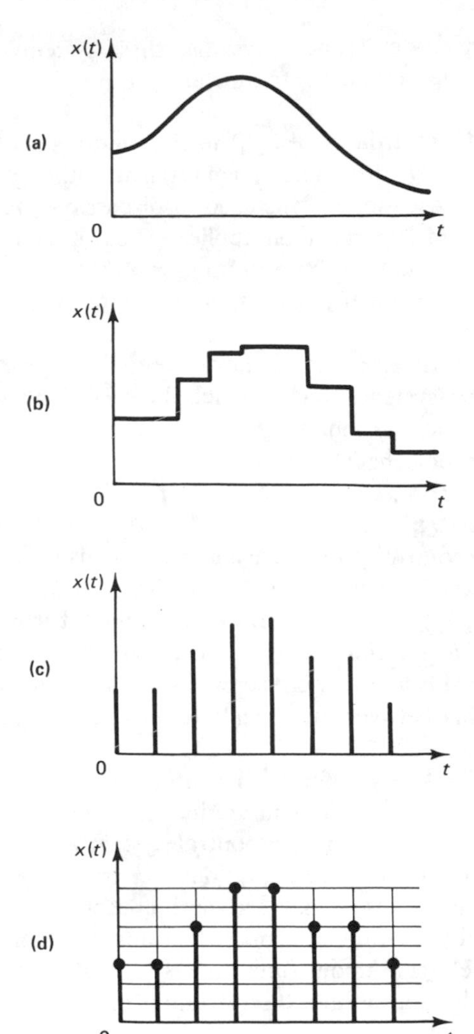

Figure 1–1 (a) Continuous-time analog signal; (b) continuous-time quantized signal; (c) sampled-data signal; (d) digital signal.

discrete-time signals exist in the system; the discrete-time signals are amplitude-modulated pulse signals. Digital control systems may include both continuous-time and discrete-time signals; here, the latter are in a numerically coded form. Both sampled-data control systems and digital control systems are discrete-time control systems.

Many industrial control systems include continuous-time signals, sampled-data signals, and digital signals. Therefore, in this book we use the term "discrete-time control systems" to describe the control systems that include some forms of sampled-data signals (amplitude-modulated pulse signals) and/or digital signals (signals in numerically coded form).

Finally, it is noted that sampled-data control systems may have analog controllers and that despite the differences existing between the components of digital control systems and those of sampled-data control systems, these systems can be analyzed and designed by means of the same discrete-time analytical techniques.

Systems dealt with in this book. The discrete-time control systems considered in this book are mostly linear and time-invariant, although nonlinear and/or time-varying systems are occasionally included in discussions. A linear system is one in which the principle of superposition applies. Thus, if y_1 is the response of the system to input x_1 and y_2 the response to input x_2, then the system is linear if and only if, for every scalar α and β, the response to input $\alpha x_1 + \beta x_2$ is $\alpha y_1 + \beta y_2$.

A linear system may be described by linear differential or linear difference equations. A time-invariant linear system is one in which the coefficients in the differential equation or difference equation do not vary with time, that is, one in which the properties of the system do not change with time.

Discrete-time control systems and continuous-time control systems. Discrete-time control systems are control systems in which one or more variables can change only at discrete instants of time. These instants, which we shall denote by kT or t_k ($k = 0, 1, 2, \ldots$), may specify the times at which some physical measurement is performed or the times at which the memory of a digital computer is read out. The time interval between two discrete instants is taken to be sufficiently short that the data for the time between them can be approximated by simple interpolation.

Discrete-time control systems differ from continuous-time control systems in that signals for a discrete-time control system are in sampled-data form or in digital form. If a digital computer is involved in a control system as a digital controller, any sampled data must be converted into digital data.

Continuous-time systems, whose signals are continuous in time, may be described by differential equations. Discrete-time systems, which involve sampled-data signals or digital signals and possibly continuous-time signals as well, may be described by difference equations after the appropriate discretization of continuous-time signals.

Sampling processes. The sampling of a continuous-time signal replaces the original continuous-time signal by a sequence of values at discrete time points. A sampling process is used whenever a control system involves a digital controller, since a sampling operation and quantization are necessary to enter data into such a controller. Also, a sampling process occurs whenever measurements necessary for control are obtained in an intermittent fashion. For example, in a radar tracking system, as the radar antenna rotates, information about azimuth and elevation is obtained once in each revolution of the antenna. Thus the scanning operation of the radar produces sampled data. In another example, a sampling process is needed whenever a large-scale controller or computer is time-shared by several plants in order to save cost. Then a control signal is sent out to each plant only periodically and thus the signal becomes a sampled-data signal.

The sampling process is usually followed by a quantization process. In the quantization process the sampled analog amplitude is replaced by a digital amplitude (represented by a binary number). Then the digital signal is processed by the computer. The output of the computer is sampled and fed to a hold circuit. The output of the hold circuit is a continuous-time signal and is fed to the actuator. We shall present details of such signal processing methods in the digital controller in Sec. 1–4.

Most modern industrial control systems are discrete-time control systems, since they invariably include sampling operations. In this book we shall present analysis and synthesis techniques for discrete-time control systems, in which signals representing the control effort are piecewise-constant and change only at discrete points in time.

The term "discretization," rather than "sampling," is frequently used in the analysis of multiple-input–multiple-output systems, although both mean basically the same thing.

It is important to note that occasionally the sampling operation or discretization is entirely fictitious and has been introduced only to simplify the analysis of control systems which actually contain only continuous-time signals. In fact, we often use a suitable discrete-time model for a continuous-time system. An example is a digital-computer simulation of a continuous-time system. Such a digital-computer-simulated system can be analyzed to yield parameters that will optimize a given performance index (see Sec. 7–1 for a definition of the optimization of a performance index).

Most of the materials presented in this book deal with control systems which can be modeled as linear time-invariant discrete-time systems. It is important to mention that many digital control systems are based on continuous-time design techniques. Since a wealth of experience has been accumulated in the design of continuous-time controllers, a thorough knowledge of them is highly valuable in designing discrete-time control systems.

1–2 DIGITAL CONTROL SYSTEMS

A control scheme in which a digital computer is included in a control loop to perform signal processing in a desired fashion is called a *direct digital control*. Direct digital control has been applied since the 1960s to large-scale process control systems; in chemical and other processes it has contributed to a great extent to improvements in productivity and the quality of products and to savings in labor costs. The impressive advances in microprocessors and microcomputers since the mid-1970s have enabled control engineers to apply direct digital control techniques to a wide variety of control systems, both large-scale and small-scale.

Direct digital control of a process or plant has the following advantages over the corresponding analog control:

1. Data processing in the digital controller is straightforward; complex control calculations can be performed easily.

2. Control programs (controller characteristics) can be changed easily if such changes are needed.
3. Digital controllers are far superior to the corresponding analog controllers from the viewpoint of internal noise and drift effects.

Digital control has some disadvantages, however, such as the following:

1. The sampling and quantizing processes tend to result in more errors, which degrade system performance.
2. Designing digital controllers to compensate for such degradation is more complex than designing analog controllers at an equivalent level of performance.

Figure 1–2 depicts a block diagram of a digital control system showing a configuration of the basic control scheme. The system includes the feedback control and the feedforward control. In designing such a control system, it should be noted that the "goodness" of the control system depends on individual circumstances. We need to choose an appropriate performance index for a given case and design a controller so as to optimize the chosen performance index.

In what follows, we shall first present definitions of the terms used in discussing digital control systems and list the types of sampling operations. Then we shall discuss

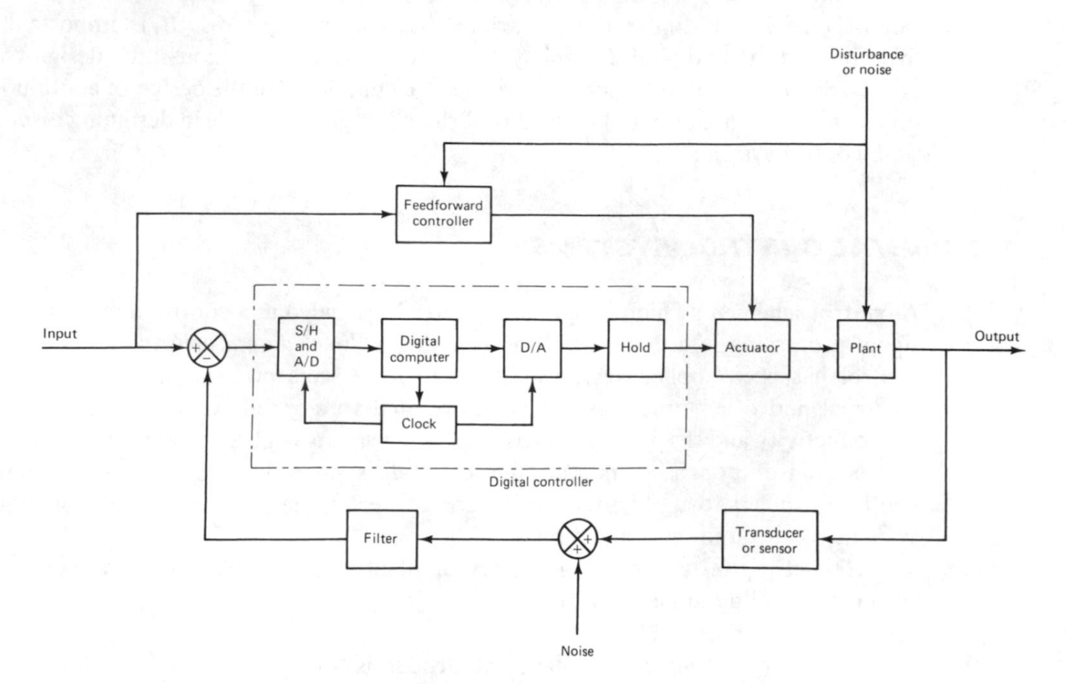

Figure 1–2 Block diagram of a digital control system.

a few examples of digital control systems, such as a numerical control system, a robot arm control system, and a robot hand grasping force control system.

A simplified digital control system. Figure 1–3 shows a block diagram of a simplified digital control system. The basic elements of the system are shown by the blocks. The controller operation is controlled by the clock. In such a digital control system, some points of the system pass signals of varying amplitude in either continuous time or discrete time, while other points pass signals in numerical code, as depicted in the figure.

The output of the plant is a continuous-time signal. The error signal is converted into digital form by the sample-and-hold circuit and the analog-to-digital converter. The conversion is done at the sampling time. The digital computer processes the sequences of numbers by means of an algorithm and produces new sequences of numbers. At every sampling instant a coded number (usually a binary number consisting of eight or more binary digits) must be converted to a physical control signal, which is usually a continuous-time or analog signal. The digital-to-analog converter and the hold circuit convert the sequence of numbers in numerical code into a piecewise continuous-time signal. The real-time clock in the computer synchronizes the events. The output of the hold circuit, a continuous-time signal, is fed to the plant, either directly or through the actuator, to control its dynamics.

The operation that transforms continuous-time signals into discrete-time data is called *sampling* or *discretization*. The reverse operation, the operation that transforms discrete-time data into a continuous-time signal, is called *data-hold*; it amounts to a reconstruction of a continuous-time signal from the sequence of discrete-time data. It is usually done using one of the many extrapolation techniques. In many cases it is done by keeping the signal constant between the successive sampling instants. (We shall discuss such extrapolation techniques in Sec. 1–4.)

The sample-and-hold (S/H) circuit and analog-to-digital (A/D) converter con-

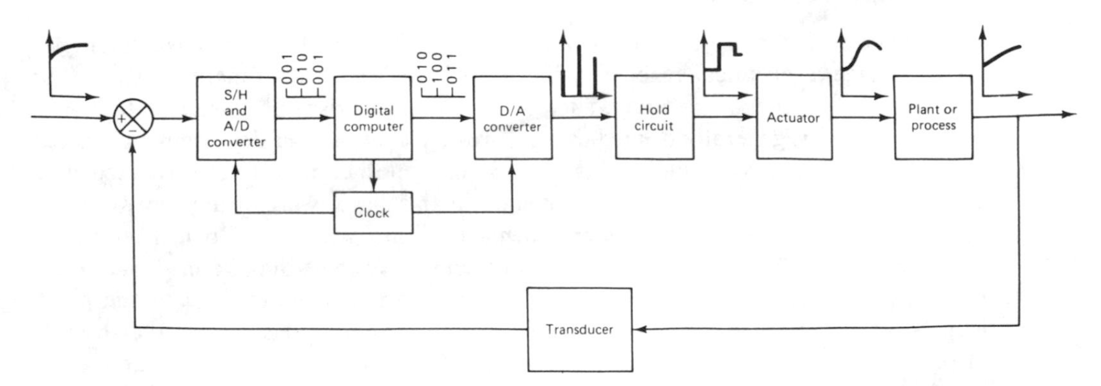

Figure 1–3 Block diagram of a digital control system showing signals in binary or graphic form.

vert the continuous-time signal into a sequence of numerically coded binary words. Such an analog-to-digital conversion process is called *coding* or *encoding*. The combination of the sample-and-hold circuit and analog-to-digital converter may be visualized as a switch that closes instantaneously at every time interval T and generates a sequence of numbers in numerical code. The digital computer operates on such numbers in numerical code and generates a desired sequence of numbers in numerical code. The digital-to-analog (D/A) conversion process is called *decoding*. We shall present more on coding and decoding in Sec. 1–3.

Definitions of terms. Before we discuss digital control systems in detail, we need to define some of the terms that appear in the block diagram of Fig. 1–3.

Sample-and-Hold (S/H). "Sample-and-hold" is a general term used for a sample-and-hold amplifier. It describes a circuit that receives an analog input signal and holds this signal at a constant value for a specified period of time. Usually the signal is electrical, but other forms are possible, such as optical and mechanical.

Analog-to-Digital Converter (A/D). An analog-to-digital converter, also called an encoder, is a device which converts an analog signal into a digital signal, usually a numerically coded signal. Such a converter is needed as an interface between an analog component and a digital component. A sample-and-hold circuit is often an integral part of a commerically available A/D converter. The conversion of an analog signal into the corresponding digital signal (binary number) is an approximation, because the analog signal can take on an infinite number of values, whereas the variety of different numbers which can be formed by a finite set of digits is limited. This approximation process is called *quantization*. (More on quantization is presented in Sec. 1–3.)

Digital-to-Analog Converter (D/A). A digital-to-analog converter, also called a decoder, is a device which converts a digital signal (numerically coded data) into an analog signal. Such a converter is needed as an interface between a digital component and an analog component.

Plant or Process. A plant is any physical object to be controlled. Examples are a furnace, a chemical reactor, and a set of machine parts functioning together to perform a particular operation, such as a servomechanism or a spacecraft.

A process is generally defined as a progressive operation or development marked by a series of gradual changes that succeed one another in a relatively fixed way and lead toward a particular result or end. In this book we call any operation to be controlled a process. Examples are chemical, economic, and biological processes.

The most difficult part in the design of control systems may lie in the accurate modeling of a physical plant or process. There are many approaches to the plant or process model, but even so, a difficulty may exist, mainly because of the absence of precise process dynamics and the presence of poorly defined random parameters in many physical plants or processes. Thus, in designing a digital controller, it is necessary to recognize the very fact that the mathematical model of a plant or process in many cases is only an approximation of the physical one. Exceptions are found in the modeling of electromechanical systems and hydraulic-mechanical systems, since these may be modeled accurately. For example, the modeling of a robot arm system may be accomplished with great accuracy.

Transducer. A transducer is a device which converts an input signal into an output signal of another form, such as a device that converts a pressure signal into a voltage output. The output signal, in general, depends on the past history of the input.

Transducers may be classified as analog transducers, sampled-data transducers, or digital transducers. An analog transducer is a transducer in which the input and output signals are continuous functions of time. The magnitudes of these signals may be any values within the physical limitations of the system. A sampled-data transducer is one in which the input and output signals occur only at discrete instants of time (usually periodic) but the magnitudes of the signals, as in the case of the analog transducer, are unquantized. A digital transducer is one in which the input and output signals occur only at discrete instants of time and the signal magnitudes are quantized (that is, they can assume only certain discrete levels).

Types of sampling operations. As stated earlier, a signal whose independent variable t is discrete is called a discrete-time signal. A sampling operation is basic in transforming a continuous-time signal into a discrete-time signal.

There are several different types of sampling operations of practical importance:

1. Periodic sampling. In this case, the sampling instants are equally spaced, or $t_k = kT$ $(k = 0, 1, 2, \ldots)$. Periodic sampling is the most conventional type of sampling operation.
2. Multiple-order sampling. The pattern of the t_k's is repeated periodically; that is, $t_{k+r} - t_k$ is constant for all k.
3. Multiple-rate sampling. In a control system having multiple loops, the largest time constant involved in one loop may be quite different from that in other loops. Hence, it may be advisable to sample slowly in a loop involving a large time constant, while in a loop involving only small time constants the sampling rate must be fast. Thus a digital control system may have different sampling periods in different feedback paths or may have multiple sampling rates.
4. Random sampling. In this case, the sampling instants are random, or t_k is a random variable.

In this book we shall treat only the case where the sampling is periodic.

Numerical control systems. Computerized numerical control systems which are direct digital control systems, are widely used in industry, since inexpensive microprocessors and microcomputers are readily available. The cost and size of numerically controlled machines have been reduced by the use of microprocessors or microcomputers for processing the necessary data.

Numerical control is a method of controlling the motions of machine components with binary numbers. The input data (consisting, for example, of the dimensions and shape of a mechanical part that is to be produced), as well as the sequential cutting orders, are stored as a sequence of binary numbers on a magnetic tape or disk. Such a tape or disk is called an *input tape* or *input disk*. The essential characteristic of a numerical control system is that binary numbers in the input data are converted into physical values (dimensions or quantities) by electrical (or other) signals that are translated into a linear or circular movement.

Figure 1–4 shows a block diagram of a numerically controlled milling machine. The system works as follows. An input tape or disk is prepared representing in binary form the desired part P. To start the system, the data in the input tape or disk are fed through the reader, which converts the binary numbers into the input pulse signal. The input pulse signal is first compared with the feedback pulse signal (usually both are frequency-modulated), and the difference signal (error) is processed in a predetermined desired fashion in the computer so as to reduce the error. The output of the computer is fed to the digital-to-analog converter. The D/A converter and the hold device convert the digital data into an analog voltage signal. This signal becomes the input to the servomotor system, as shown in the figure. (In some applications where power requirements are relatively small, stepping motors are used instead of dc servomotors.) The computer also sends out signals for controlling the sequential operation. The transducer attached to the cutterhead converts the motion into an electrical signal, which is converted into the pulse signal by the analog-to-digital converter. The precise reading of the data representing the drawing or sketch and the speedy decision making in the sequential program that are made possible by numerical control improve both product quality and productivity.

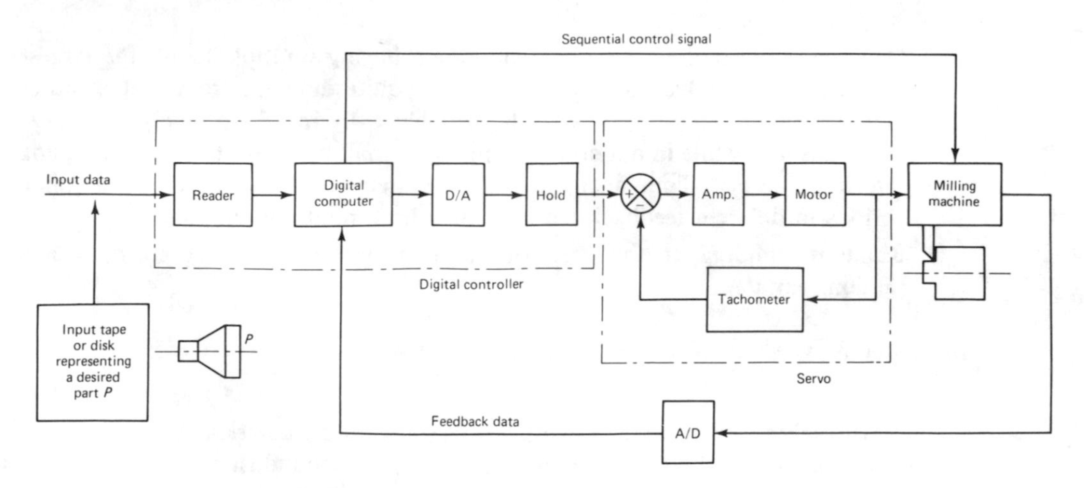

Figure 1–4 Block diagram of a numerically controlled milling machine.

Figure 1–5 shows an example of profile approximation when stepping motors are used for positioning the cutterhead. The stepping motor system that positions the cutterhead of the milling machine in the xy plane is controlled by the computer that processes the input data. The number and type of command pulses from the computer to the stepping motors determine the number and direction of motions necessary to bring the cutterhead to the correct position in x and y coordinates. The movement in the x or y direction is, in many cases, 1 micron (or 10 microns) per command pulse. For example, if 5 command pulses in the positive x direction are sent out by the computer, then the cutterhead is moved by 5 microns (or 50 microns) in the positive x direction. Thus, the cutterhead is moved in the x or y direction by 1 micron (or 10 microns) a time. This means that the cutterhead traces a given curve with an accuracy of the order of 1 micron (or 10 microns) in both the x and the y directions. Hence, the accuracy in following a given profile can be kept high. The advantage of numerical control of a milling machine is that complex parts can be produced with uniform tolerances at the maximum milling speed.

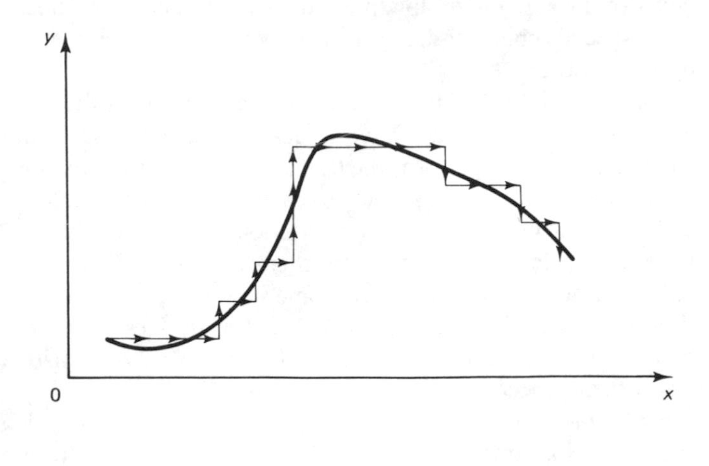

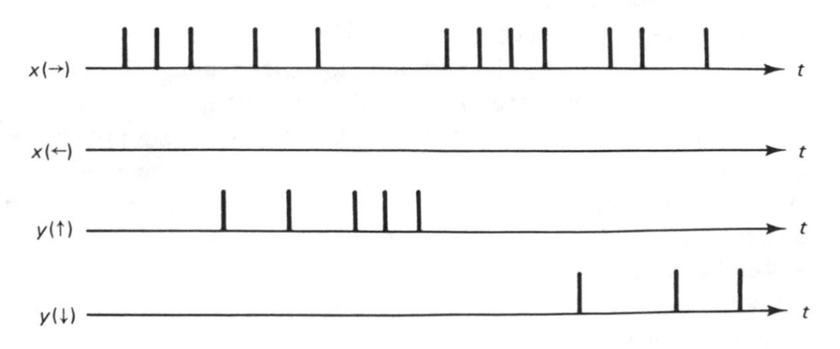

Figure 1–5 Profile approximation by a numerically controlled milling machine.

Computerized numerical control systems operate in real time. The computer must solve algebraic equations to determine the direction of motion in the xy plane and must send a pulse signal every 100 μsec or so. To speed up the computation time, not only must the microcomputer be capable of speedy data processing, but the number of steps in the computer program must also be kept to a minimum.

In addition to its low cost and compact size, the computerized numerical control system has functional advantages. There is no need to rewind the input tape every time to produce the same product. In addition, it is possible to store many different input data by increasing memory capacity and to produce many different types of products continually and thus to improve productivity.

Robot arm control system. Industrial robots are frequently used in industry to improve productivity. The robot can handle monotonous jobs as well as complex jobs without errors in operation. The robot can work in an environment intolerable to human operators. For example, it can work in extreme temperatures (both high and low) or in a high- or low-pressure environment, or under water or in space. There are special robots for firefighting, underwater exploration, and space exploration, among many others.

The industrial robot must handle mechanical parts that have particular shapes and weights. Hence, it must have at least an arm, a wrist, and a hand. It must have sufficient power to perform the task and the capability for at least limited mobility. The majority of robots in use by the early 1980s were limited in mobility to such things as motion in straight lines or motion in circular paths. But some robots of this time were able to move freely by themselves in a limited space in a factory.

The industrial robot must have some sensory devices. In low-level robots, microswitches are installed in the arms as sensory devices. The robot first touches an object and then through the microswitches confirms the existence of the object in space and then proceeds in the next step to grasp it.

In a high-level robot, an optical means (such as a television system) is used to scan the background of the object. It recognizes the pattern and determines the presence and orientation of the object. A computer is necessary to process signals in the pattern recognition process. (See Fig. 1–6.) In some applications, the computerized robot recognizes the presence and orientation of each mechanical part by a pattern recognition process that consists of reading the code numbers attached it. Then the robot picks up the part and moves it to an appropriate place for assembling, and here it assembles several parts into a component.

Figure 1–7 shows a schematic diagram for a simplified version of the robot arm control system. The diagram shows a straight-line motion control of the arm. A straight-line motion is a one-degree-of-freedom motion. The actual robot arm has 3 degrees of freedom (up-and-down motion, forward-and-backward motion, and left-and-right motion). The wrist attached to the end of the arm also has 3 degrees of freedom, and the hand has 1 degree of freedom (grasp motion). Altogether the robot arm system has 7 degrees of freedom. Additional degrees of freedom are required if the robot body must move on a plane. In general, robot hands may be interchangeable

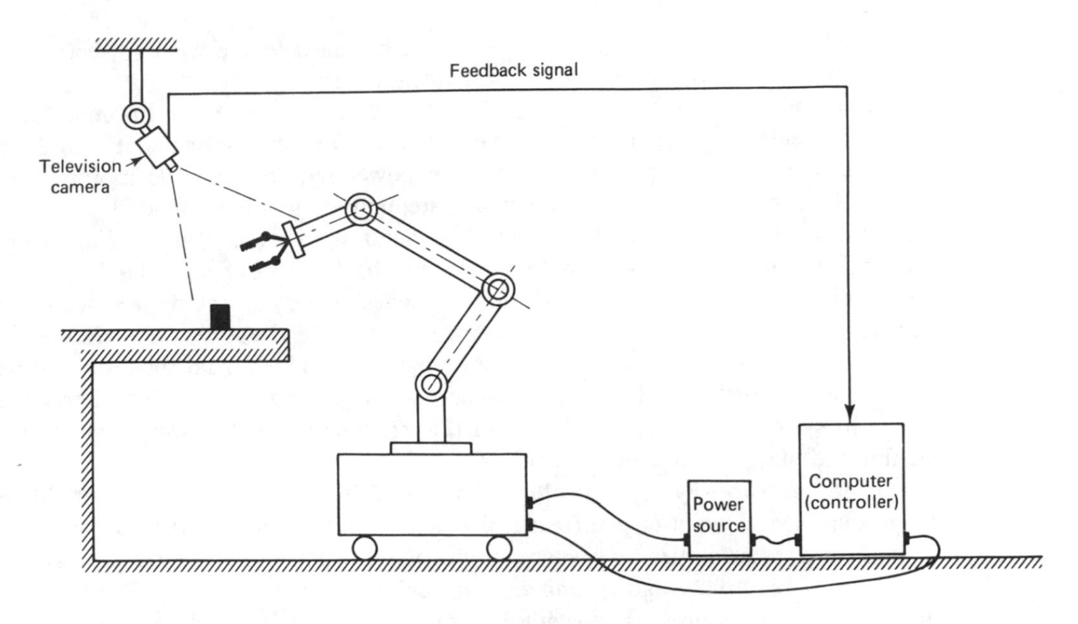

Figure 1–6 Robot using a pattern recognition process.

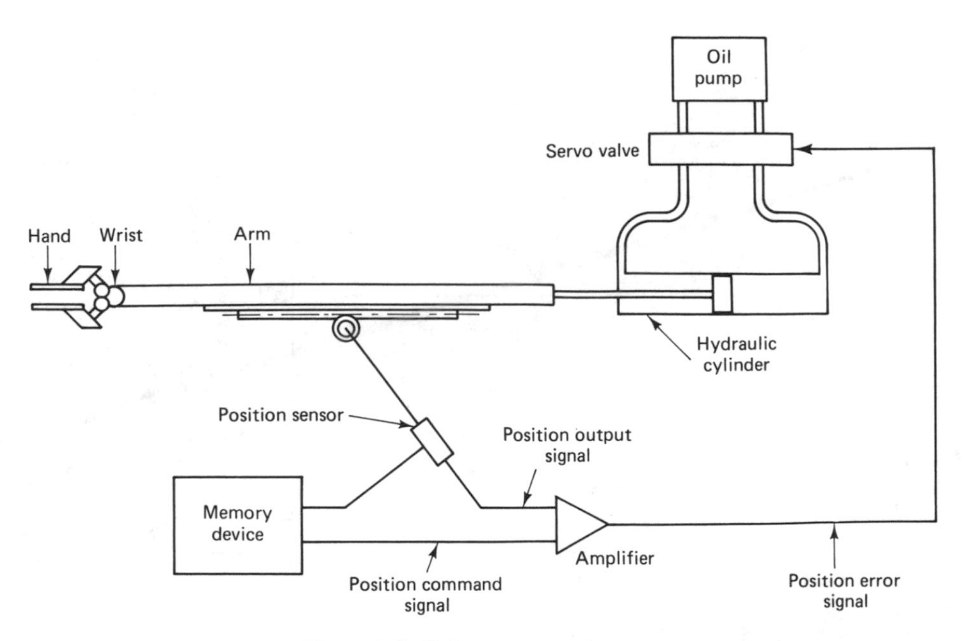

Figure 1–7 Robot arm control system.

parts: a different type of grasping device can be attached to the wrist to serve as a hand to handle each different type of mechanical object.

A servomechanism is used to position the arm and wrist. Since the robot arm motion frequently requires speed and power, hydraulic pressure or pneumatic pressure is used as the source of power. For medium power requirements, dc motors may be used. And for small power requirements, stepping motors may be used.

For control of sequential motions, command signals are stored on magnetic tapes or disks. In high-level robot systems, the "playback" mode of control is frequently used. In this mode, a human operator first "teaches" the robot the desired sequence of movements by working a lever mechanism attached to the arm; the computer in the robot memorizes the desired sequential movements. Then, from the second time on, the robot repeats faithfully the sequence of movements. Figure 1–8 shows the robot arm system shown in Fig. 1–7 with the circuits and devices that take part in teaching the playback mode of control.

There are two ways of controlling the path of motions in either two- or three-dimensional space: point-to-point control and continuous-path control (see Fig. 1–9). In moving the arm in an open space, point-to-point control may be used. However, if the arm is required to go in and out through a complex structure, continuous-path control must be used. For example, in the case of an arc welding robot in an automobile assembly plant, precise continuous-path control is necessary because the arm must move through a complex body structure. (The accuracy in moving the arm is on the order of ± 1 mm.)

Industrial robots are frequently used to perform the same task repeatedly. How-

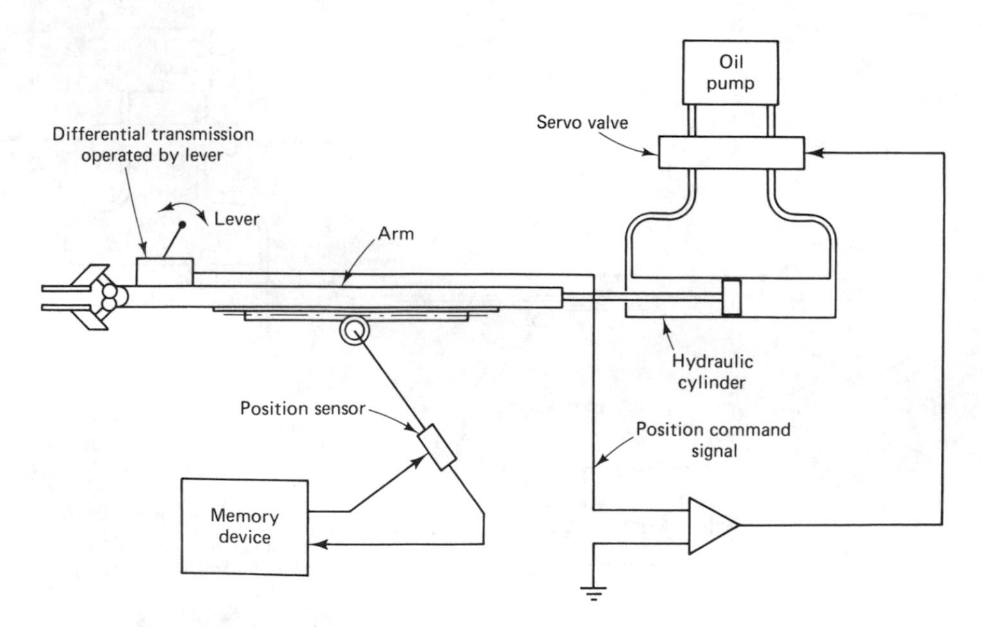

Figure 1–8 Playback mode of control of the robot arm system shown in Fig. 1–7.

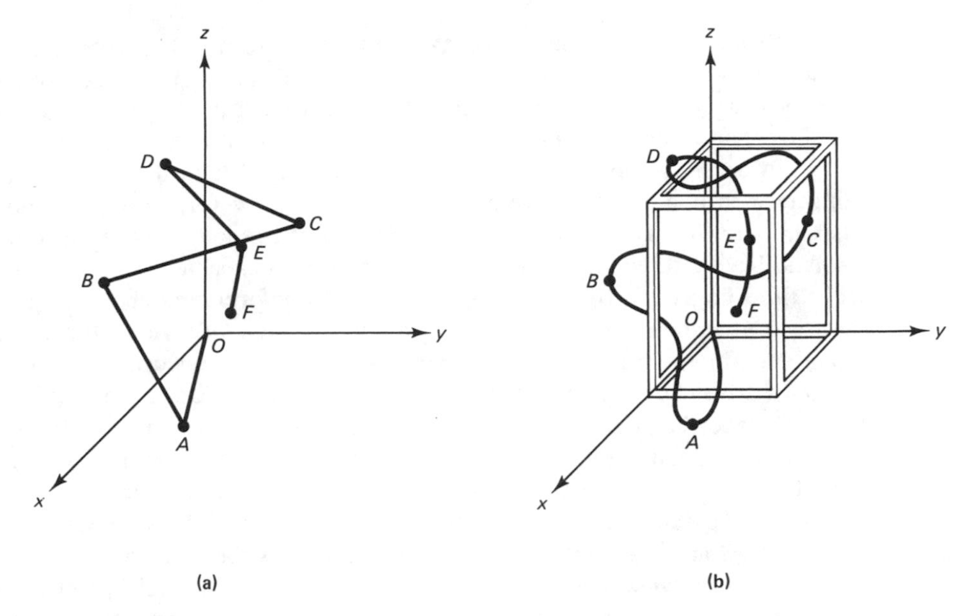

Figure 1–9 (a) Point-to-point control; (b) continuous-path control.

ever, robots with computer controls can perform many different tasks to produce a small quantity of each of a variety of products. The capability of such flexible machining is one advantage of computer-controlled robots. The flexibility comes from the fact that one robot can be programmed to provide many functions as needed: handling different parts and assembling products, even weeding out defects. In order for the robot to perform many different tasks, however, the computer must have a large memory capacity. Often it is necessary to store many thousands of step sequences in the memory. Even so, computer-controlled robots, though costly, reduce space requirements and eliminate extensive retooling procedures for the production of different products.

The high-level robot with computer control and with sufficient memory capacity controls all its motions by itself. For the majority of robots used in industry, a large-scale computer is not installed in each robot. A central computer (a large-scale computer) may be installed in a computer room, from which it may control all the motions of a group of robots or may give individual robots (which are equipped with microcomputers) only supervisory information and leave all local controls such as sequential control of the arm to them. The sensory device of each robot, however, must be continuously connected to the central computer through data communication circuits. Thus, direct on-line communication between robots and the central computer is always necessary.

If a central computer has a large capacity, it can control many individual robots. The number of robots that the central computer can handle is determined by the sampling period and the speed of signal processing. Thus, the choice of sampling period is very important.

Robot hand grasping force control system. Let us examine the hand of an industrial robot from the viewpoints of mechanisms and of control. The main points to be touched upon are the structural aspects of the hand (that is, the number of fingers and their degrees of freedom) and the type of the actuator to be used.

In order to pick up a mechanical object and move it, it may be that only two fingers with one degree of freedom are required. However, to make a more skillful hand, more than two fingers and many degrees of freedom are needed. For example, a hand having three fingers with three joints each is capable of more stable grasping. The third finger can serve as a sensor to pick up information about the object being grasped, thus increasing the operating ability of the hand. (In such a three-fingered hand a dc motor may be used as a power source. The joints can be moved by pulley systems driven by the motor. A hand designed in this way can be made quite compact.)

A robot hand actuator may include either a pneumatic or a hydraulic cylinder for linear motions and an electric motor for rotational motions. (Linear motions can be changed into rotational motions, and vice versa, by the use of appropriate linkages.) To achieve rapid finger motions, the power of the actuator must be sufficiently high and the inertia of the moving parts must be made as small as practical.

From the viewpoint of control of the robot hand, the main point to be studied is the servomechanism that controls the hand motion; it uses the feedback from sensors that act as touch feelers, as well as slip feelers.

The importance of the sensor in the robot hand cannot be overemphasized. In order to grasp a mechanical object firmly without damaging it, the robot must have a force-sensing device and a slip-sensing device. For the force-sensing device, semiconductor strain gauges which change the force into a deflection are frequently used. For the slip-sensing device, one approach is to attach a roller device to the contact surface. Slipping is measured by the rotation angle of the roller.

Note that if the grasping force is too small the robot hand will drop the mechanical object and if it is too great the hand may damage or crush the object.

Figure 1–10 shows a schematic diagram for a grasping force control system

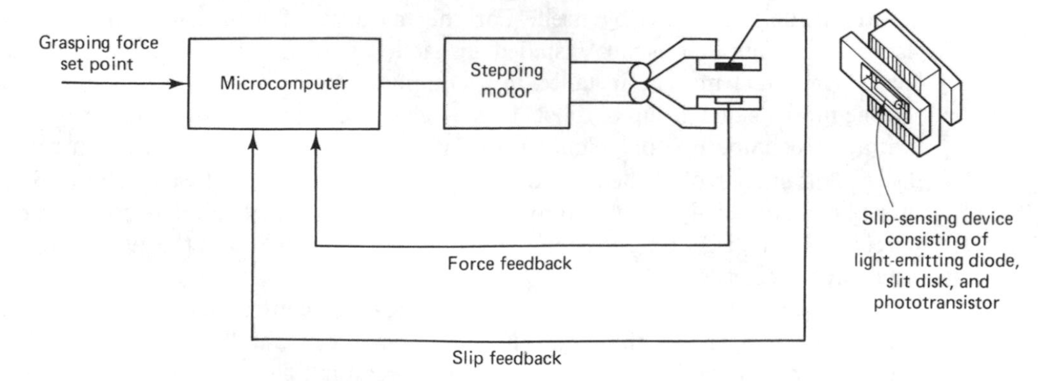

Figure 1–10 Robot hand grasping force control system.

using a force-sensing device and a slip-sensing device. In this system the grasping force is preset at a moderate level before the hand touches the mechanical object. The hand picks up and raises the object with the preset grasping force. If there is a slip in the raising motion, it will be observed by the slip-sensing device and a signal will be sent back to the controller, which will then increase the grasping force. In this way, a reasonable grasping force can be realized which prevents slipping but does not damage the mechanical object.

1-3 QUANTIZING AND CODING

The main functions involved in analog-to-digital conversion are sampling, amplitude quantizing, and coding. When the value of any sample falls between two adjacent "permitted" output states, it must be read as the permitted output state nearest the actual value of the signal. The process of representing a continuous or analog signal by a finite number of discrete states is called *amplitude quantization*. That is, "quantizing" means transforming a continuous or analog signal into a set of discrete states. (Note that quantizing occurs whenever a physical quantity is represented numerically.)

The output state of each quantized sample is then described by a numerical code. The process of representing a sample value by a numerical code (such as a binary code) is called *encoding* or *coding*. Thus, encoding is a process of assigning a digital word or code to each of the discrete states. The sampling period and quantizing levels affect the performance of digital control systems. So they must be determined carefully.

Quantizing. The standard number system used for processing digital signals is the binary number system. In this system the code group consists of n pulses each indicating either "on" (1) or "off" (0). In the case of quantizing, n "on-off" pulses can represent 2^n amplitude levels or output states.

Figure 1–11 shows the relationship between the analog input and the digital output for a unipolar 3-bit A/D converter or quantizer. Note that such a 3-bit quantizer has $2^3 = 8$ output states, since, as we have seen, an n-bit quantizer has 2^n output states.

In practice, the digital codes are standardized with data converters, as will be discussed later. The analog voltage ranges are also standardized. Most converters use standard bipolar voltages of ± 2.5, ± 5, and ± 10 V, or use standard unipolar voltages ranging from 0 to 5 V or 0 to 10 V. The horizontal axis shown in Fig. 1–11 is the analog input, ranging from 0 to 10 V. The vertical axis shown is the output binary code, in this case a 3-bit code, increasing from 000 to 111.

Referring to Fig. 1–11, notice that there are $2^n - 1$ analog decision points. In the case of the 3-bit quantizer shown in Fig. 1–11, these analog decision points are 0.625, 1.875, 3.125, 4.375, 5.625, 6.875, and 8.125 V. The center points of the output code words are 1.25, 2.50, 3.75, 5.00, 6.25, 7.50, and 8.75 V.

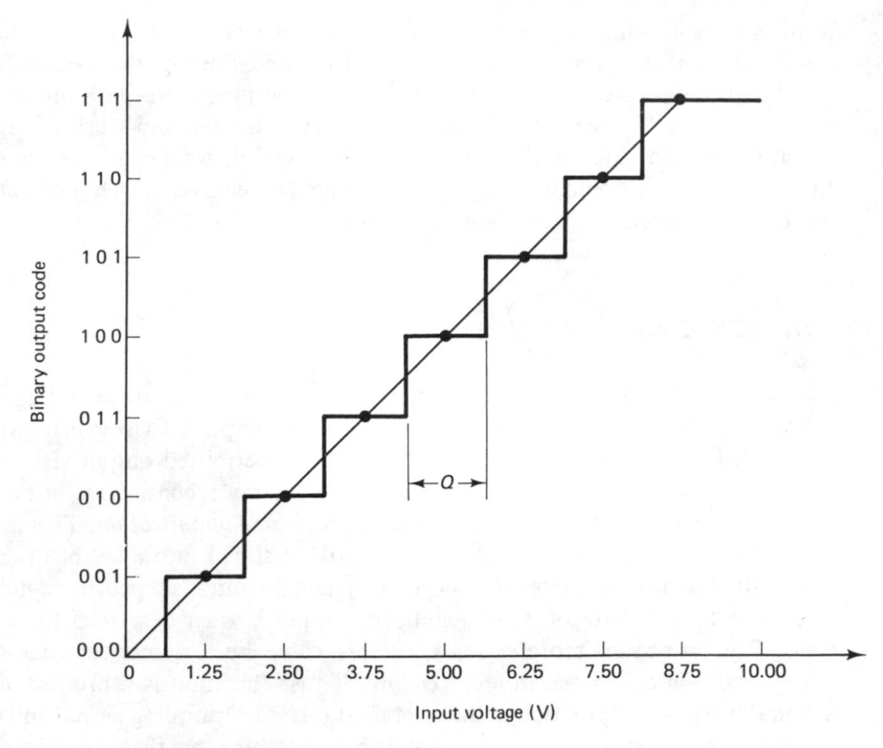

Figure 1–11 Input-output relationship of a unipolar 3-bit A/D converter.

The quantization level Q is the range between two adjacent decision points and is given by

$$Q = \frac{\text{FSR}}{2^n}$$

where the FSR is the full scale range. If the FSR is 10 V and $n = 3$, then

$$Q = \frac{10}{2^3} = \frac{10}{8} = 1.25 \text{ V}$$

Natural binary code. The natural binary code is the most popular code used in practice. Here, the binary words of the code represent fractions of the FSR, which is represented by unity. The code words are developed as follows. Any number N that lies between 0 and 1 can be approximated in natural binary code to n significant bits:

$$N = a_1 2^{-1} + a_2 2^{-2} + a_3 2^{-3} + \cdots + a_n 2^{-n}$$

where the a_i's ($i = 1, 2, \ldots, n$) are either 0 or 1.

In general, a binary fraction is written, for example, as 0.11101011. In the case of the data converter codes, the radix point (the binary point) is omitted and the code word is written simply as 1110 1011. For example, the binary code word 1111 1111 represents the decimal fraction

$$1\times\frac{1}{2}+1\times\frac{1}{2^2}+1\times\frac{1}{2^3}+1\times\frac{1}{2^4}+1\times\frac{1}{2^5}+1\times\frac{1}{2^6}+1\times\frac{1}{2^7}+1\times\frac{1}{2^8}$$
$$=0.5+0.25+0.125+0.0625+0.03125+0.015625+0.0078125$$
$$+0.00390625=0.9960937$$

Similarly, 1100 0000 represents

$$1\times\frac{1}{2}+1\times\frac{1}{2^2}=0.5+0.25=0.75$$

and 0000 0001 represents

$$1\times\frac{1}{2^8}=0.00390625$$

We derive the quantization level Q as a fraction of the FSR as follows. The leftmost bit of the binary code word has the most weight (one-half of the full scale) and is called the *most significant bit* (MSB). The rightmost bit has the least weight ($1/2^n$ times the full scale) and is called the *least significant bit* (LSB). Thus,

$$\text{LSB}=\frac{\text{FSR}}{2^n} \qquad (\text{FSR} = \text{full scale range})$$

The least significant bit is the quantization level Q. Notice that the maximum value of the digital code (binary code word 1111 1111) represents one LSB less than the full scale, that is

$$\text{FSR}(1-2^{-n})$$

For example, for $n = 8$ the maximum value of the code is 0.9960937.

Although the natural binary code is the code most commonly used in practice, other codes are also used, including binary-coded decimal, offset binary, two's complement, and others.

Table 1–1 shows binary coding for 8-bit unipolar converters and Table 1–2 shows binary-coded decimal for a 3-decimal-digit data converter. Figure 1–12 shows the relationship between the analog input and digital output of a bipolar 3-bit A/D converter.

Quantization error. Since the number of bits in the digital word is finite, A/D conversion results in a finite resolution. That is, the digital output can assume only a finite number of levels, and therefore an analog number must be rounded off to the nearest digital level. Hence, any A/D conversion involves quantization error. Such quantization error varies between 0 and $\pm\frac{1}{2}Q$. This error depends on the fineness of the quantization level and can be made as small as desired by making the quantiza-

TABLE 1–1 BINARY CODING FOR 8-BIT UNIPOLAR CONVERTERS

Fraction of FSR (analog range)	Fraction of FSR	Natural binary code
FSR − 1 LSB	0.9961	1111 1111
$\frac{7}{8}$ FSR	0.8750	1110 0000
$\frac{3}{4}$ FSR	0.7500	1100 0000
$\frac{1}{2}$ FSR	0.5000	1000 0000
$\frac{1}{4}$ FSR	0.2500	0100 0000
$\frac{1}{8}$ FSR	0.1250	0010 0000
1 LSB	0.0039	0000 0001
0	0.0000	0000 0000

tion level smaller (that is, by increasing the number of bits n). In practice, there is a maximum for the number of bits n, and so there is always some error due to quantization. The uncertainty present in the quantization process is called *quantization noise*.

In order to determine the desired size of the quantization level (or the number of output states) in a given digital control system, the engineer must have a good understanding of the relationship between the size of the quantization level and the resulting error. The variance of the quantization noise is an important measure of

TABLE 1–2 BINARY-CODED DECIMAL FOR A 3-DECIMAL-DIGIT DATA CONVERTER

Fraction of FSR (analog range)	Fraction of FSR	Binary-coded decimal
FSR − 1 LSB	0.999	1001 1001 1001
$\frac{7}{8}$ FSR	0.875	1000 0111 0101
$\frac{3}{4}$ FSR	0.750	0111 0101 0000
$\frac{1}{2}$ FSR	0.500	0101 0000 0000
$\frac{1}{4}$ FSR	0.250	0010 0101 0000
$\frac{1}{8}$ FSR	0.125	0001 0010 0101
1 LSB	0.001	0000 0000 0001
0	0.0000	0000 0000 0000

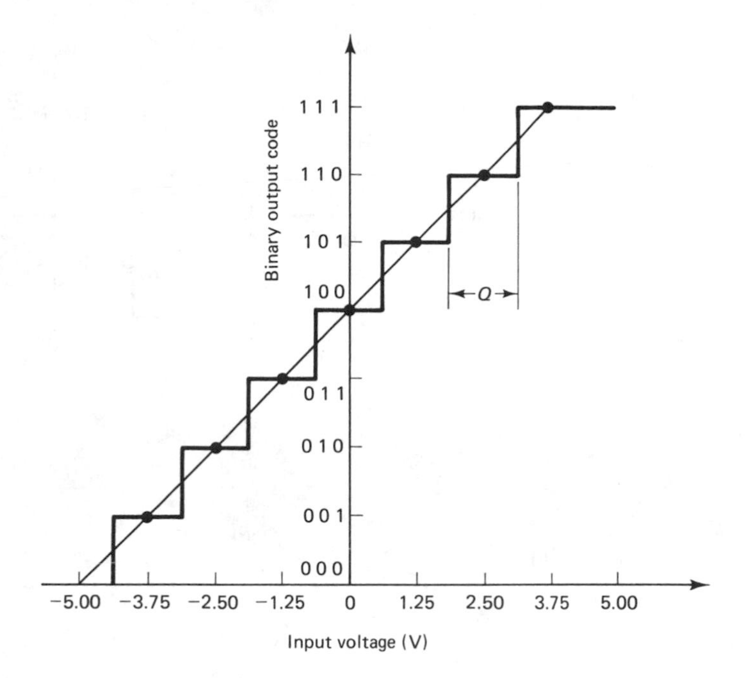

Figure 1–12 Input-output relationship of a bipolar 3-bit A/D converter.

quantization error, since the variance is proportional to the average power associated with the noise.

Figure 1–13(a) shows a block diagram of a quantizer together with its input-output characteristics. For an analog input $x(t)$ the output $y(t)$ takes on only a finite number of levels, which are integral multiples of the quantization level Q.

In numerical analysis the error resulting from neglecting the remaining digits is called the *round-off error*. Since the quantizing process is an approximating process in that the analog quantity is approximated by a finite digital number, the quantization error is a round-off error. Clearly, the finer the quantization level, the smaller the round-off error.

Figure 1–13(b) shows an analog input $x(t)$ and the discrete output $y(t)$, which is in the form of a staircase function. The quantization error $e(t)$ is the difference between the input signal and the quantized output, or

$$e(t) = x(t) - y(t)$$

Note that the magnitude of the quantized error is

$$0 \leq |e(t)| \leq \tfrac{1}{2}Q$$

For a small quantization level Q, the nature of the quantization error is similar to that of random noise. And in effect, the quantization process acts as a source of random noise. In what follows we shall obtain the variance of the quantization noise. Such variance can be obtained in terms of the quantization level Q.

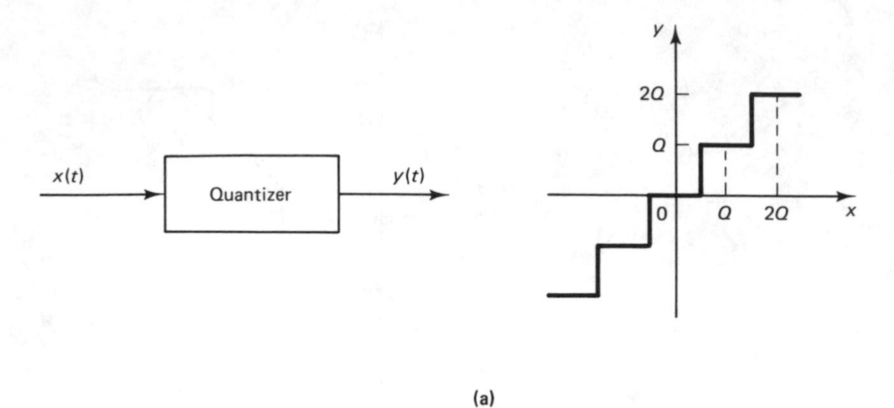

(a)

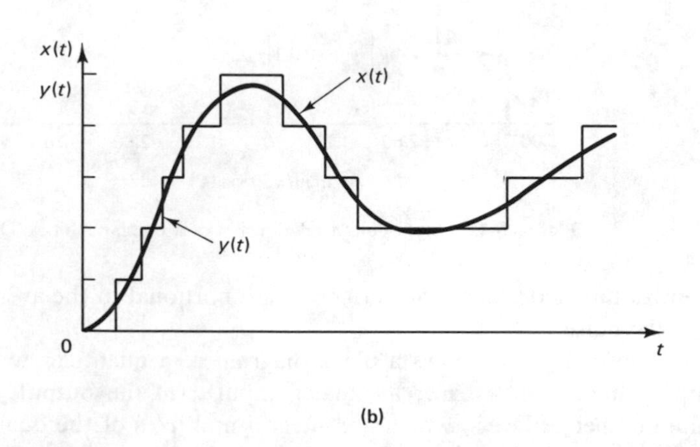

(b)

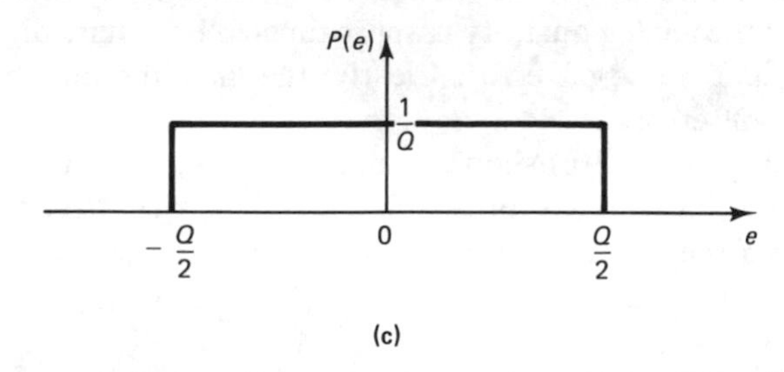

(c)

Figure 1–13 (a) Block diagram of a quantizer and its input-output characteristics; (b) analog input $x(t)$ and discrete output $y(t)$; (c) probability distribution $P(e)$ of quantization error $e(t)$.

Suppose that the quantization level Q is small and we assume that the quantization error $e(t)$ is distributed uniformly between $-\frac{1}{2}Q$ and $\frac{1}{2}Q$ and that this error acts as a white noise. [This is obviously a rather rough assumption. However, since the quantization error signal $e(t)$ is of a small amplitude, such an assumption may be acceptable as a first-order approximation.] The probability distribution $P(e)$ of signal $e(t)$ may be plotted as shown in Fig. 1–13(c). The average value of $e(t)$ is zero, or $\overline{e(t)} = 0$. Then the variance σ^2 of the quantization noise is

$$\sigma^2 = E[e(t) - \overline{e(t)}]^2 = \frac{1}{Q}\int_{-Q/2}^{Q/2} \xi^2 \, d\xi = \frac{Q^2}{12}$$

Thus, if the quantization level Q is small compared with the average amplitude of the input signal, then the variance of the quantization noise is seen to be one-twelfth of the square of the quantization level.

It is noted that when the quantization level is not small compared with the average amplitude, then the quantization phenomena may be treated as a multi-level relay system.

DATA ACQUISITION, CONVERSION, AND DISTRIBUTION SYSTEMS

With the rapid growth in the use of digital computers to perform digital control actions, both the data acquisition system and the distribution system have become an important part of the entire control system, particularly in large-scale control systems.

The signal conversion that takes place in the digital control system involves the following operations:

1. Multiplexing and demultiplexing
2. Sample and hold
3. Analog-to-digital conversion (quantizing and encoding)
4. Digital-to-analog conversion (decoding)

Figure 1–14(a) shows a block diagram of a data acquisition system, and Fig. 1–14(b) shows a block diagram of a data distribution system.

In the data acquisition system the input to the system is a physical variable such as position, velocity, acceleration, temperature, or pressure. Such a physical variable is first converted into an electrical signal (a voltage or current signal) by a suitable transducer. Once the physical variable is converted into a voltage or current signal, then the rest of the data acquisition process is done by electronic means.

In Fig. 1–14(a) the amplifier (frequently an operational amplifier) that follows the transducer performs one or more of the following functions: It amplifies the voltage output of the transducer; it converts a current signal into a voltage signal;

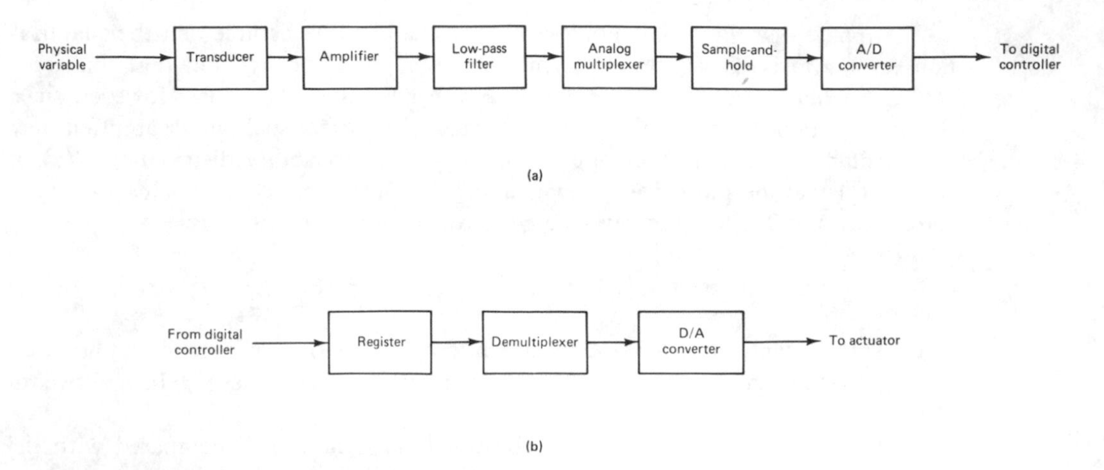

Figure 1–14 (a) Block diagram of a data acquisition system; (b) block diagram of a data distribution system.

or it buffers the signal. The low-pass filter that follows the amplifier attenuates the high-frequency signal components, such as noise signals. (Note that electronic noises are random in nature and may be reduced by low-pass filters. However, such common electrical noises as power line interference are generally periodic and may be reduced by means of notch filters.) The output of the low-pass filter is an analog signal. This signal is fed to the analog multiplexer. The output of the multiplexer is fed to the sample-and-hold circuit, whose output is, in turn, fed to the analog-to-digital converter. The output of the converter is the signal in digital form; it is fed to the digital controller.

The reverse of the data acquisition process is the data distribution process. As shown in Fig. 1–14(b) a data distribution system consists of registers, a demultiplexer, and digital-to-analog converters. It converts the signal in digital form (binary numbers) into analog form. The output of the digital-to-analog converter is fed to the analog actuator, which, in turn, directly controls the plant under consideration.

In the following, we shall discuss each of the individual components involved in the signal processing system.

Analog multiplexer. An analog-to-digital converter is the most expensive component in a data acquisition system. The analog multiplexer is a device that performs the function of time-sharing an analog-to-digital converter among many analog channels. The processing of a number of channels with a digital controller is possible because the width of each pulse representing the input signal is very narrow, so that the empty space during each sampling period may be used for other signals. If many signals are to be processed by a single digital controller, then these input signals must be fed to the controller through a multiplexer.

Figure 1–15 shows a schematic diagram of an analog multiplexer. The analog

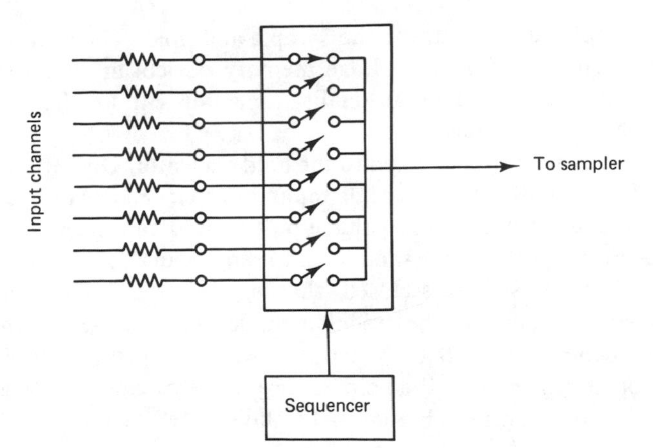

Figure 1–15 Schematic diagram of an analog multiplexer.

multiplexer is a multiple switch (usually an electronic switch) that sequentially switches among many analog input channels in some prescribed fashion. The number of channels, in many instances, is 4, 8, or 16. At a given instant of time, only one switch is in the "on" position. When the switch is on in a given input channel, the input signal is connected to the output of the multiplexer for a specified period of time. During the connection time the sample-and-hold circuit samples the signal voltage (analog signal) and holds its value while the analog-to-digital converter converts the analog value into digital data (binary numbers). Each channel is read in a sequential order, and the corresponding values are converted into digital data in the same sequence.

Demultiplexer. The demultiplexer, which is synchronized with the input sampling signal, separates the composite output digital data from the digital controller into the original channels. Each channel is connected to a D/A converter to produce the output analog signal for that channel.

Sample-and-hold circuits. A sampler in a digital system converts an analog signal into a train of amplitude-modulated pulses. The hold circuit holds the value of the sampled pulse signal over a specified period of time. The sample-and-hold is necessary in the A/D converter to produce a number which accurately represents the input signal at the sampling instant. Commercially, sample-and-hold circuits are available in a single unit, known as a sample-and-hold (S/H). Mathematically, however, the sampling operation and the holding operation are modeled separately (see Sec. 3–2). It is common practice to use a single analog-to-digital converter and multiplex many sampled analog inputs into it.

In practice, sampling duration is very short compared with the sampling period T. When the sampling duration is negligible, then the sampler may be considered an "ideal sampler." An ideal sampler enables us to obtain a relatively simple mathematical model for a sample-and-hold. (Such a mathematical model will be discussed in detail in Sec. 3–2.)

Figure 1–16 shows a simplified diagram for the sample-and-hold. The sample-and-hold circuit is an analog circuit (simply a voltage memory device) in which an input voltage is acquired and then stored on a high-quality capacitor with low leakage and low dielectric absorption characteristics.

In Fig. 1–16 the electronic switch is connected to the hold capacitor. Operational amplifier 1 is an input buffer amplifier with a high input impedance. Operational amplifier 2 is the output amplifier; it buffers the voltage on the hold capacitor.

There are two modes of operation for a sample-and-hold circuit: the tracking mode and the hold mode. When the switch is closed (that is, when the input signal is connected), then the operating mode is the tracking mode. The charge on the capacitor in the circuit tracks the input voltage. When the switch is open (the input signal is disconnected), the operating mode is the hold mode and the capacitor voltage holds constant for a specified time period. Figure 1–17 shows the tracking mode and the hold mode.

Note that, practically speaking, switching from the tracking mode to the hold mode is not instantaneous. If the hold command is given while the circuit is in the tracking mode, then the circuit will stay in the tracking mode for a short while before reacting to the hold command. The time interval during which the switching takes place (that is, the time interval when the measured amplitude is uncertain) is called the *aperture time*.

The output voltage during the hold mode may decrease slightly. The hold mode droop may be reduced by using a high-input-impedance output buffer amplifier. Such an output buffer amplifier must have very low bias current.

The sample-and-hold operation is controlled by a periodic clock.

Figure 1–16 showed a basic sample-and-hold circuit. There are several sample-and-hold circuit configurations commonly in use. Two such configurations are shown

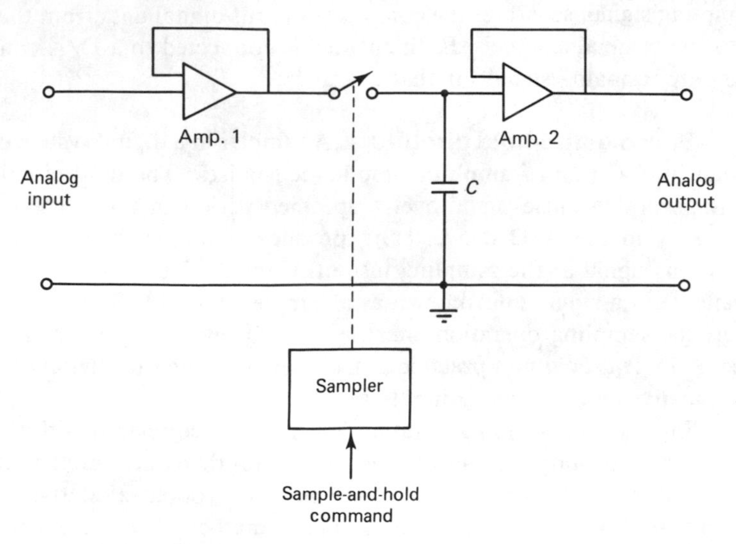

Figure 1–16 Sample-and-hold circuit.

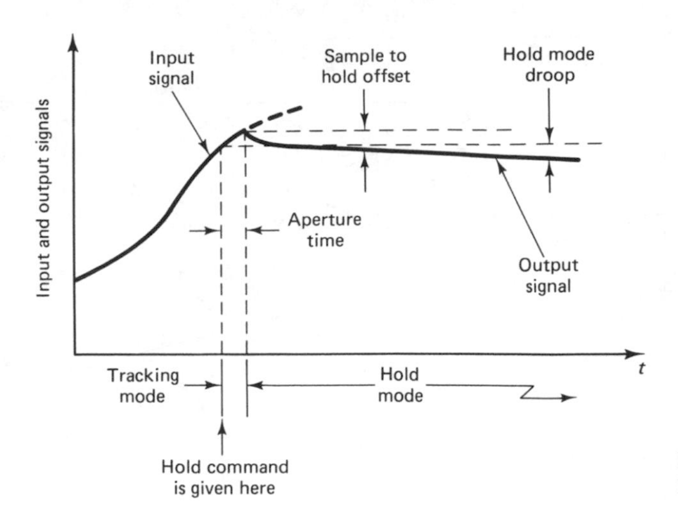

Figure 1–17 Tracking mode and hold mode.

in Fig. 1–18(a) and (b). Figure 1–18(a) shows an inverting sample-and-hold, while Fig. 1–18(b) shows a noninverting one.

Reconstructing the input signal by hold circuits. The sampling operation produces an amplitude-modulated pulse signal. The function of the hold operation is to reconstruct the analog signal that has been transmitted as a train of pulse samples. That is, the purpose of the hold operation is to fill in the spaces between sampling periods and thus roughly reconstruct the original analog input signal.

The hold circuit is designed to extrapolate the output signal between successive points according to some prescribed manner. The staircase waveform of the output shown in Fig. 1–19 is the simplest way to reconstruct the original input signal. The hold circuit that produces such a staircase waveform is called a *zero-order hold*. Because of its simplicity, the zero-order hold is commonly used in digital control systems.

There are available more sophisticated hold circuits than the zero-order hold. These are called higher-order hold circuits and include the first-order hold and the second-order hold. Higher-order hold circuits will generally reconstruct a signal more accurately than a zero-order hold, but with some disadvantages, as explained below.

The first-order hold retains the value of the previous sample as well as the present one and predicts, by extrapolation, the next sample value. This is done by generating an output slope equal to the slope of a line segment connecting previous and present samples and projecting it from the value of the present sample, as shown in Fig. 1–20.

As can easily be seen from the figure, if the slope of the original signal does not change much, the prediction is good. If, however, the original signal reverses

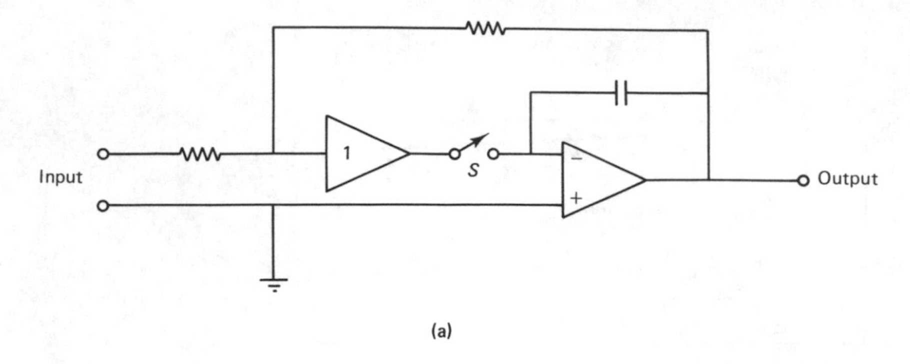

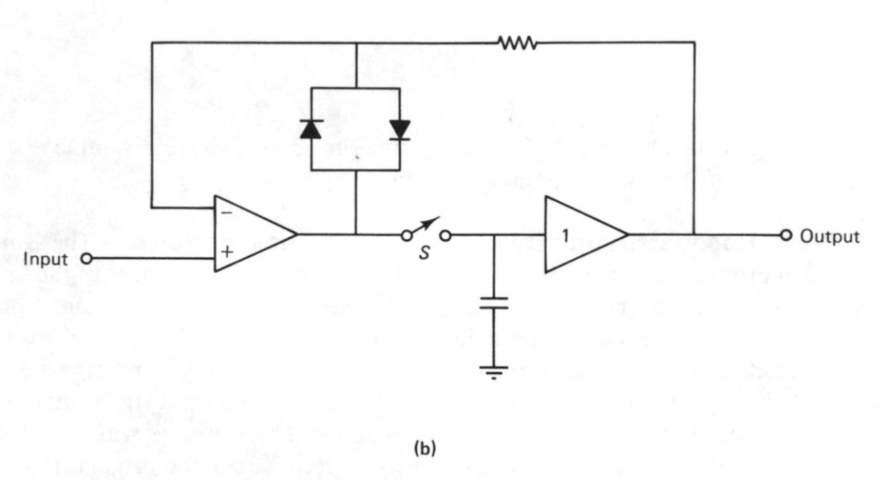

Figure 1–18 (a) Inverting sample-and-hold circuit; (b) noninverting sample-and-hold circuit.

its slope, then the prediction is wrong and the output goes in the wrong direction, thus causing a large error for the sampling period considered.

An interpolative first-order hold, also called a *polygonal* hold, reconstructs the original signal much more accurately. This hold circuit also generates a straight-line output whose slope is equal to that joining the previous sample value and the present sample value, but this time the projection is made from the current sample point with the amplitude of the previous sample. Hence, the accuracy in reconstructing the original signal is better than for other hold circuits, but there is a one-sampling-period delay, as shown in Fig. 1–21. In effect, the better accuracy is achieved at the expense of a delay of one sampling period. From the viewpoint of the stability

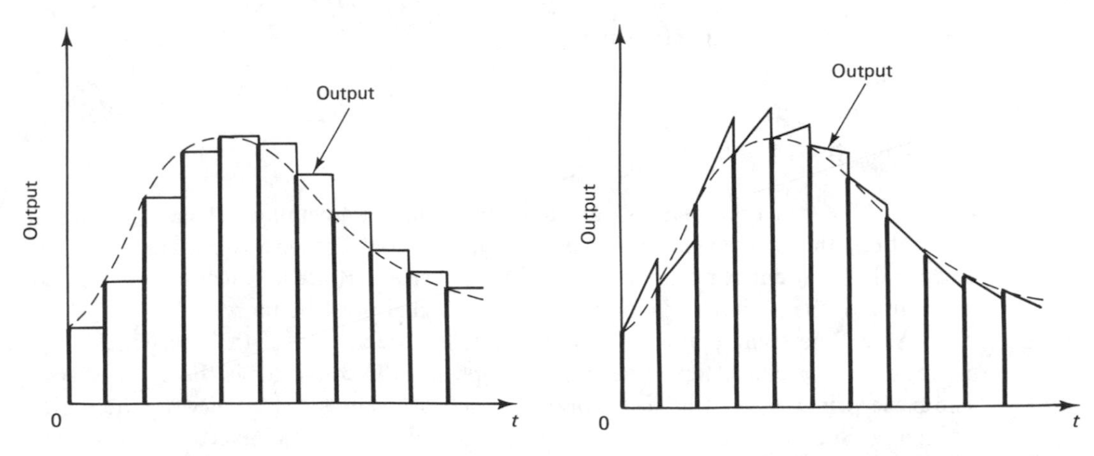

Figure 1–19 Output from a zero-order hold.

Figure 1–20 Output from a first-order hold.

of closed-loop systems, such a delay is not desirable, and so the interpolative first-order hold (polygonal hold) is not used in control system applications.

Types of analog-to-digital (A/D) converters. As stated earlier, the process by which a sampled analog signal is quantized and converted to a binary number is called analog-to-digital conversion. Thus, an analog-to-digital converter transforms an analog signal (usually in the form of a voltage or current) into a digital signal or numerically coded word. In practice, the logic is based on binary digits composed of 0s and 1s and the representation has only a finite number of digits. The analog-to-digital converter performs the operations of sample-and-hold, quantizing, and encoding. Note that in the digital system a pulse is supplied every sampling period T by a clock. The A/D converter sends a digital signal (binary number) to the digital controller each time a pulse arrives.

Among the many A/D circuits available, the following types are used most frequently:

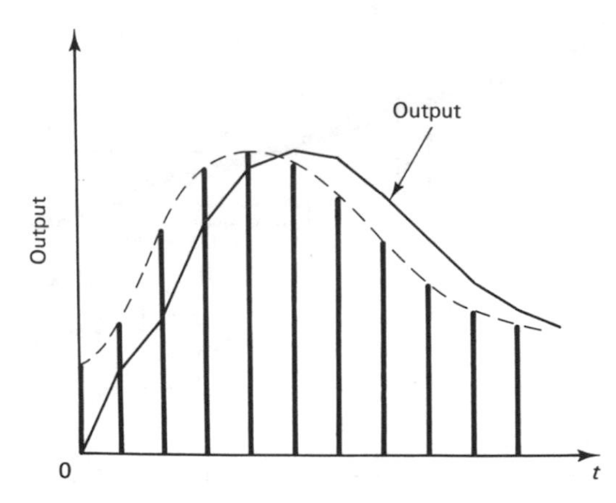

Figure 1–21 Output from an interpolative first-order hold (polygonal hold).

1. Successive-approximation type
2. Integrating type
3. Counter type
4. Parallel type

Each of these four types has its own advantages and disadvantages. In any particular application, the conversion speed, accuracy, size, and cost are the main factors to be considered in choosing the type of A/D converter. (If greater accuracy is needed, for example, the number of bits in the output signal must be increased.)

As will be seen, analog-to-digital converters use as part of their feedback loops digital-to-analog converters. The simplest type of A/D converter is the counter type. The basic principle on which it works is that clock pulses are applied to the digital counter in such a way that the output voltage of the D/A converter (that is part of the feedback loop in the A/D converter) is stepped up one least significant bit (LSB) at a time and the output voltage is compared with the analog input voltage once for each pulse. When the output voltage has reached the magnitude of the input voltage the clock pulses are stopped. The counter output voltage is then the digital output.

The successive-approximation type of A/D converter is much faster than the counter type and is the one most frequently used. Figure 1–22 shows a schematic diagram of the successive-approximation type A/D converter.

The principle of operation of this type of A/D converter is as follows. The successive-approximation register (SAR) first turns on the most significant bit (half the maximum) and compares it with the analog input. The comparator decides whether to leave the bit on or turn it off. If the analog input voltage is larger, the most significant bit is set on. The next step is to turn on bit 2 and then compare the analog input voltage with three-fourths of the maximum. After n comparisons are

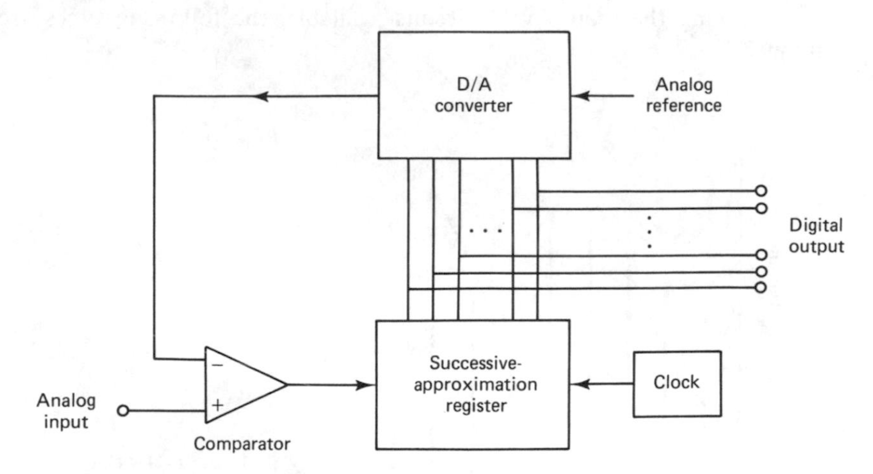

Figure 1–22 Schematic diagram of a successive-approximation type A/D converter.

completed, the digital output of the successive-approximation register indicates all those bits which remain on and produces the desired digital code. Thus, this type of A/D converter sets one bit each clock cycle, and so it requires only n clock cycles to generate n bits, where n is the resolution of the converter in bits. (The number n of bits employed determines the accuracy of conversion.) The time required for the conversion is approximately 2 μsec or less for a 12-bit conversion.

Actual analog-to-digital signal converters differ from the ideal signal converter in that the former always have some errors, such as offset error, linearity error, and gain error, the characteristics of which are shown in Fig. 1–23. Also, it is important to note that the input-output characteristics change with time and temperature.

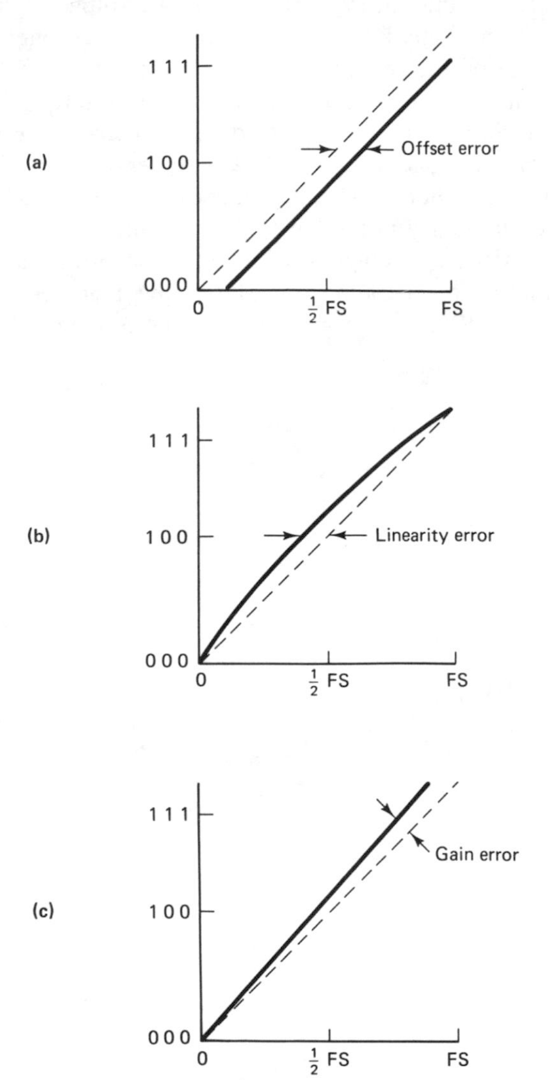

Figure 1–23 Errors in A/D converters: (a) offset error; (b) linearity error; (c) gain error.

Finally, it is noted that commercial converters are specified for three basic temperature ranges: commercial (0°C to 70°C), industrial (−25°C to 85°C), and military (−55°C to 125°C).

Digital-to-analog (D/A) converters. At the output of the digital controller the digital signal must be converted to an analog signal by the process called digital-to-analog conversion. A digital-to-analog converter is a device that transforms a digital input (binary numbers) to an analog output. Figure 1–24 shows the relationship between the digital input and the analog output for a 3-bit D/A converter. Notice that each digital input (in binary numbers) produces a single continuous-time (analog) output value. The output, in most cases, is the voltage signal.

For the full range of the digital input, there are 2^n corresponding different analog values, including 0. For the digital-to-analog conversion there is a one-to-one correspondence between the digital input and the analog output.

Two methods are commonly used for digital-to-analog conversion: the method using weighted resistors, and the one using the R-$2R$ ladder network. The former is simple in circuit configuration, but its accuracy may not be very good. The latter is a little more complicated in configuration but is more accurate.

Figure 1–25 shows a schematic diagram of a D/A converter using weighted resistors. The input resistors of the operational amplifier have their resistance values weighted in a binary fashion. When the logic circuit receives binary 1 the switch

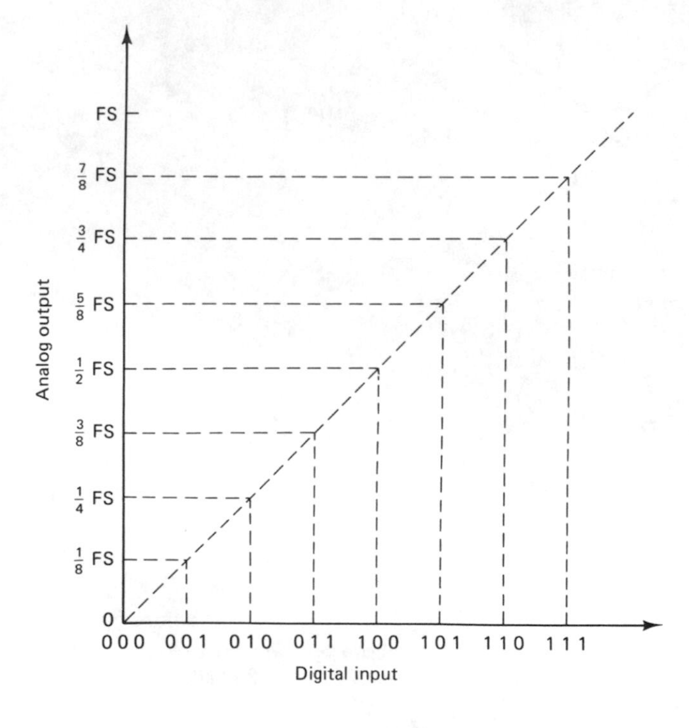

Figure 1–24 Relationship between digital input and analog output for a 3-bit D/A converter.

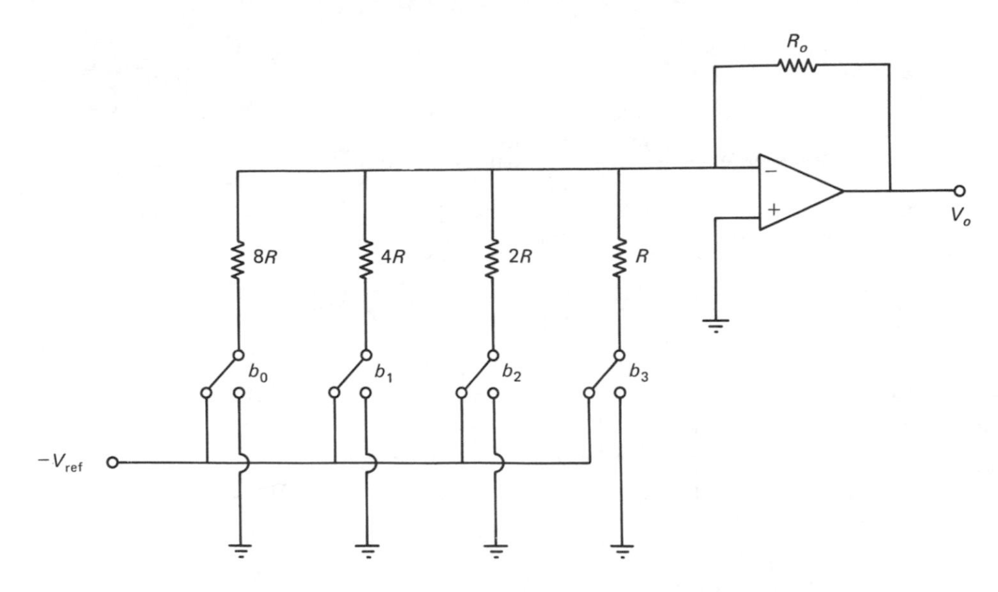

Figure 1–25 Schematic diagram of a D/A converter using weighted resistors.

(actually an electronic gate) connects the resistor to the reference voltage. When the logic circuit receives binary 0, the switch connects the resistor to ground. The digital-to-analog converters used in common practice are of the parallel type: all bits act simultaneously upon application of a digital input (binary numbers).

The D/A converter thus generates the analog output voltage corresponding to the given digital voltage. For the D/A converter shown in Fig. 1–25, if the binary number is $b_3b_2b_1b_0$, where each of the b's can be either a 0 or a 1, then the output is

$$V_o = \frac{R_o}{R}\left(b_3 + \frac{b_2}{2} + \frac{b_1}{4} + \frac{b_0}{8}\right) V_{\text{ref}}$$

Notice that as the number of bits is increased, the range of resistor values becomes large and consequently the accuracy becomes poor.

Figure 1–26 shows a schematic diagram of the 4-bit D/A converter using an R-$2R$ ladder circuit. Note that with the exception of the feedback resistor (which is $3R$) all resistors involved are either R or $2R$. This means that a high level of accuracy can be achieved.

Suppose a binary number $b_3b_2b_1b_0$ is given. If $b_3 = 1$ and $b_0 = b_1 = b_2 = 0$, then the circuit shown in Fig. 1–26 can be simplified and an equivalent circuit can be obtained as shown in Fig. 1–27(a). The output voltage is

$$V_o = 3R\frac{i_3}{2} = \frac{1}{2}V_{\text{ref}}$$

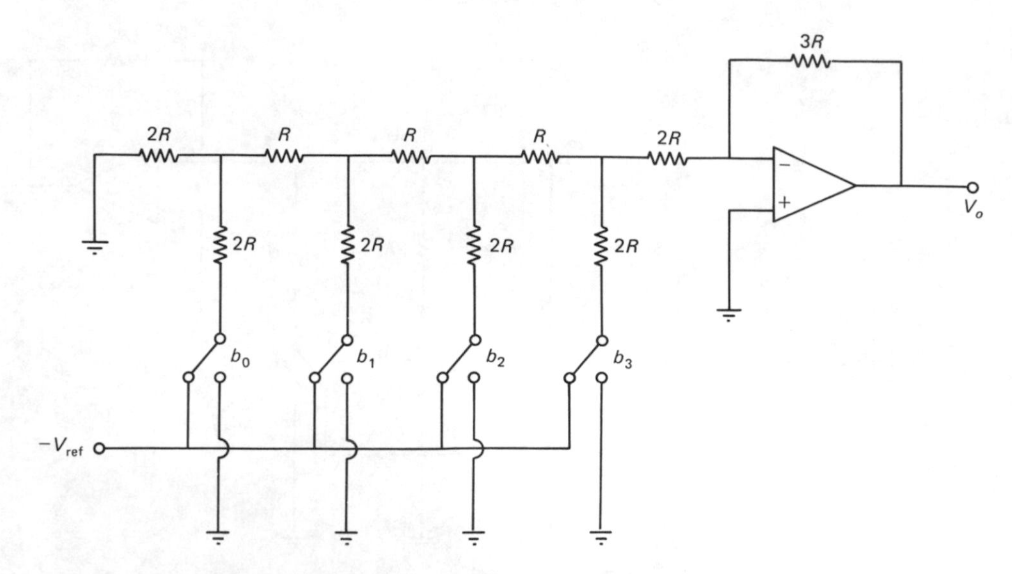

Figure 1–26 4-bit D/A converter using an R-$2R$ ladder circuit.

If $b_2 = 1$ and $b_0 = b_1 = b_3 = 0$, then the equivalent circuit is as shown in Fig. 1–27(b). The output voltage is

$$V_o = 3R\frac{i_2}{4} = \frac{1}{4}V_{\text{ref}}$$

Similarly, if $b_1 = 1$ and $b_0 = b_2 = b_3 = 0$, then the equivalent circuit of Fig. 1–27(c) can be obtained. The output voltage is

$$V_o = 3R\frac{i_1}{8} = \frac{1}{8}V_{\text{ref}}$$

Finally, the circuit shown in Fig. 1–27(d) corresponds to the case where $b_0 = 1$ and $b_1 = b_2 = b_3 = 0$. The output voltage is

$$V_o = 3R\frac{i_0}{16} = \frac{1}{16}V_{\text{ref}}$$

In this way we find that when the input data is $b_3b_2b_1b_0$ (where the b_i's are either 0 or 1), then the output voltage is

$$\begin{aligned} V_o &= (\tfrac{1}{2}b_3 + \tfrac{1}{4}b_2 + \tfrac{1}{8}b_1 + \tfrac{1}{16}b_0)V_{\text{ref}} \\ &= \tfrac{1}{2}(b_3 + \tfrac{1}{2}b_2 + \tfrac{1}{4}b_1 + \tfrac{1}{8}b_0)V_{\text{ref}} \end{aligned}$$

Figure 1–28 shows a schematic diagram of an n-bit D/A converter using an R-$2R$ ladder circuit. The output voltage in this case can be given by

$$V_o = \frac{1}{2}\left(b_{n-1} + \frac{1}{2}b_{n-2} + \cdots + \frac{1}{2^{n-1}}b_0\right)V_{\text{ref}}$$

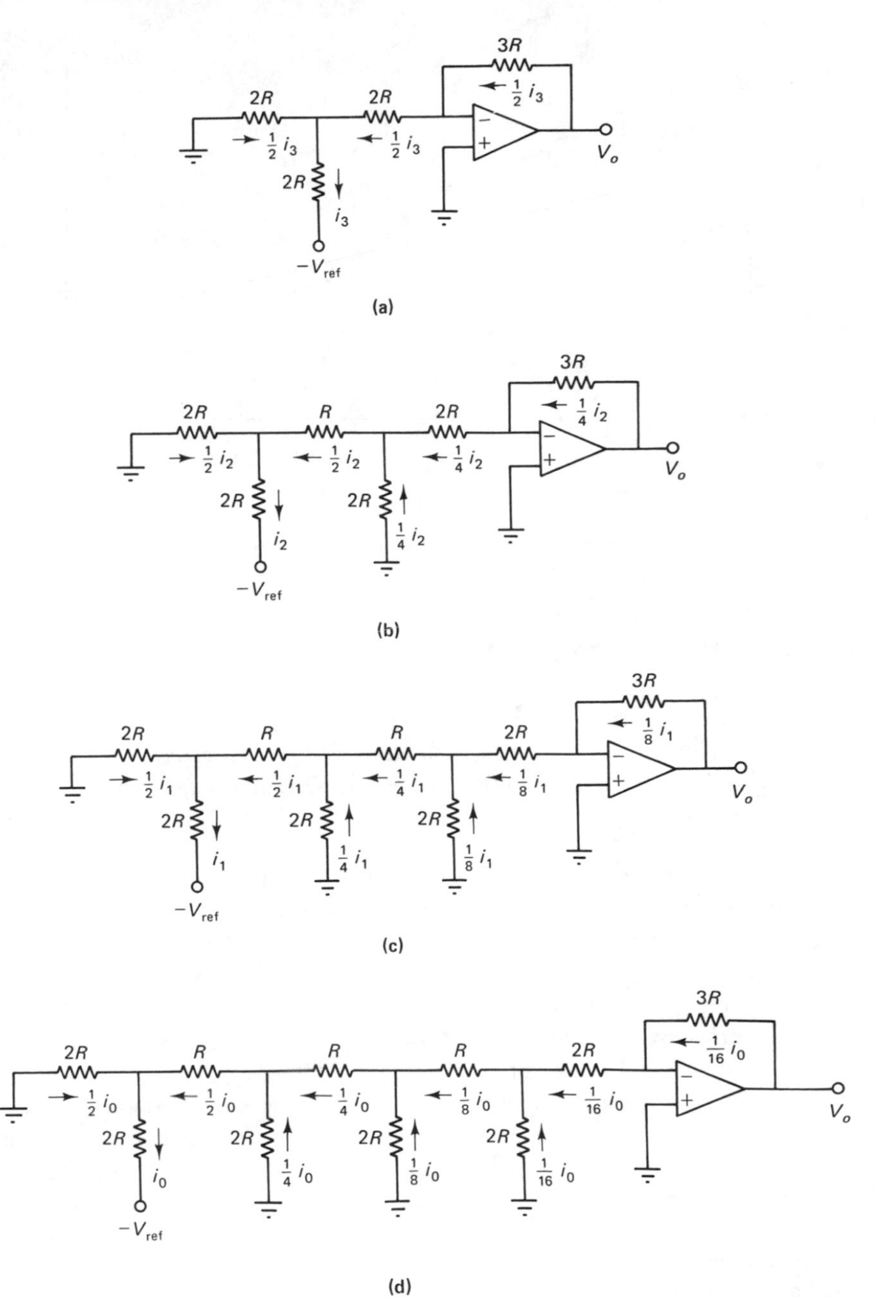

Figure 1–27 (a) Equivalent circuit of the D/A converter shown in Fig. 1–26 when $b_3 = 1$ and $b_0 = b_1 = b_2 = 0$; (b) equivalent circuit when $b_2 = 1$ and $b_0 = b_1 = b_3 = 0$; (c) equivalent circuit when $b_1 = 1$ and $b_0 = b_2 = b_3 = 0$; (d) equivalent circuit when $b_0 = 1$ and $b_1 = b_2 = b_3 = 0$.

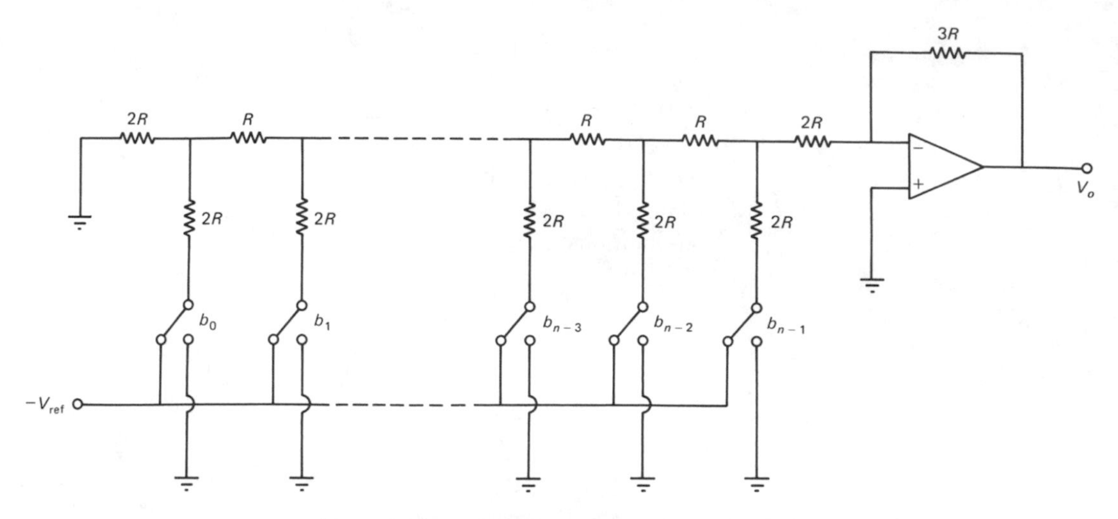

Figure 1–28 n-bit D/A converter using an R-$2R$ ladder circuit.

1–5 CONCLUDING COMMENTS

In concluding this chapter we shall compare digital controllers and analog controllers used in industrial control systems and review digital control of processes. Then we shall discuss first the selection of the sampling frequency, and then frequency folding and aliasing. Finally, we shall present an outline of the rest of the book.

Digital controllers and analog controllers. Digital controllers operate only on numbers. Decision making is one of their important functions. They are often used to solve problems involved in the optimal overall operation of industrial plants.

Digital controllers are extremely versatile. They can handle nonlinear control equations involving complicated computations or logic operations. A very much wider class of control laws can be used in digital controllers than in analog controllers. Also, in the digital controller, by merely issuing a new program the operations being performed can be changed completely. This feature is particularly important if the control system is to receive operating information or instructions from some computing center where economic analysis and optimization studies are made.

Digital controllers are capable of performing complex computations with constant accuracy at high speed and can have almost any desired degree of computational accuracy at relatively little increase in cost.

Originally, digital controllers were used as components only in large-scale control systems. At present, however, thanks to the availability of inexpensive microcomputers, digital controllers are being used in many large- and small-scale control systems. In fact, digital controllers are replacing the analog controllers that have been used in many small-scale control systems. Digital controllers are often superior in performance and lower in price than their analog counterparts.

Analog controllers represent the variables in an equation by continuous physical quantities. They can easily be designed to serve satisfactorily as non-decision-making controllers. But the cost of analog computers or analog controllers increases rapidly as the complexity of the computations increases, if constant accuracy is to be maintained.

There are additional advantages of digital controllers over analog controllers. Digital components, such as sample-and-hold circuits, A/D and D/A converters, and digital transducers, are rugged in construction, are highly reliable, and are often compact and light in weight. Moreover, digital components have high sensitivity, are often cheaper than their analog counterparts, and are less sensitive to noise signals. And, as mentioned earlier, digital controllers are flexible in allowing programming changes.

Digital control of processes. In industrial process control systems, it is generally not practical to operate for a very long time at steady state, because certain changes may occur in production requirements, raw materials, economic factors, and processing equipments and techniques. Thus the transient behavior of industrial processes must always be taken into consideration. Since there are interactions among process variables, using only one process variable for each control agent is not suitable for really complete control. By the use of a digital controller, it is possible to take into account all process variables, together with economic factors, production requirements, equipment performance, and all other needs, and thereby to accomplish optimal control of industrial processes.

Note that a system capable of controlling a process as completely as possible will have to solve complex equations. The more complete the control, the more important it is that the correct relations between operating variables be known and used. The system must be capable of accepting instructions from such varied sources as computers and human operators and must also be capable of changing its control subsystem completely in a short time. Digital controllers are most suitable in such situations. In fact, an advantage of the digital controller is flexibility, that is, ease of changing control schemes by reprogramming.

In the digital control of a complex process, the designer must have a good knowledge of the process to be controlled and must be able to obtain its mathematical model. (The mathematical model may be obtained in terms of differential equations or difference equations, or in some other form.) The designer must be familiar with the measurement technology associated with the output of the process, and other variables involved in the process. He or she must have a good working knowledge of digital computers as well as modern control theory. If the process is complicated, the designer must investigate several different approaches to the design of the control system. In this respect, a good knowledge of simulation techniques is helpful.

Selection of sampling frequency. It is necessary that the sample-and-hold operate at a sufficiently fast rate that the information contained in the input signal will not be lost through the limitations of the sample-and-hold operation. Therefore the designer of a digital control system should ask whether or not some information might be lost in the sampling process. An answer to this question is offered by the sampling theorem, which states, in effect, that if a continuous-time signal is band-limited or contains no frequency components higher than ω_c, then theoretically the original signal can be reconstructed without distortion if it is sampled at a rate of at least $2\omega_c$ radians per second. (We shall discuss details of the sampling theorem in Sec. 3–4. Note that the sampling frequency affects the stability of a closed-loop digital control system. See Sec. 3–8.) In practice, the sampling frequency is chosen to be much higher than $2\omega_c$, the minimum sampling rate.

As a matter of fact, a proper selection of the sampling frequency for a discrete-time control system is very important. The cost involved will be less if the sampling frequency is low, because with a slow sampling frequency, the computations involved can be handled by a smaller and less expensive computer. However, although slower sampling frequency is less expensive, it generally degrades the system performance. The ability of the control system to follow the input depends greatly on the sampling frequency. (The sampling frequency must be 8 to 10 times the highest-frequency component involved in the input signal.) A compromise is thus necessary between the cost involved and the system performance required. The best selection of the sampling frequency would be the slowest (least expensive) one that meets the performance requirements.

In practical situations, the sampling frequency is selected on the basis of the required closed-loop bandwidth in the frequency response or the required rise time and settling time in the transient response. A rule of thumb is to sample 8 to 10 times during a cycle of damped oscillation in the output if the system is underdamped, or to sample 8 to 10 times during the rise time in the transient response if the system is overdamped. Also, the sampling frequency may be chosen as 8 to 10 times the closed-loop bandwidth in the frequency response. (In some cases 5 samples per rise time may be acceptable, and in other cases we may require the sampling frequency to be 20 times the closed-loop bandwidth. Actual selection of the sampling frequency depends on the particular situation.)

Frequency folding and aliasing. Also in Sec. 3–4 we shall see that if the sampling frequency ω_s is not high enough, a part of the frequency spectrum of the continuous-time signal will "fold over" into the original spectrum. This phenomenon is called *frequency folding*. In the reconstruction of the original signal, the folded part of the spectrum causes distortion in the reconstructed signal. Such distortion cannot be eliminated by filtering the reconstructed signal.

In order to avoid frequency folding, one must either use a sufficiently high sampling frequency or prefilter the continuous-time signal before sampling to limit its bandwidth.

We shall also see in Sec. 3–4 that if the continuous-time signal sampled is a sinusoidal signal with a frequency higher than half the sampling frequency, then the sampled signal will exhibit an *alias frequency*. This is a frequency that may be significantly different from the original signal frequency. If aliasing occurs, then the original frequency may be mistaken for an entirely different frequency upon recovery.

Finally, it is noted that in processing a continuous-time signal we frequently prefer to represent it as a series of digital data (binary numbers) and process such data by a digital computer. This is because if the sampling frequency is fast enough to avoid aliasing, then no information is lost in the sampling operation. Also, if we choose a sufficient number of bits to represent each binary number, then the quantization error can be reduced to a negligibly small value. Thus, processing a continuous-time signal by a digital equivalent is very useful and is very common in practice.

Outline of the book. The objective of this book is to present a detailed account of the control theory that is relevant to the analysis and design of discrete-time control systems. Our emphasis is on understanding the basic concepts involved. In this book digital controllers are designed in the form of pulse transfer functions or equivalent difference equations, which can be easily implemented in the form of computer programs.

The outline of the book is as follows. Chapter 1 has presented introductory materials. Chapter 2 presents the z transform theory. This chapter includes z transforms of elementary functions, important properties and theorems of the z transform, the inverse z transform, and an introduction to the pulse transfer function and weighting sequence. Chapter 3 treats background materials for the z domain analysis of control systems. This chapter includes discussions of impulse sampling and reconstruction of original signals from sampled signals, detailed discussions of pulse transfer functions, a brief discussion of the modified z transform, and a discussion of digital controllers and digital filters. Mapping between the s plane and the z plane is discussed next. The chapter concludes with stability analysis in the z domain.

Chapter 4 deals with design techniques for discrete-time control systems via transform methods. Specifically, the chapter discusses the design technique using discrete equivalents of analog controllers, transient and steady-state response analyses, design by the root locus and frequency response methods, and finally an analytical design method.

The final three chapters present state space analysis and design of discrete-time control systems. Specifically, Chap. 5 is on the state space analysis of control systems. State space representations of discrete-time systems are obtained first, and then the solution of discrete-time state space equations is derived. The pulse-transfer-function matrix is presented next. Then continuous-time state space equations are discussed, followed by a discretization technique. Finally, Liapunov stability analysis is discussed.

Chapter 6 presents control system design in the state space. We begin the chapter with a detailed presentation of controllability and observability. We then present design techniques based on pole placement, followed by discussions of full-order state observers and minimum-order state observers. We conclude this chapter with servo systems.

The final chapter, Chap. 7, treats optimal control problems and related subjects. We begin the chapter with quadratic optimal control problems, followed by identification problems based on the least-squares method. We conclude the chapter with a discussion of Kalman filters.

The state space analysis and design of discrete-time control systems, presented in Chaps. 5, 6, and 7, make extensive use of vectors and matrices. In studying these chapters the reader may, as need arises, refer to the Appendix, which summarizes the basic materials of vector-matrix analysis.

Finally, in each chapter except Chap. 1, the main text is followed by solved problems and unsolved problems. The reader should study all solved problems carefully. Solved problems are an integral part of the text. The Appendix is also followed by solved problems. The reader who studies these solved problems will have an increased understanding of the materials presented.

REFERENCES

1–1. Cadzow, J. A., and H. R. Martens, *Discrete-Time and Computer Control Systems*. Englewood Cliffs, N.J.: Prentice-Hall, Inc., 1970.

1–2. Franklin, G. F., and J. D. Powell, *Digital Control of Dynamic Systems*. Reading, Mass.: Addison-Wesley Publishing Co., Inc., 1980.

1–3. Jury, E. I., *Sampled-Data Control Systems*. New York: John Wiley & Sons, Inc., 1958.

1–4. Jury, E. I., "Sampled-Data Systems, Revisited: Reflections, Recollections, and Reassessments," *ASME J. Dynamic Systems, Measurement, and Control*, **102** (1980), pp. 208–16.

1–5. Jury, E. I., "Sampling Schemes in Sampled-Data Control Systems," *IRE Trans. Automatic Control*, **AC-6** (1961), pp. 88–90.

1–6. Katz, P., *Digital Control Using Microprocessors*. London: Prentice-Hall International, Inc., 1981.

1–7. Kuo, B. C., *Digital Control Systems*. New York: Holt, Rinehart and Winston, Inc., 1980.

1–8. Phillips, C. L., and H. T. Nagle, Jr., *Digital Control System Analysis and Design*. Englewood Cliffs, N.J.: Prentice-Hall, Inc., 1984.

1–9. Ragazzini, J. R., and G. F. Franklin, *Sampled-Data Control Systems*. New York: McGraw-Hill Book Company, 1958.

1–10. Tou, J. T., *Digital and Sampled-Data Control Systems*. New York: McGraw-Hill Book Company, 1959.

2

The z Transform

2-1 INTRODUCTION

One of the mathematical tools commonly used for the analysis and synthesis of discrete-time control systems is the z transform. The role of the z transform in discrete-time systems is similar to that of the Laplace transform in continuous-time systems.

In a linear discrete-time control system a linear difference equation characterizes the dynamics of the system. In order to determine the system's response to a given input, such a difference equation must be solved. With the z transform method, the solutions to linear difference equations become algebraic in nature. (Just as the Laplace transformation transforms linear time-invariant differential equations into algebraic equations in s, the z transformation transforms linear time-invariant difference equations into algebraic equations in z.)

The main objective of this chapter is to present definitions of the z transform, basic theorems associated with the z transform, and methods for finding the inverse z transform. The pulse transfer function and the weighting sequence are also discussed in this chapter.

Discrete-time signals. Discrete-time signals arise if the system involves a sampling operation of continuous-time signals, or if the system involves an iterative process carried out by a digital computer.

The sequence of values (or numbers) is usually written as $x(k)$, where the argument k indicates the order in which the value (or number) occurs in the sequence: for example, $x(0)$, $x(1)$, $x(2)$, ...

If a continuous-time signal is sampled, the sampling period T becomes an important parameter. The sequence of values arising from the sampling operation is usually written as $x(kT)$. However, to simplify the notation we occasionally drop the explicit appearance of T and write $x(kT)$ as $x(k)$.

By use of the z transformation a linear discrete-time system may be represented by a transfer function called the *pulse transfer function*. Then, the z transform of the output signal may be expressed as the product of the system's pulse transfer function and the z transform of the input signal.

Only introductory material on the pulse transfer function will be discussed in this chapter. More detail will appear in Chap. 4.

Outline of the chapter. Section 2–1 has presented introductory remarks. Section 2–2 presents the definition of the z transform and associated subjects. Section 2–3 gives z transforms of elementary functions. Important properties and theorems of the z transform are presented in Sec. 2–4. Both analytical and computational methods for finding the inverse z transform are discussed in Sec. 2–5. Section 2–6 defines the pulse transfer function and the weighting sequence and then gives the solution of difference equations by the z transform method and the computational method. (BASIC computer programs for solving difference equations are presented.) Finally, Sec. 2–7 gives a summary of the chapter.

2–2 THE z TRANSFORM

The z transform method is an operational method which is very powerful in working with discrete-time systems. In what follows we shall define the z transform of a time function.

The z transform of a time function $x(t)$ where t is nonnegative, or of a sequence of values $x(kT)$ or $x(k)$ where k takes zero or positive integers and T is the sampling period, is defined by the following equation:

$$X(z) = \mathcal{Z}[x(t)] = \mathcal{Z}[x(kT)] = \mathcal{Z}[x(k)]$$

$$= \sum_{k=0}^{\infty} x(kT)z^{-k} = \sum_{k=0}^{\infty} x(k)z^{-k} \tag{2–1}$$

This z transform is referred to as the *one-sided z transform*. The symbol $\mathcal{Z}$ denotes "the z transform of." In the one-sided z transform we assume $x(t) = 0$ for $t < 0$ or $x(kT) = x(k) = 0$ for $k < 0$. Note that z is a complex variable.

The z transform of $x(t)$ where $-\infty < t < \infty$, or of $x(kT)$ or $x(k)$ where k takes integer values ($k = 0, \pm 1, \pm 2, \cdots$), is defined by the following equation:

$$X(z) = \mathcal{Z}[x(t)] = \mathcal{Z}[x(kT)] = \mathcal{Z}[x(k)]$$

$$= \sum_{k=-\infty}^{\infty} x(kT)z^{-k} = \sum_{k=-\infty}^{\infty} x(k)z^{-k}$$

This z transform is referred to as the *two-sided z transform*. In the two-sided z transform, the time function $x(t)$ is assumed to be nonzero for $t < 0$ and the time sequence $x(kT)$ or $x(k)$ is considered to have nonzero values for $k < 0$. Both the one-sided and two-sided z transforms are series in powers of z^{-1}. (The latter involves both positive and negative powers of z^{-1}.) In this book, only the one-sided z transform is considered in detail.

It is noted that since the z transform $X(z)$ is a function of z^{-1}, in the literature the notation $X(z^{-1})$ is sometimes used instead of the notation $X(z)$. In this book, however, in order to avoid any confusion in dealing with z transform theorems, we shall use the notation $X(z)$ for the z transform of $x(t)$, $x(kT)$, or $x(k)$.

For most engineering applications the one-sided z transform will have a convenient closed-form solution in its region of convergence. Note that whenever $X(z)$, an infinite series in z^{-1}, converges outside the circle $|z| = R$, where R is called the *radius of absolute convergence*, in using the z transform method for solving discrete-time problems it is not necessary each time to specify the values of z over which $X(z)$ is convergent.

Notice that expansion of the right-hand side of Eq. (2–1) gives

$$X(z) = x(0) + x(T)z^{-1} + x(2T)z^{-2} + \cdots + x(kT)z^{-k} + \cdots$$

This last equation implies that the z transform of any continuous-time function $x(t)$ may be written in the series form by inspection. The z^{-k} in this series indicates the position in time at which the amplitude $x(kT)$ occurs. Conversely, if $X(z)$ is given in the series form as above, the inverse z transform can be obtained by inspection as a sequence of the function $x(kT)$ that corresponds to the values of $x(t)$ at the respective instants of time.

If the z transform is given as a ratio of two polynomials in z, then the inverse z transform may be obtained by several different methods, such as the direct division method, the computational method, the partial-fraction-expansion method, and the inversion integral method (see Sec. 2–5 for details). Note that the partial-fraction-expansion method applicable to the z transform is similar to that applicable to the Laplace transform. The inversion integral for the z transform $X(z)$ is given by

$$\mathcal{Z}^{-1}[X(z)] = x(kT) = x(k) = \frac{1}{2\pi j} \oint_C X(z)z^{k-1}\,dz \tag{2–2}$$

where C is a circle with its center at the origin of the z plane such that all poles of $X(z)z^{k-1}$ are inside it. [For the derivation of Eq. (2–2), see Sec. 2–5.]

Poles and zeros in the *z* plane. In engineering applications of the z transform method, $X(z)$ may have the form

$$X(z) = \frac{b_0 z^m + b_1 z^{m-1} + \cdots + b_m}{z^n + a_1 z^{n-1} + \cdots + a_n} \tag{2–3}$$

or

$$X(z)=\frac{b_0(z-z_1)(z-z_2)\cdots(z-z_m)}{(z-p_1)(z-p_2)\cdots(z-p_n)}$$

where the p_i's are the poles of $X(z)$ and the z_i's the zeros of $X(z)$.

The locations of the poles and zeros of $X(z)$ determine the characteristics of $x(k)$, the sequence of values or numbers. As in the case of the s plane analysis of linear continuous-time control systems, we often use a graphical display in the z plane of the locations of the poles and zeros of $X(z)$. (For details, refer to Sec. 3–8.)

Note that in control engineering and signal processing, $X(z)$ is frequently expressed as a ratio of polynomials in z^{-1}, as follows:

$$X(z)=\frac{b_0z^{-(n-m)}+b_1z^{-(n-m+1)}+\cdots+b_mz^{-n}}{1+a_1z^{-1}+a_2z^{-2}+\cdots+a_nz^{-n}} \tag{2–4}$$

where z^{-1} is interpreted as the unit delay operator. In this chapter, where the basic properties and theorems of the z transform method are presented, $X(z)$ may be expressed in terms of powers of z, as given by Eq. (2–3), or in terms of powers of z^{-1}, as given by Eq. (2–4), depending on the circumstances.

In finding the poles and zeros of $X(z)$, it is convenient to express $X(z)$ as a ratio of polynomials in z. For example,

$$X(z)=\frac{z^2+0.5z}{z^2+3z+2}=\frac{z(z+0.5)}{(z+1)(z+2)}$$

Clearly, $X(z)$ has poles at $z=-1$ and $z=-2$ and zeros at $z=0$ and $z=-0.5$. If $X(z)$ is written as a ratio of polynomials in z^{-1}, however, the preceding $X(z)$ can be written as

$$X(z)=\frac{1+0.5z^{-1}}{1+3z^{-1}+2z^{-2}}=\frac{1+0.5z^{-1}}{(1+z^{-1})(1+2z^{-1})}$$

Although poles at $z=-1$ and $z=-2$ and a zero at $z=-0.5$ are clearly seen from the expression, a zero at $z=0$ is not explicitly shown, and so the beginner may fail to see the existence of a zero at $z=0$. Therefore, in dealing with the poles and zeros of $X(z)$, it is preferable to express $X(z)$ as a ratio of polynomials in z, rather than polynomials in z^{-1}. In addition, in obtaining the inverse z transform by use of the inversion integral method, it is desirable to express $X(z)$ as a ratio of polynomials in z, rather than polynomials in z^{-1}, in order to avoid any possible errors in determining the number of poles at the origin of function $X(z)z^{k-1}$.

2–3 *z* TRANSFORMS OF ELEMENTARY FUNCTIONS

In the following we shall present z transforms of several elementary functions. It is noted that in one-sided z transform theory, in sampling a discontinuous function $x(t)$ we assume that the function is continuous from the right; that is, if discontinuity occurs at $t = 0$, then we assume that $x(0)$ is equal to $x(0+)$ rather than to the average at the discontinuity, $[x(0-) + x(0+)]/2$.

Unit step function. Let us find the z transform of the unit step function

$$x(t) = \begin{cases} 1(t) & 0 \le t \\ 0 & t < 0 \end{cases}$$

As was just noted, in sampling a unit step function we assume that this function is continuous from the right, that is, that $1(0) = 1$. Then, referring to Eq. (2–1), we have

$$\begin{aligned} X(z) &= \mathcal{Z}[1(t)] = \sum_{k=0}^{\infty} 1z^{-k} = \sum_{k=0}^{\infty} z^{-k} \\ &= 1 + z^{-1} + z^{-2} + z^{-3} + \cdots \\ &= \frac{1}{1 - z^{-1}} = \frac{z}{z-1} \end{aligned} \tag{2–5}$$

Notice that the series converges if $|z| > 1$. In finding the z transform, the variable z acts as a dummy operator. It is not necessary to specify the region of z over which $X(z)$ is convergent. It suffices to know that such a region exists. The z transform $X(z)$ of a time function $x(t)$ obtained in this way is valid throughout the z plane except at poles of $X(z)$.

It is noted that $1(k)$ as defined by

$$1(k) = \begin{cases} 1 & k = 0, 1, 2, \ldots \\ 0 & k < 0 \end{cases}$$

is commonly called a *unit step sequence*.

Unit ramp function. Consider the unit ramp function

$$x(t) = \begin{cases} t & 0 \le t \\ 0 & t < 0 \end{cases}$$

Notice that

$$x(kT) = kT \qquad k = 0, 1, 2, \ldots$$

The z transform of the unit ramp function, therefore, can be written as

$$\begin{aligned} X(z) &= \mathcal{Z}[t] = \sum_{k=0}^{\infty} x(kT)z^{-k} = \sum_{k=0}^{\infty} kTz^{-k} = T\sum_{k=0}^{\infty} kz^{-k} \\ &= T(z^{-1} + 2z^{-2} + 3z^{-3} + \cdots) \\ &= T\frac{z^{-1}}{(1 - z^{-1})^2} \\ &= \frac{Tz}{(z-1)^2} \end{aligned}$$

Polynomial function a^k. Let us obtain the z transform of $x(k)$ as defined by

$$x(k)=\begin{cases} a^k & k=0,1,2,\dots \\ 0 & k<0 \end{cases}$$

where a is a constant. Referring to the definition of the z transform, we obtain

$$\begin{aligned} X(z) &= \mathscr{Z}[a^k] = \sum_{k=0}^{\infty} x(k)z^{-k} = \sum_{k=0}^{\infty} a^k z^{-k} \\ &= 1 + az^{-1} + a^2z^{-2} + a^3z^{-3} + \cdots \\ &= \frac{1}{1-az^{-1}} \\ &= \frac{z}{z-a} \end{aligned}$$

This geometric series converges if $|z| > |a|$. The region where this holds is the region outside a circle of radius $|a|$ centered at the origin in the complex z plane. As mentioned earlier, in using the z transform, it suffices to know that such a region exists.

Exponential function. Let us find the z transform of

$$x(t)=\begin{cases} e^{-at} & 0 \le t \\ 0 & t<0 \end{cases}$$

Since

$$x(kT) = e^{-akT} \qquad k=0,1,2,\dots$$

we have

$$\begin{aligned} X(z) &= \mathscr{Z}[e^{-at}] = \sum_{k=0}^{\infty} x(kT)z^{-k} = \sum_{k=0}^{\infty} e^{-akT}z^{-k} \\ &= 1 + e^{-aT}z^{-1} + e^{-2aT}z^{-2} + e^{-3aT}z^{-3} + \cdots \\ &= \frac{1}{1-e^{-aT}z^{-1}} \\ &= \frac{z}{z-e^{-aT}} \end{aligned}$$

Sinusoidal function. Consider the sinusoidal function

$$x(t)=\begin{cases} \sin \omega t & 0 \le t \\ 0 & t<0 \end{cases}$$

Noting that the z transform of the exponential function is

$$\mathcal{Z}[e^{-at}] = \frac{1}{1 - e^{-aT}z^{-1}}$$

and that $\sin \omega t$ can be written as

$$\sin \omega t = \frac{1}{2j}(e^{j\omega t} - e^{-j\omega t})$$

we have

$$X(z) = \mathcal{Z}[\sin \omega t] = \frac{1}{2j}\left(\frac{1}{1 - e^{j\omega T}z^{-1}} - \frac{1}{1 - e^{-j\omega T}z^{-1}}\right)$$

$$= \frac{1}{2j}\frac{(e^{j\omega T} - e^{-j\omega T})z^{-1}}{1 - (e^{j\omega T} + e^{-j\omega T})z^{-1} + z^{-2}}$$

$$= \frac{z^{-1}\sin \omega T}{1 - 2z^{-1}\cos \omega T + z^{-2}}$$

$$= \frac{z \sin \omega T}{z^2 - 2z \cos \omega T + 1}$$

Example 2–1.

Obtain the z transform of the cosine function

$$x(t) = \begin{cases} \cos \omega t & 0 \leq t \\ 0 & t < 0 \end{cases}$$

If we proceed in a manner similar to the way we treated the z transform of the sine function, we have

$$X(z) = \mathcal{Z}[\cos \omega t] = \tfrac{1}{2}\mathcal{Z}[e^{j\omega t} + e^{-j\omega t}]$$

$$= \frac{1}{2}\left(\frac{1}{1 - e^{j\omega T}z^{-1}} + \frac{1}{1 - e^{-j\omega T}z^{-1}}\right)$$

$$= \frac{1}{2}\frac{2 - (e^{-j\omega T} + e^{j\omega T})z^{-1}}{1 - (e^{j\omega T} + e^{-j\omega T})z^{-1} + z^{-2}}$$

$$= \frac{1 - z^{-1}\cos \omega T}{1 - 2z^{-1}\cos \omega T + z^{-2}}$$

$$= \frac{z^2 - z \cos \omega T}{z^2 - 2z \cos \omega T + 1}$$

Example 2–2.

Obtain z transforms of a damped sine function and a damped cosine function.

For the damped sine function defined by

$$x(t) = \begin{cases} e^{-at}\sin \omega t & 0 \leq t \\ 0 & t < 0 \end{cases}$$

we have

$$X(z) = \mathcal{Z}[e^{-at} \sin \omega t] = \frac{1}{2j} \mathcal{Z}[e^{-at}e^{j\omega t} - e^{-at}e^{-j\omega t}]$$

$$= \frac{1}{2j}\left[\frac{1}{1 - e^{-(a-j\omega)T}z^{-1}} - \frac{1}{1 - e^{-(a+j\omega)T}z^{-1}}\right]$$

$$= \frac{1}{2j} \frac{(e^{j\omega T} - e^{-j\omega T})e^{-aT}z^{-1}}{1 - (e^{j\omega T} + e^{-j\omega T})e^{-aT}z^{-1} + e^{-2aT}z^{-2}}$$

$$= \frac{e^{-aT}z^{-1} \sin \omega T}{1 - 2e^{-aT}z^{-1} \cos \omega T + e^{-2aT}z^{-2}}$$

$$= \frac{e^{-aT}z \sin \omega T}{z^2 - 2e^{-aT}z \cos \omega T + e^{-2aT}}$$

Similarly, for the damped cosine function defined by

$$x(t) = \begin{cases} e^{-at} \cos \omega t & 0 \le t \\ 0 & t < 0 \end{cases}$$

we have

$$X(z) = \mathcal{Z}[e^{-at} \cos \omega t] = \tfrac{1}{2} \mathcal{Z}[e^{-at}e^{j\omega t} + e^{-at}e^{-j\omega t}]$$

$$= \frac{1}{2}\left[\frac{1}{1 - e^{-(a-j\omega)T}z^{-1}} + \frac{1}{1 - e^{-(a+j\omega)T}z^{-1}}\right]$$

$$= \frac{1}{2} \frac{2 - (e^{-j\omega T} + e^{j\omega T})e^{-aT}z^{-1}}{1 - (e^{j\omega T} + e^{-j\omega T})e^{-aT}z^{-1} + e^{-2aT}z^{-2}}$$

$$= \frac{1 - e^{-aT}z^{-1} \cos \omega T}{1 - 2e^{-aT}z^{-1} \cos \omega T + e^{-2aT}z^{-2}}$$

$$= \frac{z^2 - e^{-aT}z \cos \omega T}{z^2 - 2e^{-aT}z \cos \omega T + e^{-2aT}}$$

Example 2–3.

Obtain the z transform of

$$X(s) = \frac{1}{s(s+1)}$$

Whenever a function in s is given, one approach for finding the corresponding z transform is to convert $X(s)$ into $x(t)$ and then find the z transform of $x(t)$. Another approach is to expand $X(s)$ into partial fractions and use a z transform table to find the z transforms of the expanded terms. Still other approaches will be discussed in Sec. 3–3.

Let us convert $X(s)$ into $x(t)$. The inverse Laplace transform of $X(s)$ is

$$x(t) = 1 - e^{-t} \qquad 0 \le t$$

Hence,

$$X(z) = \mathcal{Z}[1 - e^{-t}] = \frac{1}{1 - z^{-1}} - \frac{1}{1 - e^{-T}z^{-1}}$$

$$= \frac{(1 - e^{-T})z^{-1}}{(1 - z^{-1})(1 - e^{-T}z^{-1})}$$

$$= \frac{(1 - e^{-T})z}{(z - 1)(z - e^{-T})}$$

Comments. In this section we have presented several examples showing how to obtain the z transform of the time function $x(t)$ by directly applying the definition of the one-sided z transform. It should be noted here that two additional methods are commonly available for obtaining the z transform; they will be discussed in Sec. 3–3.

Just as in working with the Laplace transformation, a table of z transforms of commonly encountered functions is very useful for solving problems in the field of discrete-time systems. Table 2–1 is such a table.

TABLE 2–1 TABLE OF z TRANSFORMS

	$X(s)$	$x(t)$	$x(kT)$ or $x(k)$	$X(z)$
1.	—	—	Kronecker delta $\delta_0(k)$ $1 \quad k = 0$ $0 \quad k \neq 0$	1
2.	—	—	$\delta_0(n - k)$ $1 \quad n = k$ $0 \quad n \neq k$	z^{-k}
3.	$\frac{1}{s}$	$1(t)$	$1(k)$	$\frac{1}{1 - z^{-1}}$
4.	$\frac{1}{s + a}$	e^{-at}	e^{-akT}	$\frac{1}{1 - e^{-aT}z^{-1}}$
5.	$\frac{1}{s^2}$	t	kT	$\frac{Tz^{-1}}{(1 - z^{-1})^2}$
6.	$\frac{2}{s^3}$	t^2	$(kT)^2$	$\frac{T^2z^{-1}(1 + z^{-1})}{(1 - z^{-1})^3}$
7.	$\frac{6}{s^4}$	t^3	$(kT)^3$	$\frac{T^3z^{-1}(1 + 4z^{-1} + z^{-2})}{(1 - z^{-1})^4}$
8.	$\frac{a}{s(s + a)}$	$1 - e^{-at}$	$1 - e^{-akT}$	$\frac{(1 - e^{-aT})z^{-1}}{(1 - z^{-1})(1 - e^{-aT}z^{-1})}$
9.	$\frac{b - a}{(s + a)(s + b)}$	$e^{-at} - e^{-bt}$	$e^{-akT} - e^{-bkT}$	$\frac{(e^{-aT} - e^{-bT})z^{-1}}{(1 - e^{-aT}z^{-1})(1 - e^{-bT}z^{-1})}$

TABLE 2–1 *(continued)*

10.	$\frac{1}{(s+a)^2}$	te^{-at}	kTe^{-akT}	$\frac{Te^{-aT}z^{-1}}{(1-e^{-aT}z^{-1})^2}$
11.	$\frac{s}{(s+a)^2}$	$(1-at)e^{-at}$	$(1-akT)e^{-akT}$	$\frac{1-(1+aT)e^{-aT}z^{-1}}{(1-e^{-aT}z^{-1})^2}$
12.	$\frac{2}{(s+a)^3}$	t^2e^{-at}	$(kT)^2e^{-akT}$	$\frac{T^2e^{-aT}(1+e^{-aT}z^{-1})z^{-1}}{(1-e^{-aT}z^{-1})^3}$
13.	$\frac{a^2}{s^2(s+a)}$	$at-1+e^{-at}$	$akT-1+e^{-akT}$	$\frac{[(aT-1+e^{-aT})+(1-e^{-aT}-aTe^{-aT})z^{-1}]z^{-1}}{(1-z^{-1})^2(1-e^{-aT}z^{-1})}$
14.	$\frac{\omega}{s^2+\omega^2}$	$\sin \omega t$	$\sin \omega kT$	$\frac{z^{-1}\sin \omega T}{1-2z^{-1}\cos \omega T+z^{-2}}$
15.	$\frac{s}{s^2+\omega^2}$	$\cos \omega t$	$\cos \omega kT$	$\frac{1-z^{-1}\cos \omega T}{1-2z^{-1}\cos \omega T+z^{-2}}$
16.	$\frac{\omega}{(s+a)^2+\omega^2}$	$e^{-at}\sin \omega t$	$e^{-akT}\sin \omega kT$	$\frac{e^{-aT}z^{-1}\sin \omega T}{1-2e^{-aT}z^{-1}\cos \omega T+e^{-2aT}z^{-2}}$
17.	$\frac{s+a}{(s+a)^2+\omega^2}$	$e^{-at}\cos \omega t$	$e^{-akT}\cos \omega kT$	$\frac{1-e^{-aT}z^{-1}\cos \omega T}{1-2e^{-aT}z^{-1}\cos \omega T+e^{-2aT}z^{-2}}$
18.			a^k	$\frac{1}{1-az^{-1}}$
19.			a^{k-1} $k=1, 2, 3, \ldots$	$\frac{z^{-1}}{1-az^{-1}}$
20.			ka^{k-1}	$\frac{z^{-1}}{(1-az^{-1})^2}$
21.			k^2a^{k-1}	$\frac{z^{-1}(1+az^{-1})}{(1-az^{-1})^3}$
22.			k^3a^{k-1}	$\frac{z^{-1}(1+4az^{-1}+a^2z^{-2})}{(1-az^{-1})^4}$
23.			k^4a^{k-1}	$\frac{z^{-1}(1+11az^{-1}+11a^2z^{-2}+a^3z^{-3})}{(1-az^{-1})^5}$
24.			$a^k \cos k\pi$	$\frac{1}{1+az^{-1}}$

$x(t)=0$ for $t<0$.

$x(kT)=x(k)=0$ for $k<0$.

Unless otherwise noted, $k = 0, 1, 2, 3, \ldots$.

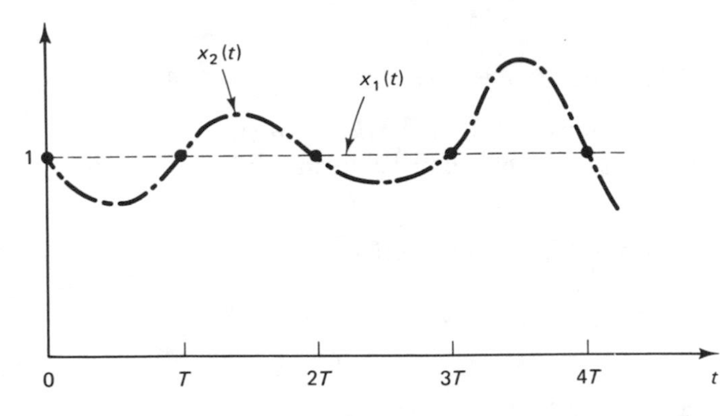

Figure 2–1 Two different continuous-time functions, $x_1(t)$ and $x_2(t)$, that have the same values at $t = 0, T, 2T, \ldots$.

It is important to note that although the z transform $X(z)$ and its inverse z transform $x(k)$ have a one-to-one correspondence, the z transform $X(z)$ and its inverse z transform $x(t)$ do not have a unique correspondence. That is, if the z transform of $x(t)$ is $X(z)$, the inverse z transform is not necessarily equal to $x(t)$. For example, the z transform of a unit step function is $1/(1 - z^{-1})$. [See Eq. (2–5).] The inverse z transform of $1/(1 - z^{-1})$ can also be any time function which has a value of unity at $t = 0, T, 2T, \ldots$ See, for example, Fig. 2–1.

2–4 IMPORTANT PROPERTIES AND THEOREMS OF THE z TRANSFORM

The use of the z transform method in the analysis of discrete-time control systems may be facilitated if theorems of the z transform are referred to. In this section we present important properties and useful theorems of the z transform. We assume that the time function $x(t)$ is z-transformable and that $x(t)$ is zero for $t < 0$.

Multiplication by a constant. If $X(z)$ is the z transform of $x(t)$, then

$$\mathcal{Z}[ax(t)] = a\,\mathcal{Z}[x(t)] = aX(z)$$

where a is a constant.

To prove this, note that by definition

$$\mathcal{Z}[ax(t)] = \sum_{k=0}^{\infty} ax(kT)z^{-k} = a\sum_{k=0}^{\infty} x(kT)z^{-k} = aX(z)$$

Linearity of the z transform. The z transform possesses an important property: linearity. This means that if $f(k)$ and $g(k)$ are z-transformable and α and β are scalars, then $x(k)$ formed by a linear combination

$$x(k) = \alpha f(k) + \beta g(k)$$

has the z transform

$$X(z) = \alpha F(z) + \beta G(z)$$

where $F(z)$ and $G(z)$ are the z transforms of $f(k)$ and $g(k)$, respectively.

The linearity property can be proved by referring to the definition of the z transform as follows:

$$\begin{aligned} X(z) &= \mathscr{Z}[x(k)] = \mathscr{Z}[\alpha f(k) + \beta g(k)] \\ &= \sum_{k=0}^{\infty} [\alpha f(k) + \beta g(k)] z^{-k} \\ &= \alpha \sum_{k=0}^{\infty} f(k) z^{-k} + \beta \sum_{k=0}^{\infty} g(k) z^{-k} \\ &= \alpha\, \mathscr{Z}[f(k)] + \beta\, \mathscr{Z}[g(k)] \\ &= \alpha F(z) + \beta G(z) \end{aligned}$$

Multiplication by a^k. If $X(z)$ is the z transform of $x(k)$, then the z transform of $a^k x(k)$ can be given by $X(a^{-1}z)$:

$$\mathscr{Z}[a^k x(k)] = X(a^{-1}z) \tag{2–6}$$

This can be proved as follows:

$$\begin{aligned} \mathscr{Z}[a^k x(k)] &= \sum_{k=0}^{\infty} a^k x(k) z^{-k} = \sum_{k=0}^{\infty} x(k)(a^{-1}z)^{-k} \\ &= X(a^{-1}z) \end{aligned}$$

Real translation theorem (shifting theorem). The real translation theorem presented below is referred to as the *shifting theorem*. If $x(t) = 0$ for $t < 0$ and $x(t)$ has the z transform $X(z)$, then

$$\mathscr{Z}[x(t - nT)] = z^{-n} X(z) \tag{2–7}$$

and

$$\mathscr{Z}[x(t + nT)] = z^n \left[X(z) - \sum_{k=0}^{n-1} x(kT) z^{-k} \right] \tag{2–8}$$

where n is zero or a positive integer.

To prove Eq. (2–7), note that

$$\begin{aligned} \mathscr{Z}[x(t - nT)] &= \sum_{k=0}^{\infty} x(kT - nT) z^{-k} \\ &= z^{-n} \sum_{k=0}^{\infty} x(kT - nT) z^{-(k-n)} \end{aligned} \tag{2–9}$$

By defining $m = k - n$, Eq. (2–9) can be written as follows:

$$\mathscr{Z}[x(t-nT)] = z^{-n} \sum_{m=-n}^{\infty} x(mT)z^{-m}$$

Since $x(mT) = 0$ for $m < 0$, we may change the lower limit of the summation from $m = -n$ to $m = 0$. Hence

$$\mathscr{Z}[x(t-nT)] = z^{-n} \sum_{m=0}^{\infty} x(mT)z^{-m} = z^{-n}X(z) \tag{2–10}$$

Thus, multiplication of a z transform by z^{-n} has the effect of delaying the time function $x(t)$ by time nT.

To prove Eq. (2–8), we note that

$$\begin{aligned}\mathscr{Z}[x(t+nT)] &= \sum_{k=0}^{\infty} x(kT+nT)z^{-k} \\ &= z^{n} \sum_{k=0}^{\infty} x(kT+nT)z^{-(k+n)} \\ &= z^{n}\left[\sum_{k=0}^{\infty} x(kT+nT)z^{-(k+n)} + \sum_{k=0}^{n-1} x(kT)z^{-k} - \sum_{k=0}^{n-1} x(kT)z^{-k}\right] \\ &= z^{n}\left[\sum_{k=0}^{\infty} x(kT)z^{-k} - \sum_{k=0}^{n-1} x(kT)z^{-k}\right] \\ &= z^{n}\left[X(z) - \sum_{k=0}^{n-1} x(kT)z^{-k}\right]\end{aligned}$$

Example 2–4.

Obtain the z transforms of $x(k+1)$, $x(k+2)$, $x(k+n)$, and $x(k-n)$. Note that $x(k+n)$ is the sequence shifted to the left by n sampling periods (in a forward time shift) and $x(k-n)$ is the sequence shifted to the right by n sampling periods (in a backward time shift).

The z transform of $x(k+1)$ is given by

$$\begin{aligned}\mathscr{Z}[x(k+1)] &= \sum_{k=0}^{\infty} x(k+1)z^{-k} \\ &= \sum_{k=1}^{\infty} x(k)z^{-k+1} \\ &= z\left[\sum_{k=0}^{\infty} x(k)z^{-k} - x(0)\right] \\ &= zX(z) - zx(0)\end{aligned} \tag{2–11}$$

As a special case, if $x(0) = 0$, then

$$\mathscr{Z}[x(k+1)] = z\,\mathscr{Z}[x(k)] \qquad \text{if } x(0) = 0 \tag{2–12}$$

It is important to note that if $x(0) = 0$, then multiplication of the z transform of a function $x(k)$ by z corresponds to a forward time shift of one period.

Equation (2–11) can easily be modified to obtain the following relationship:

$$\mathcal{Z}[x(k+2)] = z\,\mathcal{Z}[x(k+1)] - zx(1) = z^2X(z) - z^2x(0) - zx(1)$$

Similarly,

$$\mathcal{Z}[x(k+n)] = z^nX(z) - z^nx(0) - z^{n-1}x(1) - z^{n-2}x(2) - \cdots - zx(n-1) \tag{2–13}$$

where n is a positive integer.

It should be pointed out that when the difference equation is transformed into an algebraic equation in z by the z transform method, the initial data are automatically included in the algebraic representation.

The z transform of $x(k-n)$ can be obtained by referring to Eq. (2–10), as follows:

$$\mathcal{Z}[x(k-n)] = z^{-n}X(z) \tag{2–14}$$

Example 2–5.

Find the z transforms of unit step functions which are delayed by 1 sampling period and 4 sampling periods, respectively, as shown in Fig. 2–2(a) and (b).

Using the shifting theorem given by Eq. (2–7), we have

$$\mathcal{Z}[1(t-T)] = z^{-1}\,\mathcal{Z}[1(t)] = z^{-1}\frac{1}{1-z^{-1}} = \frac{z^{-1}}{1-z^{-1}}$$

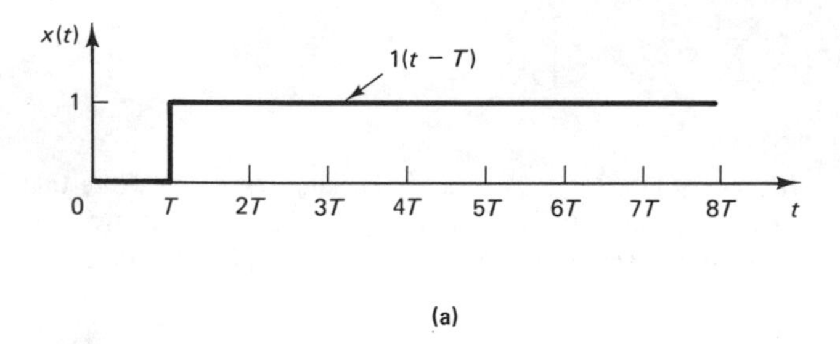

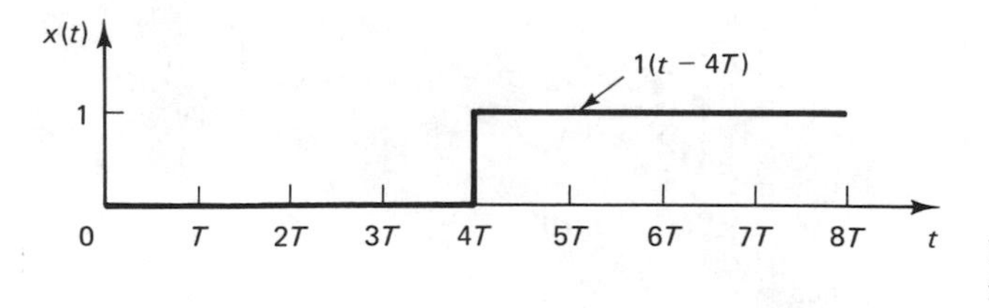

Figure 2–2 (a) Unit step function delayed by 1 sampling period; (b) unit step function delayed by 4 sampling periods.

$$\begin{aligned}\mathcal{Z}[\Delta x(k)] &= \mathcal{Z}[x(k+1)] - \mathcal{Z}[x(k)] \\ &= zX(z) - zx(0) - X(z) \\ &= (z-1)X(z) - zx(0)\end{aligned} \tag{2–18}$$

Similarly, the second forward difference is defined by

$$\begin{aligned}\Delta^2 x(k) &= \Delta[\Delta x(k)] = \Delta[x(k+1) - x(k)] \\ &= \Delta x(k+1) - \Delta x(k)\end{aligned}$$

Thus

$$\begin{aligned}\Delta^2 x(k) &= [x(k+2) - x(k+1)] - [x(k+1) - x(k)] \\ &= x(k+2) - 2x(k+1) + x(k)\end{aligned}$$

and the z transform of $\Delta^2 x(k)$ is

$$\begin{aligned}\mathcal{Z}[\Delta^2 x(k)] &= \mathcal{Z}[x(k+2) - 2x(k+1) + x(k)] \\ &= z^2 X(z) - z^2 x(0) - zx(1) - 2[zX(z) - zx(0)] + X(z) \\ &= (z-1)^2 X(z) - z(z-1)x(0) - z\,\Delta x(0)\end{aligned} \tag{2–19}$$

where $\Delta x(0) = x(1) - x(0)$. Similarly, we have

$$\Delta^3 x(k) = \Delta^2 x(k+1) - \Delta^2 x(k)$$

and the z transform of $\Delta^3 x(k)$ becomes

$$\begin{aligned}\mathcal{Z}[\Delta^3 x(k)] &= \mathcal{Z}\{\Delta[\Delta^2 x(k)]\} \\ &= \mathcal{Z}[x(k+3) - 3x(k+2) + 3x(k+1) - x(k)] \\ &= (z-1)^3 X(z) - z(z-1)^2 x(0) - z(z-1)\,\Delta x(0) - z\,\Delta^2 x(0)\end{aligned}$$

where $\Delta x(0) = x(1) - x(0)$ and $\Delta^2 x(0) = x(2) - 2x(1) + x(0)$. In general, for the mth forward difference

$$\Delta^m x(k) = \Delta^{m-1} x(k+1) - \Delta^{m-1} x(k)$$

we have

$$\mathcal{Z}[\Delta^m x(k)] = (z-1)^m X(z) - z \sum_{j=0}^{m-1} (z-1)^{m-j-1}\,\Delta^j x(0) \tag{2–20}$$

Complex translation theorem. If $x(t)$ has the z transform $X(z)$, then the z transform of $e^{-at}x(t)$ can be given by $X(ze^{aT})$. This is known as the *complex translation theorem*.

To prove this theorem, note that

$$\mathcal{Z}[e^{-at}x(t)] = \sum_{k=0}^{\infty} x(kT)e^{-akT}z^{-k} = \sum_{k=0}^{\infty} x(kT)(ze^{aT})^{-k} = X(ze^{aT}) \tag{2–21}$$

Thus, we see that replacing z in $X(z)$ by ze^{aT} gives the z transform of $e^{-at}x(t)$.

Example 2–8.

Given the z transform of $\sin \omega t$, obtain the z transform of $e^{-at} \sin \omega t$ by using the complex translation theorem.

Noting that

$$\mathcal{Z}[\sin \omega t] = \frac{z^{-1} \sin \omega T}{1 - 2z^{-1} \cos \omega T + z^{-2}}$$

we substitute ze^{aT} for z to obtain the z transform of $e^{-at} \sin \omega t$, as follows:

$$\mathcal{Z}[e^{-at} \sin \omega t] = \frac{e^{-aT}z^{-1} \sin \omega T}{1 - 2e^{-aT}z^{-1} \cos \omega T + e^{-2aT}z^{-2}}$$

Example 2–9.

Obtain the z transform of te^{-at}.

Notice that

$$\mathcal{Z}[t] = \frac{Tz^{-1}}{(1 - z^{-1})^2} = X(z)$$

Thus,

$$\mathcal{Z}[te^{-at}] = X(ze^{aT}) = \frac{Te^{-aT}z^{-1}}{(1 - e^{-aT}z^{-1})^2}$$

Initial value theorem. If $x(t)$ has the z transform $X(z)$ and if $\lim_{z \to \infty} X(z)$ exists, then the initial value $x(0)$ of $x(t)$ or $x(k)$ is given by

$$x(0) = \lim_{z \to \infty} X(z) \tag{2–22}$$

To prove this theorem, note that

$$X(z) = \sum_{k=0}^{\infty} x(k)z^{-k} = x(0) + x(1)z^{-1} + x(2)z^{-2} + \cdots$$

Letting $z \to \infty$ in this last equation, we obtain Eq. (2–22). The behavior of the signal in the neighborhood of $t = 0$ or $k = 0$ can thus be determined by the behavior of $X(z)$ at $z = \infty$.

The initial value theorem is convenient for checking z transform calculations for possible errors. Since $x(0)$ is usually known, a check of the initial value by $\lim_{z \to \infty} X(z)$ can easily spot errors in $X(z)$, if any exist.

Example 2–10.

Determine the initial value $x(0)$ if the z transform of $x(t)$ is given by

$$X(z) = \frac{(1 - e^{-T})z^{-1}}{(1 - z^{-1})(1 - e^{-T}z^{-1})}$$

By using the initial value theorem, we find

$$x(0) = \lim_{z\to\infty} \frac{(1 - e^{-T})z^{-1}}{(1 - z^{-1})(1 - e^{-T}z^{-1})} = 0$$

Referring to Example 2–3, notice that this $X(z)$ corresponds to

$$x(t) = 1 - e^{-t}$$

and thus $x(0) = 0$, which agrees with the result obtained above.

Final value theorem. Suppose that $x(k)$, where $x(k) = 0$ for $k < 0$, has the z transform $X(z)$ and that all the poles of $X(z)$ lie inside the unit circle, with the possible exception of a simple pole at $z = 1$. [This is the condition for the stability of $X(z)$, or the condition for $x(k)$ $(k = 0, 1, 2, \ldots)$ to remain finite. See Sec. 3–8.] Then the final value of $x(k)$, that is, the value of $x(k)$ as k approaches infinity, can be given by

$$\lim_{k\to\infty} x(k) = \lim_{z\to 1} [(1 - z^{-1})X(z)] \tag{2–23}$$

To prove the final value theorem, note that

$$\mathscr{Z}[x(k)] = X(z) = \sum_{k=0}^{\infty} x(k)z^{-k}$$

$$\mathscr{Z}[x(k-1)] = z^{-1}X(z) = \sum_{k=0}^{\infty} x(k-1)z^{-k}$$

Hence

$$\sum_{k=0}^{\infty} x(k)z^{-k} - \sum_{k=0}^{\infty} x(k-1)z^{-k} = X(z) - z^{-1}X(z)$$

Taking the limit as z approaches unity, we have

$$\lim_{z\to 1}\left[\sum_{k=0}^{\infty} x(k)z^{-k} - \sum_{k=0}^{\infty} x(k-1)z^{-k}\right] = \lim_{z\to 1} [(1 - z^{-1})X(z)]$$

Because of the assumed stability condition and the condition that $x(k) = 0$ for $k < 0$, the left-hand side of this last equation becomes

$$\sum_{k=0}^{\infty} [x(k) - x(k-1)] = [x(0) - x(-1)] + [x(1) - x(0)]$$
$$+ [x(2) - x(1)] + \cdots = x(\infty) = \lim_{k\to\infty} x(k)$$

Hence

$$\lim_{k\to\infty} x(k) = \lim_{z\to 1} [(1 - z^{-1})X(z)]$$

which is Eq. (2–23). The final value theorem is very useful in determining the behavior of $x(k)$ as $k \to \infty$ from its z transform $X(z)$.

Example 2–11.

Determine the final value $x(\infty)$ of

$$X(z) = \frac{1}{1 - z^{-1}} - \frac{1}{1 - e^{-aT}z^{-1}} \qquad a > 0$$

by using the final value theorem.

By applying the final value theorem to the given $X(z)$, we obtain

$$x(\infty) = \lim_{z \to 1} [(1 - z^{-1})X(z)]$$

$$= \lim_{z \to 1} \left[(1 - z^{-1}) \left(\frac{1}{1 - z^{-1}} - \frac{1}{1 - e^{-aT}z^{-1}} \right) \right]$$

$$= \lim_{z \to 1} \left(1 - \frac{1 - z^{-1}}{1 - e^{-aT}z^{-1}} \right) = 1$$

It is noted that the given $X(z)$ is actually the z transform of

$$x(t) = 1 - e^{-at}$$

By substituting $t = \infty$ in this equation, we have

$$x(\infty) = \lim_{t \to \infty} (1 - e^{-at}) = 1$$

As a matter of course the two results agree.

Example 2–12.

Consider the difference equation

$$x(k) - ax(k - 1) = 1(k) \qquad -1 < a < 1$$

where $x(k) = 0$ for $k < 0$ and $1(k)$ is the unit step sequence. Obtain the solution $x(k)$. Determine the initial and final values of $x(k)$ by using the initial and final value theorems, respectively.

Since

$$\mathcal{Z}[1(k)] = \frac{1}{1 - z^{-1}}$$

$$\mathcal{Z}[x(k) - ax(k - 1)] = X(z) - az^{-1}X(z)$$

the z transform of the given difference equation becomes

$$X(z) - az^{-1}X(z) = \frac{1}{1 - z^{-1}}$$

or

$$X(z) = \frac{1}{(1 - z^{-1})(1 - az^{-1})}$$

which can be written as

$$X(z) = \frac{1}{1-a}\left(\frac{1}{1-z^{-1}} - \frac{a}{1-az^{-1}}\right)$$

The corresponding sequence $x(k)$ is

$$x(k) = \frac{1}{1-a}(1-a^{k+1}) = \frac{1-a^{k+1}}{1-a} \tag{2–24}$$

The initial value $x(0)$ is obtained by use of the initial value theorem, as follows:

$$x(0) = \lim_{z\to\infty} X(z) = \lim_{z\to\infty} \frac{1}{(1-z^{-1})(1-az^{-1})} = 1$$

The final value $x(\infty)$ is obtained by use of the final value theorem, as follows:

$$\begin{aligned} x(\infty) &= \lim_{z\to 1}[(1-z^{-1})X(z)] \\ &= \lim_{z\to 1}\left[(1-z^{-1})\frac{1}{(1-z^{-1})(1-az^{-1})}\right] \\ &= \lim_{z\to 1}\frac{1}{1-az^{-1}} = \frac{1}{1-a} \end{aligned} \tag{2–25}$$

The initial and final values of $x(k)$ obtained from the solution, Eq. (2–24), are

$$x(0) = \frac{1-a}{1-a} = 1$$

$$x(\infty) = \frac{1-a^{\infty}}{1-a} = \frac{1}{1-a} \qquad -1 < a < 1$$

Clearly, the results agree with those obtained by use of the initial and final value theorems.

It is important to note that if $|a| \geq 1$, then $x(\infty)$ is not bounded and therefore $x(\infty)$ is not finite. Equation (2–25) does not apply to such a case, since the condition for the final value theorem is violated.

Complex differentiation. In the region of convergence a power series in z may be differentiated with respect to z any number of times to get a convergent series. The derivatives of $X(z)$ converge in the same region as $X(z)$.

Consider

$$X(z) = \sum_{k=0}^{\infty} x(k)z^{-k}$$

which converges in a certain region in the z plane. Differentiating $X(z)$ with respect to z, we obtain

$$\frac{d}{dz}X(z) = \sum_{k=0}^{\infty}(-k)x(k)z^{-k-1}$$

Multiplying both sides of this last equation by $-z$ gives

$$-z\frac{d}{dz}X(z) = \sum_{k=0}^{\infty} kx(k)z^{-k} \tag{2-26}$$

Thus we have

$$\mathcal{Z}[kx(k)] = -z\frac{d}{dz}X(z) \tag{2-27}$$

Similarly, by differentiating both sides of Eq. (2–26) with respect to z, we have

$$\frac{d}{dz}\left[-z\frac{d}{dz}X(z)\right] = \sum_{k=0}^{\infty} (-k^2)x(k)z^{-k-1}$$

Multiplying both sides of this last equation by $-z$, we obtain

$$-z\frac{d}{dz}\left[-z\frac{d}{dz}X(z)\right] = \sum_{k=0}^{\infty} k^2x(k)z^{-k}$$

or

$$\mathcal{Z}[k^2x(k)] = \left(-z\frac{d}{dz}\right)^2 X(z)$$

The operator $\left(-z\frac{d}{dz}\right)^2$ implies that we apply the operator $-z\frac{d}{dz}$ twice. Similarly, by repeating this process we have

$$\mathcal{Z}[k^mx(k)] = \left(-z\frac{d}{dz}\right)^m X(z) \tag{2-28}$$

Such complex differentiation enables us to obtain new z transform pairs from the known z transform pairs.

Example 2–13.

The z transform of the unit step sequence $1(k)$ is given by

$$\mathcal{Z}[1(k)] = \frac{1}{1-z^{-1}}$$

Obtain the z transform of the unit ramp sequence $x(k)$ where

$$x(k) = k$$

by using the complex differentiation theorem.

$$\mathcal{Z}[x(k)] = \mathcal{Z}[k] = \mathcal{Z}[k \cdot 1(k)] = -z\frac{d}{dz}\left(\frac{1}{1-z^{-1}}\right) = \frac{z^{-1}}{(1-z^{-1})^2}$$

Complex integration. Consider the sequence

$$g(k) = \frac{x(k)}{k}$$

where $x(k)/k$ is finite for $k = 0$. The z transform of $x(k)/k$ is given by

$$\mathscr{Z}\left[\frac{x(k)}{k}\right] = \int_z^\infty \frac{X(z_1)}{z_1}\,dz_1 + \lim_{k\to 0}\frac{x(k)}{k} \tag{2-29}$$

where $\mathscr{Z}[x(k)] = X(z)$.

To prove Eq. (2–29), note first that

$$\mathscr{Z}\left[\frac{x(k)}{k}\right] = G(z) = \sum_{k=0}^{\infty}\frac{x(k)}{k}z^{-k}$$

Differentiating this last equation with respect to z yields

$$\frac{d}{dz}G(z) = -\sum_{k=0}^{\infty}x(k)\,z^{-k-1} = -z^{-1}\sum_{k=0}^{\infty}x(k)z^{-k} = -\frac{X(z)}{z}$$

Integrating both sides of this last equation with respect to z from z to ∞ gives

$$\int_z^\infty \frac{d}{dz}G(z)\,dz = G(\infty) - G(z) = -\int_z^\infty \frac{X(z_1)}{z_1}\,dz_1$$

or

$$G(z) = \int_z^\infty \frac{X(z_1)}{z_1}\,dz_1 + G(\infty)$$

Noting that $G(\infty)$ is given by

$$G(\infty) = \lim_{z\to\infty} G(z) = g(0) = \lim_{k\to 0}\frac{x(k)}{k}$$

we have

$$\mathscr{Z}\left[\frac{x(k)}{k}\right] = \int_z^\infty \frac{X(z_1)}{z_1}\,dz_1 + \lim_{k\to 0}\frac{x(k)}{k}$$

Example 2–14.

Consider the sequence $x(k)$, where

$$x(k) = k^2$$

Obtain the z transform of $x(k)$ by using complex integration.

Define

$$g(k) = \frac{x(k)}{k} = k$$

Then

$$\mathscr{Z}\left[\frac{x(k)}{k}\right] = G(z) = \sum_{k=0}^{\infty}\frac{x(k)}{k}z^{-k} = \sum_{k=0}^{\infty}kz^{-k}$$

$$\frac{d}{dz}G(z) = -\sum_{k=0}^{\infty} k^2 z^{-k-1} = -z^{-1}\sum_{k=0}^{\infty} k^2 z^{-k} = -\frac{X(z)}{z}$$

Thus

$$\begin{aligned} G(z) = \mathscr{Z}\left[\frac{x(k)}{k}\right] &= \int_z^{\infty} \frac{X(z_1)}{z_1}\, dz_1 + G(\infty) \\ &= \int_z^{\infty} \sum_{k=0}^{\infty} k^2 z_1^{-k-1} dz_1 + G(\infty) \\ &= \sum_{k=0}^{\infty} k^2 \frac{z_1^{-k}}{-k}\bigg|_z^{\infty} + G(\infty) \\ &= \sum_{k=0}^{\infty} kz^{-k} + \lim_{k\to 0}\frac{x(k)}{k} \\ &= \frac{z^{-1}}{(1-z^{-1})^2} + \lim_{k\to 0} k = \frac{z^{-1}}{(1-z^{-1})^2} \end{aligned}$$

which is the correct result. This example merely illustrates the use of the complex integration technique to obtain the z transform of a function $x(k)/k$.

Partial differentiation theorem. Consider a function $x(t, a)$ or $x(kT, a)$ which is z-transformable. Here a is a constant or an independent variable. Define the z transform of $x(t, a)$ or $x(kT, a)$ as $X(z, a)$. Thus

$$\mathscr{Z}[x(t, a)] = \mathscr{Z}[x(kT, a)] = X(z, a)$$

The z transform of the partial derivative of $x(t, a)$ or $x(kT, a)$ with respect to a can be given by

$$\mathscr{Z}\left[\frac{\partial}{\partial a} x(t, a)\right] = \mathscr{Z}\left[\frac{\partial}{\partial a} x(kT, a)\right] = \frac{\partial}{\partial a} X(z, a) \tag{2–30}$$

This equation is called the partial differentiation theorem.

To prove this theorem, note that

$$\begin{aligned} \mathscr{Z}\left[\frac{\partial}{\partial a} x(t, a)\right] = \mathscr{Z}\left[\frac{\partial}{\partial a} x(kT, a)\right] &= \sum_{k=0}^{\infty} \frac{\partial}{\partial a} x(kT, a) z^{-k} \\ &= \frac{\partial}{\partial a}\sum_{k=0}^{\infty} x(kT, a) z^{-k} = \frac{\partial}{\partial a} X(z, a) \end{aligned}$$

Example 2–15.

Consider

$$x(t, a) = t^2 e^{-at}$$

Obtain the z transform of this function $x(t, a)$ by use of the partial differentiation theorem.

Notice that

$$\frac{\partial}{\partial a}(-te^{-at}) = t^2e^{-at}$$

and

$$\mathscr{Z}[te^{-at}] = \frac{Te^{-aT}z^{-1}}{(1 - e^{-aT}z^{-1})^2}$$

Then we have

$$\begin{aligned}\mathscr{Z}[x(t, a)] = \mathscr{Z}[t^2e^{-at}] &= \mathscr{Z}\left[\frac{\partial}{\partial a}(-te^{-at})\right] \\ &= \frac{\partial}{\partial a}\left[-\frac{Te^{-aT}z^{-1}}{(1 - e^{-aT}z^{-1})^2}\right] \\ &= \frac{T^2e^{-aT}(1 + e^{-aT}z^{-1})z^{-1}}{(1 - e^{-aT}z^{-1})^3}\end{aligned}$$

Real convolution theorem. Consider the functions $x_1(t)$ and $x_2(t)$, where

$$x_1(t) = 0 \qquad \text{for } t < 0$$

$$x_2(t) = 0 \qquad \text{for } t < 0$$

Assume that $x_1(t)$ and $x_2(t)$ are z-transformable and their z transforms are $X_1(z)$ and $X_2(z)$, respectively. Then

$$X_1(z)X_2(z) = \mathscr{Z}\left[\sum_{h=0}^{k} x_1(hT)x_2(kT - hT)\right] \tag{2–31}$$

This equation is called the real convolution theorem.

To prove this theorem, notice that

$$\begin{aligned}\mathscr{Z}\left[\sum_{h=0}^{k} x_1(hT)x_2(kT - hT)\right] &= \sum_{k=0}^{\infty}\sum_{h=0}^{k} x_1(hT)x_2(kT - hT)z^{-k} \\ &= \sum_{k=0}^{\infty}\sum_{h=0}^{\infty} x_1(hT)x_2(kT - hT)z^{-k}\end{aligned}$$

where we use the knowledge that $x_2(kT - hT) = 0$ for $h > k$. Now define $m = k - h$. Then

$$\mathscr{Z}\left[\sum_{h=0}^{k} x_1(hT)x_2(kT - hT)\right] = \sum_{h=0}^{\infty} x_1(hT)z^{-h}\sum_{m=-h}^{\infty} x_2(mT)z^{-m}$$

Since $x_2(mT) = 0$ for $m < 0$, this last equation becomes

$$\mathscr{Z}\left[\sum_{h=0}^{k} x_1(hT)x_2(kT - hT)\right] = \sum_{h=0}^{\infty} x_1(hT)z^{-h}\sum_{m=0}^{\infty} x_2(mT)z^{-m} = X_1(z)X_2(z)$$

Complex convolution theorem. The following, known as the complex convolution theorem, is useful in obtaining the z transform of the product of two sequences $x_1(k)$ and $x_2(k)$.

Suppose both $x_1(k)$ and $x_2(k)$ are zero for $k < 0$. Assume that

$$X_1(z) = \mathscr{Z}[x_1(k)] \qquad |z| > R_1$$

$$X_2(z) = \mathscr{Z}[x_2(k)] \qquad |z| > R_2$$

where R_1 and R_2 are the radii of absolute convergence for $x_1(k)$ and $x_2(k)$, respectively. Then the z transform of the product of $x_1(k)$ and $x_2(k)$ can be given by

$$\mathscr{Z}[x_1(k)x_2(k)] = \frac{1}{2\pi j}\oint_C \zeta^{-1}X_2(\zeta)X_1(\zeta^{-1}z)\,d\zeta \tag{2-32}$$

where $R_2 < |\zeta| < |z|/R_1$.

To prove this theorem, let us take the z transform of $x_1(k)x_2(k)$:

$$\mathscr{Z}[x_1(k)x_2(k)] = \sum_{k=0}^{\infty} x_1(k)x_2(k)z^{-k} \tag{2-33}$$

The series on the right-hand side of Eq. (2–33) coverges for $|z| > R$, where R is the radius of absolute convergence for $x_1(k)x_2(k)$. From Eq. (2–2) we have

$$\begin{aligned} x_2(k) &= \frac{1}{2\pi j}\oint_C X_2(z)z^{k-1}\,dz \\ &= \frac{1}{2\pi j}\oint_C X_2(\zeta)\zeta^{k-1}\,d\zeta \end{aligned} \tag{2-34}$$

Substituting Eq. (2–34) into Eq. (2–33), we obtain

$$\mathscr{Z}[x_1(k)x_2(k)] = \frac{1}{2\pi j}\sum_{k=0}^{\infty}\oint_C X_2(\zeta)\zeta^{k-1}x_1(k)z^{-k}\,d\zeta$$

Noting that Eq. (2–33) converges uniformly for the region $|z| > R$, we may interchange the order of summation and integration. Then

$$\mathscr{Z}[x_1(k)x_2(k)] = \frac{1}{2\pi j}\oint_C \zeta^{-1}X_2(\zeta)\sum_{k=0}^{\infty} x_1(k)(\zeta^{-1}z)^{-k}\,d\zeta$$

Since

$$\sum_{k=0}^{\infty} x_1(k)(\zeta^{-1}z)^{-k} = X_1(\zeta^{-1}z)$$

we have

$$\mathscr{Z}[x_1(k)x_2(k)] = \frac{1}{2\pi j}\oint_C \zeta^{-1}X_2(\zeta)X_1(\zeta^{-1}z)\,d\zeta \tag{2-35}$$

where C is a contour (a circle with its center at the origin) which lies in the region given by $|\zeta| > R_2$ and $|\zeta^{-1}z| > R_1$, or

$$R_2 < |\zeta| < \frac{|z|}{R_1} \tag{2-36}$$

Parseval's theorem. Suppose the z transforms of sequences $x_1(k)$ and $x_2(k)$ are such that

$$X_1(z) = \mathcal{Z}[x_1(k)] \qquad |z| > R_1 \text{ (where } R_1 < 1)$$

$$X_2(z) = \mathcal{Z}[x_2(k)] \qquad |z| > R_2$$

and inequality (2–36) is satisfied for $|z| = 1$, or

$$R_2 < |\zeta| < \frac{1}{R_1}$$

Then, by substituting $|z| = 1$ into Eq. (2–35) we obtain the following equation:

$$\mathcal{Z}[x_1(k)x_2(k)]|_{|z|=1} = \sum_{k=0}^{\infty} x_1(k)x_2(k) = \frac{1}{2\pi j}\oint_C \zeta^{-1}X_2(\zeta)X_1(\zeta^{-1})\,d\zeta$$

If we set $x_1(k) = x_2(k) = x(k)$ in this last equation, we get

$$\sum_{k=0}^{\infty} x^2(k) = \frac{1}{2\pi j}\oint_C \zeta^{-1}X(\zeta)X(\zeta^{-1})\,d\zeta$$

$$= \frac{1}{2\pi j}\oint_C z^{-1}X(z)X(z^{-1})\,dz \tag{2-37}$$

Equation (2–37) is Parseval's theorem. This theorem is useful for obtaining the summation of $x^2(k)$.

Summary. In this section we have presented important properties and theorems of the z transform that will prove to be useful in solving many z transform problems. For the purpose of convenient reference, these important properties and theorems are summarized in Table 2–2.

TABLE 2–2 IMPORTANT PROPERTIES AND THEOREMS OF THE *z* TRANSFORM

	$x(t)$ or $x(k)$	$\mathcal{Z}[x(t)]$ or $\mathcal{Z}[x(k)]$
1.	$ax(t)$	$aX(z)$
2.	$ax_1(t) + bx_2(t)$	$aX_1(z) + bX_2(z)$
3.	$x(t+T)$ or $x(k+1)$	$zX(z) - zx(0)$
4.	$x(t+2T)$	$z^2X(z) - z^2x(0) - zx(T)$

TABLE 2–2 *(continued)*

5.	$x(k+2)$	$z^2X(z) - z^2x(0) - zx(1)$
6.	$x(t+kT)$	$z^kX(z) - z^kx(0) - z^{k-1}x(T) - \cdots - zx(kT-T)$
7.	$x(t-kT)$	$z^{-k}X(z)$
8.	$x(n+k)$	$z^kX(z) - z^kx(0) - z^{k-1}x(1) - \cdots - zx(k-1)$
9.	$x(n-k)$	$z^{-k}X(z)$
10.	$tx(t)$	$-Tz\frac{d}{dz}X(z)$
11.	$kx(k)$	$-z\frac{d}{dz}X(z)$
12.	$e^{-at}x(t)$	$X(ze^{aT})$
13.	$e^{-ak}x(k)$	$X(ze^{a})$
14.	$a^kx(k)$	$X\left(\frac{z}{a}\right)$
15.	$ka^kx(k)$	$-z\frac{d}{dz}X\left(\frac{z}{a}\right)$
16.	$x(0)$	$\lim_{z\to\infty} X(z)$ if the limit exists
17.	$x(\infty)$	$\lim_{z\to 1} [(1-z^{-1})X(z)]$ if $(1-z^{-1})X(z)$ is analytic on and outside the unit circle
18.	$\nabla x(k) = x(k) - x(k-1)$	$(1-z^{-1})X(z)$
19.	$\Delta x(k) = x(k+1) - x(k)$	$(z-1)X(z) - zx(0)$
20.	$\sum_{k=0}^{n} x(k)$	$\frac{1}{1-z^{-1}}X(z)$
21.	$\frac{\partial}{\partial a}x(t, a)$	$\frac{\partial}{\partial a}X(z, a)$
22.	$k^mx(k)$	$\left(-z\frac{d}{dz}\right)^m X(z)$
23.	$\sum_{k=0}^{n} x(kT)y(nT-kT)$	$X(z)Y(z)$
24.	$\sum_{k=0}^{\infty} x(k)$	$X(1)$

2-5 THE INVERSE z TRANSFORM

The z transformation serves the same role for discrete-time control systems that the Laplace transformation serves for continuous-time control systems. For the z transform to be useful, one must be familiar with methods for finding the inverse z transform.

The notation for the inverse z transform is $\mathcal{Z}^{-1}$. The inverse z transform of $X(z)$ yields the corresponding time sequence $x(k)$.

It should be noted that only the time sequence at the sampling instants is obtained from the inverse transform. Thus, the inverse z transform of $X(z)$ yields a unique $x(k)$, but does not yield a unique $x(t)$. This means that the inverse z transform yields a time sequence which specifies the values of $x(t)$ only at discrete instants of time, $t = 0, T, 2T, \ldots$, and says nothing about the values of $x(t)$ at all other times (refer to Fig. 2–1). [In discrete-time control systems for which the sampling theorem is satisfied, the continuous-time function $x(t)$ may be roughly approximated by drawing a smooth curve through the points of $x(kT)$ ($k = 0, 1, 2, \ldots$). For details of the sampling theorem, see Sec. 3–4.]

In this book we refer to the inverse z transformation as an operation which goes from $X(z)$ to $x(k)$. An obvious method for finding the inverse z transform is to refer to a z transform table. However, unless one refers to an extensive z transform table, one may not be able to find the inverse z transform of a complicated function of z. (If one uses a less extensive table of z transforms, it is necessary to express a complex z transform as a sum of simpler z transforms. Refer to the partial-fraction-expansion method presented in this section.)

Other than referring to z transform tables, four methods for obtaining the inverse z transform are commonly available:

1. Direct division method
2. Computational method
3. Partial-fraction-expansion method
4. Inversion integral method

In obtaining the inverse z transform, we assume, as usual, that the time sequence $x(k)$ or $x(kT)$ is zero for $k < 0$.

Direct division method. In the direct division method we obtain the inverse z transform by expanding $X(z)$ into an infinite power series in z^{-1}. This method is useful when it is difficult to obtain the closed-form expression for the inverse z transform or it is desired to find only the first several terms of $x(k)$.

The direct division method stems from the definition of z transforms. If $X(z)$ is expanded into a power series in z^{-1}, that is, if

$$X(z) = \sum_{k=0}^{\infty} x(kT)z^{-k}$$

$$= x(0) + x(T)z^{-1} + x(2T)z^{-2} + \cdots + x(kT)z^{-k} + \cdots$$

then the values of $x(kT)$ can be determined by inspection.

If $X(z)$ is given in the form of a rational function, the expansion into an infinite power series in increasing powers of z^{-1} can be accomplished by simply dividing the numerator by the denominator. If the resulting series is convergent, the coefficients of the z^{-k} term in the series are the values $x(kT)$ of the time sequence. In obtaining the coefficients by long division, both the numerator and denominator of $X(z)$ must be written in increasing powers of z^{-1}.

Although the present method gives the values of $x(0)$, $x(T)$, $x(2T)$, . . . in a sequential manner, it is usually difficult to obtain an expression for the general term from a set of values of $x(kT)$.

The following formulas are sometimes useful in recognizing the closed-form expressions for finite or infinite series in z^{-1}:

$$(1 - az^{-1})^3 = 1 - 3az^{-1} + 3a^2z^{-2} - a^3z^{-3}$$

$$(1 - az^{-1})^4 = 1 - 4az^{-1} + 6a^2z^{-2} - 4a^3z^{-3} + a^4z^{-4}$$

$$(1 - az^{-1})^{-1} = 1 + az^{-1} + a^2z^{-2} + a^3z^{-3} + a^4z^{-4} + a^5z^{-5} + \cdots \qquad |z| > 1$$

$$(1 - az^{-1})^{-2} = 1 + 2az^{-1} + 3a^2z^{-2} + 4a^3z^{-3} + 5a^4z^{-4} + 6a^5z^{-5} + \cdots \qquad |z| > 1$$

$$(1 - az^{-1})^{-3} = 1 + 3az^{-1} + 6a^2z^{-2} + 10a^3z^{-3} + 15a^4z^{-4} + 21a^5z^{-5} + 28a^6z^{-6} + \cdots \qquad |z| > 1$$

$$(1 - az^{-1})^{-4} = 1 + 4az^{-1} + 10a^2z^{-2} + 20a^3z^{-3} + 35a^4z^{-4} + 56a^5z^{-5} + 84a^6z^{-6} + 120a^7z^{-7} + \cdots \qquad |z| > 1$$

For a given z transform $X(z)$, if a closed-form expression for $x(k)$ is desired, we may use the partial-fraction-expansion method or the inversion integral method discussed later in this section.

Example 2–16.

Find $x(k)$ for $k = 0, 1, 2, 3, 4$ when $X(z)$ is given by

$$X(z) = \frac{10z + 5}{(z-1)(z-0.2)}$$

First, rewrite $X(z)$ as a ratio of polynomials in z^{-1}, as follows:

$$X(z) = \frac{10z^{-1} + 5z^{-2}}{1 - 1.2z^{-1} + 0.2z^{-2}}$$

Dividing the numerator by the denominator, we have

$$
\begin{array}{r l}
 & 10z^{-1} + 17z^{-2} + 18.4z^{-3} + 18.68z^{-4} + \cdots \\
1 - 1.2z^{-1} + 0.2z^{-2} \,\big) & 10z^{-1} + 5z^{-2} \\
 & \underline{10z^{-1} - 12z^{-2} + 2z^{-3}} \\
 & \qquad 17z^{-2} - 2z^{-3} \\
 & \qquad \underline{17z^{-2} - 20.4z^{-3} + 3.4z^{-4}} \\
 & \qquad\qquad 18.4z^{-3} - 3.4z^{-4} \\
 & \qquad\qquad \underline{18.4z^{-3} - 22.08z^{-4} + 3.68z^{-5}} \\
 & \qquad\qquad\qquad 18.68z^{-4} - 3.68z^{-5} \\
 & \qquad\qquad\qquad \underline{18.68z^{-4} - 22.416z^{-5} + 3.736z^{-6}}
\end{array}
$$

Thus,

$$X(z) = 10z^{-1} + 17z^{-2} + 18.4z^{-3} + 18.68z^{-4} + \cdots$$

By comparing this infinite series expansion of $X(z)$ with $X(z) = \sum_{k=0}^{\infty} x(k)z^{-k}$, we obtain

$$x(0) = 0$$

$$x(1) = 10$$

$$x(2) = 17$$

$$x(3) = 18.4$$

$$x(4) = 18.68$$

As seen from this example, the direct division method may be carried out by hand calculations if only the first several terms of the sequence are desired. In general, the method does not yield a closed-form expression for $x(k)$, except in special cases such as those shown in Example 2–17 and Prob. A-2–18.

Example 2–17.

Obtain the inverse z transform of

$$X(z) = \frac{Te^{-aT}z}{(z - e^{-aT})^2}$$

by use of the direct division method.

We first rewrite $X(z)$ as the ratio of two polynomials in z^{-1}:

$$X(z) = \frac{Te^{-aT}z^{-1}}{1 - 2e^{-aT}z^{-1} + e^{-2aT}z^{-2}}$$

Dividing the numerator by the denominator yields

$$X(z) = Te^{-aT}z^{-1} + 2Te^{-2aT}z^{-2} + 3Te^{-3aT}z^{-3} + \cdots$$

From this expression it can be observed that

$$x(kT) = kTe^{-kaT} \qquad k = 0, 1, 2, \ldots$$

Example 2–18.

Obtain the inverse z transform of

$$X(z) = 1 + 2z^{-1} + 3z^{-2} + 4z^{-3}$$

The transform $X(z)$ is already in the form of a power series in z^{-1}. Since $X(z)$ has a finite number of terms, it corresponds to a signal of finite length. By inspection, we find

$$x(0) = 1$$

$$x(1) = 2$$

$$x(2) = 3$$

$$x(3) = 4$$

All other $x(k)$ values are zero.

Computational method. In what follows, we shall present the computational approach or computer approach to the inverse z transformation. In particular, we shall discuss the method for finding the inverse z transform by use of a microcomputer. [As a matter of course, all coefficients involved in $X(z)$ must be numerical for computer solution to be possible. We shall present the method by use of an example.]

Consider the same $X(z)$ as discussed in Example 2–16:

$$X(z) = \frac{10z + 5}{(z-1)(z-0.2)}$$

Notice that $X(z)$ can be written as

$$X(z) = \frac{10z + 5}{z^2 - 1.2z + 0.2} U(z) \tag{2–38}$$

where $U(z) = 1$. Since

$$U(z) = u(0) + u(1)z^{-1} + u(2)z^{-2} + \cdots + u(k)z^{-k} + \cdots$$

we find that $U(z) = 1$ requires that

$$u(0) = 1$$

$$u(k) = 0 \qquad k = 1, 2, 3, \ldots$$

Now let us convert Eq. (2–38) into the following difference equation:

$$x(k+2) - 1.2x(k+1) + 0.2x(k) = 10u(k+1) + 5u(k) \tag{2–39}$$

where $u(0) = 1$ and $u(k) = 0$ for $k = 1, 2, 3, \ldots$ and the initial data $x(0)$ and $x(1)$ are determined as follows. By substituting $k = -2$ into Eq. (2–39) we obtain

$$x(0) - 1.2x(-1) + 0.2x(-2) = 10u(-1) + 5u(-2)$$

Since $x(-1) = x(-2) = 0$ and $u(-1) = u(-2) = 0$, we have

$$x(0) = 0$$

Next, by substituting $k = -1$ into Eq. (2–39) we get

$$x(1) - 1.2x(0) + 0.2x(-1) = 10u(0) + 5u(-1)$$

or

$$x(1) = 10$$

Finding the inverse z transform of $X(z)$ now becomes a matter of solving the following difference equation for $x(k)$:

$$x(k+2) - 1.2x(k+1) + 0.2x(k) = 10u(k+1) + 5u(k)$$

with the initial data $x(0) = 0$, $x(1) = 10$ and $u(0) = 1$, $u(k) = 0$ for $k = 1, 2, 3, \ldots$.

On the basis of the foregoing analysis we have a BASIC computer program for finding $x(k)$, the inverse z transform of $X(z)$, as shown in Table 2–3. This program gives the values of $x(k)$ up to $k = 15$. If the values of $x(k)$ up to $k = 100$ are desired, simply modify the inequality in line 140 in Table 2–3 so that it reads $K < 101$. The results of the computer calculations are also shown in Table 2–3.

The desired inverse z transform of $X(z)$ is given by $x(k)$ in Table 2–3. From

TABLE 2–3 BASIC COMPUTER PROGRAM FOR FINDING $x(k)$, THE INVERSE z TRANSFORM OF $X(z)$, WHERE $X(z) = (10z + 5)/[(z - 1)(z - 0.2)]$

```
10  X0 = 0
20  X1 = 10
30  U0 = 1
40  U1 = 0
50  K = 0
60  X2 = 1.2 * X1 - 0.2 * X0 + 10 * U1 + 5 * U0
70  M = X0
80  X0 = X1
90  X1 = X2
100 N = U0
110 U0 = U1
120 PRINT K, M, N
130 K = K + 1
140 IF K < 16 GO TO 60
150 END
```

$k = K$	$x(k) = XK = M$	$u(k) = UK = N$
0	0	1
1	10	0
2	17	0
3	18.4	0
4	18.68	0
5	18.736	0
6	18.7472	0
7	18.7495	0
8	18.7499	0
9	18.75	0
10	18.75	0
11	18.75	0
12	18.75	0
13	18.75	0
14	18.75	0
15	18.75	0

the computer results we find the final value of $x(k)$ to be 18.75. This can be easily verified by use of the final value theorem, as follows:

$$\lim_{k\to\infty} x(k) = \lim_{z\to 1} [(1 - z^{-1})X(z)]$$

$$= \lim_{z\to 1}\left[(1 - z^{-1})\frac{10z^{-1} + 5z^{-2}}{1 - 1.2z^{-1} + 0.2z^{-2}}\right]$$

$$= \lim_{z\to 1}\frac{10z^{-1} + 5z^{-2}}{1 - 0.2z^{-1}} = 18.75$$

Partial-fraction-expansion method. The linearity of the z transform enables us to apply the partial-fraction-expansion method to obtain the inverse z transform of $X(z)$. This method is a very powerful one, as will be demonstrated in this section. The method, which is parallel to the partial-fraction-expansion method used in Laplace transformation, is widely used in routine problems involving z transforms. The method requires that all terms in the partial fraction expansion be easily recognizable in the table of z transform pairs. (If necessary, the validity of the partial-fraction-expansion terms can be checked by recombining these terms.)

To find the inverse z transform, if $X(z)$ has one or more zeros at the origin ($z = 0$), then $X(z)/z$ or $X(z)$ is expanded into a sum of simple first- or second-order terms by partial fraction expansion, and a z transform table is used to find the corresponding time function of each of the expanded terms. It is noted that the only reason that we expand $X(z)/z$ into partial fractions is that each of the expanded terms has a form that may easily be found from commonly available z transform tables. If the shifting theorem is utilized in taking inverse z transforms, however, $X(z)$, instead of $X(z)/z$, may be expanded into partial fractions, as will be shown in Example 2–19.

In applying the shifting theorem, notice that if

$$X(z) = \frac{z^{-1}}{1 - az^{-1}}$$

then by writing $zX(z)$ as $Y(z)$ we have

$$zX(z) = Y(z) = \frac{1}{1 - az^{-1}}$$

and the inverse z transform of $Y(z)$ can be obtained as follows:

$$\mathcal{Z}^{-1}[Y(z)] = y(k) = a^k$$

Hence the inverse z transform of $X(z) = z^{-1}Y(z)$ is given by

$$\mathcal{Z}^{-1}[X(z)] = x(k) = y(k - 1)$$

Since $y(k)$ is assumed to be zero for all $k < 0$, we have

$$x(k)=\begin{cases} y(k-1)=a^{k-1} & k=1,2,3,\ldots \\ 0 & k\le 0 \end{cases}$$

Now consider $X(z)$ as given by

$$X(z)=\frac{b_0z^m+b_1z^{m-1}+\cdots+b_{m-1}z+b_m}{z^n+a_1z^{n-1}+\cdots+a_{n-1}z+a_n} \qquad m\le n$$

We first factor the denominator polynomial of $X(z)$ and find the poles of $X(z)$:

$$X(z)=\frac{b_0z^m+b_1z^{m-1}+\cdots+b_{m-1}z+b_m}{(z-p_1)(z-p_2)\cdots(z-p_n)}$$

Note that if the coefficients of the denominator and numerator polynomials are real, then any complex pole or zero will always be accompanied by its complex conjugate pole or zero, respectively.

A commonly used procedure for the case where all the poles are of simple order and there is at least one zero at the origin (that is, $b_m=0$) is to divide both sides of $X(z)$ by z and then expand $X(z)/z$ into partial fractions. Once $X(z)/z$ is expanded, then it will be of the form

$$\frac{X(z)}{z}=\frac{a_1}{z-p_1}+\frac{a_2}{z-p_2}+\cdots+\frac{a_n}{z-p_n}$$

The coefficient a_i can be determined by multiplying both sides of this last equation by $z-p_i$ and setting $z=p_i$. This will result in zero for all the terms on the right-hand side except the a_i term, in which the multiplicative factor $z-p_i$ has been canceled by the denominator. Hence we have

$$a_i=\left[(z-p_i)\frac{X(z)}{z}\right]_{z=p_i}$$

Note that such determination of a_i is valid only for simple poles.

If $X(z)/z$ involves a multiple pole—for example, a double pole at $z=p_1$ and no other poles—then $X(z)/z$ will have the form

$$\frac{X(z)}{z}=\frac{c_1}{(z-p_1)^2}+\frac{c_2}{z-p_1}$$

The coefficients c_1 and c_2 are determined from

$$c_1=\left[(z-p_1)^2\frac{X(z)}{z}\right]_{z=p_1}$$

$$c_2=\left\{\frac{d}{dz}\left[(z-p_1)^2\frac{X(z)}{z}\right]\right\}_{z=p_1}$$

It is noted that if $X(z)/z$ involves a triple pole at $z=p_1$, then the partial fractions must include a term $(z+p_1)/(z-p_1)^3$. (See Prob. A-2-17.)

When expanding $X(z)/z$ or $X(z)$ into partial fractions, each of the terms must be easily recognizable in a table of z transforms. The inverse z transform of $X(z)$ is obtained as the sum of the inverse z transforms of the partial fractions. Note that in contrast to the case of the Laplace transformation, the degree of the numerator and that of denominator of $X(z)$ can be the same; see Example 2–22.

Example 2–19.

Find $x(k)$ if $X(z)$ is given by

$$X(z) = \frac{10z}{(z-1)(z-0.2)}$$

We first expand $X(z)/z$ into partial fractions as follows:

$$\frac{X(z)}{z} = \frac{10}{(z-1)(z-0.2)} = \frac{12.5}{z-1} - \frac{12.5}{z-0.2}$$

Then we obtain

$$X(z) = 12.5\left(\frac{1}{1-z^{-1}} - \frac{1}{1-0.2z^{-1}}\right)$$

From Table 2–1 we find

$$\mathscr{Z}^{-1}\left[\frac{1}{1-z^{-1}}\right] = 1, \qquad \mathscr{Z}^{-1}\left[\frac{1}{1-0.2z^{-1}}\right] = (0.2)^k \qquad k = 0, 1, 2, \ldots$$

Hence

$$x(k) = 12.5\,[1-(0.2)^k] \qquad k = 0, 1, 2, \ldots$$

or

$$\begin{aligned} x(0) &= 0 \\ x(1) &= 10 \\ x(2) &= 12 \\ x(3) &= 12.4 \\ x(4) &= 12.48 \\ &\vdots \end{aligned}$$

In this example if $X(z)$, rather than $X(z)/z$, is expanded into partial fractions, then we obtain

$$X(z) = \frac{10z}{(z-1)(z-0.2)} = \frac{12.5}{z-1} - \frac{2.5}{z-0.2} = \frac{12.5z^{-1}}{1-z^{-1}} - \frac{2.5z^{-1}}{1-0.2z^{-1}}$$

Note that the inverse transform of $z^{-1}/(1-z^{-1})$ is not available from most z transform tables. However, by use of the shifting theorem we find

$$\mathscr{Z}^{-1}\left[\frac{z^{-1}}{1-z^{-1}}\right] = \mathscr{Z}^{-1}\left[z^{-1}\left(\frac{1}{1-z^{-1}}\right)\right] = \begin{cases} 1 & k = 1, 2, 3, \ldots \\ 0 & k \le 0 \end{cases}$$

Also,

$$\mathscr{Z}^{-1}\left[\frac{z^{-1}}{1-0.2z^{-1}}\right] = \mathscr{Z}^{-1}\left[z^{-1}\left(\frac{1}{1-0.2z^{-1}}\right)\right] = \begin{cases} (0.2)^{k-1} & k = 1, 2, 3, \ldots \\ 0 & k \le 0 \end{cases}$$

Hence

$$x(k) = \begin{cases} 12.5 - 2.5(0.2)^{k-1} \\ 0 \end{cases} = \begin{cases} 12.5[1-(0.2)^k] & k = 1, 2, 3, \ldots \\ 0 & k \le 0 \end{cases}$$

which can be rewritten as

$$x(k) = 12.5[1-(0.2)^k] \qquad k = 0, 1, 2, \ldots$$

which is the same as the result obtained earlier in the same example.

Example 2–20.

Given the z transform

$$X(z) = \frac{(1-e^{-aT})z}{(z-1)(z-e^{-aT})}$$

where a is a constant and T is the sampling period, determine the inverse z transform $x(kT)$ by use of the partial-fraction-expansion method.

The partial fraction expansion of $X(z)/z$ is found to be

$$\frac{X(z)}{z} = \frac{1}{z-1} - \frac{1}{z-e^{-aT}}$$

Thus,

$$X(z) = \frac{1}{1-z^{-1}} - \frac{1}{1-e^{-aT}z^{-1}}$$

Noting that

$$\mathscr{Z}^{-1}\left[\frac{1}{1-z^{-1}}\right] = 1$$

$$\mathscr{Z}^{-1}\left[\frac{1}{1-e^{-aT}z^{-1}}\right] = e^{-akT}$$

the inverse z transform of $X(z)$ is found to be

$$x(kT) = 1 - e^{-akT} \qquad k = 0, 1, 2, \ldots$$

Example 2–21.

Obtain the inverse z transform of

$$X(z) = \frac{z+2}{(z-2)z^2}$$

Expanding $X(z)$ into partial fractions, we obtain

$$X(z) = \frac{1}{z-2} - \frac{1}{z^2} - \frac{1}{z} = \frac{z^{-1}}{1-2z^{-1}} - z^{-2} - z^{-1}$$

[Note that in this example, $X(z)$ involves a double pole at $z = 0$. Hence the partial fraction expansion must include the terms $1/(z^2)$ and $1/z$.] By referring to Table 2–1, we find the inverse z transform of each term of this last equation. That is,

$$\mathcal{Z}^{-1}\left[\frac{z^{-1}}{1-2z^{-1}}\right] = \begin{cases} 2^{k-1} & k = 1, 2, 3, \ldots \\ 0 & k \le 0 \end{cases}$$

$$\mathcal{Z}^{-1}[z^{-2}] = \begin{cases} 1 & k = 2 \\ 0 & k \ne 2 \end{cases}$$

$$\mathcal{Z}^{-1}[z^{-1}] = \begin{cases} 1 & k = 1 \\ 0 & k \ne 1 \end{cases}$$

Hence, the inverse z transform of $X(z)$ can be given by

$$x(k) = \begin{cases} 0 - 0 - 0 = 0 & k = 0 \\ 1 - 0 - 1 = 0 & k = 1 \\ 2 - 1 - 0 = 1 & k = 2 \\ 2^{k-1} - 0 - 0 = 2^{k-1} & k = 3, 4, 5, \ldots \end{cases}$$

Rewriting, we have

$$x(k) = \begin{cases} 0 & k = 0, 1 \\ 1 & k = 2 \\ 2^{k-1} & k = 3, 4, 5, \ldots \end{cases}$$

To verify this result, the direct division method may be applied to this problem. Noting that

$$\begin{aligned} X(z) &= \frac{z+2}{(z-2)z^2} = \frac{z^{-2} + 2z^{-3}}{1-2z^{-1}} \\ &= z^{-2} + 4z^{-3} + 8z^{-4} + 16z^{-5} + 32z^{-6} + \cdots \\ &= z^{-2} + (2^{3-1})z^{-3} + (2^{4-1})z^{-4} + (2^{5-1})z^{-5} + (2^{6-1})z^{-6} + \cdots \end{aligned}$$

we find

$$x(k) = \begin{cases} 0 & k = 0, 1 \\ 1 & k = 2 \\ 2^{k-1} & k = 3, 4, 5, \ldots \end{cases}$$

Example 2–22.

Consider $X(z)$, where

$$X(z) = \frac{2z^3 + z}{(z-2)^2(z-1)}$$

Obtain the inverse z transform of $X(z)$.

We shall present three different approaches in applying the partial-fraction-expansion technique to this problem.

Method 1. Expand $X(z)/z$ into partial fractions as follows:

$$\frac{X(z)}{z} = \frac{2z^2+1}{(z-2)^2(z-1)} = \frac{9}{(z-2)^2} - \frac{1}{z-2} + \frac{3}{z-1}$$

Then

$$X(z) = \frac{9z^{-1}}{(1-2z^{-1})^2} - \frac{1}{1-2z^{-1}} + \frac{3}{1-z^{-1}}$$

The inverse z transforms of the individual terms give

$$\mathcal{Z}^{-1}\left[\frac{z^{-1}}{(1-2z^{-1})^2}\right] = k(2^{k-1}) \qquad k = 0, 1, 2, \ldots$$

$$\mathcal{Z}^{-1}\left[\frac{1}{1-2z^{-1}}\right] = 2^k \qquad k = 0, 1, 2, \ldots$$

$$\mathcal{Z}^{-1}\left[\frac{1}{1-z^{-1}}\right] = 1 \qquad k = 0, 1, 2, \ldots$$

and therefore

$$x(k) = \begin{cases} 2 & k = 0 \\ 9k(2^{k-1}) - 2^k + 3 & k = 1, 2, 3, \ldots \end{cases}$$

Method 2. Since the numerator is of the same degree as the denominator we may first divide the numerator by the denominator:

$$X(z) = 2 + \frac{10z^2 - 15z + 8}{(z-2)^2(z-1)}$$

Define

$$\hat{X}(z) = \frac{10z^2 - 15z + 8}{(z-2)^2(z-1)}$$

Since the numerator involves a constant term, it is not convenient to expand $\hat{X}(z)/z$ into partial fractions, and therefore we expand $\hat{X}(z)$ into the following form:

$$\hat{X}(z) = \frac{9z}{(z-2)^2} - \frac{2}{z-2} + \frac{3}{z-1} = \frac{9z^{-1}}{(1-2z^{-1})^2} - \frac{2z^{-1}}{1-2z^{-1}} + \frac{3z^{-1}}{1-z^{-1}}$$

Then

$$X(z) = 2 + \frac{9z^{-1}}{(1-2z^{-1})^2} - \frac{2z^{-1}}{1-2z^{-1}} + \frac{3z^{-1}}{1-z^{-1}}$$

Noting that

$$\mathcal{Z}^{-1}[2] = \begin{cases} 2 & k = 0 \\ 0 & k = 1, 2, 3, \ldots \end{cases}$$

$$\mathscr{Z}^{-1}\left[\frac{z^{-1}}{(1-2z^{-1})^2}\right]=\begin{cases}k(2^{k-1}) & k=1,2,3,\dots\\ 0 & k\le 0\end{cases}$$

$$\mathscr{Z}^{-1}\left[\frac{z^{-1}}{1-2z^{-1}}\right]=\begin{cases}2^{k-1} & k=1,2,3,\dots\\ 0 & k\le 0\end{cases}$$

$$\mathscr{Z}^{-1}\left[\frac{z^{-1}}{1-z^{-1}}\right]=\begin{cases}1 & k=1,2,3,\dots\\ 0 & k\le 0\end{cases}$$

we obtain

$$x(k)=\begin{cases}2 & k=0\\ 9k(2^{k-1})-2^k+3 & k=1,2,3,\dots\end{cases}$$

which agrees with the results obtained by method 1.

Method 3. If $\hat{X}(z)$ is expanded into the form

$$\hat{X}(z)=\frac{18}{(z-2)^2}+\frac{7}{z-2}+\frac{3}{z-1}$$

then

$$X(z)=2+\frac{18z^{-2}}{(1-2z^{-1})^2}+\frac{7z^{-1}}{1-2z^{-1}}+\frac{3z^{-1}}{1-z^{-1}}$$

By referring to Prob. A-2–13, we have

$$\mathscr{Z}^{-1}\left[\frac{z^{-2}}{(1-2z^{-1})^2}\right]=\mathscr{Z}^{-1}\left[z^{-1}\frac{z^{-1}}{(1-2z^{-1})^2}\right]=\begin{cases}(k-1)2^{k-2} & k=1,2,3,\dots\\ 0 & k\le 0\end{cases}$$

The inverse z transform of $X(z)$ can then be obtained as follows:

$$x(k)=\begin{cases}2 & k=0\\ 18(k-1)2^{k-2}+7(2^{k-1})+3 & k=1,2,3,\dots\end{cases}$$

The expression $x(k)$ for $k = 1, 2, 3, \dots$ can be rewritten as

$$\begin{aligned}x(k)&=18k(2^{k-2})-18(2^{k-2})+7(2^{k-1})+3\\ &=9k(2^{k-1})-2(2^{k-1})+3\\ &=9k(2^{k-1})-2^k+3\end{aligned}$$

Hence, the complete solution is

$$x(k)=\begin{cases}2 & k=0\\ 9k(2^{k-1})-2^k+3 & k=1,2,3,\dots\end{cases}$$

Example 2–23.

Find the inverse z transform of

$$X(z)=\frac{z^2+6z}{(z^2-2z+2)(z-1)} \tag{2–40}$$

By expanding $X(z)/z$ into partial fractions, we obtain

$$\frac{X(z)}{z} = \frac{7}{z-1} - \frac{7z-8}{z^2-2z+2}$$

Hence we have

$$X(z) = \frac{7}{1-z^{-1}} - \frac{7(1-\frac{8}{7}z^{-1})}{1-2z^{-1}+2z^{-2}}$$

Note that the poles of the second term on the right-hand side of this last equation are complex conjugates. Hence, the inverse z transform of this term is a damped cosine and/or damped sine function. Referring to the formulas

$$\mathcal{Z}[e^{-akT}\cos\omega kT] = \frac{1-e^{-aT}z^{-1}\cos\omega T}{1-2e^{-aT}z^{-1}\cos\omega T + e^{-2aT}z^{-2}}$$

and

$$\mathcal{Z}[e^{-akT}\sin\omega kT] = \frac{e^{-aT}z^{-1}\sin\omega T}{1-2e^{-aT}z^{-1}\cos\omega T + e^{-2aT}z^{-2}}$$

we find the inverse z transform of

$$\hat{X}(z) = \frac{1-\frac{8}{7}z^{-1}}{1-2z^{-1}+2z^{-2}}$$

as follows. First, we rewrite this last equation as a sum of a damped cosine function and a damped sine function:

$$\hat{X}(z) = \frac{1-z^{-1}}{1-2z^{-1}+2z^{-2}} - \frac{1}{7}\frac{z^{-1}}{1-2z^{-1}+2z^{-2}}$$

and then we identify e^{-aT} and ωT as follows:

$$e^{-2aT} = 2, \qquad e^{-aT}\cos\omega T = 1$$

or

$$e^{-aT} = \sqrt{2}, \qquad \omega T = \tfrac{1}{4}\pi$$

from which we obtain

$$\cos\omega T = \frac{1}{\sqrt{2}}, \qquad \sin\omega T = \frac{1}{\sqrt{2}}$$

We can now find the inverse z transform of

$$\hat{X}(z) = \frac{1-(\sqrt{2})\dfrac{1}{\sqrt{2}}z^{-1}}{1-2(\sqrt{2})\dfrac{1}{\sqrt{2}}z^{-1}+2z^{-2}} - \frac{1}{7}\frac{(\sqrt{2})\dfrac{1}{\sqrt{2}}z^{-1}}{1-2(\sqrt{2})\dfrac{1}{\sqrt{2}}z^{-1}+2z^{-2}}$$

as follows:

$$\begin{aligned}\mathcal{Z}^{-1}[\hat{X}(z)] &= e^{-akT}\cos\omega kT - \tfrac{1}{7}e^{-akT}\sin\omega kT \\ &= (\sqrt{2})^k\cos\frac{\pi k}{4} - \frac{1}{7}(\sqrt{2})^k\sin\frac{\pi k}{4}\end{aligned}$$

Since

$$X(z) = \frac{7}{1 - z^{-1}} - 7\hat{X}(z)$$

we obtain

$$x(k) = 7 - 7(\sqrt{2})^k \cos\frac{\pi k}{4} + (\sqrt{2})^k \sin\frac{\pi k}{4} \qquad k = 0, 1, 2, \ldots$$

from which we get

$$x(0) = 0$$
$$x(1) = 1$$
$$x(2) = 9$$
$$x(3) = 23$$
$$\vdots$$

The present problem can be easily solved by the iterative method or the computational method. A BASIC computer program for finding $x(k)$, the inverse z transform of $X(z)$, is shown in Table 2–4.

TABLE 2–4 BASIC COMPUTER PROGRAM FOR FINDING $x(k)$, THE INVERSE z TRANSFORM OF $X(z)$ AS GIVEN BY EQ. (2–40)

```
10  X0 = 0
20  X1 = 1
30  X2 = 9
40  U0 = 1
50  U1 = 0
60  U2 = 0
70  K = 0
80  X3 = 3 * X2 - 4 * X1 + 2 * X0 + U2 + 6 * U1
90  M = X0
100 X0 = X1
110 X1 = X2
120 X2 = X3
130 N = U0
140 U0 = U1
150 PRINT K, M, N
160 K = K + 1
170 IF K < 16 GO TO 80
180 END
```

k = K	$x(k)$ = XK = M	$u(k)$ = UK = N
0	0	1
1	1	0
2	9	0
3	23	0
4	35	0
5	31	0
6	−1	0
7	−57	0
8	−105	0
9	−89	0
10	39	0
11	263	0
12	455	0
13	391	0
14	−121	0
15	−1017	0

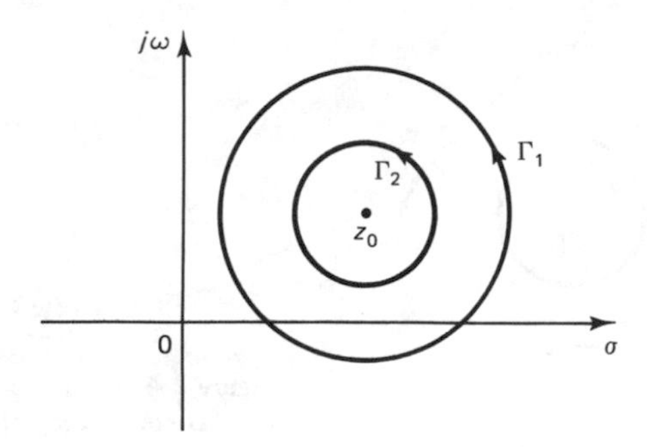

Figure 2–3 Analytic region for function $F(z)$.

Inversion integral method. The fourth method for finding the inverse z transform is to use the inversion integral. It is the most general technique for obtaining the inverse z transform. It is based on complex variable theory. (For a rigorous and complete derivation of the inversion integral, refer to a book on complex variable theory, such as Ref. 2–2.) In presenting the inversion integral formula for the z transform, we need to review the residue theorem and its associated background material.

Review of background material in deriving the inversion integral formula. Suppose z_0 is an isolated singular point (pole) of $F(z)$. It can be seen that a positive number r_1 exists such that the function $F(z)$ is analytic at every point z for which $0 < |z - z_0| \leq r_1$. Let us denote the circle with center at $z = z_0$ and radius r_1 as Γ_1. Define Γ_2 as any circle with center at $z = z_0$ and radius $|z - z_0| = r_2$ for which $r_2 \leq r_1$. Circles Γ_1 and Γ_2 are shown in Fig. 2–3. Then the Laurent series expansion of $F(z)$ about pole $z = z_0$ may be given by

$$F(z) = \sum_{n=0}^{\infty} a_n (z - z_0)^n + \sum_{n=1}^{\infty} \frac{b_n}{(z - z_0)^n}$$

where coefficients a_n and b_n are given by

$$a_n = \frac{1}{2\pi j} \oint_{\Gamma_1} \frac{F(z)}{(z - z_0)^{n+1}} dz \qquad n = 0, 1, 2, \ldots$$

$$b_n = \frac{1}{2\pi j} \oint_{\Gamma_2} \frac{F(z)}{(z - z_0)^{-n+1}} dz \qquad n = 1, 2, 3, \ldots$$

Notice that the coefficient b_1 is given by

$$b_1 = \frac{1}{2\pi j} \oint_{\Gamma_2} F(z)\, dz \tag{2–41}$$

It can be proved that the value of the integral of Eq. (2–41) is unchanged if Γ_1 is replaced by any closed curve Γ around z_0 such that $F(z)$ is analytic on and inside

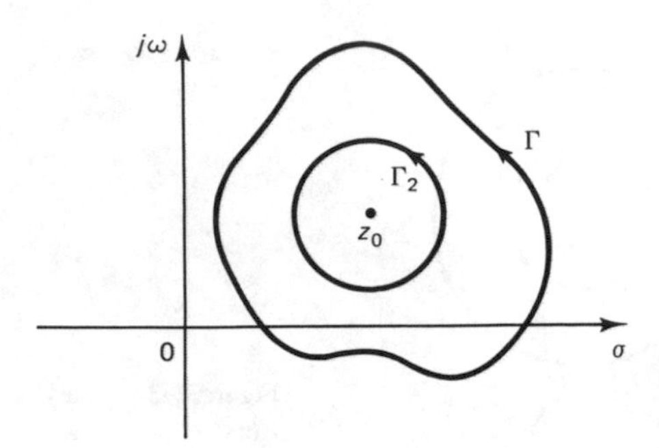

Figure 2–4 Analytic region for function $F(z)$ as bounded by closed curve Γ.

Γ except at pole $z = z_0$ (see Fig. 2–4). The closed curve Γ may extend outside the circle Γ_1. Then, by referring to the Cauchy-Goursat theorem, we have

$$\oint_{\Gamma} F(z)\, dz - \oint_{\Gamma_2} F(z)\, dz = 0$$

Thus Eq. (2–41) can be written as

$$b_1 = \frac{1}{2\pi j} \oint_{\Gamma} F(z)\, dz$$

The coefficient b_1 is called the *residue* of $F(z)$ at the pole z_0.

Next, let us assume that the closed curve Γ encloses m isolated poles $z_1, z_2, \ldots, z_m$, as shown in Fig. 2–5. Notice that the function $F(z)$ is analytic in the shaded region. According to the Cauchy-Goursat theorem the integral of $F(z)$ over the shaded region is zero. The integral over the total shaded region is

$$\oint_{\Gamma} F(z)\, dz - \oint_{\Gamma_1} F(z)\, dz - \oint_{\Gamma_2} F(z)\, dz - \cdots - \oint_{\Gamma_m} F(z)\, dz = 0$$

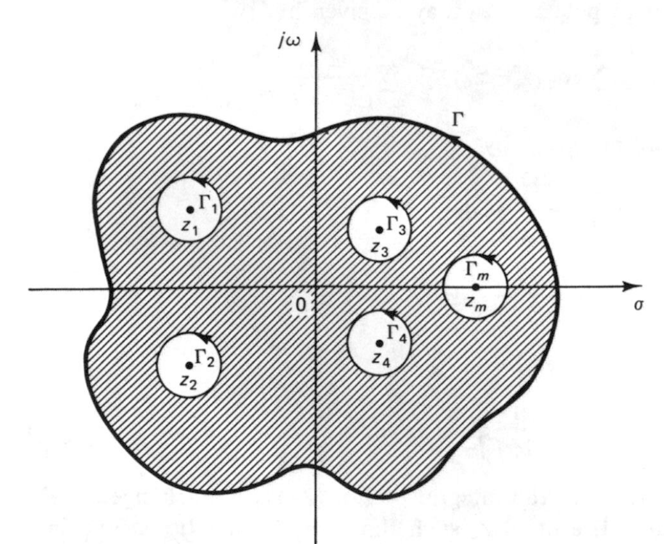

Figure 2–5 Closed curve Γ enclosing m isolated poles $z_1, z_2, \ldots, z_m$.

where $\Gamma_1, \Gamma_2, \ldots, \Gamma_m$ are closed curves around the poles $z_1, z_2, \ldots, z_m$, respectively. Hence

$$\oint_{\Gamma} F(z)\,dz = \oint_{\Gamma_1} F(z)\,dz + \oint_{\Gamma_2} F(z)\,dz + \cdots + \oint_{\Gamma_m} F(z)\,dz$$
$$= 2\pi j(b_{1_1} + b_{1_2} + \cdots + b_{1_m})$$
$$= 2\pi j(K_1 + K_2 + \cdots + K_m) \tag{2-42}$$

where $K_1 = b_{1_1}, K_2 = b_{1_2}, \ldots, K_m = b_{1_m}$ are residues of $F(z)$ at poles $z_1, z_2, \ldots, z_m$, respectively.

Equation (2–42) is known as the *residue theorem.* It states that if a function $F(z)$ is analytic within and on a closed curve Γ except at a finite number of poles $z_1, z_2, \ldots, z_m$ inside Γ, then the integral of $F(z)$ taken counterclockwise around Γ is equal to $2\pi j$ times the sum of the residues at poles $z_1, z_2, \ldots, z_m$.

Inversion integral for the z transform. We shall now use the Cauchy-Goursat theorem and the residue theorem to derive the inversion integral for the z transform.

From the definition of the z transform, we have

$$X(z) = \sum_{k=0}^{\infty} x(kT)z^{-k} = x(0) + x(T)z^{-1} + x(2T)z^{-2} + \cdots + x(kT)z^{-k} + \cdots$$

By multiplying both sides of this last equation by z^{k-1}, we obtain

$$X(z)z^{k-1} = x(0)z^{k-1} + x(T)z^{k-2} + x(2T)z^{k-3} + \cdots + x(kT)z^{-1} + \cdots \tag{2-43}$$

Notice that Eq. (2–43) is the Laurent series expansion of $X(z)z^{k-1}$ around point $z = 0$.

Consider a circle C with its center at the origin of the z plane such that all poles of $X(z)z^{k-1}$ are inside it. Noting that the coefficient $x(kT)$ associated with the term z^{-1} in Eq. (2–43) is the residue, we obtain

$$x(kT) = \frac{1}{2\pi j}\oint_C X(z)z^{k-1}\,dz \tag{2-44}$$

Equation (2–44) is the inversion integral for the z transform. The evaluation of the inversion integral can be done by using the residue theorem.

Let us define the poles of $X(z)z^{k-1}$ as $z_1, z_2, \ldots, z_m$. Since the closed curve C encloses all poles $z_1, z_2, \ldots, z_m$, then referring to Eq. (2–42) we have

$$\oint_C X(z)z^{k-1}\,dz = \oint_{C_1} X(z)z^{k-1}\,dz + \oint_{C_2} X(z)z^{k-1}\,dz + \cdots + \oint_{C_m} X(z)z^{k-1}\,dz$$
$$= 2\pi j(K_1 + K_2 + \cdots + K_m) \tag{2-45}$$

where $K_1, K_2, \ldots, K_m$ denote the residues of $X(z)z^{k-1}$ at poles $z_1, z_2, \ldots, z_m$, respectively, and $C_1, C_2, \ldots, C_m$ are small closed curves around the isolated poles $z_1, z_2, \ldots, z_m$, respectively.

Now we combine Eqs. (2–44) and (2–45) to obtain a very useful result. Since $X(z)z^{k-1}$ has m poles, namely, $z_1, z_2, \ldots, z_m$,

$$x(k) = x(kT) = K_1 + K_2 + \cdots + K_m$$

$$= \sum_{i=1}^{m} [\text{residue of } X(z)z^{k-1} \text{ at pole } z = z_i \text{ of } X(z)z^{k-1}] \tag{2–46}$$

In evaluating residues, note that if the denominator of $X(z)z^{k-1}$ contains a simple pole $z = z_i$, then the corresponding residue K is

$$K = \lim_{z \to z_i} [(z - z_i)X(z)z^{k-1}]$$

If $X(z)z^{k-1}$ contains a multiple pole z_j of order q, then the residue K is given by

$$K = \frac{1}{(q-1)!} \lim_{z \to z_j} \frac{d^{q-1}}{dz^{q-1}} [(z - z_j)^q X(z)z^{k-1}]$$

Note that in this book we treat only one-sided z transforms, as defined by Eq. (2–1). This implies that $x(k) = 0$ for $k < 0$. Hence, we restrict the values of k in Eq. (2–45) to the nonnegative integer values.

If $X(z)$ has a zero of order r at the origin, then $X(z)z^{k-1}$ in Eq. (2–45) will involve a zero of order $r + k - 1$ at the origin. If $r \geq 1$, then $r + k - 1 \geq 0$ for $k \geq 0$, and there is no pole at $z = 0$ in $X(z)z^{k-1}$. However, if $r \leq 0$, then there will be a pole at $z = 0$ for one or more nonnegative values of k. In such a case, separate inversion of Eq. (2–45) is necessary for each of such values of k. (See Prob. A-2–18.)

It should be noted that the inversion integral method, when evaluated by residues, is a very simple technique for obtaining the inverse z transform, provided that $X(z)z^{k-1}$ has no poles at the origin, $z = 0$. If, however, $X(z)z^{k-1}$ has a simple pole or a multiple pole at $z = 0$, then calculations may become cumbersome and the partial-fraction-expansion method may prove to be simpler to apply. (See Prob. A-2–18.) On the other hand, in certain problems the partial-fraction-expansion approach may become laborious. Then, the inversion integral method proves to be very convenient.

Example 2–24.

Obtain $x(kT)$ by using the inversion integral method when $X(z)$ is given by

$$X(z) = \frac{z(1 - e^{-aT})}{(z-1)(z - e^{-aT})}$$

Note that

$$X(z)z^{k-1} = \frac{(1 - e^{-aT})z^k}{(z-1)(z - e^{-aT})}$$

For $k = 0, 1, 2, \ldots, X(z)z^{k-1}$ has two simple poles, $z = z_1 = 1$ and $z = z_2 = e^{-aT}$. Hence from Eq. (2–46) we have

$$x(k) = \sum_{i=1}^{2} \left[\text{residue of } \frac{(1 - e^{-aT})z^k}{(z-1)(z-e^{-aT})} \text{ at pole } z = z_i \right]$$

$$= K_1 + K_2$$

where

$$K_1 = [\text{residue at simple pole } z = 1]$$

$$= \lim_{z \to 1} \left[(z-1) \frac{(1 - e^{-aT})z^k}{(z-1)(z-e^{-aT})} \right] = 1$$

$$K_2 = [\text{residue at simple pole } z = e^{-aT}]$$

$$= \lim_{z \to e^{-aT}} \left[(z - e^{-aT}) \frac{(1 - e^{-aT})z^k}{(z-1)(z-e^{-aT})} \right] = -e^{-akT}$$

Hence

$$x(kT) = K_1 + K_2 = 1 - e^{-akT} \qquad k = 0, 1, 2, \ldots$$

Example 2–25.

Obtain the inverse z transform of

$$X(z) = \frac{z^2}{(z-1)^2(z-e^{-aT})}$$

by using the inversion integral method.

Notice that

$$X(z)z^{k-1} = \frac{z^{k+1}}{(z-1)^2(z-e^{-aT})}$$

For $k = 0, 1, 2, \ldots, X(z)z^{k-1}$ has a simple pole at $z = z_1 = e^{-aT}$ and a double pole at $z = z_2 = 1$. Hence, from Eq. (2–46) we obtain

$$x(k) = \sum_{i=1}^{2} \left[\text{residue of } \frac{z^{k+1}}{(z-1)^2(z-e^{-aT})} \text{ at pole } z = z_i \right]$$

$$= K_1 + K_2$$

where

$$K_1 = [\text{residue at simple pole } z = e^{-aT}]$$

$$= \lim_{z \to e^{-aT}} \left[(z - e^{-aT}) \frac{z^{k+1}}{(z-1)^2(z-e^{-aT})} \right] = \frac{e^{-a(k+1)T}}{(1-e^{-aT})^2}$$

$$K_2 = [\text{residue at double pole } z = 1]$$

$$= \frac{1}{(2-1)!} \lim_{z \to 1} \frac{d}{dz} \left[(z-1)^2 \frac{z^{k+1}}{(z-1)^2(z-e^{-aT})} \right]$$

$$= \lim_{z\to 1} \frac{d}{dz}\left(\frac{z^{k+1}}{z - e^{-aT}}\right)$$

$$= \lim_{z\to 1} \frac{(k+1)z^k (z - e^{-aT}) - z^{k+1}}{(z - e^{-aT})^2}$$

$$= \frac{k}{1 - e^{-aT}} - \frac{e^{-aT}}{(1 - e^{-aT})^2}$$

Hence,

$$x(kT) = K_1 + K_2 = \frac{e^{-aT} e^{-akT}}{(1 - e^{-aT})^2} + \frac{k}{1 - e^{-aT}} - \frac{e^{-aT}}{(1 - e^{-aT})^2}$$

$$= \frac{kT}{T(1 - e^{-aT})} - \frac{e^{-aT}(1 - e^{-akT})}{(1 - e^{-aT})^2} \qquad k = 0, 1, 2, \ldots$$

Example 2–26.

Obtain the inverse z transform of

$$X(z) = \frac{10}{(z-1)(z-2)}$$

Note that

$$X(z)z^{k-1} = \frac{10z^{k-1}}{(z-1)(z-2)}$$

For $k = 0$, notice that $X(z)z^{k-1}$ becomes

$$X(z)z^{k-1} = \frac{10}{(z-1)(z-2)z} \qquad k = 0$$

Hence, for $k = 0$, $X(z)z^{k-1}$ has three simple poles, $z = z_1 = 1$, $z = z_2 = 2$, and $z = z_3 = 0$. For $k = 1, 2, 3, \ldots$, however, $X(z)z^{k-1}$ has only two simple poles, $z = z_1 = 1$ and $z = z_2 = 2$. Therefore, we must consider $x(0)$ and $x(k)$ (where $k = 1, 2, 3, \ldots$) separately.

For $k = 0$. For this case, referring to Eq. (2–46), we have

$$x(0) = \sum_{i=1}^{3} \left[\text{residue of } \frac{10}{(z-1)(z-2)z} \text{ at pole } z = z_i\right]$$

$$= K_1 + K_2 + K_3$$

where

$$K_1 = [\text{residue at simple pole } z = 1]$$

$$= \lim_{z\to 1}\left[(z-1)\frac{10}{(z-1)(z-2)z}\right] = -10$$

$$K_2 = [\text{residue at simple pole } z = 2]$$

$$= \lim_{z\to 2}\left[(z-2)\frac{10}{(z-1)(z-2)z}\right] = 5$$

$$K_3 = [\text{residue at simple pole } z = 0]$$

$$= \lim_{z \to 0} \left[z \frac{10}{(z-1)(z-2)z} \right] = 5$$

Hence

$$x(0) = K_1 + K_2 + K_3 = -10 + 5 + 5 = 0$$

For $k = 1, 2, 3, \ldots$ For this case, Eq. (2–46) becomes

$$x(k) = \sum_{i=1}^{2} \left[\text{residue of } \frac{10z^{k-1}}{(z-1)(z-2)} \text{ at pole } z = z_i \right]$$

$$= K_1 + K_2$$

where

$$K_1 = [\text{residue at simple pole } z = 1]$$

$$= \lim_{z \to 1} \left[(z-1) \frac{10z^{k-1}}{(z-1)(z-2)} \right] = -10$$

$$K_2 = [\text{residue at simple pole } z = 2]$$

$$= \lim_{z \to 2} \left[(z-2) \frac{10z^{k-1}}{(z-1)(z-2)} \right] = 10(2^{k-1})$$

Thus

$$x(k) = K_1 + K_2 = -10 + 10(2^{k-1}) = 10(2^{k-1} - 1) \qquad k = 1, 2, 3, \ldots$$

Hence, the inverse z transform of the given $X(z)$ can be written

$$x(k) = \begin{cases} 0 & k = 0 \\ 10(2^{k-1} - 1) & k = 1, 2, 3, \ldots \end{cases}$$

An alternate way to write $x(k)$ for $k \geq 0$ is

$$x(k) = 5\delta_0(k) + 10(2^{k-1} - 1) \qquad k = 0, 1, 2, \ldots$$

where $\delta_0(k)$ is the Kronecker delta and is given by

$$\delta_0(k) = \begin{cases} 1 & \text{for } k = 0 \\ 0 & \text{for } k \neq 0 \end{cases}$$

Comments on calculating residues. In obtaining the residues of a function $X(z)$, note that regardless of the way we calculate the residues, the final result is the same. Therefore, we may use any method that is convenient for a given situation. As an example, consider the following function $X(z)$:

$$X(z) = \frac{2z^2 + 5z + 6}{(z+1)^3} + \frac{4z}{(z+1)^2} + \frac{5}{z+1}$$

We shall demonstrate three methods for calculating the residue of this function $X(z)$.

Method 1. The residue of this function may be obtained as the sum of the residues of the respective terms:

$$[\text{Residue } K \text{ of } X(z) \text{ at pole } z=-1]$$

$$=\frac{1}{(3-1)!}\lim_{z\to-1}\frac{d^2}{dz^2}\left[(z+1)^3\frac{2z^2+5z+6}{(z+1)^3}\right]$$

$$+\frac{1}{(2-1)!}\lim_{z\to-1}\frac{d}{dz}\left[(z+1)^2\frac{4z}{(z+1)^2}\right]+\lim_{z\to-1}\left[(z+1)\frac{5}{z+1}\right]$$

$$=\tfrac{1}{2}\lim_{z\to-1}(4)+\lim_{z\to-1}(4)+\lim_{z\to-1}(5)=2+4+5$$

$$=11$$

Method 2. If the three terms of $X(z)$ are combined into one—

$$X(z)=\frac{2z^2+5z+6}{(z+1)^3}+\frac{4z}{(z+1)^2}+\frac{5}{z+1}=\frac{11z^2+19z+11}{(z+1)^3}$$

then the residue can be calculated as follows:

$$[\text{Residue } K \text{ of } X(z) \text{ at pole } z=-1]$$

$$=\frac{1}{(3-1)!}\lim_{z\to-1}\frac{d^2}{dz^2}\left[(z+1)^3\frac{11z^2+19z+11}{(z+1)^3}\right]$$

$$=\tfrac{1}{2}\lim_{z\to-1}(22)$$

$$=11$$

Method 3. If $X(z)$ is expanded into usual partial fractions—

$$X(z)=\frac{11z^2+19z+11}{(z+1)^3}=\frac{3}{(z+1)^3}-\frac{3}{(z+1)^2}+\frac{11}{z+1}$$

then the residue of $X(z)$ is the coefficient of the term $1/(z+1)$. Thus,

$$[\text{Residue } K \text{ of } X(z) \text{ at pole } z=-1]=11$$

2-6 PULSE TRANSFER FUNCTION AND WEIGHTING SEQUENCE

In this section we shall first define the pulse transfer function and the weighting sequence. Then we shall discuss how the z transform method is used for solving difference equations. Computer solutions of difference equations are also discussed.

Pulse transfer function and weighting sequence. Consider the linear time-invariant discrete-time system characterized by the following linear difference equation:

$$x(k) + a_1 x(k-1) + \cdots + a_n x(k-n)$$
$$= b_0 u(k) + b_1 u(k-1) + \cdots + b_n u(k-n) \tag{2–47}$$

where $u(k)$ and $x(k)$ are the system's input and output, respectively, at the kth iteration. In describing such a difference equation in the z plane, we take the z transform of each term in the equation.

Let us define

$$\mathcal{Z}[x(k)] = X(z)$$

Then, $x(k+1)$, $x(k+2)$, $x(k+3)$, . . . and $x(k-1)$, $x(k-2)$, $x(k-3)$, . . . can be expressed in terms of $X(z)$ and the initial conditions. Their exact z transforms were derived in Sec. 2–4 and are summarized in Table 2–5 for convenient reference.

Consider the difference equation given by Eq. (2–47). By referring to Table 2–5, the z transform of Eq. (2–47) becomes

$$X(z) + a_1 z^{-1} X(z) + \cdots + a_n z^{-n} X(z)$$
$$= b_0 U(z) + b_1 z^{-1} U(z) + \cdots + b_n z^{-n} U(z)$$

or

$$(1 + a_1 z^{-1} + \cdots + a_n z^{-n}) X(z) = (b_0 + b_1 z^{-1} + \cdots + b_n z^{-n}) U(z)$$

TABLE 2–5 ***z*** **TRANSFORMS OF $x(k+m)$ AND $x(k-m)$**

Discrete function	z transform
$x(k+4)$	$z^4 X(z) - z^4 x(0) - z^3 x(1) - z^2 x(2) - z x(3)$
$x(k+3)$	$z^3 X(z) - z^3 x(0) - z^2 x(1) - z x(2)$
$x(k+2)$	$z^2 X(z) - z^2 x(0) - z x(1)$
$x(k+1)$	$z X(z) - z x(0)$
$x(k)$	$X(z)$
$x(k-1)$	$z^{-1} X(z)$
$x(k-2)$	$z^{-2} X(z)$
$x(k-3)$	$z^{-3} X(z)$
$x(k-4)$	$z^{-4} X(z)$

which can be written as

$$X(z)=\frac{b_0+b_1z^{-1}+\cdots+b_nz^{-n}}{1+a_1z^{-1}+\cdots+a_nz^{-n}}U(z) \tag{2–48}$$

Define

$$G(z)=\frac{b_0+b_1z^{-1}+\cdots+b_nz^{-n}}{1+a_1z^{-1}+\cdots+a_nz^{-n}} \tag{2–49}$$

Consider the response of the linear discrete-time system given by Eq. (2–48), initially at rest, when the input $u(t)$ is the Kronecker delta function $\delta_0(kT)$:

$$\delta_0(kT)=\begin{cases}1 & \text{for } k=0\\ 0 & \text{for } k\neq 0\end{cases}$$

Since

$$\mathscr{Z}[\delta_0(kT)]=1$$

we have

$$U(z)=\mathscr{Z}[\delta_0(kT)]=1$$

and

$$X(z)=\frac{b_0+b_1z^{-1}+\cdots+b_nz^{-n}}{1+a_1z^{-1}+\cdots+a_nz^{-n}}=G(z) \tag{2–50}$$

Thus, $G(z)$ is the response of the system to the Kronecker delta input and serves in the same role as the transfer function in linear continuous-time systems. (The transfer function is the Laplace transform of the response of the linear continuous-time system to a unit impulse input when all initial conditions are assumed to be zero.) The function $G(z)$ is called the *pulse transfer function.* (We shall discuss the pulse transfer function more fully in Sec. 3–5.)

It is noted that the role of the Kronecker delta function $\delta_0(kT)$ in discrete-time control systems is similar to that of the unit impulse function $\delta(t)$ (the Dirac delta function) in continuous-time control systems.

The inverse z transform of $G(z)$ as given by Eq. (2–49),

$$g(k)=\mathscr{Z}^{-1}[G(z)]$$

is called the *weighting sequence.* To obtain the weighting sequence of a linear discrete-time system, we first obtain the pulse transfer function $G(z)$ of the system and then take its inverse z transform. Further discussion of the weighting sequence will be found in Sec. 3–5.

Example 2–27.

Consider the difference equation

$$x(k+2)+a_1x(k+1)+a_2x(k)=b_0u(k+2)+b_1u(k+1)+b_2u(k) \tag{2–51}$$

Assuming that the system is initially at rest and $u(k) = 0$ for $k < 0$, obtain the pulse transfer function $G(z) = X(z)/U(z)$.

The z transform of Eq. (2–51) gives

$$[z^2X(z) - z^2x(0) - zx(1)] + a_1[zX(z) - zx(0)] + a_2X(z)$$
$$= b_0[z^2U(z) - z^2u(0) - zu(1)] + b_1[zU(z) - zu(0)] + b_2U(z)$$

or

$$(z^2 + a_1z + a_2)X(z) = (b_0z^2 + b_1z + b_2)U(z) + z^2[x(0) - b_0u(0)]$$
$$+ z[x(1) + a_1x(0) - b_0u(1) - b_1u(0)] \tag{2–52}$$

Since the system is initially at rest, $x(k) = 0$ for $k < 0$. To determine the initial conditions $x(0)$ and $x(1)$, we substitute $k = -2$ and $k = -1$, respectively, into Eq. (2–51):

$$x(0) + a_1x(-1) + a_2x(-2) = b_0u(0) + b_1u(-1) + b_2u(-2)$$

which simplifies to

$$x(0) = b_0u(0) \tag{2–53}$$

and

$$x(1) + a_1x(0) + a_2x(-1) = b_0u(1) + b_1u(0) + b_2u(-1)$$

or

$$x(1) = -a_1x(0) + b_0u(1) + b_1u(0) \tag{2–54}$$

By substituting Eqs. (2–53) and (2–54) into Eq. (2–52), we get

$$(z^2 + a_1z + a_2)X(z) = (b_0z^2 + b_1z + b_2)U(z)$$

Thus

$$X(z) = \frac{b_0z^2 + b_1z + b_2}{z^2 + a_1z + a_2}U(z) \tag{2–55}$$

Notice that if both $x(k)$ and $u(k)$ are zero for $k < 0$, then the system's input and output are related by Eq. (2–55). The pulse transfer function $G(z) = X(z)/U(z)$ can now be obtained from Eq. (2–55), as follows:

$$G(z) = \frac{X(z)}{U(z)} = \frac{b_0z^2 + b_1z + b_2}{z^2 + a_1z + a_2} = \frac{b_0 + b_1z^{-1} + b_2z^{-2}}{1 + a_1z^{-1} + a_2z^{-2}}$$

Comments. The system described by

$$x(k) + a_1x(k-1) + \cdots + a_nx(k-n)$$
$$= b_0u(k) + b_1u(k-1) + \cdots + b_nu(k-n)$$

where $u(k)$ and $x(k)$ are the system's input and output, respectively, at the kth iteration, and the system described by

$$x(k+n) + a_1x(k+n-1) + \cdots + a_nx(k)$$
$$= b_0u(k+n) + b_1u(k+n-1) + \cdots + b_nu(k)$$

where the system is initially at rest [that is, where $x(k) = 0$ for $k < 0$] and where input $u(k) = 0$ for $k < 0$, can both be described by the same z transform equation:

$$X(z) = \frac{b_0 + b_1 z^{-1} + \cdots + b_n z^{-n}}{1 + a_1 z^{-1} + \cdots + a_n z^{-n}} U(z)$$

Example 2–28.

Obtain the weighting sequence for the discrete-time system

$$x(k) - ax(k-1) = u(k)$$

The z transform of this equation gives

$$X(z) - az^{-1}X(z) = U(z)$$

or

$$X(z) = \frac{1}{1 - az^{-1}} U(z)$$

Hence the inverse z transform of

$$G(z) = \frac{X(z)}{U(z)} = \frac{1}{1 - az^{-1}}$$

gives the weighting sequence

$$g(k) = \mathscr{Z}^{-1}[G(z)] = \begin{cases} a^k & k = 0, 1, 2, \ldots \\ 0 & k < 0 \end{cases}$$

***z* transform method for solving difference equations.** Difference equations can be solved easily by use of a digital computer, provided the numerical values of all coefficients and parameters are given. However, closed-form expressions for $x(k)$ cannot be obtained from the computer solution, except for very special cases. The usefulness of the z transform method is that it is possible to obtain the closed-form expression for $x(k)$.

In what follows, we shall demonstrate by means of examples how difference equations may be solved and the closed-form expression for $x(k)$ may be obtained by use of the z transform method. Then we shall discuss the computer solution of the difference equation. Although closed-form solutions cannot be obtained, computer solutions prove to be very convenient in saving time. [The solution $x(k)$ up to an arbitrarily large k value may be obtained in a very short time period.] In the following examples (Examples 2–29 and 2–30) we shall present BASIC computer programs for solving each of the problems so that the reader may solve each problem both ways.

Example 2–29.

Solve the following difference equation first by using the z transform method and then by using a microcomputer:

$$x(k+2) + 3x(k+1) + 2x(k) = 0, \qquad x(0) = 0, \qquad x(1) = 1$$

Taking the z transforms of both sides of this difference equation, we obtain

$$z^2X(z) - z^2x(0) - zx(1) + 3zX(z) - 3zx(0) + 2X(z) = 0$$

Substituting the initial data and simplifying gives

$$X(z) = \frac{z}{z^2 + 3z + 2} = \frac{z}{(z+1)(z+2)} = \frac{z}{z+1} - \frac{z}{z+2}$$

$$= \frac{1}{1+z^{-1}} - \frac{1}{1+2z^{-1}}$$

Noting that

$$\mathscr{Z}^{-1}\left[\frac{1}{1+z^{-1}}\right] = (-1)^k, \qquad \mathscr{Z}^{-1}\left[\frac{1}{1+2z^{-1}}\right] = (-2)^k$$

we have

$$x(k) = (-1)^k - (-2)^k \qquad k = 0, 1, 2, \ldots$$

Next, we shall present the solution of the difference equation by use of a microcomputer. A BASIC computer program for finding $x(k)$, the inverse z transform of $X(z)$, is shown in Table 2–6. The computer solution $x(k)$ for values of k up to 10 is also shown in Table 2–6.

TABLE 2–6 BASIC COMPUTER PROGRAM FOR FINDING $x(k)$, THE SOLUTION OF THE DIFFERENCE EQUATION CONSIDERED IN EXAMPLE 2–29

```
10  X0 = 0
20  X1 = 1
30  K = 0
40  X2 = -3 * X1 - 2 * X0
50  M = X0
60  X0 = X1
70  X1 = X2
80  PRINT K, M
90  K = K + 1
100 IF K < 11 GO TO 40
110 END
```

$k = \mathrm{K}$	$x(k) = \mathrm{XK} = \mathrm{M}$
0	0
1	1
2	−3
3	7
4	−15
5	31
6	−63
7	127
8	−255
9	511
10	−1023

Example 2–30.

By use of (1) the z transform method and (2) the computational method, solve the following difference equation:

$$x(k+2) - 3x(k+1) + 2x(k) = u(k) \tag{2–56}$$

where

$$x(k) = 0 \qquad \text{for } k \leq 0$$

and the input $u(k)$ is the Kronecker delta input:

$$u(k) = \delta_0(k) = \begin{cases} 1 & \text{for } k = 0 \\ 0 & \text{for } k \neq 0 \end{cases}$$

(1) We shall first solve this problem by using the z transform method. By substituting $k = -1$ into Eq. (2–56), we obtain

$$x(1) = 0$$

Taking the z transform of Eq. (2–56) with the initial data $x(0) = x(1) = 0$, we get

$$(z^2 - 3z + 2)X(z) = U(z)$$

Note that the z transform of the input function $u(k)$ is

$$U(z) = \sum_{k=0}^{\infty} u(k)z^{-k} = 1$$

Hence

$$X(z) = \frac{1}{z^2 - 3z + 2} = -\frac{1}{z-1} + \frac{1}{z-2} = -\frac{z^{-1}}{1-z^{-1}} + \frac{z^{-1}}{1-2z^{-1}} \tag{2–57}$$

Since

$$\mathcal{Z}^{-1}\left[\frac{z^{-1}}{1-z^{-1}}\right] = \begin{cases} 1 & k = 1, 2, 3, \ldots \\ 0 & k \leq 0 \end{cases}$$

and

$$\mathcal{Z}^{-1}\left[\frac{z^{-1}}{1-2z^{-1}}\right] = \begin{cases} 2^{k-1} & k = 1, 2, 3, \ldots \\ 0 & k \leq 0 \end{cases}$$

we obtain

$$x(k) = -1 + 2^{k-1} \qquad k = 1, 2, 3, \ldots$$

Note that the same solution can be obtained by first modifying Eq. (2–57) to the form

$$zX(z) = -\frac{z}{z-1} + \frac{z}{z-2} = -\frac{1}{1-z^{-1}} + \frac{1}{1-2z^{-1}}$$

and using the relationship

$$\mathcal{Z}[x(k+1)] = zX(z) - zx(0) = zX(z)$$

where we have used the condition that $x(0) = 0$, and then by taking the inverse z transforms as follows:

$$\mathcal{Z}^{-1}\left[\frac{1}{1-z^{-1}}\right] = 1$$

$$\mathcal{Z}^{-1}\left[\frac{1}{1-2z^{-1}}\right] = 2^k \qquad k = 0, 1, 2, \ldots$$

The inverse z transform of $zX(z)$ now becomes

$$x(k+1) = -1 + 2^k \qquad k = 0, 1, 2, \ldots$$

which can be modified to read

$$x(k) = -1 + 2^{k-1} \qquad k = 1, 2, 3, \ldots$$

which is exactly the same as the solution obtained earlier. Note that

$$x(0) = 0$$
$$x(1) = 0$$
$$x(2) = 1$$
$$x(3) = 3$$
$$x(4) = 7$$
$$x(5) = 15$$
$$\vdots$$

(2) Next, we shall solve this problem with a microcomputer. The BASIC computer program shown in Table 2–7 gives $x(k)$, the inverse z transform of $X(z)$. The computer solution of $x(k)$ for values of k up to 15 is also shown in Table 2–7.

TABLE 2–7 BASIC COMPUTER PROGRAM FOR FINDING $x(k)$, THE INVERSE z TRANSFORM OF $X(z)$ AS GIVEN BY EQ. (2–57)

```
10  X0 = 0
20  X1 = 0
30  U0 = 1
40  U1 = 0
50  K = 0
60  X2 = 3 * X1 - 2 * X0 + U0
70  M = X0
80  X0 = X1
90  X1 = X2
100 N = U0
110 U0 = U1
120 PRINT K, M, N
130 K = K + 1
140 IF K < 16 GO TO 60
150 END
```

k = K	$x(k)$ = XK = M	$u(k)$ = UK = N
0	0	1
1	0	0
2	1	0
3	3	0
4	7	0
5	15	0
6	31	0
7	63	0
8	127	0
9	255	0
10	511	0
11	1023	0
12	2047	0
13	4095	0
14	8191	0
15	16383	0

2-7 SUMMARY

In this chapter the basic theory of the z transform method has been presented. The z transform is seen to serve the same purpose for linear time-invariant discrete-time systems as the Laplace transform provides for linear time-invariant continuous-time systems.

The z transform of a function may be obtained in the form of an infinite series or in a closed form. For the inverse z transformation, four methods are commonly available. The direct division method is convenient for finding the first several values of the sequence. The computational method or computer method using a microcomputer provides the inverse z transformation without tedious hand calculations. The partial-fraction-expansion method is valuable in reducing a complicated z transform expression to a sum of simpler z transforms of the kind that may be found in a table of z transforms. The inversion integral method based on the residue theory is very useful in obtaining the inverse z transform when other methods cannot be applied.

Applications of the z transformation to the analysis and design of digital control systems. The z transform method, because it is very well adapted to digital analysis, is useful in the analysis and design of digital control systems. It has therefore been growing in importance as the use of computer control in control systems has increased.

The computer method of analyzing data in discrete time results in difference equations. With the z transform method, linear time-invariant difference equations can be transformed into algebraic equations. This facilitates the transient response analysis of the digital control system. Also, the z transform method allows one to use conventional analysis and design techniques available to analog (continuous-time) control systems, such as the root locus and frequency response techniques. Frequency response analysis and design can be carried out by converting the z plane into the w plane. Also, the z-transformed characteristic equation allows one to apply a simple stability test, such as the Jury stability criterion. These subjects will be discussed in detail in Chaps. 3 and 4.

REFERENCES

2–1. Aseltine, J. A., *Transform Method in Linear System Analysis*. New York: McGraw-Hill Book Company, 1958.

2–2. Churchill, R. V., and J. W. Brown, *Complex Variables and Applications*, 4th ed., New York: McGraw-Hill Book Company, 1984.

2–3. Freeman, H., *Discrete-Time Systems*. New York: John Wiley & Sons, Inc., 1965.

2–4. Jury, E. I., *Sampled-Data Control Systems*. New York: John Wiley & Sons, Inc., 1958.

2–5. Jury, E. I., *Theory and Applications of the z Transform Method*. New York: John Wiley & Sons, Inc., 1964.

2–6. Kuo, B. C., *Digital Control Systems*. New York: Holt, Rinehart and Winston, Inc., 1980.
2–7. Ogata, K., *Modern Control Engineering*. Englewood Cliffs, N.J.: Prentice-Hall, Inc., 1970.

EXAMPLE PROBLEMS AND SOLUTIONS

Problem A-2–1. Obtain the z transform of $\mathbf{G}^k$, where $\mathbf{G}$ is an $n \times n$ constant matrix.

Solution. By definition, the z transform of $\mathbf{G}^k$ is

$$\begin{aligned}\mathscr{Z}[\mathbf{G}^k] &= \sum_{k=0}^{\infty} \mathbf{G}^k z^{-k} \\ &= \mathbf{I} + \mathbf{G}z^{-1} + \mathbf{G}^2 z^{-2} + \mathbf{G}^3 z^{-3} + \cdots \\ &= (\mathbf{I} - \mathbf{G}z^{-1})^{-1} \\ &= (z\mathbf{I} - \mathbf{G})^{-1} z\end{aligned}$$

Note that $\mathbf{G}^k$ can be obtained by taking the inverse z transform of $(\mathbf{I} - \mathbf{G}z^{-1})^{-1}$ or $(z\mathbf{I} - \mathbf{G})^{-1}z$. That is,

$$\mathbf{G}^k = \mathscr{Z}^{-1}[(\mathbf{I} - \mathbf{G}z^{-1})^{-1}] = \mathscr{Z}^{-1}[(z\mathbf{I} - \mathbf{G})^{-1}z]$$

Problem A-2–2. Obtain the z transform of k by two methods.

Solution.

Method 1. By definition, the z transform of k is given by

$$\begin{aligned}\mathscr{Z}[k] &= \sum_{k=0}^{\infty} k z^{-k} = z^{-1} + 2z^{-2} + 3z^{-3} + \cdots \\ &= z^{-1}(1 + 2z^{-1} + 3z^{-2} + \cdots) \\ &= \frac{z^{-1}}{(1 - z^{-1})^2}\end{aligned}$$

Method 2: Note that the z transform of a unit step sequence $x(k) = 1(k)$ is

$$\mathscr{Z}[1(k)] = X(z) = \frac{1}{1 - z^{-1}}$$

Hence, by applying the complex differentiation theorem, the z transform of $k = k \cdot 1(k)$ is

$$\mathscr{Z}[k] = \mathscr{Z}[k \cdot 1(k)] = -z\frac{dX(z)}{dz} = -z\frac{d}{dz}\left(\frac{1}{1 - z^{-1}}\right) = \frac{z^{-1}}{(1 - z^{-1})^2}$$

Problem A-2–3. Obtain the z transform of k^2 by two methods.

Solution.

Method 1. By definition, the z transform of k^2 is

$$\mathscr{Z}[k^2] = \sum_{k=0}^{\infty} k^2 z^{-k} = z^{-1} + 4z^{-2} + 9z^{-3} + 16z^{-4} + \cdots$$

$$= z^{-1}(1 + z^{-1})(1 + 3z^{-1} + 6z^{-2} + 10z^{-3} + 15z^{-4} + \cdots)$$

$$= \frac{z^{-1}(1 + z^{-1})}{(1 - z^{-1})^3}$$

Here we have used the closed-form expression $(1 - z^{-1})^{-3}$ for the infinite series involved in the problem. (See page 70.)

Method 2. Since k^2 can be written as $k^2 = k \cdot k$, by applying the complex differentiation theorem, the z transform of k^2 is obtained as follows. First note that

$$\mathscr{Z}[k] = X(z) = \frac{z^{-1}}{(1 - z^{-1})^2}$$

Then

$$\mathscr{Z}[k^2] = \mathscr{Z}[k \cdot k] = -z\frac{dX(z)}{dz} = -z\frac{d}{dz}\left[\frac{z^{-1}}{(1 - z^{-1})^2}\right]$$

$$= -z\frac{-z^{-2}(1 - z^{-1})^2 - z^{-1}2(1 - z^{-1})z^{-2}}{(1 - z^{-1})^4}$$

$$= \frac{z^{-1}(1 + z^{-1})}{(1 - z^{-1})^3}$$

Problem A-2-4. Obtain the z transform of ka^{k-1} by three methods.

Solution.

Method 1. By definition, the z transform of ka^{k-1} is given by

$$\mathscr{Z}[ka^{k-1}] = \sum_{k=0}^{\infty} ka^{k-1}z^{-k}$$

$$= z^{-1} + 2az^{-2} + 3a^2z^{-3} + 4a^3z^{-4} + \cdots$$

$$= z^{-1}(1 + 2az^{-1} + 3a^2z^{-2} + 4a^3z^{-3} + \cdots)$$

$$= \frac{z^{-1}}{(1 - az^{-1})^2}$$

Method 2. By referring to Prob. A-2-2 (Method 1), we have

$$\mathscr{Z}[ka^{k-1}] = \sum_{k=0}^{\infty} ka^{k-1}z^{-k} = a^{-1}\sum_{k=0}^{\infty} ka^k z^{-k} = \frac{1}{a}\sum_{k=0}^{\infty} k\left(\frac{z}{a}\right)^{-k}$$

$$= \frac{1}{a}\frac{(z/a)^{-1}}{[1 - (z/a)^{-1}]^2} = \frac{z^{-1}}{(1 - az^{-1})^2}$$

Method 3. Note that

$$\mathscr{Z}[a^k] = \frac{1}{1 - az^{-1}}$$

and

$$\mathcal{Z}[a^{k-1}] = z^{-1}\frac{1}{1-az^{-1}} = \frac{z^{-1}}{1-az^{-1}}$$

Hence, by use of the complex differentiation theorem, we obtain

$$\mathcal{Z}[ka^{k-1}] = -z\frac{d}{dz}\left(\frac{z^{-1}}{1-az^{-1}}\right) = \frac{z^{-1}}{(1-az^{-1})^2}$$

Problem A-2-5. Obtain the z transform of k^2a^{k-1} by three methods.

Solution.

Method 1. By definition, the z transform of k^2a^{k-1} is

$$\begin{aligned}\mathcal{Z}[k^2a^{k-1}] &= \sum_{k=0}^{\infty} k^2a^{k-1}z^{-k} \\ &= z^{-1} + 4az^{-2} + 9a^2z^{-3} + 16a^3z^{-4} + \cdots \\ &= z^{-1}(1 + 4az^{-1} + 9a^2z^{-2} + 16a^3z^{-3} + \cdots) \\ &= z^{-1}(1+az^{-1})(1 + 3az^{-1} + 6a^2z^{-2} + 10a^3z^{-3} + \cdots) \\ &= \frac{z^{-1}(1+az^{-1})}{(1-az^{-1})^3}\end{aligned}$$

Here we have used the closed-form expression $(1 - az^{-1})^{-3}$ for the infinite series involved in the problem. (See page 70.)

Method 2. By referring to Prob. A-2-3 (Method 1), we have

$$\begin{aligned}\mathcal{Z}[k^2a^{k-1}] &= \sum_{k=0}^{\infty} k^2a^{k-1}z^{-k} = \frac{1}{a}\sum_{k=0}^{\infty} k^2a^kz^{-k} = \frac{1}{a}\sum_{k=0}^{\infty} k^2\left(\frac{z}{a}\right)^{-k} \\ &= \frac{1}{a}\frac{(z/a)^{-1}[1+(z/a)^{-1}]}{[1-(z/a)^{-1}]^3} = \frac{z^{-1}(1+az^{-1})}{(1-az^{-1})^3}\end{aligned}$$

Method 3. By referring to Prob. A-2-4, we have

$$\mathcal{Z}[ka^{k-1}] = \frac{z^{-1}}{(1-az^{-1})^2}$$

Hence, by use of the complex differentiation theorem we obtain

$$\mathcal{Z}[k^2a^{k-1}] = \mathcal{Z}[k \cdot ka^{k-1}] = -z\frac{d}{dz}\left[\frac{z^{-1}}{(1-az^{-1})^2}\right] = \frac{z^{-1}(1+az^{-1})}{(1-az^{-1})^3}$$

Problem A-2-6. Show that

$$\mathcal{Z}\left[\sum_{h=0}^{k} x(h)\right] = \frac{1}{1-z^{-1}}X(z)$$

$$\mathcal{Z}\left[\sum_{h=0}^{k-1} x(h)\right] = \frac{z^{-1}}{1-z^{-1}}X(z)$$

and

$$\sum_{k=0}^{\infty} x(k) = \lim_{z \to 1} X(z) \tag{2-58}$$

Also show that

$$\mathscr{Z}\left[\sum_{h=i}^{k} x(h)\right] = \frac{1}{1-z^{-1}}\left[X(z) - \sum_{h=0}^{i-1} x(h)z^{-h}\right] \tag{2-59}$$

where $1 \leq i \leq k - 1$.

Solution. Define

$$y(k) = \sum_{h=0}^{k} x(h) \qquad k = 0, 1, 2, \ldots$$

so that

$$\begin{aligned}
y(0) &= x(0)\\
y(1) &= x(0) + x(1)\\
y(2) &= x(0) + x(1) + x(2)\\
&\vdots\\
y(k) &= x(0) + x(1) + x(2) + \cdots + x(k)
\end{aligned}$$

Then, clearly

$$y(k) - y(k-1) = x(k)$$

By writing the z transforms of $x(k)$ and $y(k)$ as $X(z)$ and $Y(z)$, respectively, and by taking the z transform of this last equation, we have

$$Y(z) - z^{-1}Y(z) = X(z)$$

Hence

$$Y(z) = \frac{1}{1-z^{-1}} X(z)$$

or

$$\mathscr{Z}\left[\sum_{h=0}^{k} x(h)\right] = \mathscr{Z}[y(k)] = Y(z) = \frac{1}{1-z^{-1}} X(z)$$

and

$$\mathscr{Z}\left[\sum_{h=0}^{k-1} x(h)\right] = \mathscr{Z}[y(k-1)] = z^{-1}Y(z) = \frac{z^{-1}}{1-z^{-1}} X(z)$$

By using the final value theorem, we find

$$\lim_{k\to\infty} y(k) = \lim_{k\to\infty}\left[\sum_{h=0}^{k} x(h)\right] = \lim_{z\to 1}\left[(1-z^{-1})\frac{1}{1-z^{-1}}X(z)\right]$$

or

$$\sum_{h=0}^{\infty} x(h) = \sum_{k=0}^{\infty} x(k) = \lim_{z\to 1} X(z)$$

Next, to prove Eq. (2–59), first define

$$\tilde{y}(k) = \sum_{h=i}^{k} x(h) = x(i) + x(i+1) + \cdots + x(k)$$

where $1 \le i \le k-1$. Define also

$$\tilde{X}(z) = x(i)z^{-i} + x(i+1)z^{-(i+1)} + \cdots + x(k)z^{-k} + \cdots$$

Then, noting that

$$X(z) = \mathcal{Z}[x(k)] = \sum_{k=0}^{\infty} x(k)z^{-k} = x(0) + x(1)z^{-1} + x(2)z^{-2} + \cdots$$

we obtain

$$\tilde{X}(z) = X(z) - \sum_{h=0}^{i-1} x(h)z^{-h}$$

Since

$$\tilde{y}(k) - \tilde{y}(k-1) = x(k) \qquad k = i,\ i+1,\ i+2, \ldots$$

the z transform of this last equation becomes

$$\tilde{Y}(z) - z^{-1}\tilde{Y}(z) = \tilde{X}(z)$$

[Note that the z transform of $x(k)$ which begins with $k = i$ is $\tilde{X}(z)$, not $X(z)$.] Thus

$$\mathcal{Z}\left[\sum_{h=i}^{k} x(h)\right] = \tilde{Y}(z) = \frac{1}{1-z^{-1}}\tilde{X}(z) = \frac{1}{1-z^{-1}}\left[X(z) - \sum_{h=0}^{i-1} x(h)z^{-h}\right]$$

Problem A-2–7. Consider the sequence $x(k)$ given by

$$x(k) = \begin{cases} e^{-ak} & k \ge 0 \\ 0 & k < 0 \end{cases}$$

where $a > 0$. Obtain $\sum_{k=0}^{\infty} x^2(k)$ by using two methods. **(1)** Parseval's theorem and **(2)** Eq. (2–58).

Solution.

(1) Since

$$X(z) = \frac{1}{1-e^{-a}z^{-1}} = \frac{z}{z-e^{-a}} \qquad \text{for } |z| > e^{-a}$$

we have

$$X(z)X(z^{-1}) = \frac{z}{(z-e^{-a})}\frac{z^{-1}}{(z^{-1}-e^{-a})}$$

$$= \frac{z}{(z-e^{-a})(1-e^{-a}z)} \qquad \text{for } \frac{1}{e^{-a}} > |z| > e^{-a}$$

Hence from Eq. (2–37) we have

$$\sum_{k=0}^{\infty} x^2(k) = \frac{1}{2\pi j}\oint_C z^{-1}\frac{z}{(z-e^{-a})(1-e^{-a}z)}\,dz$$

$$= \frac{1}{2\pi j}\oint_C \frac{-dz}{(z-e^{-a})(ze^{-a}-1)}$$

$$= \frac{1}{2\pi j}\oint_C \frac{-e^{a}\,dz}{(z-e^{-a})(z-e^{a})}$$

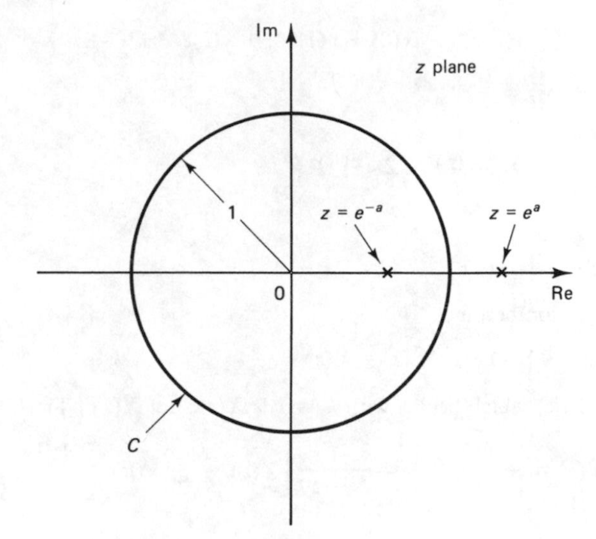

Figure 2–6 Diagram showing contour C and poles $z = e^{-a}$ and $z = e^{a}$ $(a > 0)$.

where C is taken as the unit circle centered at the origin. Referring to Fig. 2–6, we see that the pole at $z = e^{-a}$ is inside the contour C but the pole at $z = e^{a}$ is outside the contour C. Hence, we evaluate the residue at pole $z = e^{-a}$:

$$\sum_{k=0}^{\infty} x^2(k) = \frac{1}{2\pi j}\oint_C \frac{-e^{a}\,dz}{(z-e^{-a})(z-e^{a})}$$

$$= \frac{1}{2\pi j}[2\pi j(\text{residue at pole } z = e^{-a})]$$

The residue is obtained as follows:

$$[\text{Residue at pole } z = e^{-a}] = \lim_{z\to e^{-a}}\left[(z-e^{-a})\frac{-e^{a}}{(z-e^{-a})(z-e^{a})}\right]$$

$$= \frac{-e^a}{e^{-a} - e^a}$$

$$= \frac{1}{1 - e^{-2a}}$$

Thus,

$$\sum_{k=0}^{\infty} x^2(k) = \sum_{k=0}^{\infty} (e^{-ak})^2 = \frac{1}{1 - e^{-2a}}$$

(2) Note that

$$x^2(k) = (e^{-ak})^2 = e^{-2ak}$$

Define

$$x^2(k) = y(k) = e^{-2ak}$$

Then by use of Eq. (2–58) we have

$$\sum_{k=0}^{\infty} y(k) = \lim_{z \to 1} Y(z)$$

Hence

$$\sum_{k=0}^{\infty} x^2(k) = \sum_{k=0}^{\infty} y(k) = \lim_{z \to 1} Y(z) = \lim_{z \to 1} \frac{1}{1 - e^{-2a} z^{-1}} = \frac{1}{1 - e^{-2a}}$$

Problem A-2–8. Show that the first forward difference of the product of two discrete functions $x(k)$ and $y(k)$ is given by

$$\Delta[x(k)y(k)] = x(k+1)\,\Delta y(k) + y(k)\,\Delta x(k)$$

Solution.

$$\begin{aligned}
\Delta[x(k)y(k)] &= x(k+1)y(k+1) - x(k)y(k) \\
&= x(k+1)y(k+1) - x(k+1)y(k) \\
&\quad + x(k+1)y(k) - x(k)y(k) \\
&= x(k+1)[y(k+1) - y(k)] + [x(k+1) - x(k)]y(k) \\
&= x(k+1)\,\Delta y(k) + y(k)\,\Delta x(k)
\end{aligned}$$

Problem A-2–9. Obtain

$$\sum_{k=1}^{\infty} \left(\frac{1}{k}\right) z^{-k}$$

(This is similar to the z transform of $1/k$, but the k sequence begins here with 1 instead of 0.)

Solution. Since

$$\sum_{k=0}^{\infty} z^{-k} = 1 + z^{-1} + z^{-2} + \cdots = \frac{1}{1 - z^{-1}} \qquad |z| > 1$$

by multiplying both sides of this last equation by z^{-2}, we have

$$\sum_{k=0}^{\infty} z^{-k-2} = \frac{z^{-2}}{1 - z^{-1}}$$

Integrating this last equation with respect to z, we have

$$\int \sum_{k=0}^{\infty} z^{-k-2}\, dz = \int \frac{z^{-2}}{1 - z^{-1}}\, dz$$

or

$$\sum_{k=0}^{\infty} \frac{z^{-k-1}}{-k-1} = \ln (1 - z^{-1}) + \text{constant} \tag{2–60}$$

where the constant in Eq. (2–60) is zero. [To verify this, substitute ∞ for z in both sides of Eq. (2–60).] Equation (2–60) can thus be rewritten as follows:

$$\sum_{k=1}^{\infty} \frac{z^{-k}}{-k} = \ln (1 - z^{-1}) \qquad |z| > 1$$

or

$$\sum_{k=1}^{\infty} \left(\frac{1}{k}\right) z^{-k} = -\ln (1 - z^{-1}) \qquad |z| > 1$$

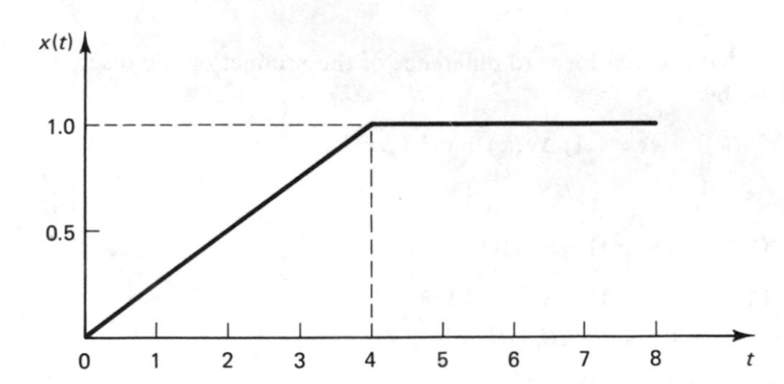

Figure 2–7 Curve $x(t)$.

Problem A-2–10. Obtain the z transform of the curve $x(t)$ shown in Fig. 2–7. Assume that the sampling period T is 1 sec.

Solution. From Fig. 2–7 we obtain

$$x(0) = 0$$
$$x(1) = 0.25$$
$$x(2) = 0.50$$
$$x(3) = 0.75$$
$$x(k) = 1 \qquad k = 4, 5, 6, \ldots$$

Then the z transform of $x(k)$ can be given by

$$X(z) = \sum_{k=0}^{\infty} x(k)z^{-k}$$

$$= 0.25z^{-1} + 0.50z^{-2} + 0.75z^{-3} + z^{-4} + z^{-5} + z^{-6} + \cdots$$

$$= 0.25(z^{-1} + 2z^{-2} + 3z^{-3}) + z^{-4}\frac{1}{1 - z^{-1}}$$

$$= \frac{z^{-1} + z^{-2} + z^{-3} + z^{-4}}{4(1 - z^{-1})}$$

$$= \frac{1}{4}\frac{z^{-1}(1 + z^{-1} + z^{-2} + z^{-3})(1 - z^{-1})}{(1 - z^{-1})^2}$$

$$= \frac{1}{4}\frac{z^{-1}(1 - z^{-4})}{(1 - z^{-1})^2}$$

Notice that the curve $x(t)$ can be written as

$$x(t) = \tfrac{1}{4}t - \tfrac{1}{4}(t - 4)1(t - 4)$$

where $1(t - 4)$ is the unit step function occurring at $t = 4$. Since the sampling period $T =$ 1 sec, the z transform of $x(t)$ can also be obtained as follows:

$$X(z) = \mathcal{Z}[x(t)] = \mathcal{Z}[\tfrac{1}{4}t] - \mathcal{Z}[\tfrac{1}{4}(t - 4)1(t - 4)]$$

$$= \frac{1}{4}\frac{z^{-1}}{(1 - z^{-1})^2} - \frac{1}{4}\frac{z^{-4}z^{-1}}{(1 - z^{-1})^2}$$

$$= \frac{1}{4}\frac{z^{-1}(1 - z^{-4})}{(1 - z^{-1})^2}$$

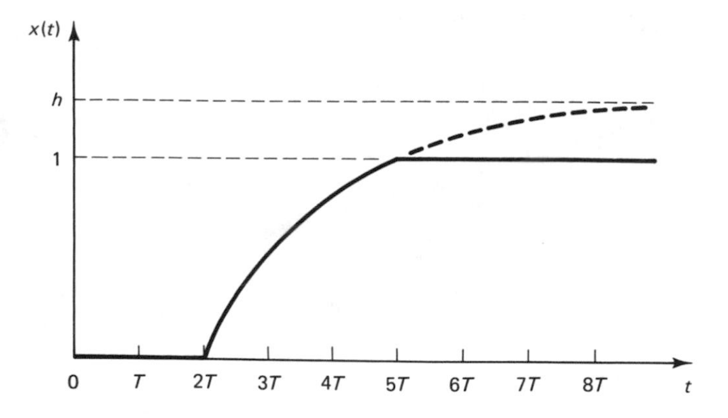

Figure 2–8 Curve $x(t)$.

Problem A-2–11. Obtain the z transform of the curve $x(t)$ shown in Fig. 2–8. The curve $x(t)$ is zero for $0 \leq t \leq 2T$, rises exponentially for $2T < t < 5T$, and reaches unity at $t = 5T$ and stays at unity for $5T < t$.

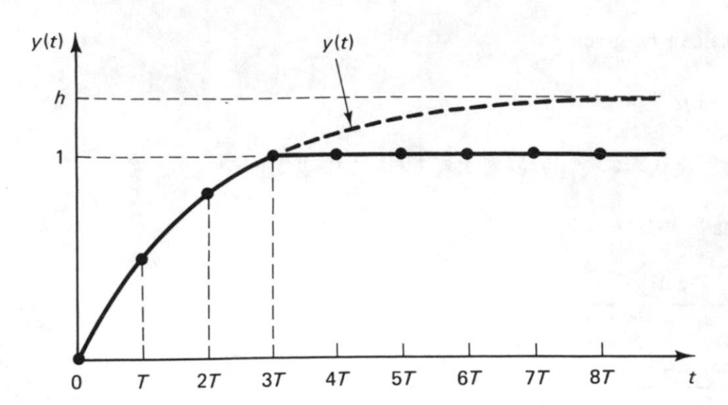

Figure 2–9 Exponential curve $y(t) = h(1 - e^{-at})$.

Solution. Referring to Fig. 2–9, the exponential curve $y(t)$ can be given by

$$y(t) = h(1 - e^{-at})$$

where a is a constant and is related to h by

$$h(1 - e^{-3aT}) = 1$$

or

$$a = \frac{1}{3T}[\ln h - \ln(h - 1)]$$

From Fig. 2–9, we have

$$y(0) = 0$$

$$y(T) = h(1 - e^{-aT})$$

$$y(2T) = h(1 - e^{-2aT})$$

$$y(3T) = h(1 - e^{-3aT}) = 1$$

Let us define

$$e^{-aT} = d$$

Then

$$h(1 - d^3) = 1$$

or

$$h = \frac{1}{1 - d^3}$$

Hence, $y(T)$ and $y(2T)$ can be written as follows:

$$y(T) = h(1 - d) = \frac{1 - d}{1 - d^3} = \frac{1}{1 + d + d^2}$$

$$y(2T) = h(1 - d^2) = \frac{1 - d^2}{1 - d^3} = \frac{1 + d}{1 + d + d^2}$$

Then, referring to Fig. 2–8, we have

$$x(0) = 0$$

$$x(T) = 0$$

$$x(2T) = 0$$

$$x(3T) = \frac{1}{1 + d + d^2}$$

$$x(4T) = \frac{1 + d}{1 + d + d^2}$$

$$x(5T) = 1$$

$$x(kT) = 1 \qquad k = 6, 7, 8, \ldots$$

The z transform of $x(kT)$ is then given by

$$\begin{aligned}
X(z) &= \mathcal{Z}[x(kT)] = \sum_{k=0}^{\infty} x(kT)z^{-k} \\
&= \frac{1}{1 + d + d^2} z^{-3} + \frac{1 + d}{1 + d + d^2} z^{-4} + z^{-5} + z^{-6} + z^{-7} + \cdots \\
&= \frac{z^{-3}(1 + z^{-1} + dz^{-1})}{1 + d + d^2} + \frac{z^{-5}}{1 - z^{-1}} \\
&= \frac{1 + dz^{-1} + d^2 z^{-2}}{1 + d + d^2} \frac{z^{-3}}{1 - z^{-1}}
\end{aligned}$$

where

$$d = \sqrt[3]{\frac{h - 1}{h}}$$

Problem A-2–12. Obtain the z transform of the curve $x(t)$ shown in Fig. 2–10, using two different approaches.

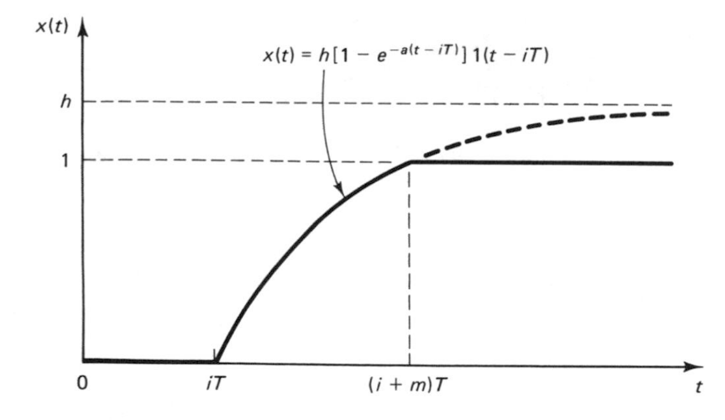

Figure 2–10 Curve $x(t)$.

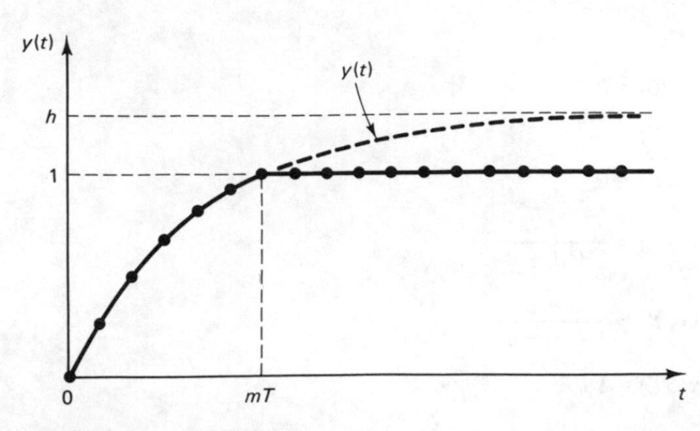

Figure 2–11 Exponential curve $y(t) = h(1 - e^{-at})$.

Solution.

Method 1. We shall use the same method as employed in solving Prob. A-2–11. First, define the exponential curve shown in Fig. 2–11 as $y(t)$. Then

$$y(t) = h(1 - e^{-at})$$

where a is a constant and is given by

$$a = \frac{1}{mT}[\ln h - \ln(h-1)]$$

Now define

$$e^{-aT} = d$$

Then we have

$$y(0) = 0$$
$$y(T) = h(1-d)$$
$$y(2T) = h(1-d^2)$$
$$\vdots$$
$$y(mT) = h(1-d^m) = 1$$

Hence

$$h = \frac{1}{1-d^m}$$

By using the foregoing expressions, we obtain from Fig. 2–10

$$x(0) = 0$$
$$x(T) = 0$$
$$\vdots$$
$$x(iT) = 0$$

$$x((i+1)T)=\frac{1-d}{1-d^m}=\frac{1}{1+d+d^2+\cdots+d^{m-1}}=\frac{1}{\Delta}$$

$$x((i+2)T)=\frac{1-d^2}{1-d^m}=\frac{1+d}{1+d+d^2+\cdots+d^{m-1}}=\frac{1+d}{\Delta}$$

$$\vdots$$

$$x((i+m-1)T)=\frac{1-d^{m-1}}{1-d^m}=\frac{1+d+d^2+\cdots+d^{m-2}}{1+d+d^2+\cdots+d^{m-1}}=\frac{1+d+d^2+\cdots+d^{m-2}}{\Delta}$$

$$x((i+m)T)=\frac{1-d^m}{1-d^m}=1$$

$$x(kT)=1 \qquad k=i+m+1,\ i+m+2,\ \ldots$$

where

$$\Delta=1+d+d^2+\cdots+d^{m-1}$$

The z transform of $x(kT)$ is then given by

$$\begin{aligned}
X(z)&=\mathscr{Z}[x(kT)]=\sum_{k=0}^{\infty}x(kT)z^{-k}\\
&=\frac{1}{\Delta}z^{-i-1}+\frac{1+d}{\Delta}z^{-i-2}+\cdots+\frac{1+d+d^2+\cdots+d^{m-2}}{\Delta}z^{-i-m+1}\\
&\quad+z^{-i-m}+z^{-i-m-1}+z^{-i-m-2}+\cdots\\
&=\frac{1}{\Delta}[1+(1+d)z^{-1}+\cdots+(1+d+d^2+\cdots+d^{m-2})z^{-m+2}]z^{-i-1}\\
&\quad+\frac{z^{-i-m}}{1-z^{-1}}\\
&=\frac{1}{\Delta(1-z^{-1})}\{[1+(1+d)z^{-1}+\cdots+(1+d+d^2+\cdots+d^{m-2})z^{-m+2}]\\
&\quad\cdot(1-z^{-1})+(1+d+d^2+\cdots+d^{m-1})z^{-m+1}\}z^{-i-1}\\
&=\frac{1+dz^{-1}+d^2z^{-2}+\cdots+d^{m-1}z^{-m+1}}{(1+d+d^2+\cdots+d^{m-1})(1-z^{-1})}z^{-i-1}
\end{aligned}$$

where $d=e^{-aT}=\sqrt[m]{(h-1)/h}$.

Since

$$\begin{aligned}
&1+dz^{-1}+d^2z^{-2}+\cdots+d^{m-1}z^{-m+1}\\
&\qquad=1+(dz^{-1})+(dz^{-1})^2+\cdots+(dz^{-1})^{m-1}\\
&\qquad=\frac{1-(dz^{-1})^m}{1-dz^{-1}}
\end{aligned}$$

and

$$1+d+d^2+\cdots+d^{m-1}=\frac{1-d^m}{1-d}$$

$X(z)$ can also be given by the following expression:

$$X(z)=\frac{1-d}{1-d^m}\frac{z^{-i-1}}{1-z^{-1}}\frac{1-(dz^{-1})^m}{1-dz^{-1}}$$

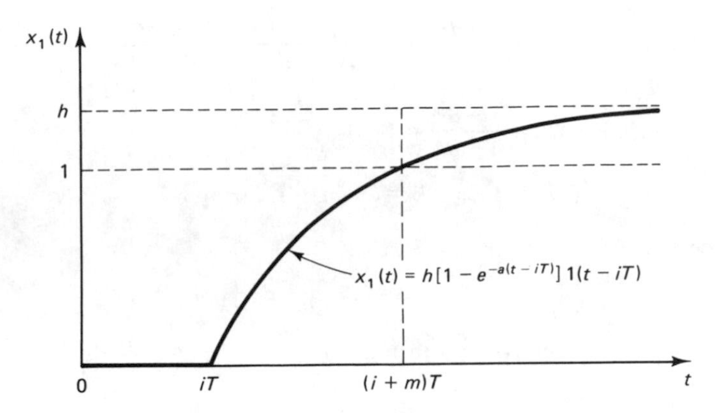

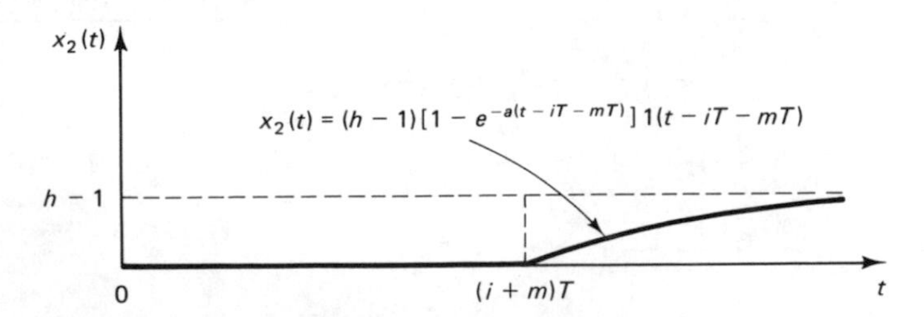

Figure 2–12 Exponential curves $x_1(t)$ and $x_2(t)$ such that $x_1(t) - x_2(t) = x(t)$, where $x(t)$ is the curve shown in Fig. 2–10.

Method 2. Referring to Fig. 2–12, we can write the curve $x(t)$ as $x_1(t) - x_2(t)$:

$$\begin{aligned}x(t)=x_1(t)-x_2(t)&=h[1-e^{-a(t-iT)}]1(t-iT)\\&\quad-(h-1)[1-e^{-a(t-iT-mT)}]1(t-iT-mT)\end{aligned}$$

Hence

$$\begin{aligned}x(kT)&=h[1-e^{-a(k-i)T}]1((k-i)T)\\&\quad-(h-1)[1-e^{-a(k-i-m)T}]1((k-i-m)T)\end{aligned}$$

Define

$$e^{-aT}=d$$

Then $x(kT)$ can be given by

$$x(kT) = h(1 - d^{k-i})1((k-i)T) - (h-1)(1 - d^{k-i-m})1((k-i-m)T)$$

The z transform of $x(kT)$ gives

$$\begin{aligned} X(z) &= h\left(z^{-i}\frac{1}{1-z^{-1}} - z^{-i}\frac{1}{1-dz^{-1}}\right) \\ &\quad - (h-1)\left(z^{-i-m}\frac{1}{1-z^{-1}} - z^{-i-m}\frac{1}{1-dz^{-1}}\right) \\ &= \left(\frac{1}{1-z^{-1}} - \frac{1}{1-dz^{-1}}\right) z^{-i}[h - (h-1)z^{-m}] \\ &= \frac{(1-d)z^{-i-1}}{(1-z^{-1})(1-dz^{-1})}[h - (h-1)z^{-m}] \end{aligned}$$

Noting that at $k = i + m$

$$x((i+m)T) = h(1 - d^m)1(mT) = 1$$

we have

$$h = \frac{1}{1-d^m}$$

Hence

$$h - 1 = \frac{d^m}{1-d^m}$$

and

$$h - (h-1)z^{-m} = \frac{1}{1-d^m}(1 - d^m z^{-m}) = \frac{1-(dz^{-1})^m}{1-d^m}$$

Thus, $X(z)$ can be given by

$$X(z) = \frac{1-d}{1-d^m}\,\frac{z^{-i-1}}{1-z^{-1}}\,\frac{1-(dz^{-1})^m}{1-dz^{-1}}$$

where $d = e^{-aT} = \sqrt[m]{(h-1)/h}$.

Problem A-2-13. Show that

$$\mathscr{Z}^{-1}\left[\frac{z^{-2}}{(1-az^{-1})^2}\right] = \begin{cases} (k-1)a^{k-2} & k = 1, 2, 3, \ldots \\ 0 & k \leq 0 \end{cases}$$

Solution. Define

$$x(k) = \begin{cases} ka^{k-1} & k = 0, 1, 2, \ldots \\ 0 & k < 0 \end{cases}$$

Then

$$\mathscr{Z}[x(k)] = X(z) = \frac{z^{-1}}{(1-az^{-1})^2}$$

Notice that

$$\mathcal{Z}[(k-1)a^{k-2}] = \mathcal{Z}[x(k-1)] = z^{-1}\frac{z^{-1}}{(1-az^{-1})^2} = \frac{z^{-2}}{(1-az^{-1})^2}$$

Hence

$$\mathcal{Z}^{-1}\left[\frac{z^{-2}}{(1-az^{-1})^2}\right] = \begin{cases} x(k-1) = (k-1)a^{k-2} & k = 1, 2, 3, \ldots \\ 0 & k \leq 0 \end{cases}$$

Problem A-2–14. Obtain the inverse z transform of

$$X(z) = \frac{1 + 2z + 3z^2 + 4z^3 + 5z^4}{z^4}$$

Solution. By dividing both numerator and denominator by z^4, we have

$$X(z) = 5 + 4z^{-1} + 3z^{-2} + 2z^{-3} + z^{-4}$$

This last equation is already in the form of a power series in z^{-1}. By inspection, we have

$$x(0) = 5$$

$$x(1) = 4$$

$$x(2) = 3$$

$$x(3) = 2$$

$$x(4) = 1$$

$$x(k) = 0 \qquad k \geq 5$$

Notice that the given $X(z)$ is the z transform of a signal of finite length.

Problem A-2–15. Obtain the inverse z transform of

$$X(z) = \frac{z(z+2)}{(z-1)^2} \tag{2–61}$$

by use of the four methods presented in Sec. 2–5.

Solution.

Method 1: Direct division method. We first rewrite $X(z)$ as a ratio of two polynomials in z^{-1}:

$$X(z) = \frac{1 + 2z^{-1}}{(1 - z^{-1})^2} = \frac{1 + 2z^{-1}}{1 - 2z^{-1} + z^{-2}}$$

Dividing the numerator by the denominator, we get

$$X(z) = 1 + 4z^{-1} + 7z^{-2} + 10z^{-3} + \cdots$$

Hence

$$x(0) = 1$$

$$x(1) = 4$$

$$x(2) = 7$$
$$x(3) = 10$$
$$\vdots$$

Method 2: Computational method. By writing the given $X(z)$ as

$$X(z) = \frac{z(z+2)}{(z-1)^2} U(z)$$

where $U(z) = 1$, the corresponding difference equation becomes

$$x(k+2) - 2x(k+1) + x(k) = u(k+2) + 2u(k+1) \tag{2–62}$$

where $u(0) = 1$ and $u(k) = 0$ for $k \neq 0$, and the initial data $x(0)$ and $x(1)$ are determined as follows. By substituting -2 for k in Eq. (2–62), we obtain

$$x(0) - 2x(-1) + x(-2) = u(0) + 2u(-1)$$

Since $x(-1) = x(-2) = 0$, $u(0) = 1$, and $u(-1) = 0$, we find

$$x(0) = 1$$

Next, by substituting -1 for k in Eq. (2–62) we get

$$x(1) - 2x(0) + x(-1) = u(1) + 2u(0)$$

or

$$x(1) = 2 + 2 = 4$$

A BASIC computer program for finding $x(k)$, the inverse z transform of $X(z)$ as given by Eq. (2–61), is shown in Table 2–8. The computation results are also given in Table 2–8.

Method 3: Partial-fraction-expansion method. We expand $X(z)$ into the following partial fractions:

$$X(z) = \frac{z(z+2)}{(z-1)^2} = 1 + \frac{3z}{(z-1)^2} + \frac{1}{z-1} = 1 + \frac{3z^{-1}}{(1-z^{-1})^2} + \frac{z^{-1}}{1-z^{-1}}$$

Then, noting that

$$\mathcal{Z}^{-1}[1] = \begin{cases} 1 & k = 0 \\ 0 & k = 1, 2, 3, \ldots \end{cases}$$

$$\mathcal{Z}^{-1}\left[\frac{z^{-1}}{(1-z^{-1})^2}\right] = k \qquad k = 0, 1, 2, \ldots$$

$$\mathcal{Z}^{-1}\left[\frac{z^{-1}}{1-z^{-1}}\right] = \begin{cases} 1 & k = 1, 2, 3, \ldots \\ 0 & k \leq 0 \end{cases}$$

we obtain

$$x(0) = 1$$
$$x(k) = 3k + 1 \qquad k = 1, 2, 3, \ldots$$

TABLE 2–8 BASIC COMPUTER PROGRAM FOR FINDING $x(k)$, THE INVERSE z TRANSFORM of $X(z)$ AS GIVEN BY EQ. (2–61)

```
10  X0 = 1
20  X1 = 4
30  U0 = 1
40  U1 = 0
50  X2 = 2 * X1 - X0 + U2 + 2 * U1
60  M = X0
70  X0 = X1
80  X1 = X2
90  N = U0
100 U0 = U1
110 PRINT K, M, N
120 K = K + 1
130 IF K < 16 GO TO 50
140 END
```

$k = \mathrm{K}$	$x(k) = \mathrm{XK} = \mathrm{M}$	$u(k) = \mathrm{UK} = \mathrm{N}$
0	1	1
1	4	0
2	7	0
3	10	0
4	13	0
5	16	0
6	19	0
7	22	0
8	25	0
9	28	0
10	31	0
11	34	0
12	37	0
13	40	0
14	43	0
15	46	0

which can be combined into one equation as follows:

$$x(k) = 3k + 1 \qquad k = 0, 1, 2, \ldots$$

Note that if we expand $X(z)$ into the following partial fractions:

$$X(z) = 1 + \frac{4}{z-1} + \frac{3}{(z-1)^2} = 1 + \frac{4z^{-1}}{1-z^{-1}} + \frac{3z^{-2}}{(1-z^{-1})^2}$$

then the inverse z transform of $X(z)$ becomes

$$x(0) = 1$$

$$x(k) = 4 + 3(k-1) = 3k + 1 \qquad k = 1, 2, 3, \ldots$$

which is the same as the result obtained by expanding $X(z)$ into the other partial fractions. [Remember that $X(z)$ can be expanded into different partial fractions but the final result for the inverse z transform is the same.]

Method 4: Inversion integral method. First, note that

$$X(z)z^{k-1} = \frac{(z+2)z^k}{(z-1)^2}$$

For $k = 0, 1, 2, \ldots$, $X(z)z^{k-1}$ has a double pole at $z = 1$. Hence, referring to Eq. (2–46), we have

$$x(k) = \left[\text{residue of } \frac{(z+2)z^k}{(z-1)^2} \text{ at double pole } z = 1\right]$$

Thus

$$x(k) = \frac{1}{(2-1)!} \lim_{z\to 1} \frac{d}{dz} \left[(z-1)^2 \frac{(z+2)z^k}{(z-1)^2} \right]$$

$$= \lim_{z\to 1} \frac{d}{dz} [(z+2)z^k]$$

$$= 3k + 1 \qquad k = 0, 1, 2, \ldots$$

Problem A-2-16. By use of the partial-fraction-expansion method, obtain the inverse z transform of

$$X(z) = \frac{z^2 + z + 2}{(z-1)(z^2 - z + 1)}$$

Solution. We first expand $X(z)$ into partial fractions:

$$X(z) = \frac{4}{z-1} + \frac{-3z+2}{z^2 - z + 1} = \frac{4z^{-1}}{1 - z^{-1}} + \frac{-3z^{-1} + 2z^{-2}}{1 - z^{-1} + z^{-2}}$$

Noting that the two poles involved in the quadratic term of this last equation are complex conjugates, we rewrite $X(z)$ as follows:

$$X(z) = \frac{4z^{-1}}{1 - z^{-1}} - 3\left(\frac{z^{-1} - 0.5z^{-2}}{1 - z^{-1} + z^{-2}}\right) + \frac{0.5z^{-2}}{1 - z^{-1} + z^{-2}}$$

$$= 4z^{-1}\frac{1}{1 - z^{-1}} - 3z^{-1}\frac{1 - 0.5z^{-1}}{1 - z^{-1} + z^{-2}} + z^{-1}\frac{0.5z^{-1}}{1 - z^{-1} + z^{-2}}$$

Since

$$\mathcal{Z}[e^{-akT} \cos \omega kT] = \frac{1 - e^{-aT}z^{-1} \cos \omega T}{1 - 2e^{-aT}z^{-1} \cos \omega T + e^{-2aT}z^{-2}}$$

$$\mathcal{Z}[e^{-akT} \sin \omega kT] = \frac{e^{-aT}z^{-1} \sin \omega T}{1 - 2e^{-aT}z^{-1} \cos \omega T + e^{-2aT}z^{-2}}$$

by identifying $e^{-2aT} = 1$ and $\cos \omega T = \frac{1}{2}$ in this case, we have $\omega T = \pi/3$ and $\sin \omega T = \sqrt{3}/2$. Hence we obtain

$$\mathcal{Z}^{-1}\left[\frac{1 - 0.5z^{-1}}{1 - z^{-1} + z^{-2}}\right] = 1^k \cos \frac{k\pi}{3}$$

and

$$\mathcal{Z}^{-1}\left[\frac{0.5z^{-1}}{1 - z^{-1} + z^{-2}}\right] = \mathcal{Z}^{-1}\left[\frac{1}{\sqrt{3}} \frac{(\sqrt{3}/2)z^{-1}}{1 - z^{-1} + z^{-2}}\right] = \frac{1}{\sqrt{3}} 1^k \sin \frac{k\pi}{3}$$

Thus, we have

$$x(k) = 4(1^{k-1}) - 3(1^{k-1}) \cos \frac{(k-1)\pi}{3} + \frac{1}{\sqrt{3}}(1^{k-1}) \sin \frac{(k-1)\pi}{3}$$

Rewriting, we have

$$x(k) = \begin{cases} 4 - 3\cos\dfrac{(k-1)\pi}{3} + \dfrac{1}{\sqrt{3}}\sin\dfrac{(k-1)\pi}{3} & k = 1, 2, 3, \ldots \\ 0 & k \leq 0 \end{cases}$$

The first several values of $x(k)$ are given by

$$\begin{aligned} x(0) &= 0 \\ x(1) &= 1 \\ x(2) &= 3 \\ x(3) &= 6 \\ x(4) &= 7 \\ x(5) &= 5 \\ &\vdots \end{aligned}$$

Note that the inverse z transform of $X(z)$ can also be obtained as follows:

$$X(z) = 4z^{-1}\frac{1}{1-z^{-1}} - 3\left(\frac{z^{-1}}{1-z^{-1}+z^{-2}}\right) + 2z^{-1}\frac{z^{-1}}{1-z^{-1}+z^{-2}}$$

Since

$$\mathcal{Z}^{-1}\left[\frac{z^{-1}}{1-z^{-1}}\right] = \begin{cases} 1 & k = 1, 2, 3, \ldots \\ 0 & k \leq 0 \end{cases}$$

and

$$\mathcal{Z}^{-1}\left[\frac{z^{-1}}{1-z^{-1}+z^{-2}}\right] = \frac{2}{\sqrt{3}}(1^k)\sin\frac{k\pi}{3}$$

we have

$$x(k) = \begin{cases} 4 - 2\sqrt{3}\sin\dfrac{k\pi}{3} + \dfrac{4}{\sqrt{3}}\sin\dfrac{(k-1)\pi}{3} & k = 1, 2, 3, \ldots \\ 0 & k \leq 0 \end{cases}$$

Note that this solution may look different from the one obtained earlier. Both solutions, however, are correct and yield the same values for $x(k)$.

Problem A-2-17. Obtain the inverse z transform of

$$X(z) = \frac{z^{-2}}{(1-z^{-1})^3}$$

Solution. The inverse z transform of $z^{-2}/(1-z^{-1})^3$ is not available from most z transform tables. It is possible, however, to write the given $X(z)$ as a sum of z transforms that are commonly available in z transform tables. Since the denominator of $X(z)$ is $(1 - z^{-1})^3$ and the z transform of k^2 is $z^{-1}(1 + z^{-1})/(1 - z^{-1})^3$, let us rewrite $X(z)$ as

$$X(z) = \frac{z^{-2}}{(1 - z^{-1})^3} = \frac{z^{-1}(1 + z^{-1})}{(1 - z^{-1})^3} - \frac{z^{-1}}{(1 - z^{-1})^3}$$

$$= \frac{z^{-1}(1 + z^{-1})}{(1 - z^{-1})^3} - \frac{z^{-1} - z^{-2} + z^{-2}}{(1 - z^{-1})^3}$$

or

$$\frac{z^{-2}}{(1 - z^{-1})^3} = \frac{z^{-1}(1 + z^{-1})}{(1 - z^{-1})^3} - \frac{z^{-1}}{(1 - z^{-1})^2} - \frac{z^{-2}}{(1 - z^{-1})^3}$$

from which we obtain the following partial fraction expansion:

$$\frac{z^{-2}}{(1 - z^{-1})^3} = \frac{1}{2}\left[\frac{z^{-1}(1 + z^{-1})}{(1 - z^{-1})^3} - \frac{z^{-1}}{(1 - z^{-1})^2}\right]$$

The z transforms of the two terms on the right-hand side of this last equation can be found from Table 2–1. Thus,

$$x(k) = \mathscr{Z}^{-1}\left[\frac{z^{-2}}{(1 - z^{-1})^3}\right] = \frac{1}{2}(k^2 - k) = \frac{1}{2}k(k - 1) \qquad k = 0, 1, 2, \ldots$$

It is noted that if the given $X(z)$ is expanded into other partial fractions, then the inverse z transform may not be obtained.

As an alternate approach, the inverse z transform of $X(z)$ may be obtained by use of the inversion integral method. First, note that

$$X(z)z^{k-1} = \frac{z^k}{(z - 1)^3}$$

Hence, for $k = 0, 1, 2, \ldots$, $X(z)z^{k-1}$ has a triple pole at $z = 1$. Referring to Eq. (2–46), we have

$$x(k) = \left[\text{residue of } \frac{z^k}{(z - 1)^3} \text{ at triple pole } z = 1\right]$$

$$= \frac{1}{(3 - 1)!}\lim_{z \to 1}\frac{d^2}{dz^2}\left[(z - 1)^3 \frac{z^k}{(z - 1)^3}\right]$$

$$= \frac{1}{2!}\lim_{z \to 1}\frac{d^2}{dz^2}(z^k)$$

$$= \frac{1}{2}\lim_{z \to 1}\frac{d}{dz}(kz^{k-1})$$

$$= \frac{1}{2}\lim_{z \to 1}[k(k - 1)z^{k-2}]$$

$$= \frac{1}{2}k(k - 1) \qquad k = 0, 1, 2, \ldots$$

TABLE 2–9 BASIC COMPUTER PROGRAM FOR FINDING $x(k)$, THE INVERSE z TRANSFORM OF $X(z)$ AS GIVEN BY EQ. (2–63)

```
10  X0 = 0
20  X1 = 0
30  X2 = 0
40  X3 = 0
50  U0 = 1
60  U1 = 0
70  X4 = 2 * X3 - X2 + U0
80  M = X0
90  X0 = X1
100 X1 = X2
110 X2 = X3
120 X3 = X4
130 N = U0
140 U0 = U1
150 PRINT K, M, N
160 K = K + 1
170 IF K < 16 GO TO 70
180 END
```

$k = \mathrm{K}$	$x(k) = \mathrm{XK} = \mathrm{M}$	$u(k) = \mathrm{UK} = \mathrm{N}$
0	0	1
1	0	0
2	0	0
3	0	0
4	1	0
5	2	0
6	3	0
7	4	0
8	5	0
9	6	0
10	7	0
11	8	0
12	9	0
13	10	0
14	11	0
15	12	0

Problem A-2–18. Obtain the inverse z transform of

$$X(z) = \frac{1}{(z-1)^2 z^2} \tag{2–63}$$

by four methods.

Solution.

Method 1: Direct division method. We first express $X(z)$ as a ratio of polynomials in z^{-1}:

$$X(z) = \frac{z^{-4}}{(1-z^{-1})^2} = \frac{z^{-4}}{1 - 2z^{-1} + z^{-2}}$$

By dividing the numerator by the denominator, we obtain

$$X(z) = z^{-4} + 2z^{-5} + 3z^{-6} + 4z^{-7} + 5z^{-8} + \cdots$$

By inspection, the inverse z transform of $X(z)$ is obtained as follows:

$$x(k) = \begin{cases} 0 & k = 0, 1, 2, 3 \\ k-3 & k = 4, 5, 6, \cdots \end{cases}$$

Method 2: Computational method. The given $X(z)$ can be written as

$$X(z) = \frac{1}{z^4 - 2z^3 + z^2} U(z)$$

where $U(z) = 1$. The corresponding difference equation is

$$x(k+4) - 2x(k+3) + x(k+2) = u(k)$$

where $u(0) = 1$, $u(k) = 0$ for $k = 1, 2, 3, \ldots$ and the initial data are easily determined as follows:

$$x(0) = 0$$
$$x(1) = 0$$
$$x(2) = 0$$
$$x(3) = 0$$

A BASIC computer program for finding $x(k)$, the inverse z transform of $X(z)$ as given by Eq. (2–63), is shown in Table 2–9. The computation results are also shown in Table 2–9.

Method 3: Partial-fraction-expansion method. By expanding $X(z)$ into partial fractions, we obtain

$$X(z) = \frac{1}{(z-1)^2 z^2} = \frac{1}{(z-1)^2} - \frac{2}{z-1} + \frac{1}{z^2} + \frac{2}{z}$$
$$= \frac{z^{-2}}{(1-z^{-1})^2} - \frac{2z^{-1}}{1-z^{-1}} + z^{-2} + 2z^{-1}$$

Note that

$$\mathcal{Z}^{-1}\left[\frac{z^{-2}}{(1-z^{-1})^2}\right] = \mathcal{Z}^{-1}\left[z^{-1}\frac{z^{-1}}{(1-z^{-1})^2}\right] = \begin{cases} k-1 & k = 1, 2, 3, \ldots \\ 0 & k \le 0 \end{cases}$$

$$\mathcal{Z}^{-1}\left[\frac{z^{-1}}{1-z^{-1}}\right] = \mathcal{Z}^{-1}\left[z^{-1}\frac{1}{1-z^{-1}}\right] = \begin{cases} 1 & k = 1, 2, 3, \ldots \\ 0 & k \le 0 \end{cases}$$

$$\mathcal{Z}^{-1}[z^{-2}] = \begin{cases} 1 & k = 2 \\ 0 & k \ne 2 \end{cases}$$

$$\mathcal{Z}^{-1}[z^{-1}] = \begin{cases} 1 & k = 1 \\ 0 & k \ne 1 \end{cases}$$

Hence the inverse z transform of $X(z)$ can be given by

$$x(k) = \begin{cases} 0 - 0 + 0 + 0 = 0 & k = 0 \\ 0 - 2 + 0 + 2 = 0 & k = 1 \\ 1 - 2 + 1 + 0 = 0 & k = 2 \\ k - 1 - 2 + 0 + 0 = k - 3 & k = 3, 4, 5, \ldots \end{cases}$$

or, by rewriting,

$$x(k) = \begin{cases} 0 & k = 0, 1, 2, 3 \\ k - 3 & k = 4, 5, 6, \ldots \end{cases}$$

Method 4: Inversion integral method. First, note that

$$X(z)z^{k-1} = \frac{z^{k-1}}{(z-1)^2 z^2} = \frac{1}{(z-1)^2 z^{3-k}}$$

Thus, $X(z)z^{k-1}$ has a double pole at $z=1$ and a $(3-k)$-multiple pole at $z=0$. The multiplicity of the pole at $z=0$ depends on the value of k.

It can be seen that the poles of $X(z)z^{k-1}$ are located as follows:

For $k=0$: double pole at $z=1$ and triple pole at $z=0$
For $k=1$: double pole at $z=1$ and double pole at $z=0$
For $k=2$: double pole at $z=1$ and simple pole at $z=0$
For $k=3, 4, 5, \ldots$: double pole at $z=1$ and no pole at $z=0$

In the following we shall consider these four cases separately.

For $k=0$. Referring to Eq. (2–46), we have

$$\begin{aligned}
x(0) &= \left[\text{residue of } \frac{1}{(z-1)^2 z^3} \text{ at double pole } z=1\right] \\
&\quad + \left[\text{residue of } \frac{1}{(z-1)^2 z^3} \text{ at triple pole } z=0\right] \\
&= \frac{1}{(2-1)!}\lim_{z\to 1}\frac{d}{dz}\left[(z-1)^2\frac{1}{(z-1)^2 z^3}\right] \\
&\quad + \frac{1}{(3-1)!}\lim_{z\to 0}\frac{d^2}{dz^2}\left[z^3\frac{1}{(z-1)^2 z^3}\right] \\
&= \lim_{z\to 1}\frac{d}{dz}\left(\frac{1}{z^3}\right) + \frac{1}{2!}\lim_{z\to 0}\frac{d^2}{dz^2}\left[\frac{1}{(z-1)^2}\right] \\
&= \lim_{z\to 1}\frac{-3}{z^4} + \lim_{z\to 0}\frac{3}{(z-1)^4} \\
&= -3+3=0
\end{aligned}$$

For $k=1$.

$$\begin{aligned}
x(1) &= \left[\text{residue of } \frac{1}{(z-1)^2 z^2} \text{ at double pole } z=1\right] \\
&\quad + \left[\text{residue of } \frac{1}{(z-1)^2 z^2} \text{ at double pole } z=0\right] \\
&= \frac{1}{(2-1)!}\lim_{z\to 1}\frac{d}{dz}\left[(z-1)^2\frac{1}{(z-1)^2 z^2}\right] \\
&\quad + \frac{1}{(2-1)!}\lim_{z\to 0}\frac{d}{dz}\left[z^2\frac{1}{(z-1)^2 z^2}\right] \\
&= \lim_{z\to 1}\frac{d}{dz}\left(\frac{1}{z^2}\right) + \lim_{z\to 0}\frac{d}{dz}\left[\frac{1}{(z-1)^2}\right] \\
&= \lim_{z\to 1}\frac{-2}{z^3} + \lim_{z\to 0}\frac{-2}{(z-1)^3} \\
&= -2+2=0
\end{aligned}$$

For $k = 2$.

$$x(2) = \left[\text{residue of } \frac{1}{(z-1)^2 z} \text{ at double pole } z = 1\right]$$

$$+ \left[\text{residue of } \frac{1}{(z-1)^2 z} \text{ at simple pole } z = 0\right]$$

$$= \frac{1}{(2-1)!} \lim_{z \to 1} \frac{d}{dz} \left[(z-1)^2 \frac{1}{(z-1)^2 z}\right] + \lim_{z \to 0} \left[z \frac{1}{(z-1)^2 z}\right]$$

$$= \lim_{z \to 1} \frac{d}{dz} \left(\frac{1}{z}\right) + \lim_{z \to 0} \frac{1}{(z-1)^2}$$

$$= \lim_{z \to 1} \frac{-1}{z^2} + \lim_{z \to 0} \frac{1}{(z-1)^2}$$

$$= -1 + 1 = 0$$

For $k = 3, 4, 5, \ldots$

$$x(k) = \left[\text{residue of } \frac{z^{k-3}}{(z-1)^2} \text{ at double pole } z = 1\right]$$

$$= \frac{1}{(2-1)!} \lim_{z \to 1} \frac{d}{dz} \left[(z-1)^2 \frac{z^{k-3}}{(z-1)^2}\right]$$

$$= \lim_{z \to 1} \frac{d}{dz} (z^{k-3})$$

$$= \lim_{z \to 1} [(k-3) z^{k-4}]$$

$$= \begin{cases} 0 & k = 3 \\ k - 3 & k = 4, 5, 6, \ldots \end{cases}$$

The complete form of the inverse z transform of $X(z)$ can now be given as follows:

$$x(k) = \begin{cases} 0 & k = 0, 1, 2, 3 \\ k - 3 & k = 4, 5, 6, \ldots \end{cases}$$

Problem A-2-19. Obtain the solution of the following difference equation in terms of $x(0)$ and $x(1)$:

$$x(k+2) + (a+b)x(k+1) + abx(k) = 0$$

where a and b are constants and $k = 0, 1, 2, \ldots$

Solution. The z transforms of $x(k+2)$, $x(k+1)$, and $x(k)$ are given, respectively, by

$$\mathscr{Z}[x(k+2)] = z^2 X(z) - z^2 x(0) - zx(1)$$

$$\mathscr{Z}[x(k+1)] = zX(z) - zx(0)$$

$$\mathscr{Z}[x(k)] = X(z)$$

Hence the z transform of the given difference equation becomes

$$[z^2X(z) - z^2x(0) - zx(1)] + (a+b)[zX(z) - zx(0)] + abX(z) = 0$$

or

$$[z^2 + (a+b)z + ab]X(z) = [z^2 + (a+b)z]x(0) + zx(1)$$

Solving this last equation for $X(z)$ gives

$$X(z) = \frac{[z^2 + (a+b)z]x(0) + zx(1)}{z^2 + (a+b)z + ab}$$

Notice that constants a and b are two roots of the characteristic equation. We shall now consider separately two cases: **(1)** $a \neq b$ and **(2)** $a = b$.

(1) For the case where $a \neq b$, expanding $X(z)/z$ into partial fractions, we obtain

$$\frac{X(z)}{z} = \frac{bx(0) + x(1)}{b - a}\,\frac{1}{z + a} + \frac{ax(0) + x(1)}{a - b}\,\frac{1}{z + b} \qquad (a \neq b)$$

from which we get

$$X(z) = \frac{bx(0) + x(1)}{b - a}\,\frac{1}{1 + az^{-1}} + \frac{ax(0) + x(1)}{a - b}\,\frac{1}{1 + bz^{-1}}$$

The inverse z transform of $X(z)$ gives

$$x(k) = \frac{bx(0) + x(1)}{b - a}(-a)^k + \frac{ax(0) + x(1)}{a - b}(-b)^k \qquad (a \neq b)$$

where $k = 0, 1, 2, \ldots$

(2) For the case where $a = b$, the z transform $X(z)$ becomes as follows:

$$\begin{aligned} X(z) &= \frac{(z^2 + 2az)x(0) + zx(1)}{z^2 + 2az + a^2} \\ &= \frac{zx(0)}{z + a} + \frac{z[ax(0) + x(1)]}{(z + a)^2} \\ &= \frac{x(0)}{1 + az^{-1}} + \frac{[ax(0) + x(1)]z^{-1}}{(1 + az^{-1})^2} \end{aligned}$$

The inverse z transform of $X(z)$ gives

$$x(k) = x(0)(-a)^k + [ax(0) + x(1)]k(-a)^{k-1}$$

where $k = 0, 1, 2, \ldots$

Problem A-2-20. Solve the following difference equation:

$$2x(k) - 2x(k-1) + x(k-2) = u(k) \tag{2-64}$$

where $x(k) = 0$ for $k < 0$ and

$$u(k) = \begin{cases} 1 & k = 0, 1, 2, \ldots \\ 0 & k < 0 \end{cases}$$

Solution. By taking the z transform of the given difference equation,

$$2X(z) - 2z^{-1}X(z) + z^{-2}X(z) = \frac{1}{1-z^{-1}}$$

Solving this last equation for $X(z)$, we obtain

$$X(z) = \frac{1}{1-z^{-1}} \frac{1}{2-2z^{-1}+z^{-2}} = \frac{z^3}{(z-1)(2z^2-2z+1)}$$

Expanding $X(z)$ into partial fractions, we get

$$X(z) = \frac{z}{z-1} + \frac{-z^2+z}{2z^2-2z+1} = \frac{1}{1-z^{-1}} + \frac{-1+z^{-1}}{2-2z^{-1}+z^{-2}}$$

Notice that the two poles involved in the quadratic term in this last equation are complex conjugates. Hence, we rewrite $X(z)$ as follows:

$$X(z) = \frac{1}{1-z^{-1}} - \frac{1}{2}\frac{1-0.5z^{-1}}{1-z^{-1}+0.5z^{-2}} + \frac{1}{2}\frac{0.5z^{-1}}{1-z^{-1}+0.5z^{-2}}$$

By referring to the formulas for the z transforms of damped cosine and damped sine functions, we identify $e^{-2aT} = 0.5$ and $\cos \omega T = 1/\sqrt{2}$ for this problem. Hence, we get $\omega T = \pi/4$, $\sin \omega T = 1/\sqrt{2}$, and $e^{-aT} = 1/\sqrt{2}$. Then the inverse z transform of $X(z)$ can be written as

$$\begin{aligned} x(k) &= 1 - \tfrac{1}{2}e^{-akT}\cos \omega kT + \tfrac{1}{2}e^{-akT}\sin \omega kT \\ &= 1 - \frac{1}{2}\left(\frac{1}{\sqrt{2}}\right)^k \cos\frac{k\pi}{4} + \frac{1}{2}\left(\frac{1}{\sqrt{2}}\right)^k \sin\frac{k\pi}{4} \qquad k = 0, 1, 2, \ldots \end{aligned}$$

from which we obtain

$$\begin{aligned} x(0) &= 0.5 \\ x(1) &= 1 \\ x(2) &= 1.25 \\ x(3) &= 1.25 \\ x(4) &= 1.125 \\ &\vdots \end{aligned}$$

A BASIC computer program for solving Eq. (2–64) and the computer solution are shown in Table 2–10.

Problem A-2–21. Consider the difference equation

$$x(k+2) - 1.3679x(k+1) + 0.3679x(k) = 0.3679u(k+1) + 0.2642u(k) \qquad (2\text{–}65)$$

where $x(k)$ is the output and $x(k) = 0$ for $k \le 0$ and where $u(k)$ is the input and is given by

TABLE 2–10 BASIC COMPUTER PROGRAM FOR SOLVING THE DIFFERENCE EQUATION GIVEN BY EQ. (2–64)

```
10  X0 = 0.5
20  X1 = 1
30  U0 = 1
40  U1 = 1
50  U2 = 1
60  X2 = X1 - 0.5 * X0 + 0.5 * U2
70  M = X0
80  X0 = X1
90  X1 = X2
100 N = U0
110 U0 = U1
120 U1 = U2
130 PRINT K, M, N
140 K = K + 1
150 IF K < 16 GO TO 60
160 END
```

$k = \mathrm{K}$	$x(k) = \mathrm{XK} = \mathrm{M}$	$u(k) = \mathrm{UK} = \mathrm{N}$
0	0.5	1
1	1	1
2	1.25	1
3	1.25	1
4	1.125	1
5	1	1
6	0.9375	1
7	0.9375	1
8	0.96875	1
9	1	1
10	1.01563	1
11	1.01563	1
12	1.00781	1
13	1	1
14	0.996094	1
15	0.996094	1

$$u(k) = 0 \qquad k < 0$$

$$u(0) = 1$$

$$u(1) = 0.2142$$

$$u(2) = -0.2142$$

$$u(k) = 0 \qquad k = 3, 4, 5, \ldots$$

Determine the output $x(k)$.

Solution. Taking the z transform of the given difference equation, we obtain

$$[z^2X(z) - z^2x(0) - zx(1)] - 1.3679[zX(z) - zx(0)] + 0.3679X(z)$$
$$= 0.3679[zU(z) - zu(0)] + 0.2642U(z) \qquad (2\text{–}66)$$

By substituting $k = -1$ into the given difference equation, we find

$$x(1) - 1.3679x(0) + 0.3679x(-1) = 0.3679u(0) + 0.2642u(-1)$$

Since $x(0) = x(-1) = 0$ and since $u(-1) = 0$ and $u(0) = 1$, we obtain

$$x(1) = 0.3679u(0) = 0.3679$$

By substituting the initial data

$$x(0) = 0, \qquad x(1) = 0.3679, \qquad u(0) = 1$$

into Eq. (2–66), we get

$$z^2X(z) - 0.3679z - 1.3679zX(z) + 0.3679X(z) = 0.3679zU(z) - 0.3679z + 0.2642U(z)$$

Solving for $X(z)$, we find

$$X(z) = \frac{0.3679z + 0.2642}{z^2 - 1.3679z + 0.3679} U(z)$$

The z transform of the input $u(k)$ is

$$U(z) = \mathscr{Z}[u(k)] = 1 + 0.2142z^{-1} - 0.2142z^{-2}$$

Hence

$$X(z) = \frac{0.3679z + 0.2642}{z^2 - 1.3679z + 0.3679}(1 + 0.2142z^{-1} - 0.2142z^{-2})$$

$$= \frac{0.3679z^{-1} + 0.3430z^{-2} - 0.02221z^{-3} - 0.05659z^{-4}}{1 - 1.3679z^{-1} + 0.3679z^{-2}}$$

$$= 0.3679z^{-1} + 0.8463z^{-2} + z^{-3} + z^{-4} + z^{-5} + \cdots$$

Thus, the inverse z transform of $X(z)$ gives

$$x(0) = 0$$

$$x(1) = 0.3679$$

$$x(2) = 0.8463$$

$$x(k) = 1 \qquad k = 3, 4, 5, \ldots$$

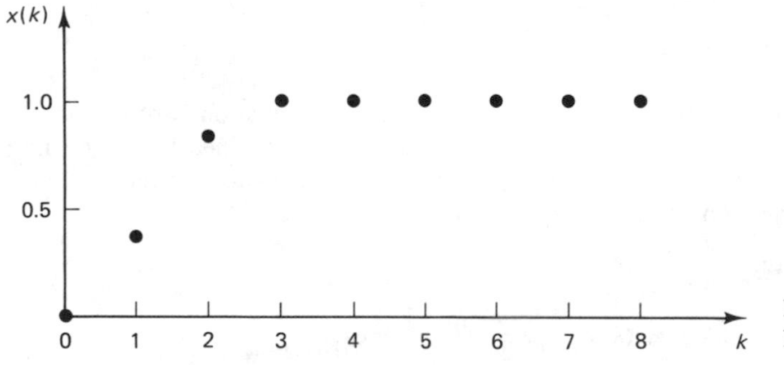

Figure 2–13 Output $x(k)$ for Prob. A-2-21.

Figure 2–13 shows the output $x(k)$.

The output $x(k)$ can also be obtained with a microcomputer. Note that the initial data $x(0)$ and $x(1)$ are as follows:

$$x(0) = 0$$

$$x(1) = 0.3679u(0) = 0.3679$$

A BASIC computer program for finding $x(k)$, the solution of Eq. (2–65), is shown in Table 2–11. The computation results are also shown in Table 2–11.

Problem A-2–22. Solve the following difference equation:

$$(k + 1)x(k + 1) - x(k) = 0$$

TABLE 2–11 BASIC COMPUTER PROGRAM FOR SOLVING THE DIFFERENCE EQUATION GIVEN BY EQ. (2–65)

```
10  X0 = 0
20  X1 = 0.3679
30  U0 = 1
40  U1 = 0.2142
50  U2 = -0.2142
60  U3 = 0
70  X2 = 1.3679 * X1 - 0.3679 * X0
         + 0.3679 * U1 + 0.2642 * U0
80  M = X0
90  X0 = X1
100 X1 = X2
110 N = U0
120 U0 = U1
130 U1 = U2
140 U2 = U3
150 PRINT K, M, N
160 K = K + 1
170 IF K < 11 GO TO 70
180 END
```

k = K	$x(k)$ = XK = M	$u(k)$ = UK = N
0	0	1
1	0.3679	0.2142
2	0.846255	−0.2142
3	1.00003	0
4	1.00001	0
5	1	0
6	1	0
7	1	0
8	1	0
9	1	0
10	1	0
11	1	0
12	1	0
13	1	0
14	1	0
15	1	0

where $x(k) = 0$ for $k < 0$ and $x(0) = 1$. Notice that this difference equation is of the time-varying kind. The solution of this type of difference equation may be obtained by use of the z transform. (It should be cautioned that, in general, the z transform approach to the solution of time-varying difference equations may not be successful.)

Solution. First, note that

$$\mathscr{Z}[kx(k)] = -z\frac{d}{dz}X(z)$$

Since the original difference equation can be written as

$$kx(k) - x(k-1) = 0$$

the z transform of this last equation can be obtained as follows:

$$-z\frac{d}{dz}X(z) - z^{-1}X(z) = 0$$

or

$$z^2\frac{d}{dz}X(z) + X(z) = 0$$

from which we have

$$\frac{dX(z)}{X(z)} = -\frac{dz}{z^2}$$

or

$$\ln X(z) = \frac{1}{z} + \ln K$$

where K is a constant. Then $X(z)$ can be found from

$$X(z) = K \exp z^{-1}$$

Since $\exp z^{-1}$ may be expanded into the series

$$\exp z^{-1} = 1 + z^{-1} + \frac{1}{2!}z^{-2} + \frac{1}{3!}z^{-3} + \cdots \qquad |z| > 0$$

we have

$$X(z) = K\left(1 + z^{-1} + \frac{1}{2!}z^{-2} + \frac{1}{3!}z^{-3} + \cdots\right)$$

from which we find the inverse z transform of $X(z)$ to be

$$x(k) = K\frac{1}{k!} \qquad k = 0, 1, 2, \ldots$$

Since $x(0)$ is given as 1, we have

$$x(0) = K = 1$$

Thus we have determined the unknown constant K. Hence the solution to the given difference equation is

$$x(k) = \frac{1}{k!} \qquad k = 0, 1, 2, \ldots$$

Problem A-2–23. Consider the difference equation

$$x(k+2) = x(k+1) + x(k)$$

where $x(0) = 0$ and $x(1) = 1$. Note that $x(2) = 1$, $x(3) = 2$, $x(4) = 3$, ... The series 0, 1, 1, 2, 3, 5, 8, 13, . . . is known as the Fibonacci series.

Obtain the general solution $x(k)$ in a closed form. Show that the limiting value of $x(k+1)/x(k)$ as k approaches infinity is $(1 + \sqrt{5})/2$, or approximately 1.6180.

Solution. The z transform of the given difference equation is

$$z^2X(z) - z^2x(0) - zx(1) = zX(z) - zx(0) + X(z)$$

Solving for $X(z)$ gives

$$X(z) = \frac{z^2x(0) + zx(1) - zx(0)}{z^2 - z - 1}$$

By substituting the initial data $x(0) = 0$ and $x(1) = 1$ into this last equation, we have

$$X(z)=\frac{z}{z^2-z-1}=\frac{1}{\sqrt{5}}\left(\frac{z}{z-\dfrac{1+\sqrt{5}}{2}}-\frac{z}{z-\dfrac{1-\sqrt{5}}{2}}\right)$$

$$=\frac{1}{\sqrt{5}}\left(\frac{1}{1-\dfrac{1+\sqrt{5}}{2}z^{-1}}-\frac{1}{1-\dfrac{1-\sqrt{5}}{2}z^{-1}}\right)$$

The inverse z transform of $X(z)$ is

$$x(k)=\frac{1}{\sqrt{5}}\left[\left(\frac{1+\sqrt{5}}{2}\right)^k-\left(\frac{1-\sqrt{5}}{2}\right)^k\right] \qquad k=0,1,2,\ldots$$

Note that although this last equation involves $\sqrt{5}$, the square roots in the right-hand side of this last equation cancel out and the values of $x(k)$ for $k = 0, 1, 2, \ldots$ turn out to be positive integers.

The limiting value of $x(k+1)/x(k)$ as k approaches infinity is obtained as follows:

$$\lim_{k\to\infty}\frac{x(k+1)}{x(k)}=\lim_{k\to\infty}\frac{\left(\dfrac{1+\sqrt{5}}{2}\right)^{k+1}-\left(\dfrac{1-\sqrt{5}}{2}\right)^{k+1}}{\left(\dfrac{1+\sqrt{5}}{2}\right)^{k}-\left(\dfrac{1-\sqrt{5}}{2}\right)^{k}}$$

Since $|(1-\sqrt{5})/2| < 1$,

$$\lim_{k\to\infty}\left(\frac{1-\sqrt{5}}{2}\right)^k \longrightarrow 0$$

Hence

$$\lim_{k\to\infty}\frac{x(k+1)}{x(k)}=\lim_{k\to\infty}\frac{\left(\dfrac{1+\sqrt{5}}{2}\right)^{k+1}}{\left(\dfrac{1+\sqrt{5}}{2}\right)^{k}}=\frac{1+\sqrt{5}}{2}=1.6180$$

Problem A-2–24. Referring to Prob. A-2–23, write a BASIC computer program to generate the Fibonacci series. Carry out the Fibonacci series to $k = 30$.

Solution. The difference equation that generates the Fibonacci series is

$$x(k+2)=x(k+1)+x(k)$$

where $x(0) = 0$ and $x(1) = 1$. The BASIC computer program based on this difference equation shown in Table 2–12 will generate the Fibonacci series. The computer solution for the Fibonacci series up to $k = 30$ is also shown in Table 2–12.

Problem A-2–25. Consider the equation

$$x=1+\cfrac{1}{1+\cfrac{1}{1+\cfrac{1}{1+\cdots}}} \tag{2–67}$$

TABLE 2–12 BASIC COMPUTER PROGRAM TO GENERATE THE FIBONACCI SERIES, TOGETHER WITH THE COMPUTER SOLUTION SHOWING THE FIBONACCI SERIES UP TO $k = 30$

```
10  X0 = 0
20  X1 = 1
30  K = 0
40  X2 = X1 + X0
50  M = X0
60  X0 = X1
70  X1 = X2
80  PRINT K, M
90  K = K + 1
100 IF K < 31 GO TO 40
110 END
```

$k = \mathrm{K}$	$x(k) = \mathrm{XK} = \mathrm{M}$
0	0
1	1
2	1
3	2
4	3
5	5
6	8
7	13
8	21
9	34
10	55
11	89
12	144
13	233
14	377
15	610
16	987
17	1597
18	2584
19	4181
20	6765
21	10946
22	17711
23	28657
24	46368
25	75025
26	121393
27	196418
28	317811
29	514229
30	832040

If we define

$$x(0) = 0$$

$$x(1) = 1$$

$$x(2) = 1 + \tfrac{1}{1} = 2$$

$$x(3) = 1 + \frac{1}{1 + \frac{1}{1}} = \frac{3}{2}$$

$$x(4) = 1 + \frac{1}{1 + \frac{1}{1 + \frac{1}{1}}} = \frac{5}{3}$$

$$\vdots$$

then Eq. (2–67) may be expressed as follows:

$$x(k+1) = 1 + \frac{1}{x(k)} \qquad k = 1, 2, 3, \ldots \tag{2–68}$$

Obtain the solution to Eq. (2–68).

Solution. In order to solve Eq. (2–68), define

$$x(k) = \frac{y(k+1)}{y(k)} \qquad k = 1, 2, 3, \ldots \tag{2–69}$$

Then Eq. (2–68) may be written as follows:

$$\frac{y(k+2)}{y(k+1)} = 1 + \frac{y(k)}{y(k+1)} \tag{2–70}$$

Notice that for $k = 0$ we have

$$\frac{y(2)}{y(1)} = 1 + \frac{y(0)}{y(1)}$$

Since $y(2)/y(1) = x(1) = 1$, we must have $y(0) = 0$ and $y(1) = y(2) = a =$ constant.

Rewriting Eq. (2–70), we have

$$y(k+2) = y(k+1) + y(k) \qquad k = 1, 2, 3, \ldots \tag{2–71}$$

Thus, Eq. (2–68) is modified into a linear difference equation. Notice that Eq. (2–71) is the difference equation that generates the Fibonacci series. (Refer to Prob. A-2–23.) The z transform of Eq. (2–71) becomes

$$z^2Y(z) - z^2y(0) - zy(1) = zY(z) - zy(0) + Y(z)$$

where $y(0) = 0$ and $y(1) = a$. Solving for $Y(z)$ gives

$$Y(z) = \frac{az}{z^2 - z - 1} = \frac{a}{\sqrt{5}}\left(\frac{1}{1 - \frac{1+\sqrt{5}}{2}z^{-1}} - \frac{1}{1 - \frac{1-\sqrt{5}}{2}z^{-1}}\right)$$

The inverse z transform of $Y(z)$ is

$$y(k) = \frac{a}{\sqrt{5}}\left[\left(\frac{1+\sqrt{5}}{2}\right)^k - \left(\frac{1-\sqrt{5}}{2}\right)^k\right] \qquad k = 0, 1, 2, \ldots$$

By referring to Eq. (2–69), we have the solution to Eq. (2–67) as follows:

$$x(k)=\frac{y(k+1)}{y(k)}=\frac{\left(\frac{1+\sqrt{5}}{2}\right)^{k+1}-\left(\frac{1-\sqrt{5}}{2}\right)^{k+1}}{\left(\frac{1+\sqrt{5}}{2}\right)^{k}-\left(\frac{1-\sqrt{5}}{2}\right)^{k}}$$

where $k = 1, 2, 3, \ldots$ Notice that as k approaches infinity, $x(k)$ approaches $(1 + \sqrt{5})/2$, or 1.6180.

Problem A-2–26. Solve the following difference equation:

$$(k+1)x(k+1)-kx(k)=k+1$$

where $x(k) = 0$ for $k \leq 0$.

Solution. First note that by substituting $k = 0$ into the given difference equation, we have

$$x(1)=1$$

Now define

$$y(k)=kx(k)$$

Then the given difference equation can be written as

$$y(k+1)-y(k)=k+1$$

Taking the z transform of this last equation, we have

$$zY(z)-zy(0)-Y(z)=\frac{z^{-1}}{(1-z^{-1})^2}+\frac{1}{1-z^{-1}}$$

Since $y(0) = 0$, we have

$$Y(z)=\frac{z^{-2}}{(1-z^{-1})^3}+\frac{z^{-1}}{(1-z^{-1})^2}$$

Referring to Prob. A-2–17, we have

$$\mathscr{Z}^{-1}\left[\frac{z^{-2}}{(1-z^{-1})^3}\right]=\frac{1}{2}(k^2-k)$$

Hence the inverse z transform of $Y(z)$ can be given by

$$y(k)=\tfrac{1}{2}(k^2-k)+k=\tfrac{1}{2}(k^2+k)$$

Then, $x(k)$ for $k = 1, 2, 3, \ldots$ is determined from

$$kx(k)=y(k)=\tfrac{1}{2}(k^2+k)$$

as follows:

$$x(k)=\tfrac{1}{2}(k+1) \qquad k=1,2,3,\ldots$$

Problem A-2-27. Solve the following difference equation:

$$kx(k+1)-(k+1)x(k)=-\frac{k+1}{n} \tag{2-72}$$

where $k = 0, 1, 2, \ldots, n$ and the final condition is specified as

$$x(n)=0$$

Plot the solution sequence when $n = 10$.

Solution. First, consider the case where $k = 0$. By substituting $k = 0$ into Eq. (2–72), we obtain

$$x(0)=\frac{1}{n}$$

By substituting $k = 1$ into Eq. (2–72), we have

$$x(2)-2x(1)=-\frac{2}{n}$$

Since $x(1)$ is unknown at this stage, $x(k)$ for increasing k cannot be found by recursion. However, since the final condition $x(n) = 0$ is specified, it is possible to start at $k = n$ and work backward.

Solving Eq. (2–72) for $x(k)$, we have

$$x(k)=\frac{k}{k+1}x(k+1)+\frac{1}{n}$$

Since the final condition is $x(n) = 0$, we obtain

$$x(n-1)=\frac{n-1}{n}x(n)+\frac{1}{n}=\frac{1}{n}$$

and

$$x(n-2)=\frac{n-2}{n-1}x(n-1)+\frac{1}{n}=\frac{1}{n}\frac{n-2}{n-1}+\frac{1}{n}$$

$$x(n-3)=\frac{n-3}{n-2}x(n-2)+\frac{1}{n}=\frac{1}{n}\left(\frac{n-3}{n-1}+\frac{n-3}{n-2}\right)+\frac{1}{n}$$

$$x(n-4)=\frac{n-4}{n-3}x(n-3)+\frac{1}{n}=\frac{1}{n}\left(\frac{n-4}{n-1}+\frac{n-4}{n-2}+\frac{n-4}{n-3}\right)+\frac{1}{n}$$

$$\vdots$$

$$x(1)=\frac{1}{n}\left(\frac{1}{n-1}+\frac{1}{n-2}+\cdots+\frac{1}{2}\right)+\frac{1}{n}$$

$$x(0)=\frac{1}{n}\left(\frac{0}{n-1}+\frac{0}{n-2}+\cdots+\frac{0}{1}\right)+\frac{1}{n}=\frac{1}{n}$$

The general expression for $x(k)$ is

$$x(k) = \frac{1}{n}\left(\frac{k}{n-1} + \frac{k}{n-2} + \cdots + \frac{k}{k+1}\right) + \frac{1}{n}$$
$$= \frac{k}{n}\left(\frac{1}{k} + \frac{1}{k+1} + \cdots + \frac{1}{n-1}\right)$$

Clearly, this solution satisfies the given difference equation, Eq. (2–72), as follows:

$$kx(k+1) - (k+1)x(k) = \frac{k(k+1)}{n}\left(\frac{1}{k+1} + \frac{1}{k+2} + \cdots + \frac{1}{n-1}\right)$$
$$- \frac{(k+1)k}{n}\left(\frac{1}{k} + \frac{1}{k+1} + \cdots + \frac{1}{n-1}\right)$$
$$= -\frac{k(k+1)}{n}\frac{1}{k} = -\frac{k+1}{n}$$

The solution to the problem can therefore be written as follows:

$$x(k) = \frac{k}{n}\sum_{j=k}^{n-1}\frac{1}{j} \qquad k = 0, 1, 2, \ldots, n-1$$

The same solution can also be obtained by a different approach. We first note that for $k = 0$,

$$x(0) = \frac{1}{n}$$

Now consider the case where $k = 1, 2, 3, \ldots, n-1$. If we divide both sides of Eq. (2–72) by $k(k+1)$, then we have

$$\frac{x(k+1)}{k+1} - \frac{x(k)}{k} = -\frac{1}{nk} \qquad k = 1, 2, 3, \ldots, n-1$$

Starting with this last equation and increasing k by 1 in each step until k equals $n-1$, we have

$$\frac{x(k+1)}{k+1} - \frac{x(k)}{k} = -\frac{1}{nk}$$
$$\frac{x(k+2)}{k+2} - \frac{x(k+1)}{k+1} = -\frac{1}{n(k+1)}$$
$$\frac{x(k+3)}{k+3} - \frac{x(k+2)}{k+2} = -\frac{1}{n(k+2)}$$
$$\vdots$$
$$\frac{x(n)}{n} - \frac{x(n-1)}{n-1} = -\frac{1}{n(n-1)}$$

Adding the terms on the left-hand side of the foregoing $n - k$ equations and equating the result with the sum of the terms on the right-hand side, we obtain

$$\frac{x(n)}{n} - \frac{x(k)}{k} = -\frac{1}{nk} - \frac{1}{n(k+1)} - \cdots - \frac{1}{n(n-1)}$$

Since $x(n) = 0$, this last equation simplifies to

$$x(k) = \frac{k}{n}\left(\frac{1}{k} + \frac{1}{k+1} + \cdots + \frac{1}{n-1}\right) \qquad k = 1, 2, 3, \ldots, n-1$$

In deriving this last equation we have excluded the case for $k = 0$. However, this last equation yields $x(0) = 1/n$, which is the valid result for $k = 0$. Hence, the general solution to the problem can be given by

$$x(k) = \frac{k}{n} \sum_{j=k}^{n-1} \frac{1}{j} \qquad k = 0, 1, 2, \ldots, n-1$$

Figure 2–14 shows the solution sequence when $n = 10$.

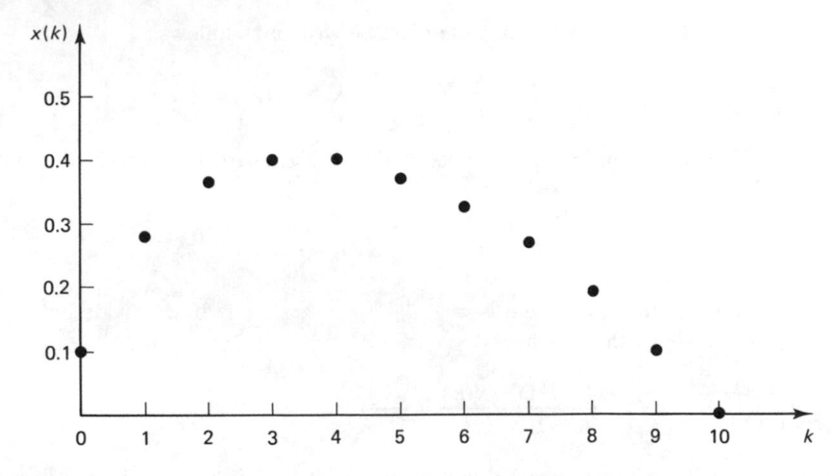

Figure 2–14 Solution sequence for Prob. A-2–27.

Problem A-2–28. Consider the difference equation

$$x(k+2) + \alpha x(k+1) + \beta x(k) = 0 \tag{2–73}$$

Find the conditions on α and β for which the solution series $x(k)$ for $k = 0, 1, 2, \ldots$, subjected to initial conditions, is finite.

Solution. Let us define

$$\alpha = a + b, \qquad \beta = ab$$

Then, referring to Prob. A-2–19, the solution $x(k)$ for $k = 0, 1, 2, \ldots$ can be given by

$$x(k) = \begin{cases} \dfrac{bx(0) + x(1)}{b-a}(-a)^k + \dfrac{ax(0) + x(1)}{a-b}(-b)^k & (a \neq b) \\ x(0)(-a)^k + [ax(0) + x(1)]k(-a)^{k-1} & (a = b) \end{cases}$$

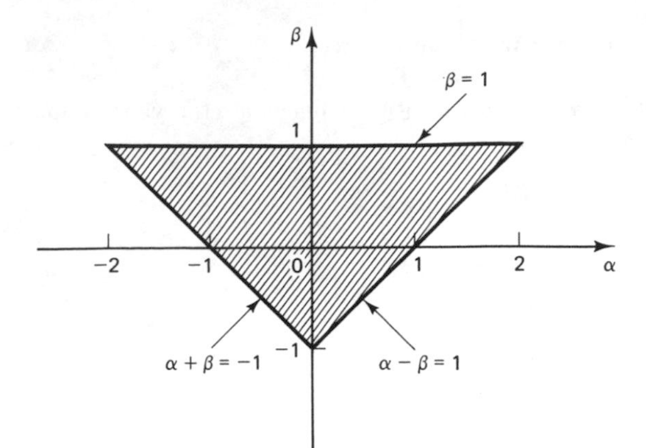

Figure 2–15 Region in the $\alpha\beta$ plane in which the solution series of Eq. (2–73), subjected to initial conditions, is finite.

The solution series $x(k)$ for $k = 0, 1, 2, \ldots$, subjected to initial conditions $x(0)$ and $x(1)$, is finite if the absolute values of a and b are less than unity. Thus, on the $\alpha\beta$ plane, three critical points can be located:

$$\alpha = 2, \qquad \beta = 1$$

$$\alpha = -2, \qquad \beta = 1$$

$$\alpha = 0, \qquad \beta = -1$$

The interior of the region bounded by lines connecting these points satisfies the condition $|a| < 1$, $|b| < 1$. The boundary lines can be given by $\beta = 1$, $\alpha - \beta = 1$, and $\alpha + \beta = -1$. See Fig. 2–15. If point (α, β) lies inside the shaded triangular region, then the solution series $x(k)$ for $k = 0, 1, 2, \ldots$, subjected to initial conditions $x(0)$ and $x(1)$, is finite.

PROBLEMS

Problem B-2–1. Obtain the z transform of

$$x(t) = \frac{1}{a}(1 - e^{-at})$$

where a is a constant.

Problem B-2–2. Obtain the z transform of k^3 by two methods.

Problem B-2–3. Obtain the z transform of $k^3 a^{k-1}$ by three methods.

Problem B-2–4. Obtain the z transform of $t^2 e^{-at}$ by two methods.

Problem B-2–5. Obtain the z transform of the following $x(k)$:

$$x(k) = 9k(2^{k-1}) - 2^k + 3 \qquad k = 0, 1, 2, \ldots$$

Assume that $x(k) = 0$ for $k < 0$.

Problem B-2–6. Obtain the z transform of $1/k!$

Problem B-2–7. Obtain the z transform of a function $x(t)$ whose Laplace transform is

$$X(s) = \frac{e^{-s}}{s^2(s+1)}$$

Problem B-2–8. Find the z transform of

$$x(k) = \sum_{h=0}^{k} a^h$$

where a is a constant.

Problem B-2–9. Using Eq. (2–30), verify the following equations:

$$\mathscr{Z}[ka^{k-1}] = \frac{z}{(z-a)^2}$$

$$\mathscr{Z}[k(k-1)a^{k-2}] = \frac{(2!)z}{(z-a)^3}$$

$$\mathscr{Z}[k(k-1)\cdots(k-h+1)a^{k-h}] = \frac{(h!)z}{(z-a)^{h+1}}$$

Problem B-2–10. Obtain the z transform of the curve $x(t)$ shown in Fig. 2–16.

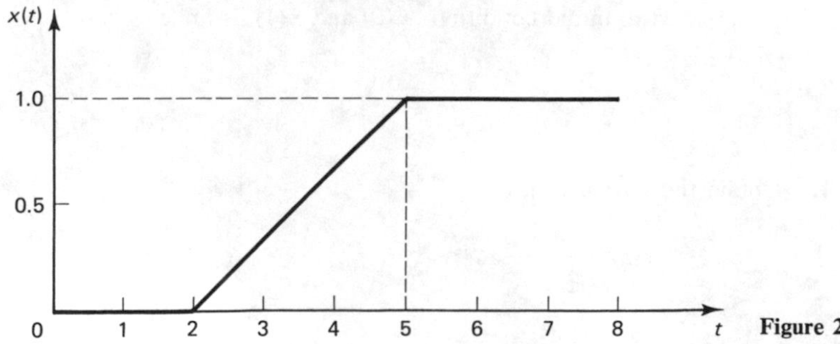

Figure 2–16 Curve $x(t)$.

Problem B-2–11. Consider the difference equation

$$x(k) = a_1 x(k-1) + a_2 x(k-2) + b_0 u(k)$$

Show that this equation can be modified to the following backward difference equation:

$$a_2 \nabla^2 x(k) - (a_1 + 2a_2) \nabla x(k) + (a_1 + a_2 - 1)x(k) = -b_0 u(k)$$

Problem B-2–12. Find the inverse z transform of

$$X(z) = \frac{z^{-1}(0.5 - z^{-1})}{(1 - 1.5z^{-1})(1 - z^{-1})^2}$$

Use (1) the partial-fraction-expansion method and (2) the computational method. Write a BASIC computer program for finding $x(k)$, the inverse z transform of $X(z)$.

Problem B-2–13. Find the inverse z transform of

$$X(z)=\frac{2+z^{-2}}{(1-0.5z^{-1})^2(1-z^{-1})}$$

Use (1) the direct division method and (2) the computational method.

Problem B-2–14. Given the z transform

$$X(z)=\frac{z^{-1}}{(1-z^{-1})(1+1.3z^{-1}+0.4z^{-2})}$$

determine the initial and final values of $x(k)$. Also find $x(k)$, the inverse z transform of $X(z)$, in a closed form.

Problem B-2–15. Obtain the inverse z transform of

$$X(z)=\frac{1+z^{-1}-z^{-2}}{1-z^{-1}}$$

Use (1) the inversion integral method and (2) the computational method.

Problem B-2–16. Obtain the inverse z transform of

$$X(z)=\frac{z^{-3}}{(1-z^{-1})(1-0.2z^{-1})}$$

in a closed form. Plot $x(k)$ versus k.

Problem B-2–17. By using the inversion integral method, obtain the inverse z transform of

$$X(z)=\frac{1+6z^{-2}+z^{-3}}{(1-z^{-1})(1-0.2z^{-1})}$$

Problem B-2–18. Find the inverse z transform of

$$X(z)=\frac{z^{-1}(1-z^{-2})}{(1+z^{-2})^2}$$

Use (1) the direct division method and (2) the computational method. Write a BASIC computer program for finding $x(k)$, the inverse z transform of $X(z)$.

Problem B-2–19. Obtain the inverse z transform of

$$X(z)=\frac{z^{-1}(1+z^{-1}+z^{-2})}{(1-z^{-1})(1-2z^{-1})(1-z^{-1}+0.5z^{-2})}$$

Write $x(k)$ in a closed form.

Problem B-2-20. Obtain the inverse z transform of

$$X(z) = \frac{0.368z^2 + 0.478z + 0.154}{(z-1)z^2}$$

by use of the inversion integral method.

Problem B-2-21. Find the solution of the following difference equation:

$$x(k+2) - 3x(k+1) + 2x(k) = u(k)$$

where $x(0) = x(1) = 0$ and $x(k) = 0$ for $k < 0$. For the input function $u(k)$ consider the following two cases:

$$u(k) = \begin{cases} 1 & k = 0, 1, 2, \ldots \\ 0 & k < 0 \end{cases}$$

and

$$u(0) = 1$$
$$u(k) = 0 \qquad k \neq 0$$

Solve this problem both analytically and computationally. (Write BASIC computer programs and carry out the computational solutions.)

Problem B-2-22. Solve the following difference equation:

$$x(k+2) - x(k+1) + 0.25x(k) = u(k+2)$$

where $x(0) = 1$ and $x(1) = 2$. The input function $u(k)$ is given by

$$u(k) = 1 \qquad k = 0, 1, 2, \ldots$$

Solve this problem both analytically and computationally.

Problem B-2-23. Consider the difference equation:

$$x(k+2) - 1.3679x(k+1) + 0.3679x(k) = 0.3679u(k+1) + 0.2642u(k)$$

where $x(k) = 0$ for $k \leq 0$. The input $u(k)$ is given by

$$u(k) = 0 \qquad k < 0$$
$$u(0) = 1.5820$$
$$u(1) = -0.5820$$
$$u(k) = 0 \qquad k = 2, 3, 4, \ldots$$

Determine the output $x(k)$. Solve this problem both analytically and computationally.

Problem B-2-24. Find the solution $x(k)$ of the following difference equation:

$$x(k+1) = (k+1)x(k)$$

where $x(0) = 1$.

Problem B-2–25. Find the solution $x(k)$ of the following difference equation:

$$x(k+1) = a(k)x(k)$$

where $x(0) = c =$ constant and $a(k)$ is a time-varying coefficient.

Problem B-2–26. Solve the following difference equation:

$$x(k+1) = x(k) + a^k$$

where $x(0) = 0$.

Problem B-2–27. Show that the nonlinear difference equation

$$x(k+1)x(k) + a(k)x(k+1) + b(k)x(k) = 0$$

can be reduced to a linear difference equation by means of the transformation

$$x(k) = \frac{y(k+1)}{y(k)} - a(k)$$

Problem B-2–28. Solve the following difference equation:

$$(k+2)x(k+2) - 2(k+1)x(k+1) + kx(k) = 1$$

where $x(k) = 0$ for $k \leq 0$.

3

Background Materials for the z Domain Analysis

3–1 INTRODUCTION

The z transform method is particularly useful for analyzing and designing single-input–single-output linear time-invariant discrete-time control systems. This chapter presents background materials necessary for the analysis and design of discrete-time control systems in the z domain. The main advantage of the z transform method is that it enables the engineer to apply conventional continuous-time design methods to discrete-time systems that may be partly discrete-time and partly continuous-time.

Applicability and limitations of the z domain analysis and design. The z domain analysis and design of discrete-time control systems are straightforward and useful. However, the z transform approach has limitations, and we must keep in mind the following facts before applying the z transform method for the analysis and design of discrete-time systems:

1. Mathematically, the sampling process is approximated by impulse sampling; that is, the sampled signal is approximated by a train of impulses whose strengths are equal to the magnitudes of the input signal at the sampling instants. This mathematical treatment is valid only if the sampling duration of the sampler is negligibly small compared with the most significant time constant involved in the system.
2. The z transform method presents accurate results only for systems in which the signals involved can be represented accurately by the values at the sampling instants. That is, the signal between any two consecutive sampling instants

must not vary very much. The inverse z transform of the system output, $X(z)$, gives the value of $x(kT)$, which is the value of $x(t)$ at the sampling instants only. No information is available concerning the behavior of the output between any two consecutive sampling instants.

A few methods, however, are available for finding the response (waveform) between any two consecutive sampling instants. For example, the modified z transform method, which is presented in Sec. 3–6, can provide response between any two consecutive sampling instants. The Laplace transform approach and the state space approach can also provide the response between any two consecutive sampling instants. (See Sec. 3–6 and Sec. 5–6, respectively.)

3. The z transform approach using pulse transfer functions is primarily useful for the analysis and synthesis of single-input–single-output linear time-invariant discrete-time control systems. For multiple-input–multiple-output linear or nonlinear discrete-time control systems, the state space approach is better suited. We shall present the state space approach to the analysis and synthesis of such control systems in Chaps. 5 through 7.

Outline of the chapter. The outline of this chapter is as follows. Section 3–1 gives introductory remarks. Section 3–2 presents a powerful method to treat the sampling operation as a mathematical representation of the operation of taking samples $x(kT)$ from a continuous-time signal $x(t)$ by impulse modulation. This section includes derivations of the transfer functions of the zero-order hold and first-order hold.

Section 3–3 deals with the convolution integral method for obtaining the z transform. Reconstructing the original continuous-time signal from the sampled signal is the main subject matter of Sec. 3–4. Based on the fact that the Laplace transform of the sampled signal is periodic, we present the sampling theorem, which states that in order to fully reconstruct a continuous-time signal from a sampled signal, it is necessary that the sampling frequency be at least twice the highest-frequency component of the original continuous-time signal. We also demonstrate that if the sampling theorem is not satisfied, aliasing will occur.

Section 3–5 presents pulse transfer functions and discusses methods for obtaining such functions. Mathematical modeling of digital controllers in terms of pulse transfer functions is also discussed. Section 3–6 presents the modified z transform method for finding the response between any two consecutive sampling periods. Section 3–7 treats the realization of digital controllers and digital filters. Section 3–8 discusses mapping between the s plane and the z plane. Finally, Sec. 3–9 presents the stability analysis of closed-loop systems in the z domain.

Remarks. In industrial control systems, time constants involved in various plants may differ widely. Hence, in practical situations, different sampling periods may be used for different loops in the system. That is, if a loop has a plant having large time constants, resulting in slowly varying signals, and the signals involved in the loop are band-limited, we may wish to sample at a low frequency. On the other hand, if a loop involves a plant having only small time constants, high-frequency signals may exist in the loop and if they exist the signals must be sampled at a high frequency. Thus the dynamics involved in the loop may suggest different sampling frequencies in different loops.

If two or more sampling rates (sampling frequencies) are involved, the digital control system is commonly referred to as a multirate digital control system. Although multirate digital control systems are important in practical situations, in this book we shall concentrate on the discussion of single-rate digital control systems. (The reader interested in multirate digital control systems may refer to Refs. 3–9 and 3–12, listed at the end of the chapter.)

Throughout this book we assume that the sampling operation is uniform, that is, that only one sampling rate exists in the system and the sampling period is constant. If a discrete-time control system involves two or more samplers in the system, we assume that all samplers are synchronized and have the same sampling rate or sampling frequency.

3–2 DISCRETE-TIME CONTROL SYSTEMS AND IMPULSE SAMPLING

Discrete-time control systems may operate partly in discrete time and partly in continuous time. Thus, in such control systems some signals appear as discrete-time functions (often in the form of a sequence of numbers or a numerical code) and other signals as continuous-time functions. In analyzing discrete-time control systems, the z transform theory plays an important role. In order to see why the z transform method is useful in the analysis of discrete-time control systems, we first introduce the concept of impulse sampling.

Impulse sampling. In a conventional sampler, a switch closes to admit an input signal every sampling period T. In practice, the sampling duration is very short in comparison with the most significant time constant of the plant. A sampler converts a continuous-time signal into a train of pulses occurring at the sampling instants $t = 0, T, 2T, \ldots$, where T is the sampling period. (Note that between any two consecutive sampling instants the sampler transmits no information. Two signals whose respective values at the sampling instants are equal will give rise to the same sampled signal.) A data hold circuit converts the sampled data into a continuous-time signal.

Data hold is a process of generating a continuous-time signal $h(t)$ from a discrete-time sequence $x(kT)$. The signal $h(t)$ during the time interval $kT \le t < (k+1)T$ may be approximated by a polynomial in τ as follows:

$$h(kT+\tau) = a_n \tau^n + a_{n-1}\tau^{n-1} + \cdots + a_1\tau + a_0 \qquad (3\text{–}1)$$

where $0 \le \tau < T$. Note that signal $h(kT)$ must equal $x(kT)$, or

$$h(kT) = x(kT)$$

Hence, Eq. (3–1) can be written as follows:

$$h(kT+\tau)=a_n\tau^n+a_{n-1}\tau^{n-1}+\cdots+a_1\tau+x(kT) \tag{3–2}$$

If the data hold circuit is an nth-order polynomial extrapolator, it is called an nth-order hold. Thus, if $n = 1$, it is called a first-order hold. [The nth-order hold uses the past $n + 1$ discrete data $x((k - n)T)$, $x((k - n + 1)T)$, . . . , $x(kT)$ to generate a signal $h(kT + \tau)$.]

Because a higher-order hold uses past samples to extrapolate a continuous-time signal between the present sampling instant and the next sampling instant, the accuracy of approximating the continuous-time signal improves as the number of past samples used is increased. However, this better accuracy is obtained at the cost of a greater time delay. In closed-loop control systems any added time delay in the loop will decrease the stability of the system and in some cases may even cause system instability.

The simplest data hold is obtained when $n = 0$ in Eq. (3–2), that is, when

$$h(kT+\tau)=x(kT) \tag{3–3}$$

where $0 \leq \tau < T$ and $k = 0, 1, 2, \ldots$. Equation (3–3) implies that the circuit holds the amplitude of the sample from one sampling instant to the next. Such a data hold is called a zero-order hold, or clamper, or staircase generator. The output of the zero-order hold is a staircase function, as shown in Fig. 3–1. In this book, unless otherwise stated, we assume that the hold circuit is of zero order.

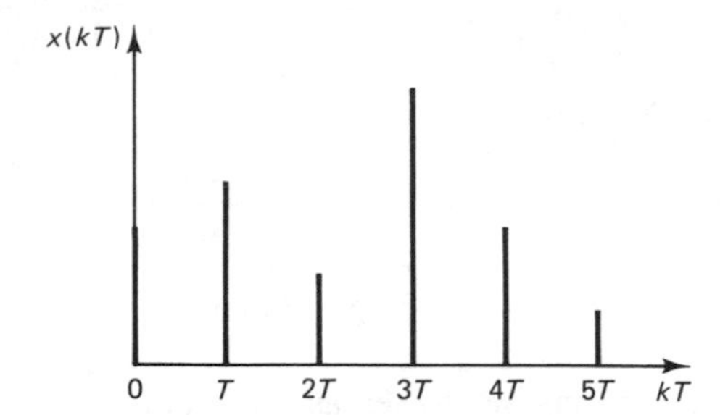

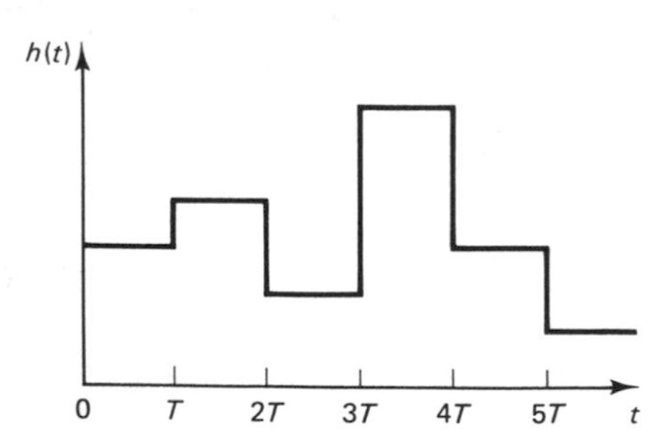

Figure 3–1 Input $x(kT)$ and output $h(t)$ of the zero-order hold.

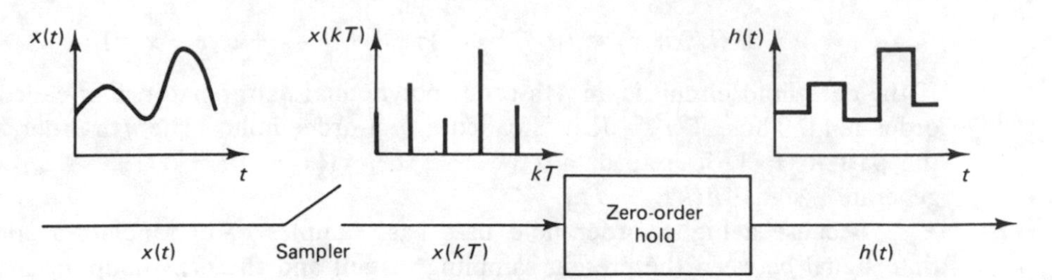

Figure 3–2 Sampler and zero-order hold.

Consider the sampler and zero-order hold shown in Fig. 3–2. Here, the output $h(t)$ is given by Eq. (3–3). Assume that the signal $x(t)$ is zero for $t < 0$. Then the output $h(t)$ is related to $x(t)$ as follows

$$\begin{aligned} h(t) &= x(0)[1(t) - 1(t-T)] + x(T)[1(t-T) - 1(t-2T)] \\ &\quad + x(2T)[1(t-2T) - 1(t-3T)] + \cdots \\ &= \sum_{k=0}^{\infty} x(kT)[1(t-kT) - 1(t-(k+1)T)] \end{aligned} \tag{3–4}$$

Since

$$\mathscr{L}[1(t-kT)] = \frac{e^{-kTs}}{s}$$

the Laplace transform of Eq. (3–4) becomes

$$\begin{aligned} \mathscr{L}[h(t)] = H(s) &= \sum_{k=0}^{\infty} x(kT) \frac{e^{-kTs} - e^{-(k+1)Ts}}{s} \\ &= \frac{1 - e^{-Ts}}{s} \sum_{k=0}^{\infty} x(kT)e^{-kTs} \end{aligned} \tag{3–5}$$

The right-hand side of Eq. (3–5) may be written as the product of two terms:

$$H(s) = G_{h0}(s)X^*(s) \tag{3–6}$$

where

$$G_{h0}(s) = \frac{1 - e^{-Ts}}{s}$$

and

$$X^*(s) = \sum_{k=0}^{\infty} x(kT)e^{-kTs} \tag{3–7}$$

The transfer function $X^*(s)$ is a function of the input signal $x(t)$. In Eq. (3–6), $G_{h0}(s)$ may be considered the transfer function between the output $H(s)$ and the input $X^*(s)$. Thus, the transfer function of a zero-order hold may be seen to be

$$G_{h0}(s) = \frac{1 - e^{-Ts}}{s}$$

Noting that the integral of an impulse function is a constant, we see that the input to the zero-order hold circuit may be regarded as a train of impulses, as shown by $x^*(t)$ in Fig. 3–3. (Since mathematically an impulse is defined as having an infinite amplitude with zero width, it is graphically represented by an arrow with an amplitude representing the strength of impulse.) Note that the zero-order hold, an integrator, resets its value to zero after one sampling period.

We may consider the continuous-time signal $x(t)$ to be represented by a sampled signal which is a sequence of impulses, with the strength of each impulse equal to the magnitude of $x(t)$ at the corresponding instant of time. [That is, at time $t = kT$, the impulse is $x(kT)\delta(t - kT)$.] The sampled signal $x^*(t)$, a train of impulses, can thus be represented by the infinite summation

$$x^*(t) = \sum_{k=-\infty}^{\infty} x(kT)\delta(t - kT)$$

[Note that $\delta(t - kT) = 0$ unless $t = kT$.]

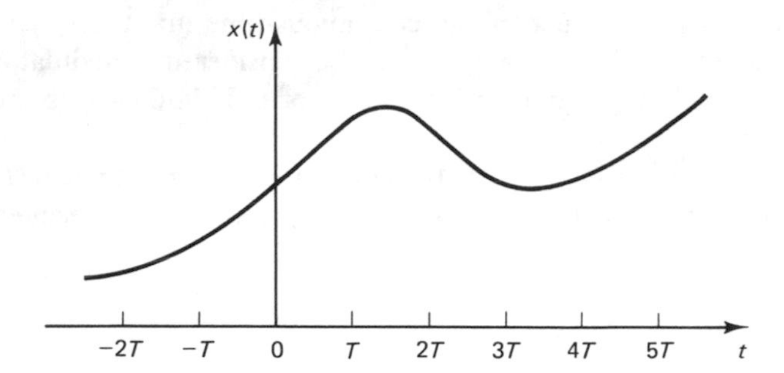

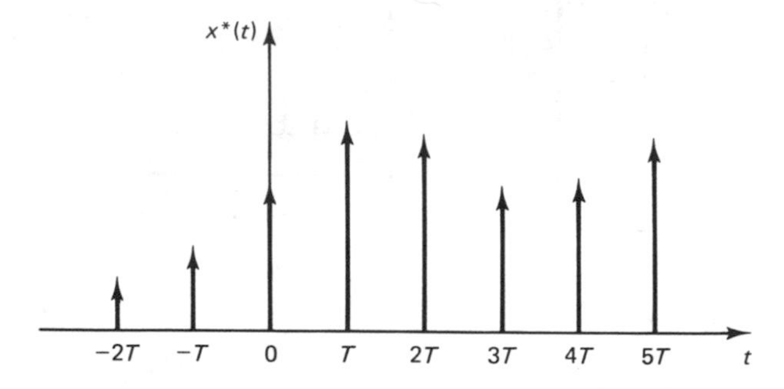

Figure 3–3 Continuous-time signal $x(t)$ and sampled signal $x^*(t)$ (train of impulses) which is the input to the zero-order hold.

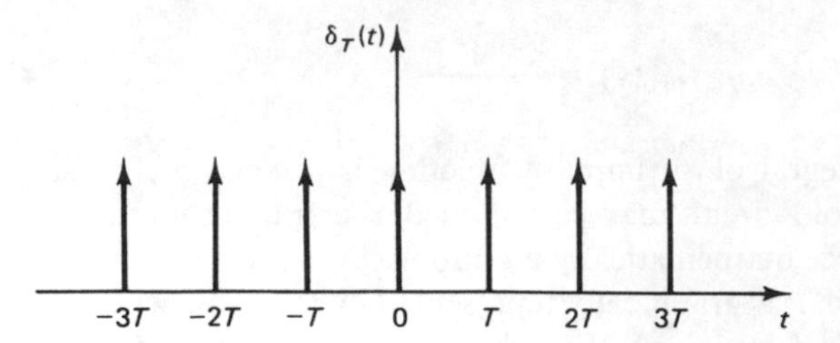

Figure 3–4 A plot of $\delta_T(t)$, a train of unit impulses.

Define $\delta_T(t)$ as a train of unit impulses, as shown in Fig. 3–4. Mathematically, the train of unit impulses $\delta_T(t)$ can be given by

$$\delta_T(t) = \sum_{k=-\infty}^{\infty} \delta(t - kT) \tag{3–8}$$

Then, the sampled signal $x^*(t)$ and the original continuous-time signal $x(t)$ can be related as follows:

$$\begin{aligned} x^*(t) &= \cdots + x(0)\delta(t) + x(T)\delta(t - T) + \cdots + x(kT)\delta(t - kT) + \cdots \\ &= \sum_{k=-\infty}^{\infty} x(kT)\delta(t - kT) \end{aligned}$$

The sampler output is equal to the product of the continuous-time input $x(t)$ and the train of unit impulses. Consequently, the sampler may be considered a modulator with the input $x(t)$ as the modulating signal and the train of unit impulses as the carrier, as shown in Fig. 3–5.

Let us summarize what we have just stated. If the continuous-time signal $x(t)$ is sampled in a periodic manner, mathematically the sampled signal may be represented by

$$x^*(t) = \sum_{k=-\infty}^{\infty} x(t)\delta(t - kT)$$

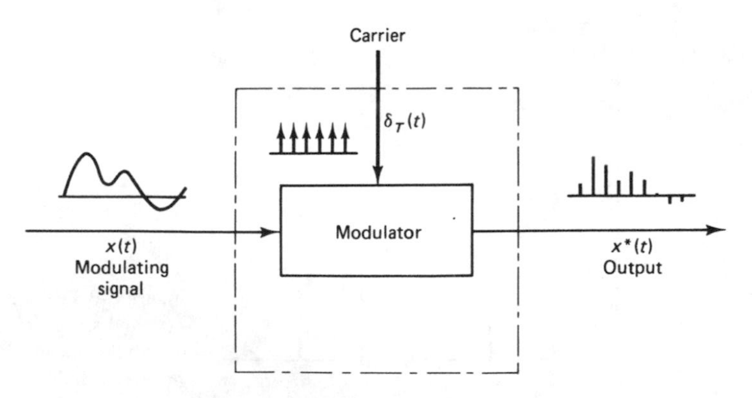

Figure 3–5 Sampler as a modulator.

or

$$x^*(t) = \sum_{k=-\infty}^{\infty} x(kT)\delta(t - kT)$$

Most of the time functions we consider in this book are zero for $t < 0$. Therefore, unless otherwise stated, we shall assume that this is the case. Then, for the signal $x(t)$, the preceding two equations for $x^*(t)$ become, respectively,

$$x^*(t) = \sum_{k=0}^{\infty} x(t)\delta(t - kT) \tag{3-9}$$

and

$$x^*(t) = \sum_{k=0}^{\infty} x(kT)\delta(t - kT) \tag{3-10}$$

The train of unit impulses $\delta_T(t)$ may then be considered to begin with $t = 0$, rather than $t = -\infty$, and to continue forever. Thus, Eq. (3–8) may be modified to

$$\delta_T(t) = \sum_{k=0}^{\infty} \delta(t - kT) \tag{3-11}$$

where we assume that the impulse $\delta(t)$ takes the full strength of unity, rather than the strength of $\frac{1}{2}$.

Figure 3–6 shows $\delta_T(t)$, $x(t)$, and $x^*(t)$. The figure depicts the amplitude modulation of the impulse train $\delta_T(t)$ by the input signal $x(t)$. The length of each arrow in the plot of the sampled signal $x^*(t)$ indicates the strength or magnitude of the associated sampled value $x(kT)$.

z transform of an impulse-sampled signal. In the analyses that follow in the rest of the book, we consider the sampled signal $x^*(t)$ to be defined by Eq. (3–10). That is, mathematically, we consider the sampler output to be a train of impulses that begins with $t = 0$, with the sampling period equal to T and the strength of each impulse equal to the sampled value of the continuous-time signal at the corresponding sampling instant.

Note that signal $x^*(t)$ may be considered to be generated by the fictitious sampler, commonly called as an *impulse sampler*, shown in Fig. 3–7. The switch may be thought of as closing instantaneously every sampling period T and generating impulses $x(kT)\delta(t - kT)$. Such a sampling process is called *impulse sampling*. The impulse sampler is introduced for mathematical convenience; it should be noted that it is a fictitious sampler and it does not exist in the real world.

Mathematically, the system shown in Fig. 3–8(a) is equivalent to the system shown in Fig. 3–8(b) from the viewpoint of the input-output relationship. That is, a real sampler and zero-order hold can be replaced by a mathematically equivalent continuous-time system which consists of an impulse sampler and a transfer function $(1 - e^{-Ts})/s$. The two sampling processes will be distinguished (as they are in Fig. 3–8) by the manner in which the sampling switches are drawn.

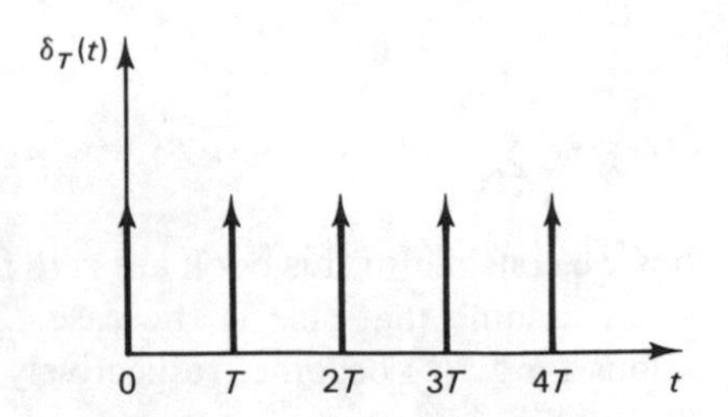

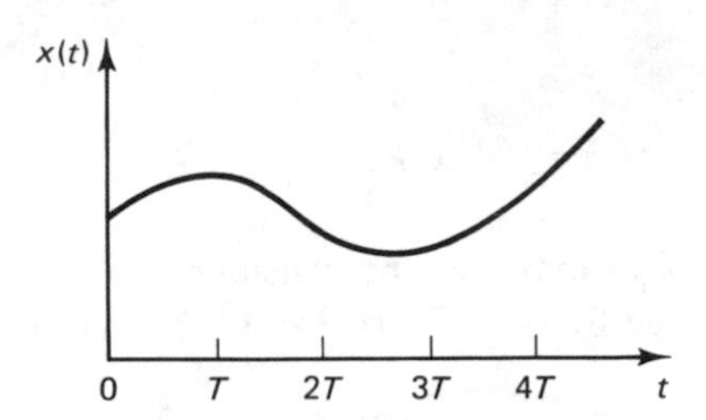

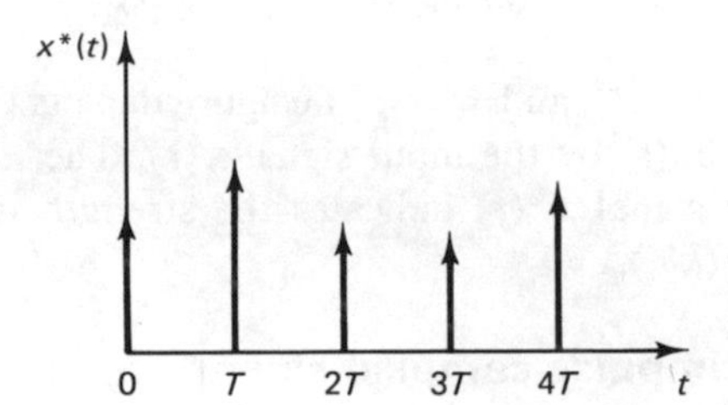

Figure 3–6 Plots of $\delta_T(t)$, $x(t)$, and $x^*(t)$.

Consider the Laplace transform of Eq. (3–10):

$$\begin{aligned} X^*(s) = \mathscr{L}[x^*(t)] &= x(0)\mathscr{L}[\delta(t)] + x(T)\mathscr{L}[\delta(t-T)] \\ &\quad + x(2T)\mathscr{L}[\delta(t-2T)] + \cdots \\ &= x(0) + x(T)e^{-Ts} + x(2T)e^{-2Ts} + \cdots \\ &= \sum_{k=0}^{\infty} x(kT)e^{-kTs} \end{aligned} \tag{3–12}$$

Notice that if we define

$$e^{Ts} = z$$

$x(t)$ / $X(s)$ → δ_T → $x^*(t)$ / $X^*(s)$

Figure 3–7 Impulse sampler.

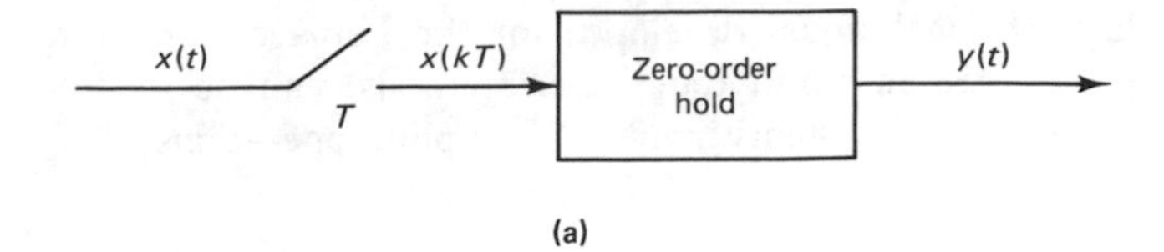

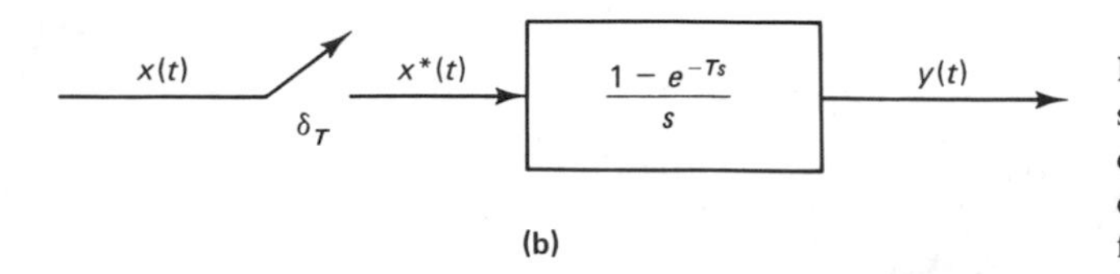

Figure 3–8 (a) System with a sampler and zero-order hold; (b) equivalent system which consists of an impulse sampler and transfer function $(1 - e^{-Ts})/s$.

or

$$s = \frac{1}{T} \ln z$$

then Eq. (3–12) becomes

$$X^*(s)\Big|_{s=\frac{1}{T}\ln z} = \sum_{k=0}^{\infty} x(kT)z^{-k} \tag{3–13}$$

The right-hand side of Eq. (3–13) is exactly the same as the right-hand side of Eq. (2–1): it is the z transform of the sequence $x(0)$, $x(T)$, $x(2T)$, . . . , generated from $x(t)$ at $t = kT$, $k = 0, 1, 2, \ldots$. Hence, we may write

$$X^*(s)\Big|_{s=\frac{1}{T}\ln z} = X(z)$$

and Eq. (3–13) becomes

$$X^*(s)\Big|_{s=\frac{1}{T}\ln z} = X^*\left(\frac{1}{T}\ln z\right) = X(z) = \sum_{k=0}^{\infty} x(kT)z^{-k} \tag{3–14}$$

Note that the variable z is a complex variable and T is the sampling period. [It should be stressed that the notation $X(z)$ does not signify $X(s)$ with s replaced by z, but rather $X^*(s = T^{-1} \ln z)$.]

When we transform e^{Ts} to z, the concept of impulse sampling (which is purely a mathematical process) enables us to analyze by the z transform method discrete-time control systems which involve samplers and hold circuits. This means that by use of the complex variable z, the techniques developed for the Laplace transform methods (such as the frequency response and root locus methods) can be readily applied to anlayze discrete-time control systems involving sampling operations.

Example 3–1.

Derive the z transform of the Kronecker delta function

$$\delta_0(kT) = \begin{cases} 1 & \text{for } k = 0 \\ 0 & \text{for } k \neq 0 \end{cases}$$

Define

$$x(t) = \delta_0(t)$$

Then the impulse-sampled signal $x^*(t)$ can be given by

$$x^*(t) = x(t)\delta_T(t) = \sum_{k=0}^{\infty} x(kT)\delta(t - kT)$$

Since $x(kT) = 0$ for $k \neq 0$, we have

$$x^*(t) = x(0)\delta(0)$$

Then, the Laplace transform of $x^*(t)$ becomes

$$X^*(s) = x(0) = \delta_0(0) = 1$$

Hence,

$$X(z) = \mathcal{Z}[\delta_0(kT)] = 1$$

Example 3–2.

Obtain the z transform of a unit step function $1(t)$.

Define

$$x(t) = 1(t)$$

The impulse-sampled signal $x^*(t)$ can be obtained as follows:

$$x^*(t) = x(t)\delta_T(t) = \sum_{k=0}^{\infty} x(kT)\delta(t - kT) = \sum_{k=0}^{\infty} \delta(t - kT)$$

Then the Laplace transform of $x^*(t)$ can be given by

$$X^*(s) = \sum_{k=0}^{\infty} e^{-kTs} = \frac{1}{1 - e^{-Ts}} \tag{3–15}$$

Hence

$$X(z) = \mathcal{Z}[1(t)] = \frac{1}{1 - z^{-1}}$$

Note that this last equation is the same as what we obtained in Eq. (2–5).

Summary. Let us summarize what we have presented so far about impulse sampling.

1. A real sampler samples the input signal periodically and produces a sequence of pulses as the output. While the sampling duration (pulse width) of the real sampler is very small (but will never become zero), the assumption of zero width, which implies that a sequence of pulses becomes a sequence of impulses whose strengths are equal to the continuous-time signal at the sampling instants, simplifies the analysis of discrete-time systems. Such an assumption is valid if the sampling duration is very small compared with the significant time constant of the system and if a hold circuit is connected to the output of the sampler.
2. The Laplace transform of the impulse-sampled signal $x^*(t)$ has been shown to be the same as the z transform of signal $x(t)$ if e^{Ts} is defined as z, or $e^{Ts} = z$.
3. In the z transform method, we consider only the values of the signal at the sampling instants. Therefore, the z transform of $x(t)$ and that of $x^*(t)$ yield the same result:

$$\mathscr{Z}[x(t)] = \mathscr{Z}[x^*(t)] = X(z) = \sum_{k=0}^{\infty} x(kT)z^{-k} \tag{3–16}$$

[It is noted that Eq. (3–16) is not the only form giving the z transform of $x^*(t)$. Other forms are available, as will be seen later in this chapter.]
4. As pointed out earlier, once the real sampler and zero-order hold are mathematically replaced by the impulse sampler and transfer function $(1 - e^{-Ts})/s$, the system becomes a continuous-time system. This simplifies the analysis of the discrete-time control system, since we may apply the techniques available to continuous-time control systems.
5. In a real sampler a switch closes and opens sufficiently fast every sampling period T to generate the sequence of numbers $x(kT)$. In a mathematical impulse sampler, the switch produces an output

$$x^*(t) = \sum_{k=0}^{\infty} x(kT)\delta(t - kT)$$

for a given input $x(t)$. The output $x^*(t)$ of the impulse sampler is a sequence of weighted impulses. In order to distinguish the real sampler (switch) and the mathematical impulse sampler, the two have been drawn differently on the diagram (Fig. 3–8). It is repeated that the impulse sampler is a fictitious sampler introduced purely for the purpose of mathematical analysis. It is not possible to physically implement such a sampler that generates impulses.

Transfer functions of hold circuits. It has been shown that the transfer function of the zero-order hold is given by

$$G_{h0}(s) = \frac{1 - e^{-Ts}}{s} \tag{3–17}$$

The transfer function of the first-order hold can be given by

$$G_{h1}(s) = \left(\frac{1 - e^{-Ts}}{s}\right)^2 \frac{Ts + 1}{T} \tag{3–18}$$

In the following, we shall derive Eq. (3–18).

Let us substitute $n = 1$ into Eq. (3–2). Then we have

$$h(kT + \tau) = a_1\tau + x(kT) \tag{3–19}$$

where $0 \le \tau < T$ and $k = 0, 1, 2, \ldots$. By applying the condition that

$$h((k-1)T) = x((k-1)T)$$

the constant a_1 can be determined as follows:

$$h((k-1)T) = -a_1T + x(kT) = x((k-1)T)$$

or

$$a_1 = \frac{x(kT) - x((k-1)T)}{T}$$

Hence, Eq. (3–19) becomes

$$h(kT + \tau) = x(kT) + \frac{x(kT) - x((k-1)T)}{T}\tau \tag{3–20}$$

where $0 \le \tau < T$. The extrapolation process of the first-order hold is based on Eq. (3–20). The continuous-time signal $h(t)$ obtained by use of the first-order hold is a piecewise-linear signal, as shown in Fig. 3–9.

Now consider the impulse sampler and the first-order hold shown in Fig. 3–10. In order to derive the transfer function of the first-order hold circuit, it is convenient to assume a simple function for $x(t)$. For example, a unit step function, a unit impulse function, and a unit ramp function are good choices for $x(t)$.

Suppose we choose a unit step function as $x(t)$. Then

$$x^*(t) = 1^*(t) = \sum_{k=0}^{\infty} 1(kT)\delta(t - kT) = \sum_{k=0}^{\infty} \delta(t - kT)$$

Noting that the output $h(t)$ of the first-order hold is the straight line that is the extrapolation of the two preceding sampled values, we get the output $h(t)$ as shown in Fig. 3–11.

From Fig. 3–11, the curve $h(t)$ may be written as follows:

$$h(t) = \left(1 + \frac{t}{T}\right)1(t) - \frac{t - T}{T}\,1(t - T) - 1(t - T)$$

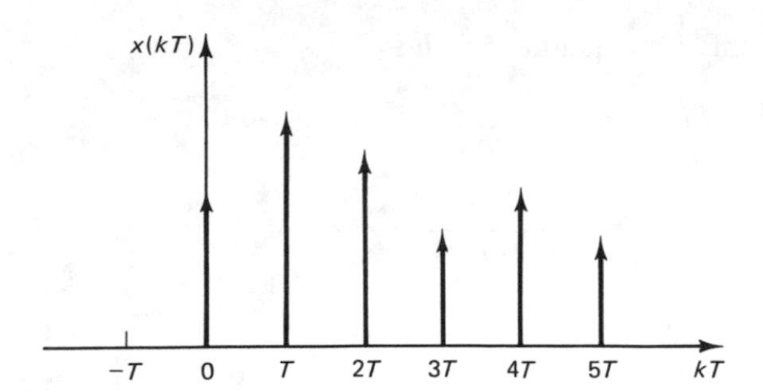

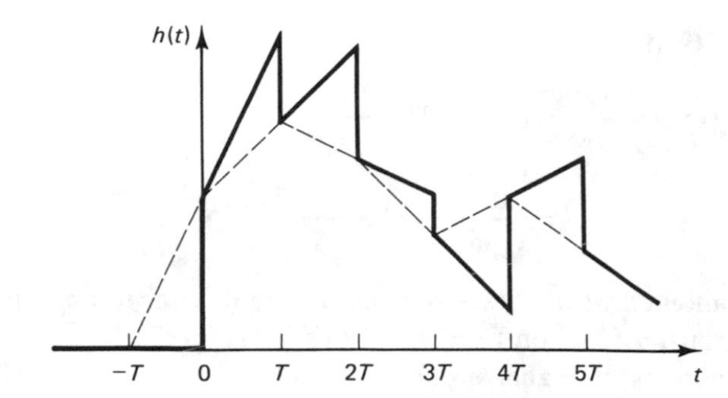

Figure 3–9 Input $x(kT)$ and output $h(t)$ of a first-order hold.

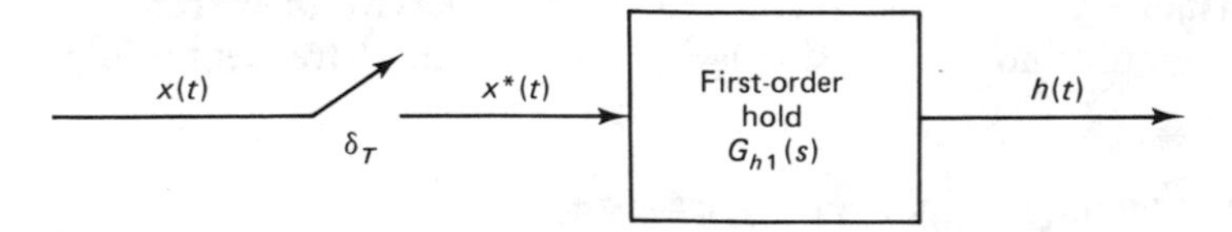

Figure 3–10 Impulse sampler and first-order hold.

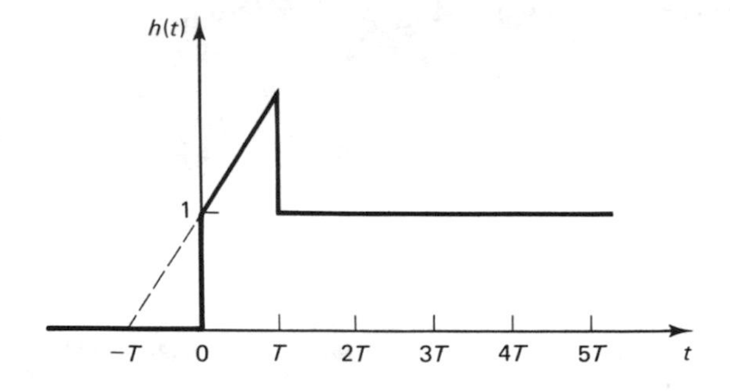

Figure 3–11 Output $h(t)$ of the first-order hold shown in Fig. 3–10 when the input $x(t)$ is a unit step function.

The Laplace transform of this last equation becomes

$$H(s)=\left(\frac{1}{s}+\frac{1}{Ts^2}\right)-\frac{1}{Ts^2}e^{-Ts}-\frac{1}{s}e^{-Ts}$$

$$=\frac{1-e^{-Ts}}{s}+\frac{1-e^{-Ts}}{Ts^2}$$

$$=(1-e^{-Ts})\frac{Ts+1}{Ts^2} \tag{3-21}$$

The Laplace transform of the unit step sequence is

$$X^*(s)=\mathscr{L}[1^*(t)]=\frac{1}{1-e^{-Ts}}$$

Hence the transfer function $G_{h1}(s)$ of the first-order hold is given by

$$G_{h1}(s)=\frac{H(s)}{X^*(s)}=(1-e^{-Ts})^2\frac{Ts+1}{Ts^2}$$

$$=\left(\frac{1-e^{-Ts}}{s}\right)^2\frac{Ts+1}{T}$$

Note that a real sampler combined with a first-order hold is equivalent to an impulse sampler combined with a transfer function $(1-e^{-Ts})^2(Ts+1)/(Ts^2)$.

Similarly, transfer functions of higher-order hold circuits may be derived by following the procedure presented above. However, since higher-order hold circuits ($n \geq 2$) are not practical from the viewpoint of delay (which may cause system instability) and noise effects, we shall not dervive their transfer functions here. (The zero-order hold is the simplest and is used most frequently in practice.)

3-3 OBTAINING THE z TRANSFORM BY THE CONVOLUTION INTEGRAL METHOD

In this section we shall obtain the z transform of $x(t)$ by using the convolution integral method.

Consider the impulse sampler shown in Fig. 3-12. The output of the impulse sampler is

$$x^*(t)=\sum_{k=0}^{\infty}x(t)\delta(t-kT)=x(t)\sum_{k=0}^{\infty}\delta(t-kT) \tag{3-22}$$

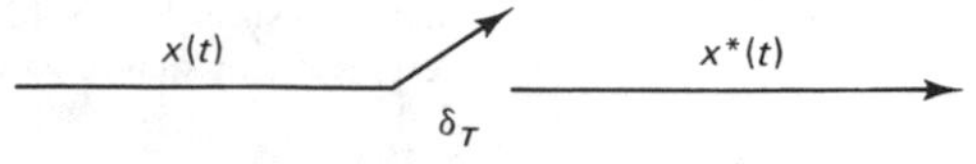

Figure 3-12 Impulse sampler.

Noting that

$$\mathscr{L}[\delta(t-kT)] = e^{-kTs}$$

we have

$$\mathscr{L}\left[\sum_{k=0}^{\infty} \delta(t-kT)\right] = 1 + e^{-Ts} + e^{-2Ts} + e^{-3Ts} + \cdots = \frac{1}{1-e^{-Ts}}$$

Since

$$X^*(s) = \mathscr{L}[x^*(t)] = \mathscr{L}\left[x(t) \sum_{k=0}^{\infty} \delta(t-kT)\right]$$

we see that $X^*(s)$ is the Laplace transform of the product of two time functions, $x(t)$ and $\sum_{k=0}^{\infty} \delta(t-kT)$. Note that it is not equal to the product of the two corresponding Laplace transforms.

The Laplace transform of the product of two Laplace-transformable functions $f(t)$ and $g(t)$ is

$$\mathscr{L}[f(t)g(t)] = \int_0^{\infty} f(t)g(t)e^{-st}\,dt \tag{3-23}$$

Note that the inversion integral is

$$f(t) = \frac{1}{2\pi j}\int_{c-j\infty}^{c+j\infty} F(s)e^{st}\,ds \qquad t > 0$$

where c is the abscissa of convergence for $F(s)$. Thus,

$$\mathscr{L}[f(t)g(t)] = \frac{1}{2\pi j}\int_0^{\infty}\int_{c-j\infty}^{c+j\infty} F(p)e^{pt}\,dp\, g(t)e^{-st}\,dt$$

Because of the uniform convergence of the integrals considered, we may invert the order of integration:

$$\mathscr{L}[f(t)g(t)] = \frac{1}{2\pi j}\int_{c-j\infty}^{c+j\infty} F(p)\,dp \int_0^{\infty} g(t)e^{-(s-p)t}\,dt$$

Noting that

$$\int_0^{\infty} g(t)e^{-(s-p)t}\,dt = G(s-p)$$

we obtain

$$\mathscr{L}[f(t)g(t)] = \frac{1}{2\pi j}\int_{c-j\infty}^{c+j\infty} F(p)G(s-p)\,dp \tag{3-24}$$

Now for $f(t)$ and $g(t)$ substitute $x(t)$ and $\sum_{k=0}^{\infty} \delta(t-kT)$, respectively. Then

$$G(s) = \mathscr{L}\left[\sum_{k=0}^{\infty} \delta(t - kT)\right] = \frac{1}{1 - e^{-Ts}}$$

and

$$G(s - p) = \frac{1}{1 - e^{-T(s-p)}}$$

Notice that the poles of $1/[1 - e^{-T(s-p)}]$ may be obtained by solving the equation

$$1 - e^{-T(s-p)} = 0$$

or

$$-T(s - p) = \pm j2\pi k \qquad k = 0, 1, 2, \ldots$$

so that the poles are

$$p = s \pm j\frac{2\pi}{T}k = s \pm j\omega_s k \qquad k = 0, 1, 2, \ldots$$

where $\omega_s = 2\pi/T$. Thus, there are infinitely many simple poles along a line parallel to the $j\omega$ axis.

The Laplace transform of $x^*(t)$ can now be written as

$$X^*(s) = \mathscr{L}\left[x(t) \sum_{k=0}^{\infty} \delta(t - kT)\right]$$

$$= \frac{1}{2\pi j}\int_{c-j\infty}^{c+j\infty} X(p)\frac{1}{1 - e^{-T(s-p)}}\,dp \qquad (3\text{–}25)$$

where the integration is along the line from $c - j\infty$ to $c + j\infty$ and this line is parallel to the imaginary axis in the p plane and separates the poles of $X(p)$ from those of $1/[1 - e^{-T(s-p)}]$. Equation (3–25) is the convolution integral. It is a well-known fact that such an integral can be evaluated in terms of residues by forming a closed contour consisting of the line from $c - j\infty$ to $c + j\infty$ and a semicircle of infinite radius in the left or right half plane, provided that the integral along the added semicircle is a constant (zero or a nonzero constant). Since there are two ways to evaluate this integral (one using an infinite semicircle in the left half plane and the other an infinite semicircle in the right half plane), we shall consider these two cases separately.

In our analysis here, we assume that the poles of $X(s)$ lie in the left half plane and that $X(s)$ can be expressed as a ratio of polynomials in s, or

$$X(s) = \frac{q(s)}{p(s)}$$

where $q(s)$ and $p(s)$ are polynomials in s. We also assume that $p(s)$ is of a higher degree in s than $q(s)$, which means that

$$\lim_{s\to\infty} X(s) = 0$$

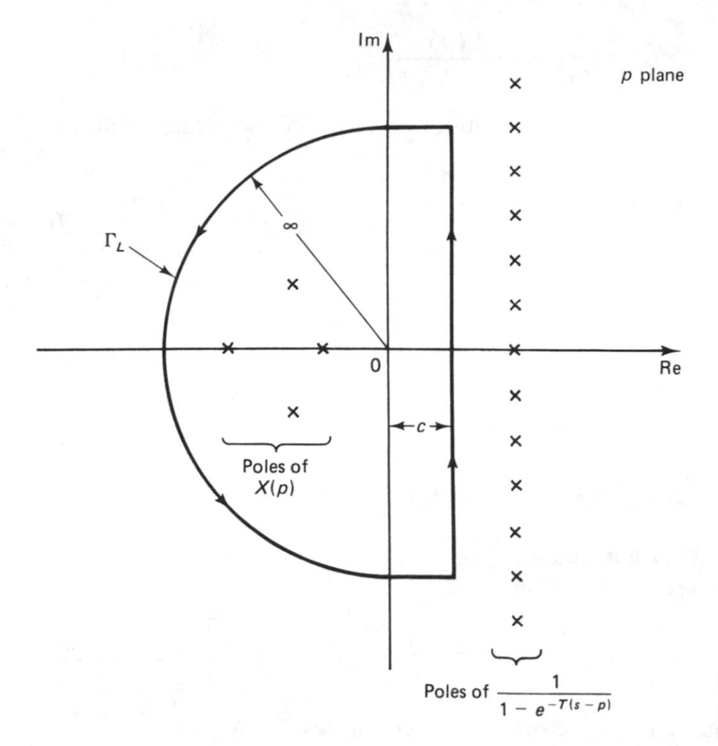

Figure 3–13 Closed contour in the left half of the p plane.

Evaluation of the convolution integral in the left half plane. We shall evaluate the convolution integral given by Eq. (3–25) using a closed contour in the left half of the p plane as shown in Fig. 3–13. Using this closed contour, Eq. (3–25) may be written as

$$X^*(s) = \frac{1}{2\pi j}\int_{c-j\infty}^{c+j\infty} X(p)\frac{1}{1-e^{-T(s-p)}}\,dp$$

$$= \frac{1}{2\pi j}\oint \frac{X(p)}{1-e^{-T(s-p)}}\,dp - \frac{1}{2\pi j}\int_{\Gamma_L}\frac{X(p)}{1-e^{-T(s-p)}}\,dp \tag{3–26}$$

where the closed contour consists of the line from $c - j\infty$ to $c + j\infty$ and Γ_L, which in turn consists of a semicircle of infinite radius and the horizontal lines at $j\infty$ and $-j\infty$ which connect the line from $c - j\infty$ to $c + j\infty$ with the semicircle in the left half of the p plane. We choose a value of c such that all the poles of $X(p)$ lie to the left of the line from $c - j\infty$ to $c + j\infty$ and all the poles of $1/[1 - e^{-T(s-p)}]$ lie to the right of this line. The closed contour encloses all poles of $X(p)$, while the poles of $1/[1 - e^{-T(s-p)}]$ are outside the closed contour.

Because we have assumed that the denominator of $X(s)$ is of a higher degree in s than the numerator, the integral along Γ_L (the infinite semicircle in the left half plane plus the horizontal lines at $j\infty$ and $-j\infty$ which connect the line from $c - j\infty$ to $c + j\infty$ with the semicircle) vanishes. Hence

$$X^*(s) = \frac{1}{2\pi j}\oint \frac{X(p)}{1 - e^{-T(s-p)}}\,dp$$

This integral is equal to the sum of the residues of $X(p)$ in the closed contour. Therefore,

$$X^*(s) = \sum \left[\text{residue of } \frac{X(p)}{1 - e^{-T(s-p)}} \text{ at pole of } X(p)\right] \tag{3–27}$$

By substituting z for e^{Ts} in Eq. (3–27) we have

$$X(z) = \sum \left[\text{residue of } \frac{X(p)z}{z - e^{Tp}} \text{ at pole of } X(p)\right]$$

By changing the complex variable notation from p to s, we obtain

$$X(z) = \sum \left[\text{residue of } \frac{X(s)z}{z - e^{Ts}} \text{ at pole of } X(s)\right] \tag{3–28}$$

Let us assume that $X(s)$ has poles $s_1, s_2, \ldots, s_m$. If a pole at $s = s_j$ is a simple pole, then the corresponding residue K_j is

$$K_j = \lim_{s \to s_j} \left[(s - s_j)\frac{X(s)z}{z - e^{Ts}}\right] \tag{3–29}$$

If a pole at $s = s_i$ is a multiple pole of order n_i, then the residue K_i is

$$K_i = \frac{1}{(n_i - 1)!} \lim_{s \to s_i} \frac{d^{n_i - 1}}{ds^{n_i - 1}} \left[(s - s_i)^{n_i} \frac{X(s)z}{z - e^{Ts}}\right] \tag{3–30}$$

Therefore, if $X(s)$ has a multiple pole s_1 of order n_1, a multiple pole s_2 of order $n_2, \ldots,$ a multiple pole s_h of order n_h, and simple poles $s_{h+1}, s_{h+2}, \ldots, s_m$, then $X(z)$ given by Eq. (3–28) can be written as

$$\begin{aligned} X(z) &= \sum \left[\text{residue of } \frac{X(s)z}{z - e^{Ts}} \text{ at pole of } X(s)\right] \\ &= \sum_{i=1}^{h} \frac{1}{(n_i - 1)!} \lim_{s \to s_i} \frac{d^{n_i - 1}}{ds^{n_i - 1}} \left[(s - s_i)^{n_i} \frac{X(s)z}{z - e^{Ts}}\right] \\ &\quad + \sum_{j=h+1}^{m} \lim_{s \to s_j} \left[(s - s_j)\frac{X(s)z}{z - e^{Ts}}\right] \end{aligned} \tag{3–31}$$

where n_i is the order of the multiple pole at $s = s_i$.

Example 3–3.

Given

$$X(s) = \frac{1}{s^2(s+1)}$$

obtain $X(z)$ by use of the convolution integral in the left half plane.

Note that $X(s)$ has a double pole at $s = 0$ and a simple pole at $s = -1$. Hence, Eq. (3–31) becomes

$$X(z) = \sum \left[\text{residue of} \frac{X(s)z}{z - e^{Ts}} \text{ at pole of } X(s)\right]$$

$$= \frac{1}{(2-1)!} \lim_{s \to 0} \frac{d}{ds} \left[s^2 \frac{1}{s^2(s+1)} \frac{z}{z - e^{Ts}}\right] + \lim_{s \to -1} \left[(s+1) \frac{1}{s^2(s+1)} \frac{z}{z - e^{Ts}}\right]$$

$$= \lim_{s \to 0} \frac{-z[z - e^{Ts} + (s+1)(-T)e^{Ts}]}{(s+1)^2(z - e^{Ts})^2} + \frac{1}{(-1)^2} \frac{z}{z - e^{-T}}$$

$$= \frac{-z(z - 1 - T)}{(z-1)^2} + \frac{z}{z - e^{-T}} = \frac{z^2(T - 1 + e^{-T}) + z(1 - e^{-T} - Te^{-T})}{(z-1)^2(z - e^{-T})}$$

$$= \frac{(T - 1 + e^{-T})z^{-1} + (1 - e^{-T} - Te^{-T})z^{-2}}{(1 - z^{-1})^2(1 - e^{-T}z^{-1})}$$

Evaluation of the convolution integral in the right half plane. Let us next evaluate the convolution integral given by Eq. (3–25) in the right half of the p plane. Let us choose the closed contour shown in Fig. 3–14, which consists of the line from $c - j\infty$ to $c + j\infty$ and Γ_R, the portion of the semicircle of infinite radius in the right half of the p plane that lies to the right of this line. The closed contour encloses all poles of $1/[1 - e^{-T(s-p)}]$, but it does not enclose any poles of $X(p)$. Now $X^*(s)$ can be written as

$$X^*(s) = \frac{1}{2\pi j} \int_{c-j\infty}^{c+j\infty} \frac{X(p)}{1 - e^{-T(s-p)}} dp$$

$$= \frac{1}{2\pi j} \oint \frac{X(p)}{1 - e^{-T(s-p)}} dp - \frac{1}{2\pi j} \int_{\Gamma_R} \frac{X(p)}{1 - e^{-T(s-p)}} dp \qquad (3\text{–}32)$$

Let us investigate the integral along Γ_R, the portion of the infinite semicircle to the right of the line from $c - j\infty$ to $c + j\infty$. Since infinitely many poles of $1/[1 - e^{-T(s-p)}]$ lie on a line parallel to the $j\omega$ axis, the evaluation of the integral along Γ_R is not as simple as in the previous case, where the closed contour enclosed a finite number of poles of $X(p)$ in the left half of the p plane.

In almost all physical control systems, as s becomes large $X(s)$ tends to zero at least as fast as $1/s$. Hence, in what follows, we consider two cases, one where the denominator of $X(s)$ is two or more degrees higher in s than the numerator, and another where the denominator of $X(s)$ is only one degree higher in s than the numerator.

Case 1: $X(s)$ Possesses at Least Two More Poles than Zeros. Referring to the theory of complex variables, it can be shown that the integral along Γ_R is zero if the degree of the denominator $p(s)$ of $X(s)$ is greater by at least 2 than the degree

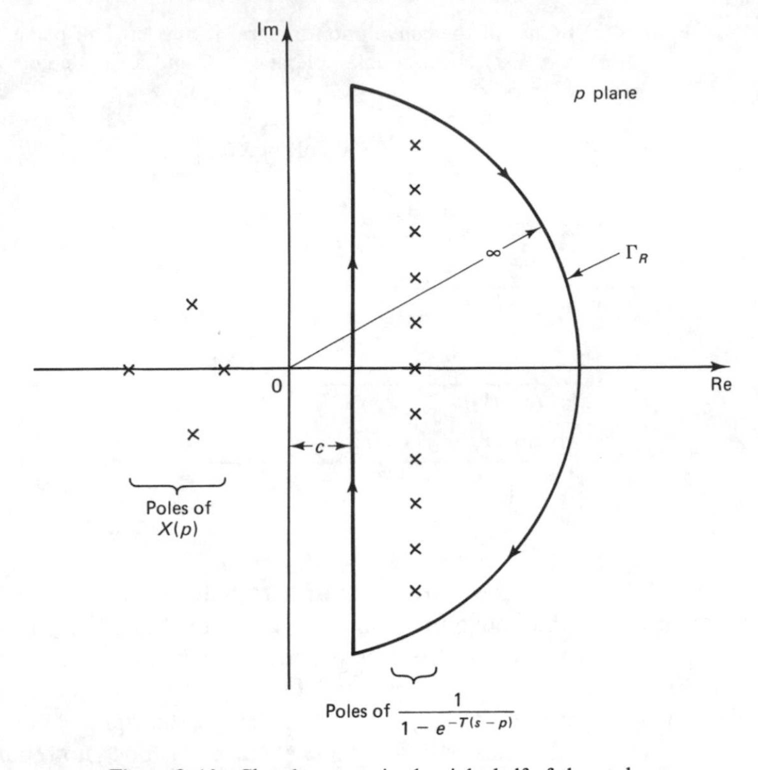

Figure 3–14 Closed contour in the right half of the p plane.

of the numerator $q(s)$; that is, if $X(s)$ possesses at least two more poles than zeros, which implies that

$$\lim_{s \to \infty} sX(s) = x(0+) = 0$$

then the integral along Γ_R is zero. Thus, in the present case

$$\frac{1}{2\pi j} \int_{\Gamma_R} \frac{X(p)}{1 - e^{-T(s-p)}} dp = 0$$

Therefore, Eq. (3–32) simplifies to

$$X^*(s) = \frac{1}{2\pi j} \oint \frac{X(p)}{1 - e^{-T(s-p)}} dp \tag{3–33}$$

The integral along the closed contour given by Eq. (3–33) can be obtained by evaluating the residues at the infinite number of poles at $p = s \pm j\omega_s k$. Thus,

$$X^*(s) = -\sum_{k=-\infty}^{\infty} \left[\lim_{p \to s + j\omega_s k} \left\{ [p - (s + j\omega_s k)] \frac{X(p)}{1 - e^{-T(s-p)}} \right\} \right]$$

The minus sign in front of the right-hand side of this last equation comes from the fact that the contour integration along the path Γ_R is taken in the clockwise direction. Using L'Hôpital's rule, we obtain

$$X^*(s) = -\sum_{k=-\infty}^{\infty} \frac{X(p)}{\dfrac{d}{dp}[1-e^{-T(s-p)}]}\Bigg|_{p=s+j\omega_s k}$$

Noting that

$$\frac{d}{dp}[1-e^{-T(s-p)}]_{p=s+j\omega_s k} = -Te^{-T(s-p)}\Big|_{p=s+j\omega_s k} = -Te^{jT\omega_s k} = -Te^{j2\pi k} = -T$$

we have

$$X^*(s) = -\sum_{k=-\infty}^{\infty} \frac{X(p)}{-T}\Bigg|_{p=s+j\omega_s k}$$

or

$$X^*(s) = \frac{1}{T}\sum_{k=-\infty}^{\infty} X(s+j\omega_s k) \tag{3–34}$$

Thus

$$X(z) = \frac{1}{T}\sum_{k=-\infty}^{\infty} X(s+j\omega_s k)\Bigg|_{s=\frac{1}{T}\ln z} \tag{3–35}$$

Note that this expression of the z transform is useful in proving the sampling theorem (see Sec. 3–4). However, it is very tedious to obtain z transform expressions of commonly encountered functions by this method.

Case 2: $X(s)$ Has a Denominator One Degree Higher in s than the Numerator. For this case $\lim_{s\to\infty} sX(s) = x(0+) \neq 0 < \infty$ and the integral along Γ_R is not zero. [The nonzero value is associated with the initial value $x(0+)$ of $x(t)$.] It can be shown that the contribution of the integral along Γ_R in Eq. (3–32) is $-\frac{1}{2}x(0+)$. That is,

$$\frac{1}{2\pi j}\int_{\Gamma_R} \frac{x(p)}{1-e^{-T(s-p)}}\,dp = -\frac{1}{2}x(0+)$$

Then the integral term on the right-hand side of Eq. (3–32) becomes

$$X^*(s) = \frac{1}{T}\sum_{k=-\infty}^{\infty} X(s+j\omega_s k) + \frac{1}{2}x(0+) \tag{3–36}$$

A different interpretation of the added term $\frac{1}{2}x(0+)$ in Eq. (3–36) is that in Eq. (3–34) only one-half of the value of $x(0+)$ is sampled. [In case 1, where Eq. (3–34) has been obtained, $x(0+) = 0$, and therefore the contribution of $\frac{1}{2}x(0+)$ is zero.]

Example 3–4.

Consider the function

$$x(t)=\begin{cases} e^{-at} & t \geq 0 \\ 0 & t<0 \end{cases}$$

Obtain $X(z)$ by using the convolution integral in the right half plane.

The Laplace transform of $x(t)$ is

$$X(s)=\frac{1}{s+a}$$

Clearly, $\lim_{s\to\infty} sX(s) = x(0+) = 1$, or the function has a jump discontinuity at $t = 0$. Hence, we must use Eq. (3–36). Referring to Eq. (3–36), we have

$$\begin{aligned} X^*(s) &= \frac{1}{T}\sum_{k=-\infty}^{\infty} X(s+j\omega_s k)+\frac{1}{2}x(0+) \\ &= \frac{1}{T}\left\{\sum_{k=1}^{\infty}[X(s+j\omega_s k)+X(s-j\omega_s k)]+X(s)\right\}+\frac{1}{2} \\ &= \frac{1}{T}\left[\sum_{k=1}^{\infty}\left(\frac{1}{s+j\omega_s k+a}+\frac{1}{s-j\omega_s k+a}\right)+\frac{1}{s+a}\right]+\frac{1}{2} \\ &= \frac{1}{T}\left[\sum_{k=1}^{\infty}\frac{2(s+a)}{(s+a)^2+(\omega_s k)^2}+\frac{1}{s+a}\right]+\frac{1}{2} \\ &= \frac{1}{2\pi}\left[\sum_{k=1}^{\infty}\frac{2(s+a)/\omega_s}{\left(\dfrac{s+a}{\omega_s}\right)^2+k^2}+\frac{\omega_s}{s+a}\right]+\frac{1}{2} \end{aligned} \tag{3–37}$$

Referring to a formula available in mathematical tables,

$$\sum_{k=1}^{\infty}\frac{2x}{x^2+k^2}+\frac{1}{x}=\pi\frac{1+e^{-2\pi x}}{1-e^{-2\pi x}}$$

and noting that

$$2\pi\frac{s+a}{\omega_s}=T(s+a)$$

we can write Eq. (3–37) in the form

$$\begin{aligned} X^*(s) &= \frac{\pi}{2\pi}\frac{1+e^{-T(s+a)}}{1-e^{-T(s+a)}}+\frac{1}{2} \\ &= \frac{1}{2}\frac{1+e^{-T(s+a)}+1-e^{-T(s+a)}}{1-e^{-T(s+a)}} \\ &= \frac{1}{2}\frac{2}{1-e^{-T(s+a)}} \\ &= \frac{1}{1-e^{-aT}e^{-Ts}} \end{aligned}$$

or

$$X(z) = \frac{1}{1 - e^{-aT}z^{-1}}$$

Thus, we have obtained $X(z)$ by using the convolution integral in the right half plane. [This process of obtaining the z transform is very tedious because an infinite series of $X(s + j\omega_s k)$ is involved. The example here is presented for demonstration purposes only. One should use other methods for obtaining the z transform.]

Example 3–5.

Show that $X^*(s)$ is periodic with period $2\pi/\omega_s$.

Referring to Eq. (3–36),

$$X^*(s) = \frac{1}{T} \sum_{h=-\infty}^{\infty} X(s + j\omega_s h) + \frac{1}{2} x(0+)$$

Hence,

$$X^*(s + j\omega_s k) = \frac{1}{T} \sum_{h=-\infty}^{\infty} X(s + j\omega_s k + j\omega_s h) + \frac{1}{2} x(0+)$$

Let $k + h = m$. Then, this last equation becomes

$$X^*(s + j\omega_s k) = \frac{1}{T} \sum_{m=-\infty}^{\infty} X(s + j\omega_s m) + \frac{1}{2} x(0+) = X^*(s)$$

Therefore, we have

$$X^*(s) = X^*(s \pm j\omega_s k) \qquad k = 0, 1, 2, \ldots$$

Thus, $X^*(s)$ is periodic, with period $2\pi/\omega_s$. This means that if a function $X(s)$ has a pole at $s = s_1$ in the s plane, then $X^*(s)$ has poles at $s = s_1 \pm j\omega_s k$ ($k = 0, 1, 2, \ldots$).

Obtaining *z* transforms of functions involving the term $(1 - e^{-Ts})/s$. We shall here consider the function $X(s)$ involving $(1 - e^{-Ts})/s$. Suppose the transfer function $G(s)$ follows the zero-order hold. Then the product of the transfer function of the zero-order hold and $G(s)$ becomes

$$X(s) = \frac{1 - e^{-Ts}}{s} G(s)$$

In what follows we shall obtain the z transform of such an $X(s)$.

Note that $X(s)$ can be written as follows:

$$X(s) = (1 - e^{-Ts}) \frac{G(s)}{s} = (1 - e^{-Ts}) G_1(s) \tag{3–38}$$

where

$$G_1(s) = \frac{G(s)}{s}$$

Consider the function

$$X_1(s) = e^{-Ts} G_1(s) \tag{3–39}$$

Since $X_1(s)$ is a product of two Laplace-transformed functions, the inverse Laplace transform of Eq. (3–39) can be given by the following convolution integral:

$$x_1(t) = \int_0^t g_0(t - \tau) g_1(\tau)\, d\tau$$

where

$$g_0(t) = \mathcal{L}^{-1}[e^{-Ts}] = \delta(t - T)$$
$$g_1(t) = \mathcal{L}^{-1}[G_1(s)]$$

Thus

$$\begin{aligned} x_1(t) &= \int_0^t \delta(t - T - \tau) g_1(\tau)\, d\tau \\ &= g_1(t - T) \end{aligned}$$

Hence, by writing

$$\mathcal{Z}[g_1(t)] = G_1(z)$$

the z transform of $x_1(t)$ becomes

$$\mathcal{Z}[x_1(t)] = \mathcal{Z}[g_1(t - T)] = z^{-1} G_1(z)$$

Referring to Eqs. (3–38) and (3–39), we have

$$\begin{aligned} X(z) &= \mathcal{Z}[G_1(s) - e^{-Ts} G_1(s)] \\ &= \mathcal{Z}[g_1(t)] - \mathcal{Z}[x_1(t)] \\ &= G_1(z) - z^{-1} G_1(z) \\ &= (1 - z^{-1}) G_1(z) \end{aligned}$$

or

$$X(z) = \mathcal{Z}[X(s)] = (1 - z^{-1})\, \mathcal{Z}\left[\frac{G(s)}{s}\right] \tag{3–40}$$

We have thus shown that if $X(s)$ involves a factor $(1 - e^{-Ts})$, then in obtaining the z transform of $X(s)$ the term $1 - e^{-Ts} = 1 - z^{-1}$ may be factored out so that $X(z)$ becomes the product of $(1 - z^{-1})$ and the z transform of the remaining term.

Similarly, if the transfer function $G(s)$ follows the first-order hold $G_{h1}(s)$, where

$$G_{h1}(s) = \left(\frac{1 - e^{-Ts}}{s}\right)^2 \frac{Ts + 1}{T}$$

then the z transform of the function

$$X(s) = \left(\frac{1 - e^{-Ts}}{s}\right)^2 \frac{Ts + 1}{T} G(s)$$

can be obtained as follows. Since

$$X(s) = (1 - e^{-Ts})^2 \frac{Ts + 1}{Ts^2} G(s)$$

by employing the same approach as used in obtaining Eq. (3–40), we have

$$\begin{aligned} X(z) &= \mathcal{Z}\left[(1 - e^{-Ts})^2 \frac{Ts + 1}{Ts^2} G(s)\right] \\ &= (1 - z^{-1})^2 \mathcal{Z}\left[\frac{Ts + 1}{Ts^2} G(s)\right] \end{aligned} \tag{3–41}$$

Equation (3–41) can be used for obtaining the z transform of the function involving the first-order hold circuit.

Example 3–6.

Obtain the z transform of

$$X(s) = \frac{1 - e^{-Ts}}{s} \frac{1}{s + 1}$$

Referring to Eq. (3–40), we have

$$\begin{aligned} X(z) &= \mathcal{Z}\left[\frac{1 - e^{-Ts}}{s} \frac{1}{s + 1}\right] \\ &= (1 - z^{-1}) \mathcal{Z}\left[\frac{1}{s(s + 1)}\right] \\ &= (1 - z^{-1}) \mathcal{Z}\left[\frac{1}{s} - \frac{1}{s + 1}\right] \\ &= (1 - z^{-1}) \left(\frac{1}{1 - z^{-1}} - \frac{1}{1 - e^{-T}z^{-1}}\right) \\ &= \frac{(1 - e^{-T})z^{-1}}{1 - e^{-T}z^{-1}} \end{aligned}$$

3–4 RECONSTRUCTING ORIGINAL SIGNALS FROM SAMPLED SIGNALS

In digital control systems, analog signals are sampled every sampling period T. In applying the control signal to the plant, in most cases, we need to convert the sampled signal to a continuous-time signal. Such conversion from the sampled signal to the continuous-time signal can be done by use of a hold circuit. Hold circuits are low-

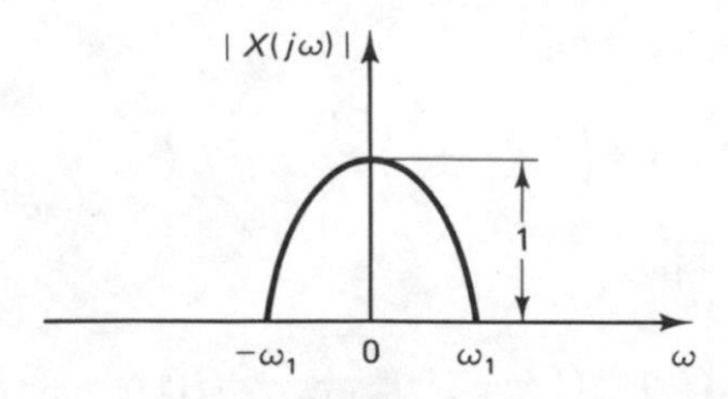

Figure 3–15 A frequency spectrum.

pass filters. They filter out high-frequency components present in the sampled signal. The ideal low-pass filter (which cannot be physically built) would exactly reconstruct the original signal. Actual low-pass filters (such as zero-order hold circuits, first-order hold circuits, and so forth) only approximate ideal low-pass filters and therefore reconstruct the original signal only approximately.

Sampling theorem. If the sampling frequency is sufficiently high compared with the highest-frequency component involved in the continuous-time signal, the amplitude characteristics of the continuous-time signal may be preserved in the envelope of the sampled signal.

In order to reconstruct the original signal from a sampled signal, there is a certain minimum frequency that the sampling operation must satisfy. Such a minimum frequency is specified in the sampling theorem. We shall assume that a continuous-time signal $x(t)$ has a frequency spectrum as shown in Fig. 3–15. This signal $x(t)$ does not contain any frequency components above ω_1 radians per second.

Sampling Theorem. If ω_s, defined as $2\pi/T$, where T is the sampling period, is greater than $2\omega_1$, or

$$\omega_s > 2\omega_1$$

where ω_1 is the highest-frequency component present in the continuous-time signal $x(t)$, then the signal $x(t)$ can be reconstructed completely from the sampled signal $x^*(t)$.

The theorem implies that if $\omega_s > 2\omega_1$, then from the knowledge of the sampled signal it is theoretically possible to reconstruct exactly the original continuous-time signal. In what follows, we shall use an intuitive graphical approach to explain the sampling theorem. For an analytical approach, see Prob. A-3–8.

To show the validity of this sampling theorem, we need to find the frequency spectrum of the sampled signal $x^*(t)$. The Laplace transform of $x^*(t)$ has been obtained in Sec. 3–3 and is given by Eq. (3–34) or (3–36), depending on whether $x(0+) = 0$ or not. In order to obtain the frequency spectrum, we substitute $j\omega$ for s in Eq. (3–34). [In discussing frequency spectra, we need not be concerned with the value of $x(0+)$.] Thus,

$$X^*(j\omega) = \frac{1}{T} \sum_{k=-\infty}^{\infty} X(j\omega + j\omega_s k)$$

$$= \cdots + \frac{1}{T}X(j(\omega - \omega_s)) + \frac{1}{T}X(j\omega) + \frac{1}{T}X(j(\omega + \omega_s)) + \cdots \qquad (3\text{–}42)$$

Equation (3–42) gives the frequency spectrum of the sampled signal $x^*(t)$. We see that the frequency spectrum of the impulse-sampled signal is reproduced an infinite number of times and is attenuated by the factor $1/T$. Thus the process of impulse modulation of the continuous-time signal is to produce a series of sidebands. Since $X^*(s)$ is periodic with period $2\pi/\omega_s$, as shown in Example 3–5, or

$$X^*(s) = X^*(s \pm j\omega_s k) \qquad k = 0, 1, 2, \ldots$$

if a function $X(s)$ has a pole at $s = s_1$, then $X^*(s)$ has poles at $s = s_1 \pm j\omega_s k$ $(k = 0, 1, 2, \ldots)$.

Figure 3–16(a) and (b) shows plots of the frequency spectra $X^*(j\omega)$ versus ω for two values of the sampling frequency ω_s. Figure 3-16(a) corresponds to $\omega_s > 2\omega_1$, while Fig. 3-16(b) corresponds to $\omega_s < 2\omega_1$. Each plot of $|X^*(j\omega)|$ versus ω consists of $|X(j\omega)|/T$ repeated every $\omega_s = 2\pi/T$ rad/sec. In the frequency spectrum of $|X^*(j\omega)|$ the component $|X(j\omega)|/T$ is called the *primary component* and the other components, $|X(j(\omega \pm \omega_s k))|/T$, are called *complementary components*.

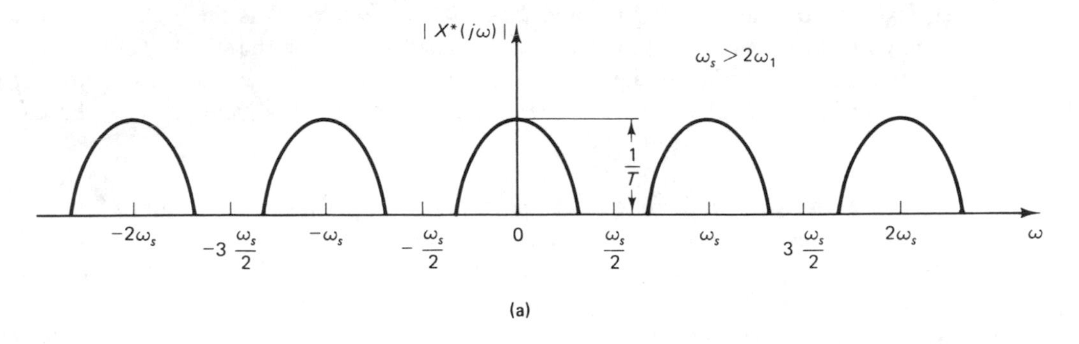

(a)

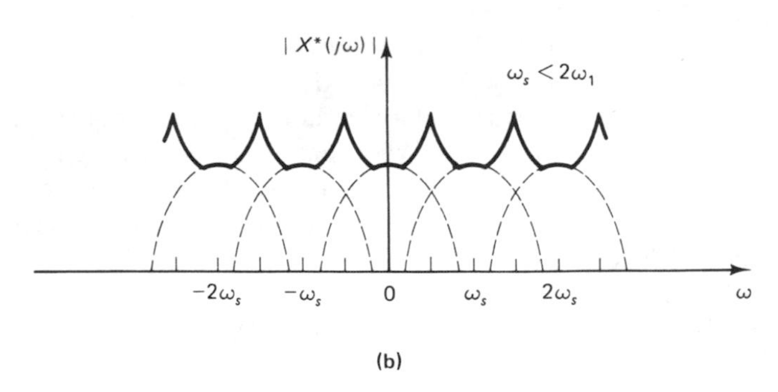

(b)

Figure 3–16 Plots of the frequency spectra $|X^*(j\omega)|$ vs. ω for two values of sampling frequency ω_s: (a) $\omega_s > 2\omega_1$; (b) $\omega_s < 2\omega_1$.

If $\omega_s > 2\omega_1$, no two components of $|X^*(j\omega)|$ will overlap, and the sampled frequency spectrum will be repeated every ω_s rad/sec.

If $\omega_s < 2\omega_1$, the original shape of $|X(j\omega)|$ no longer appears in the plot of $|X^*(j\omega)|$ versus ω because of the superposition of the spectra. Therefore, we see that the continuous-time signal $x(t)$ can be reconstructed from the impulse-sampled signal $x^*(t)$ by filtering if and only if $\omega_s > 2\omega_1$.

In summary, if the sampling frequency ω_s is higher than $2\omega_1$, then the continuous-time signal can be reconstructed by use of an ideal low-pass filter, after the signal has been sampled. Since the ideal low-pass filter has characteristics such that it passes only signals having frequencies lower than ω_1, it would be possible to obtain the frequency spectrum at the output of the filter as exactly $1/T$ times $|X(j\omega)|$ if such an ideal low-pass filter could be used.

The hold circuit following the sampler has the characteristic of a low-pass filter and thus the sampled signal is smoothed out and high-frequency components are attenuated. In practice, however, we may need an additional low-pass filter ahead of the sampler so that high-frequency components of the signal are attenuated before it is sampled.

It is noted that although the requirement on the minimum sampling frequency is specified by the sampling theorem as $\omega_s > 2\omega_1$, where ω_1 is the highest-frequency component present in the signal, practical considerations on the stability of the closed-loop system and other design considerations may make it necessary to sample at a frequency much higher than this theoretical minimum. (Frequently, ω_s is chosen to be $10\omega_1$ to $20\omega_1$.)

Aliasing. In the frequency spectra of an impulse-sampled signal $x^*(t)$ where $\omega_s < 2\omega_1$, as shown in Fig. 3–17, consider an arbitrary frequency point ω_2 which falls in the region of the overlap of the frequency spectra. The frequency spectrum at $\omega = \omega_2$ comprises two components, $|X^*(j\omega_2)|$ and $|X^*(j(\omega_s - \omega_2))|$. The latter component comes from the frequency spectrum centered at $\omega = \omega_s$. Thus, the frequency spectrum of the sampled signal at $\omega = \omega_2$ includes components not only at frequency ω_2 but also at frequency $\omega_s - \omega_2$ (in general, at $n\omega_s \pm \omega_2$, where n is an integer). When the composite spectrum is filtered by a low-pass filter, such as a zero-order hold, some higher harmonics will still be present in the output. The frequency component at $\omega = n\omega_s \pm \omega_2$ (where n is an integer) will appear in the output as if it were a frequency component at $\omega = \omega_2$. It is not possible to distinguish the frequency spectrum at $\omega = \omega_2$ from that at $\omega = n\omega_s \pm \omega_2$.

As shown in Fig. 3–17 the phenomenon that the frequency component $\omega_s - \omega_2$ (in general, $n\omega_s \pm \omega_2$, where n is an integer) shows up at frequency ω_2 when the signal $x(t)$ is sampled is called *aliasing*. This frequency $\omega_s - \omega_2$ (in general, $n\omega_s \pm \omega_2$) is called an *alias* of ω_2.

Consider, for example, signals $x(t)$ and $y(t)$, where

$$x(t) = \sin(\omega_2 t + \theta)$$

$$y(t) = \sin((\omega_2 + n\omega_s)t + \theta)$$

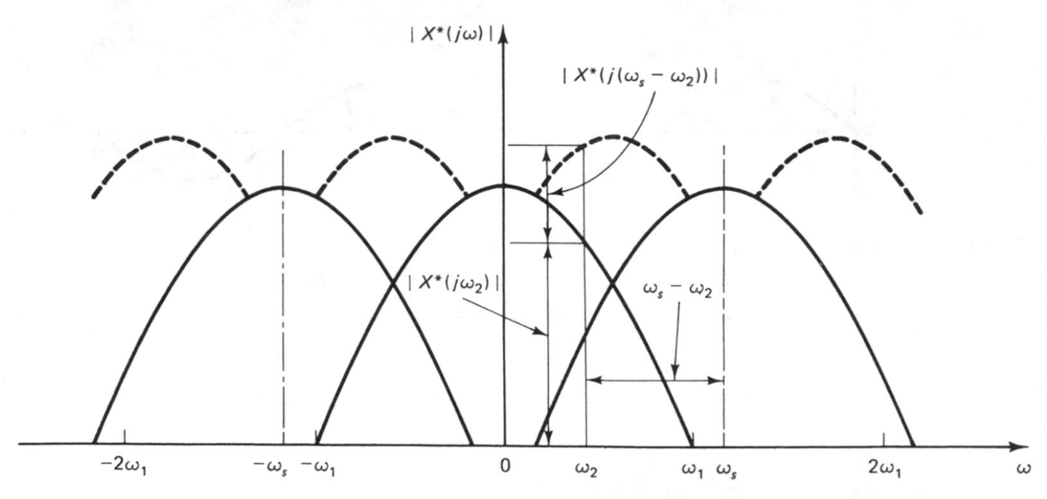

Figure 3–17 Frequency spectra of an impulse-sampled signal $x^*(t)$.

and where n is an integer. The frequencies of $x(t)$ and $y(t)$ differ from each other by an integral multiple of ω_s, the sampling frequency. If we sample these signals every sampling period T ($T = 2\pi/\omega_s$), then

$$x(kT) = \sin(\omega_2 kT + \theta) \tag{3–43}$$

and

$$y(kT) = \sin((\omega_2 + n\,\omega_s)kT + \theta)$$
$$= \sin(\omega_2 kT + 2\pi kn + \theta)$$

Since n and k are integers, we have

$$y(kT) = \sin(\omega_2 kT + \theta)$$

which is exactly the same as $x(kT)$ as given by Eq. (3–43).

Figure 3–18 shows signals $x(t) = \sin \omega_2 t = \sin t$ and $y(t) = \sin \omega_3 t = \sin 4t$ and their sampled signals $x(k)$ and $y(k)$. The sampling frequency is $\omega_s = 3$ rad/sec. Notice that we obtain exactly the same sampled values from these two signals. (Note that $\omega_3 = \omega_2 + \omega_s$.) The fact that two signals of different frequencies can have identical samples means that we cannot distinguish between them from their samples.

It is important to remember that the sampled signals are the same if the two frequencies differ by an integral multiple of the sampling frequency ω_s. If a signal is sampled at a slow frequency such that the sampling theorem is not satisfied, then high frequencies are "folded in" and appear as low frequencies. In terms of the frequency spectra for the signals $x(t)$ and $y(t)$, two frequency components at $\omega = \omega_2$ and $\omega = \omega_s + \omega_2$ cannot be distinguished.

Note that if the signal $x(t)$ is band-limited so that $X(j\omega)$ is zero for $|\omega| \geq$

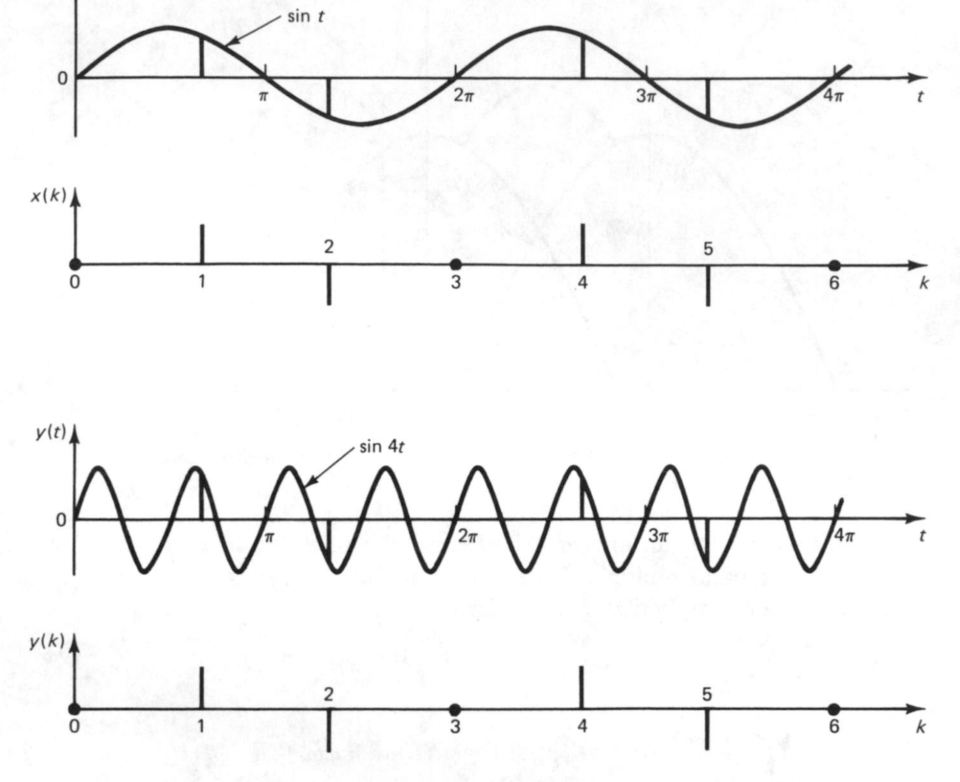

Figure 3–18 Plots of signals $x(t) = \sin t$ and $y(t) = \sin 4t$ and their sampled signals.

$\pi/T = \frac{1}{2}\omega_s$, then sampling at intervals of T will produce no aliasing and the original continuous-time signal can be reconstructed exactly from $X^*(j\omega)$.

The phenomenon of the overlap in the frequency spectra is known as *folding*. Figure 3–19 shows the regions where folding error occurs. The frequency $\frac{1}{2}\omega_s$ is called the *folding frequency* or *Nyquist frequency* ω_N. That is,

$$\omega_N = \frac{1}{2}\omega_s = \frac{\pi}{T}$$

In practice, signals in control systems have high-frequency components, and some folding effect will almost always exist. For example, in an electromechanical system some signal may be contaminated by noises. The frequency spectrum of the signal, therefore, may include low-frequency components as well as high-frequency noise components (that is, noises at 60 Hz or 400 Hz). Since sampling at frequencies higher than 400 Hz is not practical, the high frequency will be folded in and will appear as a low frequency. Remember that all signals with frequencies higher than $\frac{1}{2}\omega_s$ appear

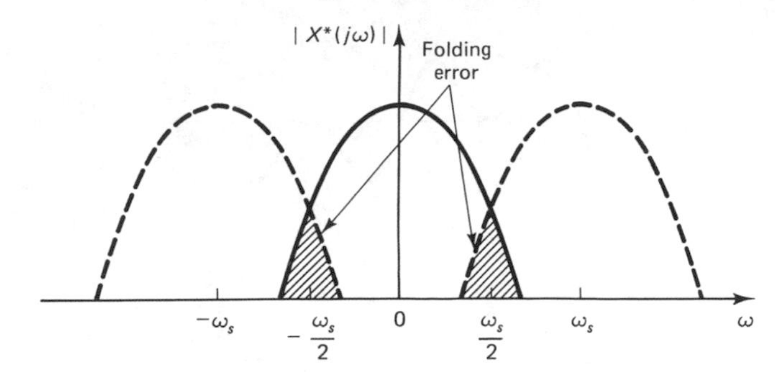

Figure 3–19 Diagram showing the regions where folding error occurs.

as signals of frequencies between 0 and $\frac{1}{2}\omega_s$. In fact, in certain cases, a signal of zero frequency may appear in the output. That is, aliasing may cause dc error in the system output. (See Prob. A-3–9 for details.)

To avoid aliasing, we must either choose the sampling frequency high enough ($\omega_s > 2\omega_1$, where ω_1 is the highest-frequency component present in the signal) or use a prefilter ahead of the sampler to reshape the frequency spectrum of the signal (so that the frequency spectrum for $\omega > \frac{1}{2}\omega_s$ is negligible) before the signal is sampled.

Hidden oscillation. It is noted that if the continuous-time signal $x(t)$ involves a frequency component equal to n times the sampling frequency ω_s (where n is an integer), then that component may not appear in the sampled signal. For example, if the signal

$$x(t) = x_1(t) + x_2(t) = \sin t + \sin 3t$$

where $x_1(t) = \sin t$ and $x_2(t) = \sin 3t$, is sampled at $t = 0, 2\pi/3, 4\pi/3, \ldots$ (the sampling frequency ω_s is 3 rad/sec), then the sampled signal will not show the frequency component with $\omega = 3$ rad/sec, the frequency equal to ω_s. (See Fig. 3–20.)

Even though the signal $x(t)$ involves an oscillation with $\omega = 3$ rad/sec [that is, the component $x_2(t) = \sin 3t$], the sampled signal does not show this oscillation. Such an oscillation existing in $x(t)$ between the sampling periods is called a *hidden oscillation*.

Ideal low-pass filter. In a discrete-time control system the high-frequency components in a signal $x^*(t)$ that are generated in the sampling operation must be attenuated prior to the application of the signal to a continuous-time element of the system.

In order to attenuate the high-frequency components, the sampled signal is usually fed to a hold circuit (see Fig. 3–21). The hold circuit, which is a low-pass filter, converts the sampled signal $x^*(t)$ back to a continuous-time signal.

An ideal low-pass filter or ideal hold circuit would convert the sampled signal $x^*(t)$ exactly back to the continuous-time signal $x(t)$. However, such an ideal filter does not exist in the real world.

In order to compare the capability of the practical zero-order hold circuit with

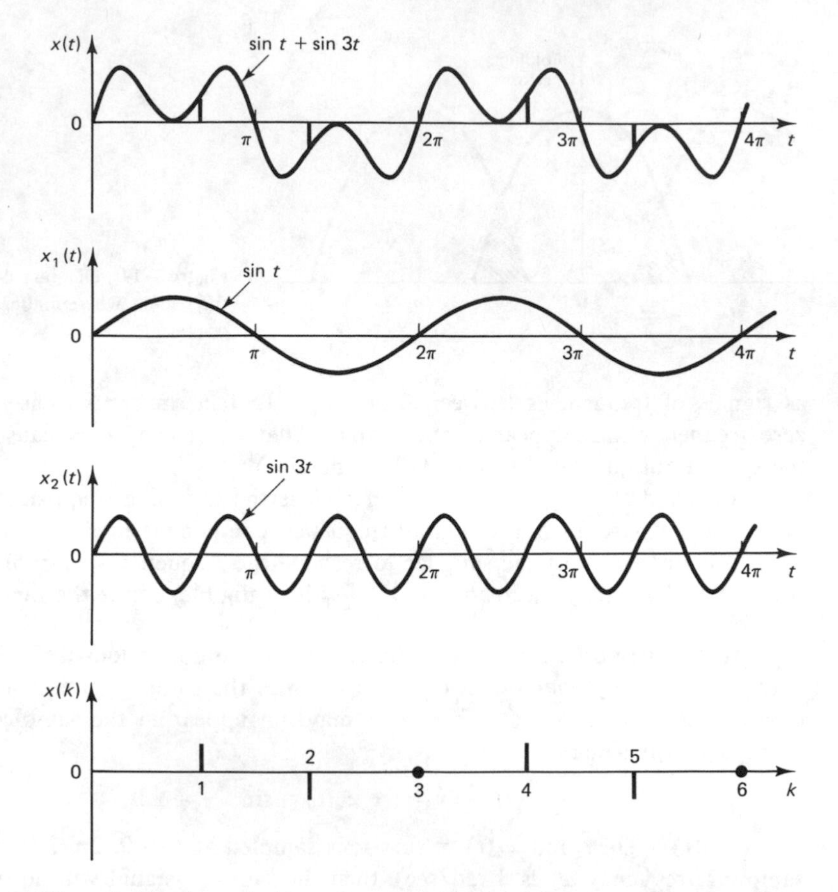

Figure 3–20 Plots of $x(t) = \sin t + \sin 3t$, $x_1(t) = \sin t$, and $x_2(t) = \sin 3t$. Sampled signal $x(k)$, where sampling frequency $\omega_s = 3$ rad/sec, does not show oscillation with frequency $\omega = 3$ rad/sec.

that of the ideal low-pass filter, let us investigate the performance of the ideal filter.

The amplitude frequency spectrum of the ideal low-pass filter $G_I(j\omega)$ is shown in Fig. 3–22. The magnitude of the ideal filter is unity over the frequency range $-\frac{1}{2}\omega_s \leq \omega \leq \frac{1}{2}\omega_s$, and is zero outside this frequency range.

The sampling process introduces an infinite number of complementary components (sideband components) in addition to the primary component. The ideal filter

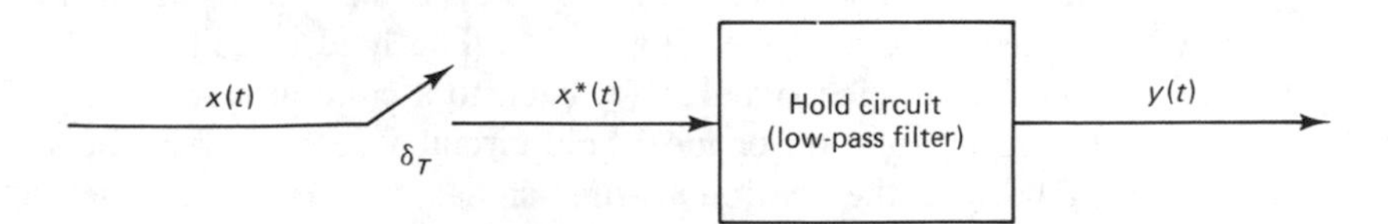

Figure 3–21 Sampler and hold circuit.

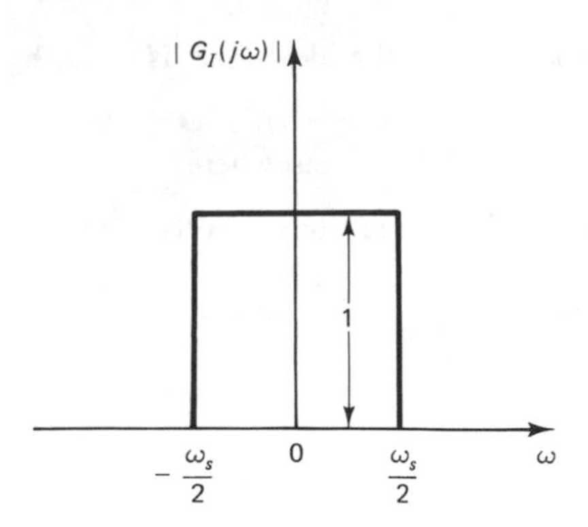

Figure 3–22 Amplitude frequency spectrum of the ideal low-pass filter.

will attenuate all such complementary components to zero and will pass only the primary component, provided the sampling frequency ω_s is greater than twice the highest-frequency component of the continuous-time signal. Such an ideal filter reconstructs the continuous-time signal represented by the samples. Figure 3–23 shows the frequency spectra of the signals before and after ideal filtering.

The frequency spectrum at the output of the ideal filter is $1/T$ times the frequency spectrum of the original continuous-time signal $x(t)$. Since the ideal filter has constant-magnitude characteristics for the frequency region $-\frac{1}{2}\omega_s \leq \omega \leq \frac{1}{2}\omega_s$, there is no distortion at any frequency within this frequency range. That is, there is no phase shift in the frequency spectrum of the ideal filter. (The phase shift of the ideal filter is zero.)

It is noted that if the sampling frequency is less than twice the highest-frequency component of the original continuous-time signal, then because of the frequency spectrum overlap of the primary component and complementary components, even the ideal filter cannot reconstruct the original continuous-time signal. (In practice, the frequency spectrum of the continuous-time signal in a control system may extend beyond $\pm\frac{1}{2}\omega_s$, even though the amplitudes at the higher frequencies are small.)

Next, let us find the impulse response function of the ideal filter. It will be shown that for the ideal filter an output is required prior to the application of the input to the filter. Thus, it is not physically realizable.

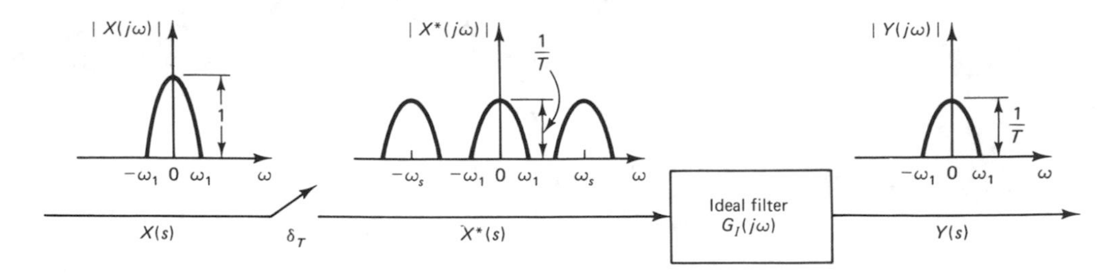

Figure 3–23 Frequency spectra of the signals before and after ideal filtering.

Since the frequency spectrum of the ideal filter is given by

$$G_I(j\omega) = \begin{cases} 1 & -\frac{1}{2}\omega_s \leq \omega \leq \frac{1}{2}\omega_s \\ 0 & \text{elsewhere} \end{cases}$$

the inverse Fourier transform of the frequency spectrum gives

$$\begin{aligned} g_I(t) &= \frac{1}{2\pi}\int_{-\infty}^{\infty} G_I(j\omega)e^{j\omega t}\,d\omega \\ &= \frac{1}{2\pi}\int_{-\omega_s/2}^{\omega_s/2} e^{j\omega t}\,d\omega \\ &= \frac{1}{2\pi jt}\left(e^{(1/2)j\omega_s t} - e^{-(1/2)j\omega_s t}\right) \\ &= \frac{1}{\pi t}\sin\frac{\omega_s t}{2} \end{aligned}$$

or

$$g_I(t) = \frac{1}{T}\frac{\sin(\omega_s t/2)}{\omega_s t/2} \tag{3–44}$$

Equation (3–44) gives the unit impulse response of the ideal filter. Figure 3–24 shows a plot of $g_I(t)$ versus t. Notice that the response extends from $t = -\infty$ to $t = \infty$. This implies that there is a response for $t < 0$ to a unit impulse applied at $t = 0$. (That is, the time response begins before an input is applied.) This cannot be true in the physical world. Hence, such an ideal filter is physically unrealizable. [In many communications systems, however, it is possible to approximate $g_I(t)$ closely by adding a phase lag, which means adding a delay to the filter. In feedback control systems, increasing phase lag is not desirable from the viewpoint of stability. Therefore, we avoid adding a phase lag to approximate the ideal filter.]

Because the ideal filter is unrealizable, and because signals in practical control systems generally have higher-frequency components and are not ideally band-limited,

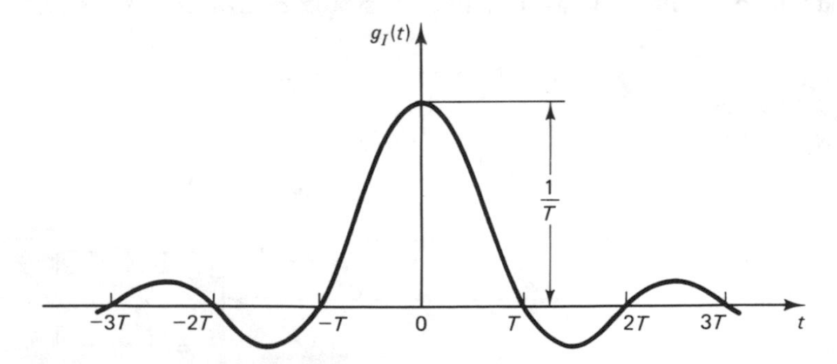

Figure 3–24 Impulse response $g_I(t)$ of ideal filter.

it is not possible, in practice, to exactly reconstruct a continuous-time signal from the sampled signal, no matter what sampling frequency is chosen. (In other words, practically speaking, it is not possible to reconstruct exactly a continuous-time signal in a practical control system once it is sampled.)

Frequency response characteristics of the zero-order hold. The transfer function of a zero-order hold is

$$G_{h0}(s) = \frac{1 - e^{-Ts}}{s} \tag{3–45}$$

In order to compare the zero-order hold with the ideal filter, we shall obtain the frequency-response characteristics of the transfer function of the zero-order hold. By substituting $j\omega$ for s in Eq. (3–45), we obtain

$$\begin{aligned} G_{h0}(j\omega) &= \frac{1 - e^{-Tj\omega}}{j\omega} \\ &= \frac{2e^{-(1/2)Tj\omega}\left(e^{(1/2)Tj\omega} - e^{-(1/2)Tj\omega}\right)}{2j\omega} \\ &= T\frac{\sin(\omega T/2)}{\omega T/2}e^{-(1/2)Tj\omega} \end{aligned}$$

The amplitude of the frequency spectrum of $G_{h0}(j\omega)$ is

$$|G_{h0}(j\omega)| = T\left|\frac{\sin(\omega T/2)}{\omega T/2}\right| \tag{3–46}$$

The magnitude becomes zero at the frequency equal to the sampling frequency and at integral multiples of the sampling frequency.

Figure 3–25(a) shows the frequency response characteristics of the zero-order hold. As can be seen from Fig. 3–25(a), there are undesirable gain peaks at frequencies of $3\omega_s/2$, $5\omega_s/2$, and so on. Notice that the magnitude is more than 3 db down ($0.637 = -3.92$ db) at frequency $\frac{1}{2}\omega_s$. Because the magnitude decreases gradually as the frequency increases, the complementary components gradually attenuate to zero. Since the magnitude characteristics of the zero-order hold are not constant, if a system is connected to the sampler and zero-order hold, distortion of the frequency spectra occurs in the system.

The phase shift characteristics of the zero-order hold can be obtained as follows. Note that $\sin(\omega T/2)$ alternates positive and negative values as ω increases from 0 to ω_s, ω_s to $2\omega_s$, $2\omega_s$ to $3\omega_s$, and so on. Thus, the phase curve [Fig. 3–25(a), bottom] is discontinuous at $\omega = k\omega_s = 2\pi k/T$, where $k = 1, 2, 3, \ldots$. Such a discontinuity or a switch from a positive value to a negative value or vice versa may be considered to be a phase shift of $\pm 180°$. In Fig. 3–25(a) phase shift is assumed to be $-180°$. (It could be assumed to be $+180°$ as well.) Thus,

$$\angle G_{h0}(j\omega) = \angle T\frac{\sin(\omega T/2)}{\omega T/2} \angle e^{-(1/2)j\omega T}$$

$$= \angle \sin\frac{\omega T}{2} + \angle e^{-(1/2)j\omega T} = \angle \sin\frac{\omega T}{2} - \frac{\omega T}{2}$$

where

$$\angle e^{-(1/2)j\omega T} = -\frac{\omega T}{2}$$

$$\angle \sin\frac{\omega T}{2} = 0^\circ \text{ or } \pm 180^\circ$$

A modification of the presentation of the frequency-response diagram of Fig. 3–25(a) is shown in Fig. 3–25(b). The diagram shown in Fig. 3–25(b) is the Bode diagram of the zero-order hold. The sampling period T is assumed to be 1 sec, or $T = 1$. Notice that the magnitude curve approaches $-\infty$ db at frequency points that are integral multiples of the sampling frequency $\omega_s = 2\pi/T = 6.28$ rad/sec.

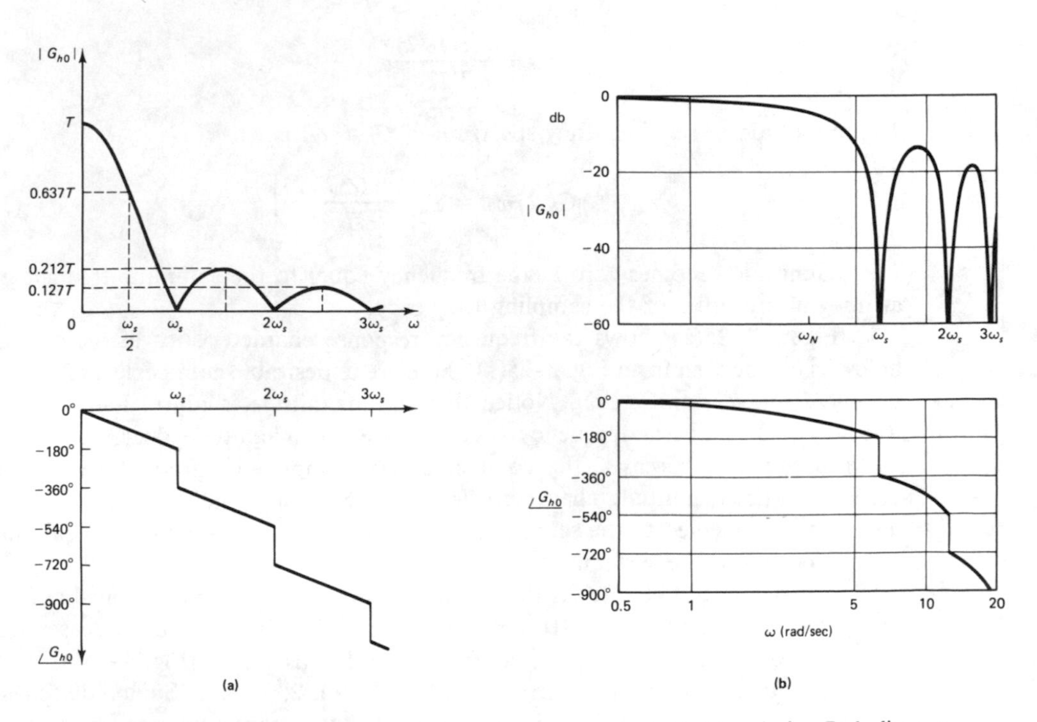

Figure 3–25 (a) Frequency response curves for the zero-order hold; (b) equivalent Bode diagram when $T = 1$ sec.

Discontinuities of the phase curve [Fig. 3–25(b), bottom] occur at these frequency points.

To summarize what we have stated, the frequency spectrum of the output of the zero-order hold includes complementary components, since the magnitude characteristics show that the magnitude of $G_{h0}(j\omega)$ is not zero for $|\omega| > \frac{1}{2}\omega_s$, except at points where $\omega = \pm\omega_s$, $\omega = \pm 2\omega_s$, $\omega = \pm 3\omega_s$, . . . In the phase curve there are phase discontinuities of $\pm 180°$ at frequency points that are multiples of ω_s. Except for these phase discontinuities, the phase characteristic is linear in ω.

Figure 3–26 shows the comparison of the ideal filter and the zero-order hold. For the sake of comparison, the magnitudes $|G(j\omega)|$ are normalized. We see that the zero-order hold is a low-pass filter, although its function is not quite good. Often, additional low-pass filtering of the signal before sampling is necessary to effectively remove frequency components higher than $\frac{1}{2}\omega_s$ so that the effect of the frequency folding can be made negligible.

The accuracy of the zero-order hold as an extrapolator depends on the sampling frequency ω_s. That is, the output of the hold may be made as close to the original continuous-time signal as possible by letting the sampling period T become as small as practically possible.

The higher-order hold, which is more sophisticated and which better approximates the ideal filter, is more complex and has more time delay than the zero-order hold. Because additional time delay in the closed-loop control system decreases the stability margin or even causes instability, higher-order holds are rarely justified in terms of improved performance, and therefore the zero-order hold is widely used in practice. In fact, the zero-order hold is satisfactory in most practical cases, provided the sampling frequency ω_s is much higher than the highest-frequency component in the continuous-time signal. (As mentioned earlier, ω_s is frequently chosen to be 10 to 20 times the highest frequency ω_1 present in the continuous-time signal.)

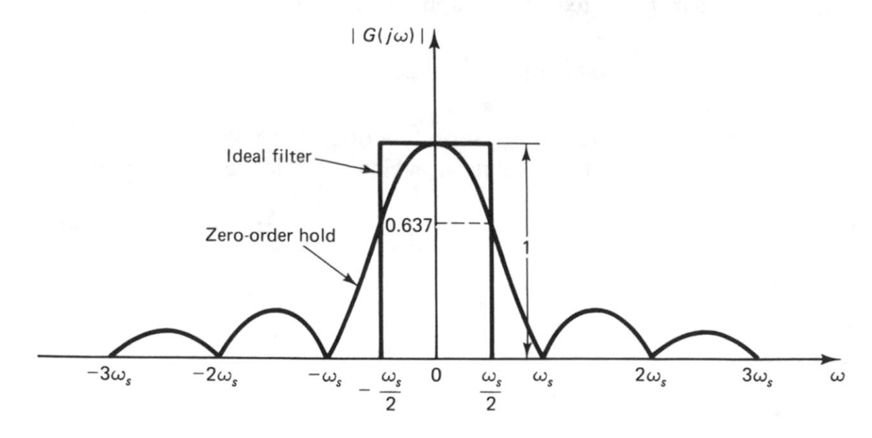

Figure 3–26 Comparison of the ideal filter and the zero-order hold.

3-5 THE PULSE TRANSFER FUNCTION

This section treats the pulse transfer function, the basic concept necessary for analyzing discrete-time control systems by the z transform method.

The transfer function for the continuous-time system relates the Laplace transform of the continuous-time output to that of the continuous-time input, while the pulse transfer function relates the z transform of the output at the sampling instants to that of the sampled input.

Before we discuss the pulse transfer function, it is appropriate to discuss convolution summation.

Convolution summation. Consider the response of a continuous-time system driven by an impulse-sampled signal (a train of impulses) as shown in Fig. 3–27. Suppose that $x(t) = 0$ for $t < 0$. The impulse-sampled signal $x^*(t)$ is the input to the continuous-time system whose transfer function is $G(s)$. The output of the system is assumed to be a continuous-time signal $y(t)$. If at the output there is another sampler, which is synchronized in phase with the input sampler and operates at the same sampling period, then the output is a train of impulses. We assume that $y(t) = 0$ for $t < 0$.

The z transform of $y(t)$ is

$$\mathcal{Z}[y(t)] = Y(z) = \sum_{k=0}^{\infty} y(kT)z^{-k} \tag{3–47}$$

In the absence of the output sampler, if we consider a fictitious sampler (which is synchronized in phase with the input sampler and operates at the same sampling period) at the output and observe the sequence of values taken by $y(t)$ only at instants $t = kT$, then the z transform of the output $y^*(t)$ can also be given by Eq. (3–47).

For the continuous-time system, it is a well-known fact that the output $y(t)$ of the system is related to the input $x(t)$ by the convolution integral, or

$$y(t) = \int_0^t g(t-\tau)x(\tau)\,d\tau = \int_0^t x(t-\tau)g(\tau)\,d\tau$$

where $g(t)$ is the impulse-response function of the system. For discrete-time systems we have a convolution summation, which is similar to the convolution integral. Since

$$x^*(t) = \sum_{k=0}^{\infty} x(t)\delta(t-kT) = \sum_{k=0}^{\infty} x(kT)\delta(t-kT)$$

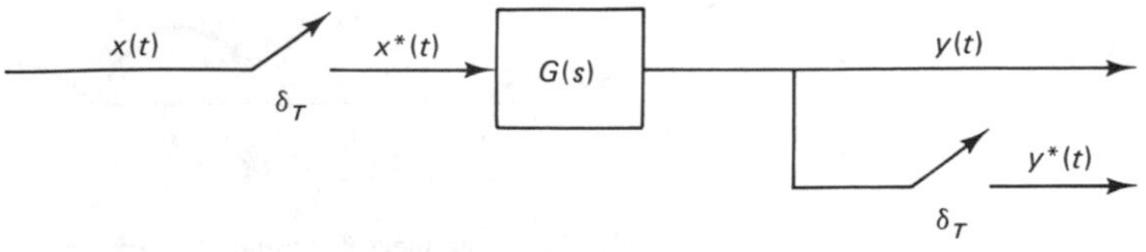

Figure 3–27 Continuous-time system $G(s)$ driven by an impulse-sampled signal.

is a train of impulses, the response $y(t)$ of the system to the input $x^*(t)$ is the sum of the individual impulse responses. Hence,

$$y(t) = \begin{cases} g(t)x(0) & 0 \le t < T \\ g(t)x(0) + g(t-T)x(T) & T \le t < 2T \\ g(t)x(0) + g(t-T)x(T) + g(t-2T)x(2T) & 2T \le t < 3T \\ \vdots & \\ g(t)x(0) + g(t-T)x(T) + \cdots + g(t-kT)x(kT) & kT \le t < (k+1)T \end{cases}$$

Noting that for a physical system a response cannot precede the input, we have $g(t) = 0$ for $t < 0$ or $g(t - kT) = 0$ for $t < kT$. Consequently, the preceding equations may be combined into one equation:

$$\begin{aligned} y(t) &= g(t)x(0) + g(t-T)x(T) + g(t-2T)x(2T) + \cdots + g(t-kT)x(kT) \\ &= \sum_{h=0}^{k} g(t-hT)x(hT) \qquad 0 \le t \le kT \end{aligned}$$

The values of the output $y(t)$ at the sampling instants $t = kT$ $(k = 0, 1, 2, \ldots)$ are given by

$$y(kT) = \sum_{h=0}^{k} g(kT - hT)x(hT) \tag{3–48}$$

$$= \sum_{h=0}^{k} x(kT - hT)g(hT) \tag{3–49}$$

where $g(kT)$ is the system's weighting sequence. The summation in Eq. (3–48) or (3–49) is called the *convolution summation*. Note that the simplified notation

$$y(kT) = x(kT)*g(kT)$$

is often used for the convolution summation.

Since we assumed that $x(t) = 0$ for $t < 0$, we have $x(kT - hT) = 0$ for $h > k$. Also, since $g(kT - hT) = 0$ for $h > k$, we may assume that the values of h in Eqs. (3–48) and (3–49) can be taken from 0 to ∞ rather than from 0 to k without changing the value of the summation. Therefore, Eqs. (3–48) and (3–49) can be rewritten as follows:

$$y(kT) = \sum_{h=0}^{\infty} g(kT - hT)x(hT) \tag{3–50}$$

$$= \sum_{h=0}^{\infty} x(kT - hT)g(hT) \tag{3–51}$$

It is noted that if $G(s)$ is a ratio of polynomials in s and if the degree of the denominator polynomial exceeds the degree of the numerator polynomial only by 1, the output $y(t)$ is discontinuous, as shown in Fig. 3–28(a). When $y(t)$ is discontinu-

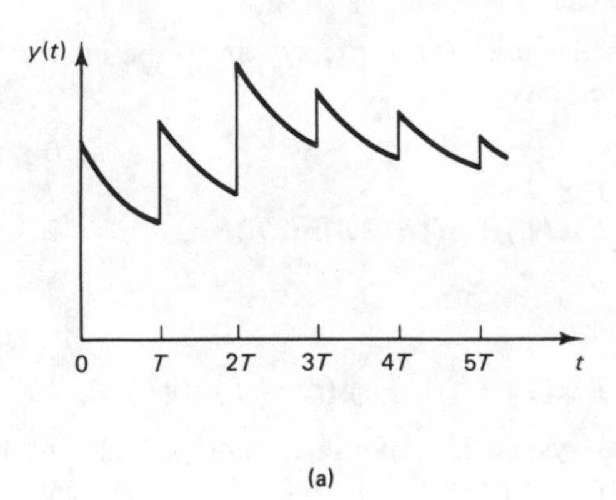

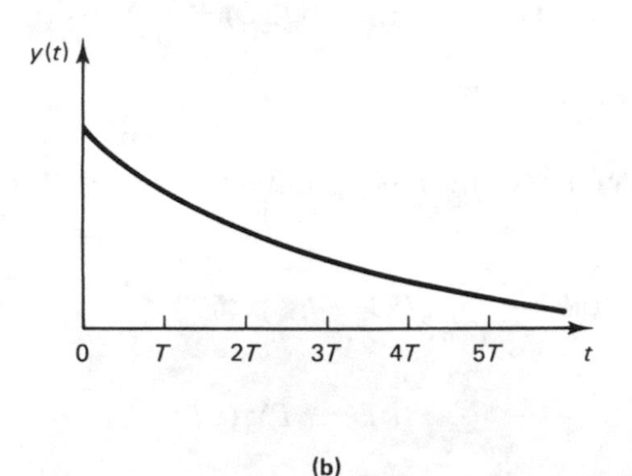

Figure 3–28 (a) Plot of output $y(t)$ (impulse response) vs. t when the degree of the denominator polynomial of $G(s)$ is higher by 1 than that of the numerator polynomial; (b) plot of output $y(t)$ vs. t when the degree of the denominator polynomial of $G(s)$ is higher by 2 or more than that of the numerator polynomial.

ous, Eqs. (3–50) and (3–51) yield the values immediately after the sampling instants, that is, $y(0+)$, $y(T+)$, . . . , $y(kT+)$. Such values do not portray the actual response curve. (See Prob. A-3–12.)

If the degree of the denominator polynomial exceeds that of the numerator polynomial by 2 or more, however, the output $y(t)$ is continuous, as shown in Fig. 3–28(b). When $y(t)$ is continuous, Eqs. (3–50) and (3–51) yield the values at the sampling instants. The values $y(k)$ in such a case portray the actual response curve.

In analyzing discrete-time control systems it is important to remember that the system response to the impulse-sampled signal may not portray the correct time-response behavior of the actual system unless the transfer function $G(s)$ of the continuous-time part of the system has at least two more poles than zeros, so that $\lim_{s\to\infty} sG(s) = 0$.

Pulse transfer function. From Eq. (3–50) we have

$$y(kT) = \sum_{h=0}^{\infty} g(kT - hT)x(hT) \qquad k = 0, 1, 2, \ldots$$

where $g(kT - hT) = 0$ for $h > k$. Hence, the z transform of $y(kT)$ becomes

$$\begin{aligned} Y(z) &= \sum_{k=0}^{\infty} y(kT)z^{-k} \\ &= \sum_{k=0}^{\infty}\sum_{h=0}^{\infty} g(kT - hT)x(hT)z^{-k} \\ &= \sum_{m=0}^{\infty}\sum_{h=0}^{\infty} g(mT)x(hT)z^{-(m+h)} \\ &= \sum_{m=0}^{\infty} g(mT)z^{-m} \sum_{h=0}^{\infty} x(hT)z^{-h} \\ &= G(z)X(z) \end{aligned} \tag{3–52}$$

where $m = k - h$ and

$$G(z) = \sum_{m=0}^{\infty} g(mT)z^{-m} = z \text{ transform of } g(t)$$

Equation (3–52) relates the pulsed output $Y(z)$ of the system to the pulse input $X(z)$. It provides a means for determining the z transform of the output sequence for any input sequence. Dividing both sides of Eq. (3–52) by $X(z)$ gives

$$G(z) = \frac{Y(z)}{X(z)} \tag{3–53}$$

As stated earlier, in Sec. 2–6, $G(z)$, given by Eq. (3–53), the ratio of the output $Y(z)$ and the input $X(z)$, is the pulse transfer function of the discrete-time system. It is the z transform of the weighting sequence. Figure 3–29 shows a block diagram for a pulse transfer function $G(z)$, together with the input $X(z)$ and the output $Y(z)$. As seen in Eq. (3–52), the z transform of the output signal can be obtained as the product of the system's pulse transfer function and the z transform of the input signal.

Also as mentioned in Sec. 2–6, $G(z)$ is also the z transform of the system's response to the Kronecker delta input:

$$x(kT) = \delta_0(kT) = \begin{cases} 1 & \text{for } k = 0 \\ 0 & \text{for } k \neq 0 \end{cases}$$

Since the z transform of the Kronecker delta input is

$$X(z) = \sum_{k=0}^{\infty} x(kT)z^{-k} = 1$$

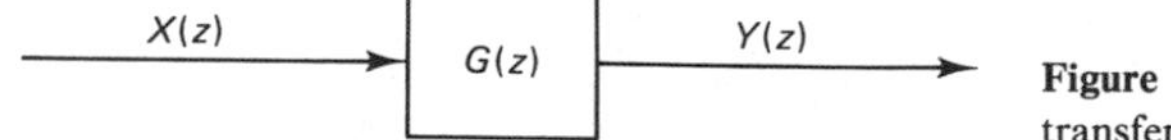

Figure 3–29 Block diagram for a pulse-transfer-function system.

then referring to Eq. (3–53), the response $Y(z)$ to the Kronecker delta input is

$$Y(z) = G(z)$$

Thus, the system's response to the Kronecker delta input is $G(z)$, the z transform of the weighting sequence. This fact is parallel to the fact that $G(s)$ is the Laplace transform of the system's weighting function, which is the system's response to the unit impulse function.

Starred Laplace transform of the signal involving both ordinary and starred Laplace transforms. In analyzing discrete-time control systems we often find that some signals in the system are starred (meaning that signals are impulse-sampled) and others are not. To obtain pulse transfer functions and to analyze discrete-time control systems, therefore, we must be able to obtain the transforms of output signals of systems which contain sampling operations in various places in the loops.

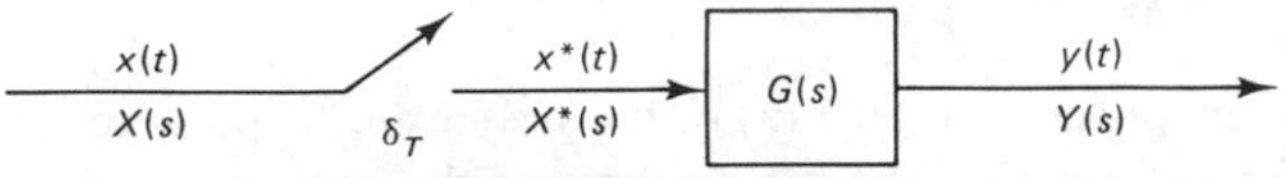

Figure 3–30 Impulse-sampled system.

Suppose the impulse sampler is followed by a linear continuous-time element whose transfer function is $G(s)$, as shown in Fig. 3–30. In the following analysis we assume that all initial conditions are zero in the system. Then the output $Y(s)$ is

$$Y(s) = G(s)X^*(s) \tag{3–54}$$

Notice that $Y(s)$ is a product of $X^*(s)$, which is periodic with period $2\pi/\omega_s$, and $G(s)$, which is not periodic. The fact that the impulse-sampled signals are periodic can be seen from the fact that

$$X^*(s) = X^*(s \pm j\omega_s k) \qquad k = 0, 1, 2, \ldots \tag{3–55}$$

(See Example 3–5.)

In the following we shall show that in taking the starred Laplace transform of Eq. (3–54) we may factor out $X^*(s)$, so that

$$Y^*(s) = [G(s)X^*(s)]^* = [G(s)]^*X^*(s) = G^*(s)X^*(s) \tag{3–56}$$

This fact is very important in deriving the pulse transfer function and also in simplifying the block diagram of the discrete-time control system.

To derive Eq. (3–56) note that the inverse Laplace transform of $Y(s)$ given by Eq. (3–54) can be written as follows:

$$y(t) = \mathscr{L}^{-1}[G(s)X^*(s)]$$

$$= \int_0^t g(t-\tau)x^*(\tau)\,d\tau$$

$$= \int_0^t g(t-\tau) \sum_{k=0}^{\infty} x(\tau)\delta(\tau - kT)\, d\tau$$

$$= \sum_{k=0}^{\infty} \int_0^t g(t-\tau)x(\tau)\delta(\tau - kT)\, d\tau$$

$$= \sum_{k=0}^{\infty} g(t-kT)x(kT)$$

Then the z transform of $y(t)$ becomes

$$Y(z) = \mathscr{Z}[y(t)] = \sum_{n=0}^{\infty} \left[\sum_{k=0}^{\infty} g(nT - kT)x(kT) \right] z^{-n}$$

$$= \sum_{m=0}^{\infty} \sum_{k=0}^{\infty} g(mT)x(kT)z^{-(k+m)}$$

where $m = n - k$. Thus

$$Y(z) = \sum_{m=0}^{\infty} g(mT)z^{-m} \sum_{k=0}^{\infty} x(kT)z^{-k}$$

$$= G(z)X(z) \tag{3–57}$$

Since the z transform can be understood as the starred Laplace transform with e^{Ts} replaced by z, the z transform may be considered to be a shorthand notation for the starred Laplace transform. Thus, Eq. (3–57) may be expressed as

$$Y^*(s) = G^*(s)X^*(s)$$

which is Eq. (3–56). We have thus shown that by taking the starred Laplace transform of both sides of Eq. (3–54), we obtain Eq. (3–56).

To summarize what we have obtained, note that Eqs. (3–54) and (3–56) state that in taking the starred Laplace transform of a product of transforms, where some are ordinary Laplace transforms and others are starred Laplace transforms, the functions already in starred transforms can be factored out of the starred Laplace transform operation.

It is noted that systems become periodic under starred Laplace transform operations. Such periodic systems are generally more complicated to analyze than the original nonperiodic ones, but the former may be analyzed without difficulty if carried out in the z domain (that is, by use of the pulse-transfer-function approach).

General procedures for obtaining pulse transfer functions. Here we shall present general procedures for obtaining the pulse transfer function of a system which has an impulse sampler at the input to the system, as shown in Fig. 3–31(a).

The pulse transfer function $G(z)$ of the system shown in Fig. 3–31(a) is

$$\frac{Y(z)}{X(z)} = G(z) = \mathscr{Z}[G(s)]$$

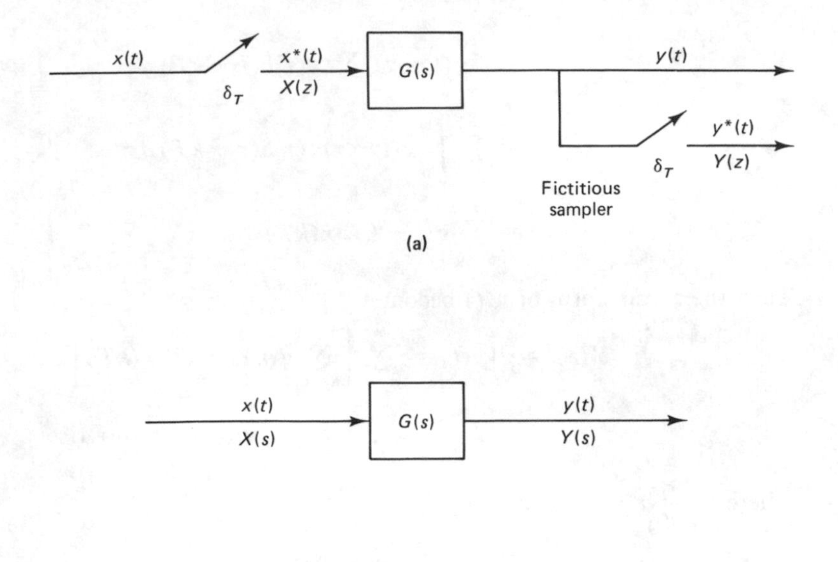

Figure 3–31 (a) Continuous-time system with an impulse sampler at the input; (b) continuous-time system.

Next, consider the system shown in Fig. 3–31(b). The transfer function $G(s)$ is given by

$$\frac{Y(s)}{X(s)} = G(s)$$

The important fact to remember is that the pulse transfer function for this system is not $\mathscr{Z}[G(s)]$, because of the absence of the input sampler.

The presence or absence of the input sampler is crucial in determining the pulse transfer function of a system, because, for example, for the system shown in Fig. 3–31(a), the Laplace transform of the output $y(t)$ is

$$Y(s) = G(s)X^*(s)$$

Hence by taking the starred Laplace transform of $Y(s)$ we have

$$Y^*(s) = G^*(s)X^*(s)$$

or, in terms of the z transform,

$$Y(z) = G(z)X(z)$$

while for the system shown in Fig. 3–31(b) the Laplace transform of the output $y(t)$ is

$$Y(s) = G(s)X(s)$$

which yields

$$Y^*(s) = [G(s)X(s)]^* = [GX(s)]^*$$

or, in terms of the z transform,

$$Y(z) = \mathcal{Z}[Y(s)] = \mathcal{Z}[G(s)X(s)] = \mathcal{Z}[GX(s)] = GX(z) \neq G(z)X(z)$$

The fact that the z transform of $G(s)X(s)$ is not equal to $G(z)X(z)$ will be discussed in detail later in this section.

In the present discussion for obtaining the pulse transfer function, we assume that there is a sampler at the input of the system. The presence or absence of a sampler at the output of the system does not affect the pulse transfer function, because if the sampler is not physically present at the output side of the system, it is always possible to assume that a fictitious sampler is present at the output. This means that although the output signal is continuous, we can consider the values of the output only at $t = kT$ $(k = 0, 1, 2, \ldots)$ and thus get sequence $y(kT)$.

In what follows we discuss general procedures for obtaining the pulse transfer function of the system shown in Fig. 3–31(a). [The procedure for obtaining an equivalent pulse transfer function for the system shown in Fig. 3–31(b) will be discussed in Chap. 4 in connection with the discretization of analog controllers.]

For the system shown in Fig. 3–31(a), where the input to the system $G(s)$ is an impulse-sampled signal, the pulse transfer function is simply

$$G(z) = \mathcal{Z}[G(s)]$$

Therefore, the procedure for obtaining the pulse transfer function boils down to the procedure for obtaining the z transform of a transfer function $G(s)$.

In this book so far we have presented four methods for obtaining the z transform of $x(t)$ or $X(s)$. These methods provide the basis for obtaining the pulse transfer function; they are listed as follows:

1. $X(z) = \mathcal{Z}[X(s) \text{ expanded into partial fractions } X_i(s)] = \sum_i \{\mathcal{Z}[X_i(s)]\}$

2. $X(z) = \sum_{k=0}^{\infty} x(kT)z^{-k}$

3. $X(z) = \sum \left[\text{residue of } \dfrac{X(s)z}{z - e^{Ts}} \text{ at pole of } X(s)\right]$

4. $X(z) = \dfrac{1}{T}\sum_{k=-\infty}^{\infty} X\left(s + j\dfrac{2\pi}{T}k\right)\Bigg|_{s=\frac{1}{T}\ln z} + \dfrac{1}{2}x(0+)$

Note that the last expression, the infinite series form, may not easily be converted into a closed form and consequently is not quite useful for the purpose of obtaining z transforms. [This equation, however, is useful in examining the properties of $X^*(s)$, as we have seen in Sec. 3–4.]

The pulse transfer function of a system may be obtained by use of any one of the above methods, except perhaps the last one, which involves the infinite series form.

Now the general procedure for obtaining the pulse transfer function may be stated as follows:

1. If a table of z transforms is readily available (such as Table 2–1) that gives both $X(z)$ and the corresponding $X(s)$ for a given $x(t)$, the process of obtaining the pulse transfer function is simple. The procedure begins with obtaining the transfer function $G(s)$ of the system. Then expand $G(s)$ into partial fractions such that the z transform of each term can be found from the z transform table. The sum of the z transforms of individual terms gives the pulse transfer function of the system. If $G(s)$ involves the term $(1 - e^{-Ts})$, then refer to Eq. (3–40).
2. The second method is to use the impulse response function. First, obtain the transfer function $G(s)$ of the system. Then obtain the impulse response function $g(t)$, where $g(t) = \mathcal{L}^{-1}[G(s)]$. Finally, evaluate

$$G(z) = \sum_{k=0}^{\infty} g(kT)z^{-k}$$

where $g(kT)$ is obtained from $g(t)$ by substituting kT for t.
3. The third method is to use the residue theorem in connection with the convolution integral in the left half plane. [Refer to Eq. (3–31).]

Examples 3–7 and 3–8 demonstrate the methods for obtaining the pulse transfer function.

Example 3–7.

Obtain the pulse transfer function $G(z)$ of the system shown in Fig. 3–31(a), where $G(s)$ is given by

$$G(s) = \frac{1}{s+a}$$

Note that there is a sampler at the input of $G(s)$ and therefore the pulse transfer function is $G(z) = \mathcal{Z}[G(s)]$.

Method 1. By referring to Table 2–1, we have

$$\mathcal{Z}\left[\frac{1}{s+a}\right] = \frac{1}{1 - e^{-aT}z^{-1}}$$

Hence

$$G(z) = \frac{1}{1 - e^{-aT}z^{-1}}$$

Method 2. The impulse response function for the system is obtained as follows:

$$g(t) = \mathscr{L}^{-1}[G(s)] = e^{-at}$$

Hence

$$g(kT) = e^{-akT} \qquad k = 0, 1, 2, \ldots$$

Therefore,

$$G(z) = \sum_{k=0}^{\infty} g(kT)z^{-k} = \sum_{k=0}^{\infty} e^{-akT}z^{-k} = \sum_{k=0}^{\infty} (e^{aT}z)^{-k}$$

$$= \frac{1}{1 - e^{-aT}z^{-1}}$$

Method 3. $G(z)$ has a simple pole at $s = -a$; thus, referring to Eq. (3–31), we obtain

$$G(z) = \lim_{s \to -a} \left[(s + a) \frac{1}{s + a} \frac{z}{z - e^{Ts}} \right] = \frac{z}{z - e^{-aT}} = \frac{1}{1 - e^{-aT}z^{-1}}$$

Example 3–8.

Obtain the pulse transfer function of the system shown in Fig. 3–31(a), where $G(s)$ is given by

$$G(s) = \frac{1 - e^{-Ts}}{s} \frac{1}{s(s + 1)}$$

Method 1. $G(s)$ involves the term $(1 - e^{-Ts})$; therefore, referring to Eq. (3–40), we obtain the pulse transfer function as follows:

$$G(z) = \mathscr{Z}[G(s)] = \mathscr{Z}\left[(1 - e^{-Ts}) \frac{1}{s^2(s + 1)} \right]$$

$$= (1 - z^{-1}) \mathscr{Z}\left[\frac{1}{s^2(s + 1)} \right]$$

$$= (1 - z^{-1}) \mathscr{Z}\left[\frac{1}{s^2} - \frac{1}{s} + \frac{1}{s + 1} \right]$$

From Table 2–1, the z transform of each of the partial-fraction-expansion terms can be found. Thus,

$$G(z) = (1 - z^{-1}) \left[\frac{Tz^{-1}}{(1 - z^{-1})^2} - \frac{1}{1 - z^{-1}} + \frac{1}{1 - e^{-T}z^{-1}} \right]$$

$$= \frac{(T - 1 + e^{-T})z^{-1} + (1 - e^{-T} - Te^{-T})z^{-2}}{(1 - z^{-1})(1 - e^{-T}z^{-1})} \tag{3–58}$$

Method 2. The given transfer function $G(s)$ can be written as follows:

$$G(s) = (1 - e^{-Ts}) \left(\frac{1}{s^2} - \frac{1}{s} + \frac{1}{s + 1} \right)$$

Therefore, by taking the inverse Laplace transform, we have the following impulse response function:

$$g(t) = (t - 1 + e^{-t})1(t) - [t - T - 1 + e^{-(t-T)}]1(t - T)$$

or

$$g(kT) = (kT - 1 + e^{-kT}) - [kT - T - 1 + e^{-(kT-T)}]1((k-1)T)$$

$$= \begin{cases} e^{-kT} + T - e^{-(kT-T)} & k = 1, 2, 3, \ldots \\ 0 & k = 0 \end{cases}$$

Then the pulse transfer function $G(z)$ can be obtained as follows:

$$G(z) = \sum_{k=0}^{\infty} g(kT)z^{-k}$$

$$= \sum_{k=0}^{\infty} [e^{-kT} + T - e^{-(kT-T)}]z^{-k} + e^{T} - 1 - T$$

$$= (1 - e^{T}) \sum_{k=0}^{\infty} e^{-kT}z^{-k} + T \sum_{k=0}^{\infty} z^{-k} + e^{T} - 1 - T$$

$$= (1 - e^{T}) \frac{1}{1 - e^{-T}z^{-1}} + \frac{T}{1 - z^{-1}} + e^{T} - 1 - T$$

$$= \frac{(T - 1 + e^{-T})z^{-1} + (1 - e^{-T} - Te^{-T})z^{-2}}{(1 - z^{-1})(1 - e^{-T}z^{-1})}$$

Method 3. Referring to Eq. (3–40), we have

$$G(z) = (1 - z^{-1})\mathcal{Z}\left[\frac{1}{s^2(s+1)}\right]$$

The z transform of $1/[s^2(s+1)]$ by use of the convolution integral in the left half plane was obtained in Example 3–3 as follows:

$$\mathcal{Z}\left[\frac{1}{s^2(s+1)}\right] = \frac{(T - 1 + e^{-T})z^{-1} + (1 - e^{-T} - Te^{-T})z^{-2}}{(1 - z^{-1})^2(1 - e^{-T}z^{-1})}$$

Hence

$$G(z) = (1 - z^{-1}) \frac{(T - 1 + e^{-T})z^{-1} + (1 - e^{-T} - Te^{-T})z^{-2}}{(1 - z^{-1})^2(1 - e^{-T}z^{-1})}$$

$$= \frac{(T - 1 + e^{-T})z^{-1} + (1 - e^{-T} - Te^{-T})z^{-2}}{(1 - z^{-1})(1 - e^{-T}z^{-1})}$$

Pulse transfer function of cascaded elements. Consider the systems shown in Fig. 3–32(a) and (b). Here we assume that the samplers are synchronized and have the same sampling period. We shall show that the pulse transfer function of the system shown in Fig. 3–32(a) is $G(z)H(z)$, while that shown in Fig. 3–32(b) is $\mathcal{Z}[G(s)H(s)] = \mathcal{Z}[GH(s)] = GH(z)$, which is different from $G(z)H(z)$.

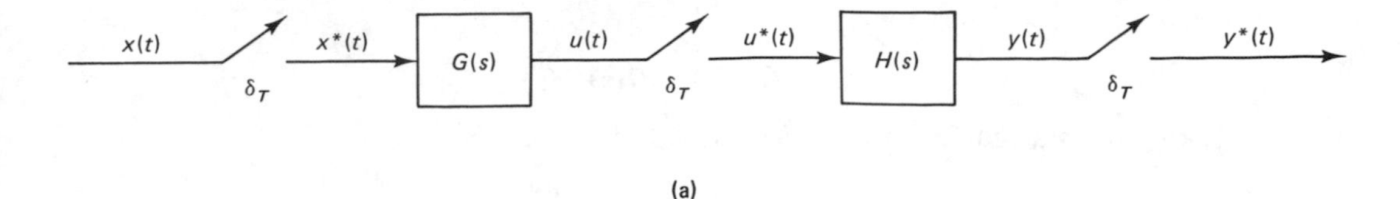

(a)

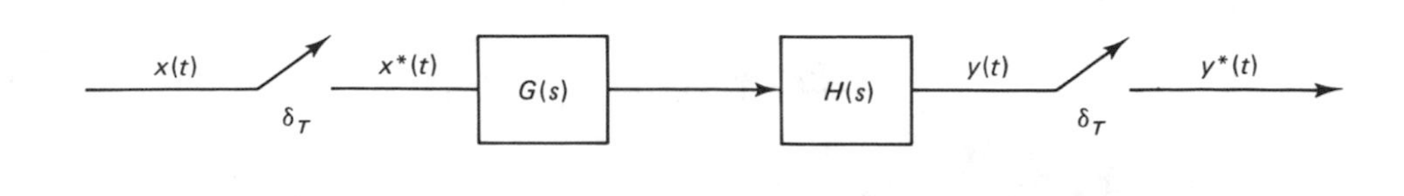

(b)

Figure 3–32 (a) Sampled system with a sampler between cascaded elements $G(s)$ and $H(s)$; (b) sampled system with no sampler between cascaded elements $G(s)$ and $H(s)$.

Consider the system shown in Fig. 3–32(a). From the diagram we obtain

$$U(s) = G(s)X^*(s)$$
$$Y(s) = H(s)U^*(s)$$

Hence, by taking the starred Laplace transform of each of these two equations, we get

$$U^*(s) = G^*(s)X^*(s)$$
$$Y^*(s) = H^*(s)U^*(s)$$

Consequently,

$$Y^*(s) = H^*(s)U^*(s) = H^*(s)G^*(s)X^*(s)$$

or

$$Y^*(s) = G^*(s)H^*(s)X^*(s)$$

In terms of the z transform notation,

$$Y(z) = G(z)H(z)X(z)$$

The pulse transfer function between the output $y^*(t)$ and input $x^*(t)$ is therefore given by

$$\frac{Y(z)}{X(z)} = G(z)H(z)$$

Next, consider the system shown in Fig. 3–32(b). From the diagram, we find

$$Y(s) = G(s)H(s)X^*(s) = GH(s)X^*(s)$$

where

$$GH(s) = G(s)H(s)$$

Taking the starred Laplace transform of $Y(s)$, we have

$$Y^*(s) = [GH(s)]^*X^*(s)$$

In terms of the z transform notation,

$$Y(z) = GH(z)X(z)$$

and the pulse transfer function between the output $y^*(t)$ and input $x^*(t)$ is

$$\frac{Y(z)}{X(z)} = GH(z) = \mathscr{Z}[GH(s)]$$

Note that

$$G(z)H(z) \neq GH(z) = \mathscr{Z}[GH(s)]$$

Hence, the pulse transfer functions of the systems shown in Fig. 3–32(a) and (b) are different. We will now verify this statement in Example 3–9.

Example 3–9.

Consider the systems shown in Fig. 3–33(a) and (b). Obtain the pulse transfer function $Y(z)/X(z)$ for each of these two systems.

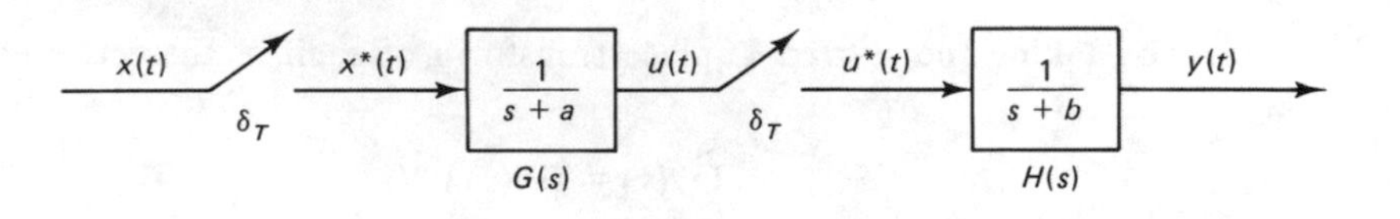

(a)

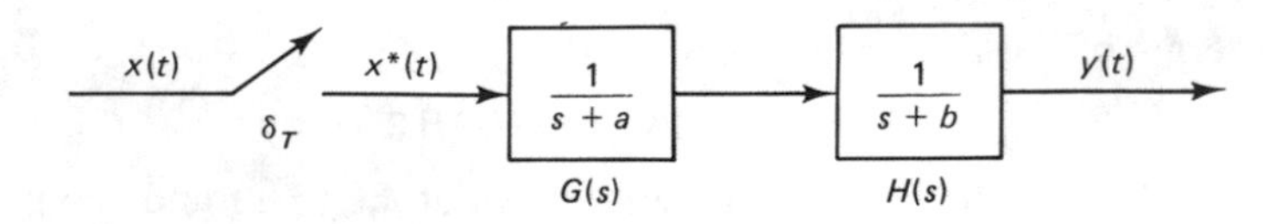

(b)

Figure 3–33 (a) Sampled system with a sampler between elements $G(s) = 1/(s + a)$ and $H(s) = 1/(s + b)$; (b) sampled system with no sampler between elements $G(s)$ and $H(s)$.

For the system of Fig. 3–33(a) the two transfer functions $G(s)$ and $H(s)$ are separated by a sampler. We assume that the two samplers shown are synchronized and have the same sampling period. The pulse transfer function for this system is

$$\frac{Y(z)}{X(z)} = \frac{Y(z)}{U(z)}\frac{U(z)}{X(z)} = H(z)G(z) = G(z)H(z)$$

Hence,

$$\frac{Y(z)}{X(z)} = G(z)H(z) = \mathscr{Z}\left[\frac{1}{s+a}\right]\mathscr{Z}\left[\frac{1}{s+b}\right] = \frac{1}{1-e^{-aT}z^{-1}}\frac{1}{1-e^{-bT}z^{-1}}$$

For the system shown in Fig. 3–33(b), the pulse transfer function $Y(z)/X(z)$ is obtained as follows:

$$\frac{Y(z)}{X(z)} = \mathscr{Z}[G(s)H(s)] = \mathscr{Z}\left[\frac{1}{s+a}\frac{1}{s+b}\right]$$

$$= \mathscr{Z}\left[\frac{1}{b-a}\left(\frac{1}{s+a} - \frac{1}{s+b}\right)\right]$$

$$= \frac{1}{b-a}\left(\frac{1}{1-e^{-aT}z^{-1}} - \frac{1}{1-e^{-bT}z^{-1}}\right)$$

Hence,

$$\frac{Y(z)}{X(z)} = GH(z) = \frac{1}{b-a}\left[\frac{(e^{-aT}-e^{-bT})z^{-1}}{(1-e^{-aT}z^{-1})(1-e^{-bT}z^{-1})}\right]$$

Clearly, we see that the pulse transfer functions of the two systems are different, that is, that

$$G(z)H(z) \neq GH(z)$$

Therefore, we must be careful to observe whether or not there is a sampler between cascaded elements.

Pulse transfer function of closed-loop systems. In a closed-loop system the existence or nonexistence of an output sampler within the loop makes a difference in the behavior of the system. (If there is an output sampler outside the loop, it will make no difference in the closed-loop operation.)

Consider the closed-loop control system shown in Fig. 3–34. In this system, the actuating error is sampled. From the block diagram,

$$E(s) = R(s) - H(s)C(s)$$

$$C(s) = G(s)E^*(s)$$

Hence,

$$E(s) = R(s) - H(s)G(s)E^*(s)$$

Then by taking the starred Laplace transform, we obtain

$$E^*(s) = R^*(s) - GH^*(s)E^*(s)$$

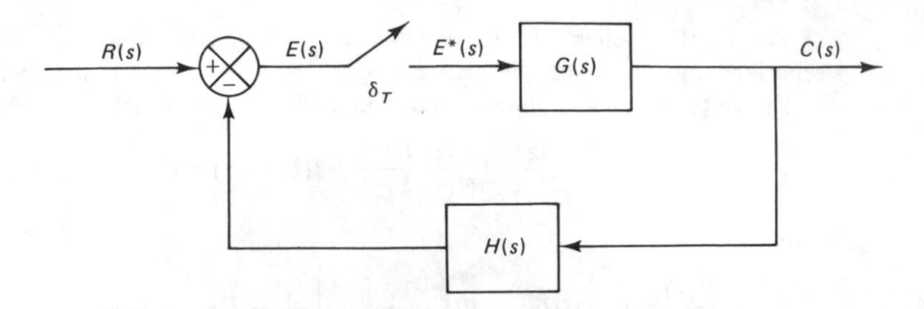

Figure 3–34 Closed-loop control system.

or

$$E^*(s) = \frac{R^*(s)}{1 + GH^*(s)}$$

Since

$$C^*(s) = G^*(s)E^*(s)$$

we obtain

$$C^*(s) = \frac{G^*(s)R^*(s)}{1 + GH^*(s)}$$

In terms of the z transform notation, the output can be given by

$$C(z) = \frac{G(z)R(z)}{1 + GH(z)}$$

The inverse z transform of this last equation gives the values of the output at the sampling instants. [Note that the actual output $c(t)$ of the system is a continuous-time signal. The inverse z transform of $C(z)$ will not give the continuous-time output $c(t)$.] The pulse transfer function for the present closed-loop system is

$$\frac{C(z)}{R(z)} = \frac{G(z)}{1 + GH(z)} \tag{3–59}$$

Table 3–1 shows five typical configurations for closed-loop discrete-time control systems. Here, the samplers are synchronized and have the same sampling period. For each configuration, the corresponding output $C(z)$ is shown. Notice that some discrete-time closed-loop control systems cannot be represented by $C(z)/R(z)$—that is, they do not have pulse transfer functions—because the input signal $R(s)$ cannot be separated from the system dynamics. Although the pulse transfer function may not exist for certain system configurations, the same techniques discussed in this chapter can still be applied for analyzing them.

TABLE 3–1 FIVE TYPICAL CONFIGURATIONS FOR CLOSED-LOOP DISCRETE-TIME CONTROL SYSTEMS

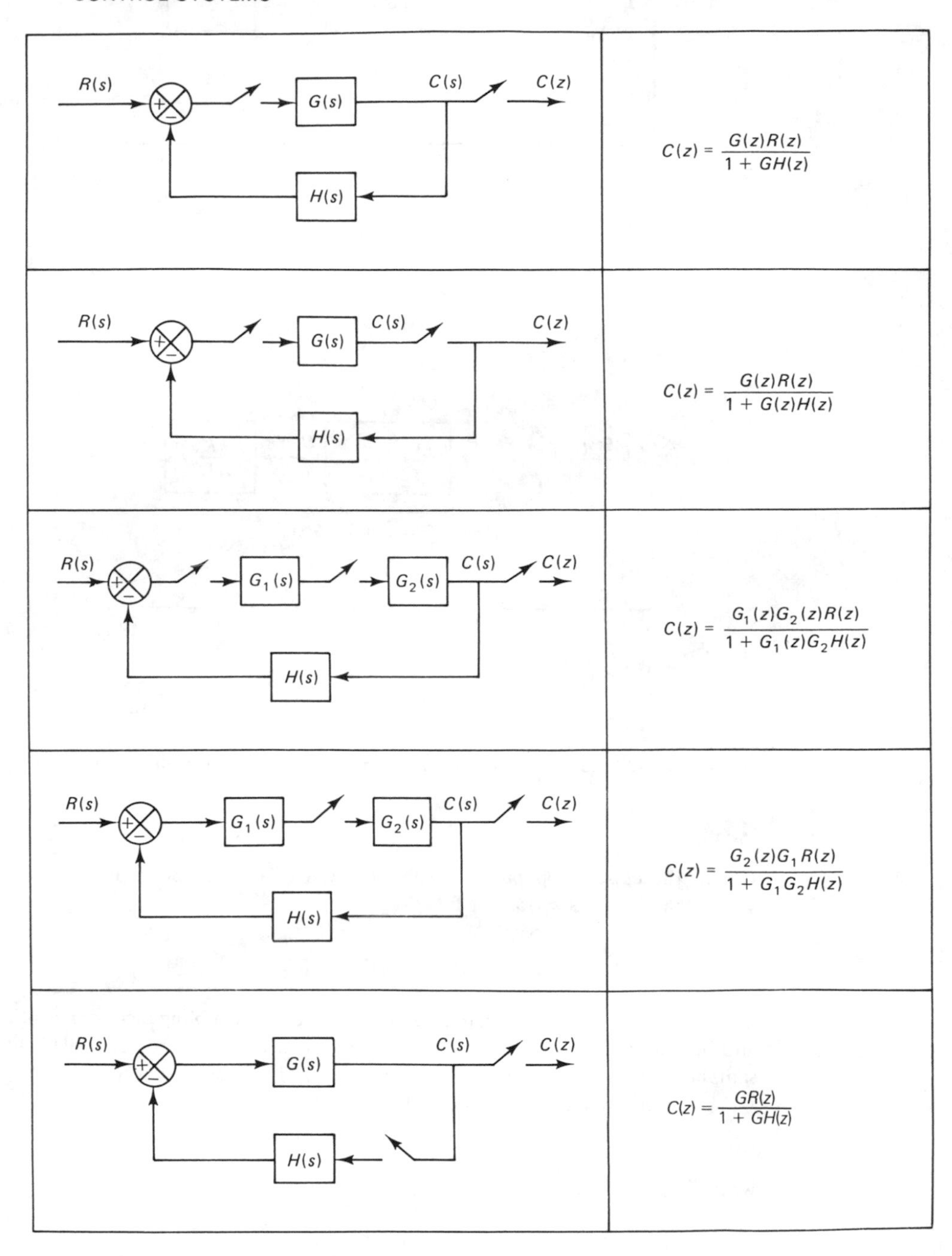

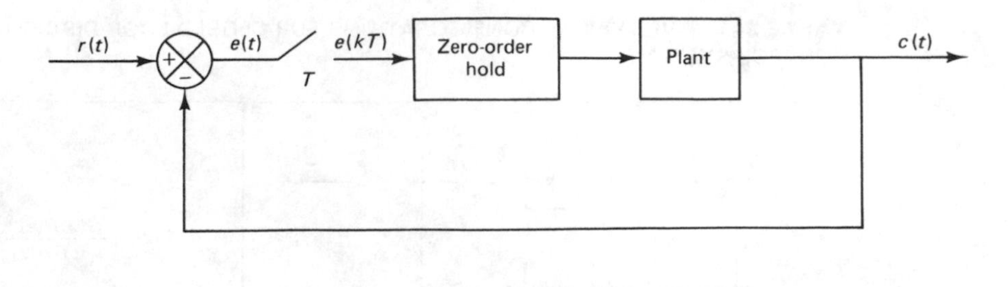

(a)

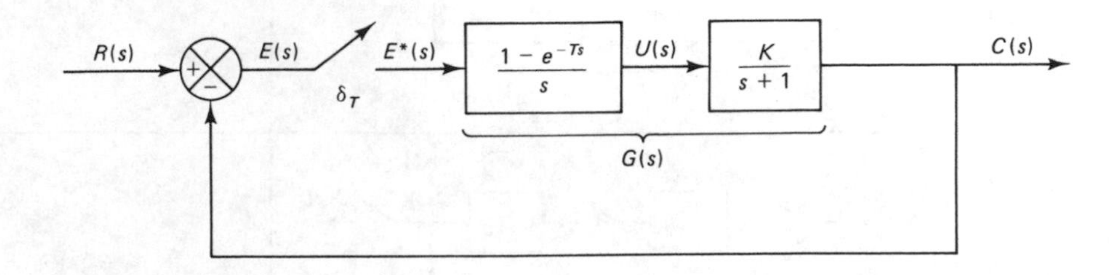

(b)

Figure 3–35 (a) Closed-loop discrete-time control system; (b) equivalent system.

Example 3–10.

Obtain the closed-loop pulse transfer function of the system shown in Fig. 3–35(a), where the plant has a transfer function

$$G_p(s) = \frac{K}{s+1}$$

As stated earlier, mathematically the process of sampling a continuous-time signal and holding it by means of a zero-order hold can be considered equivalent to impulse sampling followed by a transfer function $(1 - e^{-Ts})/s$, as shown in Fig. 3–35(b).

For the system of Fig. 3–35(b), we have

$$C(s) = G(s)E^*(s) \tag{3–60}$$

where

$$G(s) = \frac{1 - e^{-Ts}}{s} \frac{K}{s+1}$$

By taking the starred Laplace transforms of both sides of Eq. (3–60), we have

$$C^*(s) = G^*(s)E^*(s)$$

Noting that

$$E^*(s) = R^*(s) - C^*(s)$$

we have

$$\frac{C^*(s)}{R^*(s)} = \frac{G^*(s)}{1 + G^*(s)}$$

In terms of the z transform notation,

$$\frac{C(z)}{R(z)} = \frac{G(z)}{1 + G(z)}$$

Since

$$\begin{aligned} G(z) &= \mathcal{Z}[G(s)] = \mathcal{Z}\left[(1 - e^{-Ts})\frac{K}{s(s+1)}\right] \\ &= (1 - z^{-1})\,\mathcal{Z}\left[\frac{K}{s} - \frac{K}{s+1}\right] \\ &= (1 - z^{-1})\left(\frac{K}{1 - z^{-1}} - \frac{K}{1 - e^{-T}z^{-1}}\right) \\ &= \frac{K(1 - e^{-T})z^{-1}}{1 - e^{-T}z^{-1}} \end{aligned}$$

we obtain the closed-loop pulse transfer function as follows:

$$\frac{C(z)}{R(z)} = \frac{K(1 - e^{-T})z^{-1}}{1 + [K - (K+1)e^{-T}]z^{-1}}$$

Pulse transfer function of a digital controller. The pulse transfer function of a digital controller may be obtained from the required input-output characteristics of the digital controller.

Suppose the input to the digital controller is $e(k)$ and the output is $m(k)$. In general, the output $m(k)$ may be given by the following type of difference equation:

$$\begin{aligned} m(k) + a_1 m(k-1) + a_2 m(k-2) + \cdots + a_n m(k-n) \\ = b_0 e(k) + b_1 e(k-1) + \cdots + b_n e(k-n) \end{aligned} \qquad (3\text{–}61)$$

The z transform of Eq. (3–61) gives

$$\begin{aligned} M(z) + a_1 z^{-1}M(z) + a_2 z^{-2}M(z) + \cdots + a_n z^{-n}M(z) \\ = b_0 E(z) + b_1 z^{-1}E(z) + \cdots + b_n z^{-n}E(z) \end{aligned}$$

or

$$(1 + a_1 z^{-1} + a_2 z^{-2} + \cdots + a_n z^{-n}) M(z)$$
$$= (b_0 + b_1 z^{-1} + \cdots + b_n z^{-n}) E(z)$$

The pulse transfer function $G_D(z)$ of the digital controller may then be given by

$$G_D(z) = \frac{M(z)}{E(z)} = \frac{b_0 + b_1 z^{-1} + \cdots + b_n z^{-n}}{1 + a_1 z^{-1} + \cdots + a_n z^{-n}} \tag{3–62}$$

The use of the pulse transfer function $G_D(z)$ in the form of Eq. (3–62) enables us to analyze digital control systems in the z domain.

Closed-loop pulse transfer function of a digital control system. Figure 3–36(a) shows a block diagram of a digital control system. Here, the sampler, A/D converter, digital controller, zero-order hold, and D/A converter produce a continuous-time (piecewise-constant) control signal $u(t)$ to be fed to the plant. Figure 3–36(b) shows the transfer functions of blocks involved in the system.

The transfer function of the digital controller is shown as $G_D^*(s)$. In the actual system the computer (digital controller) solves a difference equation whose input-output relationship is given by the pulse transfer function $G_D(z)$.

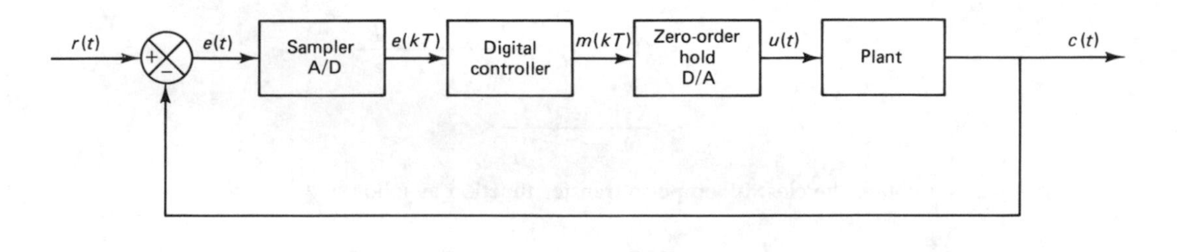

(a)

(b)

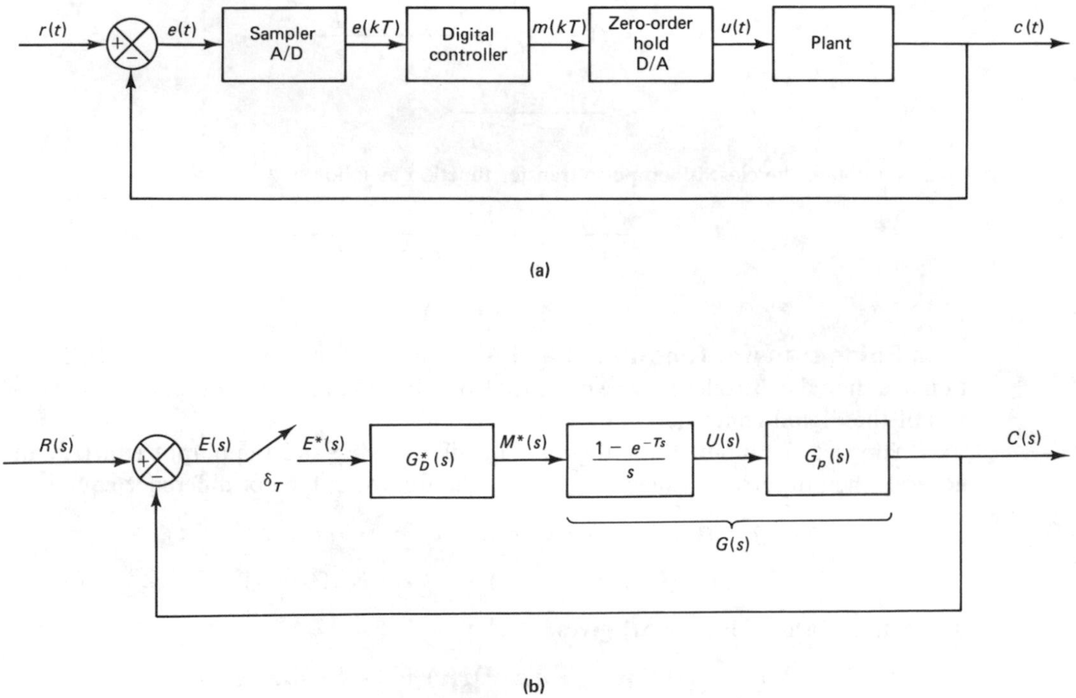

Figure 3–36 (a) Block diagram of a digital control system; (b) equivalent block diagram showing transfer functions of blocks.

In the present system the output signal $c(t)$ is fed back for comparison with the input signal $r(t)$. The error signal $e(t) = r(t) - c(t)$ is sampled and the analog signal is converted to a digital signal through the A/D device. The digital signal $e(kT)$ is fed to the digital controller, which operates on the sampled sequence $e(kT)$ in some desirable manner to produce the signal $m(kT)$.

This desirable relationship between the sequences $m(kT)$ and $e(kT)$ is specified by the pulse transfer function $G_D(z)$ of the digital controller. [By properly selecting the poles and zeros of $G_D(z)$, a number of input-output characteristics can be generated.]

Referring to Fig. 3–36(b), let us define

$$\frac{1 - e^{-Ts}}{s} G_p(s) = G(s)$$

From Fig. 3–36(b), notice that

$$C(s) = G(s)G_D^*(s)E^*(s)$$

or

$$C^*(s) = G^*(s)G_D^*(s)E^*(s)$$

In terms of the z transform notation,

$$C(z) = G(z)G_D(z)E(z)$$

Since

$$E(z) = R(z) - C(z)$$

we have

$$C(z) = G_D(z)G(z)[R(z) - C(z)]$$

and therefore,

$$\frac{C(z)}{R(z)} = \frac{G_D(z)G(z)}{1 + G_D(z)G(z)} \tag{3–63}$$

Equation (3–63) gives the closed-loop pulse transfer function of the digital control system shown in Fig. 3–36(b). The performance of such a closed-loop system can be improved by the proper choice of $G_D(z)$, the pulse transfer function of the digital controller. We shall later discuss a variety of forms for $G_D(z)$ to be used in obtaining "optimal" performance for various given performance indexes.

In the following, we shall consider only a simple case, where the pulse transfer function $G_D(z)$ is of the PID (proportional plus integral plus derivative) type.

Pulse transfer function of a digital PID controller. The analog PID control scheme has been used successfully in many industrial control systems for over half a century. The basic principle of the PID control scheme is to act upon the variable to be manipulated through a proper combination of three control actions:

proportional control action (where the control action is proportional to the actuating error signal, which is the difference between the input and the feedback signal), integral control action (where the control action is proportional to the integral of the actuating error signal), and derivative control action (where the control action is proportional to the derivative of the actuating error signal).

Where many plants are controlled directly by a single digital computer (as in a control scheme where from several loops to several hundred loops are controlled by a single digital controller), the majority of the control loops may be handled by PID control schemes.

The PID control action in analog controllers is given by

$$m(t)=K\left[e(t)+\frac{1}{T_i}\int_0^t e(t)\,dt+T_d\frac{de(t)}{dt}\right] \tag{3–64}$$

where $e(t)$ is the input to the controller (the actuating error signal), $m(t)$ is the output of the controller (the manipulating signal), K is the proportional gain, T_i is the integral time (or reset time), and T_d is the derivative time (or rate time).

In order to obtain the pulse transfer function for the digital PID controller, we may discretize Eq. (3–64). By approximating the integral term by the trapezoidal summation and the derivative term by a two-point difference form, we obtain

$$m(kT)=K\left\{e(kT)+\frac{T}{T_i}\left[\frac{e(0)+e(T)}{2}+\frac{e(T)+e(2T)}{2}+\cdots\right.\right.$$
$$\left.\left.+\frac{e((k-1)T)+e(kT)}{2}\right]+T_d\frac{e(kT)-e((k-1)T)}{T}\right\}$$

or

$$m(kT)=K\left\{e(kT)+\frac{T}{T_i}\sum_{h=1}^{k}\frac{e((h-1)T)+e(hT)}{2}\right.$$
$$\left.+\frac{T_d}{T}[e(kT)-e((k-1)T)]\right\} \tag{3–65}$$

Define

$$\frac{e((h-1)T)+e(hT)}{2}=f(hT), \qquad f(0)=0$$

Figure 3–37 shows the function $f(hT)$. Then

$$\sum_{h=1}^{k}\frac{e((h-1)T)+e(hT)}{2}=\sum_{h=1}^{k}f(hT)$$

Taking the z transform of this last equation, we obtain

$$\mathscr{Z}\left[\sum_{h=1}^{k}\frac{e((h-1)T)+e(hT)}{2}\right]=\mathscr{Z}\left[\sum_{h=1}^{k}f(hT)\right]=\frac{1}{1-z^{-1}}[F(z)-f(0)]$$
$$=\frac{1}{1-z^{-1}}F(z)$$

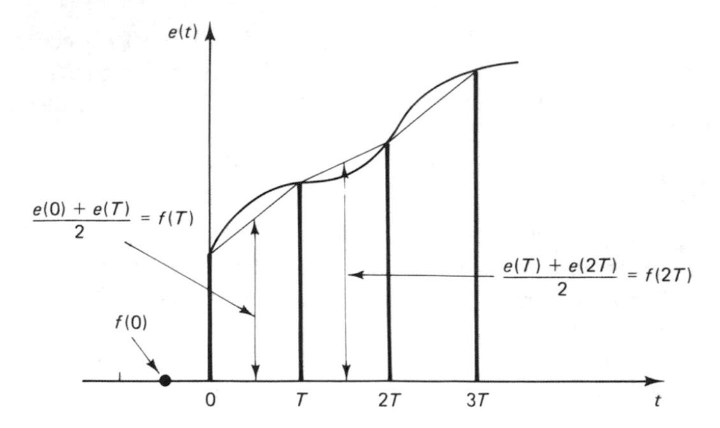

Figure 3–37 Diagram depicting function $f(hT)$.

(For the derivation of this last equation, refer to Prob. A-2–6.) Notice that

$$F(z) = \mathscr{Z}[f(hT)] = \frac{1 + z^{-1}}{2} E(z)$$

Hence

$$\mathscr{Z}\left[\sum_{h=1}^{k} \frac{e((h-1)T) + e(hT)}{2}\right] = \frac{1 + z^{-1}}{2(1 - z^{-1})} E(z)$$

Then the z transform of Eq. (3–65) gives

$$M(z) = K\left[1 + \frac{T}{2T_i}\frac{1 + z^{-1}}{1 - z^{-1}} + \frac{T_d}{T}(1 - z^{-1})\right]E(z)$$

This last equation may be rewritten as follows:

$$M(z) = K\left[1 - \frac{T}{2T_i} + \frac{T}{T_i}\frac{1}{1 - z^{-1}} + \frac{T_d}{T}(1 - z^{-1})\right]E(z)$$

$$= \left[K_P + \frac{K_I}{1 - z^{-1}} + K_D(1 - z^{-1})\right]E(z)$$

where

$$K_P = K - \frac{KT}{2T_i} = K - \frac{K_I}{2} = \text{proportional gain}$$

$$K_I = \frac{KT}{T_i} = \text{integral gain}$$

$$K_D = \frac{KT_d}{T} = \text{derivative gain}$$

Notice that the proportional gain K_P for the digital PID controller is smaller than the proportional gain K for the analog PID controller by $K_I/2$.

The pulse transfer function for the digital PID controller becomes

$$G_D(z) = \frac{M(z)}{E(z)} = K_P + \frac{K_I}{1 - z^{-1}} + K_D(1 - z^{-1}) \tag{3–66}$$

The pulse transfer function of the digital PID controller given by Eq. (3–66) is commonly referred to as the *positional form* of the PID control scheme.

The other form commonly used in the digital PID control scheme is referred to as the *velocity form*. In the velocity-form PID control scheme, we deal with $\nabla m(kT)$, the variation in the actuating signal. To derive the velocity-form PID control equation, consider the backward difference in $m(kT)$, that is, the difference between $m(kT)$ and $m((k-1)T)$:

$$\begin{aligned}\nabla m(kT) &= m(kT) - m((k-1)T)\\ &= K\left\{e(kT) - e((k-1)T) + \frac{T}{2T_i}[e(kT) + e((k-1)T)]\right.\\ &\quad \left. + \frac{T_d}{T}[e(kT) - 2e((k-1)T) + e((k-2)T)]\right\}\\ &= K_P[e(kT) - e((k-1)T)] + K_I e(kT)\\ &\quad + K_D[e(kT) - 2e((k-1)T) + e((k-2)T)]\end{aligned} \tag{3–67}$$

where we have used the relationships $K_P = K - \frac{1}{2}K_I$, $K_I = KT/T_i$, and $K_D = KT_d/T$. Note that Eq. (3–67) takes into consideration the variation of the positional form in one sampling period.

Suppose the actuating error $e(kT)$ is the difference between the input $r(kT)$ and the output $c(kT)$, or

$$e(kT) = r(kT) - c(kT)$$

By substituting this last equation into Eq. (3–67), we obtain

$$\begin{aligned}\nabla m(kT) &= K_P[r(kT) - r((k-1)T) - c(kT) + c((k-1)T)]\\ &\quad + K_I[r(kT) - c(kT)] + K_D[r(kT) - 2r((k-1)T)\\ &\quad + r((k-2)T) - c(kT) + 2c((k-1)T) - c((k-2)T)]\end{aligned} \tag{3–68}$$

The velocity-form PID control scheme given by Eq. (3–68) may be modified into a somewhat different form to cope with sudden large changes in the set point. Since the proportional and derivative control actions produce a large change in the controller output when the signal entering the controller makes a sudden large change, to suppress such a large change in the controller output, the digital proportional and derivative terms may be modified as discussed below.

If changes in the set point [input $r(kT)$] are a series of step changes, then immediately after a step change takes place the input $r(kT)$ stays constant for a while until the next step change takes place. Hence, in Eq. (3–68) we assume that

$$r(kT) = r((k-1)T) = r((k-2)T)$$

(Note that this is true if the input stays constant. But we assume that this holds true even if a step change takes place.) Then, Eq. (3–68) may be modified to

$$\begin{aligned}\nabla m(kT) = &-K_P[c(kT) - c((k-1)T)] + K_I\,[r(kT) - c(kT)] \\ &-K_D\,[c(kT) - 2c((k-1)T) + c((k-2)T)]\end{aligned} \tag{3–69}$$

The z transform of Eq. (3–69) gives

$$\begin{aligned}(1 - z^{-1})M(z) = &-K_P\,(1 - z^{-1})C(z) + K_I\,[R(z) - C(z)] \\ &-K_D\,(1 - 2z^{-1} + z^{-2})C(z)\end{aligned}$$

Simplifying, we obtain

$$M(z) = -K_P C(z) + K_I\,\frac{R(z) - C(z)}{1 - z^{-1}} - K_D\,(1 - z^{-1})C(z) \tag{3–70}$$

Equation (3–70) gives the velocity-form PID control scheme. Figure 3–38 shows the block diagram realization of the velocity-form digital PID control scheme. Notice that in Eq. (3–70) only the integral control term involves the input $R(z)$. Hence, the integral term cannot be excluded from the digital controller if the velocity form is used.

An advantage of the velocity-form PID control scheme is that initialization is not necessary when the operation is switched from manual to automatic. Thus, if there are sudden large changes in the set point or at the start of the process operation, the velocity-form PID control scheme exhibits better response characteristics than the positional-form PID control scheme. Another advantage of the velocity-form PID control scheme is that it is useful in suppressing excessive corrections in process control systems. That is, when the output is limited, integration stops automatically

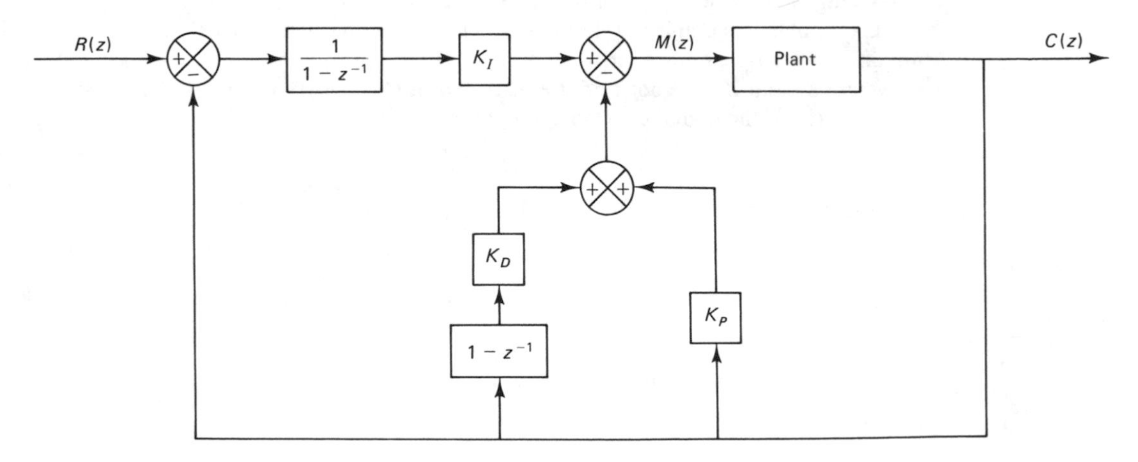

Figure 3–38 Block diagram realization of the velocity-form digital PID control scheme.

and thus integrator saturation is avoided. (Integrator saturation occurs if the output saturates and the controller, without "knowing" about the saturation, continues to integrate the error. The output of the integrator can become quite large. Once this happens, it takes some time to return to normal operation.)

Linear control laws in the form of PID control actions, in both positional form and velocity form, are basic in digital controls because they frequently give satisfactory solutions to many practical control problems—in particular, process control problems. Note that in digital controllers, control laws can be implemented by software, and therefore the hardware restrictions of analog PID controllers can be completely ignored. (For a comparison of the frequency-response characteristics of analog and digital PID controllers, see Prob. A-3-25.)

Example 3–11.

Consider the control system with a digital PID controller shown in Fig. 3–39(a). (The PID controller here is in the positional form.) The transfer function of the plant is assumed to be

$$G_p(s) = \frac{1}{s(s+1)}$$

and the sampling period T is assumed to be 1 sec. Then the transfer function of the zero-order hold becomes

$$G_h(s) = \frac{1 - e^{-s}}{s}$$

Since

$$\mathscr{Z}\left[\frac{1 - e^{-s}}{s}\frac{1}{s(s+1)}\right] = G(z) = \frac{0.3679z^{-1} + 0.2642z^{-2}}{(1 - 0.3679z^{-1})(1 - z^{-1})} \tag{3–71}$$

we may redraw the block diagram of Fig. 3–39(a) as shown in Fig. 3–39(b).

Let us obtain the unit step response of this system when the digital controller is acting in a proportional mode, or when K_I and K_D are set equal to zero. (More general cases where K_I and K_D are not zero are discussed in Chap. 4.) Then, the pulse transfer function $G_D(z)$ of the digital controller simplifies to

$$G_D(z) = K_P$$

Assume further that $K_P = 1$. Then

$$G_D(z) = 1$$

and the closed-loop pulse transfer function becomes as follows:

$$\frac{C(z)}{R(z)} = \frac{G_D(z)G(z)}{1 + G_D(z)G(z)}$$

$$= \frac{0.3679z^{-1} + 0.2642z^{-2}}{1 - z^{-1} + 0.6321z^{-2}}$$

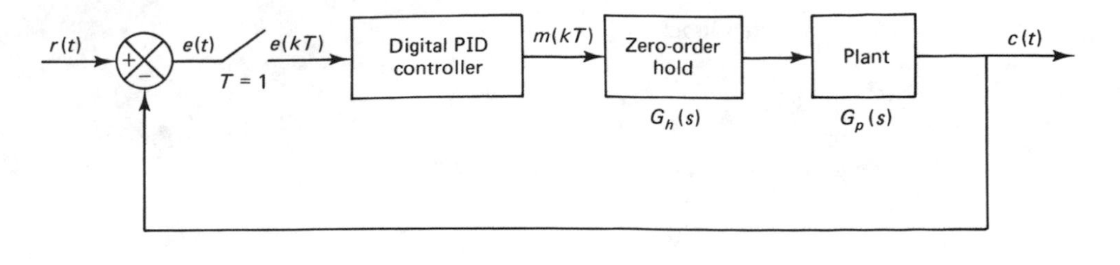

(a)

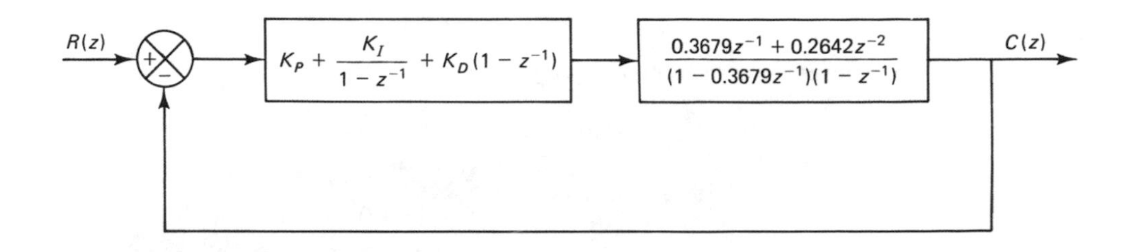

(b)

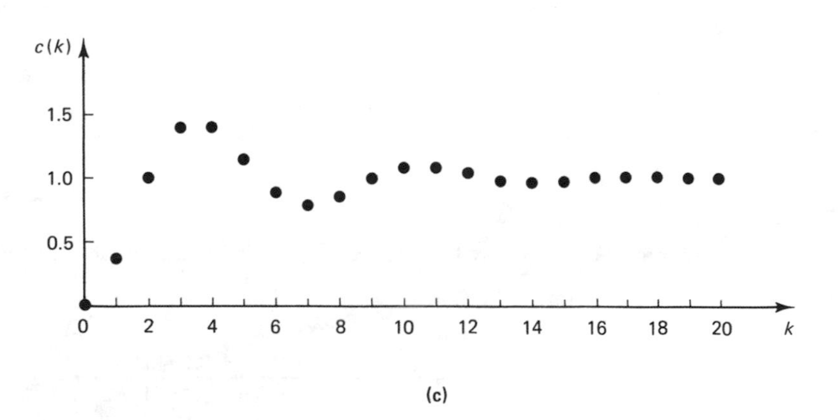

(c)

Figure 3–39 (a) Block diagram of a control sytem; (b) equivalent block diagram; (c) plot of the output $c(k)$ vs. k when the input is a unit step and when $K_P = 1$, $K_I = 0$, and $K_D = 0$.

For a unit step input,

$$R(z) = \frac{1}{1 - z^{-1}}$$

The unit step response $C(z)$ is then obtained as follows:

$$\begin{aligned}
C(z) &= \frac{0.3679z^{-1} + 0.2642z^{-2}}{(1 - z^{-1} + 0.6321z^{-2})(1 - z^{-1})} \\
&= \frac{0.3679z^{-1} + 0.2642z^{-2}}{1 - 2z^{-1} + 1.6321z^{-2} - 0.6321z^{-3}} \\
&= 0.3679z^{-1} + z^{-2} + 1.3996z^{-3} + 1.3996z^{-4} \\
&\quad + 1.1469z^{-5} + 0.8944z^{-6} + 0.8015z^{-7} + \cdots
\end{aligned}$$

The inverse z transform of $C(z)$ gives

$$\begin{aligned}
c(0) &= 0 \\
c(1) &= 0.3679 \\
c(2) &= 1.0000 \\
c(3) &= 1.3996 \\
c(4) &= 1.3996 \\
c(5) &= 1.1469 \\
c(6) &= 0.8944 \\
c(7) &= 0.8015 \\
&\vdots
\end{aligned}$$

The output $c(k)$ eventually becomes equal to unity, as can be seen as follows:

$$\begin{aligned}
c(\infty) &= \lim_{z \to 1} (1 - z^{-1})C(z) \\
&= \lim_{z \to 1} (1 - z^{-1}) \frac{0.3679z^{-1} + 0.2642z^{-2}}{(1 - z^{-1} + 0.6321z^{-2})(1 - z^{-1})} = 1
\end{aligned}$$

The output $c(k)$ is plotted in Fig. 3–39(c).

Comments. PID controllers for such process control systems as temperature systems, pressure systems, and liquid-level systems are usually tuned experimentally. In fact, in the PID control of any industrial plant, where its dynamics are not well known or defined, the controller variables (K_P, K_I, and K_D) must be determined experimentally. Such determination or tuning may be made using step changes in the reference or disturbance signal. A few established procedures are available for such a purpose. Basically, tuning (determining the constants K_P, K_I, and K_D) is accomplished by systematically varying the values until reasonably good response characteristics are obtained.

For digital PID controllers used for process control systems the sampling period must be chosen properly. Many process control systems have fairly large time constants. A rule of thumb in the selection of the sampling period (sampling period $T = 2\pi/\omega_s$, where ω_s is the sampling frequency) in process control systems is that for temperature control systems the sampling period should be 10 to 30 sec, for pressure control systems 1 to 5 sec, and for liquid-level control systems 1 to 10 sec.

3-6 OBTAINING RESPONSE BETWEEN CONSECUTIVE SAMPLING INSTANTS

The z transform analysis will not give information on the response between two consecutive sampling instants. In ordinary cases this is not serious, because if the sampling theorem is satisfied, then the output will not vary very much between any two consecutive sampling instants. In certain cases, however, we may need to find the response between consecutive sampling instants.

In this book we present three methods for providing a response between two consecutive sampling instants:

1. The Laplace transform method
2. The modified z transform method
3. The state space method

In this section we discuss the first two methods. The state space method will be discussed in Sec. 5–6.

Laplace transform method. Consider, for example, the system shown in Fig. 3–34. The output $C(s)$ can be given by

$$C(s) = G(s)E^*(s) = G(s)\frac{R^*(s)}{1 + GH^*(s)}$$

Thus,

$$c(t) = \mathscr{L}^{-1}[C(s)] = \mathscr{L}^{-1}\left[G(s)\frac{R^*(s)}{1 + GH^*(s)}\right] \tag{3–72}$$

Equation (3–72) will give the continuous-time response $c(t)$. Hence, the response at any time between two consecutive sampling instants can be calculated by the use of Eq. (3–72). [See Prob. A-3-26 for sample calculations of the right-hand side of Eq. (3–72).]

Modified *z* transform method. The modified z transform method is a modification of the z transform method. It is based on inserting a fictitious delay time at the output of the system, in addition to the insertion of the fictitious output sampler, and varying the amount of the fictitious delay time so that the output at any time between two consecutive sampling instants can be obtained.

The modified z transform method is useful not only in obtaining the response between two consecutive sampling instants, but also in obtaining the z transform of the process with pure delay or transportation lag. In addition, the modified z transform method is applicable to most sampling schemes.

Consider the system shown in Fig. 3–40(a). In this system a fictitious delay of $(1 - m)T$ seconds, where $0 \leq m \leq 1$ and T is the sampling period, is inserted at the output of the system. By varying m between 0 and 1, the output $y(t)$ at $t = kT - (1 - m)T$ (where $k = 1, 2, 3, \ldots$) may be obtained. Noting that $G^*(s)$ is given by

$$G^*(s) = \mathscr{L}[g(t)\delta_T(t)]$$

we define the modified pulse transfer function $G(z, m)$ by

$$\mathscr{Z}_m[G(s)] = G(z, m) = G^*(s, m)\Big|_{s=\frac{1}{T}\ln z}$$

$$= \mathscr{L}[g(t - (1 - m)T)\delta_T(t)]\Big|_{s=\frac{1}{T}\ln z} \tag{3–73}$$

(a)

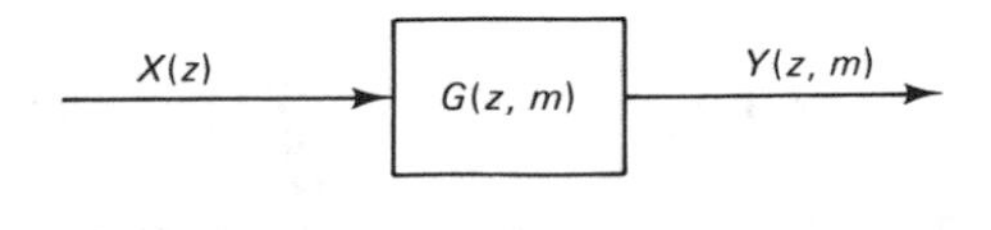

(b)

Figure 3–40 (a) System with a fictitious delay time of $(1 - m)T$ sec.; (b) modified pulse-transfer-function system with input $X(z)$ and output $Y(z, m)$.

where the notation $\mathscr{Z}_m$ signifies the modified z transform.

Noting that

$$\mathscr{L}[g(t-(1-m)T)\delta_T(t)] = \mathscr{L}[g(t-T+mT)\delta_T(t)]$$
$$= e^{-Ts}\,\mathscr{L}[g(t+mT)\delta_T(t)]$$

we have

$$G^*(s, m) = e^{-Ts}\,\mathscr{L}[g(t+mT)\delta_T(t)] \tag{3–74}$$

Since $\mathscr{L}[g(t+mT)\delta_T(t)]$ is the Laplace transform of the product of two time functions, it can be obtained as follows by referring to Eq. (3–24):

$$\mathscr{L}[g(t+mT)\delta_T(t)] = \frac{1}{2\pi j}\int_{c-j\infty}^{c+j\infty} G(p)\frac{e^{mTp}}{1-e^{-T(s-p)}}dp \tag{3–75}$$

The integration on the right-hand side of Eq. (3–75) can be carried out in a way similar to that discussed in Sec. 3–3; that is, the convolution integral can be integrated in either the left half plane or the right half plane.

Let us consider the contour integration along the infinite semicircle in the left half plane. Then,

$$\mathscr{L}[g(t+mT)\delta_T(t)] = \sum\left[\text{residue of}\,\frac{G(s)e^{mTs}z}{z-e^{Ts}}\,\text{at pole of}\ G(s)\right] \tag{3–76}$$

Hence, from Eqs. (3–73), (3–74), and (3–76) we obtain the modified z transform of $G(s)$ as follows:

$$G(z, m) = z^{-1}\sum\left[\text{residue of}\,\frac{G(s)e^{mTs}z}{z-e^{Ts}}\ \text{at pole of}\ G(s)\right] \tag{3–77}$$

Note that the modified z transform $G(z, m)$ and the z transform $G(z)$ are related as follows:

$$G(z) = \lim_{m\to 0} zG(z, m) \tag{3–78}$$

Referring to Fig. 3–40(b), the output $Y(z, m)$ is obtained as follows:

$$Y(z, m) = G(z, m)X(z) \tag{3–79}$$

As in the case of the z transform, the modified z transform $Y(z, m)$ can be expanded into an infinite series in z^{-1}, as follows:

$$Y(z, m) = y_0(m)z^{-1} + y_1(m)z^{-2} + y_2(m)z^{-3} + \cdots \tag{3–80}$$

By multiplying both sides of Eq. (3–80) by z, we have

$$zY(z, m) = y_0(m) + y_1(m)z^{-1} + y_2(m)z^{-2} + \cdots \tag{3–81}$$

where $y_k(m)$ represents the value of $y(t)$ between $t = kT$ and $t = (k+1)T$ ($k = 0, 1, 2, \ldots$), or

$$y_k(m) = y((k+m)T) \tag{3–82}$$

Note that if $y(k)$ is continuous, then

$$\lim_{m\to 1} y_{k-1}(m) = \lim_{m\to 0} y_k(m) \tag{3–83}$$

The left-hand side of Eq. (3–83) gives the values $y(0-)$, $y(T-)$, $y(2T-)$, . . . , and the right-hand side gives the values $y(0+)$, $y(T+)$, $y(2T+)$, . . . If the output $y(kT)$ is continuous, then $y(kT-) = y(kT+)$.

Example 3–12.

Obtain the modified z transform of $G(s)$ where

$$G(s) = \frac{1}{s+a}$$

Referring to Eq. (3–77), we obtain the modified z transform of $G(s)$ as follows:

$$\begin{aligned} G(z, m) &= z^{-1}\left[\text{residue of } \frac{1}{s+a}\frac{e^{mTs}z}{z-e^{Ts}} \text{ at pole } s=-a\right] \\ &= z^{-1}\left\{\lim_{s\to -a}\left[(s+a)\frac{1}{s+a}\frac{e^{mTs}z}{z-e^{Ts}}\right]\right\} \\ &= z^{-1}\frac{e^{-maT}z}{z-e^{-aT}} = \frac{e^{-maT}z^{-1}}{1-e^{-aT}z^{-1}} \end{aligned}$$

Example 3–13.

Consider the systems shown in Fig. 3–41(a) and (b). Obtain the output $Y(z, m)$ of each system.

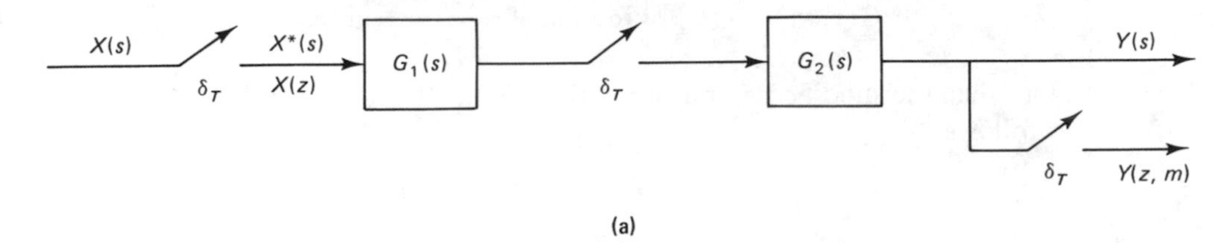

(a)

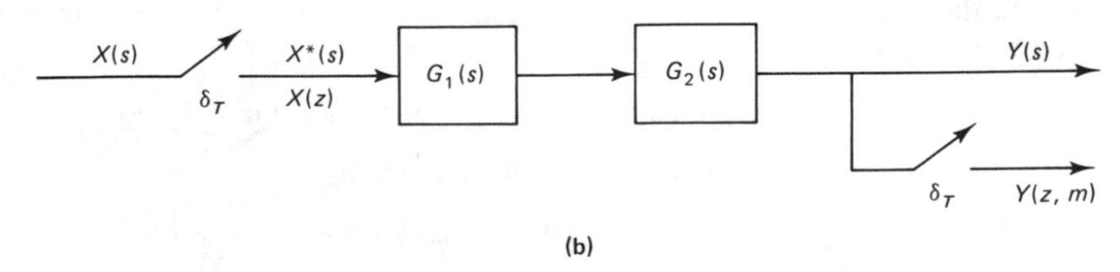

(b)

Figure 3–41 (a) System with a sampler between $G_1(s)$ and $G_2(s)$; (b) system with no sampler between $G_1(s)$ and $G_2(s)$.

For the system shown in Fig. 3–41(a), we have

$$Y(z, m) = \mathcal{Z}_m [Y(s)] = G_2(z, m)G_1(z)X(z)$$

Note that

$$Y(z) = \mathcal{Z}[Y(s)] = G_2(z)G_1(z)X(z)$$

For the system shown in Fig. 3–41(b), we have

$$Y(z, m) = \mathcal{Z}_m [Y(s)] = G_1G_2(z, m)X(z)$$

where

$$G_1G_2(z, m) = \mathcal{Z}_m [G_1(s)G_2(s)]$$

Note that

$$Y(z) = G_1G_2(z)X(z)$$

Example 3–14.

Consider the system shown in Fig. 3–34. Obtain the modified z transform of $C(s)$.

Referring to Eq. (3–59), the output $C(z)$ is given by

$$C(z) = \frac{G(z)}{1 + GH(z)} R(z)$$

The modified z transform of $C(z)$ is given by

$$C(z, m) = \frac{G(z, m)}{1 + GH(z)} R(z) \tag{3–84}$$

Example 3–15.

Consider the system shown in Fig. 3–42. The sampling period T is 1 sec, or $T = 1$. Suppose the system is subjected to a unit step input. Obtain $c_k(m)$ for $m = 0.5$ and $k = 0, 1, 2, \ldots, 9$.

First, referring to Example 3–8, we have

$$G(z) = \mathcal{Z}[G(s)] = \mathcal{Z}\left[\frac{1 - e^{-s}}{s} \frac{1}{s(s+1)}\right]$$

$$= \frac{(T - 1 + e^{-T})z^{-1} + (1 - e^{-T} - Te^{-T})z^{-2}}{(1 - z^{-1})(1 - e^{-T}z^{-1})}$$

$$= \frac{0.3679z^{-1} + 0.2642z^{-2}}{(1 - z^{-1})(1 - 0.3679z^{-1})}$$

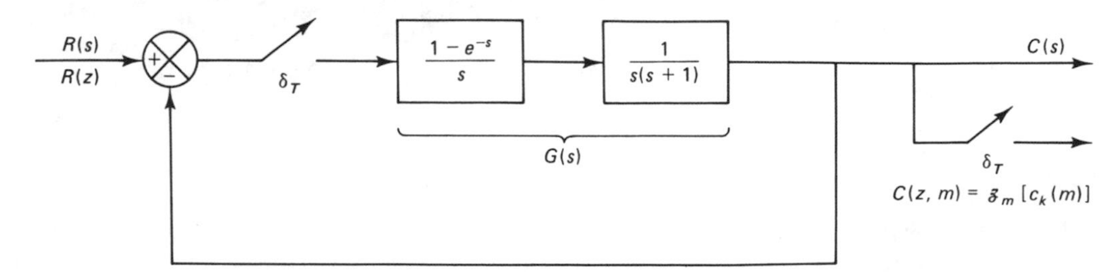

Figure 3–42 Closed-loop discrete-time control system for Example 3–15.

The modified z transform of $G(s)$ is obtained from Eq. (3–77) as follows:

$$G(z, m) = z^{-1} \sum \left[\text{residue of } \frac{G(s)e^{ms}z}{z - e^s} \text{ at pole of } G(s) \right]$$

$$= z^{-1}(1 - z^{-1}) \left\{ \left[\text{residue of } \frac{1}{s^2(s+1)} \frac{e^{ms}z}{z - e^s} \text{ at double pole } s = 0 \right] + \left[\text{residue of } \frac{1}{s^2(s+1)} \frac{e^{ms}z}{z - e^s} \text{ at simple pole } s = -1 \right] \right\}$$

$$= z^{-1}(1 - z^{-1}) \left\{ \frac{1}{(2-1)!} \lim_{s \to 0} \frac{d}{ds} \left[s^2 \frac{1}{s^2(s+1)} \frac{e^{ms}z}{z - e^s} \right] + \lim_{s \to -1} \left[(s+1) \frac{1}{s^2(s+1)} \frac{e^{ms}z}{z - e^s} \right] \right\}$$

$$= z^{-1}(1 - z^{-1}) \left[\frac{mz^2 - mz - z^2 + 2z}{(z-1)^2} + \frac{e^{-m}z}{z - e^{-1}} \right]$$

$$= \frac{(m-1)z^{-1} + (2-m)z^{-2}}{1 - z^{-1}} + \frac{e^{-m}z^{-1}(1 - z^{-1})}{1 - e^{-1}z^{-1}}$$

$$= \frac{(m - 1 + e^{-m})z^{-1} + (2.3679 - 1.3679m - 2e^{-m})z^{-2} + [-0.3679(2-m) + e^{-m}]z^{-3}}{(1 - z^{-1})(1 - 0.3679z^{-1})}$$

Referring to Eq. (3–84) and noting that $R(z) = 1/(1 - z^{-1})$, we have

$$C(z, m) = \frac{G(z, m)}{1 + G(z)} \frac{1}{1 - z^{-1}}$$

$$= \frac{(m - 1 + e^{-m})z^{-1} + (2.3679 - 1.3679m - 2e^{-m})z^{-2} + (-0.7358 + 0.3679m + e^{-m})z^{-3}}{1 - 2z^{-1} + 1.6321z^{-2} - 0.6321z^{-3}}$$

Hence, for $m = 0.5$ we have

$$C(z, 0.5) = \frac{0.1065z^{-1} + 0.4709z^{-2} + 0.05468z^{-3}}{1 - 2z^{-1} + 1.6321z^{-2} - 0.6321z^{-3}} \tag{3–85}$$

By referring to Eq. (3–81), Eq. (3–85) can be expanded into an infinite series in z^{-1}, as follows:

$$C(z, 0.5) = c_0(0.5)z^{-1} + c_1(0.5)z^{-2} + c_2(0.5)z^{-3} + \cdots$$

or

$$zC(z, 0.5) = c_0(0.5) + c_1(0.5)z^{-1} + c_2(0.5)z^{-2} + \cdots$$

where $c_k(0.5) = c((k + 0.5)T) = c(k + 0.5)$ and $k = 0, 1, 2, \ldots$ The values of $c_k(0.5)$ can easily be obtained with a microcomputer. (Refer to Sec. 2–6.) The computer solution for $k = 0, 1, 2, \ldots, 9$ is as follows:

$$c_0(0.5) = c(0.5) = 0.1065$$
$$c_1(0.5) = c(1.5) = 0.6839$$
$$c_2(0.5) = c(2.5) = 1.2487$$
$$c_3(0.5) = c(3.5) = 1.4485$$
$$c_4(0.5) = c(4.5) = 1.2913$$
$$c_5(0.5) = c(5.5) = 1.0078$$
$$c_6(0.5) = c(6.5) = 0.8236$$
$$c_7(0.5) = c(7.5) = 0.8187$$
$$c_8(0.5) = c(8.5) = 0.9302$$
$$c_9(0.5) = c(9.5) = 1.0447$$

These values give the response at the midpoints between pairs of consecutive sampling points. Note that by varying the value of m between 0 and 1, it is possible to find the response at any point between two consecutive sampling points, such as $c(1.2)$ and $c(2.8)$.

Summary. The main purpose of this section has been to present two methods (the Laplace transform and modified z transform methods) for finding the response for any time between two consecutive sampling instants. The modified z transform method can be used not only for such a purpose, but also for dealing with multirate sampling schemes. The reader who is interested in a more detailed account of the modified z transform method should refer, for instance, to Ref. 3–9 listed at the end of the chapter.

3–7 REALIZATION OF DIGITAL CONTROLLERS AND DIGITAL FILTERS

In this section we discuss realization methods for pulse transfer functions that represent digital controllers and digital filters. Realization of digital controllers and digital filters may involve either software or hardware or both. In general, "realization" of a pulse transfer function means determining the physical layout for the appropriate combination of arithmetic and storage operations.

In a software realization we obtain computer programs for the digital computer involved. In a hardware realization, we build a special-purpose processor using such circuitry as digital adders, multipliers, and delay elements (shift registers with a sampling period T as a unit time delay) that can be used for our purpose.

In the field of digital signal processing, a *digital filter* is a computational algorithm which converts an input sequence of numbers into an output sequence in such a way that the characteristics of the signal are changed in some prescribed fashion. That is, a digital filter processes a digital signal by passing desirable frequency components of the digital input signal and rejecting undesirable ones. In general terms, a digital controller is a form of digital filter.

Note that for the purpose of mathematical analysis a sequence of numbers which appears at the output of a digital filter (or digital controller) can be considered to represent weights of an impulse train. The weighted impulse train is fed to a hold device to produce a continuous-time signal that is fed in turn to a plant.

A digital filter (whether in the form of a hardware device or of software, such as a digital computer program) may be used to replace a continuous-time or analog filter. Approximating continuous-time filters by discrete-time equivalents is an important subject in the field of digital signal processing. There are important differences between the digital signal processing used in communications and that used in control. In digital control, the processing of signals must be done in real time. In communications, signal processing need not be done in real time, and therefore delays can be tolerated in the processing to improve accuracy. (We shall discuss details of approximating continuous-time or analog filters by discrete-time equivalents in Chap. 4.)

The software realization of digital controllers and filters is more flexible than the hardware realization; it is achieved through numerical calculations carried out by the appropriate digital computer programs designed for real-time control computers. Software realization of digital filters has become very widespread in many industrial control systems; its increased popularity is due mainly to a rapid increase in the ability of digital computers to process digital signals, to rapid technical advances in the field of A/D and D/A converters, and to the low cost of having microprocessors and microcomputers in control systems.

This section deals with the block diagram realization of digital filters using delay elements, adders, and multipliers. Here several different structures of block diagram realizations will be discussed. We shall discuss in detail block diagram realizations which provide the layout of the signal processing schemes. Such block diagram realizations can be used as a basis for a software or a hardware design. In fact, once the block diagram realization is completed, the physical realization in hardware or software is straightforward. Note that in a block diagram realization, a pulse transfer function of z^{-1} represents a delay of one time unit (see Fig. 3–43). (Note also that in the s plane, z^{-1} corresponds to a pure delay e^{-Ts}.)

In what follows we shall deal with the digital filters which are used for filtering and control purposes. The general form of the pulse transfer function between the output $Y(z)$ and input $X(z)$ is given by

$$G(z) = \frac{Y(z)}{X(z)} = \frac{b_0 + b_1 z^{-1} + b_2 z^{-2} + \cdots + b_m z^{-m}}{1 + a_1 z^{-1} + a_2 z^{-2} + \cdots + a_n z^{-n}} \qquad n \geq m \qquad (3\text{–}86)$$

where the a_i's and b_i's are real coefficients (some of them may be zero). The pulse transfer function is in this form for many digital controllers. For example, the pulse

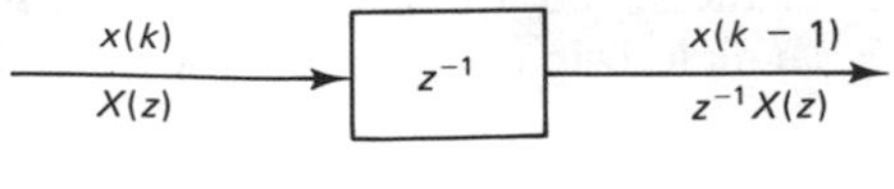

$x(k)$, $X(z)$ → z^{-1} → $x(k-1)$, $z^{-1}X(z)$

Figure 3–43 Pulse transfer function showing a delay of one time unit.

transfer function of the PID controller given by Eq. (3–66) can be expressed in the form of Eq. (3–86), as follows:

$$G_D(z) = \frac{(K_P + K_I + K_D) - (K_P + 2K_D)z^{-1} + K_D z^{-2}}{1 - z^{-1}}$$

$$= \frac{b_0 + b_1 z^{-1} + b_2 z^{-2}}{1 + a_1 z^{-1} + a_2 z^{-2}}$$

where

$$a_1 = -1$$

$$a_2 = 0$$

$$b_0 = K_P + K_I + K_D$$

$$b_1 = -(K_P + 2K_D)$$

$$b_2 = K_D$$

We shall now discuss the direct programming and the standard programming of digital filters. In these programmings, coefficients a_i and b_i (which are real quantities) appear as multipliers in the block diagram realization. Those block diagram schemes where the coefficients a_i and b_i appear directly as multipliers are called *direct structures*.

Direct programming. Consider the digital filter given by Eq. (3–86). Notice that the pulse transfer function has n poles and m zeros. Figure 3–44 shows a block

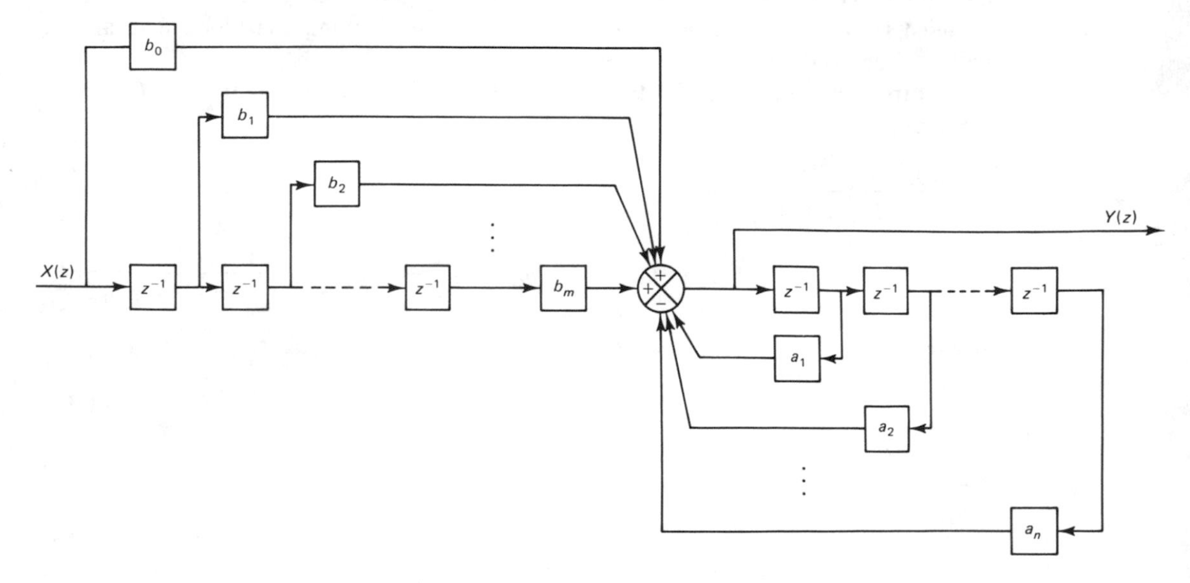

Figure 3–44 Block diagram realization of a filter showing direct programming.

diagram realization of the filter. The fact that this block diagram represents Eq. (3–86) can be seen easily, since from the diagram we have

$$Y(z) = -a_1 z^{-1} Y(z) - a_2 z^{-2} Y(z) - \cdots - a_n z^{-n} Y(z) + b_0 X(z) + b_1 z^{-1} X(z) + \cdots + b_m z^{-m} X(z)$$

Rearranging this last equation yields Eq. (3–86).

The type of realization here is called direct programming. Direct programming means that we realize the numerator and denominator of the pulse transfer function using separate sets of delay elements. The numerator uses a set of m delay elements and the denominator uses a different set of n delay elements. Thus, the total number of delay elements used in direct programming is $m + n$.

The number of delay elements used in direct programming can be reduced. In fact, the number of delay elements can be reduced from $n + m$ to n (if $n \geq m$) or to m (if $m \geq n$). The programming method that uses a minimum possible number of delay elements is called standard programming.

In practice, we try to use the minimum number of delay elements in realizing a given pulse transfer function. Therefore, the direct programming that requires more than the minimum number of delay elements is more or less of academic value rather than of practical value. It is noted that if either the denominator or numerator is constant, both the direct programming and the standard programming use the same number of delay elements and yield the same block diagram.

Standard programming As previously stated the number of delay elements required in direct programming can be reduced. In fact, the number of delay elements used in realizing the pulse transfer function given by Eq. (3–86) can be reduced from $n + m$ to n (where $n \geq m$) by rearranging the block diagram, as will be discussed here.

First, rewrite the pulse transfer function $Y(z)/X(z)$ given by Eq. (3–86) as follows:

$$\frac{Y(z)}{X(z)} = \frac{Y(z)}{H(z)} \frac{H(z)}{X(z)}$$

$$= (b_0 + b_1 z^{-1} + b_2 z^{-2} + \cdots + b_m z^{-m}) \frac{1}{1 + a_1 z^{-1} + a_2 z^{-2} + \cdots + a_n z^{-n}}$$

where

$$\frac{Y(z)}{H(z)} = b_0 + b_1 z^{-1} + b_2 z^{-2} + \cdots + b_m z^{-m} \tag{3–87}$$

and

$$\frac{H(z)}{X(z)} = \frac{1}{1 + a_1 z^{-1} + a_2 z^{-2} + \cdots + a_n z^{-n}} \tag{3–88}$$

Then, draw block diagrams for the systems given by Eqs. (3–87) and (3–88), respectively. To draw the block diagrams, we may rewrite Eq. (3–87) as

$$Y(z) = b_0 H(z) + b_1 z^{-1} H(z) + \cdots + b_m z^{-m} H(z) \tag{3–89}$$

and Eq. (3–88) as

$$H(z) = X(z) - a_1 z^{-1} H(z) - a_2 z^{-2} H(z) - \cdots - a_n z^{-n} H(z) \tag{3–90}$$

Then from Eq. (3–89) we obtain Fig. 3–45(a). Similarly, we get Fig. 3–45(b) from Eq. (3–90). The combination of these two block diagrams gives the block diagram for the digital filter $G(z)$, as shown in Fig. 3–45(c). The block diagram realization as presented here is based on the standard programming. Notice that we use only n delay elements. The coefficients $a_1, a_2, \ldots, a_n$ appear as feedback elements and the coefficients $b_0, b_1, \ldots, b_m$ appear as feedforward elements.

The block diagrams in Figs. 3–44 and 3–45(c) are equivalent, but the latter uses n delay elements, while the former uses $n + m$ delay elements. Obviously, the latter, which uses a smaller number of delay elements, is preferred.

Comments. Note first that the use of a minimal number of delay elements saves memory space in digital controllers. Also, the use of a minimal number of summing points is desirable.

In realizing digital controllers or digital filters, it is important to have a good level of accuracy. Basically, there are three sources of errors that affect the accuracy:

1. The error due to the quantization of the input signal into a finite number of discrete levels. (In Chap. 1 we discussed this type of error, which may be considered an additive source of noise, called quantization noise. The quantization noise may be considered white noise; the variance of the noise is $\sigma^2 = Q^2/12$.)
2. The error due to the accumulation of round-off errors in the arithmetic operations in the digital system.
3. The error due to quantization of the coefficients a_i and b_i of the pulse transfer function. This error may become large as the order of the pulse transfer function is increased. That is, in a higher-order digital filter in direct structure, small errors in the coefficients a_i and b_i cause large errors in the locations of the poles and zeros of the filter.

These three errors arise because of the practical limitations of the number of bits that represent various signal samples and coefficients. Note that the third type of error listed may be reduced by mathematically decomposing a higher-order pulse transfer function into a combination of lower-order pulse transfer functions. In this way, the system may be made less sensitive to coefficient inaccuracies.

For decomposing higher-order pulse transfer functions in order to avoid the coefficient sensitivity problem, the following three approaches are commonly used.

1. Series programming
2. Parallel programming
3. Ladder programming

We shall discuss these three programmings in what follows.

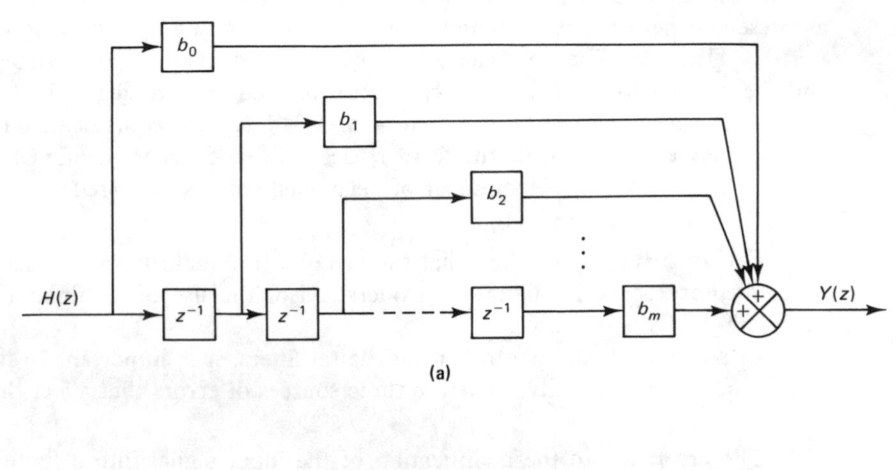

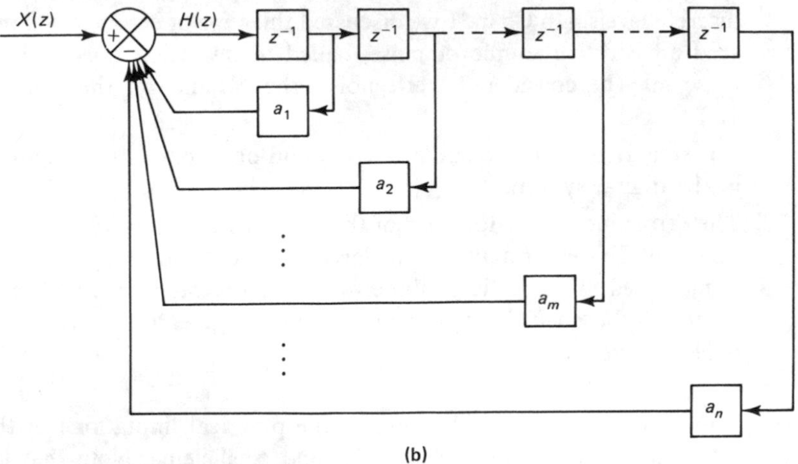

Figure 3–45 (a) Block diagram realization of Eq. (3–89); (b) block diagram realization of Eq. (3–90); (c) block diagram realization of the digital filter given by Eq. (3–86) by standard programming.

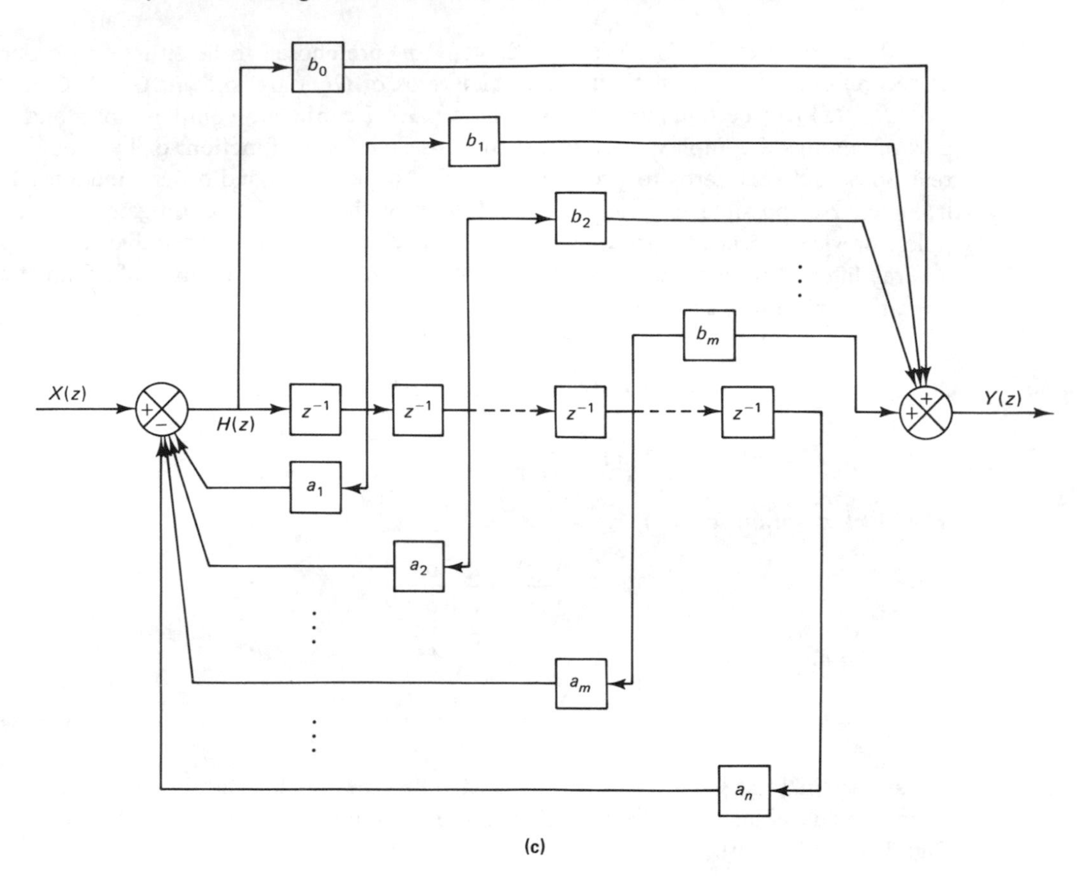

(c)

Series programming. The first approach used to avoid the sensitivity problem is to implement the pulse transfer function $G(z)$ as a series connection of first-order and/or second-order pulse transfer functions. If $G(z)$ can be written as a product of pulse transfer functions $G_1(z), G_2(z), \ldots, G_p(z)$, or

$$G(z) = G_1(z)G_2(z) \cdots G_p(z)$$

then the digital filter for $G(z)$ may be given as a series connection of the component digital filters $G_1(z), G_2(z), \ldots, G_p(z)$, as shown in Fig. 3–46.

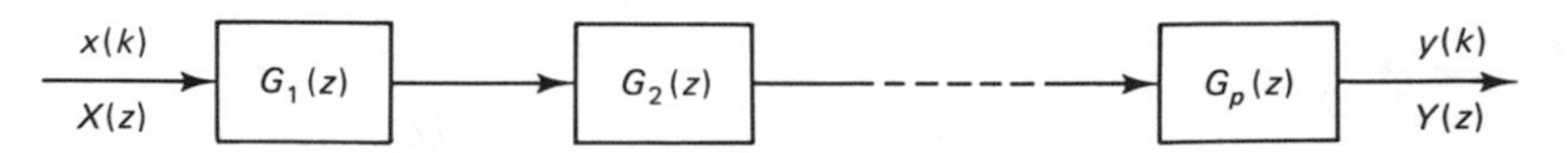

Figure 3–46 Digital filter $G(z)$ decomposed into a series connection of $G_1(z), G_2(z), \ldots, G_p(z)$.

In most cases the $G_i(z)$ ($i = 1, 2, \ldots, p$) are chosen to be either first-order or second-order functions. If the poles and zeros of $G(z)$ are known, $G_1(z), G_2(z), \ldots, G_p(z)$ can be obtained by grouping a pair of conjugate complex poles and a pair of conjugate complex zeros to produce a second-order function, or by grouping real poles and real zeros to produce either first-order or second-order functions. It is, of course, possible to group two real zeros with a pair of conjugate complex poles, or vice versa. The grouping is, in a sense, arbitrary. It is desirable to group several different ways to see which one is best with respect to the number of arithmetic operations required, the range of coefficients, and so forth.

To summarize, $G(z)$ may be decomposed as follows:

$$G(z) = G_1(z)G_2(z) \cdots G_p(z)$$
$$= \prod_{i=1}^{j} \frac{1 + b_i z^{-1}}{1 + a_i z^{-1}} \prod_{i=j+1}^{p} \frac{1 + e_i z^{-1} + f_i z^{-2}}{1 + c_i z^{-1} + d_i z^{-2}}$$

The block diagram for

$$\frac{Y(z)}{X(z)} = \frac{1 + b_i z^{-1}}{1 + a_i z^{-1}} \tag{3–91}$$

and that for

$$\frac{Y(z)}{X(z)} = \frac{1 + e_i z^{-1} + f_i z^{-2}}{1 + c_i z^{-1} + d_i z^{-2}} \tag{3–92}$$

are shown in Fig. 3–47(a) and (b), respectively. The block diagram for the digital filter $G(z)$ is a series connection of p component digital filters such as shown in Fig. 3–47(a) and (b).

Parallel programming. The second approach to avoiding the coefficient sensitivity problem is to expand the pulse transfer function $G(z)$ into partial fractions. If $G(z)$ is expanded as a sum of A, $G_1(z), G_2(z), \ldots, G_q(z)$, or so that

$$G(z) = A + G_1(z) + G_2(z) + \cdots + G_q(z)$$

where A is simply a constant, then the block diagram for the digital filter $G(z)$ can be obtained as a parallel connection of $q + 1$ digital filters, as shown in Fig. 3–48.

Because of the presence of the constant term A, the first-order and second-order functions can be chosen in simpler forms. That is, $G(z)$ may be expressed as follows:

$$G(z) = A + G_1(z) + G_2(z) + \cdots + G_q(z)$$
$$= A + \sum_{i=1}^{j} G_i(z) + \sum_{i=j+1}^{q} G_i(z)$$
$$= A + \sum_{i=1}^{j} \frac{b_i}{1 + a_i z^{-1}} + \sum_{i=j+1}^{q} \frac{e_i + f_i z^{-1}}{1 + c_i z^{-1} + d_i z^{-2}}$$

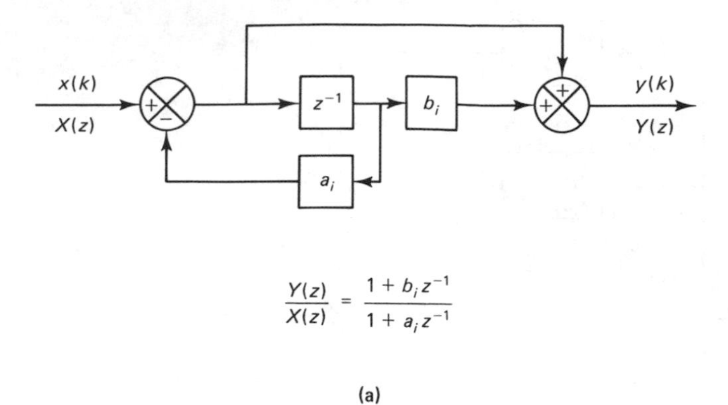

(a)

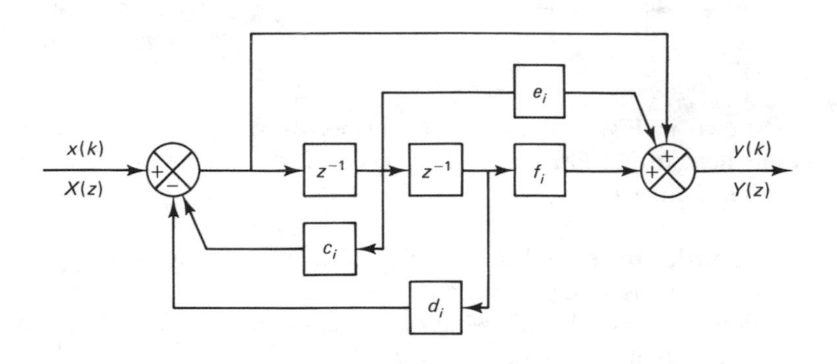

(b)

Figure 3–47 (a) Block diagram representation of Eq. (3–91); (b) block diagram representation of Eq. (3–92).

The block diagram for

$$\frac{Y(z)}{X(z)} = \frac{b_i}{1 + a_i z^{-1}} \tag{3–93}$$

and that for

$$\frac{Y(z)}{X(z)} = \frac{e_i + f_i z^{-1}}{1 + c_i z^{-1} + d_i z^{-2}} \tag{3–94}$$

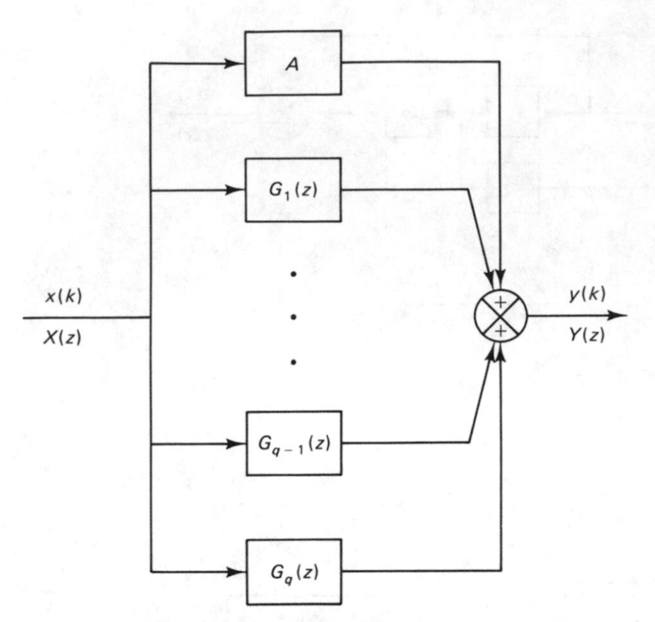

Figure 3–48 Digital filter $G(z)$ decomposed as a parallel connection of A, $G_1(z)$, $G_2(z)$, . . . , $G_q(z)$.

are shown in Fig. 3–49(a) and (b), respectively. The parallel connection of $q + 1$ component digital filters as shown in Fig. 3–48 will produce the block diagram for the digital filter $G(z)$.

Ladder programming. The third approach to avoiding the coefficient sensitivity problem is to implement a ladder structure, that is, to expand the pulse transfer function $G(z)$ into the following continued-fraction form and to program according to this equation:

$$G(z) = A_0 + \cfrac{1}{B_1 z + \cfrac{1}{A_1 + \cfrac{1}{B_2 z + \cfrac{1}{\ddots \; A_{n-1} + \cfrac{1}{B_n z + \cfrac{1}{A_n}}}}}} \tag{3–95}$$

The programming method based on this scheme is called *ladder programming*. Let us define

$$G_i^{(B)}(z) = \frac{1}{B_i z + G_i^{(A)}(z)} \qquad i = 1, 2, \ldots, n-1$$

$$G_i^{(A)}(z) = \frac{1}{A_i + G_{i+1}^{(B)}(z)} \qquad i = 1, 2, \ldots, n-1$$

$$G_n^{(B)}(z) = \frac{1}{B_n z + \dfrac{1}{A_n}}$$

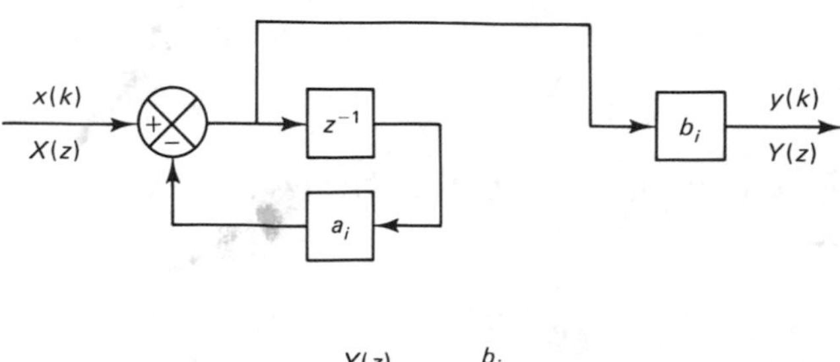

$$\frac{Y(z)}{X(z)} = \frac{b_i}{1 + a_i z^{-1}}$$

(a)

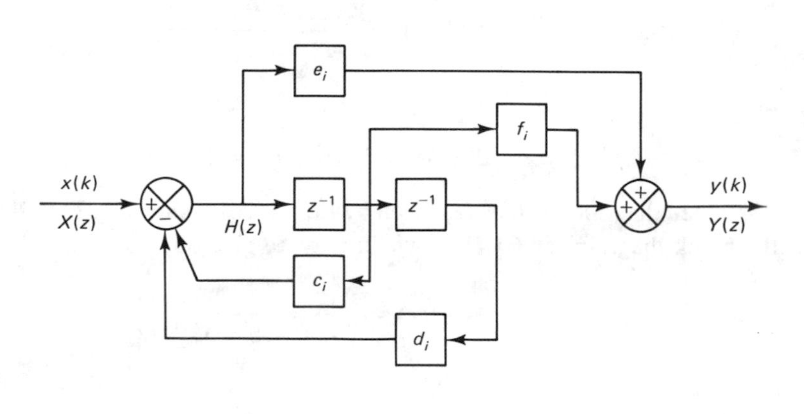

$$\frac{Y(z)}{X(z)} = \frac{e_i + f_i z^{-1}}{1 + c_i z^{-1} + d_i z^{-2}}$$

(b)

Figure 3–49 (a) Block diagram representation of Eq. (3–93); (b) block diagram representation of Eq. (3–94).

Then $G(z)$ may be written as follows:

$$G(z) = A_0 + G_1^{(B)}(z)$$

We shall explain this programming method by using a simple example where $n = 2$. That is,

$$G(z) = A_0 + \cfrac{1}{B_1 z + \cfrac{1}{A_1 + \cfrac{1}{B_2 z + \cfrac{1}{A_2}}}}$$

By the use of the functions $G_1^{(A)}(z)$, $G_1^{(B)}(z)$, and $G_2^{(B)}(z)$, the transfer function $G(z)$ may be written as follows:

$$\begin{aligned} G(z) &= A_0 + \cfrac{1}{B_1 z + \cfrac{1}{A_1 + G_2^{(B)}(z)}} \\ &= A_0 + \frac{1}{B_1 z + G_1^{(A)}(z)} \\ &= A_0 + G_1^{(B)}(z) \end{aligned}$$

Notice that $G_i^{(B)}(z)$ may be written as

$$G_i^{(B)}(z) = \frac{Y_i(z)}{X_i(z)} = \frac{1}{B_i z + G_i^{(A)}(z)} \tag{3–96}$$

or

$$X_i(z) - G_i^{(A)}(z) Y_i(z) = B_i z Y_i(z)$$

The block diagram for $G_i^{(B)}(z)$ given by Eq. (3–96) is shown in Fig. 3–50(a). Similarly, the block diagram for $G_i^{(A)}(z)$, which may be given by

$$G_i^{(A)}(z) = \frac{Y_i(z)}{X_i(z)} = \frac{1}{A_i + G_{i+1}^{(B)}(z)} \tag{3–97}$$

or

$$X_i(z) - G_{i+1}^{(B)}(z) Y_i(z) = A_i Y_i(z)$$

may be drawn as shown in Fig. 3–50(b). Note that

$$G_n^{(A)}(z) = \frac{1}{A_n}$$

By combining component digital filters as shown in Fig. 3–51(a), it is possible to draw the block diagram of the digital filter $G(z)$ as shown in Fig. 3–51(b). [Note that Fig. 3–51(a) and (b) corresponds to the case where $n = 2$.]

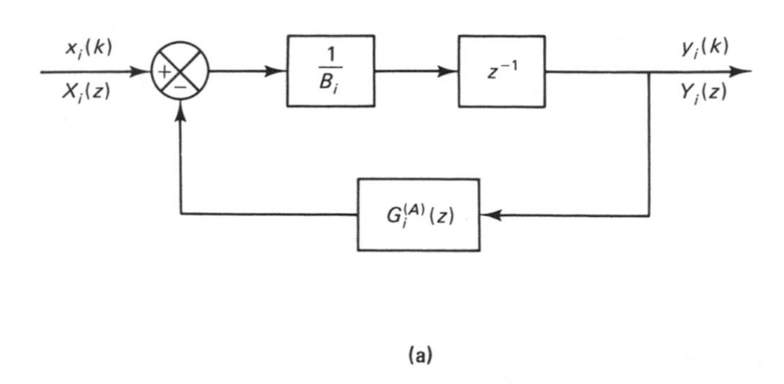

(a)

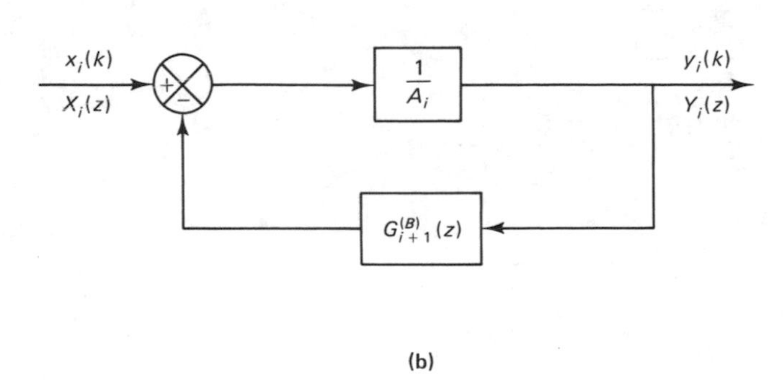

(b)

Figure 3–50 (a) Block diagram for $G_i^{(B)}(z)$ given by Eq. (3–96); (b) block diagram for $G_i^{(A)}(z)$ given by Eq. (3–97).

Comments. Digital filters based on ladder programming have advantages with respect to coefficient sensitivity and accuracy. Realization of the ladder structure is achieved by expanding $G(z)$ into continued fractions around the origin.

It is noted that the continued-fraction expansion given by Eq. (3–95) is not the only way possible. There are a few different ways to construct the ladder structure. For example, a digital filter $G(z)$ may be structured as a continued-fraction expansion form around the origin in terms of z^{-1}, as follows:

$$G(z)=\hat{A}_0+\cfrac{1}{\hat{B}_1z^{-1}+\cfrac{1}{\hat{A}_1+\cfrac{1}{\hat{B}_2z^{-1}+\cfrac{1}{\begin{matrix}\cdot\\ \cdot\\ \cdot\\ \hat{A}_{n-1}+\cfrac{1}{\hat{B}_nz^{-1}+\cfrac{1}{\hat{A}_n}}\end{matrix}}}}}$$

Also, instead of $G(z)$, its inverse $1/G(z)$ may be expanded into continued-fraction forms in terms of z or z^{-1} in order to carry out the ladder programming.

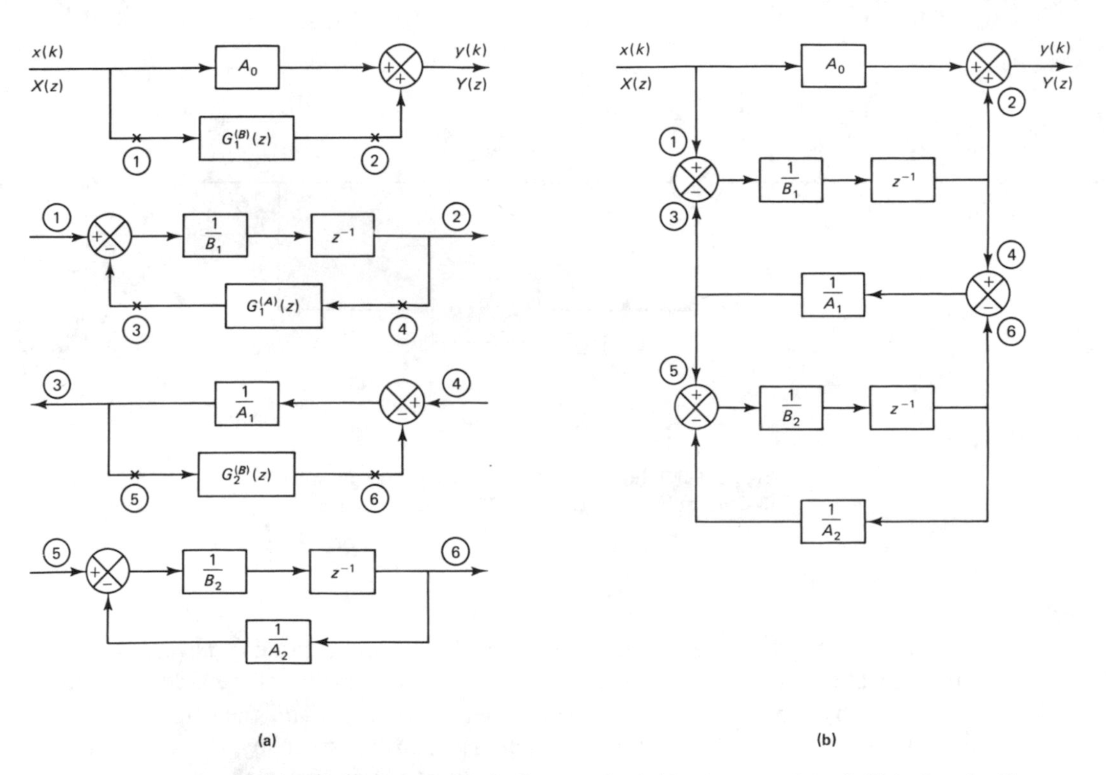

Figure 3–51 (a) Component block diagrams for ladder programming of $G(z)$ given by Eq. (3–95) when $n = 2$; (b) combination of component block diagrams showing ladder programming of $G(z)$.

Example 3–16.

Obtain the block diagrams for the following pulse-transfer-function system (a digital filter) by (1) direct programming, (2) standard programming, and (3) ladder programming:

$$\frac{Y(z)}{X(z)} = G(z) = \frac{2 - 0.6z^{-1}}{1 + 0.5z^{-1}}$$

1. *Direct programming.* Since the given pulse transfer function can be written as

$$Y(z) = -0.5z^{-1}Y(z) + 2X(z) - 0.6z^{-1}X(z)$$

direct programming yields the block diagram shown in Fig. 3–52. Notice that we need two delay elements.

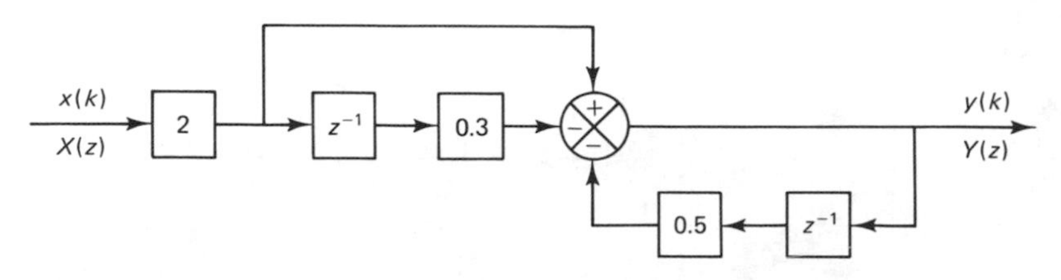

Figure 3–52 Block diagram realization of $Y(z)/X(z) = (2 - 0.6z^{-1})/(1 + 0.5z^{-1})$ (direct programming).

2. *Standard programming.* We shall first rewrite the pulse transfer function as follows:

$$\frac{Y(z)}{X(z)} = \frac{Y(z)}{H(z)}\frac{H(z)}{X(z)} = (1 - 0.3z^{-1})\frac{2}{1 + 0.5z^{-1}}$$

where

$$\frac{Y(z)}{H(z)} = 1 - 0.3z^{-1}$$

and

$$\frac{H(z)}{X(z)} = \frac{2}{1 + 0.5z^{-1}}$$

Block diagram realizations of these last two equations are shown in Fig. 3–53(a) and (b), respectively. If we combine these two diagrams, we obtain the block diagram for the digital filter $Y(z)/X(z)$, as shown in Fig. 3–53(c). Notice that the number of delay elements required has been reduced to 1 by the standard programming.

3. *Ladder programming.* We shall first rewrite the given $Y(z)/X(z)$ in the ladder form as follows:

$$\frac{Y(z)}{X(z)} = G(z) = \frac{2z - 0.6}{z + 0.5} = 2 + \frac{-1.6}{z + 0.5} = 2 + \cfrac{1}{-0.625z + \cfrac{1}{-3.2}}$$

Thus, $A_0 = 2$ and

$$G_1^{(B)}(z) = \frac{1}{-0.625z + \dfrac{1}{-3.2}} = \frac{1}{-\dfrac{1}{1.6}z - 0.3125}$$

Hence, we obtain

$$Y(z) = 2X(z) + G_1^{(B)}(z)X(z)$$

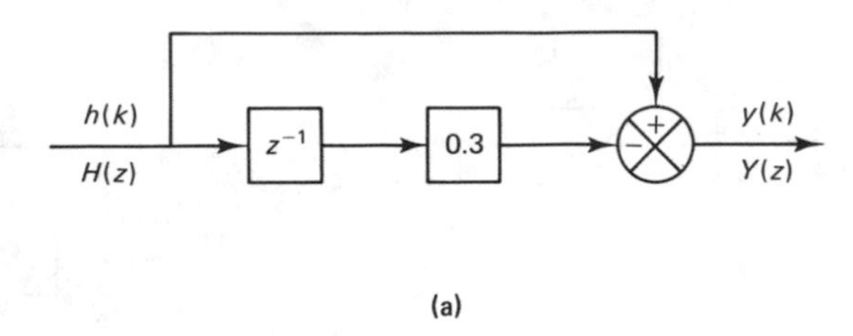

(a)

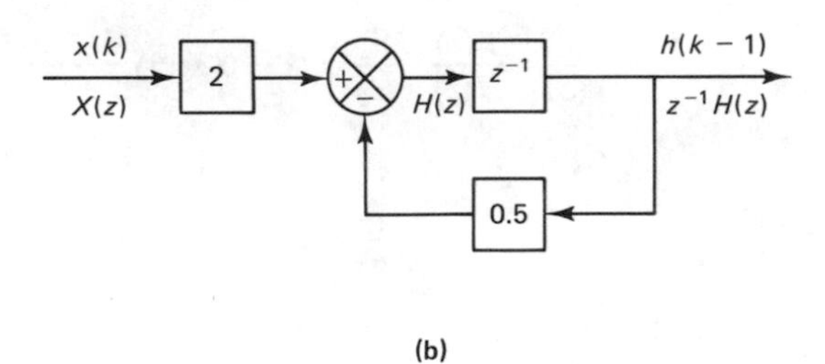

(b)

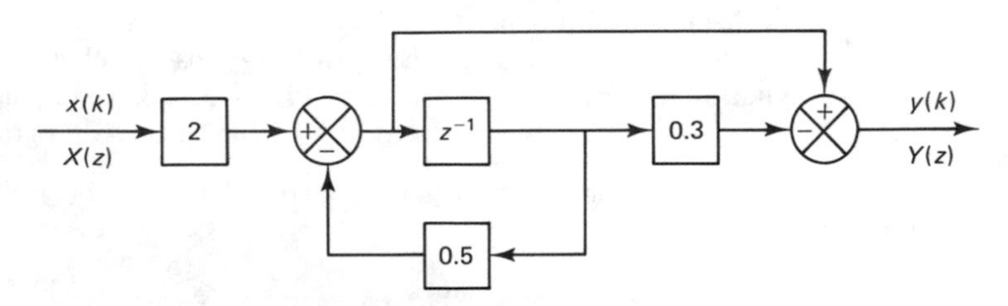

Figure 3–53 (a) Block diagram realization of $Y(z)/H(z) = 1 - 0.3z^{-1}$; (b) block diagram realization of $H(z)/X(z) = 2/(1 + 0.5z^{-1})$; (c) combination of block diagrams in parts (a) and (b) (standard programming).

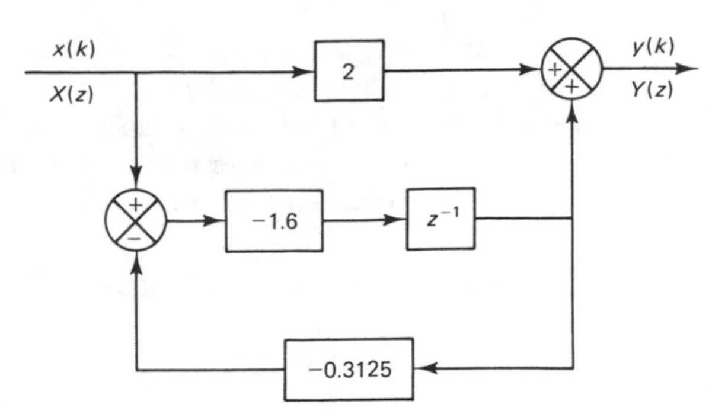

Figure 3–54 Block diagram realization of $Y(z)/X(z) = (2 - 0.6z^{-1})/(1 + 0.5z^{-1})$ (ladder programming).

Referring to Fig. 3–50(a) for the block diagram of $G_1^{(B)}(z)$, we obtain the block diagram of the digital filter $Y(z)/X(z)$ as shown in Fig. 3–54. Notice that we need only one delay element.

Infinite–impulse response filter and finite–impulse response filter. Digital filters may be classified according to the duration of the impulse response. Consider a digital filter defined by the following pulse transfer function:

$$\frac{Y(z)}{X(z)} = \frac{b_0 + b_1 z^{-1} + \cdots + b_m z^{-m}}{1 + a_1 z^{-1} + a_2 z^{-2} + \cdots + a_n z^{-n}} \tag{3–98}$$

where $n \geq m$. In terms of the difference equation,

$$\begin{aligned} y(k) = &-a_1 y(k-1) - a_2 y(k-2) - \cdots - a_n y(k-n) \\ &+ b_0 x(k) + b_1 x(k-1) + \cdots + b_m x(k-m) \end{aligned}$$

The impulse response of the digital filter defined by Eq. (3–98), where we assume not all a_i's are zero, has an infinite number of nonzero samples, although their magnitudes may become negligibly small as k increases. This type of digital filter is called an *infinite–impulse response filter*. Such a digital filter is also called a *recursive filter*, because the previous values of the output together with the present and past values of the input are used in processing the signal to obtain the current output $y(k)$. Because of the recursive nature, errors in previous outputs may accumulate. A recursive filter may be recognized by the presence of both a_i and b_i in the block diagram realization.

Next, consider a digital filter where the coefficients a_i are all zero, or where

$$\frac{Y(z)}{X(z)} = b_0 + b_1 z^{-1} + b_2 z^{-2} + \cdots + b_m z^{-m} \tag{3–99}$$

In terms of the difference equation,

$$y(k) = b_0 x(k) + b_1 x(k-1) + \cdots + b_m x(k-m)$$

The impulse response of the digital filter defined by Eq. (3–99) is limited to a finite number of samples defined over a finite range of time intervals; that is, the impulse response sequence is finite. This type of digital filter is called a *finite–impulse response filter*. It is also called a *nonrecursive filter* or a *moving average filter*.

In a nonrecursive realization, the present value of the output depends only on the present and past values of the input. The finite–impulse response filter can be recognized by the absence of the a_i's in the block diagram realization.

Realization of a finite–impulse response filter. We shall next consider the realization of the finite–impulse response filter.

Let us define the finite–impulse response sequence (weighting sequence) of the digital filter as $g(kT)$. If the input $x(kT)$ is applied to this filter, then the output $y(kT)$ can be given by

$$\begin{aligned} y(kT) &= \sum_{h=0}^{k} g(hT)x(kT - hT) \\ &= g(0)x(kT) + g(T)x((k-1)T) + \cdots + g(kT)x(0) \end{aligned} \tag{3–100}$$

The output $y(kT)$ is a convolution summation of the input signal and the impulse response sequence. The right-hand side of Eq. (3–100) consists of $k + 1$ terms. Thus, the output $y(kT)$ is given in terms of the past k inputs $x(0)$, $x(T)$, . . . , $x((k-1)T)$ and the current input $x(kT)$. Notice that as k increases, it is physically not possible to process all past values of input to produce the current output. We need to limit the number of the past values of the input to process.

Suppose we decide to employ the N immediate past values of the input $x((k-1)T)$, $x((k-2)T)$, . . . , $x((k-N)T)$ and the current input $x(kT)$. This is equivalent to approximating the right-hand side of Eq. (3–100) by the $N + 1$ most recent input values including the current one, or

$$y(kT) = g(0)x(kT) + g(T)x((k-1)T) + \cdots + g(NT)x((k-N)T) \tag{3–101}$$

Since Eq. (3–101) is a difference equation, the corresponding digital filter in the z plane can be obtained as follows. By taking the z transform of Eq. (3–101), we have

$$Y(z) = g(0)X(z) + g(T)z^{-1}X(z) + \cdots + g(NT)z^{-N}X(z) \tag{3–102}$$

Figure 3–55 shows the block diagram realization of this filter.

The characteristics of the finite–impulse response filter can be summarized as follows:

1. The finite–impulse response filter is nonrecursive. Thus, because of the lack of feedback, the accumulation of errors in past outputs can be avoided in the processing of the signal.
2. Implementation of the finite–impulse response filter does not require feedback, so the direct programming and standard programming are identical. Also, implementation may be achieved by high-speed convolution using the fast Fourier transform.

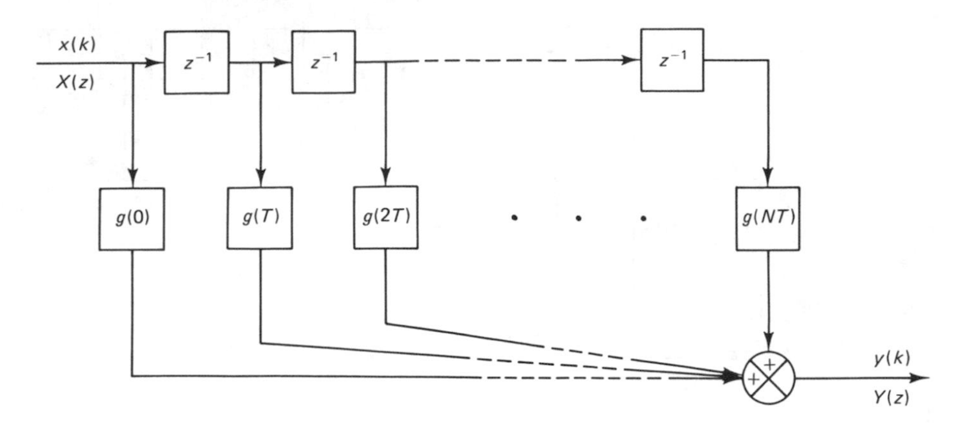

Figure 3–55 Block diagram realization of Eq. (3–102).

3. The poles of the pulse transfer function of the finite–impulse response filter are at the origin, and therefore it is always stable.
4. If the input signal involves high-frequency components, then the number of delay elements needed in the finite–impulse response filter increases and the amount of time delay becomes large. (This is a disadvantage of the finite–impulse response filter compared with the infinite–impulse response filter.)

Example 3–17.

The digital filter discussed in Example 3–16 is a recursive filter. Modify this digital filter and realize it as a nonrecursive filter. Then, obtain the response of this nonrecursive filter to a Kronecker delta input.

By dividing the numerator of the recursive filter $G(z)$ by the denominator, we obtain

$$G(z) = \frac{2 - 0.6z^{-1}}{1 + 0.5z^{-1}}$$

$$= 2 - 1.6z^{-1} + 0.8z^{-2} - 0.4z^{-3} + 0.2z^{-4} - 0.1z^{-5} + 0.05z^{-6} - 0.025z^{-7} + \cdots$$

By arbitrarily truncating this series at z^{-7}, we obtain the desired nonrecursive filter, as follows:

$$\frac{Y(z)}{X(z)} = 2 - 1.6z^{-1} + 0.8z^{-2} - 0.4z^{-3} + 0.2z^{-4} - 0.1z^{-5} + 0.05z^{-6} - 0.025z^{-7} \qquad (3\text{–}103)$$

Figure 3–56 shows the block digram for this nonrecursive digital filter. Notice that we need a large number of delay elements to obtain a good level of accuracy.

Noting that the digital filter is the z transform of the impulse response sequence, the inverse z transform of the digital filter gives the impulse response sequence. By taking the inverse z transform of the nonrecursive filter given by Eq. (3–103), we obtain

$$y(kT) = 2x(kT) - 1.6x((k-1)T) + 0.8x((k-2)T) - 0.4x((k-3)T)$$
$$+ 0.2x((k-4)T) - 0.1x((k-5)T) + 0.05x((k-6)T) - 0.025x((k-7)T)$$

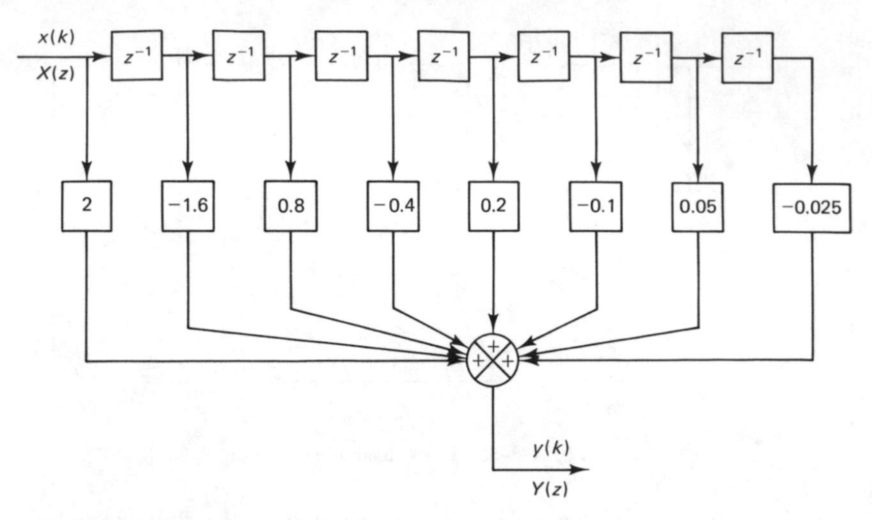

Figure 3–56 Block diagram for the digital filter given by Eq. (3–103) (nonrecursive form).

For the Kronecker delta input, where $x(0) = 1$ and $x(kT) = 0$ for $k \neq 0$, this last equation gives

$$y(0) = 2$$
$$y(T) = -1.6$$
$$y(2T) = 0.8$$
$$y(3T) = -0.4$$
$$y(4T) = 0.2$$
$$y(5T) = -0.1$$
$$y(6T) = 0.05$$
$$y(7T) = -0.025$$

The impulse response sequence for this digital filter is shown in Fig. 3–57.

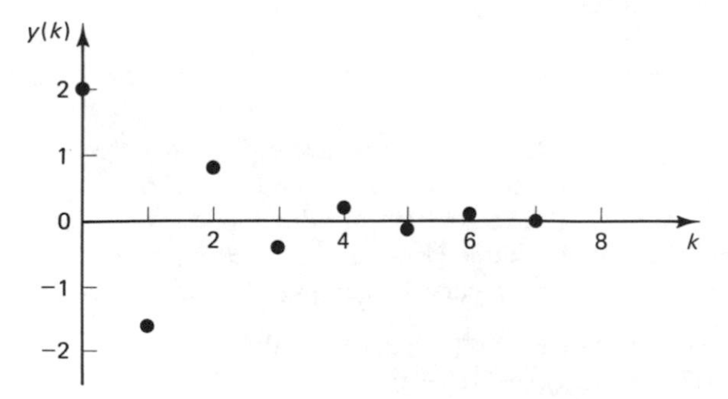

Figure 3–57 Impulse response sequence for the digital filter given by Eq. (3–103).

3-8 MAPPING BETWEEN THE s PLANE AND THE z PLANE

The absolute stability and relative stability of the linear time-invariant continuous-time closed-loop control system are determined by the locations of the closed-loop poles in the s plane. For example, complex closed-loop poles in the left half of the s plane near the $j\omega$ axis will exhibit oscillatory behavior, and closed-loop poles on the negative real axis will exhibit exponential decay.

Since the complex variables z and s are related by $z = e^{Ts}$, the pole and zero locations in the z plane are related to the pole and zero locations in the s plane. Therefore, the stability of the linear time-invariant discrete-time closed-loop system can be determined in terms of the locations of the poles of the closed-loop pulse transfer function. It is noted that the dynamic behavior of the discrete-time control system depends on the sampling period T. In terms of poles and zeros in the z plane, their locations depend on the sampling period T. In other words, a change in the sampling period T modifies the pole and zero locations in the z plane and causes the response behavior to change.

Mapping of the left half of the *s* plane into the *z* plane. In the design of a continuous-time control system, the locations of the poles and zeros in the s plane are very important in predicting the dynamic behavior of the system. Similarly, in designing discrete-time control systems, the locations of the poles and zeros in the z plane are very important. In the following paragraphs we shall investigate how the locations of the poles and zeros in the s plane compare with the locations of the poles and zeros in the z plane.

When impulse-sampling is incorporated into the process, the complex variables z and s are related by the equation

$$z = e^{Ts}$$

This means that a pole in the s plane can be located in the z plane through the transformation $z = e^{Ts}$. Since the complex variable s has real part σ and imaginary part ω, we have

$$s = \sigma + j\omega$$

and

$$z = e^{T(\sigma + j\omega)} = e^{T\sigma}e^{jT\omega} = e^{T\sigma}e^{j(T\omega + 2\pi k)}$$

From this last equation we see that poles and zeros in the s plane, where frequencies differ in integral multiples of the sampling frequency $2\pi/T$, are mapped into the same locations in the z plane. This means that there are infinitely many values of s for each value of z.

Since σ is negative in the left half of the s plane, the left half of the s plane corresponds to

$$|z| = e^{T\sigma} < 1$$

The $j\omega$ axis in the s plane corresponds to $|z| = 1$. That is, the imaginary axis in the s plane (the line $\sigma = 0$) corresponds to the unit circle in the z plane, and the interior of the unit circle corresponds to the left half of the s plane.

Since the linear time-invariant continuous-time system is stable if all poles of the transfer function lie in the left half of the s plane, we find that in the z plane the system is stable if all poles of the pulse transfer function lie in the unit circle centered at the origin. The locations of the poles and zeros of the pulse transfer function determine the nature of the response of the discrete-time system.

Primary strip and complementary strips. Note that since $\angle z = \omega T$, the angle of z varies from $-\infty$ to ∞ as ω varies from $-\infty$ to ∞. Consider a representative point on the $j\omega$ axis in the s plane. As this point moves from $-j\frac{1}{2}\omega_s$ to $j\frac{1}{2}\omega_s$ on the $j\omega$ axis, where ω_s is the sampling frequency, we have $|z| = 1$, and $\angle z$ varies from $-\pi$ to π in the counterclockwise direction in the z plane. As the representative point moves from $j\frac{1}{2}\omega_s$ to $j\frac{3}{2}\omega_s$ on the $j\omega$ axis, the corresponding point in the z plane traces out the unit circle once in the counterclockwise direction. Thus, as the point in the s plane moves from $-\infty$ to ∞ on the $j\omega$ axis, we trace the unit circle in the z plane an infinite number of times. From this analysis, it is clear that each strip of width ω_s in the left half of the s plane maps into the inside of the unit circle in the z plane. This implies that the s plane may be divided into an infinite number of periodic strips as shown in Fig. 3–58. The primary strip extends from $j\omega = -j\frac{1}{2}\omega_s$ to $j\frac{1}{2}\omega_s$. The complementary strips extend from $j\frac{1}{2}\omega_s$ to $j\frac{3}{2}\omega_s$, $j\frac{3}{2}\omega_s$ to $j\frac{5}{2}\omega_s$, . . . , and from $-j\frac{1}{2}\omega_s$ to $-j\frac{3}{2}\omega_s$, $-j\frac{3}{2}\omega_s$ to $-j\frac{5}{2}\omega_s$, ...

In the primary strip, if we trace the sequence of points 1–2–3–4–5–1 in the s plane as shown by the circled numbers in Fig. 3–59(a), then this path is mapped into the unit circle centered at the origin of the z plane, as shown in Fig. 3–59(b). The corresponding points 1, 2, 3, 4, and 5 in the z plane are shown by the circled numbers in Fig. 3–59(b).

The area enclosed by any of the complementary strips is mapped into the same unit circle in the z plane. This means that the correspondence between the z plane and the s plane is not unique. A point in the z plane corresponds to an infinite number of points in the s plane, although a point in the s plane corresponds to a single point in the z plane.

Since the entire left half of the s plane is mapped into the interior of the unit circle in the z plane, the entire right half of the s plane is mapped into the exterior of the unit circle in the z plane. As mentioned earlier, the $j\omega$ axis in the s plane maps into the unit circle in the z plane. Note that if the sampling frequency is at least twice as fast as the highest-frequency component involved in the system, then every point in the unit circle in the z plane represents frequencies between $-\frac{1}{2}\omega_s$ and $\frac{1}{2}\omega_s$.

In what follows we shall investigate the mapping of some of the commonly used contours in the s plane into the z plane. Specifically, we shall map constant-attenuation loci, constant-frequency loci, and constant-damping-ratio loci.

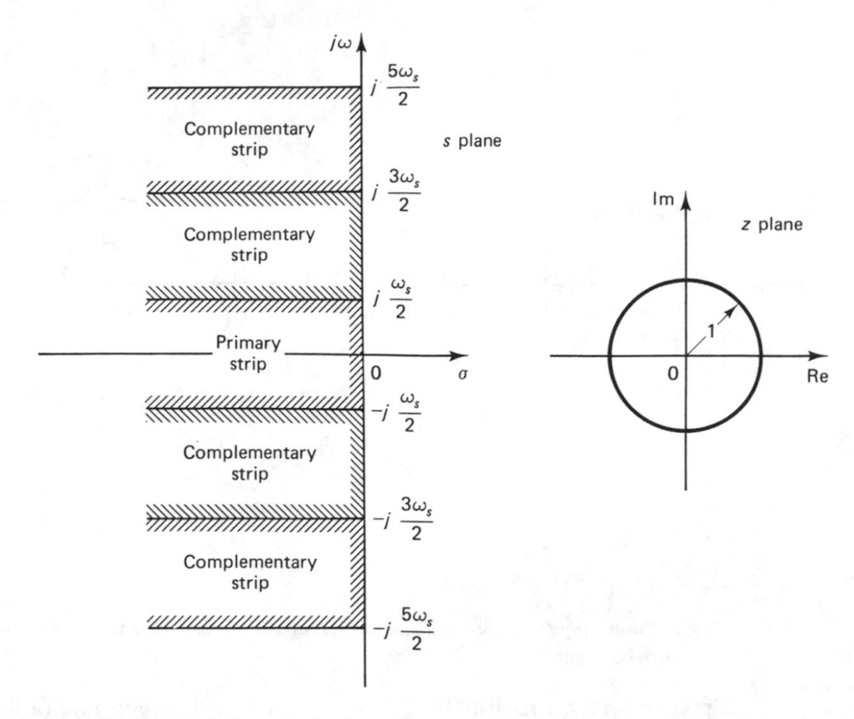

Figure 3–58 Periodic strips in the s plane and the corresponding region (unit circle centered at the origin) in the z plane.

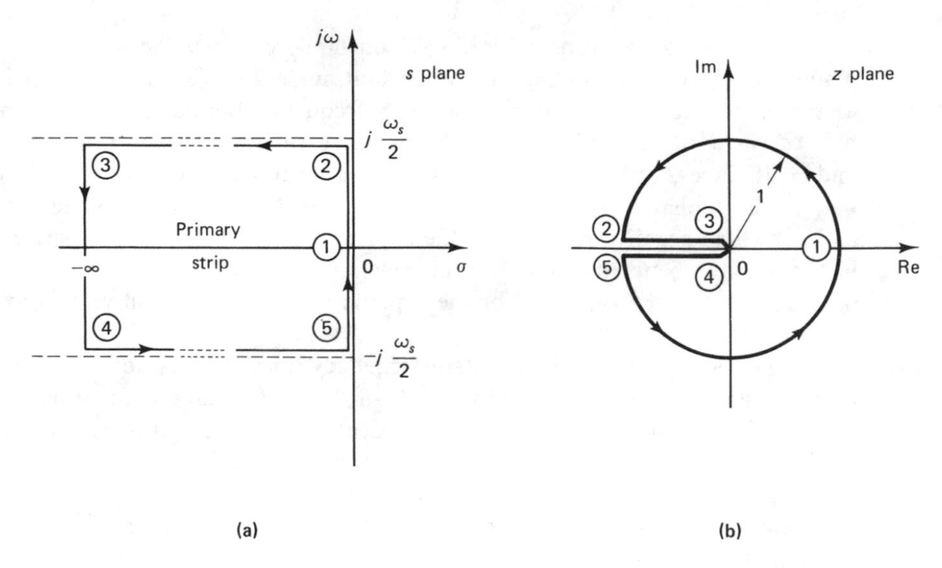

Figure 3–59 Diagrams showing the correspondence between the primary strip in the s plane and the unit circle in the z plane: (a) a path in the s plane; (b) the corresponding path in the z plane.

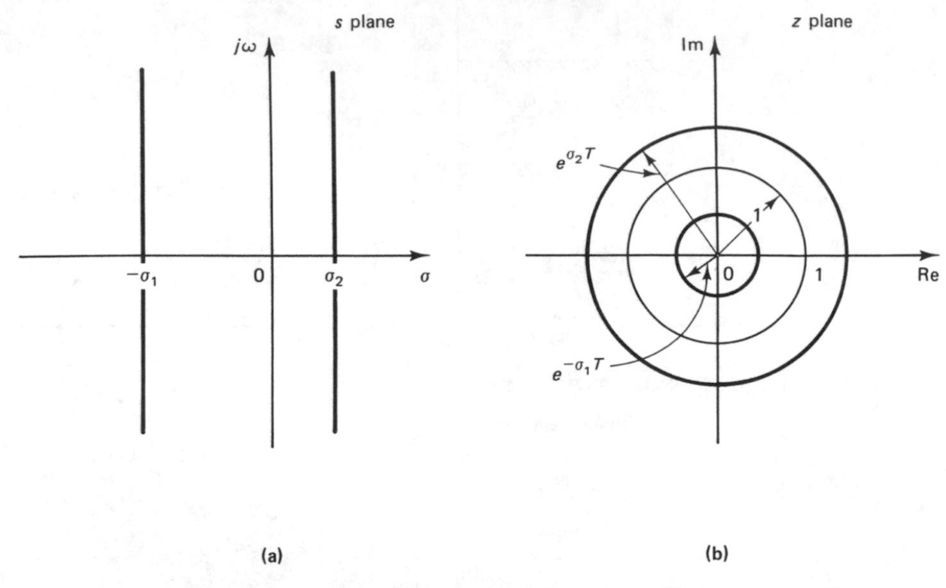

Figure 3–60 (a) Constant-attenuation lines in the s plane; (b) the corresponding loci in the z plane.

Constant-attenuation loci. A constant-attenuation line (a line plotted as σ = constant) in the s plane maps into a circle of radius $z = e^{T\sigma}$ centered at the origin in the z plane, as shown in Fig. 3–60.

Constant-frequency loci. A constant-frequency locus $\omega = \omega_1$ in the s plane is mapped into a radial line of constant angle $T\omega_1$ (in radians) in the z plane, as shown in Fig. 3–61. Note that constant-frequency lines at $\omega = \pm\frac{1}{2}\omega_s$ in the left half of the s plane correspond to the negative real axis in the z plane between 0 and $-$ 1, since $T(\pm\frac{1}{2}\omega_s) = \pm\pi$. Constant-frequency lines at $\omega = \pm\frac{1}{2}\omega_s$ in the right half of the s plane correspond to the negative real axis in the z plane between -1 and $-\infty$. The negative real axis in the s plane corresponds to the positive real axis in the z plane between 0 and 1. And constant frequency lines at $\omega = \pm n\,\omega_s$ (n = 0, 1, 2, . . .) in the right half of the s plane map into the positive real axis in the z plane between 1 and ∞.

The region bounded by constant-frequency lines $\omega = \omega_1$ and $\omega = -\omega_2$ (where both ω_1 and ω_2 lie between $-\frac{1}{2}\omega_s$ and $\frac{1}{2}\omega_s$) and constant-attenuation lines $\sigma = -\sigma_1$ and $\sigma = -\sigma_2$, as shown in Fig. 3–62(a), is mapped into a region bounded by two radial lines and two circular arcs, as shown in Fig. 3–62(b).

Constant-damping-ratio loci. A constant-damping-ratio line (a radial line) in the s plane is mapped into a spiral in the z plane. This can be seen as follows. In the s plane a constant-damping-ratio line can be given by

$$s = -\zeta\omega_n + j\omega_n\sqrt{1-\zeta^2} = -\zeta\omega_n + j\omega_d$$

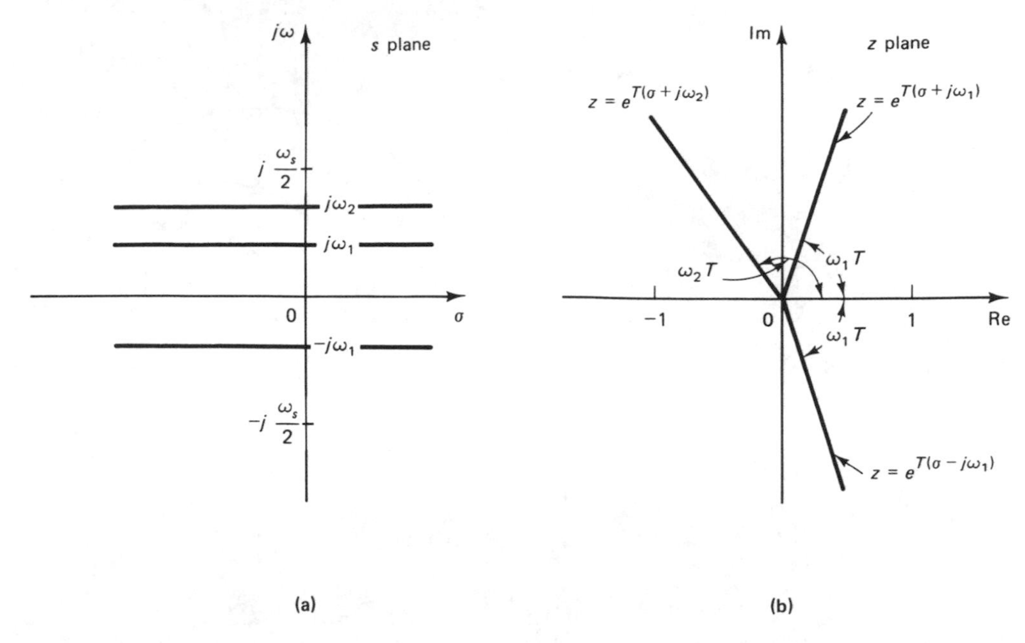

Figure 3–61 (a) Constant-frequency loci in the s plane; (b) the corresponding loci in the z plane.

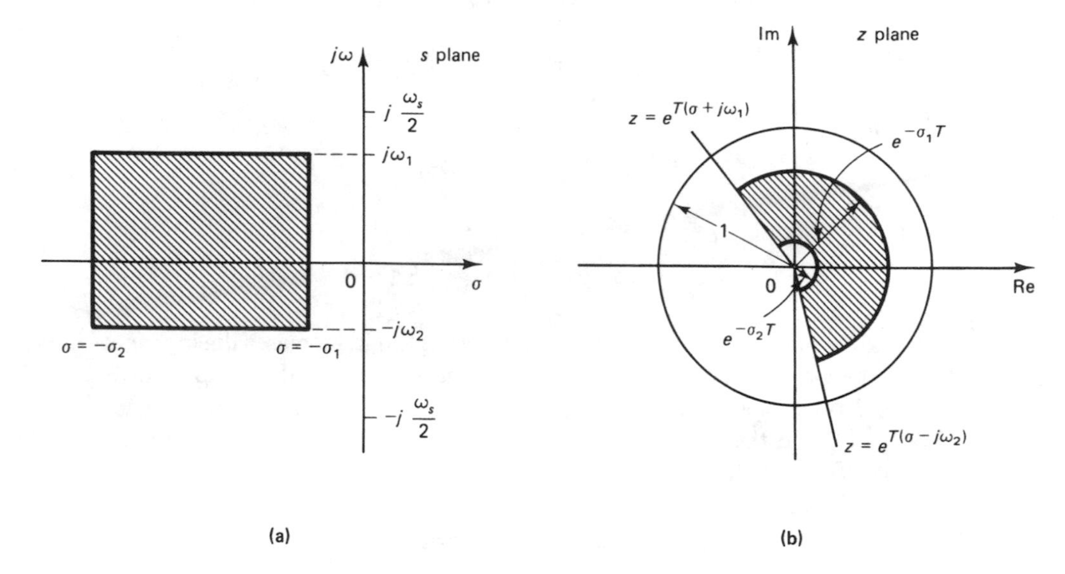

Figure 3–62 (a) Region bounded by lines $\omega = \omega_1$, $\omega = -\omega_2$, $\sigma = -\sigma_1$, and $\sigma = -\sigma_2$ in the s plane; (b) the corresponding region in the z plane.

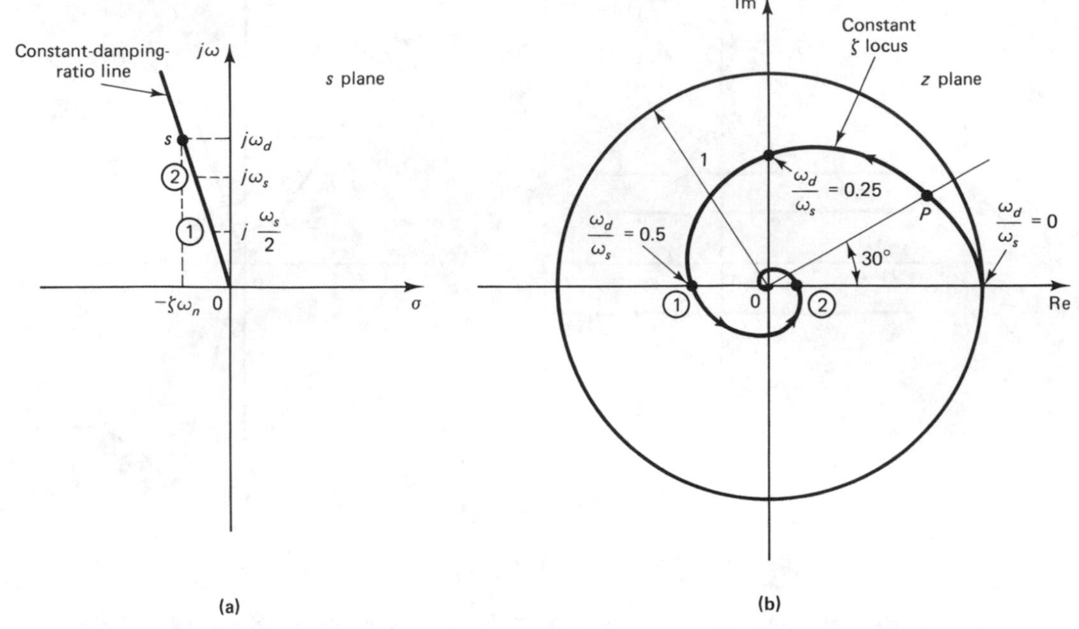

Figure 3–63 (a) Constant-damping-ratio line in the s plane; (b) the corresponding locus in the z plane.

where $\omega_d = \omega_n\sqrt{1-\zeta^2}$ [see Fig. 3–63(a)]. In the z plane this line becomes

$$z = e^{Ts} = \exp(-\zeta\omega_n T + j\omega_d T)$$
$$= \exp\left(-\frac{2\pi\zeta}{\sqrt{1-\zeta^2}}\frac{\omega_d}{\omega_s} + j2\pi\frac{\omega_d}{\omega_s}\right)$$

Hence

$$|z| = \exp\left(-\frac{2\pi\zeta}{\sqrt{1-\zeta^2}}\frac{\omega_d}{\omega_s}\right) \tag{3–104}$$

and

$$\angle z = 2\pi\frac{\omega_d}{\omega_s} \tag{3–105}$$

Thus, the magnitude of z decreases and the angle of z increases linearly as ω_d increases, and the locus in the z plane becomes a logarithmic spiral, as shown in Fig. 3–63(b).

Notice that for a given ratio of ω_d/ω_s, the magnitude $|z|$ becomes a function only of ζ and the angle of z becomes a constant. For example, if the damping ratio is specified as 0.3, or $\zeta = 0.3$, then for $\omega_d = 0.25\omega_s$, we have

$$|z| = \exp\left(-\frac{2\pi \times 0.3}{\sqrt{1-0.3^2}} \times 0.25\right) = 0.610$$

$$\angle z = 2\pi \times 0.25 = 0.5\pi = 90°$$

For $\omega_d = 0.5\omega_s$,

$$|z| = \exp\left(-\frac{2\pi \times 0.3}{\sqrt{1-0.3^2}} \times 0.5\right) = 0.3725$$

$$\underline{/z} = 2\pi \times 0.5 = \pi = 180°$$

Thus, the spiral can be graduated in terms of a normalized frequency ω_d/ω_s [see Fig. 3–63(b)]. Once the sampling frequency ω_s is specified, the numerical value of ω_d at any point on the spiral can be determined. For example, at point P in Fig. 3–63(b), ω_d can be determined as follows. If, for example, the sampling frequency is specified as $\omega_s = 10\pi$ rad/sec, then at point P

$$\underline{/z} = \frac{\pi}{6} = 2\pi \frac{\omega_d}{\omega_s}$$

Hence, ω_d at point P is

$$\omega_d = \tfrac{1}{12}\omega_s = \tfrac{5}{6}\pi \text{ rad/sec}$$

Note that if a constant-damping-ratio line is in the second or third quadrant in the s plane, then the spiral decays within the unit circle in the z plane. However, if a constant-damping-ratio line is in the first or fourth quadrant in the s plane (which corresponds to negative damping), then the spiral grows outside the unit circle. Figure 3–64 shows constant-damping-ratio loci for $\zeta = 0$, $\zeta = 0.2$, $\zeta = 0.4$, $\zeta = 0.6$, $\zeta = 0.8$, and $\zeta = 1$. The $\zeta = 1$ locus is a horizontal line between points $z = 0$ and $z = 1$. (Note that Fig. 3–64 shows only the loci in the upper half of the z plane, which correspond to $0 \le \omega \le \frac{1}{2}\omega_s$. The loci corresponding to $-\frac{1}{2}\omega_s \le \omega \le 0$ are the mirror images of the loci in the upper half of the z plane about the horizontal axis.)

Notice that the constant ζ loci are normal to the constant ω_n loci in the s plane, as shown in Fig. 3–65(a). In the z plane mapping, constant ω_n loci intersect

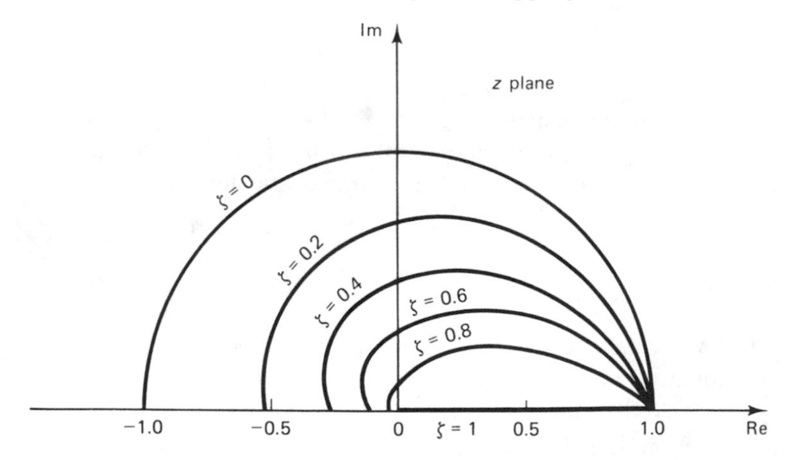

Figure 3–64 Constant-damping-ratio loci in the z plane.

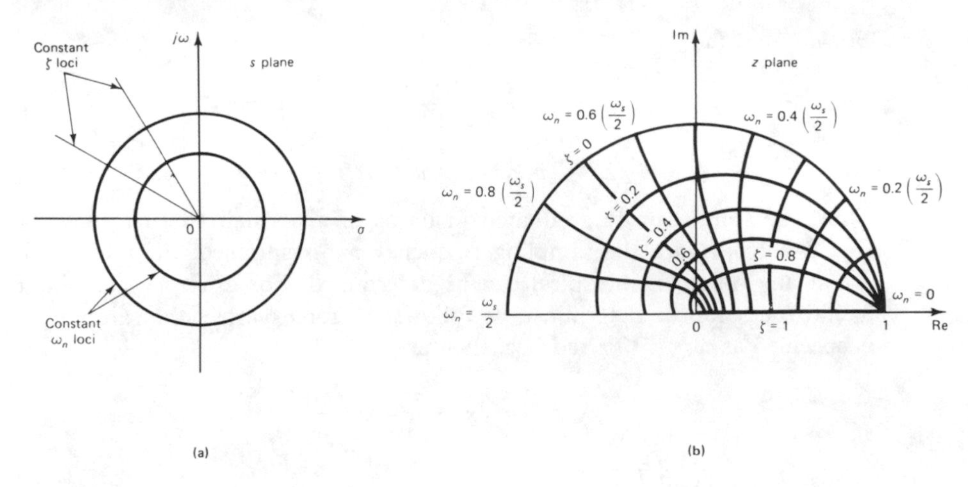

Figure 3–65 (a) Diagram showing orthogonality of the constant ζ loci and constant ω_n loci in the s plane; (b) the corresponding diagram in the z plane.

constant ζ spirals at right angles, as shown in Fig. 3–65(b). A mapping such as this which preserves both the size and the sense of angles is called a *conformal mapping*.

3-9 STABILITY ANALYSIS OF CLOSED-LOOP SYSTEMS IN THE z DOMAIN

Stability analysis of a closed-loop system. In this section we shall discuss the stability of linear time-invariant single-input–single-output discrete-time control systems. Consider the following closed-loop pulse-transfer function system:

$$\frac{C(z)}{R(z)} = \frac{G(z)}{1 + GH(z)} \tag{3–106}$$

The stability of the system defined by Eq. (3–106), as well as of other types of discrete-time control systems, may be determined from the locations of the closed-loop poles in the z plane, or the roots of the characteristic equation

$$P(z) = 1 + GH(z) = 0$$

as follows:

1. For the system to be stable, the closed-loop poles or the roots of the characteristic equation must lie within the unit circle in the z plane. Any closed-loop pole outside the unit circle makes the system unstable.
2. If a simple pole lies at $z = 1$ or $z = -1$ (or if a simple pole lies at $z = 1$ and another simple pole at $z = -1$), then the system becomes critically stable.

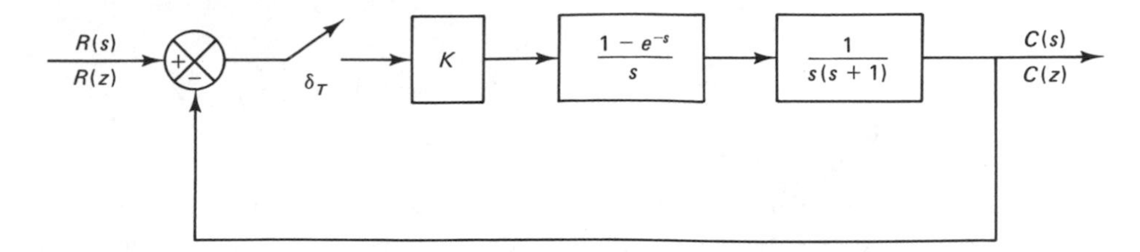

Figure 3–66 Closed-loop control system of Example 3–18.

Also, the system becomes critically stable if a single pair of conjugate complex poles lie on the unit circle in the z plane. Any multiple closed-loop pole on the unit circle makes the system unstable.

3. Closed-loop zeros do not affect the absolute stability and therefore may be located anywhere in the z plane.

Thus, a linear time-invariant single-input–single-output discrete-time closed-loop control system becomes unstable if any of the closed-loop poles lies outside the unit circle and/or any multiple closed-loop pole lies on the unit circle in the z plane.

Example 3–18.

Consider the closed-loop control system shown in Fig. 3-66. Determine the stability of the system when $K = 1$. The open-loop transfer function $G(s)$ of the system is

$$G(s) = \frac{1 - e^{-s}}{s} \frac{1}{s(s+1)}$$

Referring to Eq. (3–71), the z transform of $G(s)$ is

$$G(z) = \mathscr{Z}\left[\frac{1 - e^{-s}}{s} \frac{1}{s(s+1)}\right] = \frac{0.3679z + 0.2642}{(z - 0.3679)(z - 1)} \tag{3–107}$$

Since the closed-loop pulse transfer function for the system is

$$\frac{C(z)}{R(z)} = \frac{G(z)}{1 + G(z)}$$

the characteristic equation is

$$1 + G(z) = 0$$

which becomes

$$(z - 0.3679)(z - 1) + 0.3679z + 0.2642 = 0$$

or

$$z^2 - z + 0.6321 = 0$$

The roots of the characteristic equation are found to be

$$z_1 = 0.5 + j0.6181, \qquad z_2 = 0.5 - j0.6181$$

Since

$$|z_1| = |z_2| < 1$$

the system is stable.

It is important to note that in the absence of the sampler, a second-order system is always stable. In the presence of the sampler, however, a second-order system such as this can become unstable for large values of gain. In fact, it can be shown that the second-order system shown in Fig. 3–66 will become unstable if K > 2.3925. (See Example 3–23.)

Methods for testing absolute stability. There are available three stability tests which can be applied directly to the characteristic equation $P(z) = 0$ without solving for the roots. Two of them are the Schur-Cohn stability test and the Jury stability test. These two tests reveal the existence of any unstable roots (the roots that lie outside the unit circle in the z plane). However, these tests neither give the locations of unstable roots nor indicate the effects of parameter changes on the system stability, except for the simple case of low-order systems. (See Example 3–23.) The third method is based on the bilinear transformation coupled with the Routh stability criterion, which will be outlined later in this section.

Both the Schur-Cohn stability test and the Jury stability test may be applied to polynomial equations with real or complex coefficients. The computations required in the Jury test, when the polynomial equation involves only real coefficients, are

TABLE 3–2 GENERAL FORM OF THE JURY STABILITY TABLE

Row	z^0	z^1	z^2	z^3	. . .	z^{n-2}	z^{n-1}	z^n
1	a_n	a_{n-1}	a_{n-2}	a_{n-3}	. . .	a_2	a_1	a_0
2	a_0	a_1	a_2	a_3	. . .	a_{n-2}	a_{n-1}	a_n
3	b_{n-1}	b_{n-2}	b_{n-3}	b_{n-4}	. . .	b_1	b_0	
4	b_0	b_1	b_2	b_3	. . .	b_{n-2}	b_{n-1}	
5	c_{n-2}	c_{n-3}	c_{n-4}	c_{n-5}	. . .	c_0		
6	c_0	c_1	c_2	c_3	. . .	c_{n-2}		
⋮	⋮							
$2n-5$	p_3	p_2	p_1	p_0				
$2n-4$	p_0	p_1	p_2	p_3				
$2n-3$	q_2	q_1	q_0					

much simpler than those required in the Schur-Cohn test. Since the coefficients of the characteristic equations corresponding to physically realizable systems are always real, the Jury test is preferred to the Schur-Cohn test.

The Jury stability test. In applying the Jury stability test to a given characteristic equation $P(z) = 0$, we construct a table whose elements are based on the coefficients of $P(z)$. Assume that the characteristic equation $P(z)$ is a polynomial in z as follows:

$$P(z) = a_0 z^n + a_1 z^{n-1} + \cdots + a_{n-1} z + a_n \tag{3-108}$$

where $a_0 > 0$. Then the Jury table becomes as given in Table 3–2.

Notice that the elements in the first row consist of the coefficients in $P(z)$ arranged in the ascending order of powers of z. The elements in the second row consist of the coefficients of $P(z)$ arranged in the descending order of powers of z. The elements for rows 3 through $2n - 3$ are given by the following determinants:

$$b_k = \begin{vmatrix} a_n & a_{n-1-k} \\ a_0 & a_{k+1} \end{vmatrix} \qquad k = 0, 1, 2, \ldots, n-1$$

$$c_k = \begin{vmatrix} b_{n-1} & b_{n-2-k} \\ b_0 & b_{k+1} \end{vmatrix} \qquad k = 0, 1, 2, \ldots, n-2$$

$$\vdots$$

$$q_k = \begin{vmatrix} p_3 & p_{2-k} \\ p_0 & p_{k+1} \end{vmatrix} \qquad k = 0, 1, 2$$

Note that the last row in the table consists of three elements. (For second-order systems, $2n - 3 = 1$ and the Jury table consists only of one row containing three elements.) Notice that the elements in any even-numbered row are simply the reverse of the immediately preceding odd-numbered row.

Stability criterion by the Jury test. A system with the characteristic equation $P(z) = 0$ given by Eq. (3–108) is stable if the following conditions are all satisfied:

1. $|a_n| < a_0$
2. $P(z)|_{z=1} > 0$
3. $P(z)|_{z=-1} \begin{cases} > 0 \text{ for } n \text{ even} \\ < 0 \text{ for } n \text{ odd} \end{cases}$
4. $|b_{n-1}| > |b_0|$

   $|c_{n-2}| > |c_0|$

   $\vdots$

   $|q_2| > |q_0|$

TABLE 3–3 JURY STABILITY TABLE FOR THE FOURTH-ORDER SYSTEM

Row	z^0	z^1	z^2	z^3	z^4	
	a_4 a_0				a_0 a_4	$= b_3$
	a_4 a_0			a_1 a_3		$= b_2$
	a_4 a_0		a_2 a_2			$= b_1$
1 2	a_4 a_0	a_3 a_1				$= b_0$
	b_3 b_0			b_0 b_3		$= c_2$
	b_3 b_0		b_1 b_2			$= c_1$
3 4	b_3 b_0	b_2 b_1				$= c_0$
5	c_2	c_1	c_0			

Example 3–19.

Construct the Jury stability table for the following characteristic equation:

$$P(z) = a_0 z^4 + a_1 z^3 + a_2 z^2 + a_3 z + a_4$$

where $a_0 > 0$. Write the stability conditions.

Referring to the general case of the Jury stability table given by Table 3–2, a Jury stability table for the fourth-order system may be constructed as shown in Table 3–3. This table is slightly modified from the standard form and is convenient for the computations of the b's and c's. The determinant given in the left-hand side of each row gives the value of b or c written in the right hand side of the same row.

The stability conditions are

1. $|a_4| < a_0$
2. $P(1) = a_0 + a_1 + a_2 + a_3 + a_4 > 0$
3. $P(-1) = a_0 - a_1 + a_2 - a_3 + a_4 > 0 \qquad n = 4 = \text{even}$
4. $|b_3| > |b_0|$
   $|c_2| > |c_0|$

It is noted that the value of c_1 (or, in the case of the nth-order system, the value of q_1) is not used in the stability test and therefore the computation of c_1 (or q_1) may be omitted.

Example 3–20.

Examine the stability of the following characteristic equation:

$$P(z) = z^4 - 1.2z^3 + 0.07z^2 + 0.3z - 0.08 = 0$$

Notice that for this characteristic equation

$$a_0 = 1$$

$$a_1 = -1.2$$

$$a_2 = 0.07$$

$$a_3 = 0.3$$

$$a_4 = -0.08$$

Clearly, the first condition, $|a_4| < a_0$, is satisfied. Let us examine the second condition for stability:

$$P(1) = 1 - 1.2 + 0.07 + 0.3 - 0.08 = 0.09 > 0$$

The second condition is satisfied. The third condition for stability becomes

$$P(-1) = 1 + 1.2 + 0.07 - 0.3 - 0.08 = 1.89 > 0 \qquad n = 4 = \text{even}$$

Hence the third condition is satisfied.

We now construct the Jury stability table. Referring to Example 3–19, we compute the values of b_3, b_2, b_1, and b_0 and c_2 and c_0. The result is shown in Table 3–4. (Although the value of c_1 is shown in the table, c_1 is not needed in the stability test and therefore need not be computed.) From this table, we get

$$|b_3| = 0.994 > 0.204 = |b_0|$$

$$|c_2| = 0.946 > 0.315 = |c_0|$$

Thus the two parts of the fourth condition in the Jury stability test are satisfied. Since all conditions for stability are satisfied, the given characteristic equation is stable, or all roots lie inside the unit circle in the z plane.

As a matter of fact, the given characteristic equation $P(z)$ can be factored as follows:

$$P(z) = (z - 0.8)(z + 0.5)(z - 0.5)(z - 0.4)$$

As a matter of course, the result obtained above agrees with the fact that all roots are within the unit circle in the z plane.

Example 3–21.

Examine the stability of the characteristic equation given by

$$P(z) = z^3 - 1.1z^2 - 0.1z + 0.2 = 0$$

TABLE 3–4. JURY STABILITY TABLE FOR THE SYSTEM OF EXAMPLE 3–20

Row	z^0	z^1	z^2	z^3	z^4	
	-0.08				1	$= b_3 = -0.994$
	1				-0.08	
	-0.08			-1.2		$= b_2 = 1.176$
	1			0.3		
	-0.08		0.07			$= b_1 = -0.0756$
	1		0.07			
1	-0.08	0.3				$= b_0 = -0.204$
2	1	-1.2				
	-0.994			-0.204		$= c_2 = 0.946$
	-0.204			-0.994		
	-0.994		-0.0756			$= c_1 = -1.184$
	-0.204		1.176			
3	-0.994	1.176				$= c_0 = 0.315$
4	-0.204	-0.0756				
5	0.946	-1.184	0.315			

First we identify the coefficients:

$$a_0 = 1$$

$$a_1 = -1.1$$

$$a_2 = -0.1$$

$$a_3 = 0.2$$

The conditions for stability in the Jury test for the third-order system are

1. $|a_3| < a_0$
2. $P(1) > 0$
3. $P(-1) < 0 \qquad n = 3 = \text{odd}$
4. $|b_2| > |b_0|$

The first condition, $|a_3| < a_0$, is clearly satisfied. Now we examine the second condition of the Jury stability test:

$$P(1) = 1 - 1.1 - 0.1 + 0.2 = 0$$

This indicates that at least one root is at $z = 1$. Therefore, the system is at best critically stable. The remaining tests determine whether the system is critically stable or unstable. (If the given characteristic equation represents a control system, critical stability will not be desired. The stability test may be stopped at this point.)

The third condition of the Jury test gives

$$P(-1) = -1 - 1.1 + 0.1 + 0.2 = -1.8 < 0 \qquad n = 3 = \text{odd}$$

The third condition is satisfied. Now we examine the fourth condition of the Jury test. Simple computations give $b_2 = -0.96$ and $b_0 = -0.12$. Hence

$$|b_2| > |b_0|$$

The fourth condition of the Jury test is satisfied.

From the above analysis we conclude that the given characteristic equation has one root on the unit circle ($z = 1$) and its other two roots within the unit circle in the z plane. Hence the system is critically stable.

Example 3–22.

A control system has the following characteristic equation:

$$P(z) = z^3 - 1.3z^2 - 0.08z + 0.24 = 0$$

Determine the stability of the system.

We first identify the coefficients:

$$a_0 = 1$$

$$a_1 = -1.3$$

$$a_2 = -0.08$$

$$a_3 = 0.24$$

Clearly, the first condition for stability, $|a_3| < a_0$, is satisfied. Next, we examine the second condition for stability:

$$P(1) = 1 - 1.3 - 0.08 + 0.24 = -0.14 < 0$$

The test indicates that the second condition for stability is violated. The system is therefore unstable. We may stop the test here.

Example 3–23.

Consider the discrete-time unity-feedback control system (with sampling period $T = 1$ sec) whose open-loop pulse transfer function is given by

$$G(z) = \frac{K(0.3679z + 0.2642)}{(z - 0.3679)(z - 1)}$$

Determine the range of gain K for stability by use of the Jury stability test.

The closed-loop pulse transfer function becomes

$$\frac{C(z)}{R(z)} = \frac{K(0.3679z + 0.2642)}{z^2 + (0.3679K - 1.3679)z + 0.3679 + 0.2642K}$$

Thus the characteristic equation for the system is

$$P(z) = z^2 + (0.3679K - 1.3679)z + 0.3679 + 0.2642K = 0$$

Since this is a second-order system, the Jury stability conditions may be written as follows:

1. $|a_2| < a_0$
2. $P(1) > 0$
3. $P(-1) > 0 \qquad n = 2 = \text{even}$

We shall now apply the first condition for stability. Since $a_2 = 0.3679 + 0.2642K$ and $a_0 = 1$, the first condition for stability becomes

$$|0.3679 + 0.2642K| < 1$$

or

$$2.3925 > K > -5.1775 \qquad (3\text{–}109)$$

The second condition for stability becomes

$$P(1) = 1 + (0.3679K - 1.3679) + 0.3679 + 0.2642K = 0.6321K > 0$$

which gives

$$K > 0 \qquad (3\text{–}110)$$

The third condition for stability gives

$$P(-1) = 1 - (0.3679K - 1.3679) + 0.3679 + 0.2642K = 2.7358 - 0.1037K > 0$$

which yields

$$26.382 > K \qquad (3\text{–}111)$$

For stability, gain constant K must satisfy inequalities (3–109) (3–110), and (3–111). Hence

$$2.3925 > K > 0$$

The range of gain constant K for stability is between 0 and 2.3925.

If gain K is set equal to 2.3925, then the system becomes critically stable (meaning that sustained oscillations exist at the output). The frequency of the sustained oscillations can be determined if 2.3925 is substituted for K in the characteristic equation and the resulting equation is investigated. With $K = 2.3925$ the characteristic equation becomes

$$z^2 - 0.4877z + 1 = 0$$

The characteristic roots are at $z = 0.2439 \pm j0.9698$. Noting that the sampling period T is equal to 1 sec, from Eq. (3–105) we have

$$\omega_d = \frac{\omega_s}{2\pi} \angle z = \frac{2\pi}{2\pi} \angle z = \tan^{-1} \frac{0.9698}{0.2439} = 1.3244 \text{ rad/sec}$$

The frequency of the sustained oscillations is 1.3244 rad/sec.

Comments. When the characteristic polynomial involves only real coefficients, the Jury stability test is very simple to apply. As was noted earlier, the Jury stability test is not limited to polynomials with real coefficients. In applying this test to polynomials with complex coefficients, we insert the complex conjugate coefficients of the first row into the second row and do the same for the third and fourth rows, and so on. Then the resulting table is applicable for the stability test of the polynomial with complex coefficients. The stability conditions, however, become more complicated. We shall not discuss this case in any detail because the characteristic polynomials for physically realizable systems involve only real coefficients.

Stability analysis by use of the bilinear transformation and Routh stability criterion. Another method frequently used in the stability analysis of discrete-time control systems is to use the bilinear transformation coupled with the Routh stability criterion. The method requires transformation from the z plane to another complex plane, the w plane. Those who are familiar with the Routh-Hurwitz stability criterion will find the method simple and straightforward. However, the amount of computation required is much more than that required in the Jury stability criterion.

The bilinear transformation defined by

$$z = \frac{w+1}{w-1}$$

which, when solved for w, gives

$$w = \frac{z+1}{z-1}$$

maps the inside of the unit circle in the z plane into the left half of the w plane. This can be seen as follows. Let the real part of w be called σ and the imaginary part $j\omega$, so that

$$w = \sigma + j\omega$$

Since the inside of the unit circle in the z plane is

$$|z| = \left|\frac{w+1}{w-1}\right| = \left|\frac{\sigma + j\omega + 1}{\sigma + j\omega - 1}\right| < 1$$

or

$$\frac{(\sigma+1)^2 + \omega^2}{(\sigma-1)^2 + \omega^2} < 1$$

we get

$$(\sigma+1)^2 + \omega^2 < (\sigma-1)^2 + \omega^2$$

which yields

$$\sigma < 0$$

Thus the inside of the unit circle in the z plane ($|z| < 1$) corresponds to the left half of the w plane. The unit circle in the z plane is mapped into the imaginary axis in the w plane, and the outside of the unit circle in the z plane is mapped into the right half of the w plane. (It is pointed out that although the w plane is similar to the s plane in that it maps the inside of the unit circle to the left half plane, it is by no means quantitatively equivalent to the s plane. Therefore, estimating the relative stability of the system from the pole locations in the w plane is difficult.)

In the stability analysis using the bilinear transformation coupled with the Routh criterion, we first substitute $(w + 1)/(w - 1)$ for z in the characteristic equation

$$P(z) = a_0 z^n + a_1 z^{n-1} + \cdots + a_{n-1} z + a_n = 0$$

as follows:

$$a_0 \left(\frac{w+1}{w-1}\right)^n + a_1 \left(\frac{w+1}{w-1}\right)^{n-1} + \cdots + a_{n-1} \frac{w+1}{w-1} + a_n = 0$$

Then, clearing the fractions by multiplying both sides of this last equation by $(w - 1)^n$, we obtain

$$Q(w) = b_0 w^n + b_1 w^{n-1} + \cdots + b_{n-1} w + b_n = 0$$

Once we transform $P(z) = 0$ into $Q(w) = 0$, it is possible to apply the Routh stability criterion in the same manner as in continuous-time systems.

It is noted that the bilinear transformation coupled with the Routh stability criterion will indicate exactly how many roots of the characteristic equation lie in the right half of the w plane and how many lie on the imaginary axis. However, such information about the exact number of unstable poles is usually not needed in control systems design, because unstable or critically stable control systems are not desired. As mentioned earlier, the amount of computation required in this approach is much more than that required in the Jury stability test. Therefore, we shall not go any further on this subject here. We refer the reader to Prob. A-3–34, where the present method is used for stability analysis.

A few comments on the stability of closed-loop control systems.

1. If one is interested in the effect of a system parameter on the stability of a closed-loop control system, a root locus diagram may prove to be useful. A digital computer may be employed to compute and plot a root locus diagram.
2. It is noted that in testing the stability of a characteristic equation it may be simpler, in some cases, to find the roots of the characteristic equation directly by use of a digital computer.
3. It is important to point out that stability has nothing to do with the system's ability to follow a particular input. The error signal in a closed-loop control system may increase without bound, even if the system is stable. (Refer to Sec. 4–4 for a discussion of error constants.)

REFERENCES

3–1. Antoniou, A., *Digital Filters: Analysis and Design*. New York: McGraw-Hill Book Company, 1979.
3–2. Bristol, E. H., "Design and Programming Control Algorithms for DDC Systems," *Control Engineering*, **24**, Jan. 1977, pp. 24–26.
3–3. Cadzow, J. A., and H. R. Martens, *Discrete-Time and Computer Control Systems*. Englewood Cliffs, N.J.: Prentice-Hall, Inc., 1970.
3–4. Franklin, G. F., and J. D. Powell, *Digital Control of Dynamic Systems*. Reading, Mass: Addison-Wesley Publishing Co., Inc., 1980.
3–5. Freeman, H., *Discrete-Time Systems*. New York: John Wiley & Sons, Inc., 1965.
3–6. Jerri, A. J., "The Shannon Sampling Theorem—Its Various Extensions and Applications: A Tutorial Review," *Proc. IEEE*, **65** (1977), pp. 1565–95.
3–7. Jury, E. I., "A General z-Transform Formula for Sampled-Data Systems," *IEEE Trans. Automatic Control*, AC-**12** (1967), pp. 606–8.
3–8. Jury, E. I., "Hidden Oscillations in Sampled-Data Control Systems," *AIEE Trans. part II*, **75** (1956), pp. 391–95.
3–9. Jury, E. I., *Sampled-Data Control Systems*. New York: John Wiley & Sons, Inc., 1958.
3–10. Jury E. I., and J. Blanchard, "A Stability Test for Linear Discrete-Time Systems in Table Forms," *Proc. IRE*, **49** (1961), pp. 1947–48.
3–11. Katz, P., *Digital Control Using Microprocessors*. London: Prentice-Hall International, Inc., 1981.
3–12. Kuo, B. C., *Digital Control Systems*. New York: Holt, Rinehart and Winston, Inc., 1980.
3–13. Li, Y. T., J. L. Meiry, and R. E. Curry, "On the Ideal Sampler Approximation," *IEEE Trans. Automatic Control*, AC-**17** (1972), pp. 167–68.
3–14. Linvill, W. K., "Sampled-Data Control Systems Studied through Comparison of Sampling with Amplitude Modulation," *AIEE Trans. part II*, **70** (1951), pp. 1779–88.
3–15. Mitra, S. K., and R. J. Sherwood, "Canonic Realizations of Digital Filters Using the Continued Fraction Expansion," *IEEE Trans. Audio and Electroacoustics*, AU-**20** (1972), pp. 185–94.
3–16. Mitra, S. K., and R. J. Sherwood, "Digital Ladder Networks," *IEEE Trans. Audio and Electroacoustics*, AU-**21** (1973), pp. 30–36.
3–17. Neuman, C. P., and C. S. Baradello, "Digital Transfer Functions for Microcomputer Control," *IEEE Trans. Systems, Man, and Cybernetics*, SMC-**9** (1979), pp. 856–60.
3–18. Ogata, K., *Modern Control Engineering*. Englewood Cliffs, N.J.: Prentice-Hall, Inc., 1970.
3–19. Phillips, C. L., and H. T. Nagle, Jr., *Digital Control System Analysis and Design*. Englewood Cliffs, N.J.: Prentice-Hall, Inc., 1984.
3–20. Ragazzini, J. R., and L. A. Zadeh, "The Analysis of Sampled-Data Systems," *AIEE Trans. part II*, **71** (1952), pp. 225–34.

EXAMPLE PROBLEMS AND SOLUTIONS

Problem A-3–1. Consider a zero-order hold preceded by a sampler. Figure 3–67 shows the input $x(t)$ to the sampler and the output $y(t)$ of the zero-order hold. In the zero-order hold the value of the last sample is retained until the next sample is taken.

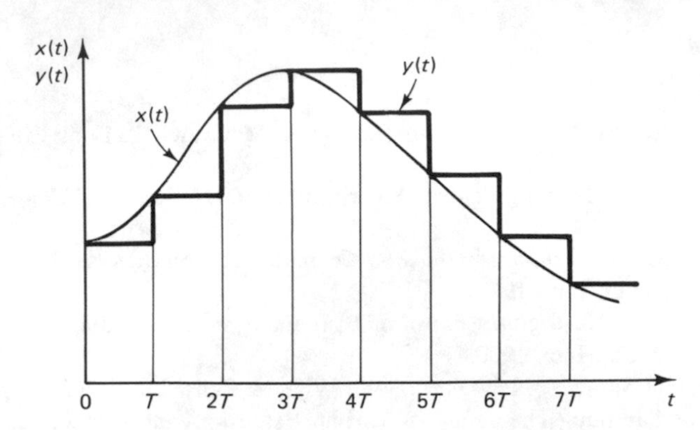

Figure 3–67 Input and output curves for a zero-order hold.

Obtain the expression for $y(t)$. Then find $Y(s)$ and obtain the transfer function of the zero-order hold.

Solution. From Fig. 3–67 we obtain

$$y(t) = x(0)[1(t) - 1(t-T)] + x(T)[1(t-T) - 1(t-2T)]$$
$$+ x(2T)[1(t-2T) - 1(t-3T)] + \cdots$$

The Laplace transform of $y(t)$ is

$$Y(s) = x(0)\left(\frac{1}{s} - \frac{e^{-Ts}}{s}\right) + x(T)\left(\frac{e^{-Ts}}{s} - \frac{e^{-2Ts}}{s}\right)$$
$$+ x(2T)\left(\frac{e^{-2Ts}}{s} - \frac{e^{-3Ts}}{s}\right) + \cdots$$
$$= \frac{1-e^{-Ts}}{s}[x(0) + x(T)e^{-Ts} + x(2T)e^{-2Ts} + \cdots]$$
$$= \frac{1-e^{-Ts}}{s} X^*(s)$$

where

$$X^*(s) = \sum_{k=0}^{\infty} x(kT)e^{-kTs} = \mathscr{L}\left[\sum_{k=0}^{\infty} x(kT)\delta(t-kT)\right]$$

The transfer function of the zero-order hold is thus

$$G_{h0} = \frac{Y(s)}{X^*(s)} = \frac{1-e^{-Ts}}{s}$$

Problem A-3–2. Consider a first-order hold preceded by a sampler. The input to the sampler is $x(t)$ and the output of the first-order hold is $y(t)$. In the first-order hold the output $y(t)$ for $kT \leq t < (k+1)T$ is the straight line that is the extrapolation of the two

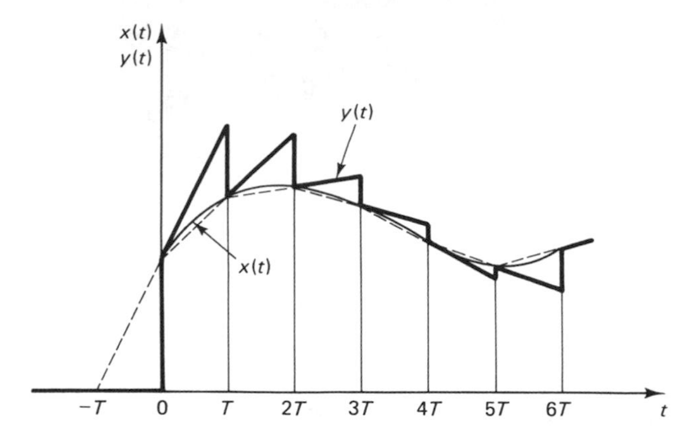

Figure 3–68 Input and output curves for a first-order hold.

preceding sampled values, $x((k-1)T)$ and $x(kT)$, as shown in Fig. 3–68. The equation for the output $y(t)$ is

$$y(t)=\frac{t-kT}{T}[x(kT)-x((k-1)T)]+x(kT) \qquad kT \le t < (k+1)T \qquad (3\text{–}112)$$

Obtain the transfer function of the first-order hold, assuming a simple function such as an impulse function at $t = 0$ as the input $x(t)$.

Solution. For an impulse input of magnitude $x(0)$ such that $x^*(t) = x(0)\delta(t)$, the output $y(t)$ given by Eq. (3–112) becomes as shown in Fig. 3–69. The mathematical expression for $y(t)$ becomes as follows:

$$y(t)=x(0)\left(1+\frac{t}{T}\right)1(t)-\left[2x(0)+2x(0)\frac{t-T}{T}\right]1(t-T)$$

$$+\left[x(0)+x(0)\frac{t-2T}{T}\right]1(t-2T)$$

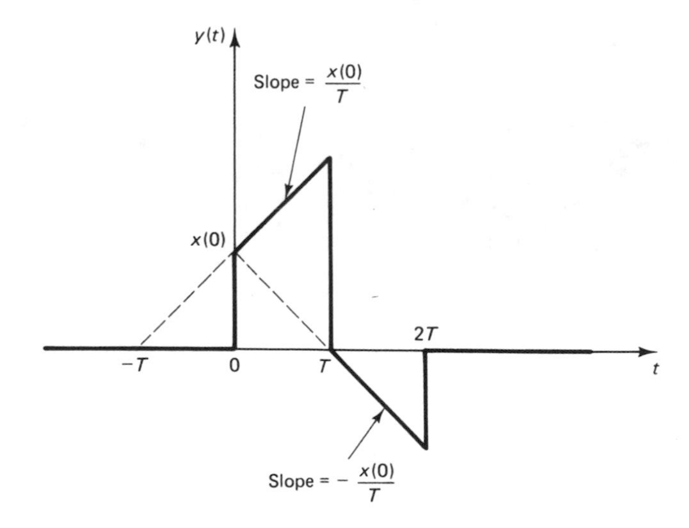

Figure 3–69 Output curve of the first-order hold when the input is a unit impulse function.

Hence,

$$Y(s) = x(0)\left(\frac{1}{s} + \frac{1}{Ts^2}\right) - 2x(0)\left(\frac{e^{-Ts}}{s} + \frac{e^{-Ts}}{Ts^2}\right) + x(0)\left(\frac{e^{-2Ts}}{s} + \frac{e^{-2Ts}}{Ts^2}\right)$$

$$= x(0)(1 - 2e^{-Ts} + e^{-2Ts})\left(\frac{1}{s} + \frac{1}{Ts^2}\right)$$

$$= x(0)\frac{Ts+1}{Ts^2}(1 - e^{-Ts})^2$$

Since

$$X^*(s) = \mathscr{L}[x^*(t)] = \mathscr{L}[x(0)\delta(t)] = x(0)$$

the transfer function of the first-order hold is obtained as follows:

$$G_{h1}(s) = \frac{Y(s)}{X^*(s)} = \frac{Ts+1}{T}\left(\frac{1 - e^{-Ts}}{s}\right)^2$$

Problem A-3-3. Consider the impulse sampler and first-order hold shown in Fig. 3-70. Derive the transfer function of the first-order hold, assuming a unit ramp function as the input $x(t)$ to the sampler.

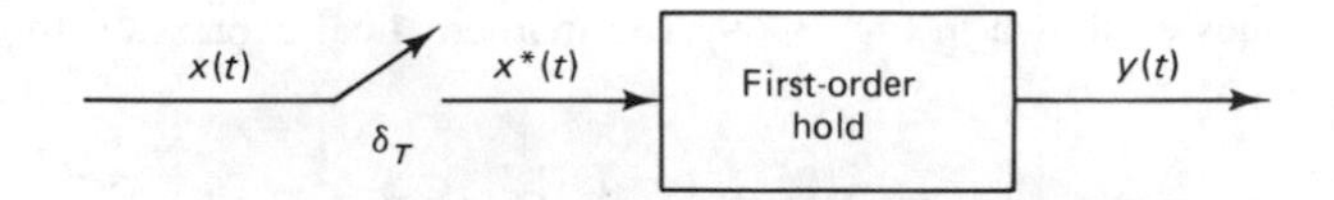

Figure 3-70 Impulse sampler and first-order hold.

Solution. For the unit ramp input $x(t)$, the output $y(t)$ of the first-order hold can be sketched as shown in Fig. 3-71. The equation for the curve $y(t)$ is

$$y(t) = (t - T)1(t - T) + T1(t - T)$$

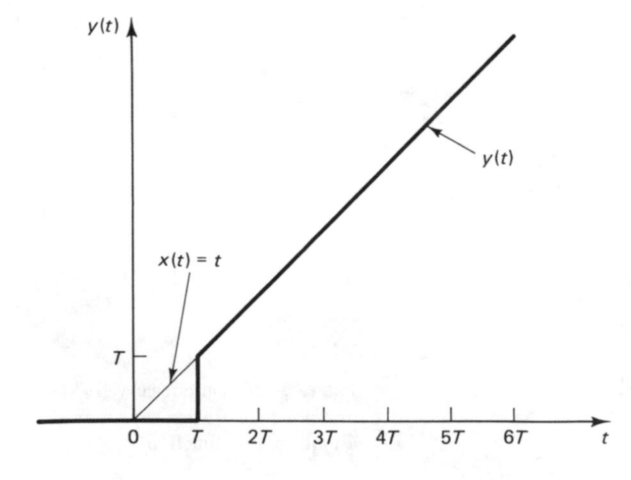

Figure 3-71 Output curve of the first-order hold when the input is a unit ramp function.

The Laplace transform of this last equation gives

$$Y(s)=\frac{1}{s^2}e^{-Ts}+\frac{1}{s}Te^{-Ts}=e^{-Ts}\frac{1+Ts}{s^2}$$

The Laplace transform of the pulsed unit ramp function

$$x^*(t)=\sum_{k=0}^{\infty} kT\delta(t-kT)$$

is

$$X^*(s)=\sum_{k=0}^{\infty} kTe^{-kTs}=Te^{-Ts}+2Te^{-2Ts}+3Te^{-3Ts}+\cdots$$

$$=\frac{Te^{-Ts}}{(1-e^{-Ts})^2}$$

Hence the transfer function of the first-order hold is

$$G_{h1}(s)=\frac{Y(s)}{X^*(s)}=\frac{e^{-Ts}(Ts+1)/s^2}{Te^{-Ts}/(1-e^{-Ts})^2}=\frac{Ts+1}{T}\left(\frac{1-e^{-Ts}}{s}\right)^2$$

Problem A-3-4. Consider the function

$$X(s)=\frac{1-e^{-Ts}}{s}$$

Show that $s=0$ is not a pole of $X(s)$. Show also that

$$Y(s)=\frac{1-e^{-Ts}}{s^2}$$

has a simple pole at $s=0$.

Solution. If a transfer function involves a transcendental term e^{-Ts}, then it may be replaced by a series valid in the vicinity of the pole in question.

For the function

$$X(s)=\frac{1-e^{-Ts}}{s} \tag{3-113}$$

let us obtain the Laurent series expansion about the pole at the origin. Since in the vicinity of the origin, e^{-Ts} may be replaced by

$$e^{-Ts}=1-Ts+\frac{(Ts)^2}{2!}-\frac{(Ts)^3}{3!}+\cdots \tag{3-114}$$

substitution of Eq. (3-114) into Eq. (3-113) gives

$$X(s)=\frac{1}{s}\left[Ts-\frac{(Ts)^2}{2!}+\frac{(Ts)^3}{3!}-\cdots\right]$$

$$=T-\frac{T^2s}{2!}+\frac{T^3s^2}{3!}-\cdots$$

which is the Laurent series expansion of $X(s)$. From this last equation we see that $s = 0$ is not a pole of $X(s)$.

Next, consider $Y(s)$. Since

$$Y(s) = \frac{1 - e^{-Ts}}{s^2}$$

it may be expanded into the Laurent series as

$$Y(s) = \frac{T}{s} - \frac{T^2}{2!} + \frac{T^3 s}{3!} - \cdots$$

We see that pole at the origin ($s = 0$) is of order 1, or is a simple pole.

Problem A-3–5. Obtain the z transform of

$$X(s) = \frac{s}{(s+1)^2(s+2)}$$

by using (1) the partial-fraction-expansion method and (2) the residue method.

Solution.

1. *Partial-fraction-expansion method*. Since $X(s)$ can be expanded into the form

$$X(s) = \frac{2}{s+1} - \frac{1}{(s+1)^2} - \frac{2}{s+2}$$

we have

$$X(z) = 2\left(\frac{1}{1 - e^{-T}z^{-1}}\right) - \frac{Te^{-T}z^{-1}}{(1 - e^{-T}z^{-1})^2} - 2\left(\frac{1}{1 - e^{-2T}z^{-1}}\right)$$

$$= \frac{2 - 2e^{-T}z^{-1} - Te^{-T}z^{-1}}{(1 - e^{-T}z^{-1})^2} - \frac{2}{1 - e^{-2T}z^{-1}}$$

2. *Residue method*. Referring to Eq. (3–31) and noting that $X(s)$ has a double pole at $s = -1$ and a simple pole at $s = -2$, we have

$$X(z) = \frac{1}{(2-1)!} \lim_{s \to -1} \frac{d}{ds}\left[(s+1)^2 \frac{s}{(s+1)^2(s+2)} \frac{z}{z - e^{Ts}}\right]$$

$$+ \lim_{s \to -2}\left[(s+2)\frac{s}{(s+1)^2(s+2)} \frac{z}{z - e^{Ts}}\right]$$

$$= \frac{2z^2 - 2ze^{-T} - Tze^{-T}}{(z - e^{-T})^2} - \frac{2z}{z - e^{-2T}}$$

$$= \frac{2 - 2e^{-T}z^{-1} - Te^{-T}z^{-1}}{(1 - e^{-T}z^{-1})^2} - \frac{2}{1 - e^{-2T}z^{-1}}$$

Problem A-3–6. Consider the function

$$X(s) = \frac{11s^2 + 19s + 11}{(s+1)^3}$$

Obtain $X(z)$ by using two methods: (1) the partial-fraction-expansion method and (2) the residue method.

Solution.

1. *Partial-fraction-expansion method*. Notice that $X(s)$ can be expanded as follows:

$$X(s) = \frac{3}{(s+1)^3} - \frac{3}{(s+1)^2} + \frac{11}{s+1}$$

Hence

$$X(z) = \frac{3}{2}\frac{T^2e^{-T}z(z+e^{-T})}{(z-e^{-T})^3} - \frac{3Tze^{-T}}{(z-e^{-T})^2} + \frac{11z}{z-e^{-T}}$$

$$= \frac{3T^2e^{-2T}z^{-2}}{(1-e^{-T}z^{-1})^3} + \frac{1.5Te^{-T}(T-2)z^{-1}}{(1-e^{-T}z^{-1})^2} + \frac{11}{1-e^{-T}z^{-1}}$$

2. *Residue method*.

$$X(z) = \left[\text{residue of } \frac{X(s)z}{z-e^{Ts}} \text{ at the pole of } X(s)\right]$$

$$= \frac{1}{(3-1)!}\lim_{s\to-1}\frac{d^2}{ds^2}\left[(s+1)^3X(s)\frac{z}{z-e^{Ts}}\right]$$

$$= \frac{1}{2}\lim_{s\to-1}\frac{d^2}{ds^2}\left[(11s^2+19s+11)\frac{z}{z-e^{Ts}}\right]$$

$$= \frac{1}{2}\lim_{s\to-1}\frac{d}{ds}\left[\frac{(22s+19)z(z-e^{Ts}) - (11s^2+19s+11)z(-Te^{Ts})}{(z-e^{Ts})^2}\right]$$

$$= \frac{1}{2}\frac{22z^2 - 22ze^{-T} + 3zT^2e^{-T} - 6zTe^{-T}}{(z-e^{-T})^2} + \frac{1}{2}\frac{6zT^2e^{-2T}}{(z-e^{-T})^3}$$

$$= \frac{3zT^2e^{-2T}}{(z-e^{-T})^3} + \frac{1.5zTe^{-T}(T-2)}{(z-e^{-T})^2} + \frac{11z}{z-e^{-T}}$$

$$= \frac{3T^2e^{-2T}z^{-2}}{(1-e^{-T}z^{-1})^3} + \frac{1.5Te^{-T}(T-2)z^{-1}}{(1-e^{-T}z^{-1})^2} + \frac{11}{1-e^{-T}z^{-1}}$$

Problem A-3-7. Assume that a sampled signal $X^*(s)$ is applied to a system $G(s)$. Assume also that the output of $G(s)$ is $Y(s)$ and $y(0+) = 0$.

$$Y(s) = G(s)X^*(s)$$

Using the relationship

$$Y^*(s) = \frac{1}{T}\sum_{k=-\infty}^{\infty} Y(s + j\omega_s k) \tag{3-115}$$

show that

$$Y^*(s) = G^*(s)X^*(s)$$

Solution. Referring to Eq. (3–115), we have

$$Y^*(s) = \frac{1}{T} \sum_{k=-\infty}^{\infty} G(s + j\omega_s k) X^*(s + j\omega_s k) \tag{3–116}$$

Note that from Eq. (3–36)

$$X^*(s) = \frac{1}{T} \sum_{h=-\infty}^{\infty} X(s + j\omega_s h) + \frac{1}{2} x(0+)$$

Hence

$$X^*(s + j\omega_s k) = \frac{1}{T} \sum_{h=-\infty}^{\infty} X(s + j\omega_s h + j\omega_s k) + \frac{1}{2} x(0+)$$

By letting $h + k = m$, we have

$$X^*(s + j\omega_s k) = \frac{1}{T} \sum_{m=-\infty}^{\infty} X(s + j\omega_s m) + \frac{1}{2} x(0+) = X^*(s) \tag{3–117}$$

Substitution of Eq. (3–117) into Eq. (3–116) yields

$$Y^*(s) = \frac{1}{T} \sum_{k=-\infty}^{\infty} G(s + j\omega_s k) X^*(s)$$

Since from Eq. (3–115) we have

$$G^*(s) = \frac{1}{T} \sum_{k=-\infty}^{\infty} G(s + j\omega_s k)$$

we obtain

$$Y^*(s) = G^*(s) X^*(s)$$

Problem A-3–8. Consider a continuous-time signal $x(t)$ with frequency spectrum limited to between $-\omega_1$ and ω_1. That is,

$$X(j\omega) = 0 \qquad \text{for} \qquad \omega < -\omega_1 \text{ and } \omega_1 < \omega$$

Prove that if this signal is sampled with frequency $\omega_s \geq 2\omega_1$, then the Fourier transform of $x(t)$ is uniquely determined by $x(kT)$, $k = \ldots, -2, -1, 0, 1, 2, \ldots$, and the original continuous-time signal $x(t)$ can be given by a sum of an infinite series of weighted sampled values $x(kT)$ as follows:

$$x(t) = \sum_{k=-\infty}^{\infty} x(kT) \frac{\sin[\omega_s(t - kT)/2]}{\omega_s(t - kT)/2}$$

(This is Shannon's sampling theorem.)

Solution. The Fourier transform of $x(t)$ is given by

$$X(j\omega) = \int_{-\infty}^{\infty} e^{-j\omega t} x(t)\, dt$$

and the inverse Fourier transform is given by

$$x(t) = \frac{1}{2\pi}\int_{-\infty}^{\infty} e^{j\omega t}X(j\omega)d\omega$$

Define the sampled version of $x(t)$ as $x^*(t)$. Then, $x^*(t)$ can be given by

$$x^*(t) = \cdots + x(-T)\delta(t+T) + x(0)\delta(t) + x(T)\delta(t-T) + \cdots$$

$$= \sum_{k=-\infty}^{\infty} x(kT)\delta(t-kT)$$

The Fourier transform of $x^*(t)$ is

$$X^*(j\omega) = \int_{-\infty}^{\infty} e^{-j\omega t}x^*(t)\,dt = \int_{-\infty}^{\infty} e^{-j\omega t}\left[\sum_{k=-\infty}^{\infty} x(kT)\delta(t-kT)\right]dt$$

$$= \sum_{k=-\infty}^{\infty} x(kT)e^{-j\omega kT}$$

Thus, $X^*(j\omega)$ is uniquely determined by $x(kT)$, $k = \ldots, -2, -1, 0, 1, 2, \ldots$.

Referring to Eq. (3–42), the Fourier transform of $x^*(t)$ can be given by

$$X^*(j\omega) = \frac{1}{T}\sum_{k=-\infty}^{\infty} X(j\omega + j\omega_s k)$$

Since the frequency spectrum of the original continuous-time signal $x(t)$ is limited to between $-\omega_1$ and ω_1, we have

$$X(j\omega) = 0 \qquad \text{for} \qquad \omega < -\omega_1 \text{ and } \omega_1 < \omega$$

Since the sampling frequency ω_s is greater than $2\omega_1$, we have

$$X(j\omega) = 0 \qquad \text{for} \qquad \omega < -\tfrac{1}{2}\omega_s \text{ and } \tfrac{1}{2}\omega_s < \omega$$

Hence

$$X^*(j\omega) = \frac{1}{T}[\cdots + X(j\omega + j\omega_s) + X(j\omega) + X(j\omega - j\omega_s) + \cdots]$$

$$= \frac{1}{T}X(j\omega)$$

Thus, we obtain

$$X(j\omega) = \begin{cases} TX^*(j\omega) & -\tfrac{1}{2}\omega_s \le \omega \le \tfrac{1}{2}\omega_s \\ 0 & \omega < -\tfrac{1}{2}\omega_s, \qquad \tfrac{1}{2}\omega_s < \omega \end{cases}$$

The inverse Fourier transform of $X(j\omega)$ gives

$$x(t) = \frac{1}{2\pi}\int_{-\infty}^{\infty} e^{j\omega t}X(j\omega)\,d\omega$$

$$= \frac{T}{2\pi}\int_{-\omega_s/2}^{\omega_s/2} e^{j\omega t}X^*(j\omega)\,d\omega$$

$$= \frac{1}{\omega_s}\int_{-\omega_s/2}^{\omega_s/2} e^{j\omega t}\left[\sum_{k=-\infty}^{\infty} x(kT)e^{-j\omega kT}\right]d\omega$$

$$= \frac{1}{\omega_s} \sum_{k=-\infty}^{\infty} x(kT) \int_{-\omega_s/2}^{\omega_s/2} e^{j\omega(t-kT)}\, d\omega$$

$$= \frac{1}{\omega_s} \sum_{k=-\infty}^{\infty} x(kT) \frac{e^{j\omega(t-kT)}}{j(t-kT)} \Bigg|_{-\omega_s/2}^{\omega_s/2}$$

$$= \sum_{k=-\infty}^{\infty} x(kT) \frac{\sin [\omega_s (t-kT)/2]}{\omega_s (t-kT)/2}$$

Hence we have shown that the original continuous-time signal $x(t)$ can be reconstructed from the sampled data $x(kT)$. [Note that unless $X(j\omega) = 0$ for $\omega < -\omega_1$ and $\omega_1 < \omega$, the continuous-time signal $x(t)$ cannot be determined by sampled data $x(kT)$, $k = \ldots, -2, -1, 0, 1, 2, \ldots$.]

Problem A-3-9. It is known that aliasing may cause a dc error in the system output. Suppose that the sampling frequency is 50 rad/sec, or $\omega_s = 50$ rad/sec, and the noise signal $x(t)$ entering into the system is $\cos 50t$. The frequency spectrum $|X(j\omega)|$ of the noise signal is shown in Fig. 3–72(a). Show that in the present case aliasing will cause a dc component in the system output. Such a dc component will show up as a steady-state error (a dc error) in the output.

Solution. The alias frequency spectrum is shown in Fig. 3–72(b). From this diagram it can be seen that there is a dc component, which will show up in the output as a steady-state error (a dc error).

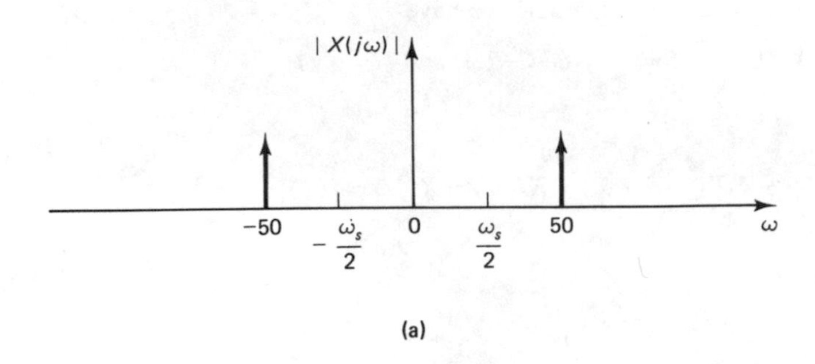

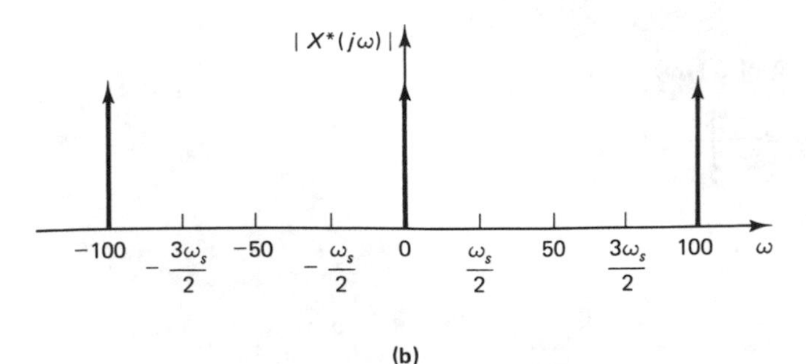

Figure 3–72 (a) Frequency spectrum of noise signal $\cos 50t$; (b) alias frequency spectrum of noise signal $\cos 50t$ when the sampling frequency ω_s is 50 rad/sec.

The dc component caused by aliasing can also be seen from the z transform analysis. Since the z transform of $\cos 50t$ sampled every T sec, where $T = 2\pi/\omega_s = 2\pi/50$, is

$$\mathcal{Z}[\cos 50t] = \mathcal{Z}[\cos 50kT] = \mathcal{Z}\left[\cos\left(50k\frac{2\pi}{50}\right)\right] = \mathcal{Z}[\cos 2k\pi]$$

$$= \frac{1}{1 - z^{-1}} = 1 + z^{-1} + z^{-2} + \cdots$$

it can be seen that the noise signal $\cos 50t$ sampled at every $2\pi/50$ sec becomes unity at every sampling instant. Thus, the noise signal considered here can generate an undesirable dc component in the output.

Problem A-3–10. Draw the magnitude and phase curves of the first-order hold. Then compare the magnitude and phase characteristics of the first-order hold with those of the zero-order hold.

Solution. The transfer function of the first-order hold is

$$G_{h1}(s) = \frac{Ts + 1}{T}\left(\frac{1 - e^{-Ts}}{s}\right)^2$$

By substituting $j\omega$ for s in $G_{h1}(s)$, we obtain

$$\begin{aligned} G_{h1}(j\omega) &= \frac{Tj\omega + 1}{T}\left(\frac{1 - e^{-Tj\omega}}{j\omega}\right)^2 \\ &= \frac{Tj\omega + 1}{T}\left[e^{-j(1/2)T\omega}\,\frac{e^{j(1/2)T\omega} - e^{-j(1/2)T\omega}}{j\omega}\right]^2 \\ &= \frac{Tj\omega + 1}{T}\,e^{-jT\omega}\left[\frac{2j\sin(T\omega/2)}{j\omega}\right]^2 \\ &= \frac{Tj\omega + 1}{T}\,e^{-jT\omega}\,\frac{4\sin^2(T\omega/2)}{\omega^2} \end{aligned}$$

Hence

$$|G_{h1}(j\omega)| = T\sqrt{1 + T^2\omega^2}\left[\frac{\sin(T\omega/2)}{T\omega/2}\right]^2$$

$$\begin{aligned} \underline{/G_{h1}(j\omega)} &= \underline{/Tj\omega + 1} + \underline{/e^{-jT\omega}} \\ &= \tan^{-1}T\omega - T\omega \\ &= \tan^{-1}\frac{2\pi\omega}{\omega_s} - \frac{2\pi\omega}{\omega_s} \end{aligned}$$

where we have used the relationship $T = 2\pi/\omega_s$.

At a few selected values of ω, we have

$$|G_{h1}(j0)| = T \qquad \underline{/G_{h1}(j0)} = 0°$$

$$\left|G_{h1}\left(j\frac{\pi}{T}\right)\right| = 1.336T \qquad \underline{/G_{h1}\left(j\frac{\pi}{T}\right)} = -107.7°$$

$$\left|G_{h1}\left(j\frac{2\pi}{T}\right)\right| = 0 \qquad \underline{/G_{h1}\left(j\frac{2\pi}{T}\right)} = -279.0°$$

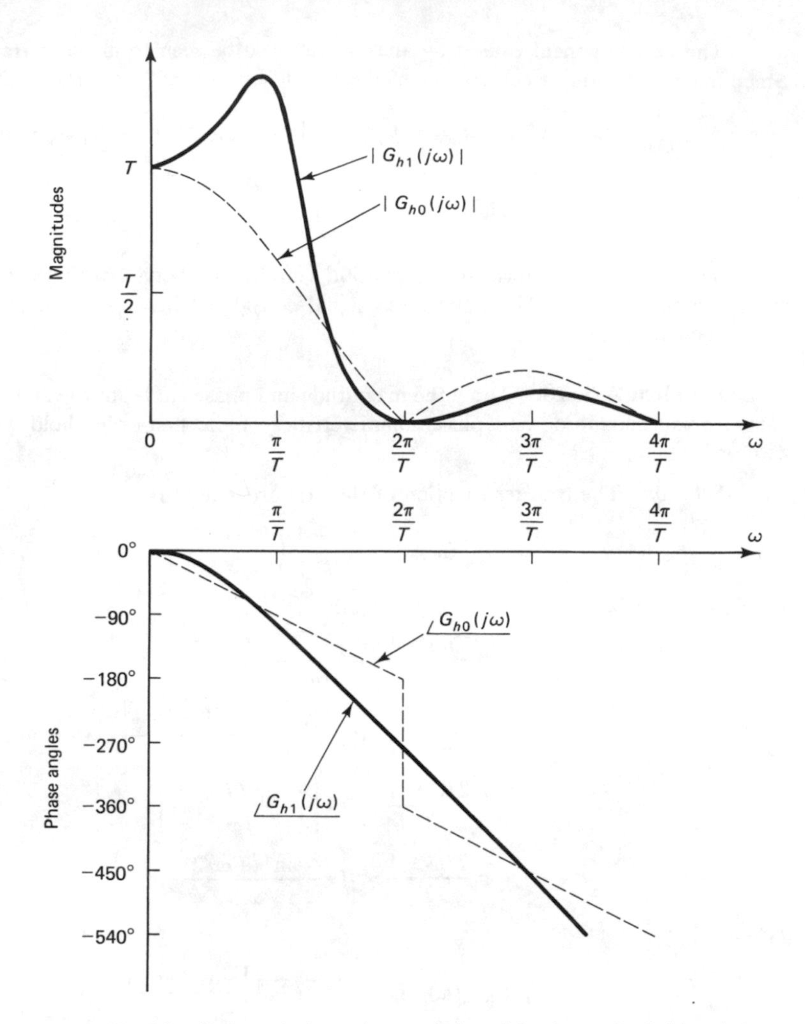

Figure 3–73 Magnitude and phase characteristics of the first-order hold and those of the zero-order hold.

Figure 3–73 shows plots of the magnitude and phase characteristics of the first-order hold and those of the zero-order hold. From Fig. 3–73 it is seen that both the zero-order hold and the first-order hold are not quite satisfactory low-pass filters. They allow significant transmission above the Nyquist frequency, $\omega_N = \pi/T$. It is important, therefore, that the signal be low-pass-filtered before the sampling operation so that the frequency components above the Nyquist frequency are negligible.

Problem A-3–11. Consider the zero-order hold shown in Fig. 3–74. From the diagram we have

$$Y(s) = G(s)X^*(s) = \frac{1 - e^{-Ts}}{s} X^*(s) \tag{3–118}$$

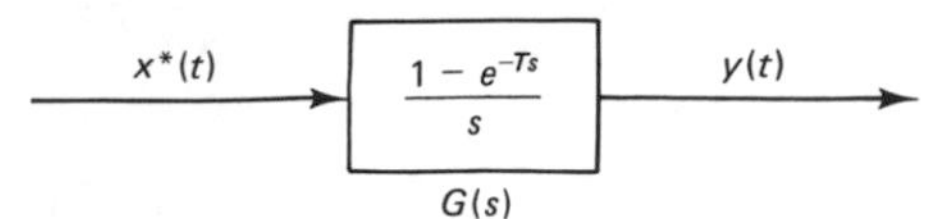

Figure 3–74 Zero-order hold for Prob. A-3–11.

Show that

$$Y^*(s) = X^*(s)$$

Solution. By taking the starred Laplace transform of Eq. (3–118), we have

$$Y^*(s) = \left(\frac{1 - e^{-Ts}}{s}\right)^* X^*(s)$$

In terms of the z transform notation, we have

$$Y(z) = \mathscr{Z}\left[\frac{1 - e^{-Ts}}{s}\right] X(z)$$

where

$$\mathscr{Z}\left[\frac{1 - e^{-Ts}}{s}\right] = (1 - z^{-1})\mathscr{Z}\left[\frac{1}{s}\right] = (1 - z^{-1})\frac{1}{1 - z^{-1}} = 1$$

Hence

$$Y(z) = X(z)$$

In terms of the starred Laplace transform notation, this last equation can be written as

$$Y^*(s) = X^*(s)$$

Problem A-3–12. Consider the system shown in Fig. 3–75(a), where the impulse-sampled signal is fed directly to the system $G(s)$. Obtain the response $y_1(t)$ to the exponential input $x(t) = e^{-t}$. Also, obtain $y_1(kT)$, where the sampling period T is 1 sec.

Next, obtain the response $y_2(kT)$ of the same system with a zero-order hold inserted as shown in Fig. 3–75(b). The input is again $x(t) = e^{-t}$. Then compare $y_1(kT)$ and $y_2(kT)$ with $y_3(kT)$, the response at $t = kT$ of the system shown in Fig. 3–75(c), where again the input is $x(t) = e^{-t}$.

Solution. The response $y_1(t)$ of the system shown in Fig. 3–75(a) can be obtained as follows:

$$y_1(t) = \int_0^t g(t - \tau)x^*(\tau)\, d\tau$$

Since

$$x^*(t) = \sum_{k=0}^{\infty} x(t)\delta(t - kT) = \sum_{k=0}^{\infty} x(kT)\delta(t - kT)$$

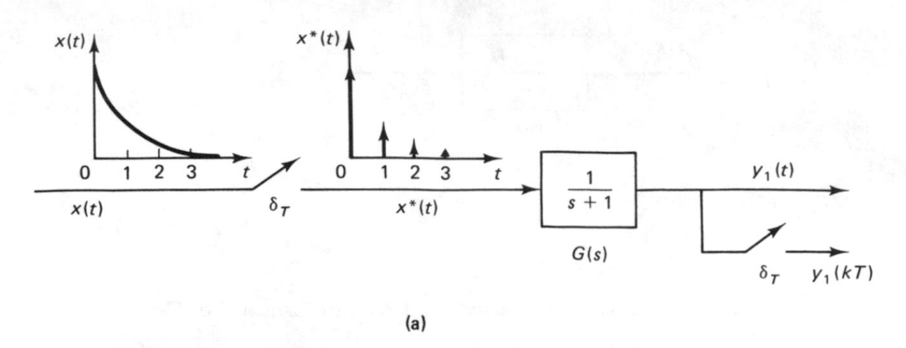

(a)

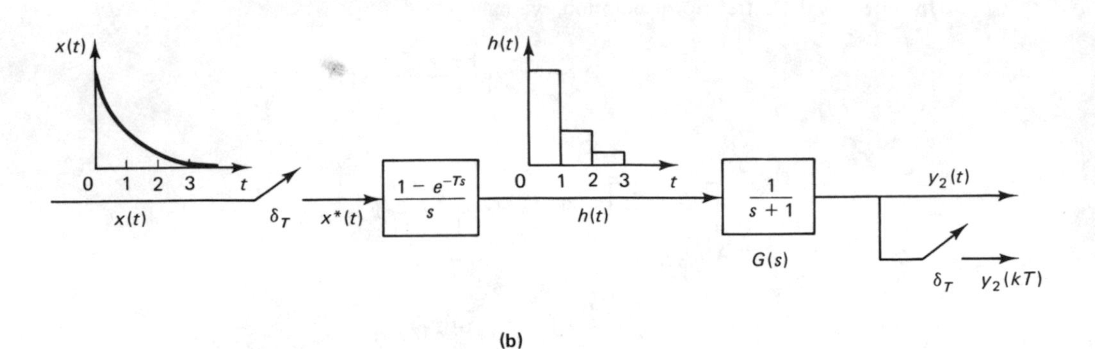

(b)

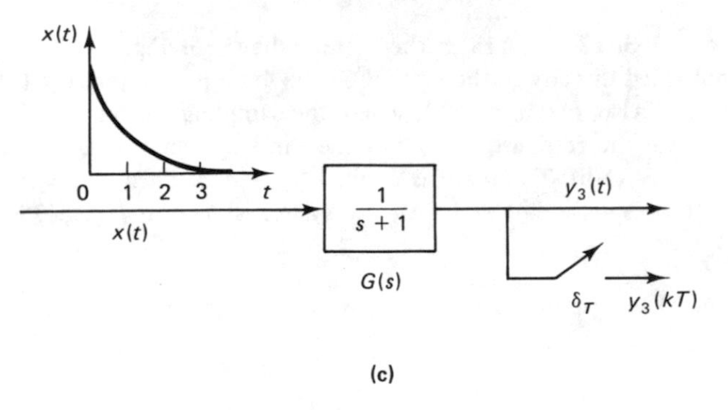

(c)

Figure 3–75 (a) Impulse-sampled system; (b) system with an impulse sampler followed by a zero-order hold; (c) continuous-time system.

is a train of impulses, the response $y_1(t)$ to the input $x^*(t)$ is the sum of the individual impulse responses:

$$y_1(t) = \begin{cases} g(t)x(0) & 0 \le t < T \\ g(t)x(0) + g(t-T)x(T) & T \le t < 2T \\ \cdot \\ \cdot \\ \cdot \\ g(t)x(0) + g(t-T)x(T) + \cdots + g(t-kT)x(kT) & kT \le t < (k+1)T \end{cases}$$

Since $g(t) = e^{-t}$ and $x(t) = e^{-t}$, the preceding $(k + 1)$ equations can be rewritten as follows:

$$y_1(t) = \begin{cases} e^{-t} & 0 \le t < T \\ e^{-t} + e^{-(t-T)}e^{-T} = 2e^{-t} & T \le t < 2T \\ \cdot \\ \cdot \\ \cdot \\ e^{-t} + e^{-(t-T)}e^{-T} + \cdots + e^{-(t-kT)}e^{-kT} = (k+1)e^{-t} & kT \le t < (k+1)T \end{cases}$$

The response $y_1(t)$ when $T = 1$ sec is shown in Fig. 3–76. It is discontinuous.

The response $y_1(kT)$ can be obtained by substituting $t = 0$, $t = T$, $t = 2T$, . . . , $t = kT$ in the preceding $(k + 1)$ equations, or

$$y_1(0) = 1$$
$$y_1(T) = 2e^{-T}$$
$$\cdot$$
$$\cdot$$
$$\cdot$$
$$y_1(kT) = (k + 1)e^{-kT}$$

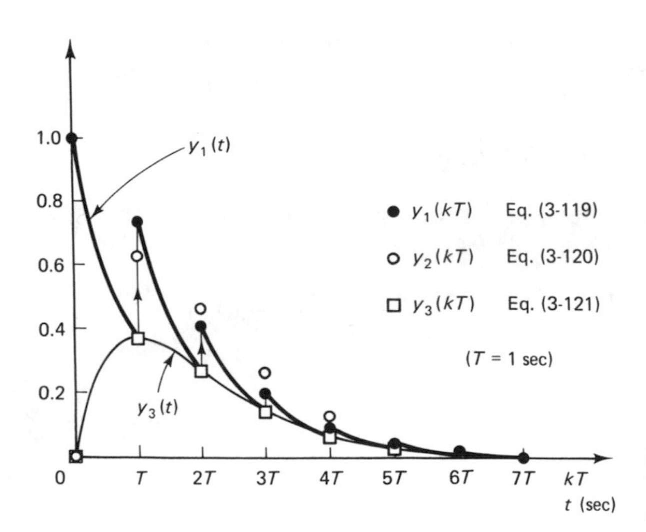

Figure 3–76 Plots of $y_1(t)$ and $y_3(t)$ vs. t and plots of $y_1(kT)$, $y_2(kT)$, and $y_3(kT)$ vs. kT, where $T = 1$ sec.

The same result can also be obtained by the z transform method. Since

$$Y_1(s) = G(s)X^*(s)$$

we have

$$Y_1^*(s) = G^*(s)X^*(s)$$

or

$$Y_1(z) = G(z)X(z)$$

Noting that

$$g(t) = e^{-t} \qquad \text{and} \qquad x(t) = e^{-t}$$

we have

$$G(z) = \frac{1}{1 - e^{-T}z^{-1}} \qquad \text{and} \qquad X(z) = \frac{1}{1 - e^{-T}z^{-1}}$$

Hence

$$Y_1(z) = \frac{1}{1 - e^{-T}z^{-1}} \frac{1}{1 - e^{-T}z^{-1}} = \frac{1}{(1 - e^{-T}z^{-1})^2}$$

or

$$z^{-1}Y_1(z) = \frac{z^{-1}}{(1 - e^{-T}z^{-1})^2}$$

The inverse z transform of this last equation gives

$$y_1((k-1)T) = ke^{-(k-1)T}$$

or

$$y_1(kT) = (k+1)e^{-kT} \qquad k = 0, 1, 2, \ldots \tag{3-119}$$

The response $y_1(kT)$ when $T = 1$ sec is also shown in Fig. 3–76. It is noted that the reason that $y_1(kT)$ does not portray the response $y_1(t)$ is that $\lim_{s\to\infty} sG(s)$ does not vanish, or

$$\lim_{s\to\infty} sG(s) = \lim_{s\to\infty}\left(s\frac{1}{s+1}\right) = 1 \neq 0$$

Next, consider the response $y_2(kT)$ of the system shown in Fig. 3–75(b). Since $Y_2(s)$ can be obtained in the form

$$Y_2(s) = \frac{1 - e^{-Ts}}{s} G(s)X^*(s) = \frac{1 - e^{-Ts}}{s} \frac{1}{s+1} X^*(s)$$

we get

$$Y_2^*(s) = \left(\frac{1 - e^{-Ts}}{s} \frac{1}{s+1}\right)^* X^*(s)$$

or, in terms of the z transform,

$$Y_2(z) = (1 - z^{-1})\,\mathscr{Z}\left[\frac{1}{s(s+1)}\right]\frac{1}{1 - e^{-T}z^{-1}}$$

$$= \frac{1 - e^{-T}}{e^{-T}}\,\frac{e^{-T}z^{-1}}{(1 - e^{-T}z^{-1})^2}$$

The inverse z transform of $Y_2(z)$ gives

$$y_2(kT) = \frac{1 - e^{-T}}{e^{-T}}\,ke^{-kT}$$

Since the sampling period T is 1 sec, we have

$$y_2(kT) = 1.7181ke^{-kT} \qquad T = 1;\ k = 0, 1, 2, \ldots \tag{3–120}$$

The response sequence $y_2(kT)$ when $T = 1$ sec is also shown in Fig. 3–76.

Finally, consider the response of the system shown in Fig. 3–75(c). For the input $x(t) = e^{-t}$, we have

$$X(s) = \frac{1}{s+1}$$

Hence

$$Y_3(s) = G(s)X(s) = \frac{1}{s+1}\,\frac{1}{s+1} = \frac{1}{(s+1)^2}$$

The inverse Laplace transform of $Y_3(s)$ gives

$$y_3(t) = te^{-t}$$

Hence

$$y_3(kT) = kTe^{-kT} \tag{3–121}$$

The response $y_3(t)$ and response sequence $y_3(kT)$ when $T = 1$ sec are shown in Fig. 3–76.

Notice that the values of $y_1(kT)$, $y_2(kT)$, and $y_3(kT)$ for the same k and T values may differ considerably. This is because the net input energy to $G(s)$ is different in each case.

Problem A-3–13. An impulse-sampled system is shown in Fig. 3–77(a). Obtain the response $y_1(t)$ to the exponential input $x(t) = e^{-t}$. Also obtain the response sequence $y_1(kT)$. Figure 3–77(b) shows the same system with an impulse sampler followed by a zero-order hold. Obtain the response sequence $y_2(kT)$ to the exponential input $x(t) = e^{-t}$.

Solution. Referring to the system shown in Fig. 3–77(a), we have

$$g(t) = \mathscr{L}^{-1}[G(s)] = \mathscr{L}^{-1}\left[\frac{1}{s(s+1)}\right] = \mathscr{L}^{-1}\left[\frac{1}{s} - \frac{1}{s+1}\right] = 1 - e^{-t}$$

The response $y_1(t)$ can be obtained as follows:

$$y_1(t) = \int_0^t g(t-\tau)x^*(\tau)\,d\tau$$

$$= \int_0^t g(t-\tau)\sum_{k=0}^{\infty} x(kT)\delta(\tau - kT)\,d\tau$$

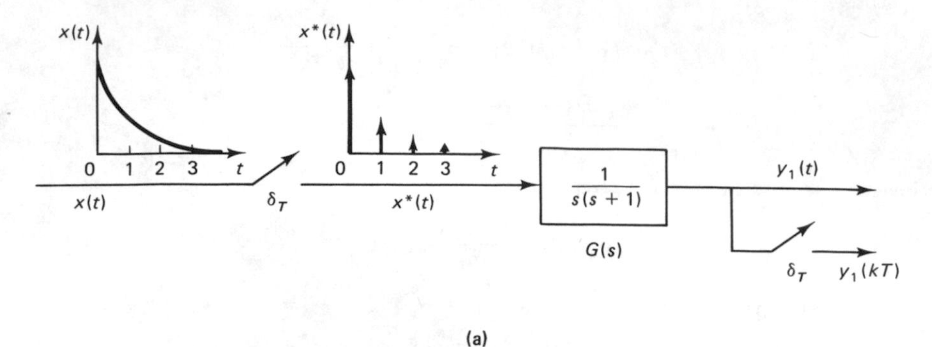

(a)

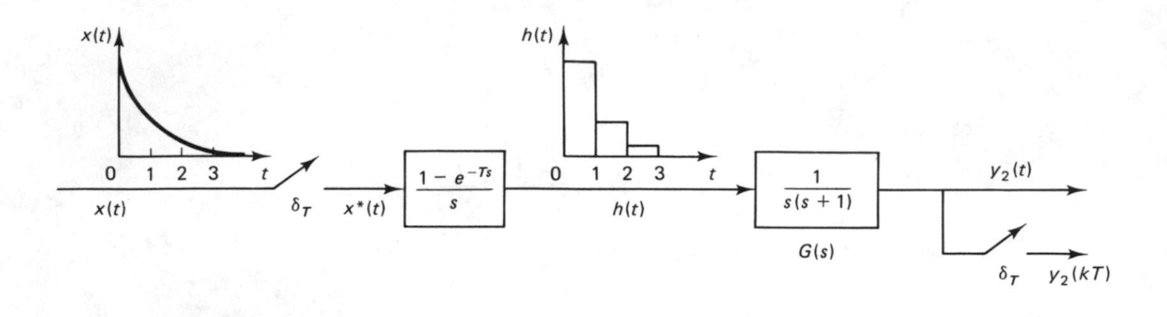

(b)

Figure 3–77 (a) Impulse-sampled system; (b) system with an impulse sampler followed by a zero-order hold (Prob. A-3–13).

or

$$y_1(t) = \begin{cases} g(t)x(0) & 0 \le t < T \\ g(t)x(0) + g(t-T)x(T) & T \le t < 2T \\ \quad \vdots & \\ g(t)x(0) + g(t-T)x(T) + \cdots + g(t-kT)x(kT) & kT \le t < (k+1)T \end{cases}$$

Noting that $x(t) = e^{-t}$, the above $(k + 1)$ equations can be rewritten as follows:

$$y_1(t) = \begin{cases} 1 - e^{-t} & 0 \le t < T \\ (1 - e^{-t}) + [1 - e^{-(t-T)}]e^{-T} = 1 + e^{-T} - 2e^{-t} & T \le t < 2T \\ \quad \vdots & \\ (1 - e^{-t}) + [1 - e^{-(t-T)}]e^{-T} + \cdots + [1 - e^{-(t-kT)}]e^{-kT} & \\ = 1 + e^{-T} + e^{-2T} + \cdots + e^{-(k-1)T} + e^{-kT} - e^{-t} - ke^{-t} & kT \le t < (k+1)T \end{cases}$$

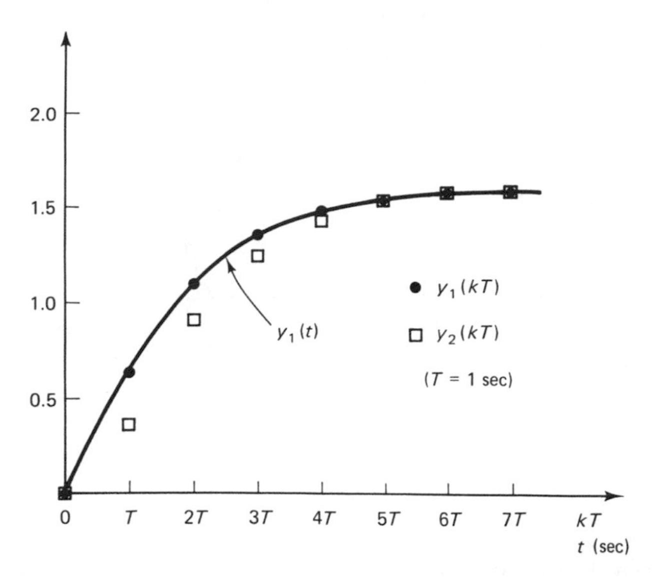

Figure 3–78 Plots of $y_1(t)$ vs. t, $y_1(kT)$ vs. kT, and $y_2(kT)$ vs. kT, where $T = 1$ sec.

Notice that the curve $y_1(t)$ is continuous, since $y_1(T-) = y_1(T+)$, $y_1(2T-) = y_1(2T+)$, . . . , $y_1(kT-) = y_1(kT+)$. The curve $y_1(t)$ versus t is shown in Fig. 3–78.

The values of $y_1(kT)$ can be obtained from $y_1(t)$ by substituting kT for t ($k = 0, 1, 2, \ldots$):

$$y_1(0) = 0$$

$$y_1(T) = 1 - e^{-T}$$

$$y_1(2T) = 1 + e^{-T} - 2e^{-2T}$$

$$\vdots$$

$$y_1(kT) = 1 + e^{-T} + e^{-2T} + \cdots + e^{-(k-1)T} - ke^{-kT}$$

A plot of $y_1(kT)$ versus kT when $T = 1$ sec is also shown in Fig. 3–78.

It is noted that the response $y_1(kT)$ exactly portrays the response $y_1(t)$. This is because

$$\lim_{s\to\infty} sG(s) = \lim_{s\to\infty}\left[s\frac{1}{s(s+1)}\right] = 0$$

and the response $y_1(t)$ is a continuous curve.

To obtain the closed-form expression for $y_1(kT)$, it is convenient to find $Y_1(z)$ by the z transform method and to obtain the inverse z transform of $Y_1(z)$. From Fig. 3–77(a) we find

$$Y_1(s) = G(s)X^*(s)$$

Hence

$$Y_1^*(s) = G^*(s)X^*(s)$$

or

$$\begin{aligned}
Y_1(z) &= G(z)X(z) \\
&= \mathscr{Z}\left[\frac{1}{s(s+1)}\right]\mathscr{Z}[e^{-t}] \\
&= \frac{(1-e^{-T})z^{-1}}{(1-z^{-1})(1-e^{-T}z^{-1})}\frac{1}{1-e^{-T}z^{-1}} \\
&= \frac{1}{1-e^{-T}}\frac{1}{1-z^{-1}} - \frac{e^{-T}z^{-1}}{(1-e^{-T}z^{-1})^2} - \frac{1}{1-e^{-T}}\frac{1}{1-e^{-T}z^{-1}}
\end{aligned}$$

The inverse z transform of $Y_1(z)$ gives

$$\begin{aligned}
y_1(kT) &= \frac{1}{1-e^{-T}} - ke^{-kT} - \frac{1}{1-e^{-T}}e^{-kT} \\
&= \frac{1}{1-e^{-T}} - \left(\frac{1}{1-e^{-T}} + k\right)e^{-kT} \qquad (3\text{–}122)
\end{aligned}$$

Equation (3–122) gives the closed form expression for $y_1(kT)$. To verify that the previously obtained values of $y_1(kT)$ $(k = 0, 1, 2, \ldots)$ are equivalent to Eq. (3–122), note that

$$\begin{aligned}
y_1(0) &= 0 \\
y_1(T) &= \frac{1}{1-e^{-T}} - \left(\frac{1}{1-e^{-T}} + 1\right)e^{-T} = 1 - e^{-T} \\
y_1(2T) &= \frac{1}{1-e^{-T}} - \left(\frac{1}{1-e^{-T}} + 2\right)e^{-2T} = 1 + e^{-T} - 2e^{-2T} \\
&\ \vdots \\
y_1(kT) &= \frac{1}{1-e^{-T}} - \left(\frac{1}{1-e^{-T}} + k\right)e^{-kT} \\
&= 1 + e^{-T} + e^{-2T} + \cdots + e^{-(k-1)T} - ke^{-kT}
\end{aligned}$$

For the system shown in Fig. 3–77(b), we have

$$Y_2(s) = \frac{1-e^{-Ts}}{s}G(s)X^*(s)$$

Hence

$$Y_2^*(s) = \left[\frac{1-e^{-Ts}}{s}\frac{1}{s(s+1)}\right]^* X^*(s)$$

In terms of the z transform,

$$Y_2(z) = (1-z^{-1})\,\mathscr{Z}\left[\frac{1}{s^2(s+1)}\right]X(z)$$

$$= (1 - z^{-1})\, \mathscr{Z}\left[\frac{1}{s^2} - \frac{1}{s} + \frac{1}{s+1}\right] \frac{1}{1 - e^{-T}z^{-1}}$$

$$= (1 - z^{-1}) \left[\frac{Tz^{-1}}{(1 - z^{-1})^2} - \frac{1}{1 - z^{-1}} + \frac{1}{1 - e^{-T}z^{-1}}\right] \frac{1}{1 - e^{-T}z^{-1}}$$

$$= \frac{T}{1 - e^{-T}} \left(\frac{1}{1 - z^{-1}} - \frac{1}{1 - e^{-T}z^{-1}}\right) - \frac{1 - e^{-T}}{e^{-T}} \frac{e^{-T}z^{-1}}{(1 - e^{-T}z^{-1})^2}$$

The inverse z transform of $Y_2(z)$ gives

$$y_2(kT) = \frac{T}{1 - e^{-T}} (1 - e^{-kT}) - \frac{1 - e^{-T}}{e^{-T}} ke^{-kT} \tag{3–123}$$

For $T = 1$ sec, we have

$$y_2(kT) = 1.5820(1 - e^{-kT}) - 1.7181ke^{-kT} \qquad T = 1 \text{ sec}$$

A plot of $y_2(kT)$ versus kT when $T = 1$ sec is also shown in Fig. 3–78.

Problem A-3–14. By direct calculation show that if

$$y(kT) = \sum_{n=0}^{k} g(kT - nT)x(nT)$$

then

$$Y(z) = G(z)X(z)$$

where $G(z) = \mathscr{Z}[g(kT)]$ and $X(z) = \mathscr{Z}[x(kT)]$. Note that $g(kT - nT) = 0$ for $k < n$.

Solution. Since $g(kT - nT) = 0$ for $k < n$, we have

$$\begin{aligned}
G(z)X(z) &= [g(0) + g(T)z^{-1} + g(2T)z^{-2} + g(3T)z^{-3} + \cdots] \\
&\quad \times [x(0) + x(T)z^{-1} + x(2T)z^{-2} + x(3T)z^{-3} + \cdots] \\
&= g(0)x(0) + [g(T)x(0) + g(0)x(T)]z^{-1} \\
&\quad + [g(2T)x(0) + g(T)x(T) + g(0)x(2T)]z^{-2} + \cdots \\
&= \sum_{n=0}^{0} g(0 - nT)x(nT) + \sum_{n=0}^{1} g(T - nT)x(nT)z^{-1} \\
&\quad + \sum_{n=0}^{2} g(2T - nT)x(nT)z^{-2} + \cdots \\
&= y(0) + y(T)z^{-1} + y(2T)z^{-2} + \cdots \\
&= \sum_{k=0}^{\infty} y(kT)z^{-k} \\
&= Y(z)
\end{aligned}$$

Thus, we have shown that

$$Y(z) = G(z)X(z)$$

Problem A-3–15. The Laplace transform of the impulse-sampled signal $x^*(t)$ is given by

$$X^*(s) = \sum_{k=0}^{\infty} x(kT)e^{-kTs} \tag{3–124}$$

Using Eq. (3–124), show that

$$X^*(s) = X^*(s + j\omega_s n) \qquad n = \pm 1, \pm 2, \pm 3, \ldots$$

where $\omega_s = 2\pi/T$.

Solution. Notice that from Eq. (3–124) we have

$$X^*(s + j\omega_s n) = \sum_{k=0}^{\infty} x(kT)e^{-kT(s+j\omega_s n)}$$

$$= \sum_{k=0}^{\infty} x(kT)e^{-kTs}e^{-jknT\omega_s}$$

Since $\omega_s = 2\pi/T$, we have

$$e^{-jknT\omega_s} = e^{-jkn2\pi} = 1 \qquad n = \pm 1, \pm 2, \pm 3, \ldots$$

Hence

$$X^*(s + j\omega_s n) = \sum_{k=0}^{\infty} x(kT)e^{-kTs} = X^*(s) \qquad n = \pm 1, \pm 2, \pm 3, \ldots$$

Problem A-3–16. Consider the system shown in Fig. 3–79. If the impulse response sequence of $G(s)$ is known, then the response $y(kT)$ due to any input $x(kT)$ can be expressed in terms of the convolution summation as follows [refer to Eqs. (3–48) and (3–49)]:

$$y(kT) = \sum_{h=0}^{k} g(kT - hT)x(hT)$$

$$= \sum_{h=0}^{k} x(kT - hT)g(hT)$$

Suppose that the impulse response sequence $g(kT)$ is given as shown in Fig. 3–80(a). Assuming that the input sequence $x(kT)$ is as shown in Fig. 3–80(b), determine the response $y(kT)$ by the convolution summation. Graphically show how to calculate $y(3T)$.

Solution. Figure 3–80(c) shows a plot of $x(3T - hT)$ versus hT and Fig. 3–80(d) depicts the product of $x(3T - hT)$ and $g(hT)$, where $h = 0, 1, 2,$ and 3. The output $y(3T)$ is then given by

$$y(3T) = x(3T)g(0) + x(2T)g(T) + x(T)g(2T) + x(0)g(3T)$$

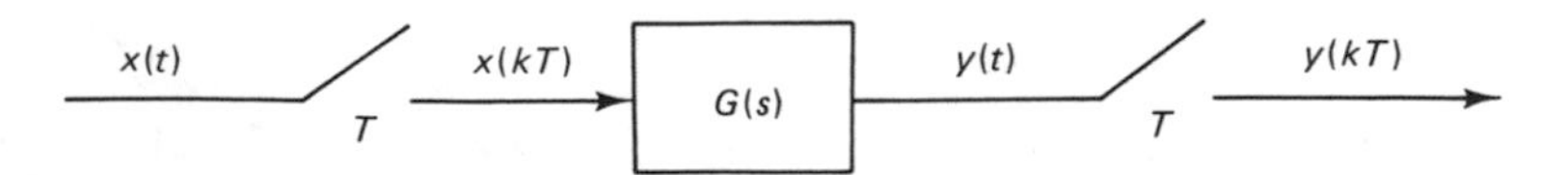

Figure 3–79 Discrete-time system in Prob. A-3–16.

(a)

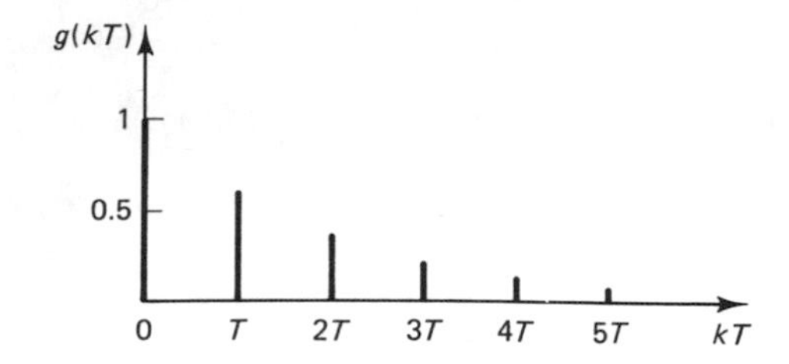

(b)

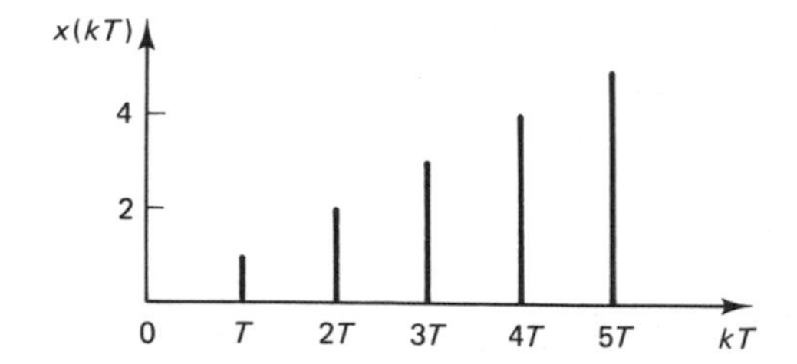

(c)

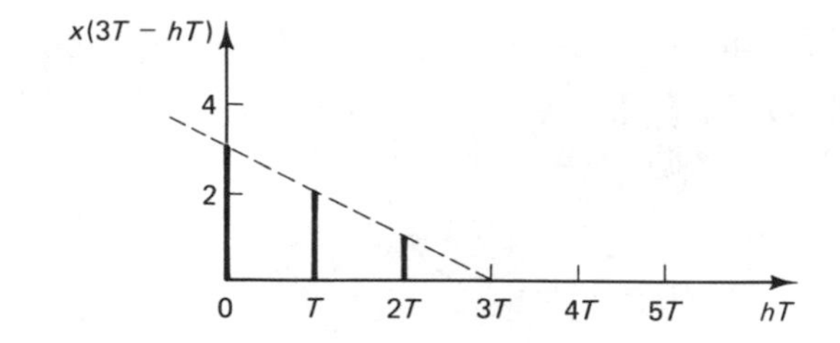

(d)

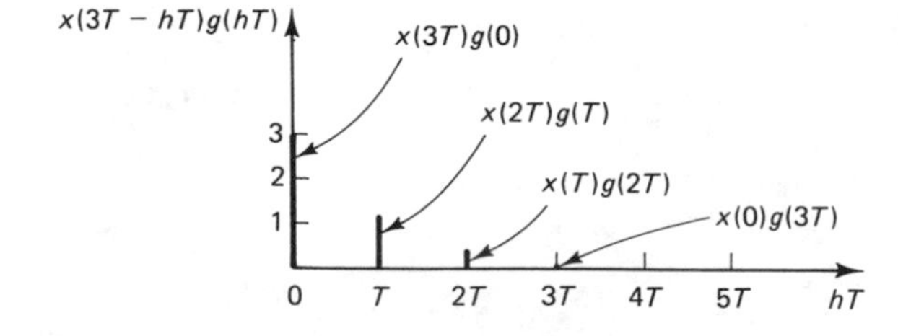

Figure 3–80 (a) Plot of impulse response sequence $g(kT)$ vs. kT; (b) plot of input sequence $x(kT)$ vs. kT; (c) plot of $x(3T - hT)$ vs. hT; (d) plot of $x(3T - hT)g(hT)$ vs. hT for $h = 0, 1, 2,$ and 3.

Note that if the impulse response is relatively short in duration, the direct convolution approach to signal processing can be used. In fact, if the values of $g(kT)$ are stored in the system memory, then as the sampled input enters the system the convolution summation operation can be performed to yield the successive output samples. If the impulse response lasts long, however, the fast Fourier transform method for calculating the convolution summation is preferred in order to avoid the excessive number of operations that will result if the direct convolution approach is used.

Problem A-3–17. Consider the system

$$\frac{Y(z)}{U(z)} = G(z) = \frac{0.8253z^2 + 0.3494z - 0.1747}{z^3}$$

Using the convolution summation

$$y(k) = \sum_{h=0}^{k} g(k-h)u(h)$$

obtain the response $y(k)$ to a unit step sequence input $u(k)$.

Solution. Note that

$$G(z) = 0.8253z^{-1} + 0.3494z^{-2} - 0.1747z^{-3}$$

Hence

$$g(0) = 0$$
$$g(1) = 0.8253$$
$$g(2) = 0.3494$$
$$g(3) = -0.1747$$
$$g(k) = 0 \qquad k = 4, 5, 6, \ldots$$

Since for a unit step sequence input we have $u(k) = 1$ for $k = 0, 1, 2, \ldots$, the output $y(k)$ can be obtained from the convolution summation equation as follows:

$$y(0) = g(0)u(0) = 0$$
$$y(1) = g(1)u(0) + g(0)u(1) = 0.8253$$
$$y(2) = g(2)u(0) + g(1)u(1) + g(0)u(2) = 0.3494 + 0.8253 = 1.1747$$
$$\begin{aligned} y(3) &= g(3)u(0) + g(2)u(1) + g(1)u(2) + g(0)u(3) \\ &= -0.1747 + 0.3494 + 0.8253 + 0 = 1.0000 \end{aligned}$$
$$\begin{aligned} y(k) &= g(k)u(0) + g(k-1)u(1) + \cdots + g(1)u(k-1) + g(0)u(k) \\ &= 1.0000 \qquad k = 4, 5, 6, \ldots \end{aligned}$$

Rewriting, we have

$$y(0) = 0$$
$$y(1) = 0.8253$$
$$y(2) = 1.1747$$
$$y(k) = 1.0000 \qquad k = 3, 4, 5, \ldots$$

Problem A-3–18. Consider the system

$$y(k) - ay(k-1) = x(k) \tag{3–125}$$

where $-1 < a < 1$ and $y(-1) = 0$. Assume that $x(k)$ is the Kronecker delta input, so that

$$x(0) = 1$$

$$x(k) = 0 \qquad k \neq 0$$

From Eq. (3–125) we have

$$y(k) = ay(k-1) + x(k)$$

Hence

$$y(0) = ay(-1) + x(0) = 1$$

$$y(1) = ay(0) + x(1) = a$$

$$y(2) = ay(1) + x(2) = a^2$$

$$\vdots$$

$$y(k) = a^k$$

The sequence $1, a, a^2, a^3, \ldots$ is the weighting sequence for the given system. Once the weighting sequence is obtained, the response of the system to any input can be determined by use of the convolution summation.

By using the convolution summation approach, determine the response $y(k)$ of the system when the input $x(k)$ is the unit step sequence.

Solution. The convolution summation is given by Eq. (3–48) or (3–49):

$$y(k) = \sum_{h=0}^{k} g(k-h)x(h)$$

$$= \sum_{h=0}^{k} x(k-h)g(h)$$

For the unit step sequence input, we have

$$x(k) = \begin{cases} 1 & k = 0, 1, 2, \ldots \\ 0 & k < 0 \end{cases}$$

Since the weighting sequence is

$$g(h) = a^h$$

we obtain the response to the unit step sequence as follows:

$$y(k) = \sum_{h=0}^{k} x(k-h)a^h = \sum_{h=0}^{k} a^h$$

$$= 1 + a + a^2 + \cdots + a^k$$

$$= \frac{1 - a^{k+1}}{1 - a}$$

Problem A-3–19. Obtain the weighting sequence of the system defined by

$$G_n(z)=\frac{1}{(1+az^{-1})^n}$$

for $n = 1$, 2, and 3, respectively.

Solution. For $n = 1$, we have

$$G_1(z)=\frac{1}{1+az^{-1}}=1-az^{-1}+a^2z^{-2}-a^3z^{-3}+\cdots$$

Hence, the weighting sequence $g_1(k)$ is found to be

$$g_1(k)=(-a)^k$$

For $n = 2$, we obtain

$$G_2(z)=\frac{1}{(1+az^{-1})^2}=\frac{1-az^{-1}+a^2z^{-2}-a^3z^{-3}+\cdots}{1+az^{-1}}$$
$$=1-2az^{-1}+3a^2z^{-2}-4a^3z^{-3}+\cdots$$

Hence, the weighting sequence $g_2(k)$ is

$$g_2(k)=(k+1)(-a)^k$$

For $n = 3$, we get

$$G_3(z)=\frac{1}{(1+az^{-1})^3}=\frac{1-2az^{-1}+3a^2z^{-2}-4a^3z^{-3}+\cdots}{1+az^{-1}}$$
$$=1-3az^{-1}+6a^2z^{-2}-10a^3z^{-3}+\cdots$$

Hence, the weighting sequence $g_3(k)$ is

$$g_3(k)=\frac{(k+2)(k+1)}{2}(-a)^k$$

Problem A-3–20. Consider the system shown in Fig. 3–81(a), where two discrete-time systems whose weighting sequences are $g_1(k)$ and $g_2(k)$, respectively, are connected in parallel. Obtain the equivalent weighting sequence $g(k)$ of the system.

Consider next the case where two discrete-time systems are connected in series, as shown in Fig. 3–81(b). Obtain the equivalent weighting sequence $g(k)$ of the system.

Solution. For the system of Fig. 3–81(a) we have

$$c_1(k)=\sum_{h=0}^{k}g_1(k-h)r(h)$$

and

$$c_2(k)=\sum_{h=0}^{k}g_2(k-h)r(h)$$

Hence

$$c(k) = c_1(k) + c_2(k)$$
$$= \sum_{h=0}^{k} [g_1(k-h) + g_2(k-h)]r(h)$$

Since $c(k)$ can be given by

$$c(k) = \sum_{h=0}^{k} g(k-h)r(h)$$

we find

$$g(k-h) = g_1(k-h) + g_2(k-h)$$

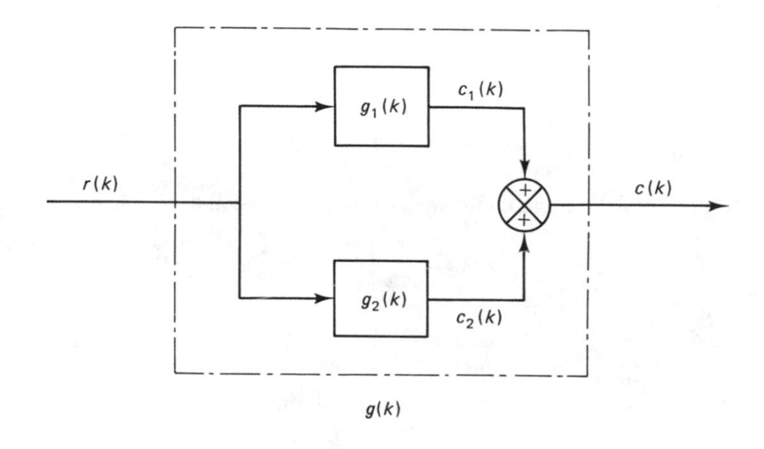

(a)

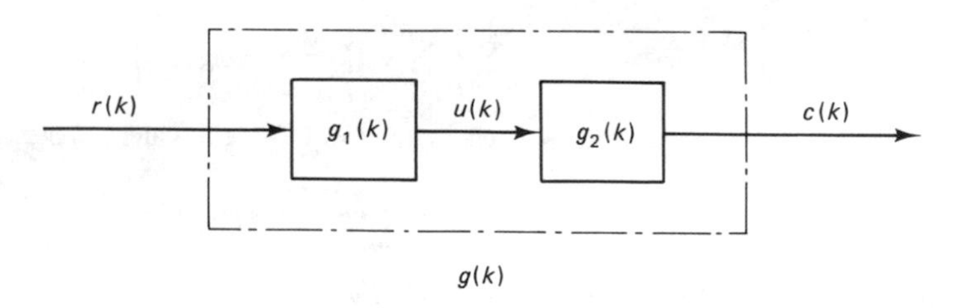

(b)

Figure 3–81 (a) Parallel-connected system; (b) series-connected system.

The equivalent weighting sequence $g(k)$ is thus given by

$$g(k) = g_1(k) + g_2(k)$$

For the system of Fig. 3–81(b), we have

$$c(k) = \sum_{n=0}^{k} g_2(k-n)u(n)$$

and

$$u(n) = \sum_{h=0}^{n} g_1(n-h)r(h)$$

Hence

$$c(k) = \sum_{n=0}^{k} g_2(k-n) \sum_{h=0}^{n} g_1(n-h)r(h)$$

$$= \sum_{n=0}^{k} \left[\sum_{h=0}^{n} g_1(n-h)r(h) \right] g_2(k-n)$$

Define $n = k - i$ and eliminate n from this last equation. Then

$$c(k) = \sum_{i=k}^{0} \left[\sum_{h=0}^{k-i} g_1(k-i-h)r(h) \right] g_2(i)$$

$$= \sum_{i=0}^{k} \left[\sum_{h=0}^{k-i} g_1(k-i-h)r(h) \right] g_2(i)$$

Note that the upper limit of the second summation can be changed from $k - i$ to k, since $g_1(j) = 0$ for $j < 0$. Consequently,

$$c(k) = \sum_{i=0}^{k} \left[\sum_{h=0}^{k} g_1(k-i-h)r(h) \right] g_2(i)$$

$$= \sum_{h=0}^{k} \left[\sum_{i=0}^{k} g_1(k-h-i)g_2(i) \right] r(h)$$

Similarly, the upper limit of the second summation can be changed from k to $k - h$ without changing the value of the summation. Hence

$$c(k) = \sum_{h=0}^{k} \left[\sum_{i=0}^{k-h} g_1(k-h-i)g_2(i) \right] r(h) \tag{3–126}$$

Notice that $c(k)$ can also be written in terms of the equivalent weighting sequence $g(k)$ as

$$c(k) = \sum_{h=0}^{k} g(k-h)r(h) \tag{3–127}$$

Hence, comparing Eqs. (3–126) and (3–127) we obtain

$$g(k-h) = \sum_{i=0}^{k-h} g_1(k-h-i)g_2(i)$$

or

$$g(k) = \sum_{i=0}^{k} g_1(k-i)g_2(i)$$

This last equation gives the equivalent weighting sequence $g(k)$ of the system shown in Fig. 3–81(b).

Problem A-3–21. Given the system equation

$$y(k) = ay(k-1) + bx(k-1) + cx(k)$$

obtain $y(k)$ in terms of $x(0)$, $x(1)$, . . . , $x(k)$. Assume that $y(k)$ and $x(k)$ are zero for negative k.

Solution. Notice that

$$y(0) = cx(0)$$

$$\begin{aligned} y(1) &= ay(0) + bx(0) + cx(1) \\ &= (ac+b)x(0) + cx(1) \end{aligned}$$

$$\begin{aligned} y(2) &= ay(1) + bx(1) + cx(2) \\ &= a[(ac+b)x(0) + cx(1)] + bx(1) + cx(2) \\ &= (ac+b)[ax(0) + x(1)] + cx(2) \end{aligned}$$

$$\begin{aligned} y(3) &= ay(2) + bx(2) + cx(3) \\ &= a\{(ac+b)[ax(0) + x(1)] + cx(2)\} + bx(2) + cx(3) \\ &= (ac+b)[a^2x(0) + ax(1) + x(2)] + cx(3) \end{aligned}$$

In this way we get

$$y(k) = (ac+b)[a^{k-1}x(0) + a^{k-2}x(1) + \cdots + x(k-1)] + cx(k)$$

Problem A-3–22. Obtain the discrete-time output $C(z)$ of the closed-loop control system shown in Fig. 3–82. Also, obtain the continuous-time output $C(s)$.

Solution. From the diagram we have

$$C(s) = G_2(s)M^*(s)$$

$$M(s) = G_1(s)E(s)$$

$$E(s) = R(s) - H(s)C(s)$$

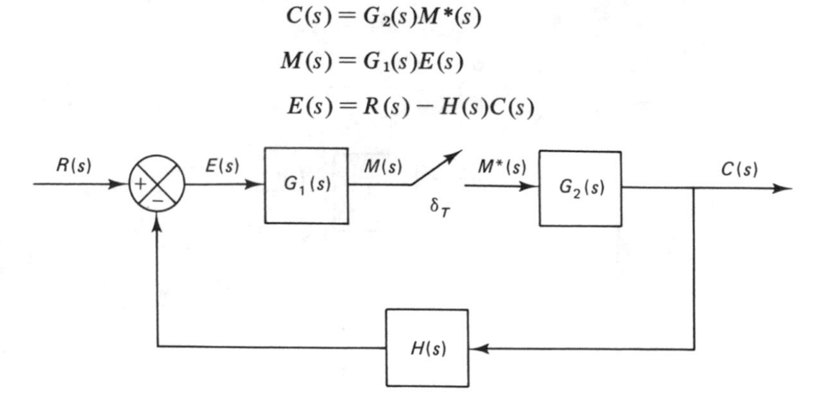

Figure 3–82 Discrete-time control system.

Hence

$$M(s) = G_1(s)[R(s) - H(s)C(s)]$$
$$= G_1(s)R(s) - G_1(s)H(s)G_2(s)M^*(s)$$

Taking the starred Laplace transform of this last equation, we obtain

$$M^*(s) = [G_1R(s)]^* - [G_1G_2H(s)]^* M^*(s)$$

or

$$M^*(s) = \frac{[G_1R(s)]^*}{1 + [G_1G_2H(s)]^*}$$

Since $C(s) = G_2(s)M^*(s)$, we have

$$C^*(s) = G_2^*(s)M^*(s) = \frac{G_2^*(s)[G_1R(s)]^*}{1 + [G_1G_2H(s)]^*}$$

In terms of the z transform notation,

$$C(z) = \frac{G_2(z)G_1R(z)}{1 + G_1G_2H(z)}$$

This last equation gives the discrete-time output $C(z)$.

The continuous-time output $C(s)$ can be obtained from the following equation:

$$C(s) = G_2(s)M^*(s) = G_2(s)\frac{[G_1R(s)]^*}{1 + [G_1G_2H(s)]^*}$$

Notice that $[G_1R(s)]^*/\{1 + [G_1G_2H(s)]^*\}$ is a series of impulses. The continuous-time output $C(s)$ is the response of $G_2(s)$ to the sequence of such impulses. [See Prob. A-3-26 for details of determining the continuous-time output $c(t)$, the inverse Laplace transform of $C(s)$.]

Problem A-3-23. Consider the system shown in Fig. 3-83. Obtain the closed-loop pulse transfer function $C(z)/R(z)$. Also, obtain the expression for $C(s)$.

Solution. From the diagram we have

$$C(s) = G_2(s)M^*(s)$$
$$M(s) = G_1(s)E^*(s)$$
$$E(s) = R(s) - H(s)C(s) = R(s) - H(s)G_2(s)M^*(s)$$

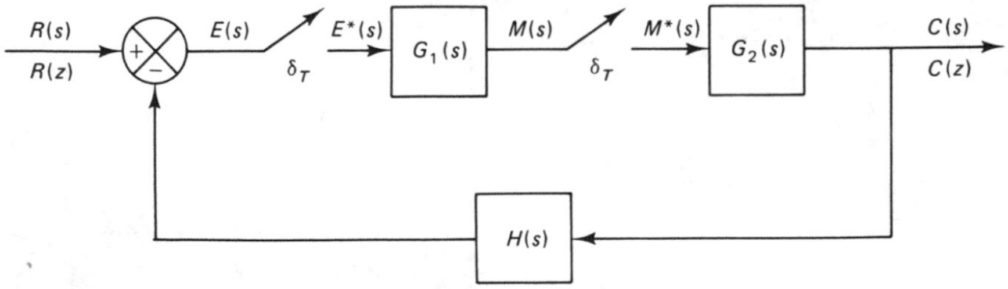

Figure 3-83 Discrete-time control system.

Taking the starred Laplace transforms of both sides of the last three equations gives

$$C^*(s) = G_2^*(s)M^*(s)$$

$$M^*(s) = G_1^*(s)E^*(s)$$

$$E^*(s) = R^*(s) - HG_2^*(s)M^*(s)$$

Solving for $C^*(s)$ gives

$$C^*(s) = G_2^*(s)G_1^*(s)[R^*(s) - HG_2^*(s)M^*(s)]$$

or

$$\begin{aligned} C^*(s) &= G_1^*(s)G_2^*(s)R^*(s) - G_1^*(s)G_2^*(s)HG_2^*(s)M^*(s) \\ &= G_1^*(s)G_2^*(s)R^*(s) - G_1^*(s)HG_2^*(s)C^*(s) \end{aligned}$$

Thus

$$C^*(s)[1 + G_1^*(s)HG_2^*(s)] = G_1^*(s)G_2^*(s)R^*(s)$$

or

$$\frac{C^*(s)}{R^*(s)} = \frac{G_1^*(s)G_2^*(s)}{1 + G_1^*(s)HG_2^*(s)}$$

In terms of the z transform notation, we have

$$\frac{C(z)}{R(z)} = \frac{G_1(z)G_2(z)}{1 + G_1(z)HG_2(z)}$$

The continuous-time output $C(s)$ can be obtained from the following equation:

$$C(s) = G_2(s)M^*(s) = G_2(s)\frac{G_1^*(s)R^*(s)}{1 + G_1^*(s)HG_2^*(s)}$$

Problem A-3–24. Obtain the pulse transfer function $C(z)/R(z)$ of the closed-loop control system shown in Fig. 3–84. Also, obtain the pulse transfer function between $X(z)$ and $R(z)$.

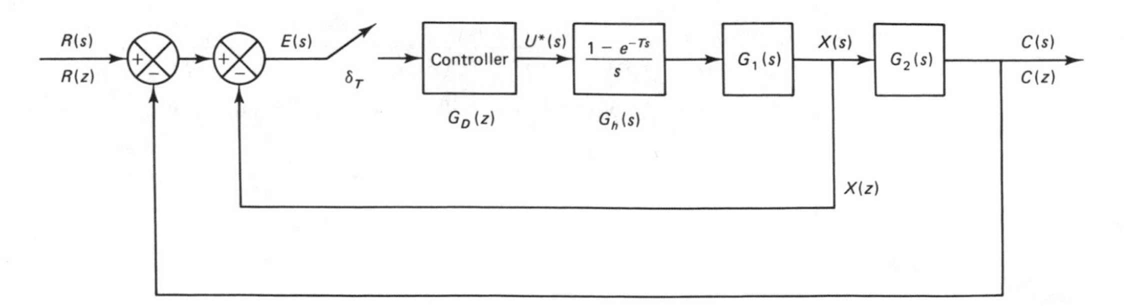

Figure 3–84 Discrete-time control system.

Solution. From the block diagram of Fig. 3–84, we have

$$C(s) = G_h(s)G_1(s)G_2(s)U^*(s)$$

$$X(s) = G_h(s)G_1(s)U^*(s)$$

$$E(s) = R(s) - X(s) - C(s)$$

Taking the starred Laplace transform of the preceding three equations, we obtain

$$C^*(s) = [G_h(s)G_1(s)G_2(s)]^*U^*(s) = [G_hG_1G_2(s)]^*U^*(s)$$

$$X^*(s) = [G_h(s)G_1(s)]^*U^*(s) = [G_hG_1(s)]^*U^*(s)$$

$$E^*(s) = R^*(s) - X^*(s) - C^*(s)$$

In terms of z transform notation, we have

$$C(z) = G_hG_1G_2(z)U(z)$$

$$X(z) = G_hG_1(z)U(z)$$

$$E(z) = R(z) - X(z) - C(z)$$

Also, from the given block diagram we have

$$U(z) = G_D(z)E(z)$$

Hence

$$\begin{aligned} U(z) &= G_D(z)[R(z) - X(z) - C(z)] \\ &= G_D(z)[R(z) - G_hG_1(z)U(z) - G_hG_1G_2(z)U(z)] \end{aligned}$$

or

$$U(z)[1 + G_D(z)G_hG_1(z) + G_D(z)G_hG_1G_2(z)] = G_D(z)R(z)$$

The closed-loop pulse transfer function is then obtained from

$$\begin{aligned} C(z) &= G_hG_1G_2(z)U(z) \\ &= \frac{G_hG_1G_2(z)G_D(z)R(z)}{1 + G_D(z)G_hG_1(z) + G_D(z)G_hG_1G_2(z)} \end{aligned}$$

as follows:

$$\frac{C(z)}{R(z)} = \frac{G_D(z)G_hG_1G_2(z)}{1 + G_D(z)[G_hG_1(z) + G_hG_1G_2(z)]}$$

The pulse transfer function between $X(z)$ and $R(z)$ is obtained from

$$\begin{aligned} X(z) &= G_hG_1(z)U(z) \\ &= \frac{G_hG_1(z)G_D(z)R(z)}{1 + G_D(z)G_hG_1(z) + G_D(z)G_hG_1G_2(z)} \end{aligned}$$

as follows:

$$\frac{X(z)}{R(z)} = \frac{G_D(z)G_hG_1(z)}{1 + G_D(z)[G_hG_1(z) + G_hG_1G_2(z)]}$$

Problem A-3-25. Consider the analog PID controller and the digital PID controller. The equation for the analog PID controller is

$$m(t) = K\left[e(t) + \frac{1}{T_i}\int_0^t e(t)\,dt + T_d\frac{de(t)}{dt}\right]$$

where $e(t)$ is the input to the controller and $m(t)$ is the output of the controller. The transfer function of the analog PID controller is

$$G(s) = \frac{M(s)}{E(s)} = K\left(1 + \frac{1}{T_i s} + T_d s\right)$$

The pulse transfer function of the digital PID controller in the positional form is as given by Eq. (3–66):

$$G_D(z) = \frac{M(z)}{E(z)} = K_P + \frac{K_I}{1 - z^{-1}} + K_D(1 - z^{-1})$$

where $K_P = K - \frac{1}{2}K_I$.

Compare the polar plots (frequency response characteristics) of the analog PID controller with those of the digital PID controller.

Solution. For the analog PID controller, the frequency response characteristics can be obtained by substituting $j\omega$ for s in $G(s)$. Thus,

$$G(j\omega) = K\left(1 + \frac{1}{T_i j\omega} + T_d j\omega\right)$$

$$= K\left(1 - j\frac{1}{T_i\,\omega} + T_d j\omega\right) \qquad (3\text{–}128)$$

For the digital PID controller, the frequency response characteristics can be obtained by substituting $z = e^{j\omega T}$ into $G_D(z)$:

$$G_D(e^{j\omega T}) = K_P + \frac{K_I}{1 - e^{-j\omega T}} + K_D(1 - e^{-j\omega T})$$

$$= K_P + \frac{K_I}{1 - \cos\omega T + j\sin\omega T} + K_D(1 - \cos\omega T + j\sin\omega T)$$

$$= K_P + \frac{K_I}{2}\left(1 - j\frac{\sin\omega T}{1 - \cos\omega T}\right) + K_D(1 - \cos\omega T + j\sin\omega T) \qquad (3\text{–}129)$$

We shall first compare separately the P action, the I action, and the D action of the analog controller with their counterparts in the digital controller. Notice that in the proportional action (P action) the digital controller has a gain $K_I/2$ less than the corresponding gain in the analog controller, since $K_P = K - \frac{1}{2}K_I$. See Fig. 3–85(a).

For the integral action (I action) the real parts of the polar plots of the analog controller and digital controller differ by $K_I/2$, as shown in Fig. 3–85(b).

When the proportional action and integral action are combined, then the real parts of the polar plots for the analog PI action and the digital PI action become the same, as shown in Fig. 3–85(c).

The polar plots of the derivative action (D action) for the analog controller and the

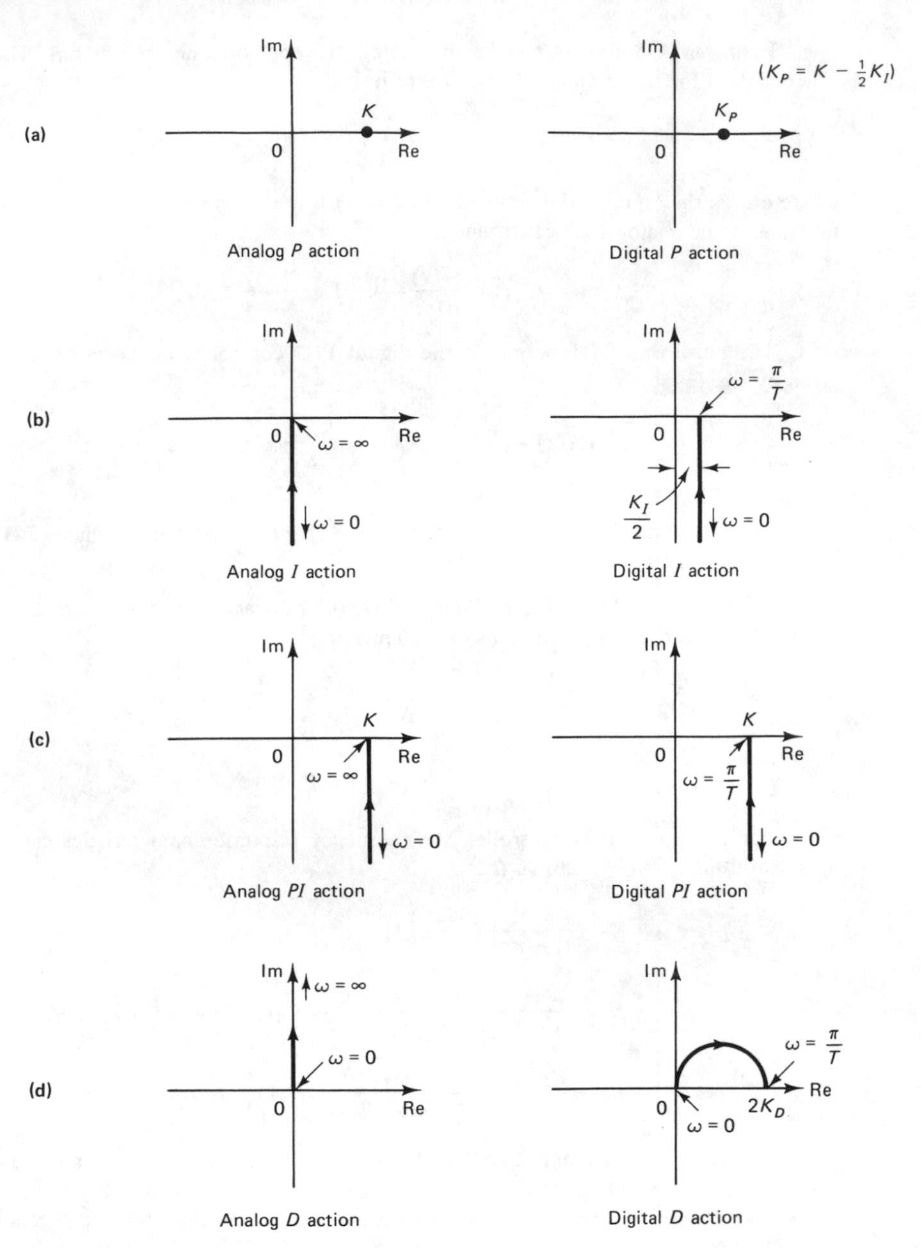

Figure 3–85 Polar plots of analog and digital controllers with (a) proportional action, (b) integral action, (c) proportional plus integral action, and (d) derivative action.

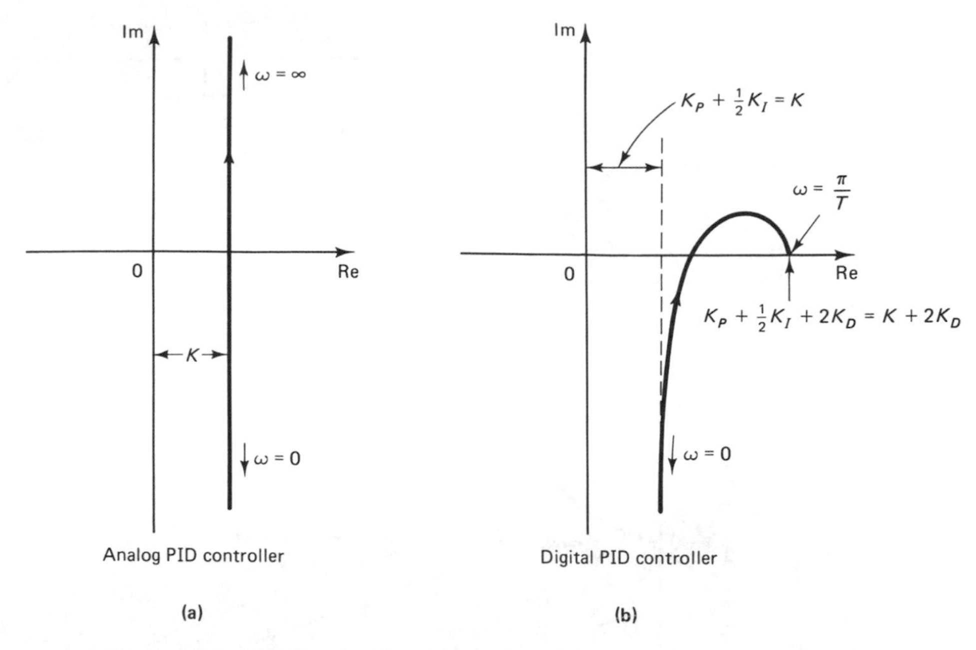

Figure 3–86 (a) Polar plot of analog PID controller; (b) polar plot of digital PID controller.

digital controller differ very much, as shown in Fig. 3–85(d). Hence, there are considerable differences in the analog D action and the digital D action.

The qualitative polar plot of the analog PID controller can be obtained from Eq. (3–128) by varying ω from 0 to ∞, as shown in Fig. 3–86(a). Similarly, the qualitative polar plot of the digital PID controller can be obtained from Eq. (3–129) by varying ω from 0 to π/T, as shown in Fig. 3–86(b).

Note that although the polar plots of the analog PI controller and the digital PI controller are similar, there are significant differences between the polar plots of the analog PID controller and the digital PID controller.

Problem A-3–26. In Example 3–11 we obtained the unit step response sequence $c(k)$ for the system shown in Fig. 3–87(a). (The sampling period $T = 1$ sec.) The output $c(k)$ was obtained as the inverse z transform of $C(z)$.

It is desired to obtain the continuous-time output $c(t)$ so that the output between any two consecutive sampling instants can be determined. Find the expression for the continuous-time output $c(t)$.

Solution. For the system shown in Fig. 3–87(a), we have

$$C(s) = G(s)E^*(s)$$

$$E(s) = R(s) - C(s)$$

Hence

$$E^*(s) = R^*(s) - C^*(s) = R^*(s) - G^*(s)E^*(s)$$

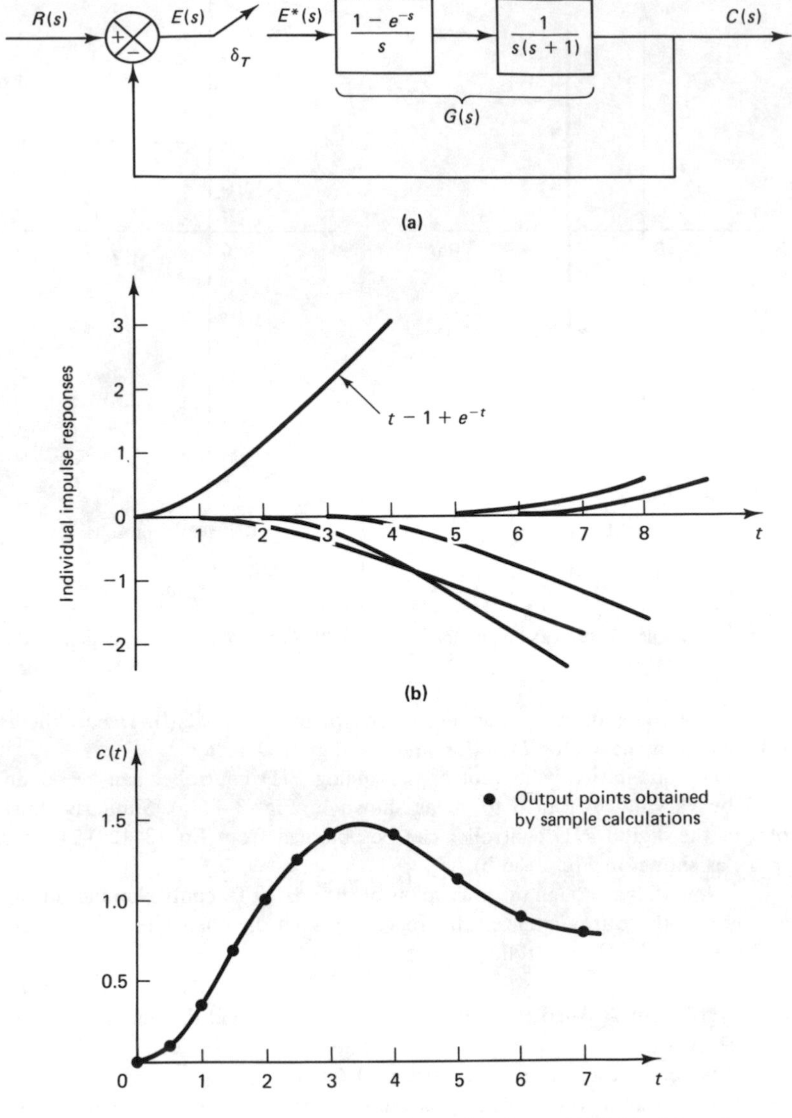

Figure 3–87 (a) Discrete-time control system; (b) plots of individual impulse responses; (c) plot of continuous-time output $c(t)$ vs. t.

or

$$E^*(s) = \frac{R^*(s)}{1 + G^*(s)}$$

Thus

$$C(s) = G(s)\frac{R^*(s)}{1 + G^*(s)}$$

The continuous-time output $c(t)$ can therefore be obtained as the inverse Laplace transform of $C(s)$:

$$c(t) = \mathscr{L}^{-1}[C(s)] = \mathscr{L}^{-1}\left[G(s)\frac{R^*(s)}{1+G^*(s)}\right]$$

For the present system,

$$G(s) = \frac{1-e^{-s}}{s}\frac{1}{s(s+1)}$$

Hence

$$c(t) = \mathscr{L}^{-1}\left[\frac{1-e^{-s}}{s}\frac{1}{s(s+1)}\frac{R^*(s)}{1+G^*(s)}\right]$$

Let us define

$$X^*(s) = (1-e^{-s})\frac{R^*(s)}{1+G^*(s)}$$

Then the z transform expression for this last equation is

$$X(z) = (1-z^{-1})\frac{R(z)}{1+G(z)}$$

Referring to Eq. (3–71) for the z transform of $G(s)$, we obtain

$$X(z) = (1-z^{-1})\frac{\dfrac{1}{1-z^{-1}}}{1+\dfrac{0.3679z^{-1}+0.2642z^{-2}}{(1-0.3679z^{-1})(1-z^{-1})}}$$

$$= \frac{1-1.3679z^{-1}+0.3679z^{-2}}{1-z^{-1}+0.6321z^{-2}}$$

Hence, noting that the sampling period T is 1 sec or $T = 1$, we have

$$X^*(s) = \frac{1-1.3679e^{-s}+0.3679e^{-2s}}{1-e^{-s}+0.6321e^{-2s}}$$

Therefore,

$$c(t) = \mathscr{L}^{-1}\left[\frac{1}{s^2(s+1)}\frac{1-1.3679e^{-s}+0.3679e^{-2s}}{1-e^{-s}+0.6321e^{-2s}}\right]$$

$$= \mathscr{L}^{-1}\left[\frac{1}{s^2(s+1)}(1-0.3679e^{-s}-0.6321e^{-2s}-0.3996e^{-3s}\right.$$

$$\left.+0e^{-4s}+0.2526e^{-5s}+0.2526e^{-6s}+\cdots)\right]$$

Since

$$\frac{1}{s^2(s+1)} = \frac{1}{s^2}-\frac{1}{s}+\frac{1}{s+1}$$

the inverse Laplace transform of this last equation is

$$\mathscr{L}^{-1}\left[\frac{1}{s^2(s+1)}\right] = t - 1 + e^{-t}$$

Hence we obtain

$$\begin{aligned} c(t) = (t - 1 + e^{-t}) &- 0.3679[(t-1) - 1 + e^{-(t-1)}]1(t-1) \\ &- 0.6321[(t-2) - 1 + e^{-(t-2)}]1(t-2) \\ &- 0.3996[(t-3) - 1 + e^{-(t-3)}]1(t-3) \\ &+ 0.0000[(t-4) - 1 + e^{-(t-4)}]1(t-4) \\ &+ 0.2526[(t-5) - 1 + e^{-(t-5)}]1(t-5) \\ &+ 0.2526[(t-6) - 1 + e^{-(t-6)}]1(t-6) \\ &+ \cdots \end{aligned} \tag{3-130}$$

Figure 3–87(b) shows plots of individual impulse responses given by Eq. (3–130). [Observe that $c(t)$ consists of the sum of impulse responses which occur at $t = 0$, $t = 1$, $t = 2$, . . . with weighting factors 1, −0.3679, −0.6321, ...]

From Eq. (3–130) we see that for time intervals $0 \leq t < 1$, $1 \leq t < 2$, $2 \leq t < 3$, . . . the output $c(t)$ is the sum of impulse responses as follows:

$$c(t) = \begin{cases} t - 1 + e^{-t} & 0 \leq t < 1 \\ (t - 1 + e^{-t}) - 0.3679[(t-1) - 1 + e^{-(t-1)}]1(t-1) & 1 \leq t < 2 \\ (t - 1 + e^{-t}) - 0.3679[(t-1) - 1 + e^{-(t-1)}]1(t-1) & \\ \quad - 0.6321[(t-2) - 1 + e^{-(t-2)}]1(t-2) & 2 \leq t < 3 \\ \vdots & \end{cases}$$

Sample calculations for several values of t follow.

$$c(0) = 0 - 1 + 1 = 0$$

$$c(0.5) = 0.5 - 1 + 0.6065 = 0.1065$$

$$c(1.0) = 0.3679 - 0.3679 \times 0 = 0.3679$$

$$c(1.5) = 0.7231 - 0.3679 \times 0.1065 = 0.6839$$

$$c(2.0) = 1.1353 - 0.3679 \times 0.3679 = 1.0000$$

$$c(2.5) = 1.5821 - 0.3679 \times 0.7231 - 0.6321 \times 0.1065 = 1.2487$$

$$c(3.0) = 2.0498 - 0.3679 \times 1.1353 - 0.6321 \times 0.3679 = 1.3996$$

$$c(4.0) = 3.0183 - 0.3679 \times 2.0498 - 0.6321 \times 1.1353 - 0.3996 \times 0.3679 = 1.3996$$

$$\begin{aligned} c(5.0) = 4.0067 &- 0.3679 \times 3.0183 - 0.6321 \times 2.0498 - 0.3996 \times 1.1353 \\ &+ 0 \times 0.3679 = 1.1469 \end{aligned}$$

$$c(6.0) = 5.0025 - 0.3679 \times 4.0067 - 0.6321 \times 3.0183 - 0.3996 \times 2.0498$$
$$+ 0 \times 1.1353 + 0.2526 \times 0.3679 = 0.8944$$

$$\vdots$$

(Note that such calculations can easily be carried out by a microcomputer.) The continuous-time output $c(t)$ thus obtained is plotted in Fig. 3–87(c).

It is noted that the same problem was solved in Examples 3–11 and 3–15. As a matter of course, the $c(k)$ values for $k = 0, 1, 2, \ldots$ obtained in Example 3–11 agree with the values of $c(t)$ for $t = 0, 1, 2, \ldots$ obtained here, and the $c(k)$ values for $k = 0.5, 1.5, 2.5, \ldots$ obtained in Example 3–15 agree with the corresponding values obtained here.

Problem A-3–27. Consider the system shown in Fig. 3–88. The sampling period is 2 sec, or $T = 2$. The input $x(t)$ is a Kronecker delta function $\delta_0(t)$; that is,

$$\delta_0(k) = \begin{cases} 1 & k = 0 \\ 0 & k \neq 0 \end{cases}$$

Obtain the response every 0.5 sec by using the modified z transform method.

Solution. Since the input $x(t)$ is a Kronecker delta function, we have

$$X(z) = 1$$

The modified pulse transfer function $G(z, m)$ is obtained as follows. Referring to Eq. (3–77),

$$G(z, m) = z^{-1}\left[\text{residue of } \frac{1}{s + 0.6931}\frac{e^{mTs}z}{z - e^{Ts}} \text{ at pole } s = -0.6931\right]$$

Noting that $T = 2$, we obtain

$$G(z, m) = z^{-1}\left\{\lim_{s \to -0.6931}\left[(s + 0.6931)\frac{1}{s + 0.6931}\frac{e^{2ms}z}{z - e^{2s}}\right]\right\}$$

$$= z^{-1}\frac{(e^{-1.3862})^m z}{z - e^{-1.3862}} = \frac{4^{-m}}{z - 0.25}$$

Hence, the output $Y(z, m)$ can be obtained as follows:

$$Y(z, m) = G(z, m)X(z) = \frac{4^{-m}}{z - 0.25}$$

Referring to Eq. (3–81), we have

$$zY(z, m) = y_0(m) + y_1(m)z^{-1} + y_2(m)z^{-2} + \cdots$$

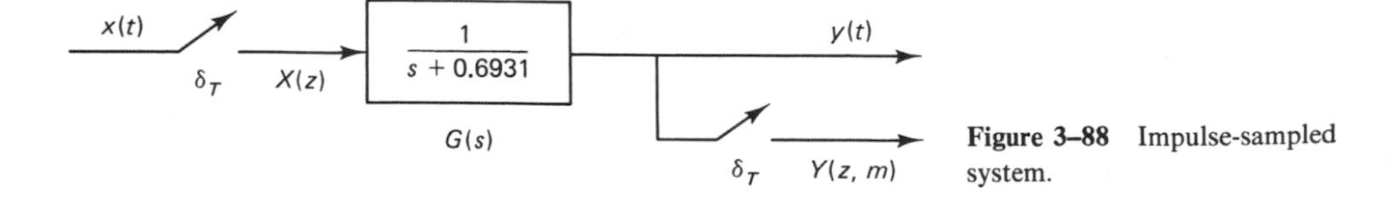

Figure 3–88 Impulse-sampled system.

where $y_k(m) = y((k + m)T) = y(2k + 2m)$. In this problem $zY(z, m)$ can be expanded into an infinite series in z^{-1} as follows:

$$zY(z, m) = \frac{4^{-m}}{1 - 0.25z^{-1}}$$
$$= 4^{-m} + 4^{-m-1}z^{-1} + 4^{-m-2}z^{-2} + 4^{-m-3}z^{-3} + \cdots$$

Hence,

$$y_0(m) = 4^{-m}$$
$$y_1(m) = 4^{-m-1}$$
$$y_2(m) = 4^{-m-2}$$
$$y_3(m) = 4^{-m-3}$$
$$\vdots$$

To obtain the system output every 0.5 sec, we set $m = 0, 0.25, 0.50$, and 0.75. For $m = 0.25$, we obtain

$$y_0(0.25) = y(0.5) = 4^{-0.25} = 0.7071$$
$$y_1(0.25) = y(2.5) = 4^{-1.25} = 0.1768$$
$$y_2(0.25) = y(4.5) = 4^{-2.25} = 0.04419$$
$$y_3(0.25) = y(6.5) = 4^{-3.25} = 0.01105$$
$$\vdots$$

Similarly, the values of $y_k(m)$ for $m = 0, 0.5$, and 0.75 can be calculated. The result is shown in Fig. 3–89 as a plot of $y_k(m)$ versus k.

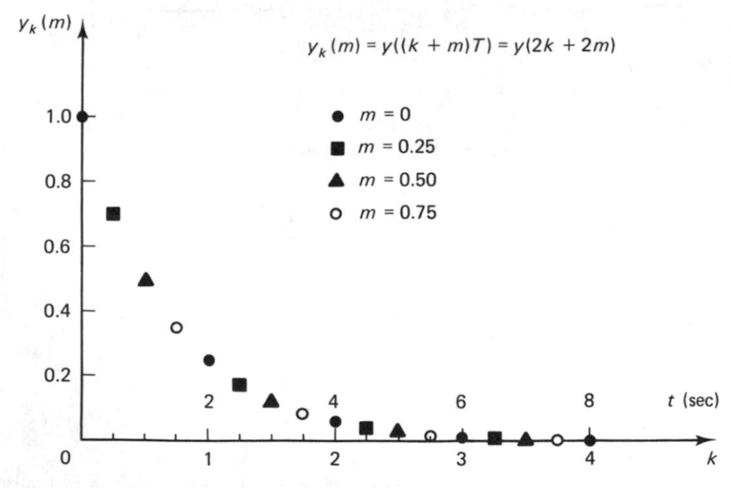

Figure 3–89 Plot of $y_k(m)$ vs. k for the system considered in Prob. A-3–27.

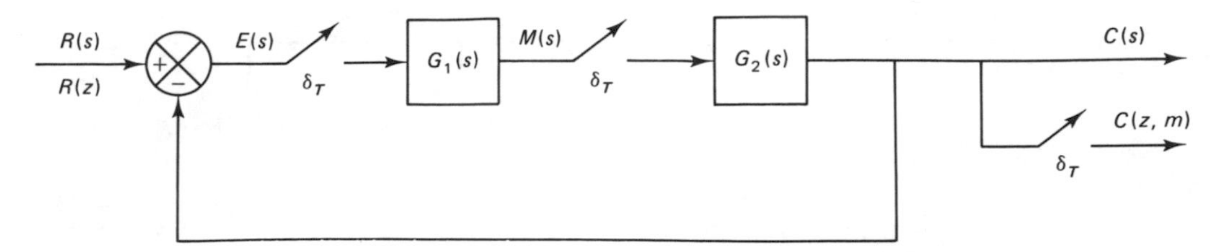

Figure 3–90 Closed-loop discrete-time control system.

Problem A-3–28. Obtain $C(z, m)$, the modified z transform of the output, of the system shown in Fig. 3–90.

Solution. From Fig. 3–90 we have

$$E(s) = R(s) - C(s)$$

$$M(s) = G_1(s)E^*(s)$$

$$C(s) = G_2(s)M^*(s)$$

Hence

$$M^*(s) = G_1^*(s)E^*(s)$$

or

$$M(z) = G_1(z)E(z)$$

Also,

$$E^*(s) = R^*(s) - C^*(s) = R^*(s) - G_2^*(s)M^*(s)$$

or

$$E(z) = R(z) - G_2(z)M(z)$$

Therefore,

$$M(z) = G_1(z)[R(z) - G_2(z)M(z)]$$

from which we obtain

$$M(z) = \frac{G_1(z)R(z)}{1 + G_1(z)G_2(z)}$$

Since $C(z, m)$ can be given by $G_2(z, m)M(z)$, we have

$$C(z, m) = G_2(z, m)M(z) = \frac{G_1(z)G_2(z, m)}{1 + G_1(z)G_2(z)} R(z)$$

Problem A-3–29. Consider the digital filter system shown in Fig. 3–91. Obtain the pulse transfer function $Y(z)/X(z)$.

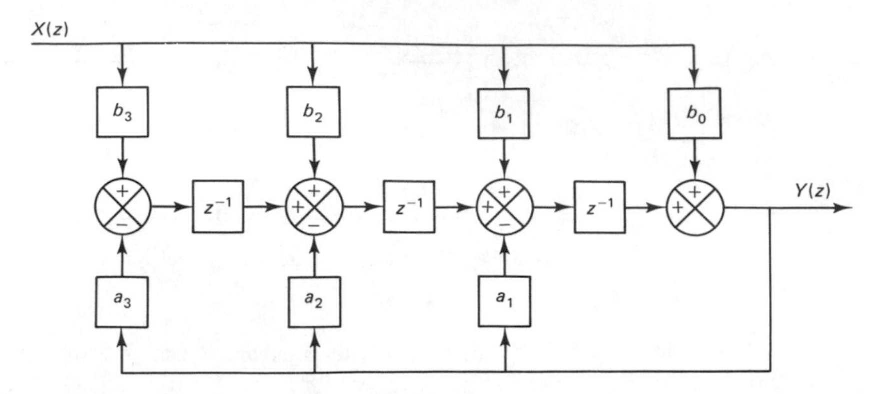

Figure 3–91 Digital filter system.

Solution. We may first simplify the diagram to the form shown in Fig. 3–92(a). Then by eliminating one feedback loop we obtain Fig. 3–92(b). Further simplification results in Fig. 3–92(c). The pulse transfer function $Y(z)/X(z)$ can now be obtained as follows:

$$\frac{Y(z)}{X(z)} = \frac{(b_0 z^3 + b_1 z^2 + b_2 z + b_3)z^{-3}}{1 + a_1 z^{-1} + a_2 z^{-2} + a_3 z^{-3}}$$

$$= \frac{b_0 + b_1 z^{-1} + b_2 z^{-2} + b_3 z^{-3}}{1 + a_1 z^{-1} + a_2 z^{-2} + a_3 z^{-3}}$$

Problem A-3–30. Consider the digital filter defined by

$$G(z) = \frac{Y(z)}{X(z)} = \frac{4(z-1)(z^2+1.2z+1)}{(z+0.1)(z^2-0.3z+0.8)}$$

Draw a series realization diagram and a parallel realization diagram. (Use one first-order section and one second-order section.)

Solution. We shall first consider the series realization scheme. In order to limit the coefficients to real quantities, we group the second-order term in the numerator (which has complex zeros) and the second-order term in the denominator (which has complex poles). Hence, we group $G(z)$ as follows:

$$G(z) = 4\frac{z-1}{z+0.1}\,\frac{z^2+1.2z+1}{z^2-0.3z+0.8}$$

$$= 4\frac{1-z^{-1}}{1+0.1z^{-1}}\,\frac{1+1.2z^{-1}+z^{-2}}{1-0.3z^{-1}+0.8z^{-2}}$$

Figure 3–93(a) shows a series realization diagram.

Next, we shall consider the parallel realization scheme. Expansion of $G(z)/z$ into partial fractions gives

$$\frac{G(z)}{z} = \frac{4(z-1)(z^2+1.2z+1)}{z(z+0.1)(z^2-0.3z+0.8)}$$

$$= -\frac{50}{z} + \frac{\frac{979}{21}}{z + 0.1} + \frac{\frac{155}{21}z + \frac{85}{21}}{z^2 - 0.3z + 0.8}$$

Then $G(z)$ can be written as follows:

$$G(z) = -50 + \frac{46.61905}{1 + 0.1z^{-1}} + \frac{7.38095 + 4.04762z^{-1}}{1 - 0.3z^{-1} + 0.8z^{-2}}$$

Figure 3–93(b) shows a parallel realization diagram.

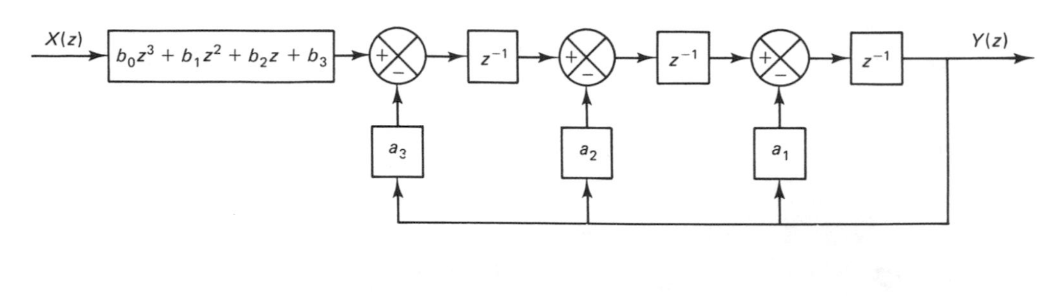

(a)

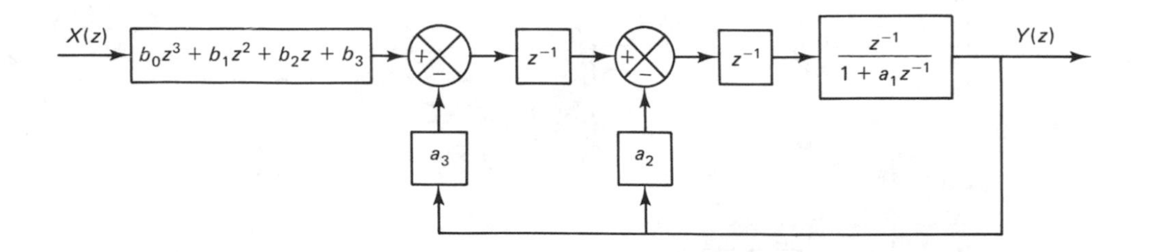

(b)

X(z) $b_0z^3 + b_1z^2 + b_2z + b_3$ $\frac{z^{-3}}{1 + a_1z^{-1} + a_2z^{-2} + a_3z^{-3}}$ Y(z)

(c)

Figure 3–92 Successive simplifications of the block diagram shown in Fig. 3–91: (a) feedforward elements combined, (b) one feedback loop eliminated, and (c) all feedback loops eliminated.

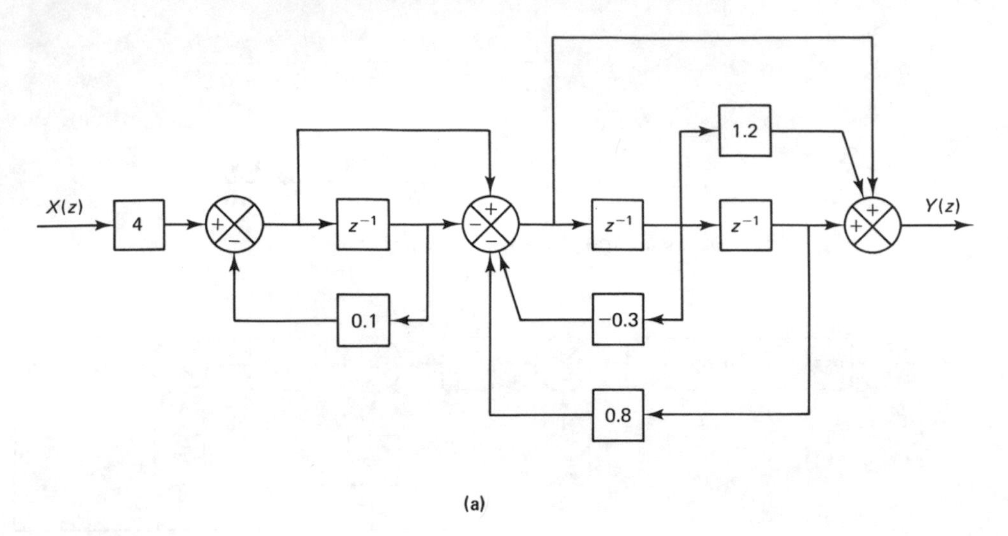

(a)

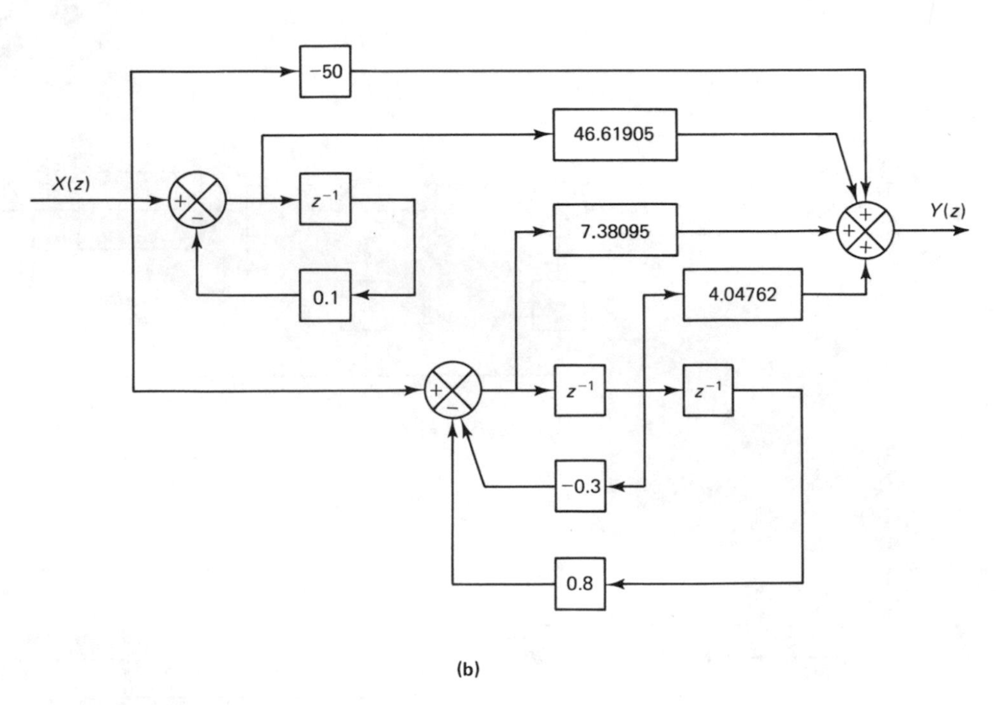

(b)

Figure 3–93 Block diagram realizations of the digital filter considered in Prob. A-3-30. (a) Series realization; (b) parallel realization.

Problem A-3-31. Show that geometrically the patterns of the poles near $z = 1$ in the z plane are similar to the patterns of poles in the s plane near the origin.

Solution. Note that

$$z = e^{Ts}$$

Near the origin of the s plane,

$$z = e^{Ts} = 1 + Ts + \tfrac{1}{2}T^2s^2 + \cdots$$

or

$$z - 1 \cong Ts$$

Thus, geometrical patterns of the poles near $z = 1$ in the z plane are similar to the patterns of poles in the s plane near the origin.

Problem A-3-32. State the conditions for stability, instability, and critical stability in terms of the weighting sequence $g(kT)$ of a linear time-invariant discrete-time control system.

Solution.

Stability: The system is stable if the weighting sequence $g(kT)$ vanishes for sufficiently large k.

Instability: The system is unstable if $g(kT)$ grows without bound as k increases indefinitely.

Critical stability: The system is critically stable if $g(kT)$ approaches a constant nonzero value or a bounded oscillation for large values of k.

Problem A-3-33. Consider the system shown in Fig. 3-94. Determine the stability range for K (where $K > 0$) by using the Jury stability criterion. The sampling period is assumed to be T seconds.

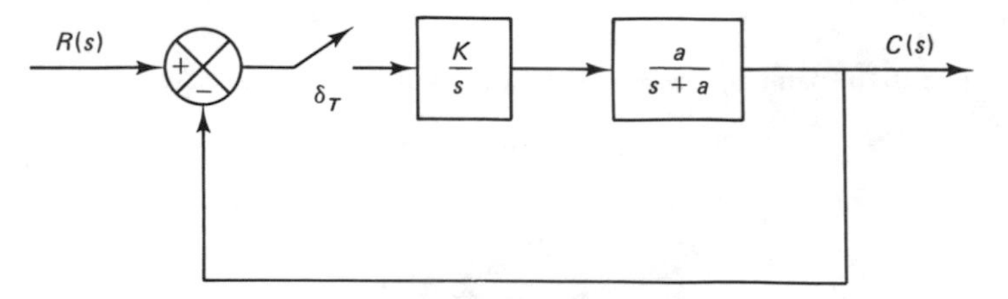

Figure 3-94 Control system.

Solution. Define

$$G(s) = \frac{K}{s}\frac{a}{s+a}$$

Then

$$G(z) = \mathscr{Z}[G(s)] = \mathscr{Z}\left[\frac{Ka}{s(s+a)}\right]$$

$$= \mathscr{Z}\left[\frac{K}{s} - \frac{K}{s+a}\right]$$

$$= \frac{K}{1-z^{-1}} - \frac{K}{1-e^{-aT}z^{-1}}$$

$$= \frac{K(1-e^{-aT})z^{-1}}{(1-z^{-1})(1-e^{-aT}z^{-1})}$$

$$= \frac{K(1-e^{-aT})z}{(z-1)(z-e^{-aT})}$$

Thus the characteristic equation for the system is

$$z^2 - (1 + e^{-aT} - K + Ke^{-aT})z + e^{-aT} = 0$$

Note that for the characteristic equation

$$P(z) = a_0 z^2 + a_1 z + a_2 = 0$$

where $a_0 > 0$, the Jury stability criterion is met if the following conditions are satisfied:

$$|a_2| < a_0$$

$$P(z)|_{z=1} > 0$$

$$P(z)|_{z=-1} > 0 \qquad n = 2 = \text{even}$$

For the present problem the first condition is satisfied, since

$$|e^{-aT}| < 1 \qquad aT > 0$$

The second condition becomes

$$1 - (1 + e^{-aT} - K + Ke^{-aT}) + e^{-aT} > 0$$

or

$$e^{-aT} < 1$$

which is the same as the first condition and is clearly satisfied. The third condition becomes

$$1 + (1 + e^{-aT} - K + Ke^{-aT}) + e^{-aT} > 0$$

or

$$K < \frac{2(1+e^{-aT})}{1-e^{-aT}}$$

Hence, the stability range for K is found to be

$$0 < K < \frac{2(1+e^{-aT})}{1-e^{-aT}} = 2\coth\frac{aT}{2}$$

Problem A-3–34. Consider the following characteristic equation:

$$P(z) = z^3 - 1.3z^2 - 0.08z + 0.24 = 0 \tag{3–131}$$

Determine whether or not any of the roots of the characteristic equation lie outside the unit circle in the z plane. Use the bilinear transformation and the Routh stability criterion.

Solution. Let us substitute $(w + 1)/(w - 1)$ for z in the given characteristic equation, resulting in

$$\left(\frac{w+1}{w-1}\right)^3 - 1.3\left(\frac{w+1}{w-1}\right)^2 - 0.08\frac{w+1}{w-1} + 0.24 = 0$$

Clearing the fractions by multiplying both sides of this last equation by $(w - 1)^3$, we get

$$-0.14w^3 + 1.06w^2 + 5.10w + 1.98 = 0$$

By dividing both sides of this last equation by -0.14, we obtain

$$w^3 - 7.571w^2 - 36.43w - 14.14 = 0 \qquad (3\text{–}132)$$

The Routh array for Eq. (3–132) becomes as follows:

one sign	w^3	1	−36.43
change	w^2	−7.571	−14.14
	w^1	−38.30	0
	w^0	−14.14	

Routh stability criterion states that the number of roots with positive real parts is equal to the number of changes in sign of the coefficients of the first column of the array. In the present Routh array there is one sign change for the coefficients in the first column. Hence, there is one root in the right half of the w plane. This means that the original characteristic equation given by Eq. (3–131) has one root outside the unit circle in the z plane. (Compare the amount of computation needed in the present method and that needed in the Jury stability test. See in particular Example 3–22.)

Problem A-3–35. Consider the plant governed by the equation

$$x(k+1) = 0.3679x(k) + 0.6321u(k) \qquad (3\text{–}133)$$

It is desired to have a stable equilibrium state at x_d. This may be accomplished by applying a suitable control function $u(k)$.

Show that if the control function

$$u(k) = (1+a)x_d - ax(k) \qquad (3\text{–}134)$$

is applied to the plant, then it is possible to have a stable equilibrium state at x_d, provided the constant a lies within a certain range. Determine such a range for the constant a.

Solution. By substituting Eq. (3–134) into Eq. (3–133), we have

$$\begin{aligned} x(k+1) &= 0.3679x(k) + 0.6321[(1+a)x_d - ax(k)] \\ &= (0.3679 - 0.6321a)x(k) + 0.6321(1+a)x_d \end{aligned} \qquad (3\text{–}135)$$

In order for the system to have a stable equilibrium state, it must be a stable system. For stability, we require that

$$|0.3679 - 0.6321a| < 1$$

or

$$-1 < 0.3679 - 0.6321a < 1$$

which can be reduced to

$$-1 < a < 2.1641$$

If the value of a lies between -1 and 2.1641, the system is stable.

Let us assume that the stable equilibrium state is x_e. Then at steady state,

$$x(k+1) = x(k) = x_e$$

At steady state, Eq. (3–135) becomes

$$x_e = (0.3679 - 0.6321a)x_e + 0.6321(1+a)x_d$$

or

$$0.6321(1+a)x_e = 0.6321(1+a)x_d$$

Hence

$$x_e = x_d$$

Thus, we have shown that the stable equilibrium state, under control $u(k)$ given by Eq. (3–134), is x_d.

PROBLEMS

Problem B-3–1. Show that the circuit shown in Fig. 3–95 acts as a zero-order hold.

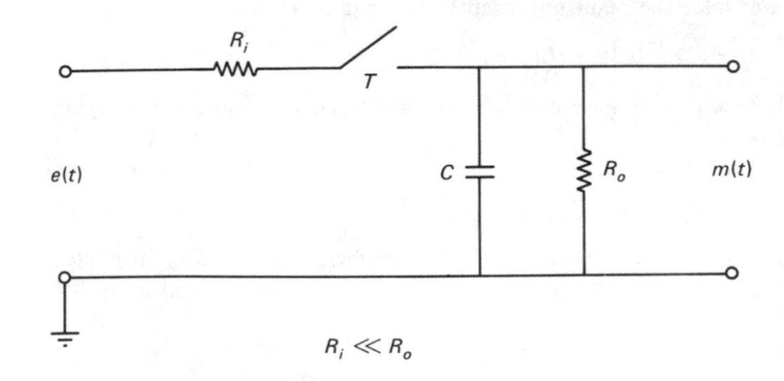

Figure 3–95 Circuit approximating a zero-order hold.

Problem B-3–2. Consider the circuit shown in Fig. 3–96. Derive a difference equation describing the system dynamics when the input voltage applied is piecewise-constant, or

$$e(t) = e(kT) \qquad kT \le t < (k+1)T$$

(Derive first a differential equation and then discretize it to obtain a difference equation.)

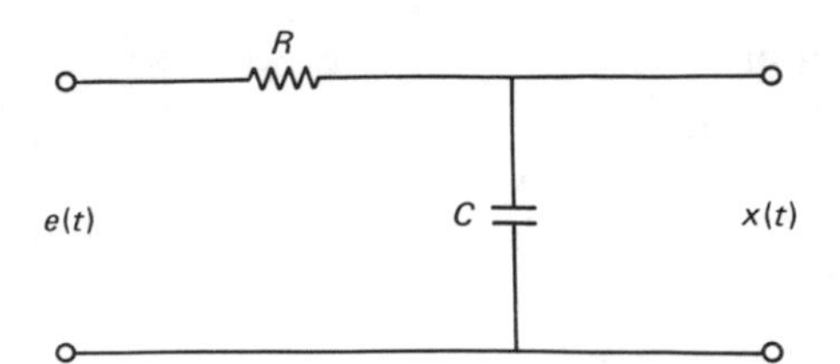

Figure 3–96 *RC* circuit for Prob. B-3–2.

Problem B-3–3. Consider a transfer function system

$$X(s) = \frac{s+3}{(s+1)(s+2)}$$

Obtain the pulse transfer function by two different methods.

Problem B-3–4. Obtain the z transform of

$$X(s) = \frac{K}{(s+a)(s+b)}$$

Use the residue method and the method based on the impulse response function.

Problem B-3–5. Obtain the z transform of

$$X(s) = \frac{1-e^{-Ts}}{s} \frac{1}{(s+a)^2}$$

Problem B-3–6. Obtain the z transform of

$$X(s) = \frac{1-e^{-Ts}}{s} \frac{s+1}{(s+2)(s+10)}$$

Problem B-3–7. Consider the difference equation system

$$y(k+1) + 0.5y(k) = x(k)$$

where $y(0) = 0$. Obtain the response $y(k)$ when the input $x(k)$ is a unit step sequence. Also, write a BASIC computer program for solving this problem.

Problem B-3–8. Consider the difference equation system

$$y(k+2) + y(k) = x(k)$$

where $y(k) = 0$ for $k < 0$. Obtain the response $y(k)$ when the input $x(k)$ is a unit step sequence. Also, write a BASIC computer program for solving this problem.

Problem B-3–9. Consider the difference equation system

$$y(k) = ay(k-1) + x(k) \qquad -1 < a < 1$$

Obtain the response sequence $y(k)$ for the following two inputs:

1. Kronecker delta input, or $x(0) = 1$ and $x(k) = 0$ for $k \neq 0$.
2. Unit step sequence, or $x(k) = 1$ for $k = 0, 1, 2, \ldots$.

Problem B-3–10. Obtain the weighting sequence $g(k)$ of the system described by the difference equation

$$y(k) - ay(k-1) = x(k) \qquad -1 < a < 1$$

If two systems described by this last equation are connected in series, what is the weighting sequence of the resulting system?

Problem B-3–11. Consider the system described by

$$y(k) - y(k-1) + 0.24y(k-2) = x(k) + x(k-1)$$

where $x(k)$ is the input and $y(k)$ is the output of the system.

Determine the weighting sequence of the system. Assuming that $y(k) = 0$ for $k < 0$, determine the response $y(k)$ when the input $x(k)$ is a unit step sequence. Also, write a BASIC computer program for solving this problem.

Problem B-3–12. Consider the system

$$G(z) = \frac{1 - 0.5z^{-1}}{(1 - 0.3z^{-1})(1 + 0.7z^{-1})}$$

Obtain the response of this system to a unit step sequence input. Also, write a BASIC computer program for solving this problem.

Problem B-3–13. Obtain the response $y(kT)$ of the following system:

$$\frac{Y(s)}{X^*(s)} = \frac{1}{(s+1)(s+2)}$$

where $x(t)$ is the unit step function and $x^*(t)$ is its impulse-sampled version. Assume that the sampling period T is 0.1 sec.

Problem B-3–14. Consider a system whose pulse transfer function is

$$\frac{Y(z)}{X(z)} = \frac{10z - 2}{(z - 0.5)(z + 0.5)}$$

Obtain the response $y(k)$ to a Kronecker delta input $x(k)$.

Problem B-3–15. Consider the system defined by

$$\frac{Y(z)}{U(z)} = H(z) = \frac{0.5z^3 + 0.4127z^2 + 0.1747z - 0.0874}{z^3}$$

Using the convolution equation

$$y(k) = \sum_{j=0}^{k} h(k-j)u(j)$$

obtain the response $y(k)$ to a unit step sequence input $u(k)$.

Problem B-3–16. Obtain the closed-loop pulse transfer function of the system shown in Fig. 3–97.

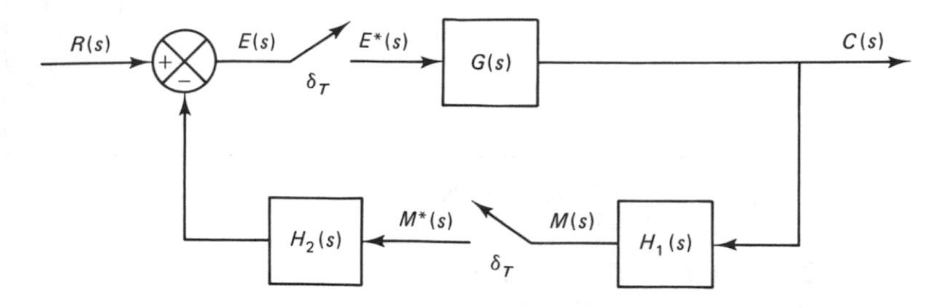

Figure 3–97 Discrete-time control system.

Problem B-3–17. Obtain the closed-loop pulse transfer function of the system shown in Fig. 3–98.

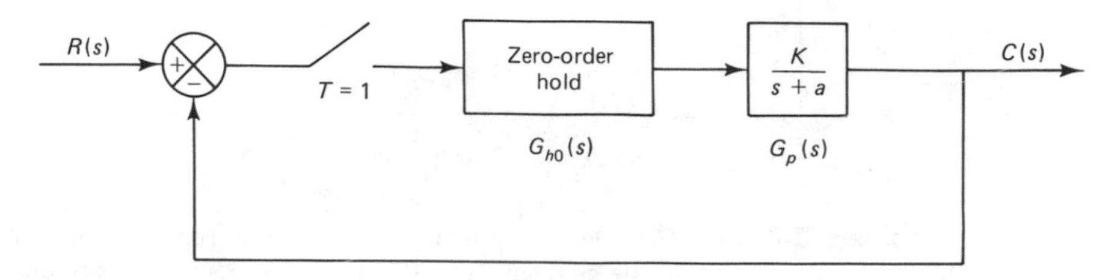

Figure 3–98 Discrete-time control system.

Problem B-3–18. Consider the discrete-time control system shown in Fig. 3–99. Obtain the discrete-time output $C(z)$ and the continuous-time output $C(s)$ in terms of the input and the transfer functions of the blocks.

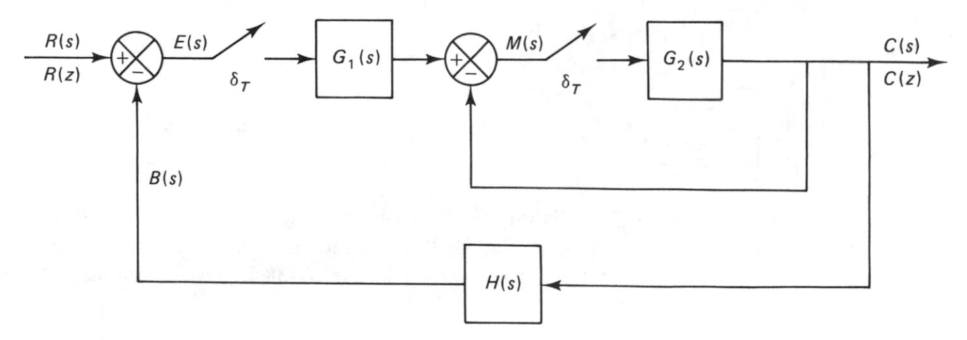

Figure 3–99 Discrete-time control system.

Problem B-3–19. Consider the discrete-time control system shown in Fig. 3–100. Obtain the output sequence $c(kT)$ of the system when it is subjected to a unit step input. Assume that the sampling period T is 1 sec. Also, obtain the continuous-time output $c(t)$.

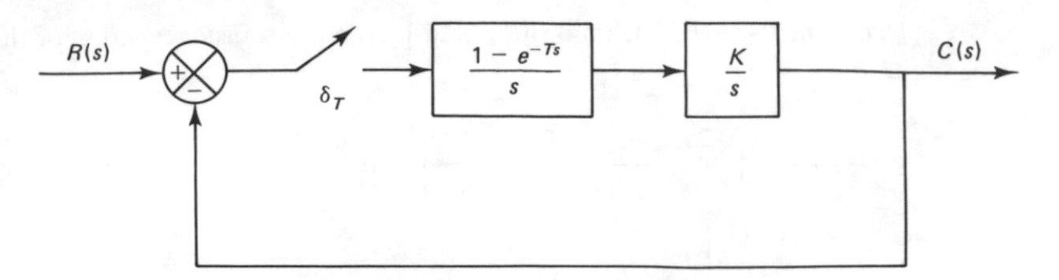

Figure 3–100 Discrete-time control system.

Problem B-3–20. Obtain in a closed form the response sequence $c(kT)$ of the system shown in Fig. 3–101 when it is subjected to a Kronecker delta input $r(k)$. Assume that the sampling period T is 1 sec.

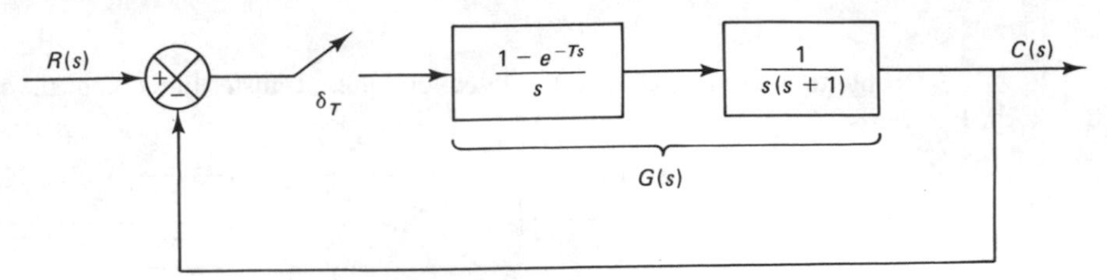

Figure 3–101 Discrete-time control system.

Problem B-3–21. Consider the system shown in Fig. 3–102. Assuming that the sampling period T is 0.2 sec and the gain constant K is unity, determine the response $c(kT)$ for $k = 0, 1, 2, 3,$ and 4 when the input $r(t)$ is a unit step function. Also, determine the final value $c(\infty)$.

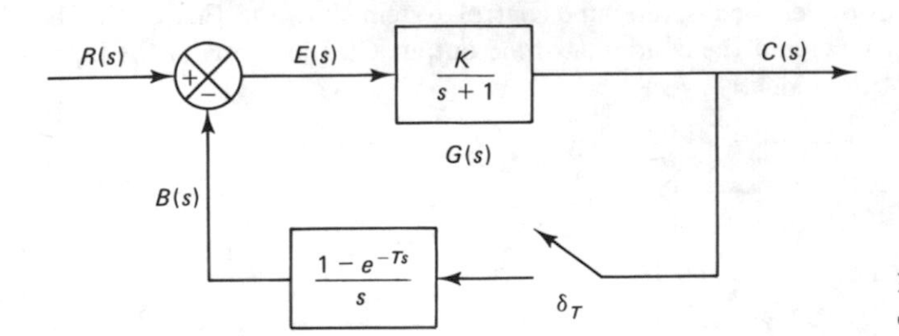

Figure 3–102 Discrete-time control system.

Problem B-3–22. Obtain the closed-loop pulse transfer function $C(z)/R(z)$ of the digital control system shown in Fig. 3–38. Assume that the pulse transfer function of the plant is $G(z)$. (Note that the system shown in Fig. 3–38 is a velocity-form PID control of the plant.)

Problem B-3–23. Obtain the modified z transform of each of the following transfer functions:

1. $$G_1(s) = \frac{\omega}{(s+a)^2 + \omega^2}$$

2. $$G_2(s) = \frac{s+a}{(s+a)^2 + \omega^2}$$

Problem B-3–24. Obtain $C(z, m)$, the modified z transform of the output, of the system shown in Fig. 3–103.

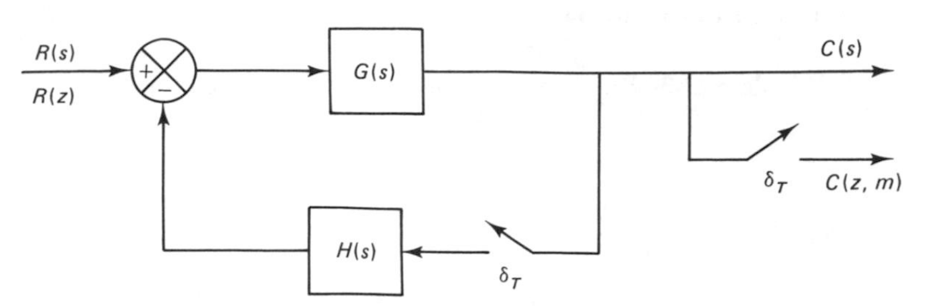

Figure 3–103 Closed-loop discrete-time control system.

Problem B-3–25. Referring to the system considered in Prob. A-3–27, suppose that the sampling period is 1 sec, or $T = 1$, and the input $x(t)$ is a unit step. Obtain the output every 0.2 sec by the modified z transform method.

Problem B-3–26. Consider the digital filter system shown in Fig. 3–104. Derive the pulse transfer function for the system.

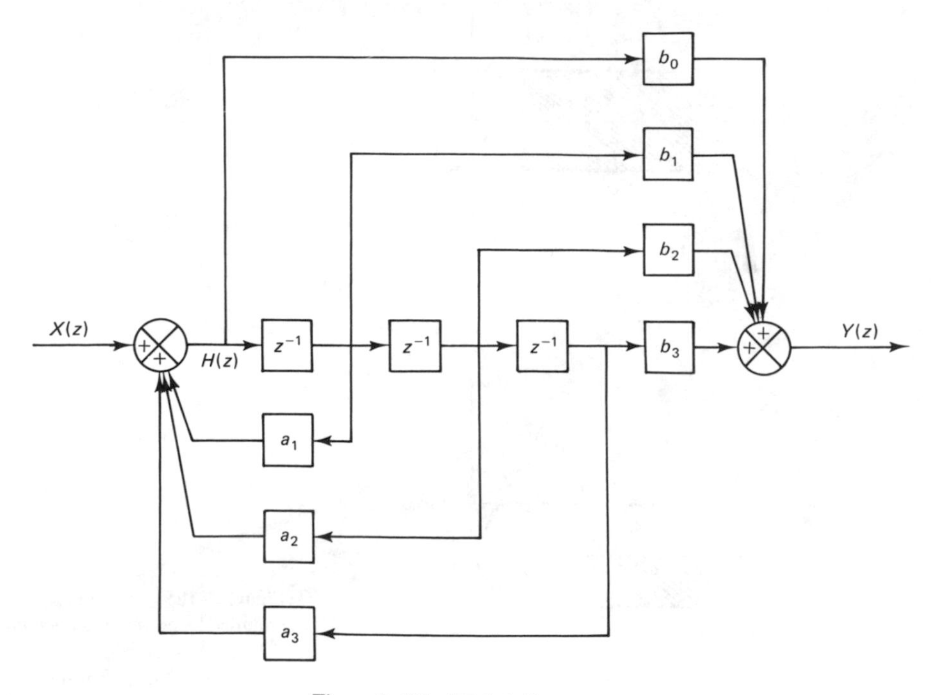

Figure 3–104 Digital filter system.

Problem B-3–27. Assume that a digital filter is given by the following difference equation:

$$y(k) + a_1 y(k-1) + a_2 y(k-2) = b_1 x(k) + b_2 x(k-1)$$

Draw block diagrams for the filter using (1) direct programming, (2) standard programming, and (3) ladder programming.

Problem B-3–28. Consider the digital filter defined by

$$G(z) = \frac{2 + 2.2z^{-1} + 0.2z^{-2}}{1 + 0.4z^{-1} - 0.12z^{-2}}$$

Realize this digital filter in the series scheme, the parallel scheme, and the ladder scheme.

Problem B-3–29. Consider the regions in the s plane shown in Fig. 3–105(a) and (b). Draw the corresponding regions in the z plane. The sampling period T is assumed to be 0.3 sec. (The sampling frequency is $\omega_s = 2\pi/T = 2\pi/0.3 = 20.9$ rad/sec.)

(a)

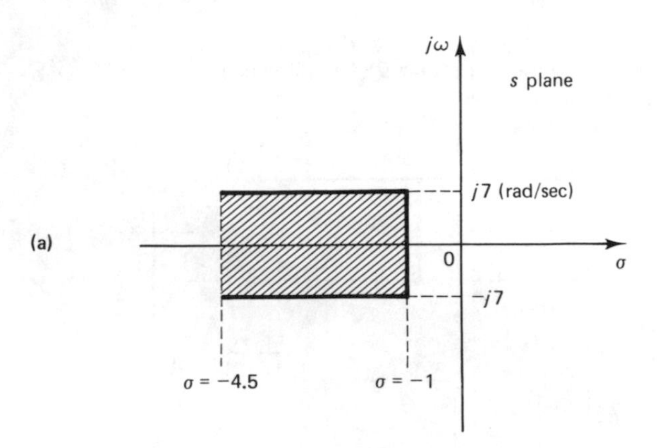

(b)

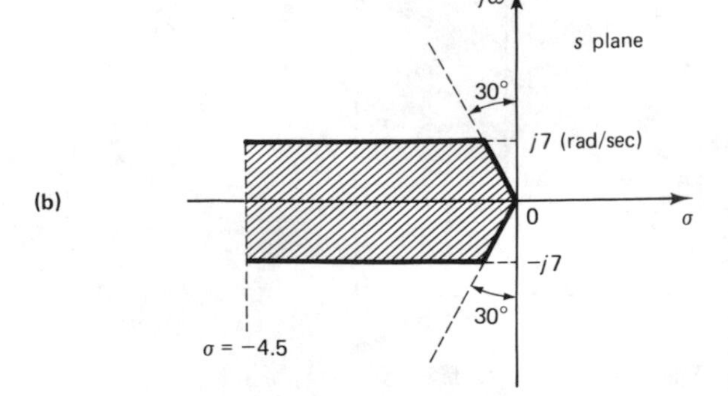

Figure 3–105 (a) Region in the s plane bounded by constant ω lines and constant σ lines; (b) region in the s plane bounded by constant ζ lines, constant ω lines, and a constant σ line.

Problem B-3–30. Consider the following characteristic equation:

$$z^3 + 2.1z^2 + 1.44z + 0.32 = 0$$

Determine whether or not any of the roots of the characteristic equation lie outside the unit circle centered at the origin of the z plane.

Problem B-3–31. Consider the discrete-time closed-loop control system shown in Fig. 3–66. Determine the range of gain K for stability by use of the Jury stability criterion.

Problem B-3–32. Solve Prob. B-3–31 by using the bilinear transformation coupled with the Routh stability criterion.

Problem B-3–33. Consider the system described by

$$y(k) - 0.6y(k-1) - 0.81y(k-2) + 0.67y(k-3) - 0.12y(k-4) = x(k)$$

where $x(k)$ is the input and $y(k)$ is the output of the system. Determine the stability of the system.

Problem B-3–34. Determine the stability of the following discrete-time system:

$$\frac{Y(z)}{X(z)} = \frac{z^{-3}}{1 + 0.5z^{-1} - 1.34z^{-2} + 0.24z^{-3}}$$

Problem B-3–35. Consider the system

$$\frac{Y(z)}{X(z)} = G(z) = \frac{b_0 + b_1 z^{-1} + \cdots + b_n z^{-n}}{1 + a_1 z^{-1} + \cdots + a_n z^{-n}}$$

Suppose that the input sequence $\{x(k)\}$ is bounded, that is, that

$$|x(k)| \le M_1 = \text{constant} \qquad k = 0, 1, 2, \ldots$$

Show that if all poles of $G(z)$ lie inside the unit circle in the z plane, then the output $y(k)$ is also bounded, that is, that

$$|y(k)| \le M_2 = \text{constant} \qquad k = 0, 1, 2, \ldots$$

4

Design of Discrete-Time Control Systems via Transform Methods

4–1 INTRODUCTION

In Chap. 3 we discussed methods for analyzing discrete-time control systems, including the pulse-transfer-function formulation of discrete-time models of continuous-time plants. In this chapter we present four different design methods for single-input–single-output discrete-time or digital control systems. The first method is based on discrete equivalents of analog controllers (Sec. 4–3). The second method is based on the root locus technique using pole-zero configurations in the z plane (Sec. 4–5). The third method is based on the frequency response method in the w plane (Sec. 4–6). The fourth method is an analytical method in which we attempt to obtain a desired behavior of the closed-loop system by manipulating the pulse transfer function of the digital controller (Sec. 4–7). Also, we include a summary of transient and steady-state response analyses in this chapter.

Design techniques for continuous-time control systems based on conventional transform methods (the root locus and frequency response methods) have become well established since the 1950s. So it is advantageous to utilize such techniques to design discrete-time control systems or digital control systems. In our discussions in this chapter, we assume that the reader has an adequate background in design methods for continuous-time control systems.

As to the four design methods to be discussed in this chapter, the first method is an *indirect* method. In it, we carry out a conventional continuous-time design and then discretize the resulting controller. For example, consider the case where we want to design a digital controller for an industrial plant. (In most cases the controlled plant is a continuous-time system.) Since techniques for designing analog controllers for continuous-time control systems are well established, we may assume that the entire control system in consideration is continuous-time and design a suitable analog controller to satisfy the given performance specifications. Then we discretize the well-designed analog controller and use the discretized version of the analog controller as a digital controller. This is a very useful design technique available for discrete-time control systems. In fact, in practice, the control engineer may have more experience in designing analog controllers and therefore may wish to design analog controllers first and then convert them into digital controllers, or it may be desired to change from an existing analog controller to a digital one at the time the analog controller is to be replaced.

The other three methods (the root locus method, the frequency response method in the w plane, and the analytical method) are called *direct* methods. Here we discretize the plant model and carry out the design in the z plane using discrete-time techniques.

Conventional design techniques based on the root locus and frequency response methods are especially useful for designing industrial control systems. In fact, in the past, many industrial digital control systems were successfully designed on the basis of conventional transform methods. Both familiarity with the root locus and frequency response techniques and experience gained in the design of analog filters are immensely valuable in designing discrete-time control systems. It is important to point out that discretizing the continuous-time control system creates new phenomena not present in the original continuous-time control system. However, in general, digital controllers are superior to analog controllers with respect to reliability and electronic noises.

Familiarity with various discretization techniques is necessary for the analysis and design of discrete-time control systems. These techniques are also useful for simulating continuous-time control systems on the digital computer. One of the simplest ways of discretizing or approximating a continuous-time plant by a discrete-time model is to insert fictitious sample-and-hold devices in the loop as needed for discretization purposes. (See the discussion of the step-invariance method in Sec. 4–2.) Then the plant can be described in terms of a pulse transfer function. There are other methods for discretizing continuous-time filters or systems in order to obtain discrete-time models or digital models. They are based on numerical integration methods or mapping methods (see Sec. 4–2) or on the state space method (see Chap. 5).

If the responses of the continuous-time control system and the discrete-time control system (whether obtained by a discretization process or obtained otherwise) for typical inputs (such as step and sinusoidal inputs) are in good agreement with each other, then we may consider the discrete-time control system to be equivalent to the continuous-time control system.

It is noted that the sampling operation has the effect of modifying the transient response characteristics and adversely affects system stability. Even if a continuous-time control system is stable, the discretized version of it may become unstable if the sampling period is not small enough. Since the discrete-time equivalent of the continuous-time control system needs to be stable, it is necessary to investigate the stability of the discretized control system.

Another fact that is important to note is that when the z transform method is used for the analysis or design of the discrete-time control system, the output of the system is obtained only at the sampling instants. Although the values of the system response at the sampling instants may be correct, the system response obtained by the z transform method may not portray the correct time response behavior of the actual system unless the sampling frequency is sufficiently high to satisfy the sampling theorem and unless the transfer function of the continuous-time part of the system has at least two more poles than zeros in the s plane. (Note that by the use of the modified z transform method, it is possible to obtain the response between sampling instants.)

Outline of the chapter. Section 4–1 has presented introductory materials. Section 4–2 discusses several methods for obtaining discrete-time equivalents of continuous-time filters or controllers. Section 4–3 treats a design problem utilizing a discrete-time equivalent of a continuous-time controller. Section 4–4 summarizes transient and steady-state response characteristics of both continuous-time and discrete-time control systems. The design technique based on the root locus method is presented in Sec. 4–5. Section 4–6 first reviews the frequency response method and then presents frequency response techniques using the w transformation for designing discrete-time control systems. Section 4–7 treats an analytical design method.

4–2 OBTAINING DISCRETE-TIME EQUIVALENTS OF CONTINUOUS-TIME FILTERS

With the rapid advances in digital computer technology and A/D and D/A converters, it is a simple matter to realize an equivalent digital controller for a given analog controller. Figure 4–1(a) shows an analog controller and Fig. 4–1(b) an equivalent digital controller. As we shall see later, many different digital controllers are possible which are equivalent to a given analog controller, depending on how we approximate the response between the sampling points.

In the equivalent digital controller shown in Fig. 4–1(b), the continuous-time input signal is converted to a digital signal, which is then processed by a digital controller. The output of the digital controller is converted to a continuous-time signal. The computer is programmed in such a way that the overall dynamic characteristics of the digital controller are equivalent to those of the original analog controller. Note that controllers are a class of filters in a general sense.

There are several methods commonly available for obtaining discrete-time equivalents of continuous-time filters (or analog filters). A discrete-time equivalent of a continuous-time filter must have approximately the same dynamic characteristics as the original continuous-time filter. That is, in obtaining a discrete-time equivalent

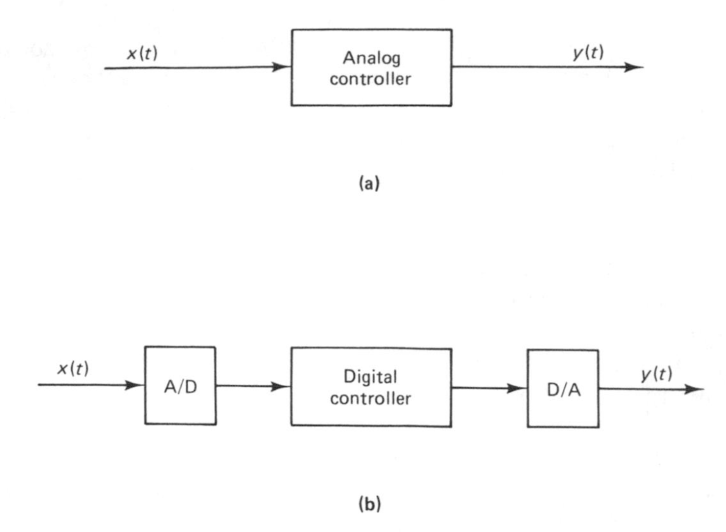

Figure 4–1 (a) Analog controller; (b) equivalent digital controller.

of a continuous-time filter, it is desired to have transient and frequency response characteristics as close as possible to those of the original continuous-time filter. However, this may not be possible to accomplish. In fact, in using a given discretization technique, it may be possible to obtain the same or almost the same impulse response characteristics while it is not possible to get reasonably good fidelity in the frequency response characteristics, or vice versa. In general, the characteristics of the discrete-time filter depend on the sampling frequency and on the particular method of discretization. By lowering the sampling frequency, the fidelity of the discrete-time filter is decreased.

In most practical situations, in discretizing a continuous-time filter (or analog filter) the designer may wish to preserve one or more of the following properties, among others: the number of poles and zeros, impulse response characteristics or step response characteristics, dc gain, phase and gain margins, or bandwidth. (Not all of these properties will be preserved once a continuous-time filter is discretized.) Some of the discretized filters are in good agreement in impulse response characteristics with the original continuous-time filter, but others are not. In particular, in most cases, it is difficult to match the frequency response characteristics of a discrete-time filter to those of the original continuous-time filter. In discretizing the continuous-time filter, the designer must decide what dynamic characteristics (impulse response characteristics, frequency response characteristics, and so forth) are most important in a particular situation. Then the goal must be to preserve those important characteristics by choosing a suitable discretization method. With respect to the frequency response characteristics, it is important to point out that if the continuous-time filter is discretized, the resulting filter may exhibit an undesirable phenomenon (namely, frequency folding) not present in the original continuous-time filter if the sampling frequency is too low to satisfy the sampling theorem.

Remember that in discretizing continuous-time filters, not only the particular method used, but also the sampling frequency chosen, will affect the dynamic characteristics of the resulting filter. If the sampling frequency is sufficiently high, the equivalent discrete-time filter will yield a good approximation to a given continuous-time filter. However, a slow sampling frequency always results in a poor approximation.

Once a continuous-time filter is discretized and quantized, we have obtained an equivalent digital filter. Such a digital filter may be realized by means of software (through a computer program). But the digital controller must operate in real time; care must be exercised in regard to problems associated with finite word length, for example, and in regard to sampling frequency and programming of the algorithm.

In what follows we discuss several discretizing methods which are based on numerical integration methods, the z transform method, and a mapping method.

Discretizing a simple continuous-time filter. Consider the continuous-time filter (analog filter) shown in Fig. 4–2. The transfer function for this filter is

$$\frac{Y(s)}{X(s)} = \frac{1}{RCs + 1} = \frac{a}{s + a} \tag{4–1}$$

where $a = 1/RC$. We shall obtain discrete-time equivalents to this continuous-time filter by using several methods.

Discrete-time equivalents of the continuous-time filter may be derived directly from the transfer function given by Eq. (4–1) or its equivalent differential equation

$$\frac{dy}{dt} + ay = ax \tag{4–2}$$

If Eq. (4–2) is used as the dynamic description of the continuous-time filter, we need to derive a difference equation whose solution approximates the solution of the differential equation. (There are various methods of obtaining equivalent difference equations.) Once we have a difference equation, then it is a simple matter to obtain an equivalent discrete-time filter, or an equivalent digital filter if the amplitude is quantized. For convenience we may occasionally call a discrete-time filter a discrete filter or a digital filter. (Precisely speaking, the discrete-time filter becomes a digital filter only if the output is quantized. "Discrete filter" is an abbreviation of "discrete-time filter.")

In discretizing a continuous-time filter, it is important to note that the fidelity

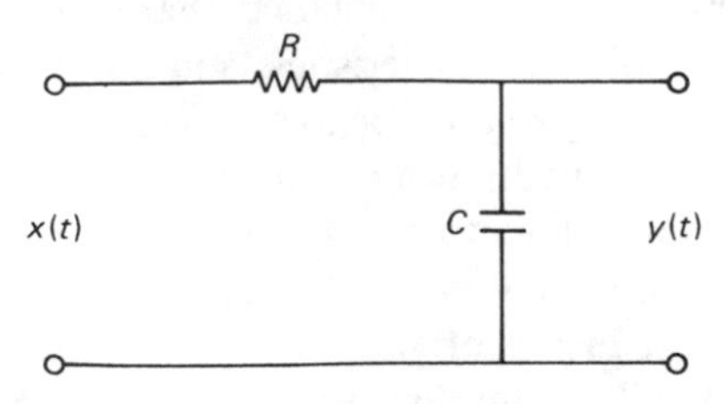

Figure 4–2 Continuous-time filter.

of the impulse response alone is not sufficient to evaluate the behavior of the discrete-time filter. The fidelity of frequency response is also important.

In what follows, we shall discuss the following methods for obtaining discrete-time equivalents of continuous-time filters:

1. Backward difference method (a numerical integration method).
2. Forward difference method (a numerical integration method). This method may lead to an unstable system and therefore cannot be used in practice.
3. Bilinear transformation method (a numerical integration method based on the trapezoidal integration rule).
4. Bilinear transformation method with frequency prewarping (an improved version of the bilinear transformation method).
5. Impulse-invariance method (z transform method).
6. Step-invariance method (impulse-invariance method with sample-and-hold—the z transform method coupled with a fictitious sample-and-hold).
7. Matched pole-zero mapping method.

For a given continuous-time filter, each of these seven methods will yield a different discrete-time filter.

Backward difference method. Equation (4–2) may be written as follows:

$$\frac{dy}{dt} = -ay + ax \tag{4–3}$$

Let us integrate both sides of Eq. (4–3) from 0 to t:

$$\int_0^t \frac{dy(t)}{dt}\, dt = -a \int_0^t y(t)\, dt + a \int_0^t x(t)\, dt$$

Suppose we determine the value of $y(t)$ every sampling period T. Then by substituting kT for t in this last equation, we obtain

$$\int_0^{kT} \frac{dy(t)}{dt}\, dt = -a \int_0^{kT} y(t)\, dt + a \int_0^{kT} x(t)\, dt$$

or

$$y(kT) - y(0) = -a \int_0^{kT} y(t)\, dt + a \int_0^{kT} x(t)\, dt \tag{4–4}$$

Similarly, by changing kT to $(k - 1)T$ in Eq. (4–4), we obtain

$$y((k-1)T) - y(0) = -a \int_0^{(k-1)T} y(t)\, dt + a \int_0^{(k-1)T} x(t)\, dt \tag{4–5}$$

By subtracting Eq. (4–5) from Eq. (4–4), we get

$$y(kT) - y((k-1)T) = -a \int_{(k-1)T}^{kT} y(t)\,dt + a \int_{(k-1)T}^{kT} x(t)\,dt \qquad (4\text{–}6)$$

The terms on the right-hand side of Eq. (4–6) can be numerically integrated by various methods. Here we shall use the backward difference method.

Integration by the backward difference method means that we approximate the areas

$$\int_{(k-1)T}^{kT} y(t)\,dt \qquad \text{and} \qquad \int_{(k-1)T}^{kT} x(t)\,dt$$

by $y(kT)T$ and $x(kT)T$, respectively (see Fig. 4–3). Thus, Eq. (4–6) can be written as follows:

$$y(kT) = y((k-1)T) - aT[y(kT) - x(kT)] \qquad (4\text{–}7)$$

The z transform of Eq. (4–7) is

$$Y(z) = z^{-1}Y(z) - aT[Y(z) - X(z)]$$

from which we get

$$\frac{Y(z)}{X(z)} = G_D(z) = \frac{aT}{1 - z^{-1} + aT} = \frac{a}{\dfrac{1 - z^{-1}}{T} + a} \qquad (4\text{–}8)$$

By comparing Eq. (4–1) with Eq. (4–8), we note that the right-hand sides of these two equations become identical if we let

$$s = \frac{1 - z^{-1}}{T} \qquad (4\text{–}9)$$

Equation (4–9) is the mapping from the s plane to the z plane when the backward difference method is used to discretize Eq. (4–2).

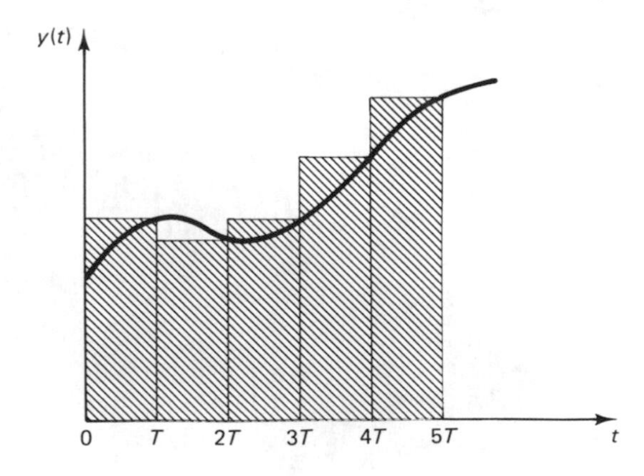

Figure 4–3 Area approximation by the backward difference method.

Note that Eq. (4–7) can also be derived by using Eq. (4–9), as follows. By using the substitution

$$\frac{dy}{dt} = \frac{y(kT) - y((k-1)T)}{T}$$

in Eq. (4–2) and taking $x = x(kT)$ and $y = y(kT)$, we obtain

$$\frac{y(kT) - y((k-1)T)}{T} + ay(kT) = ax(kT)$$

or

$$y(kT) = y((k-1)T) - aT[y(kT) - x(kT)]$$

which is identical to Eq. (4–7).

The stability region in the s plane can be mapped by Eq. (4–9) into the z plane as follows. Noting that the stable region in the s plane is given by $\text{Re}(s) < 0$ and referring to Eq. (4–9), we may write the stability region in the z plane as

$$\text{Re}\left(\frac{1 - z^{-1}}{T}\right) = \text{Re}\left(\frac{z - 1}{Tz}\right) < 0$$

Noting that T is positive and writing z as $\sigma + j\omega$, we may write this last inequality as

$$\text{Re}\left(\frac{\sigma + j\omega - 1}{\sigma + j\omega}\right) < 0$$

or

$$\begin{aligned} &\text{Re}\left[\frac{(\sigma + j\omega - 1)(\sigma - j\omega)}{(\sigma + j\omega)(\sigma - j\omega)}\right] \\ &= \text{Re}\left(\frac{\sigma^2 - \sigma + \omega^2 + j\omega}{\sigma^2 + \omega^2}\right) \\ &= \frac{\sigma^2 - \sigma + \omega^2}{\sigma^2 + \omega^2} < 0 \end{aligned}$$

which can be written as

$$(\sigma - \tfrac{1}{2})^2 + \omega^2 < (\tfrac{1}{2})^2$$

The stable region can thus be mapped into a circle with center at $\sigma = \frac{1}{2}$, $\omega = 0$ and radius equal to $\frac{1}{2}$, as shown in Fig. 4–4.

The backward difference method is simple and will produce a stable discrete-time filter for a stable continuous-time filter. (Note that some unstable continuous-time filters may be mapped into stable discrete-time filters.) However, because the stable region is mapped into a circle within the unit circle, there is considerable distortion in the transient and frequency response characteristics of the discrete-time filter obtained in this way when they are compared with the characteristics of the

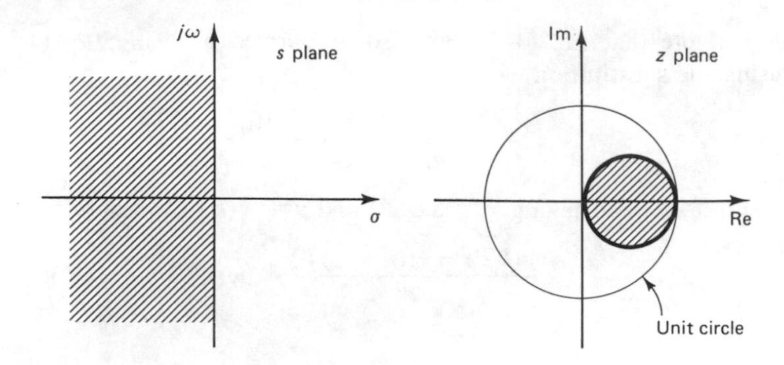

Figure 4–4 Mapping of the left half of the s plane into the z plane by $s = (1 - z^{-1})/T$ (backward difference method).

original continuous-time filter. To reduce the distortion, we need to use a faster sampling frequency, that is, a smaller sampling period T.

Forward difference method. It is interesting to investigate the discretization process by use of the forward difference method. In this method we approximate the areas

$$\int_{(k-1)T}^{kT} y(t)\,dt \qquad \text{and} \qquad \int_{(k-1)T}^{kT} x(t)\,dt$$

by $y((k-1)T)T$ and $x((k-1)T)T$, respectively (see Fig. 4–5). Thus Eq. (4–6) becomes

$$y(kT) = y((k-1)T) - aT[y((k-1)T) - x((k-1)T)]$$

or

$$y(kT) = (1 - aT)y((k-1)T) + aTx((k-1)T)$$

The z transform of this last equation is

$$Y(z) = (1 - aT)z^{-1}Y(z) + aTz^{-1}X(z)$$

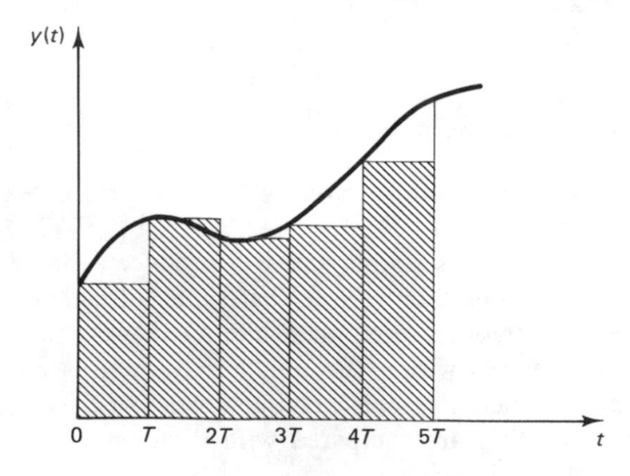

Figure 4–5 Area approximation by the forward difference method.

and the pulse transfer function of the filter becomes

$$\frac{Y(z)}{X(z)} = G_D(z) = \frac{aTz^{-1}}{1-(1-aT)z^{-1}} = \frac{a}{\dfrac{1-z^{-1}}{Tz^{-1}} + a} \tag{4–10}$$

Now, if we compare Eqs. (4–1) and (4–10), we see that the right-hand sides of these two equations become identical if we let

$$s = \frac{1-z^{-1}}{Tz^{-1}} \tag{4–11}$$

That is, if we substitute Eq. (4–11) into the right-hand side of Eq. (4–1) we obtain the right-hand side of Eq. (4–10). We may consider Eq. (4–11) to be the mapping from the s plane to the z plane when the forward difference method is used to discretize Eq. (4–2).

There is one serious problem in this method, however, regarding the stability of the resulting discrete-time filter, as seen from the following analysis. From Eq. (4–11), the left half of the s plane is mapped into the region $\text{Re}\,[(1 - z^{-1})/(Tz^{-1})] = \text{Re}\,[(z-1)/T] < 0$, or $\text{Re}(z) < 1$, as shown in Fig. 4–6. This mapping shows that the poles in the left half of the s plane may be mapped outside the unit circle in the z plane. Hence, the discrete-time filter obtained by this method may become unstable. Consequently, the forward difference method is not acceptable as a discretizing method and cannot be used in practice.

Bilinear transformation method. The bilinear transformation method is also called the *trapezoidal integration method* or *Tustin transformation method*. By this method we approximate the areas

$$\int_{(k-1)T}^{kT} y(t)\,dt \qquad \text{and} \qquad \int_{(k-1)T}^{kT} x(t)\,dt$$

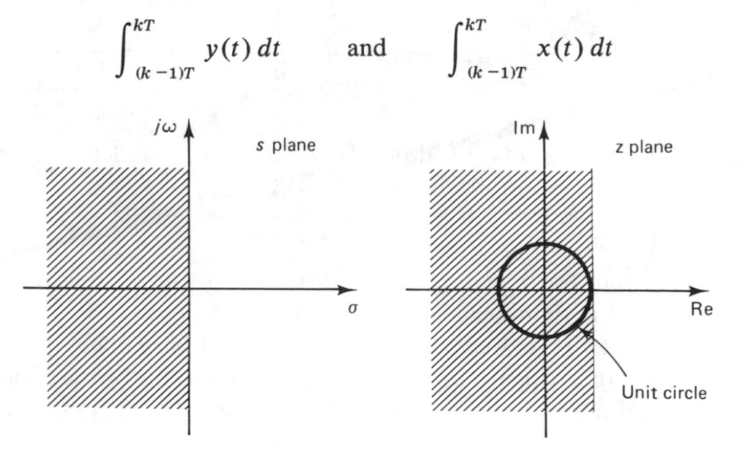

Figure 4–6 Mapping of the left half of the s plane into the z plane by $s = (1 - z^{-1})/(Tz^{-1})$ (forward difference method).

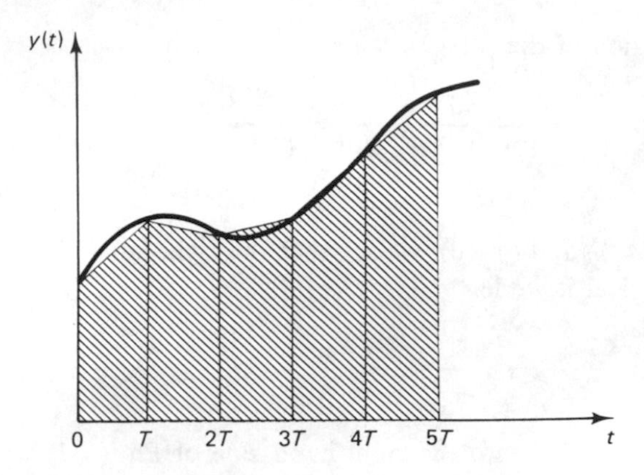

Figure 4–7 Area approximation by the bilinear transformation (trapezoidal integration) method.

by $\frac{1}{2}[y(kT) + y((k-1)T)]T$ and $\frac{1}{2}[x(kT) + x((k-1)T)]T$, respectively.

Notice that the integration scheme assumes that the variation between successive sampling points is linear, as shown in Fig. 4–7. Hence, by use of the bilinear transformation method, Eq. (4–6) can be written as follows:

$$y(kT) = y((k-1)T) - \frac{aT}{2}[y(kT) + y((k-1)T)] + \frac{aT}{2}[x(kT) + x((k-1)T)] \tag{4–12}$$

The z transform of Eq. (4–12) is

$$Y(z) = z^{-1}Y(z) - \frac{aT}{2}[Y(z) + z^{-1}Y(z)] + \frac{aT}{2}[X(z) + z^{-1}X(z)]$$

or

$$\frac{Y(z)}{X(z)} = G_D(z) = \frac{\frac{aT}{2}(1 + z^{-1})}{(1 - z^{-1}) + \frac{aT}{2}(1 + z^{-1})} = \frac{a}{\frac{2}{T}\frac{1 - z^{-1}}{1 + z^{-1}} + a} \tag{4–13}$$

By comparing Eqs. (4–1) and (4–13), we see that if we let

$$s = \frac{2}{T}\frac{1 - z^{-1}}{1 + z^{-1}} \tag{4–14}$$

and substitute for s in Eq. (4–1), then the right-hand side of Eq. (4–1) becomes identical to the right-hand side of Eq. (4–13).

Equation (4–14) is called the *bilinear transformation*. We may consider Eq. (4–14) to be the mapping from the s plane to the z plane by the bilinear transformation (trapezoidal integration) method. Note that since $G_D(z)$ is obtained by the equation

$$G_D(z) = G(s)\Big|_{s = \frac{2}{T}\frac{1 - z^{-1}}{1 + z^{-1}}}$$

the number of poles and that of zeros are equal in $G_D(z)$. Also, the order of the discrete-time filter (the number of poles of the discrete-time filter) is the same as that of the original continuous-time filter.

By means of Eq. (4–14), the left half of the s plane [$\mathrm{Re}(s) < 0$] is mapped into the region

$$\mathrm{Re}\left(\frac{2}{T}\frac{1-z^{-1}}{1+z^{-1}}\right) = \mathrm{Re}\left(\frac{2}{T}\frac{z-1}{z+1}\right) < 0$$

Since $T > 0$, this last inequality can be simplified to

$$\mathrm{Re}\left(\frac{z-1}{z+1}\right) < 0$$

By taking $z = \sigma + j\omega$, this last inequality becomes

$$\begin{aligned}\mathrm{Re}\left(\frac{z-1}{z+1}\right) &= \mathrm{Re}\left(\frac{\sigma+j\omega-1}{\sigma+j\omega+1}\right) \\ &= \mathrm{Re}\left[\frac{(\sigma-1+j\omega)(\sigma+1-j\omega)}{(\sigma+1+j\omega)(\sigma+1-j\omega)}\right] \\ &= \mathrm{Re}\left[\frac{\sigma^2-1+\omega^2+j2\omega}{(\sigma+1)^2+\omega^2}\right] < 0\end{aligned}$$

which is equivalent to

$$\sigma^2 - 1 + \omega^2 < 0$$

or

$$\sigma^2 + \omega^2 < 1^2$$

which corresponds to the inside of the unit circle in the z plane. Hence the bilinear transformation maps the entire left half of the s plane into the unit circle with center at the origin of the z plane (which is the stable region in the z plane). The bilinear transformation thus produces a stable discrete-time filter for a stable continuous-time filter.

Notice that by means of the bilinear transformation given by Eq. (4–14) the entire $j\omega$ axis of the s plane is mapped into one complete revolution of the unit circle in the z plane. (This means that there is no frequency folding.) This mapping coincides with that obtained by the mapping $z = e^{Ts}$, which maps the entire $j\omega$ axis of the s plane into an infinite number of revolutions of the unit circle in the z plane. Although the bilinear transformation and the z transformation appear similar to each other in that both transformations map the left half of the s plane into the unit circle in the z plane, there are considerable differences between them in their effects on the transient and frequency response characteristics of the discrete-time filter. There are noticeable distortions in the transient response characteristics of the discrete-time filter obtained by the bilinear transformation method when they are compared with those of the original continuous-time filter. Also, there are frequency response distortions. The degree of the frequency response distortions can be reduced by modifying the bilinear transformation method with frequency prewarping, as will be discussed in the following.

Bilinear transformation method with frequency prewarping. Consider the continuous-time filter given by

$$\frac{Y(s)}{X(s)} = G(s) = \frac{a}{s+a}$$

Let us define

$$G_D(z) = \frac{a}{s+a}\bigg|_{s=\frac{2}{T}\frac{1-z^{-1}}{1+z^{-1}}} = \frac{a}{\frac{2}{T}\frac{1-z^{-1}}{1+z^{-1}}+a}$$

and examine the frequency response of $G(s)$ and that of $G_D(z)$. Note that the frequency response of $G(s)$ is given by $G(j\omega)$, while the frequency response of $G_D(z)$ can be given by $G_D(e^{j\omega T})$.

In order to compare the frequency responses of $G(j\omega)$ and $G_D(e^{j\omega T})$, let us first use the substitutions

$$s = j\omega_A \qquad \text{and} \qquad z = e^{j\omega_D T}$$

in Eq. (4–14), which was

$$s = \frac{2}{T}\frac{1-z^{-1}}{1+z^{-1}}$$

Then we obtain

$$\begin{aligned} j\omega_A &= \frac{2}{T}\frac{1-e^{-j\omega_D T}}{1+e^{-j\omega_D T}} = \frac{2}{T}\frac{e^{j(1/2)\omega_D T} - e^{-j(1/2)\omega_D T}}{e^{j(1/2)\omega_D T} + e^{-j(1/2)\omega_D T}} \\ &= \frac{2}{T}\frac{2j\sin(\omega_D T/2)}{2\cos(\omega_D T/2)} = j\frac{2}{T}\tan\frac{\omega_D T}{2} \end{aligned}$$

or

$$\omega_A = \frac{2}{T}\tan\frac{\omega_D T}{2} \tag{4–15}$$

Equation (4–15) relates the frequencies ω_A and ω_D and gives a way of measuring the frequency distortion. By using the relationship given by Eq. (4–15), we can relate $G(j\omega_A)$ and $G_D(e^{j\omega_D T})$ as follows:

$$G(j\omega_A) = G_D(e^{j\omega_D T}) \tag{4–16}$$

Equation (4–16) states that $G(j\omega_A)$ at frequency ω_A is equal to $G_D(e^{j\omega_D T})$ at frequency ω_D, where $\omega_A = (2/T)\tan(\omega_D T/2)$.

It is noted that if $\omega_D T$ is small, then Eq. (4–15) can be simplified to

$$\omega_A \cong \frac{2}{T}\frac{\omega_D T}{2} = \omega_D$$

This means that for relatively low frequencies (compared with the folding frequency $\frac{1}{2}\omega_s$) the relationship between ω_D and ω_A is approximately linear. The low-frequency behavior of the equivalent discrete-time filter is approximately the same as that of the original continuous-time filter and thus the frequency distortion is small for small $\omega_D T$. However, as the frequency ω_D approaches the folding frequency $\frac{1}{2}\omega_s$, tan $(\omega_D T/2)$ approaches tan $(\pi/2)$ and the frequency ω_A of the continuous-time filter increases rapidly to infinity. The frequency distortion thus becomes stronger as ω_D approaches $\frac{1}{2}\omega_s$.

In using the bilinear transformation method, however, it is possible to take into account the possible distortion of the frequency response and to minimize the distortion at least near the frequency range of interest. Specifically, the frequency prewarping technique utilizing Eq. (4–15) can accomplish this purpose.

In discretizing a continuous-time filter using the frequency prewarping technique, we adjust the corner frequency (where the magnitude is −3 db) to a new value before applying the bilinear transformation method. If the bilinear transformation is applied to the adjusted continuous-time filter, it will bring the −3 db point in the ω_D domain to a desired frequency point. The following procedure will accomplish the objective.

For a Low-Pass Filter. Consider the low-pass filter

$$G(s) = \frac{a}{s+a} \tag{4–17}$$

We first carry out the adjustment of the frequency scale warping before transforming $G(s)$ into the z domain. Thus we replace a by $(2/T)\tan(aT/2)$, so that we get the expression

$$\frac{\frac{2}{T}\tan\frac{aT}{2}}{s + \frac{2}{T}\tan\frac{aT}{2}}$$

Then, we apply the bilinear transformation to this adjusted filter. That is, we substitute $(2/T)(1-z^{-1})/(1+z^{-1})$ for s in this last expression:

$$G_D(z) = \left.\frac{\frac{2}{T}\tan\frac{aT}{2}}{s + \frac{2}{T}\tan\frac{aT}{2}}\right|_{s=\frac{2}{T}\frac{1-z^{-1}}{1+z^{-1}}}$$

$$= \frac{\dfrac{2}{T}\tan\dfrac{aT}{2}}{\dfrac{2}{T}\dfrac{1-z^{-1}}{1+z^{-1}}+\dfrac{2}{T}\tan\dfrac{aT}{2}} = \frac{\tan\dfrac{aT}{2}}{\dfrac{1-z^{-1}}{1+z^{-1}}+\tan\dfrac{aT}{2}} \tag{4-18}$$

Equation (4–18) gives the discretized filter for $G(s)$ when the bilinear transformation technique is used with frequency prewarping.

For a High-Pass Filter. Similarly, for a high-pass filter

$$H(s) = \frac{s}{s+a}$$

we replace a by $(2/T)\tan(aT/2)$ and then substitute $(2/T)(1-z^{-1})/(1+z^{-1})$ for s in $H(s)$. The result is

$$H_D(z) = \frac{\dfrac{2}{T}\dfrac{1-z^{-1}}{1+z^{-1}}}{\dfrac{2}{T}\dfrac{1-z^{-1}}{1+z^{-1}}+\dfrac{2}{T}\tan\dfrac{aT}{2}} = \frac{\dfrac{1-z^{-1}}{1+z^{-1}}}{\dfrac{1-z^{-1}}{1+z^{-1}}+\tan\dfrac{aT}{2}} \tag{4-19}$$

Equation (4–19) gives the discretized high-pass filter for $H(s)$ when the frequency prewarping technique is used.

Note that the bilinear transformation with frequency prewarping maps the left half of the s plane into a unit circle in the z plane and therefore produces a stable discrete-time filter for a stable continuous-time filter. Also, there is no frequency folding, because the frequency range $0 < \omega_A < \infty$ in the s plane is compressed to $0 < \omega_D < \pi/T$ in the z plane.

The frequency response characteristics of the equivalent discrete-time filter obtained in this way will agree with those of the original continuous-time filter, in the low-frequency region for a low-pass filter or in the high-frequency region for a high-pass filter, and in the region near the corner frequency or cutoff frequency. There is considerable distortion in the phase characteristics, however, since the frequency response at $\omega_A = \infty$ is compressed to $\omega_D = \pi/T$. Also, the impulse response characteristics are considerably distorted.

To summarize, the bilinear transformation method with frequency prewarping is useful if a relatively simple design procedure is desired for an equivalent discrete-time filter and if it is required to have frequency response characteristics in the frequency range of interest reasonably similar to those of the given continuous-time filter.

Example 4–1.

Consider the continuous-time filter given by Eq. (4–17) and the discrete-time equivalent given by Eq. (4–18). Show that $|G(ja)|$ and $|G_D(e^{jaT})|$ are equal.

Notice that the magnitude of $G(j\omega)$ at $\omega = a$ is

$$|G(ja)| = \frac{a}{\sqrt{a^2 + a^2}} = \frac{a}{\sqrt{2}a} = 0.707 = -3 \text{ db}$$

Since $G_D(z)$ at frequency ω is given by

$$G_D(e^{j\omega T}) = \frac{\tan \dfrac{aT}{2}}{j \tan \dfrac{\omega T}{2} + \tan \dfrac{aT}{2}}$$

the magnitude of $G_D(e^{j\omega T})$ at $\omega = a$ is

$$|G_D(e^{jaT})| = \frac{\tan \dfrac{aT}{2}}{\sqrt{2} \tan \dfrac{aT}{2}} = 0.707 = -3 \text{ db}$$

Hence

$$|G(ja)| = |G_D(e^{jaT})|$$

Example 4–2.

Consider the continuous-time filter

$$G(s) = \frac{10}{s + 10}$$

Obtain an equivalent discrete-time filter using the bilinear transformation method with frequency prewarping. It is desired to have the same magnitude at $\omega = 10$ rad/sec for both the continuous-time and discrete-time filters. Assume that the sampling period T is given as 0.2 sec, or $T = 0.2$.

We replace 10 by $(2/T) \tan (10T/2)$ in the continuous-time filter $G(s)$ and then substitute $(2/T)(1 - z^{-1})/(1 + z^{-1})$ for s in the resulting equation. Thus,

$$G_D(z) = \left. \frac{\dfrac{2}{T} \tan \dfrac{10T}{2}}{s + \dfrac{2}{T} \tan \dfrac{10T}{2}} \right|_{s = \frac{2}{T}\frac{1 - z^{-1}}{1 + z^{-1}}}$$

$$= \frac{\dfrac{2}{T} \tan \dfrac{10T}{2}}{\dfrac{2}{T}\dfrac{1 - z^{-1}}{1 + z^{-1}} + \dfrac{2}{T} \tan \dfrac{10T}{2}} = \frac{\tan \dfrac{2}{2}}{\dfrac{1 - z^{-1}}{1 + z^{-1}} + \tan \dfrac{2}{2}}$$

$$= \frac{1.5574}{\dfrac{1 - z^{-1}}{1 + z^{-1}} + 1.5574} = \frac{0.6090(1 + z^{-1})}{1 + 0.2180z^{-1}}$$

This last equation gives the desired discrete-time filter. (Refer to Prob. A-4–3 for more detailed discussions on the bilinear transformation method with frequency prewarping.)

Impulse-invariance method. Consider the continuous-time filter $G(s)$ given by

$$G(s)=\frac{a}{s+a}$$

The equivalent discrete-time filter $G_D(z)$ obtained by the impulse-invariance method is the one whose impulse response $g_D(kT)=\mathscr{Z}^{-1}[G_D(z)]$ is T times the corresponding impulse response $g(t)=\mathscr{L}^{-1}[G(s)]$ of the continuous-time filter at the sampled instants, where T is the sampling period. That is,

$$g_D(kT)=Tg(t)|_{t=kT}$$

Thus, the impulse-invariance method preserves the shape of the impulse response. The z transform of the preceding equation gives

$$G_D(z)=\mathscr{Z}[g_D(kT)]=T\mathscr{Z}[g(t)]=T\mathscr{Z}[G(s)]=TG(z) \tag{4–20}$$

Thus, for the given continuous-time filter we have

$$G_D(z)=TG(z)=\frac{Ta}{1-e^{-aT}z^{-1}}$$

Because the equivalent discrete-time filter $G_D(z)$ is proportional to the z transform of the continuous-time filter $G(s)$, the impulse-invariance method is also called the z transform method.

The frequency response of $G_D(z)$ is the frequency response of $G(s)$ plus an infinite number of complementary frequency response components centered at integral multiples of the sampling frequency. The most serious problem associated with the discrete-time filter under discussion is the possibility of frequency folding, the folding of high frequencies into the low-frequency range of interest. If frequency folding occurs, the frequency response of the discrete-time filter will be greatly distorted. The occurrence of frequency folding may be avoided by two approaches: (1) cascading a low-pass filter ahead of the sampler to attenuate high-frequency components and (2) using a sufficiently high sampling frequency. Both approaches have some disadvantages. Adding a low-pass guard filter increases the time lag in a closed-loop; increased time lag in the loop always decreases the system's stability margin. Using a very high sampling frequency in relation to the cutoff frequency may result in an impractical hardware realization and may be an expensive way to solve the frequency folding problem. Therefore, the impulse-invariance method may be used only if the continuous-time filter $G(s)$ has sharp attenuation characteristics and the signals involved are band-limited. The sampling frequency must be sufficiently high that the effect of frequency folding will be minimal. In this case, the frequency response characteristics of $G_D(z)$ will be very close to those of the original continuous-time filter $G(s)$.

As for the stability, notice that since the z transformation always maps a stable $G(s)$ into a stable $G_D(z)$, there is no stability problem if this method is used for obtaining the equivalent discrete-time filter for a stable continuous-time filter.

It should be pointed out, however, that the determination of the equivalent discrete-time filter by this method may not be simple from a computational viewpoint, since finding the z transform of a relatively complex filter $G(s)$ is a laborious process. [If $G(s)$ is of higher order, then it must be expanded into partial fractions.]

Step-invariance method. Consider a continuous-time filter $G(s)$, where

$$G(s) = \frac{a}{s+a} \tag{4–21}$$

The equivalent discrete-time filter $G_D(z)$ of the continuous-time filter $G(s)$ obtained by the step-invariance method is the one whose step response is the same as that of the original continuous-time filter at the sampling instants. That is,

$$\mathscr{Z}^{-1}\left[G_D(z)\frac{1}{1-z^{-1}}\right] = \mathscr{L}^{-1}\left[G(s)\frac{1}{s}\right]_{t=kT}$$

where $G_D(z)[1/(1-z^{-1})]$ represents the step response of $G_D(z)$ and $G(s)/s$ represents the step response of $G(s)$. By taking the z transform of this last equation we obtain

$$G_D(z)\frac{1}{1-z^{-1}} = \mathscr{Z}\left\{\mathscr{L}^{-1}\left[\frac{G(s)}{s}\right]\right\} = \mathscr{Z}\left[\frac{G(s)}{s}\right]$$

or

$$G_D(z) = (1-z^{-1})\,\mathscr{Z}\left[\frac{G(s)}{s}\right]$$

Notice that this last equation can be rewritten as follows:

$$G_D(z) = \mathscr{Z}\left[\frac{1-e^{-Ts}}{s}G(s)\right] \tag{4–22}$$

The right-hand side of this last equation can be seen as the z transform of the filter $G(s)$ preceded by an impulse sampler and a hold device. Therefore, we may first assume that a continuous-time input signal $x(t)$ is fed to a fictitious sample-and-hold device as shown in Fig. 4–8. Then we assume that the output of the hold is fed to the continuous-time filter $G(s)$. The ratio of the z transforms of the output

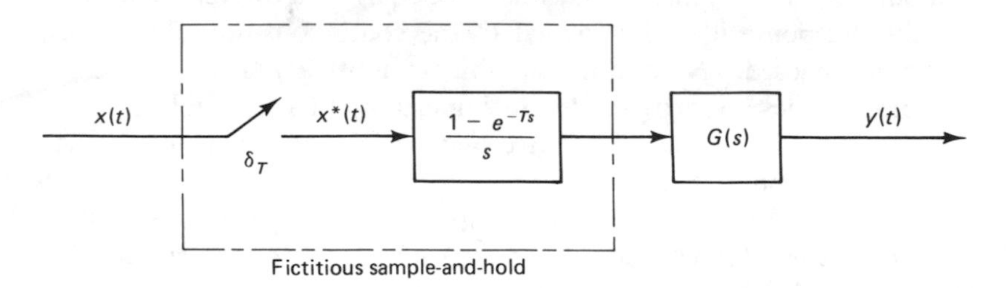

Figure 4–8 $G(s)$ preceded by a fictitious sample-and-hold device.

$y(t)$ and the input $x(t)$ gives the discrete-time filter $G_D(z)$. It gives the same step response (at $t = kT$, where $k = 0, 1, 2, \ldots$) as that of the original continuous-time filter $G(s)$.

Because the equivalent discrete-time filter obtained by the step-invariance method has the form given by Eq. (4–22), the method is also called the z transform method with sample-and-hold. The discrete-time system obtained in this way is called the zero-order hold equivalent of the continuous-time system. It should be noted that in the present method the sampler and hold are fictitious and are used only as an analytical part of the method. They do not exist as physical hardware. The step-invariance method is commonly used to obtain discrete equivalents of plants and other components in control systems. That is, the dynamics of plants and other components in digital control systems are very often approximated by this method.

For the continuous-time filter given by Eq. (4–21), the equivalent discrete-time filter $G_D(z)$ is obtained as follows:

$$G_D(z) = (1 - z^{-1}) \mathscr{Z}\left[\frac{G(s)}{s}\right] = (1 - z^{-1}) \mathscr{Z}\left[\frac{a}{s(s+a)}\right]$$

$$= (1 - z^{-1}) \mathscr{Z}\left[\frac{1}{s} - \frac{1}{s+a}\right] = (1 - z^{-1})\left(\frac{1}{1 - z^{-1}} - \frac{1}{1 - e^{-aT}z^{-1}}\right)$$

$$= \frac{(1 - e^{-aT})z^{-1}}{1 - e^{-aT}z^{-1}}$$

Equivalent discrete-time filters obtained by the step-invariance method may exhibit frequency folding phenomena and may present the same kind of aliasing errors as filters designed by the impulse-invariance method. Notice, however, that the presence of the $1/s$ term in $G(s)/s$ adds high-frequency attenuation. Consequently, the equivalent discrete-time filter designed by the step-invariance method will exhibit smaller aliasing errors than that designed by the impulse-invariance method. In applying the step-invariance method, the signals must be band-limited.

As for stability, the equivalent discrete-time filter obtained by the step-invariance method is stable if the original continuous-time filter is a stable one.

In a particular situation where the transient response characteristics to step inputs are most important and where the equivalent discrete-time filter needs to exhibit a step response identical to that of the continuous-time filter, the present method should be used. (Note, however, that in most applications, although the response characteristics to step inputs are important, they are not the only characteristics that need to be considered. There may be other types of inputs that are important, and the response characteristics to those inputs must also be considered.)

Although both the impulse response and frequency response characteristics of the equivalent discrete-time filter designed by the step-invariance method are distorted, this method is convenient for digital computer implementation.

Matched pole-zero mapping method. Roughly speaking, in the matched pole-zero mapping method, we consider separately the numerator and denominator of the transfer function $G(s)$ of the continuous-time filter and map the poles of $G(s)$ to the poles of the discrete-time filter and the zeros of $G(s)$ to the zeros of the discrete-time filter. Consider the continuous-time filter

$$G(s) = \frac{a}{s + a}$$

The z transform of $G(s)$ is

$$G(z) = \frac{a}{1 - e^{-aT}z^{-1}} = \frac{az}{z - e^{-aT}}$$

Notice that if we look at only the denominator of $G(z)$, the pole of $G(s)$ is related to the pole of $G(z)$ by the relationship $z = e^{sT}$. An extension of this very fact enables us to obtain a discrete equivalent to a continuous-time filter.

The present continuous-time filter $G(s) = a/(s + a)$ has no finite zero but has one infinite zero. If $G(s)$ has a finite zero at $s = -b$, then we simply assume that the equivalent discrete-time filter $G_D(z)$ has a zero at $z = e^{-bT}$. For an infinite zero, we assume that there is a zero at $z = -1$ in the equivalent discrete-time filter. The reason for placing a zero at $z = -1$ may be explained as follows.

First, notice that the $j\omega$ axis from $\omega = 0$ to $\omega = \frac{1}{2}\omega_s = \pi/T$ in the s plane maps into a unit semicircle from $z = 1$ to $z = -1$ in the z plane. If we choose the sampling frequency ω_s to satisfy the sampling theorem, then $\omega = \frac{1}{2}\omega_s$ (instead of $\omega = \infty$) can be considered the possible highest frequency. Since the present continuous-time filter $G(s) = a/(s + a)$ is a low-pass filter we may say that $G(j\omega)$ approaches zero as ω approaches $\frac{1}{2}\omega_s$. [Precisely speaking, $G(j\omega)$ approaches zero as ω approaches ∞.] This is equivalent to saying that $G_D(z)$, the discrete-time equivalent of $G(s)$, approaches zero as z approaches -1. (Point $z = -1$ corresponds to the possible highest frequency point $\omega = \pi/T$ in the s plane.) This is why we place a zero at $z = -1$ for each infinite zero in the continuous-time filter.

Note that the frequency response of the discrete-time filter for $0 \leqslant \omega \leqslant \frac{1}{2}\omega_s$ corresponds to the frequency response of the continuous-time filter for $0 \leqslant \omega \leqslant \infty$. Therefore, the matched pole-zero mapping method involves no aliasing. This method preserves the general shape of the frequency response.

The matched pole-zero mapping method follows the following procedure:

1. First, $G(s)$ must be in a factored form before the method can be applied. Then the poles of $G(s)$ are mapped to the z plane poles according to the relation $z = e^{sT}$. For example, a pole of $G(s)$ at $s = -a$ is mapped to a pole at $z = e^{-aT}$.
2. The finite zeros of $G(s)$ are mapped to the z plane zeros according to the relation $z = e^{sT}$. That is, a finite zero of $G(s)$ at $s = -b$ is mapped to a zero at $z = e^{-bT}$ in $G_D(z)$.

3. The infinite zeros (zeros at infinity) are mapped to the point $z = -1$. Thus, for each infinite zero we have a factor $z + 1$ in the numerator of the discrete-time filter. (The number of such infinite zeros is equal to the number of excess poles in the transfer function of the continuous-time filter.) Similarly, infinite poles (poles at infinity), if any, are mapped to the point $z = -1$. Thus, for each infinite pole we have a factor $z + 1$ in the denominator of the discrete-time filter.
4. Adjust the gain of the discrete-time filter to match the gain of the continuous-time filter. For low-pass filters, the gain of the discrete-time filter at $z = 1$ should be the same as the gain of the continuous-time filter at $s = 0$. Similarly, for high-pass filters, the gains should be matched at $z = -1$ and $s = \infty$, respectively.

If $G(s)$ involves conjugate complex poles or zeros, then it is usually advantageous to treat a set of conjugate complex poles or zeros as a single unit rather than treat them separately. (See Example 4–6.)

Let us now obtain a discrete-time equivalent of the low-pass filter given by

$$G(s) = \frac{a}{s + a}$$

by the matched pole-zero mapping method.

First, note that there is no finite zero in $G(s)$ but there is one infinite zero. The infinite zero is mapped to $z = -1$. The finite pole at $s = -a$ is mapped to a pole at $z = e^{-aT}$. Hence, the discrete-time equivalent $G_D(z)$ of the given $G(s)$ is

$$G_D(z) = K\frac{a(z + 1)}{z - e^{-aT}}$$

where the gain K is adjusted so that the gains at low frequencies are the same for both the discrete-time and continuous-time filters. That is, we equate $G_D(1)$ and $G(0)$, or

$$G_D(1) = K\frac{2a}{1 - e^{-aT}} = G(0) = 1$$

from which the constant K is determined to be

$$K = \frac{1 - e^{-aT}}{2a}$$

Thus, the desired $G_D(z)$ is obtained as follows:

$$\frac{Y(z)}{X(z)} = G_D(z) = \frac{(1 - e^{-aT})(z + 1)}{2(z - e^{-aT})} = \frac{(1 - e^{-aT})(1 + z^{-1})}{2(1 - e^{-aT}z^{-1})} \tag{4–23}$$

Notice that the degrees of the numerator and denominator polynomials are equal. Thus, in the equivalent difference equation

$$y(kT) = e^{-aT}y((k-1)T) + \tfrac{1}{2}(1 - e^{-aT})[x(kT) + x((k-1)T)]$$

the output $y(kT)$ at time kT will require a sampled input at time kT or $x(kT)$.

Example 4–3.

Consider the continuous-time filter

$$G(s) = \frac{s+b}{s+a}$$

where our interest is in the low-frequency range. Obtain an equivalent discrete-time filter by using the matched pole-zero mapping method.

The zero at $s = -b$ is mapped to $z = e^{-bT}$, while the pole at $s = -a$ is mapped to $z = e^{-aT}$. Hence, the equivalent discrete-time filter may be obtained as

$$G_D(z) = K\frac{z - e^{-bT}}{z - e^{-aT}}$$

where gain K is determined so that the low-frequency gains of the discrete-time and continuous-time filters are the same. Thus

$$G_D(1) = K\frac{1 - e^{-bT}}{1 - e^{-aT}} = G(0) = \frac{b}{a}$$

from which gain K is determined as follows:

$$K = \frac{b}{a}\frac{1 - e^{-aT}}{1 - e^{-bT}}$$

The equivalent discrete-time filter may therefore be given by

$$G_D(z) = \frac{b}{a}\frac{1 - e^{-aT}}{1 - e^{-bT}}\frac{z - e^{-bT}}{z - e^{-aT}}$$

$$= \frac{b}{a}\frac{1 - e^{-aT}}{1 - e^{-bT}}\frac{1 - e^{-bT}z^{-1}}{1 - e^{-aT}z^{-1}}$$

Example 4–4.

Consider the high-pass filter

$$G(s) = \frac{s}{s+a}$$

Obtain an equivalent discrete-time filter by the matched pole-zero mapping method.

The zero $s = 0$ is mapped to $z = e^{-0T} = 1$ and the pole $s = -a$ is mapped to $z = e^{-aT}$. Consequently,

$$G_D(z) = K\frac{z - 1}{z - e^{-aT}}$$

where gain K is determined so that the high-frequency gains of the discrete-time and continuous-time filters are the same. Thus,

$$G_D(-1) = K\frac{-1 - 1}{-1 - e^{-aT}} = G(\infty) = 1$$

from which we obtain

$$K = \frac{1 + e^{-aT}}{2}$$

and the equivalent discrete-time filter can be given by

$$G_D(z) = \frac{1 + e^{-aT}}{2} \frac{z - 1}{z - e^{-aT}}$$

$$= \frac{1 + e^{-aT}}{2} \frac{1 - z^{-1}}{1 - e^{-aT}z^{-1}}$$

Example 4–5.

Consider the continuous-time filter

$$G(s) = s + a$$

where we are interested in the low-frequency range. Obtain an equivalent discrete-time filter by the matched pole-zero mapping method.

The continuous-time filter $G(s)$ has a zero at $s = -a$. $G(s)$ has no finite pole, but it has an infinite pole. The zero $s = -a$ is mapped to $z = e^{-aT}$, and the infinite pole is mapped to $z = -1$. Hence, the equivalent discrete-time filter may be given by

$$G_D(z) = K \frac{z - e^{-aT}}{z + 1}$$

where gain K is determined so that the low-frequency gains of the discrete-time and continuous-time filters are the same. Thus

$$G_D(1) = K \frac{1 - e^{-aT}}{1 + 1} = G(0) = a$$

Hence

$$K = \frac{2a}{1 - e^{-aT}}$$

The equivalent discrete-time filter may therefore be given by

$$G_D(z) = \frac{2a}{1 - e^{-aT}} \frac{z - e^{-aT}}{z + 1} = \frac{2a}{1 - e^{-aT}} \frac{1 - e^{-aT}z^{-1}}{1 + z^{-1}}$$

Example 4–6.

Given the continuous-time filter

$$G(s) = \frac{1}{(s + a)^2 + b^2} = \frac{1}{(s + a + jb)(s + a - jb)}$$

obtain an equivalent discrete-time filter by the matched pole-zero mapping method.

The conjugate poles in the s plane are mapped to conjugate poles in the z plane. For the given $G(s)$ there is no finite zero, but there are two infinite zeros. The infinite zeros are mapped to $z = -1$. Hence the equivalent discrete-time filter may be given by

$$G_D(z) = K \frac{(z+1)^2}{z^2 - 2ze^{-aT}\cos bT + e^{-2aT}}$$

where gain K is determined so that the low-frequency gains of $G_D(z)$ and $G(s)$ are the same. By equating $G_D(1)$ with $G(0)$ and solving for K, we obtain

$$K = \frac{1 - 2e^{-aT}\cos bT + e^{-2aT}}{4(a^2 + b^2)}$$

Hence the equivalent discrete-time filter is given by

$$G_D(z) = \frac{1 - 2e^{-aT}\cos bT + e^{-2aT}}{4(a^2+b^2)} \frac{(1+z^{-1})^2}{1 - 2e^{-aT}z^{-1}\cos bT + e^{-2aT}z^{-2}}$$

Comments. In the preceding discussions we have presented several forms of an equivalent discrete-time filter $G_D(z)$ for a continuous-time filter $G(s) = a/(s+a)$. Among the methods yielding stable filters, the step-invariance method (the z transform method with sample-and-hold) yielded $G_D(z)$ in the form

$$G_D(z) = \frac{Y(z)}{X(z)} = \frac{\alpha z^{-1}}{1 + \beta z^{-1}} \tag{4-24}$$

where α and β are constants; whereas other methods yielded $G_D(z)$ in the form

$$G_D(z) = \frac{Y(z)}{X(z)} = K\frac{1 + \alpha z^{-1}}{1 + \beta z^{-1}} \tag{4-25}$$

where K, α, and β are constants.

The difference equation corresponding to Eq. (4–24) is

$$y(kT) = -\beta y((k-1)T) + \alpha x((k-1)T)$$

To produce the output $y(kT)$ the digital computer requires the input from the previous sampling point, $x((k-1)T)$, and the previous output, $y((k-1)T)$.

On the other hand, Eq. (4–25) yields the following difference equation:

$$y(kT) = -\beta y((k-1)T) + Kx(kT) + \alpha K x((k-1)T)$$

In this case, in order to produce the output $y(kT)$ the digital computer requires the input at the same sampling point, $x(kT)$, and the input and output at the previous sampling point, $x((k-1)T)$ and $y((k-1)T)$. If, in a particular situation, it is not possible to include the current input $x(kT)$ to produce the output $y(kT)$, then an equivalent discrete-time filter of the form given by Eq. (4–24) (the step-invariance method) should be used.

Summary. In this section we have presented several methods for obtaining discrete-time equivalents for continuous-time filters. (One of the methods, the forward difference method, may produce unstable discrete-time filters and therefore cannot be used in practice.)

The response between sampling points is different for each discretization method used. Furthermore, none of these equivalent discrete-time filters can have complete fidelity; the actual (continuous-time) response between any two consecutive sampling points is always different from the response between the same two consecutive sampling points that is taking place in each equivalent discrete-time filter, no matter what method of discretization is used.

It is noted that at or near the frequency $\omega = \frac{1}{2}\omega_s = \pi/T$ there are considerable differences in frequency response between the original continuous-time filter and its discrete-time equivalents. ($\frac{1}{2}\omega_s$ is the highest frequency that we consider in the response of the discrete-time or digital control system.) If the sampling frequency satisfies the sampling theorem, then in most cases the frequency $\frac{1}{2}\omega_s$ is outside the range of interest. And if the corner frequency of any transfer function is well below $\frac{1}{4}\omega_s$,

TABLE 4–1 EQUIVALENT DISCRETE-TIME FILTERS FOR A CONTINUOUS-TIME FILTER $G(s) = a/(s + a)$

Mapping method	Mapping equation	Equivalent discrete-time filter for $G(s) = \frac{a}{s+a}$
Backward difference method	$s = \frac{1 - z^{-1}}{T}$	$G_D(z) = \frac{a}{\frac{1 - z^{-1}}{T} + a}$
Forward difference method	$s = \frac{1 - z^{-1}}{Tz^{-1}}$	This method is not recommended, because the discrete-time equivalent may become unstable.
Bilinear transformation method	$s = \frac{2}{T}\frac{1 - z^{-1}}{1 + z^{-1}}$	$G_D(z) = \frac{a}{\frac{2}{T}\frac{1 - z^{-1}}{1 + z^{-1}} + a}$
Bilinear transformation method with frequency prewarping	$s = \frac{2}{T}\frac{1 - z^{-1}}{1 + z^{-1}}$ $\left(\omega_A = \frac{2}{T}\tan\frac{\omega_D T}{2}\right)$	$G_D(z) = \frac{\tan\frac{aT}{2}}{\frac{1 - z^{-1}}{1 + z^{-1}} + \tan\frac{aT}{2}}$
Impulse-invariance method	$G_D(z) = T\mathcal{Z}[G(s)]$	$G_D(z) = \frac{Ta}{1 - e^{-aT}z^{-1}}$
Step-invariance method	$G_D(z) = \mathcal{Z}\left[\frac{1 - e^{-Ts}}{s}G(s)\right]$	$G_D(z) = \frac{(1 - e^{-aT})z^{-1}}{1 - e^{-aT}z^{-1}}$
Matched pole-zero mapping method	A pole or zero at $s = -a$ is mapped to $z = e^{-aT}$. An infinite pole or zero is mapped to $z = -1$.	$G_D(z) = \frac{1 - e^{-aT}}{2}\frac{1 + z^{-1}}{1 - e^{-aT}z^{-1}}$

then the frequency response characteristics near the corner frequency are accurately reproduced.

Table 4–1 summarizes the seven different forms of the equivalent discrete-time filter for the same continuous-time filter $G(s) = a/(s + a)$. It is not possible to say which equivalent discrete-time filter is best for any given system, since the degree of distortions in transient response and frequency response characteristics depends on the sampling frequency, the cutoff frequency, the highest-frequency component involved in the system, transportation lag present in the system, etc.

If the design of a discrete-time control system is attempted through continuous-time design techniques and an equivalent discrete-time filter is to be used in place of a continuous-time (analog) filter, then it may be advisable for the designer to try a few alternate forms of the equivalent discrete-time filter before the design is completed. (The final design should be reached only after a complete simulation on a digital computer yields satisfactory test results.) Since the matched pole-zero mapping method, the bilinear transformation method, and the bilinear transformation method with frequency prewarping generally give satisfactory results, one of these methods may be used as a first trial in the design process. (The matched pole-zero mapping method requires simpler algebra than the other methods mentioned here. For this reason, it is frequently used for the first trial.)

Finally, it is noted again that if quantization effects are ignored, discrete-time filters are the same as digital filters. In the remaining sections of this chapter, we ignore the quantization effects and use "discrete-time controller" and "digital controller" interchangeably. Likewise in our discussions we shall frequently substitute "digital control systems" for "discrete-time control systems" and vice versa.

4–3 DESIGN PRINCIPLES BASED ON A DISCRETE-TIME EQUIVALENT OF AN ANALOG CONTROLLER

In this section we design a digital control system by using a discrete-time equivalent of an analog controller. Our approach here is to first assume that the entire control system under consideration is to be continuous-time and therefore to design first a suitable analog controller in the s plane by using conventional transform methods (such as the root locus and frequency response methods), and then to discretize the analog controller and use the discretized version of the analog controller as a digital controller.

Consider the control system shown in Fig. 4–9. We assume that the plant $G_p(s)$ is continuous-time, that its dynamic characteristics are given, and that the controller that satisfies the given performance specifications is of the analog type. Our objective here is to transform the analog controller into an equivalent digital controller so that we have a digital control system that satisfies the given performance specifications.

In replacing the analog controller by a digital controller we need to add a sampler between the summing point and the digital controller and a hold circuit between the digital controller and the plant, as shown in the block diagram of Fig.

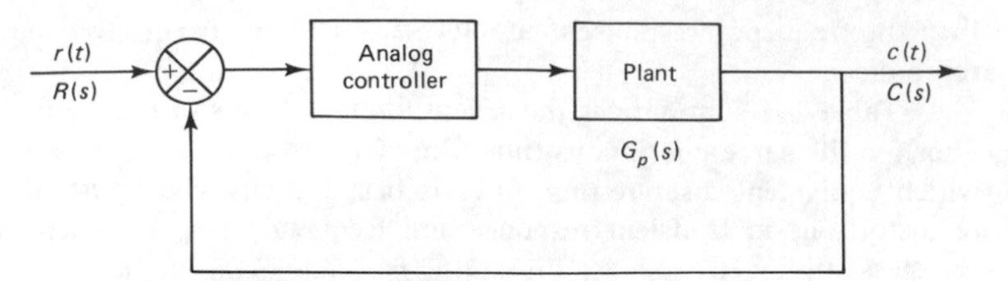

Figure 4–9 Continuous-time control system.

4–10. Also, although we do not explicitly show a sampler between the digital controller and the hold, the output of the digital controller is always sampled before being fed to the hold circuit. In fact, the hold shown in the diagram is a sample-and-hold circuit. (It is a common practice to denote a sample-and-hold circuit by simply a hold.)

It is important to note that the hold circuit produces a time lag in the system; this time lag is inevitably introduced into the loop whenever we convert an analog controller to a digital controller. The time lag produces phase lag and reduces the stability margin in a closed-loop system; therefore if analog-to-digital conversion is attempted, it is necessary to allow for the time lag in the closed loop at the time the original analog controller is being designed.

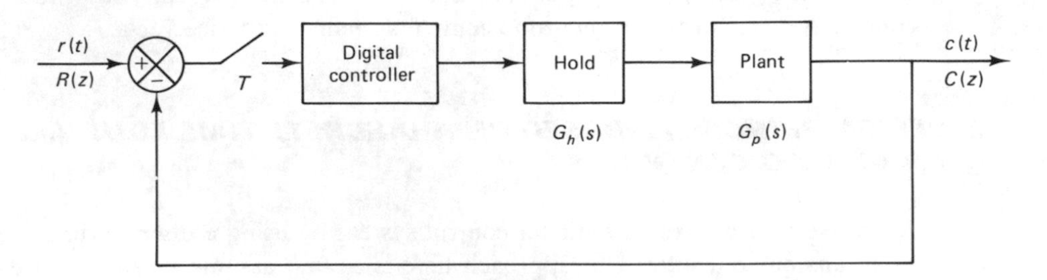

Figure 4–10 Digital control system.

Let us assume that we use a zero-order hold, $(1 - e^{-Ts})/s$, in our digital control system. The hold may be approximated by a transfer function which is a ratio of polynomials in s. Notice that e^{-Ts} may be written in the form of a ratio of polynomials in Ts as follows:

$$e^{-Ts} = \frac{1 - \dfrac{Ts}{2} + \dfrac{(Ts)^2}{8} - \cdots}{1 + \dfrac{Ts}{2} + \dfrac{(Ts)^2}{8} + \cdots} \cong \frac{1 - \dfrac{Ts}{2}}{1 + \dfrac{Ts}{2}}$$

The zero-order hold may therefore be approximated by

$$\frac{1-e^{-Ts}}{s} = \frac{1}{s}\left(1 - \frac{1-\dfrac{Ts}{2}}{1+\dfrac{Ts}{2}}\right) = \frac{T}{\dfrac{T}{2}s+1}$$

The time lag that will be introduced in the closed loop by the zero-order hold can thus be approximated by the time lag of $T/(\frac{1}{2}Ts + 1)$. (The sampling period T should be chosen to satisfy the sampling theorem.)

What we are concerned with is the time lag or phase lag introduced by the hold. Since the overall system dc gain will be determined at the final stage of the design, here we introduce, instead of $T/(\frac{1}{2}Ts + 1)$, a transfer function

$$G_h(s) = \frac{1}{\dfrac{T}{2}s+1} \tag{4–26}$$

into the continuous-time control system in anticipation of the conversion from an analog controller to a digital controller. [$G_h(s)$ has the property that it approximates the phase characteristics of the hold but has unity dc gain.] Figure 4–11 shows the modified version of the continuous-time control system shown in Fig. 4–9.

Once the analog controller for the system of Fig. 4–11 is properly designed, then we discretize the analog controller $G_c(s)$ and obtain an equivalent digital controller $G_D(z)$ by using one of the techniques presented in Sec. 4–2. This completes the tentative determination of the digital controller.

The next step is to make sure that the system so designed will behave as expected. To test the performance of the designed system, we must analyze the system response. To analyze the behavior of the system, we need to determine the pulse transfer function $G(z)$ of the continuous-time plant $G_p(s)$ when it is preceded by a zero-order hold. Then we test the response to various input signals of the system whose block diagram is shown in Fig. 4–12. If the results are satisfactory, then the analytical part of the design is complete.

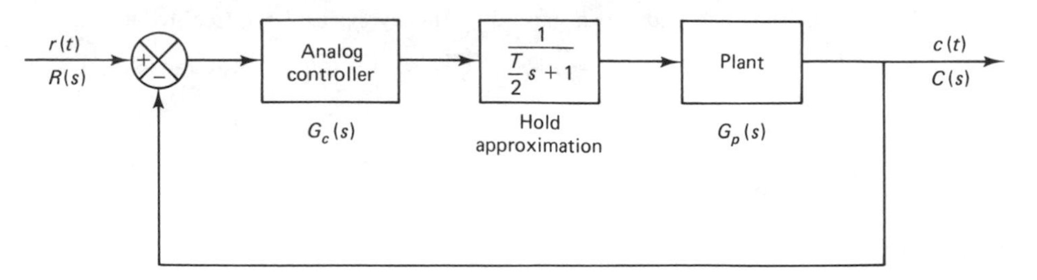

Figure 4–11 Continuous-time control system shown in Fig. 4–9 modified to allow for time lag.

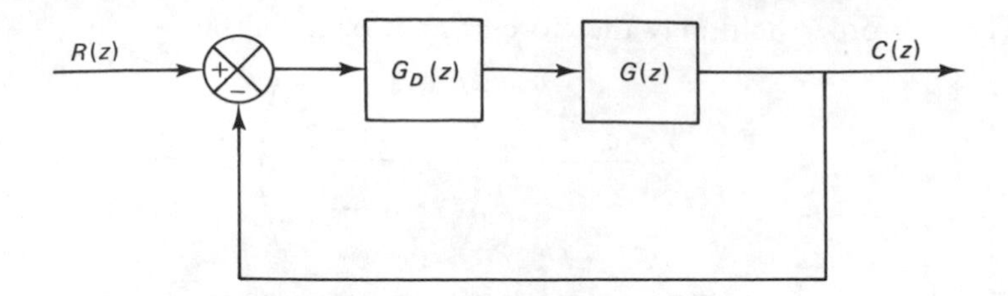

Figure 4–12 Digital control system obtained by discretizing the analog control system shown in Fig. 4–11.

Finally, we convert the digital controller, which is specified in the form of a ratio of polynomials in z or a pulse transfer function, to a numerical algorithm and solve it in real time by a digital computer.

It is important to point out that an increase in the sampling period T modifies the system dynamics and may destabilize the closed-loop system. Therefore, it is necessary to choose a sufficiently small sampling period T. (Barely satisfying the sampling theorem is not enough.)

Example 4–7, which follows, shows the details of the design process just presented.

Example 4–7.

Consider the continuous-time control system shown in Fig. 4–13. The specifications for the system are that the damping ratio ζ of the pair of dominant closed-loop poles be 0.5 and the settling time be 2 sec (the settling time is defined as $4/\zeta\omega_n$). The given specifications can be translated into terms of the unit step response as follows: the maximum overshoot be approximately 16.3 percent and the undamped natural frequency be 4 rad/sec. A well-designed analog controller will satisfy the given specifications.

We wish to convert the analog control system to a digital control system that will exhibit a response similar to that of the analog control system. That is, the digital control system should exhibit approximately 16.3 percent overshoot in the unit step response and should have a settling time of approximately 2 sec.

The sampling period T for the digital control system must be decided upon before the present design process can be started. Noting that the damping ratio ζ is specified as 0.5 and the undamped natural frequency ω_n as 4 rad/sec, we have

$$\omega_d = \omega_n\sqrt{1-\zeta^2} = 4\sqrt{1-0.5^2} = 3.464 \text{ rad/sec}$$

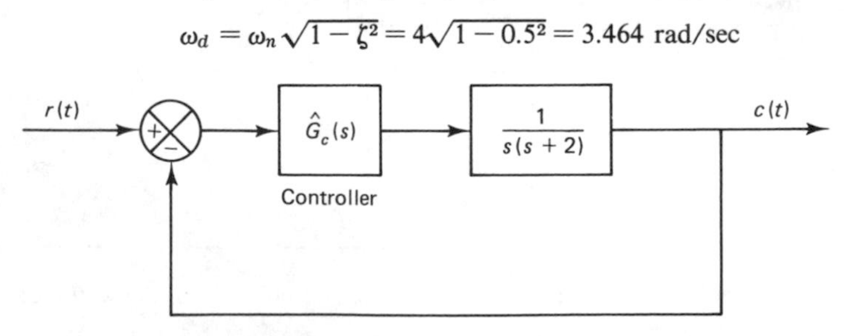

Figure 4–13 Control system considered in Example 4–7.

Hence the response of the continuous-time or analog control system to a step input will exhibit a damped oscillation of period $2\pi/\omega_d$, or 1.814 sec. It is desirable to have at least eight samples per period. (This is a rule of thumb. In some systems we may choose a sampling period T of approximately one-tenth to one-half, depending on the circumstances, of the smallest significant time constant involved in the plant.) Here, we may choose a sampling period T of 0.2 sec. In addition to the requirement that we should sample at least eight times per period, the choice of the sampling period depends on other factors such as the highest-frequency component involved in the input and disturbances (noises) to which the system may be subjected. Here, however, for the convenience of demonstration of the design method, we simply choose $T = 0.2$ sec.

Let us assume that we use a zero-order hold in the digital control system. The effect of such a hold is to introduce a time lag, as mentioned earlier.

We shall first design an analog controller that allows for the time lag to be introduced by the hold. Once a suitable analog controller is designed, it is possible to obtain an equivalent digital controller with one or more of the techniques presented in Sec. 4–2.

The time lag due to the hold circuit produces a phase lag. The amount of this phase lag can be approximated by $G_h(s)$ given by Eq. (4–26), or

$$G_h(s) = \frac{1}{(T/2)s + 1} = \frac{1}{0.1s + 1} = \frac{10}{s + 10}$$

The inclusion of $G_h(s)$ in the control system modifies the block diagram to the form shown in Fig. 4–14.

A suitable analog controller can easily be designed by use of a conventional method. (Any of the design methods using the root locus or frequency response technique may be applied to obtain a satisfactory controller.) For the modified continuous-time or analog control system we may determine the transfer function $G_c(s)$ of a suitable analog controller to be

$$G_c(s) = 20.25\left(\frac{s + 2}{s + 6.66}\right)$$

The controller zero at $s = -2$ will cancel the plant pole at $s = -2$. Thus, the controller $G_c(s)$ effectively replaces the open-loop pole at $s = -2$ by the open-loop pole at $s = -6.66$. Figure 4–15 shows the block diagram of the analog control system so designed. (Because of the difficulty in building pure analog differentiators, a pole is usually included in the analog controller. In the present case, the transfer function of the analog controller includes a pole at $s = -6.66$. In fact, having such a pole in the controller is advantageous

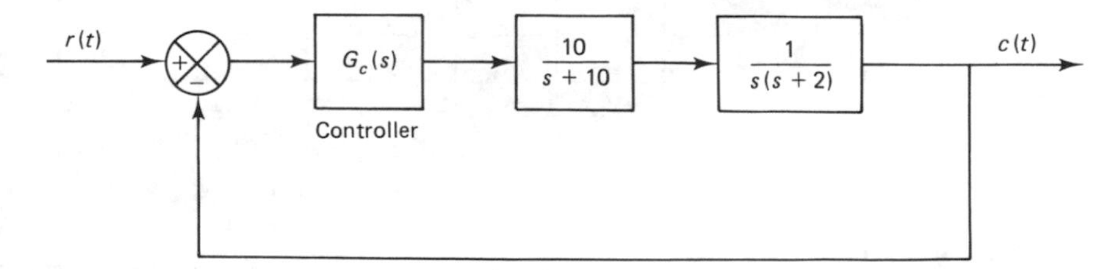

Figure 4–14 Control system of Fig. 4–13 modified to include anticipated phase lag due to the hold device when the analog controller is converted to the digital controller.

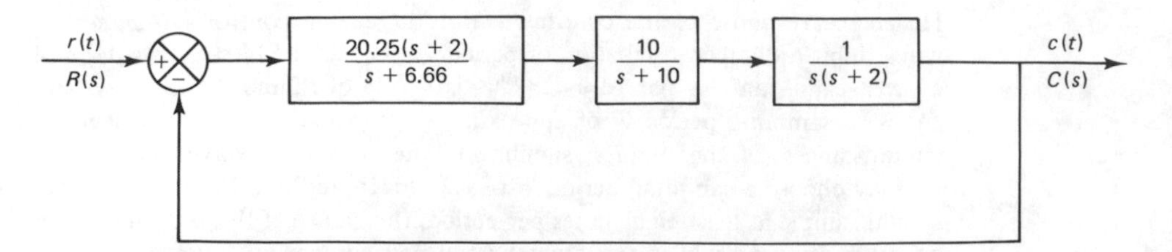

Figure 4–15 Block diagram for the analog control system designed in Example 4–7.

in attenuating noises because it has a smoothing effect.) The closed-loop transfer function is

$$\frac{C(s)}{R(s)} = \frac{\dfrac{202.5}{(s+6.66)(s+10)s}}{1+\dfrac{202.5}{(s+6.66)(s+10)s}}$$

$$= \frac{202.5}{(s+2+j2\sqrt{3})(s+2-j2\sqrt{3})(s+12.66)}$$

The designed system has closed-loop poles at $s = -2 + j2\sqrt{3}$, $s = -2 - j2\sqrt{3}$, and $s = -12.66$. Since the third pole at $s = -12.66$ is far away from the origin (that is, since the real part of this third pole is over 6 times the real part of the conjugate complex poles), the response of this system can be approximated by two dominant closed-loop poles at $s = -2 \pm j2\sqrt{3}$. Notice that the damping ratio ζ and the undamped natural frequency ω_n of the dominant closed-loop poles are 0.5 and 4 rad/sec, respectively.

Once we have designed the analog controller $G_c(s)$, it is an easy matter to discretize it and to obtain an equivalent digital controller $G_D(z)$. Figure 4–16 shows the block diagram of the discretized version of the control system of Fig. 4–15.

Now we discretize $G_c(s)$. Since our analog controller has been designed to cancel the undesired pole at $s = -2$ with the zero of the controller, it is convenient to use the matched pole-zero mapping method. By this method we are able to cancel the pole at $z = e^{-2T}$ (that is, $s = -2$) with the zero of the equivalent digital controller. Since our analog controller $G_c(s)$ is

$$G_c(s) = 20.25\left(\frac{s+2}{s+6.66}\right)$$

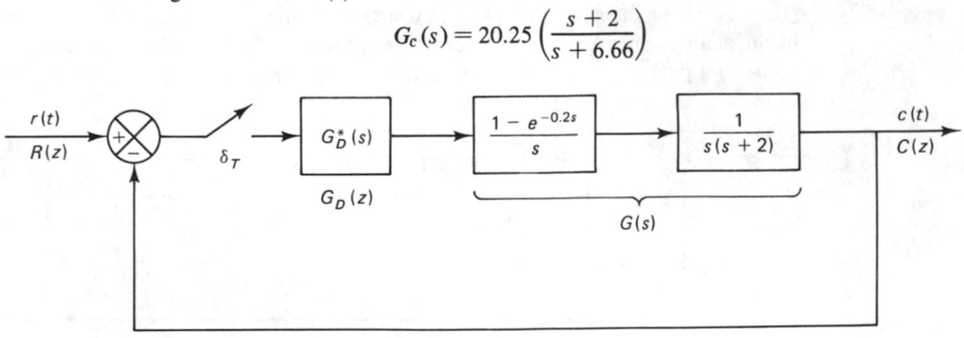

Figure 4–16 Block diagram of the discretized version of the control system shown in Fig. 4–15.

by use of the matched pole-zero mapping method the pole at $s = -6.66$ is mapped to a pole at $z = e^{-6.66T} = e^{-6.66\times 0.2} = e^{-1.332} = 0.2644$ and the zero at $s = -2$ is mapped to a zero at $z = e^{-2T} = e^{-2\times 0.2} = e^{-0.4} = 0.6703$. Hence the equivalent digital controller will have the form

$$G_D(z) = K\frac{z - 0.6703}{z - 0.2644}$$

The gain constant K is determined so that the low-frequency gains are the same for both $G_D(z)$ and $G_c(s)$. Hence we substitute 1 for z in $G_D(z)$ and equate $G_D(1)$ with $G_c(0)$:

$$G_D(1) = K\frac{1 - 0.6703}{1 - 0.2644} = G_c(0) = 20.25\left(\frac{0 + 2}{0 + 6.66}\right)$$

which yields

$$K = 13.57$$

The equivalent digital controller is thus determined to be

$$G_D(z) = 13.57\left(\frac{z - 0.6703}{z - 0.2644}\right) \tag{4–27}$$

Equation (4–27) can be converted to a difference equation which will be solved by the digital controller. This completes the demonstration of how to design an equivalent digital controller from an analog controller.

The following analysis is to check the response of the digital control system thus designed. To analyze the digital control system we must obtain the pulse transfer function $G(z)$ of the plant which is preceded by a zero-order hold. Noting that $T = 0.2$ sec, we have

$$\begin{aligned}
G(z) &= \mathscr{Z}\left[\frac{1 - e^{-0.2s}}{s}\frac{1}{s(s + 2)}\right] \\
&= (1 - z^{-1})\,\mathscr{Z}\left[\frac{1}{s^2(s + 2)}\right] \\
&= (1 - z^{-1})\,\mathscr{Z}\left[\frac{0.5}{s^2} - \frac{0.25}{s} + \frac{0.25}{s + 2}\right] \\
&= \frac{z - 1}{z}\left[\frac{0.1z}{(z - 1)^2} - \frac{0.25z}{z - 1} + \frac{0.25z}{z - e^{-0.4}}\right] \\
&= \frac{0.01758(z + 0.8760)}{(z - 1)(z - 0.6703)}
\end{aligned}$$

Figure 4–17(a) shows the block diagram of the designed digital control system.

Notice that the feedforward pulse transfer function can be simplified to

$$\begin{aligned}
13.57\left(\frac{z - 0.6703}{z - 0.2644}\right)\frac{0.01758(z + 0.8760)}{(z - 1)(z - 0.6703)} &= 0.2385\left[\frac{z + 0.8760}{(z - 0.2644)(z - 1)}\right] \\
&= 0.2385\left[\frac{(1 + 0.8760z^{-1})z^{-1}}{(1 - 0.2644z^{-1})(1 - z^{-1})}\right]
\end{aligned}$$

Figure 4–17(b) shows the simplified block diagram of the designed digital control system. The closed-loop pulse transfer function is then obtained as follows:

$$\frac{C(z)}{R(z)} = \frac{0.2385(1 + 0.8760z^{-1})z^{-1}}{(1 - 0.2644z^{-1})(1 - z^{-1}) + 0.2385(1 + 0.8760z^{-1})z^{-1}}$$

$$= \frac{0.2385z^{-1} + 0.2089z^{-2}}{1 - 1.0259z^{-1} + 0.4733z^{-2}}$$

A BASIC computer program for finding the response to a unit step sequence input $u(k) = 1$ ($k = 0, 1, 2, \ldots$) is shown in Table 4–2. The computation results are also shown in Table 4–2.

Figure 4–18(a) shows a plot of the unit step response of the analog control system, and Fig. 4–18(b) shows a plot of that of the equivalent digital control system. The unit step response plots show a fairly good agreement between the analog control system and the equivalent digital control system. Careful examination shows that the maximum overshoot is approximately 16.5 percent and the settling time is approximately 2 sec for the analog control system, whereas the maximum overshoot is approximately 19

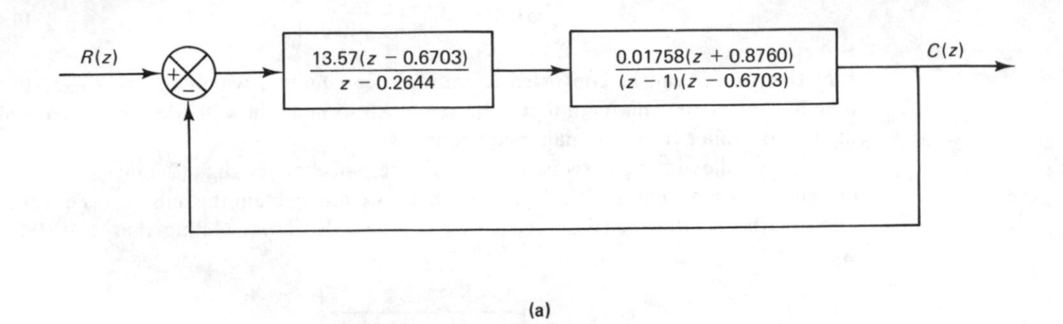

(a)

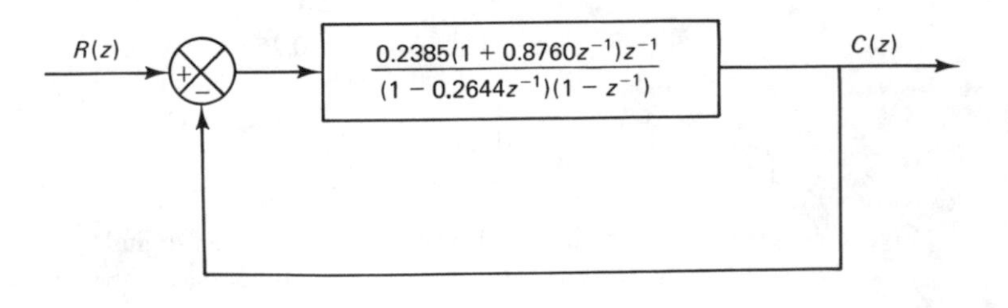

(b)

Figure 4–17 (a) Block diagram of the digital control system equivalent to the analog control system shown in Fig. 4–15; (b) simplified block diagram.

TABLE 4–2 BASIC COMPUTER PROGRAM FOR FINDING THE UNIT STEP RESPONSE OF THE SYSTEM SHOWN IN FIG. 4–17(b)

```
10  C0 = 0
20  C1 = 0.2385
30  R0 = 1
40  R1 = 1
50  C2 = 1.0259 * C1 - 0.4733 * C0
         + 0.2385 * R1 + 0.2089 * R0
60  M = C0
70  C0 = C1
80  C1 = C2
90  N = R0
100 R0 = R1
110 PRINT K, M, N
120 K = K + 1
130 IF K < 21 GO TO 50
140 END
```

k = K	$c(k)$ = CK = M	$r(k)$ = RK = N
0	0	1
1	0.2385	1
2	0.692077	1
3	1.04452	1
4	1.19141	1
5	1.1753	1
6	1.08924	1
7	1.00859	1
8	0.96657	1
9	0.96164	1
10	0.976469	1
11	0.994016	1
12	1.005	1
13	1.00796	1
14	1.0058	1
15	1.00218	1
16	0.999495	1
17	0.998449	1
18	0.998648	1
19	0.999347	1
20	0.99997	1

percent and the settling time is approximately 2 sec for the digital control system. It is noted that the maximum overshoot for the digital control system can be reduced toward 16.5 percent by making the sampling period T smaller.

The analog control system shown in Fig. 4–15 is a third-order system having three closed-loop poles (because the zero at $s = -2$ and the pole at $s = -2$ cancel each other), whereas the digital control system shown in Fig. 4–17(b) is a second-order system having two closed-loop poles. The difference in system order may be explained as follows. In the analog control system the zero-order hold is approximated by a first-order lag. Thus, the zero-order hold increases the system order by 1. The analog control system becomes one of the fourth order, but because of the cancellation of one pole and one zero the system is reduced to a third-order one. In the digital control system, however, the z transform of the second-order plant preceded by the zero-order hold results in a second-order system. The digital control system is now of the third order, but the cancellation of one pole of the plant and the zero of the digital controller has resulted in the reduction of the system order by 1. Thus, the pulse transfer function of the digital control system is of the second order.

Comments. If the sampling frequency ω_s is found to be low (less than 5 or 6 times the damped natural frequency), then it is advisable to verify the design

1.4
1.2
1.0
0.8
c(t)
0.6
0.4
0.2
0
0.4
0.8
1.2
1.6
2.0
2.4
2.8
3.2
3.6
4.0
t (sec)

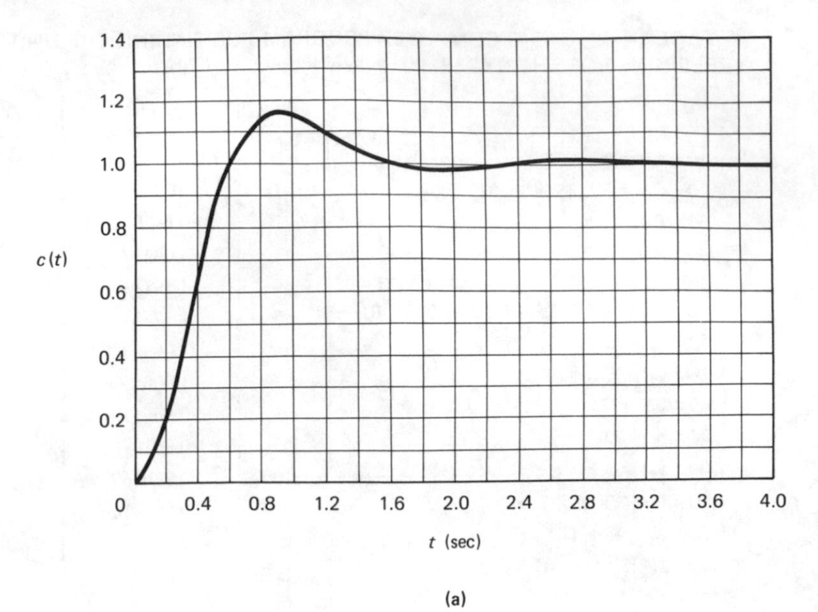

(a)

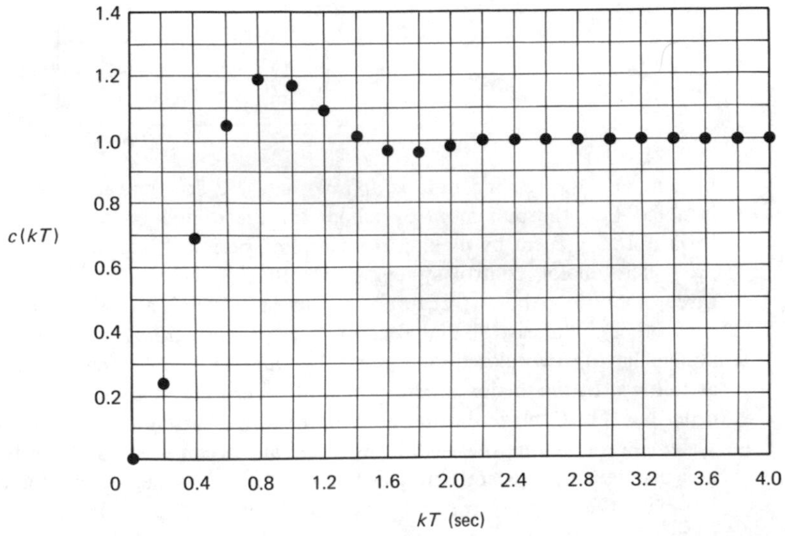

1.4
1.2
1.0
0.8
c(kT)
0.6
0.4
0.2
0
0.4
0.8
1.2
1.6
2.0
2.4
2.8
3.2
3.6
4.0
kT (sec)

(b)

Figure 4–18 (a) Unit step response of the analog control system shown in Fig. 4–15; (b) unit step response of the equivalent digital control system shown in Fig. 4–17(b).

by an exact discrete-time analysis, because the equivalent digital controller may not exactly portray the analog controller. (This means that the design may not be satisfactory and it will be necessary either to redesign by use of the same design approach but using a smaller sampling period T or to use an exact discrete-time method.)

On the other hand, if a discrete-time control system designed via a discrete-time equivalent of an analog controller has resulted in a very high sampling frequency, then it is desirable to redesign the system using an exact discrete-time method so that the sampling frequency can be made the lowest that will satisfy the performance requirements.

4-4 TRANSIENT AND STEADY-STATE RESPONSE ANALYSIS

Absolute stability is a basic requirement of all control systems. In addition, good relative stability and steady-state accuracy are also required of any control system, whether continuous-time or discrete-time.

In this section we shall discuss transient response and steady-state response characteristics of closed-loop control systems. The transient response refers to that portion of the response due to the closed-loop poles of the system, and the steady-state response refers to that portion of the response due to the poles of the input or forcing function.

Discrete-time control systems are very frequently analyzed with "standard" inputs such as impulse inputs, step inputs, ramp inputs, or sinusoidal inputs. This is because the system's response to any arbitrary input may be estimated from its response to such standard inputs. In this section, we shall consider the response of the discrete-time control system to time-domain inputs such as impulse, step, ramp, and acceleration inputs.

Transient response specifications. In many practical cases, the desired performance characteristics of control systems, whether they are continuous-time or discrete-time, are specified in terms of time-domain quantities. This is because systems with energy storage cannot respond instantaneously and will always exhibit transient response whenever they are subjected to inputs or disturbances.

Frequently, the performance characteristics of a control system are specified in terms of the transient response to a unit step input, since the unit step input is easy to generate and is sufficiently drastic to provide useful information on both the transient response and the steady-state response characteristics of the system.

The transient response of a system to a unit step input depends on the initial conditions. For convenience in comparing transient responses of various systems, it is a common practice to use the standard initial condition: that the system is at rest initially and the output and all its time derivatives are zero. The response characteristics can then be easily compared.

The transient response of a practical control system, where the output signal is continuous-time, often exhibits damped oscillations before reaching the steady state.

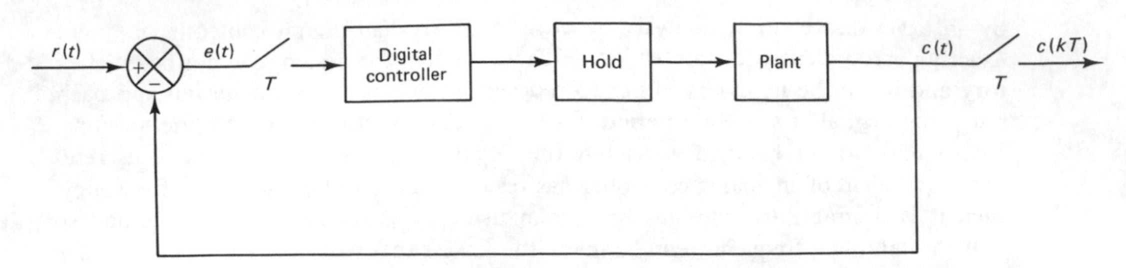

Figure 4–19 A digital control system.

(This is true for the majority of discrete-time or digital control systems because the plants to be controlled are in most cases continuous-time and, therefore, the output signals are continuous-time.)

Consider, for example, the digital control system shown in Fig. 4–19. The output $c(t)$ of such a system to a unit step input may exhibit damped oscillations as shown in Fig. 4–20(a). Figure 4–20(b) shows the discrete-time output $c(kT)$.

Just as in the case of continuous-time control systems, the transient response of a digital control system may be characterized not only by the damping ratio and

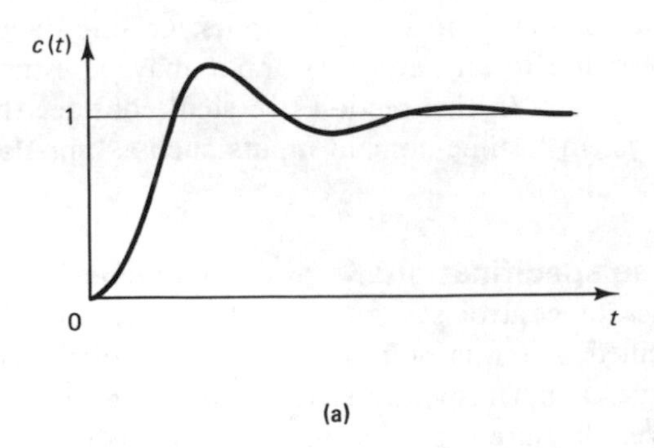

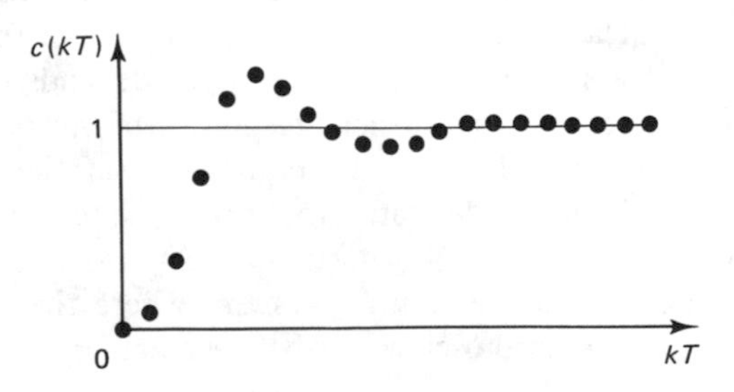

Figure 4–20 (a) Unit step response of the system shown in Fig. 4–19; (b) discrete-time output in the unit step response.

damped natural frequency, but also by the rise time, maximum overshoot, settling time, and so forth, in response to a step input. In fact, in specifying such transient response characteristics, it is common to specify the following quantities:

TRANSIENT RESPONSE SPECIFICATIONS

1. Delay time t_d
2. Rise time t_r
3. Peak time t_p
4. Maximum overshoot M_p
5. Settling time t_s

The aforementioned transient response specifications in the unit step response are defined in what follows and are shown graphically in Fig. 4–21.

1. Delay time t_d. The delay time is the time required for the response to reach half the final value the very first time.
2. Rise time t_r. The rise time is the time required for the response to rise from 10 to 90 percent, or 5 to 95 percent, or 0 to 100 percent of its final value, depending on the situation. For underdamped second-order systems, the 0 to 100 percent rise time is commonly used. For overdamped systems and systems with transportation lags, the 10 to 90 percent rise time is commonly used.
3. Peak time t_p. The peak time is the time required for the response to reach the first peak of the overshoot.

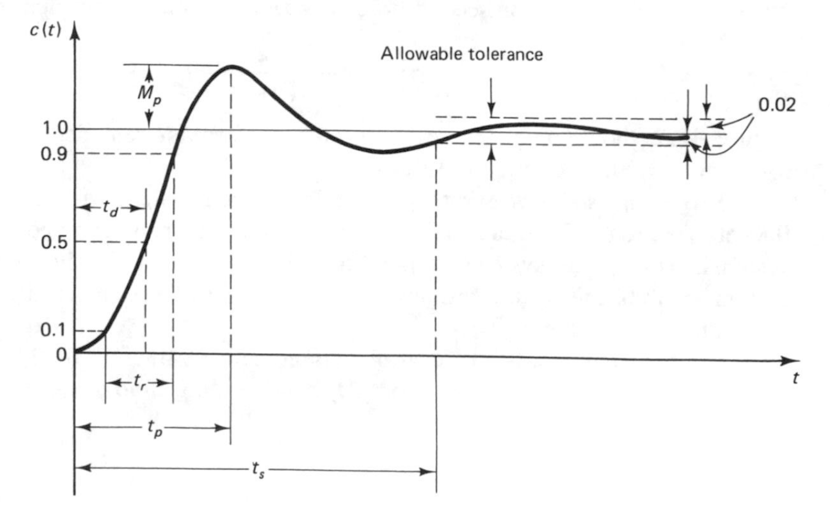

Figure 4–21 Unit step response curve showing transient response specifications t_d, t_r, t_p, M_p, and t_s.

4. Maximum overshoot M_p. The maximum overshoot is the maximum peak value of the response curve measured from unity. If the final steady-state value of the response differs from unity, then it is common to use the maximum percent overshoot. It is defined by the relation

$$\text{Maximum percent overshoot} = \frac{c(t_p) - c(\infty)}{c(\infty)} \times 100\%$$

The amount of the maximum (percent) overshoot directly indicates the relative stability of the system.

5. Settling time t_s. The settling time is the time required for the response curve to reach and stay within a range about the final value of a size specified as an absolute percentage of the final value, usually 2 percent. The settling time is related to the largest time constant of the control system.

The time-domain specifications just given are quite important since most control systems are time-domain systems; that is, they must exhibit acceptable time responses. (This means that the control system being designed must be modified until the transient response is satisfactory.)

Not all the specifications we have just defined necessarily apply to any given case. For example, for an overdamped system, the peak time and maximum overshoot terms do not apply. On the other hand, other specifications may be involved: for systems which yield steady-state errors for step inputs, the error must be kept within a specified percentage level. (Detailed discussions of steady-state errors will be given later in this section.)

Transient response specifications for second-order continuous-time systems. Consider the second-order continuous-time system defined by

$$\frac{C(s)}{R(s)} = \frac{\omega_n^2}{s^2 + 2\zeta\omega_n s + \omega_n^2} \tag{4-28}$$

Figure 4–22 shows unit step response curves of the system defined by Eq. (4–28) for various values of damping ratio ζ.

From Fig. 4–22 we see that an underdamped system with ζ between 0.4 and 0.8 gets close to the final value more rapidly than the critically damped and overdamped systems. Among the systems responding without oscillation, the critically damped system exhibits the fastest response. The overdamped system is always sluggish in responding to any input.

For the continuous-time control system defined by Eq. (4–28), the rise time t_r, peak time t_p, maximum overshoot M_p, and settling time t_s can be given in terms of ζ and ω_n, as follows:

RISE TIME t_r:

$$t_r = \frac{1}{\omega_d} \tan^{-1} \frac{\omega_d}{-\sigma} = \frac{\pi - \beta}{\omega_d} \tag{4-29}$$

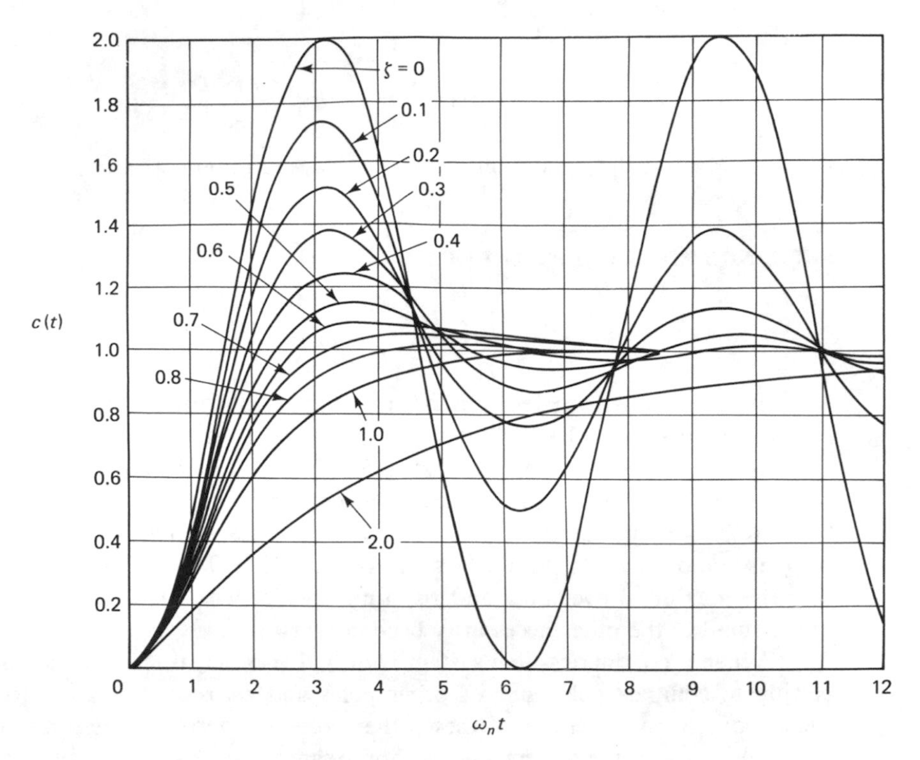

Figure 4–22 Unit step response curves of the system defined by Eq. (4–28) for various values of damping ratio ζ.

where $\omega_d = \omega_n\sqrt{1-\zeta^2}$ and angle β is as defined in Fig. 4–23.

PEAK TIME t_p:

$$t_p = \frac{\pi}{\omega_d} \tag{4–30}$$

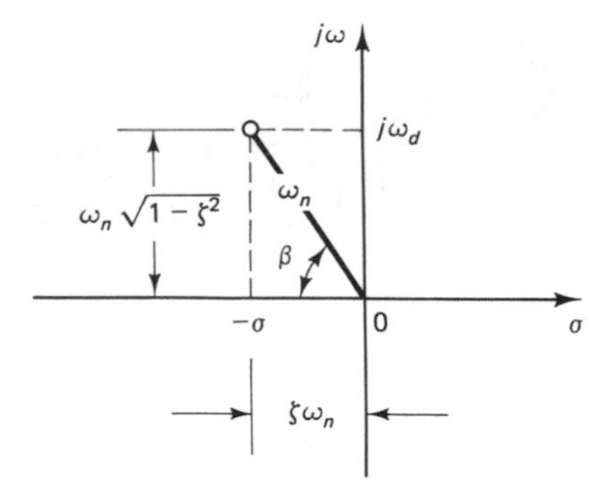

Figure 4–23 Definition of the angle β.

MAXIMUM OVERSHOOT M_p:

$$M_p = \exp\left(-\frac{\sigma\pi}{\omega_d}\right) = \exp\left(-\frac{\zeta\pi}{\sqrt{1-\zeta^2}}\right) \tag{4–31}$$

A curve relating the maximum overshoot M_p and damping ratio ζ is shown in Fig. 4–24.

SETTLING TIME (TO 2 PERCENT OF FINAL VALUE) t_s:

$$t_s = \frac{4}{\sigma} = \frac{4}{\zeta\omega_n} \tag{4–32}$$

A few comments on transient response characteristics. Except for certain applications where oscillations cannot be tolerated, it is desirable that the transient response be sufficiently fast and sufficiently damped. Thus, for a desirable transient response for a second-order system, the damping ratio must be between 0.4 and 0.8. Small values of ζ ($\zeta < 0.4$) yield excessive overshoot in the transient response, and a system with a large value of ζ ($\zeta > 0.8$) responds sluggishly. Note that the maximum overshoot and rise time conflict with each other. If either one is made smaller, the other necessarily becomes larger.

When a continuous-time system is of higher-order, if it has a pair of dominant conjugate complex poles and its other poles and zeros are far away to the left in the s plane, then the effects of these other poles and zeros on the transient response are negligible and the system may be approximated by a second-order model.

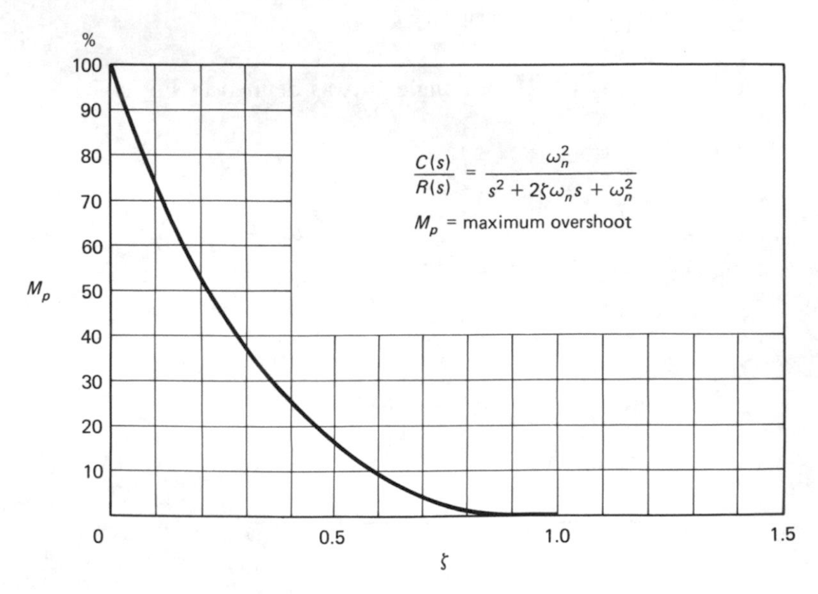

Figure 4–24 Curve relating the maximum overshoot M_p and damping ratio ζ for the second-order system defined by Eq. (4–28).

Mapping from the s plane to the z plane. Absolute stability requires that all closed-loop poles of a discrete-time control system lie inside the unit circle in the z plane. Relative stability may be predicted from the closed-loop pole and zero locations in the z plane.

Let us examine how the specifications for the rise time, maximum overshoot, settling time, and so forth, may be interpreted in the z plane.

Note that if the continuous-time control system can be approximated by the standard second-order system defined by Eq. (4–28), the maximum overshoot is directly related to the damping ratio ζ. Also, the rise time is related to the damped natural frequency ω_d and the damping ratio ζ.

Damping ratio ζ. In the s plane a constant damping ratio may be represented by a straight line from the origin. As was pointed out in Sec. 3–8, a constant-damping-ratio locus (for $0 < \zeta < 1$) in the z plane is a logarithmic spiral. Figure 4–25(a) shows constant ζ loci in both the s plane and the z plane. Note that the logarithmic spirals shown correspond to the primary strip in the s plane. (If the sampling theorem is satisfied, then we need to consider only the primary strip in the s plane.)

If all the poles in the s plane are specified as having a damping ratio not less than a specified value ζ_1, then the poles must lie to the left of the constant-damping ratio line in the s plane (the shaded region). In the z plane, the poles must lie in the region bounded by logarithmic spirals corresponding to $\zeta = \zeta_1$ (the shaded region).

Damped natural frequency ω_d. The rise time or the speed of response depends on the damped natural frequency ω_d and the damping ratio ζ of the (dominant) conjugate complex closed-loop poles. As stated in Sec. 3–8, in the s plane, constant ω_d loci are horizontal lines, while in the z plane they are straight lines emanating from the origin, as shown in Fig. 4–25(b).

Settling time t_s. The settling time is determined by the value of attenuation σ of the dominant closed-loop poles. If the settling time is specified, it is possible to draw a line $\sigma = -\sigma_1$ in the s plane corresponding to a given settling time. The region to the left of the line $\sigma = -\sigma_1$ in the s plane corresponds to the inside of a circle with radius $e^{-\sigma_1 T}$ in the z plane, as shown in Fig. 4–25(c).

Example 4–8.

Specify the region in the z plane that corresponds to a desirable region (shaded region) in the s plane bounded by lines $\omega = \pm\omega_1$, lines $\zeta = \zeta_1$, and a line $\sigma = -\sigma_1$, as shown in Fig. 4–26(a).

On the basis of the preceding discussions on mapping from the s plane to the z plane, the desirable region can be mapped to the z plane as in Fig. 4–26(b).

Note that if the dominant closed-loop poles of the continuous-time control system are required to be in the desirable region specified in the s plane, then the dominant closed-loop poles of the equivalent discrete-time control system must lie inside the region in the z plane that corresponds to the desirable region in the s plane. Once the discrete-

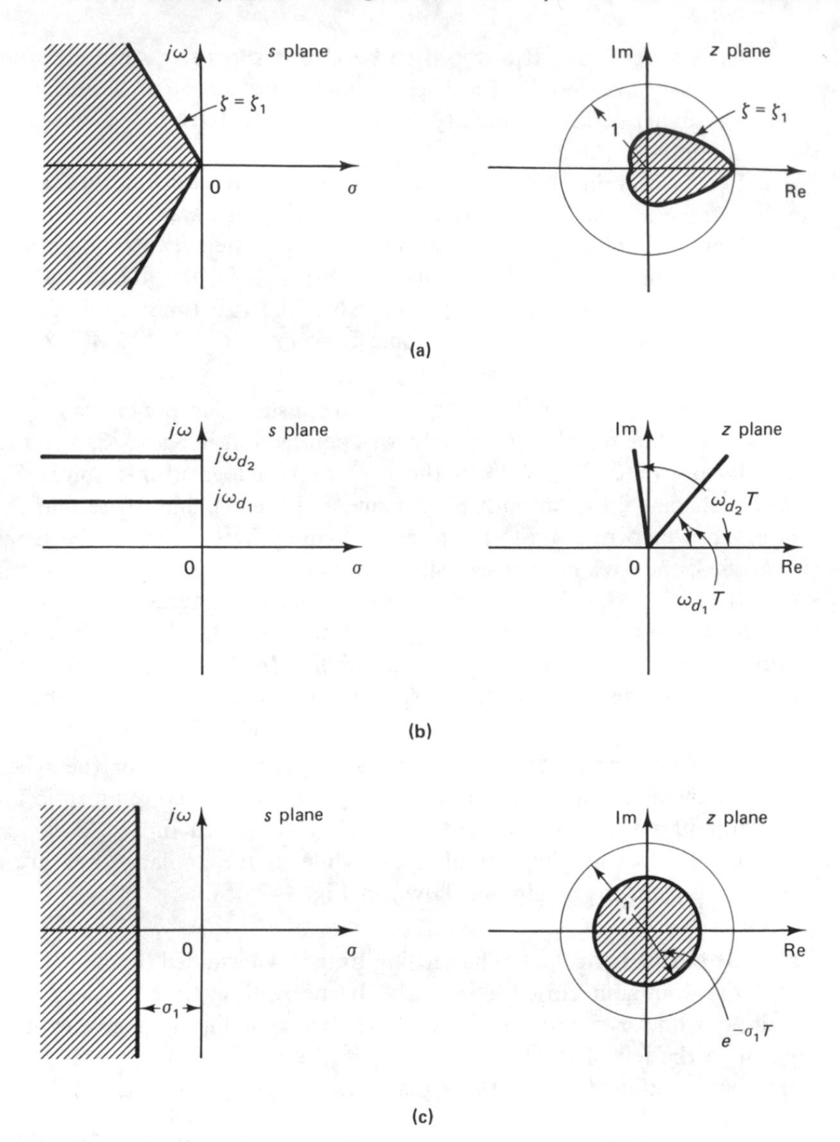

Figure 4–25 (a) Constant ζ loci in the s plane and z plane; (b) constant ω_d loci in the s plane and z plane; (c) constant σ loci in the s plane and z plane.

time control system is designed, the system response characteristics must be checked by experiments or simulation. If the response characteristics are not satisfactory, then closed-loop pole and zero locations must be modified until satisfactory results are obtained.

Relationship between z plane pole and zero locations and transient response. For continuous-time control systems the relationship between s plane pole and zero locations and transient response characteristics are well documented.

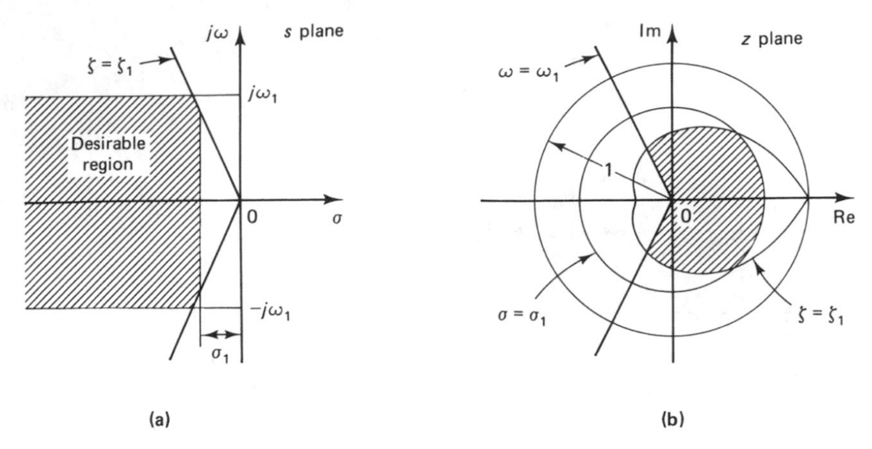

Figure 4–26 (a) A desirable region in the s plane for closed-loop pole locations; (b) corresponding region in the z plane.

For discrete-time control systems, it is necessary to pay particular attention to the sampling period T. This is because if the sampling period is too long and the sampling theorem is not satisfied, then frequency folding occurs and the effective pole and zero locations will be changed.

Suppose a continuous-time control system has closed-loop poles at $s = -\sigma_1 \pm j\omega_1$ in the s plane. If the sampling operation is involved in this system and if $\omega_1 > \frac{1}{2}\omega_s$, where ω_s is the sampling frequency, then frequency folding occurs and the system behaves as if it had poles at $s = -\sigma_1 \pm j(\omega_1 \pm n\omega_s)$, where $n = 1, 2, 3, \ldots$. This means that the sampling operation folds the poles outside the primary strip back into the primary strip and the poles will appear at $s = -\sigma_1 \pm j(\omega_1 - \omega_s)$; see Fig. 4–27(a). On the z plane those poles are mapped into one pair of conjugate complex poles, as shown in Fig. 4–27(b). When frequency folding occurs, oscillations with frequency $\omega_s - \omega_1$, rather than frequency ω_1, are observed.

Let us assume that the sampling theorem is satisfied and no frequency folding occurs. The nature of the transient response of a discrete-time control system to a given input depends on the actual locations of the closed-loop poles and zeros in the z plane. Consider the discrete-time control system defined by

$$\frac{C(z)}{R(z)} = \frac{b_0 z^n + b_1 z^{n-1} + \cdots + b_n}{z^n + a_1 z^{n-1} + \cdots + a_n} \tag{4–33}$$

where $R(z)$ is the z transform of the input and $C(z)$ is the z transform of the output. The output $C(z)$ can be written as follows:

$$C(z) = \frac{b_0 z^n + b_1 z^{n-1} + \cdots + b_n}{z^n + a_1 z^{n-1} + \cdots + a_n} R(z)$$

$$= \frac{b_0 z^n + b_1 z^{n-1} + \cdots + b_n}{(z - p_1)(z - p_2) \cdots (z - p_n)} R(z)$$

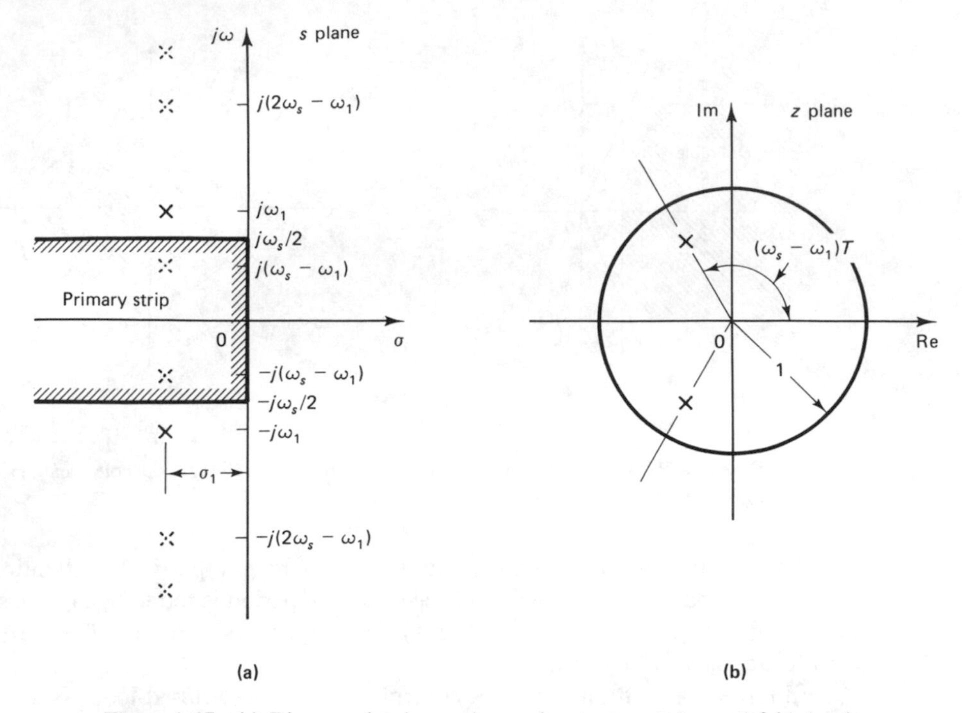

Figure 4–27 (a) Diagram showing s plane poles at $-\sigma_1 \pm j\omega_1$ and folded poles appearing at $-\sigma_1 \pm j(\omega_1 \pm \omega_s)$, $-\sigma_1 \pm j(\omega_1 \pm 2\omega_s)$, . . . ; (b) z plane mapping of s plane poles at $-\sigma_1 \pm j\omega_1$, $-\sigma_1 \pm j(\omega_1 \pm \omega_s)$, $-\sigma_1 \pm j(\omega_1 \pm 2\omega_s)$, ...

where multiple poles may be involved. Thus, this last equation may be expanded into partial fractions as follows:

$$C(z) = \alpha_0 + \sum_h \frac{\alpha_h z}{z - p_h} + \sum_i \frac{\beta_i z e^{-a_i T} \sin \omega_i T}{z^2 - 2e^{-a_i T} z \cos \omega_i T + e^{-2a_i T}}$$

$$+ \sum_j \frac{\gamma_j z (z - e^{-a_j T} \cos \omega_j T)}{z^2 - 2e^{-a_j T} z \cos \omega_j T + e^{-2a_j T}}$$

$$+ \left[\text{terms due to multiple poles such as } \frac{z}{(z-p)^2}, \frac{z(z+p)}{(z-p)^3}, \text{etc.} \right] \qquad (4\text{–}34)$$

Note that the poles of $C(z)$ are determined by the coefficients $a_1, a_2, \ldots, a_n$ in Eq. (4–33) and the poles of the input $R(z)$. The coefficients α_0, α_h, β_i, and γ_j are determined by $a_1, a_2, \ldots, a_n$; $b_0, b_1, \ldots, b_n$; and the poles and zeros of the input $R(z)$.

As can be seen clearly from Eq. (4–34) the response $c(k)$, the inverse z transform of $C(z)$, is a sum of the contributions of all the components in the partial fraction

TABLE 4–3 GRAPHICAL REPRESENTATIONS OF INVERSE *z* TRANSFORM OF $z/(z-p)$

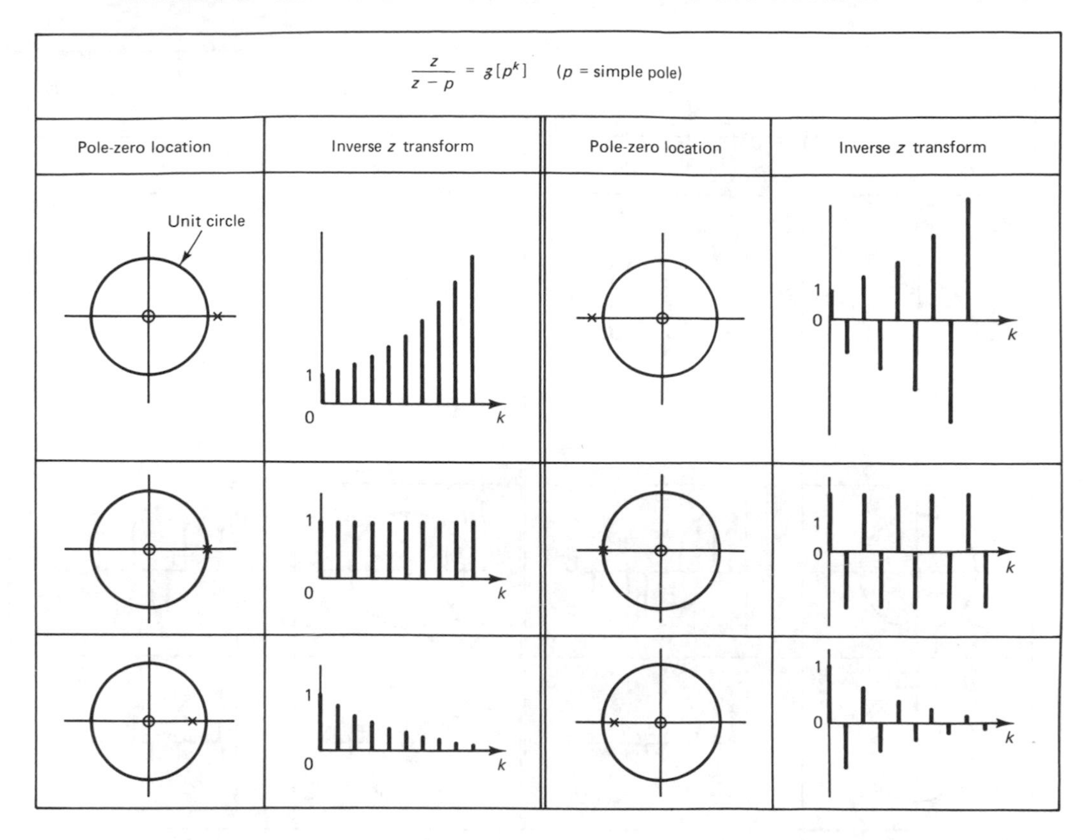

terms. Tables 4–3, 4–4, and 4–5 show graphical representations of inverse z transforms of partial-fraction-expansion components appearing in Eq. (4–34).

Steady-state error analysis (continuous-time control systems). An important feature associated with transient response is steady-state error. The steady-state performance of a stable control system is generally judged by the steady-state error due to step, ramp, and acceleration inputs. In what follows we shall investigate a type of steady-state error which is caused by the inability of a system to follow particular types of inputs. (It should be noted that besides this type of steady-state error, there are errors which can be attributed to other causes, such as imperfections in system components, static friction, backlash, or deterioration or aging of components. In this section, however, we shall not discuss steady-state error due to such causes.)

Any physical control system inherently suffers steady-state error in response to certain types of inputs. That is, a system may have no steady-state error with step inputs, but the same system may exhibit nonzero steady-state error in response

TABLE 4–4 GRAPHICAL REPRESENTATIONS OF INVERSE *z* TRANSFORM OF $ze^{-aT} \sin \omega T/[z^2 - 2e^{-aT} z \cos \omega T + e^{-2aT}]$

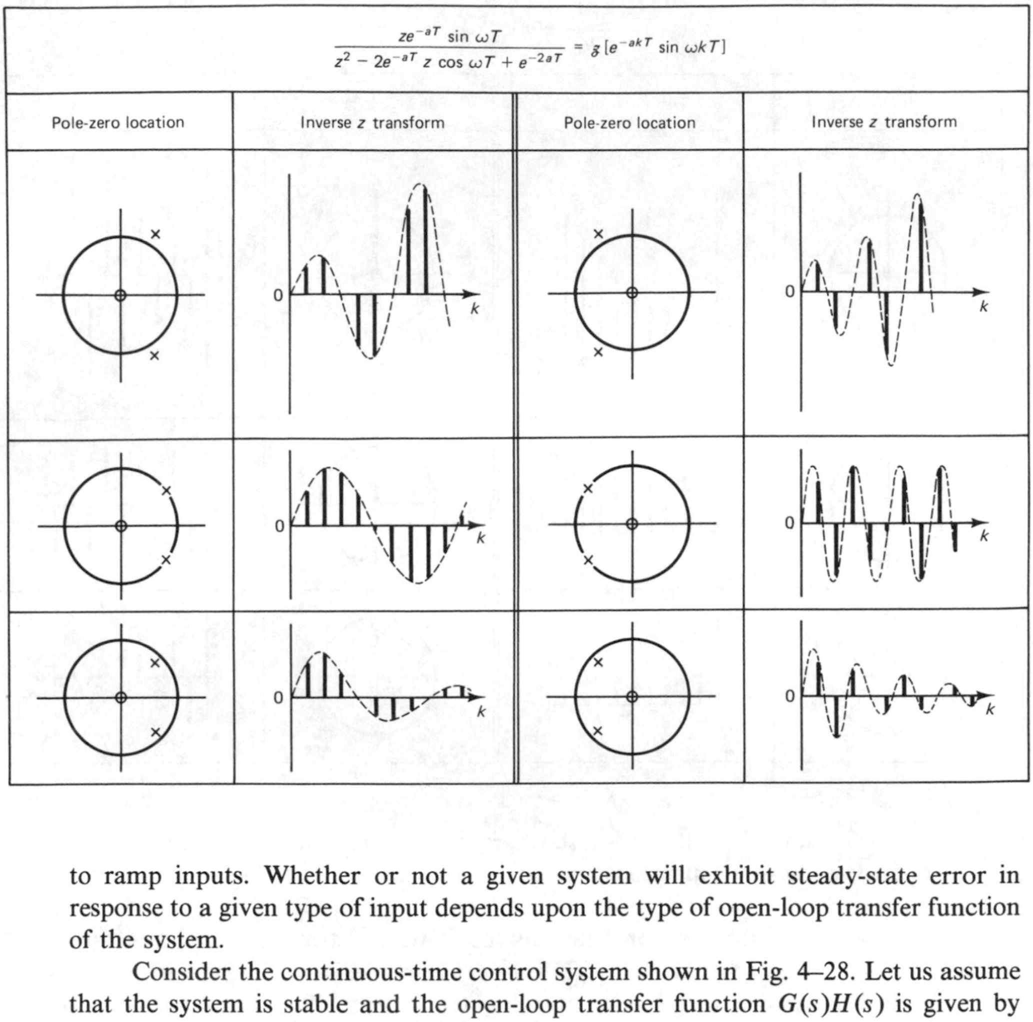

$$\frac{ze^{-aT} \sin \omega T}{z^2 - 2e^{-aT} z \cos \omega T + e^{-2aT}} = \mathcal{Z}[e^{-akT} \sin \omega kT]$$

Pole-zero location	Inverse z transform	Pole-zero location	Inverse z transform

to ramp inputs. Whether or not a given system will exhibit steady-state error in response to a given type of input depends upon the type of open-loop transfer function of the system.

Consider the continuous-time control system shown in Fig. 4–28. Let us assume that the system is stable and the open-loop transfer function $G(s)H(s)$ is given by

$$G(s)H(s) = \frac{K(T_a s + 1)(T_b s + 1) \cdots (T_m s + 1)}{s^N(T_1 s + 1)(T_2 s + 1) \cdots (T_p s + 1)}$$

The term s^N in the denominator represents a pole of multiplicity N at the origin. It is customary to classify the system according to the number of integrators in the open-loop transfer function.

A system is said to be of type 0, type 1, type 2, . . . , if $N = 0$, $N = 1$, $N = 2$, . . . , respectively. Type 0 systems will exhibit finite steady-state errors in response to step inputs and infinite errors in response to ramp and higher-order

TABLE 4–5 GRAPHICAL REPRESENTATIONS OF INVERSE z TRANSFORM OF $z(z - e^{-aT}\cos\omega T)/[z^2 - 2e^{-aT} z \cos\omega T + e^{-2aT}]$

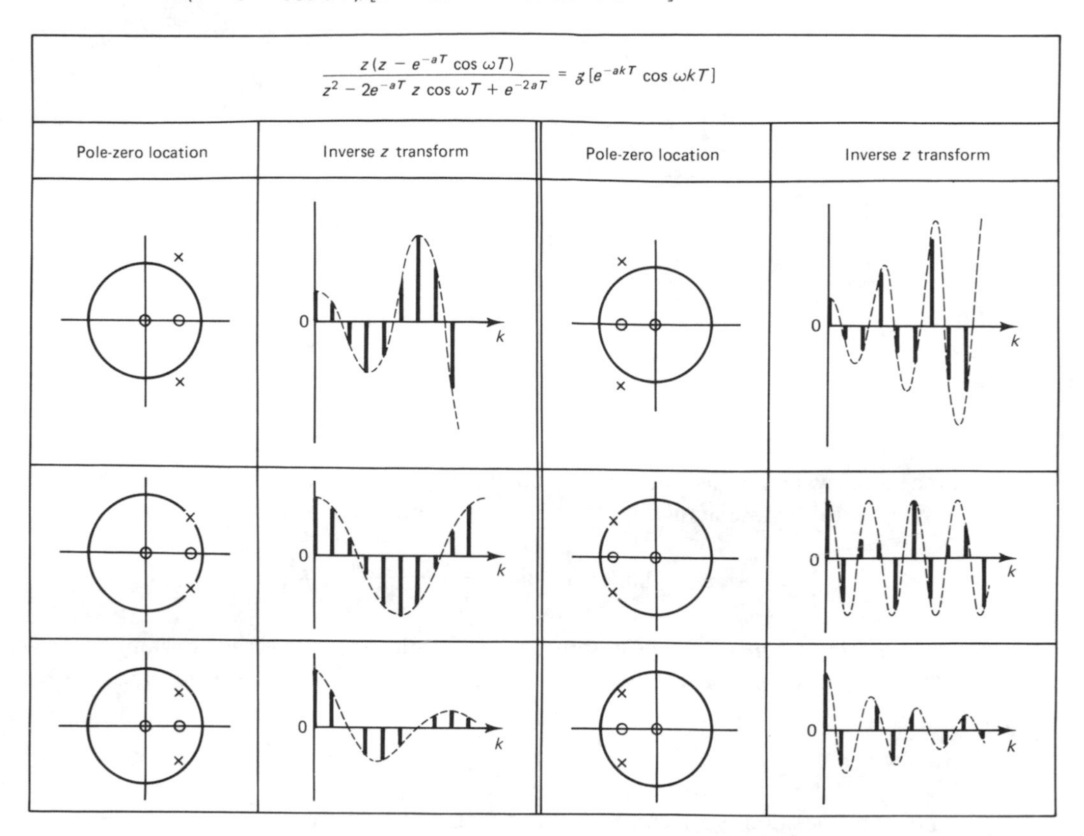

inputs. Type 1 systems will exhibit no steady-state error in response to step inputs, finite steady-state errors in response to ramp inputs, and infinite steady-state errors in response to acceleration and higher-order inputs. As the type number is increased, accuracy is improved. However, increasing the type number aggravates the stability problem. A compromise between steady-state accuracy and relative stability (transient response characteristics) is always necessary.

Consider the system shown in Fig. 4–28. The closed-loop transfer function can be given by

$$\frac{C(s)}{R(s)} = \frac{G(s)}{1 + G(s)H(s)}$$

Hence

$$\frac{E(s)}{R(s)} = 1 - \frac{C(s)H(s)}{R(s)} = \frac{1}{1 + G(s)H(s)}$$

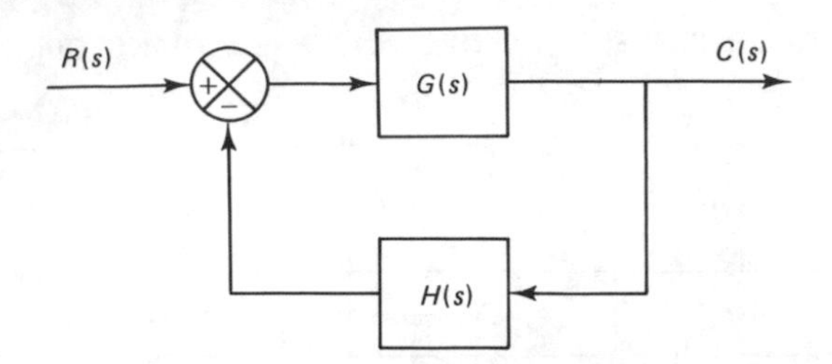

$$G(s)H(s) = \frac{K(T_a s + 1)(T_b s + 1) \ldots (T_m s + 1)}{s^N (T_1 s + 1)(T_2 s + 1) \ldots (T_p s + 1)}$$

Figure 4–28 Continuous-time control system.

and

$$E(s) = \frac{R(s)}{1 + G(s)H(s)}$$

Since the system is assumed to be stable, the final value theorem can be applied and the steady-state actuating error e_{ss} can be obtained as follows:

$$e_{ss} = \lim_{t \to \infty} e(t) = \lim_{s \to 0} \left[\frac{s}{1 + G(s)H(s)} R(s) \right]$$

Static Position Error Constant. The static position error constant K_p is an error constant defined when the input is a unit step. For a unit step input we have

$$e_{ss} = \lim_{s \to 0} \left[\frac{s}{1 + G(s)H(s)} \frac{1}{s} \right] = \frac{1}{1 + G(0)H(0)}$$

The static position error constant K_p is defined as follows:

$$K_p = \lim_{s \to 0} G(s)H(s) = G(0)H(0)$$

The steady-state error for a unit step input is

$$e_{ss} = \frac{1}{1 + K_p}$$

Static Velocity Error Constant. The static velocity error constant K_v is defined when the input is a unit ramp. For a unit ramp input,

$$e_{ss} = \lim_{s \to 0} \left[\frac{s}{1 + G(s)H(s)} \frac{1}{s^2} \right] = \lim_{s \to 0} \frac{1}{sG(s)H(s)}$$

The static velocity error constant K_v is defined by the equation

$$K_v = \lim_{s \to 0} sG(s)H(s)$$

The steady-state actuating error for a unit ramp input is

$$e_{ss} = \frac{1}{K_v}$$

Static Acceleration Error Constant. For a unit acceleration input $r(t) = \frac{1}{2}t^2 1(t)$, the steady-state error is

$$e_{ss} = \lim_{s\to 0}\left[\frac{s}{1 + G(s)H(s)}\frac{1}{s^3}\right] = \lim_{s\to 0}\frac{1}{s^2 G(s)H(s)}$$

The static acceleration error constant K_a is defined by

$$K_a = \lim_{s\to 0} s^2 G(s)H(s)$$

The steady-state actuating error for a unit acceleration input is

$$e_{ss} = \frac{1}{K_a}$$

Summary. From the foregoing analysis we see that a type 0 system will exhibit a constant steady-state error in response to a step input and an infinite error in response to ramp, acceleration, or higher-order inputs. A type 1 system will exhibit a zero steady-state error in response to a step input, a constant steady-state error in response to a ramp input, and an infinite steady-state error in response to acceleration or higher-order inputs.

Table 4–6 lists system types and the corresponding steady-state errors in response to step, ramp, and acceleration inputs for the continuous-time control system of the configuration shown in Fig. 4–28.

TABLE 4–6 SYSTEM TYPES AND THE CORRESPONDING STEADY-STATE ERRORS IN RESPONSE TO STEP, RAMP, AND ACCELERATION INPUTS FOR THE CONTINUOUS-TIME CONTROL SYSTEM SHOWN IN FIG. 4–28

System	Steady-state errors in response to		
	Step input $r(t) = 1$	Ramp input $r(t) = t$	Acceleration input $r(t) = \frac{1}{2}t^2$
Type 0 System	$\frac{1}{1 + K_p}$	∞	∞
Type 1 System	0	$\frac{1}{K_v}$	∞
Type 2 System	0	0	$\frac{1}{K_a}$

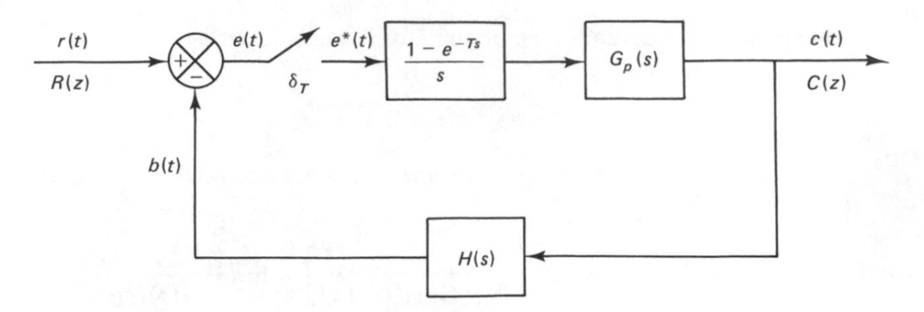

Figure 4–29 Discrete-time control system.

It is important to note that the terms "position error," "velocity error," and "acceleration error" mean steady-state deviations in the output position. A finite velocity error implies that after transients have died out, the input and output move at the same velocity but have a finite position difference.

The concepts of static error constants can be extended to the discrete-time control system, as discussed in the next subsection.

Steady-state error analysis (discrete-time control systems). Consider the discrete-time control system shown in Fig. 4–29. We assume that the system is stable so that the final value theorem can be applied to find the steady-state values. From the diagram we have the actuating error

$$e(t) = r(t) - b(t)$$

We shall consider the steady-state error at the sampling instants. Noting that the sampled actuating error is $e^*(t)$, we have

$$e_{ss}^* = \lim_{t\to\infty} e^*(t) = \lim_{k\to\infty} e(kT) = \lim_{z\to 1} [(1 - z^{-1})E(z)] \tag{4–35}$$

For the system shown in Fig. 4–29, define

$$G(z) = (1 - z^{-1})\, \mathcal{Z}\left[\frac{G_p(s)}{s}\right]$$

and

$$GH(z) = (1 - z^{-1})\, \mathcal{Z}\left[\frac{G_p(s)H(s)}{s}\right]$$

Then we have

$$\frac{C(z)}{R(z)} = \frac{G(z)}{1 + GH(z)}$$

and

$$E(z) = R(z) - B(z) = R(z) - GH(z)E(z)$$

or

$$E(z) = \frac{1}{1 + GH(z)} R(z) \tag{4–36}$$

By substituting Eq. (4–36) into Eq. (4–35), we obtain

$$e_{ss}^* = \lim_{z \to 1} \left[(1 - z^{-1}) \frac{1}{1 + GH(z)} R(z) \right] \tag{4–37}$$

As in the case of the continuous-time control system, we consider three types of inputs: unit step, unit ramp, and unit acceleration inputs.

Static Position Error Constant. For a unit step input $r(t) = 1(t)$ we have

$$R(z) = \frac{1}{1 - z^{-1}}$$

By substituting this last equation into Eq. (4–37) the steady-state error in response to a unit step input can be obtained as follows:

$$e_{ss}^* = \lim_{z \to 1} \left[(1 - z^{-1}) \frac{1}{1 + GH(z)} \frac{1}{1 - z^{-1}} \right] = \lim_{z \to 1} \frac{1}{1 + GH(z)}$$

We define the static position error constant K_p as follows:

$$K_p = \lim_{z \to 1} GH(z) \tag{4–38}$$

Then, the steady-state error in response to a unit step input can be obtained from the equation

$$e_{ss}^* = \frac{1}{1 + K_p} \tag{4–39}$$

The steady-state error in response to a unit step input becomes zero if $K_p = \infty$, which requires that $GH(z)$ have at least one pole at $z = 1$.

Static Velocity Error Constant. For a unit ramp input $r(t) = t\,1(t)$ we have

$$R(z) = \frac{Tz^{-1}}{(1 - z^{-1})^2}$$

By substituting this last equation into Eq. (4–37) we have

$$e_{ss}^* = \lim_{z \to 1} \left[(1 - z^{-1}) \frac{1}{1 + GH(z)} \frac{Tz^{-1}}{(1 - z^{-1})^2} \right] = \lim_{z \to 1} \frac{T}{(1 - z^{-1})GH(z)}$$

Now we define the static velocity error constant K_v as follows:

$$K_v = \lim_{z \to 1} \frac{(1 - z^{-1})GH(z)}{T} \tag{4–40}$$

Then the steady-state error in response to a unit ramp input can be given by

$$e_{ss}^* = \frac{1}{K_v} \tag{4-41}$$

If $K_v = \infty$, then the steady-state error in response to a unit ramp input is zero. This requires $GH(z)$ to possess a double pole at $z = 1$.

Static Acceleration Error Constant. For a unit acceleration input $r(t) = \frac{1}{2}t^2 1(t)$, we have

$$R(z) = \frac{T^2(1 + z^{-1})z^{-1}}{2(1 - z^{-1})^3}$$

By substituting this last equation into Eq. (4–37), we obtain

$$e_{ss}^* = \lim_{z \to 1} \left[(1 - z^{-1}) \frac{1}{1 + GH(z)} \frac{T^2(1 + z^{-1})z^{-1}}{2(1 - z^{-1})^3} \right] = \lim_{z \to 1} \frac{T^2}{(1 - z^{-1})^2 GH(z)}$$

We define the static acceleration error constant K_a as follows:

$$K_a = \lim_{z \to 1} \frac{(1 - z^{-1})^2 GH(z)}{T^2} \tag{4-42}$$

Then the steady-state error becomes

$$e_{ss}^* = \frac{1}{K_a} \tag{4-43}$$

The steady-state error in response to a unit acceleration input becomes zero if $K_a = \infty$. This requires $GH(z)$ to possess a triple pole at $z = 1$.

Equations (4–39), (4–41), and (4–43) give the expressions for steady-state errors of the discrete-time control system shown in Fig. 4–29 at the sampling instants for a unit step, unit ramp, and unit acceleration input, respectively. If we compare these equations with those corresponding to the continuous-time control system, then we immediately see that they are of exactly the same form.

The steady-state error analysis just presented applies to the closed-loop discrete-time control system shown in Fig. 4–29. For a different closed-loop configuration, it is noted that if the closed-loop discrete-time control system has a closed-loop pulse transfer function, then the static error constants can be determined by an analysis similar to the one just presented. Table 4–7 lists the static error constants for typical closed-loop configurations of discrete-time control systems. If the closed-loop discrete-time control system does not have a closed-loop pulse transfer function, however, the static error constants cannot be defined, because the input signal cannot be separated from the system dynamics.

Discrete-time control systems that have closed-loop pulse transfer functions, such as those shown in Table 4–7, can be classified according to the number of

TABLE 4–7 STATIC ERROR CONSTANTS FOR TYPICAL CLOSED-LOOP CONFIGURATIONS OF DISCRETE-TIME CONTROL SYSTEMS

Closed-loop configuration	Values of K_p, K_v, and K_a
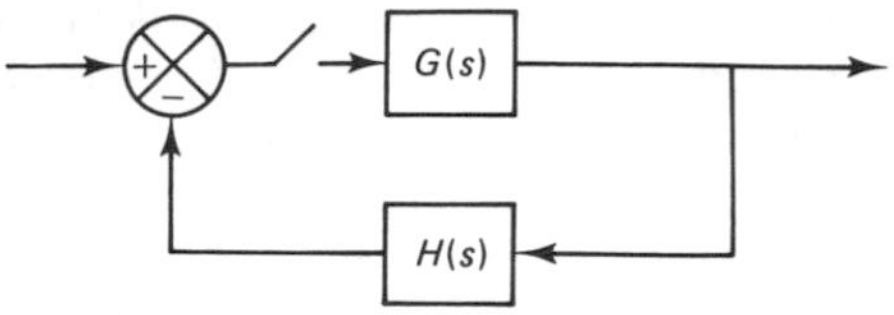 $G(s)$, $H(s)$	$K_p = \lim_{z\to 1} GH(z)$ $K_v = \lim_{z\to 1} \frac{(1-z^{-1})GH(z)}{T}$ $K_a = \lim_{z\to 1} \frac{(1-z^{-1})^2 GH(z)}{T^2}$
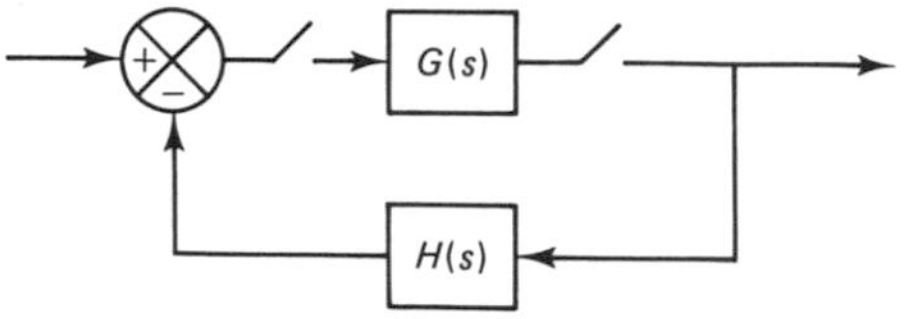 $G(s)$, $H(s)$	$K_p = \lim_{z\to 1} G(z)H(z)$ $K_v = \lim_{z\to 1} \frac{(1-z^{-1})G(z)H(z)}{T}$ $K_a = \lim_{z\to 1} \frac{(1-z^{-1})^2 G(z)H(z)}{T^2}$
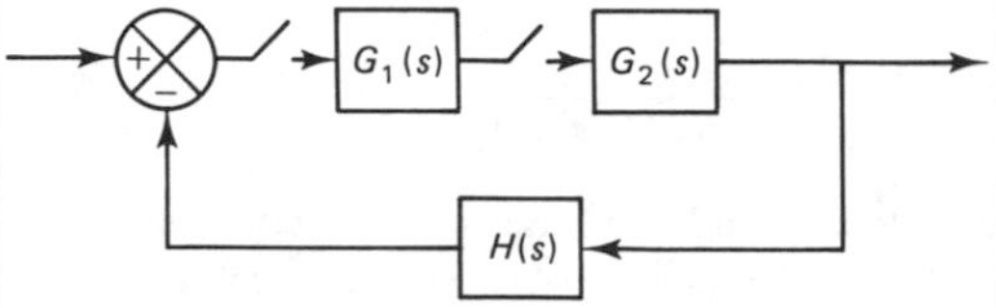 $G_1(s)$, $G_2(s)$, $H(s)$	$K_p = \lim_{z\to 1} G_1(z)HG_2(z)$ $K_v = \lim_{z\to 1} \frac{(1-z^{-1})G_1(z)HG_2(z)}{T}$ $K_a = \lim_{z\to 1} \frac{(1-z^{-1})^2 G_1(z)HG_2(z)}{T^2}$
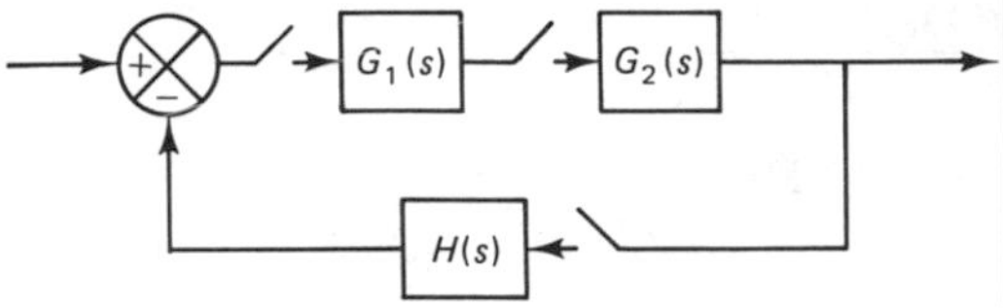 $G_1(s)$, $G_2(s)$, $H(s)$	$K_p = \lim_{z\to 1} G_1(z)G_2(z)H(z)$ $K_v = \lim_{z\to 1} \frac{(1-z^{-1})G_1(z)G_2(z)H(z)}{T}$ $K_a = \lim_{z\to 1} \frac{(1-z^{-1})^2 G_1(z)G_2(z)H(z)}{T^2}$

open-loop poles at $z = 1$. (An open-loop pole at $z = 1$ corresponds to an integrator in the loop.) Suppose the open-loop pulse transfer function is given by the equation

$$\text{Open-loop pulse transfer function} = \frac{1}{(z-1)^N}\frac{A(z)}{B(z)}$$

where $A(z)/B(z)$ contains neither a pole nor a zero at $z = 1$. Then the system can be classified as a type 0 system, a type 1 system, or a type 2 system according to whether $N = 0$, $N = 1$, or $N = 2$, respectively. The system type specifies the steady-state characteristics or steady-state accuracy.

The physical meaning of the static error constants for discrete-time control systems is the same as that for continuous-time control systems, except that the former transmit information only at the sampling instants.

Finally, it is noted that when an analog controller in a control system is replaced by an equivalent digital controller, the static error constants for the analog and the equivalent digital control systems must agree.

Example 4–9.

In Example 4–7 we considered the problem of converting an analog control system to a digital control system. Consider the analog control system shown in Fig. 4–15 and the equivalent digital control system obtained in Example 4–7 and shown in Fig. 4–17(b).

Compare the static error constants for the analog control system with those for the equivalent digital control system and verify that the respective static error constants are the same for both systems.

The static position, velocity, and acceleration error constants for the analog control system shown in Fig. 4–15 are

$$K_p = \lim_{s\to 0}\left[\frac{20.25(s+2)}{s+6.66}\,\frac{10}{s+10}\,\frac{1}{s(s+2)}\right] = \infty$$

$$K_v = \lim_{s\to 0} s\left[\frac{20.25(s+2)}{s+6.66}\,\frac{10}{s+10}\,\frac{1}{s(s+2)}\right] = 3.041$$

$$K_a = \lim_{s\to 0} s^2\left[\frac{20.25(s+2)}{s+6.66}\,\frac{10}{s+10}\,\frac{1}{s(s+2)}\right] = 0$$

The static position, velocity, and acceleration error constants for the equivalent digital control system (with the sampling period $T = 0.2$ sec) shown in Fig. 4–17(b) are

$$K_p = \lim_{z\to 1}\left[\frac{0.2385(1+0.8760z^{-1})z^{-1}}{(1-0.2644z^{-1})(1-z^{-1})}\right] = \infty$$

$$K_v = \lim_{z\to 1}\left[\frac{1-z^{-1}}{0.2}\,\frac{0.2385(1+0.8760z^{-1})z^{-1}}{(1-0.2644z^{-1})(1-z^{-1})}\right] = 3.041$$

$$K_a = \lim_{z\to 1}\left[\frac{(1-z^{-1})^2}{0.2^2}\,\frac{0.2385(1+0.8760z^{-1})z^{-1}}{(1-0.2644z^{-1})(1-z^{-1})}\right] = 0$$

For the analog control system and its equivalent digital control system, the static error constants must agree, as we have just demonstrated. Therefore, in converting an analog control system to a digital control system, it is a good idea to check to see whether the static error constants of the two systems agree. If they do not, there must be errors in the discretization process and/or computations.

Response to disturbances. In examining transient response characteristics and steady-state errors, it is important to note that the effects of disturbances, in addition to those of reference inputs, must be explored.

For the system shown in Fig. 4–30(a), let us assume that the reference input is zero, or $R(z) = 0$, but the system is subjected to disturbance $N(z)$. For this case the system block diagram can be redrawn as shown in Fig. 4–30(b). Then the response $C(z)$ to disturbance $N(z)$ can be found from the closed-loop pulse transfer function

$$\frac{C(z)}{N(z)} = \frac{G(z)}{1 + G_D(z)G(z)}$$

If $|G_D(z)G(z)| \gg 1$, then we find

$$\frac{C(z)}{N(z)} \cong \frac{1}{G_D(z)}$$

Since the system error is

$$E(z) = R(z) - C(z) = -C(z)$$

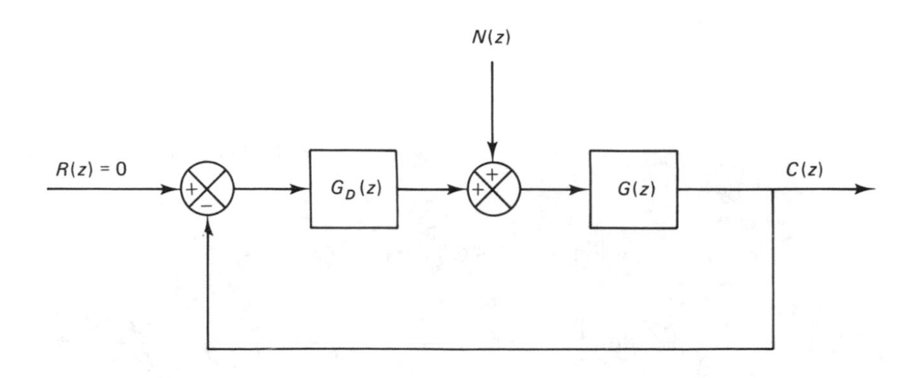

(a)

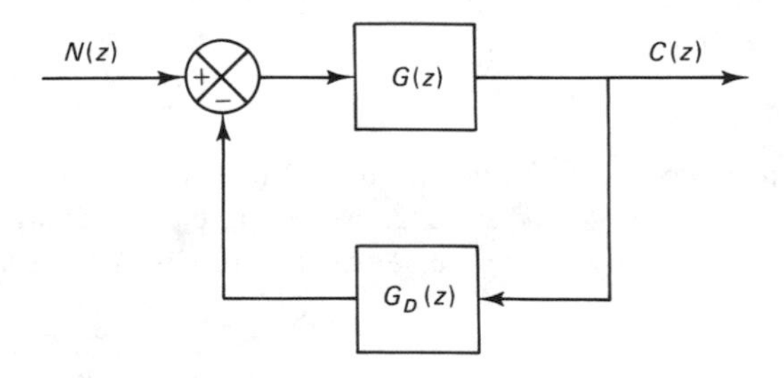

(b)

Figure 4–30 (a) Digital closed-loop control system subjected to reference input and disturbance input; (b) modified block diagram where the disturbance input is considered the input to the system.

we find the error $E(z)$ due to the disturbance $N(z)$ to be

$$E(z) = -\frac{1}{G_D(z)} N(z)$$

Thus the larger the gain of $G_D(z)$, the smaller the error $E(z)$. If $G_D(z)$ includes an integrator [which means that $G_D(z)$ has a pole at $z = 1$], then the steady-state error due to a constant disturbance is zero. This may be seen as follows. Since for a constant disturbance of magnitude N we have

$$N(z) = \frac{N}{1 - z^{-1}}$$

if $G_D(z)$ involves a pole at $z = 1$, then it may be written as

$$G_D(z) = \frac{\hat{G}_D(z)}{z - 1} = \frac{\hat{G}_D(z) z^{-1}}{1 - z^{-1}}$$

where $\hat{G}_D(z)$ does not involve any zeros at $z = 1$. Then the steady-state error can be given by

$$e_{ss}^* = \lim_{z \to 1} \left[(1 - z^{-1}) E(z) \right] = \lim_{z \to 1} \left[(1 - z^{-1}) \frac{-N(z)}{G_D(z)} \right]$$

$$= -\lim_{z \to 1} \left[(1 - z^{-1}) \frac{N}{1 - z^{-1}} \frac{1}{G_D(z)} \right] = -\lim_{z \to 1} \frac{(1 - z^{-1}) N}{\hat{G}_D(z) z^{-1}} = 0$$

If a linear system is subjected to both the reference input and a disturbance input, then the resulting error is the sum of the errors due to the reference input and the disturbance input. The total error must be kept within acceptable limits.

Note that the point where the disturbance enters the system is very important in adjusting the gain of $G_D(z)G(z)$. For example, consider the system shown in Fig. 4–31(a). The closed-loop pulse transfer function for the disturbance is

$$\frac{C(z)}{N(z)} = -\frac{E(z)}{N(z)} = \frac{1}{1 + G_D(z)G(z)}$$

To minimize the effects of disturbance $N(z)$ on the system error $E(z)$, the gain of $G_D(z)G(z)$ must be made as large as possible. However, for the system shown in Fig. 4–31(b), the closed-loop pulse transfer function for the disturbance is

$$\frac{C(z)}{N(z)} = -\frac{E(z)}{N(z)} = \frac{G_D(z)G(z)}{1 + G_D(z)G(z)}$$

and to minimize the effects of disturbance $N(z)$ on the system error $E(z)$, the gain of $G_D(z)G(z)$ must be made as small as possible.

Therefore, it is advantageous to obtain the expression for $E(z)/N(z)$ before concluding whether the gain of $G_D(z)G(z)$ should be large or small to minimize

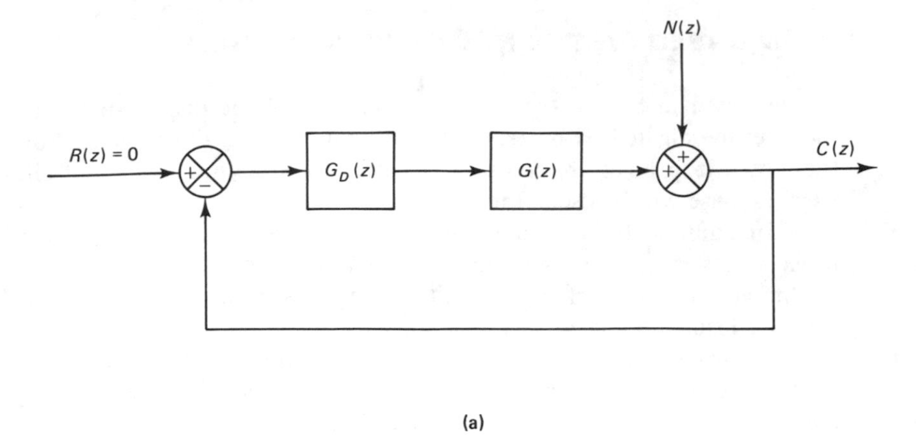

(a)

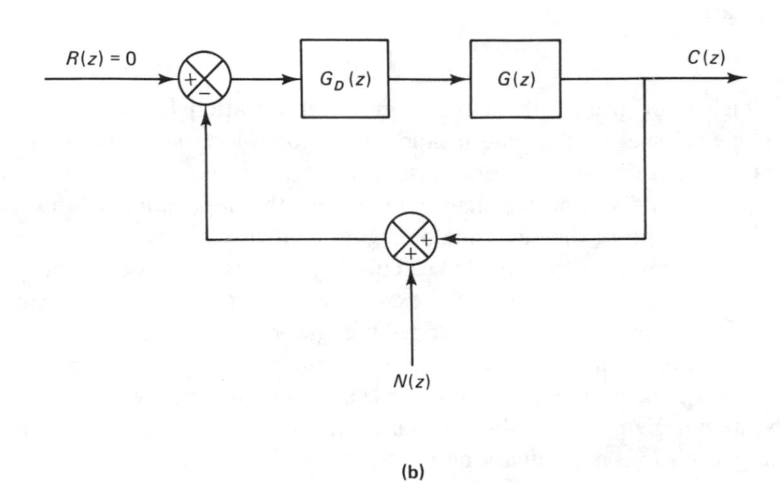

(b)

Figure 4–31 (a) Digital closed-loop control system subjected to reference input and disturbance input; (b) digital closed-loop control system where the disturbance enters the feedback loop.

the error due to disturbances. It is important to remember, however, that the magnitude of the gain cannot be determined solely from the disturbance considerations. It must be determined by considering the responses to both reference and disturbance inputs. If the frequency regions for the reference input and disturbance input are sufficiently apart, a suitable filter may be inserted in the system. If the frequency regions overlap, then modification of the block diagram configuration may become necessary to get acceptable responses to both reference and disturbance inputs.

4-5 DESIGN BASED ON THE ROOT LOCUS METHOD

As discussed in Sec. 4-4, the relative stability of the discrete-time control system may be investigated with respect to the unit circle in the z plane. For example, if the closed-loop poles are complex conjugates and lie inside the unit circle, the unit step response will be oscillatory.

In addition to the transient response characteristics of a given system, it is often necessary to investigate the effects of the system gain and/or sampling period on the absolute and relative stability of the closed-loop system. For such purposes the root locus method proves to be very useful.

The root locus method developed for continuous-time systems can be extended to discrete-time systems without modifications, except that the stability boundary is changed from the $j\omega$ axis in the s plane to the unit circle in the z plane. The reason the root locus method can be extended to discrete-time systems is that the characteristic equation for the discrete-time system is of the same form as that for the root locus in the s plane. For example, for the system shown in Fig. 4-32 the characteristic equation is

$$1 + G(z)H(z) = 0$$

which is of exactly the same form as the equation for root locus analysis in the s plane. However, the pole locations for closed-loop systems in the z plane must be interpreted differently from those in the s plane.

In this section we shall demonstrate the application of the root locus method to the design of discrete-time or digital control systems.

Computer programs for calculating and tracing root loci are available for most computer systems. Therefore, exact plotting of the root loci can be done on the computer and may not require tedious graphical plotting procedures. However, skill in plotting root loci is an advantage, since it will enable the control engineer to make quick graphical plots for given problems to speed up preliminary stages of system design. In fact, the experienced control engineer frequently uses the root locus approach to a preliminary design to locate the dominant closed-loop poles at desired positions in the z plane and then uses a digital simulation to improve the closed-loop performance.

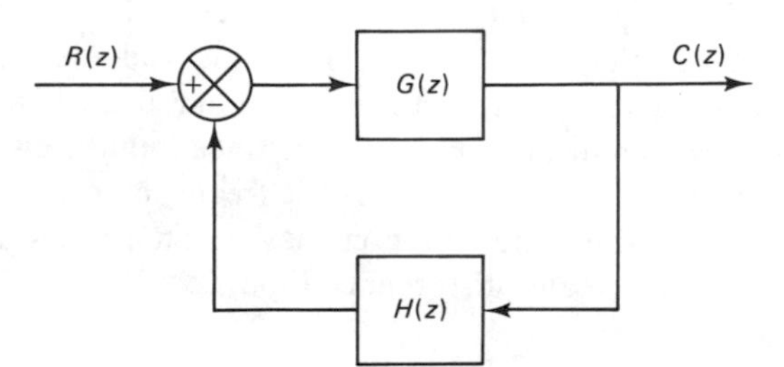

Figure 4-32 Closed-loop control system.

Angle and magnitude conditions. In many linear time-invariant discrete-time control systems, the characteristic equation may have either of the following two forms:

$$1 + G(z)H(z) = 0$$

and

$$1 + GH(z) = 0$$

To combine these two forms into one, let us define the characteristic equation as

$$1 + F(z) = 0 \tag{4-44}$$

where

$$F(z) = G(z)H(z) \qquad \text{or} \qquad F(z) = GH(z)$$

Note that $F(z)$ is the open-loop pulse transfer function. The characteristic equation given by Eq. (4-44) can then be written as

$$F(z) = -1$$

Since $F(z)$ is a complex quantity, this last equation can be split into two equations by equating first the angles and then the magnitudes of the two sides to obtain

ANGLE CONDITION:

$$\angle F(z) = \pm 180^\circ(2k + 1) \qquad k = 0, 1, 2, \ldots$$

MAGNITUDE CONDITION:

$$|F(z)| = 1$$

The values of z that fulfill both the angle and the magnitude conditions are the roots of the characteristic equation, or the closed-loop poles.

A plot of the points in the complex plane satisfying the angle condition alone is the root locus. The roots of the characteristic equation (the closed-loop poles) corresponding to a given value of the gain can be located on the root loci by use of the magnitude condition. The details of applying the angle and magnitude conditions to obtain the closed-loop poles are presented in the following.

General procedure for constructing root loci. For a complicated system with many open-loop poles and zeros, constructing a root locus plot may seem complicated, but actually it is not difficult if the established rules for constructing root loci are applied.

By locating particular points and asymptotes and by computing angles of departure from complex poles and angles of arrival at complex zeros, it is possible to construct root loci without difficulty. Note that while root loci may be conveniently drawn with a digital computer, if manual construction of the root locus plot is attempted, we essentially proceed on a trial-and-error basis. But the number of trials required can be greatly reduced if the established rules are used.

Because the open-loop conjugate complex poles and conjugate complex zeros, if any, are always located symmetrically about the real axis, the root loci are always symmetric with respect to the real axis. Hence, we need only construct the upper half of the root loci and draw the mirror image of the upper half in the lower half of the z plane. Remember that the angles of the complex quantities originating from the open-loop poles and open-loop zeros and drawn to a test point z are measured in the counterclockwise direction. We shall now present the general rules and procedures for constructing root loci.

General rules for constructing root loci.

1. First obtain the characteristic equation

$$1 + F(z) = 0$$

and then rearrange this equation so that the parameter of interest, such as gain K, appears as the multiplying factor in the form

$$1 + \frac{K(z + z_1)(z + z_2)\cdots(z + z_m)}{(z + p_1)(z + p_2)\cdots(z + p_n)} = 0$$

In the present discussion, we assume that the parameter of interest is gain K, where $K > 0$. From the factored form of the open-loop pulse transfer function, locate the open-loop poles and zeros in the z plane. [Note that if $F(z) = G(z)H(z)$, then the open-loop zeros are zeros of $G(z)H(z)$, while the closed-loop zeros consist of the zeros of $G(z)$ and the poles of $H(z)$.]

2. Find the starting points and terminating points of the root loci. Find also the number of separate branches of the root loci. The points on the root loci corresponding to $K = 0$ are open-loop poles and those corresponding to $K = \infty$ are open-loop zeros. Hence, as K is increased from zero to infinity a root locus starts from an open-loop pole and terminates at a finite open-loop zero or an open-loop zero at infinity. This means that a root locus plot will have just as many branches as there are roots of the characteristic equation. [If the zeros at infinity are included in the count, $F(z)$ has the same number of zeros as poles.]

If the number n of closed-loop poles is the same as the number of open-loop poles, then the number of individual root locus branches terminating at finite open-loop zeros is equal to the number m of the open-loop zeros. The remaining $n - m$ branches terminate at infinity (at $n - m$ implicit zeros at infinity) along asymptotes.

3. Determine the root loci on the real axis. Root loci on the real axis are determined by open-loop poles and zeros lying on it. The conjugate complex poles and zeros of the open-loop pulse transfer function have no effect on the location

of the root loci on the real axis because the angle contribution of a pair of conjugate complex poles or zeros is 360° on the real axis. Each portion of the root locus on the real axis extends over a range from a pole or zero to another pole or zero.

In constructing the root loci on the real axis, choose a test point on it. If the total number of real poles and real zeros to the right of this test point is odd, then this point lies on a root locus. The root locus and its complement form alternate segments along the real axis.

4. Determine the asymptotes of the root loci. If the test point z is located far from the origin, then the angles of all the complex quantities may be considered the same. One open-loop zero and one open-loop pole then each cancel the effects of the other.

   Therefore, the root loci for very large values of z must be asymptotic to straight lines whose angles are given as follows:

$$\text{Angle of asymptote} = \frac{\pm 180°(2N+1)}{n-m} \qquad N = 0, 1, 2, \ldots$$

where

$$n = \text{number of finite poles of } F(z)$$

$$m = \text{number of finite zeros of } F(z)$$

Here, $N = 0$ corresponds to the asymptote with the smallest angle with the real axis. Although N assumes an infinite number of values, the angle repeats itself, as N is increased, and the number of distinct asymptotes is $n - m$.

All the asymptotes intersect on the real axis. The point at which they do so is obtained as follows. Since

$$F(z) = \frac{K[z^m + (z_1 + z_2 + \cdots + z_m)z^{m-1} + \cdots + z_1 z_2 \cdots z_m]}{z^n + (p_1 + p_2 + \cdots + p_n)z^{n-1} + \cdots + p_1 p_2 \cdots p_n}$$

$$= \frac{K}{z^{n-m} + [(p_1 + p_2 + \cdots + p_n) - (z_1 + z_2 + \cdots + z_m)]z^{n-m-1} + \cdots}$$

for a large value of z this last equation may be approximated as follows:

$$F(z) \cong \frac{K}{\left[z + \dfrac{(p_1 + p_2 + \cdots + p_n) - (z_1 + z_2 + \cdots + z_m)}{n-m}\right]^{n-m}}$$

If the abscissa of the intersection of the asymptotes and the real axis is denoted by $-\sigma_a$, then

$$-\sigma_a = -\frac{(p_1 + p_2 + \cdots + p_n) - (z_1 + z_2 + \cdots + z_m)}{n-m} \qquad (4\text{–}45)$$

Because all the complex poles and zeros occur in conjugate pairs, $-\sigma_a$ given by Eq. (4–45) is always a real quantity.

Once the intersection of the asymptotes and the real axis is found, the asymptotes can be readily drawn in the complex z plane.

5. Find the breakaway and break-in points. Because of the conjugate symmetry of the root loci, the breakaway points and break-in points either lie on the real axis or occur in conjugate complex pairs.

   If a root locus lies between two adjacent open-loop poles on the real axis, then there exists at least one breakaway point between the two poles. Similarly, if the root locus lies between two adjacent zeros (one zero may be located at $-\infty$) on the real axis, then there always exists at least one break-in point between the two zeros.

   If the root locus lies between an open-loop pole and a zero (finite or infinite) on the real axis, then there may exist no breakaway or break-in points or there may exist both breakaway and break-in points.

   If the characteristic equation

$$1 + F(z) = 0$$

is written as

$$1 + \frac{KB(z)}{A(z)} = 0$$

where $KB(z)/A(z) = F(z)$, then

$$K = -\frac{A(z)}{B(z)} \tag{4–46}$$

and the breakaway and break-in points (which correspond to double roots) can be determined from the roots of

$$\frac{dK}{dz} = -\frac{A'(z)B(z) - A(z)B'(z)}{B^2(z)} = 0 \tag{4–47}$$

where the prime indicates differentiation with respect to z.

   If the value of K corresponding to a root $z = z_0$ of $dK/dz = 0$ is positive, point $z = z_0$ is an actual breakaway or break-in point. Since K is assumed to be nonnegative, if the value of K thus obtained is negative, then point $z = z_0$ is neither a breakaway nor a break-in point.

   Note that this approach can be used when there are complex poles and/or complex zeros.

6. Determine the angle of departure (or angle of arrival) of the root loci from the complex poles (or at the complex zeros).

   In order to sketch the root loci with reasonable accuracy, we must find the direction of the root loci near the complex poles and zeros. The angle of

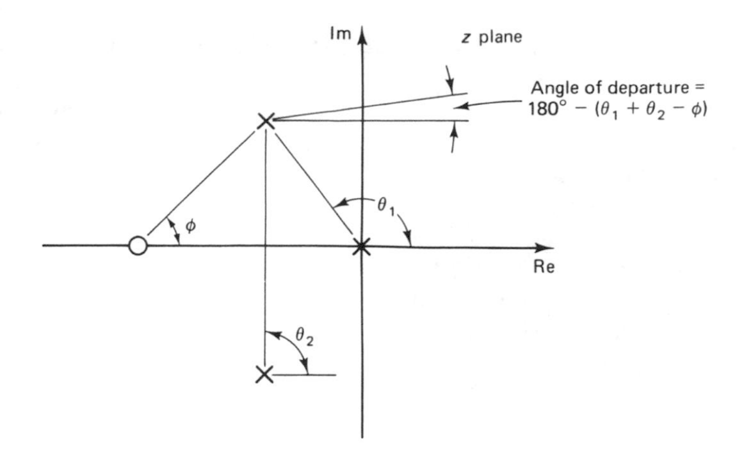

Figure 4–33 Diagram showing angle of departure.

departure (or angle of arrival) of the root locus from a complex pole (or at a complex zero) can be found by subtracting from 180° the sum of all the angles of lines (complex quantities) from all other poles and zeros to the complex pole (or complex zero) in question, with appropriate signs included. The angle of departure is shown in Fig. 4–33.

7. Find the points where the root loci cross the imaginary axis. The points where the root loci intersect the imaginary axis can be found by setting $z = j\nu$ in the characteristic equation (which involves undetermined gain K), equating both the real part and the imaginary part to zero, and solving for ν and K. The values of ν and K thus found give the location at which the root loci cross the imaginary axis and the value of the corresponding gain K, respectively.
8. Any point on the root loci is a possible closed-loop pole. A particular point will be a closed-loop pole when the value of gain K satisfies the magnitude condition. Conversely, the magnitude condition enables us to determine the value of gain K at any specific root location on the locus. The magnitude condition is

$$|F(z)| = 1$$

or

$$\left|\frac{(z+z_1)(z+z_2)\cdots(z+z_m)}{(z+p_1)(z+p_2)\cdots(z+p_n)}\right| = \frac{1}{K} \tag{4–48}$$

If gain K of the open-loop pulse transfer function is given in the problem, then by applying the magnitude condition, Eq. (4–48), it is possible to locate the closed-loop poles for a given K on each branch of the root loci by a trial-and-error method.

Cancellation of poles of $G(z)$ with zeros of $H(z)$. It is important to note that if $F(z) = G(z)H(z)$ and the denominator of $G(z)$ and the numerator of $H(z)$ involve common factors, then the corresponding open-loop poles and zeros will cancel each other, reducing the degree of the characteristic equation by one or more. The root locus plot of $G(z)H(z)$ will not show all the roots of the characteristic equation, but only the roots of the reduced equation.

To obtain the complete set of closed-loop poles, we must add the canceled pole or poles of $G(z)H(z)$ to those closed-loop poles obtained from the root locus plot of $G(z)H(z)$. The important thing to remember is that a canceled pole of $G(z)H(z)$ is a closed-loop pole of the system.

As an example, consider the case where $G(z)$ and $H(z)$ of the system shown in Fig. 4–32 are given by

$$G(z) = \frac{z+c}{(z+a)(z+b)}$$

and

$$H(z) = \frac{z+a}{z+d}$$

Then, clearly the pole $z = -a$ of $G(z)$ and the zero $z = -a$ of $H(z)$ cancel each other, resulting in

$$G(z)H(z) = \frac{z+c}{(z+a)(z+b)}\frac{z+a}{z+d} = \frac{z+c}{(z+b)(z+d)}$$

However, the closed-loop pulse transfer function of the system is

$$\frac{C(z)}{R(z)} = \frac{G(z)}{1+G(z)H(z)} = \frac{(z+c)(z+d)}{(z+a)[(z+b)(z+d)+z+c]}$$

and we see that $z = -a$, the canceled pole of $G(z)H(z)$, is a closed-loop pole of the closed-loop system.

Note, however, that if pole-zero cancellation occurs in the feedforward pulse transfer function, then the same pole-zero cancellation occurs in the closed-loop pulse transfer function. Consider again the system shown in Fig. 4–32, where we assume

$$G(z) = G_D(z)G_1(z), \qquad H(z) = 1$$

Suppose pole-zero cancellation occurs in $G_D(z)G_1(z)$. For example, suppose

$$G_D(z)G_1(z) = \frac{z+b}{z+a}\frac{z+d}{(z+b)(z+c)} = \frac{z+d}{(z+a)(z+c)}$$

Then the closed-loop pulse transfer function becomes

$$\frac{C(z)}{R(z)}=\frac{G_D(z)G_1(z)}{1+G_D(z)G_1(z)}=\frac{(z+b)(z+d)}{(z+b)[(z+a)(z+c)+z+d]}$$

$$=\frac{z+d}{(z+a)(z+c)+z+d}$$

Because of the pole-zero cancellation, the third-order system becomes one of the second order.

It is important to summarize that the effect of pole-zero cancellation in $G(z)$ and $H(z)$ is different from that of pole-zero cancellation in the feedforward pulse transfer function (such as pole-zero cancellation in the digital controller and the plant.) In the former, the canceled pole is still a pole of the closed-loop system, whereas in the latter the canceled pole does not appear as a pole of the closed-loop system (in the latter, the order of the system is reduced by the number of canceled poles).

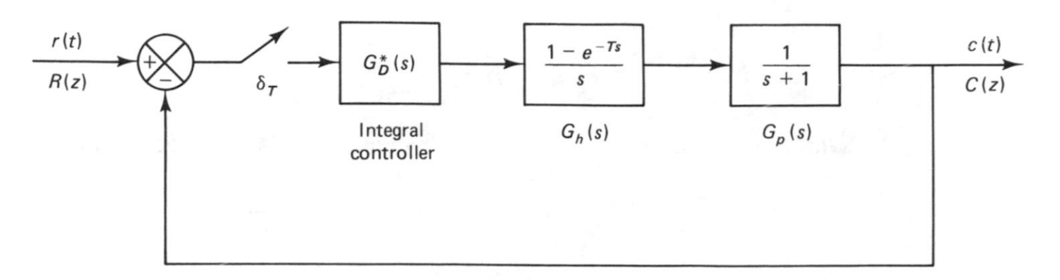

Figure 4–34 Closed-loop digital control system.

Root locus diagrams of digital control systems. In what follows we shall investigate the effects of gain K and sampling period T on the relative stability of the closed-loop control system. Consider the system shown in Fig. 4–34. Assume that the digital controller is of the integral type, or that

$$G_D(z)=\frac{K}{1-z^{-1}}=K\frac{z}{z-1}$$

Let us draw root locus diagrams for the system for three values of the sampling period T: 0.5 sec, 1 sec, and 2 sec. Let us also determine the critical value of K for each case. And finally let us locate the closed-loop poles corresponding to $K = 2$ for each of the three cases.

We shall first obtain the z transform of $G_h(s)G_p(s)$:

$$\mathscr{Z}[G_h(s)G_p(s)]=\mathscr{Z}\left[\frac{1-e^{-Ts}}{s}\frac{1}{s+1}\right]$$

$$=(1-z^{-1})\,\mathscr{Z}\left[\frac{1}{s(s+1)}\right]$$

$$= (1 - z^{-1}) \mathscr{Z}\left[\frac{1}{s} - \frac{1}{s+1}\right]$$

$$= \frac{z-1}{z}\left(\frac{z}{z-1} - \frac{z}{z - e^{-T}}\right)$$

$$= \frac{1 - e^{-T}}{z - e^{-T}}$$

The feedforward pulse transfer function becomes

$$G(z) = G_D(z) \mathscr{Z}[G_h(s)G_p(s)] = \frac{Kz}{z-1}\frac{1 - e^{-T}}{z - e^{-T}} \tag{4–49}$$

The characteristic equation is

$$1 + G(z) = 0$$

or

$$1 + \frac{Kz(1 - e^{-T})}{(z-1)(z - e^{-T})} = 0 \tag{4–50}$$

1. *Sampling period* $T = 0.5$ sec: For this case, Eq. (4–49) becomes as follows:

$$G(z) = \frac{0.3935Kz}{(z-1)(z-0.6065)}$$

Notice that $G(z)$ has poles at $z = 1$ and $z = 0.6065$ and a zero at $z = 0$.

In order to draw a root locus diagram, we first locate the poles and zero in the z plane and then find the breakaway point and break-in point. Notice that this open-loop pulse transfer function with two poles and one zero results in a circular root locus centered at the zero. The breakaway point and break-in point are determined by writing the characteristic equation in the form of Eq. (4–46):

$$K = -\frac{(z-1)(z-0.6065)}{0.3935z} \tag{4–51}$$

and differentiating K with respect to z and equating the result to zero:

$$\frac{dK}{dz} = -\frac{z^2 - 0.6065}{0.3935z^2} = 0$$

Hence

$$z^2 = 0.6065$$

or

$$z = 0.7788 \qquad \text{and} \qquad z = -0.7788$$

Notice that substitution of 0.7788 for z in Eq. (4–51) yields $K = 0.1244$, while letting $z = -0.7788$ yields $K = 8.041$. Since both K values are positive, $z = 0.7788$ is the actual breakaway point and $z = -0.7788$ is the actual break-in point.

Figure 4–35(a) shows the root locus diagram when $T = 0.5$ sec. The critical value of gain K for this case is obtained by use of the magnitude condition, which can be obtained from Eq. (4–50) as follows:

$$\left|\frac{z(1 - e^{-T})}{(z - 1)(z - e^{-T})}\right| = \frac{1}{K}$$

For the present case, $T = 0.5$ and this last equation becomes

$$\left|\frac{0.3935z}{(z - 1)(z - 0.6065)}\right| = \frac{1}{K} \tag{4–52}$$

Since the critical gain K_c corresponds to point $z = -1$, we substitute -1 for z in Eq. (4–52):

$$\left|\frac{0.3935(-1)}{(-2)(-1.6065)}\right| = \frac{1}{K}$$

or

$$K = 8.165$$

The critical gain K_c is thus 8.165.

The closed-loop poles corresponding to $K = 2$ can be found to be

$$z_1 = 0.4098 + j0.6623 \qquad \text{and} \qquad z_2 = 0.4098 - j0.6623$$

These closed-loop poles are indicated by dots in the root locus diagram.

2. *Sampling period* $T = 1$ sec: For this case, Eq. (4–49) becomes as follows:

$$G(z) = \frac{0.6321Kz}{(z - 1)(z - 0.3679)}$$

Hence, $G(z)$ has poles at $z = 1$ and $z = 0.3679$ and a zero at $z = 0$.

The breakaway point and break-in point are found to be $z = 0.6065$ and $z = -0.6065$, respectively. The corresponding gain values are $K = 0.2449$ and $K = 4.083$, respectively.

Figure 4–35(b) shows the root locus diagram when $T = 1$ sec. The critical value of gain K is 4.328. The closed-loop poles corresponding to $K = 2$ are found to be

$$z_1 = 0.05185 + j0.6043 \qquad \text{and} \qquad z_2 = 0.05185 - j0.6043$$

and are shown in the root locus diagram by dots.

3. *Sampling period* $T = 2$ sec: For this case, Eq. (4–49) becomes

$$G(z) = \frac{0.8647Kz}{(z - 1)(z - 0.1353)}$$

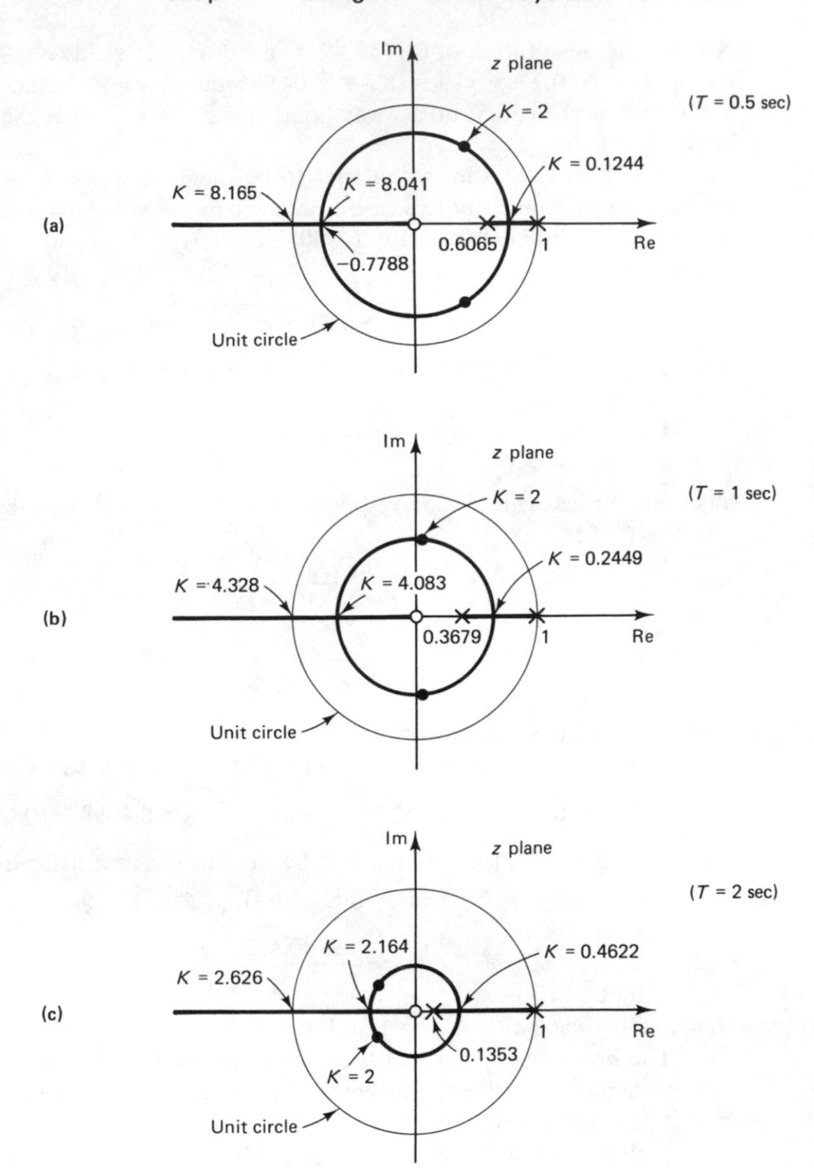

Figure 4–35 (a) Root locus diagram for the system shown in Fig. 4–34 when $T = 0.5$ sec; (b) root locus diagram when $T = 1$ sec; (c) root locus diagram when $T = 2$ sec.

We see that $G(z)$ has poles at $z = 1$ and $z = 0.1353$ and a zero at $z = 0$.

The breakaway point and break-in point are found to be $z = 0.3678$ and $z = -0.3678$, with corresponding gain values $K = 0.4622$ and $K = 2.164$, respectively. The critical value of gain K for this case is 2.626.

Figure 4–35(c) shows the root locus diagram when $T = 2$ sec. The closed-loop poles corresponding to $K = 2$ are found to be

$$z_1 = -0.2971 + j0.2169 \qquad \text{and} \qquad z_2 = -0.2971 - j0.2169$$

These closed-loop poles are shown by dots in the root locus diagram.

Effects of sampling period T on transient response characteristics. The transient response characteristics of the discrete-time control system depend on the sampling period T. A large sampling period has detrimental effects on the relative stability of the system. A rule of thumb is to sample eight to ten times during a cycle of the damped sinusoidal oscillations of the output of the closed-loop system, if it is underdamped. For overdamped systems, sample eight to ten times during the rise time in the step response.

As seen from the preceding analysis, for a given value of gain K, increasing the sampling period T will make the discrete-time control system less stable and eventually will make it unstable. Conversely, making the sampling period T shorter allows the critical value of gain K for stability to be larger. In fact, making the sampling period shorter and shorter tends to make the system behave more like the continuous-time system. (For the continuous-time second-order control system, the critical gain for stability is infinity, or $K = \infty$.)

For the system shown in Fig. 4–34, the damping ratio ζ for the closed-loop poles for $K = 2$ for each of the preceding three cases can be found from Fig. 4–36. Graphically, the damping ratios for the closed-loop poles corresponding to $T = 0.5$, $T = 1$, and $T = 2$ are determined approximately as $\zeta = 0.24$, $\zeta = 0.32$, and $\zeta = 0.37$, respectively.

The damping ratio ζ of a closed-loop pole can be analytically determined from the location of the closed-loop pole in the z plane. If the damping ratio of a closed-loop pole is ζ, then in the s plane the closed-loop pole location (in the upper half plane) can be given by

$$s = -\zeta\omega_n + j\omega_n\sqrt{1-\zeta^2}$$

Since $z = e^{Ts}$, the corresponding point in the z plane is

$$z = \exp\left[T\left(-\zeta\omega_n + j\omega_n\sqrt{1-\zeta^2}\right)\right]$$

from which we get

$$|z| = e^{-T\zeta\omega_n} \tag{4–53}$$

and

$$\angle z = T\omega_n\sqrt{1-\zeta^2} = \theta \qquad \text{(rad)} \tag{4–54}$$

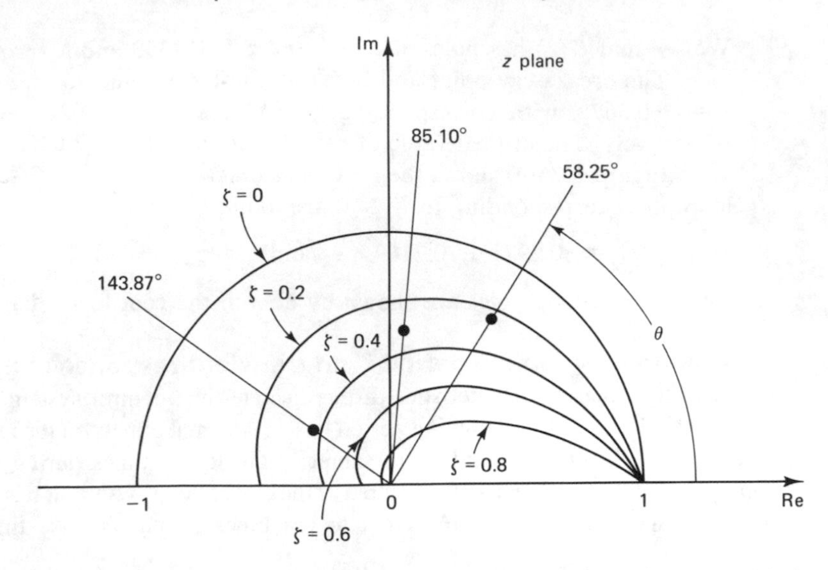

Figure 4–36 Closed-loop pole locations in the z plane shown with constant ζ loci.

From Eqs. (4–53) and (4–54) the value of ζ can be calculated. For example, in the case where the sampling period T is 0.5 sec, we have the closed-loop pole for $K = 2$ at $z = 0.4098 + j0.6623$. Hence

$$|z| = \sqrt{0.4098^2 + 0.6623^2} = 0.7788$$

By solving

$$|z| = e^{-T\zeta\omega_n} = 0.7788$$

for the exponent, we find

$$T\zeta\omega_n = 0.25 \tag{4–55}$$

Also

$$\underline{/z} = \tan^{-1}\frac{0.6623}{0.4098} = 58.25° = 1.0167 \text{ rad}$$

Hence

$$\underline{/z} = T\omega_n\sqrt{1-\zeta^2} = 1.0167 \text{ rad} \tag{4–56}$$

From Eqs. (4–55) and (4–56) we obtain

$$\frac{T\zeta\omega_n}{T\omega_n\sqrt{1-\zeta^2}} = \frac{0.25}{1.0167}$$

or

$$\frac{\zeta}{\sqrt{1-\zeta^2}} = 0.2459$$

which yields

$$\zeta = 0.2388$$

[From Fig. 4–36 we graphically obtained 0.24 for ζ, which is very close to the actual ζ value of 0.2388.]

It is important to point out that in the second-order system the damping ratio ζ is indicative of the relative stability (for example, in respect to the maximum overshoot in the unit step response) only if the sampling frequency is sufficiently high (so that there are 8 to 10 samplings in a cycle of oscillation). If the sampling frequency is not high enough, the maximum overshoot in the unit step response will be much higher than would be predicted by the damping ratio ζ.

In order to compare the effects of different sampling periods T on the transient response, we shall compare the unit step response sequences for the three values of T considered in the preceding analysis.

The closed-loop pulse transfer function for the system of Fig. 4–34, whose feedforward pulse transfer function $G(z)$ is given by Eq. (4–49), is

$$\frac{C(z)}{R(z)} = \frac{G(z)}{1 + G(z)} = \frac{Kz(1 - e^{-T})}{(z - 1)(z - e^{-T}) + Kz(1 - e^{-T})}$$

For $T = 0.5$ sec and $K = 2$, the unit step response can be given by

$$C(z) = \frac{0.3935 \times 2z}{(z - 1)(z - 0.6065) + 0.3935 \times 2z} R(z)$$

$$= \frac{0.7870z^{-1}}{1 - 0.8195z^{-1} + 0.6065z^{-2}} \frac{1}{1 - z^{-1}}$$

from which we obtain the unit step response sequence $c(kT)$ versus kT shown in Fig. 4–37(a). A BASIC computer program for finding the response is shown in Table 4–8, together with the computation results.

From Fig. 4–36 we see that the angle θ of the line connecting the origin and the dominant closed-loop pole at $z = 0.4098 + j0.6623$ (this line is a constant ω line in the s plane) is approximately 58.25°. The angle θ of the dominant closed-loop poles determines the number of samples per cycle of sinusoidal oscillation. Note that

$$\cos \theta k = \cos \theta \left(k + \frac{360°}{\theta}\right)$$

Hence, for $\theta = 58.25°$, we have $360°/\theta = 360°/58.25° = 6.18$ samples per cycle of damped oscillation, as seen from Fig. 4–37(a).

Similarly, for $T = 1$ sec and $K = 2$, the unit step response is given by

$$C(z) = \frac{1.2642z^{-1}}{1 - 0.1037z^{-1} + 0.3679z^{-2}} \frac{1}{1 - z^{-1}}$$

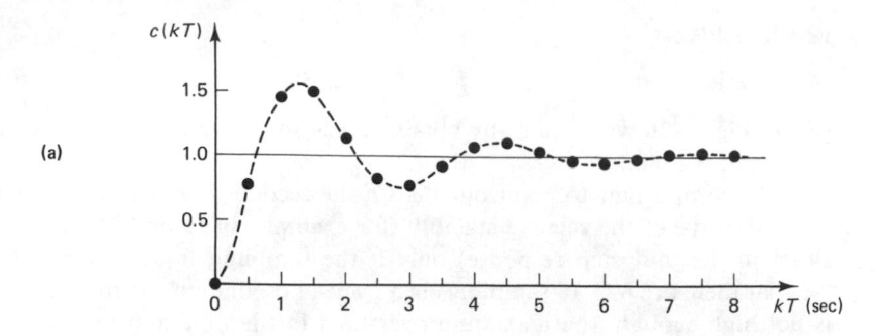

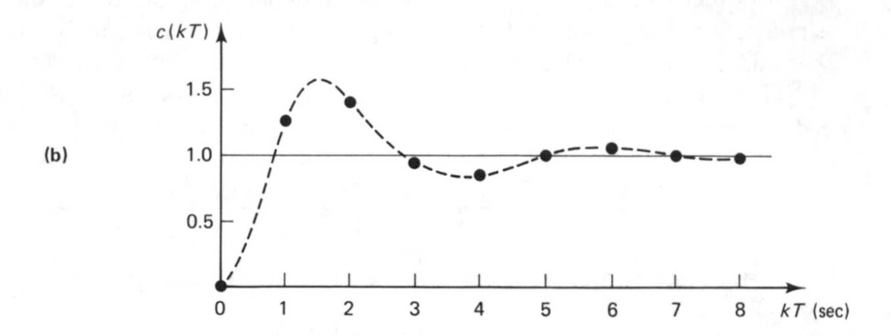

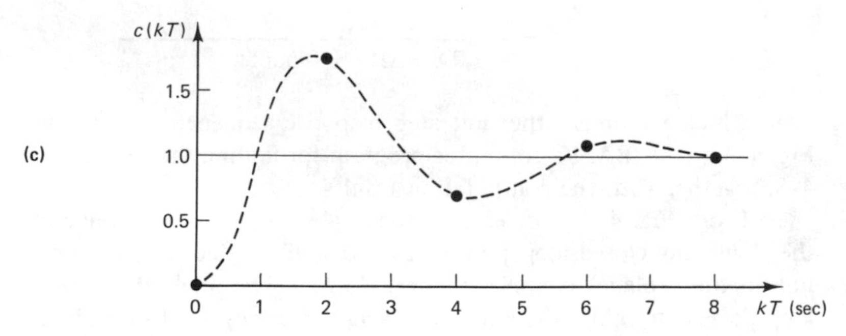

Figure 4–37 (a) Unit step response sequence of the system shown in Fig. 4–34 when $T = 0.5$ sec and $K = 2$; (b) unit step response sequence when $T = 1$ sec and $K = 2$; (c) unit step response sequence when $T = 2$ sec and $K = 2$.

The unit step response sequence $c(kT)$ versus kT is shown in Fig. 4–37(b). Since the angle of the line connecting the origin and the closed-loop pole for the present case is 85.10°, as shown in Fig. 4–36, we have approximately $360°/85.10° = 4.23$ samples per cycle, which is very much less than what we normally recommend. (We recommend 8 or more samples per cycle of damped sinusoidal oscillation.)

Finally, for $T = 2$ sec and $K = 2$, the unit step response is given by

$$C(z) = \frac{1.7294z^{-1}}{1 + 0.5941z^{-1} + 0.1353z^{-2}} \frac{1}{1 - z^{-1}}$$

The unit step response sequence $c(kT)$ versus kT is shown in Fig. 4–37(c). From Fig. 4–36 the angle of the line connecting the origin and the closed-loop pole for the present case is 143.87°, and consequently we have 360°/143.87° = 2.50 samples per cycle, as seen from Fig. 4–37(c). (Note that a slow sampling frequency such as 2.50 samples per cycle is unacceptable.)

Figure 4–37 has shown three different plots of the unit step response sequence $c(kT)$ versus kT. As can be seen from these plots, if the sampling period is small, a plot of $c(kT)$ versus kT will give a fairly accurate portrait of the response $c(t)$. However, if the sampling period is not sufficiently small, then the plot of $c(kT)$ versus kT will not portray an accurate result. It is very important to choose an adequate sampling period based on the satisfaction of the sampling theorem, system dynamics, and actual hardware considerations. Note that barely satisfying the sampling

TABLE 4–8 BASIC COMPUTER PROGRAM FOR FINDING THE UNIT STEP RESPONSE $c(k)$ WHERE $C(z)$ IS GIVEN BY

$$C(z) = \frac{0.7870z^{-1}}{1 - 0.8195z^{-1} + 0.6065z^{-2}} R(z)$$

```
10  C0 = 0
20  C1 = 0.7870
30  R0 = 1
40  R1 = 1
50  C2 = 0.8195 * C1 - 0.6065 * C0
         + 0.7870 * R1
60  M = C0
70  C0 = C1
80  C1 = C2
90  N = R0
100 R0 = R1
110 PRINT K, M, N
120 K = K + 1
130 IF K < 21 GO TO 50
140 END
```

k = K	$c(k)$ = CK = M	$r(k)$ = RK = N
0	0	1
1	0.7870	1
2	1.43195	1
3	1.48316	1
4	1.13398	1
5	0.816755	1
6	0.768574	1
7	0.921484	1
8	1.07602	1
9	1.10992	1
10	1.04397	1
11	0.969371	1
12	0.948231	1
13	0.976152	1
14	1.01185	1
15	1.02418	1
16	1.01262	1
17	0.995682	1
18	0.988804	1
19	0.993444	1
20	1.00142	1

theorem is not sufficient. An acceptable rule of thumb is to have 8 to 10 samples per cycle (6 samples per cycle is marginal) if the system is underdamped and exhibits oscillation in the response.

Next, let us investigate the effect of the sampling period T on the steady-state accuracy. We shall consider the unit ramp response for each of the three cases.

For the case where the sampling period T is 0.5 sec and gain K is 2, the open-loop pulse transfer function is

$$G(z) = \frac{0.7870z}{(z-1)(z-0.6065)}$$

and the static velocity error constant K_v is given by

$$K_v = \lim_{z \to 1} \frac{(1-z^{-1})G(z)}{T}$$

$$= \lim_{z \to 1} \left[\frac{z-1}{0.5z} \frac{0.7870z}{(z-1)(z-0.6065)} \right]$$

$$= 4$$

Thus the steady-state error in response to a unit ramp input is

$$e_{ss}^* = \frac{1}{K_v} = \frac{1}{4} = 0.25$$

Similarly, for the case where $T = 1$ sec and $K = 2$, the open-loop pulse transfer function is

$$G(z) = \frac{1.2642z}{(z-1)(z-0.3679)}$$

and the static velocity error constant K_v is given by

$$K_v = \lim_{z \to 1} \frac{(1-z^{-1})G(z)}{T}$$

$$= \lim_{z \to 1} \left[\frac{z-1}{z} \frac{1.2642z}{(z-1)(z-0.3679)} \right]$$

$$= 2$$

and the steady-state error in response to a unit ramp input is

$$e_{ss}^* = \frac{1}{K_v} = \frac{1}{2} = 0.5$$

Finally, for the case where $T = 2$ sec and $K = 2$, the open-loop pulse transfer function is

$$G(z) = \frac{1.7294z}{(z-1)(z-0.1353)}$$

and the static velocity error constant K_v and the steady-state error in response to a unit ramp input are obtained, respectively, as

$$K_v = 1$$

and

$$e_{ss}^* = \frac{1}{K_v} = 1$$

Parts (a), (b), and (c) of Fig. 4–38 show, respectively, the plots of the unit ramp response sequence $c(kT)$ versus kT for the three cases considered.

In the preceding analysis we considered the discrete-time control system shown in Fig. 4–34. Let us now investigate the dynamic characteristics of an equivalent analog control system. Figure 4–39 shows the block diagram of such an equivalent analog control system. Notice that the analog controller is represented by

$$G_c(s) = \frac{K_a}{s}$$

The discrete equivalent of this analog controller (obtained by the backward difference method or the impulse-invariance method) is

$$G_D(z) = \frac{TK_a}{1-z^{-1}}$$

In the preceding analysis the digital controller was assumed to be

$$G_D(z) = \frac{K}{1-z^{-1}}$$

Hence

$$K = TK_a$$

Consider the case where the digital control system has the sampling period $T = 0.5$ sec and gain $K = 2$. The gain K_a of this equivalent analog controller is

$$K_a = \frac{K}{T} = 4$$

The closed-loop transfer function of the equivalent analog control system is thus

$$\frac{C(s)}{R(s)} = \frac{\dfrac{4}{s(s+1)}}{1 + \dfrac{4}{s(s+1)}} = \frac{4}{s^2+s+4}$$

(a)

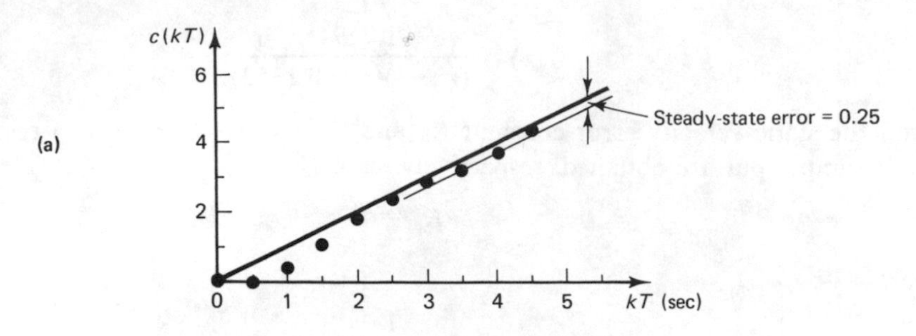

(b)

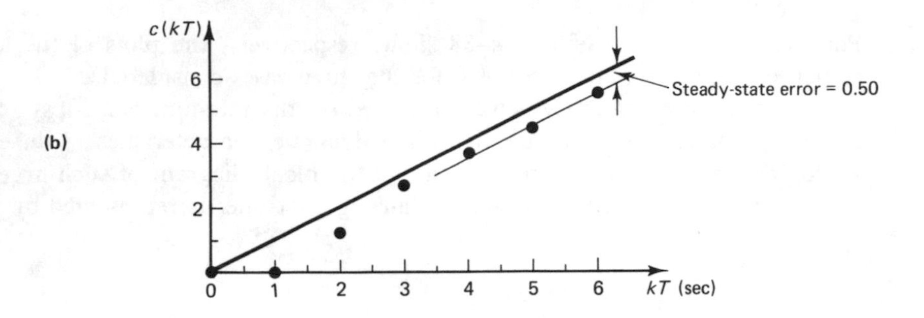

(c)

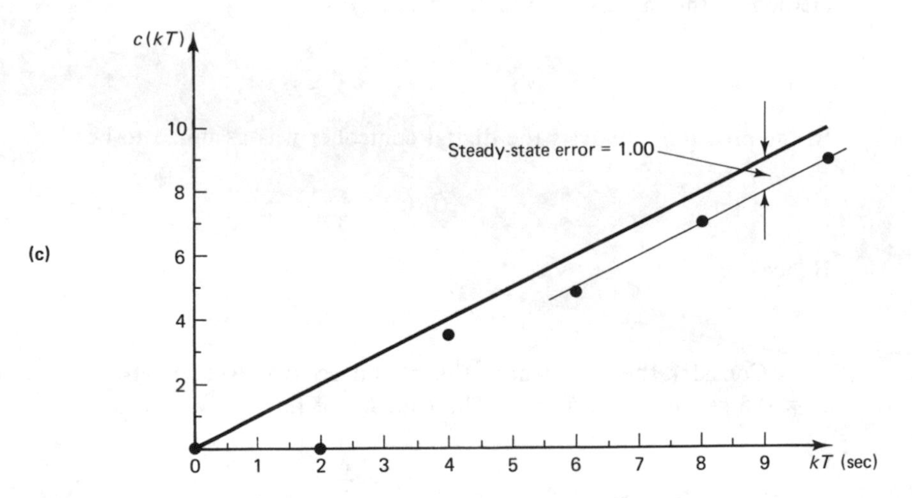

Figure 4–38 (a) Unit ramp response sequence of the system shown in Fig. 4–34 when $T = 0.5$ sec and $K = 2$; (b) unit ramp response sequence when $T = 1$ sec and $K = 2$; (c) unit ramp response sequence when $T = 2$ sec and $K = 2$.

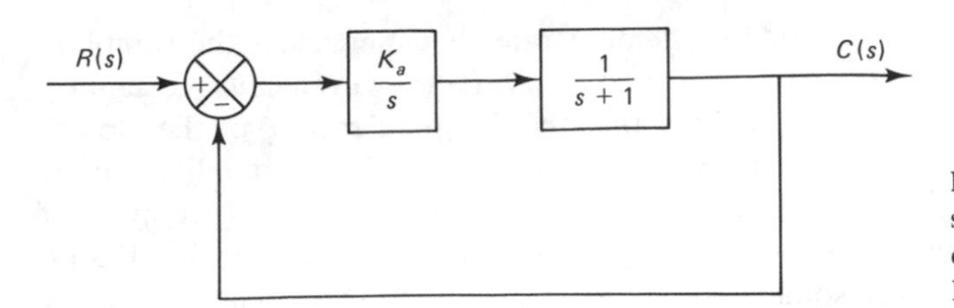

Figure 4–39 Analog control system which is equivalent to the digital control system shown in Fig. 4–34.

There are two conjugate complex closed-loop poles. The damping ratio ζ of the closed-loop poles is 0.25. The maximum overshoot in the unit step response is found from Fig. 4–24 to be 45 percent. The digital control system exhibited approximately 55 percent overshoot in the unit step response. [The amount of overshoot is measured from the envelope of the unit step response curve; see Fig. 4–37(a).] Thus, the digital control system exhibited considerably higher overshoot. (Note that if the sampling period T had been smaller so that there were 8 to 10 samples during a cycle of oscillation, then the maximum overshoot would have been smaller and closer to 45 percent.)

The closed-loop transfer function of the equivalent analog control system for the case where the digital control system has the sampling period $T = 1$ sec and gain $K = 2$ can be obtained as follows:

$$\frac{C(s)}{R(s)} = \frac{\dfrac{2}{s(s+1)}}{1 + \dfrac{2}{s(s+1)}} = \frac{2}{s^2 + s + 2}$$

The damping ratio ζ of the closed-loop poles is 0.354. The maximum overshoot in the unit step response of the equivalent analog control system is 30 percent. The digital control system exhibited approximately 57 percent overshoot. [See Fig. 4–37(b).] The large maximum overshoot is the result of a relatively large sampling period.

Finally, the closed-loop transfer function of the equivalent analog control system for the case where the digital control system has the sampling period $T = 2$ sec and gain $K = 2$ can be given by the equation

$$\frac{C(s)}{R(s)} = \frac{\dfrac{1}{s(s+1)}}{1 + \dfrac{1}{s(s+1)}} = \frac{1}{s^2 + s + 1}$$

The damping ratio ζ of the closed-loop poles is 0.5. The maximum overshoot in the unit step response of the equivalent analog control system is 17 percent. The digital control system exhibited approximately 76 percent overshoot, as seen from Fig. 4–37(c). This very large overshoot is the result of the unacceptably slow sampling rate.

The three cases we have considered demonstrate that increasing the sampling period T adversely affects the system's relative stability. (It may even cause instability in some cases.) It is important to remember that the damping ratio ζ of the closed-loop poles of the digital control system is indicative of the relative stability only if the sampling frequency is sufficiently high (that is, 8 or more samples per cycle of damped sinusoidal oscillation). If the sampling frequency is low (that is, less than 6 samples per cycle of damped sinusoidal oscillation), then predicting the relative stability from the damping ratio value is erroneous.

Example 4–10.

Consider the digital control system shown in Fig. 4–40. In the z plane, design a digital controller such that the dominant closed-loop poles have a damping ratio of 0.5 and a settling time of 2 sec. The sampling period is assumed to be 0.2 sec, or $T = 0.2$. (Notice that the same design problem was discussed in Example 4–7.) Obtain the response of the designed digital control system to a unit step input. Also, obtain the static velocity error constant K_v of the system.

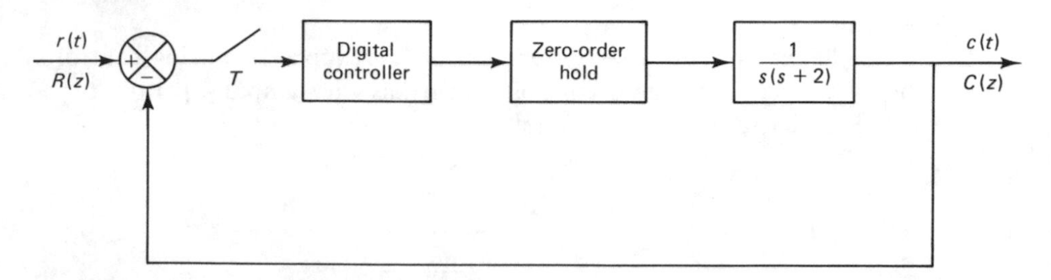

Figure 4–40 Digital control system for Example 4–10.

For the standard second-order system having a pair of dominant closed-loop poles, the settling time of 2 sec means that

$$\text{Settling time} = \frac{4}{\zeta\omega_n} = \frac{4}{0.5\omega_n} = 2$$

which gives the undamped natural frequency ω_n of the dominant closed-loop poles as

$$\omega_n = 4$$

The damped natural frequency ω_d is determined to be

$$\omega_d = \omega_n\sqrt{1-\zeta^2} = 4\sqrt{1-0.5^2} = 3.464$$

Since the sampling period T is 0.2 sec, we have

$$\omega_s = \frac{2\pi}{T} = \frac{2\pi}{0.2} = 10\pi = 31.42$$

(Notice that there are approximately 9 samples per cycle of damped oscillation. Thus, the sampling period of 0.2 sec is satisfactory.)

We shall first locate the desired dominant closed-loop poles in the z plane. Referring to Eqs. (3–104) and (3–105), for a constant-damping-ratio locus we have

$$|z| = e^{-T\zeta\omega_n} = \exp\left(-\frac{2\pi\zeta}{\sqrt{1-\zeta^2}}\frac{\omega_d}{\omega_s}\right)$$

and

$$\underline{/z} = T\omega_d = 2\pi\frac{\omega_d}{\omega_s}$$

From the given specifications ($\zeta = 0.5$ and $\omega_d = 3.464$) the magnitude and angle of the dominant closed-loop pole in the upper half of the z plane are determined as follows:

$$|z| = \exp\left(-\frac{2\pi \times 0.5}{\sqrt{1-0.5^2}}\frac{3.464}{31.42}\right) = e^{-0.400} = 0.6703$$

and

$$\underline{/z} = 2\pi\frac{3.464}{31.42} = 0.6927 \text{ rad} = 39.69°$$

We can now locate the desired dominant closed-loop pole in the upper half of the z plane, shown in Fig. 4–41 as point P. Note that at point P

$$z = 0.6703\underline{/39.69°} = 0.5158 + j0.4281$$

Noting that the sampling period T is 0.2 sec, the pulse transfer function $G(z)$ of the plant preceded by the zero-order hold can be obtained as follows:

$$G(z) = \mathscr{Z}\left[\frac{1-e^{-0.2s}}{s}\frac{1}{s(s+2)}\right] = (1-z^{-1})\,\mathscr{Z}\left[\frac{1}{s^2(s+2)}\right]$$

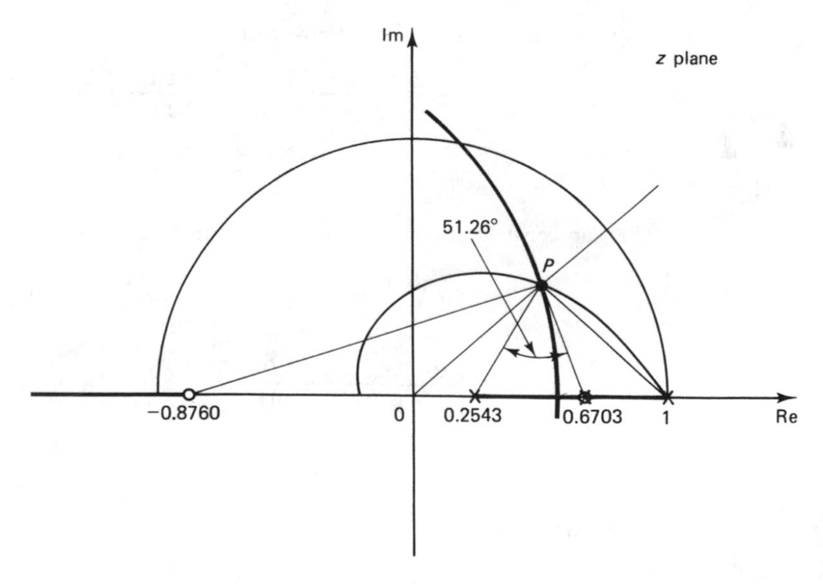

Figure 4–41 Root locus diagram of the system considered in Example 4–10.

Referring to Example 4–7, this last equation can be given by

$$G(z) = \frac{0.01758(z + 0.8760)}{(z - 1)(z - 0.6703)}$$

Next, we locate the poles ($z = 1$ and $z = 0.6703$) and zero ($z = -0.8760$) of $G(z)$ on the z plane, as shown in Fig. 4–41. If point P is to be the location for the desired dominant closed-loop pole in the upper half of the z plane, then the sum of the angles at point P must be equal to $\pm 180°$. However, the sum of the angle contributions at point P is

$$17.10° - 138.52° - 109.84° = -231.26°$$

Hence the angle deficiency is

$$-231.26° + 180° = -51.26°$$

The controller pulse transfer function must provide $+51.26°$. The pulse transfer function for the controller may be assumed to be

$$G_D(z) = K \frac{z + \alpha}{z + \beta}$$

where K is the gain constant of the controller.

If we decide to cancel the pole at $z = 0.6703$ by the zero of the controller at $z = -\alpha$, then the pole of the controller can be determined (from the condition that the controller must provide $+51.26°$) as a point at $z = 0.2543$ ($\beta = -0.2543$). Thus, the pulse transfer function for the controller may be determined as

$$G_D(z) = K \frac{z - 0.6703}{z - 0.2543}$$

The open-loop pulse transfer function now becomes

$$G_D(z)G(z) = K \frac{z - 0.6703}{z - 0.2543} \frac{0.01758(z + 0.8760)}{(z - 1)(z - 0.6703)}$$

$$= K \frac{0.01758(z + 0.8760)}{(z - 0.2543)(z - 1)}$$

The gain constant K can be determined from the following magnitude condition:

$$|G_D(z)G(z)|_{z = 0.5158 + j0.4281} = 1$$

Hence

$$K \left| \frac{0.01758(z + 0.8760)}{(z - 0.2543)(z - 1)} \right|_{z = 0.5158 + j0.4281} = 1$$

which gives

$$K = 12.67$$

The designed digital controller is

$$G_D(z) = 12.67 \frac{z - 0.6703}{z - 0.2543} \tag{4–57}$$

We shall now compare this result with that designed in Example 4–7 as a discrete equivalent of an analog controller. Notice that $G_D(z)$ obtained in Example 4–7 [given by Eq. (4–27)] is slightly different from that given by Eq. (4–57). This difference is due to the fact that in the design based on the discrete equivalent of an analog controller, we used an approximation for the zero-order hold. That is, in obtaining Eq. (4–27) we approximated the zero-order hold by

$$\frac{1-e^{-0.2s}}{s} \cong \frac{1}{(0.2/2)s+1} = \frac{10}{s+10}$$

whereas in the present root locus method we have used the exact form for the zero-order hold.

The open-loop pulse transfer function for the present system is

$$G_D(z)G(z) = \frac{12.67 \times 0.01758(z+0.8760)}{(z-0.2543)(z-1)} = \frac{0.2227(z+0.8760)}{(z-0.2543)(z-1)}$$

Hence the closed-loop pulse transfer function is

$$\frac{C(z)}{R(z)} = \frac{G_D(z)G(z)}{1+G_D(z)G(z)} = \frac{0.2227z+0.1951}{z^2-1.0316z+0.4494}$$

The response to the unit step input $R(z) = 1/(1-z^{-1})$ can be obtained from

$$C(z) = \frac{0.2227z+0.1951}{z^2-1.0316z+0.4494}\,\frac{1}{1-z^{-1}}$$

$$= \frac{0.2227z^{-1}+0.1951z^{-2}}{1-1.0316z^{-1}+0.4494z^{-2}}\,\frac{1}{1-z^{-1}}$$

A BASIC computer program and the result of the computation are shown in Table 4–9.

Figure 4–42 shows the unit step response sequence $c(kT)$ versus kT. The plot shows that the maximum overshoot is approximately 16 percent (which means that the damping ratio is approximately 0.5) and the settling time is approximately 2 sec. The digital controller just designed satisfies the given specifications and is satisfactory.

The static velocity error constant K_v of the system is given by

$$K_v = \lim_{z\to 1}\left[\frac{1-z^{-1}}{T}G_D(z)G(z)\right]$$

$$= \lim_{z\to 1}\left[\frac{z-1}{0.2z}\,\frac{0.2227(z+0.8760)}{(z-0.2543)(z-1)}\right]$$

$$= 2.801$$

If it is required to have a large value of K_v, then we may include a lag compensator. For example, adding a zero at $z = 0.94$ and a pole at $z = 0.98$ would raise the K_v value 3 times, since $(1-0.94)/(1-0.98) = 3$. A lag compensator, which has a pole and a zero very close to each other, does not significantly change the root locus near the dominant closed-loop poles. The effect of a lag compensator on the transient response is to introduce a small but slowly decreasing transient component. Such a small but

TABLE 4–9 BASIC COMPUTER PROGRAM FOR FINDING THE UNIT STEP RESPONSE $c(k)$ WHERE $C(z)$ IS GIVEN BY

$$C(z) = \frac{0.2227z^{-1} + 0.1951z^{-2}}{1 - 1.0316z^{-1} + 0.4494z^{-2}} R(z)$$

```
10  C0 = 0
20  C1 = 0.2227
30  R0 = 1
40  R1 = 1
50  C2 = 1.0316 * C1 - 0.4494 * C0
        + 0.2227 * R1 + 0.1951 * R0
60  M = C0
70  C0 = C1
80  C1 = C2
90  N = R0
100 R0 = R1
110 PRINT K, M, N
120 K = K + 1
130 IF K < 21 GO TO 50
140 END
```

k = K	$c(k)$ = CK = M	$r(k)$ = RK = N
0	0	1
1	0.222700	1
2	0.647537	1
3	0.985718	1
4	1.14366	1
5	1.15462	1
6	1.09495	1
7	1.02846	1
8	0.986689	1
9	0.973480	1
10	0.978623	1
11	0.989866	1
12	0.999153	1
13	1.00368	1
14	1.00418	1
15	1.00266	1
16	1.00086	1
17	0.999696	1
18	0.999299	1
19	0.999414	1
20	0.99971	1

slow transient, however, is not desirable from the viewpoint of disturbance or noise attenuation, since the response to disturbances would not attenuate promptly.

Finally, it is noted that although the designed system is of the third order, it acts as a second-order system, since one pole of the plant has been canceled by the zero of the controller. Because of this, the present system has only two closed-loop poles. The dominant closed-loop poles are the only closed-loop poles in this case. If a pole and a zero do not cancel each other, then the system will be of the third order and there will be a pair of conjugate complex closed-loop poles (which will be dominant) and a third pole (which will be real).

Comments. It is important to note that the poles of the closed-loop pulse transfer function determine the natural modes of the system. The transient response and frequency response behaviors, however, are strongly influenced by the zeros of the closed-loop pulse transfer function.

Familiarity with the relationship between the z plane pole and zero locations and the time response characteristics is useful in designing discrete-time control sys-

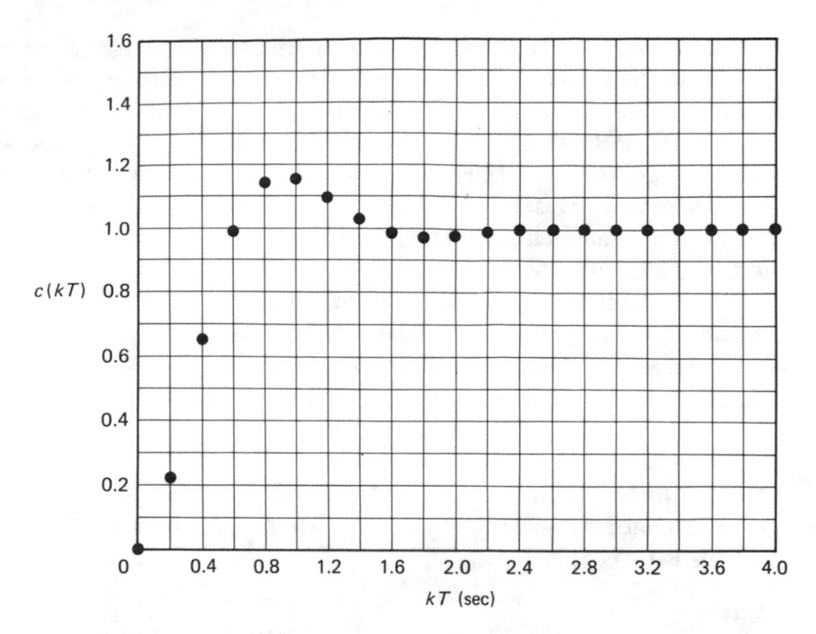

Figure 4–42 Unit step response sequence of the system designed in Example 4–10.

tems. It is important to note that in the s plane adding a zero on the negative real axis near the origin increases the maximum overshoot in response to a step input. Such a zero in the s plane is mapped to a zero on the positive real axis in the z plane between 0 and 1. Therefore, in the z plane, adding a zero on the positive real axis between 0 and 1 increases the maximum overshoot. In fact, moving a zero toward point $z = 1$ will greatly increase the maximum overshoot.

Similarly, in the s plane a closed-loop pole on the negative real axis near the origin increases the settling time. In the z plane, such a closed-loop pole is mapped to a closed-loop pole on the positive real axis between 0 and 1. Thus, a closed-loop pole in the z plane between 0 and 1 (in particular, near point $z = 1$) increases the settling time. The presence of a closed-loop pole or zero on the negative real axis between 0 and -1 in the z plane, however, affects the transient response only slightly.

4–6 DESIGN BASED ON THE FREQUENCY RESPONSE METHOD

The frequency response concept plays the same powerful role in digital control systems as it does in analog or continuous-time control systems. As stated earlier, it is assumed in this book that the reader is familiar with conventional frequency response design techniques for analog or continuous-time control systems. In fact, familiarity with Bode diagrams (logarithmic plots) is necessary in the extension of the conventional frequency response techniques to the analysis and design of digital or discrete-time control systems.

Frequency response methods, which are basically trial-and-error methods, have very frequently been used in the compensator design. The basic reason is the simplicity of the methods. In performing frequency response tests on a discrete-time system, it is important that the system have a low-pass filter before the sampler so that sidebands are filtered out. Then the response of the linear time-invariant system to a sinusoidal input preserves the frequency and modifies only the amplitude and phase of the input signal. Thus, the amplitude and phase are the only two quantities that must be dealt with.

In the following, we shall analyze the response of the linear time-invariant discrete-time system to a sinusoidal input; this analysis will be followed by the definition of the sinusoidal pulse transfer function. Then we discuss the design of a discrete-time control system in the w plane by use of a Bode diagram.

Response of a linear time-invariant discrete-time system to a sinusoidal input. Earlier in this book we stated that the frequency response of $G(z)$ can be obtained by substituting $z = e^{j\omega T}$ into $G(z)$. In what follows we shall show that this is indeed true.

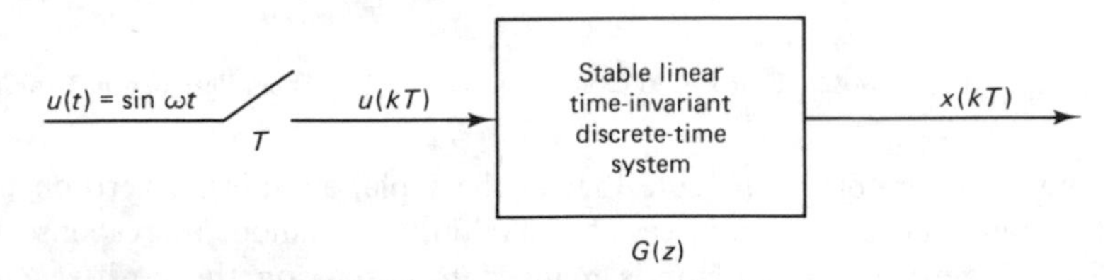

Figure 4–43 Stable linear time-invariant discrete-time system.

Consider the stable linear time-invariant discrete-time system shown in Fig. 4–43. The input to the system $G(z)$ before sampling is

$$u(t) = \sin \omega t$$

The sampled signal $u(k)$ is

$$u(k) = u(kT) = \sin k\omega T$$

The z transform of the sampled input is

$$U(z) = \mathcal{Z}[\sin k\omega T] = \frac{z \sin \omega T}{(z - e^{j\omega T})(z - e^{-j\omega T})}$$

The response of the system is given by

$$X(z) = G(z)U(z) = G(z)\frac{z \sin \omega T}{(z - e^{j\omega T})(z - e^{-j\omega T})}$$

$$= \frac{az}{z - e^{j\omega T}} + \frac{\bar{a}z}{z - e^{-j\omega T}} + [\text{terms due to poles of } G(z)] \qquad (4\text{–}58)$$

Multiplying both sides of Eq. (4–58) by $(z - e^{j\omega T})/z$, we obtain

$$G(z)\frac{\sin \omega T}{z - e^{-j\omega T}} = a + \frac{\bar{a}(z - e^{j\omega T})}{z - e^{-j\omega T}} + \frac{z - e^{j\omega T}}{z} \text{[terms due to poles of } G(z)]$$

The second term on the right-hand side of this last equation approaches zero as z approaches $e^{j\omega T}$. Since the system considered here is a stable one, the third term on the right-hand side also approaches zero as z approaches $e^{j\omega T}$. Hence, by letting z approach $e^{j\omega T}$, we have

$$a = G(z)\frac{\sin \omega T}{z - e^{-j\omega T}}\bigg|_{z=e^{j\omega T}} = \frac{G(e^{j\omega T})}{2j}$$

The coefficient $\bar{a}$, the complex conjugate of a, is then obtained as follows:

$$\bar{a} = -\frac{G(e^{-j\omega T})}{2j}$$

Let us define

$$G(e^{j\omega T}) = Me^{j\theta}$$

Then

$$G(e^{-j\omega T}) = Me^{-j\theta}$$

Equation (4–58) can now be written as

$$X(z) = \frac{Me^{j\theta}}{2j}\frac{z}{z - e^{j\omega T}} - \frac{Me^{-j\theta}}{2j}\frac{z}{z - e^{-j\omega T}} + \text{[terms due to poles of } G(z)]$$

or

$$X(z) = \frac{M}{2j}\left(\frac{e^{j\theta}z}{z - e^{j\omega T}} - \frac{e^{-j\theta}z}{z - e^{-j\omega T}}\right) + \text{[terms due to poles of } G(z)]$$

The inverse z transform of this last equation is

$$x(kT) = \frac{M}{2j}(e^{jk\omega T}e^{j\theta} - e^{-jk\omega T}e^{-j\theta}) + \mathscr{Z}^{-1}\text{[terms due to poles of } G(z)] \tag{4–59}$$

The last term on the right-hand side of Eq. (4–59) represents the transient response. Since the system $G(z)$ has been assumed to be stable, all transient response terms will disappear at steady state and we will get the following steady-state response $x_{ss}(kT)$:

$$x_{ss}(kT) = \frac{M}{2j}[e^{j(k\omega T+\theta)} - e^{-j(k\omega T+\theta)}] = M \sin(k\omega T + \theta) \tag{4–60}$$

where M, the gain of the discrete-time system when subjected to a sinusoidal input, is given by

$$M = M(\omega) = |G(e^{j\omega T})|$$

and θ, the phase angle, is given by

$$\theta = \theta(\omega) = \underline{/G(e^{j\omega T})}$$

In terms of $G(e^{j\omega T})$, Eq. (4–60) can be written as follows:

$$x_{ss}(kT) = |G(e^{j\omega T})| \sin (k\omega T + \underline{/G(e^{j\omega T})})$$

We have shown that $G(e^{j\omega T})$ indeed gives the magnitude and phase of the frequency response of $G(z)$. Thus, to obtain the frequency response of $G(z)$, we need only to substitute $e^{j\omega T}$ for z in $G(z)$. The function $G(e^{j\omega T})$ is commonly called the *sinusoidal pulse transfer function*. Noting that

$$e^{j\left(\omega + \frac{2\pi}{T}\right)T} = e^{j\omega T}e^{j2\pi} = e^{j\omega T}$$

we find that the sinusoidal pulse transfer function $G(e^{j\omega T})$ is periodic, with the period equal to T.

Example 4–11.

Consider the system defined by

$$x(kT) = u(kT) + ax((k-1)T) \qquad 0 < a < 1$$

where $u(kT)$ is the input and $x(kT)$ the output. Obtain the steady-state output $x(kT)$ when the input $u(kT)$ is the sampled sinusoid, or $u(kT) = A \sin k\omega T$.

The z transform of the system equation is

$$X(z) = U(z) + az^{-1}X(z)$$

By defining $G(z) = X(z)/U(z)$, we have

$$G(z) = \frac{X(z)}{U(z)} = \frac{1}{1 - az^{-1}}$$

Let us substitute $e^{j\omega T}$ for z in $G(z)$. Then the sinusoidal pulse transfer function $G(e^{j\omega T})$ can be obtained as

$$G(e^{j\omega T}) = \frac{1}{1 - ae^{-j\omega T}} = \frac{1}{1 - a\cos\omega T + ja\sin\omega T}$$

The amplitude of $G(e^{j\omega T})$ is

$$|G(e^{j\omega T})| = M = \frac{1}{\sqrt{1 + a^2 - 2a\cos\omega T}}$$

and the phase angle of $G(e^{j\omega T})$ is

$$\underline{/G(e^{j\omega T})} = \theta = -\tan^{-1}\frac{a\sin\omega T}{1 - a\cos\omega T}$$

Then the steady-state output $x_{ss}(kT)$ can be written as follows:

$$\begin{aligned} x_{ss}(kT) &= AM \sin(k\omega T + \theta) \\ &= \frac{A}{\sqrt{1 + a^2 - 2a\cos\omega T}} \sin\left(k\omega T - \tan^{-1}\frac{a\sin\omega T}{1 - a\cos\omega T}\right) \end{aligned}$$

Bilinear transformation and the *w* plane. Before we can advantageously apply our well-developed frequency response methods to the analysis and design of discrete-time control systems, certain modifications in the z plane approach are necessary. Since in the z plane the frequency appears as $z = e^{j\omega T}$, if we treat frequency response in the z plane, the simplicity of the logarithmic plots will be completely lost. Thus, the direct application of frequency response methods is not worthy of consideration. In fact, since the z transformation maps the primary and complementary strips of the left half of the s plane into the unit circle in the z plane, conventional frequency response methods, which deal with the entire left half plane, do not apply to the z plane.

The difficulty, however, can be overcome by transforming the pulse transfer function in the z plane into that in the w plane. The transformation, commonly called the w transformation, a bilinear transformation, is defined by

$$z = \frac{1 + (T/2)w}{1 - (T/2)w} \tag{4–61}$$

where T is the sampling period involved in the discrete-time control system under consideration. By converting a given pulse transfer function in the z plane into a rational function of w, the frequency response methods can be extended to discrete-time control systems. By solving Eq. (4–61) for w, we obtain the inverse relationship

$$w = \frac{2}{T}\frac{z - 1}{z + 1} \tag{4–62}$$

(Note that this transformation has the same form as the bilinear transformation discussed in Sec. 4–2.)

Through the z transformation and the w transformation, the primary strip of the left half of the s plane is first mapped into the inside of the unit circle in the z plane and then mapped into the entire left half of the w plane. The two mapping processes are depicted in Fig. 4–44. (Note that in the s plane we consider only the primary strip.) Notice that the origin of the z plane is mapped into the point $w = -2/T$ in the w plane. Notice also that as s varies from 0 to $j\omega_s/2$ along the $j\omega$ axis in the s plane, z varies from 1 to -1 along the unit circle in the z plane and w varies from 0 to ∞ along the imaginary axis in the w plane.

Although the left half of the w plane corresponds to the left half of the s plane and the imaginary axis of the w plane corresponds to the imaginary axis of the s plane, there are differences between the two planes. The chief difference is that the behavior in the s plane over the frequency range $-\frac{1}{2}\omega_s \leqslant \omega \leqslant \frac{1}{2}\omega_s$ maps to the range $-\infty < \nu < \infty$, where ν is the fictitious frequency in the w plane. This means that although the frequency response characteristics of the analog filter will be reproduced in the discrete or digital filter, the frequency scale on which the response occurs will be compressed from an infinite interval in the analog filter to a finite interval in the digital filter.

Once the pulse transfer function $G(z)$ is transformed into $G(w)$ by means of

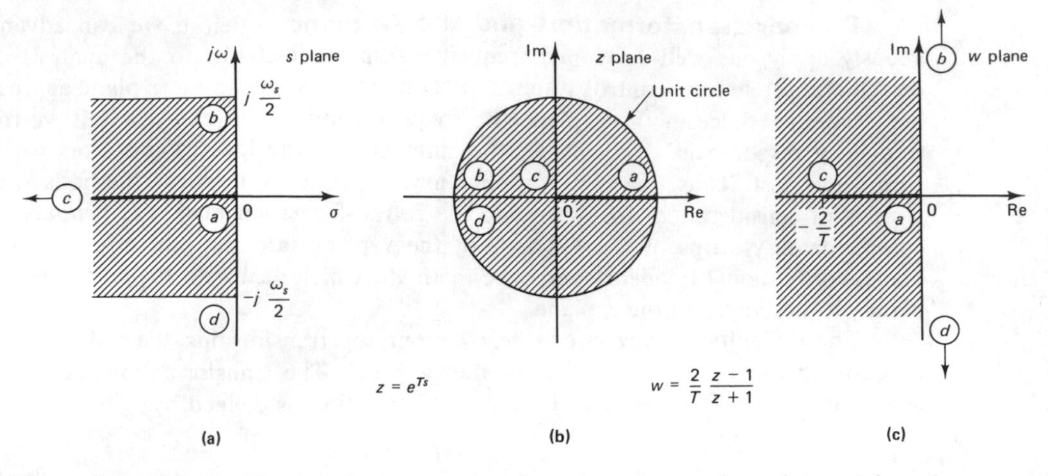

Figure 4–44 Diagrams showing mappings from the s plane to the z plane and from the z plane to the w plane. (a) Primary strip in the left half of the s plane; (b) z plane mapping of the primary strip in the s plane; (c) w plane mapping of the unit circle in the z plane.

the w transformation, it may be treated as a conventional transfer function in w. Conventional frequency response techniques can then be used in the w plane and so the well-established frequency response design techniques can be applied to the design of discrete-time control systems.

As noted earlier, ν represents the fictitious frequency. By replacing w by $j\nu$, conventional frequency response techniques may be used to draw the Bode diagram for the transfer function in w. (In the brief review of the Bode diagrams in this section, we shall use the fictitious frequency ν as the variable.)

Although the w plane resembles the s plane geometrically, the frequency axis in the w plane is distorted. The fictitious frequency ν and the actual frequency ω are related as follows:

$$w\Big|_{w=j\nu} = j\nu = \frac{2}{T}\frac{z-1}{z+1}\Big|_{z=e^{j\omega T}} = \frac{2}{T}\frac{e^{j\omega T}-1}{e^{j\omega T}+1}$$

$$= \frac{2}{T}\frac{e^{j(1/2)\omega T} - e^{-j(1/2)\omega T}}{e^{j(1/2)\omega T} + e^{-j(1/2)\omega T}} = \frac{2}{T}\, j \tan\frac{\omega T}{2}$$

or

$$\nu = \frac{2}{T}\tan\frac{\omega T}{2} \qquad (4\text{–}63)$$

Equation (4–63) gives the relationship between the actual frequency ω and the fictitious frequency ν. Note that as the actual frequency ω moves from $-\frac{1}{2}\omega_s$ to 0, the fictitious frequency ν moves from $-\infty$ to 0, and as ω moves from 0 to $\frac{1}{2}\omega_s$, ν moves from 0 to ∞.

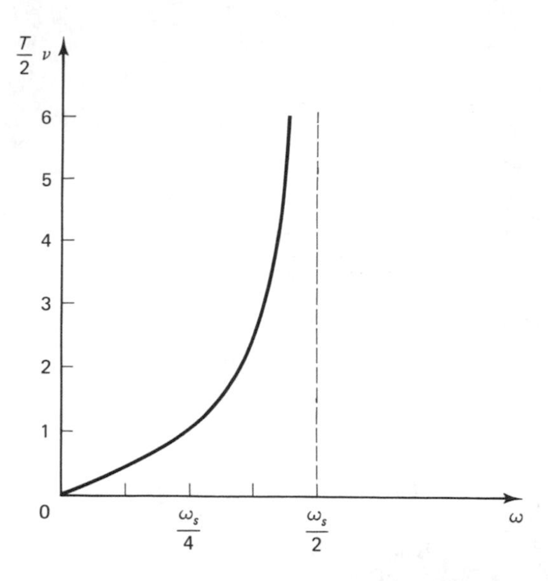

Figure 4–45 Relationship between the fictitious frequency ν and the actual frequency ω for the frequency range between 0 and $\frac{1}{2}\omega_s$.

Referring to Eq. (4–63), the actual frequency ω can be translated into the fictitious frequency ν. For example, if the bandwidth is specified as ω_b, then the corresponding bandwidth in the w plane is $(2/T)\tan(\omega_b T/2)$. Similarly, $G(j\nu_1)$ corresponds to $G(j\omega_1)$, where $\omega_1 = (2/T)\tan^{-1}(\nu_1 T/2)$. Figure 4–45 shows the relationship between the fictitious frequency ν times $\frac{1}{2}T$ and the actual frequency ω for the frequency range between 0 and $\frac{1}{2}\omega_s$.

Notice that in Eq. (4–63) if ωT is small, then

$$\nu \cong \omega$$

This means that for small ωT the transfer functions $G(s)$ and $G(w)$ resemble each other. Note that this is the direct result of the inclusion of the scale factor $2/T$ in Eq. (4–62). The presence of this scale factor in the transformation enables us to maintain the same error constants before and after the w transformation. (This means that the transfer function in the w plane will approach that in the s plane as T approaches zero. See Example 4–12, which follows.)

Example 4–12.

Consider the transfer-function system shown in Fig. 4–46. The sampling period T is assumed to be 0.1 sec. Obtain $G(w)$.

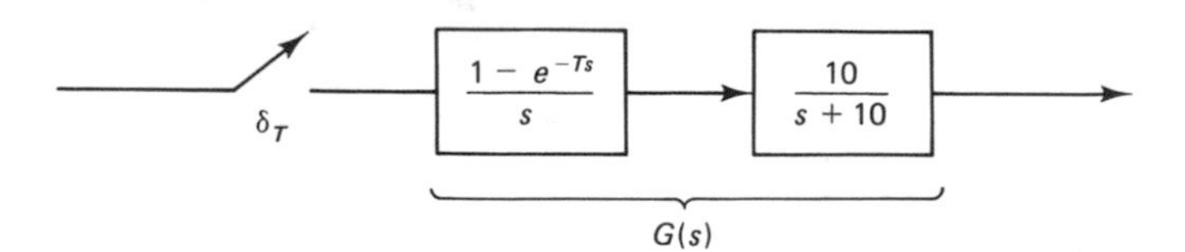

Figure 4–46 Transfer-function system of Example 4–12.

The z transform of $G(s)$ is

$$G(z) = \mathcal{Z}\left[\frac{1-e^{-Ts}}{s}\frac{10}{s+10}\right]$$

$$= (1-z^{-1})\,\mathcal{Z}\left[\frac{10}{s(s+10)}\right]$$

$$= \frac{0.6321}{z-0.3679}$$

By use of the bilinear transformation given by Eq. (4–61), or

$$z = \frac{1+(T/2)w}{1-(T/2)w} = \frac{1+0.05w}{1-0.05w}$$

$G(z)$ can be transformed into $G(w)$ as follows:

$$G(w) = \frac{0.6321}{\dfrac{1+0.05w}{1-0.05w} - 0.3679} = \frac{0.6321(1-0.05w)}{0.6321+0.06840w}$$

$$= 9.241\,\frac{1-0.05w}{w+9.241}$$

Notice that the location of the pole of the plant is $s = -10$ and that of the pole in the w plane is $w = -9.241$. The gain value in the s plane is 10 and that in the w plane is 9.241. (Thus, both the pole locations and the gain values are similar in the s plane and the w plane.) However, $G(w)$ has a zero at $w = 2/T = 20$, although the plant does not have any zero. As the sampling period T becomes smaller, the w plane zero at $w = 2/T$ approaches infinity in the right half of the w plane. Note that we have

$$\lim_{w\to 0} G(w) = \lim_{s\to 0} \frac{10}{s+10}$$

This fact is very useful in checking the numerical calculations in transforming $G(s)$ into $G(w)$.

To summarize, the w transformation, a bilinear transformation, maps the inside of the unit circle of the z plane into the left half of the w plane. The overall result due to the transformations from the s plane into the z plane and from the z plane into the w plane is that the w plane and the s plane are similar over the region of interest in the s plane. This is because some of the distortions caused by the transformation from the s plane into the z plane are partly compensated for by the transformation from the z plane into the w plane.

Note that if

$$G(z) = \frac{b_0 z^m + b_1 z^{m-1} + \cdots + b_m}{z^n + a_1 z^{n-1} + \cdots + a_n} \qquad m \le n$$

where the a_i's and the b_i's are constants, is transformed into the w plane by the transformation

$$z = \frac{1 + (T/2)w}{1 - (T/2)w}$$

then $G(w)$ takes the form

$$G(w) = \frac{\beta_0 w^n + \beta_1 w^{n-1} + \cdots + \beta_n}{\alpha_0 w^n + \alpha_1 w^{n-1} + \cdots + \alpha_n}$$

where the α_i's and the β_i's are constants (some of them may be zero). Thus, $G(w)$ is a ratio of polynominals in w, where the degrees of the numerator and denominator may or may not be equal. (For example, β_0 may be zero while α_0 is nonzero, and vice versa. Then the degree of the numerator becomes different from that of the denominator.) Since $G(j\nu)$ is a rational function of ν, the Nyquist stability criterion can be applied to $G(j\nu)$. In terms of the Bode diagram the conventional straight-line approximation to the magnitude curve as well as the concept of the phase margin and gain margin apply to $G(j\nu)$.

Bode diagrams. Design by means of Bode diagrams has been widely used in dealing with single-input–single-output continuous-time control systems. In particular, if the transfer function is in a factored form, the simplicity and ease with which the asymptotic Bode diagram can be drawn and reshaped is well known.

As stated earlier, the conventional frequency response methods apply to the transfer functions in the w plane. Recall that the Bode diagram consists of two separate plots, the logarithmic magnitude $|G(j\nu)|$ versus log ν, and the phase angle $\underline{/G(j\nu)}$ versus log ν. Sketching of the logarithmic magnitude is based on the factoring of $G(j\nu)$, so that it works on the principle of adding the individual factored terms instead of multiplying individual terms. Familiar asymptotic plotting techniques can be applied, and therefore the magnitude curve can be quickly drawn by using straight-line asymptotes. Using the Bode diagram, a digital compensator or digital controller may be designed with conventional design techniques.

It is important to note that there may be a difference in the high-frequency magnitudes for $G(j\omega)$ and $G(j\nu)$. The high-frequency asymptote of the logarithmic magnitude curve for $G(j\nu)$ may be a constant-db line (that is, a horizontal line). On the other hand, if $\lim_{s \to \infty} G(s) = 0$, then the magnitude of $G(j\omega)$ always approaches zero ($-\infty$ db) as ω approaches infinity. For example, referring to Example 4–12, we obtained $G(w)$ for $G(s)$ as follows:

$$G(w) = 9.241 \left(\frac{1 - 0.05w}{w + 9.241} \right)$$

The high-frequency magnitude of $G(j\nu)$ is

$$\lim_{\nu \to \infty} |G(j\nu)| = \lim_{\nu \to \infty} \left| 9.241 \left(\frac{1 - 0.05j\nu}{j\nu + 9.241} \right) \right| = 0.4621$$

while the high-frequency magnitude of the plant is

$$\lim_{\omega \to \infty} \left| \frac{10}{j\omega + 10} \right| = 0$$

The difference in the Bode diagrams at the high-frequency end can be explained as follows. First recall that we are interested only in the frequency range $0 \leqslant \omega \leqslant \frac{1}{2}\omega_s$, which corresponds to $0 \leqslant \nu \leqslant \infty$. Then, noting that $\nu = \infty$ in the w plane corresponds to $\omega = \frac{1}{2}\omega_s$ in the s plane, it can be said that $\lim_{\nu \to \infty} |G(j\nu)|$ corresponds to $\lim_{\omega \to \omega_s/2} |10/(j\omega + 10)|$, which is a constant. (It is important to note that these two values are generally not equal to each other.) From the pole-zero point of view, it can be said that when $|G(j\nu)|$ is a nonzero constant at $\nu = \infty$ it is implied that $G(w)$ contains the same number of poles and zeros.

In general, one or more zeros of $G(w)$ lie in the right half of the w plane. The presence of a zero in the right half of the w plane means that $G(w)$ is a nonminimum phase transfer function. Therefore, we must be careful in drawing the phase angle curve in the Bode diagram.

Advantages of the Bode diagram approach to the design. The Bode diagram approach to the analysis and design of control systems is particularly useful for the following reasons:

1. In the Bode diagram the low-frequency asymptote of the magnitude curve is indicative of one of the static error constants K_p, K_v, or K_a.
2. Specifications of the transient response can be translated into those of the frequency response in terms of the phase margin, gain margin, bandwidth, and so forth. These specifications can easily be handled in the Bode diagram. In particular, the phase and gain margins can be read directly from the Bode diagram.
3. The design of a digital compensator (or digital controller) to satisfy the given specifications (in terms of the phase margin and gain margin) can be carried out in the Bode diagram in a simple and straightforward manner.

Phase lead, phase lag, and phase lag-lead compensation. Before we discuss design procedures in the w plane, let us review the phase lead, phase lag, and phase lag-lead compensation techniques.

Phase lead compensation is commonly used for improving stability margins. The phase lead compensation increases the system bandwidth. Thus, the system has a faster speed to respond. However, such a system using phase lead compensation may be subjected to high-frequency noise problems due to its increased high-frequency gains.

Phase lag compensation reduces the system gain at higher frequencies without reducing the system gain at lower frequencies. The system bandwidth is reduced and thus the system has a slower speed to respond. Because of the reduced high-frequency gain the total system gain can be increased and thereby low-frequency gain can be increased and the steady-state accuracy can be improved. Also, any high-frequency noises involved in the system can be attenuated.

In some applications, a phase lag compensator is cascaded with a phase lead compensator. The cascaded compensator is known as a *phase lag-lead* compensator. By use of the lag-lead compensator, the low-frequency gain can be increased (which means an improvement in steady-state accuracy) while at the same time the system bandwith and stability margins can be increased.

Note that the PID controller is a special case of a phase lag-lead controller. The PD control action, which affects the high-frequency region, increases the phase lead angle and improves system stability as well as increasing the system bandwidth (and thus increasing the speed of response). That is, the PD controller behaves in much the same way as a phase lead compensator. The PI control action affects the low-frequency portion and, in fact, increases the low-frequency gain and improves steady-state accuracy. Therefore, the PI controller acts as a phase lag compensator. The PID control action is a combination of the PI and PD control actions. The design techniques for PID controllers basically follow those of phase lag-lead compensators. (In industrial control systems, however, each of the PID control actions in the PID controller may be adjusted experimentally.)

Some remarks on the coefficient quantization problem. From the viewpoint of microprocessor implementation of the phase lead, phase lag, and phase lag-lead compensators, phase lead compensators present no coefficient quantization problem, because the locations of poles and zeros are widely separated. However, in the case of phase lag compensators and phase lag-lead compensators, the phase lag network presents a coefficient quantization problem because the locations of poles and zeros are close to each other. (They are near the point $z = 1$.)

Since the filter coefficients must be realized by binary words that use limited numbers of bits, if the number of bits employed is insufficient, the pole and zero locations of the filter may not be realized exactly as desired and the resulting filter will not behave as it is expected.

Since small deviations in the pole and zero locations from the desired locations can have significant effects on the frequency response characteristics of the compensator, the digital version of the compensator may not perform as expected. To minimize the effect of the coefficient quantization problem it is necessary to structure the filter so that it is least subject to coefficient inaccuracies due to quantization.

Because the sensitivity of the roots of polynomials to the parameter variations becomes severe as the order of the polynomial increases, direct realization of a higher-order filter is not desirable. It is preferable to place lower-order elements in cascade or in parallel, as discussed in Sec. 3–7. As a matter of course, from the outset if we choose poles and zeros of the digital compensator from allowable discrete points, then the coefficient quantization problem can be avoided.

In the analog compensator, the poles and zeros of the compensator can be placed with an arbitrary accuracy. In converting an analog compensator to a digital compensator the digital version of the lag compensator may involve considerable inaccuracies in the locations of poles and zeros. (The important thing to remember is that the poles and zeros of the filter in the z plane must lie on a finite number of allowable discrete points.)

Design procedure in the *w* plane. Referring to the digital control system shown in Fig. 4–47, the design procedure in the w plane may be stated as follows:

1. First, obtain $G(z)$, the z transform of the plant preceded by a hold. Then, transform $G(z)$ into a transfer function $G(w)$ through the bilinear transformation given by Eq. (4–61):

$$z = \frac{1 + (T/2)w}{1 - (T/2)w}$$

   That is,

$$G(w) = G(z)\Big|_{z = \frac{1 + (T/2)w}{1 - (T/2)w}}$$

   It is important that the sampling period T be chosen properly. A rule of thumb is to sample at the frequency 10 times that of the bandwidth of the closed-loop system. (Although digital controls and signal processing use similar approaches in sampling continuous-time signals, the sampling frequencies involved are very different. In the field of signal processing, sampling frequencies are generally very high, while in the field of digital control systems, the sampling frequencies used are generally low. Such a difference in the sampling frequencies is mainly due to the different dynamics involved and the different trade-offs in these two fields.)
2. Substitute $w = j\nu$ into $G(w)$ and plot the Bode diagram for $G(j\nu)$.
3. Read from the Bode diagram the static error constants, the phase margin, and the gain margin.

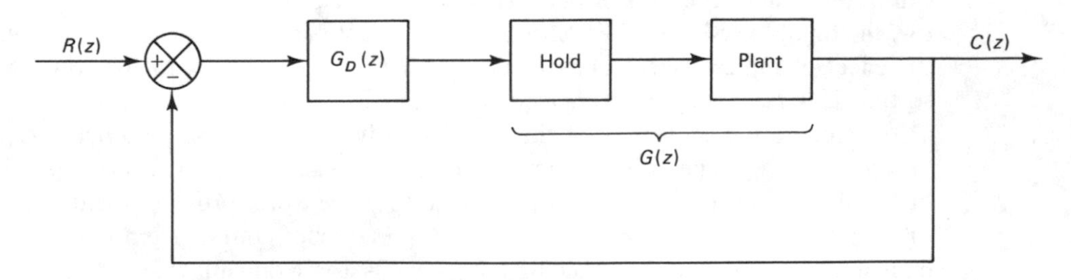

Figure 4–47 Digital control system.

4. By assuming that the low-frequency gain of the discrete-time controller (or digital controller) transfer function $G_D(w)$ is unity, determine the system gain by satisfying the requirement for a given static error constant. Then, by using conventional design techniques for continuous-time control systems, determine the pole(s) and zero(s) of the digital controller transfer function. [$G_D(w)$ is a ratio of two polynomials in w.] Then the open-loop transfer function of the designed system is given by $G_D(w)G(w)$.
5. Transform the controller transfer function $G_D(w)$ into $G_D(z)$ through the bilinear transformation given by Eq. (4–62):

$$w = \frac{2}{T}\frac{z-1}{z+1}$$

Then

$$G_D(z) = G_D(w)\Big|_{w=\frac{2}{T}\frac{z-1}{z+1}}$$

is the pulse transfer function of the digital controller.
6. Realize the pulse transfer function $G_D(z)$ by a computational algorithm.

In following the design procedure just given, it is important to note the following:

1. The transfer function $G(w)$ is a nonminimum phase transfer function. Hence, the phase angle curve is different from that for the more typical minimum phase transfer function. It is necessary to make sure that the phase angle curve is drawn correctly by taking into consideration the nonminimum phase term.
2. The frequency axis in the w plane is distorted. The relationship between the fictitious frequency ν and the actual frequency ω is

$$\nu = \frac{2}{T}\tan\frac{\omega T}{2}$$

If, for example, a bandwidth ω_b is specified, we need to design the system for a bandwidth ν_b, where

$$\nu_b = \frac{2}{T}\tan\frac{\omega_b T}{2}$$

Example 4–13.

Consider the digital control system shown in Fig. 4–48. Design a digital controller in the w plane such that the phase margin is 50°, the gain margin is at least 10 db (which corresponds to a damping ratio ζ of the dominant closed-loop poles of about 0.5), and the static velocity error constant K_v is 2 sec^{-1}. Assume that the sampling period is 0.2 sec, or $T = 0.2$.

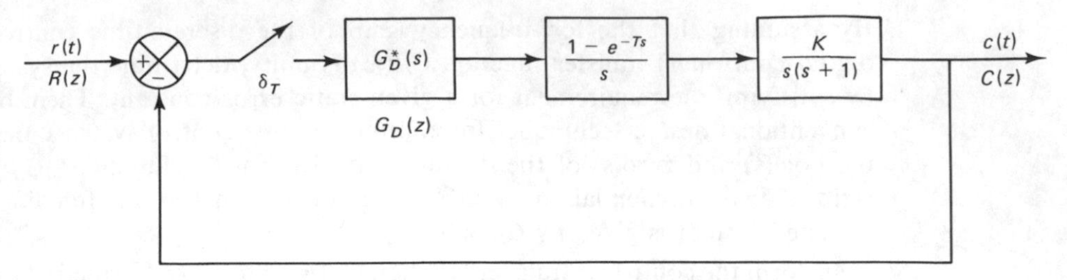

Figure 4–48 Digital control system of Example 4–13.

First, we obtain the pulse transfer function $G(z)$ of the plant which is preceded by the zero-order hold:

$$G(z) = \mathcal{Z}\left[\frac{1-e^{-0.2s}}{s}\frac{K}{s(s+1)}\right]$$

$$= (1-z^{-1})\,\mathcal{Z}\left[\frac{K}{s^2(s+1)}\right]$$

$$= 0.01873\left[\frac{K(z+0.9356)}{(z-1)(z-0.8187)}\right]$$

(Refer to Example 3–8 for the derivation of this result.)

Next, we transform the pulse transfer function $G(z)$ into a transfer function $G(w)$ by means of the bilinear transformation given by Eq. (4–61):

$$z = \frac{1+(T/2)w}{1-(T/2)w} = \frac{1+0.1w}{1-0.1w}$$

Thus,

$$G(w) = \frac{0.01873K\left(\dfrac{1+0.1w}{1-0.1w}+0.9356\right)}{\left(\dfrac{1+0.1w}{1-0.1w}-1\right)\left(\dfrac{1+0.1w}{1-0.1w}-0.8187\right)}$$

$$= 0.003316\left[\frac{K(300.6+w)(1-0.1w)}{w(0.997+w)}\right]$$

$$= \frac{K\left(\dfrac{w}{300.6}+1\right)\left(1-\dfrac{w}{10}\right)}{w\left(\dfrac{w}{0.997}+1\right)}$$

Now let us assume that the transfer function of the digital controller $G_D(w)$ has unity gain for the low-frequency range and has the following form:

$$G_D(w) = \frac{1+(w/\alpha)}{1+(w/\beta)}$$

(This is one of the simplest forms for the digital controller transfer function. Other forms may be assumed as well.) The open-loop transfer function is

$$G_D(w)G(w) = \frac{1+\dfrac{w}{\alpha}}{1+\dfrac{w}{\beta}} \frac{K\left(\dfrac{w}{300.6}+1\right)\left(1-\dfrac{w}{10}\right)}{w\left(\dfrac{w}{0.997}+1\right)}$$

The static velocity error constant K_v is specified as 2 sec^{-1}. Hence

$$K_v = \lim_{w\to 0} wG_D(w)G(w)$$

$$= \lim_{w\to 0}\left[w\frac{1+\dfrac{w}{\alpha}}{1+\dfrac{w}{\beta}} \frac{K\left(\dfrac{w}{300.6}+1\right)\left(1-\dfrac{w}{10}\right)}{w\left(\dfrac{w}{0.997}+1\right)}\right]$$

$$= K = 2$$

The gain K is thus determined to be 2.

Figure 4–49 shows the Bode diagram for the system considered. For the magnitude curves we have used straight-line asymptotes. The magnitude and phase angle of $G(j\nu)$ are shown by dashed curves. (Note that the zero at $\nu = 10$ which lies in the right half

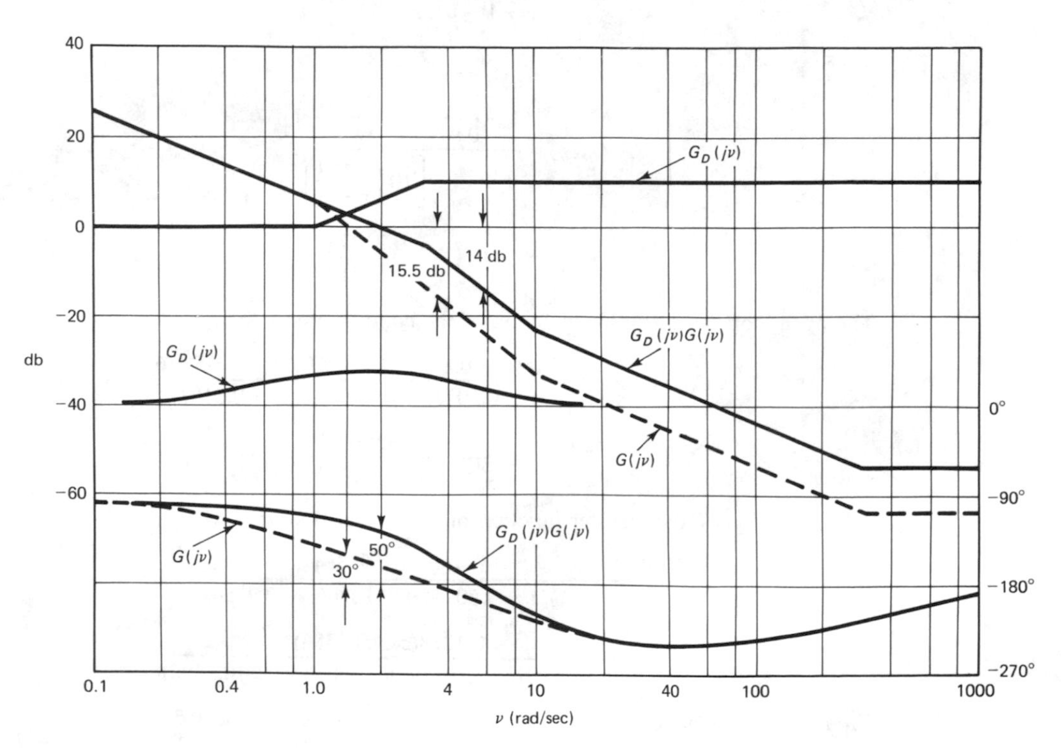

Figure 4–49 Bode diagram for the system designed in Example 4–13.

of the w plane gives phase lag.) The phase margin can be read from the Bode diagram (dashed curves) as 30° and the gain margin as 15.5 db.

The given specifications require, in addition to $K_v = 2$, the phase margin of 50° and a gain margin of at least 10 db. Let us design a digital controller to satisfy these specifications. Using a conventional design technique, we choose the zero of the controller at 0.997. Then the new gain crossover frequency is at $\nu = 2$ and the phase angle of $G(j2)$ is approximately $-162°$. To obtain the phase margin of 50°, we need to add approximately 32° at $\nu = 2$. By using a conventional frequency response design technique, we choose the pole of the controller at 3.27. Thus, the controller transfer function is

$$G_D(w) = \frac{1 + \dfrac{w}{0.997}}{1 + \dfrac{w}{3.27}} \tag{4–64}$$

The magnitude and phase angle curves for $G_D(j\nu)$ and the magnitude and phase angle curves of the compensated open-loop transfer function $G_D(j\nu)G(j\nu)$ are shown by solid curves in Fig. 4–49. From the Bode diagram we see that the phase margin is 50° and the gain margin is 14 db.

The controller transfer function given by Eq. (4–64) will now be transformed back to the z plane by the bilinear transformation given by Eq. (4–62):

$$w = \frac{2}{T}\frac{z-1}{z+1} = \frac{2}{0.2}\frac{z-1}{z+1} = 10\frac{z-1}{z+1}$$

Thus,

$$G_D(z) = \frac{1 + \dfrac{1}{0.997}\left[10\left(\dfrac{z-1}{z+1}\right)\right]}{1 + \dfrac{1}{3.27}\left[10\left(\dfrac{z-1}{z+1}\right)\right]}$$

$$= 2.718\left(\frac{z - 0.8187}{z - 0.5071}\right)$$

The open-loop pulse transfer function of the compensated system is

$$G_D(z)G(z) = \frac{2.718(z - 0.8187)}{z - 0.5071}\,\frac{2 \times 0.01873(z + 0.9356)}{(z-1)(z - 0.8187)}$$

$$= 0.1018\frac{z + 0.9356}{(z-1)(z - 0.5071)}$$

The closed-loop pulse transfer function of the designed system is

$$\frac{C(z)}{R(z)} = \frac{0.1018(z + 0.9356)}{(z-1)(z - 0.5071) + 0.1018(z + 0.9356)}$$

$$= \frac{0.1018(z + 0.9356)}{(z - 0.7026 + j0.3296)(z - 0.7026 - j0.3296)}$$

From this last equation we see that the closed-loop poles are located at

$$z = 0.7026 \pm j0.3296$$

Notice that the closed-loop pulse transfer function involves a zero at $z = -0.9356$. The effect of this zero on the transient and frequency responses is very small, however, since it is located on the negative real axis of the z plane between 0 and -1 and is close to point $z = -1$.

Figure 4–50 shows the root locus plot for the designed system. Notice that the closed-loop poles for the system are located on the $\zeta = 0.5$ locus. Hence, the closed-loop poles of the system have a damping ratio ζ of 0.5.

The designed system satisfies the given specifications that the phase margin be 50°, the gain margin be at least 10 db, and the static velocity error constant K_v be 2 sec^{-1}. Therefore, the system just designed is satisfactory.

Note that for the present system the sampling period T is chosen as 0.2 sec. From Fig. 4–50, it can be seen that the number of samples per cycle of sinusoidal oscillation is $360°/25.15° = 14.3$. This means that the sampling frequency ω_s is 14.3 times the damped natural frequency ω_d. Thus, the sampling period is satisfactory under normal circumstances.

Comments. The advantage of the w transform method is that the conventional frequency response method using Bode diagrams can be used for the design of discrete-time control systems. In applying this method, one must carefully choose a reasonable sampling frequency. (A rule of thumb is to choose a sampling frequency

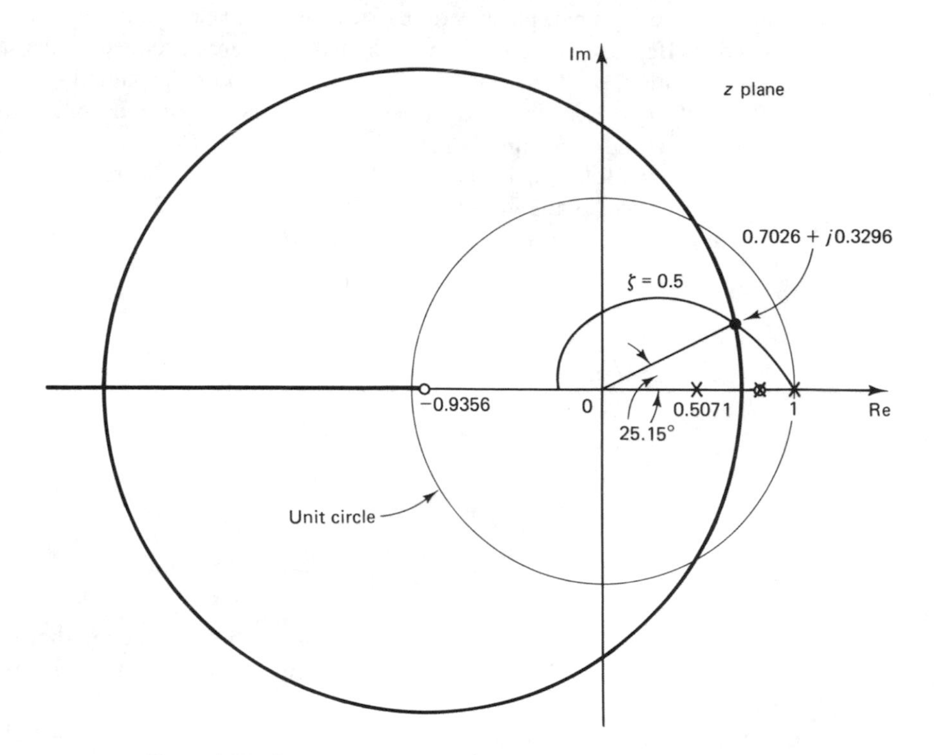

Figure 4–50 Root locus plot for the system designed in Example 4–13.

10 times the closed-loop bandwidth.) Before we conclude this section, we summarize the important facts about design in the w plane.

1. The magnitude and phase angle of $G(j\nu)$ are the magnitude and phase angle of $G(z)$ as z moves on the unit circle from $z = 1$ to $z = -1$. Since $z = e^{j\omega T}$, the ω value varies from 0 to $\frac{1}{2}\omega_s$. The fictitious frequency ν varies from 0 to ∞, since $\nu = (2/T)\tan(\omega T/2)$. Thus, the frequency response of the digital control system for $0 \le \omega \le \frac{1}{2}\omega_s$ is similar to the frequency response of the corresponding analog control system for $0 \le \nu \le \infty$.
2. Since $G(j\nu)$ is a rational function of ν, it is basically the same as $G(j\omega)$. In determining possible unstable zeros of the characteristic equation, the Nyquist stability criterion can be applied. Therefore, both the conventional straight-line approximation to the magnitude curve in the Bode diagram and the concept of phase margin and gain margin apply to $G(j\nu)$.
3. Compare transfer functions $G(w)$ and $G(s)$. As we mentioned earlier, because of the presence of the scale factor $2/T$ in the w transformation, the corresponding static error constants for $G(w)$ and $G(s)$ become identical. (Without the scale factor $2/T$, this will not be true.)
4. The w transformation may generate one or more right half plane zeros in $G(w)$. If one or more right half plane zeros exist, then $G(w)$ is a nonminimum phase transfer function. Because the zeros in the right half plane are generated by the sample-and-hold operation, the locations of these zeros depend on the sampling period T. The effects of these zeros in the right half plane on the response become smaller as the sampling period T becomes smaller.

   In what follows, let us consider the effects on the response of the zero in the right half plane at $w = 2/T$. The zero at $w = 2/T$ causes distortion in the frequency response as ν approaches $2/T$. Since

$$\nu = \frac{2}{T}\tan\frac{\omega T}{2}$$

then as ν approaches $2/T$, $\tan(\omega T/2)$ approaches 1, or

$$\frac{\omega T}{2} = \frac{\pi}{4}$$

and thus

$$\omega = \frac{\pi}{2T}$$

As we stated earlier, $\omega = \frac{1}{2}\omega_s = \pi/T$ is the highest frequency that we consider in the response of the discrete-time or digital control system. Therefore, $\omega = \omega_s/4 = \pi/2T$, which is one-half the highest frequency considered, is well within the frequency range of interest. Thus the zero at $w = 2/T$, which is in the right half of the w plane, will seriously affect the response.

5. It should be noted that the Bode diagram method in the w plane is frequently used in practice and many successful digital control systems have been designed by this approach.

4-7 ANALYTICAL DESIGN METHOD

In analog control systems, PID controllers, among others, have long been used to obtain generally satisfactory response characteristics. The adjustments in PID control actions are limited to proportional gain, integral gain, and derivative gain. In digital control systems, the algorithms for systems control are not limited to any particular control schemes such as PID controls and, in fact, digital controllers can produce an infinite variety of control actions.

The main reason why the control actions of analog controllers are limited is that there are physical limitations in pneumatic, hydraulic, and electronic components. Such limitations may be completely ignored in designing digital controllers. Thus, many control schemes that have been impossible with analog controls are possible with digital controls. In fact, optimal control schemes that are not possible with analog controllers are made possible by digital control schemes.

In this section we specifically present an analytical design method for digital controllers that will force the error sequence, when subjected to a specific type of time-domain input, to become zero after a finite number of sampling periods and, in fact, to become zero and stay zero after the minimum possible number of sampling periods.

If the response of a closed-loop control system to a step input exhibits the minimum possible settling time (that is, the output reaches the final value in the minimum time and stays there), no steady-state error, and no ripples between the sampling instants, then this type of response is commonly called a *deadbeat response*. The deadbeat response will be discussed in this section. (We shall treat the deadbeat response again in Chap. 6, where we discuss the pole placement technique and the design of state observers.)

The discussions that follow are limited to the determination of the control algorithms or pulse transfer functions of digital controllers for single-input–single-output systems, given desired optimal response characteristics. For optimal control of multiple-input–multiple-output systems, see Chap. 7, where the state space approach is used.

Design of digital controllers for minimum settling time with zero steady-state error. Consider the digital control system shown in Fig. 4–51(a). The error signal $e(t)$, which is the difference between the input $r(t)$ and the output $c(t)$, is sampled every time interval T. The input to the digital controller is the error signal $e(kT)$. The output of the digital controller is the control signal $u(kT)$. The control signal $u(kT)$ is fed to the zero-order hold, and the output of the hold, $u(t)$, which is a piecewise continuous-time signal, is fed to the plant. (Although the

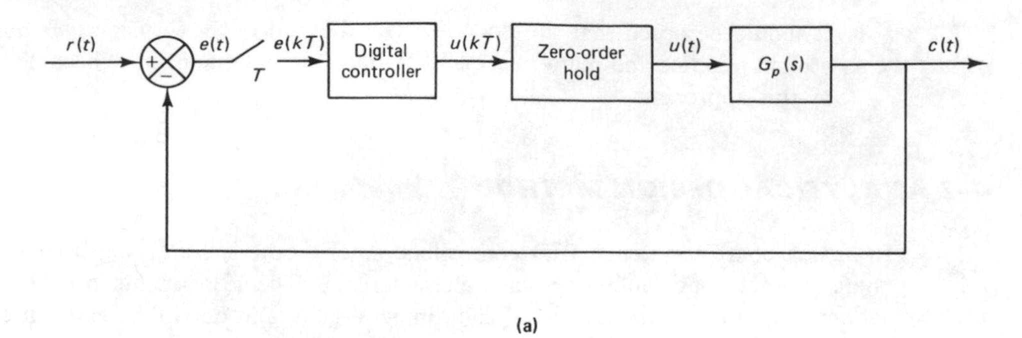

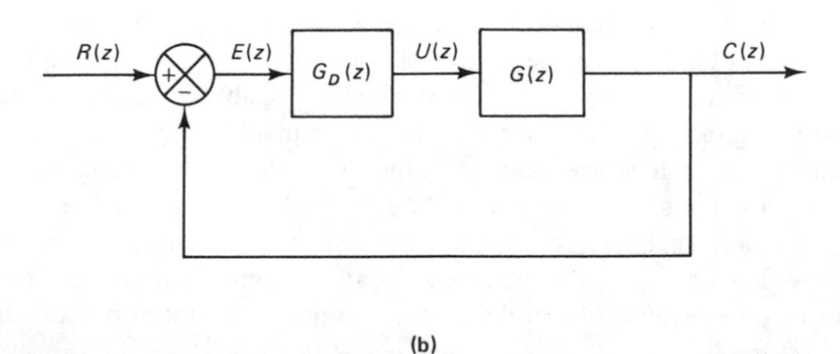

Figure 4–51 (a) A digital control system; (b) diagram showing equivalent digital control system.

sampler at the input of the zero-order hold is not shown, the signal $u(kT)$ is first sampled and fed to the zero-order hold. As mentioned earlier, the zero-order hold shown in the diagram is a sample-and-hold device.) It is desired to design a digital controller $G_D(z)$ such that the closed-loop control system will exhibit the minimum possible settling time with zero steady-state error in response to a step, a ramp, or an acceleration input. It is required that the output not exhibit intersampling ripples after the steady state is reached. The system must satisfy any other specifications, if required, such as a specification for the static velocity error constant.

Let us define the z transform of the plant which is preceded by the zero-order hold as $G(z)$, or

$$G(z) = \mathscr{Z}\left[\frac{1 - e^{-Ts}}{s} G_p(s)\right]$$

Then, the open-loop pulse transfer function becomes $G_D(z)G(z)$, as shown in Fig. 4–51(b). Next, define the desired closed-loop pulse transfer function as $F(z)$:

$$\frac{C(z)}{R(z)} = \frac{G_D(z)G(z)}{1 + G_D(z)G(z)} = F(z) \tag{4–65}$$

Since it is required that the system exhibit a finite settling time with zero steady-state error, the system must exhibit a finite impulse response. Hence, the desired closed-loop pulse transfer function must be of the following form:

$$F(z) = \frac{a_0 z^N + a_1 z^{N-1} + \cdots + a_N}{z^N}$$

or

$$F(z) = a_0 + a_1 z^{-1} + \cdots + a_N z^{-N} \tag{4–66}$$

where $N \geq n$ and n is the order of the system. [Note that $F(z)$ must not contain any terms with positive powers in z, since such terms in the series expansion of $F(z)$ imply that the output precedes the input, which is not possible for a physically realizable system.] In our design approach, we solve the closed-loop pulse transfer function for the digital controller pulse transfer function $G_D(z)$. That is, we find the pulse transfer function $G_D(z)$ that will satisfy Eq. (4–65). Solving Eq. (4–65) for $G_D(z)$, we obtain

$$G_D(z) = \frac{F(z)}{G(z)[1 - F(z)]} \tag{4–67}$$

The designed system must be physically realizable. The conditions for physical realizability place certain constraints on the closed-loop pulse transfer function $F(z)$ and the digital controller pulse transfer function $G_D(z)$. The conditions for physical realizability may be stated as follows:

1. The order of the numerator of $G_D(z)$ must be equal to or lower than the order of the denominator. (Otherwise, the controller requires future input data to produce the current output.)
2. If the plant $G_p(s)$ involves a transportation lag e^{-Ls}, then the designed closed-loop system must involve at least the same magnitude of the transportation lag. (Otherwise, the closed-loop system would have to respond before an input was given, which is impossible for a physically realizable system.)
3. If $G(z)$ is expanded into a series in z^{-1}, the lowest-power term of the series expansion of $F(z)$ in z^{-1} must be at least as large as that of $G(z)$. For example, if an expansion of $G(z)$ into a series in z^{-1} begins with the z^{-1} term, then the first term of $F(z)$ given by Eq. (4–66) must be zero, or a_0 must equal 0; that is, the expansion has to be of the form

$$F(z) = a_1 z^{-1} + a_2 z^{-2} + \cdots + a_N z^{-N}$$

where $N \geq n$ and n is the order of the system. This means that the plant cannot respond instantaneously when a control signal of finite magnitude is applied: the response comes at a delay of at least one sampling period if the series expansion of $G(z)$ begins with a term in z^{-1}.

In addition to the physical realizability conditions, we must pay attention to the stability aspects of the system. Specifically, we must avoid canceling an unstable pole of the plant by a zero of the digital controller. If such a cancellation is attempted, any error in the pole-zero cancellation will diverge as time elapses and the system will become unstable. Similarly, the digital controller pulse transfer function should not involve unstable poles to cancel plant zeros which lie outside the unit circle.

Next, let us investigate what will happen to the closed-loop pulse transfer function $F(z)$ if $G(z)$ involves an unstable (or critically stable) pole, that is, a pole $z = \alpha$ outside (or on) the unit circle. [Note that the following argument applies equally, if $G(z)$ involves two or more unstable—or critically stable—poles.] Let us define

$$G(z) = \frac{G_1(z)}{z - \alpha}$$

where $G_1(z)$ does not include a term that cancels with $z = \alpha$. Then the closed-loop pulse transfer function becomes

$$\frac{C(z)}{R(z)} = \frac{G_D(z)G(z)}{1 + G_D(z)G(z)} = \frac{G_D(z)\dfrac{G_1(z)}{z - \alpha}}{1 + G_D(z)\dfrac{G_1(z)}{z - \alpha}} = F(z) \tag{4–68}$$

Since we require that no zero of $G_D(z)$ cancel the unstable pole of $G(z)$ at $z = \alpha$, we must have

$$1 - F(z) = \frac{1}{1 + G_D(z)\dfrac{G_1(z)}{z - \alpha}} = \frac{z - \alpha}{z - \alpha + G_D(z)G_1(z)}$$

that is, $1 - F(z)$ must have $z = \alpha$ as a zero. Also, notice that from Eq. (4–68) if zeros of $G(z)$ do not cancel poles of $G_D(z)$, the zeros of $G(z)$ become zeros of $F(z)$. [$F(z)$ may involve additional zeros.]

Let us summarize what we have stated concerning stability.

1. Since the digital controller $G_D(z)$ should not cancel unstable (or critically stable) poles of $G(z)$, all unstable (or critically stable) poles of $G(z)$ must be included in $1 - F(z)$ as zeros.
2. Zeros of $G(z)$ that lie inside the unit circle may be canceled with poles of $G_D(z)$. However, zeros of $G(z)$ that lie on or outside the unit circle must not be canceled with poles of $G_D(z)$. Hence, all zeros of $G(z)$ that lie on or outside the unit circle must be included in $F(z)$ as zeros.

Now, we shall proceed with the design. Since $e(kT) = r(kT) - c(kT)$, referring to Eq. (4–65) we have

$$E(z) = R(z) - C(z) = R(z)[1 - F(z)] \tag{4–69}$$

Note that for a unit step input $r(t) = 1(t)$,

$$R(z) = \frac{1}{1 - z^{-1}}$$

For a unit ramp input $r(t) = t\,1(t)$,

$$R(z) = \frac{Tz^{-1}}{(1 - z^{-1})^2}$$

And for a unit acceleration input $r(t) = \frac{1}{2}t^2 1(t)$,

$$R(z) = \frac{T^2 z^{-1}(1 + z^{-1})}{2(1 - z^{-1})^3}$$

Thus, in general, z transforms of such time-domain polynomial inputs may be written as

$$R(z) = \frac{P(z)}{(1 - z^{-1})^{q+1}} \tag{4–70}$$

where $P(z)$ is a polynomial in z^{-1}. Notice that for a unit step input, $P(z) = 1$ and $q = 0$; for a unit ramp input, $P(z) = Tz^{-1}$ and $q = 1$; and for a unit acceleration input, $P(z) = \frac{1}{2}T^2 z^{-1}(1 + z^{-1})$ and $q = 2$.

By substituting Eq. (4–70) into Eq. (4–69) we obtain

$$E(z) = \frac{P(z)[1 - F(z)]}{(1 - z^{-1})^{q+1}} \tag{4–71}$$

To ensure that the system reaches steady state in a finite number of sampling periods and maintains zero steady-state error, $E(z)$ must be a polynomial in z^{-1} with a finite number of terms. Then, by referring to Eq. (4–71), we choose the function $1 - F(z)$ to be of the form

$$1 - F(z) = (1 - z^{-1})^{q+1} N(z) \tag{4–72}$$

where $N(z)$ is a polynomial in z^{-1} with a finite number of terms. Then

$$E(z) = P(z)N(z) \tag{4–73}$$

which is a polynomial in z^{-1} with a finite number of terms. This means that the error signal becomes zero in a finite number of sampling periods.

From the preceding analysis, the pulse transfer function of the digital controller can be determined as follows. By first letting $F(z)$ satisfy the physical realizability and stability conditions and then substituting Eq. (4–72) into Eq. (4–67) we obtain

$$G_D(z) = \frac{F(z)}{G(z)(1 - z^{-1})^{q+1} N(z)} \tag{4–74}$$

Equation (4–74) gives the pulse transfer function of the digital controller that will produce zero steady-state error after a finite number of sampling periods.

For a stable plant $G_p(s)$, the condition that the output not exhibit intersampling ripples after the settling time is reached may be written as follows:

$$c(t \geq nT) = \text{constant} \qquad \text{for step inputs}$$

$$\dot{c}(t \geq nT) = \text{constant} \qquad \text{for ramp inputs}$$

$$\ddot{c}(t \geq nT) = \text{constant} \qquad \text{for acceleration inputs}$$

The applicable condition must be satisfied when the system is designed. In designing the system, the condition on $c(t)$, $\dot{c}(t)$, or $\ddot{c}(t)$ must be interpreted in terms of $u(t)$. Note that the plant is continuous-time and the input to the plant is $u(t)$, a continuous-time function; therefore in order to have no ripples in the output $c(t)$, the control signal $u(t)$ at steady state must be either constant or monotonically increasing (or monotonically decreasing) for step, ramp, or acceleration inputs.

Comments.

1. Since the closed-loop pulse transfer function $F(z)$ is a polynomial in z^{-1}, all the closed-loop poles are at the origin or at $z = 0$. The multiple closed-loop pole at the origin is very sensitive to system parameter variations.
2. Although a digital control system designed to exhibit minimum settling time with zero steady-state error in response to a specific type of input has excellent transient response characteristics for the input it is designed for, it may exhibit inferior or sometimes unacceptable transient response characteristics for other types of input. (This is always true in optimal control systems. An optimal control system will exhibit the best response characteristics for the type of input it is designed for, but will not exhibit optimal response characteristics for other types of input.)
3. In the case in which an analog controller is discretized, an increase in the sampling period changes the system dynamics and may lead to system instability. On the other hand, the behavior of the digital control system we are designing in this section does not depend on the choice of the sampling period. Since the inputs $r(t)$ considered here are time-domain inputs (such as step inputs, ramp inputs, and acceleration inputs), the sampling period T can be chosen arbitrarily. For a smaller sampling period, the response time (which is an integral multiple of the sampling period T) becomes smaller. However, for a very small sampling period T, the magnitude of the control signal will become excessively large, with the result that saturation phenomena will take place in the system and the design method presented in this section will no longer apply. Hence, the sampling period T should not be too small. On the other hand, if the sampling period T is chosen too large, the system may behave unsatisfactorily or may even become unstable when it is subjected to sufficiently time-varying inputs (such as frequency-domain inputs). Thus a compromise is necessary. A

rule of thumb is to choose the smallest sampling period T such that no saturation phenomena occur in the control signal.

Example 4–14.

Consider the digital control system shown in Fig. 4–51(a), where the plant transfer function $G_p(s)$ is given by

$$G_p(s) = \frac{1}{s(s+1)}$$

Design a digital controller $G_D(z)$ such that the closed-loop system will exhibit a deadbeat response to a unit step input. (In a deadbeat response the system should not exhibit intersampling ripples in the output after the settling time is reached.) The sampling period T is assumed to be 1 sec. Then, using the digital controller $G_D(z)$ so designed, investigate the response of this system to a unit ramp input.

The first step in the design is to determine the z transform of the plant which is preceded by the zero-order hold:

$$\begin{aligned} G(z) &= \mathscr{Z}\left[\frac{1-e^{-Ts}}{s}\frac{1}{s(s+1)}\right] \\ &= (1-z^{-1})\,\mathscr{Z}\left[\frac{1}{s^2(s+1)}\right] \\ &= (1-z^{-1})\left[\frac{z^{-1}}{(1-z^{-1})^2} - \frac{1}{1-z^{-1}} + \frac{1}{1-0.3679z^{-1}}\right] \\ &= \frac{0.3679(1+0.7181z^{-1})z^{-1}}{(1-z^{-1})(1-0.3679z^{-1})} \end{aligned} \tag{4–75}$$

Now redraw the block diagram of the system as shown in Fig. 4–51(b). Define the closed-loop pulse transfer function as $F(z)$, or

$$\frac{C(z)}{R(z)} = \frac{G_D(z)G(z)}{1+G_D(z)G(z)} = F(z)$$

Notice that if $G(z)$ is expanded into a series in z^{-1}, then the first term will be $0.3679z^{-1}$. Hence $F(z)$ must begin with a term in z^{-1}.

Referring to Eq. (4–66) and noting that the system is of the second order ($n = 2$), we assume $F(z)$ to be of the following form:

$$F(z) = a_1 z^{-1} + a_2 z^{-2} \tag{4–76}$$

Since the input is a step function, from Eq. (4–72) we require that

$$1 - F(z) = (1-z^{-1})N(z) \tag{4–77}$$

Since $G(z)$ has a critically stable pole at $z = 1$, the stability requirement states that $1 - F(z)$ must have a zero at $z = 1$. However, the function $1 - F(z)$ already has a term $1 - z^{-1}$ and therefore satisfies the requirement.

Since the system should not exhibit intersampling ripples and the input is a step function, we require $c(t \geq 2T)$ to be constant. Noting that $u(t)$, the output of the zero-order hold, is a continuous-time function, a constant $c(t \geq 2T)$ requires that $u(t)$ also be constant for $t \geq 2T$. In terms of the z transform, $U(z)$ must be of the following type of series in z^{-1}:

$$U(z) = b_0 + b_1 z^{-1} + b(z^{-2} + z^{-3} + z^{-4} + \cdots)$$

where b is a constant. Because the plant transfer function $G_p(s)$ involves an integrator, b must be zero. (Otherwise the output cannot stay constant.) Consequently, we have

$$U(z) = b_0 + b_1 z^{-1}$$

From Fig. 4–51(b), $U(z)$ can be given as follows:

$$\begin{aligned} U(z) &= \frac{C(z)}{G(z)} = \frac{C(z)}{R(z)} \frac{R(z)}{G(z)} = F(z) \frac{R(z)}{G(z)} \\ &= F(z) \frac{1}{1 - z^{-1}} \frac{(1 - z^{-1})(1 - 0.3679z^{-1})}{0.3679(1 + 0.7181z^{-1})z^{-1}} \\ &= F(z) \frac{1 - 0.3679z^{-1}}{0.3679(1 + 0.7181z^{-1})z^{-1}} \end{aligned}$$

In order for $U(z)$ to be a series in z^{-1} with only two terms, $F(z)$ must be of the following form:

$$F(z) = (1 + 0.7181z^{-1})z^{-1}F_1 \tag{4–78}$$

where F_1 is a constant. Then, $U(z)$ can be written as follows:

$$U(z) = 2.7181(1 - 0.3679z^{-1})F_1 \tag{4–79}$$

Equation (4–79) gives $U(z)$ in terms of F_1. Once constant F_1 is determined, $U(z)$ can be given as a series in z^{-1} with only two terms.

Now we shall determine $N(z)$, $F(z)$, and F_1. By substituting Eq. (4–76) into Eq. (4–77), we obtain

$$1 - a_1 z^{-1} - a_2 z^{-2} = (1 - z^{-1})N(z)$$

The left-hand side of this last equation must be divisible by $1 - z^{-1}$. If we divide the left-hand side by $1 - z^{-1}$, the quotient is $1 + (1 - a_1)z^{-1}$ and the remainder is $(1 - a_1 - a_2)z^{-2}$. Hence, $N(z)$ is determined as

$$N(z) = 1 + (1 - a_1)z^{-1} \tag{4–80}$$

and the remainder must be zero. This requires that

$$1 - a_1 - a_2 = 0 \tag{4–81}$$

Also, from Eqs. (4–76) and (4–78) we have

$$F(z) = a_1 z^{-1} + a_2 z^{-2} = (1 + 0.7181z^{-1})z^{-1}F_1$$

Hence

$$a_1 + a_2 z^{-1} = (1 + 0.7181z^{-1})F_1$$

Division of the left-hand side of this last equation by $1 + 0.7181z^{-1}$ yields the quotient a_1 and the remainder $(a_2 - 0.7181a_1)z^{-1}$. By equating the quotient with F_1 and the remainder with zero, we obtain

$$F_1 = a_1$$

and

$$a_2 - 0.7181a_1 = 0 \tag{4-82}$$

Solving Eqs. (4–81) and (4–82) for a_1 and a_2 gives

$$a_1 = 0.5820, \qquad a_2 = 0.4180$$

Thus, $F(z)$ is determined as

$$F(z) = 0.5820z^{-1} + 0.4180z^{-2} \tag{4-83}$$

and

$$F_1 = 0.5820$$

Equation (4–80) gives

$$N(z) = 1 + 0.4180z^{-1} \tag{4-84}$$

The digital controller pulse transfer function $G_D(z)$ is then determined from Eq. (4–74), as follows. By referring to Eqs. (4–75), (4–83), and (4–84),

$$\begin{aligned} G_D(z) &= \frac{F(z)}{G(z)(1 - z^{-1})N(z)} \\ &= \frac{(1 + 0.7181z^{-1})z^{-1}(0.5820)}{\dfrac{0.3679(1 + 0.7181z^{-1})z^{-1}}{(1 - z^{-1})(1 - 0.3679z^{-1})}(1 - z^{-1})(1 + 0.4180z^{-1})} \\ &= \frac{1.5820 - 0.5820z^{-1}}{1 + 0.4180z^{-1}} \end{aligned}$$

With the digital controller thus designed, the system output in response to a unit step input $r(t) = 1$ can be obtained as follows:

$$\begin{aligned} C(z) &= F(z)R(z) \\ &= (0.5820z^{-1} + 0.4180z^{-2})\frac{1}{1 - z^{-1}} \\ &= 0.5820z^{-1} + z^{-2} + z^{-3} + z^{-4} + \cdots \end{aligned}$$

Hence

$$c(0) = 0$$

$$c(1) = 0.5820$$

$$c(k) = 1 \qquad k = 2, 3, 4, \cdots$$

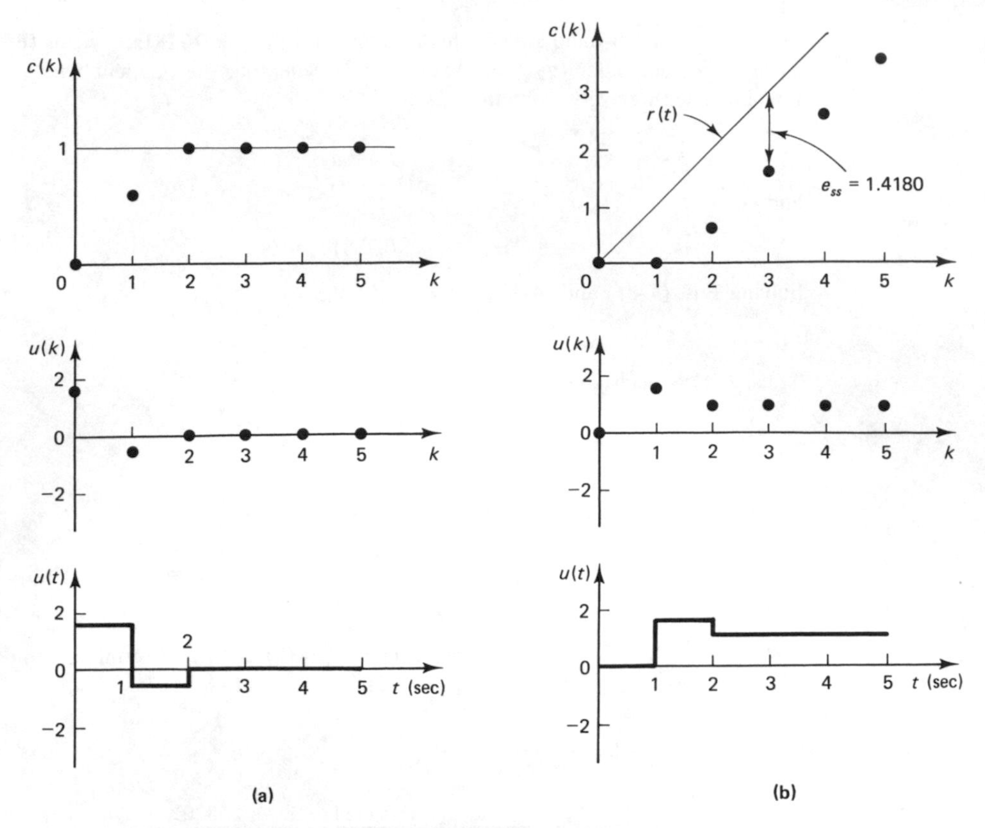

Figure 4–52 Responses of the system designed in Example 4–14. (a) Plots of $c(k)$ vs. k, $u(k)$ vs. k, and $u(t)$ vs. t in the unit step response; (b) plots of $c(k)$ vs. k, $u(k)$ vs. k, and $u(t)$ vs. t in the unit ramp response.

Notice that substitution of 0.5820 for F_1 in Eq. (4–79) yields

$$U(z) = 2.7181(1 - 0.3679z^{-1})(0.5820)$$
$$= 1.5820 - 0.5820z^{-1}$$

Thus, the control signal $u(k)$ becomes zero for $k \geq 2$, as required. There is no intersampling ripple in the output after the settling time is reached. Figure 4–52(a) shows plots of $c(k)$ versus k, $u(k)$ versus k, and $u(t)$ versus t in the unit step response.

Next, let us investigate the response of this system to a unit ramp input:

$$C(z) = F(z)R(z)$$
$$= (0.5820z^{-1} + 0.4180z^{-2})\frac{z^{-1}}{(1 - z^{-1})^2}$$
$$= 0.5820z^{-2} + 1.5820z^{-3} + 2.5820z^{-4} + 3.5820z^{-5} + \cdots$$

For the unit ramp response, the control signal $U(z)$ is obtained as follows. Referring to Eqs. (4–75) and (4–83),

$$U(z)=\frac{C(z)}{G(z)}=\frac{F(z)}{G(z)}R(z)=\frac{F(z)}{G(z)}\frac{z^{-1}}{(1-z^{-1})^2}$$

$$=(1.5820-0.5820z^{-1})\frac{z^{-1}}{1-z^{-1}}$$

$$=1.5820z^{-1}+z^{-2}+z^{-3}+z^{-4}+\cdots$$

The signal $u(k)$ becomes constant ($b = 1$) for $k \geq 2$. Hence, the system output will not exhibit intersampling ripples. Figure 4–52(b) shows plots of $c(k)$ versus k, $u(k)$ versus k, and $u(t)$ versus t in the unit ramp response.

Note that the static velocity error constant K_v for the present system is

$$K_v=\lim_{z\to 1}\left[\frac{1-z^{-1}}{T}G_D(z)G(z)\right]$$

$$=\lim_{z\to 1}\left[(1-z^{-1})\frac{F(z)}{(1-z^{-1})N(z)}\right]$$

$$=\lim_{z\to 1}\frac{0.5820z^{-1}+0.4180z^{-2}}{1+0.4180z^{-1}}=0.7052$$

Thus, the steady-state error in the unit ramp response is

$$e_{ss}^{*}=\frac{1}{K_v}=1.4180$$

which is indicated in Fig. 4–52(b).

In the present design problem, we have required that in response to a step input the system exhibit the minimum settling time with no steady-state error and no ripples in the output after the settling time is reached. If one or more additional constraints are present in the design problem (for example, if the value of the static velocity error constant K_v is arbitrarily specified), then the number of sampling periods required before reaching the steady state must be increased. For example, the second-order system may require three or more sampling periods before the steady state is reached, depending on the additional constraints imposed. See Example 4–15, which follows.

Example 4–15.

Consider a design problem the same as that of Example 4–14 except that the static velocity error constant K_v is specified. (Because of this additional constraint, the settling time will be longer than 2 sec.) The block diagram of the digital control system is shown in Fig. 4–51(a). The plant transfer function $G_p(s)$ under consideration is

$$G_p(s)=\frac{1}{s(s+1)}$$

The design specifications are (1) that the closed-loop system is to exhibit a finite settling time with zero steady-state error in the unit step response, (2) that the output is not to exhibit intersampling ripples after the settling time is reached, (3) that the static velocity error constant K_v is to be 1 sec^{-1}, and (4) that the settling time is to be the minimum possible that will satisfy all these specifications. The sampling period T is assumed to be 1 sec. Design a digital controller $G_D(z)$ that satisfies the given specifications. After the controller is designed, investigate the response of the system to a unit ramp input.

The z transform of the plant which is preceded by the zero-order hold was obtained in Example 4–14 as

$$G(z) = \mathscr{Z}\left[\frac{1-e^{-Ts}}{s}\frac{1}{s(s+1)}\right]$$

$$= \frac{0.3679(1+0.7181z^{-1})z^{-1}}{(1-z^{-1})(1-0.3679z^{-1})}$$

Define the closed-loop pulse transfer function as $F(z)$:

$$\frac{C(z)}{R(z)} = \frac{G_D(z)G(z)}{1+G_D(z)G(z)} = F(z)$$

Since the first term in the expansion of $G(z)$ is $0.3679z^{-1}$, $F(z)$ must begin with a term in z^{-1}:

$$F(z) = a_1 z^{-1} + a_2 z^{-2} + \cdots + a_N z^{-N}$$

where $N \geq n$ and n is the order of the system (that is, $n = 2$ in the present case). Because of the added constraint, we may assume $N > 2$. We shall try $N = 3$. Thus, we assume

$$F(z) = a_1 z^{-1} + a_2 z^{-2} + a_3 z^{-3} \tag{4–85}$$

(If a satisfactory result is not obtained, we must assume $N > 3$.) Since the input is a step function, from Eq. (4–72) we require that

$$1 - F(z) = (1 - z^{-1})N(z) \tag{4–86}$$

Note that the presence of a critically stable pole at $z = 1$ in the plant pulse transfer function $G(z)$ requires $1 - F(z)$ to have a zero at $z = 1$. However, the function $1 - F(z)$ already has a term $1 - z^{-1}$ and therefore satisfies the stability requirement.

The requirement that the static velocity error constant be 1 sec^{-1} can be written as follows:

$$K_v = \lim_{z\to 1}\left[\frac{1-z^{-1}}{T} G_D(z)G(z)\right]$$

$$= \lim_{z\to 1}\left[(1-z^{-1})\frac{F(z)}{(1-z^{-1})N(z)}\right]$$

$$= \frac{F(1)}{N(1)} = 1$$

Notice that from Eq. (4–86) we have $F(1) = 1$. Hence, K_v can be written as follows:

$$K_v = \frac{1}{N(1)} = 1 \tag{4-87}$$

Since the system output should not exhibit intersampling ripples after the settling time is reached, we require $U(z)$ to be of the following form:

$$U(z) = b_0 + b_1 z^{-1} + b_2 z^{-2} + b(z^{-3} + z^{-4} + z^{-5} + \cdots)$$

Because the plant transfer function $G_p(s)$ involves an integrator, b must be zero. Consequently, we have

$$U(z) = b_0 + b_1 z^{-1} + b_2 z^{-2}$$

Also, from Fig. 4–51(b), $U(z)$ can be given by

$$U(z) = \frac{C(z)}{G(z)} = \frac{C(z)}{R(z)}\frac{R(z)}{G(z)} = F(z)\frac{R(z)}{G(z)}$$

$$= F(z)\frac{1 - 0.3679z^{-1}}{0.3679(1 + 0.7181z^{-1})z^{-1}}$$

In order for $U(z)$ to be a series in z^{-1} with three terms, $F(z)$ must be of the following form:

$$F(z) = (1 + 0.7181z^{-1})z^{-1}F_1(z) \tag{4-88}$$

where $F_1(z)$ is a first-degree polynomial in z^{-1}. Then, $U(z)$ can be written as follows:

$$U(z) = 2.7181(1 - 0.3679z^{-1})F_1(z) \tag{4-89}$$

From Eqs. (4–85) and (4–86), we have

$$1 - F(z) = 1 - a_1 z^{-1} - a_2 z^{-2} - a_3 z^{-3} = (1 - z^{-1})N(z)$$

If we divide $1 - a_1 z^{-1} - a_2 z^{-2} - a_3 z^{-3}$ by $1 - z^{-1}$, the quotient is $1 + (1 - a_1)z^{-1} + (1 - a_1 - a_2)z^{-2}$ and the remainder is $(1 - a_1 - a_2 - a_3)z^{-3}$. Hence, $N(z)$ is determined as

$$N(z) = 1 + (1 - a_1)z^{-1} + (1 - a_1 - a_2)z^{-2} \tag{4-90}$$

and the remainder must be zero, so that

$$1 - a_1 - a_2 - a_3 = 0 \tag{4-91}$$

Note that from Eq. (4–87) we require $N(1) = 1$. Therefore, by substituting $z^{-1} = 1$ into Eq. (4–90) we obtain

$$2a_1 + a_2 = 2 \tag{4-92}$$

Also, Eq. (4–88) can be rewritten as

$$F(z) = a_1 z^{-1} + a_2 z^{-2} + a_3 z^{-3} = (1 + 0.7181z^{-1})z^{-1}F_1(z)$$

Hence

$$a_1 + a_2 z^{-1} + a_3 z^{-2} = (1 + 0.7181z^{-1})F_1(z)$$

Division of the left-hand side of this last equation by $1 + 0.7181z^{-1}$ yields the quotient $[a_1 + (a_2 - 0.7181a_1)z^{-1}]$ and the remainder $[a_3 - 0.7181(a_2 - 0.7181a_1)]z^{-2}$. By equating the quotient with $F_1(z)$ and the remainder with zero, we obtain

$$F_1(z) = a_1 + (a_2 - 0.7181a_1)z^{-1}$$

and

$$a_3 - 0.7181(a_2 - 0.7181a_1) = 0 \tag{4–93}$$

Solving Eqs. (4–91), (4–92), and (4–93) for a_1, a_2, and a_3 gives

$$a_1 = 0.8253, \qquad a_2 = 0.3494, \qquad a_3 = -0.1747$$

Thus $F(z)$ is determined as

$$F(z) = 0.8253z^{-1} + 0.3494z^{-2} - 0.1747z^{-3}$$

and

$$F_1(z) = 0.8253 - 0.2432z^{-1}$$

Equation (4–90) gives

$$N(z) = 1 + 0.1747z^{-1} - 0.1747z^{-2}$$

The digital controller pulse transfer function $G_D(z)$ is then determined from Eq. (4–74) as follows:

$$\begin{aligned} G_D(z) &= \frac{F(z)}{G(z)(1 - z^{-1})N(z)} \\ &= \frac{(1 + 0.7181z^{-1})z^{-1}(0.8253 - 0.2432z^{-1})}{\dfrac{0.3679(1 + 0.7181z^{-1})z^{-1}}{(1 - z^{-1})(1 - 0.3679z^{-1})}(1 - z^{-1})(1 + 0.1747z^{-1} - 0.1747z^{-2})} \\ &= (2.2433)\frac{(1 - 0.2947z^{-1})(1 - 0.3679z^{-1})}{(1 + 0.5144z^{-1})(1 - 0.3397z^{-1})} \end{aligned}$$

With the digital controller thus designed, the system output in response to a unit step input $r(t) = 1$ is obtained as follows:

$$\begin{aligned} C(z) &= F(z)R(z) \\ &= (0.8253z^{-1} + 0.3494z^{-2} - 0.1747z^{-3})\frac{1}{1 - z^{-1}} \\ &= 0.8253z^{-1} + 1.1747z^{-2} + z^{-3} + z^{-4} + \cdots \end{aligned}$$

Hence

$$\begin{aligned} c(0) &= 0 \\ c(1) &= 0.8253 \\ c(2) &= 1.1747 \\ c(k) &= 1 \qquad k = 3, 4, 5, \ldots \end{aligned}$$

The unit step response sequence has a maximum overshoot of approximately 17.5 percent. The settling time is 3 sec.

Notice that from Eq. (4–89) we have

$$U(z) = 2.7181(1 - 0.3679z^{-1})(0.8253 - 0.2432z^{-1})$$

$$= 2.2432 - 1.4863z^{-1} + 0.2432z^{-2}$$

Thus the control signal $u(k)$ becomes zero for $k \geq 3$. Consequently, there are no intersampling ripples in the response. Figure 4–53(a) shows plots of $c(k)$ versus k, $u(k)$ versus k, and $u(t)$ versus t in the unit step response. Notice that the assumption of $N = 3$, that is, the assumption of $F(z)$ as given by Eq. (4–85) is satisfactory.

Next, let us investigate the response of this system to a unit ramp input:

$$C(z) = F(z)R(z)$$

$$= (0.8253z^{-1} + 0.3494z^{-2} - 0.1747z^{-3})\frac{z^{-1}}{(1 - z^{-1})^2}$$

$$= 0.8253z^{-2} + 2z^{-3} + 3z^{-4} + 4z^{-5} + \cdots$$

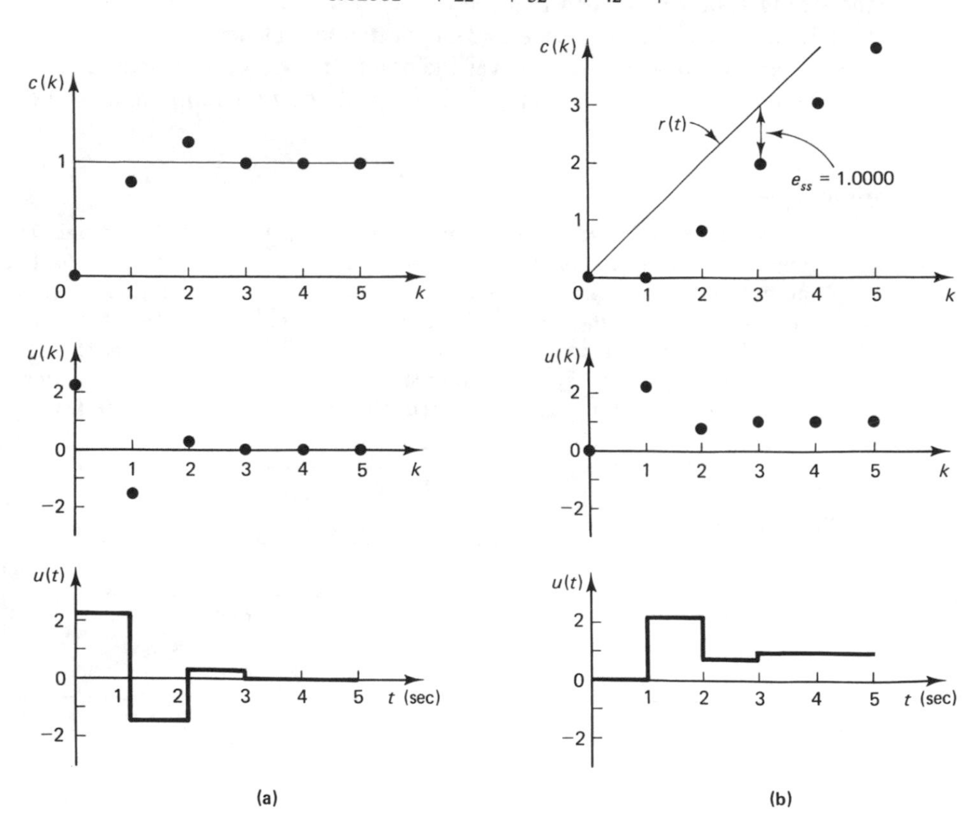

Figure 4–53 Responses of the system designed in Example 4–15. (a) Plots of $c(k)$ vs. k, $u(k)$ vs. k, and $u(t)$ vs. t in the unit step response; (b) plots of $c(k)$ vs. k, $u(k)$ vs. k, and $u(t)$ vs. t in the unit ramp response.

In the unit ramp response, the control signal $U(z)$ is obtained as follows:

$$U(z)=\frac{C(z)}{G(z)}=\frac{F(z)}{G(z)}R(z)=\frac{F(z)}{G(z)}\frac{1}{1-z^{-1}}\frac{z^{-1}}{1-z^{-1}}$$

$$=(2.2432-1.4863z^{-1}+0.2432z^{-2})\frac{z^{-1}}{1-z^{-1}}$$

$$=2.2432z^{-1}+0.7569z^{-2}+z^{-3}+z^{-4}+z^{-5}+\cdots$$

The signal $u(k)$ becomes constant ($b = 1$) for $k \geq 3$. Hence, the system output will not exhibit intersampling ripples. Figure 4–53(b) shows plots of $c(k)$ versus k, $u(k)$ versus k, and $u(t)$ versus t in the unit ramp response. Notice that the steady-state error in the unit ramp response is $e_{ss} = 1/K_v = 1$, as indicated in Fig. 4–53(b).

Comparing the digital control systems designed in Examples 4–14 and 4–15, we note that the latter improves the ramp response characteristics at the expense of the settling time. (The latter system requires one extra sampling period to reach the steady state.) Note also that the former has better step response characteristics, that is, a shorter settling time and no overshoot. Depending on the objectives of the system, we may choose one over the other. If good ramp response characteristics are required, then the system should be designed using the ramp input as the reference input, rather than the step input.

Example 4–16.

Consider the digital control system shown in Fig. 4–54. The plant transfer function involves a transportation lag e^{-5s}. The delay time is 5 sec, or $L = 5$. The desired output $c(t)$ in response to a unit step input is as shown in Fig. 4–55(a). The curve rises from zero to the final value in 10 sec (measured from $t = 5$ to $t = 15$) and there is neither overshoot nor steady-state error. The settling time is 15 sec (measured from $t = 0$ to $t = 15$). It is required that there be no intersampling ripples in the output after the settling time is reached. Design a digital controller $G_D(z)$.

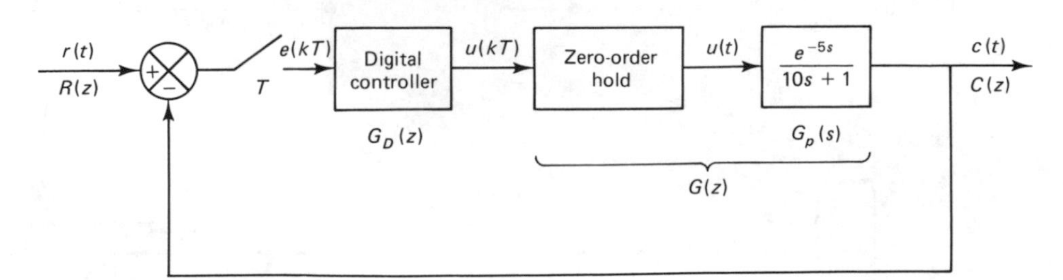

Figure 4–54 Digital control system of Example 4–16.

Let us choose the sampling period to be 5 sec, or $T = 5$ sec. (We may, of course, choose the sampling period to be 2.5 sec, 1 sec, or another value. In this example, however, in order to simplify our presentation, we set the sampling period at 5 sec.) The z transform of the plant which is preceded by the zero-order hold is

$$G(z) = \mathscr{Z}\left[\frac{1 - e^{-Ts}}{s} \frac{e^{-5s}}{10s + 1}\right]$$

$$= (1 - z^{-1})z^{-1} \mathscr{Z}\left[\frac{1}{s(10s + 1)}\right]$$

$$= \frac{0.3935z^{-2}}{1 - 0.6065z^{-1}}$$

Notice that there is no unstable or critically stable pole involved in $G(z)$. Therefore there is no stability problem involved in this case.

Let us define the closed-loop pulse transfer function as $F(z)$:

$$\frac{C(z)}{R(z)} = \frac{G_D(z)G(z)}{1 + G_D(z)G(z)} = F(z) \tag{4–94}$$

In the present case the output $c(t)$ in the unit step response is specified as shown in Fig. 4–55(a). Since $h[1 - e^{-0.1(15-5)}] = h(1 - e^{-1}) = 1$, we have $h = 1.5820$. From the deadbeat response curve shown in Fig. 4–55(a), we obtain

$$c(0) = 0$$

$$c(1) = 0$$

$$c(2) = h(1 - e^{-0.5}) = 1.5820 \times 0.3935 = 0.6225$$

$$c(k) = 1 \qquad k = 3, 4, 5, \ldots$$

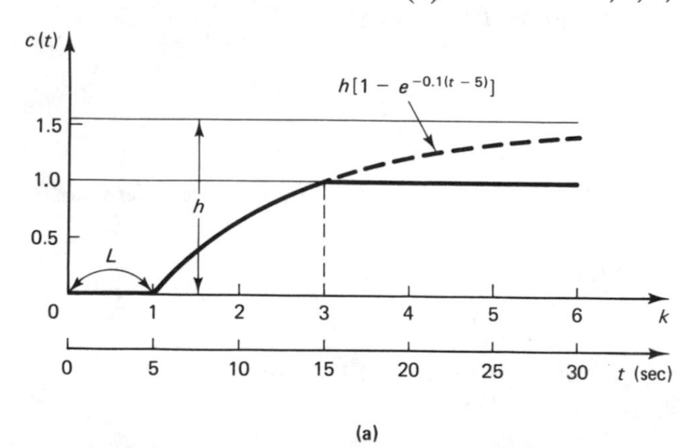

(a)

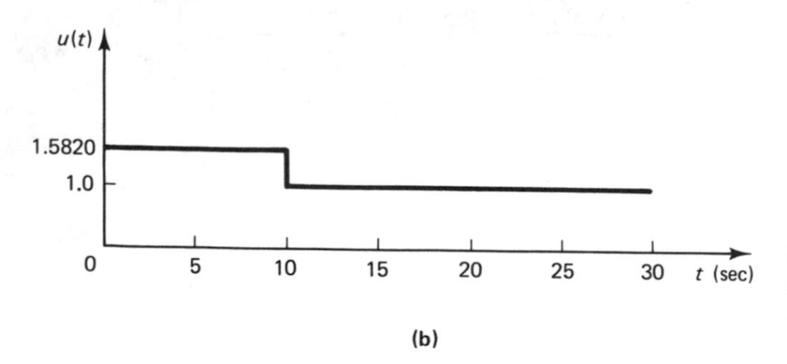

(b)

Figure 4–55 (a) Desired output $c(t)$ in response to a unit step input; (b) plot of $u(t)$ vs. t.

from which we get

$$C(z) = 0.6225z^{-2} + z^{-3} + z^{-4} + z^{-5} + \cdots$$

$$= 0.6225z^{-2} + z^{-3}\frac{1}{1 - z^{-1}}$$

$$= \frac{0.6225z^{-2} + 0.3775z^{-3}}{1 - z^{-1}}$$

Noting that

$$C(z) = F(z)R(z) = F(z)\frac{1}{1 - z^{-1}} = \frac{0.6225z^{-2} + 0.3775z^{-3}}{1 - z^{-1}}$$

we obtain

$$F(z) = 0.6225z^{-2} + 0.3775z^{-3} = 0.6225(1 + 0.6065z^{-1})z^{-2}$$

Once $F(z)$ is determined, the pulse transfer function of the digital controller can be obtained from Eq. (4–94):

$$G_D(z) = \frac{F(z)}{G(z)[1 - F(z)]}$$

Notice that from Eq. (4–72) we have

$$1 - F(z) = (1 - z^{-1})N(z)$$

or

$$1 - 0.6225z^{-2} - 0.3775z^{-3} = (1 - z^{-1})N(z)$$

By dividing $(1 - 0.6225z^{-2} - 0.3775z^{-3})$ by $(1 - z^{-1})$, $N(z)$ can be determined as follows:

$$N(z) = 1 + z^{-1} + 0.3775z^{-2}$$

Consequently,

$$1 - F(z) = (1 - z^{-1})(1 + z^{-1} + 0.3775z^{-2})$$

and

$$G_D(z) = \frac{0.6225(1 + 0.6065z^{-1})z^{-2}}{\dfrac{0.3935z^{-2}}{1 - 0.6065z^{-1}}(1 - z^{-1})(1 + z^{-1} + 0.3775z^{-2})}$$

$$= \frac{1.5820(1 - 0.3678z^{-2})}{(1 - z^{-1})(1 + z^{-1} + 0.3775z^{-2})}$$

This last equation gives the pulse transfer function of the digital controller. Since $c(t)$ must be unity at steady state, $u(t)$, a continuous-time signal, must be constant after the steady state is reached.

Let us determine $U(z)$:

$$U(z) = \frac{C(z)}{G(z)} = \frac{0.6225z^{-2} + 0.3775z^{-3}}{(1 - z^{-1})\dfrac{0.3935z^{-2}}{1 - 0.6065z^{-1}}} = 1.5820\left(\frac{1 - 0.3678z^{-2}}{1 - z^{-1}}\right)$$

$$= 1.5820 + 1.5820z^{-1} + z^{-2} + z^{-3} + z^{-4} + \cdots$$

Taking the inverse z transform of $U(z)$, we find that $u(k)$ is constant for $k \geq 2$. Thus, there are no intersampling ripples in the output after the settling time is reached. The signal $u(t)$ versus t is plotted in Fig. 4–55(b).

For the general case of designing a digital controller where the output curve $c(t)$ is specified for the unit step input and we have many sampling points between the time the response starts and the settling time, see Prob. A-4–17.

Also, for the same plant as that in Example 4–16, if we require only a minimum settling time with no steady-state error and no intersampling ripples [instead of specifying the desired response curve $c(t)$ in the unit step response], then the resulting digital controller will be different from that designed in Example 4–16, although the two response curves will look quite similar. For details, refer to Prob. A-4–15.

REFERENCES

4–1. Evans, W. R., "Control System Synthesis by Root Locus Method," *AIEE Trans. part II*, **69** (1950), pp. 66–69.
4–2. Franklin, G. F., and J. D. Powell, *Digital Control of Dynamic Systems*. Reading, Mass: Addison-Wesley Publishing Co., Inc., 1980.
4–3. Jury, E. I., *Sampled-Data Control Systems*. New York: John Wiley & Sons, Inc., 1958.
4–4. Katz, P., *Digital Control Using Microprocessors*. London: Prentice-Hall International, 1981.
4–5. Kuo, B. C., *Digital Control Systems*. New York: Holt, Rinehart and Winston, Inc., 1980.
4–6. Ogata, K., *Modern Control Engineering*. Englewood Cliffs, N.J.: Prentice-Hall, Inc., 1970.
4–7. Phillips, C. L., and H. T. Nagle, Jr., *Digital Control System Analysis and Design*. Englewood Cliffs, N.J.: Prentice-Hall, Inc., 1984.
4–8. Ragazzini, J. R., and G. F. Franklin, *Sampled-Data Control Systems*. New York: McGraw-Hill Book Company, 1958.
4–9. Tou, J. T., *Digital and Sampled-Data Control Systems*. New York: McGraw-Hill Book Company, 1959.

EXAMPLE PROBLEMS AND SOLUTIONS

Problem A-4–1. Obtain an equivalent digital filter $G_D(z)$ for the following analog filter system by the backward difference method:

$$\ddot{y} + a_1\dot{y} + a_2 y = \dot{x} \tag{4–95}$$

Solution. By the backward difference method, the derivative terms can be written as

$$\frac{dy}{dt} = \frac{y(kT) - y((k-1)T)}{T}$$

$$\frac{d^2y}{dt^2} = \frac{y(kT) - 2y((k-1)T) + y((k-2)T)}{T^2}$$

and

$$\frac{dx}{dt} = \frac{x(kT) - x((k-1)T)}{T}$$

By substituting these three equations into Eq. (4–95), we obtain

$$\frac{y(kT) - 2y((k-1)T) + y((k-2)T)}{T^2} + a_1 \frac{y(kT) - y((k-1)T)}{T} + a_2 y(kT) = \frac{x(kT) - x((k-1)T)}{T}$$

or

$$\left(\frac{1}{T^2} + \frac{a_1}{T} + a_2\right) y(kT) - \left(\frac{2}{T^2} + \frac{a_1}{T}\right) y((k-1)T) + \frac{1}{T^2} y((k-2)T) = \frac{1}{T} x(kT) - \frac{1}{T} x((k-1)T)$$

Taking the z transform of this last equation, we obtain

$$G_D(z) = \frac{Y(z)}{X(z)} = \frac{\frac{1}{T}(1 - z^{-1})}{\left(\frac{1}{T^2} + \frac{a_1}{T} + a_2\right) - \left(\frac{2}{T^2} + \frac{a_1}{T}\right) z^{-1} + \frac{1}{T^2} z^{-2}}$$

$$= \frac{T - Tz^{-1}}{1 + a_1 T + a_2 T^2 - (2 + a_1 T) z^{-1} + z^{-2}}$$

The same result can be obtained much more easily if we first convert Eq. (4–95) into the transfer function

$$G(s) = \frac{Y(s)}{X(s)} = \frac{s}{s^2 + a_1 s + a_2} \tag{4–96}$$

and then use the transformation given by Eq. (4–9),

$$s = \frac{1 - z^{-1}}{T}$$

By substituting this last equation into Eq. (4–96), we obtain

$$G_D(z) = \left.\frac{s}{s^2 + a_1 s + a_2}\right|_{s = (1 - z^{-1})/T}$$

$$= \frac{\dfrac{1-z^{-1}}{T}}{\left(\dfrac{1-z^{-1}}{T}\right)^2 + a_1 \dfrac{1-z^{-1}}{T} + a_2}$$

$$= \frac{T - Tz^{-1}}{1 + a_1 T + a_2 T^2 - (2 + a_1 T)z^{-1} + z^{-2}}$$

which is the result already obtained.

Problem A-4–2. Obtain the discrete-time equivalent (or digital equivalent) of the following continuous-time (or analog) filter $G(s)$ by the bilinear transformation method:

$$G(s) = \frac{2}{(s+1)(s+2)}$$

The sampling period T is assumed to be 0.1 sec.

Solution. From Eq. (4–14), the bilinear transformation is given by

$$s = \frac{2}{T}\frac{1-z^{-1}}{1+z^{-1}} = \frac{2}{0.1}\left(\frac{1-z^{-1}}{1+z^{-1}}\right) = 20\left(\frac{1-z^{-1}}{1+z^{-1}}\right)$$

Hence

$$G_D(z) = \frac{2}{(s+1)(s+2)}\Bigg|_{s=20\left(\frac{1-z^{-1}}{1+z^{-1}}\right)}$$

$$= \frac{2}{\left[20\left(\dfrac{1-z^{-1}}{1+z^{-1}}\right)+1\right]\left[20\left(\dfrac{1-z^{-1}}{1+z^{-1}}\right)+2\right]}$$

$$= \frac{0.004329(1+z^{-1})^2}{(1-0.9048z^{-1})(1-0.8182z^{-1})}$$

Problem A-4–3. Using the bilinear transformation method and the bilinear transformation method with frequency prewarping, design low-pass digital filters that have frequency response characteristics similar to those of the following analog filter:

$$G(s) = \frac{10}{s+10} \tag{4–97}$$

The frequency region of interest is $0 \leq \omega \leq 10$ rad/sec. The sampling period is 0.2 sec, or $T = 0.2$.

Solution. In the bilinear transformation method, the complex variables s and z are related by the equation

$$s = \frac{2}{T}\frac{1-z^{-1}}{1+z^{-1}} \tag{4–98}$$

By substituting Eq. (4–98) into Eq. (4–97), we obtain the following equivalent digital filter $G(z)$:

$$G(z) = \frac{10}{\frac{2}{T}\frac{1-z^{-1}}{1+z^{-1}} + 10}$$

By substituting $z = e^{j\omega_D T}$ into $G(z)$, we obtain

$$G(e^{j\omega_D T}) = \frac{10}{\frac{2}{T}\frac{1-e^{-j\omega_D T}}{1+e^{-j\omega_D T}} + 10}$$

Noting that

$$\frac{2}{T}\frac{1-e^{-j\omega_D T}}{1+e^{-j\omega_D T}} = \frac{2}{T}\frac{e^{j(1/2)\omega_D T} - e^{-j(1/2)\omega_D T}}{e^{j(1/2)\omega_D T} + e^{-j(1/2)\omega_D T}} = \frac{2}{T} j \tan \frac{\omega_D T}{2}$$

we get

$$G(e^{j\omega_D T}) = \frac{10}{\frac{2}{T} j \tan \frac{\omega_D T}{2} + 10}$$

Since $T = 0.2$ sec, $G(e^{j\omega_D T})$ can be simplified to

$$G(e^{j0.2\omega_D}) = \frac{10}{j10 \tan 0.1\omega_D + 10}$$

The Bode diagrams (magnitude curves only) of the analog filter $G(j\omega) = 10/(j\omega + 10)$ and the digital equivalent $G(e^{j\omega_D T})$ obtained by the bilinear transformation method are shown in Fig. 4–56 as curves (a) and (b), respectively. Notice that in the frequency region of interest (0 to 10 rad/sec) there is a significant difference between the two magnitude curves, (a) and (b).

In this problem the desired digital filter must exhibit frequency response characteristics similar to $G(j\omega) = 10/(j\omega + 10)$. Since the frequency region of interest is $0 \leq \omega \leq 10$ rad/sec, we require that at $\omega = 10$, the magnitude be equal to -3.010 db. Since ω_A and ω_D are related by the equation

$$\omega_A = \frac{2}{T} \tan \frac{\omega_D T}{2} = 10 \tan 0.1\, \omega_D$$

if we require the magnitude of -3.010 db at $\omega_D = 10$ in the desired digital filter, the corresponding analog filter must have the magnitude of -3.010 db at $\omega_A = 10 \tan 1 = 15.574$. Hence we must first modify the original analog filter to

$$G_D(s) = \frac{15.574}{s + 15.574} \tag{4–99}$$

and then obtain the equivalent digital filter based on the modified analog filter given by Eq. (4–99).

By substituting Eq. (4–98) into Eq. (4–99) we obtain

$$G_D(z) = \frac{15.574}{\frac{2}{T}\frac{1-z^{-1}}{1+z^{-1}} + 15.574} = \frac{0.6090(1+z^{-1})}{1+0.2180z^{-1}} \tag{4–100}$$

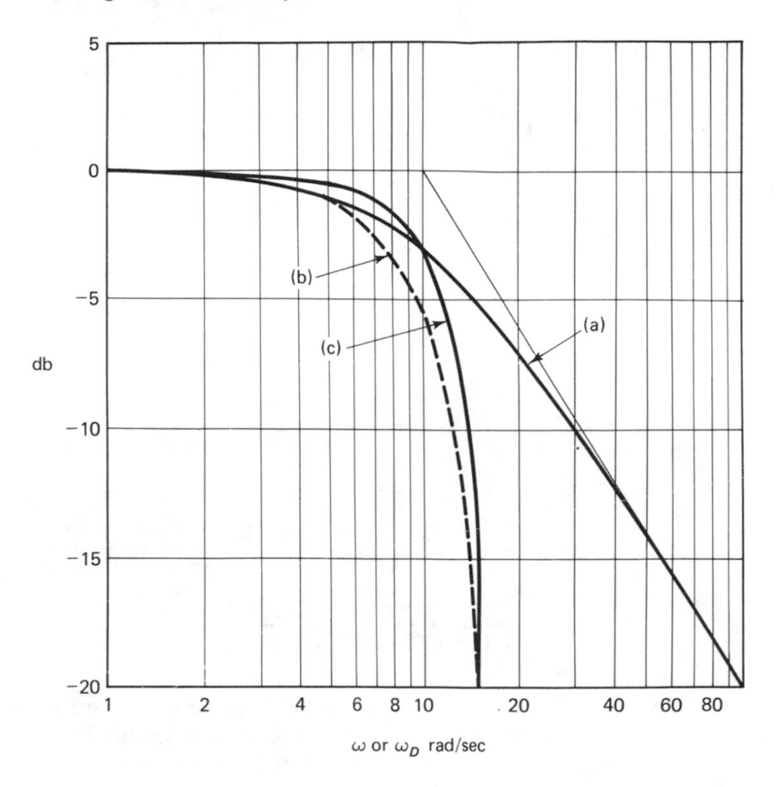

Figure 4–56 Bode diagrams for the analog filter $G(j\omega) = 10/(j\omega + 10)$ [shown by curve (a)], the digital equivalent obtained by the bilinear transformation method [shown by curve (b)], and the digital equivalent obtained by the bilinear transformation method with frequency prewarping [shown by curve (c)].

This is the equivalent digital filter obtained by the bilinear transformation method with frequency prewarping. The Bode diagram (magnitude curve only) of this digital filter is shown in Fig. 4–56 as curve (c).

Notice that in the frequency range of interest (0 to 10 rad/sec), the frequency response characteristic of the designed digital filter given by Eq. (4–100) [curve (c)] and that of the original analog filter given by Eq. (4–97) [curve (a)] are generally in good agreement.

Problem A-4–4. Obtain the discrete-time equivalent (or digital equivalent) of the following continuous-time (or analog) filter $G(s)$ by the step-invariance method:

$$G(s) = \frac{2}{(s+1)(s+2)}$$

Assume that the sampling period T is 0.1 sec.

Solution. From Eq. (4–22), the discrete-time equivalent $G_D(z)$ is given by

$$G_D(z) = \mathcal{Z}\left[\frac{1-e^{-Ts}}{s}G(s)\right] = (1-z^{-1})\,\mathcal{Z}\left[\frac{G(s)}{s}\right]$$

The z transform of $G(s)/s$ is given as follows:

$$\mathscr{Z}\left[\frac{G(s)}{s}\right] = \mathscr{Z}\left[\frac{2}{s(s+1)(s+2)}\right]$$

$$= \mathscr{Z}\left[\frac{1}{s} - \frac{2}{s+1} + \frac{1}{s+2}\right]$$

$$= \frac{1}{1-z^{-1}} - \frac{2}{1-e^{-T}z^{-1}} + \frac{1}{1-e^{-2T}z^{-1}}$$

$$= \frac{1}{1-z^{-1}} - \frac{2}{1-0.9048z^{-1}} + \frac{1}{1-0.8187z^{-1}}$$

$$= \frac{0.0091z^{-1} + 0.0082z^{-2}}{(1-z^{-1})(1-0.9048z^{-1})(1-0.8187z^{-1})}$$

Hence

$$G_D(z) = (1-z^{-1})\frac{0.0091z^{-1} + 0.0082z^{-2}}{(1-z^{-1})(1-0.9048z^{-1})(1-0.8187z^{-1})}$$

$$= \frac{0.0091z^{-1} + 0.0082z^{-2}}{(1-0.9048z^{-1})(1-0.8187z^{-1})}$$

Problem A-4-5. The Butterworth filter is frequently used to obtain flat amplitude in the passband and sharp cutoff characteristics in the stopband in frequency response. It is a good approximation of the ideal low-pass filter. The Butterworth filter is characterized by the following magnitude-squared equation:

$$G^2(\omega) = \frac{1}{1+(\omega/\omega_c)^{2n}}$$

where n denotes the order of the Butterworth transfer function. A Bode diagram (magnitude curve only) for a fourth-order Butterworth filter is shown in Fig. 4-57. The frequency ω_c is the cutoff frequency. The magnitude of the Butterworth filter decreases monotonically with ω. As the order n is increased, the magnitude curve becomes much flatter in the passband and attenuation becomes sharper in the stopband. The high-frequency asymptote has a slope of $-20n$ db/decade.

Obtain the transfer function of a fourth-order Butterworth low-pass filter in the s domain. Assume that the cutoff frequency ω_c is 1 rad/sec. Then using the matched pole-zero mapping method, obtain the pulse transfer function of the fourth-order Butterworth low-pass filter.

Solution. For a fourth-order Butterworth filter with $\omega_c = 1$, the amplitude-squared function becomes

$$G^2(\omega) = \frac{1}{1+\omega^8}$$

In the s plane expression, this last equation can be written as

$$G(s)G(-s) = \frac{1}{1+s^8}$$

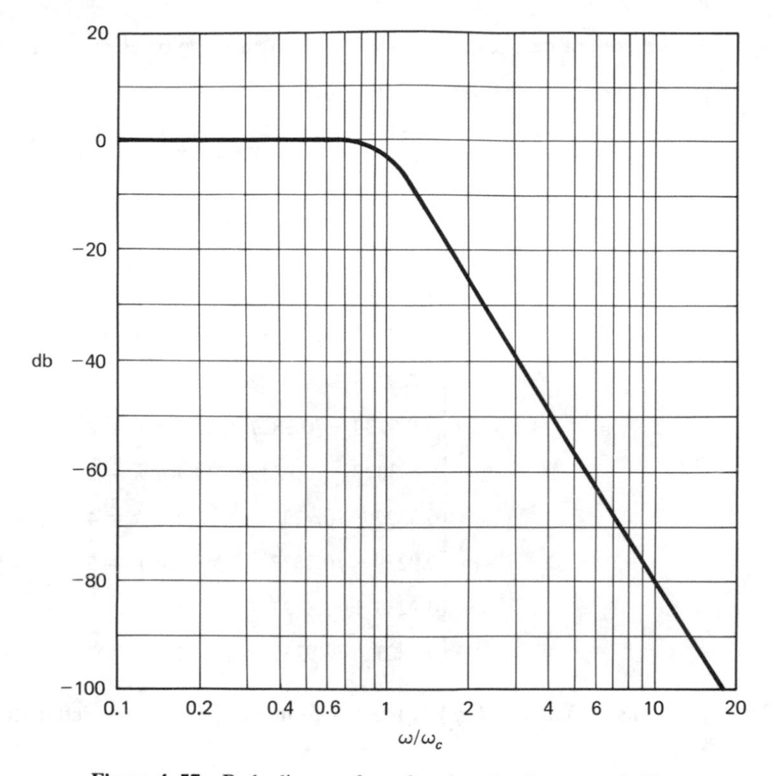

Figure 4–57 Bode diagram for a fourth-order Butterworth filter.

Note that the poles of $G(s)$ consist of left–half plane poles and those of $G(-s)$ consist of right–half plane poles. The poles of $G(s)G(-s)$ can be found as follows. Let us consider the general case first. In the nth-order Butterworth transfer function

$$G(s)G(-s) = \frac{1}{1 + \left(\dfrac{s}{j\omega_c}\right)^{2n}}$$

the poles of $G(s)G(-s)$ can be found by solving

$$\left(\frac{s}{j\omega_c}\right)^{2n} = -1 = e^{j\pi}e^{j2\pi k} \qquad k = 0, 1, 2, \ldots$$

By taking the $2n$th root, we have

$$\frac{s}{j\omega_c} = e^{j\pi/2n}e^{j\pi k/n}$$

or

$$s = e^{j\pi/2}\omega_c e^{j\pi/2n}e^{j\pi k/n} = \omega_c e^{j(n+1)\pi/2n}e^{j\pi k/n}$$

Since $|s| = |\omega_c|$, $2n$ poles of $G(s)G(-s)$ lie on a circle of radius ω_c in the s plane. Notice also that the poles are equally spaced at angle π/n rad. (All complex poles are in conjugate pairs.)

In the present case, $\omega_c = 1$ and $n = 4$. Hence, the poles are located at

$$\begin{aligned} s &= e^{j5\pi/8}e^{j\pi k/4} \\ &= \left(\cos\frac{5\pi}{8} + j\sin\frac{5\pi}{8}\right)\left(\cos\frac{\pi k}{4} + j\sin\frac{\pi k}{4}\right) \\ &= (-0.3827 + j0.9239)\left(\cos\frac{\pi k}{4} + j\sin\frac{\pi k}{4}\right) \end{aligned}$$

where $k = 0, 1, 2, 3, 4, 5, 6, 7$. That is,

$$s = \begin{cases} -0.3827 + j0.9239 & \text{for } k = 0 \\ -0.9239 + j0.3827 & \text{for } k = 1 \\ -0.9239 - j0.3827 & \text{for } k = 2 \\ -0.3827 - j0.9239 & \text{for } k = 3 \\ 0.3827 - j0.9239 & \text{for } k = 4 \\ 0.9239 - j0.3827 & \text{for } k = 5 \\ 0.9239 + j0.3827 & \text{for } k = 6 \\ 0.3827 + j0.9239 & \text{for } k = 7 \end{cases}$$

Then the transfer function $G(s)$ is formed from the poles in the left half of the s plane as follows:

$$\begin{aligned} G(s) &= \frac{1}{(s + 0.3827 + j0.9239)(s + 0.3827 - j0.9239)} \\ &\quad \times \frac{1}{(s + 0.9239 + j0.3827)(s + 0.9239 - j0.3827)} \end{aligned}$$

This is the fourth-order Butterworth filter defined in the s plane. The pole locations in the s plane are shown in Fig. 4–58(a).

Using the matched pole-zero mapping method, the equivalent Butterworth digital filter can be obtained as follows:

$$\begin{aligned} G_D(z) &= K\frac{(z+1)^2}{(z - e^{-0.3827-j0.9239})(z - e^{-0.3827+j0.9239})} \\ &\quad \times \frac{(z+1)^2}{(z - e^{-0.9239-j0.3827})(z - e^{-0.9239+j0.3827})} \\ &= K\frac{(z+1)^2}{(z - 0.4112 + j0.5443)(z - 0.4112 - j0.5443)} \\ &\quad \times \frac{(z+1)^2}{(z - 0.3683 + j0.1482)(z - 0.3683 - j0.1482)} \end{aligned}$$

where gain constant K is determined by equating $G(s)|_{s=0}$ and $G_D(z)|_{z=1}$. Notice that

$$G(s)|_{s=0} = 1$$

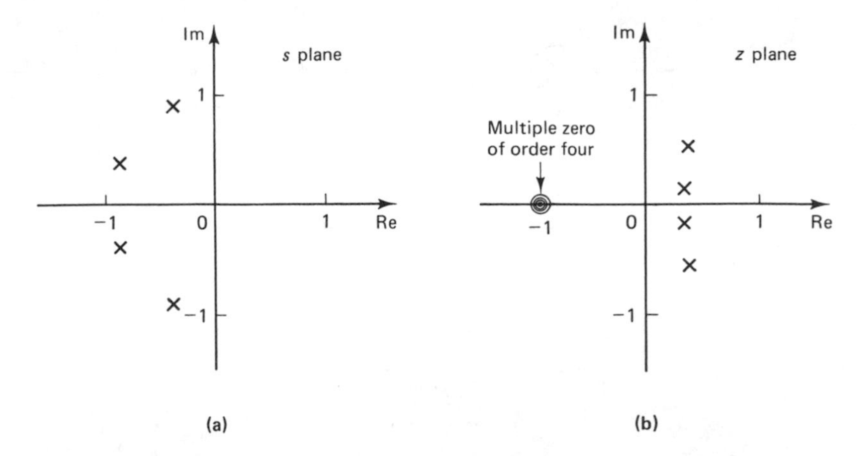

Figure 4–58 (a) Pole locations for a fourth-order Butterworth analog filter; (b) pole and zero locations for the equivalent fourth-order Butterworth digital filter.

and $G_D(z)|_{z=1}$ can be calculated as follows:

$$G_D(z)|_{z=1} = K\frac{(1+1)^2}{(0.5888+j0.5443)(0.5888-j0.5443)}$$

$$\times\frac{(1+1)^2}{(0.6317+j0.1482)(0.6317-j0.1482)}$$

$$= 59.11K$$

Hence, the gain constant is determined as follows:

$$K = \frac{1}{59.11} = 0.01692$$

The pulse transfer function of the fourth-order Butterworth filter thus becomes

$$G_D(z) = 0.01692\frac{(z+1)^2}{(z-0.4112+j0.5443)(z-0.4112-j0.5443)}$$

$$\times\frac{(z+1)^2}{(z-0.3683+j0.1482)(z-0.3683-j0.1482)}$$

$$= \frac{0.01692\,(z+1)^4}{(z^2-0.8224z+0.4653)(z^2-0.7366z+0.1576)}$$

The pole and zero locations of this fourth-order Butterworth digital filter are shown in Fig. 4–58(b). Notice that a multiple zero of order 4 is located at $z = -1$. It is noted that Butterworth low-pass filters in general give excellent low-pass filtering characteristics and may be used for filtering undesirable high-frequency components above the cutoff frequency.

Problem A-4–6. Consider the analog control system shown in Fig. 4–59(a), where the first-order plant $a/(s+a)$ is controlled by the analog integral controller K/s. Obtain the static velocity error constant for the system and the steady-state error in the response to a unit ramp input.

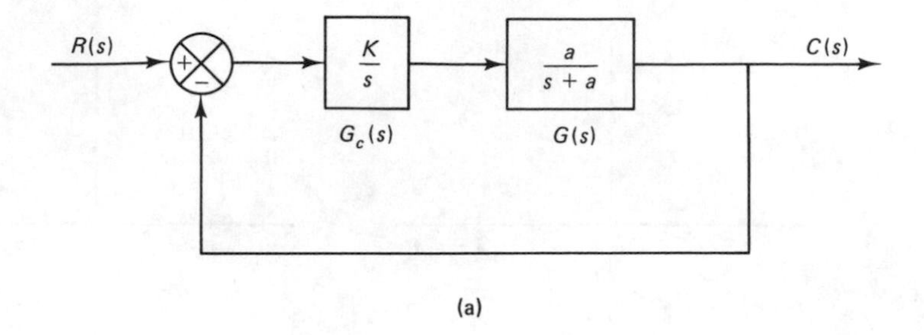

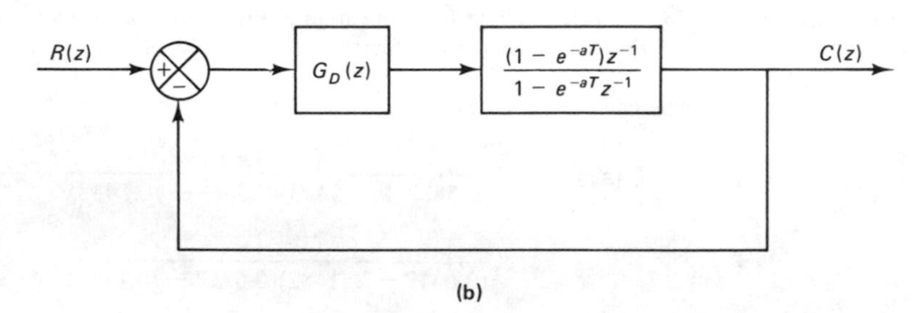

Figure 4–59 (a) Analog control system of Prob. A-4–6; (b) equivalent digital control system.

Next, consider equivalent digital control systems. Using equivalent digital controllers obtained by (1) the backward difference method and (2) the impulse-invariance method, show that the static velocity error constants for the equivalent digital control systems are the same as that for the analog control system.

Noting the requirement that the static error constants for equivalent analog and digital control systems must remain the same, determine an equivalent digital integral controller by using the matched pole-zero mapping method.

Solution. For the system shown in Fig. 4–59(a) the static velocity error constant K_v is

$$K_v = \lim_{s\to 0}\left(s\frac{K}{s}\frac{a}{s+a}\right) = K$$

and the steady-state error in the unit ramp response is

$$e_{ss} = \frac{1}{K}$$

Next, consider the digital control of the given plant. The pulse transfer function of the plant preceded by a zero-order hold becomes as follows:

$$G(z) = \mathscr{Z}\left[\frac{1-e^{-Ts}}{s}\frac{a}{s+a}\right]$$

$$= (1 - z^{-1}) \mathcal{Z}\left[\frac{a}{s(s + a)}\right]$$

$$= (1 - z^{-1}) \left(\frac{1}{1 - z^{-1}} - \frac{1}{1 - e^{-aT}z^{-1}}\right)$$

$$= \frac{(1 - e^{-aT})z^{-1}}{1 - e^{-aT}z^{-1}}$$

Figure 4–59(b) shows an equivalent digital control system.

The equivalent digital controller based on the backward difference method is

$$G_D(z) = \frac{K}{s}\bigg|_{s=(1-z^{-1})/T} = \frac{KT}{1 - z^{-1}}$$

The equivalent digital controller based on the impulse-invariance method is

$$G_D(z) = T\,\mathcal{Z}\left[\frac{K}{s}\right] = \frac{KT}{1 - z^{-1}}$$

Notice that the backward difference method and the impulse-invariance method yield the same equivalent digital controller in this case. The static velocity error constant becomes

$$K_v = \lim_{z \to 1}\left[\frac{1 - z^{-1}}{T} G_D(z)G(z)\right]$$

$$= \lim_{z \to 1}\left[\frac{1 - z^{-1}}{T}\,\frac{KT}{1 - z^{-1}}\,\frac{(1 - e^{-aT})z^{-1}}{1 - e^{-aT}z^{-1}}\right] = K$$

Thus, the static velocity error constants are the same for the analog and digital control systems. The steady-state error in the unit ramp response is

$$e_{ss}^* = \frac{1}{K}$$

The equivalent digital controller based on the matched pole-zero mapping method is

$$G_D(z) = \hat{K}\frac{z + 1}{z - 1} = \hat{K}\frac{1 + z^{-1}}{1 - z^{-1}}$$

where $\hat{K}$ is a constant. For a low-pass filter, such a constant is normally determined by the requirement that $G_D(1)$ and $G_c(0)$ be equal. In this particular case, both $G_D(1)$ and $G_c(0)$ become infinity and it is not possible to determine the unique value for $\hat{K}$. However, the constant $\hat{K}$ can be determined by the requirement that the analog control system and the equivalent digital control system have the same static velocity error constant. Thus, equating the static velocity error constant to K, we obtain

$$K_v = \lim_{z \to 1}\left[\frac{1 - z^{-1}}{T}\,\hat{K}\,\frac{1 + z^{-1}}{1 - z^{-1}}\,\frac{(1 - e^{-aT})z^{-1}}{1 - e^{-aT}z^{-1}}\right] = K$$

or

$$\hat{K} = \frac{KT}{2}$$

Thus we have determined the constant $\hat{K}$. The equivalent digital controller based on the matched pole-zero mapping method is

$$G_D(z) = \frac{KT}{2} \frac{1 + z^{-1}}{1 - z^{-1}}$$

Problem A-4-7. Consider the analog control system shown in Fig. 4–60(a). We wish to convert this analog control system into a digital control system. Design a suitable digital

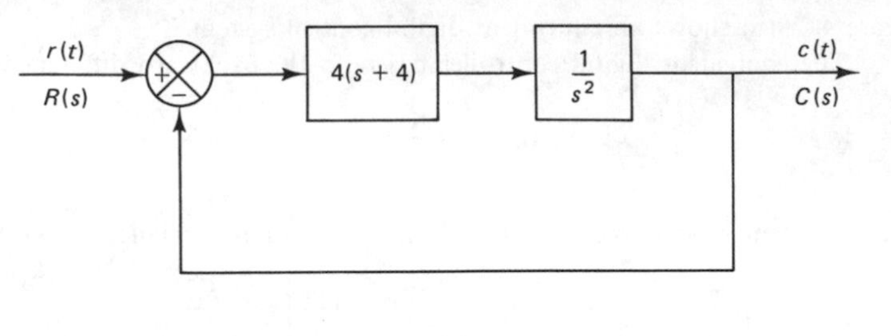

(a)

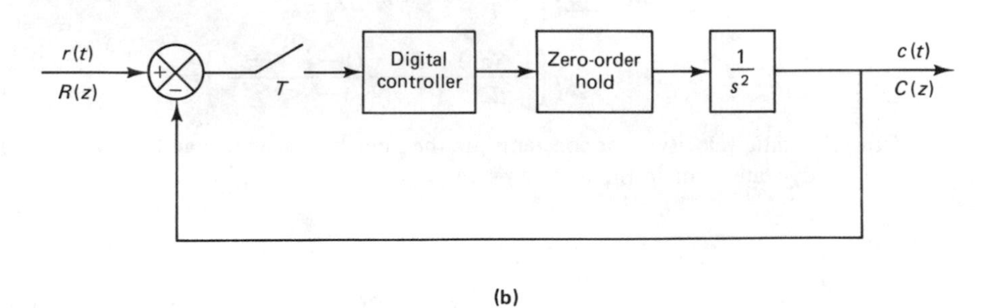

(b)

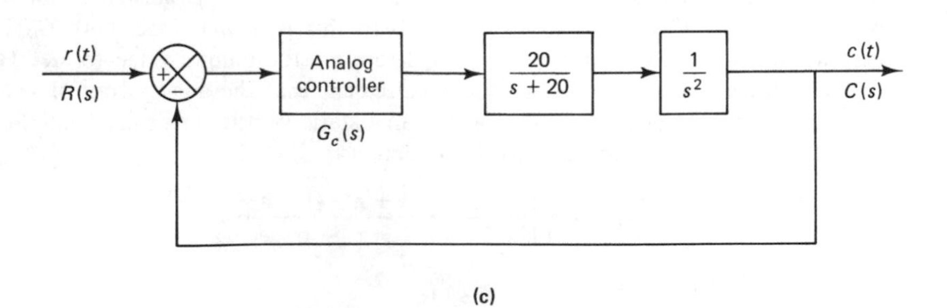

(c)

Figure 4–60 (a) Analog control system of Prob. A-4–7; (b) digital control system; (c) modified analog control system.

controller by the matched pole-zero mapping method. The sampling period T is assumed to be 0.1 sec. Also, obtain the response of the digital control system thus designed to a unit step input, and plot $c(kT)$ versus kT.

Solution. First, we examine the damping ratio ζ and the undamped natural frequency ω_n of the closed-loop poles of the analog control system. Since

$$\frac{C(s)}{R(s)} = \frac{4(s+4)}{s^2 + 4s + 16}$$

the closed-loop poles are located at

$$s = -2 \pm j2\sqrt{3}$$

and have a damping ratio ζ of 0.5 and an undamped natural frequency ω_n of 4 rad/sec. Although the damping ratio is 0.5, because of the presence of a zero at $s = -4$ the maximum overshoot in the unit step response is approximately 30 percent.

To convert the analog controller into a digital controller, we need to insert a sampler and a zero-order hold in the system, as shown in Fig. 4–60(b). As mentioned in Sec. 4–3, the hold introduces a time lag, which produces a phase lag. For a zero-order hold, such a phase lag may be approximated by

$$G_h(s) = \frac{1}{(T/2)s + 1}$$

Since $T = 0.1$ in the present case, we have

$$G_h(s) = \frac{1}{(0.1/2)s + 1} = \frac{20}{s + 20}$$

It is necessary to insert $G_h(s)$ into the analog control system and redesign the analog controller to compensate for the phase lag that will be introduced by the addition of the zero-order hold. Figure 4–60(c) shows the block diagram for the modified analog control system.

Now it is necessary to redesign the analog controller to include the effect of $G_h(s)$. By use of a conventional design method, such as the root locus method, a suitable analog controller can easily be designed. An example is

$$G_c(s) = 4(s + 3.2)$$

Let us use this analog controller in the system of Fig. 4–60(c). Then the redesigned system has three closed-loop poles: $s_1 = -2 + j2\sqrt{3}$, $s_2 = -2 - j2\sqrt{3}$, and $s_3 = -16$. The dominant closed-loop poles have a damping ratio ζ of 0.5 and an undamped natural frequency ω_n of 4 rad/sec. Since the third pole is located far to the left, its effect on the transient response is very small.

In what follows, we design the equivalent digital controller by the matched pole-zero mapping method. Notice that $G_c(s)$ has a zero at $s = -3.2$ and a pole at infinity. The zero at $s = -3.2$ is mapped to $z = e^{-3.2 \times 0.1}$ and the infinite pole is mapped to $z = -1$. Hence the equivalent digital controller is

$$G_D(z) = K\frac{z - e^{-0.32}}{z + 1} = K\frac{z - 0.7261}{z + 1}$$

where the gain K is determined so that the low-frequency gains of the digital and analog controllers are the same. [Note that although the controller transfer function $4(s + 3.2)$ is a high-pass filter, we are interested in the low-frequency range and therefore the analog and digital controllers must have the same low-frequency characteristics.] Now equating $G_D(1)$ and $G_c(0)$, we have

$$G_D(1) = K\frac{1-0.7261}{1+1} = G_c(0) = 4 \times 3.2$$

from which we obtain

$$K = 93.465$$

Thus

$$G_D(z) = 93.465\left(\frac{z-0.7261}{z+1}\right) = 93.465\left(\frac{1-0.7261z^{-1}}{1+z^{-1}}\right)$$

The pulse transfer function of the plant which is preceded by the zero-order hold is

$$G(z) = \mathscr{Z}\left[\frac{1-e^{-0.1s}}{s}\frac{1}{s^2}\right] = (1-z^{-1})\,\mathscr{Z}\left[\frac{1}{s^3}\right]$$

$$= \frac{0.005(1+z^{-1})z^{-1}}{(1-z^{-1})^2}$$

Hence, the open-loop pulse transfer function becomes

$$G_D(z)G(z) = 93.465\left[\frac{1-0.7261z^{-1}}{1+z^{-1}}\,\frac{0.005(1+z^{-1})z^{-1}}{(1-z^{-1})^2}\right]$$

$$= (0.4673)\frac{(1-0.7261z^{-1})z^{-1}}{(1-z^{-1})^2}$$

Figure 4–61 shows the block diagram of the digital control system just designed.

The closed-loop pulse transfer function is

$$\frac{C(z)}{R(z)} = \frac{0.4673(1-0.7261z^{-1})z^{-1}}{(1-z^{-1})^2 + 0.4673(1-0.7261z^{-1})z^{-1}}$$

$$= \frac{0.4673z^{-1} - 0.3393z^{-2}}{1 - 1.5327z^{-1} + 0.6607z^{-2}}$$

Hence the unit step response sequence is obtained from the equivalent difference equation:

$$c(k) - 1.5327c(k-1) + 0.6607c(k-2) = 0.4673r(k-1) - 0.3393r(k-2)$$

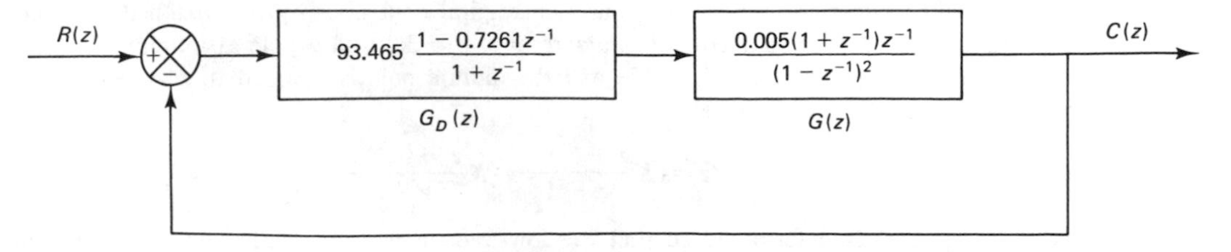

Figure 4–61 Digital control system designed in Prob. A-4–7.

where

$$r(k) = \begin{cases} 0 & k < 0 \\ 1 & k = 0, 1, 2, \ldots \end{cases}$$

A BASIC computer program for solving this difference equation is shown in Table 4–10, together with the computation results.

Figure 4–62 shows a plot of $c(kT)$ versus kT, together with the unit step response curve for the original analog control system. The maximum overshoot in the unit step response of the digital control system is approximately 35 percent.

Problem A-4–8. Consider a unity feedback system with the closed-loop pulse transfer function

$$\frac{C(z)}{R(z)} = \frac{K(1 + cz^{-1})z^{-1}}{1 + az^{-1} + bz^{-2}}$$

TABLE 4–10 BASIC COMPUTER PROGRAM FOR SOLVING DIFFERENCE EQUATION:

$$c(k) - 1.5327c(k-1) + 0.6607c(k-2) = 0.4673r(k-1) - 0.3393r(k-2)$$

where $r(k) = 0$ for $k < 0$ and $r(k) = 1$ for $k = 0, 1, 2, \ldots$

```
10  C0 = 0
20  C1 = 0.4673
30  R0 = 1
40  R1 = 1
50  C2 = 1.5327 * C1 - 0.6607 * C0
         + 0.4673 * R1 - 0.3393 * R0
60  M = C0
70  C0 = C1
80  C1 = C2
90  N = R0
100 R0 = R1
110 PRINT K, M, N
120 K = K + 1
130 IF K < 21 GO TO 50
140 END
```

k = K	$c(k)$ = CK = M	$r(k)$ = RK = N
0	0	1
1	0.4673	1
2	0.844231	1
3	1.11321	1
4	1.27643	1
5	1.34889	1
6	1.3521	1
7	1.30916	1
8	1.24121	1
9	1.16544	1
10	1.09421	1
11	1.03508	1
12	0.991529	1
13	0.963837	1
14	0.950169	1
15	0.947518	1
16	0.952483	1
17	0.961846	1
18	0.972916	1
19	0.983696	1
20	0.992906	1

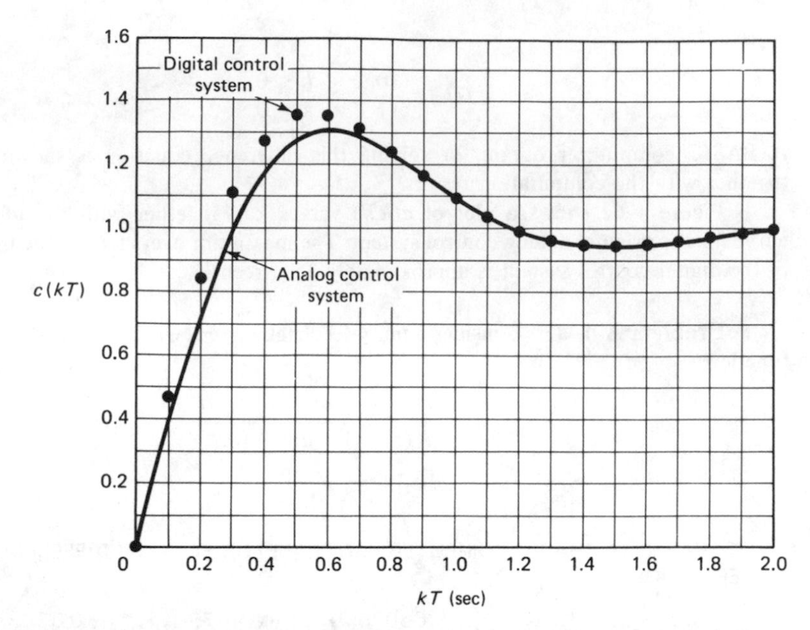

Figure 4–62 Plot of $c(kT)$ vs. kT of the digital control system designed in Prob. A-4–7 and a plot of the unit step response curve for the original analog control system shown in Fig. 4–60(a).

Assume that the open-loop pulse transfer function involves an integrator, meaning that it has a simple pole at $z = 1$. (The system is of type 1.) Show that the steady-state error in the unit ramp response is given by

$$e_{ss}^* = \frac{1}{K_v} = T\left(\frac{2+a}{1+a+b} - \frac{1}{1+c}\right)$$

where T is the sampling period.

Solution. Let us define the open-loop pulse transfer function of the system as $G(z)$. Then

$$\frac{C(z)}{R(z)} = \frac{K(1+cz^{-1})z^{-1}}{1+az^{-1}+bz^{-2}} = \frac{G(z)}{1+G(z)}$$

from which we get

$$G(z) = \frac{K(1+cz^{-1})z^{-1}}{(1+az^{-1}+bz^{-2}) - K(1+cz^{-1})z^{-1}} \tag{4–101}$$

Since the system is of type 1, the static position error constant K_p is infinity. That is,

$$K_p = \lim_{z\to 1} G(z) = \lim_{z\to 1} \frac{K(1+cz^{-1})z^{-1}}{(1+az^{-1}+bz^{-2}) - K(1+cz^{-1})z^{-1}}$$

$$= \frac{K(1+c)}{(1+a+b) - K(1+c)} = \infty$$

which gives

$$(1 + a + b) - K(1 + c) = 0$$

or

$$K = \frac{1 + a + b}{1 + c} \tag{4–102}$$

Hence by substituting Eq. (4–102) into Eq. (4–101) we obtain

$$G(z) = \frac{(1 + a + b)(1 + cz^{-1})z^{-1}}{(1 + az^{-1} + bz^{-2})(1 + c) - (1 + a + b)(1 + cz^{-1})z^{-1}}$$

$$= \frac{(1 + a + b)(1 + cz^{-1})z^{-1}}{1 + c + (ac - b - 1)z^{-1} + (b - c - ac)z^{-2}}$$

$$= \frac{(1 + a + b)(1 + cz^{-1})z^{-1}}{(1 - z^{-1})[1 + c + (ac - b + c)z^{-1}]}$$

Clearly, $G(z)$ has a term $(1 - z^{-1})$ in the denominator, or a pole at $z = 1$.

The static velocity error constant K_v is

$$K_v = \lim_{z \to 1} \left[\frac{1 - z^{-1}}{T} G(z) \right]$$

$$= \lim_{z \to 1} \left\{ \frac{1 - z^{-1}}{T} \frac{(1 + a + b)(1 + cz^{-1})z^{-1}}{(1 - z^{-1})[1 + c + (ac - b + c)z^{-1}]} \right\}$$

$$= \frac{1}{T} \frac{(1 + a + b)(1 + c)}{1 + 2c + ac - b}$$

Hence the steady-state error in the unit ramp response is

$$e_{ss}^* = \frac{1}{K_v} = T \frac{1 + 2c + ac - b}{(1 + a + b)(1 + c)}$$

$$= T \frac{(1 + c)(2 + a) - (1 + a + b)}{(1 + a + b)(1 + c)}$$

$$= T \left(\frac{2 + a}{1 + a + b} - \frac{1}{1 + c} \right)$$

Problem A-4–9. Draw root locus diagrams in the z plane for the system shown in Fig. 4–63 for the following three sampling periods: $T = 1$ sec, $T = 2$ sec, and $T = 4$ sec.

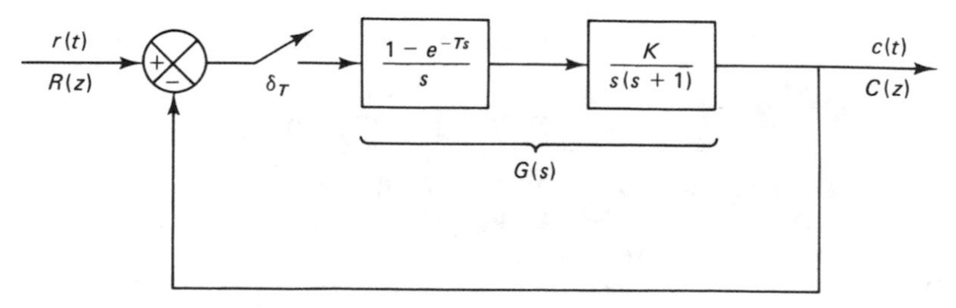

Figure 4–63 Control system of Prob. A-4–9.

Solution. We first obtain the z transform of $G(s)$. Referring to Example 3–8, we get

$$G(z) = \mathscr{Z}\left[\frac{1 - e^{-Ts}}{s}\frac{K}{s(s+1)}\right]$$

$$= (1 - z^{-1})\,\mathscr{Z}\left[\frac{K}{s^2(s+1)}\right]$$

$$= \frac{K[(T - 1 + e^{-T})z^{-1} + (1 - e^{-T} - Te^{-T})z^{-2}]}{(1 - z^{-1})(1 - e^{-T}z^{-1})} \qquad (4\text{–}103)$$

Next we construct root locus diagrams for the three cases considered.

1. *Sampling period $T = 1$:* For $T = 1$, Eq. (4–103) becomes

$$G(z) = \frac{K[(1 - 1 + e^{-1})z^{-1} + (1 - e^{-1} - e^{-1})z^{-2}]}{(1 - z^{-1})(1 - e^{-1}z^{-1})}$$

$$= \frac{0.3679K(z + 0.7181)}{(z - 1)(z - 0.3679)}$$

Notice that $G(z)$ possesses a zero at $z = -0.7181$ and poles at $z = 1$ and $z = 0.3679$. The breakaway point is at $z = 0.6479$, and the break-in point is at $z = -2.0841$. The root locus diagram for this case is shown in Fig. 4–64(a). The value of gain K of any point on the root loci can be determined from the magnitude condition

$$K = \left|\frac{(z - 1)(z - 0.3679)}{0.3679(z + 0.7181)}\right|$$

If we choose a point z on the root loci, the value of K at that point can be calculated by substituting the value of z into this last equation. (This means that with this value of K, that particular point becomes a closed-loop pole.) The critical gain is found to be $K = 2.3925$.

2. *Sampling period $T = 2$.* For the sampling period $T = 2$, we have from Eq. (4–103)

$$G(z) = \frac{1.1353K(z + 0.5232)}{(z - 1)(z - 0.1353)}$$

The pulse transfer function $G(z)$ in this case possesses a zero at $z = -0.5232$ and poles at $z = 1$ and $z = 0.1353$. The breakaway point is at $z = 0.4783$, and the break-in point is at $z = -1.5247$. The root locus diagram for this case is shown in Fig. 4–64(b). The critical gain K for stability is $K = 1.4557$.

3. *Sampling period $T = 4$.* For the case of $T = 4$, Eq. (4–103) gives

$$G(z) = \frac{3.0183K(z + 0.3010)}{(z - 1)(z - 0.0183)}$$

The breakaway point is at $z = 0.3435$, and the break-in point is at $z = -0.9455$. The root locus diagram is shown in Fig. 4–64(c). The critical gain for stability is $K = 0.9653$.

From the three cases considered, notice that the smaller the sampling period, the larger the critical gain K for stability.

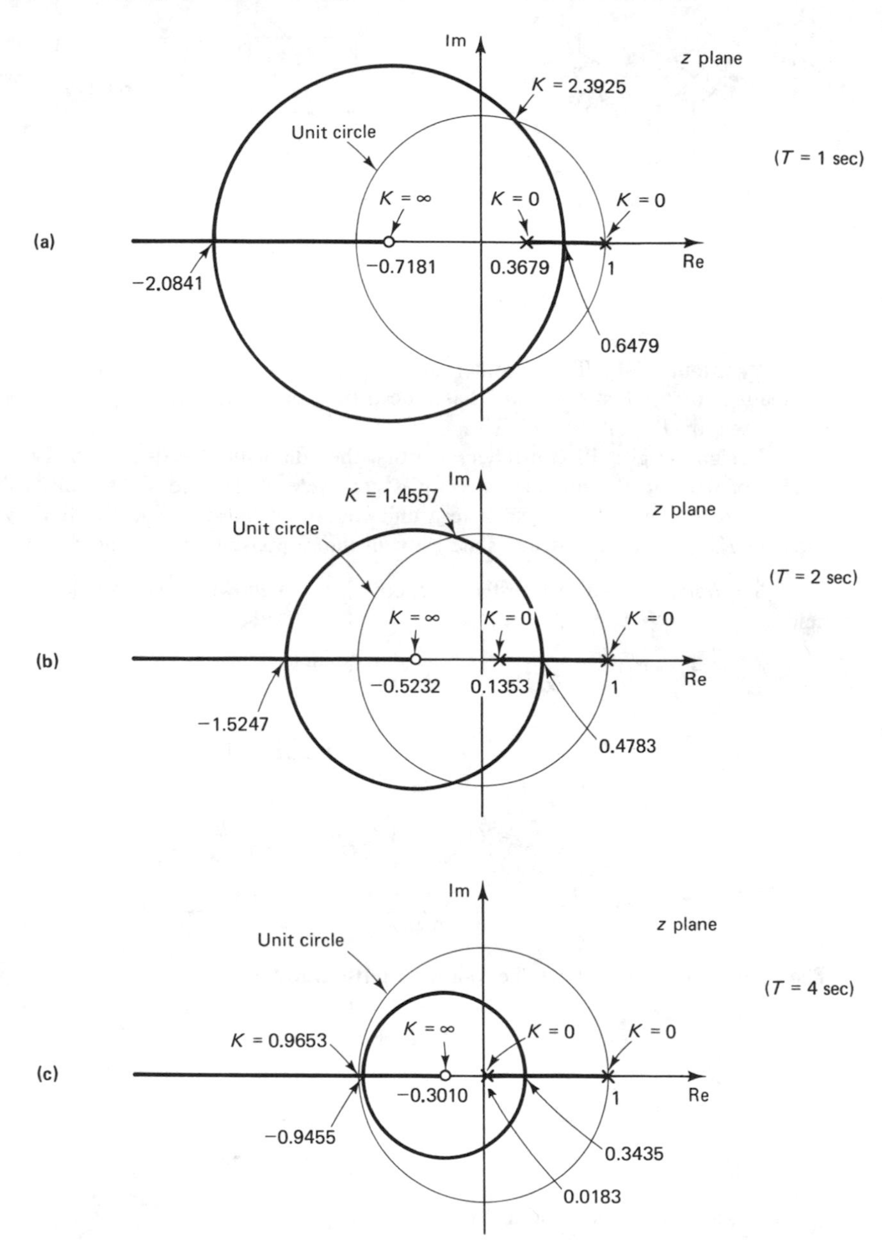

Figure 4–64 Root locus diagrams for the system shown in Fig. 4–63; (a) when $T = 1$; (b) when $T = 2$; (c) when $T = 4$.

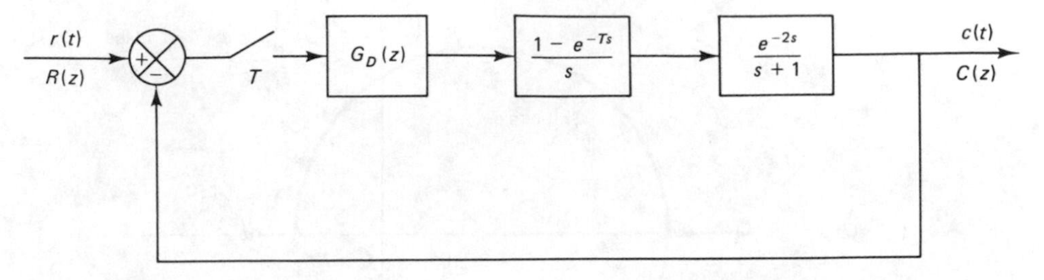

Figure 4–65 Digital control system of Prob. A-4–10.

Problem A-4–10. Consider the digital control system shown in Fig. 4–65, where the plant is of the first order and has a dead time of 2 sec. The sampling period is assumed to be 1 sec, or $T = 1$.

Design a digital PI controller such that the dominant closed-loop poles have a damping ratio ζ of 0.5 and the number of samples per cycle of damped sinusoidal oscillation is 10. Obtain the response of the system to a unit step input. Also, obtain the static velocity error constant K_v and find the steady-state error in the response to a unit ramp input.

Solution. The pulse transfer function of the plant which is preceded by a zero-order hold is

$$G(z) = \mathcal{Z}\left[\frac{1 - e^{-Ts}}{s} \frac{e^{-2s}}{s+1}\right]$$

$$= (1 - z^{-1})z^{-2} \mathcal{Z}\left[\frac{1}{s(s+1)}\right]$$

$$= (1 - z^{-1})z^{-2} \frac{(1 - e^{-1})z^{-1}}{(1 - z^{-1})(1 - e^{-1}z^{-1})}$$

$$= \frac{0.6321z^{-3}}{1 - 0.3679z^{-1}} = \frac{0.6321}{z^2(z - 0.3679)}$$

The digital PI controller has the following pulse transfer function:

$$G_D(z) = K_p + K_i \frac{1}{1 - z^{-1}}$$

$$= (K_p + K_i) \frac{z - \dfrac{K_p}{K_p + K_i}}{z - 1}$$

The open-loop pulse transfer function becomes

$$G_D(z)G(z) = \frac{(K_p + K_i)\left(z - \dfrac{K_p}{K_p + K_i}\right)}{z - 1} \frac{0.6321}{z^2(z - 0.3679)}$$

We locate the open-loop poles in the z plane as shown in Fig. 4–66. There is one open-loop zero involved in this case, but its location is unknown at this point.

Since it is required to have 10 samples per cycle of damped sinusoidal oscillation, the

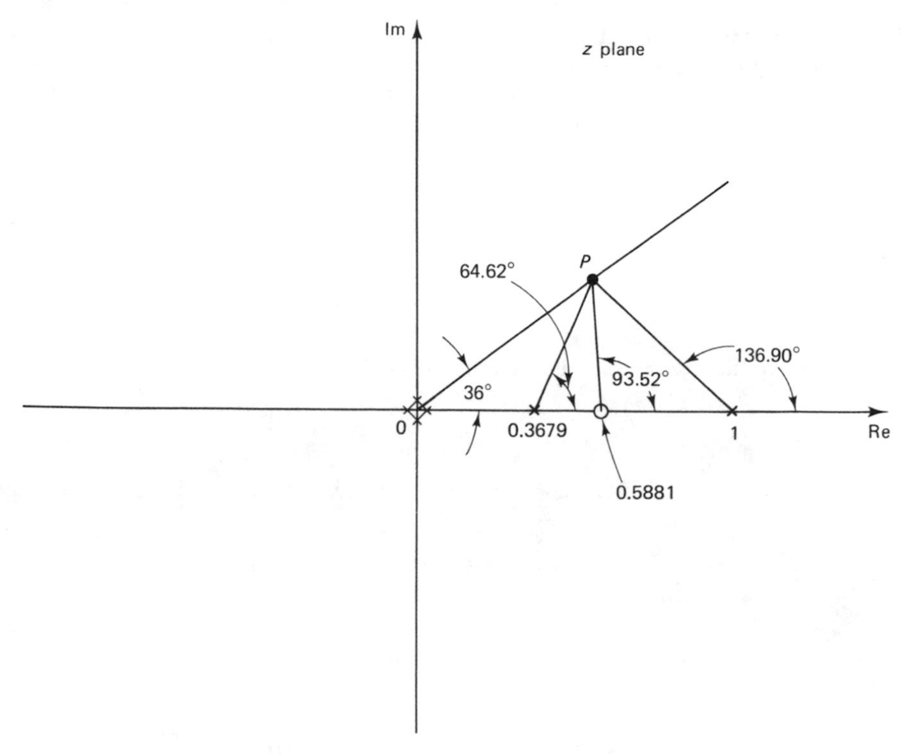

Figure 4–66 Pole and zero locations in the z plane of the system considered in Prob. A-4–10.

dominant closed-loop pole in the upper half of the z plane must lie on a line from the origin having an angle of $360°/10 = 36°$. From Eqs. (3–104) and (3–105), rewritten as

$$|z| = \exp\left(-\frac{2\pi\zeta}{\sqrt{1-\zeta^2}}\frac{\omega_d}{\omega_s}\right)$$

$$\underline{/z} = 2\pi\frac{\omega_d}{\omega_s}$$

the desired closed-loop pole location can be determined as follows. Noting that $\underline{/z} = 36°$, we have

$$2\pi\frac{\omega_d}{\omega_s} = \frac{2\pi}{10}$$

or $\omega_d/\omega_s = 0.1$. Since ζ is specified as 0.5, we have

$$|z| = \exp\left(-\frac{2\pi \times 0.5}{\sqrt{1-0.5^2}}\frac{1}{10}\right) = e^{-0.3628} = 0.6958$$

The closed-loop pole is located at point P in Fig. 4–66, where (at point P)

$$\begin{aligned} z &= 0.6958\,\underline{/36°} \\ &= 0.5629 + j0.4090 \end{aligned}$$

(Note that this point is the intersection of the $\zeta = 0.5$ locus and the line from the origin having an angle of 36°.)

If point P is to be the closed-loop pole location in the upper half of the z plane, then the angle deficiency at point P is

$$-36° - 36° - 136.90° - 64.62° + 180° = -93.52°$$

The controller zero must contribute +93.52°. This means that the zero of the digital controller must be located at $z = 0.5881$. Therefore

$$\frac{K_p}{K_p + K_i} = 0.5881 \tag{4–104}$$

Hence the PI controller is determined as follows:

$$G_D(z) = K\frac{z - 0.5881}{z - 1}$$

where $K = K_p + K_i$. Gain constant K is determined from the magnitude condition:

$$K\left|\frac{z - 0.5881}{z - 1}\frac{0.6321}{z^2(z - 0.3679)}\right|_{z = 0.5629 + j0.4090} = 1$$

or

$$K = 0.5070$$

Thus,

$$K_p + K_i = 0.5070 \tag{4–105}$$

From Eqs. (4–104) and (4–105) we find that

$$K_p = 0.2982 \qquad \text{and} \qquad K_i = 0.2088$$

Hence the PI controller just designed can be given by

$$G_D(z) = 0.5070\frac{1 - 0.5881z^{-1}}{1 - z^{-1}}$$

Finally, the open-loop pulse transfer function becomes

$$G_D(z)G(z) = 0.5070\left(\frac{1 - 0.5881z^{-1}}{1 - z^{-1}}\frac{0.6321z^{-3}}{1 - 0.3679z^{-1}}\right)$$

$$= \frac{0.3205(1 - 0.5881z^{-1})z^{-3}}{(1 - z^{-1})(1 - 0.3679z^{-1})}$$

The closed-loop pulse transfer function becomes

$$\frac{C(z)}{R(z)} = \frac{0.3205z^{-3} - 0.1885z^{-4}}{1 - 1.3679z^{-1} + 0.3679z^{-2} + 0.3205z^{-3} - 0.1885z^{-4}}$$

The response $c(k)$ to the unit step input is obtained from the following difference equation:

$$c(k) - 1.3679c(k-1) + 0.3679c(k-2) + 0.3205c(k-3) - 0.1885c(k-4)$$
$$= 0.3205r(k-3) - 0.1885r(k-4)$$

where

$$r(k) = \begin{cases} 0 & k < 0 \\ 1 & k = 0, 1, 2, \ldots \end{cases}$$

A BASIC computer program for solving this last equation is shown in Table 4–11, together with the computational results. The unit step response sequence $c(kT)$ versus kT is plotted in Fig. 4–67.

The static velocity error constant K_v is

$$K_v = \lim_{z \to 1} \left[\frac{1 - z^{-1}}{T} G_D(z) G(z) \right]$$

$$= \lim_{z \to 1} \left[\frac{1 - z^{-1}}{1} \frac{0.3205(1 - 0.5881z^{-1})z^{-3}}{(1 - z^{-1})(1 - 0.3679z^{-1})} \right]$$

$$= 0.2088$$

TABLE 4–11 BASIC COMPUTER PROGRAM FOR SOLVING DIFFERENCE EQUATION:

$$c(k) - 1.3679c(k-1) + 0.3679c(k-2) + 0.3205c(k-3) - 0.1885c(k-4) = 0.3205r(k-3) - 0.1885r(k-4)$$

where $r(k) = 0$ for $k < 0$ and $r(k) = 1$ for $k = 0, 1, 2, \ldots$

```
10  C0 = 0
20  C1 = 0
30  C2 = 0
40  C3 = 0.3205
50  R0 = 1
60  R1 = 1
70  C4 = 1.3679 * C3 - 0.3679 * C2 - 0.3205 * C1
       + 0.1885 * C0 + 0.3205 * R1 - 0.1885 * R0
80  M = C0
90  C0 = C1
100 C1 = C2
110 C2 = C3
120 C3 = C4
130 N = R0
140 R0 = R1
150 PRINT K, M, N
160 K = K + 1
170 IF K < 21 GO TO 70
180 END
```

k = K	$c(k)$ = CK = M	$r(k)$ = RK = N
0	0	1
1	0	1
2	0	1
3	0.3205	1
4	0.570412	1
5	0.794355	1
6	0.906023	1
7	0.956703	1
8	0.960281	1
9	0.952952	1
10	0.946418	1
11	0.948583	1
12	0.956972	1
13	0.968362	1
14	0.978932	1
15	0.986919	1
16	0.991887	1
17	0.994503	1
18	0.995687	1
19	0.996257	1
20	0.9967	1

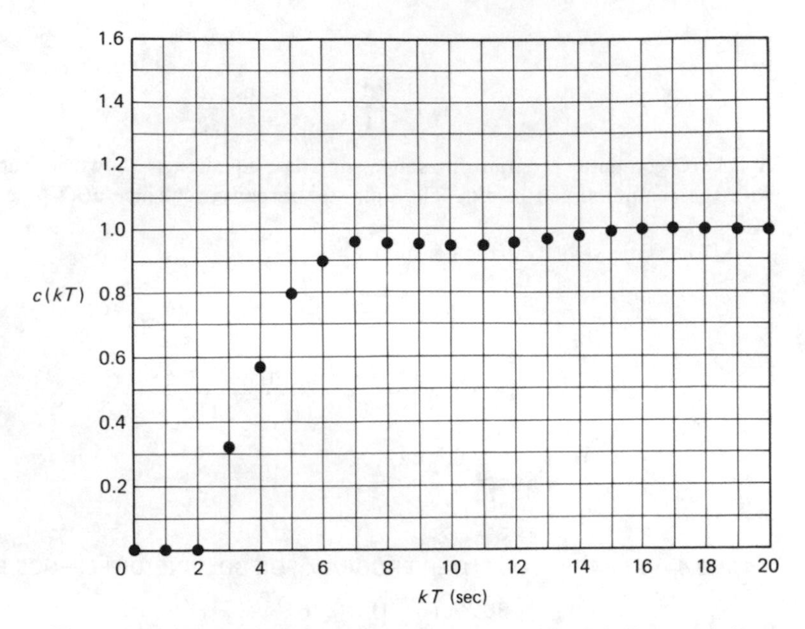

Figure 4–67 Plot of $c(kT)$ vs. kT for the system designed in Prob. A-4–10.

Thus, for a unit ramp input the steady-state error is

$$e_{ss}^* = \frac{1}{K_v} = 4.7881$$

Problem A-4–11. Consider the system shown in Fig. 4–68. We wish to design a digital controller such that the dominant closed-loop poles of the system will have a damping ratio ζ of 0.5. We also want the number of samples per cycle of damped sinusoidal oscillation to be 8. Assume that the sampling period T is 0.2 sec.

Using the root locus method in the z plane, determine the pulse transfer function of the digital controller. Obtain the response of the designed system to a unit step input. Also obtain the static velocity error constant K_v.

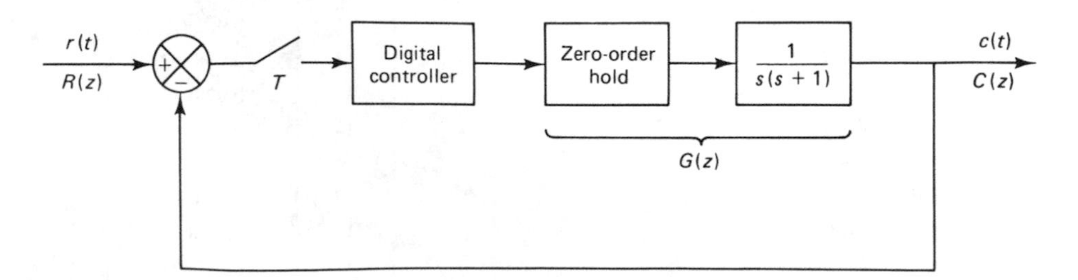

Figure 4–68 Digital control system of Prob. A-4–11.

Solution. We shall first locate the desired closed-loop poles in the z plane. Referring to Eqs. (3–104) and (3–105), for a constant-damping-ratio locus we have

$$|z| = e^{-\zeta T \omega_n} = \exp\left(-\frac{2\pi\zeta}{\sqrt{1-\zeta^2}} \frac{\omega_d}{\omega_s}\right) \tag{4–106}$$

and

$$\angle z = T\omega_d = 2\pi \frac{\omega_d}{\omega_s} = \theta$$

Since we require 8 samples per cycle of damped sinusoidal oscillation, the dominant closed-loop pole in the upper half of the z plane must be located on a line having an angle of 45° and passing through the origin as shown in Fig. 4–69. (Note that the number of samples per cycle is $360°/\theta$. Hence, 8 samples per cycle requires $\theta = 360°/8 = 45°$.) Thus,

$$\angle z = 45° = \frac{\pi}{4} = 2\pi \frac{\omega_d}{\omega_s}$$

which gives

$$\frac{\omega_d}{\omega_s} = \frac{1}{8} \tag{4–107}$$

Since the sampling period T is specified as 0.2 sec, we have

$$\omega_s = \frac{2\pi}{T} = \frac{2\pi}{0.2} = 10\pi$$

Therefore,

$$\omega_d = \frac{1}{8}\omega_s = \frac{10\pi}{8} = 3.9270$$

By letting $\zeta = 0.5$ and substituting Eq. (4–107) into Eq. (4–106), we obtain

$$|z| = e^{-0.4535} = 0.6354$$

Hence, we can locate the desired closed-loop pole in the upper half of the z plane, as shown by point P in Fig. 4–69. Note that at point P

$$|z| \angle z = 0.6354 \angle 45° = 0.4493 + j0.4493$$

Next, we obtain the pulse transfer function $G(z)$ of the plant which is preceded by a zero-order hold:

$$\begin{aligned} G(z) &= \mathscr{Z}\left[\frac{1-e^{-Ts}}{s} \frac{1}{s(s+1)}\right] \\ &= (1-z^{-1})\,\mathscr{Z}\left[\frac{1}{s^2(s+1)}\right] \\ &= (1-z^{-1})\left[\frac{0.2z^{-1}}{(1-z^{-1})^2} - \frac{1}{1-z^{-1}} + \frac{1}{1-e^{-0.2}z^{-1}}\right] \\ &= \frac{0.01873(1+0.9356z^{-1})z^{-1}}{(1-z^{-1})(1-0.8187z^{-1})} = \frac{0.01873(z+0.9356)}{(z-1)(z-0.8187)} \end{aligned}$$

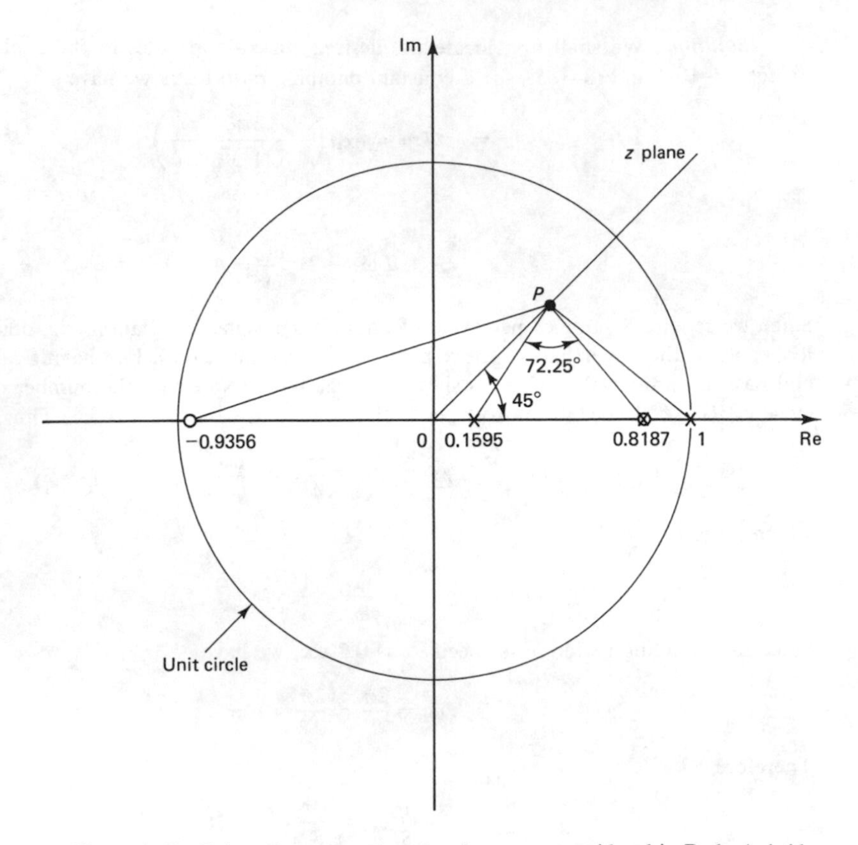

Figure 4–69 Pole and zero locations for the system considered in Prob. A-4–11.

We can now locate the open-loop poles and a zero on the z plane as shown in Fig. 4–69. Since point P is the location of the desired closed-loop pole, the angle deficiency at point P can be calculated easily as follows:

$$-140.79° - 129.43° + 17.97° + 180° = -72.25°$$

The controller pulse transfer function must contribute 72.25°.

Let us choose the controller pulse transfer function to be

$$G_D(z) = K\frac{z + \alpha}{z + \beta}$$

and choose the zero of the controller to cancel the pole at $z = 0.8187$. Then, the pole of the controller can be determined easily from the angle condition as $z = 0.1595$. Thus, we have

$$G_D(z) = K\frac{1 - 0.8187z^{-1}}{1 - 0.1595z^{-1}}$$

The open-loop pulse transfer function of the system is therefore obtained as follows:

$$G_D(z)G(z) = K\frac{1-0.8187z^{-1}}{1-0.1595z^{-1}}\frac{0.01873(1+0.9356z^{-1})z^{-1}}{(1-z^{-1})(1-0.8187z^{-1})}$$

$$= K\frac{0.01873(1+0.9356z^{-1})z^{-1}}{(1-0.1595z^{-1})(1-z^{-1})}$$

The gain constant K can be determined from the magnitude condition:

$$K\left|\frac{0.01873(z+0.9356)}{(z-0.1595)(z-1)}\right|_{z=0.4493+j\,0.4493} = 1$$

or

$$K = 13.934$$

Hence, we have determined the pulse transfer function of the digital controller to be

$$G_D(z) = 13.934\left(\frac{1-0.8187z^{-1}}{1-0.1595z^{-1}}\right)$$

The open-loop pulse transfer function is

$$G_D(z)G(z) = \frac{0.2610(1+0.9356z^{-1})z^{-1}}{(1-0.1595z^{-1})(1-z^{-1})}$$

The closed-loop pulse transfer function is

$$\frac{C(z)}{R(z)} = \frac{G_D(z)G(z)}{1+G_D(z)G(z)}$$

$$= \frac{0.2610z^{-1}+0.2442z^{-2}}{1-0.8985z^{-1}+0.4037z^{-2}}$$

Because of the cancellation of a pole of the plant and the zero of the controller, the order of the system is reduced from third to second. The system has only a pair of conjugate complex closed-loop poles.

We shall now obtain the unit step response. Although a computer solution can be obtained easily, here we demonstrate the solution by long division. Noting that $R(z) = 1/(1 - z^{-1})$, we have

$$C(z) = \frac{0.2610z^{-1}+0.2442z^{-2}}{1-0.8985z^{-1}+0.4037z^{-2}}\frac{1}{1-z^{-1}}$$

$$= \frac{0.2610z^{-1}+0.2442z^{-2}}{1-1.8985z^{-1}+1.3022z^{-2}-0.4037z^{-3}}$$

$$= 0.2610z^{-1} + 0.7397z^{-2} + 1.0645z^{-3} + 1.1630z^{-4}$$
$$+ 1.1204z^{-5} + 1.0424z^{-6} + 0.9895z^{-7} + 0.9734z^{-8}$$
$$+ 0.9804z^{-9} + 0.9931z^{-10} + 1.0017z^{-11} + 1.0043z^{-12}$$
$$+ 1.0032z^{-13} + 1.0011z^{-14} + 0.9997z^{-15} + 0.9993z^{-16} + \cdots$$

Figure 4–70 shows the unit step response sequence $c(kT)$ versus kT. The plot shows the maximum overshoot to be approximately 16.5 percent.

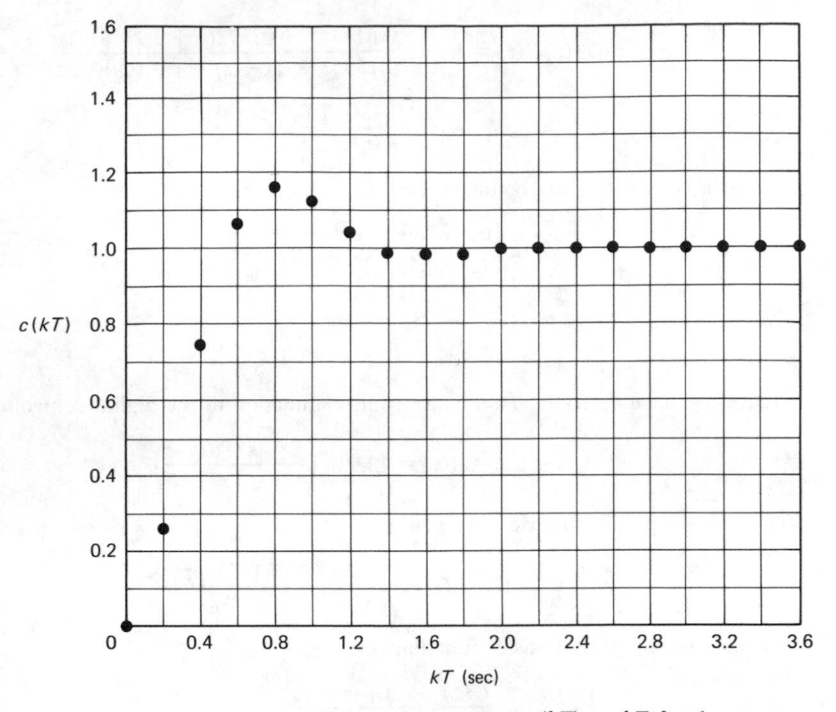

Figure 4–70 Plot of the unit step response sequence $c(kT)$ vs. kT for the system designed in Prob. A-4–11.

Finally, the static velocity error constant K_v is determined as follows:

$$K_v = \lim_{z\to 1}\left[\frac{1-z^{-1}}{T} G_D(z)G(z)\right]$$

$$= \lim_{z\to 1}\left[\frac{1-z^{-1}}{0.2}\frac{0.2610(1+0.9356z^{-1})z^{-1}}{(1-0.1595z^{-1})(1-z^{-1})}\right]$$

$$= 3.005$$

Problem A-4–12. Consider the frequency response of the following difference-equation system:

$$y(k) = 0.8y(k-1) + x(k) + 2x(k-1)$$

where $x(k)$ is the input and $y(k)$ is the output of the system. Determine the sinusoidal pulse transfer function.

Solution. The pulse transfer function of the system is

$$\frac{Y(z)}{X(z)} = G(z) = \frac{1+2z^{-1}}{1-0.8z^{-1}} = \frac{z+2}{z-0.8}$$

The sinusoidal pulse transfer function is

$$G(e^{j\omega T}) = \frac{e^{j\omega T} + 2}{e^{j\omega T} - 0.8} = \frac{2 + \cos \omega T + j \sin \omega T}{-0.8 + \cos \omega T + j \sin \omega T}$$

Note that the magnitude of $G(e^{j\omega T})$ is

$$|G(e^{j\omega T})| = \frac{\sqrt{(2 + \cos \omega T)^2 + \sin^2 \omega T}}{\sqrt{(-0.8 + \cos \omega T)^2 + \sin^2 \omega T}}$$

$$= \sqrt{\frac{5 + 4 \cos \omega T}{1.64 - 1.6 \cos \omega T}}$$

and the phase angle of $G(e^{j\omega T})$ is

$$\underline{/G(e^{j\omega T})} = \tan^{-1}\left(\frac{\sin \omega T}{2 + \cos \omega T}\right) - \tan^{-1}\left(\frac{\sin \omega T}{-0.8 + \cos \omega T}\right)$$

Problem A-4–13. Design a digital controller for the system shown in Fig. 4-71. Use the Bode diagram approach in the w domain. The design specifications are that the phase margin be 55°, the gain margin be at least 10 db, and the static velocity error constant be 5 sec^{-1}. The sampling period is specified as 0.1 sec, or $T = 0.1$. After the controller is designed, draw a root locus diagram. Locate the closed-loop poles on the diagram and find the number of samples per cycle of damped sinusoidal oscillation.

Solution. The z transform of the plant which is preceded by a zero-order hold is

$$G(z) = \mathscr{Z}\left[\frac{1 - e^{-Ts}}{s} \frac{1}{s(s+2)}\right]$$

$$= (1 - z^{-1})\, \mathscr{Z}\left[\frac{1}{s^2(s+2)}\right]$$

$$= 0.004683z^{-1} \frac{1 + 0.9355z^{-1}}{(1 - z^{-1})(1 - 0.8187z^{-1})}$$

$$= (0.004683) \frac{z + 0.9355}{(z-1)(z-0.8187)}$$

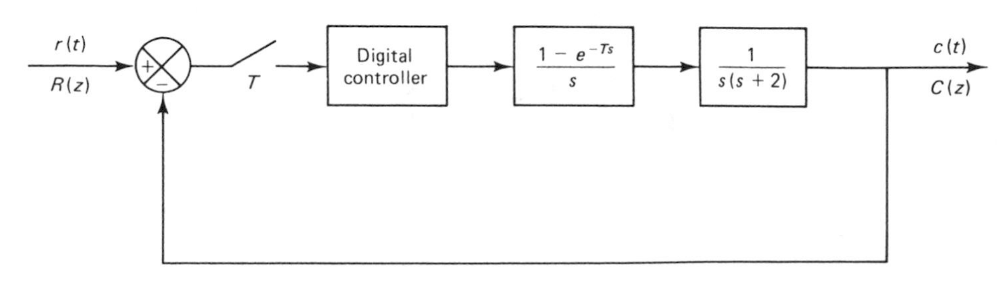

Figure 4–71 Digital control system of Prob. A-4–13.

Let us transform $G(z)$ into $G(w)$ by using the following bilinear transformation:

$$z = \frac{1 + (Tw/2)}{1 - (Tw/2)} = \frac{1 + 0.05w}{1 - 0.05w}$$

By substituting this last equation into $G(z)$, we obtain

$$G(w) = \frac{0.004683\left(\dfrac{1 + 0.05w}{1 - 0.05w} + 0.9355\right)}{\left(\dfrac{1 + 0.05w}{1 - 0.05w} - 1\right)\left(\dfrac{1 + 0.05w}{1 - 0.05w} - 0.8187\right)}$$

$$= \frac{0.5(1 + 0.001666w)(1 - 0.05w)}{w(1 + 0.5016w)}$$

The Bode diagram of $G(j\nu)$ is shown in Fig. 4–72.

We shall now choose the controller transfer function to be of the form

$$G_D(w) = K\frac{1 + (w/\alpha)}{1 + (w/\beta)}$$

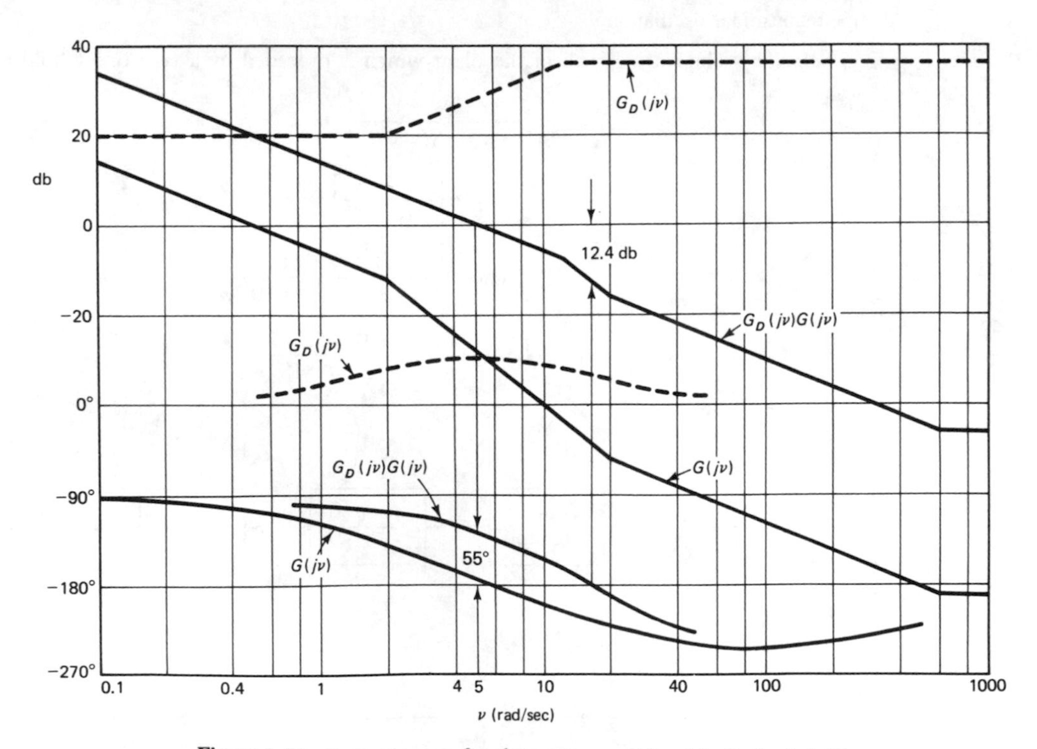

Figure 4–72 Bode diagram for the system considered in Prob. A-4–13.

where K, α, and β are constants. The open-loop transfer function is

$$G_D(w)G(w) = K\frac{1 + (w/\alpha)}{1 + (w/\beta)}\frac{0.5(1 + 0.001666w)(1 - 0.05w)}{w(1 + 0.5016w)}$$

The required static velocity error constant K_v is 5 $\sec^{-1}$. Hence

$$K_v = \lim_{w \to 0}[wG_D(w)G(w)] = 0.5K = 5$$

from which we determine that

$$K = 10$$

Using a conventional design technique, the digital controller transfer function is determined as

$$G_D(w) = 10\left(\frac{1 + \dfrac{w}{1.994}}{1 + \dfrac{w}{12.5}}\right)$$

Therefore, the open-loop transfer function becomes

$$G_D(w)G(w) = 10\left(\frac{1 + \dfrac{w}{1.994}}{1 + \dfrac{w}{12.5}}\right)\frac{0.5(1 + 0.001666w)(1 - 0.05w)}{w(1 + 0.5016w)}$$

This open-loop transfer function gives the phase margin of approximately 55° and the gain margin of approximately 12.4 db. The static velocity error constant K_v is 5 $\sec^{-1}$. Hence, all requirements are satisfied and the designed controller transfer function $G_D(w)$ is satisfactory.

Next, we transform $G_D(w)$ into $G_D(z)$. The following bilinear transformation should be used:

$$w = \frac{2}{T}\frac{z - 1}{z + 1} = \frac{2}{0.1}\frac{z - 1}{z + 1} = 20\left(\frac{z - 1}{z + 1}\right)$$

Then

$$G_D(z) = 10\left(\frac{1 + \dfrac{20}{1.994}\dfrac{z - 1}{z + 1}}{1 + \dfrac{20}{12.5}\dfrac{z - 1}{z + 1}}\right)$$

$$= 42.423\left(\frac{z - 0.8187}{z - 0.2308}\right) = 42.423\left(\frac{1 - 0.8187z^{-1}}{1 - 0.2308z^{-1}}\right)$$

The open-loop pulse transfer function now becomes

$$G_D(z)G(z) = \frac{0.1987(z + 0.9355)}{(z - 0.2308)(z - 1)}$$

Figure 4-73 shows the root locus diagram for the system. Using the magnitude condition we find that the closed-loop poles are located at $z = 0.516 \pm j0.388$. On the root locus diagram,

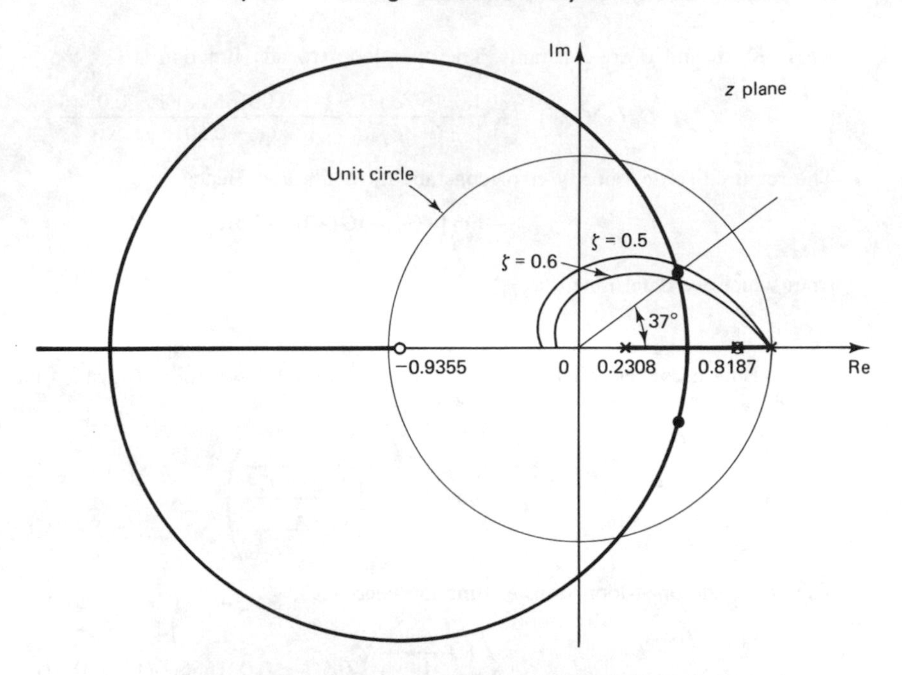

Figure 4–73 Root locus diagram for the system designed in Prob. A-4–13.

constant ζ loci for $\zeta = 0.5$ and 0.6 are superimposed. From the diagram it can be seen that the damping ratio ζ of the closed-loop poles is approximately 0.55.

The line connecting the closed-loop pole in the upper half of the z plane and the origin has an angle of 37°. Hence the number of samples per cycle of damped sinusoidal oscillation is $360°/37° = 9.73$.

Problem A-4–14. Consider the control system shown in Fig. 4–74, where the plant transfer function is $1/s^2$. Design a digital controller in the w plane such that the phase margin is 50° and the gain margin is at least 10 db. The sampling period is 0.1 sec, or $T = 0.1$. After designing the controller, obtain the static velocity error constant K_v. Also, obtain the response of the designed system to a unit step input.

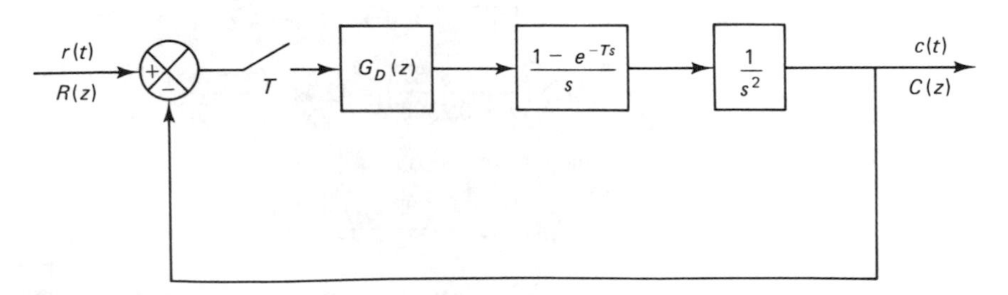

Figure 4–74 Digital control system of Prob. A-4–14.

Solution. We shall first obtain the z transform of the plant which is preceded by the zero-order hold:

$$G(z) = \mathscr{Z}\left[\frac{1-e^{-Ts}}{s}\frac{1}{s^2}\right] = (1-z^{-1})\,\mathscr{Z}\left[\frac{1}{s^3}\right]$$

$$= (1-z^{-1})\frac{T^2(1+z^{-1})z^{-1}}{2(1-z^{-1})^3}$$

$$= \frac{0.005(1+z^{-1})z^{-1}}{(1-z^{-1})^2} = \frac{0.005(z+1)}{(z-1)^2}$$

Next, using the bilinear transformation given by

$$z = \frac{1+(Tw/2)}{1-(Tw/2)} = \frac{1+0.05w}{1-0.05w}$$

we transform $G(z)$ into $G(w)$:

$$G(w) = \frac{0.005\left(\dfrac{1+0.05w}{1-0.05w}+1\right)}{\left(\dfrac{1+0.05w}{1-0.05w}-1\right)^2} = \frac{1-0.05w}{w^2}$$

Thus

$$G(j\nu) = \frac{1-0.05j\nu}{(j\nu)^2}$$

Figure 4–75 shows the Bode diagram of $G(j\nu)$ thus obtained. Notice that the phase margin is $-2°$. It is necessary to add a lead network to give the required phase margin and gain margin. By applying a conventional design technique, it can be seen that the following lead network will satisfy the requirements:

$$G_D(w) = 64\left(\frac{w+1}{w+16}\right)$$

The addition of this lead network modifies the Bode diagram. The gain crossover frequency is shifted to $\nu = 4$. Note that the maximum phase lead ϕ_m this lead network can produce is 61.93°, since

$$\phi_m = \sin^{-1}\frac{1-\frac{1}{16}}{1+\frac{1}{16}} = \sin^{-1} 0.8824 = 61.93°$$

At the gain crossover frequency $\nu = 4$, the phase angle of $G_D(j\nu)G(j\nu)$ becomes $-191.31° + 61.93° = -129.38°$. Thus, the phase margin is 50.62°. The gain margin is found to be approximately 13 db. Hence, the given design specifications are satisfied.

We now transform the controller transfer function $G_D(w)$ into $G_D(z)$. By using the bilinear transformation

$$w = \frac{2}{T}\frac{z-1}{z+1} = \frac{2}{0.1}\frac{z-1}{z+1} = 20\left(\frac{z-1}{z+1}\right)$$

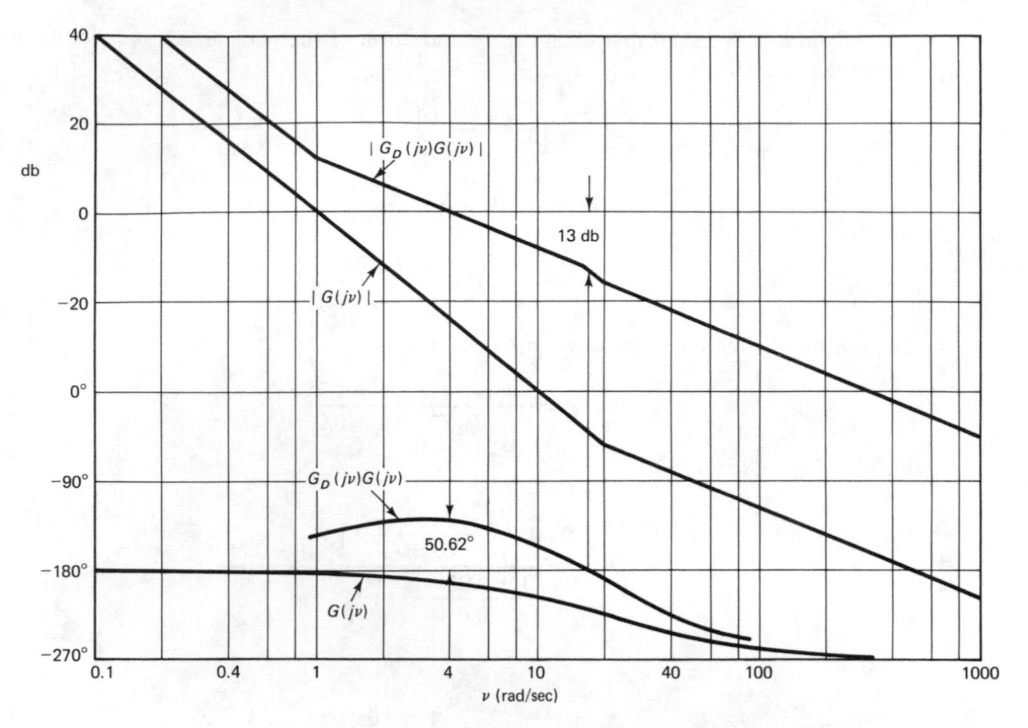

Figure 4–75 Bode diagram for the system considered in Prob. A-4–14.

we obtain

$$G_D(z) = 64 \frac{20\left(\dfrac{z-1}{z+1}\right)+1}{20\left(\dfrac{z-1}{z+1}\right)+16} = 37.333\left(\frac{z-0.9048}{z-0.1111}\right)$$

Hence, the open-loop pulse tranfer function becomes

$$G_D(z)G(z) = 37.333\left(\frac{z-0.9048}{z-0.1111}\right)\frac{0.005(z+1)}{(z-1)^2}$$

$$= \frac{0.1867(1-0.9048z^{-1})(1+z^{-1})z^{-1}}{(1-0.1111z^{-1})(1-z^{-1})^2}$$

The static velocity error constant K_v is obtained as follows:

$$K_v = \lim_{z\to 1}\left[\frac{1-z^{-1}}{T} G_D(z)G(z)\right]$$

$$= \lim_{z\to 1}\left[\frac{1-z^{-1}}{0.1}\frac{0.1867(1-0.9048z^{-1})(1+z^{-1})z^{-1}}{(1-0.1111z^{-1})(1-z^{-1})^2}\right] = \infty$$

Thus, the static velocity error constant K_v is infinity. There is no steady-state error in the ramp response.

Next we shall obtain the unit step response. The closed-loop pulse transfer function of the system is

$$\frac{C(z)}{R(z)} = \frac{0.1867z^{-1} + 0.0178z^{-2} - 0.1689z^{-3}}{1 - 1.9244z^{-1} + 1.2400z^{-2} - 0.2800z^{-3}}$$

A BASIC computer program for finding the unit step response sequence is shown in Table 4–12. The computational results are also shown in Table 4–12. The output $c(kT)$ versus kT is plotted in Fig. 4–76. Notice that the zero of the digital controller at $z = 0.9048$ is close to the double pole at $z = 1$. A zero-pole pair near point $z = 1$ creates a long tail with small amplitude in the response.

TABLE 4–12 BASIC COMPUTER PROGRAM FOR FINDING THE UNIT STEP RESPONSE OF THE SYSTEM GIVEN BY

$$\frac{C(z)}{R(z)} = \frac{0.1867z^{-1} + 0.0178z^{-2} - 0.1689z^{-3}}{1 - 1.9244z^{-1} + 1.2400z^{-2} - 0.2800z^{-3}}$$

```
10  C0 = 0
20  C1 = 0.1867
30  C2 = 0.5638
40  R0 = 1
50  R1 = 1
60  R2 = 1
70  C3 = 1.9244 * C2 - 1.2400 * C1 + 0.2800 * C0
         + 0.1867 * R2 + 0.0178 * R1 - 0.1689 * R0
80  M = C0
90  C0 = C1
100 C1 = C2
110 C2 = C3
120 N = R0
130 R0 = R1
140 R1 = R2
150 PRINT K, M, N
160 K = K + 1
170 IF K < 25 GO TO 70
180 END
```

k = K	$c(k)$ = CK = M	$r(k)$ = RK = N
0	0	1
1	0.1867	1
2	0.5638	1
3	0.889069	1
4	1.09969	1
5	1.20726	1
6	1.24417	1
7	1.2408	1
8	1.21865	1
9	1.19055	1
10	1.16299	1
11	1.1386	1
12	1.11797	1
13	1.10079	1
14	1.08649	1
15	1.07449	1
16	1.06432	1
17	1.05563	1
18	1.04816	1
19	1.0417	1
20	1.03611	1
21	1.03126	1
22	1.02706	1
23	1.02343	1
24	1.02028	1

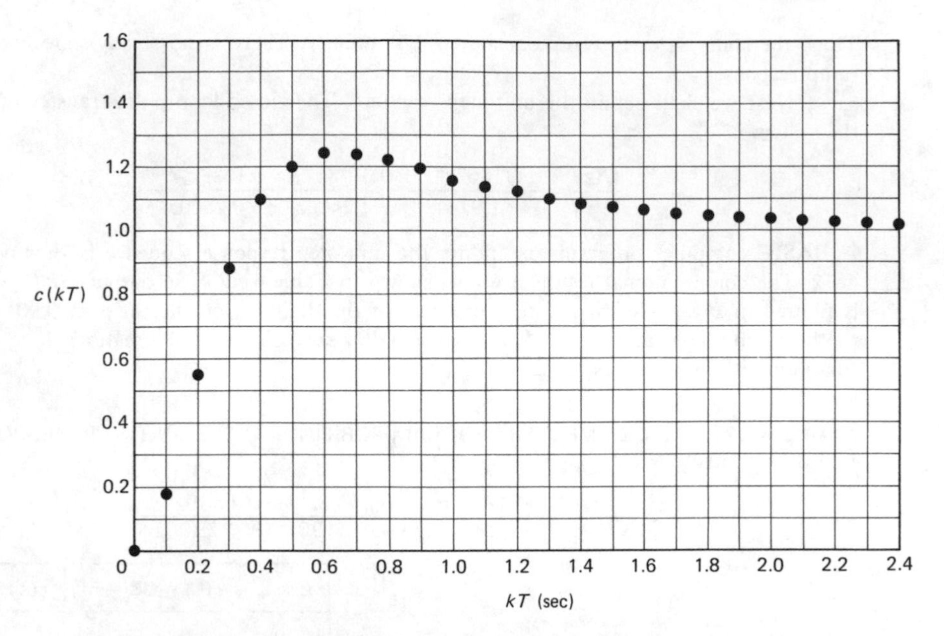

Figure 4–76 Plot of $c(kT)$ vs. kT for the system designed in Prob. A-4–14.

Problem A-4–15. Consider the digital control system shown in Fig. 4–54. It is desired that the system exhibit a deadbeat response to a unit step input. (That is, the settling time should be the minimum possible and the steady-state error must be zero. Also, the system output should not exhibit intersampling ripples after the settling time is reached.) The sampling period T is assumed to be 5 sec, or $T = 5$. Design the digital controller $G_D(z)$.

Solution. We shall first obtain the z transform of the plant which is preceded by the zero-order hold. Referring to Example 4–16, we have

$$G(z) = \mathscr{Z}\left[\frac{1 - e^{-Ts}}{s} G_p(s)\right] = \mathscr{Z}\left[\frac{1 - e^{-Ts}}{s} \frac{e^{-5s}}{10s + 1}\right]$$

$$= \frac{0.3935z^{-2}}{1 - 0.6065z^{-1}}$$

Let us define the closed-loop pulse transfer function as $F(z)$:

$$\frac{C(z)}{R(z)} = \frac{G_D(z)G(z)}{1 + G_D(z)G(z)} = F(z)$$

Notice that if $G(z)$ is expanded into a series in z^{-1}, then the first term is $0.3935z^{-2}$. Hence, $F(z)$ must begin with a term in z^{-2}:

$$F(z) = a_2z^{-2} + a_3z^{-3} + \cdots + a_Nz^{-N}$$

Since the input is a unit step, Eq. (4–72) gives

$$1 - F(z) = (1 - z^{-1})N(z) \tag{4–108}$$

Note that because $G(z)$ involves only a stable pole there is no stability requirement on $F(z)$.

Since the system should not exhibit intersampling ripples after the settling time is reached, we require $U(z)$ to be of the following type of series in z^{-1}:

$$U(z) = b_0 + b_1 z^{-1} + b_2 z^{-2} + \cdots + b_{N-1} z^{-N+1} + b(z^{-N} + z^{-N-1} + z^{-N-2} + \cdots)$$

where b must be a nonzero constant, since the plant does not have an integrator. Thus, $U(z)$ is an infinite series in z^{-1}. From Fig. 4–54, $U(z)$ can be given by

$$U(z) = \frac{C(z)}{G(z)} = \frac{C(z)}{R(z)}\frac{R(z)}{G(z)} = F(z)\frac{R(z)}{G(z)}$$

$$= F(z)\frac{1}{1 - z^{-1}}\frac{1 - 0.6065z^{-1}}{0.3935z^{-2}}$$

$$= F(z)\frac{2.5413(1 - 0.6065z^{-1})}{(1 - z^{-1})z^{-2}} \tag{4–109}$$

In order for $U(z)$ to be an infinite series in z^{-1}, $F(z)$ should not be divisible by $1 - z^{-1}$.

Since there is no other requirement on $F(z)$, referring to Eq. (4–108) we may choose $F(z)$ to be $a_2 z^{-2}$. (With this choice, the response will have the minimum settling time.) Thus,

$$F(z) = a_2 z^{-2}$$

From Eq. (4–108) we have

$$1 - F(z) = 1 - a_2 z^{-2} = (1 - z^{-1})N(z)$$

Hence we choose

$$a_2 = 1$$

and

$$N(z) = 1 + z^{-1}$$

We have thus determined that

$$F(z) = z^{-2}$$

The digital controller pulse transfer function $G_D(z)$ is then determined from Eq. (4–74) as follows:

$$G_D(z) = \frac{F(z)}{G(z)(1 - z^{-1})N(z)}$$

$$= \frac{2.5413(1 - 0.6065z^{-1})}{(1 - z^{-1})(1 + z^{-1})}$$

With the digital controller thus designed, the system output in response to a unit step input is obtained as follows:

$$C(z) = F(z)R(z) = z^{-2}\frac{1}{1 - z^{-1}}$$

$$= z^{-2} + z^{-3} + z^{-4} + \cdots$$

Hence

$$c(0) = 0$$
$$c(1) = 0$$
$$c(k) = 1 \qquad k = 2, 3, 4, \ldots$$

Notice that from Eq. (4–109) we have

$$U(z) = z^{-2}\frac{2.5413(1 - 0.6065z^{-1})}{(1 - z^{-1})z^{-2}} = 2.5413\left(\frac{1 - 0.6065z^{-1}}{1 - z^{-1}}\right)$$
$$= 2.5413 + z^{-1} + z^{-2} + z^{-3} + \cdots$$

Thus the control signal $u(k)$ becomes unity for $k \geq 1$. Consequently, there are no intersampling ripples in the output after the settling time is reached. Figure 4–77 shows plots of $c(k)$ versus k, $u(k)$ versus k, and $u(t)$ versus t in the unit step response.

Problem A-4–16. Consider the digital control system shown in Fig. 4–78. Design a digital controller $G_D(z)$ such that the closed-loop system will exhibit the minimum settling time with zero steady-state error in a unit ramp response. The system should not exhibit intersampling ripples at steady state. The sampling period T is assumed to be 1 sec. After

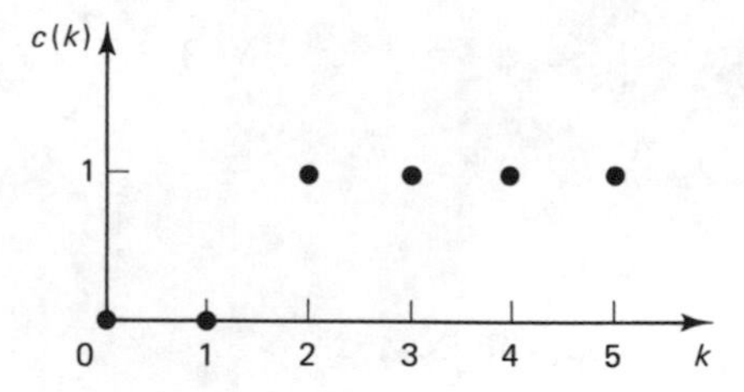

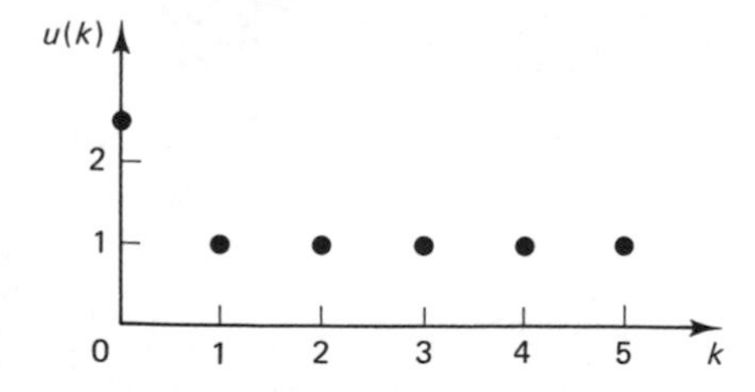

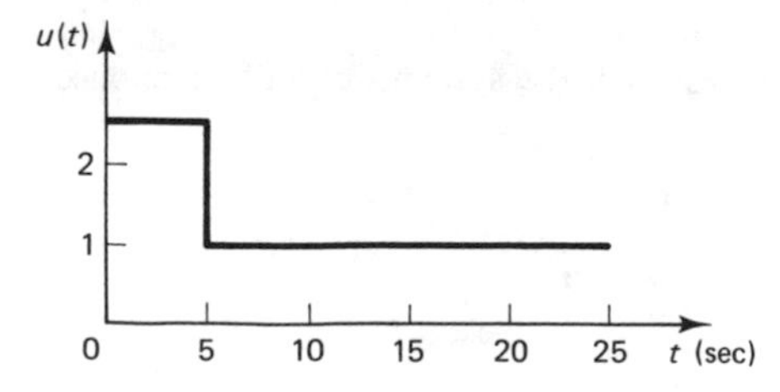

Figure 4–77 Plots of $c(k)$ vs. k, $u(k)$ vs. k, and $u(t)$ vs. t in the unit step response of the system designed in Prob. A-4–15.

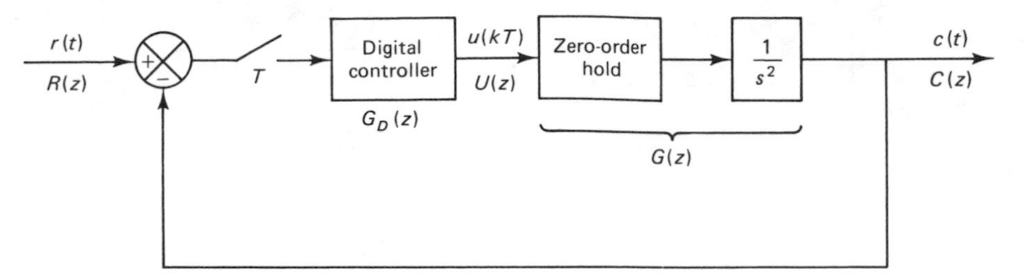

Figure 4–78 Digital control system of Prob. A-4–16.

the controller is designed, investigate the response of the system to a Kronecker delta input and a unit step input.

Solution. The first step in the design is to determine the z transform of the plant which is preceded by the zero-order hold:

$$G(z) = \mathscr{Z}\left[\frac{1-e^{-Ts}}{s}\frac{1}{s^2}\right] = (1-z^{-1})\,\mathscr{Z}\left[\frac{1}{s^3}\right]$$

$$= \frac{(1+z^{-1})z^{-1}}{2(1-z^{-1})^2}$$

Now define the closed-loop pulse transfer function as $F(z)$:

$$\frac{C(z)}{R(z)} = \frac{G_D(z)G(z)}{1+G_D(z)G(z)} = F(z)$$

Notice that if $G(z)$ is expanded into a series in z^{-1}, then the first term will be $0.5z^{-1}$. Hence, $F(z)$ must begin with a term in z^{-1}:

$$F(z) = a_1 z^{-1} + a_2 z^{-2} + \cdots + a_N z^{-N}$$

where $N \geq n$ and n is the order of the system. Since the system here is of the second order, $n = 2$.

Since the input is a unit ramp, from Eq. (4–72) we require that

$$1 - F(z) = (1-z^{-1})^2 N(z) \tag{4–110}$$

Notice that $G(z)$ has a critically stable double pole at $z = 1$. Therefore, from the stability requirement, $1 - F(z)$ must have a double zero at $z = 1$. However, the function $1 - F(z)$ already involves a term $(1 - z^{-1})^2$, and therefore it satisfies the stability requirement.

Since the system should not exhibit intersampling ripples at steady state, we require $U(z)$ to be of the following type of series in z^{-1}:

$$U(z) = b_0 + b_1 z^{-1} + b_2 z^{-2} + \cdots + b_{N-1} z^{-N+1} + b(z^{-N} + z^{-N-1} + z^{-N-2} + \cdots)$$

Because the plant transfer function $G_p(s)$ involves a double integrator, b must be zero. (Otherwise the output increases parabolically, instead of linearly.) Consequently, we have

$$U(z) = b_0 + b_1 z^{-1} + \cdots + b_{N-1} z^{-N+1}$$

From Fig. 4–78, $U(z)$ can be given by

$$U(z) = \frac{C(z)}{G(z)} = \frac{C(z)}{R(z)} \frac{R(z)}{G(z)} = F(z) \frac{R(z)}{G(z)}$$

$$= F(z) \frac{z^{-1}}{(1 - z^{-1})^2} \frac{2(1 - z^{-1})^2}{(1 + z^{-1})z^{-1}}$$

$$= F(z) \frac{2}{1 + z^{-1}}$$

For $U(z)$ to be a series in z^{-1} with a finite number of terms, $F(z)$ must be divisible by $1 + z^{-1}$:

$$F(z) = (1 + z^{-1})F_1(z) \tag{4–111}$$

Then, $U(z)$ can be written as follows:

$$U(z) = 2F_1(z) \tag{4–112}$$

where $F_1(z)$ is a polynomial in z^{-1} with a finite number of terms.

By comparing Eqs. (4–110) and (4–111) and by making a simple analysis, we see that $F(z)$ must involve a term with at least z^{-3}. Hence we assume

$$F(z) = a_1 z^{-1} + a_2 z^{-2} + a_3 z^{-3}$$

This assumed form of $F(z)$ involves the minimum number of terms; the transient response will settle in three sampling periods.

We shall now determine constants a_1, a_2, and a_3. From Eq. (4–110) we have

$$1 - a_1 z^{-1} - a_2 z^{-2} - a_3 z^{-3} = (1 - z^{-1})^2 N(z)$$

If we divide the left-hand side of this last equation by $(1 - z^{-1})^2$, the quotient is $1 + (2 - a_1)z^{-1}$. The remainder is $[2(2 - a_1) - (1 + a_2)]z^{-2} - [(2 - a_1) + a_3]z^{-3}$. Hence, $N(z)$ is determined as follows:

$$N(z) = 1 + (2 - a_1)z^{-1}$$

and the remainder is set equal to zero:

$$[2(2 - a_1) - (1 + a_2)]z^{-2} - (2 - a_1 + a_3)z^{-3} = 0$$

To satisfy this last equation, we require that

$$2(2 - a_1) - (1 + a_2) = 0 \tag{4–113}$$

$$2 - a_1 + a_3 = 0 \tag{4–114}$$

From Eq. (4–111) we have

$$a_1 z^{-1} + a_2 z^{-2} + a_3 z^{-3} = (1 + z^{-1})F_1(z)$$

If we divide the left-hand side of this last equation by $1 + z^{-1}$, the quotient is $a_1 z^{-1} + (a_2 - a_1)z^{-2}$. The remainder is $(a_1 - a_2 + a_3)z^{-3}$. Hence,

$$F_1(z) = a_1 z^{-1} + (a_2 - a_1)z^{-2}$$

and the remainder is set equal to zero:

$$a_1 - a_2 + a_3 = 0 \tag{4–115}$$

By solving Eqs. (4–113), (4–114), and (4–115) for a_1, a_2, and a_3, we obtain

$$a_1 = 1.25, \qquad a_2 = 0.5, \qquad a_3 = -0.75$$

Hence,

$$N(z) = 1 + 0.75z^{-1}$$

and

$$F_1(z) = 1.25z^{-1} - 0.75z^{-2} = 1.25z^{-1}(1 - 0.6z^{-1})$$

and $F(z)$ is determined as follows:

$$\begin{aligned} F(z) &= 1.25z^{-1} + 0.5z^{-2} - 0.75z^{-3} \\ &= 1.25z^{-1}(1 + z^{-1})(1 - 0.6z^{-1}) \end{aligned}$$

The digital controller $G_D(z)$ is then determined from Eq. (4–74):

$$\begin{aligned} G_D(z) &= \frac{F(z)}{G(z)(1 - z^{-1})^2 N(z)} \\ &= \frac{1.25z^{-1}(1 + z^{-1})(1 - 0.6z^{-1})}{\dfrac{(1 + z^{-1})z^{-1}}{2(1 - z^{-1})^2}(1 - z^{-1})^2 (1 + 0.75z^{-1})} \\ &= \frac{2.5(1 - 0.6z^{-1})}{1 + 0.75z^{-1}} \end{aligned}$$

With the digital controller thus designed, the system output in response to a unit ramp input is obtained as follows:

$$\begin{aligned} C(z) &= F(z)R(z) \\ &= (1.25z^{-1} + 0.5z^{-2} - 0.75z^{-3})\frac{z^{-1}}{(1 - z^{-1})^2} \\ &= 1.25z^{-2} + 3z^{-3} + 4z^{-4} + 5z^{-5} + \cdots \end{aligned}$$

Hence

$$\begin{aligned} c(0) &= 0 \\ c(1) &= 0 \\ c(2) &= 1.25 \\ c(k) &= k \qquad k = 3, 4, 5, \ldots \end{aligned}$$

Notice that from Eq. (4–112) we have

$$\begin{aligned} U(z) &= 2F_1(z) \\ &= 2(1.25z^{-1})(1 - 0.6z^{-1}) \\ &= 2.5z^{-1} - 1.5z^{-2} \end{aligned}$$

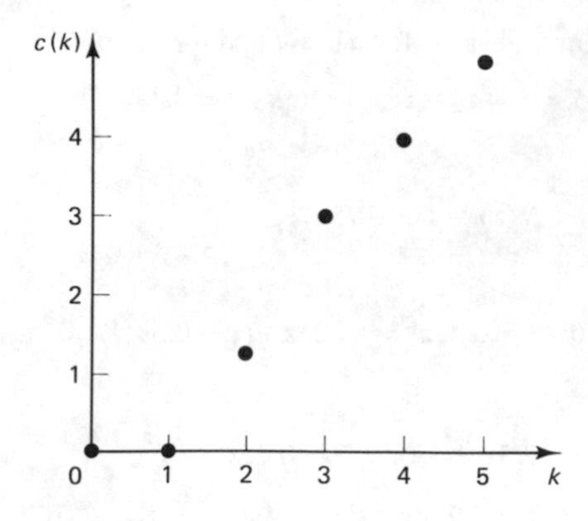

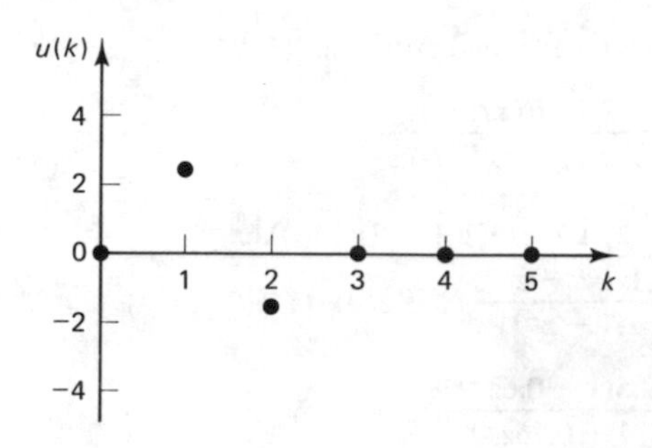

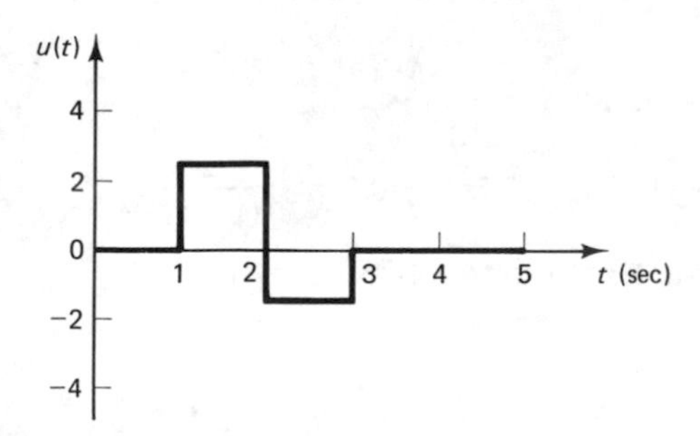

Figure 4–79 Plots of $c(k)$ vs. k, $u(k)$ vs. k, and $u(t)$ vs. t in the unit ramp response of the system designed in Prob. A-4–16.

Thus, the control signal $u(k)$ becomes zero for $k \geq 3$. Consequently, there are no intersampling ripples in the response at steady state. Figure 4–79 shows plots of $c(k)$ versus k, $u(k)$ versus k, and $u(t)$ versus t in the unit ramp response.

Next, let us investigate the response of this system to a Kronecker delta input and a unit step input. For a Kronecker delta input,

$$C(z) = F(z)R(z) = F(z) = 1.25z^{-1} + 0.5z^{-2} - 0.75z^{-3}$$

Notice that $U(z)$ in this case becomes as follows:

$$U(z) = F(z)\frac{R(z)}{G(z)} = \frac{1.25z^{-1}(1 + z^{-1})(1 - 0.6z^{-1})}{(1 + z^{-1})z^{-1}/[2(1 - z^{-1})^2]}$$

$$= 2.5(1 - 0.6z^{-1})(1 - z^{-1})^2$$

$$= 2.5 - 6.5z^{-1} + 5.5z^{-2} - 1.5z^{-3}$$

The control signal $u(k)$ becomes zero for $k \geq 4$. Hence, there are no intersampling ripples after $t \geq 4T = 4$.

For the unit step input,

$$C(z) = F(z)R(z) = (1.25z^{-1} + 0.5z^{-2} - 0.75z^{-3})\frac{1}{1 - z^{-1}}$$

$$= 1.25z^{-1} + 1.75z^{-2} + z^{-3} + z^{-4} + z^{-5} + \cdots$$

The maximum overshoot is 75 percent in the unit step response. Notice that

$$U(z) = F(z)\frac{R(z)}{G(z)} = \frac{1.25z^{-1}(1 + z^{-1})(1 - 0.6z^{-1})}{\dfrac{(1 + z^{-1})z^{-1}}{2(1 - z^{-1})^2}(1 - z^{-1})}$$

$$= 1.25(1 - 0.6z^{-1})(2)(1 - z^{-1})$$

$$= 2.5 - 4z^{-1} + 1.5z^{-2}$$

The control signal $u(k)$ becomes zero for $k \geq 3$. Consequently, there are no intersampling ripples after the settling time is reached. Figure 4–80(a) shows plots of $c(k)$ versus k, $u(k)$ versus k, and $u(t)$ versus t in the response to the Kronecker delta input. Figure 4–80(b) shows similar plots in the unit step response. Notice that when the system is designed for the ramp input, the response to a step input is no longer deadbeat.

Problem A-4–17. Consider the digital control system shown in Fig. 4–81(a). The plant transfer function is in the form of a first-order term with a delay. That is,

$$G_p(s) = \frac{Ke^{-Ls}}{\tau s + 1} = Ke^{-Ls}\frac{a}{s + a}$$

where K is the plant gain, L is the delay time, and $\tau = 1/a$ is the time constant of the plant.

It is assumed that the response to a unit step input is specified as shown in Fig. 4–81(b). The system output exhibits a finite settling time with no steady-state error and no intersampling ripples after the settling time is reached. The characteristic of this output curve is that after a delay equal to L, the output rises exponentially and stays at the desired value, $c(t) = 1$.

Design a digital controller $G_D(z)$ such that the output follows the specified response curve when it is subjected to a unit step input.

Solution. First, we must choose the sampling period T. A simple way to determine the sampling period T is to choose it so that

$$T = \frac{L}{i} = \frac{T_s}{m}$$

where T_s is the transient response time, or the time it takes for the output to move from zero to unity, and i and m are positive integers. In the following analysis, we assume that we can find such positive integers i and m.

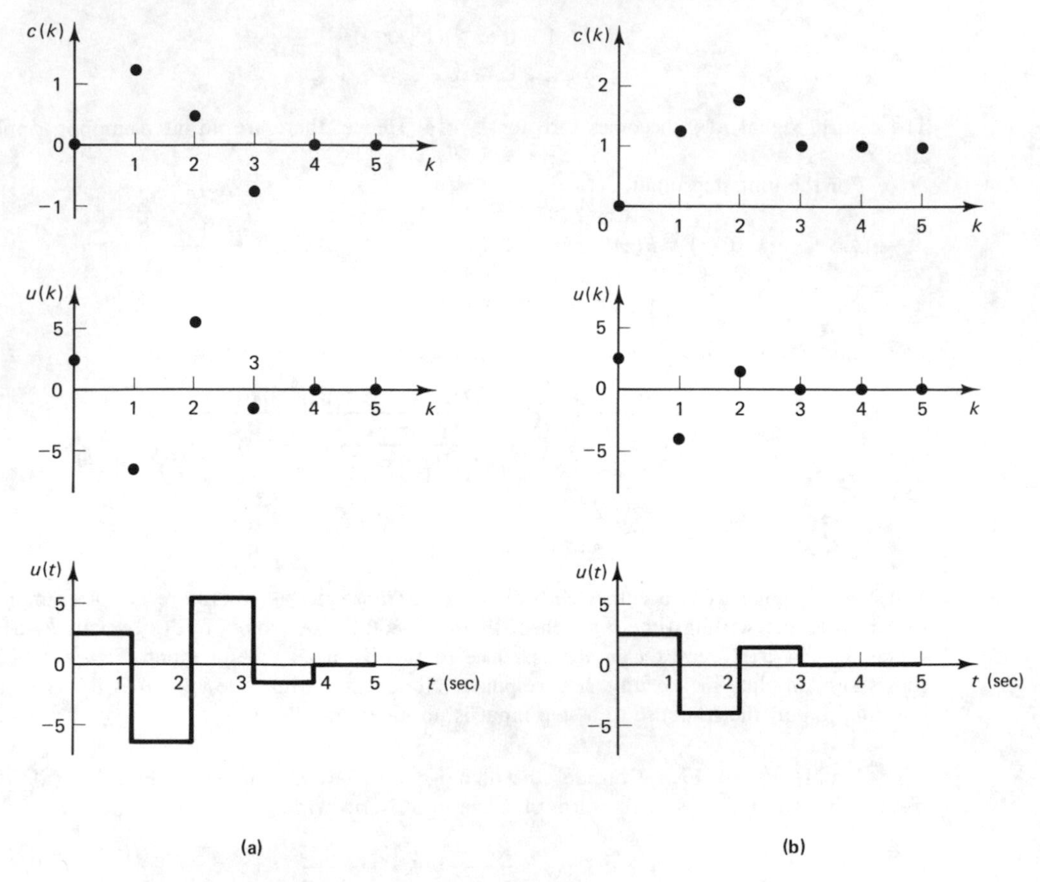

Figure 4–80 (a) Plots of $c(k)$ vs. k, $u(k)$ vs. k, and $u(t)$ vs. t in the response to the Kronecker delta input of the system designed in Prob. A-4–16; (b) plots of $c(k)$ vs. k, $u(k)$ vs. k, and $u(t)$ vs. t in the unit step response of the same system.

Next, we determine the z transform of the plant which is preceded by the zero-order hold. Since

$$Ke^{-Ls}\frac{a}{s+a} = Ke^{-iTs}\frac{a}{s+a}$$

we have

$$\begin{aligned} G(z) &= \mathscr{Z}\left[\frac{1-e^{-Ts}}{s}Ke^{-iTs}\frac{a}{s+a}\right] \\ &= (1-z^{-1})Kz^{-i}\,\mathscr{Z}\left[\frac{a}{s(s+a)}\right] \\ &= Kz^{-i-1}\frac{1-e^{-aT}}{1-e^{-aT}z^{-1}} \end{aligned}$$

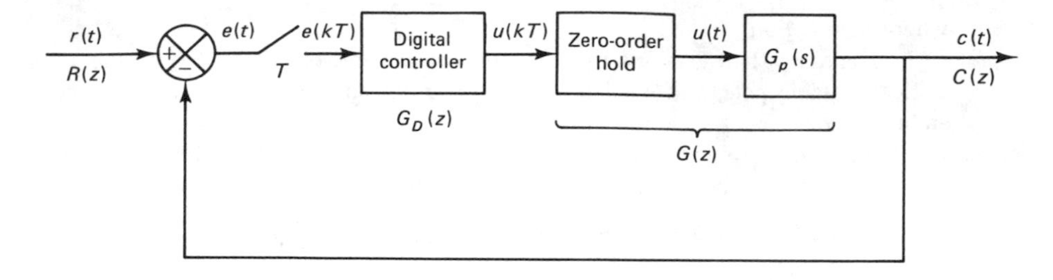

(a)

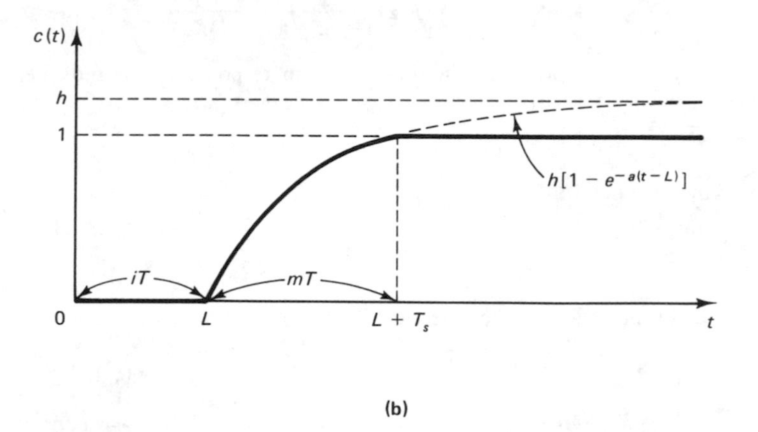

(b)

Figure 4–81 (a) Digital control system of Prob. A-4–17; (b) specified unit step response curve.

Let us define

$$e^{-aT} = d$$

Then, $G(z)$ can be written in the form

$$G(z) = Kz^{-i-1}\frac{1-d}{1-dz^{-1}} \tag{4–116}$$

Now define the closed-loop pulse transfer function as $F(z)$:

$$\frac{C(z)}{R(z)} = \frac{G_D(z)G(z)}{1+G_D(z)G(z)} = F(z) \tag{4–117}$$

Notice that if $G(z)$ is expanded into a series in z^{-1}, then the first term will be Kz^{-i-1}. Hence, $F(z)$ must begin with a term in z^{-i-1}:

$$F(z) = a_{i+1}z^{-i-1} + a_{i+2}z^{-i-2} + \cdots + a_n z^{-n}$$

where $n = i + m$.

Next, we shall examine the stability requirement. Notice that $G(z)$ has no unstable or critically stable pole, and so there is no constraint on $1 - F(z)$ from the standpoint of the stability requirement.

Since the output should not exhibit intersampling ripples after the settling time is reached, we require $U(z)$ to be a series in z^{-1} of the following form:

$$U(z) = b_0 + b_1 z^{-1} + b_2 z^{-2} + \cdots + b_{m-1} z^{-m+1} + b(z^{-m} + z^{-m-1} + \cdots)$$

Because the plant transfer function $G_p(s)$ does not have an integrator, b must be a nonzero constant.

From Fig. 4–81(a) and Eq. (4–116), $U(z)$ may be obtained as follows:

$$\begin{aligned} U(z) &= \frac{C(z)}{G(z)} = \frac{C(z)}{R(z)}\frac{R(z)}{G(z)} = F(z)\frac{R(z)}{G(z)} \\ &= F(z)\frac{1 - dz^{-1}}{(1 - z^{-1})(1 - d)Kz^{-i-1}} \end{aligned} \tag{4–118}$$

In the present problem, the output $c(t)$ in response to the unit step input is specified as shown in Fig. 4–81(b). From the curve, we have

$$c(n) = h(1 - e^{-amT}) = h(1 - d^m) = 1$$

from which we obtain

$$h = \frac{1}{1 - d^m}$$

Hence the output $c(k)$ can be given as follows:

$$c(k) = \begin{cases} 0 & k = 0, 1, 2, \ldots, i \\ h[1 - e^{-a(k-i)T}] = h(1 - d^{k-i}) = \dfrac{1 - d^{k-i}}{1 - d^m} & k = i + 1, i + 2, \ldots, n - 1 \\ 1 & k = n, n + 1, n + 2, \ldots \end{cases}$$

Referring to Prob. A-2–12 and noting that $n = i + m$, we can rewrite $c(k)$ for $k = i + 1, i + 2, \ldots, n - 1$ as follows:

$$c(i + 1) = \frac{1 - d}{1 - d^m} = \frac{1}{\Delta}$$

$$c(i + 2) = \frac{1 - d^2}{1 - d^m} = \frac{1 + d}{\Delta}$$

$$\vdots$$

$$c(n - 1) = \frac{1 - d^{n-1-i}}{1 - d^m} = \frac{1 - d^{m-1}}{1 - d^m} = \frac{1 + d + d^2 + \cdots + d^{m-2}}{\Delta}$$

$$c(n) = 1$$

where

$$\Delta = 1 + d + d^2 + \cdots + d^{m-1}$$

Then, $C(z)$ can be obtained as follows:

$$\begin{aligned}
C(z) &= F(z)R(z) = F(z)\frac{1}{1 - z^{-1}} \\
&= \frac{1}{\Delta}z^{-i-1} + \frac{1+d}{\Delta}z^{-i-2} + \cdots \\
&\quad + \frac{1 + d + d^2 + \cdots + d^{m-2}}{\Delta}z^{-n+1} + z^{-n} + z^{-n-1} + z^{-n-2} + \cdots \\
&= \frac{1}{\Delta(1 - z^{-1})}\{[1 + (1+d)z^{-1} + \cdots \\
&\quad + (1 + d + d^2 + \cdots + d^{m-2})z^{-m+2}](1 - z^{-1}) \\
&\quad + (1 + d + d^2 + \cdots + d^{m-1})z^{-m+1}\}z^{-i-1} \\
&= \frac{(1 + dz^{-1} + d^2z^{-2} + \cdots + d^{m-1}z^{-m+1})z^{-i-1}}{(1 + d + d^2 + \cdots + d^{m-1})(1 - z^{-1})}
\end{aligned}$$

Hence $F(z)$ is determined as follows:

$$\begin{aligned}
F(z) &= \frac{(1 + dz^{-1} + d^2z^{-2} + \cdots + d^{m-1}z^{-m+1})z^{-i-1}}{1 + d + d^2 + \cdots + d^{m-1}} \\
&= \frac{1-d}{1-d^m}\frac{[1 - (dz^{-1})^m]z^{-i-1}}{1 - dz^{-1}}
\end{aligned}$$

Once $F(z)$ is determined, then by referring to Eq. (4–117) the pulse transfer function of the digital controller can be obtained as follows:

$$G_D(z) = \frac{F(z)}{G(z)[1 - F(z)]}$$

Since

$$\begin{aligned}
1 - F(z) &= 1 - \frac{(1 + dz^{-1} + d^2z^{-2} + \cdots + d^{m-1}z^{-m+1})z^{-i-1}}{1 + d + d^2 + \cdots + d^{m-1}} \\
&= \frac{(1 - z^{-i-1}) + d(1 - z^{-i-2}) + \cdots + d^{m-1}(1 - z^{-i-m})}{1 + d + d^2 + \cdots + d^{m-1}} \\
&= (1 - z^{-1})\frac{\sum_{k=i}^{n-1} d^{k-i}(1 + z^{-1} + \cdots + z^{-k})}{1 + d + d^2 + \cdots + d^{m-1}}
\end{aligned}$$

we have

$$G_D(z) = \frac{\dfrac{1-d}{1-d^m}\dfrac{1 - (dz^{-1})^m}{1 - dz^{-1}}z^{-i-1}(1 + d + d^2 + \cdots + d^{m-1})}{Kz^{-i-1}\dfrac{1-d}{1 - dz^{-1}}(1 - z^{-1})\sum_{k=i}^{n-1} d^{k-i}(1 + z^{-1} + \cdots + z^{-k})}$$

$$= \frac{1-(dz^{-1})^m}{(1-d)K(1-z^{-1})\sum_{k=i}^{n-1} d^{k-i}(1+z^{-1}+\cdots+z^{-k})}$$

This last equation gives the pulse transfer function for the desired digital controller. For example, if $i = 1$, $m = 2$, $n = i + m = 3$, $K = 1$, and $d = 0.6065$, then the pulse transfer function of the digital controller becomes as follows:

$$G_D(z) = \frac{1 - 0.3678z^{-2}}{0.3935(1 - z^{-1})[1 + z^{-1} + 0.6065(1 + z^{-1} + z^{-2})]}$$

$$= \frac{1.5820(1-0.3678z^{-2})}{(1-z^{-1})(1+z^{-1}+0.3775z^{-2})}$$

which is the same as that for the digital controller designed in Example 4–16.

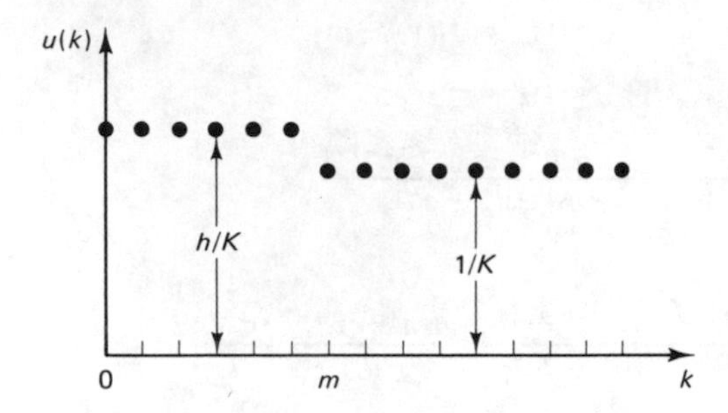

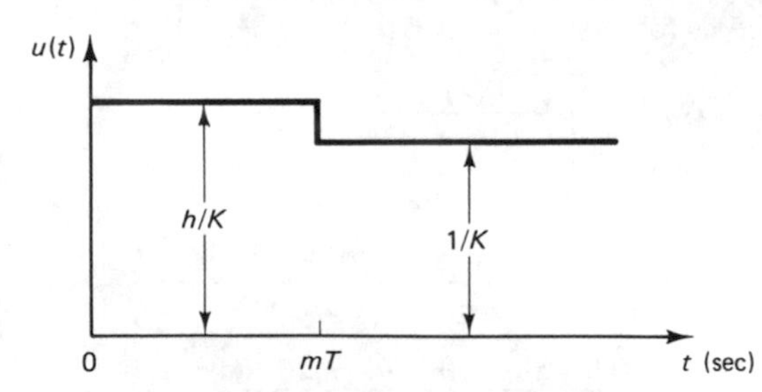

Figure 4–82 Plots of $u(k)$ vs. k and $u(t)$ vs. t for the system designed in Prob. A-4–17.

Now let us examine $U(z)$. From Eq. (4–118), $U(z)$ is obtained as follows:

$$U(z) = F(z)\frac{1-dz^{-1}}{(1-z^{-1})(1-d)Kz^{-i-1}}$$

$$= \frac{1-d}{1-d^m}\frac{[1-(dz^{-1})^m]z^{-i-1}}{1-dz^{-1}}\frac{1-dz^{-1}}{(1-z^{-1})(1-d)Kz^{-i-1}}$$

$$= \frac{1-(dz^{-1})^m}{K(1-d^m)(1-z^{-1})} = \frac{h-(h-1)z^{-m}}{K(1-z^{-1})} = \frac{h(1-z^{-m})+z^{-m}}{K(1-z^{-1})}$$

$$= \frac{h}{K}(1 + z^{-1} + z^{-2} + \cdots + z^{-m+1})$$

$$+ \frac{1}{K}(z^{-m} + z^{-m-1} + z^{-m-2} + \cdots)$$

Since $u(k)$ becomes constant for $k \geq m$, there are no intersampling ripples after the settling time is reached.

Figure 4–82 shows plots of $u(k)$ versus k and $u(t)$ versus t.

PROBLEMS

Problem B-4–1. Obtain the discrete-time equivalent of the following continuous-time filter by use of (1) the backward difference method and (2) the impulse-invariance method:

$$G(s) = \frac{2}{(s+1)(s+2)}$$

Assume that the sampling period is 0.1 sec, or $T = 0.1$.

Problem B-4–2. Obtain the equivalent discrete-time filter of the following continuous-time filter by the matched pole-zero mapping method:

$$G(s) = \frac{s+a}{s(s+b)}$$

The sampling period is T.

Problem B-4–3. Consider the analog control system shown in Fig. 4–83(a). We wish to replace the analog controller by a digital controller, as shown in Fig. 4–83(b). First modify the analog controller to take into account the effect of the hold that must be included in the equivalent digital control system. Then by using the matched pole-zero mapping method, determine the equivalent digital controller. The sampling period is assumed to be 0.2 sec, or $T = 0.2$. Compare the static velocity error constants of the original analog control system and the equivalent digital control system.

Problem B-4–4. In the analog control system shown in Fig. 4–84, we wish to replace the analog controller by a digital controller. Design an equivalent digital controller. The sampling period is assumed to be 0.1 sec, or $T = 0.1$. Then, obtain the unit step response and plot $c(kT)$ versus kT.

Problem B-4–5. Consider the control system shown in Fig. 4–85. Plot the root loci as the gain K is varied from 0 to ∞. Determine the critical value of gain K for stability. The sampling period is 0.1 sec, or $T = 0.1$. What value of gain K will yield a damping ratio ζ of the closed-loop poles equal to 0.5? With gain K set to yield $\zeta = 0.5$, determine the damped natural frequency ω_d and the number of samples per cycle of damped sinusoidal oscillation.

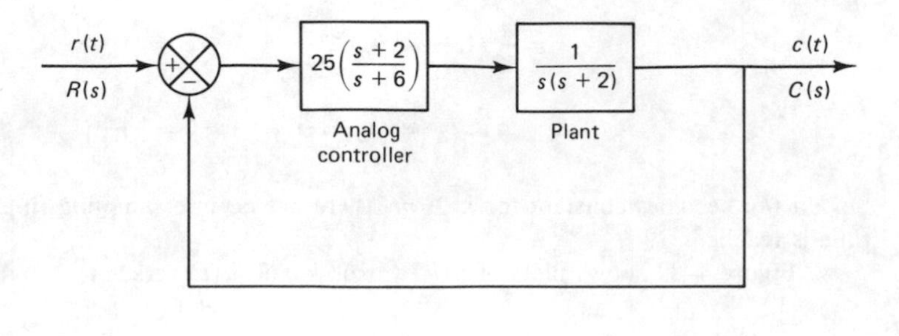

(a)

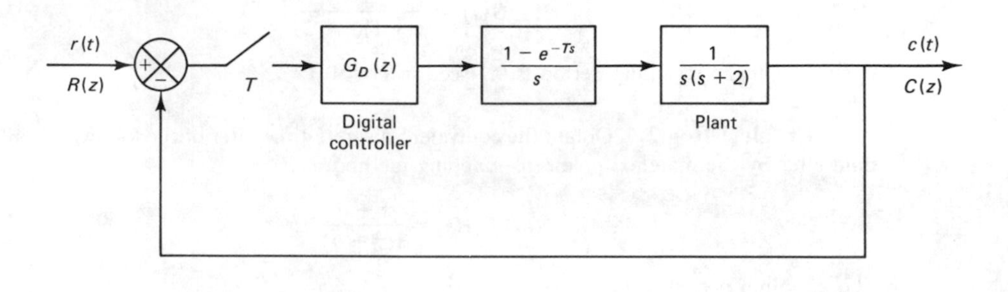

(b)

Figure 4–83 (a) Analog control system of Prob. B-4–3; (b) equivalent digital control system.

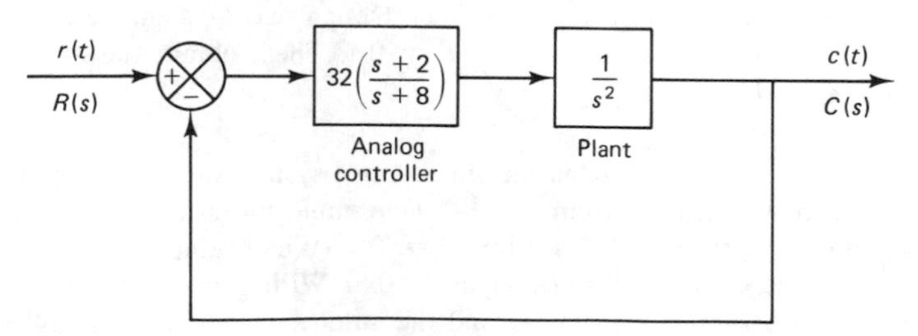

Figure 4–84 Analog control system for Prob. B-4–4.

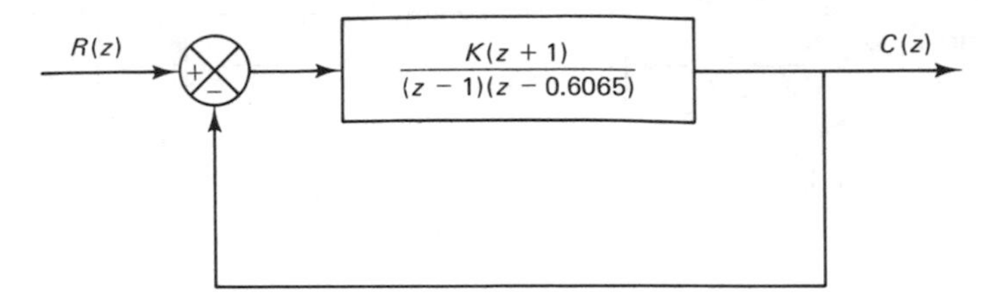

Figure 4–85 Control system for Prob. B-4–5.

Problem B-4–6. Referring to the digital control system shown in Fig. 4–86, design a digital controller $G_D(z)$ such that the damping ratio ζ of the dominant closed-loop poles is 0.5 and the number of samples per cycle of damped sinusoidal oscillation is 8. Assume that the sampling period is 0.1 sec, or $T = 0.1$. Determine the static velocity error constant. Also, determine the response of the designed system to a unit step input.

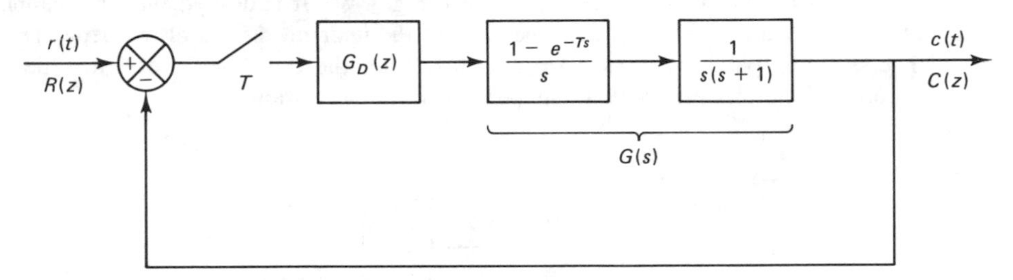

Figure 4–86 Digital control system for Prob. B-4–6.

Problem B-4–7. Consider the control system shown in Fig. 4–87. Design a suitable digital controller that includes an integral control action. The design specifications are that the damping ratio ζ of the dominant closed-loop poles be 0.5 and that there be at least 8 samples per cycle of damped sinusoidal oscillation. The sampling period is assumed to be 0.2 sec, or $T = 0.2$. After the digital controller is designed, determine the static velocity error constant K_v.

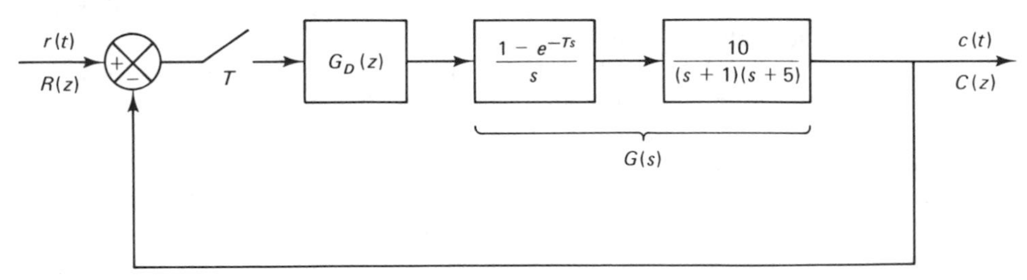

Figure 4–87 Digital control system for Prob. B-4–7.

Problem B-4–8. Consider the digital control system shown in Fig. 4–88, where the plant is of the first order and has a dead time of 5 sec. By choosing a reasonable sampling period T, design a digital PI controller such that the dominant closed-loop poles have a damping

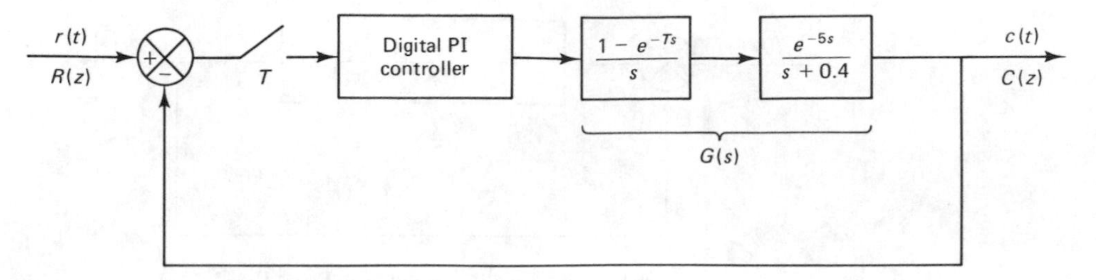

Figure 4–88 Digital control system for Prob. B-4–8.

ratio ζ of 0.5 and the number of samples per cycle of damped sinusoidal oscillation is 10. After the controller is designed, determine the response of the system to a unit step input.

Problem B-4–9. Design a digital proportional-plus-derivative controller for the plant whose transfer function is $1/s^2$, as shown in Fig. 4-89. It is desired that the damping ratio ζ of the dominant closed-loop poles be 0.5 and the undamped natural frequency be 4 rad/sec. The sampling period is 0.1 sec, or $T = 0.1$. After the controller is designed, determine the number of samples per cycle of damped sinusoidal oscillation.

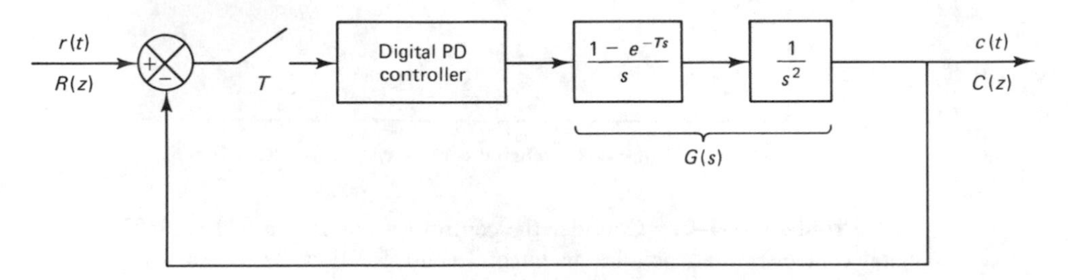

Figure 4–89 Digital control system for Prob. B-4–9.

Problem B-4–10. Referring to the system considered in Prob. A-4–11, redesign the digital controller so that the static velocity error constant K_v is 12 sec^{-1}, without appreciably changing the locations of the dominant closed-loop poles in the z plane. The sampling period is assumed to be 0.2 sec, or $T = 0.2$. After the controller is redesigned, obtain the unit step response and unit ramp response of the digital control system.

Problem B-4–11. Consider the system defined by

$$G(z) = \frac{1}{1 - e^{-aT}z^{-1}}$$

Determine the steady-state amplitude and phase angle of $G(z)$ when it is subjected to sinusoidal inputs.

Problem B-4–12. Consider the control system shown in Fig. 4–90. Draw a Bode diagram in the w domain. Set the gain K so that the phase margin becomes equal to 50°.

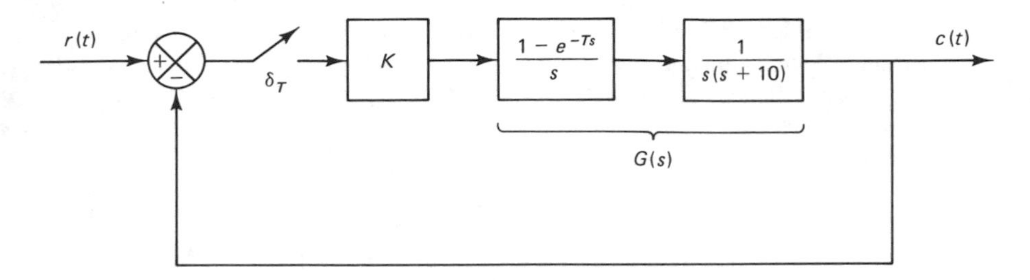

Figure 4–90 Control system for Prob. B-4–12.

With the gain K so set, determine the gain margin and the static velocity error constant K_v. The sampling period is assumed to be 0.1 sec, or $T = 0.1$.

Problem B-4–13. Using the Bode diagram approach in the w domain, design a digital controller for the system shown in Fig. 4–91. The design specifications are that the phase margin be 50° and that the gain margin be at least 10 db, and that the static velocity error constant K_v be 20 sec^{-1}. The sampling period is assumed to be 0.1 sec, or $T = 0.1$. After the controller is designed, calculate the number of samples per cycle of damped sinusoidal oscillation.

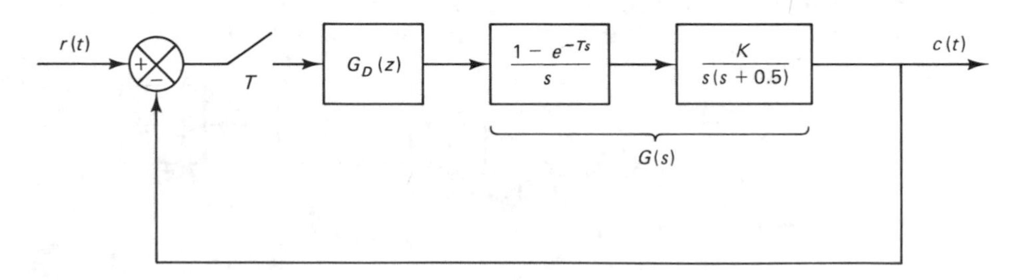

Figure 4–91 Digital control system for Prob. B-4–13.

Problem B-4–14. Consider the digital control system shown in Fig. 4–92. Using the Bode diagram approach in the w domain, design a digital controller such that the phase margin is 60°, the gain margin is 12 db or more, and the static velocity error constant is 5 sec^{-1}. The sampling period is assumed to be 0.1 sec, or $T = 0.1$.

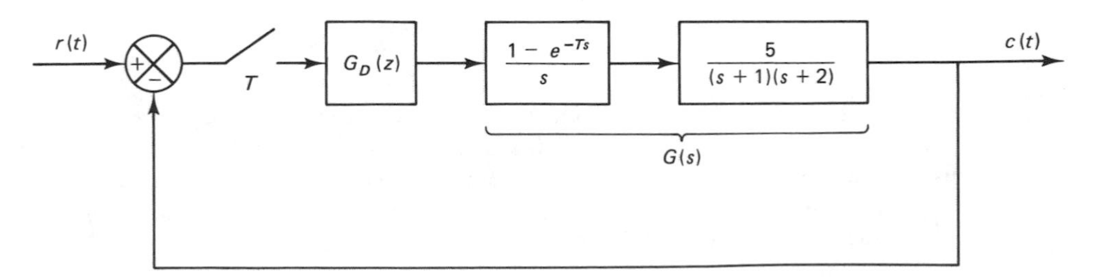

Figure 4–92 Digital control system for Prob. B-4–14.

Problem B-4-15. Consider the system shown in Fig. 4-93. Design a digital controller using a Bode diagram in the w domain such that the phase margin is 50° and the gain margin is at least 10 db. It is desired that the static velocity error constant K_v be 10 sec^{-1}. The sampling period is specified as 0.1 sec, or $T = 0.1$. After the controller is designed, determine the number of samples per cycle of damped sinusoidal oscillation.

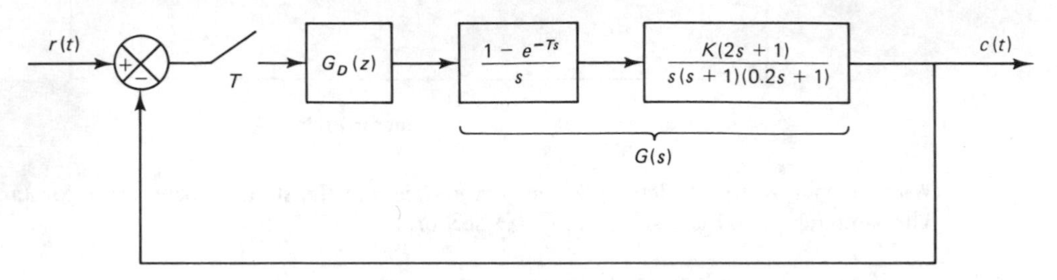

Figure 4-93 Digital control system for Prob. B-4-15.

Problem B-4-16. Consider the digital control system shown in Fig. 4-94. Design a digital controller $G_D(z)$ such that the system output will exhibit a deadbeat response to a unit step input (that is, the settling time will be the minimum possible and the steady-state error will be zero; also, the system output will not exhibit intersampling ripples after the settling time is reached). The sampling period T is assumed to be 1 sec, or $T = 1$.

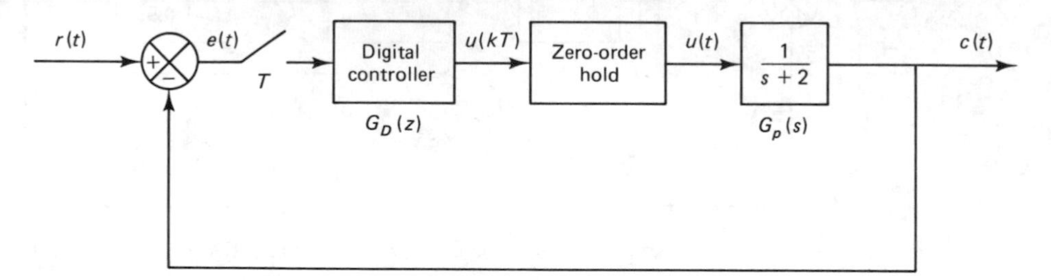

Figure 4-94 Digital control system for Prob. B-4-16.

Problem B-4-17. Consider the digital control system that was shown in Fig. 4-81(a). Assume that the plant transfer function is given by

$$G_p(s) = \frac{e^{-10s}}{100s + 1}$$

It is desired to have the output to a unit step input as given by the curve shown in Fig. 4-81(b), where we assume that $L = 10$ sec and $T_s = 20$ sec. (Note that the system output exhibits a finite settling time with no steady-state error and no intersampling ripples after the settling time is reached.) Choosing a suitable sampling period T, design a digital controller $G_D(z)$ such that the output follows the desired response curve.

5

State Space Analysis

5–1 INTRODUCTION

In Chaps. 3 and 4 we were concerned with so-called conventional methods for the analysis and design of control systems. Conventional methods such as the root locus and frequency response methods are useful for dealing with single-input–single-output systems. Conventional methods are conceptually simple and require only a reasonable number of computations, but they are applicable only to linear time-invariant systems having a single input and a single output. They are based on the input-output relationship of the system, that is, the transfer function or the pulse transfer function. They do not apply to nonlinear systems except in simple cases. Also, the conventional methods do not apply to the design of optimal and adaptive control systems, which are mostly time-varying and/or nonlinear.

A modern control system may have many inputs and many outputs, and these may be interrelated in a complicated manner. To analyze such systems and design optimal controllers, it is essential to reduce the complexity of the mathematical expressions, as well as to resort to computers for most of the tedious computations necessary in the analysis. Modern control systems are increasingly becoming digital control systems. The state space methods for the analysis and synthesis of control systems to be presented in this and succeeding chapters are best suited for dealing with multiple-input–multiple-output systems that are required to be optimal in some sense.

Concept of the state space method. The state space method is based on the description of system equations in terms of n first-order difference equations or differential equations, which may be combined into a first-order vector-matrix difference equation or differential equation. The use of the vector-matrix notation greatly simplifies the mathematical representation of the systems of equations.

System design by use of the state space concept enables the engineer to design control systems with respect to given performance indexes. In addition, design in the state space can be carried out for a *class* of inputs, instead of a specific input function such as the impulse function, step function, or sinusoidal function. Also, state space methods enable the engineer to include initial conditions in the design.

This is a very convenient and useful feature that is not possible in the conventional design methods.

We shall now first define state, state variable, state vector, and state space, and then we shall present state space equations.

State. The state of a dynamic system is the smallest set of variables (called *state variables*) such that the knowledge of these variables at $t = t_0$, together with the knowledge of the input for $t \geq t_0$, completely determines the behavior of the system for any time $t \geq t_0$. Note that the concept of state is by no means limited to physical systems. It is applicable to biological systems, economic systems, social systems, and others.

State variables. The state variables of a dynamic system are the variables making up the smallest set of variables which determine the state of the dynamic system. If at least n variables $x_1, x_2, \ldots, x_n$ are needed to completely describe the behavior of a dynamic system (so that once the input is given for $t \geq t_0$ and the initial state at $t = t_0$ is specified, the future state of the system is completely determined), then such n variables are a set of state variables.

Note that state variables need not be physically measurable or observable quantities. Variables which do not represent physical quantities and those which are neither measurable nor observable can be chosen as state variables. Such freedom in choosing state variables is an advantage of the state space methods. Practically speaking, however, it is convenient to choose easily measurable quantities for the state variables, if this is possible at all, because optimal control laws will require the feedback of all state variables with suitable weighting.

State vector. If n state variables are needed to completely describe the behavior of a given system, then these n state variables can be considered the n components of a vector $\mathbf{x}$. Such a vector is called a *state vector*. A state vector is thus a vector which determines uniquely the system state $\mathbf{x}(t)$ for any time $t \geq t_0$, once the state at $t = t_0$ is given and the input $\mathbf{u}(t)$ for $t \geq t_0$ is specified.

State space. The n-dimensional space whose coordinate axes consist of the x_1 axis, x_2 axis, $\ldots$, x_n axis is called a *state space*. Any state can be represented by a point in the state space.

State space equations. In state space analysis we are concerned with three types of variables that are involved in the modeling of dynamic systems: input variables, output variables, and state variables. As we shall see in Sec. 5–2, the state space representation for a given system is not unique, except that the number of state variables is the same for any of the different state space representations of the same system.

For time-varying (linear or nonlinear) discrete-time systems, the state equation may be written as

$$\mathbf{x}(k+1) = \mathbf{f}[\mathbf{x}(k), \mathbf{u}(k), k]$$

and the output equation as

$$\mathbf{y}(k) = \mathbf{g}[\mathbf{x}(k), \mathbf{u}(k), k]$$

For linear time-varying discrete-time systems, the state equation and output equation may be simplified to

$$\mathbf{x}(k+1) = \mathbf{G}(k)\mathbf{x}(k) + \mathbf{H}(k)\mathbf{u}(k)$$

$$\mathbf{y}(k) = \mathbf{C}(k)\mathbf{x}(k) + \mathbf{D}(k)\mathbf{u}(k)$$

where

$\mathbf{x}(k) = n$-vector (state vector)

$\mathbf{y}(k) = m$-vector (output vector)

$\mathbf{u}(k) = r$-vector (input vector)

$\mathbf{G}(k) = n \times n$ matrix (state matrix)

$\mathbf{H}(k) = n \times r$ matrix (input matrix)

$\mathbf{C}(k) = m \times n$ matrix (output matrix)

$\mathbf{D}(k) = m \times r$ matrix (direct transmission matrix)

The appearance of the variable k in the arguments of matrices $\mathbf{G}(k)$, $\mathbf{H}(k)$, $\mathbf{C}(k)$, and $\mathbf{D}(k)$ implies that these matrices are time-varying. If the variable k does not appear explicitly in the matrices, they are assumed to be time-invariant, or constant. That is, if the system is time-invariant, then the last two equations can be simplified to

$$\mathbf{x}(k+1) = \mathbf{G}\mathbf{x}(k) + \mathbf{H}\mathbf{u}(k) \tag{5–1}$$

$$\mathbf{y}(k) = \mathbf{C}\mathbf{x}(k) + \mathbf{D}\mathbf{u}(k) \tag{5–2}$$

As in the discrete-time case, continuous-time (linear or nonlinear) systems may be represented by the following state equation and output equation:

$$\dot{\mathbf{x}}(t) = \mathbf{f}[\mathbf{x}(t), \mathbf{u}(t), t]$$

$$\mathbf{y}(t) = \mathbf{g}[\mathbf{x}(t), \mathbf{u}(t), t]$$

For linear time-varying continuous-time systems, the state equation and output equation are given by

$$\dot{\mathbf{x}}(t) = \mathbf{A}(t)\mathbf{x}(t) + \mathbf{B}(t)\mathbf{u}(t)$$

$$\mathbf{y}(t) = \mathbf{C}(t)\mathbf{x}(t) + \mathbf{D}(t)\mathbf{u}(t)$$

If the system is time-invariant, then the last two equations are simplified to

$$\dot{\mathbf{x}}(t) = \mathbf{A}\mathbf{x}(t) + \mathbf{B}\mathbf{u}(t) \tag{5–3}$$

$$\mathbf{y}(t) = \mathbf{C}\mathbf{x}(t) + \mathbf{D}\mathbf{u}(t) \tag{5–4}$$

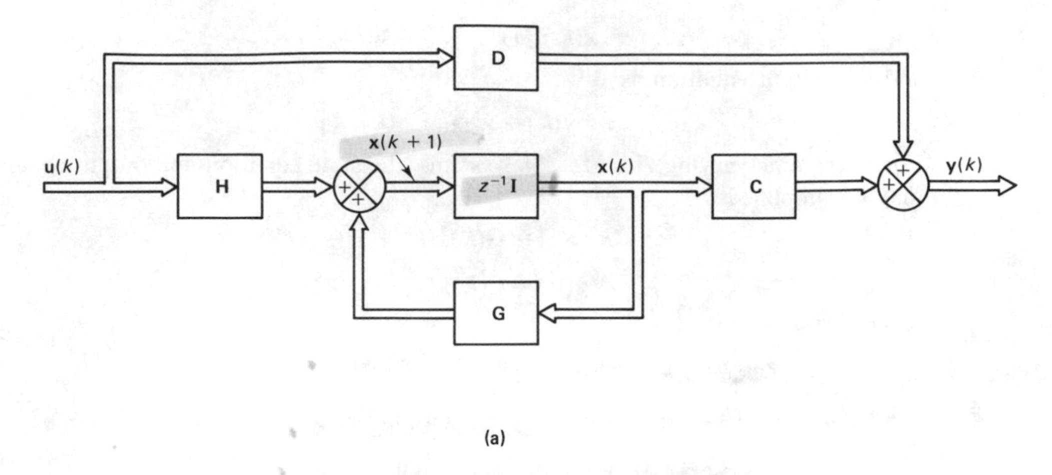

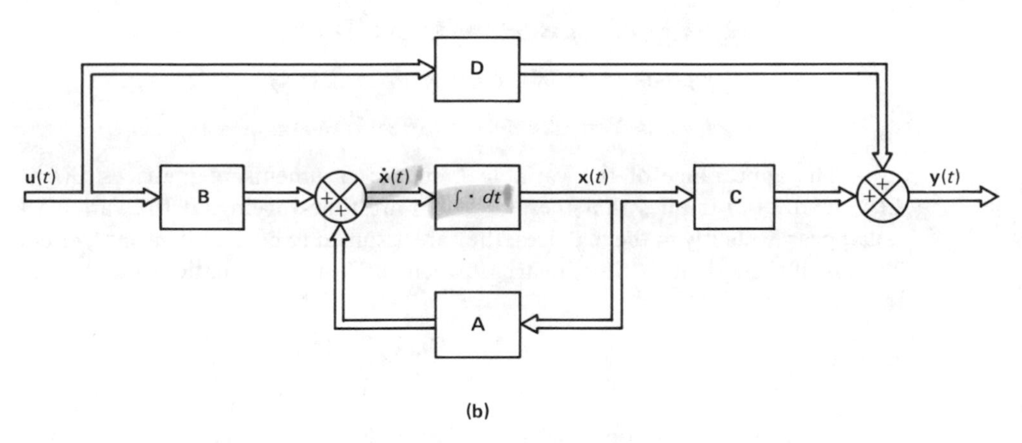

Figure 5–1 (a) Block diagram of the linear time-invariant discrete-time control system represented in state space; (b) block diagram of the linear time-invariant continuous-time system represented in state space.

Figure 5–1(a) shows the block diagram representation of the discrete-time control system defined by Eqs. (5–1) and (5–2), and Fig. 5–1(b) shows the continuous-time control system defined by Eqs. (5–3) and (5–4). Notice that the basic configurations of the discrete-time and continuous-time systems are the same.

Outline of the chapter. Section 5–1 has introduced the state space method and defined some basic terms. Section 5–2 presents various state space representations of linear time-invariant discrete-time systems. Section 5–3 first treats the solution of the linear time-invariant discrete-time state equation by the recursion procedure and by the z transform approach. Then it presents a method for computing $(z\mathbf{I} - \mathbf{G})^{-1}$.

Finally in Sec. 5–3, the solution of the linear time-varying discrete-time state equation is discussed and the properties of the state transition matrix are presented. Section 5–4 deals with the pulse-transfer-function matrix and the weighting sequence matrix. Section 5–5 gives the solution of continuous-time state equations. It begins with the solution of the linear time-invariant continuous-time homogeneous state equation, which is followed by discussion of the matrix exponential. Then it presents the solution of nonhomogeneous state equations, followed by a brief discussion of the transfer function matrix. Methods for computing $e^{\mathbf{A}t}$ are discussed next. The section concludes by presenting the solution of the linear time-varying continuous-time state equation. Section 5–6 treats the discretization of linear continuous-time state space equations. Both the time-invariant case and the time-varying case are discussed. The final section, Sec. 5–7, presents Liapunov stability analysis. It begins with discussions of the Liapunov function and definitions of stability in dynamic systems. Then it presents Liapunov's main stability theorem, followed by its application to linear time-invariant continuous-time systems. Finally, Liapunov stability analysis is applied to discrete-time systems.

5–2 STATE SPACE REPRESENTATIONS OF DISCRETE-TIME SYSTEMS

There are many techniques available for obtaining state space representations of discrete-time systems. Consider the discrete-time system described by

$$\begin{aligned} y(k) + a_1 y(k-1) + a_2 y(k-2) + \cdots + a_n y(k-n) \\ = b_0 u(k) + b_1 u(k-1) + \cdots + b_n u(k-n) \end{aligned} \tag{5–5}$$

where $u(k)$ is the input and $y(k)$ is the output of the system at the kth sampling instant. Note that some of the coefficients a_i $(i = 1, 2, \ldots, n)$ and b_j $(j = 0, 1, 2, \ldots, n)$ may be zero. Equation (5–5) can be written in the form of the pulse transfer function as

$$G(z) = \frac{Y(z)}{U(z)} = \frac{b_0 + b_1 z^{-1} + \cdots + b_n z^{-n}}{1 + a_1 z^{-1} + \cdots + a_n z^{-n}} \tag{5–6}$$

or

$$G(z) = \frac{Y(z)}{U(z)} = \frac{b_0 z^n + b_1 z^{n-1} + \cdots + b_n}{z^n + a_1 z^{n-1} + \cdots + a_n} \tag{5–7}$$

In what follows, we shall present three methods, from among the several methods available, for obtaining state space representation of discrete-time systems as described by Eq. (5–5), Eq. (5–6), or Eq. (5–7). The three methods to be presented are

1. Direct programming method
2. Nested programming method
3. Partial-fraction-expansion programming method

Direct programming method. In applying the direct programming method to obtain a state space representation of a pulse-transfer-function system, it is not necessary that the denominator be factored.

Consider the pulse-transfer-function system defined by Eq. (5–6):

$$G(z)=\frac{Y(z)}{U(z)}=\frac{b_0+b_1z^{-1}+b_2z^{-2}+\cdots+b_nz^{-n}}{1+a_1z^{-1}+a_2z^{-2}+\cdots+a_nz^{-n}}$$

$$=b_0+\frac{(b_1-a_1b_0)z^{-1}+(b_2-a_2b_0)z^{-2}+\cdots+(b_n-a_nb_0)z^{-n}}{1+a_1z^{-1}+a_2z^{-2}+\cdots+a_nz^{-n}}$$

This last equation can be written as follows:

$$Y(z)=b_0U(z)+\frac{(b_1-a_1b_0)z^{-1}+(b_2-a_2b_0)z^{-2}+\cdots+(b_n-a_nb_0)z^{-n}}{1+a_1z^{-1}+a_2z^{-2}+\cdots+a_nz^{-n}}U(z) \tag{5–8}$$

Let us define

$$\tilde{Y}(z)=\frac{(b_1-a_1b_0)z^{-1}+(b_2-a_2b_0)z^{-2}+\cdots+(b_n-a_nb_0)z^{-n}}{1+a_1z^{-1}+a_2z^{-2}+\cdots+a_nz^{-n}}U(z) \tag{5–9}$$

Then, Eq. (5–8) becomes

$$Y(z)=b_0U(z)+\tilde{Y}(z) \tag{5–10}$$

Let us rewrite Eq. (5–9) in the following form:

$$\frac{\tilde{Y}(z)}{(b_1-a_1b_0)z^{-1}+(b_2-a_2b_0)z^{-2}+\cdots+(b_n-a_nb_0)z^{-n}}$$

$$=\frac{U(z)}{1+a_1z^{-1}+a_2z^{-2}+\cdots+a_nz^{-n}}=Q(z)$$

From this last equation the following two equations may be obtained:

$$Q(z)=-a_1z^{-1}Q(z)-a_2z^{-2}Q(z)-\cdots-a_nz^{-n}Q(z)+U(z) \tag{5–11}$$

and

$$\tilde{Y}(z)=(b_1-a_1b_0)z^{-1}Q(z)+(b_2-a_2b_0)z^{-2}Q(z)+\cdots+(b_n-a_nb_0)z^{-n}Q(z) \tag{5–12}$$

Now we define the state variables as follows:

$$X_1(z)=z^{-n}Q(z)$$

$$X_2(z)=z^{-n+1}Q(z)$$

$$\vdots \tag{5–13}$$

$$X_{n-1}(z) = z^{-2}Q(z)$$
$$X_n(z) = z^{-1}Q(z)$$

Then clearly we have

$$zX_1(z) = X_2(z)$$
$$zX_2(z) = X_3(z)$$
$$\vdots$$
$$zX_{n-1}(z) = X_n(z)$$

In terms of difference equations, the preceding $n-1$ equations become

$$\begin{aligned} x_1(k+1) &= x_2(k) \\ x_2(k+1) &= x_3(k) \\ &\vdots \\ x_{n-1}(k+1) &= x_n(k) \end{aligned} \tag{5–14}$$

By substituting Eq. (5–13) into Eq. (5–11) we obtain

$$zX_n(z) = -a_1X_n(z) - a_2X_{n-1}(z) - \cdots - a_nX_1(z) + U(z)$$

which may be transformed into a difference equation:

$$x_n(k+1) = -a_nx_1(k) - a_{n-1}x_2(k) - \cdots - a_1x_n(k) + u(k) \tag{5–15}$$

Also, Eq. (5–12) can be rewritten as follows:

$$\tilde{Y}(z) = (b_1 - a_1b_0)X_n(z) + (b_2 - a_2b_0)X_{n-1}(z) + \cdots + (b_n - a_nb_0)X_1(z)$$

By use of this last equation, Eq. (5–10) can be written in the form

$$\begin{aligned} y(k) = (b_n - a_nb_0)x_1(k) &+ (b_{n-1} - a_{n-1}b_0)x_2(k) \\ &+ \cdots + (b_1 - a_1b_0)x_n(k) + b_0u(k) \end{aligned} \tag{5–16}$$

Combining Eqs. (5–14) and (5–15) results in the following state equation:

$$\begin{bmatrix} x_1(k+1) \\ x_2(k+1) \\ \cdot \\ \cdot \\ \cdot \\ x_{n-1}(k+1) \\ x_n(k+1) \end{bmatrix} = \begin{bmatrix} 0 & 1 & 0 & \cdots & 0 \\ 0 & 0 & 1 & \cdots & 0 \\ \cdot & \cdot & \cdot & & \cdot \\ \cdot & \cdot & \cdot & & \cdot \\ \cdot & \cdot & \cdot & & \cdot \\ 0 & 0 & 0 & \cdots & 1 \\ -a_n & -a_{n-1} & -a_{n-2} & \cdots & -a_1 \end{bmatrix} \begin{bmatrix} x_1(k) \\ x_2(k) \\ \cdot \\ \cdot \\ \cdot \\ x_{n-1}(k) \\ x_n(k) \end{bmatrix} + \begin{bmatrix} 0 \\ 0 \\ \cdot \\ \cdot \\ \cdot \\ 0 \\ 1 \end{bmatrix} u(k) \tag{5–17}$$

The output equation, Eq. (5–16), can be rewritten as follows:

$$y(k) = [b_n - a_n b_0 \vdots b_{n-1} - a_{n-1} b_0 \vdots \cdots \vdots b_1 - a_1 b_0] \begin{bmatrix} x_1(k) \\ x_2(k) \\ \cdot \\ \cdot \\ \cdot \\ x_n(k) \end{bmatrix} + b_0 u(k) \qquad (5\text{–}18)$$

The state space representation given by Eqs. (5–17) and (5–18) is commonly called a *controllable canonical form*. (For the definition of "controllability," see Sec. 6–2.) Figure 5–2(a) shows the block diagram representation of the system given by Eqs. (5–17) and (5–18).

Note that if we reverse the order of the state variables, that is, if instead of defining the state variables by Eq. (5–13) we define them according to the fashion

$$\hat{X}_1(z) = z^{-1}Q(z)$$
$$\hat{X}_2(z) = z^{-2}Q(z)$$
$$\cdot$$
$$\cdot$$
$$\cdot$$
$$\hat{X}_n(z) = z^{-n}Q(z)$$

then we obtain

$$z\hat{X}_2(z) = \hat{X}_1(z)$$
$$z\hat{X}_3(z) = \hat{X}_2(z)$$
$$\cdot$$
$$\cdot$$
$$\cdot$$
$$z\hat{X}_n(z) = \hat{X}_{n-1}(z)$$

and Eq. (5–11) can be written as follows:

$$z\hat{X}_1(z) = -a_1\hat{X}_1(z) - a_2\hat{X}_2 - \cdots - a_n\hat{X}_n(z) + U(z)$$

Hence

$$\hat{x}_1(k+1) = -a_1\hat{x}_1(k) - a_2\hat{x}_2(k) - \cdots - a_n\hat{x}_n(k) + u(k)$$
$$\hat{x}_2(k+1) = \hat{x}_1(k)$$
$$\hat{x}_3(k+1) = \hat{x}_2(k)$$
$$\cdot$$
$$\cdot$$
$$\cdot$$
$$\hat{x}_n(k+1) = \hat{x}_{n-1}(k)$$

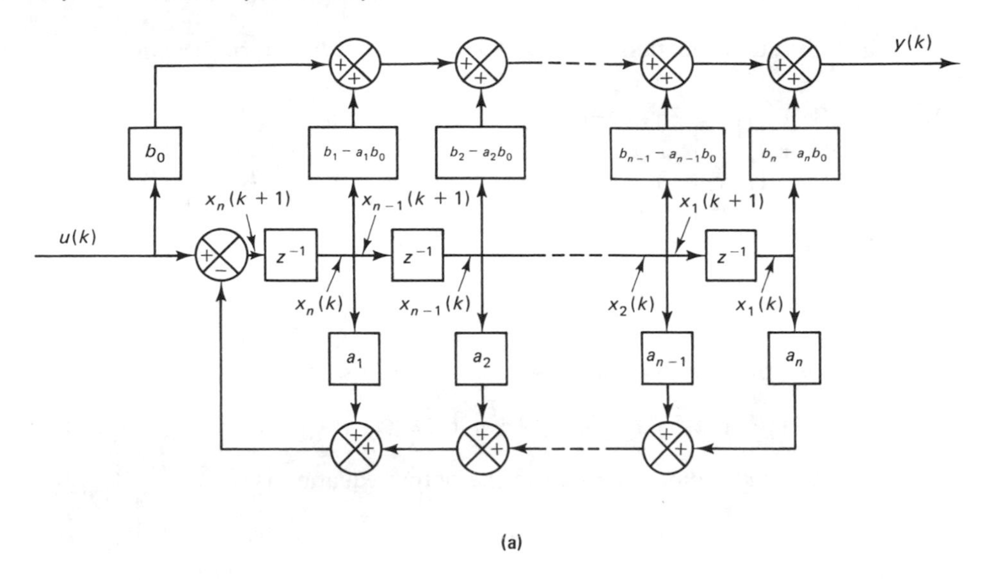

(a)

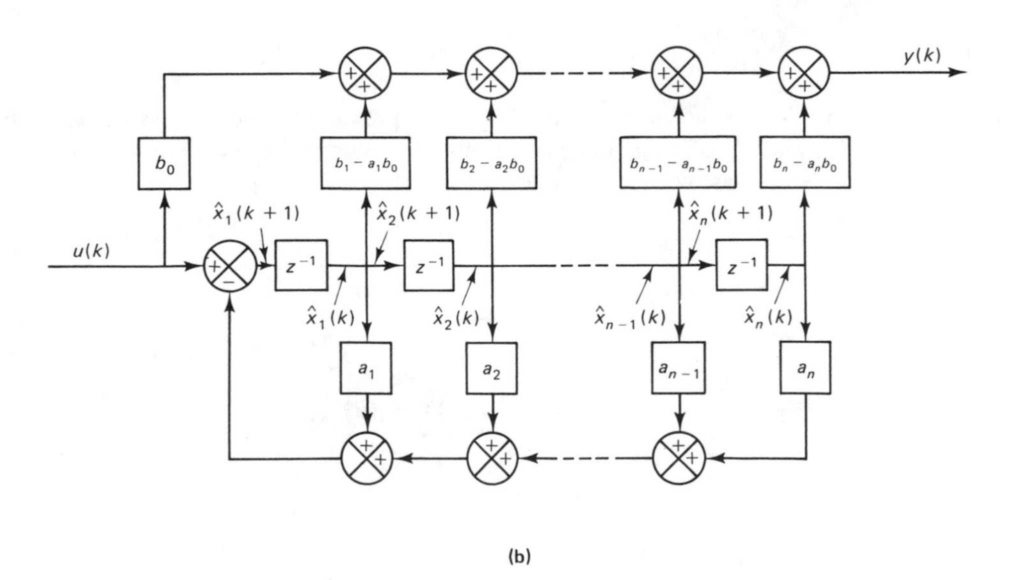

(b)

Figure 5–2 (a) Block diagram representation of the system given by Eqs. (5–17) and (5–18); (b) block diagram representation of the system given by Eqs. (5–19) and (5–20). Both representations are in the controllable canonical form.

and the state equation in the standard vector-matrix form becomes

$$\begin{bmatrix} \hat{x}_1(k+1) \\ \hat{x}_2(k+1) \\ \hat{x}_3(k+1) \\ \cdot \\ \cdot \\ \cdot \\ \hat{x}_n(k+1) \end{bmatrix} = \begin{bmatrix} -a_1 & -a_2 & \cdots & -a_{n-1} & -a_n \\ 1 & 0 & \cdots & 0 & 0 \\ 0 & 1 & \cdots & 0 & 0 \\ \cdot & \cdot & & \cdot & \cdot \\ \cdot & \cdot & & \cdot & \cdot \\ \cdot & \cdot & & \cdot & \cdot \\ 0 & 0 & \cdots & 1 & 0 \end{bmatrix} \begin{bmatrix} \hat{x}_1(k) \\ \hat{x}_2(k) \\ \hat{x}_3(k) \\ \cdot \\ \cdot \\ \cdot \\ \hat{x}_n(k) \end{bmatrix} + \begin{bmatrix} 1 \\ 0 \\ 0 \\ \cdot \\ \cdot \\ \cdot \\ 0 \end{bmatrix} u(k) \qquad (5\text{-}19)$$

The output equation can be obtained from Eqs. (5–10) and (5–12) as follows:

$$Y(z) = (b_1 - a_1 b_0)\hat{X}_1(z) + (b_2 - a_2 b_0)\hat{X}_2(z) + \cdots + (b_n - a_n b_0)\hat{X}_n(z) + b_0 U(z)$$

Thus, in the vector-matrix form the output equation is

$$y(k) = [b_1 - a_1 b_0 \vdots b_2 - a_2 b_0 \vdots \cdots \vdots b_n - a_n b_0] \begin{bmatrix} \hat{x}_1(k) \\ \hat{x}_2(k) \\ \cdot \\ \cdot \\ \cdot \\ \hat{x}_n(k) \end{bmatrix} + b_0 u(k) \qquad (5\text{-}20)$$

Figure 5–2(b) shows the block diagram representation of the system given by Eqs. (5–19) and (5–20). The system is shown in a controllable canonical form.

Note that two sets of state variables are related by

$$x_1(k) = \hat{x}_n(k)$$
$$x_2(k) = \hat{x}_{n-1}(k)$$
$$\cdot$$
$$\cdot$$
$$\cdot$$
$$x_n(k) = \hat{x}_1(k)$$

or

$$\begin{bmatrix} x_1(k) \\ x_2(k) \\ \cdot \\ \cdot \\ \cdot \\ x_n(k) \end{bmatrix} = \begin{bmatrix} 0 & 0 & \cdots & 0 & 1 \\ 0 & 0 & \cdots & 1 & 0 \\ \cdot & \cdot & & \cdot & \cdot \\ \cdot & \cdot & & \cdot & \cdot \\ \cdot & \cdot & & \cdot & \cdot \\ 1 & 0 & \cdots & 0 & 0 \end{bmatrix} \begin{bmatrix} \hat{x}_1(k) \\ \hat{x}_2(k) \\ \cdot \\ \cdot \\ \cdot \\ \hat{x}_n(k) \end{bmatrix}$$

If we define

$$\mathbf{T} = \begin{bmatrix} 0 & 0 & \cdots & 0 & 1 \\ 0 & 0 & \cdots & 1 & 0 \\ \cdot & \cdot & & \cdot & \cdot \\ \cdot & \cdot & & \cdot & \cdot \\ \cdot & \cdot & & \cdot & \cdot \\ 0 & 1 & & 0 & 0 \\ 1 & 0 & \cdots & 0 & 0 \end{bmatrix}$$

then it can be seen that

$$\mathbf{T}^T = \mathbf{T} \qquad \text{and} \qquad \mathbf{T}\mathbf{T}^T = \mathbf{I}$$

A matrix that satisfies the condition that $\mathbf{T}\mathbf{T}^T = \mathbf{T}^T\mathbf{T} = \mathbf{I}$, where $\mathbf{T}$ is a real matrix, is called an *orthogonal matrix*. In the orthogonal matrix, the inverse is exactly the same as the transpose, or

$$\mathbf{T}^{-1} = \mathbf{T}^T$$

In the present case

$$\mathbf{x} = \mathbf{T}\hat{\mathbf{x}}$$

and

$$\hat{\mathbf{x}} = \mathbf{T}\mathbf{x}$$

The state vectors $\mathbf{x}$ and $\hat{\mathbf{x}}$ are related by the orthogonal matrix $\mathbf{T}$.

Nested programming method. The nested programming method also does not require the denominator of the pulse transfer function to be in a factored form.

Consider the pulse transfer function system defined by Eq. (5–6),

$$G(z) = \frac{Y(z)}{U(z)} = \frac{b_0 + b_1 z^{-1} + \cdots + b_n z^{-n}}{1 + a_1 z^{-1} + \cdots + a_n z^{-n}}$$

which may be modified to

$$Y(z) - b_0 U(z) + z^{-1}[a_1 Y(z) - b_1 U(z)] \\ + z^{-2}[a_2 Y(z) - b_2 U(z)] + \cdots + z^{-n}[a_n Y(z) - b_n U(z)] = 0$$

or

$$Y(z) = b_0 U(z) + z^{-1}\big(b_1 U(z) - a_1 Y(z) + z^{-1}\{b_2 U(z) - a_2 Y(z) \\ + z^{-1}[b_3 U(z) - a_3 Y(z) + \cdots]\}\big) \qquad (5\text{–}21)$$

Now define the state variables as follows:

$$
\begin{aligned}
X_n(z) &= z^{-1}[b_1U(z) - a_1Y(z) + X_{n-1}(z)] \\
X_{n-1}(z) &= z^{-1}[b_2U(z) - a_2Y(z) + X_{n-2}(z)] \\
&\cdot \\
&\cdot \\
&\cdot \\
X_2(z) &= z^{-1}[b_{n-1}U(z) - a_{n-1}Y(z) + X_1(z)] \\
X_1(z) &= z^{-1}[b_nU(z) - a_nY(z)]
\end{aligned}
\tag{5–22}
$$

Then Eq. (5–21) can be written in the form

$$
Y(z) = b_0U(z) + X_n(z) \tag{5–23}
$$

By substituting Eq. (5–23) into Eq. (5–22) and multiplying both sides of the equations by z, we obtain

$$
\begin{aligned}
zX_n(z) &= X_{n-1}(z) - a_1X_n(z) + (b_1 - a_1b_0)U(z) \\
zX_{n-1}(z) &= X_{n-2}(z) - a_2X_n(z) + (b_2 - a_2b_0)U(z) \\
&\cdot \\
&\cdot \\
&\cdot \\
zX_2(z) &= X_1(z) - a_{n-1}X_n(z) + (b_{n-1} - a_{n-1}b_0)U(z) \\
zX_1(z) &= -a_nX_n(z) + (b_n - a_nb_0)U(z)
\end{aligned}
$$

The inverse z transforms of the preceding n equations taken in the reverse order give

$$
\begin{aligned}
x_1(k+1) &= -a_nx_n(k) + (b_n - a_nb_0)u(k) \\
x_2(k+1) &= x_1(k) - a_{n-1}x_n(k) + (b_{n-1} - a_{n-1}b_0)u(k) \\
&\cdot \\
&\cdot \\
&\cdot \\
x_{n-1}(k+1) &= x_{n-2}(k) - a_2x_n(k) + (b_2 - a_2b_0)u(k) \\
x_n(k+1) &= x_{n-1}(k) - a_1x_n(k) + (b_1 - a_1b_0)u(k)
\end{aligned}
$$

Also, the inverse z transform of Eq. (5–23) yields

$$
y(k) = x_n(k) + b_0u(k)
$$

Rewriting the state equation and the output equation in the standard vector-matrix forms gives

$$\begin{bmatrix} x_1(k+1) \\ x_2(k+1) \\ \cdot \\ \cdot \\ \cdot \\ x_{n-1}(k+1) \\ x_n(k+1) \end{bmatrix} = \begin{bmatrix} 0 & 0 & \cdots & 0 & 0 & -a_n \\ 1 & 0 & \cdots & 0 & 0 & -a_{n-1} \\ \cdot & \cdot & & \cdot & \cdot & \cdot \\ \cdot & \cdot & & \cdot & \cdot & \cdot \\ \cdot & \cdot & & \cdot & \cdot & \cdot \\ 0 & 0 & \cdots & 1 & 0 & -a_2 \\ 0 & 0 & \cdots & 0 & 1 & -a_1 \end{bmatrix} \begin{bmatrix} x_1(k) \\ x_2(k) \\ \cdot \\ \cdot \\ \cdot \\ x_{n-1}(k) \\ x_n(k) \end{bmatrix} + \begin{bmatrix} b_n - a_n b_0 \\ b_{n-1} - a_{n-1} b_0 \\ \cdot \\ \cdot \\ \cdot \\ b_2 - a_2 b_0 \\ b_1 - a_1 b_0 \end{bmatrix} u(k) \tag{5–24}$$

$$y(k) = [0 \;\; 0 \;\; \cdots \;\; 0 \;\; 1] \begin{bmatrix} x_1(k) \\ x_2(k) \\ \cdot \\ \cdot \\ \cdot \\ x_{n-1}(k) \\ x_n(k) \end{bmatrix} + b_0 u(k) \tag{5–25}$$

The state space representation given by Eqs. (5–24) and (5–25) is called an *observable canonical form*. (For the definition of "observability," see Sec. 6–3.) Notice that the $n \times n$ state matrix of the state equation given by Eq. (5–24) is the transpose of that of the state equation defined by Eq. (5–17). Figure 5–3(a) shows the block diagram representation of the system given by Eqs. (5–24) and (5–25).

Note that if we reverse the order of the state variables, that is, if we define

$$\hat{x}_1(k) = x_n(k)$$
$$\hat{x}_2(k) = x_{n-1}(k)$$
$$\cdot$$
$$\cdot$$
$$\cdot$$
$$\hat{x}_n(k) = x_1(k)$$

then the state equation and the output equation become as follows:

$$\begin{bmatrix} \hat{x}_1(k+1) \\ \hat{x}_2(k+1) \\ \cdot \\ \cdot \\ \cdot \\ \hat{x}_{n-1}(k+1) \\ \hat{x}_n(k+1) \end{bmatrix} = \begin{bmatrix} -a_1 & 1 & 0 & \cdots & 0 & 0 \\ -a_2 & 0 & 1 & \cdots & 0 & 0 \\ \cdot & \cdot & \cdot & & \cdot & \cdot \\ \cdot & \cdot & \cdot & & \cdot & \cdot \\ \cdot & \cdot & \cdot & & \cdot & \cdot \\ -a_{n-1} & 0 & 0 & \cdots & 0 & 1 \\ -a_n & 0 & 0 & \cdots & 0 & 0 \end{bmatrix} \begin{bmatrix} \hat{x}_1(k) \\ \hat{x}_2(k) \\ \cdot \\ \cdot \\ \cdot \\ \hat{x}_{n-1}(k) \\ \hat{x}_n(k) \end{bmatrix} + \begin{bmatrix} b_1 - a_1 b_0 \\ b_2 - a_2 b_0 \\ \cdot \\ \cdot \\ \cdot \\ b_{n-1} - a_{n-1} b_0 \\ b_n - a_n b_0 \end{bmatrix} u(k) \tag{5–26}$$

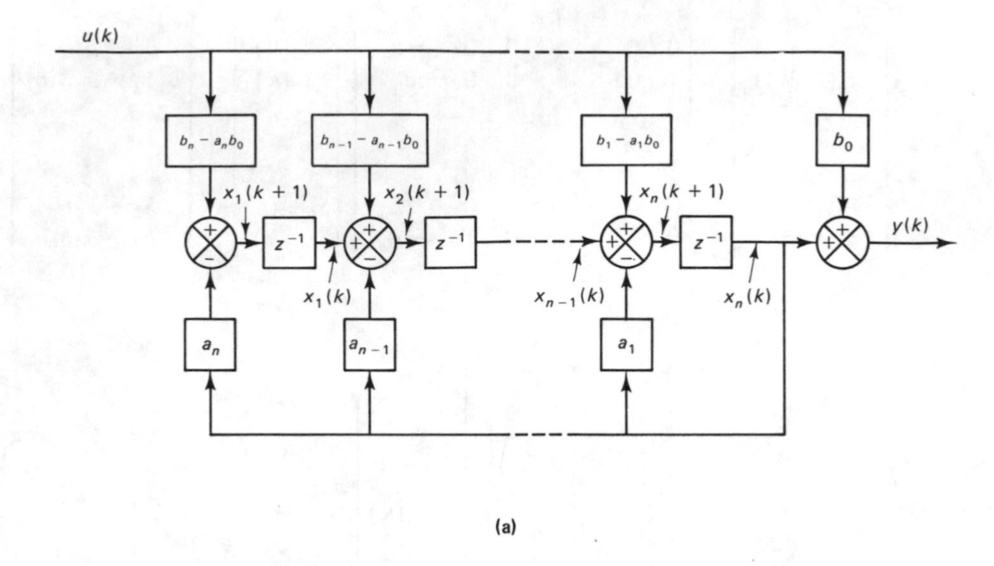

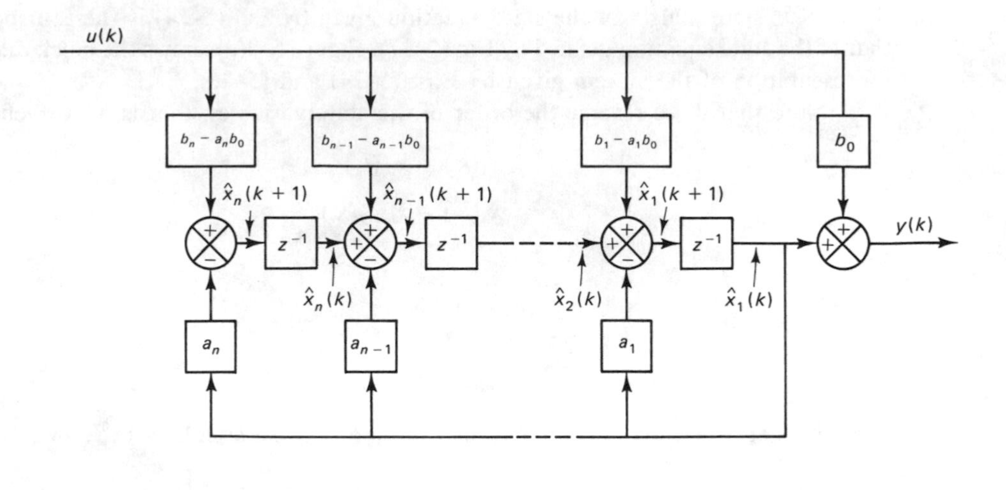

Figure 5–3 (a) Block diagram representation of the system given by Eqs. (5–24) and (5–25); (b) block diagram representation of the system given by Eqs. (5–26) and (5–27). Both representations are in the observable canonical form.

$$y(k) = [1 \quad 0 \quad \cdots \quad 0 \quad 0] \begin{bmatrix} \hat{x}_1(k) \\ \hat{x}_2(k) \\ \cdot \\ \cdot \\ \cdot \\ \hat{x}_{n-1}(k) \\ \hat{x}_n(k) \end{bmatrix} + b_0 u(k) \tag{5–27}$$

Figure 5–3(b) shows the block diagram representation of the system given by Eqs. (5–26) and (5–27). It is in an observable canonical form.

Partial-fraction-expansion programming method. The programming method using partial fraction expansion applies when the denominator of the pulse transfer function is in a factored form.

Consider the system defined by

$$G(z) = \frac{Y(z)}{U(z)} = \frac{b_0 z^n + b_1 z^{n-1} + \cdots + b_n}{z^n + a_1 z^{n-1} + \cdots + a_n}$$

$$= b_0 + \frac{(b_1 - a_1 b_0) z^{n-1} + (b_2 - a_2 b_0) z^{n-2} + \cdots + (b_n - a_n b_0)}{(z - p_1)(z - p_2) \cdots (z - p_n)} \tag{5–28}$$

We shall first discuss the case where all poles are distinct. Then we shall consider the case where multiple poles are involved.

Case 1: G(z) Involves Distinct Poles Only. In this case Eq. (5–28) can be expanded into partial fractions as follows:

$$\frac{Y(z)}{U(z)} = b_0 + \frac{c_1}{z - p_1} + \frac{c_2}{z - p_2} + \cdots + \frac{c_n}{z - p_n} \tag{5–29}$$

where

$$c_i = \lim_{z \to p_i} \left[\frac{Y(z)}{U(z)} (z - p_i) \right]$$

Equation (5–29) can be written in the form

$$Y(z) = b_0 U(z) + \frac{c_1}{z - p_1} U(z) + \frac{c_2}{z - p_2} U(z) + \cdots + \frac{c_n}{z - p_n} U(z) \tag{5–30}$$

Let us define the state variables as follows:

$$\begin{aligned} X_1(z) &= \frac{1}{z - p_1} U(z) \\ X_2(z) &= \frac{1}{z - p_2} U(z) \\ &\cdot \\ &\cdot \\ &\cdot \end{aligned} \tag{5–31}$$

$$X_n(z) = \frac{1}{z - p_n} U(z)$$

Then Eq. (5–31) can be rewritten as

$$\begin{aligned} zX_1(z) &= p_1 X_1(z) + U(z) \\ zX_2(z) &= p_2 X_2(z) + U(z) \\ &\cdot \\ &\cdot \\ &\cdot \\ zX_n(z) &= p_n X_n(z) + U(z) \end{aligned} \tag{5–32}$$

Also, Eq. (5–30) can be written as

$$Y(z) = b_0 U(z) + c_1 X_1(z) + c_2 X_2(z) + \cdots + c_n X_n(z) \tag{5–33}$$

The inverse z transformation of Eqs. (5–32) and (5–33) gives

$$\begin{aligned} x_1(k+1) &= p_1 x_1(k) + u(k) \\ x_2(k+1) &= p_2 x_2(k) + u(k) \\ &\cdot \\ &\cdot \\ &\cdot \\ x_n(k+1) &= p_n x_n(k) + u(k) \end{aligned} \tag{5–34}$$

and

$$y(k) = c_1 x_1(k) + c_2 x_2(k) + \cdots + c_n x_n(k) + b_0 u(k) \tag{5–35}$$

Rewriting the state equation and the output equation in the form of vector-matrix equations, we obtain

$$\begin{bmatrix} x_1(k+1) \\ x_2(k+1) \\ \cdot \\ \cdot \\ \cdot \\ x_n(k+1) \end{bmatrix} = \begin{bmatrix} p_1 & 0 & \cdots & 0 \\ 0 & p_2 & \cdots & 0 \\ \cdot & \cdot & & \cdot \\ \cdot & \cdot & & \cdot \\ \cdot & \cdot & & \cdot \\ 0 & 0 & \cdots & p_n \end{bmatrix} \begin{bmatrix} x_1(k) \\ x_2(k) \\ \cdot \\ \cdot \\ \cdot \\ x_n(k) \end{bmatrix} + \begin{bmatrix} 1 \\ 1 \\ \cdot \\ \cdot \\ \cdot \\ 1 \end{bmatrix} u(k) \tag{5–36}$$

and

$$y(k) = [c_1 \quad c_2 \quad \cdots \quad c_n] \begin{bmatrix} x_1(k) \\ x_2(k) \\ \cdot \\ \cdot \\ \cdot \\ x_n(k) \end{bmatrix} + b_0 u(k) \tag{5–37}$$

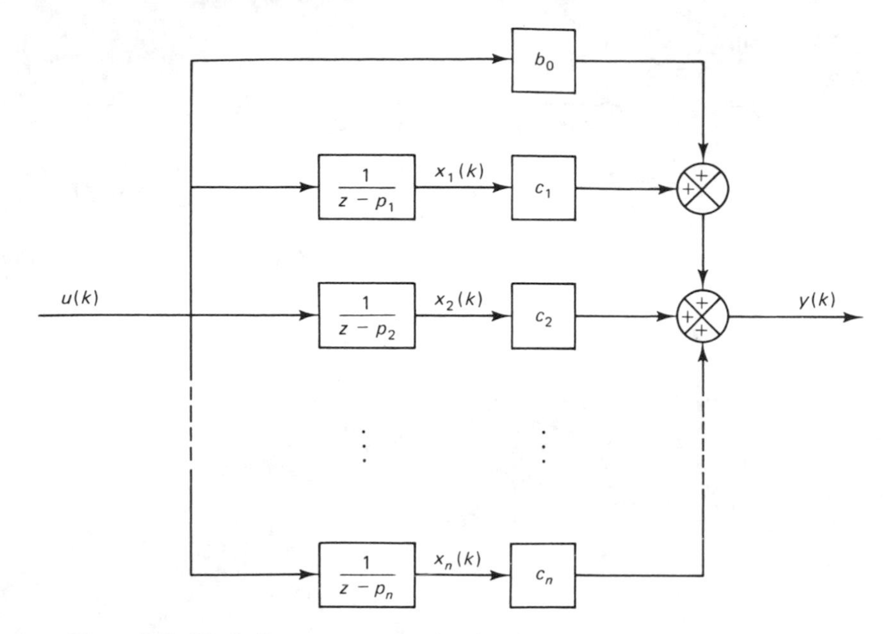

Figure 5–4 Block diagram representation for the system (involving distinct poles only) defined by Eq. (5–28), obtained by the partial-fraction-expansion programming method.

Notice that the $n \times n$ state matrix is a diagonal matrix. The diagonal elements are the poles of $G(z) = Y(z)/U(z)$.

Figure 5–4 shows the block diagram representation of the system defined by Eq. (5–28) according to the partial-fraction-expansion programming method. Note that this diagram corresponds to the case where all poles of $G(z)$ are distinct.

Case 2: G(z) Involves Multiple Poles. In the discussion that follows, we assume that $G(z)$ involves a multiple pole of order m at $z = p_1$ and that all other poles are distinct.

Consider the system defined by Eq. (5–7). Since $G(z)$ can be written in the form

$$G(z) = \frac{Y(z)}{U(z)} = \frac{b_0 z^n + b_1 z^{n-1} + \cdots + b_{n-1} z + b_n}{(z - p_1)^m (z - p_{m+1})(z - p_{m+2}) \cdots (z - p_n)}$$

$$= b_0 + \frac{(b_1 - a_1 b_0) z^{n-1} + (b_2 - a_2 b_0) z^{n-2} + \cdots + (b_n - a_n b_0)}{(z - p_1)^m (z - p_{m+1}) \cdots (z - p_n)}$$

$$= b_0 + \frac{c_1}{(z - p_1)^m} + \frac{c_2}{(z - p_1)^{m-1}} + \cdots + \frac{c_m}{z - p_1}$$

$$+ \frac{c_{m+1}}{z - p_{m+1}} + \frac{c_{m+2}}{z - p_{m+2}} + \cdots + \frac{c_n}{z - p_n} \tag{5–38}$$

we obtain

$$Y(z) = b_0 U(z) + \frac{c_1}{(z-p_1)^m} U(z) + \frac{c_2}{(z-p_1)^{m-1}} U(z) + \cdots + \frac{c_m}{z-p_1} U(z)$$
$$+ \frac{c_{m+1}}{z-p_{m+1}} U(z) + \frac{c_{m+2}}{z-p_{m+2}} U(z) + \cdots + \frac{c_n}{z-p_n} U(z) \tag{5-39}$$

Let us define the first m state variables $X_1(z), X_2(z), \cdots, X_m(z)$ by the equations:

$$\begin{aligned} X_1(z) &= \frac{1}{(z-p_1)^m} U(z) \\ X_2(z) &= \frac{1}{(z-p_1)^{m-1}} U(z) \\ &\vdots \\ X_m(z) &= \frac{1}{z-p_1} U(z) \end{aligned} \tag{5-40}$$

and the remaining $n - m$ state variables $X_{m+1}(z), X_{m+2}(z), \cdots, X_n(z)$ by the equations

$$\begin{aligned} X_{m+1}(z) &= \frac{1}{z-p_{m+1}} U(z) \\ X_{m+2}(z) &= \frac{1}{z-p_{m+2}} U(z) \\ &\vdots \\ X_n(z) &= \frac{1}{z-p_n} U(z) \end{aligned} \tag{5-41}$$

Notice that the m state variables defined by Eq. (5–40) are related each to the next by the following equations:

$$\begin{aligned} \frac{X_1(z)}{X_2(z)} &= \frac{1}{z-p_1} \\ \frac{X_2(z)}{X_3(z)} &= \frac{1}{z-p_1} \\ &\vdots \end{aligned} \tag{5-42}$$

$$\frac{X_{m-1}(z)}{X_m(z)} = \frac{1}{z - p_1}$$

By taking the inverse z transforms of all of Eq. (5–42), the last equation in Eq. (5–40), and all of Eq. (5–41), we obtain

$$\begin{aligned}
x_1(k+1) &= p_1 x_1(k) + x_2(k) \\
x_2(k+1) &= p_1 x_2(k) + x_3(k) \\
&\vdots \\
x_{m-1}(k+1) &= p_1 x_{m-1}(k) + x_m(k) \\
x_m(k+1) &= p_1 x_m(k) + u(k) \\
x_{m+1}(k+1) &= p_{m+1} x_{m+1}(k) + u(k) \\
&\vdots \\
x_n(k+1) &= p_n x_n(k) + u(k)
\end{aligned} \tag{5–43}$$

The output equation given by Eq. (5–39) can be rewritten as follows:

$$\begin{aligned}
Y(z) = c_1 X_1(z) + c_2 X_2(z) + \cdots + c_m X_m(z) + c_{m+1} X_{m+1}(z) \\
+ c_{m+2} X_{m+2}(z) + \cdots + c_n X_n(z) + b_0 U(z)
\end{aligned}$$

By taking the inverse z transform of this last equation, we get

$$\begin{aligned}
y(k) = c_1 x_1(k) + c_2 x_2(k) + \cdots + c_m x_m(k) + c_{m+1} x_{m+1}(k) \\
+ c_{m+2} x_{m+2}(k) + \cdots + c_n x_n(k) + b_0 u(k)
\end{aligned} \tag{5–44}$$

Rewriting the state equation and the output equation in vector-matrix form, we obtain

$$\begin{bmatrix} x_1(k+1) \\ x_2(k+1) \\ \vdots \\ x_m(k+1) \\ \hline x_{m+1}(k+1) \\ \vdots \\ x_n(k+1) \end{bmatrix} = \left[\begin{array}{cccccc|ccc} p_1 & 1 & 0 & \cdots & 0 & & 0 & \cdots & 0 \\ 0 & p_1 & 1 & \cdots & 0 & & 0 & \cdots & 0 \\ \vdots & \vdots & \vdots & & \vdots & & \vdots & & \vdots \\ 0 & 0 & 0 & \cdots & p_1 & & 0 & \cdots & 0 \\ \hline 0 & 0 & 0 & \cdots & 0 & & p_{m+1} & \cdots & 0 \\ \vdots & \vdots & \vdots & & \vdots & & \vdots & & \vdots \\ 0 & 0 & 0 & \cdots & 0 & & 0 & \cdots & p_n \end{array}\right] \begin{bmatrix} x_1(k) \\ x_2(k) \\ \vdots \\ x_m(k) \\ \hline x_{m+1}(k) \\ \vdots \\ x_n(k) \end{bmatrix} + \begin{bmatrix} 0 \\ 0 \\ \vdots \\ 1 \\ \hline 1 \\ \vdots \\ 1 \end{bmatrix} u(k) \tag{5–45}$$

and

$$y(k) = [c_1 \quad c_2 \quad \cdots \quad c_n] \begin{bmatrix} x_1(k) \\ x_2(k) \\ \cdot \\ \cdot \\ \cdot \\ x_n(k) \end{bmatrix} + b_0 u(k) \tag{5-46}$$

The $n \times n$ state matrix in this case is in a Jordan canonical form. Figure 5–5 shows the block diagram representation of the system defined by Eq. (5–7) or Eq. (5–38), where $G(z)$ involves a multiple pole of order m at $z = p_1$ and all other poles are distinct. (For details of Jordan canonical forms, see the Appendix, Sec. 6.)

Nonuniqueness of state space representations. For a given pulse-transfer-function system the state space representation is not unique. We have demonstrated that different state space representations for a given pulse-transfer-function

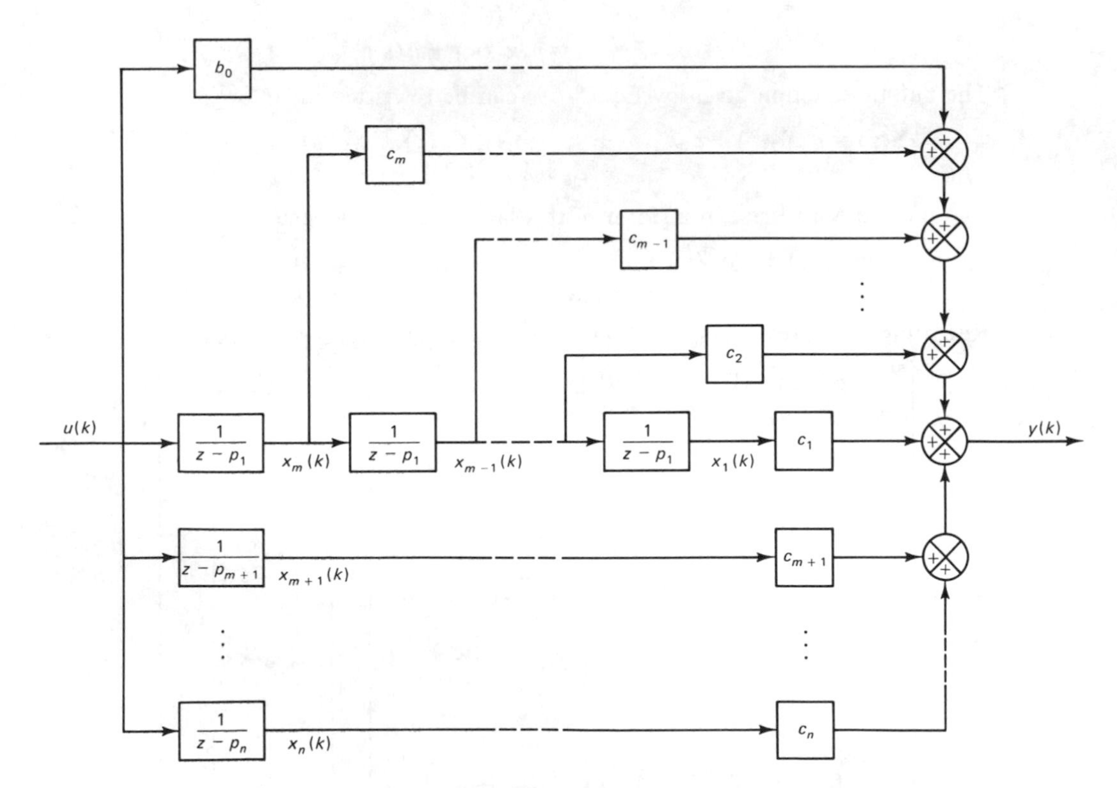

Figure 5–5 Block diagram representation for the system (involving multiple poles) defined by Eq. (5–38), obtained by the partial-fraction-expansion programming method.

system are possible. The state equations, however, are related to each other by the similarity transformation.

Consider the system defined by

$$\mathbf{x}(k+1) = \mathbf{G}\mathbf{x}(k) + \mathbf{H}\mathbf{u}(k) \tag{5–47}$$

$$\mathbf{y}(k) = \mathbf{C}\mathbf{x}(k) + \mathbf{D}\mathbf{u}(k) \tag{5–48}$$

Let us define a new state vector $\hat{\mathbf{x}}(k)$ by

$$\mathbf{x}(k) = \mathbf{P}\hat{\mathbf{x}}(k) \tag{5–49}$$

where $\mathbf{P}$ is a nonsingular matrix. [Note that since both $\mathbf{x}(k)$ and $\hat{\mathbf{x}}(k)$ are n-dimensional vectors, they are related to each other by a nonsingular matrix.]

Then, by substituting Eq. (5–49) into Eq. (5–47), we obtain

$$\mathbf{P}\hat{\mathbf{x}}(k+1) = \mathbf{G}\mathbf{P}\hat{\mathbf{x}}(k) + \mathbf{H}\mathbf{u}(k) \tag{5–50}$$

Premultiplying both sides of Eq. (5–50) by $\mathbf{P}^{-1}$ yields

$$\hat{\mathbf{x}}(k+1) = \mathbf{P}^{-1}\mathbf{G}\mathbf{P}\hat{\mathbf{x}}(k) + \mathbf{P}^{-1}\mathbf{H}\mathbf{u}(k) \tag{5–51}$$

Let us define

$$\mathbf{P}^{-1}\mathbf{G}\mathbf{P} = \hat{\mathbf{G}}, \qquad \mathbf{P}^{-1}\mathbf{H} = \hat{\mathbf{H}}$$

Then Eq. (5–51) can be written as follows:

$$\hat{\mathbf{x}}(k+1) = \hat{\mathbf{G}}\hat{\mathbf{x}}(k) + \hat{\mathbf{H}}\mathbf{u}(k) \tag{5–52}$$

Similarly, by substituting Eq. (5-49) into Eq. (5–48), we obtain

$$\mathbf{y}(k) = \mathbf{C}\mathbf{P}\hat{\mathbf{x}}(k) + \mathbf{D}\mathbf{u}(k) \tag{5–53}$$

By defining

$$\mathbf{C}\mathbf{P} = \hat{\mathbf{C}}, \qquad \mathbf{D} = \hat{\mathbf{D}}$$

we can write Eq. (5–53) as

$$\mathbf{y}(k) = \hat{\mathbf{C}}\hat{\mathbf{x}}(k) + \hat{\mathbf{D}}\mathbf{u}(k) \tag{5–54}$$

We have thus shown that the state space representation given by Eqs. (5–47) and (5–48),

$$\mathbf{x}(k+1) = \mathbf{G}\mathbf{x}(k) + \mathbf{H}\mathbf{u}(k)$$

$$\mathbf{y}(k) = \mathbf{C}\mathbf{x}(k) + \mathbf{D}\mathbf{u}(k)$$

is equivalent to the state space representation given by Eqs. (5–52) and (5–54),

$$\hat{\mathbf{x}}(k+1) = \hat{\mathbf{G}}\hat{\mathbf{x}}(k) + \hat{\mathbf{H}}\mathbf{u}(k)$$

$$\mathbf{y}(k) = \hat{\mathbf{C}}\hat{\mathbf{x}}(k) + \hat{\mathbf{D}}\mathbf{u}(k)$$

The state vectors $\mathbf{x}(k)$ and $\hat{\mathbf{x}}(k)$ are related to each other by Eq. (5–49). Since matrix $\mathbf{P}$ can be any nonsingular $n \times n$ matrix, there are infinitely many state space representations for a given system.

In some applications, we may desire to diagonalize the state matrix $\mathbf{G}$. This may be done by properly choosing a matrix $\mathbf{P}$ such that

$$\mathbf{P}^{-1}\mathbf{G}\mathbf{P} = \text{diagonal matrix}$$

In the case where diagonalization is not possible, $\mathbf{P}^{-1}\mathbf{G}\mathbf{P}$ may be transformed into a Jordan canonical form:

$$\mathbf{P}^{-1}\mathbf{G}\mathbf{P} = \mathbf{J} = \text{Jordan canonical form}$$

For methods for transforming matrix $\mathbf{G}$ into a diagonal matrix or into a matrix in a Jordan canonical form, refer to the Appendix. [Note that if the partial-fraction-expansion programming method is used, the state matrix becomes diagonal if all poles involved are distinct, and it becomes a Jordan canonical form if multiple poles are involved in $Y(z)/U(z)$.]

5–3 SOLVING DISCRETE-TIME STATE SPACE EQUATIONS

In this section, we first present the solution of the linear time-invariant discrete-time state equation

$$\mathbf{x}(k+1) = \mathbf{G}\mathbf{x}(k) + \mathbf{H}\mathbf{u}(k)$$

by a recursion procedure and then by the z transform method. Then we discuss methods for computing $(z\mathbf{I} - \mathbf{G})^{-1}$. Finally, we treat the solution of the linear time-varying discrete-time state equation

$$\mathbf{x}(k+1) = \mathbf{G}(k)\mathbf{x}(k) + \mathbf{H}(k)\mathbf{u}(k)$$

Solution of the linear time-invariant discrete-time state equation. In general, discrete-time equations are easier to solve than differential equations because the former can be solved easily by means of a recursion procedure. The recursion procedure is quite simple and convenient for digital computations.

Consider the following state equation and output equation:

$$\mathbf{x}(k+1) = \mathbf{G}\mathbf{x}(k) + \mathbf{H}\mathbf{u}(k) \tag{5–55}$$

$$\mathbf{y}(k) = \mathbf{C}\mathbf{x}(k) + \mathbf{D}\mathbf{u}(k) \tag{5–56}$$

The solution of Eq. (5–55) for any positive integer k may be obtained directly by recursion, as follows:

$$\mathbf{x}(1) = \mathbf{G}\mathbf{x}(0) + \mathbf{H}\mathbf{u}(0)$$

$$\mathbf{x}(2) = \mathbf{G}\mathbf{x}(1) + \mathbf{H}\mathbf{u}(1) = \mathbf{G}^2\mathbf{x}(0) + \mathbf{G}\mathbf{H}\mathbf{u}(0) + \mathbf{H}\mathbf{u}(1)$$

$$\mathbf{x}(3) = \mathbf{G}\mathbf{x}(2) + \mathbf{H}\mathbf{u}(2) = \mathbf{G}^3\mathbf{x}(0) + \mathbf{G}^2\mathbf{H}\mathbf{u}(0) + \mathbf{G}\mathbf{H}\mathbf{u}(1) + \mathbf{H}\mathbf{u}(2)$$

$$\vdots$$

By repeating this procedure, we obtain

$$\mathbf{x}(k) = \mathbf{G}^k\mathbf{x}(0) + \sum_{j=0}^{k-1} \mathbf{G}^{k-j-1}\mathbf{H}\mathbf{u}(j) \qquad k = 1, 2, 3, \ldots \tag{5–57}$$

Clearly, $\mathbf{x}(k)$ consists of two parts, one representing the contribution of the initial state $\mathbf{x}(0)$, and the other the contribution of the input $\mathbf{u}(j)$, where $j = 0, 1, 2, \ldots, k - 1$. The output $\mathbf{y}(k)$ is given by

$$\mathbf{y}(k) = \mathbf{C}\mathbf{G}^k\mathbf{x}(0) + \mathbf{C}\sum_{j=0}^{k-1} \mathbf{G}^{k-j-1}\mathbf{H}\mathbf{u}(j) + \mathbf{D}\mathbf{u}(k) \tag{5–58}$$

State transition matrix. Notice that it is possible to write the solution of the homogeneous state equation

$$\mathbf{x}(k + 1) = \mathbf{G}\mathbf{x}(k) \tag{5–59}$$

as

$$\mathbf{x}(k) = \mathbf{\Psi}(k)\mathbf{x}(0) \tag{5–60}$$

where $\mathbf{\Psi}(k)$ is a unique $n \times n$ matrix satisfying the conditions

$$\mathbf{\Psi}(k + 1) = \mathbf{G}\mathbf{\Psi}(k), \qquad \mathbf{\Psi}(0) = \mathbf{I} \tag{5–61}$$

Clearly, $\mathbf{\Psi}(k)$ can be given by

$$\mathbf{\Psi}(k) = \mathbf{G}^k \tag{5–62}$$

From Eq. (5–60), we see that the solution of Eq. (5–59) is simply a transformation of the initial state. Therefore, the unique matrix $\mathbf{\Psi}(k)$ is called the *state transition matrix*. It is also called the *fundamental matrix*. The state transition matrix contains all the information about the free motions of the system defined by Eq. (5–59).

In terms of the state transition matrix $\mathbf{\Psi}(k)$, Eq. (5–57) can be written in the form

$$\mathbf{x}(k) = \mathbf{\Psi}(k)\mathbf{x}(0) + \sum_{j=0}^{k-1} \mathbf{\Psi}(k - j - 1)\mathbf{H}\mathbf{u}(j) \tag{5–63}$$

$$= \mathbf{\Psi}(k)\mathbf{x}(0) + \sum_{j=0}^{k-1} \mathbf{\Psi}(j)\mathbf{H}\mathbf{u}(k - j - 1) \tag{5–64}$$

Substituting Eq. (5–63) or Eq. (5–64) into Eq. (5–56), the following output equation can be obtained:

$$\mathbf{y}(k) = \mathbf{C}\mathbf{\Psi}(k)\mathbf{x}(0) + \mathbf{C}\sum_{j=0}^{k-1} \mathbf{\Psi}(k - j - 1)\mathbf{H}\mathbf{u}(j) + \mathbf{D}\mathbf{u}(k) \tag{5–65}$$

$$= \mathbf{C}\mathbf{\Psi}(k)\mathbf{x}(0) + \mathbf{C}\sum_{j=0}^{k-1}\mathbf{\Psi}(j)\mathbf{H}\mathbf{u}(k-j-1) + \mathbf{D}\mathbf{u}(k) \tag{5-66}$$

z transform approach to the solution of discrete-time state equations. We next present the solution of a discrete-time state equation by the z transform method. Consider the discrete-time system described by Eq. (5–55):

$$\mathbf{x}(k+1) = \mathbf{G}\mathbf{x}(k) + \mathbf{H}\mathbf{u}(k) \tag{5-67}$$

Taking the z transform of both sides of Eq. (5–67) gives

$$z\mathbf{X}(z) - z\mathbf{x}(0) = \mathbf{G}\mathbf{X}(z) + \mathbf{H}\mathbf{U}(z)$$

where $\mathbf{X}(z) = \mathcal{Z}[\mathbf{x}(k)]$ and $\mathbf{U}(z) = \mathcal{Z}[\mathbf{u}(k)]$. Then

$$(z\mathbf{I} - \mathbf{G})\mathbf{X}(z) = z\mathbf{x}(0) + \mathbf{H}\mathbf{U}(z)$$

Premultiplying both sides of this last equation by $(z\mathbf{I} - \mathbf{G})^{-1}$, we obtain

$$\mathbf{X}(z) = (z\mathbf{I} - \mathbf{G})^{-1}z\mathbf{x}(0) + (z\mathbf{I} - \mathbf{G})^{-1}\mathbf{H}\mathbf{U}(z) \tag{5-68}$$

Taking the inverse z transform of both sides of Eq. (5–68) gives

$$\mathbf{x}(k) = \mathcal{Z}^{-1}[(z\mathbf{I} - \mathbf{G})^{-1}z]\mathbf{x}(0) + \mathcal{Z}^{-1}[(z\mathbf{I} - \mathbf{G})^{-1}\mathbf{H}\mathbf{U}(z)] \tag{5-69}$$

Comparing Eq. (5–57) with Eq. (5–69), we obtain

$$\mathbf{G}^k = \mathcal{Z}^{-1}[(z\mathbf{I} - \mathbf{G})^{-1}z] \tag{5-70}$$

and

$$\sum_{j=0}^{k-1}\mathbf{G}^{k-j-1}\mathbf{H}\mathbf{u}(j) = \mathcal{Z}^{-1}[(z\mathbf{I} - \mathbf{G})^{-1}\mathbf{H}\mathbf{U}(z)] \tag{5-71}$$

where $k = 1, 2, 3, \ldots$.

Notice that the solution by the z transform method involves the process of inverting the matrix $(z\mathbf{I} - \mathbf{G})$, which may be accomplished by analytical means or by use of a computer routine. The solution also requires the inverse z transforms of $(z\mathbf{I} - \mathbf{G})^{-1}z$ and $(z\mathbf{I} - \mathbf{G})^{-1}\mathbf{H}\mathbf{U}(z)$.

Example 5–1.

Obtain the state transition matrix of the following discrete-time system:

$$\mathbf{x}(k+1) = \mathbf{G}\mathbf{x}(k) + \mathbf{H}u(k)$$

$$y(k) = \mathbf{C}\mathbf{x}(k)$$

where

$$\mathbf{G} = \begin{bmatrix} 0 & 1 \\ -0.16 & -1 \end{bmatrix}, \qquad \mathbf{H} = \begin{bmatrix} 1 \\ 1 \end{bmatrix}, \qquad \mathbf{C} = [1 \quad 0]$$

Then obtain the state $\mathbf{x}(k)$ and the output $y(k)$ when the input $u(k) = 1$ for $k = 0, 1, 2, \ldots$. Assume that the initial state is given by

$$\mathbf{x}(0) = \begin{bmatrix} x_1(0) \\ x_2(0) \end{bmatrix} = \begin{bmatrix} 1 \\ -1 \end{bmatrix}$$

From Eqs. (5–62) and (5–70) the state transition matrix $\mathbf{\Psi}(k)$ is

$$\mathbf{\Psi}(k) = \mathbf{G}^k = \mathscr{Z}^{-1}[(z\mathbf{I} - \mathbf{G})^{-1}z]$$

Therefore, we first obtain $(z\mathbf{I} - \mathbf{G})^{-1}$:

$$(z\mathbf{I} - \mathbf{G})^{-1} = \begin{bmatrix} z & -1 \\ 0.16 & z+1 \end{bmatrix}^{-1}$$

$$= \begin{bmatrix} \dfrac{z+1}{(z+0.2)(z+0.8)} & \dfrac{1}{(z+0.2)(z+0.8)} \\[2ex] \dfrac{-0.16}{(z+0.2)(z+0.8)} & \dfrac{z}{(z+0.2)(z+0.8)} \end{bmatrix}$$

$$= \begin{bmatrix} \dfrac{\frac{4}{3}}{z+0.2} + \dfrac{-\frac{1}{3}}{z+0.8} & \dfrac{\frac{5}{3}}{z+0.2} + \dfrac{-\frac{5}{3}}{z+0.8} \\[2ex] \dfrac{-\frac{0.8}{3}}{z+0.2} + \dfrac{\frac{0.8}{3}}{z+0.8} & \dfrac{-\frac{1}{3}}{z+0.2} + \dfrac{\frac{4}{3}}{z+0.8} \end{bmatrix}$$

The state transition matrix $\mathbf{\Psi}(k)$ is now obtained as follows:

$$\mathbf{\Psi}(k) = \mathbf{G}^k = \mathscr{Z}^{-1}[(z\mathbf{I} - \mathbf{G})^{-1}z]$$

$$= \mathscr{Z}^{-1}\begin{bmatrix} \dfrac{4}{3}\dfrac{z}{z+0.2} - \dfrac{1}{3}\dfrac{z}{z+0.8} & \dfrac{5}{3}\dfrac{z}{z+0.2} - \dfrac{5}{3}\dfrac{z}{z+0.8} \\[2ex] -\dfrac{0.8}{3}\dfrac{z}{z+0.2} + \dfrac{0.8}{3}\dfrac{z}{z+0.8} & -\dfrac{1}{3}\dfrac{z}{z+0.2} + \dfrac{4}{3}\dfrac{z}{z+0.8} \end{bmatrix}$$

$$= \begin{bmatrix} \frac{4}{3}(-0.2)^k - \frac{1}{3}(-0.8)^k & \frac{5}{3}(-0.2)^k - \frac{5}{3}(-0.8)^k \\[2ex] -\frac{0.8}{3}(-0.2)^k + \frac{0.8}{3}(-0.8)^k & -\frac{1}{3}(-0.2)^k + \frac{4}{3}(-0.8)^k \end{bmatrix} \tag{5–72}$$

Equation (5–72) gives the state transition matrix.

Next, compute $\mathbf{x}(k)$. The z transform of $\mathbf{x}(k)$ is given by

$$\mathscr{Z}[\mathbf{x}(k)] = \mathbf{X}(z) = (z\mathbf{I} - \mathbf{G})^{-1}z\mathbf{x}(0) + (z\mathbf{I} - \mathbf{G})^{-1}\mathbf{H}U(z)$$
$$= (z\mathbf{I} - \mathbf{G})^{-1}[z\mathbf{x}(0) + \mathbf{H}U(z)]$$

Since

$$U(z) = \frac{1}{1 - z^{-1}} = \frac{z}{z-1}$$

we obtain

$$z\mathbf{x}(0) + \mathbf{H}U(z) = \begin{bmatrix} z \\ -z \end{bmatrix} + \begin{bmatrix} \dfrac{z}{z-1} \\[2ex] \dfrac{z}{z-1} \end{bmatrix} = \begin{bmatrix} \dfrac{z^2}{z-1} \\[2ex] \dfrac{-z^2+2z}{z-1} \end{bmatrix}$$

Hence

$$\mathbf{X}(z) = (z\mathbf{I} - \mathbf{G})^{-1}[z\mathbf{x}(0) + \mathbf{H}U(z)]$$

$$= \begin{bmatrix} \dfrac{(z^2+2)z}{(z+0.2)(z+0.8)(z-1)} \\ \dfrac{(-z^2+1.84z)z}{(z+0.2)(z+0.8)(z-1)} \end{bmatrix}$$

$$= \begin{bmatrix} \dfrac{-\frac{17}{6}z}{z+0.2} + \dfrac{\frac{22}{9}z}{z+0.8} + \dfrac{\frac{25}{18}z}{z-1} \\ \dfrac{\frac{3.4}{6}z}{z+0.2} + \dfrac{-\frac{17.6}{9}z}{z+0.8} + \dfrac{\frac{7}{18}z}{z-1} \end{bmatrix}$$

Thus, the state vector $\mathbf{x}(k)$ is given by

$$\mathbf{x}(k) = \mathcal{Z}^{-1}[\mathbf{X}(z)] = \begin{bmatrix} -\frac{17}{6}(-0.2)^k + \frac{22}{9}(-0.8)^k + \frac{25}{18} \\ \frac{3.4}{6}(-0.2)^k - \frac{17.6}{9}(-0.8)^k + \frac{7}{18} \end{bmatrix}$$

Finally, the output $y(k)$ is obtained as follows:

$$y(k) = \mathbf{C}\mathbf{x}(k) = [1 \quad 0]\begin{bmatrix} -\frac{17}{6}(-0.2)^k + \frac{22}{9}(-0.8)^k + \frac{25}{18} \\ \frac{3.4}{6}(-0.2)^k - \frac{17.6}{9}(-0.8)^k + \frac{7}{18} \end{bmatrix}$$

$$= -\tfrac{17}{6}(-0.2)^k + \tfrac{22}{9}(-0.8)^k + \tfrac{25}{18}$$

Computation of $(z\mathbf{I} - \mathbf{G})^{-1}$. The solution of the state equation given by Eq. (5–55) by the z transform method requires the computation of $(z\mathbf{I} - \mathbf{G})^{-1}$. Computing $(z\mathbf{I} - \mathbf{G})^{-1}$ is, except in simple cases, generally a time-consuming task. There are both analytical and computational methods available for computing $(z\mathbf{I} - \mathbf{G})^{-1}$. We shall present one method here and two other methods later. (See Probs. A-5–14, A-5–15, and A-5–16.)

Method for computing $(z\mathbf{I} - \mathbf{G})^{-1}$. The method presented here is based on the expansion of the adjoint of $(z\mathbf{I} - \mathbf{G})$. The inverse of $(z\mathbf{I} - \mathbf{G})$ can be written in terms of the adjoint of $(z\mathbf{I} - \mathbf{G})$, as follows:

$$(z\mathbf{I} - \mathbf{G})^{-1} = \frac{\operatorname{adj}(z\mathbf{I} - \mathbf{G})}{|z\mathbf{I} - \mathbf{G}|} \tag{5–73}$$

Note that the determinant $|z\mathbf{I} - \mathbf{G}|$ may be written as follows:

$$|z\mathbf{I} - \mathbf{G}| = z^n + a_1 z^{n-1} + a_2 z^{n-2} + \cdots + a_n \tag{5–74}$$

It can be shown (see Prob. A-5–14) that $\operatorname{adj}(z\mathbf{I} - \mathbf{G})$ may be given by

$$\operatorname{adj}(z\mathbf{I} - \mathbf{G}) = \mathbf{I}z^{n-1} + \mathbf{H}_1 z^{n-2} + \mathbf{H}_2 z^{n-3} + \cdots + \mathbf{H}_{n-1} \tag{5–75}$$

where

$$
\begin{aligned}
\mathbf{H}_1 &= \mathbf{G} + a_1\mathbf{I} \\
\mathbf{H}_2 &= \mathbf{G}\mathbf{H}_1 + a_2\mathbf{I} \\
&\cdot \\
&\cdot \\
&\cdot \\
\mathbf{H}_{n-1} &= \mathbf{G}\mathbf{H}_{n-2} + a_{n-1}\mathbf{I} \\
\mathbf{H}_n &= \mathbf{G}\mathbf{H}_{n-1} + a_n\mathbf{I} = \mathbf{0}
\end{aligned}
\tag{5-76}
$$

Note that $a_1, a_2, \ldots, a_n$ are the coefficients appearing in the determinant given by Eq. (5–74). The a_i's can also be given (see Prob. A-5–14) by use of the trace, as follows:

$$
\begin{aligned}
a_1 &= -\operatorname{tr}\mathbf{G} \\
a_2 &= -\tfrac{1}{2}\operatorname{tr}\mathbf{G}\mathbf{H}_1 \\
a_3 &= -\tfrac{1}{3}\operatorname{tr}\mathbf{G}\mathbf{H}_2 \\
&\cdot \\
&\cdot \\
&\cdot \\
a_n &= -\frac{1}{n}\operatorname{tr}\mathbf{G}\mathbf{H}_{n-1}
\end{aligned}
\tag{5-77}
$$

(The trace of an $n \times n$ matrix is the sum of its diagonal elements.)

For a higher-order determinant ($n > 3$), expanding the determinant $|z\mathbf{I} - \mathbf{G}|$ into the form given by Eq. (5–74) may be time-consuming; in this case, using Eq. (5–77) to compute the a_i's proves to be useful, since a_1, $\mathbf{H}_1$, a_2, $\mathbf{H}_2, \ldots, a_{n-1}$, $\mathbf{H}_{n-1}$ can easily be computed sequentially.

By substituting Eq. (5–76) into Eq. (5–75) and substituting the resulting equation into Eq. (5–73), we obtain the inverse of $(z\mathbf{I} - \mathbf{G})$. The present method is convenient for computer solution; a standard program is available.

Example 5–2.

Determine the inverse of the matrix $(z\mathbf{I} - \mathbf{G})$, where

$$
\mathbf{G} = \begin{bmatrix} 0.1 & 0.1 & 0 \\ 0.3 & -0.1 & -0.2 \\ 0 & 0 & -0.3 \end{bmatrix}
$$

Also, obtain $\mathbf{G}^k$.

From Eq. (5–73), we have

$$
(z\mathbf{I} - \mathbf{G})^{-1} = \frac{\operatorname{adj}(z\mathbf{I} - \mathbf{G})}{|z\mathbf{I} - \mathbf{G}|}
$$

Although the determinant $|z\mathbf{I} - \mathbf{G}|$ can be expanded easily, here for demonstration purposes let us use Eq. (5–77) to compute a_1, a_2, and a_3. First, notice that

$$a_1 = -\text{tr}\,\mathbf{G} = -\text{tr}\begin{bmatrix} 0.1 & 0.1 & 0 \\ 0.3 & -0.1 & -0.2 \\ 0 & 0 & -0.3 \end{bmatrix} = 0.3$$

Then, from Eq. (5–76) we obtain

$$\mathbf{H}_1 = \mathbf{G} + a_1\mathbf{I} = \begin{bmatrix} 0.1 & 0.1 & 0 \\ 0.3 & -0.1 & -0.2 \\ 0 & 0 & -0.3 \end{bmatrix} + \begin{bmatrix} 0.3 & 0 & 0 \\ 0 & 0.3 & 0 \\ 0 & 0 & 0.3 \end{bmatrix}$$

$$= \begin{bmatrix} 0.4 & 0.1 & 0 \\ 0.3 & 0.2 & -0.2 \\ 0 & 0 & 0 \end{bmatrix}$$

Hence

$$a_2 = -\tfrac{1}{2}\,\text{tr}\,\mathbf{GH}_1 = -\tfrac{1}{2}\,\text{tr}\left\{\begin{bmatrix} 0.1 & 0.1 & 0 \\ 0.3 & -0.1 & -0.2 \\ 0 & 0 & -0.3 \end{bmatrix}\begin{bmatrix} 0.4 & 0.1 & 0 \\ 0.3 & 0.2 & -0.2 \\ 0 & 0 & 0 \end{bmatrix}\right\}$$

$$= -\tfrac{1}{2}\,\text{tr}\begin{bmatrix} 0.07 & 0.03 & -0.02 \\ 0.09 & 0.01 & 0.02 \\ 0 & 0 & 0 \end{bmatrix} = -0.04$$

By substituting the matrix $\mathbf{H}_1$ and the value of a_2 just obtained into Eq. (5–76), we get

$$\mathbf{H}_2 = \mathbf{GH}_1 + a_2\mathbf{I} = \begin{bmatrix} 0.03 & 0.03 & -0.02 \\ 0.09 & -0.03 & 0.02 \\ 0 & 0 & -0.04 \end{bmatrix}$$

and

$$a_3 = -\tfrac{1}{3}\,\text{tr}\,\mathbf{GH}_2 = -\tfrac{1}{3}\,\text{tr}\begin{bmatrix} 0.012 & 0 & 0 \\ 0 & 0.012 & 0 \\ 0 & 0 & 0.012 \end{bmatrix} = -0.012$$

Notice that

$$\mathbf{H}_3 = \mathbf{GH}_2 - 0.012\mathbf{I} = \mathbf{0}$$

The adjoint of $(z\mathbf{I} - \mathbf{G})$ can now be given by Eq. (5–75), or

$$\text{adj}(z\mathbf{I} - \mathbf{G}) = \mathbf{I}z^2 + \mathbf{H}_1 z + \mathbf{H}_2$$

$$= \begin{bmatrix} 1 & 0 & 0 \\ 0 & 1 & 0 \\ 0 & 0 & 1 \end{bmatrix} z^2 + \begin{bmatrix} 0.4 & 0.1 & 0 \\ 0.3 & 0.2 & -0.2 \\ 0 & 0 & 0 \end{bmatrix} z + \begin{bmatrix} 0.03 & 0.03 & -0.02 \\ 0.09 & -0.03 & 0.02 \\ 0 & 0 & -0.04 \end{bmatrix}$$

$$= \begin{bmatrix} z^2 + 0.4z + 0.03 & 0.1z + 0.03 & -0.02 \\ 0.3z + 0.09 & z^2 + 0.2z - 0.03 & -0.2z + 0.02 \\ 0 & 0 & z^2 - 0.04 \end{bmatrix}$$

Also,

$$|z\mathbf{I}-\mathbf{G}| = z^3 + a_1 z^2 + a_2 z + a_3 = z^3 + 0.3z^2 - 0.04z - 0.012$$
$$= (z+0.2)(z-0.2)(z+0.3)$$

Hence

$$(z\mathbf{I}-\mathbf{G})^{-1} = \frac{\text{adj}\,(z\mathbf{I}-\mathbf{G})}{|z\mathbf{I}-\mathbf{G}|}$$
$$= \begin{bmatrix} \frac{z+0.1}{(z+0.2)(z-0.2)} & \frac{0.1}{(z+0.2)(z-0.2)} & \frac{-0.02}{(z+0.2)(z-0.2)(z+0.3)} \\ \frac{0.3}{(z+0.2)(z-0.2)} & \frac{z-0.1}{(z+0.2)(z-0.2)} & \frac{-0.2(z-0.1)}{(z+0.2)(z-0.2)(z+0.3)} \\ 0 & 0 & \frac{1}{z+0.3} \end{bmatrix}$$

This last equation gives the inverse of $(z\mathbf{I}-\mathbf{G})$.

Next, we shall obtain $\mathbf{G}^k$. From Eq. (5–70), we have

$$\mathbf{G}^k = \mathcal{Z}^{-1}[(z\mathbf{I}-\mathbf{G})^{-1}z]$$
$$= \mathcal{Z}^{-1}\begin{bmatrix} \frac{0.25z}{z+0.2} + \frac{0.75z}{z-0.2} & -\frac{0.25z}{z+0.2} + \frac{0.25z}{z-0.2} & \frac{0.5z}{z+0.2} - \frac{0.1z}{z-0.2} - \frac{0.4z}{z+0.3} \\ -\frac{0.75z}{z+0.2} + \frac{0.75z}{z-0.2} & \frac{0.75z}{z+0.2} + \frac{0.25z}{z-0.2} & -\frac{1.5z}{z+0.2} - \frac{0.1z}{z-0.2} + \frac{1.6z}{z+0.3} \\ 0 & 0 & \frac{z}{z+0.3} \end{bmatrix}$$
$$= \begin{bmatrix} 0.25(-0.2)^k + 0.75(0.2)^k & -0.25(-0.2)^k + 0.25(0.2)^k & 0.5(-0.2)^k - 0.1(0.2)^k - 0.4(-0.3)^k \\ -0.75(-0.2)^k + 0.75(0.2)^k & 0.75(-0.2)^k + 0.25(0.2)^k & -1.5(-0.2)^k - 0.1(0.2)^k + 1.6(-0.3)^k \\ 0 & 0 & (-0.3)^k \end{bmatrix} \quad (5\text{–}78)$$

Solution of linear time-varying discrete-time state equations. Consider the following linear time-varying discrete-time state equation and output equation:

$$\mathbf{x}(k+1) = \mathbf{G}(k)\mathbf{x}(k) + \mathbf{H}(k)\mathbf{u}(k) \quad (5\text{–}79)$$

$$\mathbf{y}(k) = \mathbf{C}(k)\mathbf{x}(k) + \mathbf{D}(k)\mathbf{u}(k) \quad (5\text{–}80)$$

The solution of Eq. (5–79) may be found easily by recursion, as follows:

$$\mathbf{x}(h+1) = \mathbf{G}(h)\mathbf{x}(h) + \mathbf{H}(h)\mathbf{u}(h)$$
$$\mathbf{x}(h+2) = \mathbf{G}(h+1)\mathbf{x}(h+1) + \mathbf{H}(h+1)\mathbf{u}(h+1)$$
$$= \mathbf{G}(h+1)\mathbf{G}(h)\mathbf{x}(h) + \mathbf{G}(h+1)\mathbf{H}(h)\mathbf{u}(h) + \mathbf{H}(h+1)\mathbf{u}(h+1)$$
$$\vdots$$

Let us define the state transition matrix (fundamental matrix) for the system defined by Eq. (5–79) as $\mathbf{\Psi}(k, h)$. It is a unique matrix satisfying the conditions

$$\mathbf{\Psi}(k+1, h) = \mathbf{G}(k)\mathbf{\Psi}(k, h), \qquad \mathbf{\Psi}(h, h) = \mathbf{I}$$

where $k = h, h + 1, h + 2, \ldots$. It can be seen that the state transition matrix $\mathbf{\Psi}(k, h)$ is given by the equation

$$\mathbf{\Psi}(k, h) = \mathbf{G}(k-1)\mathbf{G}(k-2) \cdots \mathbf{G}(h) \qquad k > h \tag{5–81}$$

Using $\mathbf{\Psi}(k, h)$, the solution of Eq. (5–79) becomes

$$\mathbf{x}(k) = \mathbf{\Psi}(k, h)\mathbf{x}(h) + \sum_{j=h}^{k-1} \mathbf{\Psi}(k, j+1)\mathbf{H}(j)\mathbf{u}(j) \qquad k > h \tag{5–82}$$

Notice that the first term on the right-hand side of Eq. (5–82) is the contribution of the initial state $\mathbf{x}(h)$ to the current state $\mathbf{x}(k)$ and that the second term is the contribution of the input $\mathbf{u}(h)$, $\mathbf{u}(h+1), \ldots, \mathbf{u}(k-1)$.

Equation (5–82) can be verified easily. Referring to Eq. (5–81), we have

$$\mathbf{\Psi}(k+1, h) = \mathbf{G}(k)\mathbf{G}(k-1) \cdots \mathbf{G}(h) = \mathbf{G}(k)\mathbf{\Psi}(k, h) \tag{5–83}$$

If we substitute Eq. (5–83) into

$$\mathbf{x}(k+1) = \mathbf{\Psi}(k+1, h)\mathbf{x}(h) + \sum_{j=h}^{k} \mathbf{\Psi}(k+1, j+1)\mathbf{H}(j)\mathbf{u}(j)$$

we obtain

$$\begin{aligned}
\mathbf{x}(k+1) &= \mathbf{G}(k)\mathbf{\Psi}(k, h)\mathbf{x}(h) + \sum_{j=h}^{k-1} \mathbf{\Psi}(k+1, j+1)\mathbf{H}(j)\mathbf{u}(j) \\
&\quad + \mathbf{\Psi}(k+1, k+1)\mathbf{H}(k)\mathbf{u}(k) \\
&= \mathbf{G}(k)\left[\mathbf{\Psi}(k, h)\mathbf{x}(h) + \sum_{j=h}^{k-1} \mathbf{\Psi}(k, j+1)\mathbf{H}(j)\mathbf{u}(j)\right] + \mathbf{H}(k)\mathbf{u}(k) \\
&= \mathbf{G}(k)\mathbf{x}(k) + \mathbf{H}(k)\mathbf{u}(k)
\end{aligned}$$

Thus, we have shown that Eq. (5–82) is the solution of Eq. (5–79).

Once we get the solution $\mathbf{x}(k)$, the output equation, Eq. (5–80), becomes as follows:

$$\mathbf{y}(k) = \mathbf{C}(k)\mathbf{\Psi}(k, h)\mathbf{x}(h) + \sum_{j=h}^{k-1} \mathbf{C}(k)\mathbf{\Psi}(k, j+1)\mathbf{H}(j)\mathbf{u}(j) + \mathbf{D}(k)\mathbf{u}(k) \qquad k > h$$

If $\mathbf{G}(k)$ is nonsingular for all k values considered, so that the inverse of $\mathbf{\Psi}(k, h)$ exists, then the inverse of $\mathbf{\Psi}(k, h)$, denoted by $\mathbf{\Psi}(h, k)$, is given as follows:

$$\begin{aligned}
\mathbf{\Psi}^{-1}(k, h) &= \mathbf{\Psi}(h, k) \\
&= [\mathbf{G}(k-1)\mathbf{G}(k-2) \cdots \mathbf{G}(h)]^{-1} \\
&= \mathbf{G}^{-1}(h)\mathbf{G}^{-1}(h+1) \cdots \mathbf{G}^{-1}(k-1)
\end{aligned} \tag{5–84}$$

Summary on $\Psi(k, h)$. A summary on the state transition matrix $\Psi(k, h)$ gives the following:

1. $\Psi(k, k) = \mathbf{I}$
2. $\Psi(k, h) = \mathbf{G}(k-1)\mathbf{G}(k-2) \cdots \mathbf{G}(h) \qquad k > h$
3. If the inverse of $\Psi(k, h)$ exists, then
$$\Psi^{-1}(k, h) = \Psi(h, k)$$
4. If $\mathbf{G}(k)$ is nonsingular for all k values considered, then
$$\Psi(k, i) = \Psi(k, j)\Psi(j, i) \qquad \text{for any } i, j, k$$
If $\mathbf{G}(k)$ is singular for any value of k, then
$$\Psi(k, i) = \Psi(k, j)\Psi(j, i) \qquad \text{for } k > j > i$$

(For details, see Prob. A-5–21.)

5–4 THE PULSE-TRANSFER-FUNCTION MATRIX

A single-input–single-output discrete-time system may be modeled by a pulse transfer function. Extension of the pulse-transfer-function concept to a multiple-input–multiple-output discrete-time system gives us the pulse-transfer-function matrix. In this section we shall investigate the relationship between state space representation and representation by the pulse-transfer-function matrix.

Pulse-transfer-function matrix. The state space representation of an nth-order linear time-invariant discrete-time system with r inputs and m outputs can be given by

$$\mathbf{x}(k+1) = \mathbf{G}\mathbf{x}(k) + \mathbf{H}\mathbf{u}(k) \tag{5–85}$$

$$\mathbf{y}(k) = \mathbf{C}\mathbf{x}(k) + \mathbf{D}\mathbf{u}(k) \tag{5–86}$$

where $\mathbf{x}(k)$ is an n-vector, $\mathbf{u}(k)$ is an r-vector, $\mathbf{y}(k)$ is an m-vector, $\mathbf{G}$ is an $n \times n$ matrix, $\mathbf{H}$ is an $n \times r$ matrix, $\mathbf{C}$ is an $m \times n$ matrix, and $\mathbf{D}$ is an $m \times r$ matrix. Taking the z transforms of Eqs. (5–85) and (5–86), we obtain

$$z\mathbf{X}(z) - z\mathbf{x}(0) = \mathbf{G}\mathbf{X}(z) + \mathbf{H}\mathbf{U}(z)$$

$$\mathbf{Y}(z) = \mathbf{C}\mathbf{X}(z) + \mathbf{D}\mathbf{U}(z)$$

Noting that the definition of the pulse transfer function calls for the assumption of zero initial conditions, here we also assume that the initial state $\mathbf{x}(0)$ is zero. Then we obtain

$$\mathbf{X}(z) = (z\mathbf{I} - \mathbf{G})^{-1}\mathbf{H}\mathbf{U}(z)$$

and

$$\mathbf{Y}(z) = [\mathbf{C}(z\mathbf{I} - \mathbf{G})^{-1}\mathbf{H} + \mathbf{D}]\mathbf{U}(z) = \mathbf{F}(z)\mathbf{U}(z) \tag{5-87}$$

where

$$\mathbf{F}(z) = \mathbf{C}(z\mathbf{I} - \mathbf{G})^{-1}\mathbf{H} + \mathbf{D} \tag{5-88}$$

$\mathbf{F}(z)$ is called the *pulse-transfer-function matrix*. It is an $m \times r$ matrix. The pulse transfer function matrix $\mathbf{F}(z)$ characterizes the input-output dynamics of the given discrete-time system.

Since the inverse of matrix $(z\mathbf{I} - \mathbf{G})$ can be written as

$$(z\mathbf{I} - \mathbf{G})^{-1} = \frac{\operatorname{adj}(z\mathbf{I} - \mathbf{G})}{|z\mathbf{I} - \mathbf{G}|}$$

the pulse transfer function matrix $\mathbf{F}(z)$ can be given by the equation

$$\mathbf{F}(z) = \frac{\mathbf{C} \operatorname{adj}(z\mathbf{I} - \mathbf{G})\,\mathbf{H}}{|z\mathbf{I} - \mathbf{G}|} + \mathbf{D}$$

Clearly, the poles of $\mathbf{F}(z)$ are the zeros of $|z\mathbf{I} - \mathbf{G}| = 0$. This means that the characteristic equation of the discrete-time system is given by

$$|z\mathbf{I} - \mathbf{G}| = 0$$

or

$$z^n + a_1 z^{n-1} + a_2 z^{n-2} + \cdots + a_{n-1} z + a_n = 0$$

where the coefficients a_i depend on the elements of $\mathbf{G}$. (In Chap. 3 it was shown that the discrete-time system is stable if the roots of the characteristic equation are less than unity in their absolute values. Noting that the roots of the characteristic equation are the eigenvalues of the matrix $\mathbf{G}$, for stability the eigenvalues of $\mathbf{G}$ must be less than unity in their absolute values.)

For the system defined by Eqs. (5–85) and (5–86), the $(n + m) \times (n + r)$ matrix

$$\mathbf{E}(z) = \begin{bmatrix} \mathbf{G} - z\mathbf{I} & \mathbf{H} \\ \mathbf{C} & \mathbf{D} \end{bmatrix} \tag{5-89}$$

is called the *system matrix*. The values of z that make

$$\operatorname{rank} \mathbf{E}(z) < n + \min(m, r)$$

are called the zeros of the system.

For example, if $r = 1$ and $m = 1$, that is, if both $u(k)$ and $y(k)$ are scalar, then

$$\mathbf{E}(z) = \begin{bmatrix} \mathbf{G} - z\mathbf{I} & \mathbf{H} \\ \mathbf{C} & d \end{bmatrix} \tag{5-90}$$

which is an $(n + 1) \times (n + 1)$ matrix. Referring to the Appendix, Eq. (3), the determinant of $\mathbf{E}(z)$ can be obtained as follows:

$$
\begin{aligned}
|\mathbf{E}(z)| &= \begin{vmatrix} \mathbf{G} - z\mathbf{I} & \mathbf{H} \\ \mathbf{C} & d \end{vmatrix} \\
&= |\mathbf{G} - z\mathbf{I}|\,|d - \mathbf{C}(\mathbf{G} - z\mathbf{I})^{-1}\mathbf{H}| \\
&= (-1)^n |z\mathbf{I} - \mathbf{G}|\,|d + \mathbf{C}(z\mathbf{I} - \mathbf{G})^{-1}\mathbf{H}| \\
&= (-1)^n |z\mathbf{I} - \mathbf{G}| \left| d + \mathbf{C}\frac{\operatorname{adj}(z\mathbf{I} - \mathbf{G})}{|z\mathbf{I} - \mathbf{G}|}\mathbf{H} \right| \\
&= (-1)^n [d\,|z\mathbf{I} - \mathbf{G}| + \mathbf{C}\operatorname{adj}(z\mathbf{I} - \mathbf{G})\,\mathbf{H}]
\end{aligned}
$$

The values of z that make the rank of $\mathbf{E}(z)$ given by Eq. (5–90) less than $n + 1$, that is, the values of z that make $|\mathbf{E}(z)| = 0$, are the zeros of the system. Thus, the roots of the equation

$$|\mathbf{E}(z)| = 0$$

or

$$d\,|z\mathbf{I} - \mathbf{G}| + \mathbf{C}\operatorname{adj}(z\mathbf{I} - \mathbf{G})\,\mathbf{H} = 0$$

are the zeros of the system when the input $u(k)$ and output $y(k)$ are scalar.

Weighting sequence matrix. By rewriting Eq. (5–87) as the equation

$$
\begin{bmatrix} Y_1(z) \\ Y_2(z) \\ \cdot \\ \cdot \\ \cdot \\ Y_m(z) \end{bmatrix}
=
\begin{bmatrix} F_{11}(z) & F_{12}(z) & \cdots & F_{1r}(z) \\ F_{21}(z) & F_{22}(z) & \cdots & F_{2r}(z) \\ \cdot & \cdot & & \cdot \\ \cdot & \cdot & & \cdot \\ \cdot & \cdot & & \cdot \\ F_{m1}(z) & F_{m2}(z) & \cdots & F_{mr}(z) \end{bmatrix}
\begin{bmatrix} U_1(z) \\ U_2(z) \\ \cdot \\ \cdot \\ \cdot \\ U_r(z) \end{bmatrix}
$$

we see that the ith output $Y_i(z)$ in response to the r inputs $U_1(z)$, $U_2(z)$, $\cdots$, $U_r(z)$ is given by

$$Y_i(z) = \sum_{j=1}^{r} F_{ij}(z) U_j(z) \qquad i = 1, 2, \ldots, m$$

Notice that the element $F_{ij}(z)$ of the pulse-transfer-function matrix $\mathbf{F}(z)$ is the z transform of the ith output in response to a Kronecker delta function applied to the jth input.

The weighting sequence matrix is the extension of the concept of the weighting function for the single-input–single-output system to the multiple-input–multiple-output system. It is the system's response sequence matrix to Kronecker delta inputs when the system is initially at rest, or when $\mathbf{x}(0) = \mathbf{0}$. Thus, it is the inverse z transform of $\mathbf{F}(z)$.

Since $(z\mathbf{I} - \mathbf{G})^{-1}$ may be expanded into the infinite series

$$(z\mathbf{I} - \mathbf{G})^{-1} = \mathbf{I}z^{-1} + \mathbf{G}z^{-2} + \mathbf{G}^2 z^{-3} + \cdots$$

it is possible to write Eq. (5–88) as follows:

$$\begin{aligned} \mathbf{F}(z) &= \mathbf{C}[\mathbf{I}z^{-1} + \mathbf{G}z^{-2} + \mathbf{G}^2 z^{-3} + \cdots]\mathbf{H} + \mathbf{D} \\ &= \mathbf{D} + \mathbf{CH}z^{-1} + \mathbf{CGH}z^{-2} + \mathbf{CG}^2\mathbf{H}z^{-3} + \cdots + \mathbf{CG}^{k-1}\mathbf{H}z^{-k} + \cdots \end{aligned} \tag{5–91}$$

Then the weighting sequence matrix $\mathbf{F}(k)$ is given by

$$\mathbf{F}(k) = \mathscr{Z}^{-1}[\mathbf{F}(z)] \tag{5–92}$$

Referring to the definition of the z transform, we may write $\mathbf{F}(z)$ as follows:

$$\begin{aligned} \mathbf{F}(z) &= \sum_{k=0}^{\infty} \mathbf{F}(k) z^{-k} \\ &= \mathbf{F}(0) + \mathbf{F}(1)z^{-1} + \mathbf{F}(2)z^{-2} + \cdots + \mathbf{F}(k)z^{-k} + \cdots \end{aligned} \tag{5–93}$$

By comparing the coefficients of the z^{-k} $(k = 0, 1, 2, \ldots)$ terms of Eqs. (5–91) and (5–93), we obtain the weighting sequence matrices as follows:

$$\begin{aligned} \mathbf{F}(0) &= \mathbf{D} \\ \mathbf{F}(1) &= \mathbf{CH} \\ \mathbf{F}(2) &= \mathbf{CGH} \\ &\vdots \\ \mathbf{F}(k) &= \mathbf{CG}^{k-1}\mathbf{H} \end{aligned}$$

Thus, the weighting sequence matrix $\mathbf{F}(k)$ may be given as follows:

$$\mathbf{F}(k) = \begin{cases} \mathbf{0} & k < 0 \\ \mathbf{D} & k = 0 \\ \mathbf{CG}^{k-1}\mathbf{H} & k = 1, 2, 3, \ldots \end{cases} \tag{5–94}$$

Since the output $\mathbf{Y}(z)$ is the product of the pulse-transfer-function matrix $\mathbf{F}(z)$ and the input $\mathbf{U}(z)$, the output $\mathbf{y}(k)$ may be given by the convolution summation of the weighting sequence matrix and the input as follows:

$$\mathbf{y}(k) = \sum_{j=0}^{k} \mathbf{F}(k - j)\mathbf{u}(j) \qquad k = 0, 1, 2, \ldots$$

Example 5–3.

Determine the pulse-transfer-function matrix and weighting sequence matrix for the system given by the equations

$$\begin{bmatrix} x_1(k+1) \\ x_2(k+1) \end{bmatrix} = \begin{bmatrix} 0 & 1 \\ -0.16 & -1 \end{bmatrix} \begin{bmatrix} x_1(k) \\ x_2(k) \end{bmatrix} + \begin{bmatrix} 0 \\ 1 \end{bmatrix} u(k)$$

$$\begin{bmatrix} y_1(k) \\ y_2(k) \end{bmatrix} = \begin{bmatrix} 1 & 1 \\ 0 & 1 \end{bmatrix} \begin{bmatrix} x_1(k) \\ x_2(k) \end{bmatrix}$$

From Eq. (5–88), the pulse-transfer-function matrix $\mathbf{F}(z)$ is given as follows:

$$\begin{aligned}
\mathbf{F}(z) &= \mathbf{C}(z\mathbf{I} - \mathbf{G})^{-1}\mathbf{H} + \mathbf{D} \\
&= \begin{bmatrix} 1 & 1 \\ 0 & 1 \end{bmatrix} \begin{bmatrix} z & -1 \\ 0.16 & z+1 \end{bmatrix}^{-1} \begin{bmatrix} 0 \\ 1 \end{bmatrix} + \begin{bmatrix} 0 \\ 0 \end{bmatrix} \\
&= \begin{bmatrix} 1 & 1 \\ 0 & 1 \end{bmatrix} \begin{bmatrix} \frac{4}{3}\frac{1}{z+0.2} - \frac{1}{3}\frac{1}{z+0.8} & \frac{5}{3}\frac{1}{z+0.2} - \frac{5}{3}\frac{1}{z+0.8} \\ -\frac{0.8}{3}\frac{1}{z+0.2} + \frac{0.8}{3}\frac{1}{z+0.8} & -\frac{1}{3}\frac{1}{z+0.2} + \frac{4}{3}\frac{1}{z+0.8} \end{bmatrix} \begin{bmatrix} 0 \\ 1 \end{bmatrix} \\
&= \begin{bmatrix} \frac{4}{3}\frac{1}{z+0.2} - \frac{1}{3}\frac{1}{z+0.8} \\ -\frac{1}{3}\frac{1}{z+0.2} + \frac{4}{3}\frac{1}{z+0.8} \end{bmatrix}
\end{aligned}$$

The weighting sequence matrix $\mathbf{F}(k)$ is given by the equation

$$\mathbf{F}(k) = \mathscr{Z}^{-1}[\mathbf{F}(z)] = \begin{bmatrix} \frac{4}{3}(-0.2)^{k-1} - \frac{1}{3}(-0.8)^{k-1} \\ -\frac{1}{3}(-0.2)^{k-1} + \frac{4}{3}(-0.8)^{k-1} \end{bmatrix}$$

where $k = 1, 2, 3, \ldots$. The same result can, of course, be obtained by evaluating

$$\mathbf{F}(k) = \mathbf{C}\mathbf{G}^{k-1}\mathbf{H} \qquad k = 1, 2, 3, \ldots$$

Since $\mathbf{G}^k$ can be given [see Eq. (5–72)] by the equation

$$\begin{aligned}
\mathbf{G}^k &= \mathscr{Z}^{-1}[(z\mathbf{I} - \mathbf{G})^{-1}z] \\
&= \begin{bmatrix} \frac{4}{3}(-0.2)^k - \frac{1}{3}(-0.8)^k & \frac{5}{3}(-0.2)^k - \frac{5}{3}(-0.8)^k \\ -\frac{0.8}{3}(-0.2)^k + \frac{0.8}{3}(-0.8)^k & -\frac{1}{3}(-0.2)^k + \frac{4}{3}(-0.8)^k \end{bmatrix}
\end{aligned}$$

we obtain

$$\begin{aligned}
\mathbf{F}(k) &= \begin{bmatrix} 1 & 1 \\ 0 & 1 \end{bmatrix} \begin{bmatrix} \frac{4}{3}(-0.2)^{k-1} - \frac{1}{3}(-0.8)^{k-1} & \frac{5}{3}(-0.2)^{k-1} - \frac{5}{3}(-0.8)^{k-1} \\ -\frac{0.8}{3}(-0.2)^{k-1} + \frac{0.8}{3}(-0.8)^{k-1} & -\frac{1}{3}(-0.2)^{k-1} + \frac{4}{3}(-0.8)^{k-1} \end{bmatrix} \begin{bmatrix} 0 \\ 1 \end{bmatrix} \\
&= \begin{bmatrix} \frac{4}{3}(-0.2)^{k-1} - \frac{1}{3}(-0.8)^{k-1} \\ -\frac{1}{3}(-0.2)^{k-1} + \frac{4}{3}(-0.8)^{k-1} \end{bmatrix} \qquad k = 1, 2, 3, \ldots
\end{aligned}$$

which is the same expression for the weighting sequence matrix we have already obtained.

Similarity transformation. We have shown that for the system defined by

$$\mathbf{x}(k+1) = \mathbf{G}\mathbf{x}(k) + \mathbf{H}\mathbf{u}(k)$$

$$\mathbf{y}(k) = \mathbf{C}\mathbf{x}(k) + \mathbf{D}\mathbf{u}(k)$$

the pulse-transfer-function matrix is

$$\mathbf{F}(z) = \mathbf{C}(z\mathbf{I} - \mathbf{G})^{-1}\mathbf{H} + \mathbf{D}$$

In Sec. 5–2 we showed that various different state space representations for a given system are related by the similarity transformation. By defining a new state vector $\hat{\mathbf{x}}(k)$ by using a similarity transformation matrix $\mathbf{P}$, or

$$\mathbf{x}(k) = \mathbf{P}\hat{\mathbf{x}}(k)$$

where $\mathbf{P}$ is a nonsingular $n \times n$ matrix, we have

$$\hat{\mathbf{x}}(k+1) = \hat{\mathbf{G}}\hat{\mathbf{x}}(k) + \hat{\mathbf{H}}\mathbf{u}(k) \tag{5–95}$$

$$\mathbf{y}(k) = \hat{\mathbf{C}}\hat{\mathbf{x}}(k) + \hat{\mathbf{D}}\mathbf{u}(k) \tag{5–96}$$

where $\mathbf{G}$, $\mathbf{H}$, $\mathbf{C}$, $\mathbf{D}$ and $\hat{\mathbf{G}}$, $\hat{\mathbf{H}}$, $\hat{\mathbf{C}}$, $\hat{\mathbf{D}}$ are related, respectively, by

$$\mathbf{P}^{-1}\mathbf{G}\mathbf{P} = \hat{\mathbf{G}}$$

$$\mathbf{P}^{-1}\mathbf{H} = \hat{\mathbf{H}}$$

$$\mathbf{C}\mathbf{P} = \hat{\mathbf{C}}$$

$$\mathbf{D} = \hat{\mathbf{D}}$$

The pulse-transfer-function matrix $\hat{\mathbf{F}}(z)$ for the system defined by Eqs. (5–95) and (5–96) is

$$\hat{\mathbf{F}}(z) = \hat{\mathbf{C}}(z\mathbf{I} - \hat{\mathbf{G}})^{-1}\hat{\mathbf{H}} + \hat{\mathbf{D}}$$

Notice that the pulse-transfer-function matrices $\mathbf{F}(z)$ and $\hat{\mathbf{F}}(z)$ are the same, since

$$\begin{aligned}\hat{\mathbf{F}}(z) &= \hat{\mathbf{C}}(z\mathbf{I} - \hat{\mathbf{G}})^{-1}\hat{\mathbf{H}} + \hat{\mathbf{D}} = \mathbf{C}\mathbf{P}(z\mathbf{I} - \mathbf{P}^{-1}\mathbf{G}\mathbf{P})^{-1}\mathbf{P}^{-1}\mathbf{H} + \mathbf{D} \\ &= \mathbf{C}\mathbf{P}(z\mathbf{P} - \mathbf{G}\mathbf{P})^{-1}\mathbf{H} + \mathbf{D} = \mathbf{C}(z\mathbf{P}\mathbf{P}^{-1} - \mathbf{G}\mathbf{P}\mathbf{P}^{-1})^{-1}\mathbf{H} + \mathbf{D} \\ &= \mathbf{C}(z\mathbf{I} - \mathbf{G})^{-1}\mathbf{H} + \mathbf{D} = \mathbf{F}(z)\end{aligned}$$

Thus, the pulse-transfer-function matrix is invariant under similarity transformation. That is, it does not depend on the particular state vector $\mathbf{x}(k)$ chosen for the system representation.

The characteristic equation $|z\mathbf{I} - \mathbf{G}| = 0$ is also invariant under similarity transformation, since

$$|z\mathbf{I} - \mathbf{G}| = |\mathbf{P}^{-1}||z\mathbf{I} - \mathbf{G}||\mathbf{P}| = |z\mathbf{I} - \mathbf{P}^{-1}\mathbf{G}\mathbf{P}| = |z\mathbf{I} - \hat{\mathbf{G}}|$$

Thus, the eigenvalues of $\mathbf{G}$ are invariant under similarity transformation.

Similarly, the $(n+m) \times (n+r)$ system matrix given by Eq. (5–89),

$$\mathbf{E}(z) = \begin{bmatrix} \mathbf{G} - z\mathbf{I} & \mathbf{H} \\ \mathbf{C} & \mathbf{D} \end{bmatrix}$$

is invariant under similarity transformation:

$$\begin{bmatrix} \mathbf{P}^{-1} & \mathbf{0} \\ \mathbf{0} & \mathbf{I}_m \end{bmatrix}\begin{bmatrix} \mathbf{G} - z\mathbf{I} & \mathbf{H} \\ \mathbf{C} & \mathbf{D} \end{bmatrix}\begin{bmatrix} \mathbf{P} & \mathbf{0} \\ \mathbf{0} & \mathbf{I}_r \end{bmatrix} = \begin{bmatrix} \mathbf{P}^{-1}(\mathbf{G} - z\mathbf{I})\mathbf{P} & \mathbf{P}^{-1}\mathbf{H} \\ \mathbf{C}\mathbf{P} & \mathbf{D} \end{bmatrix} = \begin{bmatrix} \hat{\mathbf{G}} - z\mathbf{I} & \hat{\mathbf{H}} \\ \hat{\mathbf{C}} & \hat{\mathbf{D}} \end{bmatrix}$$

Hence

$$\begin{vmatrix} \mathbf{G} - z\mathbf{I} & \mathbf{H} \\ \mathbf{C} & \mathbf{D} \end{vmatrix} = \begin{vmatrix} \hat{\mathbf{G}} - z\mathbf{I} & \hat{\mathbf{H}} \\ \hat{\mathbf{C}} & \hat{\mathbf{D}} \end{vmatrix}$$

5-5 CONTINUOUS-TIME STATE SPACE EQUATIONS

In this section we shall first present state space representations for continuous-time systems and then discuss the general solution of the linear continuous-time state space equation. At the beginning of this section, we shall consider time-invariant systems. Later, however, we include discussions of time-varying systems. We shall begin this section by presenting state space equations in canonical forms.

Canonical forms for continuous-time state space equations. Consider the transfer function system

$$G(s) = \frac{Y(s)}{U(s)} = \frac{b_0 s^n + b_1 s^{n-1} + \cdots + b_{n-1}s + b_n}{s^n + a_1 s^{n-1} + \cdots + a_{n-1}s + a_n} \tag{5-97}$$

where $U(s)$ is the Laplace transform of the input and $Y(s)$ is the Laplace transform of the output. There are many ways to realize state space representations for the system. And as in the case of discrete-time control systems, we can have state space representations for this continuous-time system in controllable canonical form, observable canonical form, and diagonal canonical form (or Jordan canonical form). (For the meaning of the terms "controllable" and "observable," see Secs. 6–2 and 6–3.)

The derivation of the canonical forms in state space for the continuous-time case is the same as for the discrete-time case. [All we need to do is to change z to s and $X(z)$ to $X(s)$.] Therefore, we shall not give the derivations here but only the results.

Controllable Canonical Form. The state space representation of the transfer function system given by Eq. (5–97) in the controllable canonical form is given by the following equations:

$$\begin{bmatrix} \dot{x}_1(t) \\ \dot{x}_2(t) \\ \cdot \\ \cdot \\ \cdot \\ \dot{x}_{n-1}(t) \\ \dot{x}_n(t) \end{bmatrix} = \begin{bmatrix} 0 & 1 & 0 & \cdots & 0 \\ 0 & 0 & 1 & \cdots & 0 \\ \cdot & \cdot & \cdot & & \cdot \\ \cdot & \cdot & \cdot & & \cdot \\ \cdot & \cdot & \cdot & & \cdot \\ 0 & 0 & 0 & \cdots & 1 \\ -a_n & -a_{n-1} & -a_{n-2} & \cdots & -a_1 \end{bmatrix} \begin{bmatrix} x_1(t) \\ x_2(t) \\ \cdot \\ \cdot \\ \cdot \\ x_{n-1}(t) \\ x_n(t) \end{bmatrix} + \begin{bmatrix} 0 \\ 0 \\ \cdot \\ \cdot \\ \cdot \\ 0 \\ 1 \end{bmatrix} u(t)$$

$$y(t) = [b_n - a_n b_0 \vdots b_{n-1} - a_{n-1} b_0 \vdots \cdots \vdots b_1 - a_1 b_0] \begin{bmatrix} x_1(t) \\ x_2(t) \\ \cdot \\ \cdot \\ \cdot \\ x_n(t) \end{bmatrix} + b_0 u(t)$$

Observable Canonical Form. The state space representation of the transfer function system given by Eq. (5–97) in the observable canonical form is as follows:

$$\begin{bmatrix} \dot{x}_1(t) \\ \dot{x}_2(t) \\ \cdot \\ \cdot \\ \cdot \\ \dot{x}_n(t) \end{bmatrix} = \begin{bmatrix} 0 & 0 & \cdots & 0 & -a_n \\ 1 & 0 & \cdots & 0 & -a_{n-1} \\ \cdot & \cdot & & \cdot & \cdot \\ \cdot & \cdot & & \cdot & \cdot \\ \cdot & \cdot & & \cdot & \cdot \\ 0 & 0 & \cdots & 1 & -a_1 \end{bmatrix} \begin{bmatrix} x_1(t) \\ x_2(t) \\ \cdot \\ \cdot \\ \cdot \\ x_n(t) \end{bmatrix} + \begin{bmatrix} b_n - a_n b_0 \\ b_{n-1} - a_{n-1} b_0 \\ \cdot \\ \cdot \\ \cdot \\ b_1 - a_1 b_0 \end{bmatrix} u(t)$$

$$y(t) = [0 \quad 0 \quad \cdots \quad 0 \quad 1] \begin{bmatrix} x_1(t) \\ x_2(t) \\ \cdot \\ \cdot \\ \cdot \\ x_{n-1}(t) \\ x_n(t) \end{bmatrix} + b_0 u(t)$$

Diagonal Canonical Form. The transfer function system given by Eq. (5–97) may be factored as follows:

$$G(s) = \frac{Y(s)}{U(s)} = \frac{b_0 s^n + b_1 s^{n-1} + \cdots + b_{n-1} s + b_n}{(s - p_1)(s - p_2) \cdots (s - p_n)}$$

If all poles of $G(s)$ are distinct, then $G(s)$ can be expanded into the following form:

$$\frac{Y(s)}{U(s)} = b_0 + \frac{c_1}{s - p_1} + \frac{c_2}{s - p_2} + \cdots + \frac{c_n}{s - p_n}$$

The state space representation in the diagonal canonical form is

$$\begin{bmatrix} \dot{x}_1(t) \\ \dot{x}_2(t) \\ \cdot \\ \cdot \\ \cdot \\ \dot{x}_n(t) \end{bmatrix} = \begin{bmatrix} p_1 & & & & 0 \\ & p_2 & & & \\ & & \cdot & & \\ & & & \cdot & \\ & & & & \cdot \\ 0 & & & & p_n \end{bmatrix} \begin{bmatrix} x_1(t) \\ x_2(t) \\ \cdot \\ \cdot \\ \cdot \\ x_n(t) \end{bmatrix} + \begin{bmatrix} 1 \\ 1 \\ \cdot \\ \cdot \\ \cdot \\ 1 \end{bmatrix} u(t)$$

$$y(t)=[c_1 \quad c_2 \quad \cdots \quad c_n]\begin{bmatrix} x_1(t) \\ x_2(t) \\ \cdot \\ \cdot \\ \cdot \\ x_n(t) \end{bmatrix}+b_0 u(t)$$

Jordan Canonical Form. If $G(s)$ involves multiple poles, then the preceding diagonal canonical form must be modified into the Jordan canonical form. Suppose, for example, that the p_i's are different from one another, except that the first three p_i's are equal, or $p_1 = p_2 = p_3$. Then the factored form of $G(s)$ is

$$G(s)=\frac{b_0 s^n + b_1 s^{n-1} + \cdots + b_{n-1}s + b_n}{(s-p_1)^3(s-p_4)(s-p_5)\cdots(s-p_n)}$$

The partial fraction expansion of $G(s)$ gives

$$G(s)=b_0+\frac{c_1}{(s-p_1)^3}+\frac{c_2}{(s-p_1)^2}+\frac{c_3}{s-p_1}+\frac{c_4}{s-p_4}+\cdots+\frac{c_n}{s-p_n}$$

Then, the state space representation in the Jordan canonical form becomes as follows:

$$\begin{bmatrix} \dot{x}_1(t) \\ \dot{x}_2(t) \\ \dot{x}_3(t) \\ \dot{x}_4(t) \\ \cdot \\ \cdot \\ \cdot \\ \dot{x}_n(t) \end{bmatrix} = \left[\begin{array}{ccc|cccc} p_1 & 1 & 0 & & & & 0 \\ 0 & p_1 & 1 & & & & \\ 0 & 0 & p_1 & & & & \\ \hline & & & p_4 & & & 0 \\ & & & & \cdot & & \\ & & & & & \cdot & \\ & & & & & & \cdot \\ 0 & & & 0 & & & p_n \end{array}\right] \begin{bmatrix} x_1(t) \\ x_2(t) \\ x_3(t) \\ x_4(t) \\ \cdot \\ \cdot \\ \cdot \\ x_n(t) \end{bmatrix} + \begin{bmatrix} 0 \\ 0 \\ 1 \\ 1 \\ \cdot \\ \cdot \\ \cdot \\ 1 \end{bmatrix} u(t)$$

$$y(t)=[c_1 \quad c_2 \quad \cdots \quad c_n]\begin{bmatrix} x_1(t) \\ x_2(t) \\ \cdot \\ \cdot \\ \cdot \\ x_n(t) \end{bmatrix}+b_0 u(t)$$

As an example, consider the following system:

$$G(s)=\frac{Y(s)}{U(s)}=\frac{s+3}{s^2+3s+2}$$

The state space representations in the controllable canonical form, observable canonical form, and diagonal canonical form become as follows:

CONTROLLABLE CANONICAL FORM:

$$\begin{bmatrix} \dot{x}_1(t) \\ \dot{x}_2(t) \end{bmatrix} = \begin{bmatrix} 0 & 1 \\ -2 & -3 \end{bmatrix} \begin{bmatrix} x_1(t) \\ x_2(t) \end{bmatrix} + \begin{bmatrix} 0 \\ 1 \end{bmatrix} u(t)$$

$$y(t) = [3 \quad 1] \begin{bmatrix} x_1(t) \\ x_2(t) \end{bmatrix}$$

OBSERVABLE CANONICAL FORM:

$$\begin{bmatrix} \dot{x}_1(t) \\ \dot{x}_2(t) \end{bmatrix} = \begin{bmatrix} 0 & -2 \\ 1 & -3 \end{bmatrix} \begin{bmatrix} x_1(t) \\ x_2(t) \end{bmatrix} + \begin{bmatrix} 3 \\ 1 \end{bmatrix} u(t)$$

$$y(t) = [0 \quad 1] \begin{bmatrix} x_1(t) \\ x_2(t) \end{bmatrix}$$

DIAGONAL CANONICAL FORM:

The given $G(s)$ can be expanded as follows:

$$G(s) = \frac{s+3}{(s+1)(s+2)} = \frac{2}{s+1} - \frac{1}{s+2}$$

Hence

$$\begin{bmatrix} \dot{x}_1(t) \\ \dot{x}_2(t) \end{bmatrix} = \begin{bmatrix} -1 & 0 \\ 0 & -2 \end{bmatrix} \begin{bmatrix} x_1(t) \\ x_2(t) \end{bmatrix} + \begin{bmatrix} 1 \\ 1 \end{bmatrix} u(t)$$

$$y(t) = [2 \quad -1] \begin{bmatrix} x_1(t) \\ x_2(t) \end{bmatrix}$$

As seen from the preceding discussions, the state space representation of the continuous-time control system given by Eq. (5–97) is

$$\dot{\mathbf{x}}(t) = \mathbf{A}\mathbf{x}(t) + \mathbf{B}u(t)$$

$$y(t) = \mathbf{C}\mathbf{x}(t) + Du(t)$$

where D may be zero. In general, for a multiple-input–multiple-output system [such that $\mathbf{x}(t)$ is an n-vector, $\mathbf{y}(t)$ is an m-vector, and $\mathbf{u}(t)$ is an r-vector], the state space representation is

$$\dot{\mathbf{x}}(t) = \mathbf{A}\mathbf{x}(t) + \mathbf{B}\mathbf{u}(t)$$

$$\mathbf{y}(t) = \mathbf{C}\mathbf{x}(t) + \mathbf{D}\mathbf{u}(t)$$

Solution of homogeneous linear time-invariant state equations. Consider the state equation

$$\dot{\mathbf{x}} = \mathbf{A}\mathbf{x} \tag{5–98}$$

where $\mathbf{x}$ is the state vector (n-vector) and $\mathbf{A}$ is an $n \times n$ constant matrix. Let us assume that the solution of Eq. (5–98) is in the form of a vector power series in t, or

$$\mathbf{x}(t) = \mathbf{b}_0 + \mathbf{b}_1 t + \mathbf{b}_2 t^2 + \cdots + \mathbf{b}_k t^k + \cdots \tag{5–99}$$

By substituting Eq. (5–99) into Eq. (5–98), we obtain

$$\begin{aligned} \mathbf{b}_1 + 2\mathbf{b}_2 t + 3\mathbf{b}_3 t^2 + \cdots + k\mathbf{b}_k t^{k-1} + \cdots \\ = \mathbf{A}(\mathbf{b}_0 + \mathbf{b}_1 t + \mathbf{b}_2 t^2 + \cdots + k\mathbf{b}_k t^k + \cdots) \end{aligned}$$

If the assumed solution is to be the true solution, this last equation must hold for all t. Thus, we require that the coefficients of the equal powers of t be identical, or that

$$\begin{aligned} \mathbf{b}_1 &= \mathbf{A}\mathbf{b}_0 \\ \mathbf{b}_2 &= \frac{1}{2}\mathbf{A}\mathbf{b}_1 = \frac{1}{2}\mathbf{A}^2\mathbf{b}_0 = \frac{1}{2!}\mathbf{A}^2\mathbf{b}_0 \\ \mathbf{b}_3 &= \frac{1}{3}\mathbf{A}\mathbf{b}_2 = \frac{1}{3 \times 2}\mathbf{A}^3\mathbf{b}_0 = \frac{1}{3!}\mathbf{A}^3\mathbf{b}_0 \\ &\ \ \vdots \\ \mathbf{b}_k &= \frac{1}{k!}\mathbf{A}^k\mathbf{b}_0 \end{aligned}$$

By substituting 0 for t in Eq. (5–99), we obtain

$$\mathbf{x}(0) = \mathbf{b}_0$$

Thus the solution $\mathbf{x}(t)$ can be written as follows:

$$\mathbf{x}(t) = \left(\mathbf{I} + \mathbf{A}t + \frac{1}{2!}\mathbf{A}^2 t^2 + \cdots + \frac{1}{k!}\mathbf{A}^k t^k + \cdots\right)\mathbf{x}(0)$$

The expression in parentheses in the right-hand side of this last equation is an $n \times n$ matrix. Because of its similarity to the infinite power series for a scalar exponential, we call it the *matrix exponential* and write

$$\mathbf{I} + \mathbf{A}t + \frac{1}{2!}\mathbf{A}^2 t^2 + \cdots + \frac{1}{k!}\mathbf{A}^k t^k + \cdots = e^{\mathbf{A}t}$$

In terms of the matrix exponential, the solution to Eq. (5–98) can be written as follows:

$$\mathbf{x}(t) = e^{\mathbf{A}t}\mathbf{x}(0) \tag{5–100}$$

Since the matrix exponential is very important in the state space analysis of linear systems, we shall examine the properties of the matrix exponential in the following.

Matrix exponential. It can be proved that the matrix exponential of an $n \times n$ matrix $\mathbf{A}$,

$$e^{\mathbf{A}t} = \sum_{k=0}^{\infty} \frac{\mathbf{A}^k t^k}{k!}$$

converges absolutely for all finite t. (Hence computer computations for evaluating the elements of $e^{\mathbf{A}t}$ by using the series expansion can easily be carried out.)

Because of the convergence of the infinite series $\sum_{k=0}^{\infty} \mathbf{A}^k t^k / k!$, the series can be differentiated term by term to give

$$\frac{d}{dt} e^{\mathbf{A}t} = \mathbf{A} + \mathbf{A}^2 t + \frac{\mathbf{A}^3 t^2}{2!} + \cdots + \frac{\mathbf{A}^k t^{k-1}}{(k-1)!} + \cdots$$

$$= \mathbf{A}\left[\mathbf{I} + \mathbf{A}t + \frac{\mathbf{A}^2 t^2}{2!} + \cdots + \frac{\mathbf{A}^{k-1} t^{k-1}}{(k-1)!} + \cdots\right] = \mathbf{A}e^{\mathbf{A}t}$$

$$= \left[\mathbf{I} + \mathbf{A}t + \frac{\mathbf{A}^2 t^2}{2!} + \cdots + \frac{\mathbf{A}^{k-1} t^{k-1}}{(k-1)!} + \cdots\right]\mathbf{A} = e^{\mathbf{A}t}\mathbf{A}$$

The matrix exponential has the property that

$$e^{\mathbf{A}(t+s)} = e^{\mathbf{A}t} e^{\mathbf{A}s}$$

This can be proved as follows:

$$e^{\mathbf{A}t} e^{\mathbf{A}s} = \left(\sum_{k=0}^{\infty} \frac{\mathbf{A}^k t^k}{k!}\right)\left(\sum_{k=0}^{\infty} \frac{\mathbf{A}^k s^k}{k!}\right) = \sum_{k=0}^{\infty} \mathbf{A}^k \left[\sum_{i=0}^{k} \frac{t^i s^{k-i}}{i!(k-i)!}\right]$$

$$= \sum_{k=0}^{\infty} \mathbf{A}^k \frac{(t+s)^k}{k!} = e^{\mathbf{A}(t+s)}$$

In particular, if $s = -t$, then

$$e^{\mathbf{A}t} e^{-\mathbf{A}t} = e^{-\mathbf{A}t} e^{\mathbf{A}t} = e^{\mathbf{A}(t-t)} = \mathbf{I}$$

Thus, the inverse of $e^{\mathbf{A}t}$ is $e^{-\mathbf{A}t}$. Since the inverse of $e^{\mathbf{A}t}$ always exists, $e^{\mathbf{A}t}$ is nonsingular.

It is important to point out that

$$e^{(\mathbf{A}+\mathbf{B})t} = e^{\mathbf{A}t} e^{\mathbf{B}t} \quad \text{if} \quad \mathbf{AB} = \mathbf{BA}$$

$$e^{(\mathbf{A}+\mathbf{B})t} \neq e^{\mathbf{A}t} e^{\mathbf{B}t} \quad \text{if} \quad \mathbf{AB} \neq \mathbf{BA}$$

To prove this, note that

$$e^{(\mathbf{A}+\mathbf{B})t} = \mathbf{I} + (\mathbf{A}+\mathbf{B})t + \frac{(\mathbf{A}+\mathbf{B})^2}{2!} t^2 + \frac{(\mathbf{A}+\mathbf{B})^3}{3!} t^3 + \cdots$$

$$e^{\mathbf{A}t}e^{\mathbf{B}t} = \left(\mathbf{I} + \mathbf{A}t + \frac{\mathbf{A}^2t^2}{2!} + \frac{\mathbf{A}^3t^3}{3!} + \cdots\right)\left(\mathbf{I} + \mathbf{B}t + \frac{\mathbf{B}^2t^2}{2!} + \frac{\mathbf{B}^3t^3}{3!} + \cdots\right)$$

$$= \mathbf{I} + (\mathbf{A} + \mathbf{B})t + \frac{\mathbf{A}^2t^2}{2!} + \mathbf{AB}t^2 + \frac{\mathbf{B}^2t^2}{2!} + \frac{\mathbf{A}^3t^3}{3!} + \frac{\mathbf{A}^2\mathbf{B}t^3}{2!}$$

$$+ \frac{\mathbf{AB}^2t^3}{2!} + \frac{\mathbf{B}^3t^3}{3!} + \cdots$$

Hence

$$e^{(\mathbf{A}+\mathbf{B})t} - e^{\mathbf{A}t}e^{\mathbf{B}t} = \frac{\mathbf{BA} - \mathbf{AB}}{2!}t^2$$

$$+ \frac{\mathbf{BA}^2 + \mathbf{ABA} + \mathbf{B}^2\mathbf{A} + \mathbf{BAB} - 2\mathbf{A}^2\mathbf{B} - 2\mathbf{AB}^2}{3!}t^3 + \cdots$$

The difference between $e^{(\mathbf{A}+\mathbf{B})t}$ and $e^{\mathbf{A}t}e^{\mathbf{B}t}$ vanishes if $\mathbf{A}$ and $\mathbf{B}$ commute.

Laplace transform approach to the solution of homogeneous linear time-invariant state equations. We shall next consider the Laplace transform approach to the solution of Eq. (5–98),

$$\dot{\mathbf{x}} = \mathbf{Ax}$$

Taking the Laplace transform of both sides of this last equation, we obtain

$$s\mathbf{X}(s) - \mathbf{x}(0) = \mathbf{AX}(s)$$

where $\mathbf{X}(s) = \mathscr{L}[\mathbf{x}]$. Hence

$$(s\mathbf{I} - \mathbf{A})\mathbf{X}(s) = \mathbf{x}(0)$$

Premultiplying both sides of this last equation by $(s\mathbf{I} - \mathbf{A})^{-1}$, we obtain

$$\mathbf{X}(s) = (s\mathbf{I} - \mathbf{A})^{-1}\mathbf{x}(0)$$

The inverse Laplace transform of $\mathbf{X}(s)$ gives the solution $\mathbf{x}(t)$. Thus

$$\mathbf{x}(t) = \mathscr{L}^{-1}[(s\mathbf{I} - \mathbf{A})^{-1}]\mathbf{x}(0) \tag{5–101}$$

Note that

$$(s\mathbf{I} - \mathbf{A})^{-1} = \frac{\mathbf{I}}{s} + \frac{\mathbf{A}}{s^2} + \frac{\mathbf{A}^2}{s^3} + \cdots$$

Hence the inverse Laplace transform of $(s\mathbf{I} - \mathbf{A})^{-1}$ is

$$\mathscr{L}^{-1}[(s\mathbf{I} - \mathbf{A})^{-1}] = \mathbf{I} + \mathbf{A}t + \frac{\mathbf{A}^2t^2}{2!} + \frac{\mathbf{A}^3t^3}{3!} + \cdots = e^{\mathbf{A}t} \tag{5–102}$$

(The inverse Laplace transform of a matrix is the matrix consisting of the inverse Laplace transforms of all elements.) From Eqs. (5–101) and (5–102), the solution of Eq. (5–98) is

$$\mathbf{x}(t) = e^{\mathbf{A}t}\mathbf{x}(0)$$

The importance of Eq. (5–102) lies in the fact that it provides a convenient means for finding the closed-form solution for the matrix exponential.

State transition matrix. The solution of the homogeneous state equation

$$\dot{\mathbf{x}} = \mathbf{A}\mathbf{x} \tag{5–103}$$

can be written as follows:

$$\mathbf{x}(t) = \mathbf{\Phi}(t)\mathbf{x}(0) \tag{5–104}$$

where $\mathbf{\Phi}(t)$ is the state transition matrix (fundamental matrix). It is the unique solution of

$$\dot{\mathbf{\Phi}}(t) = \mathbf{A}\mathbf{\Phi}(t), \qquad \mathbf{\Phi}(0) = \mathbf{I}$$

To verify this, note that

$$\mathbf{x}(0) = \mathbf{\Phi}(0)\mathbf{x}(0) = \mathbf{I}\mathbf{x}(0)$$

and

$$\dot{\mathbf{x}}(t) = \dot{\mathbf{\Phi}}(t)\mathbf{x}(0) = \mathbf{A}\mathbf{\Phi}(t)\mathbf{x}(0) = \mathbf{A}\mathbf{x}(t)$$

We thus confirm that Eq. (5–104) is the solution of Eq. (5–103).

From Eqs. (5–100), (5–101), and (5–104), we obtain

$$\mathbf{\Phi}(t) = e^{\mathbf{A}t} = \mathscr{L}^{-1}[(s\mathbf{I} - \mathbf{A})^{-1}]$$

Note that

$$\mathbf{\Phi}^{-1}(t) = e^{-\mathbf{A}t} = \mathbf{\Phi}(-t)$$

From Eq. (5–104) we see that the solution of Eq. (5–103) is simply a transformation of the initial state. The state transition matrix contains all the information about the free motions of the system defined by Eq. (5–103).

Note that if the eigenvalues $\lambda_1, \lambda_2, \ldots, \lambda_n$ of the matrix $\mathbf{A}$ are all distinct, then $\mathbf{\Phi}(t)$ will contain the n exponentials

$$e^{\lambda_1 t}, e^{\lambda_2 t}, \ldots, e^{\lambda_n t}$$

In particular, if the matrix $\mathbf{A}$ is diagonal, then

$$\mathbf{\Phi}(t) = e^{\mathbf{A}t} = \begin{bmatrix} e^{\lambda_1 t} & & & & 0 \\ & e^{\lambda_2 t} & & & \\ & & \cdot & & \\ & & & \cdot & \\ & & & & \cdot \\ 0 & & & & e^{\lambda_n t} \end{bmatrix} \qquad (\mathbf{A}\text{: diagonal})$$

If there is a multiplicity in the eigenvalues, for example, if the eigenvalues of $\mathbf{A}$ are

$$\lambda_1, \lambda_1, \lambda_1, \lambda_4, \lambda_5, \ldots, \lambda_n$$

then $\mathbf{\Phi}(t)$ will contain, in addition to the exponentials $e^{\lambda_1 t}, e^{\lambda_4 t}, e^{\lambda_5 t}, \ldots, e^{\lambda_n t}$, terms like $te^{\lambda_1 t}$ and $t^2 e^{\lambda_1 t}$. Note that the exponential terms vanish as t approaches infinity if $\lambda_i < 0$, become unbounded if $\lambda_i > 0$, and become bounded if $\lambda_i = 0$. The terms like $te^{\lambda_i t}$, $t^2 e^{\lambda_i t}$, and so forth, approach zero as t approaches infinity if $\lambda_i < 0$ and are unbounded if $\lambda_i \geq 0$.

Properties of the state transition matrix $\mathbf{\Phi}(t)$. We shall summarize the important properties of the state transition matrix $\mathbf{\Phi}(t)$. For the time-invariant system

$$\dot{\mathbf{x}} = \mathbf{A}\mathbf{x}$$

for which

$$\mathbf{\Phi}(t) = e^{\mathbf{A}t}$$

we have

1. $\mathbf{\Phi}(0) = e^{\mathbf{A}0} = \mathbf{I}$
2. $\mathbf{\Phi}(t) = e^{\mathbf{A}t} = (e^{-\mathbf{A}t})^{-1} = [\mathbf{\Phi}(-t)]^{-1}$ or $\mathbf{\Phi}^{-1}(t) = \mathbf{\Phi}(-t)$
3. $\mathbf{\Phi}(t_1 + t_2) = e^{\mathbf{A}(t_1+t_2)} = e^{\mathbf{A}t_1} e^{\mathbf{A}t_2} = \mathbf{\Phi}(t_1)\mathbf{\Phi}(t_2) = \mathbf{\Phi}(t_2)\mathbf{\Phi}(t_1)$
4. $[\mathbf{\Phi}(t)]^n = \mathbf{\Phi}(nt)$
5. $\mathbf{\Phi}(t_2 - t_1)\mathbf{\Phi}(t_1 - t_0) = \mathbf{\Phi}(t_2 - t_0) = \mathbf{\Phi}(t_1 - t_0)\mathbf{\Phi}(t_2 - t_1)$

Example 5–4.

Obtain the state transition matrix $\mathbf{\Phi}(t)$ of the following system:

$$\begin{bmatrix} \dot{x}_1 \\ \dot{x}_2 \end{bmatrix} = \begin{bmatrix} 0 & 1 \\ -2 & -3 \end{bmatrix} \begin{bmatrix} x_1 \\ x_2 \end{bmatrix}$$

Obtain also the inverse of the state transition matrix, $\mathbf{\Phi}^{-1}(t)$.

For this system

$$\mathbf{A} = \begin{bmatrix} 0 & 1 \\ -2 & -3 \end{bmatrix}$$

The state transition matrix $\mathbf{\Phi}(t)$ is given as follows:

$$\mathbf{\Phi}(t) = e^{\mathbf{A}t} = \mathscr{L}^{-1}[(s\mathbf{I} - \mathbf{A})^{-1}]$$

Since

$$s\mathbf{I} - \mathbf{A} = \begin{bmatrix} s & 0 \\ 0 & s \end{bmatrix} - \begin{bmatrix} 0 & 1 \\ -2 & -3 \end{bmatrix} = \begin{bmatrix} s & -1 \\ 2 & s+3 \end{bmatrix}$$

the inverse of $(s\mathbf{I} - \mathbf{A})$ is

$$(s\mathbf{I} - \mathbf{A})^{-1} = \frac{1}{(s+1)(s+2)}\begin{bmatrix} s+3 & 1 \\ -2 & s \end{bmatrix}$$

$$= \begin{bmatrix} \dfrac{s+3}{(s+1)(s+2)} & \dfrac{1}{(s+1)(s+2)} \\ \dfrac{-2}{(s+1)(s+2)} & \dfrac{s}{(s+1)(s+2)} \end{bmatrix}$$

Hence

$$\mathbf{\Phi}(t) = e^{\mathbf{A}t} = \mathscr{L}^{-1}[(s\mathbf{I} - \mathbf{A})^{-1}] = \begin{bmatrix} 2e^{-t} - e^{-2t} & e^{-t} - e^{-2t} \\ -2e^{-t} + 2e^{-2t} & -e^{-t} + 2e^{-2t} \end{bmatrix}$$

Noting that $\mathbf{\Phi}^{-1}(t) = \mathbf{\Phi}(-t)$, we obtain the inverse of the state transition matrix as follows:

$$\mathbf{\Phi}^{-1}(t) = e^{-\mathbf{A}t} = \begin{bmatrix} 2e^{t} - e^{2t} & e^{t} - e^{2t} \\ -2e^{t} + 2e^{2t} & -e^{t} + 2e^{2t} \end{bmatrix}$$

Solution of nonhomogeneous linear time-invariant state equations. We shall next consider the nonhomogeneous state equation described by

$$\dot{\mathbf{x}} = \mathbf{A}\mathbf{x} + \mathbf{B}\mathbf{u} \tag{5–105}$$

where $\mathbf{x}$ is the state vector (n-vector), $\mathbf{u}$ the input vector (r-vector), $\mathbf{A}$ an $n \times n$ constant matrix, and $\mathbf{B}$ an $n \times r$ constant matrix.

By writing Eq. (5–105) as

$$\dot{\mathbf{x}}(t) - \mathbf{A}\mathbf{x}(t) = \mathbf{B}\mathbf{u}(t)$$

and premultiplying both sides of this last equation by $e^{-\mathbf{A}t}$, we obtain

$$e^{-\mathbf{A}t}[\dot{\mathbf{x}}(t) - \mathbf{A}\mathbf{x}(t)] = \frac{d}{dt}[e^{-\mathbf{A}t}\mathbf{x}(t)] = e^{-\mathbf{A}t}\mathbf{B}\mathbf{u}(t)$$

Integrating the preceding equation between 0 and t gives

$$e^{-\mathbf{A}t}\mathbf{x}(t) = \mathbf{x}(0) + \int_0^t e^{-\mathbf{A}\tau}\mathbf{B}\mathbf{u}(\tau)\,d\tau$$

or

$$\mathbf{x}(t) = e^{\mathbf{A}t}\mathbf{x}(0) + \int_0^t e^{\mathbf{A}(t-\tau)}\mathbf{B}\mathbf{u}(\tau)\,d\tau \tag{5–106}$$

Equation (5–106) can also be written as follows:

$$\mathbf{x}(t) = \mathbf{\Phi}(t)\mathbf{x}(0) + \int_0^t \mathbf{\Phi}(t-\tau)\mathbf{B}\mathbf{u}(\tau)\,d\tau \tag{5–107}$$

where

$$\mathbf{\Phi}(t) = e^{\mathbf{A}t}$$

Equation (5–106) or (5–107) is the solution of Eq. (5–105). The solution $\mathbf{x}(t)$ is clearly the sum of a term consisting of the transition of the initial state and a term arising from the input vector.

Laplace transform approach to the solution of nonhomogeneous linear time-invariant state equations. The solution of the nonhomogeneous state equation

$$\dot{\mathbf{x}} = \mathbf{A}\mathbf{x} + \mathbf{B}\mathbf{u}$$

can also be obtained by the Laplace transform approach. The Laplace transform of this state equation gives

$$s\mathbf{X}(s) - \mathbf{x}(0) = \mathbf{A}\mathbf{X}(s) + \mathbf{B}\mathbf{U}(s)$$

or

$$(s\mathbf{I} - \mathbf{A})\mathbf{X}(s) = \mathbf{x}(0) + \mathbf{B}\mathbf{U}(s)$$

Premultiplying both sides of this last equation by $(s\mathbf{I} - \mathbf{A})^{-1}$, we obtain

$$\mathbf{X}(s) = (s\mathbf{I} - \mathbf{A})^{-1}\mathbf{x}(0) + (s\mathbf{I} - \mathbf{A})^{-1}\mathbf{B}\mathbf{U}(s)$$

The relationship given by Eq. (5–102) gives

$$\mathbf{X}(s) = \mathscr{L}[e^{\mathbf{A}t}]\mathbf{x}(0) + \mathscr{L}[e^{\mathbf{A}t}]\,\mathbf{B}\mathbf{U}(s)$$

The inverse Laplace transform of this last equation can be obtained by use of the convolution integral, as follows:

$$\mathbf{x}(t) = e^{\mathbf{A}t}\mathbf{x}(0) + \int_0^t e^{\mathbf{A}(t-\tau)}\,\mathbf{B}\mathbf{u}(\tau)\,d\tau$$

Solution in terms of $\mathbf{x}(t_0)$. Thus far we have assumed the initial time to be zero. If, however, the initial time is given by t_0 instead of 0, then the solution to Eq. (5–105) must be modified to

$$\mathbf{x}(t) = e^{\mathbf{A}(t-t_0)}\mathbf{x}(t_0) + \int_{t_0}^t e^{\mathbf{A}(t-\tau)}\mathbf{B}\mathbf{u}(\tau)\,d\tau \tag{5–108}$$

Example 5–5.

Obtain the time response of the following system:

$$\begin{bmatrix} \dot{x}_1 \\ \dot{x}_2 \end{bmatrix} = \begin{bmatrix} 0 & 1 \\ -2 & -3 \end{bmatrix} \begin{bmatrix} x_1 \\ x_2 \end{bmatrix} + \begin{bmatrix} 0 \\ 1 \end{bmatrix} u$$

where $u(t)$ is the unit step function occurring at $t = 0$, or

$$u(t) = 1(t)$$

For this system

$$\mathbf{A} = \begin{bmatrix} 0 & 1 \\ -2 & -3 \end{bmatrix}, \qquad \mathbf{B} = \begin{bmatrix} 0 \\ 1 \end{bmatrix}$$

The state transition matrix $\Phi(t) = e^{\mathbf{A}t}$ was obtained in Example 5–4 as follows:

$$\Phi(t) = e^{\mathbf{A}t} = \begin{bmatrix} 2e^{-t} - e^{-2t} & e^{-t} - e^{-2t} \\ -2e^{-t} + 2e^{-2t} & -e^{-t} + 2e^{-2t} \end{bmatrix}$$

The response to the unit step input is then obtained from

$$\mathbf{x}(t) = e^{\mathbf{A}t}\mathbf{x}(0) + \int_0^t \begin{bmatrix} 2e^{-(t-\tau)} - e^{-2(t-\tau)} & e^{-(t-\tau)} - e^{-2(t-\tau)} \\ -2e^{-(t-\tau)} + 2e^{-2(t-\tau)} & -e^{-(t-\tau)} + 2e^{-2(t-\tau)} \end{bmatrix} \begin{bmatrix} 0 \\ 1 \end{bmatrix} [1]\, d\tau$$

or

$$\begin{bmatrix} x_1(t) \\ x_2(t) \end{bmatrix} = \begin{bmatrix} 2e^{-t} - e^{-2t} & e^{-t} - e^{-2t} \\ -2e^{-t} + 2e^{-2t} & -e^{-t} + 2e^{-2t} \end{bmatrix} \begin{bmatrix} x_1(0) \\ x_2(0) \end{bmatrix} + \begin{bmatrix} \frac{1}{2} - e^{-t} + \frac{1}{2}e^{-2t} \\ e^{-t} - e^{-2t} \end{bmatrix}$$

Transfer-function matrix. In the following we shall discuss the transfer-function matrix, which is similar to the pulse-transfer-function matrix discussed in Sec. 5–4. Consider the system defined by the equations

$$\dot{\mathbf{x}} = \mathbf{A}\mathbf{x} + \mathbf{B}\mathbf{u} \tag{5–109}$$

$$\mathbf{y} = \mathbf{C}\mathbf{x} + \mathbf{D}\mathbf{u} \tag{5–110}$$

where $\mathbf{x}$ is the state vector, $\mathbf{u}$ is the input vector, and $\mathbf{y}$ is the output vector. The Laplace transforms of Eqs. (5–109) and (5–110) are given by

$$s\mathbf{X}(s) - \mathbf{x}(0) = \mathbf{A}\mathbf{X}(s) + \mathbf{B}\mathbf{U}(s) \tag{5–111}$$

$$\mathbf{Y}(s) = \mathbf{C}\mathbf{X}(s) + \mathbf{D}\mathbf{U}(s) \tag{5–112}$$

By substituting $\mathbf{0}$ for $\mathbf{x}(0)$ in Eq. (5–111), we obtain

$$s\mathbf{X}(s) = \mathbf{A}\mathbf{X}(s) + \mathbf{B}\mathbf{U}(s)$$

or

$$\mathbf{X}(s) = (s\mathbf{I} - \mathbf{A})^{-1}\mathbf{B}\mathbf{U}(s)$$

By substituting this last equation into Eq. (5–112), we get

$$\mathbf{Y}(s) = [\mathbf{C}(s\mathbf{I} - \mathbf{A})^{-1}\mathbf{B} + \mathbf{D}]\mathbf{U}(s) = \mathbf{F}(s)\mathbf{U}(s)$$

where

$$\mathbf{F}(s) = \mathbf{C}(s\mathbf{I} - \mathbf{A})^{-1}\mathbf{B} + \mathbf{D} \tag{5–113}$$

$\mathbf{F}(s)$ is the transfer-function matrix of the system defined by Eqs. (5–109) and (5–110).

Machine computation of $e^{\mathbf{A}t}$. The solution of the linear time-invariant continuous-time state equation involves the matrix exponential $e^{\mathbf{A}t}$. There are several methods available for computing $e^{\mathbf{A}t}$. If the number of rows of the square matrix becomes 4 or larger, hand computation becomes exceedingly tedious and machine computation becomes necessary. (There are standard computational schemes for $e^{\mathbf{A}t}$.)

Conceptually, the simplest way to compute $e^{\mathbf{A}t}$ is to expand $e^{\mathbf{A}t}$ into a power series in t. (The matrix exponential $e^{\mathbf{A}t}$ is uniformly convergent for finite values of t.) For example, $e^{\mathbf{A}t}$ may be expanded into a power series as follows:

$$e^{\mathbf{A}t} = \mathbf{I} + (\mathbf{A}t) + \frac{\mathbf{A}t}{2}\left(\frac{\mathbf{A}t}{1!}\right) + \frac{\mathbf{A}t}{3}\left(\frac{\mathbf{A}^2t^2}{2!}\right) + \cdots + \frac{\mathbf{A}t}{n+1}\left(\frac{\mathbf{A}^nt^n}{n!}\right) + \cdots$$

Notice that each term in parentheses is equal to the entire preceding term. This provides a convenient recursion scheme. The computation is carried out to only enough terms that additional terms are negligible by comparison with the partial sum to that point. In this approach, a norm of the matrix may be used as a check to stop computation. A norm is a scalar quantity that gives a means of determining the absolute magnitude of the n^2 elements of an $n \times n$ matrix. There are several different forms of norms commonly used. Any one of them may be used. One example is

$$\text{Norm of } \mathbf{M} = \|\mathbf{M}\| = \sum_{\substack{i=1\\ j=1}}^{n} |m_{ij}|$$

where the m_{ij}'s are the elements of matrix $\mathbf{M}$. In machine computation, in addition to computing and adding terms in the series, the computer keeps a tab on the norm and stops the computation when the norm reaches a prescribed limit value.

The virtues of the series expansion technique are its simplicity and the ease of programming it. It is not necessary to find the eigenvalues of $\mathbf{A}$. There are, however, some computational disadvantages to the series expansion method, stemming from the convergence requirements for the series $e^{\mathbf{A}t}$. The Jordan canonical form requires considerably more programming than the series expansion method but the program will run in a fraction of the time needed for the series solution. (A standard program is available for computing $e^{\mathbf{A}t}$ by use of the Jordan canonical form.) For details of the analytical aspects of the Jordan canonical form, see the Appendix.

There are several analytical approaches for obtaining $e^{\mathbf{A}t}$. In the following, we shall discuss three of them.

Computation of $e^{\mathbf{A}t}$: Method 1. Transform matrix $\mathbf{A}$ into the diagonal form or into a Jordan canonical form. We shall first consider the case where matrix $\mathbf{A}$ involves only distinct eigenvectors and therefore can be transformed into the diagonal form. We shall next consider the case where matrix $\mathbf{A}$ involves multiple eigenvectors and therefore cannot be diagonalized.

Consider the state equation

$$\dot{\mathbf{x}} = \mathbf{A}\mathbf{x}$$

Let $\mathbf{P}$ be a diagonalizing matrix, and let us define

$$\mathbf{x} = \mathbf{P}\hat{\mathbf{x}}$$

Then

$$\dot{\hat{\mathbf{x}}} = \mathbf{P}^{-1}\mathbf{A}\mathbf{P}\hat{\mathbf{x}} = \mathbf{D}\hat{\mathbf{x}}$$

where $\mathbf{D}$ is a diagonal matrix. The solution of this last equation is

$$\hat{\mathbf{x}}(t) = e^{\mathbf{D}t}\hat{\mathbf{x}}(0)$$

Hence

$$\mathbf{x}(t) = \mathbf{P}\hat{\mathbf{x}}(t) = \mathbf{P}e^{\mathbf{D}t}\mathbf{P}^{-1}\mathbf{x}(0)$$

Noting that $\mathbf{x}(t)$ can also be given by the equation

$$\mathbf{x}(t) = e^{\mathbf{A}t}\mathbf{x}(0)$$

we obtain

$$e^{\mathbf{A}t} = \mathbf{P}e^{\mathbf{D}t}\mathbf{P}^{-1}$$

Thus, once matrix $\mathbf{A}$ is given and if it can be diagonalized, then a necessary transformation matrix $\mathbf{P}$ can be obtained by a standard method presented in the Appendix. Then, as derived here, $e^{\mathbf{A}t}$ can be given as follows:

$$e^{\mathbf{A}t} = \mathbf{P}e^{\mathbf{D}t}\mathbf{P}^{-1} = \mathbf{P}\begin{bmatrix} e^{\lambda_1 t} & & & & 0 \\ & e^{\lambda_2 t} & & & \\ & & \cdot & & \\ & & & \cdot & \\ & & & & \cdot \\ 0 & & & & e^{\lambda_n t} \end{bmatrix}\mathbf{P}^{-1} \tag{5–114}$$

Next, we shall consider the case where matrix $\mathbf{A}$ may be transformed into a Jordan canonical form. Consider again the state equation

$$\dot{\mathbf{x}} = \mathbf{A}\mathbf{x}$$

First obtain a transformation matrix $\mathbf{S}$ that will transform matrix $\mathbf{A}$ into a Jordan canonical form (for details, see the Appendix), so that

$$\mathbf{S}^{-1}\mathbf{A}\mathbf{S} = \mathbf{J}$$

where $\mathbf{J}$ is a matrix in a Jordan canonical form. Now define

$$\mathbf{x} = \mathbf{S}\hat{\mathbf{x}}$$

Then

$$\dot{\hat{\mathbf{x}}} = \mathbf{S}^{-1}\mathbf{A}\mathbf{S}\hat{\mathbf{x}} = \mathbf{J}\hat{\mathbf{x}}$$

The solution of this last equation is

$$\hat{\mathbf{x}}(t) = e^{\mathbf{J}t}\hat{\mathbf{x}}(0)$$

Hence

$$\mathbf{x}(t) = \mathbf{S}\hat{\mathbf{x}}(t) = \mathbf{S}e^{\mathbf{J}t}\mathbf{S}^{-1}\mathbf{x}(0)$$

Since the solution $\mathbf{x}(t)$ can also be given by the equation

$$\mathbf{x}(t) = e^{\mathbf{A}t}\mathbf{x}(0)$$

we obtain

$$e^{\mathbf{A}t} = \mathbf{S}e^{\mathbf{J}t}\mathbf{S}^{-1}$$

Note that $e^{\mathbf{J}t}$ is a triangular matrix (which means that the elements below the principal diagonal line are zeros) whose elements are $e^{\lambda t}$, $te^{\lambda t}$, $\frac{1}{2}t^2e^{\lambda t}$, and so forth. For example, if matrix $\mathbf{J}$ has the following Jordan canonical form:

$$\mathbf{J} = \begin{bmatrix} \lambda_1 & 1 & 0 \\ 0 & \lambda_1 & 1 \\ 0 & 0 & \lambda_1 \end{bmatrix}$$

then

$$e^{\mathbf{J}t} = \begin{bmatrix} e^{\lambda_1 t} & te^{\lambda_1 t} & \frac{1}{2}t^2e^{\lambda_1 t} \\ 0 & e^{\lambda_1 t} & te^{\lambda_1 t} \\ 0 & 0 & e^{\lambda_1 t} \end{bmatrix}$$

Similarly, if

$$\mathbf{J} = \left[\begin{array}{ccc|cc|cc} \lambda_1 & 1 & 0 & & & & \\ 0 & \lambda_1 & 1 & & & & 0 \\ 0 & 0 & \lambda_1 & & & & \\ \hline & & & \lambda_4 & 1 & & \\ & & & 0 & \lambda_4 & & \\ \hline & & & & & \lambda_6 & \\ 0 & & & & & & \lambda_7 \end{array}\right]$$

then

$$e^{\mathbf{J}t} = \left[\begin{array}{ccc|cc|cc} e^{\lambda_1 t} & te^{\lambda_1 t} & \frac{1}{2}t^2e^{\lambda_1 t} & & & & \\ 0 & e^{\lambda_1 t} & te^{\lambda_1 t} & & & & 0 \\ 0 & 0 & e^{\lambda_1 t} & & & & \\ \hline & & & e^{\lambda_4 t} & te^{\lambda_4 t} & & \\ & & & 0 & e^{\lambda_4 t} & & \\ \hline & & & & & e^{\lambda_6 t} & 0 \\ 0 & & & & & 0 & e^{\lambda_7 t} \end{array}\right]$$

As an example, consider the following matrix $\mathbf{A}$:

$$\mathbf{A} = \begin{bmatrix} 0 & 1 & 0 \\ 0 & 0 & 1 \\ 1 & -3 & 3 \end{bmatrix}$$

The characteristic equation is

$$|\lambda\mathbf{I} - \mathbf{A}| = \lambda^3 - 3\lambda^2 + 3\lambda - 1 = (\lambda - 1)^3$$

Thus matrix **A** has a multiple eigenvalue of order 3 at $\lambda = 1$. It can be shown that matrix **A** has a multiple eigenvector of order 3. The transformation matrix that will transform matrix **A** into a Jordan canonical form can be given by

$$\mathbf{S} = \begin{bmatrix} 1 & 0 & 0 \\ 1 & 1 & 0 \\ 1 & 2 & 1 \end{bmatrix}$$

Thus

$$\mathbf{S}^{-1} = \begin{bmatrix} 1 & 0 & 0 \\ -1 & 1 & 0 \\ 1 & -2 & 1 \end{bmatrix}$$

Then, it can be seen that

$$\mathbf{S}^{-1}\mathbf{A}\mathbf{S} = \begin{bmatrix} 1 & 0 & 0 \\ -1 & 1 & 0 \\ 1 & -2 & 1 \end{bmatrix} \begin{bmatrix} 0 & 1 & 0 \\ 0 & 0 & 1 \\ 1 & -3 & 3 \end{bmatrix} \begin{bmatrix} 1 & 0 & 0 \\ 1 & 1 & 0 \\ 1 & 2 & 1 \end{bmatrix}$$

$$= \begin{bmatrix} 1 & 1 & 0 \\ 0 & 1 & 1 \\ 0 & 0 & 1 \end{bmatrix} = \mathbf{J}$$

Noting that

$$e^{\mathbf{J}t} = \begin{bmatrix} e^t & te^t & \frac{1}{2}t^2e^t \\ 0 & e^t & te^t \\ 0 & 0 & e^t \end{bmatrix}$$

we find

$$e^{\mathbf{A}t} = \mathbf{S}e^{\mathbf{J}t}\mathbf{S}^{-1}$$

$$= \begin{bmatrix} 1 & 0 & 0 \\ 1 & 1 & 0 \\ 1 & 2 & 1 \end{bmatrix} \begin{bmatrix} e^t & te^t & \frac{1}{2}t^2e^t \\ 0 & e^t & te^t \\ 0 & 0 & e^t \end{bmatrix} \begin{bmatrix} 1 & 0 & 0 \\ -1 & 1 & 0 \\ 1 & -2 & 1 \end{bmatrix}$$

$$= \begin{bmatrix} e^t - te^t + \frac{1}{2}t^2e^t & te^t - t^2e^t & \frac{1}{2}t^2e^t \\ \frac{1}{2}t^2e^t & e^t - te^t - t^2e^t & te^t + \frac{1}{2}t^2e^t \\ te^t + \frac{1}{2}t^2e^t & -3te^t - t^2e^t & e^t + 2te^t + \frac{1}{2}t^2e^t \end{bmatrix}$$

Computation of $e^{\mathbf{A}t}$: Method 2. The second method of computing $e^{\mathbf{A}t}$ uses the Laplace transform approach. Referring to Eq. (5–102), $e^{\mathbf{A}t}$ can be given as follows:

$$e^{\mathbf{A}t} = \mathcal{L}^{-1}[(s\mathbf{I} - \mathbf{A})^{-1}]$$

Thus, to obtain $e^{\mathbf{A}t}$, first invert the matrix $(s\mathbf{I} - \mathbf{A})$. This results in a matrix whose elements are rational functions of s. Then, take the inverse Laplace transform of each element of the matrix. Note that $(s\mathbf{I} - \mathbf{A})^{-1}$ can be obtained by the methods discussed in Sec. 5–3 for computing $(z\mathbf{I} - \mathbf{G})^{-1}$.

Computation of $e^{\mathbf{A}t}$: Method 3. The third method of computing $e^{\mathbf{A}t}$ uses Sylvester's interpolation formulas. We shall first consider the case where the roots of the minimal polynomial $\phi(\lambda)$ of $\mathbf{A}$ are distinct. (For the definition of the minimal polynomial, see Prob. A-5–11.) Then we shall deal with the case of multiple roots. (For details of Sylvester's interpolation formulas, see Probs. A-5–16 and A-5–17.)

Case 1: Minimal Polynomial of **A** *Involves Only Distinct Roots.* We shall assume that the degree of the minimal polynomial of $\mathbf{A}$ is m. By using Sylvester's interpolation formula, it can be shown that $e^{\mathbf{A}t}$ can be obtained by solving the following determinant equation:

$$\begin{vmatrix} 1 & \lambda_1 & \lambda_1^2 & \cdots & \lambda_1^{m-1} & e^{\lambda_1 t} \\ 1 & \lambda_2 & \lambda_2^2 & \cdots & \lambda_2^{m-1} & e^{\lambda_2 t} \\ \cdot & \cdot & \cdot & & \cdot & \cdot \\ \cdot & \cdot & \cdot & & \cdot & \cdot \\ \cdot & \cdot & \cdot & & \cdot & \cdot \\ 1 & \lambda_m & \lambda_m^2 & \cdots & \lambda_m^{m-1} & e^{\lambda_m t} \\ \mathbf{I} & \mathbf{A} & \mathbf{A}^2 & \cdots & \mathbf{A}^{m-1} & e^{\mathbf{A}t} \end{vmatrix} = \mathbf{0} \tag{5–115}$$

By solving Eq. (5–115) for $e^{\mathbf{A}t}$, $e^{\mathbf{A}t}$ can be obtained in terms of the $\mathbf{A}^k$ ($k = 0, 1, 2, \ldots, m - 1$) and the $e^{\lambda_i t}$ ($i = 1, 2, 3, \ldots, m$). [Equation (5–115) may be expanded, for example, about the last column.]

Notice that solving Eq. (5–115) for $e^{\mathbf{A}t}$ is the same as writing

$$e^{\mathbf{A}t} = \alpha_0\mathbf{I} + \alpha_1\mathbf{A} + \alpha_2\mathbf{A}^2 + \cdots + \alpha_{m-1}\mathbf{A}^{m-1} \tag{5–116}$$

and determining the α_k ($k = 0, 1, 2, \ldots, m - 1$) by solving the following set of m equations for the α_k:

$$\begin{aligned} \alpha_0 + \alpha_1\lambda_1 + \alpha_2\lambda_1^2 + \cdots + \alpha_{m-1}\lambda_1^{m-1} &= e^{\lambda_1 t} \\ \alpha_0 + \alpha_1\lambda_2 + \alpha_2\lambda_2^2 + \cdots + \alpha_{m-1}\lambda_2^{m-1} &= e^{\lambda_2 t} \\ &\cdot \\ &\cdot \\ &\cdot \\ \alpha_0 + \alpha_1\lambda_m + \alpha_2\lambda_m^2 + \cdots + \alpha_{m-1}\lambda_m^{m-1} &= e^{\lambda_m t} \end{aligned}$$

If $\mathbf{A}$ is an $n \times n$ matrix and has distinct eigenvalues, then the number of α_k's to be determined is $m = n$. If $\mathbf{A}$ involves multiple eigenvalues but its minimal polynomial has only simple roots, however, then the number m of α_k's to be determined is less than n.

Case 2: Minimal Polynomial of* A *Involves Multiple Roots. As an example, consider the case where the minimal polynomial of **A** involves three equal roots ($\lambda_1 = \lambda_2 = \lambda_3$) and has other roots ($\lambda_4, \lambda_5, \ldots, \lambda_m$) that are all distinct. By applying Sylvester's interpolation formula, it can be shown that $e^{\mathbf{A}t}$ can be obtained from the following determinant equation (see Prob. A-5–17):

$$\begin{vmatrix} 0 & 0 & 1 & 3\lambda_1 & \cdots & \frac{(m-1)(m-2)}{2}\lambda_1^{m-3} & \frac{t^2}{2}e^{\lambda_1 t} \\ 0 & 1 & 2\lambda_1 & 3\lambda_1^2 & \cdots & (m-1)\lambda_1^{m-2} & te^{\lambda_1 t} \\ 1 & \lambda_1 & \lambda_1^2 & \lambda_1^3 & \cdots & \lambda_1^{m-1} & e^{\lambda_1 t} \\ 1 & \lambda_4 & \lambda_4^2 & \lambda_4^3 & \cdots & \lambda_4^{m-1} & e^{\lambda_4 t} \\ \cdot & \cdot & \cdot & \cdot & & \cdot & \cdot \\ \cdot & \cdot & \cdot & \cdot & & \cdot & \cdot \\ \cdot & \cdot & \cdot & \cdot & & \cdot & \cdot \\ 1 & \lambda_m & \lambda_m^2 & \lambda_m^3 & \cdots & \lambda_m^{m-1} & e^{\lambda_m t} \\ \mathbf{I} & \mathbf{A} & \mathbf{A}^2 & \mathbf{A}^3 & \cdots & \mathbf{A}^{m-1} & e^{\mathbf{A}t} \end{vmatrix} = \mathbf{0} \qquad (5\text{–}117)$$

Equation (5–117) can be solved for $e^{\mathbf{A}t}$ by expanding it about the last column.

It is noted that just as in case 1, solving Eq. (5–117) for $e^{\mathbf{A}t}$ is the same as writing

$$e^{\mathbf{A}t} = \alpha_0\mathbf{I} + \alpha_1\mathbf{A} + \alpha_2\mathbf{A}^2 + \cdots + \alpha_{m-1}\mathbf{A}^{m-1} \qquad (5\text{–}118)$$

and determining the α_k's ($k = 0, 1, 2, \ldots, m-1$) from

$$\begin{aligned} \alpha_2 + 3\alpha_3\lambda_1 + \cdots + \frac{(m-1)(m-2)}{2}\alpha_{m-1}\lambda_1^{m-3} &= \frac{t^2}{2}e^{\lambda_1 t} \\ \alpha_1 + 2\alpha_2\lambda_1 + 3\alpha_3\lambda_1^2 + \cdots + (m-1)\alpha_{m-1}\lambda_1^{m-2} &= te^{\lambda_1 t} \\ \alpha_0 + \alpha_1\lambda_1 + \alpha_2\lambda_1^2 + \cdots + \alpha_{m-1}\lambda_1^{m-1} &= e^{\lambda_1 t} \\ \alpha_0 + \alpha_1\lambda_4 + \alpha_2\lambda_4^2 + \cdots + \alpha_{m-1}\lambda_4^{m-1} &= e^{\lambda_4 t} \\ &\cdot \\ &\cdot \\ &\cdot \\ \alpha_0 + \alpha_1\lambda_m + \alpha_2\lambda_m^2 + \cdots + \alpha_{m-1}\lambda_m^{m-1} &= e^{\lambda_m t} \end{aligned}$$

The extension to other cases where, for example, there are two or more sets of multiple roots will be apparent. Note that if the minimal polynomial of **A** is not found, it is possible to substitute the characteristic polynomial for the minimal polynomial. The number of computations may, of course, be increased.

Example 5–6.

Consider the matrix

$$\mathbf{A} = \begin{bmatrix} 0 & 1 \\ 0 & -2 \end{bmatrix}$$

Compute $e^{\mathbf{A}t}$ as a sum of an infinite series.

The matrix exponential $e^{\mathbf{A}t}$ may always be expanded into a series of matrices, which can then be added together into a closed form. In the present case,

$$e^{\mathbf{A}t} = \mathbf{I} + \sum_{k=1}^{\infty} \frac{\mathbf{A}^k t^k}{k!}$$

$$= \begin{bmatrix} 1 & 0 \\ 0 & 1 \end{bmatrix} + \begin{bmatrix} 0 & 1 \\ 0 & -2 \end{bmatrix} t + \begin{bmatrix} 0 & 1 \\ 0 & -2 \end{bmatrix}^2 \frac{t^2}{2!} + \cdots$$

$$= \begin{bmatrix} 1 & \frac{1}{2} - \frac{1}{2}\left[1 - 2t + \frac{(2t)^2}{2!} - \frac{(2t)^3}{3!} + \cdots\right] \\ 0 & 1 - 2t + \frac{(2t)^2}{2!} - \frac{(2t)^3}{3!} + \frac{(2t)^4}{4!} - \cdots \end{bmatrix}$$

$$= \begin{bmatrix} 1 & \frac{1}{2}(1 - e^{-2t}) \\ 0 & e^{-2t} \end{bmatrix}$$

Example 5–7.

Consider the same matrix as in Example 5–6:

$$\mathbf{A} = \begin{bmatrix} 0 & 1 \\ 0 & -2 \end{bmatrix}$$

Compute $e^{\mathbf{A}t}$ by use of the three analytical methods presented in this section.

Method 1. The eigenvalues of $\mathbf{A}$ are 0 and -2 ($\lambda_1 = 0$, $\lambda_2 = -2$). A necessary transformation matrix $\mathbf{P}$ may be obtained (see the Appendix for details) as follows:

$$\mathbf{P} = \begin{bmatrix} 1 & 1 \\ 0 & -2 \end{bmatrix}$$

Then, from Eq. (5–114), $e^{\mathbf{A}t}$ is obtained as follows:

$$e^{\mathbf{A}t} = \begin{bmatrix} 1 & 1 \\ 0 & -2 \end{bmatrix} \begin{bmatrix} e^0 & 0 \\ 0 & e^{-2t} \end{bmatrix} \begin{bmatrix} 1 & \frac{1}{2} \\ 0 & -\frac{1}{2} \end{bmatrix} = \begin{bmatrix} 1 & \frac{1}{2}(1 - e^{-2t}) \\ 0 & e^{-2t} \end{bmatrix}$$

Method 2. Since

$$s\mathbf{I} - \mathbf{A} = \begin{bmatrix} s & 0 \\ 0 & s \end{bmatrix} - \begin{bmatrix} 0 & 1 \\ 0 & -2 \end{bmatrix} = \begin{bmatrix} s & -1 \\ 0 & s+2 \end{bmatrix}$$

we obtain

$$(s\mathbf{I} - \mathbf{A})^{-1} = \begin{bmatrix} \frac{1}{s} & \frac{1}{s(s+2)} \\ 0 & \frac{1}{s+2} \end{bmatrix}$$

Hence

$$e^{\mathbf{A}t} = \mathcal{L}^{-1}[(s\mathbf{I} - \mathbf{A})^{-1}] = \begin{bmatrix} 1 & \frac{1}{2}(1 - e^{-2t}) \\ 0 & e^{-2t} \end{bmatrix}$$

Method 3. From Eq. (5–115) we get

$$\begin{vmatrix} 1 & \lambda_1 & e^{\lambda_1 t} \\ 1 & \lambda_2 & e^{\lambda_2 t} \\ \mathbf{I} & \mathbf{A} & e^{\mathbf{A}t} \end{vmatrix} = \mathbf{0}$$

Substituting 0 for λ_1 and -2 for λ_2 in this last equation, we obtain

$$\begin{vmatrix} 1 & 0 & 1 \\ 1 & -2 & e^{-2t} \\ \mathbf{I} & \mathbf{A} & e^{\mathbf{A}t} \end{vmatrix} = \mathbf{0}$$

Expanding the determinant, we obtain

$$-2e^{\mathbf{A}t} + \mathbf{A} + 2\mathbf{I} - \mathbf{A}e^{-2t} = \mathbf{0}$$

or

$$\begin{aligned} e^{\mathbf{A}t} &= \tfrac{1}{2}(\mathbf{A} + 2\mathbf{I} - \mathbf{A}e^{-2t}) \\ &= \frac{1}{2}\left\{\begin{bmatrix} 0 & 1 \\ 0 & -2 \end{bmatrix} + \begin{bmatrix} 2 & 0 \\ 0 & 2 \end{bmatrix} - \begin{bmatrix} 0 & 1 \\ 0 & -2 \end{bmatrix} e^{-2t}\right\} \\ &= \begin{bmatrix} 1 & \frac{1}{2}(1 - e^{-2t}) \\ 0 & e^{-2t} \end{bmatrix} \end{aligned}$$

An alternate approach is to solve the following equations for α_0 and α_1:

$$\alpha_0 + \alpha_1\lambda_1 = e^{\lambda_1 t}$$

$$\alpha_0 + \alpha_1\lambda_2 = e^{\lambda_2 t}$$

Since $\lambda_1 = 0$ and $\lambda_2 = -2$, the last two equations become as follows:

$$\alpha_0 = 1$$

$$\alpha_0 - 2\alpha_1 = e^{-2t}$$

Solving for α_0 and α_1 gives

$$\alpha_0 = 1, \qquad \alpha_1 = \tfrac{1}{2}(1 - e^{-2t})$$

Then, $e^{\mathbf{A}t}$ can be written as follows:

$$\begin{aligned} e^{\mathbf{A}t} &= \alpha_0\mathbf{I} + \alpha_1\mathbf{A} = \mathbf{I} + \tfrac{1}{2}(1 - e^{-2t})\mathbf{A} \\ &= \begin{bmatrix} 1 & \frac{1}{2}(1 - e^{-2t}) \\ 0 & e^{-2t} \end{bmatrix} \end{aligned}$$

Linear time-varying systems. An advantage of the state-space approach to control system analysis is that it can easily be extended to linear time-varying systems. Most of the results obtained in this section so far carry over to linear time-varying systems by changing the state transition matrix $\mathbf{\Phi}(t)$ to $\mathbf{\Phi}(t, t_0)$. (This is because for time-varying systems the state transition matrix depends upon both t and t_0 and not the difference $t - t_0$. Also, it is not always possible to set the initial time equal to zero.) It is important to point out, however, that there is an important difference between the time-invariant case and the time-varying case: The state transition matrix for the time-varying system cannot, in general, be given as a matrix exponential.

Solution of homogeneous time-varying state equations. For a scalar differential equation

$$\dot{x} = a(t)x$$

the solution can be given by

$$x(t) = \exp\left[\int_{t_0}^{t} a(\tau)\,d\tau\right] x(t_0)$$

and the state transition function is given by

$$\phi(t, t_0) = \exp\left[\int_{t_0}^{t} a(\tau)\,d\tau\right]$$

The same result, however, does not carry over to the vector-matrix differential equation, except in special cases.

Consider the state equation

$$\dot{\mathbf{x}} = \mathbf{A}(t)\mathbf{x} \tag{5–119}$$

where

$\mathbf{x}(t) = n$-vector

$\mathbf{A}(t) = n \times n$ matrix whose elements are piecewise-continuous functions of t in the interval $t_0 \leq t \leq t_1$.

The solution to Eq. (5–119) is given by

$$\mathbf{x}(t) = \mathbf{\Phi}(t, t_0)\mathbf{x}(t_0) \tag{5–120}$$

where $\mathbf{\Phi}(t, t_0)$ is the $n \times n$ nonsingular matrix satisfying the following matrix differential equation:

$$\dot{\mathbf{\Phi}}(t, t_0) = \mathbf{A}(t)\mathbf{\Phi}(t, t_0), \qquad \mathbf{\Phi}(t_0, t_0) = \mathbf{I} \tag{5–121}$$

The fact that Eq. (5–120) is the solution of Eq. (5–119) can be verified easily, since

$$\mathbf{x}(t_0) = \mathbf{\Phi}(t_0, t_0)\mathbf{x}(t_0) = \mathbf{I}\mathbf{x}(t_0)$$

and

$$\dot{\mathbf{x}}(t) = \frac{d}{dt}\left[\mathbf{\Phi}(t, t_0)\mathbf{x}(t_0)\right] = \dot{\mathbf{\Phi}}(t, t_0)\mathbf{x}(t_0)$$

$$= \mathbf{A}(t)\mathbf{\Phi}(t, t_0)\mathbf{x}(t_0) = \mathbf{A}(t)\mathbf{x}(t)$$

We see that the solution of Eq. (5–119) is simply a transformation of the initial state. The matrix $\mathbf{\Phi}(t, t_0)$ is the state transition matrix for the time-varying system described by Eq. (5–119).

State transition matrix for the time-varying case. It is important to note that the state transition matrix $\mathbf{\Phi}(t, t_0)$ is given by a matrix exponential if and only if $\mathbf{A}(t)$ and $\int_{t_0}^{t} \mathbf{A}(\tau)\, d\tau$ commute. That is,

$$\mathbf{\Phi}(t, t_0) = \exp\left[\int_{t_0}^{t} \mathbf{A}(\tau)\, d\tau\right] \qquad \text{if and only if } \mathbf{A}(t) \text{ and } \int_{t_0}^{t} \mathbf{A}(\tau)\, d\tau \text{ commute}$$

Note that if $\mathbf{A}(t)$ is a constant matrix or a diagonal matrix, $\mathbf{A}(t)$ and $\int_{t_0}^{t} \mathbf{A}(\tau)\, d\tau$ commute. If $\mathbf{A}(t)$ and $\int_{t_0}^{t} \mathbf{A}(\tau)\, d\tau$ do not commute, there is no simple way to compute the state transition matrix $\mathbf{\Phi}(t, t_0)$.

To compute $\mathbf{\Phi}(t, t_0)$ numerically, we may use the following series expansion for $\mathbf{\Phi}(t, t_0)$:

$$\mathbf{\Phi}(t, t_0) = \mathbf{I} + \int_{t_0}^{t} \mathbf{A}(\tau)\, d\tau + \int_{t_0}^{t} \mathbf{A}(\tau_1)\left[\int_{t_0}^{\tau_1} \mathbf{A}(\tau_2)\, d\tau_2\right] d\tau_1 + \cdots \tag{5–122}$$

This, in general, will not give $\mathbf{\Phi}(t, t_0)$ in a closed form.

Example 5–8.

Obtain $\mathbf{\Phi}(t, 0)$ for the following time-varying system:

$$\begin{bmatrix} \dot{x}_1 \\ \dot{x}_2 \end{bmatrix} = \begin{bmatrix} 0 & 1 \\ 0 & t \end{bmatrix} \begin{bmatrix} x_1 \\ x_2 \end{bmatrix}$$

To obtain $\mathbf{\Phi}(t, 0)$, let us use Eq. (5–122). Since

$$\int_0^t \mathbf{A}(\tau)\, d\tau = \int_0^t \begin{bmatrix} 0 & 1 \\ 0 & \tau \end{bmatrix} d\tau = \begin{bmatrix} 0 & t \\ 0 & \frac{1}{2}t^2 \end{bmatrix}$$

$$\int_0^t \begin{bmatrix} 0 & 1 \\ 0 & \tau_1 \end{bmatrix} \left\{\int_0^{\tau_1} \begin{bmatrix} 0 & 1 \\ 0 & \tau_2 \end{bmatrix} d\tau_2\right\} d\tau_1 = \int_0^t \begin{bmatrix} 0 & 1 \\ 0 & \tau_1 \end{bmatrix} \begin{bmatrix} 0 & \tau_1 \\ 0 & \frac{1}{2}\tau_1^2 \end{bmatrix} d\tau_1 = \begin{bmatrix} 0 & \frac{1}{6}t^3 \\ 0 & \frac{1}{8}t^4 \end{bmatrix}$$

we obtain

$$\begin{aligned} \mathbf{\Phi}(t, 0) &= \begin{bmatrix} 1 & 0 \\ 0 & 1 \end{bmatrix} + \begin{bmatrix} 0 & t \\ 0 & \frac{1}{2}t^2 \end{bmatrix} + \begin{bmatrix} 0 & \frac{1}{6}t^3 \\ 0 & \frac{1}{8}t^4 \end{bmatrix} + \cdots \\ &= \begin{bmatrix} 1 & t + \frac{1}{6}t^3 + \cdots \\ 0 & 1 + \frac{1}{2}t^2 + \frac{1}{8}t^4 + \cdots \end{bmatrix} \end{aligned}$$

Properties of the state transition matrix $\mathbf{\Phi}(t, t_0)$. We shall list two important properties of the state transition matrix $\mathbf{\Phi}(t, t_0)$:

1. $$\mathbf{\Phi}(t_2, t_1)\mathbf{\Phi}(t_1, t_0) = \mathbf{\Phi}(t_2, t_0) \tag{5–123}$$
2. $$\mathbf{\Phi}(t_1, t_0) = \mathbf{\Phi}^{-1}(t_0, t_1) \tag{5–124}$$

To prove Eq. (5–123), note that

$$\mathbf{x}(t_1) = \mathbf{\Phi}(t_1, t_0)\mathbf{x}(t_0)$$

$$\mathbf{x}(t_2) = \mathbf{\Phi}(t_2, t_0)\mathbf{x}(t_0)$$

Also

$$\mathbf{x}(t_2) = \mathbf{\Phi}(t_2, t_1)\mathbf{x}(t_1)$$

Hence

$$\mathbf{x}(t_2) = \mathbf{\Phi}(t_2, t_1)\mathbf{\Phi}(t_1, t_0)\mathbf{x}(t_0) = \mathbf{\Phi}(t_2, t_0)\mathbf{x}(t_0)$$

Thus

$$\mathbf{\Phi}(t_2, t_1)\mathbf{\Phi}(t_1, t_0) = \mathbf{\Phi}(t_2, t_0)$$

To prove Eq. (5–124), note that

$$\mathbf{\Phi}(t_1, t_0) = \mathbf{\Phi}^{-1}(t_2, t_1)\mathbf{\Phi}(t_2, t_0)$$

If we let $t_2 = t_0$ in this last equation, then

$$\mathbf{\Phi}(t_1, t_0) = \mathbf{\Phi}^{-1}(t_0, t_1)\mathbf{\Phi}(t_0, t_0) = \mathbf{\Phi}^{-1}(t_0, t_1)$$

Solution of a nonhomogeneous time-varying state equation. Consider the state equation

$$\dot{\mathbf{x}} = \mathbf{A}(t)\mathbf{x} + \mathbf{B}(t)\mathbf{u} \tag{5–125}$$

where the elements of $\mathbf{A}(t)$ and $\mathbf{B}(t)$ are piecewise-continuous functions of t in the interval $t_0 \leq t \leq t_1$. In the following, we shall obtain the solution of this state equation.

Let us put

$$\mathbf{x}(t) = \mathbf{\Phi}(t, t_0)\boldsymbol{\xi}(t)$$

where t_0 is the initial time (constant) and $\mathbf{\Phi}(t, t_0)$ is the unique matrix satisfying the following equation:

$$\dot{\mathbf{\Phi}}(t, t_0) = \mathbf{A}(t)\mathbf{\Phi}(t, t_0), \qquad \mathbf{\Phi}(t_0, t_0) = \mathbf{I}$$

Then

$$\dot{\mathbf{x}}(t) = \frac{d}{dt}[\mathbf{\Phi}(t, t_0)\boldsymbol{\xi}(t)] = \dot{\mathbf{\Phi}}(t, t_0)\boldsymbol{\xi}(t) + \mathbf{\Phi}(t, t_0)\dot{\boldsymbol{\xi}}(t)$$

$$= \mathbf{A}(t)\mathbf{\Phi}(t, t_0)\boldsymbol{\xi}(t) + \mathbf{\Phi}(t, t_0)\dot{\boldsymbol{\xi}}(t)$$

$$= \mathbf{A}(t)\mathbf{\Phi}(t, t_0)\boldsymbol{\xi}(t) + \mathbf{B}(t)\mathbf{u}(t)$$

From this last equation we find that

$$\mathbf{\Phi}(t, t_0)\dot{\boldsymbol{\xi}}(t) = \mathbf{B}(t)\mathbf{u}(t)$$

or

$$\dot{\boldsymbol{\xi}}(t) = \boldsymbol{\Phi}^{-1}(t, t_0)\mathbf{B}(t)\mathbf{u}(t)$$

Hence

$$\boldsymbol{\xi}(t) = \boldsymbol{\xi}(t_0) + \int_{t_0}^{t} \boldsymbol{\Phi}^{-1}(\tau, t_0)\mathbf{B}(\tau)\mathbf{u}(\tau)\, d\tau$$

Since

$$\boldsymbol{\xi}(t_0) = \boldsymbol{\Phi}^{-1}(t_0, t_0)\mathbf{x}(t_0) = \mathbf{x}(t_0)$$

the solution of Eq. (5–125) is obtained as follows:

$$\mathbf{x}(t) = \boldsymbol{\Phi}(t, t_0)\mathbf{x}(t_0) + \boldsymbol{\Phi}(t, t_0)\int_{t_0}^{t} \boldsymbol{\Phi}^{-1}(\tau, t_0)\mathbf{B}(\tau)\mathbf{u}(\tau)\, d\tau$$

$$= \boldsymbol{\Phi}(t, t_0)\mathbf{x}(t_0) + \int_{t_0}^{t} \boldsymbol{\Phi}(t, \tau)\mathbf{B}(\tau)\mathbf{u}(\tau)\, d\tau \qquad (5\text{–}126)$$

Evaluating the right-hand side of Eq. (5–126) for practical cases will require a machine computation.

5-6 DISCRETIZATION OF CONTINUOUS-TIME STATE SPACE EQUATIONS

In digital control of continuous-time plants, we need to convert continuous-time state space equations into discrete-time state space equations. Such conversion can be done by introducing fictitious samplers and fictitious holding devices into continuous-time systems. The error introduced by discretization may be made negligible by using a sufficiently small sampling period compared with the significant time constant of the system. (The discrete-time solutions must be valid at equally spaced sampling instants.)

Discretization of continuous-time state space equations: time-invariant case. In what follows we shall present a procedure for discretizing continuous-time state space equations. We assume that the input vector $\mathbf{u}(t)$ changes only at equally spaced sampling instants. Note that the sampling operation here is fictitious. We shall derive the discrete-time state equation and output equation which yield the exact values at $t = kT$, where $k = 0, 1, 2, \ldots$.

Consider the continuous-time state equation and output equation

$$\dot{\mathbf{x}} = \mathbf{A}\mathbf{x} + \mathbf{B}\mathbf{u} \qquad (5\text{–}127)$$

$$\mathbf{y} = \mathbf{C}\mathbf{x} + \mathbf{D}\mathbf{u} \qquad (5\text{–}128)$$

In the following analysis, in order to clarify the presentation, we use the notation kT and $(k + 1)T$ instead of k and $k + 1$. The discrete-time representation of Eq. (5–127) will take the form

$$\mathbf{x}((k+1)T) = \mathbf{G}(T)\mathbf{x}(kT) + \mathbf{H}(T)\mathbf{u}(kT) \tag{5–129}$$

Note that the matrices **G** and **H** depend upon the sampling period T. Once the sampling period T is fixed, **G** and **H** are constant matrices.

In order to determine $\mathbf{G}(T)$ and $\mathbf{H}(T)$, we use the solution of Eq. (5–127), or

$$\mathbf{x}(t) = e^{\mathbf{A}t}\mathbf{x}(0) + e^{\mathbf{A}t}\int_0^t e^{-\mathbf{A}\tau}\mathbf{B}\mathbf{u}(\tau)\,d\tau$$

We assume that the input $\mathbf{u}(t)$ is sampled and fed to a zero-order hold so that all the components of $\mathbf{u}(t)$ are constant over the interval between any two consecutive sampling instants, or

$$\mathbf{u}(t) = \mathbf{u}(kT) \qquad \text{for} \qquad kT \le t < kT + T \tag{5–130}$$

Since

$$\mathbf{x}((k+1)T) = e^{\mathbf{A}(k+1)T}\mathbf{x}(0) + e^{\mathbf{A}(k+1)T}\int_0^{(k+1)T} e^{-\mathbf{A}\tau}\mathbf{B}\mathbf{u}(\tau)\,d\tau \tag{5–131}$$

and

$$\mathbf{x}(kT) = e^{\mathbf{A}kT}\mathbf{x}(0) + e^{\mathbf{A}kT}\int_0^{kT} e^{-\mathbf{A}\tau}\mathbf{B}\mathbf{u}(\tau)\,d\tau \tag{5–132}$$

multiplying Eq. (5–132) by $e^{\mathbf{A}T}$ and subtracting it from Eq. (5–131) gives us

$$\mathbf{x}((k+1)T) = e^{\mathbf{A}T}\mathbf{x}(kT) + e^{\mathbf{A}(k+1)T}\int_{kT}^{(k+1)T} e^{-\mathbf{A}\tau}\mathbf{B}\mathbf{u}(\tau)\,d\tau$$

Since from Eq. (5–130) $\mathbf{u}(t) = \mathbf{u}(kT)$ for $kT \le t < kT + T$, we may substitute $\mathbf{u}(\tau) = \mathbf{u}(kT) =$ constant in this last equation. [Note that $\mathbf{u}(t)$ may jump at $t = kT + T$ and thus $\mathbf{u}(kT + T)$ may be different from $\mathbf{u}(kT)$. Such a jump in $\mathbf{u}(\tau)$ at $\tau = kT + T$, the upper limit of integration, does not affect the value of the integral in this last equation, because the integrand does not involve impulse functions.] Hence, we may write

$$\mathbf{x}((k+1)T) = e^{\mathbf{A}T}\mathbf{x}(kT) + e^{\mathbf{A}T}\int_0^T e^{-\mathbf{A}t}\mathbf{B}\mathbf{u}(kT)\,dt$$

$$= e^{\mathbf{A}T}\mathbf{x}(kT) + \int_0^T e^{\mathbf{A}\lambda}\mathbf{B}\mathbf{u}(kT)\,d\lambda \tag{5–133}$$

where $\lambda = T - t$. If we define

$$\mathbf{G}(T) = e^{\mathbf{A}T} \tag{5–134}$$

$$\mathbf{H}(T) = \left(\int_0^T e^{\mathbf{A}\lambda}\,d\lambda\right)\mathbf{B} \tag{5–135}$$

then Eq. (5–133) becomes

$$\mathbf{x}((k+1)T) = \mathbf{G}(T)\mathbf{x}(kT) + \mathbf{H}(T)\mathbf{u}(kT) \tag{5–136}$$

which is Eq. (5–129). Thus, Eqs. (5–134) and (5–135) give the desired matrices $\mathbf{G}(T)$ and $\mathbf{H}(T)$. Note that $\mathbf{G}(T)$ and $\mathbf{H}(T)$ depend upon the sampling period T. Referring to Eq. (5–128), the output equation becomes

$$\mathbf{y}(kT) = \mathbf{C}\mathbf{x}(kT) + \mathbf{D}\mathbf{u}(kT) \tag{5–137}$$

where matrices $\mathbf{C}$ and $\mathbf{D}$ are constant matrices and do not depend on the sampling period T.

If matrix $\mathbf{A}$ is nonsingular, then $\mathbf{H}(T)$ given by Eq. (5–135) can be simplified to

$$\mathbf{H}(T) = \left(\int_0^T e^{\mathbf{A}\lambda}\, d\lambda\right)\mathbf{B} = \mathbf{A}^{-1}(e^{\mathbf{A}T} - \mathbf{I})\mathbf{B} = (e^{\mathbf{A}T} - \mathbf{I})\mathbf{A}^{-1}\mathbf{B}$$

Comments.

1. In the state space approach, notice that by assuming the control vector $\mathbf{u}(t)$ to be constant between any two consecutive sampling instants, the discrete-time model can be obtained simply by integrating the continuous-time state equation over one sampling period. The discrete-time state equation given by Eq. (5–129) is called the *zero-order hold equivalent* of the continuous-time state equation given by Eq. (5–127).
2. In general, in converting the continuous-time system equation into a discrete-time system equation, some sort of approximation is necessary. It is important to point out that Eq. (5–136) involves no approximation, provided the control vector $\mathbf{u}(t)$ is constant between any two consecutive sampling instants, as assumed in the derivation.
3. Notice that for $T \ll 1$, $\mathbf{G}(T) \cong \mathbf{G}(0) = e^{\mathbf{A}0} = \mathbf{I}$. Thus, as the sampling period T becomes very small, $\mathbf{G}(T)$ approaches the identity matrix.

Example 5–9.

Consider the continuous-time system given by

$$G(s) = \frac{Y(s)}{U(s)} = \frac{1}{s+a}$$

Obtain the continuous-time state space representation of the system. Then, discretize the state equation and output equation and obtain the discrete-time state space representation of the system. Also, obtain the pulse transfer function for the system by using Eq. (5–88).

The continuous-time state space representation of the system is simply

$$\dot{x} = -ax + u$$

$$y = x$$

Now we discretize the state equation and the output equation. Referring to Eqs. (5–134) and (5–135), we have

$$G(T) = e^{-aT}$$

$$H(T) = \int_0^T e^{-a\lambda}\, d\lambda = \frac{1 - e^{-aT}}{a}$$

Hence the discretized version of the system equations is

$$x(k+1) = e^{-aT}x(k) + \frac{1 - e^{-aT}}{a}\, u(k)$$

$$y(k) = x(k)$$

Referring to Eq. (5–88), the pulse transfer function for this system is

$$F(z) = C(zI - G)^{-1}H$$

$$= (z - e^{-aT})^{-1}\frac{1 - e^{-aT}}{a} = \frac{(1 - e^{-aT})z^{-1}}{a(1 - e^{-aT}z^{-1})}$$

This result, of course, agrees with the z transform of $G(s)$ where it is preceded by a sampler and zero-order hold [that is, where the signal $u(t)$ is sampled and fed to a zero-order hold before being applied to $G(s)$]:

$$G(z) = \mathscr{Z}\left[\frac{1 - e^{-Ts}}{s}\,\frac{1}{s+a}\right] = (1 - z^{-1})\,\mathscr{Z}\left[\frac{1}{s(s+a)}\right]$$

$$= \frac{(1 - e^{-aT})z^{-1}}{a(1 - e^{-aT}z^{-1})}$$

Example 5–10.

Obtain the discrete-time state and output equations and the pulse transfer function (when the sampling period $T = 1$) of the following continuous-time system:

$$G(s) = \frac{Y(s)}{U(s)} = \frac{1}{s(s+2)}$$

which may be represented in state space by the equations

$$\begin{bmatrix} \dot{x}_1 \\ \dot{x}_2 \end{bmatrix} = \begin{bmatrix} 0 & 1 \\ 0 & -2 \end{bmatrix}\begin{bmatrix} x_1 \\ x_2 \end{bmatrix} + \begin{bmatrix} 0 \\ 1 \end{bmatrix} u$$

$$y = [1 \quad 0]\begin{bmatrix} x_1 \\ x_2 \end{bmatrix}$$

The desired discrete-time state equation will have the form

$$\mathbf{x}((k+1)T) = \mathbf{G}(T)\mathbf{x}(kT) + \mathbf{H}(T)u(kT)$$

where matrices $\mathbf{G}(T)$ and $\mathbf{H}(T)$ are obtained from Eqs. (5–134) and (5–135) as follows:

$$\mathbf{G}(T) = e^{\mathbf{A}T} = \begin{bmatrix} 1 & \frac{1}{2}(1 - e^{-2T}) \\ 0 & e^{-2T} \end{bmatrix}$$

$$\mathbf{H}(T) = \left(\int_0^T e^{\mathbf{A}t}\, dt\right)\mathbf{B} = \left\{\int_0^T \begin{bmatrix} 1 & \frac{1}{2}(1 - e^{-2t}) \\ 0 & e^{-2t} \end{bmatrix} dt\right\}\begin{bmatrix} 0 \\ 1 \end{bmatrix}$$

$$= \begin{bmatrix} \frac{1}{2}\left(T + \frac{e^{-2T} - 1}{2}\right) \\ \frac{1}{2}(1 - e^{-2T}) \end{bmatrix}$$

Thus

$$\begin{bmatrix} x_1((k+1)T) \\ x_2((k+1)T) \end{bmatrix} = \begin{bmatrix} 1 & \frac{1}{2}(1 - e^{-2T}) \\ 0 & e^{-2T} \end{bmatrix} \begin{bmatrix} x_1(kT) \\ x_2(kT) \end{bmatrix} + \begin{bmatrix} \frac{1}{2}\left(T + \frac{e^{-2T} - 1}{2}\right) \\ \frac{1}{2}(1 - e^{-2T}) \end{bmatrix} u(kT)$$

The output equation becomes

$$y(kT) = [1 \quad 0] \begin{bmatrix} x_1(kT) \\ x_2(kT) \end{bmatrix}$$

When the sampling period is 1 sec, or $T = 1$, the discrete-time state equation and the output equation become, respectively,

$$\begin{bmatrix} x_1(k+1) \\ x_2(k+1) \end{bmatrix} = \begin{bmatrix} 1 & 0.4323 \\ 0 & 0.1353 \end{bmatrix} \begin{bmatrix} x_1(k) \\ x_2(k) \end{bmatrix} + \begin{bmatrix} 0.2838 \\ 0.4323 \end{bmatrix} u(k)$$

and

$$y(k) = [1 \quad 0] \begin{bmatrix} x_1(k) \\ x_2(k) \end{bmatrix}$$

The pulse-transfer-function representation of this system can be obtained from Eq. (5–88), as follows:

$$\begin{aligned} F(z) &= \mathbf{C}(z\mathbf{I} - \mathbf{G})^{-1}\mathbf{H} + D \\ &= [1 \quad 0] \begin{bmatrix} z - 1 & -0.4323 \\ 0 & z - 0.1353 \end{bmatrix}^{-1} \begin{bmatrix} 0.2838 \\ 0.4323 \end{bmatrix} + 0 \\ &= [1 \quad 0] \begin{bmatrix} \frac{1}{z-1} & \frac{0.4323}{(z-1)(z-0.1353)} \\ 0 & \frac{1}{z - 0.1353} \end{bmatrix} \begin{bmatrix} 0.2838 \\ 0.4323 \end{bmatrix} \\ &= \frac{0.2838z + 0.1485}{(z-1)(z-0.1353)} \\ &= \frac{0.2838z^{-1} + 0.1485z^{-2}}{(1 - z^{-1})(1 - 0.1353z^{-1})} \end{aligned}$$

Note that the same pulse transfer function can be obtained by taking the z transform of $G(s)$ when it is preceded by a sampler and zero-order hold. Assuming $T = 1$, we obtain

$$\begin{aligned} G(z) &= \mathscr{Z}\left[\frac{1 - e^{-Ts}}{s} \frac{1}{s(s+2)}\right] = (1 - z^{-1})\,\mathscr{Z}\left[\frac{1}{s^2(s+2)}\right] \\ &= (1 - z^{-1})\,\mathscr{Z}\left[\frac{0.5}{s^2} - \frac{0.25}{s} + \frac{0.25}{s+2}\right] \end{aligned}$$

$$= (1 - z^{-1}) \left[\frac{0.5z^{-1}}{(1 - z^{-1})^2} - \frac{0.25}{1 - z^{-1}} + \frac{0.25}{1 - 0.1353z^{-1}} \right]$$

$$= \frac{0.2838z^{-1} + 0.1485z^{-2}}{(1 - z^{-1})(1 - 0.1353z^{-1})}$$

Discretization of continuous-time state space equations: time-varying case. Consider the following continuous-time state equation and output equation:

$$\dot{\mathbf{x}} = \mathbf{A}(t)\mathbf{x} + \mathbf{B}(t)\mathbf{u} \tag{5–138}$$

$$\mathbf{y} = \mathbf{C}(t)\mathbf{x} + \mathbf{D}(t)\mathbf{u} \tag{5–139}$$

The solution of Eq. (5–138) can be given by

$$\mathbf{x}(t) = \mathbf{\Phi}(t, t_0)\mathbf{x}(t_0) + \int_{t_0}^{t} \mathbf{\Phi}(t, \tau)\mathbf{B}(\tau)\mathbf{u}(\tau)\, d\tau$$

$$= \mathbf{\Phi}(t, t_0)\mathbf{x}(t_0) + \mathbf{\Phi}(t, t_0) \int_{t_0}^{t} \mathbf{\Phi}^{-1}(\tau, t_0)\mathbf{B}(\tau)\mathbf{u}(\tau)\, d\tau \tag{5–140}$$

In the following analysis, we assume, as in the case of time-invariant state equations, that vector $\mathbf{u}(t)$ is sampled and held constant by a zero-order hold so that it is constant between any two consecutive sampling instants:

$$\mathbf{u}(t) = \mathbf{u}(kT) \qquad \text{for} \qquad kT \le t < kT + T$$

By substituting $(k+1)T$ for t and kT for t_0 in Eq. (5–140), we obtain

$$\mathbf{x}((k+1)T) = \mathbf{\Phi}((k+1)T, kT)\mathbf{x}(kT)$$

$$+ \mathbf{\Phi}((k+1)T, kT) \int_{kT}^{(k+1)T} \mathbf{\Phi}^{-1}(\tau, kT)\mathbf{B}(\tau)\mathbf{u}(kT)\, d\tau$$

Let us define

$$\mathbf{G}(kT) = \mathbf{\Phi}((k+1)T, kT) \tag{5–141}$$

$$\mathbf{H}(kT) = \mathbf{\Phi}((k+1)T, kT) \int_{kT}^{(k+1)T} \mathbf{\Phi}^{-1}(\tau, kT)\mathbf{B}(\tau)\, d\tau \tag{5–142}$$

Then, the discrete-time state equation becomes

$$\mathbf{x}((k+1)T) = \mathbf{G}(kT)\mathbf{x}(kT) + \mathbf{H}(kT)\mathbf{u}(kT)$$

and the output equation given by Eq. (5–139) becomes

$$\mathbf{y}(kT) = \mathbf{C}(kT)\mathbf{x}(kT) + \mathbf{D}(kT)\mathbf{u}(kT)$$

Note that matrices $\mathbf{G}(kT)$, $\mathbf{H}(kT)$, $\mathbf{C}(kT)$, and $\mathbf{D}(kT)$ depend upon both k and T. [In the case of the time-invariant system, matrices $\mathbf{G}(T)$ and $\mathbf{H}(T)$ are independent of k, and $\mathbf{C}$ and $\mathbf{D}$ are constant matrices.]

Time response between two consecutive sampling instants. In a sampled continuous-time system the output is continuous in time. As seen in Chaps. 3 and 4, the z transform solution of the discrete-time system equation gives the output response only at the sampling instants. In practice, we may wish to determine the output between two consecutive sampling instants. There are a few methods available for finding the response (output) between two consecutive sampling instants. In Chap. 3 we discussed two such methods (the Laplace transform method and the modified z transform method). Here we shall show that the state space method can be easily modified to obtain the output between any two consecutive sampling instants. In what follows, we shall demonstrate this.

Consider the time-invariant continuous-time system defined by

$$\dot{\mathbf{x}} = \mathbf{A}\mathbf{x} + \mathbf{B}\mathbf{u}$$

$$\mathbf{y} = \mathbf{C}\mathbf{x} + \mathbf{D}\mathbf{u}$$

Let us assume that the input $\mathbf{u}$ is sampled and fed to a zero-order hold. Then $\mathbf{u}(\tau) = \mathbf{u}(kT)$ for $kT \leq \tau < kT + T$. Referring to Eq. (5–108), the solution of the state equation starting with the initial state $\mathbf{x}(t_0)$ is

$$\mathbf{x}(t) = e^{\mathbf{A}(t-t_0)}\mathbf{x}(t_0) + \int_{t_0}^{t} e^{\mathbf{A}(t-\tau)}\mathbf{B}\mathbf{u}(\tau)\, d\tau$$

In order to obtain the response of the sampled system at $t = kT + \Delta T$, where $0 < \Delta T < T$, we put $t = kT + \Delta T$, $t_0 = kT$, and $\mathbf{u}(\tau) = \mathbf{u}(kT)$ in the solution $\mathbf{x}(t)$. Then

$$\begin{aligned}\mathbf{x}(kT + \Delta T) &= e^{\mathbf{A}\Delta T}\mathbf{x}(kT) + \int_{kT}^{kT+\Delta T} e^{\mathbf{A}(kT+\Delta T-\tau)}\mathbf{B}\mathbf{u}(kT)\, d\tau \\ &= e^{\mathbf{A}\Delta T}\mathbf{x}(kT) + \int_{0}^{\Delta T} e^{\mathbf{A}\lambda}\mathbf{B}\mathbf{u}(kT)\, d\lambda\end{aligned}$$

where $\lambda = kT + \Delta T - \tau$. Let us define

$$\mathbf{G}(\Delta T) = e^{\mathbf{A}\Delta T} \tag{5–143}$$

$$\mathbf{H}(\Delta T) = \left(\int_{0}^{\Delta T} e^{\mathbf{A}\lambda}\, d\lambda\right)\mathbf{B} \tag{5–144}$$

Then, we obtain

$$\mathbf{x}(kT + \Delta T) = \mathbf{G}(\Delta T)\mathbf{x}(kT) + \mathbf{H}(\Delta T)\mathbf{u}(kT) \tag{5–145}$$

The output $\mathbf{y}(kT + \Delta T)$ can be given by

$$\begin{aligned}\mathbf{y}(kT + \Delta T) &= \mathbf{C}\mathbf{x}(kT + \Delta T) + \mathbf{D}\mathbf{u}(kT) \\ &= \mathbf{C}\mathbf{G}(\Delta T)\mathbf{x}(kT) + [\mathbf{C}\mathbf{H}(\Delta T) + \mathbf{D}]\mathbf{u}(kT)\end{aligned} \tag{5–146}$$

Thus, the values of $\mathbf{x}(kT + \Delta T)$ and $\mathbf{y}(kT + \Delta T)$ between any two consecutive sampling instants can be obtained by computing $\mathbf{G}(\Delta T)$ and $\mathbf{H}(\Delta T)$ for various values of ΔT, where $0 < \Delta T < T$, and substituting the computed values into Eqs. (5–145) and (5–146). (Such computations can easily be programmed for digital computer calculation.)

Example 5–11.

Consider the system discussed in Example 5–10. Obtain the discrete-time state equation and the output equation at $t = kT + \Delta T$. Also, obtain the specific expressions for the state equation and output equation when $T = 1$ sec and $\Delta T = 0.5$ sec.

In Example 5–10, the matrices $\mathbf{G}(T)$ and $\mathbf{H}(T)$ were obtained as follows:

$$\mathbf{G}(T) = \begin{bmatrix} 1 & \frac{1}{2}(1 - e^{-2T}) \\ 0 & e^{-2T} \end{bmatrix}$$

$$\mathbf{H}(T) = \begin{bmatrix} \frac{1}{2}\left(T + \frac{e^{-2T} - 1}{2}\right) \\ \frac{1}{2}(1 - e^{-2T}) \end{bmatrix}$$

To obtain the state equation and the output equation at $t = kT + \Delta T$, where $0 < \Delta T < T$, we first convert $\mathbf{G}(T)$ to $\mathbf{G}(\Delta T)$ and $\mathbf{H}(T)$ to $\mathbf{H}(\Delta T)$ and then substitute $\mathbf{G}(\Delta T)$ and $\mathbf{H}(\Delta T)$ into Eqs. (5–145) and (5–146), as follows:

$$\begin{bmatrix} x_1(kT + \Delta T) \\ x_2(kT + \Delta T) \end{bmatrix} = \begin{bmatrix} 1 & \frac{1}{2}(1 - e^{-2\Delta T}) \\ 0 & e^{-2\Delta T} \end{bmatrix} \begin{bmatrix} x_1(kT) \\ x_2(kT) \end{bmatrix} + \begin{bmatrix} \frac{1}{2}\left(\Delta T + \frac{e^{-2\Delta T} - 1}{2}\right) \\ \frac{1}{2}(1 - e^{-2\Delta T}) \end{bmatrix} u(kT)$$

$$y(kT + \Delta T) = [1 \quad 0] \begin{bmatrix} 1 & \frac{1}{2}(1 - e^{-2\Delta T}) \\ 0 & e^{-2\Delta T} \end{bmatrix} \begin{bmatrix} x_1(kT) \\ x_2(kT) \end{bmatrix}$$

$$+ [1 \quad 0] \begin{bmatrix} \frac{1}{2}\left(\Delta T + \frac{e^{-2\Delta T} - 1}{2}\right) \\ \frac{1}{2}(1 - e^{-2\Delta T}) \end{bmatrix} u(kT)$$

For $T = 1$ and $\Delta T = 0.5$ we obtain the state equation and output equation as follows:

$$\begin{bmatrix} x_1(k + 0.5) \\ x_2(k + 0.5) \end{bmatrix} = \begin{bmatrix} 1 & 0.3161 \\ 0 & 0.3679 \end{bmatrix} \begin{bmatrix} x_1(k) \\ x_2(k) \end{bmatrix} + \begin{bmatrix} 0.0920 \\ 0.3161 \end{bmatrix} u(k)$$

$$y(k + 0.5) = [1 \quad 0.3161] \begin{bmatrix} x_1(k) \\ x_2(k) \end{bmatrix} + (0.0920)u(k)$$

5–7 LIAPUNOV STABILITY ANALYSIS

Liapunov stability analysis plays an important role in the stability analysis of control systems described by state space equations. There are two methods of stability analysis due to Liapunov, called the *first method* and the *second method*; both apply to the determination of the stability of dynamic systems described by ordinary differential or difference equations. The first method consists entirely of procedures in which the explicit forms of the solutions of the differential equations or difference equations are used for the analysis. The second method, on the other hand, does not require the solutions of the differential or difference equations. This is the reason the second method is so useful in practice.

Although there are many powerful stability criteria available for control systems, such as the Jury stability criterion and the Routh-Hurwitz stability criteria, they are limited to linear time-invariant systems. The second method of Liapunov, on the other hand, is not limited to linear time-invariant systems: it is applicable to both linear and nonlinear systems, time-invariant or time-varying. In particular, we find that the second method of Liapunov is indispensable for the stability analysis of nonlinear systems for which exact solutions may be unobtainable. (It is cautioned, however, that although the second method of Liapunov is applicable to any nonlinear system, obtaining successful results may not be an easy task. Experience and imagination may be necessary to carry out the stability analysis of most nonlinear systems.)

The second method of Liapunov is also called the *direct method* of Liapunov.

Second method of Liapunov. From the classical theory of mechanics, we know that a vibratory system is stable if its total energy is continually decreasing until an equilibrium state is reached.

The second method of Liapunov is based on a generalization of this fact: If the system has an asymptotically stable equilibrium state, then the stored energy of the system displaced within a domain of attraction decays with increasing time until it finally assumes its minimum value at the equilibrium state. For purely mathematical systems, however, there is no simple way of defining an "energy function." In order to circumvent this difficulty, Liapunov introduced the so-called Liapunov function, a fictitious energy function. This idea is, however, more general than that of energy and is more widely applicable. In fact, any scalar function satisfying the hypotheses of Liapunov's stability theorems (see Theorems 5–1 through 5–6) can serve as a Liapunov function.

Before we discuss the Liapunov function further, it is necessary to define the positive definiteness of scalar functions.

Positive definiteness of scalar functions. A scalar function $V(\mathbf{x})$ is said to be *positive definite* in a region Ω (which includes the origin of the state space) if $V(\mathbf{x}) > 0$ for all nonzero states $\mathbf{x}$ in the region Ω and if $V(\mathbf{0}) = 0$.

A time-varying function $V(\mathbf{x}, t)$ is said to be positive definite in a region Ω (which includes the origin of the state space) if it is bounded from below by a time-invariant positive definite function, that is, if there exists a positive definite function $V(\mathbf{x})$ such that

$$\begin{aligned} V(\mathbf{x}, t) &> V(\mathbf{x}) \qquad \text{for all } t \geq t_0 \\ V(\mathbf{0}, t) &= 0 \qquad \text{for all } t \geq t_0 \end{aligned}$$

Negative definiteness of scalar functions. A scalar function $V(\mathbf{x})$ is said to be *negative definite* if $-V(\mathbf{x})$ is positive definite.

Positive semidefiniteness of scalar functions. A scalar function $V(\mathbf{x})$ is said to be *positive semidefinite* if it is positive at all states in the region Ω except at the origin and at certain other states, where it is zero.

Negative semidefiniteness of scalar functions. A scalar function $V(\mathbf{x})$ is said to be *negative semidefinite* if $-V(\mathbf{x})$ is positive semidefinite.

Indefiniteness of scalar functions. A scalar function $V(\mathbf{x})$ is said to be *indefinite* if in the region Ω it assumes both positive and negative values, no matter how small the region Ω is.

Example 5–12.

In this example, we give several scalar functions and their classifications according to the foregoing definitions. Here we assume $\mathbf{x}$ to be a two-dimensional vector.

1. $V(\mathbf{x}) = x_1^2 + x_2^2$ — positive definite
2. $V(\mathbf{x}) = x_1^2 + \dfrac{x_2^2}{1 + x_2^2}$ — positive definite
3. $V(\mathbf{x}) = (x_1 + x_2)^2$ — positive semidefinite
4. $V(\mathbf{x}) = -x_1^2 - (x_1 + x_2)^2$ — negative definite
5. $V(\mathbf{x}) = x_1 x_2 + x_2^2$ — indefinite

Liapunov functions. The Liapunov function, a scalar function, is a positive definite function, and it is continuous together with its first partial derivatives (with respect to its arguments) in the region Ω about the origin and has a time derivative which, when taken along the trajectory, is negative definite (or negative semidefinite). Liapunov functions involve $x_1, x_2, \ldots, x_n$, and possibly t. We denote them by $V(x_1, x_2, \ldots, x_n, t)$, or simply by $V(\mathbf{x}, t)$. If Liapunov functions do not include t explicitly, then we denote them by $V(x_1, x_2, \ldots, x_n)$, or $V(\mathbf{x})$.

Notice that $\dot{V}(\mathbf{x}, t)$ is actually the total derivative of $V(\mathbf{x}, t)$ with respect to t along a solution of the system. Hence $\dot{V}(\mathbf{x}, t) < 0$ implies that $V(\mathbf{x}, t)$ is a decreasing function of t. A Liapunov function is not unique for a given system. (For this reason, the second method of Liapunov is a more powerful tool than conventional energy considerations. Note that a system whose energy E decreases on the average but not necessarily at each instant is stable but that E is not a Liapunov function.)

Later in this section we shall show that in the second method of Liapunov, the sign behavior of $V(\mathbf{x}, t)$ and that of its time derivative $\dot{V}(\mathbf{x}, t) = dV(\mathbf{x}, t)/dt$ give information about the stability of an equilibrium state without having the solution.

Note that the simplest positive definite function is of a quadratic form:

$$V(\mathbf{x}) = \sum_{i=1}^{n} \sum_{j=1}^{n} q_{ij} x_i x_j \qquad i, j = 1, 2, \ldots, n$$

In general, Liapunov functions may not be of a simple quadratic form. For any Liapunov function, however, the lowest-degree terms in V must be even. This can be seen as follows. If we define

$$\frac{x_1}{x_n} = \hat{x}_1, \frac{x_2}{x_n} = \hat{x}_2, \ldots, \frac{x_{n-1}}{x_n} = \hat{x}_{n-1}$$

then in the neighborhood of the origin the lowest-degree terms alone will become dominant and we can write $V(\mathbf{x})$ as

$$V(\mathbf{x}) = x_n^p \; V(\hat{x}_1, \hat{x}_2, \ldots, \hat{x}_{n-1}, 1)$$

If we keep the $\hat{x}_i$'s fixed, $V(\hat{x}_1, \hat{x}_2, \ldots, \hat{x}_{n-1}, 1)$ is a fixed quantity. For p odd, x_n^p can assume both positive and negative values near the origin, which means that $V(\mathbf{x})$ is not positive definite. Hence, p must be even.

In what follows, we give definitions of a system, an equilibrium state, stability, asymptotic stability, and instability.

System. The system we consider here is defined by

$$\dot{\mathbf{x}} = \mathbf{f}(\mathbf{x}, t) \tag{5–147}$$

where $\mathbf{x}$ is a state vector (an n-vector) and $\mathbf{f}(\mathbf{x}, t)$ is an n-vector whose elements are functions of $x_1, x_2, \ldots, x_n$, and t. (Note that we use the continuous-time system as a model to present basic materials on stability analysis by the second method of Liapunov. Then we extend the results obtained to the discrete-time system.) We assume that the system of Eq. (5–147) has a unique solution starting at the given initial condition. We shall denote the solution of Eq. (5–147) as $\boldsymbol{\phi}(t; \mathbf{x}_0, t_0)$, where $\mathbf{x} = \mathbf{x}_0$ at $t = t_0$ and t is the observed time. Thus,

$$\boldsymbol{\phi}(t_0; \mathbf{x}_0, t_0) = \mathbf{x}_0$$

Equilibrium state. In the system of Eq. (5–147), a state $\mathbf{x}_e$ where

$$\mathbf{f}(\mathbf{x}_e, t) = \mathbf{0} \qquad \text{for all } t \tag{5–148}$$

is called an *equilibrium state* of the system. If the system is linear and time-invariant, that is, if $\mathbf{f}(\mathbf{x}, t) = \mathbf{A}\mathbf{x}$, then there exists only one equilibrium state if $\mathbf{A}$ is nonsingular, and there exist infinitely many equilibrium states if $\mathbf{A}$ is singular. For nonlinear systems, there may be one or more equilibrium states. These states correspond to the constant solutions of the system ($\mathbf{x} = \mathbf{x}_e$ for all t). Determination of the equilibrium states does not involve the solution of the differential equation of the system, Eq. (5–147), but only the solution of Eq. (5–148).

Any isolated equilibrium state (that is, where isolated from each other) can be shifted to the origin of the coordinates, or $\mathbf{f}(\mathbf{0}, t) = \mathbf{0}$, by a translation of coordinates. In this section, we shall treat the stability analysis only of such states.

Stability in the sense of Liapunov. In the following, we shall denote a spherical region of radius r about an equilibrium state $\mathbf{x}_e$ as

$$\|\mathbf{x} - \mathbf{x}_e\| \leq r$$

where $\|\mathbf{x} - \mathbf{x}_e\|$ is called the *Euclidean norm* and is defined as follows:

$$\|\mathbf{x} - \mathbf{x}_e\| = [(x_1 - x_{1e})^2 + (x_2 - x_{2e})^2 + \cdots + (x_n - x_{ne})^2]^{1/2}$$

Let $S(\delta)$ consist of all points such that

$$\|\mathbf{x}_0 - \mathbf{x}_e\| \leq \delta$$

and let $S(\epsilon)$ consist of all points such that

$$\|\boldsymbol{\phi}(t; \mathbf{x}_0, t_0) - \mathbf{x}_e\| \leq \epsilon \qquad \text{for all } t \geq t_0$$

An equilibrium state $\mathbf{x}_e$ of the system of Eq. (5–147) is said to be *stable in the sense of Liapunov* if, corresponding to each $S(\epsilon)$, there is an $S(\delta)$ such that trajectories starting in $S(\delta)$ do not leave $S(\epsilon)$ as t increases indefinitely. The real number δ depends on ϵ and, in general, also depends on t_0. If δ does not depend on t_0, the equilibrium state is said to be *uniformly stable*.

What we have stated here is that we first choose the region $S(\epsilon)$, and for each $S(\epsilon)$, there must be a region $S(\delta)$ such that trajectories starting within $S(\delta)$ do not leave $S(\epsilon)$ as t increases indefinitely.

Asymptotic stability. An equilibrium state $\mathbf{x}_e$ of the system of Eq. (5–147) is said to be *asymptotically stable* if it is stable in the sense of Liapunov and if every solution starting within $S(\delta)$ converges, without leaving $S(\epsilon)$, to $\mathbf{x}_e$ as t increases indefinitely.

In practice, asymptotic stability is more important than mere stability. Also, since asymptotic stability is a local concept, simply to establish asymptotic stability does not necessarily mean that the system will operate properly. Some knowledge of the size of the largest region of asymptotic stability is usually necessary. This region is called the *domain of attraction*. It is that part of the state space in which asymptotically stable trajectories originate. In other words, every trajectory originating in the domain of attraction is asymptotically stable.

Asymptotic stability in the large. If asymptotic stability holds for all states (all points in the state space) from which trajectories originate, the equilibrium state is said to be *asymptotically stable in the large*. That is, the equilibrium state $\mathbf{x}_e$ of the system given by Eq. (5–147) is said to be asymptotically stable in the large if it is stable and if every solution converges to $\mathbf{x}_e$ as t increases indefinitely. Obviously a necessary condition for asymptotic stability in the large is that there be only one equilibrium state in the whole state space.

In control engineering problems, asymptotic stability in the large is a desirable feature. If the equilibrium state is not asymptotically stable in the large, then the problem becomes one of determining the largest region of asymptotic stability. This is usually very difficult. For all practical purposes, however, it is sufficient to determine a region of asymptotic stability large enough that no disturbance will exceed it.

Instability. An equilibrium state $\mathbf{x}_e$ is said to be unstable if for some real number $\epsilon > 0$ and any real number $\delta > 0$, no matter how small, there is always a state $\mathbf{x}_0$ in $S(\delta)$ such that the trajectory starting at this state leaves $S(\epsilon)$.

Graphical representation of stability, asymptotic stability, and instability. A graphical representation of the foregoing definitions will clarify their meanings.

Let us consider the two-dimensional case. Parts (a), (b), and (c) of Fig. 5–6 show equilibrium states and typical trajectories corresponding to stability, asymptotic stability, and instability, respectively. In Fig. 5–6(a), (b), or (c), the region $S(\delta)$ bounds the initial state $\mathbf{x}_0$, and the region $S(\epsilon)$ corresponds to the boundary for the trajectory starting from any initial state $\mathbf{x}_0$ in the region $S(\delta)$.

Note that the foregoing definitions do not specify the exact region of allowable initial conditions. Thus the definitions apply to the neighborhood of the equilibrium state, unless $S(\epsilon)$ corresponds to the entire state plane.

Note that in Fig. 5–6(c), the trajectory leaves $S(\epsilon)$ and therefore the equilibrium state is unstable. We cannot, however, say that the trajectory will go to infinity, since it may approach a limit cycle outside the region $S(\epsilon)$. (If a linear time-invariant system is unstable, trajectories starting even near the unstable equilibrium state go to infinity. But in the case of nonlinear systems this is not necessarily true.)

It is important to point out that the definitions presented above are not the only ones defining the concepts of the stability of an equilibrium state. In fact, various other ways to define stability are available in the literature. For example, in conventional control theory, only systems that are asymptotically stable are called stable systems, and those systems that are stable in the sense of Liapunov but are not asymptotically stable are called unstable. Another example is BIBO stability. A linear time-invariant system is called bounded-input–bounded-output stable (BIBO stable) if the output starting from an arbitrary initial state is bounded when the input is bounded. It is noted, however, that in this book, when the word "stability" is casually used, it normally means asymptotic stability in the sense of Liapunov.

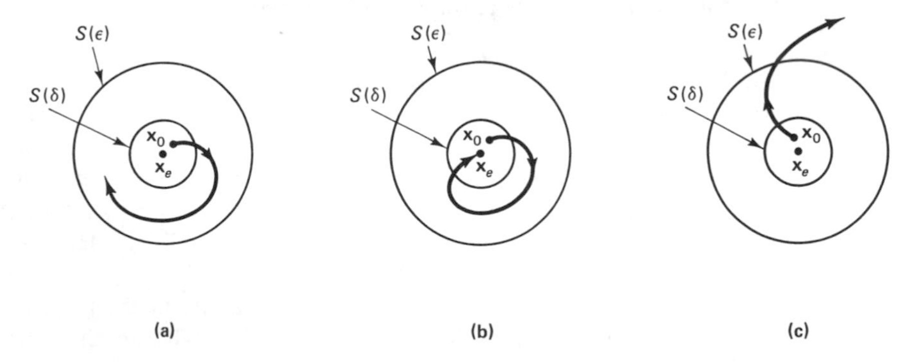

Figure 5–6 (a) Stable equilibrium state and a representative trajectory; (b) asymptotically stable equilibrium state and a representative trajectory; (c) unstable equilibrium state and a representative trajectory.

Liapunov theorem on asymptotic stability. It can be shown that if a scalar function $V(\mathbf{x})$, where $\mathbf{x}$ is an n-vector, is positive definite, then the states $\mathbf{x}$ which satisfy

$$V(\mathbf{x}) = C$$

where C is a positive constant, lie on a closed hypersurface in the n-dimensional state space, at least in the neighborhood of the origin. If $V(\mathbf{x}) \longrightarrow \infty$ as $\|\mathbf{x}\| \longrightarrow \infty$, then such closed surfaces extend over the entire state space. The hypersurface $V(\mathbf{x}) = C_1$ lies entirely inside the hypersurface $V(\mathbf{x}) = C_2$ if $C_1 < C_2$.

For a given system, if a positive definite scalar function $V(\mathbf{x})$ can be found such that its time derivative taken along a trajectory is always negative, then as time increases, $V(\mathbf{x})$ takes smaller and smaller values of C. As time increases, $V(\mathbf{x})$ finally shrinks to zero, and therefore $\mathbf{x}$ also shrinks to zero. This implies the asymptotic stability of the origin of the state space. Liapunov's main stability theorem, which is a generalization of the foregoing, provides a sufficient condition for asymptotic stability. This theorem may be stated as follows.

Theorem 5–1. Suppose a system is described by

$$\dot{\mathbf{x}} = \mathbf{f}(\mathbf{x}, t)$$

where

$$\mathbf{f}(\mathbf{0}, t) = \mathbf{0} \qquad \text{for all } t$$

If there exists a scalar function $V(\mathbf{x}, t)$ having continuous first partial derivatives and satisfying the conditions

1. $V(\mathbf{x}, t)$ is positive definite.
2. $\dot{V}(\mathbf{x}, t)$ is negative definite.

then the equilibrium state at the origin is uniformly asymptotically stable.

If, in addition, $V(\mathbf{x}, t) \longrightarrow \infty$ as $\|\mathbf{x}\| \longrightarrow \infty$, then the equilibrium state at the origin is uniformly asymptotically stable in the large. (For the proof of this theorem, see Prob. A-5–31.)

The conditions of this theorem may be modified as follows:

1′. $V(\mathbf{x}, t)$ is positive definite.

2′. $\dot{V}(\mathbf{x}, t)$ is negative semidefinite.

3′. $\dot{V}(\boldsymbol{\phi}(t; \mathbf{x}_0, t_0), t)$ does not vanish identically in $t \geq t_0$ for any t_0 and any $\mathbf{x}_0 \neq \mathbf{0}$, where $\boldsymbol{\phi}(t; \mathbf{x}_0, t_0)$ denotes the solution starting from $\mathbf{x}_0$ at t_0.

Then the origin of the system is uniformly asymptotically stable in the large.

The equivalence of condition 2 in the theorem and the modified conditions 2′ and 3′ may be seen as follows. If $\dot{V}(\mathbf{x}, t)$ is not negative definite but only negative semidefinite, then the trajectory of the representative point can become tangent to some particular surface $V(\mathbf{x}, t) = C$. Since $\dot{V}(\boldsymbol{\phi}(t; \mathbf{x}_0, t_0), t)$ does not vanish identically in $t \geq t_0$ for any t_0 and any $\mathbf{x}_0 \neq \mathbf{0}$, the representative point cannot remain at the tangent point [the point which corresponds to $\dot{V}(\mathbf{x}, t) = 0$] and therefore must move toward the origin.

Liapunov theorem on stability. To prove stability (but not asymptotic stability) of the origin of the system defined by Eq. (5–147), the following theorem may be applied.

Theorem 5–2. Suppose a system is described by

$$\dot{\mathbf{x}} = \mathbf{f}(\mathbf{x}, t)$$

where $\mathbf{f}(\mathbf{0}, t) = \mathbf{0}$ for all t. If there exists a scalar function $V(\mathbf{x}, t)$ having continuous first partial derivatives and satisfying the conditions

1. $V(\mathbf{x}, t)$ is positive definite.
2. $\dot{V}(\mathbf{x}, t)$ is negative semidefinite.

then the equilibrium state at the origin is uniformly stable.

It should be noted that the negative semidefiniteness of $\dot{V}(\mathbf{x}, t)$ [$\dot{V}(\mathbf{x}, t) \leq 0$ along the trajectories] means that the origin is uniformly stable but not necessarily uniformly asymptotically stable. Hence, in this case the system may exhibit a limit cycle operation.

Instability. If an equilibrium state $\mathbf{x} = \mathbf{0}$ of a system is unstable, then there exists a scalar function $W(\mathbf{x}, t)$ which determines the instability of the equilibrium state. We shall present a theorem on instability in the following.

Theorem 5–3. Suppose a system is described by

$$\dot{\mathbf{x}} = \mathbf{f}(\mathbf{x}, t)$$

where

$$\mathbf{f}(\mathbf{0}, t) = \mathbf{0} \qquad \text{for all } t \geq t_0$$

If there exists a scalar function $W(\mathbf{x}, t)$ having continuous first partial derivatives and satisfying the conditions

1. $W(\mathbf{x}, t)$ is positive definite in some region about the origin.
2. $\dot{W}(\mathbf{x}, t)$ is positive definite in the same region.

then the equilibrium state at the origin is unstable.

Remarks. A few comments are in order when the Liapunov stability analysis is applied to nonlinear systems.

1. In applying Liapunov stability theorems to a nonlinear system, the stability conditions obtained from a particular Liapunov function are sufficient conditions but are not necessary conditions.
2. A Liapunov function for a particular system is not unique. Therefore, it is important to note that failure in finding a suitable Liapunov function to show stability or asymptotic stability or instability of the equilibrium state under consideration can give no information on stability.
3. Although a particular Liapunov function may prove that the equilibrium state under consideration is stable or asymptotically stable in the region Ω which includes this equilibrium state, it does not necessarily mean that the motions are unstable outside the region Ω.
4. For a stable or asymptotically stable equilibrium state, a Liapunov function with the required properties always exists.

Stability analysis of linear time-invariant systems. There are many approaches to the investigation of the asymptotic stability of linear time-invariant systems. For example, for a continuous-time system described by the equation

$$\dot{\mathbf{x}} = \mathbf{A}\mathbf{x}$$

it can be stated that a necessary and sufficient condition for the asymptotic stability of the origin of the system is that all eigenvalues of $\mathbf{A}$ have negative real parts, or that the zeros of the characteristic polynomial

$$|s\mathbf{I} - \mathbf{A}| = s^n + a_1 s^{n-1} + \cdots + a_{n-1}s + a_n$$

have negative real parts.

Similarly, for a discrete-time system represented by the equation

$$\mathbf{x}(k+1) = \mathbf{G}\mathbf{x}(k)$$

a necessary and sufficient condition that can be stated for the asymptotic stability of the origin is that all eigenvalues of $\mathbf{G}$ be less than unity in their magnitudes, or that the zeros of the characteristic polynomial

$$|z\mathbf{I} - \mathbf{G}| = z^n + a_1 z^{n-1} + \cdots + a_{n-1}z + a_n$$

lie within the unit circle centered at the origin of the z plane.

Finding the eigenvalues, however, may become difficult in the case of higher-order systems or in the case where some of the coefficients of the characteristic polynomial are nonnumerical. In such a case, the Jury stability criterion or the Routh-Hurwitz stability criteria may be applied. The Liapunov approach, which provides an alternate approach to the stability analysis of linear time-invariant systems, is algebraic and does not require factoring of the characteristic polynomial, as will be seen later. It is important to note that for linear time-invariant systems the second method of Liapunov gives not just sufficient conditions but the necessary and sufficient conditions for stability or asymptotic stability.

In the following stability analysis of linear time-invariant systems, it is assumed that if an eigenvalue λ_i of matrix **A** is a complex quantity, then **A** must have $\bar{\lambda}_i$, the complex conjugate of λ_i, as its eigenvalue. Thus any complex eigenvalues of **A** will appear as conjugate complex pairs. Also, in the following discussions on stability, we shall use the conjugate transpose expression, rather than the transpose expression, of matrix **A**, since the elements of matrix **A** may include complex conjugates. The conjugate transpose of **A** is denoted by **A***. It is a conjugate of the transpose:

$$\mathbf{A}^* = \bar{\mathbf{A}}^T$$

Liapunov stability analysis of linear time-invariant continuous-time systems. Consider the following linear time-invariant system:

$$\dot{\mathbf{x}} = \mathbf{A}\mathbf{x} \tag{5–149}$$

where **x** is a state vector (an n-vector) and **A** is an $n \times n$ constant matrix. We assume that **A** is nonsingular. Then the only equilibrium state is the origin, $\mathbf{x} = \mathbf{0}$. The stability of the equilibrium state of the linear time-invariant system can be investigated easily with the second method of Liapunov.

For the system defined by Eq. (5–149), let us choose as a possible Liapunov function

$$V(\mathbf{x}) = \mathbf{x}^*\mathbf{P}\mathbf{x}$$

where **P** is a positive definite Hermitian matrix. (If **x** is a real vector, then **P** can be chosen to be a positive definite real symmetric matrix.) The time derivative of $V(\mathbf{x})$ along any trajectory is

$$\begin{aligned}\dot{V}(\mathbf{x}) &= \dot{\mathbf{x}}^*\mathbf{P}\mathbf{x} + \mathbf{x}^*\mathbf{P}\dot{\mathbf{x}} \\ &= (\mathbf{A}\mathbf{x})^*\mathbf{P}\mathbf{x} + \mathbf{x}^*\mathbf{P}\mathbf{A}\mathbf{x} \\ &= \mathbf{x}^*\mathbf{A}^*\mathbf{P}\mathbf{x} + \mathbf{x}^*\mathbf{P}\mathbf{A}\mathbf{x} \\ &= \mathbf{x}^*(\mathbf{A}^*\mathbf{P} + \mathbf{P}\mathbf{A})\mathbf{x}\end{aligned}$$

Since $V(\mathbf{x})$ was chosen to be positive definite, we require, for asymptotic stability, that $\dot{V}(\mathbf{x})$ be negative definite. Therefore, we require that

$$\dot{V}(\mathbf{x}) = -\mathbf{x}^*\mathbf{Q}\mathbf{x}$$

where

$$\mathbf{Q} = -(\mathbf{A}^*\mathbf{P} + \mathbf{P}\mathbf{A}) = \text{positive definite}$$

Hence, for the asymptotic stability of the system of Eq. (5–149), it is sufficient that **Q** be positive definite.

For a test of positive definiteness of an $n \times n$ matrix, we apply Sylvester's criterion, which states that a necessary and sufficient condition for the matrix to be positive definite is that the determinants of all the successive principal minors of the matrix be positive. Consider, for example, the following $n \times n$ Hermitian matrix

P (if the elements of **P** are all real, then the Hermitian matrix becomes a real symmetric matrix):

$$\mathbf{P} = \begin{bmatrix} p_{11} & p_{12} & \cdots & p_{1n} \\ \bar{p}_{12} & p_{22} & \cdots & p_{2n} \\ \cdot & \cdot & & \cdot \\ \cdot & \cdot & & \cdot \\ \cdot & \cdot & & \cdot \\ \bar{p}_{1n} & \bar{p}_{2n} & \cdots & p_{nn} \end{bmatrix}$$

where $\bar{p}_{ij}$ denotes the complex conjugate of p_{ij}. The matrix **P** is positive definite if all the successive principal minors are positive, that is, if

$$p_{11} > 0, \qquad \begin{vmatrix} p_{11} & p_{12} \\ \bar{p}_{12} & p_{22} \end{vmatrix} > 0, \ \cdots, \qquad \begin{vmatrix} p_{11} & p_{12} & \cdots & p_{1n} \\ \bar{p}_{12} & p_{22} & \cdots & p_{2n} \\ \cdot & \cdot & & \cdot \\ \cdot & \cdot & & \cdot \\ \cdot & \cdot & & \cdot \\ \bar{p}_{1n} & \bar{p}_{2n} & \cdots & p_{nn} \end{vmatrix} > 0$$

Instead of first specifying a positive definite matrix **P** and examining whether or not **Q** is positive definite, it is convenient to specify a positive definite matrix **Q** first and then examine whether or not **P** determined from

$$\mathbf{A}^*\mathbf{P} + \mathbf{P}\mathbf{A} = -\mathbf{Q}$$

is positive definite. Note that positive definite **P** is a necessary and sufficient condition. We shall summarize what we have just stated in the form of a theorem.

Theorem 5–4. Consider the system described by

$$\dot{\mathbf{x}} = \mathbf{A}\mathbf{x}$$

where $\mathbf{x}$ is a state vector (an n-vector) and **A** is an $n \times n$ constant nonsingular matrix. A necessary and sufficient condition for the equilibrium state $\mathbf{x} = \mathbf{0}$ to be asymptotically stable in the large is that, given any positive definite Hermitian (or any positive definite real symmetric) matrix **Q**, there exists a positive definite Hermitian (or a positive definite real symmetric) matrix **P** such that

$$\mathbf{A}^*\mathbf{P} + \mathbf{P}\mathbf{A} = -\mathbf{Q} \tag{5–150}$$

The scalar function $\mathbf{x}^*\mathbf{P}\mathbf{x}$ is a Liapunov function for this system. [Note that in the linear system considered, if the equilibrium state (the origin) is asymptotically stable, then it is asymptotically stable in the large.]

Remarks. In applying Theorem 5–4 to the stability analysis of linear time-invariant continuous-time systems, several important remarks may be made.

1. If $\dot{V}(\mathbf{x}) = -\mathbf{x}^*\mathbf{Q}\mathbf{x}$ does not vanish identically along any trajectory, then **Q** may be chosen to be positive semidefinite.

2. If an arbitrary positive definite matrix is chosen for **Q** [or an arbitrary positive-semidefinite matrix if $\dot{V}(\mathbf{x})$ does not vanish identically along any trajectory] and the matrix equation

$$\mathbf{A}^*\mathbf{P} + \mathbf{P}\mathbf{A} = -\mathbf{Q}$$

is solved to determine **P**, then the positive definiteness of **P** is a necessary and sufficient condition for the asymptotic stability of the equilibrium state $\mathbf{x} = \mathbf{0}$.

3. The final result does not depend on the particular **Q** matrix chosen so long as **Q** is positive definite (or positive semidefinite, as the case may be).

4. To determine the elements of the **P** matrix, we equate the matrices $\mathbf{A}^*\mathbf{P} + \mathbf{P}\mathbf{A}$ and $-\mathbf{Q}$ element by element. This results in $n(n+1)/2$ linear equations for the determination of the elements $p_{ij} = \bar{p}_{ji}$ of **P**. If we denote the eigenvalues of **A** by $\lambda_1, \lambda_2, \ldots, \lambda_n$, each repeated a number of times equal to its multiplicity as a root of the characteristic equation, and if for every sum of two roots

$$\lambda_j + \lambda_k \neq 0$$

then the elements of **P** are uniquely determined. (Note that for a stable matrix **A**, the sum $\lambda_j + \lambda_k$ is always nonzero.)

5. In determining whether or not there exists a positive definite Hermitian or positive definite real symmetric matrix **P**, it is convenient to choose $\mathbf{Q} = \mathbf{I}$, where **I** is the identity matrix. Then the elements of **P** are determined from

$$\mathbf{A}^*\mathbf{P} + \mathbf{P}\mathbf{A} = -\mathbf{I}$$

and the matrix **P** is tested for positive definiteness.

Example 5–13.

Determine the stability of the equilibrium state of the following system:

$$\dot{x}_1 = -x_1 - 2x_2$$

$$\dot{x}_2 = x_1 - 4x_2$$

The system has only one equilibrium state at the origin. By choosing $\mathbf{Q} = \mathbf{I}$ and substituting **I** into Eq. (5–150), we have

$$\mathbf{A}^*\mathbf{P} + \mathbf{P}\mathbf{A} = -\mathbf{I}$$

Noting that **A** is a real matrix, **P** must be a real symmetric matrix. This last equation may then be written as follows:

$$\begin{bmatrix} -1 & 1 \\ -2 & -4 \end{bmatrix} \begin{bmatrix} p_{11} & p_{12} \\ p_{12} & p_{22} \end{bmatrix} + \begin{bmatrix} p_{11} & p_{12} \\ p_{12} & p_{22} \end{bmatrix} \begin{bmatrix} -1 & -2 \\ 1 & -4 \end{bmatrix} = -\begin{bmatrix} 1 & 0 \\ 0 & 1 \end{bmatrix} \tag{5–151}$$

where we have noted that $p_{21} = p_{12}$ and made the appropriate substitution. If the matrix **P** turns out to be positive definite, then $\mathbf{x}^*\mathbf{P}\mathbf{x}$ is a Liapunov function and the origin is asymptotically stable.

Equation (5–151) yields the following three equations:

$$-2p_{11}+2p_{12}=-1$$
$$-2p_{11}-5p_{12}+p_{22}=0$$
$$-4p_{12}-8p_{22}=-1$$

Solving for the p's, we obtain

$$p_{11}=\tfrac{23}{60}, \qquad p_{12}=-\tfrac{7}{60}, \qquad p_{22}=\tfrac{11}{60}$$

Hence

$$\mathbf{P}=\begin{bmatrix} \frac{23}{60} & -\frac{7}{60} \\ -\frac{7}{60} & \frac{11}{60} \end{bmatrix}$$

By Sylvester's criterion, this matrix is positive definite. Hence, we conclude that the origin of the system is asymptotically stable in the large.

It is noted that a Liapunov function for this system is

$$V(\mathbf{x})=\mathbf{x}^*\mathbf{P}\mathbf{x}=[x_1 \quad x_2]\begin{bmatrix} \frac{23}{60} & -\frac{7}{60} \\ -\frac{7}{60} & \frac{11}{60} \end{bmatrix}\begin{bmatrix} x_1 \\ x_2 \end{bmatrix}$$
$$=\tfrac{1}{60}(23x_1^2-14x_1x_2+11x_2^2)$$

and $\dot{V}(\mathbf{x})$ is given by

$$\dot{V}(\mathbf{x})=-x_1^2-x_2^2$$

Liapunov stability analysis of discrete-time systems. In what follows, we extend the Liapunov stability analysis presented thus far in this section to discrete-time systems. As in the case of continuous-time systems, asymptotic stability is the most important concept in the stability of equilibrium states of discrete-time systems.

We shall now present a stability theorem for linear or nonlinear time-invariant discrete-time systems based on the second method of Liapunov. It is noted that for discrete-time systems, instead of $\dot{V}(\mathbf{x})$, we use the forward difference $V(\mathbf{x}(k+1)T)-V(\mathbf{x}(kT))$, or

$$\Delta V(\mathbf{x}(kT))=V(\mathbf{x}(k+1)T)-V(\mathbf{x}(kT)) \tag{5–152}$$

Theorem 5–5. Consider the discrete-time system

$$\mathbf{x}((k+1)T)=\mathbf{f}(\mathbf{x}(kT)) \tag{5–153}$$

where

$$\mathbf{x}=n\text{-vector}$$
$$\mathbf{f}(\mathbf{x})=n\text{-vector with property that } \mathbf{f}(\mathbf{0})=\mathbf{0}$$
$$T=\text{sampling period}$$

Suppose there exists a scalar function $V(\mathbf{x})$ continuous in $\mathbf{x}$ such that

1. $V(\mathbf{x}) > 0$ for $\mathbf{x} \neq \mathbf{0}$.
2. $\Delta V(\mathbf{x}) < 0$ for $\mathbf{x} \neq \mathbf{0}$, where

$$\Delta V(\mathbf{x}(kT)) = V(\mathbf{x}(k+1)T) - V(\mathbf{x}(kT)) = V(\mathbf{f}(\mathbf{x}(kT))) - V(\mathbf{x}(kT))$$

3. $V(\mathbf{0}) = 0$.
4. $V(\mathbf{x}) \to \infty$ as $\|\mathbf{x}\| \to \infty$.

Then the equilibrium state $\mathbf{x} = \mathbf{0}$ is asymptotically stable in the large and $V(\mathbf{x})$ is a Liapunov function.

Note that in this theorem condition 2 may be replaced by

2′. $\Delta V(\mathbf{x}) \leq 0$ for all $\mathbf{x}$, and
$\Delta V(\mathbf{x})$ does not vanish identically for any solution sequence $\{\mathbf{x}(kT)\}$ satisfying Eq. (5–153).

This means that $\Delta V(\mathbf{x})$ need not be negative definite if it does not vanish identically on any solution sequence of the difference equation.

Liapunov stability analysis of linear time-invariant discrete-time systems. Consider the discrete-time system described by

$$\mathbf{x}(k+1) = \mathbf{G}\mathbf{x}(k) \tag{5–154}$$

where $\mathbf{x}$ is a state vector (an n-vector) and $\mathbf{G}$ is an $n \times n$ constant nonsingular matrix. The origin $\mathbf{x} = \mathbf{0}$ is the equilibrium state. We shall investigate the stability of this state by use of the second method of Liapunov.

Let us choose as a possible Liapunov function

$$V(\mathbf{x}(k)) = \mathbf{x}^*(k)\mathbf{P}\mathbf{x}(k)$$

where $\mathbf{P}$ is a positive definite Hermitian (or a positive definite real symmetric) matrix. Then

$$\begin{aligned}\Delta V(\mathbf{x}(k)) &= V(\mathbf{x}(k+1)) - V(\mathbf{x}(k)) \\ &= \mathbf{x}^*(k+1)\mathbf{P}\mathbf{x}(k+1) - \mathbf{x}^*(k)\mathbf{P}\mathbf{x}(k) \\ &= [\mathbf{G}\mathbf{x}(k)]^*\mathbf{P}[\mathbf{G}\mathbf{x}(k)] - \mathbf{x}^*(k)\mathbf{P}\mathbf{x}(k) \\ &= \mathbf{x}^*(k)\mathbf{G}^*\mathbf{P}\mathbf{G}\mathbf{x}(k) - \mathbf{x}^*(k)\mathbf{P}\mathbf{x}(k) \\ &= \mathbf{x}^*(k)(\mathbf{G}^*\mathbf{P}\mathbf{G} - \mathbf{P})\mathbf{x}(k)\end{aligned}$$

Since $V(\mathbf{x}(k))$ is chosen to be positive definite, we require, for asymptotic stability, that $\Delta V(\mathbf{x}(k))$ be negative definite. Therefore,

$$\Delta V(\mathbf{x}(k)) = -\mathbf{x}^*(k)\mathbf{Q}\mathbf{x}(k)$$

where

$$\mathbf{Q} = -(\mathbf{G}^*\mathbf{P}\mathbf{G} - \mathbf{P}) = \text{positive definite}$$

Hence, for the asymptotic stability of the discrete-time system of Eq. (5–154), it is sufficient that $\mathbf{Q}$ be positive-definite.

As in the case of linear continuous-time systems, it is convenient to specify first a positive definite Hermitian (or a positive definite real symmetric) matrix $\mathbf{Q}$ and then to see whether or not the $\mathbf{P}$ matrix determined from

$$\mathbf{G}^*\mathbf{P}\mathbf{G} - \mathbf{P} = -\mathbf{Q}$$

is positive definite. Note that a positive definite $\mathbf{P}$ is a necessary and sufficient condition. We shall summarize in a theorem what we have stated here.

Theorem 5–6. Consider the discrete-time system

$$\mathbf{x}(k+1) = \mathbf{G}\mathbf{x}(k)$$

where $\mathbf{x}$ is a state vector (an n-vector) and $\mathbf{G}$ is an $n \times n$ constant nonsingular matrix. A necessary and sufficient condition for the equilibrium state $\mathbf{x} = \mathbf{0}$ to be asymptotically stable in the large is that, given any positive-definite Hermitian (or any positive definite real symmetric) matrix $\mathbf{Q}$, there exists a positive definite Hermitian (or a positive definite real symmetric) matrix $\mathbf{P}$ such that

$$\mathbf{G}^*\mathbf{P}\mathbf{G} - \mathbf{P} = -\mathbf{Q} \tag{5–155}$$

The scalar function $\mathbf{x}^*\mathbf{P}\mathbf{x}$ is a Liapunov function for this system.

If $\Delta V(\mathbf{x}(k)) = -\mathbf{x}^*(k)\mathbf{Q}\mathbf{x}(k)$ does not vanish identically along any solution series, then $\mathbf{Q}$ may be chosen to be positive semidefinite.

Stability of a discrete-time system obtained by discretizing a continuous-time system. If the system is described in terms of state space equations, the asymptotic stability of an equilibrium state of a discrete-time system obtained by discretizing a continuous-time system is equivalent to that of the original continuous-time system.

Consider a continuous-time system

$$\dot{\mathbf{x}} = \mathbf{A}\mathbf{x}$$

and the corresponding discrete-time system

$$\mathbf{x}((k+1)T) = \mathbf{G}\mathbf{x}(kT)$$

where

$$\mathbf{G} = e^{\mathbf{A}T}$$

If the continuous-time system is asymptotically stable, that is, if all the eigenvalues of the matrix $\mathbf{A}$ have negative real parts, then

$$\|\mathbf{G}^n\| \rightarrow 0 \qquad \text{as } n \rightarrow \infty$$

and the discretized system is also asymptotically stable. This is because if the λ_i's are the eigenvalues of $\mathbf{A}$, then the $e^{\lambda_i T}$'s are the eigenvalues of $\mathbf{G}$. (Note that $|e^{\lambda_i T}| < 1$ if $\lambda_i T$ is negative.)

It should be noted here that if a continuous-time system having complex poles is discretized, then in certain exceptional cases, hidden instability may occur, depending on the choice of the sampling period T. That is, in some cases where a continuous-time system is not asymptotically stable, the equivalent discretized system may seem to be asymptotically stable if one looks at the values of the output only at the sampling instants. This phenomenon occurs only at certain values of the sampling period T. If the value of T is varied, then such hidden instability shows up as explicit instability. See Prob. A-5-24.

Contraction. A norm of $\mathbf{x}$ denoted by $\|\mathbf{x}\|$ may be thought of as a measure of the length of the vector. There are several different definitions of a norm. Any norm, however, has the following properties:

$$\|\mathbf{x}\| = 0 \qquad \text{for } \mathbf{x} = \mathbf{0}$$

$$\|\mathbf{x}\| > 0 \qquad \text{for } \mathbf{x} \neq \mathbf{0}$$

$$\|\mathbf{x} + \mathbf{y}\| \leq \|\mathbf{x}\| + \|\mathbf{y}\| \qquad \text{for all } \mathbf{x} \text{ and } \mathbf{y}$$

$$\|k\mathbf{x}\| = |k|\,\|\mathbf{x}\| \qquad \text{for all } \mathbf{x} \text{ and real constant } k$$

A function $f(\mathbf{x})$ is said to be a contraction if $\mathbf{f}(\mathbf{0}) = \mathbf{0}$ and

$$\|\mathbf{f}(\mathbf{x})\| < \|\mathbf{x}\|$$

for some set of values of $\mathbf{x} \neq \mathbf{0}$ and some norm.

For discrete-time systems a norm $\|\mathbf{x}\|$ may be used as a Liapunov function. Consider the following discrete-time system:

$$\mathbf{x}(k+1) = \mathbf{f}(\mathbf{x}(k)), \qquad \mathbf{f}(\mathbf{0}) = \mathbf{0} \tag{5–156}$$

where $\mathbf{x}$ is an n-vector and $\mathbf{f}(\mathbf{x})$ is also an n-vector. Assume that $\mathbf{f}(\mathbf{x})$ is a contraction for all $\mathbf{x}$ and some norm. Then the origin of the system of Eq. (5–156) is asymptotically stable in the large and one of its Liapunov functions is

$$V(\mathbf{x}) = \|\mathbf{x}\|$$

This can be seen as follows. Since $V(\mathbf{x}) = \|\mathbf{x}\|$ is positive definite and

$$\Delta V(\mathbf{x}(k)) = V(\mathbf{f}(\mathbf{x}(k))) - V(\mathbf{x}(k)) = \|\mathbf{f}(\mathbf{x})\| - \|\mathbf{x}\|$$

is negative definite because $\mathbf{f}(\mathbf{x})$ is a contraction for all $\mathbf{x}$, we find $V(\mathbf{x}) = \|\mathbf{x}\|$ is a Liapunov function and by Theorem 5–5 the origin of the system is asymptotically stable in the large. (See Prob. A-5–34.)

Example 5–14.

Consider the following system:

$$\begin{bmatrix} x_1(k+1) \\ x_2(k+1) \end{bmatrix} = \begin{bmatrix} 0 & 1 \\ -0.5 & -1 \end{bmatrix} \begin{bmatrix} x_1(k) \\ x_2(k) \end{bmatrix}$$

Determine the stability of the origin of the system.

Let us choose $\mathbf{Q}$ to be $\mathbf{I}$. Then, referring to Eq. (5–155), the Liapunov stability equation becomes

$$\begin{bmatrix} 0 & -0.5 \\ 1 & -1 \end{bmatrix}\begin{bmatrix} p_{11} & p_{12} \\ p_{12} & p_{22} \end{bmatrix}\begin{bmatrix} 0 & 1 \\ -0.5 & -1 \end{bmatrix} - \begin{bmatrix} p_{11} & p_{12} \\ p_{12} & p_{22} \end{bmatrix} = -\begin{bmatrix} 1 & 0 \\ 0 & 1 \end{bmatrix} \tag{5–157}$$

If matrix $\mathbf{P}$ is found to be positive definite, then the origin $\mathbf{x} = \mathbf{0}$ is asymptotically stable in the large.

From Eq. (5–157) we obtain the following three equations:

$$0.25p_{22} - p_{11} = -1$$

$$0.5(-p_{12} + p_{22}) - p_{12} = 0$$

$$p_{11} - 2p_{12} = -1$$

from which we get

$$p_{11} = \tfrac{11}{5}, \qquad p_{12} = \tfrac{8}{5}, \qquad p_{22} = \tfrac{24}{5}$$

Consequently,

$$\mathbf{P} = \begin{bmatrix} \frac{11}{5} & \frac{8}{5} \\ \frac{8}{5} & \frac{24}{5} \end{bmatrix}$$

By applying Sylvester's criterion for the positive definiteness of matrix $\mathbf{P}$, we find $\mathbf{P}$ is positive definite. Hence, the equilibrium state, the origin $\mathbf{x} = \mathbf{0}$, is asymptotically stable in the large.

Note that instead of choosing $\mathbf{Q}$ to be $\mathbf{I}$, we could choose $\mathbf{Q}$ to be a positive-semidefinite matrix, such as

$$\mathbf{Q} = \begin{bmatrix} 0 & 0 \\ 0 & 1 \end{bmatrix}$$

as long as $\Delta V(\mathbf{x}) = -\mathbf{x}^*(k)\,\mathbf{Q}\mathbf{x}(k)$ does not vanish identically along any solution series. For the positive semidefinite matrix $\mathbf{Q}$ just given, we have

$$\Delta V(\mathbf{x}) = -x_2^2(k)$$

For the present system, $x_2(k)$ identically zero implies that $x_1(k)$ is identically zero. Hence $\Delta V(\mathbf{x})$ does not vanish identically along any solution series, except at the origin. Therefore, we can choose this positive semidefinite matrix $\mathbf{Q}$ for the determination of matrix $\mathbf{P}$ of the Liapunov stability equation. The Liapunov stability equation in this case becomes

$$\begin{bmatrix} 0 & -0.5 \\ 1 & -1 \end{bmatrix}\begin{bmatrix} p_{11} & p_{12} \\ p_{12} & p_{22} \end{bmatrix}\begin{bmatrix} 0 & 1 \\ -0.5 & -1 \end{bmatrix} - \begin{bmatrix} p_{11} & p_{12} \\ p_{12} & p_{22} \end{bmatrix} = -\begin{bmatrix} 0 & 0 \\ 0 & 1 \end{bmatrix}$$

By solving this last equation, we obtain

$$\mathbf{P} = \begin{bmatrix} \frac{3}{5} & \frac{4}{5} \\ \frac{4}{5} & \frac{12}{5} \end{bmatrix}$$

By applying the Sylvester criterion, we find $\mathbf{P}$ to be positive definite. Hence, we get the same conclusion as before: The origin is asymptotically stable in the large.

REFERENCES

5–1. Bellman, R., *Introduction to Matrix Analysis*. New York: McGraw-Hill Book Company, 1960.

5–2. Hahn, W., *Theory and Application of Liapunov's Direct Method*. Englewood Cliffs, N.J.: Prentice-Hall, Inc., 1963.

5–3. Jury, E. I., "Hidden Oscillations in Sampled-Data Control Systems," *AIEE Trans. part II*, **75** (1956), pp. 391–95.

5–4. Kalman, R. E., and J. E. Bertram, "Control System Analysis and Design via the Second Method of Lyapunov: I. Continuous-Time Systems; II. Discrete-Time Systems," *ASME J. Basic Engineering, ser.* **D, 82** (1960), pp. 371–93, 394–400.

5–5. LaSalle, J. P., and S. Lefschetz, *Stability by Liapunov's Direct Method with Applications*. New York: Academic Press, Inc., 1961.

5–6. Ogata, K., *Modern Control Engineering*. Englewood Cliffs, N.J.: Prentice-Hall, Inc., 1970.

5–7. Ogata, K., *State Space Analysis of Control Systems*. Englewood Cliffs, N.J.: Prentice-Hall, Inc., 1967.

EXAMPLE PROBLEMS AND SOLUTIONS

Problem A-5–1. Obtain the state equation and output equation for the system shown in Fig. 5–7.

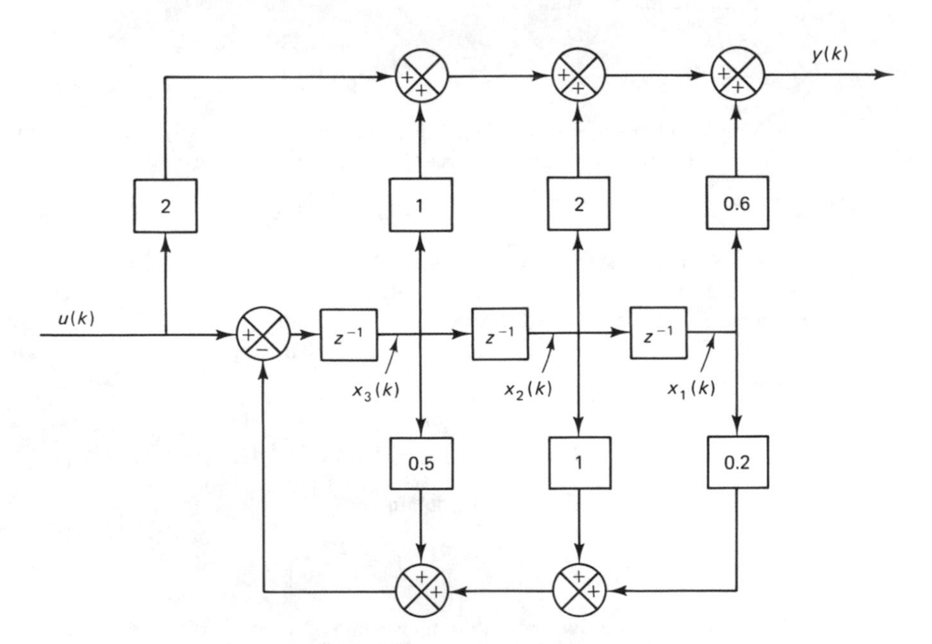

Figure 5–7 Block diagram of the control system of Prob. A-5–1.

Solution. From the block diagram we obtain

$$x_1(k+1) = x_2(k)$$

$$x_2(k+1) = x_3(k)$$

$$x_3(k+1) = -0.2x_1(k) - x_2(k) - 0.5x_3(k) + u(k)$$

and

$$y(k) = 0.6x_1(k) + 2x_2(k) + x_3(k) + 2u(k)$$

Hence

$$\begin{bmatrix} x_1(k+1) \\ x_2(k+1) \\ x_3(k+1) \end{bmatrix} = \begin{bmatrix} 0 & 1 & 0 \\ 0 & 0 & 1 \\ -0.2 & -1 & -0.5 \end{bmatrix} \begin{bmatrix} x_1(k) \\ x_2(k) \\ x_3(k) \end{bmatrix} + \begin{bmatrix} 0 \\ 0 \\ 1 \end{bmatrix} u(k)$$

$$y(k) = [0.6 \quad 2 \quad 1] \begin{bmatrix} x_1(k) \\ x_2(k) \\ x_3(k) \end{bmatrix} + 2u(k)$$

(This is in a controllable canonical form.)

Problem A-5–2. Using the nested programming method, obtain the state equation and output equation for the system defined by

$$\frac{Y(z)}{U(z)} = \frac{z^{-1} + 5z^{-2}}{1 + 4z^{-1} + 3z^{-2}}$$

Then, draw a block diagram for the system showing all state variables.

Solution. The given pulse transfer function can be written as

$$Y(z) = z^{-1}\{U(z) - 4Y(z) + z^{-1}[5U(z) - 3Y(z)]\}$$

Define

$$X_1(z) = z^{-1}[U(z) - 4Y(z) + X_2(z)]$$

$$X_2(z) = z^{-1}[5U(z) - 3Y(z)]$$

$$Y(z) = X_1(z)$$

Then, we obtain

$$zX_1(z) = -4X_1(z) + X_2(z) + U(z)$$

$$zX_2(z) = -3X_1(z) + 5U(z)$$

The state equation can therefore be given by

$$\begin{bmatrix} x_1(k+1) \\ x_2(k+1) \end{bmatrix} = \begin{bmatrix} -4 & 1 \\ -3 & 0 \end{bmatrix} \begin{bmatrix} x_1(k) \\ x_2(k) \end{bmatrix} + \begin{bmatrix} 1 \\ 5 \end{bmatrix} u(k)$$

and the output equation becomes

$$y(k) = [1 \quad 0] \begin{bmatrix} x_1(k) \\ x_2(k) \end{bmatrix}$$

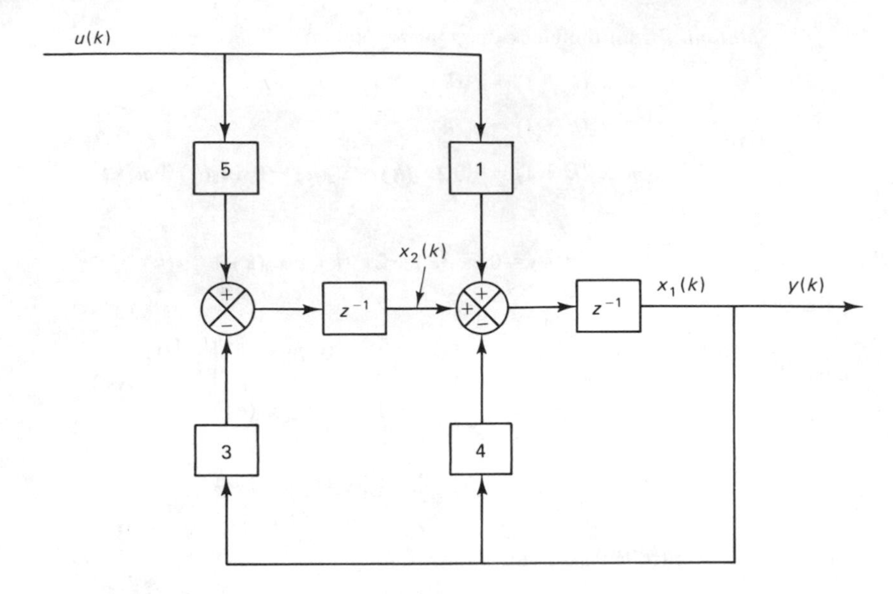

Figure 5–8 Block diagram for the system considered in Prob. A-5–2.

Figure 5–8 shows the block diagram for the system defined by the state space equations. The output of each delay element constitutes a state variable.

Problem A-5–3. Obtain a state space representation for the system shown in Fig. 5–9.

Solution. Let us choose the state variables as shown in the diagram. (Note that the output of a first-order lag term can always be chosen as a state variable.) From the block diagram we obtain the following equations:

$$\frac{X_1(s)}{U_1(s) - Y(s)} = \frac{5}{s+1}$$

$$\frac{X_2(s)}{X_1(s)} = \frac{1}{2s+1}$$

$$\frac{X_3(s)}{U_2(s)} = \frac{2}{s+3}$$

$$Y(s) = 10X_1(s) + X_2(s) + X_3(s)$$

or

$$\dot{x}_1 + x_1 = 5u_1 - 5y$$

$$2\dot{x}_2 + x_2 = x_1$$

$$\dot{x}_3 + 3x_3 = 2u_2$$

$$y = 10x_1 + x_2 + x_3$$

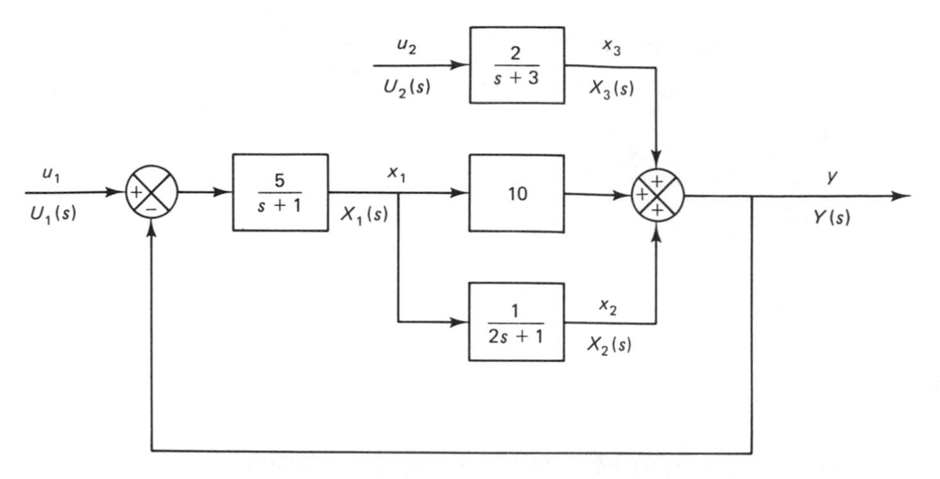

Figure 5–9 Block diagram of the control system of Prob. A-5–3.

Hence

$$\dot{x}_1 = -51x_1 - 5x_2 - 5x_3 + 5u_1$$
$$\dot{x}_2 = 0.5x_1 - 0.5x_2$$
$$\dot{x}_3 = -3x_3 + 2u_2$$
$$y = 10x_1 + x_2 + x_3$$

or

$$\begin{bmatrix} \dot{x}_1 \\ \dot{x}_2 \\ \dot{x}_3 \end{bmatrix} = \begin{bmatrix} -51 & -5 & -5 \\ 0.5 & -0.5 & 0 \\ 0 & 0 & -3 \end{bmatrix} \begin{bmatrix} x_1 \\ x_2 \\ x_3 \end{bmatrix} + \begin{bmatrix} 5 & 0 \\ 0 & 0 \\ 0 & 2 \end{bmatrix} \begin{bmatrix} u_1 \\ u_2 \end{bmatrix}$$

$$y = [10 \quad 1 \quad 1] \begin{bmatrix} x_1 \\ x_2 \\ x_3 \end{bmatrix}$$

Problem A-5–4. Obtain the state space representation of the system shown in Fig. 5–10.

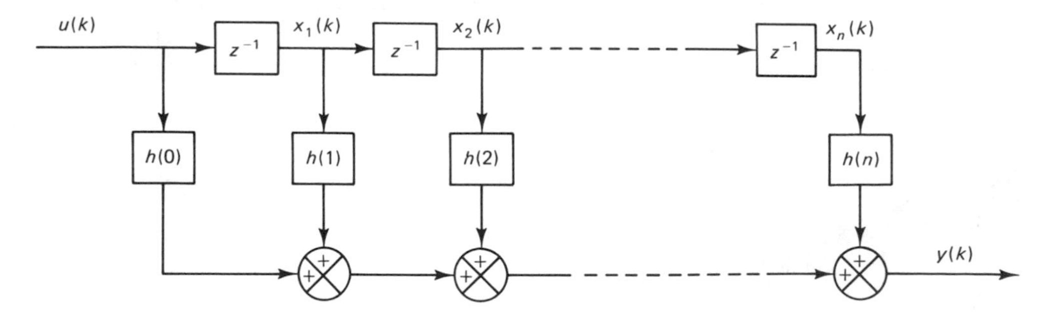

Figure 5–10 Block diagram of the control system of Prob. A-5–4.

Solution. From the block diagram we obtain

$$y(k) = h(0)u(k) + h(1)x_1(k) + h(2)x_2(k) + \cdots + h(n)x_n(k)$$

and

$$x_1(k+1) = u(k)$$
$$x_2(k+1) = x_1(k)$$
$$x_3(k+1) = x_2(k)$$
$$\vdots$$
$$x_n(k+1) = x_{n-1}(k)$$

Thus, the state space representation for the system becomes

$$\begin{bmatrix} x_1(k+1) \\ x_2(k+1) \\ \cdot \\ \cdot \\ \cdot \\ x_n(k+1) \end{bmatrix} = \begin{bmatrix} 0 & 0 & \cdots & 0 & 0 \\ 1 & 0 & \cdots & 0 & 0 \\ \cdot & \cdot & & \cdot & \cdot \\ \cdot & \cdot & & \cdot & \cdot \\ \cdot & \cdot & & \cdot & \cdot \\ 0 & 0 & \cdots & 1 & 0 \end{bmatrix} \begin{bmatrix} x_1(k) \\ x_2(k) \\ \cdot \\ \cdot \\ \cdot \\ x_n(k) \end{bmatrix} + \begin{bmatrix} 1 \\ 0 \\ \cdot \\ \cdot \\ \cdot \\ 0 \end{bmatrix} u(k)$$

$$y(k) = [h(1) \quad h(2) \cdots h(n)] \begin{bmatrix} x_1(k) \\ x_2(k) \\ \cdot \\ \cdot \\ \cdot \\ x_n(k) \end{bmatrix} + h(0)u(k)$$

Problem A-5-5. Obtain a state space representation of the system shown in Fig. 5-11.

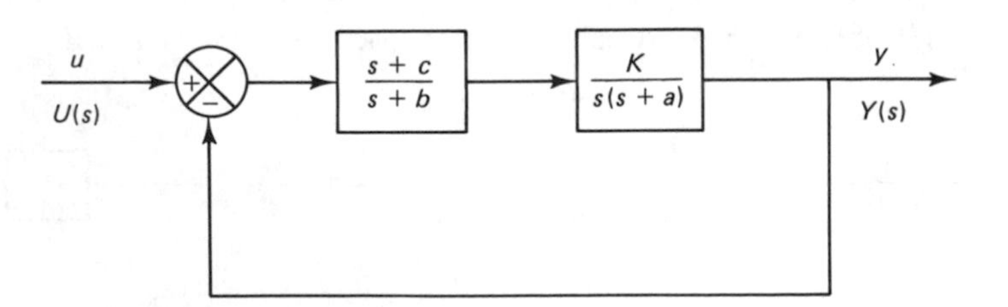

Figure 5-11 Block diagram of the control system of Prob. A-5-5.

Solution. Let us first expand $(s + c)/(s + b)$ into partial fractions:

$$\frac{s+c}{s+b} = 1 + \frac{c-b}{s+b}$$

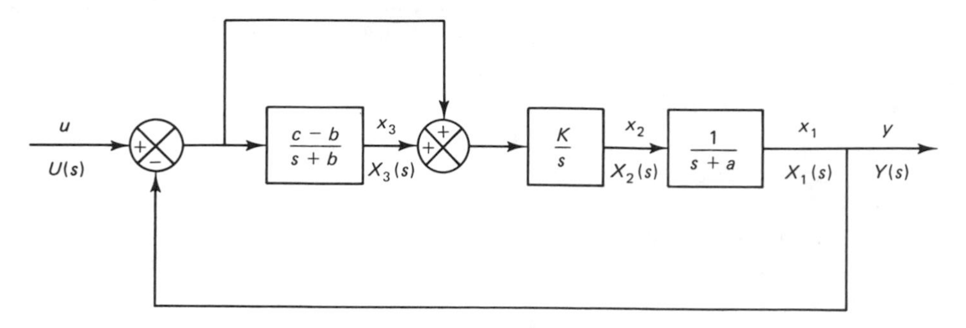

Figure 5–12 Modification of the block diagram shown in Fig. 5–11.

Next, convert $K/[s(s + a)]$ into the product of K/s and $1/(s + a)$. Then redraw the block diagram as in Fig. 5–12. The output of an integral block or the output of a first-order lag block always qualifies as a state variable. Therefore, defining a set of state variables as shown in Fig. 5–12; we obtain the following equations:

$$\dot{x}_1 = -ax_1 + x_2$$

$$\dot{x}_2 = -Kx_1 + Kx_3 + Ku$$

$$\dot{x}_3 = -(c - b)x_1 - bx_3 + (c - b)u$$

$$y = x_1$$

Rewriting gives

$$\begin{bmatrix} \dot{x}_1 \\ \dot{x}_2 \\ \dot{x}_3 \end{bmatrix} = \begin{bmatrix} -a & 1 & 0 \\ -K & 0 & K \\ -(c-b) & 0 & -b \end{bmatrix} \begin{bmatrix} x_1 \\ x_2 \\ x_3 \end{bmatrix} + \begin{bmatrix} 0 \\ K \\ c - b \end{bmatrix} u$$

$$y = [1 \quad 0 \quad 0] \begin{bmatrix} x_1 \\ x_2 \\ x_3 \end{bmatrix}$$

Problem A-5–6. Obtain a state space representation of the system shown in Fig. 5–13. The sampling period T is 1 sec.

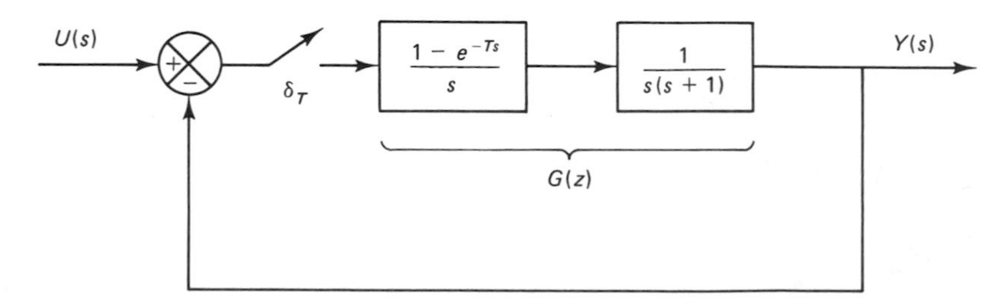

Figure 5–13 Block diagram of the control system of Prob. A-5–6.

Solution. We shall first obtain the z transform of the feedforward transfer function:

$$G(z) = \mathscr{Z}\left[\frac{1-e^{-s}}{s}\frac{1}{s(s+1)}\right] = (1-z^{-1})\,\mathscr{Z}\left[\frac{1}{s^2(s+1)}\right]$$

$$= \frac{0.3679(z+0.7181)}{(z-1)(z-0.3679)}$$

Then the closed-loop pulse transfer function can be obtained easily. There are many ways to obtain the state space representation for such a system, as discussed in Sec. 5–2. In this problem, we shall demonstrate another approach, based on block diagram modification.

Let us expand $G(z)$ into partial fractions:

$$G(z) = \frac{1}{z-1} - \frac{0.6321}{z-0.3679} = \frac{z^{-1}}{1-z^{-1}} - \frac{0.6321z^{-1}}{1-0.3679z^{-1}}$$

Figure 5–14 shows the block diagram for the system. Let us choose the output of each unit delay element as a state variable, as shown in Fig. 5–14. Then we obtain

$$zX_1(z) = X_1(z) - [X_1(z) - 0.6321X_2(z)] + U(z)$$

$$zX_2(z) = 0.3679X_2(z) - [X_1(z) - 0.6321X_2(z)] + U(z)$$

$$Y(z) = X_1(z) - 0.6321X_2(z)$$

from which we get

$$x_1(k+1) = 0.6321x_2(k) + u(k)$$

$$x_2(k+1) = -x_1(k) + x_2(k) + u(k)$$

$$y(k) = x_1(k) - 0.6321x_2(k)$$

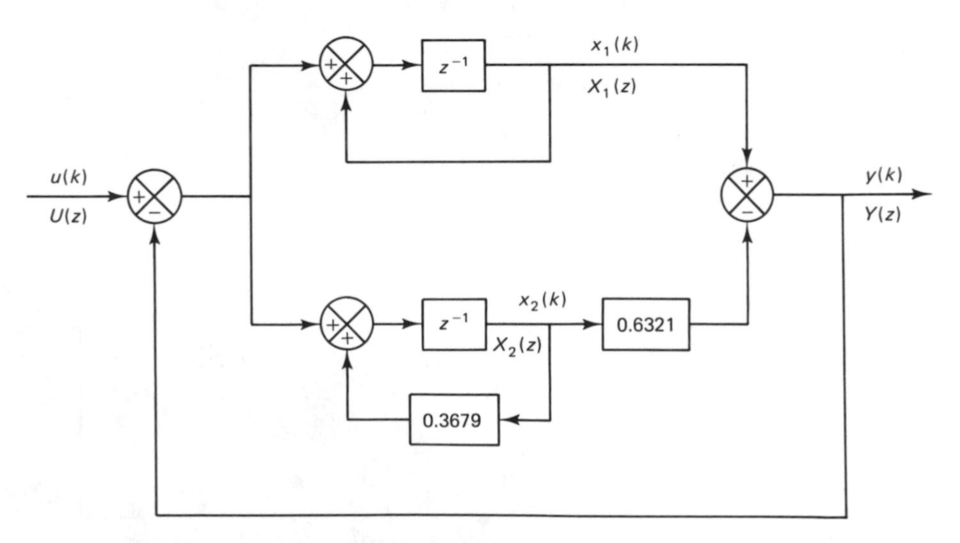

Figure 5–14 Modified block diagram for the system shown in Fig. 5–13.

or

$$\begin{bmatrix} x_1(k+1) \\ x_2(k+1) \end{bmatrix} = \begin{bmatrix} 0 & 0.6321 \\ -1 & 1 \end{bmatrix} \begin{bmatrix} x_1(k) \\ x_2(k) \end{bmatrix} + \begin{bmatrix} 1 \\ 1 \end{bmatrix} u(k)$$

$$y(k) = \begin{bmatrix} 1 & -0.6321 \end{bmatrix} \begin{bmatrix} x_1(k) \\ x_2(k) \end{bmatrix}$$

Problem A-5-7. Figure 5-15 shows a block diagram of a discrete-time multiple-input-multiple-output system. Obtain state space equations for the system by considering $x_1(k)$, $x_2(k)$, and $x_3(k)$ as shown in the diagram to be state variables. Then, define new state variables such that the state matrix becomes a diagonal matrix.

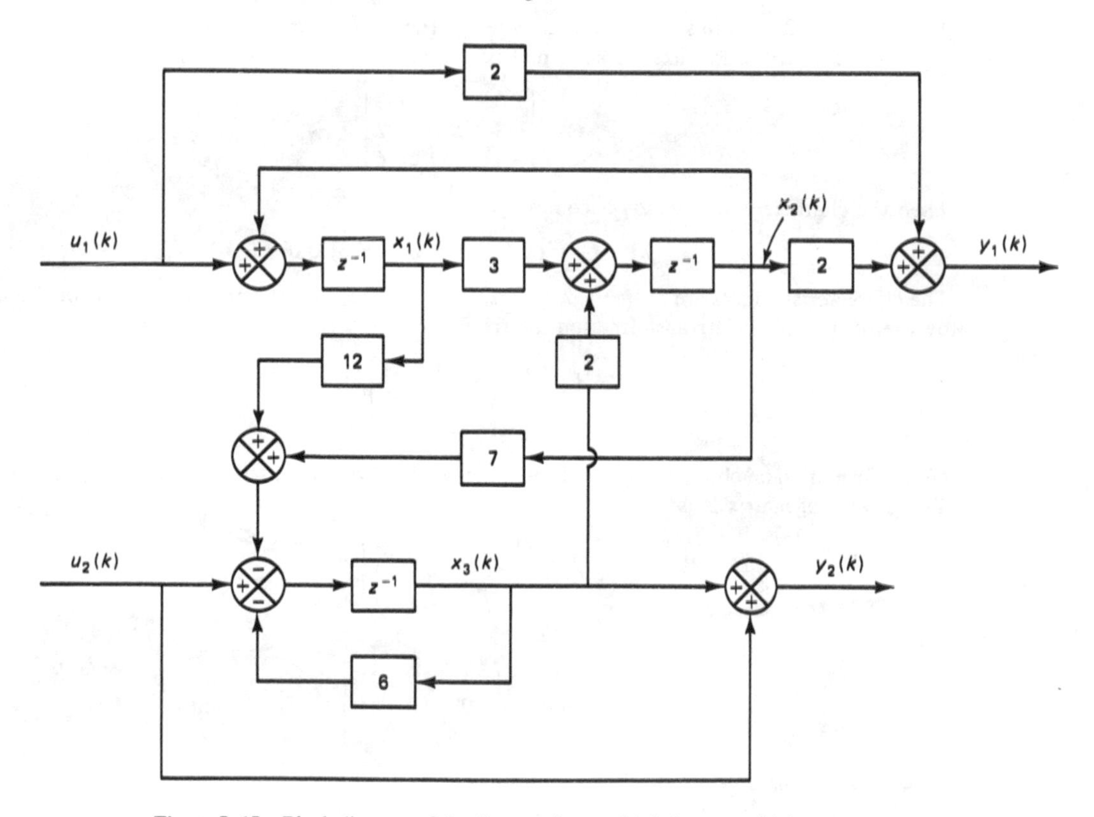

Figure 5-15 Block diagram of the discrete-time multiple-input-multiple-output system of Prob. A-5-7.

Solution. From Fig. 5-15 we obtain the following discrete-time state space equations:

$$x_1(k+1) = x_2(k) + u_1(k)$$

$$x_2(k+1) = 3x_1(k) + 2x_3(k)$$

$$x_3(k+1) = -12x_1(k) - 7x_2(k) - 6x_3(k) + u_2(k)$$

$$y_1(k) = 2x_2(k) + 2u_1(k)$$

$$y_2(k) = x_3(k) + u_2(k)$$

Rewriting in the form of vector-matrix equations, we obtain

$$\begin{bmatrix} x_1(k+1) \\ x_2(k+1) \\ x_3(k+1) \end{bmatrix} = \begin{bmatrix} 0 & 1 & 0 \\ 3 & 0 & 2 \\ -12 & -7 & -6 \end{bmatrix} \begin{bmatrix} x_1(k) \\ x_2(k) \\ x_3(k) \end{bmatrix} + \begin{bmatrix} 1 & 0 \\ 0 & 0 \\ 0 & 1 \end{bmatrix} \begin{bmatrix} u_1(k) \\ u_2(k) \end{bmatrix}$$

$$\begin{bmatrix} y_1(k) \\ y_2(k) \end{bmatrix} = \begin{bmatrix} 0 & 2 & 0 \\ 0 & 0 & 1 \end{bmatrix} \begin{bmatrix} x_1(k) \\ x_2(k) \\ x_3(k) \end{bmatrix} + \begin{bmatrix} 2 & 0 \\ 0 & 1 \end{bmatrix} \begin{bmatrix} u_1(k) \\ u_2(k) \end{bmatrix}$$

These two equations are state space equations for the system being considered.

In order to diagonalize the state matrix, let us define

$$\mathbf{G} = \begin{bmatrix} 0 & 1 & 0 \\ 3 & 0 & 2 \\ -12 & -7 & -6 \end{bmatrix}$$

Then the characteristic equation becomes

$$|\lambda\mathbf{I} - \mathbf{G}| = (\lambda + 1)(\lambda + 2)(\lambda + 3) = 0$$

The characteristic roots are $\lambda_1 = -1$, $\lambda_2 = -2$, and $\lambda_3 = -3$. The matrix $\mathbf{G}$ can be diagonalized by use of the following transformation matrix $\mathbf{P}$:

$$\mathbf{P} = \begin{bmatrix} 1 & 2 & 1 \\ -1 & -4 & -3 \\ -1 & 1 & 3 \end{bmatrix}$$

(For information on obtaining such a diagonalizing transformation matrix $\mathbf{P}$, see the Appendix.) The inverse of matrix $\mathbf{P}$ is

$$\mathbf{P}^{-1} = \begin{bmatrix} \frac{9}{2} & \frac{5}{2} & 1 \\ -3 & -2 & -1 \\ \frac{5}{2} & \frac{3}{2} & 1 \end{bmatrix}$$

Thus,

$$\mathbf{P}^{-1}\mathbf{G}\mathbf{P} = \begin{bmatrix} -1 & 0 & 0 \\ 0 & -2 & 0 \\ 0 & 0 & -3 \end{bmatrix}$$

Now let us define a new state vector $\hat{\mathbf{x}}$ as follows:

$$\mathbf{x} = \mathbf{P}\hat{\mathbf{x}}$$

Then, in terms of the new state vector, state space equations can be written as follows:

$$\begin{bmatrix} \hat{x}_1(k+1) \\ \hat{x}_2(k+1) \\ \hat{x}_3(k+1) \end{bmatrix} = \begin{bmatrix} -1 & 0 & 0 \\ 0 & -2 & 0 \\ 0 & 0 & -3 \end{bmatrix} \begin{bmatrix} \hat{x}_1(k) \\ \hat{x}_2(k) \\ \hat{x}_3(k) \end{bmatrix} + \begin{bmatrix} \frac{9}{2} & 1 \\ -3 & -1 \\ \frac{5}{2} & 1 \end{bmatrix} \begin{bmatrix} u_1(k) \\ u_2(k) \end{bmatrix}$$

$$\begin{bmatrix} y_1(k) \\ y_2(k) \end{bmatrix} = \begin{bmatrix} -2 & -8 & -6 \\ -1 & 1 & 3 \end{bmatrix} \begin{bmatrix} \hat{x}_1(k) \\ \hat{x}_2(k) \\ \hat{x}_3(k) \end{bmatrix} + \begin{bmatrix} 2 & 0 \\ 0 & 1 \end{bmatrix} \begin{bmatrix} u_1(k) \\ u_2(k) \end{bmatrix}$$

Notice that the initial data $\hat{x}_1(0)$, $\hat{x}_2(0)$, and $\hat{x}_3(0)$ are obtained from

$$\begin{bmatrix} \hat{x}_1(0) \\ \hat{x}_2(0) \\ \hat{x}_3(0) \end{bmatrix} = \mathbf{P}^{-1}\mathbf{x}(0) = \begin{bmatrix} \frac{9}{2} & \frac{5}{2} & 1 \\ -3 & -2 & -1 \\ \frac{5}{2} & \frac{3}{2} & 1 \end{bmatrix} \begin{bmatrix} x_1(0) \\ x_2(0) \\ x_3(0) \end{bmatrix}$$

Problem A-5-8. Obtain a state space representation of the following pulse-transfer-function system:

$$\frac{Y(z)}{U(z)} = \frac{5}{(z+1)^2(z+2)}$$

Use the partial-fraction-expansion programming method. Also, obtain the initial values of the state variables in terms of $y(0)$, $y(1)$, and $y(2)$. Then draw a block diagram for the system.

Solution. Because we need the initial values of the state variables in terms of $y(0)$, $y(1)$, and $y(2)$, we slightly modify the partial-fraction-expansion programming method presented in Sec. 5-2. Let us expand $Y(z)/U(z)$, $zY(z)/U(z)$, and $z^2Y(z)/U(z)$ into partial fractions as follows:

$$\frac{Y(z)}{U(z)} = \frac{5}{(z+1)^2} - \frac{5}{z+1} + \frac{5}{z+2}$$

$$\frac{zY(z)}{U(z)} = -\frac{5}{(z+1)^2} + \frac{10}{z+1} - \frac{10}{z+2}$$

$$\frac{z^2Y(z)}{U(z)} = \frac{5}{(z+1)^2} - \frac{15}{z+1} + \frac{20}{z+2}$$

Then we have

$$\begin{bmatrix} \dfrac{Y(z)}{U(z)} \\[2ex] \dfrac{zY(z)}{U(z)} \\[2ex] \dfrac{z^2Y(z)}{U(z)} \end{bmatrix} = \begin{bmatrix} 5 & -5 & 5 \\ -5 & 10 & -10 \\ 5 & -15 & 20 \end{bmatrix} \begin{bmatrix} \dfrac{1}{(z+1)^2} \\[2ex] \dfrac{1}{z+1} \\[2ex] \dfrac{1}{z+2} \end{bmatrix}$$

Now let us define the state variables by the following equation:

$$\begin{bmatrix} \dfrac{X_1(z)}{U(z)} \\[2ex] \dfrac{X_2(z)}{U(z)} \\[2ex] \dfrac{X_3(z)}{U(z)} \end{bmatrix} = \begin{bmatrix} \dfrac{1}{(z+1)^2} \\[2ex] \dfrac{1}{z+1} \\[2ex] \dfrac{1}{z+2} \end{bmatrix} \tag{5-158}$$

Then the state variables $X_1(z)$, $X_2(z)$, and $X_3(z)$ are related to $Y(z)$, $zY(z)$, and $z^2Y(z)$ as follows:

$$\begin{bmatrix} Y(z) \\ zY(z) \\ z^2Y(z) \end{bmatrix} = \begin{bmatrix} 5 & -5 & 5 \\ -5 & 10 & -10 \\ 5 & -15 & 20 \end{bmatrix} \begin{bmatrix} X_1(z) \\ X_2(z) \\ X_3(z) \end{bmatrix} \tag{5-159}$$

From Eq. (5–158), we obtain

$$(z+1)^2X_1(z) = U(z)$$

$$(z+1)X_2(z) = U(z)$$

$$(z+2)X_3(z) = U(z)$$

Noting that

$$(z+1)X_1(z) = X_2(z)$$

we get

$$zX_1(z) = -X_1(z) + X_2(z)$$

$$zX_2(z) = -X_2(z) + U(z)$$

$$zX_3(z) = -2X_3(z) + U(z)$$

The output $Y(z)$ is given by the equation

$$Y(z) = 5X_1(z) - 5X_2(z) + 5X_3(z)$$

Consequently, we have the state space equations as follows:

$$x_1(k+1) = -x_1(k) + x_2(k)$$

$$x_2(k+1) = -x_2(k) + u(k)$$

$$x_3(k+1) = -2x_3(k) + u(k)$$

$$y(k) = 5x_1(k) - 5x_2(k) + 5x_3(k)$$

or

$$\begin{bmatrix} x_1(k+1) \\ x_2(k+1) \\ x_3(k+1) \end{bmatrix} = \begin{bmatrix} -1 & 1 & 0 \\ 0 & -1 & 0 \\ 0 & 0 & -2 \end{bmatrix} \begin{bmatrix} x_1(k) \\ x_2(k) \\ x_3(k) \end{bmatrix} + \begin{bmatrix} 0 \\ 1 \\ 1 \end{bmatrix} u(k)$$

$$y(k) = [5 \quad -5 \quad 5] \begin{bmatrix} x_1(k) \\ x_2(k) \\ x_3(k) \end{bmatrix}$$

The initial data are obtained by use of Eq. (5–159), as follows:

$$\begin{bmatrix} x_1(0) \\ x_2(0) \\ x_3(0) \end{bmatrix} = \begin{bmatrix} 5 & -5 & 5 \\ -5 & 10 & -10 \\ 5 & -15 & 20 \end{bmatrix}^{-1} \begin{bmatrix} y(0) \\ y(1) \\ y(2) \end{bmatrix}$$

$$= \begin{bmatrix} \frac{2}{5} & \frac{1}{5} & 0 \\ \frac{2}{5} & \frac{3}{5} & \frac{1}{5} \\ \frac{1}{5} & \frac{2}{5} & \frac{1}{5} \end{bmatrix} \begin{bmatrix} y(0) \\ y(1) \\ y(2) \end{bmatrix}$$

The block diagram for this system is shown in Fig. 5–16.

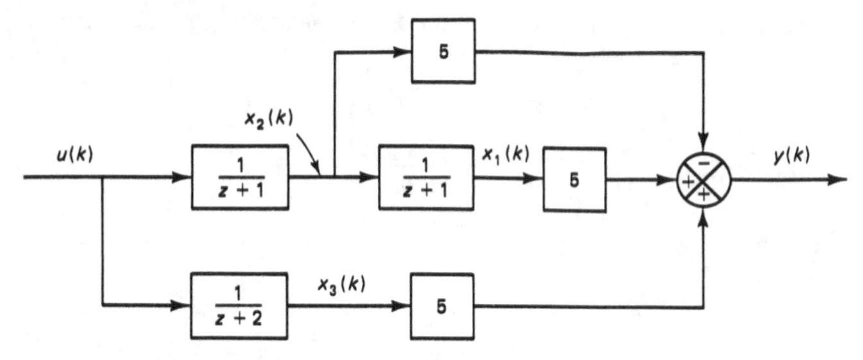

Figure 5–16 Block diagram for the system considered in Prob. A-5–8.

Problem A-5–9. Obtain a state space representation of the following pulse-transfer-function system such that the state matrix is diagonal:

$$\frac{Y(z)}{U(z)} = \frac{z^3 + 8z^2 + 17z + 8}{(z+1)(z+2)(z+3)}$$

Then, obtain the initial state $\mathbf{x}(0)$ in terms of $y(0)$, $y(1)$, $y(2)$ and $u(0)$, $u(1)$, $u(2)$.

Solution. Let us first divide the numerators of the right-hand sides of $Y(z)/U(z)$, $zY(z)/U(z)$, and $z^2Y(z)/U(z)$ by the respective denominators and expand the remaining terms into partial fractions, as follows:

$$\frac{Y(z)}{U(z)} = 1 - \frac{1}{z+1} + \frac{2}{z+2} + \frac{1}{z+3}$$

$$\frac{zY(z)}{U(z)} = z + 2 + \frac{1}{z+1} - \frac{4}{z+2} - \frac{3}{z+3}$$

$$\frac{z^2Y(z)}{U(z)} = z^2 + 2z - 6 - \frac{1}{z+1} + \frac{8}{z+2} + \frac{9}{z+3}$$

Rewriting, we have

$$\frac{Y(z) - U(z)}{U(z)} = -\frac{1}{z+1} + \frac{2}{z+2} + \frac{1}{z+3}$$

$$\frac{zY(z) - zU(z) - 2U(z)}{U(z)} = \frac{1}{z+1} - \frac{4}{z+2} - \frac{3}{z+3}$$

$$\frac{z^2Y(z) - z^2U(z) - 2zU(z) + 6U(z)}{U(z)} = -\frac{1}{z+1} + \frac{8}{z+2} + \frac{9}{z+3}$$

or

$$\begin{bmatrix} \dfrac{Y(z) - U(z)}{U(z)} \\ \dfrac{zY(z) - zU(z) - 2U(z)}{U(z)} \\ \dfrac{z^2Y(z) - z^2U(z) - 2zU(z) + 6U(z)}{U(z)} \end{bmatrix} = \begin{bmatrix} -1 & 2 & 1 \\ 1 & -4 & -3 \\ -1 & 8 & 9 \end{bmatrix} \begin{bmatrix} \dfrac{1}{z+1} \\ \dfrac{1}{z+2} \\ \dfrac{1}{z+3} \end{bmatrix}$$

Let us define the state variables $X_1(z)$, $X_2(z)$, and $X_3(z)$ as follows:

$$\begin{bmatrix} \dfrac{X_1(z)}{U(z)} \\ \dfrac{X_2(z)}{U(z)} \\ \dfrac{X_3(z)}{U(z)} \end{bmatrix} = \begin{bmatrix} \dfrac{1}{z+1} \\ \dfrac{1}{z+2} \\ \dfrac{1}{z+3} \end{bmatrix} \tag{5–160}$$

Then, we have

$$\begin{bmatrix} Y(z) - U(z) \\ zY(z) - zU(z) - 2U(z) \\ z^2Y(z) - z^2U(z) - 2zU(z) + 6U(z) \end{bmatrix} = \begin{bmatrix} -1 & 2 & 1 \\ 1 & -4 & -3 \\ -1 & 8 & 9 \end{bmatrix} \begin{bmatrix} X_1(z) \\ X_2(z) \\ X_3(z) \end{bmatrix} \tag{5–161}$$

Notice that Eq. (5–160) can be written as

$$zX_1(z) = -X_1(z) + U(z)$$

$$zX_2(z) = -2X_2(z) + U(z)$$

$$zX_3(z) = -3X_3(z) + U(z)$$

from which we obtain

$$x_1(k+1) = -x_1(k) + u(k)$$

$$x_2(k+1) = -2x_2(k) + u(k)$$

$$x_3(k+1) = -3x_3(k) + u(k)$$

The output $Y(z)$ is given by

$$Y(z) = -X_1(z) + 2X_2(z) + X_3(z) + U(z)$$

or

$$y(k) = -x_1(k) + 2x_2(k) + x_3(k) + u(k)$$

In vector-matrix notation, the state space equations become

$$\begin{bmatrix} x_1(k+1) \\ x_2(k+1) \\ x_3(k+1) \end{bmatrix} = \begin{bmatrix} -1 & 0 & 0 \\ 0 & -2 & 0 \\ 0 & 0 & -3 \end{bmatrix} \begin{bmatrix} x_1(k) \\ x_2(k) \\ x_3(k) \end{bmatrix} + \begin{bmatrix} 1 \\ 1 \\ 1 \end{bmatrix} u(k)$$

$$y(k) = [-1 \quad 2 \quad 1] \begin{bmatrix} x_1(k) \\ x_2(k) \\ x_3(k) \end{bmatrix} + u(k)$$

The initial data are obtained from Eq. (5–161) as follows:

$$\begin{bmatrix} x_1(0) \\ x_2(0) \\ x_3(0) \end{bmatrix} = \begin{bmatrix} -1 & 2 & 1 \\ 1 & -4 & -3 \\ -1 & 8 & 9 \end{bmatrix}^{-1} \begin{bmatrix} y(0) - u(0) \\ y(1) - u(1) - 2u(0) \\ y(2) - u(2) - 2u(1) + 6u(0) \end{bmatrix}$$

$$= \begin{bmatrix} -3 & -\frac{5}{2} & -\frac{1}{2} \\ -\frac{3}{2} & -2 & -\frac{1}{2} \\ 1 & \frac{3}{2} & \frac{1}{2} \end{bmatrix} \begin{bmatrix} y(0) - u(0) \\ y(1) - u(1) - 2u(0) \\ y(2) - u(2) - 2u(1) + 6u(0) \end{bmatrix}$$

Figure 5–17 shows the block diagram for the present system.

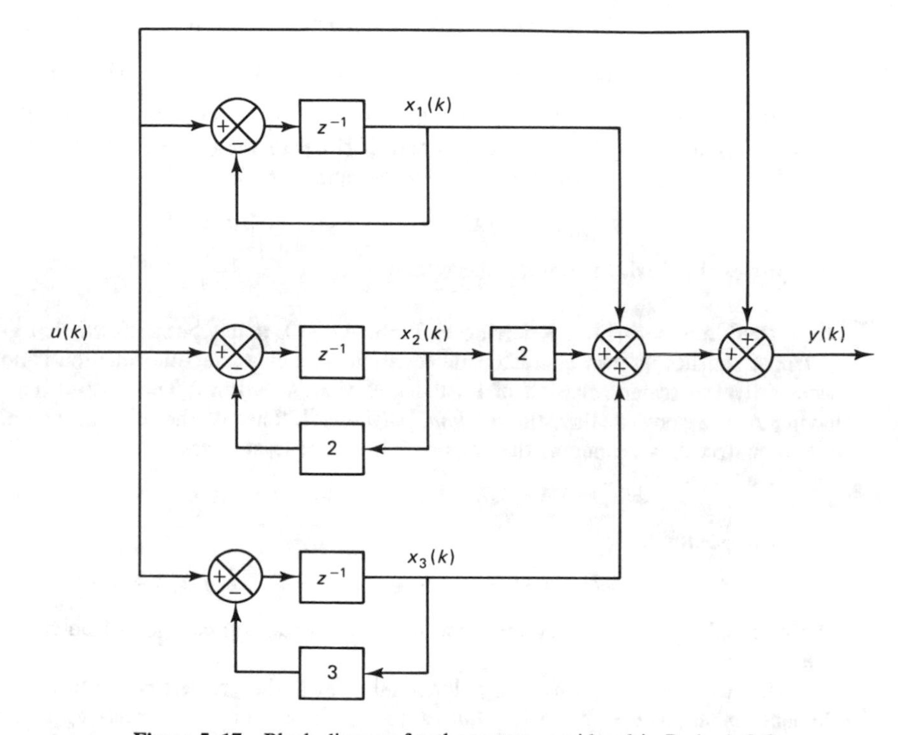

Figure 5–17 Block diagram for the system considered in Prob. A-5-9.

Problem A-5-10. Let **A** be an $n \times n$ matrix, and let its characteristic equation be

$$|\lambda\mathbf{I} - \mathbf{A}| = \lambda^n + a_1\lambda^{n-1} + \cdots + a_{n-1}\lambda + a_n = 0$$

Show that the matrix **A** satisfies its characteristic equation, or that

$$\mathbf{A}^n + a_1\mathbf{A}^{n-1} + \cdots + a_{n-1}\mathbf{A} + a_n\mathbf{I} = \mathbf{0}$$

(This is the Cayley-Hamilton theorem.)

Solution. We first note that adj $(\lambda\mathbf{I} - \mathbf{A})$ is a polynomial in λ of degree $n - 1$. That is,

$$\text{adj}\,(\lambda\mathbf{I} - \mathbf{A}) = \mathbf{B}_1\lambda^{n-1} + \mathbf{B}_2\lambda^{n-2} + \cdots + \mathbf{B}_{n-1}\lambda + \mathbf{B}_n$$

where

$$\mathbf{B}_1 = \mathbf{I}$$

Note also that

$$(\lambda\mathbf{I}-\mathbf{A})\,\mathrm{adj}\,(\lambda\mathbf{I}-\mathbf{A}) = [\mathrm{adj}\,(\lambda\mathbf{I}-\mathbf{A})](\lambda\mathbf{I}-\mathbf{A}) = |\lambda\mathbf{I}-\mathbf{A}|\mathbf{I}$$

Hence we obtain

$$\begin{aligned}|\lambda\mathbf{I}-\mathbf{A}|\mathbf{I} &= \mathbf{I}\lambda^n + a_1\mathbf{I}\lambda^{n-1} + \cdots + a_{n-1}\mathbf{I}\lambda + a_n\mathbf{I}\\ &= (\lambda\mathbf{I}-\mathbf{A})(\mathbf{B}_1\lambda^{n-1} + \mathbf{B}_2\lambda^{n-2} + \cdots + \mathbf{B}_{n-1}\lambda + \mathbf{B}_n)\\ &= (\mathbf{B}_1\lambda^{n-1} + \mathbf{B}_2\lambda^{n-2} + \cdots + \mathbf{B}_{n-1}\lambda + \mathbf{B}_n)(\lambda\mathbf{I}-\mathbf{A})\end{aligned}$$

From this equation we see that $\mathbf{A}$ and $\mathbf{B}_i$ $(i = 1, 2, \ldots, n)$ commute. Hence the product of $(\lambda\mathbf{I} - \mathbf{A})$ and $\mathrm{adj}\,(\lambda\mathbf{I} - \mathbf{A})$ becomes zero if either of these is zero. If $\mathbf{A}$ is substituted for λ in this last equation, then clearly $\lambda\mathbf{I} - \mathbf{A}$ becomes zero. Hence

$$\mathbf{A}^n + a_1\mathbf{A}^{n-1} + \cdots + a_{n-1}\mathbf{A} + a_n\mathbf{I} = \mathbf{0}$$

This proves the Cayley-Hamilton theorem.

Problem A-5-11. Referring to Prob. A-5-10, it has been shown that every $n \times n$ matrix $\mathbf{A}$ satisfies its own characteristic equation. The characteristic equation is not, however, necessarily the scalar equation of least degree that $\mathbf{A}$ satisfies. The least-degree polynomial having $\mathbf{A}$ as a root is called the *minimal polynomial*. That is, the minimal polynomial of an $n \times n$ matrix $\mathbf{A}$ is defined as the polynomial $\phi(\lambda)$ of least degree

$$\phi(\lambda) = \lambda^m + a_1\lambda^{m-1} + \cdots + a_{m-1}\lambda + a_m \qquad m \le n$$

such that $\phi(\mathbf{A}) = \mathbf{0}$, or

$$\phi(\mathbf{A}) = \mathbf{A}^m + a_1\mathbf{A}^{m-1} + \cdots + a_{m-1}\mathbf{A} + a_m\mathbf{I} = \mathbf{0}$$

The minimal polynomial plays an important role in the computation of polynomials in an $n \times n$ matrix.

Let us suppose that $d(\lambda)$, a polynomial in λ, is the greatest common divisor of all the elements of $\mathrm{adj}\,(\lambda\mathbf{I} - \mathbf{A})$. Show that if the coefficient of the highest-degree term in λ of $d(\lambda)$ is chosen as 1, then the minimal polynomial $\phi(\lambda)$ is given by

$$\phi(\lambda) = \frac{|\lambda\mathbf{I}-\mathbf{A}|}{d(\lambda)}$$

Solution. By assumption, the greatest common divisor of the matrix $\mathrm{adj}\,(\lambda\mathbf{I} - \mathbf{A})$ is $d(\lambda)$. Therefore,

$$\mathrm{adj}\,(\lambda\mathbf{I}-\mathbf{A}) = d(\lambda)\mathbf{B}(\lambda)$$

where the greatest common divisor of the n^2 elements (which are functions of λ) of $\mathbf{B}(\lambda)$ is unity. Since

$$(\lambda\mathbf{I}-\mathbf{A})\,\mathrm{adj}\,(\lambda\mathbf{I}-\mathbf{A}) = |\lambda\mathbf{I}-\mathbf{A}|\mathbf{I}$$

we obtain

$$d(\lambda)(\lambda\mathbf{I}-\mathbf{A})\mathbf{B}(\lambda) = |\lambda\mathbf{I}-\mathbf{A}|\mathbf{I} \tag{5-162}$$

from which we find that $|\lambda\mathbf{I} - \mathbf{A}|$ is divisible by $d(\lambda)$. Let us put

$$|\lambda\mathbf{I}-\mathbf{A}| = d(\lambda)\psi(\lambda) \tag{5-163}$$

Then the coefficient of the highest-degree term in λ of $\psi(\lambda)$ is unity. From Eqs. (5–162) and (5–163) we have

$$(\lambda\mathbf{I} - \mathbf{A})\mathbf{B}(\lambda) = \psi(\lambda)\mathbf{I}$$

Hence

$$\psi(\mathbf{A}) = \mathbf{0}$$

Note that $\psi(\lambda)$ can be written as follows:

$$\psi(\lambda) = g(\lambda)\phi(\lambda) + \alpha(\lambda)$$

where $\alpha(\lambda)$ is of lower degree than $\phi(\lambda)$. Since $\psi(\mathbf{A}) = \mathbf{0}$ and $\phi(\mathbf{A}) = \mathbf{0}$, we must have $\alpha(\mathbf{A}) = \mathbf{0}$. Since $\phi(\lambda)$ is the minimal polynomial, $\alpha(\lambda)$ must be identically zero, or

$$\psi(\lambda) = g(\lambda)\phi(\lambda)$$

Note that because $\phi(\mathbf{A}) = \mathbf{0}$, we can write

$$\phi(\lambda)\mathbf{I} = (\lambda\mathbf{I} - \mathbf{A})\mathbf{C}(\lambda)$$

Hence

$$\psi(\lambda)\mathbf{I} = g(\lambda)\phi(\lambda)\mathbf{I} = g(\lambda)(\lambda\mathbf{I} - \mathbf{A})\mathbf{C}(\lambda)$$

and we obtain

$$\mathbf{B}(\lambda) = g(\lambda)\mathbf{C}(\lambda)$$

Note that the greatest common divisor of the n^2 elements of $\mathbf{B}(\lambda)$ is unity. Hence

$$g(\lambda) = 1$$

Therefore,

$$\psi(\lambda) = \phi(\lambda)$$

Then, from this last equation and Eq. (5–163) we obtain

$$\phi(\lambda) = \frac{|\lambda\mathbf{I} - \mathbf{A}|}{d(\lambda)}$$

It is noted that the minimal polynomial $\phi(\lambda)$ of an $n \times n$ matrix $\mathbf{A}$ can be determined by the following procedure:

1. Form adj $(\lambda\mathbf{I} - \mathbf{A})$ and write the elements of adj $(\lambda\mathbf{I} - \mathbf{A})$ as factored polynomials in λ.
2. Determine $d(\lambda)$ as the greatest common divisor of all the elements of adj $(\lambda\mathbf{I} - \mathbf{A})$. Choose the coefficient of the highest-degree term in λ of $d(\lambda)$ to be 1. If there is no common divisor, $d(\lambda) = 1$.
3. The minimal polynomial $\phi(\lambda)$ is then given as $|\lambda\mathbf{I} - \mathbf{A}|$ divided by $d(\lambda)$.

Problem A-5–12. If an $n \times n$ matrix $\mathbf{A}$ has n distinct eigenvalues, then the minimal polynomial of $\mathbf{A}$ is identical with the characteristic polynomial. Also, if the multiple eigenvalues of $\mathbf{A}$ are linked in a Jordan chain, the minimal polynomial and the characteristic polynomial are identical. If, however, the multiple eigenvalues of $\mathbf{A}$ are not linked in a Jordan chain, the minimal polynomial is of lower degree than the characteristic polynomial.

Using the following matrices $\mathbf{A}$ and $\mathbf{B}$ as examples, verify the foregoing statements about the minimal polynomial when multiple eigenvalues are involved.

$$\mathbf{A} = \begin{bmatrix} 2 & 1 & 4 \\ 0 & 2 & 0 \\ 0 & 3 & 1 \end{bmatrix}, \qquad \mathbf{B} = \begin{bmatrix} 2 & 0 & 0 \\ 0 & 2 & 0 \\ 0 & 3 & 1 \end{bmatrix}$$

Solution. First, consider the matrix $\mathbf{A}$. The characteristic polynomial is given by

$$|\lambda\mathbf{I} - \mathbf{A}| = \begin{vmatrix} \lambda - 2 & -1 & -4 \\ 0 & \lambda - 2 & 0 \\ 0 & -3 & \lambda - 1 \end{vmatrix} = (\lambda - 2)^2(\lambda - 1)$$

Thus the eigenvalues of $\mathbf{A}$ are 2, 2, and 1. It can be shown that the Jordan canonical form of $\mathbf{A}$ is

$$\begin{bmatrix} 2 & 1 & 0 \\ 0 & 2 & 0 \\ 0 & 0 & 1 \end{bmatrix}$$

and the multiple eigenvalues are linked in the Jordan chain as shown. (For the procedure for deriving the Jordan canonical form of $\mathbf{A}$, refer to the Appendix.)

To determine the minimal polynomial let us first obtain adj $(\lambda\mathbf{I} - \mathbf{A})$. It is given by

$$\text{adj}\,(\lambda\mathbf{I} - \mathbf{A}) = \begin{bmatrix} (\lambda - 2)(\lambda - 1) & (\lambda + 11) & 4(\lambda - 2) \\ 0 & (\lambda - 2)(\lambda - 1) & 0 \\ 0 & 3(\lambda - 2) & (\lambda - 2)^2 \end{bmatrix}$$

Notice that there is no common divisor of all the elements of adj $(\lambda\mathbf{I} - \mathbf{A})$. Hence $d(\lambda) = 1$. Thus, the minimal polynomial $\phi(\lambda)$ is identical with the characteristic polynomial, or

$$\begin{aligned} \phi(\lambda) &= |\lambda\mathbf{I} - \mathbf{A}| = (\lambda - 2)^2(\lambda - 1) \\ &= \lambda^3 - 5\lambda^2 + 8\lambda - 4 \end{aligned}$$

A simple calculation proves that

$$\mathbf{A}^3 - 5\mathbf{A}^2 + 8\mathbf{A} - 4\mathbf{I} = \mathbf{0}$$

but

$$\mathbf{A}^2 - 3\mathbf{A} + 2\mathbf{I} \neq \mathbf{0}$$

Thus, we have shown that the minimal polynomial and the characteristic polynomial of this matrix $\mathbf{A}$ are the same.

Next, consider the matrix $\mathbf{B}$. The characteristic polynomial is given by

$$|\lambda\mathbf{I} - \mathbf{B}| = \begin{vmatrix} \lambda - 2 & 0 & 0 \\ 0 & \lambda - 2 & 0 \\ 0 & -3 & \lambda - 1 \end{vmatrix} = (\lambda - 2)^2(\lambda - 1)$$

A simple computation reveals that matrix $\mathbf{B}$ has three eigenvectors, and the Jordan canonical form of $\mathbf{B}$ is given by

$$\begin{bmatrix} 2 & 0 & 0 \\ 0 & 2 & 0 \\ 0 & 0 & 1 \end{bmatrix}$$

Thus, the multiple eigenvalues are not linked. To obtain the minimal polynomial we first compute adj $(\lambda\mathbf{I} - \mathbf{B})$:

$$\text{adj}\,(\lambda\mathbf{I} - \mathbf{B}) = \begin{bmatrix} (\lambda-2)(\lambda-1) & 0 & 0 \\ 0 & (\lambda-2)(\lambda-1) & 0 \\ 0 & 3(\lambda-2) & (\lambda-2)^2 \end{bmatrix}$$

from which it is evident that

$$d(\lambda) = \lambda - 2$$

Hence

$$\phi(\lambda) = \frac{|\lambda\mathbf{I} - \mathbf{B}|}{d(\lambda)} = \frac{(\lambda-2)^2(\lambda-1)}{\lambda-2} = \lambda^2 - 3\lambda + 2$$

As a check, let us compute $\phi(\mathbf{B})$:

$$\phi(\mathbf{B}) = \mathbf{B}^2 - 3\mathbf{B} + 2\mathbf{I} = \begin{bmatrix} 4 & 0 & 0 \\ 0 & 4 & 0 \\ 0 & 9 & 1 \end{bmatrix} - 3\begin{bmatrix} 2 & 0 & 0 \\ 0 & 2 & 0 \\ 0 & 3 & 1 \end{bmatrix} + 2\begin{bmatrix} 1 & 0 & 0 \\ 0 & 1 & 0 \\ 0 & 0 & 1 \end{bmatrix} = \begin{bmatrix} 0 & 0 & 0 \\ 0 & 0 & 0 \\ 0 & 0 & 0 \end{bmatrix}$$

For the given matrix $\mathbf{B}$ the degree of the minimal polynomial is lower by 1 than that of the characteristic polynomial. As shown here, if the multiple eigenvalues of an $n \times n$ matrix are not linked in a Jordan chain, the minimal polynomial is of lower degree than the characteristic polynomial.

Problem A-5-13. Show that by use of the minimal polynomial, the inverse of a nonsingular matrix $\mathbf{A}$ can be expressed as a polynomial in $\mathbf{A}$ with scalar coefficients as follows:

$$\mathbf{A}^{-1} = -\frac{1}{a_m}(\mathbf{A}^{m-1} + a_1\mathbf{A}^{m-2} + \cdots + a_{m-2}\mathbf{A} + a_{m-1}\mathbf{I}) \tag{5–164}$$

where $a_1, a_2, \ldots, a_m$ are coefficients of the minimal polynomial

$$\phi(\lambda) = \lambda^m + a_1\lambda^{m-1} + \cdots + a_{m-1}\lambda + a_m$$

Then, obtain the inverse of the following matrix $\mathbf{A}$:

$$\mathbf{A} = \begin{bmatrix} 1 & 2 & 0 \\ 3 & -1 & -2 \\ 1 & 0 & -3 \end{bmatrix}$$

Solution. For a nonsingular matrix $\mathbf{A}$, its minimal polynomial $\phi(\mathbf{A})$ can be written as follows:

$$\phi(\mathbf{A}) = \mathbf{A}^m + a_1\mathbf{A}^{m-1} + \cdots + a_{m-1}\mathbf{A} + a_m\mathbf{I} = \mathbf{0}$$

where $a_m \neq 0$. Hence,

$$\mathbf{I} = -\frac{1}{a_m}(\mathbf{A}^m + a_1\mathbf{A}^{m-1} + \cdots + a_{m-2}\mathbf{A}^2 + a_{m-1}\mathbf{A})$$

Premultiplying by $\mathbf{A}^{-1}$, we obtain

$$\mathbf{A}^{-1} = -\frac{1}{a_m}(\mathbf{A}^{m-1} + a_1\mathbf{A}^{m-2} + \cdots + a_{m-2}\mathbf{A} + a_{m-1}\mathbf{I})$$

which is Eq. (5–164).

For the given matrix $\mathbf{A}$, adj $(\lambda\mathbf{I} - \mathbf{A})$ can be given as follows:

$$\text{adj}\,(\lambda\mathbf{I} - \mathbf{A}) = \begin{bmatrix} \lambda^2 + 4\lambda + 3 & 2\lambda + 6 & -4 \\ 3\lambda + 7 & \lambda^2 + 2\lambda - 3 & -2\lambda + 2 \\ \lambda + 1 & 2 & \lambda^2 - 7 \end{bmatrix}$$

Clearly, there is no common divisor $d(\lambda)$ of all elements of adj $(\lambda\mathbf{I} - \mathbf{A})$. Hence, $d(\lambda) = 1$. Consequently, the minimal polynomial $\phi(\lambda)$ is given by the equation

$$\phi(\lambda) = \frac{|\lambda\mathbf{I} - \mathbf{A}|}{d(\lambda)} = |\lambda\mathbf{I} - \mathbf{A}|$$

Thus, the minimal polynomial $\phi(\lambda)$ is the same as the characteristic polynomial.

Since the characteristic equation is

$$|\lambda\mathbf{I} - \mathbf{A}| = \lambda^3 + 3\lambda^2 - 7\lambda - 17 = 0$$

we obtain

$$\phi(\lambda) = \lambda^3 + 3\lambda^2 - 7\lambda - 17$$

By identifying the coefficients a_i of the minimal polynomial (which is the same as the characteristic polynomial in this case), we have

$$a_1 = 3, \qquad a_2 = -7, \qquad a_3 = -17$$

The inverse of $\mathbf{A}$ can then be obtained from Eq. (5–164) as follows:

$$\mathbf{A}^{-1} = -\frac{1}{a_3}(\mathbf{A}^2 + a_1\mathbf{A} + a_2\mathbf{I}) = \frac{1}{17}(\mathbf{A}^2 + 3\mathbf{A} - 7\mathbf{I})$$

$$= \frac{1}{17}\left\{\begin{bmatrix} 7 & 0 & -4 \\ -2 & 7 & 8 \\ -2 & 2 & 9 \end{bmatrix} + 3\begin{bmatrix} 1 & 2 & 0 \\ 3 & -1 & -2 \\ 1 & 0 & -3 \end{bmatrix} - 7\begin{bmatrix} 1 & 0 & 0 \\ 0 & 1 & 0 \\ 0 & 0 & 1 \end{bmatrix}\right\}$$

$$= \frac{1}{17}\begin{bmatrix} 3 & 6 & -4 \\ 7 & -3 & 2 \\ 1 & 2 & -7 \end{bmatrix}$$

$$= \begin{bmatrix} \frac{3}{17} & \frac{6}{17} & -\frac{4}{17} \\ \frac{7}{17} & -\frac{3}{17} & \frac{2}{17} \\ \frac{1}{17} & \frac{2}{17} & -\frac{7}{17} \end{bmatrix}$$

Problem A-5–14. Show that the inverse of $z\mathbf{I} - \mathbf{G}$ can be given by the equation

$$(z\mathbf{I} - \mathbf{G})^{-1} = \frac{\text{adj}\,(z\mathbf{I} - \mathbf{G})}{|z\mathbf{I} - \mathbf{G}|}$$

$$= \frac{\mathbf{I}z^{n-1} + \mathbf{H}_1 z^{n-2} + \mathbf{H}_2 z^{n-3} + \cdots + \mathbf{H}_{n-1}}{|z\mathbf{I} - \mathbf{G}|} \qquad (5\text{–}165)$$

where

$$\mathbf{H}_1 = \mathbf{G} + a_1\mathbf{I}$$
$$\mathbf{H}_2 = \mathbf{G}\mathbf{H}_1 + a_2\mathbf{I}$$
$$\vdots$$
$$\mathbf{H}_{n-1} = \mathbf{G}\mathbf{H}_{n-2} + a_{n-1}\mathbf{I}$$
$$\mathbf{H}_n = \mathbf{G}\mathbf{H}_{n-1} + a_n\mathbf{I} = \mathbf{0}$$

and $a_1, a_2, \ldots, a_n$ are the coefficients appearing in the characteristic polynomial given by

$$|z\mathbf{I} - \mathbf{G}| = z^n + a_1 z^{n-1} + a_2 z^{n-2} + \cdots + a_n$$

Show also that

$$a_1 = -\operatorname{tr}\mathbf{G}$$
$$a_2 = -\tfrac{1}{2}\operatorname{tr}\mathbf{G}\mathbf{H}_1$$
$$\vdots$$
$$a_n = -\frac{1}{n}\operatorname{tr}\mathbf{G}\mathbf{H}_{n-1}$$

In order to simplify the derivation, assume that $n = 3$. (The derivation can be easily extended to the case of arbitrary n.)

Solution. Note that

$$\begin{aligned}(z\mathbf{I} - \mathbf{G})(\mathbf{I}z^2 + \mathbf{H}_1 z + \mathbf{H}_2) &= z^3\mathbf{I} - z^2\mathbf{G} + z^2\mathbf{H}_1 - z\mathbf{G}\mathbf{H}_1 + z\mathbf{H}_2 - \mathbf{G}\mathbf{H}_2 \\ &= z^3\mathbf{I} - z^2\mathbf{G} + z^2(\mathbf{G} + a_1\mathbf{I}) - z\mathbf{G}(\mathbf{G} + a_1\mathbf{I}) + z[\mathbf{G}(\mathbf{G} + a_1\mathbf{I}) + a_2\mathbf{I}] \\ &\quad - \mathbf{G}[\mathbf{G}(\mathbf{G} + a_1\mathbf{I}) + a_2\mathbf{I}] \\ &= z^3\mathbf{I} + a_1 z^2\mathbf{I} + a_2 z\mathbf{I} + a_3\mathbf{I} - \mathbf{G}^3 - a_1\mathbf{G}^2 - a_2\mathbf{G} - a_3\mathbf{I} \end{aligned} \tag{5–166}$$

The Cayley-Hamilton theorem (see Prob. A-5–10) states that an $n \times n$ matrix $\mathbf{G}$ satisfies its own characteristic equation. Since $n = 3$ in the present case, the characteristic equation is

$$|z\mathbf{I} - \mathbf{G}| = z^3 + a_1 z^2 + a_2 z + a_3 = 0$$

and $\mathbf{G}$ satisfies the following equation:

$$\mathbf{G}^3 + a_1\mathbf{G}^2 + a_2\mathbf{G} + a_3\mathbf{I} = \mathbf{0}$$

Hence Eq. (5–166) simplifies to

$$(z\mathbf{I} - \mathbf{G})(\mathbf{I}z^2 + \mathbf{H}_1 z + \mathbf{H}_2) = (z^3 + a_1 z^2 + a_2 z + a_3)\mathbf{I} = |z\mathbf{I} - \mathbf{G}|\mathbf{I}$$

Consequently,

$$\mathbf{I} = \frac{(z\mathbf{I} - \mathbf{G})(\mathbf{I}z^2 + \mathbf{H}_1 z + \mathbf{H}_2)}{|z\mathbf{I} - \mathbf{G}|}$$

or

$$(z\mathbf{I}-\mathbf{G})^{-1}=\frac{\mathbf{I}z^2+\mathbf{H}_1z+\mathbf{H}_2}{|z\mathbf{I}-\mathbf{G}|}$$

which is Eq. (5–165) when $n = 3$.

Next, we shall show that

$$a_1=-\operatorname{tr}\mathbf{G}$$

$$a_2=-\tfrac{1}{2}\operatorname{tr}\mathbf{G}\mathbf{H}_1$$

$$a_3=-\tfrac{1}{3}\operatorname{tr}\mathbf{G}\mathbf{H}_2$$

We shall transform **G** into a diagonal matrix if **G** involves n linearly independent eigenvectors (where $n = 3$ in the present case) or into a matrix in a Jordan canonical form if **G** involves fewer than n linearly independent eigenvectors. That is,

$$\mathbf{P}^{-1}\mathbf{G}\mathbf{P}=\mathbf{D}=\text{matrix in diagonal form}$$

or

$$\mathbf{S}^{-1}\mathbf{G}\mathbf{S}=\mathbf{J}=\text{matrix in a Jordan canonical form}$$

where matrices **P** and **S** are nonsingular transformation matrices.

Since the following derivation applies regardless of whether matrix **G** can be transformed into a diagonal matrix or into a matrix in a Jordan canonical form, we shall use the notation

$$\mathbf{T}^{-1}\mathbf{G}\mathbf{T}=\hat{\mathbf{D}}$$

where $\hat{\mathbf{D}}$ represents either a diagonal matrix or a matrix in a Jordan canonical form, as the case may be.

In the following we shall first show that

$$\operatorname{tr}\mathbf{G}=\operatorname{tr}\hat{\mathbf{D}}$$

$$\operatorname{tr}\mathbf{G}\mathbf{H}_1=\operatorname{tr}\hat{\mathbf{D}}\hat{\mathbf{H}}_1$$

$$\operatorname{tr}\mathbf{G}\mathbf{H}_2=\operatorname{tr}\hat{\mathbf{D}}\hat{\mathbf{H}}_2$$

where

$$\hat{\mathbf{H}}_1=\hat{\mathbf{D}}+a_1\mathbf{I}$$

$$\hat{\mathbf{H}}_2=\hat{\mathbf{D}}\hat{\mathbf{H}}_1+a_2\mathbf{I}$$

Then we shall show that

$$a_1=-\operatorname{tr}\hat{\mathbf{D}}$$

$$a_2=-\tfrac{1}{2}\operatorname{tr}\hat{\mathbf{D}}\hat{\mathbf{H}}_1$$

$$a_3=-\tfrac{1}{3}\operatorname{tr}\hat{\mathbf{D}}\hat{\mathbf{H}}_2$$

Notice that since

$$\operatorname{tr}\mathbf{A}\mathbf{B}=\operatorname{tr}\mathbf{B}\mathbf{A}$$

we have

$$\text{tr}\,\mathbf{T}\hat{\mathbf{D}}\mathbf{T}^{-1} = \text{tr}\,(\mathbf{T}\hat{\mathbf{D}})(\mathbf{T}^{-1}) = \text{tr}\,(\mathbf{T}^{-1})(\mathbf{T}\hat{\mathbf{D}}) = \text{tr}\,\hat{\mathbf{D}}$$

Notice also that

$$\text{tr}\,(\mathbf{A} + \mathbf{B}) = \text{tr}\,\mathbf{A} + \text{tr}\,\mathbf{B}$$

Now we have

$$\begin{aligned}
\text{tr}\,\mathbf{G} &= \text{tr}\,\mathbf{T}\hat{\mathbf{D}}\mathbf{T}^{-1} = \text{tr}\,\hat{\mathbf{D}} \\
\text{tr}\,\mathbf{G}\mathbf{H}_1 &= \text{tr}\,\mathbf{G}(\mathbf{G} + a_1\mathbf{I}) = \text{tr}\,\mathbf{G}^2 + \text{tr}\,a_1\mathbf{G} \\
&= \text{tr}\,\mathbf{T}\hat{\mathbf{D}}^2\mathbf{T}^{-1} + \text{tr}\,a_1\mathbf{T}\hat{\mathbf{D}}\mathbf{T}^{-1} \\
&= \text{tr}\,\hat{\mathbf{D}}^2 + \text{tr}\,a_1\hat{\mathbf{D}} = \text{tr}\,(\hat{\mathbf{D}}^2 + a_1\hat{\mathbf{D}}) = \text{tr}\,\hat{\mathbf{D}}\hat{\mathbf{H}}_1 \\
\text{tr}\,\mathbf{G}\mathbf{H}_2 &= \text{tr}\,\mathbf{G}(\mathbf{G}\mathbf{H}_1 + a_2\mathbf{I}) = \text{tr}\,\mathbf{G}[\mathbf{G}(\mathbf{G} + a_1\mathbf{I}) + a_2\mathbf{I}] \\
&= \text{tr}\,(\mathbf{G}^3 + a_1\mathbf{G}^2 + a_2\mathbf{G}) \\
&= \text{tr}\,\mathbf{T}\hat{\mathbf{D}}^3\mathbf{T}^{-1} + \text{tr}\,a_1\mathbf{T}\hat{\mathbf{D}}^2\mathbf{T}^{-1} + \text{tr}\,a_2\mathbf{T}\hat{\mathbf{D}}\mathbf{T}^{-1} \\
&= \text{tr}\,\hat{\mathbf{D}}^3 + \text{tr}\,a_1\hat{\mathbf{D}}^2 + \text{tr}\,a_2\hat{\mathbf{D}} \\
&= \text{tr}\,(\hat{\mathbf{D}}^3 + a_1\hat{\mathbf{D}}^2 + a_2\hat{\mathbf{D}}) = \text{tr}\,\hat{\mathbf{D}}\hat{\mathbf{H}}_2
\end{aligned}$$

Let us write

$$\mathbf{T}^{-1}\mathbf{G}\mathbf{T} = \hat{\mathbf{D}} = \begin{bmatrix} p_1 & * & 0 \\ 0 & p_2 & * \\ 0 & 0 & p_3 \end{bmatrix}$$

where an asterisk denotes "either 0 or 1." Then

$$\begin{aligned}
|z\mathbf{I} - \hat{\mathbf{D}}| &= z^3 - (p_1 + p_2 + p_3)z^2 + (p_1p_2 + p_2p_3 + p_3p_1)z - p_1p_2p_3 \\
&= z^3 + a_1z^2 + a_2z + a_3
\end{aligned}$$

where

$$\begin{aligned}
a_1 &= -(p_1 + p_2 + p_3) \\
a_2 &= p_1p_2 + p_2p_3 + p_3p_1 \\
a_3 &= -p_1p_2p_3
\end{aligned}$$

Notice that

$$\begin{aligned}
\text{tr}\,\hat{\mathbf{D}} &= p_1 + p_2 + p_3 = -a_1 \\
\text{tr}\,\hat{\mathbf{D}}\hat{\mathbf{H}}_1 &= \text{tr}\,\hat{\mathbf{D}}(\hat{\mathbf{D}} + a_1\mathbf{I}) = \text{tr}\,\hat{\mathbf{D}}^2 + \text{tr}\,a_1\hat{\mathbf{D}} \\
&= p_1^2 + p_2^2 + p_3^2 - (p_1 + p_2 + p_3)(p_1 + p_2 + p_3) \\
&= -2(p_1p_2 + p_2p_3 + p_3p_1) = -2a_2 \\
\text{tr}\,\hat{\mathbf{D}}\hat{\mathbf{H}}_2 &= \text{tr}\,\hat{\mathbf{D}}(\hat{\mathbf{D}}\hat{\mathbf{H}}_1 + a_2\mathbf{I}) = \text{tr}\,(\hat{\mathbf{D}}^3 + a_1\hat{\mathbf{D}}^2 + a_2\hat{\mathbf{D}}) \\
&= \text{tr}\,\hat{\mathbf{D}}^3 + \text{tr}\,a_1\hat{\mathbf{D}}^2 + \text{tr}\,a_2\hat{\mathbf{D}}
\end{aligned}$$

$$
\begin{aligned}
&= (p_1^3 + p_2^3 + p_3^3) - (p_1 + p_2 + p_3)(p_1^2 + p_2^2 + p_3^2) \\
&\quad + (p_1 p_2 + p_2 p_3 + p_3 p_1)(p_1 + p_2 + p_3) \\
&= 3 p_1 p_2 p_3 = -3 a_3
\end{aligned}
$$

Thus we have shown that

$$
\begin{aligned}
a_1 &= -\mathrm{tr}\, \hat{\mathbf{D}} = -\mathrm{tr}\, \mathbf{G} \\
a_2 &= -\tfrac{1}{2}\, \mathrm{tr}\, \hat{\mathbf{D}}\hat{\mathbf{H}}_1 = -\tfrac{1}{2}\, \mathrm{tr}\, \mathbf{G}\mathbf{H}_1 \\
a_3 &= -\tfrac{1}{3}\, \mathrm{tr}\, \hat{\mathbf{D}}\hat{\mathbf{H}}_2 = -\tfrac{1}{3}\, \mathrm{tr}\, \mathbf{G}\mathbf{H}_2
\end{aligned}
$$

Problem A-5-15. Show that the inverse of $z\mathbf{I} - \mathbf{G}$ can be given by the equation

$$
(z\mathbf{I} - \mathbf{G})^{-1} = \frac{\sum_{j=0}^{m-1} z^j \sum_{i=1+j}^{m} a_{m-i}\, \mathbf{G}^{i-j-1}}{\sum_{i=0}^{m} a_{m-i} z^i} \tag{5-167}
$$

where the a_i's are the coefficients appearing in the following minimal polynomial of an $n \times n$ matrix $\mathbf{G}$:

$$
a_0 \mathbf{G}^m + a_1 \mathbf{G}^{m-1} + \cdots + a_{m-1}\mathbf{G} + a_m \mathbf{I} = \mathbf{0}
$$

where $a_0 = 1$. Note that $m \le n$.

Solution. Let us define

$$
(z\mathbf{I} - \mathbf{G})^{-1} = \mathbf{P}
$$

Then

$$
\mathbf{I} = (z\mathbf{I} - \mathbf{G})\mathbf{P}
$$

or

$$
z\mathbf{P} = \mathbf{G}\mathbf{P} + \mathbf{I}
$$

By premultiplying both sides of this last equation by $(z\mathbf{I} + \mathbf{G})$, we obtain

$$
(z\mathbf{I} + \mathbf{G})z\mathbf{P} = (z\mathbf{I} + \mathbf{G})(\mathbf{G}\mathbf{P} + \mathbf{I})
$$

or

$$
z^2\mathbf{P} = \mathbf{G}^2\mathbf{P} + \mathbf{G} + z\mathbf{I}
$$

Similarly, by premultiplying both sides of this last equation by $(z\mathbf{I} + \mathbf{G})$ and simplifying, we obtain

$$
z^3\mathbf{P} = \mathbf{G}^3\mathbf{P} + \mathbf{G}^2 + z\mathbf{G} + z^2\mathbf{I}
$$

By repeating this process we obtain the following set of equations:

$$
\begin{aligned}
\mathbf{P} &= \mathbf{P} \\
z\mathbf{P} &= \mathbf{G}\mathbf{P} + \mathbf{I}
\end{aligned}
$$

$$z^2\mathbf{P} = \mathbf{G}^2\mathbf{P} + \mathbf{G} + z\mathbf{I}$$

$$z^3\mathbf{P} = \mathbf{G}^3\mathbf{P} + \mathbf{G}^2 + z\mathbf{G} + z^2\mathbf{I}$$

$$\vdots$$

$$z^m\mathbf{P} = \mathbf{G}^m\mathbf{P} + \mathbf{G}^{m-1} + z\mathbf{G}^{m-2} + \cdots + z^{m-2}\mathbf{G} + z^{m-1}\mathbf{I}$$

where m is the degree of the minimal polynomial of $\mathbf{G}$. Then, by multiplying the $z^i\mathbf{P}$'s by a_{m-i} (where $i = 0, 1, 2, \ldots, m$) in the above $m + 1$ equations and adding the products together, we get

$$a_m\mathbf{P} + a_{m-1}z\mathbf{P} + a_{m-2}z^2\mathbf{P} + \cdots + a_0z^m\mathbf{P} = \sum_{i=0}^{m} a_{m-i}\mathbf{G}^i\mathbf{P} + \sum_{i=1}^{m} a_{m-i}\mathbf{G}^{i-1}$$
$$+ z\sum_{i=2}^{m} a_{m-i}\mathbf{G}^{i-2} + \cdots + z^{m-2}\sum_{i=m-1}^{m} a_{m-i}\mathbf{G}^{i-m+1} + z^{m-1}a_0\mathbf{I} \qquad (5\text{–}168)$$

Noting that

$$\sum_{i=0}^{m} a_{m-i}\mathbf{G}^i\mathbf{P} = (a_0\mathbf{G}^m + a_1\mathbf{G}^{m-1} + \cdots + a_{m-1}\mathbf{G} + a_m\mathbf{I})\mathbf{P} = \mathbf{0}$$

we can simplify Eq. (5–168) to

$$\sum_{i=0}^{m} a_{m-i}z^i\mathbf{P} = \sum_{j=0}^{m-1} z^j \sum_{i=1+j}^{m} a_{m-i}\mathbf{G}^{i-j-1}$$

Therefore

$$(z\mathbf{I} - \mathbf{G})^{-1} = \mathbf{P} = \frac{\sum_{j=0}^{m-1} z^j \sum_{i=1+j}^{m} a_{m-i}\mathbf{G}^{i-j-1}}{\sum_{i=0}^{m} a_{m-i}z^i}$$

If the minimal polynomial and the characteristic polynomial of $\mathbf{G}$ are identical, then $m = n$, and in this case

$$(z\mathbf{I} - \mathbf{G})^{-1} = \frac{\sum_{j=0}^{n-1} z^j \sum_{i=1+j}^{n} a_{n-i}\mathbf{G}^{i-j-1}}{|z\mathbf{I} - \mathbf{G}|}$$

where

$$|z\mathbf{I} - \mathbf{G}| = a_0z^n + a_1z^{n-1} + \cdots + a_{n-1}z + a_n$$
$$= \sum_{i=0}^{n} a_{n-i}z^i = \sum_{i=0}^{n} a_iz^{n-i}$$

Note that $a_0 = 1$.

Problem A-5-16. Consider the following polynomial of degree $m - 1$, where we assume $\lambda_1, \lambda_2, \ldots, \lambda_m$ to be distinct:

$$p_k(\lambda) = \frac{(\lambda - \lambda_1)\cdots(\lambda - \lambda_{k-1})(\lambda - \lambda_{k+1})\cdots(\lambda - \lambda_m)}{(\lambda_k - \lambda_1)\cdots(\lambda_k - \lambda_{k-1})(\lambda_k - \lambda_{k+1})\cdots(\lambda_k - \lambda_m)}$$

where $k = 1, 2, \ldots, m$. Notice that

$$p_k(\lambda_i) = \begin{cases} 1 & \text{if } i = k \\ 0 & \text{if } i \neq k \end{cases}$$

Then the polynomial $f(\lambda)$ of degree $m - 1$

$$\begin{aligned} f(\lambda) &= \sum_{k=1}^{m} f(\lambda_k) p_k(\lambda) \\ &= \sum_{k=1}^{m} f(\lambda_k) \frac{(\lambda - \lambda_1)\cdots(\lambda - \lambda_{k-1})(\lambda - \lambda_{k+1})\cdots(\lambda - \lambda_m)}{(\lambda_k - \lambda_1)\cdots(\lambda_k - \lambda_{k-1})(\lambda_k - \lambda_{k+1})\cdots(\lambda_k - \lambda_m)} \end{aligned}$$

takes on the values $f(\lambda_k)$ at the points λ_k. This last equation is commonly called Lagrange's interpolation formula. The polynomial $f(\lambda)$ of degree $m - 1$ is determined from m independent data $f(\lambda_1), f(\lambda_2), \ldots, f(\lambda_m)$. That is, the polynomial $f(\lambda)$ passes through m points $f(\lambda_1)$, $f(\lambda_2) \ldots, f(\lambda_m)$. Since $f(\lambda)$ is a polynomial of degree $m - 1$, it is uniquely determined. Any other representations of the polynomial of degree $m - 1$ can be reduced to the Lagrange polynomial $f(\lambda)$.

Assuming that the eigenvalues of an $n \times n$ matrix **A** are distinct, substitute **A** for λ in the polynomial $p_k(\lambda)$. Then we get

$$p_k(\mathbf{A}) = \frac{(\mathbf{A} - \lambda_1\mathbf{I})\cdots(\mathbf{A} - \lambda_{k-1}\mathbf{I})(\mathbf{A} - \lambda_{k+1}\mathbf{I})\cdots(\mathbf{A} - \lambda_m\mathbf{I})}{(\lambda_k - \lambda_1)\cdots(\lambda_k - \lambda_{k-1})(\lambda_k - \lambda_{k+1})\cdots(\lambda_k - \lambda_m)}$$

Notice that $p_k(\mathbf{A})$ is a polynomial in **A** of degree $m - 1$. Notice also that

$$p_k(\lambda_i\mathbf{I}) = \begin{cases} \mathbf{I} & \text{if } i = k \\ \mathbf{0} & \text{if } i \neq k \end{cases}$$

Now define

$$\begin{aligned} f(\mathbf{A}) &= \sum_{k=1}^{m} f(\lambda_k) p_k(\mathbf{A}) \\ &= \sum_{k=1}^{m} f(\lambda_k) \frac{(\mathbf{A} - \lambda_1\mathbf{I})\cdots(\mathbf{A} - \lambda_{k-1}\mathbf{I})(\mathbf{A} - \lambda_{k+1}\mathbf{I})\cdots(\mathbf{A} - \lambda_m\mathbf{I})}{(\lambda_k - \lambda_1)\cdots(\lambda_k - \lambda_{k-1})(\lambda_k - \lambda_{k+1})\cdots(\lambda_k - \lambda_m)} \end{aligned} \tag{5–169}$$

Equation (5–169) is known as Sylvester's interpolation formula. Equation (5–169) is equivalent to the following equation:

$$\begin{vmatrix} 1 & 1 & \cdots & 1 & \mathbf{I} \\ \lambda_1 & \lambda_2 & \cdots & \lambda_m & \mathbf{A} \\ \lambda_1^2 & \lambda_2^2 & \cdots & \lambda_m^2 & \mathbf{A}^2 \\ \vdots & \vdots & & \vdots & \vdots \\ \lambda_1^{m-1} & \lambda_2^{m-1} & \cdots & \lambda_m^{m-1} & \mathbf{A}^{m-1} \\ f(\lambda_1) & f(\lambda_2) & \cdots & f(\lambda_m) & f(\mathbf{A}) \end{vmatrix} = \mathbf{0} \tag{5–170}$$

Equations (5–169) and (5–170) are frequently used for evaluating functions $f(\mathbf{A})$ of matrix $\mathbf{A}$, for example, $(\lambda\mathbf{I} - \mathbf{A})^{-1}$, $e^{\mathbf{A}t}$, and so forth. Note that Eq. (5–170) can also be written as follows:

$$\begin{vmatrix} 1 & \lambda_1 & \lambda_1^2 & \cdots & \lambda_1^{m-1} & f(\lambda_1) \\ 1 & \lambda_2 & \lambda_2^2 & \cdots & \lambda_2^{m-1} & f(\lambda_2) \\ \cdot & \cdot & \cdot & & \cdot & \cdot \\ \cdot & \cdot & \cdot & & \cdot & \cdot \\ \cdot & \cdot & \cdot & & \cdot & \cdot \\ 1 & \lambda_m & \lambda_m^2 & \cdots & \lambda_m^{m-1} & f(\lambda_m) \\ \mathbf{I} & \mathbf{A} & \mathbf{A}^2 & \cdots & \mathbf{A}^{m-1} & f(\mathbf{A}) \end{vmatrix} = \mathbf{0} \tag{5–171}$$

Show that Eqs. (5–169) and (5–170) are equivalent. In order to simplify the arguments, assume that $m = 4$.

Solution. Equation (5–170), when $m = 4$, can be expanded as follows:

$$\Delta = \begin{vmatrix} 1 & 1 & 1 & 1 & \mathbf{I} \\ \lambda_1 & \lambda_2 & \lambda_3 & \lambda_4 & \mathbf{A} \\ \lambda_1^2 & \lambda_2^2 & \lambda_3^2 & \lambda_4^2 & \mathbf{A}^2 \\ \lambda_1^3 & \lambda_2^3 & \lambda_3^3 & \lambda_4^3 & \mathbf{A}^3 \\ f(\lambda_1) & f(\lambda_2) & f(\lambda_3) & f(\lambda_4) & f(\mathbf{A}) \end{vmatrix}$$

$$= f(\mathbf{A})\begin{vmatrix} 1 & 1 & 1 & 1 \\ \lambda_1 & \lambda_2 & \lambda_3 & \lambda_4 \\ \lambda_1^2 & \lambda_2^2 & \lambda_3^2 & \lambda_4^2 \\ \lambda_1^3 & \lambda_2^3 & \lambda_3^3 & \lambda_4^3 \end{vmatrix} - f(\lambda_4)\begin{vmatrix} 1 & 1 & 1 & \mathbf{I} \\ \lambda_1 & \lambda_2 & \lambda_3 & \mathbf{A} \\ \lambda_1^2 & \lambda_2^2 & \lambda_3^2 & \mathbf{A}^2 \\ \lambda_1^3 & \lambda_2^3 & \lambda_3^3 & \mathbf{A}^3 \end{vmatrix}$$

$$+ f(\lambda_3)\begin{vmatrix} 1 & 1 & 1 & \mathbf{I} \\ \lambda_1 & \lambda_2 & \lambda_4 & \mathbf{A} \\ \lambda_1^2 & \lambda_2^2 & \lambda_4^2 & \mathbf{A}^2 \\ \lambda_1^3 & \lambda_2^3 & \lambda_4^3 & \mathbf{A}^3 \end{vmatrix} - f(\lambda_2)\begin{vmatrix} 1 & 1 & 1 & \mathbf{I} \\ \lambda_1 & \lambda_3 & \lambda_4 & \mathbf{A} \\ \lambda_1^2 & \lambda_3^2 & \lambda_4^2 & \mathbf{A}^2 \\ \lambda_1^3 & \lambda_3^3 & \lambda_4^3 & \mathbf{A}^3 \end{vmatrix}$$

$$+ f(\lambda_1)\begin{vmatrix} 1 & 1 & 1 & \mathbf{I} \\ \lambda_2 & \lambda_3 & \lambda_4 & \mathbf{A} \\ \lambda_2^2 & \lambda_3^2 & \lambda_4^2 & \mathbf{A}^2 \\ \lambda_2^3 & \lambda_3^3 & \lambda_4^3 & \mathbf{A}^3 \end{vmatrix}$$

Since

$$\begin{vmatrix} 1 & 1 & 1 & 1 \\ \lambda_1 & \lambda_2 & \lambda_3 & \lambda_4 \\ \lambda_1^2 & \lambda_2^2 & \lambda_3^2 & \lambda_4^2 \\ \lambda_1^3 & \lambda_2^3 & \lambda_3^3 & \lambda_4^3 \end{vmatrix} = (\lambda_4 - \lambda_3)(\lambda_4 - \lambda_2)(\lambda_4 - \lambda_1)(\lambda_3 - \lambda_2)(\lambda_3 - \lambda_1)(\lambda_2 - \lambda_1)$$

and

$$\begin{vmatrix} 1 & 1 & 1 & \mathbf{I} \\ \lambda_i & \lambda_j & \lambda_k & \mathbf{A} \\ \lambda_i^2 & \lambda_j^2 & \lambda_k^2 & \mathbf{A}^2 \\ \lambda_i^3 & \lambda_j^3 & \lambda_k^3 & \mathbf{A}^3 \end{vmatrix} = (\mathbf{A} - \lambda_k\mathbf{I})(\mathbf{A} - \lambda_j\mathbf{I})(\mathbf{A} - \lambda_i\mathbf{I})(\lambda_k - \lambda_j)(\lambda_k - \lambda_i)(\lambda_j - \lambda_i)$$

we obtain

$$\begin{aligned}\Delta = f(\mathbf{A})[(\lambda_4-\lambda_3)(\lambda_4-\lambda_2)(\lambda_4-\lambda_1)(\lambda_3-\lambda_2)(\lambda_3-\lambda_1)(\lambda_2-\lambda_1)] \\ -f(\lambda_4)[(\mathbf{A}-\lambda_3\mathbf{I})(\mathbf{A}-\lambda_2\mathbf{I})(\mathbf{A}-\lambda_1\mathbf{I})(\lambda_3-\lambda_2)(\lambda_3-\lambda_1)(\lambda_2-\lambda_1)] \\ +f(\lambda_3)[(\mathbf{A}-\lambda_4\mathbf{I})(\mathbf{A}-\lambda_2\mathbf{I})(\mathbf{A}-\lambda_1\mathbf{I})(\lambda_4-\lambda_2)(\lambda_4-\lambda_1)(\lambda_2-\lambda_1)] \\ -f(\lambda_2)[(\mathbf{A}-\lambda_4\mathbf{I})(\mathbf{A}-\lambda_3\mathbf{I})(\mathbf{A}-\lambda_1\mathbf{I})(\lambda_4-\lambda_3)(\lambda_4-\lambda_1)(\lambda_3-\lambda_1)] \\ +f(\lambda_1)[(\mathbf{A}-\lambda_4\mathbf{I})(\mathbf{A}-\lambda_3\mathbf{I})(\mathbf{A}-\lambda_2\mathbf{I})(\lambda_4-\lambda_3)(\lambda_4-\lambda_2)(\lambda_3-\lambda_2)] \\ = 0\end{aligned}$$

Solving this last equation for $\mathbf{f}(\mathbf{A})$, we obtain

$$\begin{aligned}f(\mathbf{A}) &= f(\lambda_1)\frac{(\mathbf{A}-\lambda_2\mathbf{I})(\mathbf{A}-\lambda_3\mathbf{I})(\mathbf{A}-\lambda_4\mathbf{I})}{(\lambda_1-\lambda_2)(\lambda_1-\lambda_3)(\lambda_1-\lambda_4)} \\ &+ f(\lambda_2)\frac{(\mathbf{A}-\lambda_1\mathbf{I})(\mathbf{A}-\lambda_3\mathbf{I})(\mathbf{A}-\lambda_4\mathbf{I})}{(\lambda_2-\lambda_1)(\lambda_2-\lambda_3)(\lambda_2-\lambda_4)} \\ &+ f(\lambda_3)\frac{(\mathbf{A}-\lambda_1\mathbf{I})(\mathbf{A}-\lambda_2\mathbf{I})(\mathbf{A}-\lambda_4\mathbf{I})}{(\lambda_3-\lambda_1)(\lambda_3-\lambda_2)(\lambda_3-\lambda_4)} \\ &+ f(\lambda_4)\frac{(\mathbf{A}-\lambda_1\mathbf{I})(\mathbf{A}-\lambda_2\mathbf{I})(\mathbf{A}-\lambda_3\mathbf{I})}{(\lambda_4-\lambda_1)(\lambda_4-\lambda_2)(\lambda_4-\lambda_3)} \\ &= \sum_{k=1}^{m} f(\lambda_k)\frac{(\mathbf{A}-\lambda_1\mathbf{I})\cdots(\mathbf{A}-\lambda_{k-1}\mathbf{I})(\mathbf{A}-\lambda_{k+1}\mathbf{I})\cdots(\mathbf{A}-\lambda_m\mathbf{I})}{(\lambda_k-\lambda_1)\cdots(\lambda_k-\lambda_{k-1})(\lambda_k-\lambda_{k+1})\cdots(\lambda_k-\lambda_m)}\end{aligned}$$

where $m = 4$. Thus we have shown the equivalence of Eqs. (5–169) and (5–170). Although we assumed $m = 4$, the entire argument can be extended to an arbitrary positive integer m. (For the case where the matrix $\mathbf{A}$ involves multiple eigenvalues, refer to Prob. A-5–17.)

Problem A-5–17. Consider Sylvester's interpolation formula in the form given by Eq. (5–171):

$$\begin{vmatrix} 1 & \lambda_1 & \lambda_1^2 & \cdots & \lambda_1^{m-1} & f(\lambda_1) \\ 1 & \lambda_2 & \lambda_2^2 & \cdots & \lambda_2^{m-1} & f(\lambda_2) \\ \cdot & \cdot & \cdot & & \cdot & \cdot \\ \cdot & \cdot & \cdot & & \cdot & \cdot \\ \cdot & \cdot & \cdot & & \cdot & \cdot \\ 1 & \lambda_m & \lambda_m^2 & \cdots & \lambda_m^{m-1} & f(\lambda_m) \\ \mathbf{I} & \mathbf{A} & \mathbf{A}^2 & \cdots & \mathbf{A}^{m-1} & f(\mathbf{A}) \end{vmatrix} = \mathbf{0}$$

This formula for the determination of $f(\mathbf{A})$ applies to the case where the minimal polynomial of $\mathbf{A}$ involves only distinct roots.

Suppose the minimal polynomial of $\mathbf{A}$ involves multiple roots. Then the rows in the determinant that correspond to the multiple roots become identical, and therefore modification of the determinant in Eq. (5–171) becomes necessary.

Modify the form of Sylvester's interpolation formula given by Eq. (5–171) when the minimal polynomial of $\mathbf{A}$ involves multiple roots. In deriving a modified determinant equation, assume that there are three equal roots ($\lambda_1 = \lambda_2 = \lambda_3$) in the minimal polynomial of $\mathbf{A}$ and that there are other roots ($\lambda_4, \lambda_5, \ldots, \lambda_m$) that are distinct.

Solution. Since the minimal polynomial of **A** involves three equal roots, the minimal polynomial $\phi(\lambda)$ can be written as follows:

$$\phi(\lambda) = \lambda^m + a_1\lambda^{m-1} + \cdots + a_{m-1}\lambda + a_m$$
$$= (\lambda - \lambda_1)^3(\lambda - \lambda_4)(\lambda - \lambda_5)\cdots(\lambda - \lambda_m)$$

An arbitrary function $f(\mathbf{A})$ of an $n \times n$ matrix **A** can be written as

$$f(\mathbf{A}) = g(\mathbf{A})\phi(\mathbf{A}) + \alpha(\mathbf{A})$$

where the minimal polynomial $\phi(\mathbf{A})$ is of degree m and $\alpha(\mathbf{A})$ is a polynomial in **A** of degree $m - 1$ or less. Hence we have

$$f(\lambda) = g(\lambda)\phi(\lambda) + \alpha(\lambda)$$

where $\alpha(\lambda)$ is a polynomial in λ of degree $m - 1$ or less which can thus be written as follows:

$$\alpha(\lambda) = \alpha_0 + \alpha_1\lambda + \alpha_2\lambda^2 + \cdots + \alpha_{m-1}\lambda^{m-1} \qquad (5\text{–}172)$$

In the present case we have

$$f(\lambda) = g(\lambda)\phi(\lambda) + \alpha(\lambda)$$
$$= g(\lambda)[(\lambda - \lambda_1)^3(\lambda - \lambda_4)\cdots(\lambda - \lambda_m)] + \alpha(\lambda) \qquad (5\text{–}173)$$

By substituting $\lambda_1, \lambda_4, \ldots, \lambda_m$ for λ in Eq. (5–173), we obtain the following $m - 2$ equations:

$$\begin{aligned} f(\lambda_1) &= \alpha(\lambda_1) \\ f(\lambda_4) &= \alpha(\lambda_4) \\ &\ \ \vdots \\ f(\lambda_m) &= \alpha(\lambda_m) \end{aligned} \qquad (5\text{–}174)$$

By differentiating Eq. (5–173) with respect to λ, we obtain

$$\frac{d}{d\lambda} f(\lambda) = (\lambda - \lambda_1)^2 h(\lambda) + \frac{d}{d\lambda}\alpha(\lambda) \qquad (5\text{–}175)$$

where

$$(\lambda - \lambda_1)^2 h(\lambda) = \frac{d}{d\lambda}[g(\lambda)(\lambda - \lambda_1)^3(\lambda - \lambda_4)\cdots(\lambda - \lambda_m)]$$

Substitution of λ_1 for λ in Eq. (5–175) gives

$$\frac{d}{d\lambda} f(\lambda)\bigg|_{\lambda=\lambda_1} = f'(\lambda_1) = \frac{d}{d\lambda}\alpha(\lambda)\bigg|_{\lambda=\lambda_1}$$

Referring to Eq. (5–172), this last equation becomes

$$f'(\lambda_1) = \alpha_1 + 2\alpha_2\lambda_1 + \cdots + (m - 1)\alpha_{m-1}\lambda_1^{m-2} \qquad (5\text{–}176)$$

Similarly, differentiating Eq. (5–173) twice with respect to λ and substituting λ_1 for λ, we obtain

$$\left.\frac{d^2}{d\lambda^2}f(\lambda)\right|_{\lambda=\lambda_1}=f''(\lambda_1)=\left.\frac{d^2}{d\lambda^2}\alpha(\lambda)\right|_{\lambda=\lambda_1}$$

This last equation can be written as follows:

$$f''(\lambda_1)=2\alpha_2+6\alpha_3\lambda_1+\cdots+(m-1)(m-2)\alpha_{m-1}\lambda_1^{m-3} \tag{5–177}$$

Rewriting Eqs. (5–177), (5–176), and (5–174), we get

$$\begin{aligned}
\alpha_2+3\alpha_3\lambda_1+\cdots+\frac{(m-1)(m-2)}{2}\alpha_{m-1}\lambda_1^{m-3}&=\frac{f''(\lambda_1)}{2}\\
\alpha_1+2\alpha_2\lambda_1+\cdots+(m-1)\alpha_{m-1}\lambda_1^{m-2}&=f'(\lambda_1)\\
\alpha_0+\alpha_1\lambda_1+\alpha_2\lambda_1^2+\cdots+\alpha_{m-1}\lambda_1^{m-1}&=f(\lambda_1)\\
\alpha_0+\alpha_1\lambda_4+\alpha_2\lambda_4^2+\cdots+\alpha_{m-1}\lambda_4^{m-1}&=f(\lambda_4)\\
&\;\;\vdots\\
\alpha_0+\alpha_1\lambda_m+\alpha_2\lambda_m^2+\cdots+\alpha_{m-1}\lambda_m^{m-1}&=f(\lambda_m)
\end{aligned} \tag{5–178}$$

These m simultaneous equations determine the α_k values (where $k=0, 1, 2, \ldots, m-1$). Noting that $\phi(\mathbf{A})=\mathbf{0}$ because it is a minimal polynomial, we have $f(\mathbf{A})$ as follows:

$$f(\mathbf{A})=g(\mathbf{A})\phi(\mathbf{A})+\alpha(\mathbf{A})=\alpha(\mathbf{A})$$

Hence, referring to Eq. (5–172) we have

$$f(\mathbf{A})=\alpha(\mathbf{A})=\alpha_0\mathbf{I}+\alpha_1\mathbf{A}+\alpha_2\mathbf{A}^2+\cdots+\alpha_{m-1}\mathbf{A}^{m-1} \tag{5–179}$$

where the α_k values are given in terms of $f(\lambda_1)$, $f'(\lambda_1)$, $f''(\lambda_1)$, $f(\lambda_4)$, $f(\lambda_5)$, $\ldots$, $f(\lambda_m)$. In terms of the determinant equation, $f(\mathbf{A})$ can be obtained by solving the following equation:

$$\begin{vmatrix}
0 & 0 & 1 & 3\lambda_1 & \cdots & \frac{(m-1)(m-2)}{2}\lambda_1^{m-3} & \frac{f''(\lambda_1)}{2}\\
0 & 1 & 2\lambda_1 & 3\lambda_1^2 & \cdots & (m-1)\lambda_1^{m-2} & f'(\lambda_1)\\
1 & \lambda_1 & \lambda_1^2 & \lambda_1^3 & \cdots & \lambda_1^{m-1} & f(\lambda_1)\\
1 & \lambda_4 & \lambda_4^2 & \lambda_4^3 & \cdots & \lambda_4^{m-1} & f(\lambda_4)\\
\vdots & \vdots & \vdots & \vdots & & \vdots & \vdots\\
1 & \lambda_m & \lambda_m^2 & \lambda_m^3 & \cdots & \lambda_m^{m-1} & f(\lambda_m)\\
\mathbf{I} & \mathbf{A} & \mathbf{A}^2 & \mathbf{A}^3 & \cdots & \mathbf{A}^{m-1} & f(\mathbf{A})
\end{vmatrix}=\mathbf{0} \tag{5–180}$$

Equation (5–180) shows the desired modification in the form of the determinant. This equation gives the form of Sylvester's interpolation formula when the minimal polynomial of $\mathbf{A}$ involves three equal roots. (The necessary modification of the form of the determinant for other cases will be apparent.)

Problem A-5–18. Using a form of Sylvester's interpolation formula (refer to Sec. 5–5 and Probs. A-5–16 and A-5–17), compute $e^{\mathbf{A}t}$, where

$$\mathbf{A} = \begin{bmatrix} 2 & 1 & 4 \\ 0 & 2 & 0 \\ 0 & 3 & 1 \end{bmatrix}$$

Solution. It can easily be shown that the characteristic polynomial and the minimal polynomial are the same for this **A**. The minimal polynomial (characteristic polynomial) is given by

$$\phi(\lambda) = (\lambda - 2)^2(\lambda - 1)$$

Note that $\lambda_1 = \lambda_2 = 2$ and $\lambda_3 = 1$. By referring to Sylvester's interpolation formula as presented in Sec. 5–5 and Prob. A-5–17, $e^{\mathbf{A}t}$ can be given by

$$e^{\mathbf{A}t} = \alpha_0 \mathbf{I} + \alpha_1 \mathbf{A} + \alpha_2 \mathbf{A}^2$$

where α_0, α_1, and α_2 are determined from the equations

$$\alpha_1 + 2\alpha_2\lambda_1 = te^{\lambda_1 t}$$

$$\alpha_0 + \alpha_1\lambda_1 + \alpha_2\lambda_1^2 = e^{\lambda_1 t}$$

$$\alpha_0 + \alpha_1\lambda_3 + \alpha_2\lambda_3^2 = e^{\lambda_3 t}$$

Substituting $\lambda_1 = 2$, and $\lambda_3 = 1$ into these three equations gives

$$\alpha_1 + 4\alpha_2 = te^{2t}$$

$$\alpha_0 + 2\alpha_1 + 4\alpha_2 = e^{2t}$$

$$\alpha_0 + \alpha_1 + \alpha_2 = e^t$$

Solving for α_0, α_1, and α_2, we obtain

$$\alpha_0 = 4e^t - 3e^{2t} + 2te^{2t}$$

$$\alpha_1 = -4e^t + 4e^{2t} - 3te^{2t}$$

$$\alpha_2 = e^t - e^{2t} + te^{2t}$$

Hence,

$$\begin{aligned} e^{\mathbf{A}t} &= (4e^t - 3e^{2t} + 2te^{2t}) \begin{bmatrix} 1 & 0 & 0 \\ 0 & 1 & 0 \\ 0 & 0 & 1 \end{bmatrix} \\ &\quad + (-4e^t + 4e^{2t} - 3te^{2t}) \begin{bmatrix} 2 & 1 & 4 \\ 0 & 2 & 0 \\ 0 & 3 & 1 \end{bmatrix} \\ &\quad + (e^t - e^{2t} + te^{2t}) \begin{bmatrix} 4 & 16 & 12 \\ 0 & 4 & 0 \\ 0 & 9 & 1 \end{bmatrix} \\ &= \begin{bmatrix} e^{2t} & 12e^t - 12e^{2t} + 13te^{2t} & -4e^t + 4e^{2t} \\ 0 & e^{2t} & 0 \\ 0 & -3e^t + 3e^{2t} & e^t \end{bmatrix} \end{aligned}$$

Problem A-5-19. Consider the system

$$\dot{\mathbf{x}} = \mathbf{A}\mathbf{x}$$

where the eigenvalues $\lambda_1, \lambda_2, \ldots, \lambda_n$ of matrix $\mathbf{A}$ are distinct. Let $\mathbf{P}$ be a diagonalizing matrix and define the state vector $\hat{\mathbf{x}}$ by

$$\mathbf{x} = \mathbf{P}\hat{\mathbf{x}}$$

Then

$$\dot{\hat{\mathbf{x}}} = \mathbf{P}^{-1}\mathbf{A}\mathbf{P}\hat{\mathbf{x}} = \mathbf{D}\hat{\mathbf{x}}$$

where

$$\mathbf{D} = \mathbf{P}^{-1}\mathbf{A}\mathbf{P} = \text{diagonal matrix} = \begin{bmatrix} \lambda_1 & & & & 0 \\ & \lambda_2 & & & \\ & & \cdot & & \\ & & & \cdot & \\ & & & & \cdot \\ 0 & & & & \lambda_n \end{bmatrix}$$

Show that

$$\mathbf{P}^{-1}e^{\mathbf{A}t}\mathbf{P} = e^{\mathbf{D}t} = \begin{bmatrix} e^{\lambda_1 t} & & & & 0 \\ & e^{\lambda_2 t} & & & \\ & & \cdot & & \\ & & & \cdot & \\ & & & & \cdot \\ 0 & & & & e^{\lambda_n t} \end{bmatrix} \tag{5-181}$$

Next, define

$$\mathbf{G} = e^{\mathbf{A}T}$$

Show also that

$$|z\mathbf{I} - \mathbf{G}| = (z - e^{\lambda_1 T})(z - e^{\lambda_2 T}) \ldots (z - e^{\lambda_n T}) \tag{5-182}$$

Solution. To derive Eq. (5-181), notice that

$$\mathbf{P}^{-1}e^{\mathbf{A}t}\mathbf{P} = \mathbf{P}^{-1}(\mathbf{I} + \mathbf{A}t + \frac{1}{2!}\mathbf{A}^2t^2 + \cdots)\mathbf{P}$$

$$= \mathbf{I} + \mathbf{P}^{-1}\mathbf{A}\mathbf{P}t + \frac{1}{2!}\mathbf{P}^{-1}\mathbf{A}^2\mathbf{P}t^2 + \cdots$$

$$= \begin{bmatrix} 1 & & & & 0 \\ & 1 & & & \\ & & \cdot & & \\ & & & \cdot & \\ & & & & \cdot \\ 0 & & & & 1 \end{bmatrix} + \begin{bmatrix} \lambda_1 t & & & & 0 \\ & \lambda_2 t & & & \\ & & \cdot & & \\ & & & \cdot & \\ & & & & \cdot \\ 0 & & & & \lambda_n t \end{bmatrix} + \frac{1}{2!}\begin{bmatrix} \lambda_1^2 t^2 & & & & 0 \\ & \lambda_2^2 t^2 & & & \\ & & \cdot & & \\ & & & \cdot & \\ & & & & \cdot \\ 0 & & & & \lambda_n^2 t^2 \end{bmatrix}$$

$$+ \cdots$$

$$= \begin{bmatrix} e^{\lambda_1 t} & & & & 0 \\ & e^{\lambda_2 t} & & & \\ & & \cdot & & \\ & & & \cdot & \\ & & & & \cdot \\ 0 & & & & e^{\lambda_n t} \end{bmatrix}$$

Next, we derive Eq. (5–182):

$$|z\mathbf{I} - \mathbf{G}| = |\mathbf{P}^{-1}||z\mathbf{I} - e^{\mathbf{A}T}||\mathbf{P}| = |z\mathbf{I} - \mathbf{P}^{-1}e^{\mathbf{A}T}\mathbf{P}|$$

$$= \left| z\mathbf{I} - \begin{bmatrix} e^{\lambda_1 T} & & & & 0 \\ & e^{\lambda_2 T} & & & \\ & & \cdot & & \\ & & & \cdot & \\ & & & & \cdot \\ 0 & & & & e^{\lambda_n T} \end{bmatrix} \right|$$

$$= \begin{vmatrix} z - e^{\lambda_1 T} & & & & 0 \\ & z - e^{\lambda_2 T} & & & \\ & & \cdot & & \\ & & & \cdot & \\ & & & & \cdot \\ 0 & & & & z - e^{\lambda_n T} \end{vmatrix}$$

$$= (z - e^{\lambda_1 T})(z - e^{\lambda_2 T}) \cdots (z - e^{\lambda_n T})$$

Notice that the eigenvalues of **G** are given by $z_i = e^{\lambda_i T}$, where $i = 1, 2, \ldots, n$. We see the equivalence of Re $\lambda_i < 0$ and $|z_i| < 1$. Thus, the discrete-time system obtained by discretizing an asymptotically stable continuous-time system is also asymptotically stable.

Problem A-5–20. Consider the following matrix **A**:

$$\mathbf{A} = \begin{bmatrix} 5 & 4 & 0 \\ 0 & 1 & 0 \\ -4 & 4 & 1 \end{bmatrix}$$

The characteristic equation for this matrix is

$$|\lambda\mathbf{I} - \mathbf{A}| = (\lambda - 1)^2(\lambda - 5)$$

The following transformation matrix **S**:

$$\mathbf{S} = \begin{bmatrix} 0 & 4 & 16 \\ 0 & -4 & 0 \\ -32 & 4 & -16 \end{bmatrix}$$

will transform matrix **A** into a Jordan canonical form:

$$\mathbf{S}^{-1}\mathbf{A}\mathbf{S} = \mathbf{J} = \begin{bmatrix} 1 & 1 & 0 \\ 0 & 1 & 0 \\ 0 & 0 & 5 \end{bmatrix}$$

(For the derivation of this transformation matrix **S**, see the Appendix.)

Obtain $e^{\mathbf{A}t}$ by using the relationship

$$e^{\mathbf{A}t} = \mathbf{S}e^{\mathbf{J}t}\mathbf{S}^{-1}$$

Solution. First note that the inverse of matrix **S** is

$$\mathbf{S}^{-1} = \begin{bmatrix} -\frac{1}{32} & -\frac{1}{16} & -\frac{1}{32} \\ 0 & -\frac{1}{4} & 0 \\ \frac{1}{16} & \frac{1}{16} & 0 \end{bmatrix}$$

For matrix **J**, where

$$\mathbf{J} = \begin{bmatrix} 1 & 1 & 0 \\ 0 & 1 & 0 \\ 0 & 0 & 5 \end{bmatrix}$$

we have

$$e^{\mathbf{J}t} = \begin{bmatrix} e^t & te^t & 0 \\ 0 & e^t & 0 \\ 0 & 0 & e^{5t} \end{bmatrix}$$

Hence

$$\begin{aligned} e^{\mathbf{A}t} &= \mathbf{S}e^{\mathbf{J}t}\mathbf{S}^{-1} \\ &= \begin{bmatrix} 0 & 4 & 16 \\ 0 & -4 & 0 \\ -32 & 4 & -16 \end{bmatrix} \begin{bmatrix} e^t & te^t & 0 \\ 0 & e^t & 0 \\ 0 & 0 & e^{5t} \end{bmatrix} \begin{bmatrix} -\frac{1}{32} & -\frac{1}{16} & -\frac{1}{32} \\ 0 & -\frac{1}{4} & 0 \\ \frac{1}{16} & \frac{1}{16} & 0 \end{bmatrix} \\ &= \begin{bmatrix} e^{5t} & -e^t + e^{5t} & 0 \\ 0 & e^t & 0 \\ e^t - e^{5t} & e^t + 8te^t - e^{5t} & e^t \end{bmatrix} \end{aligned}$$

Problem A-5–21. Consider the state equation

$$\mathbf{x}(k+1) = \mathbf{G}(k)\mathbf{x}(k)$$

The state transition matrix $\mathbf{\Psi}(k, h)$ is defined as follows:

$$\mathbf{\Psi}(k+1, h) = \mathbf{G}(k)\mathbf{\Psi}(k, h), \qquad \mathbf{\Psi}(h, h) = \mathbf{I}$$

where $k = h, h+1, h+2, \ldots$. Assume that $\mathbf{G}(k)$ is nonsingular for all k values considered, so that the inverse of $\mathbf{\Psi}(k, h)$ exists.

Show that for any positive integer values i, j, k

$$\mathbf{\Psi}(k, i) = \mathbf{\Psi}(k, j)\mathbf{\Psi}(j, i)$$

Solution. Noting that

$$\mathbf{\Psi}(h, k) = \mathbf{\Psi}^{-1}(k, h) \qquad k > h$$

we have for $k > i > j$

$$\begin{aligned} \mathbf{\Psi}(k, i) &= \mathbf{G}(k-1)\mathbf{G}(k-2) \cdots \mathbf{G}(i) \\ &= \mathbf{G}(k-1)\mathbf{G}(k-2) \cdots \mathbf{G}(i) \cdots \mathbf{G}(j)[\mathbf{G}^{-1}(j)\mathbf{G}^{-1}(j+1) \cdots \mathbf{G}^{-1}(i-1)] \end{aligned}$$

$$= [\mathbf{G}(k-1)\mathbf{G}(k-2) \cdots \mathbf{G}(j)][\mathbf{G}(i-1)\mathbf{G}(i-2) \cdots \mathbf{G}(j)]^{-1}$$
$$= \mathbf{\Psi}(k,j)\mathbf{\Psi}^{-1}(i,j)$$
$$= \mathbf{\Psi}(k,j)\mathbf{\Psi}(j,i)$$

For $k > j > i$, we have

$$\mathbf{\Psi}(k,i) = [\mathbf{G}(k-1)\mathbf{G}(k-2) \cdots \mathbf{G}(j)][\mathbf{G}(j-1)\mathbf{G}(j-2) \cdots \mathbf{G}(i)]$$
$$= \mathbf{\Psi}(k,j)\mathbf{\Psi}(j,i)$$

Note that if $k < j < i$, then

$$\mathbf{\Psi}(k,i) = \mathbf{\Psi}^{-1}(i,k) = [\mathbf{G}(i-1)\mathbf{G}(i-2) \cdots \mathbf{G}(k)]^{-1}$$

and

$$\mathbf{\Psi}^{-1}(j,k)\mathbf{\Psi}^{-1}(i,j) = [\mathbf{G}(j-1)\mathbf{G}(j-2) \cdots \mathbf{G}(k)]^{-1}[\mathbf{G}(i-1)\mathbf{G}(i-2) \cdots \mathbf{G}(j)]^{-1}$$
$$= [\mathbf{G}^{-1}(k)\mathbf{G}^{-1}(k+1) \cdots \mathbf{G}^{-1}(j-1)][\mathbf{G}^{-1}(j)\mathbf{G}^{-1}(j+1) \cdots \mathbf{G}^{-1}(i-1)]$$
$$= \mathbf{G}^{-1}(k)\mathbf{G}^{-1}(k+1) \cdots \mathbf{G}^{-1}(i-1)$$
$$= [\mathbf{G}(i-1)\mathbf{G}(i-2) \cdots \mathbf{G}(k)]^{-1}$$

Hence

$$\mathbf{\Psi}^{-1}(i,k) = \mathbf{\Psi}^{-1}(j,k)\mathbf{\Psi}^{-1}(i,j) = [\mathbf{\Psi}(i,j)\mathbf{\Psi}(j,k)]^{-1}$$

or

$$\mathbf{\Psi}(i,k) = \mathbf{\Psi}(i,j)\mathbf{\Psi}(j,k)$$

Similarly, we obtain for $k < i < j$

$$\mathbf{\Psi}(i,k) = \mathbf{\Psi}(i,j)\mathbf{\Psi}(j,k)$$

In this way, for any positive integer values of k, j, and i, we have

$$\mathbf{\Psi}(k,i) = \mathbf{\Psi}(k,j)\mathbf{\Psi}(j,i)$$

If $\mathbf{G}(k)$ is singular, then the inverse of $\mathbf{\Psi}(k,i)$ does not exist. Hence an equation such as

$$\mathbf{\Psi}(k,i) = \mathbf{\Psi}(k,j)\mathbf{\Psi}^{-1}(i,j)$$

does not apply. However, for $k > j > i$, we have

$$\mathbf{\Psi}(k,i) = \mathbf{G}(k-1)\mathbf{G}(k-2) \cdots \mathbf{G}(j)\mathbf{G}(j-1) \cdots \mathbf{G}(i)$$
$$= \mathbf{\Psi}(k,j)\mathbf{\Psi}(j,i)$$

Problem A-5-22. Find the pulse transfer function of the system defined by

$$\mathbf{x}(k+1) = \mathbf{G}\mathbf{x}(k) + \mathbf{H}u(k)$$
$$y(k) = \mathbf{C}\mathbf{x}(k) + \mathbf{D}u(k)$$

where

$$\mathbf{G}=\begin{bmatrix}-a_1 & 1 & 0\\ -a_2 & 0 & 1\\ -a_3 & 0 & 0\end{bmatrix},\qquad \mathbf{H}=\begin{bmatrix}h_1\\ h_2\\ h_3\end{bmatrix}$$

$$\mathbf{C}=[1 \quad 0 \quad 0],\qquad D=b_0$$

Solution. From Eq. (5–88), the pulse transfer function $F(z)$ is given by the equation

$$\begin{aligned}F(z)&=\mathbf{C}(z\mathbf{I}-\mathbf{G})^{-1}\mathbf{H}+D\\ &=[1 \quad 0 \quad 0]\begin{bmatrix}z+a_1 & -1 & 0\\ a_2 & z & -1\\ a_3 & 0 & z\end{bmatrix}^{-1}\begin{bmatrix}h_1\\ h_2\\ h_3\end{bmatrix}+b_0\end{aligned}$$

First note that

$$|z\mathbf{I}-\mathbf{G}|=z^3+a_1z^2+a_2z+a_3$$

and

$$\begin{bmatrix}z+a_1 & -1 & 0\\ a_2 & z & -1\\ a_3 & 0 & z\end{bmatrix}^{-1}=\frac{1}{|z\mathbf{I}-\mathbf{G}|}\begin{bmatrix}z^2 & z & 1\\ -(a_2z+a_3) & (z+a_1)z & z+a_1\\ -a_3z & -a_3 & z(z+a_1)+a_2\end{bmatrix}$$

Then, $F(z)$ can be written as follows:

$$\begin{aligned}F(z)&=\frac{1}{|z\mathbf{I}-\mathbf{G}|}[z^2 \quad z \quad 1]\begin{bmatrix}h_1\\ h_2\\ h_3\end{bmatrix}+b_0\\ &=\frac{1}{|z\mathbf{I}-\mathbf{G}|}(h_1z^2+h_2z+h_3)+b_0\\ &=\frac{h_1z^2+h_2z+h_3}{z^3+a_1z^2+a_2z+a_3}+b_0\end{aligned}$$

Thus, the pulse transfer function for the system is

$$\begin{aligned}F(z)&=\frac{b_0z^3+(a_1b_0+h_1)z^2+(a_2b_0+h_2)z+a_3b_0+h_3}{z^3+a_1z^2+a_2z+a_3}\\ &=\frac{b_0+(a_1b_0+h_1)z^{-1}+(a_2b_0+h_2)z^{-2}+(a_3b_0+h_3)z^{-3}}{1+a_1z^{-1}+a_2z^{-2}+a_3z^{-3}}\end{aligned}$$

Problem A-5-23. Consider the following oscillator system:

$$\frac{Y(s)}{U(s)}=\frac{\omega^2}{s^2+\omega^2}$$

Obtain the continuous-time state space representation of the system. Then discretize the system and obtain the discrete-time state space representation. Also obtain the pulse transfer function of the discretized system.

Solution. From the given transfer function, we have

$$\ddot{y} + \omega^2 y = \omega^2 u$$

Define

$$x_1 = y$$

$$x_2 = \frac{1}{\omega}\dot{y}$$

Then we obtain the following continuous-time state space representation of the system:

$$\begin{bmatrix} \dot{x}_1 \\ \dot{x}_2 \end{bmatrix} = \begin{bmatrix} 0 & \omega \\ -\omega & 0 \end{bmatrix} \begin{bmatrix} x_1 \\ x_2 \end{bmatrix} + \begin{bmatrix} 0 \\ \omega \end{bmatrix} u$$

$$y = [1 \quad 0] \begin{bmatrix} x_1 \\ x_2 \end{bmatrix}$$

The discrete-time state space representation of the system is obtained as follows. Noting that

$$\mathbf{A} = \begin{bmatrix} 0 & \omega \\ -\omega & 0 \end{bmatrix}, \qquad \mathbf{B} = \begin{bmatrix} 0 \\ \omega \end{bmatrix}$$

we have

$$\mathbf{G} = e^{\mathbf{A}T} = \mathscr{L}^{-1}[(s\mathbf{I} - \mathbf{A})^{-1}] = \mathscr{L}^{-1}\left\{\begin{bmatrix} s & -\omega \\ \omega & s \end{bmatrix}^{-1}\right\}$$

$$= \mathscr{L}^{-1} \begin{bmatrix} \dfrac{s}{s^2 + \omega^2} & \dfrac{\omega}{s^2 + \omega^2} \\ \dfrac{-\omega}{s^2 + \omega^2} & \dfrac{s}{s^2 + \omega^2} \end{bmatrix} = \begin{bmatrix} \cos \omega T & \sin \omega T \\ -\sin \omega T & \cos \omega T \end{bmatrix}$$

and

$$\mathbf{H} = \left(\int_0^T e^{\mathbf{A}\lambda}\, d\lambda\right) \mathbf{B} = \left(\int_0^T \begin{bmatrix} \cos \omega\lambda & \sin \omega\lambda \\ -\sin \omega\lambda & \cos \omega\lambda \end{bmatrix} d\lambda\right) \begin{bmatrix} 0 \\ \omega \end{bmatrix}$$

$$= \begin{bmatrix} 1 - \cos \omega T \\ \sin \omega T \end{bmatrix}$$

Hence the discrete-time state space representation of the oscillator system becomes as follows:

$$\begin{bmatrix} x_1((k+1)T) \\ x_2((k+1)T) \end{bmatrix} = \begin{bmatrix} \cos \omega T & \sin \omega T \\ -\sin \omega T & \cos \omega T \end{bmatrix} \begin{bmatrix} x_1(kT) \\ x_2(kT) \end{bmatrix} + \begin{bmatrix} 1 - \cos \omega T \\ \sin \omega T \end{bmatrix} u(kT)$$

$$y(kT) = [1 \quad 0] \begin{bmatrix} x_1(kT) \\ x_2(kT) \end{bmatrix}$$

The pulse transfer function of the discretized system can be obtained from Eq. (5–88):

$$F(z) = \mathbf{C}(z\mathbf{I} - \mathbf{G})^{-1}\mathbf{H} + D$$

Noting that D is zero, we have

$$F(z) = [1 \quad 0] \begin{bmatrix} z - \cos \omega T & -\sin \omega T \\ \sin \omega T & z - \cos \omega T \end{bmatrix}^{-1} \begin{bmatrix} 1 - \cos \omega T \\ \sin \omega T \end{bmatrix}$$

$$= \frac{1}{z^2 - 2z\cos\omega T + 1}[1 \quad 0]\begin{bmatrix} z - \cos\omega T & \sin\omega T \\ -\sin\omega T & z - \cos\omega T \end{bmatrix}\begin{bmatrix} 1 - \cos\omega T \\ \sin\omega T \end{bmatrix}$$

$$= \frac{(1 - \cos\omega T)(z + 1)}{z^2 - 2z\cos\omega T + 1}$$

Hence

$$\frac{Y(z)}{U(z)} = F(z) = \frac{(1 - \cos\omega T)(1 + z^{-1})z^{-1}}{1 - 2z^{-1}\cos\omega T + z^{-2}}$$

Note that the pulse transfer function obtained in this way is the same as that obtained by taking the z transform of the system which is preceded by a zero-order hold. That is,

$$\frac{Y(z)}{U(z)} = \mathcal{Z}\left[\frac{1 - e^{-Ts}}{s}\frac{\omega^2}{s^2 + \omega^2}\right] = (1 - z^{-1})\,\mathcal{Z}\left[\frac{1}{s} - \frac{s}{s^2 + \omega^2}\right]$$

$$= (1 - z^{-1})\left(\frac{1}{1 - z^{-1}} - \frac{1 - z^{-1}\cos\omega T}{1 - 2z^{-1}\cos\omega T + z^{-2}}\right)$$

$$= \frac{(1 - \cos\omega T)(1 + z^{-1})z^{-1}}{1 - 2z^{-1}\cos\omega T + z^{-2}}$$

Thus, we get the same expression for the pulse transfer function. The reason for this is that discretization in the state space yields the zero-order hold equivalent of the continuous-time system.

Problem A-5-24. Consider the system shown in Fig. 5-18(a). This system involves complex poles. It is stable but not asymptotically stable in the sense of Liapunov. Figure 5-18(b) shows a discretized version of the continuous-time system. The discretized system is also stable but not asymptotically stable.

Assuming a unit step input, show that the discretized system may exhibit hidden oscillations when the sampling period T assumes a certain value.

Solution. The unit step response of the continuous-time system shown in Fig. 5-18(a) is

$$Y(s) = \frac{s^2}{s^2 + 4}\frac{1}{s} = \frac{s}{s^2 + 4}$$

Hence

$$y(t) = \cos 2t$$

[Notice that the average value of the output $y(t)$ is zero, not unity.] The response $y(t)$ versus t is shown in Fig. 5-19(a).

The pulse transfer function of the discretized system shown in Fig. 5-18(b) is

$$\frac{Y(z)}{U(z)} = \mathcal{Z}\left[\frac{1 - e^{-Ts}}{s}\frac{s^2}{s^2 + 4}\right] = (1 - z^{-1})\,\mathcal{Z}\left[\frac{s}{s^2 + 4}\right]$$

$$= (1 - z^{-1})\frac{1 - z^{-1}\cos 2T}{1 - 2z^{-1}\cos 2T + z^{-2}}$$

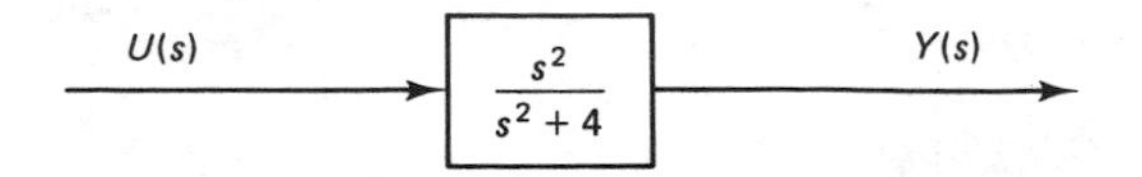

(a)

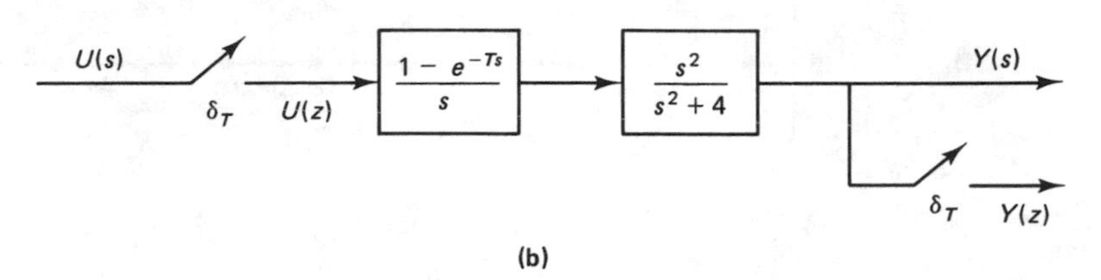

(b)

Figure 5–18 (a) Continuous-time system of Prob. A-5–24; (b) discretized version of the system.

Hence the unit step response is obtained as follows:

$$Y(z) = \frac{(1 - z^{-1})(1 - z^{-1}\cos 2T)}{1 - 2z^{-1}\cos 2T + z^{-2}} \frac{1}{1 - z^{-1}}$$

$$= \frac{1 - z^{-1}\cos 2T}{1 - 2z^{-1}\cos 2T + z^{-2}}$$

The response $y(kT)$ becomes oscillatory if $T \neq n\pi$ sec ($n = 1, 2, 3, \ldots$). For example, the response of the discretized system when $T = \frac{1}{4}\pi$ sec becomes as follows:

$$Y(z) = \frac{1}{1 + z^{-2}} = 1 - z^{-2} + z^{-4} - z^{-6} + \cdots$$

Hence

$$y(0) = 1$$
$$y(T) = 0$$
$$y(2T) = -1$$
$$y(3T) = 0$$
$$y(4T) = 1$$
$$\vdots$$

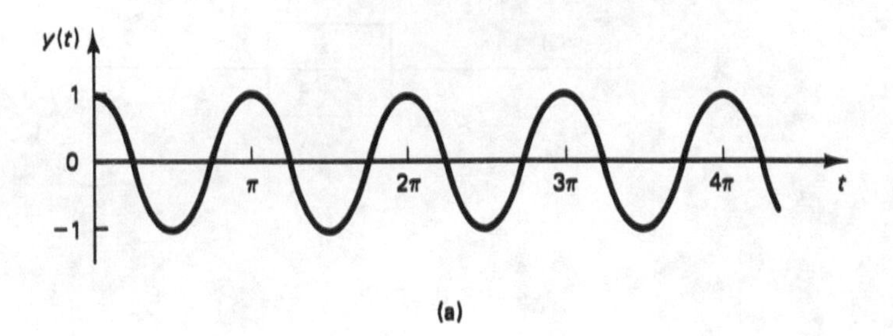

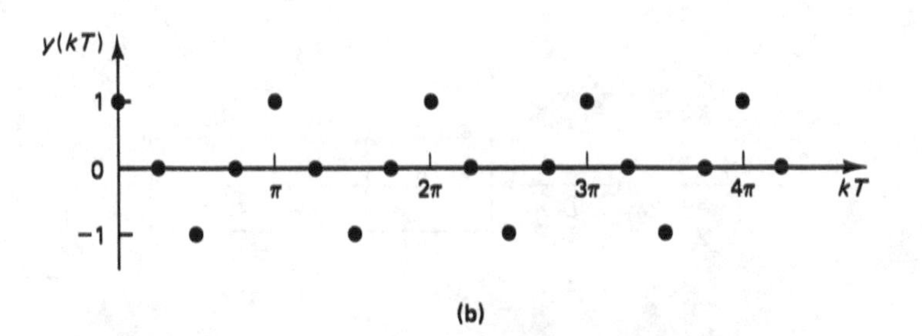

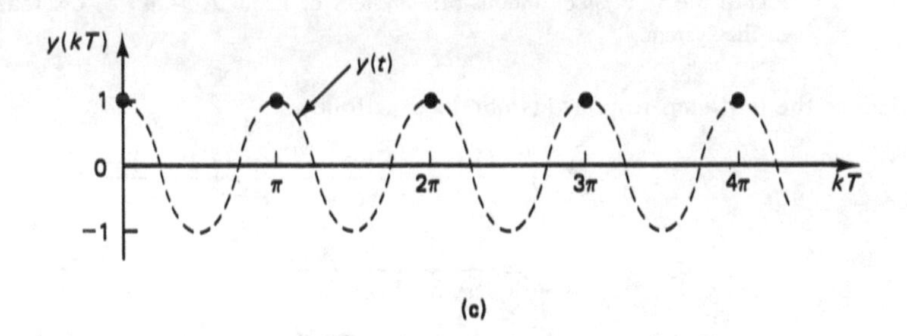

Figure 5–19 (a) Unit step response $y(t)$ of the continuous-time system shown in Fig. 5–18(a); (b) plot of $y(kT)$ vs. kT of the discretized system shown in Fig. 5–18(b) when $T = \frac{1}{4}\pi$ sec; (c) plot of $y(kT)$ vs. kT of the discretized system when $T = \pi$ sec. (Hidden oscillations are shown in the diagram.)

A plot of $y(kT)$ versus kT when $T = \frac{1}{4}\pi$ sec is shown in Fig. 5–19(b). Clearly, the response is oscillatory. If, however, the sampling period T is π sec, or $T = \pi$, then

$$Y(z) = \frac{(1 - z^{-1})(1 - z^{-1})}{1 - 2z^{-1} + z^{-2}} \frac{1}{1 - z^{-1}} = \frac{1}{1 - z^{-1}}$$

$$= 1 + z^{-1} + z^{-2} + z^{-3} + \cdots$$

The response $y(kT)$ for $k = 0, 1, 2, \ldots$ is constant at unity. A plot of $y(kT)$ versus kT when $T = \pi$ is shown in Fig. 5–19(c).

Notice that if $T = \pi$ sec (in fact, if $T = n\pi$ sec, where $n = 1, 2, 3, \ldots$) the unit step response sequence stays at unity. Such a response may give us an impression that $y(t)$

is constant. The actual response is not unity but oscillates between 1 and -1. Thus, the output of the discretized system when $T = \pi$ sec (or when $T = n\pi$ sec, where $n = 1, 2, 3, \ldots$) exhibits hidden oscillations.

Note that such hidden oscillations (hidden instability) occur only at certain particular values of the sampling period T. If the value of T is varied, such hidden oscillations (hidden instability) show up in the output as explicit oscillations (explicit instability).

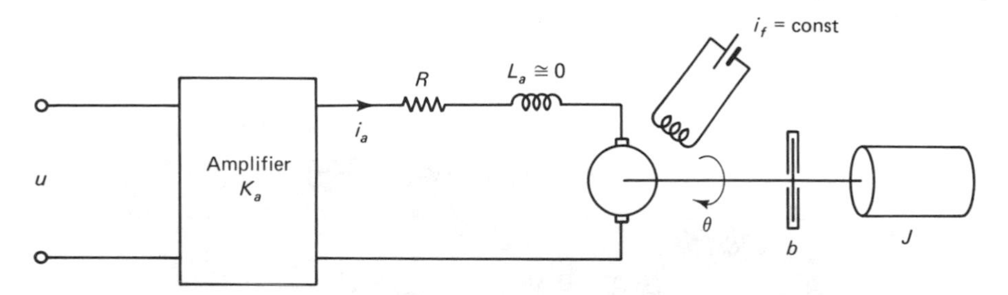

Figure 5–20 Schematic diagram of a dc servomotor system.

Problem A-5–25. Consider the dc servomotor system shown in Fig. 5–20. In the diagram J is the moment of inertia of the rotor and load, b is the viscous friction coefficient, R is the armature winding resistance, i_a is the armature winding current, K_a is the power amplifier gain, and i_f is the field current (constant). We assume that the armature winding inductance L_a is negligible. The input to the system is u, the voltage applied to the amplifier, and the output is θ, the angular displacement of the rotor and load.

Obtain the continuous-time state space representation and the transfer function of the dc servomotor system. Then, obtain the discrete-time state space representation of the servomotor system. Also, obtain the pulse transfer function.

Solution. For the system shown in the diagram, we have

$$Ri_a + K_b\dot{\theta} = K_a u \tag{5–183}$$

$$J\ddot{\theta} + b\dot{\theta} = T = K_1 i_f K_2 i_a = K i_a \tag{5–184}$$

where K_b is the back emf constant and $K = K_1 i_f K_2$ is the motor torque constant. By eliminating i_a from Eqs. (5–183) and (5–184) we have

$$J\ddot{\theta} + b\dot{\theta} + \frac{KK_b}{R}\dot{\theta} = KK_a\frac{u}{R}$$

Now define the state variables x_1 and x_2 and the output variable y as follows:

$$x_1 = \theta$$

$$x_2 = \dot{\theta}$$

$$y = x_1$$

Then we obtain

$$\dot{x}_1 = x_2$$

$$\dot{x}_2 = -\frac{1}{J}\left(b + \frac{KK_b}{R}\right)x_2 + \frac{KK_a}{JR}u$$

Hence the state space representation of the system is

$$\begin{bmatrix} \dot{x}_1 \\ \dot{x}_2 \end{bmatrix} = \begin{bmatrix} 0 & 1 \\ 0 & -\frac{1}{J}\left(b + \frac{KK_b}{R}\right) \end{bmatrix} \begin{bmatrix} x_1 \\ x_2 \end{bmatrix} + \begin{bmatrix} 0 \\ \frac{KK_a}{JR} \end{bmatrix} u$$

$$y = [1 \quad 0] \begin{bmatrix} x_1 \\ x_2 \end{bmatrix}$$

The transfer function $F(s)$ for the system can be obtained from Eq. (5–113):

$$F(s) = \mathbf{C}(s\mathbf{I} - \mathbf{A})^{-1}\mathbf{B} + D$$

or

$$F(s) = \frac{\Theta(s)}{U(s)} = [1 \quad 0] \begin{bmatrix} s & -1 \\ 0 & s + \frac{1}{J}\left(b + \frac{KK_b}{R}\right) \end{bmatrix}^{-1} \begin{bmatrix} 0 \\ \frac{KK_a}{JR} \end{bmatrix} + 0$$

$$= \frac{1}{s\left[s + \frac{1}{J}\left(b + \frac{KK_b}{R}\right)\right]} [1 \quad 0] \begin{bmatrix} s + \frac{1}{J}\left(b + \frac{KK_b}{R}\right) & 1 \\ 0 & s \end{bmatrix} \begin{bmatrix} 0 \\ \frac{KK_a}{JR} \end{bmatrix}$$

$$= \frac{\frac{KK_a}{JR}}{s\left[s + \frac{1}{J}\left(b + \frac{KK_b}{R}\right)\right]} = \frac{K_a K_m}{s(T_m s + 1)}$$

where

$$K_m = \frac{K}{bR + KK_b} = \text{motor gain constant}$$

$$T_m = \frac{JR}{bR + KK_b} = \text{motor time constant}$$

The transfer function $\Theta(s)/U(s)$ can, of course, be obtained by taking the Laplace transforms of Eqs. (5–183) and (5–184) and eliminating i_a from the two equations.

The discrete-time state space representation of the servomotor system is given by

$$\mathbf{x}((k+1)T) = \mathbf{G}\mathbf{x}(kT) + \mathbf{H}\mathbf{u}(kT)$$

$$y(kT) = \mathbf{C}\mathbf{x}(kT)$$

Let us define

$$\alpha = \frac{1}{J}\left(b + \frac{KK_b}{R}\right), \qquad \beta = \frac{KK_a}{JR}$$

and

$$\mathbf{A} = \begin{bmatrix} 0 & 1 \\ 0 & -\alpha \end{bmatrix}, \qquad \mathbf{B} = \begin{bmatrix} 0 \\ \beta \end{bmatrix}$$

Then

$$\mathbf{G} = e^{\mathbf{A}T} = \mathscr{L}^{-1}[(s\mathbf{I} - \mathbf{A})^{-1}] = \mathscr{L}^{-1}\left\{\begin{bmatrix} s & -1 \\ 0 & s+\alpha \end{bmatrix}^{-1}\right\}$$

$$= \mathscr{L}^{-1}\begin{bmatrix} \frac{1}{s} & \frac{1}{s(s+\alpha)} \\ 0 & \frac{1}{s+\alpha} \end{bmatrix} = \begin{bmatrix} 1 & \frac{1}{\alpha}(1 - e^{-\alpha T}) \\ 0 & e^{-\alpha T} \end{bmatrix}$$

$$\mathbf{H} = \left(\int_0^T e^{\mathbf{A}\lambda}\, d\lambda\right)\mathbf{B} = \int_0^T \begin{bmatrix} 1 & \frac{1}{\alpha}(1 - e^{-\alpha\lambda}) \\ 0 & e^{-\alpha\lambda} \end{bmatrix} d\lambda \begin{bmatrix} 0 \\ \beta \end{bmatrix}$$

$$= \begin{bmatrix} \frac{\beta}{\alpha}\left(T + \frac{e^{-\alpha T} - 1}{\alpha}\right) \\ \frac{\beta}{\alpha}(1 - e^{-\alpha T}) \end{bmatrix}$$

Hence, the discrete-time state equation and output equation become

$$\begin{bmatrix} x_1((k+1)T) \\ x_2((k+1)T) \end{bmatrix} = \begin{bmatrix} 1 & \frac{1}{\alpha}(1 - e^{-\alpha T}) \\ 0 & e^{-\alpha T} \end{bmatrix} \begin{bmatrix} x_1(kT) \\ x_2(kT) \end{bmatrix} + \begin{bmatrix} \frac{\beta}{\alpha}\left(T + \frac{e^{-\alpha T} - 1}{\alpha}\right) \\ \frac{\beta}{\alpha}(1 - e^{-\alpha T}) \end{bmatrix} u(kT)$$

$$y(kT) = [1 \quad 0] \begin{bmatrix} x_1(kT) \\ x_2(kT) \end{bmatrix}$$

where

$$\alpha = \frac{1}{J}\left(b + \frac{KK_b}{R}\right), \qquad \beta = \frac{KK_a}{JR}$$

Problem A-5–26. Consider a linear system described by

$$\dot{\mathbf{x}} = \mathbf{A}\mathbf{x}$$

where $\mathbf{x}$ is an n vector and $\mathbf{A}$ is an $n \times n$ constant matrix. Utilizing the method of successive approximations, an approximate solution of the system equation can be given by

$$\mathbf{x}_k(t) = \int_0^t \mathbf{A}\mathbf{x}_{k-1}(t)\, dt + \mathbf{x}(0)$$

With the initial approximation $\mathbf{x}_0(t) = \mathbf{x}(0)$, obtain the kth approximation to the solution. Show that as k is increased to infinity the solution approaches $e^{\mathbf{A}t}\mathbf{x}(0)$.

Solution. The successive approximations are

$$\mathbf{x}_0(t) = \mathbf{x}(0)$$

$$\mathbf{x}_1(t) = (\mathbf{I} + \mathbf{A}t)\mathbf{x}(0)$$

$$\mathbf{x}_2(t) = \left(\mathbf{I} + \mathbf{A}t + \frac{1}{2!}\mathbf{A}^2t^2\right)\mathbf{x}(0)$$

$$\vdots$$

$$\mathbf{x}_k(t) = \left(\mathbf{I} + \mathbf{A}t + \frac{1}{2!}\mathbf{A}^2t^2 + \cdots + \frac{1}{k!}\mathbf{A}^kt^k\right)\mathbf{x}(0)$$

As k approaches infinity,

$$\begin{aligned}\mathbf{x}(t) &= \lim_{k\to\infty} \mathbf{x}_k(t) \\ &= \lim_{k\to\infty}\left(\mathbf{I} + \mathbf{A}t + \frac{1}{2!}\mathbf{A}^2t^2 + \cdots + \frac{1}{k!}\mathbf{A}^kt^k\right)\mathbf{x}(0) \\ &= \sum_{k=0}^{\infty}\frac{1}{k!}\mathbf{A}^kt^k\,\mathbf{x}(0) = e^{\mathbf{A}t}\,\mathbf{x}(0)\end{aligned}$$

Problem A-5–27. Discretize the continuous-time state equation

$$\dot{\mathbf{x}} = \mathbf{A}\mathbf{x} + \mathbf{B}\mathbf{u} + \mathbf{v}$$

and obtain the discrete-time state equation in the following form:

$$\mathbf{x}((k+1)T) = \mathbf{G}(T)\mathbf{x}(kT) + \mathbf{H}(T)\mathbf{u}(kT) + \mathbf{N}(T)\mathbf{v}(kT)$$

Determine $\mathbf{G}(T)$, $\mathbf{H}(T)$, and $\mathbf{N}(T)$. Assume that $\mathbf{A}$ and $\mathbf{B}$ are constant matrices and $\mathbf{u}$ and $\mathbf{v}$ are constant vectors between any two consecutive sampling instants.

Solution. The solution of the given continuous-time state equation is

$$\mathbf{x}(t) = e^{\mathbf{A}t}\mathbf{x}(0) + e^{\mathbf{A}t}\int_0^t e^{-\mathbf{A}\tau}[\mathbf{B}\mathbf{u}(\tau) + \mathbf{v}(\tau)]\,d\tau$$

At $t = (k+1)T$ and $t = kT$, we have

$$\mathbf{x}((k+1)T) = e^{\mathbf{A}(k+1)T}\mathbf{x}(0) + e^{\mathbf{A}(k+1)T}\int_0^{(k+1)T} e^{-\mathbf{A}\tau}[\mathbf{B}\mathbf{u}(\tau) + \mathbf{v}(\tau)]\,d\tau \qquad (5\text{–}185)$$

and

$$x(kT) = e^{\mathbf{A}kT}\mathbf{x}(0) + e^{\mathbf{A}kT}\int_0^{kT} e^{-\mathbf{A}\tau}[\mathbf{B}\mathbf{u}(\tau) + \mathbf{v}(\tau)]\,d\tau \qquad (5\text{–}186)$$

Multiplying Eq. (5–186) by $e^{\mathbf{A}T}$ and subtracting from Eq. (5–185), we obtain

$$\begin{aligned}\mathbf{x}((k+1)T) &= e^{\mathbf{A}T}\mathbf{x}(kT) + e^{\mathbf{A}(k+1)T}\int_{kT}^{(k+1)T} e^{-\mathbf{A}\tau}[\mathbf{B}\mathbf{u}(\tau) + \mathbf{v}(\tau)]\,d\tau \\ &= e^{\mathbf{A}T}\mathbf{x}(kT) + e^{\mathbf{A}T}\int_0^T e^{-\mathbf{A}t}\mathbf{B}\,\mathbf{u}(kT)\,dt + e^{\mathbf{A}T}\int_0^T e^{-\mathbf{A}t}\,\mathbf{v}(kT)\,dt \\ &= \mathbf{G}(T)\mathbf{x}(kT) + \mathbf{H}(T)\mathbf{u}(kT) + \mathbf{N}(T)\mathbf{v}(kT)\end{aligned}$$

where we are using the relationships

$$\mathbf{u}(\tau) = \mathbf{u}(kT) \qquad \text{for} \qquad kT \le \tau < (k+1)T$$

$$\mathbf{v}(\tau) = \mathbf{v}(kT) \qquad \text{for} \qquad kT \le \tau < (k+1)T$$

and also the relationship $\tau = t + kT$, where $0 \le t < T$. Thus, $\mathbf{G}(T)$, $\mathbf{H}(T)$, and $\mathbf{N}(T)$ are given as follows:

$$\mathbf{G}(T) = e^{\mathbf{A}T}$$

$$\mathbf{H}(T) = e^{\mathbf{A}T} \int_0^T e^{-\mathbf{A}t}\mathbf{B}\, dt = \int_0^T e^{\mathbf{A}\tau}\mathbf{B}\, d\tau$$

$$\mathbf{N}(T) = e^{\mathbf{A}T} \int_0^T e^{-\mathbf{A}t}\, dt = \int_0^T e^{\mathbf{A}\tau}\, d\tau$$

Problem A-5–28. Obtain the response of the following time-varying continuous-time system:

$$\dot{x} + tx = u$$

where u is an arbitrary input.

Solution. The solution of the given system can be given by either

$$x(t) = \phi(t, t_0)x(t_0) + \phi(t, t_0) \int_{t_0}^{t} \phi^{-1}(\tau, t_0)u(\tau)\, d\tau$$

or

$$x(t) = \phi(t, t_0)x(t_0) + \int_{t_0}^{t} h(t, \tau)u(\tau)\, d\tau$$

Note that

$$\phi(t, t_0)\phi^{-1}(\tau, t_0) = h(t, \tau)$$

In this system, since t and $\int_{t_0}^{t} \tau\, d\tau$ commute for all t, $\phi(t, t_0)$ can be expressed as follows:

$$\phi(t, t_0) = \exp\left(-\int_{t_0}^{t} \tau\, d\tau\right) = \exp\left[-\tfrac{1}{2}(t^2 - t_0^2)\right]$$

Hence

$$\phi^{-1}(\tau, t_0) = \exp\left[\tfrac{1}{2}(\tau^2 - t_0^2)\right]$$

Therefore,

$$h(t, \tau) = \begin{cases} \exp\left[-\tfrac{1}{2}(t^2 - \tau^2)\right] & t \ge \tau \\ 0 & t < \tau \end{cases}$$

The output $x(t)$ is thus given by

$$x(t) = \exp\left[-\tfrac{1}{2}(t^2 - t_0^2)\right] x(t_0) + \exp\left(-\tfrac{1}{2}t^2\right) \int_{t_0}^{t} \exp\left(\tfrac{1}{2}\tau^2\right) u(\tau)\, d\tau$$

Problem A-5–29. Even though the double integrator system is dynamically simple, it represents an important class of systems. An example of double integrator systems is a satellite attitude control system, which can be described by

$$J\ddot{\theta} = u + v$$

where J is the moment of inertia, θ is the attitude angle, u is the control torque, and v is the disturbance torque.

Consider the double integrator system in the absence of disturbance input. Define $J\theta = y$. Then the system equation becomes

$$\ddot{y} = u$$

Obtain a continuous-time state space representation of the system. Then obtain a discrete-time equivalent. Also obtain the pulse transfer function for the discrete-time system.

***Solution*.** Define

$$x_1 = y$$

$$x_2 = \dot{y}$$

Then the continuous-time state equation and output equation become

$$\begin{bmatrix} \dot{x}_1 \\ \dot{x}_2 \end{bmatrix} = \begin{bmatrix} 0 & 1 \\ 0 & 0 \end{bmatrix} \begin{bmatrix} x_1 \\ x_2 \end{bmatrix} + \begin{bmatrix} 0 \\ 1 \end{bmatrix} u$$

$$y = [1 \quad 0] \begin{bmatrix} x_1 \\ x_2 \end{bmatrix}$$

The discrete-time equivalent of this system can be given by

$$\mathbf{x}((k+1)T) = \mathbf{G}\mathbf{x}(kT) + \mathbf{H}u(kT)$$

$$y(kT) = \mathbf{C}\mathbf{x}(kT)$$

Matrices $\mathbf{G}$ and $\mathbf{H}$ are obtained from Eqs. (5–134) and (5–135). Noting that

$$\mathbf{A} = \begin{bmatrix} 0 & 1 \\ 0 & 0 \end{bmatrix}, \qquad \mathbf{B} = \begin{bmatrix} 0 \\ 1 \end{bmatrix}$$

we have

$$\mathbf{G} = e^{\mathbf{A}T} = \begin{bmatrix} 1 & T \\ 0 & 1 \end{bmatrix}$$

and

$$\mathbf{H} = \left(\int_0^T e^{\mathbf{A}\lambda}\, d\lambda \right) \mathbf{B} = \left(\int_0^T \begin{bmatrix} 1 & \lambda \\ 0 & 1 \end{bmatrix} d\lambda \right) \begin{bmatrix} 0 \\ 1 \end{bmatrix} = \begin{bmatrix} \frac{T^2}{2} \\ T \end{bmatrix}$$

Hence, the discrete-time state equation and output equation become

$$\begin{bmatrix} x_1((k+1)T) \\ x_2((k+1)T) \end{bmatrix} = \begin{bmatrix} 1 & T \\ 0 & 1 \end{bmatrix} \begin{bmatrix} x_1(kT) \\ x_2(kT) \end{bmatrix} + \begin{bmatrix} \frac{T^2}{2} \\ T \end{bmatrix} u(kT)$$

$$y(kT) = [1 \quad 0]\begin{bmatrix} x_1(kT) \\ x_2(kT) \end{bmatrix}$$

The pulse transfer function of the discrete-time system is obtained from Eq. (5–88), as follows:

$$\frac{Y(z)}{U(z)} = F(z) = \mathbf{C}(z\mathbf{I} - \mathbf{G})^{-1}\mathbf{H} + D$$

$$= [1 \quad 0]\begin{bmatrix} z-1 & -T \\ 0 & z-1 \end{bmatrix}^{-1}\begin{bmatrix} \dfrac{T^2}{2} \\ T \end{bmatrix} + 0$$

$$= \frac{T^2(z+1)}{2(z-1)^2} = \frac{T^2 z^{-1}(1+z^{-1})}{2(1-z^{-1})^2}$$

Problem A-5–30. Show that the following quadratic form is positive definite:

$$V(\mathbf{x}) = 10x_1^2 + 4x_2^2 + x_3^2 + 2x_1x_2 - 2x_2x_3 - 4x_1x_3$$

***Solution*.** The quadratic form $V(\mathbf{x})$ can be written as follows:

$$V(\mathbf{x}) = \mathbf{x}^T\mathbf{P}\mathbf{x} = [x_1 \quad x_2 \quad x_3]\begin{bmatrix} 10 & 1 & -2 \\ 1 & 4 & -1 \\ -2 & -1 & 1 \end{bmatrix}\begin{bmatrix} x_1 \\ x_2 \\ x_3 \end{bmatrix}$$

Applying Sylvester's criterion, we obtain

$$10 > 0, \qquad \begin{vmatrix} 10 & 1 \\ 1 & 4 \end{vmatrix} > 0, \qquad \begin{vmatrix} 10 & 1 & -2 \\ 1 & 4 & -1 \\ -2 & -1 & 1 \end{vmatrix} > 0$$

Since all the successive principal minors of the matrix $\mathbf{P}$ are positive, $V(\mathbf{x})$ is positive definite.

Problem A-5–31. Consider the system defined by

$$\dot{\mathbf{x}} = \mathbf{f}(\mathbf{x}, t)$$

Suppose that

$$\mathbf{f}(\mathbf{0}, t) = \mathbf{0} \qquad \text{for all } t$$

Suppose also that there exists a scalar function $V(\mathbf{x}, t)$ which has continuous first partial derivatives. If $V(\mathbf{x}, t)$ satisfies the conditions

1. $V(\mathbf{x}, t)$ is positive definite. That is, $V(\mathbf{0}, t) = 0$ and $V(\mathbf{x}, t) \geq \alpha\,(\|\mathbf{x}\|) > 0$ for all $\mathbf{x} \neq \mathbf{0}$ and all t, where α is a continuous nondecreasing scalar function such that $\alpha(0) = 0$.
2. The total derivative $\dot{V}$ is negative for all $\mathbf{x} \neq \mathbf{0}$ and all t, or $\dot{V}(\mathbf{x}, t) \leq -\gamma(\|\mathbf{x}\|) < 0$ for all $\mathbf{x} \neq \mathbf{0}$ and all t, where γ is a continuous nondecreasing scalar function such that $\gamma(0) = 0$.
3. There exists a continuous nondecreasing scalar function β such that $\beta(0) = 0$ and, for all t, $V(\mathbf{x}, t) \leq \beta(\|\mathbf{x}\|)$.

4. $\alpha(\|\mathbf{x}\|)$ approaches infinity as $\|\mathbf{x}\|$ increases indefinitely, or

$$\alpha(\|\mathbf{x}\|) \to \infty \qquad \text{as } \|\mathbf{x}\| \to \infty$$

then the origin of the system, $\mathbf{x} = \mathbf{0}$, is uniformly asymptotically stable in the large. (This is Liapunov's main stability theorem.)

Prove this theorem.

Solution. To prove uniform asymptotic stability in the large, we need to prove the following:

1. The origin is uniformly stable.
2. Every solution is uniformly bounded.
3. Every solution converges to the origin when $t \to \infty$ uniformly in t_0 and $\|\mathbf{x}_0\| \le \delta$, where δ is fixed but arbitrarily large. That is, given two real numbers $\delta > 0$ and $\mu > 0$ there is a real number $T(\mu, \delta)$ such that

$$\|\mathbf{x}_0\| \le \delta$$

implies

$$\|\boldsymbol{\phi}(t; \mathbf{x}_0, t_0)\| \le \mu \qquad \text{for all } t \ge t_0 + T(\mu, \delta)$$

where $\boldsymbol{\phi}(t; \mathbf{x}_0, t_0)$ is the solution to the given differential equation.

Since β is continuous and $\beta(0) = 0$, we can take a $\delta(\epsilon) > 0$ such that $\beta(\delta) < \alpha(\epsilon)$ for any $\epsilon > 0$. Figure 5–21 shows the curves $\alpha(\|\mathbf{x}\|)$, $\beta(\|\mathbf{x}\|)$, and $V(\mathbf{x}, t)$. Noting that

$$V(\boldsymbol{\phi}(t; \mathbf{x}_0, t_0), t) - V(\mathbf{x}_0, t_0) = \int_{t_0}^{t} \dot{V}(\boldsymbol{\phi}(\tau; \mathbf{x}_0, t_0), \tau)\, d\tau < 0 \qquad t > t_0$$

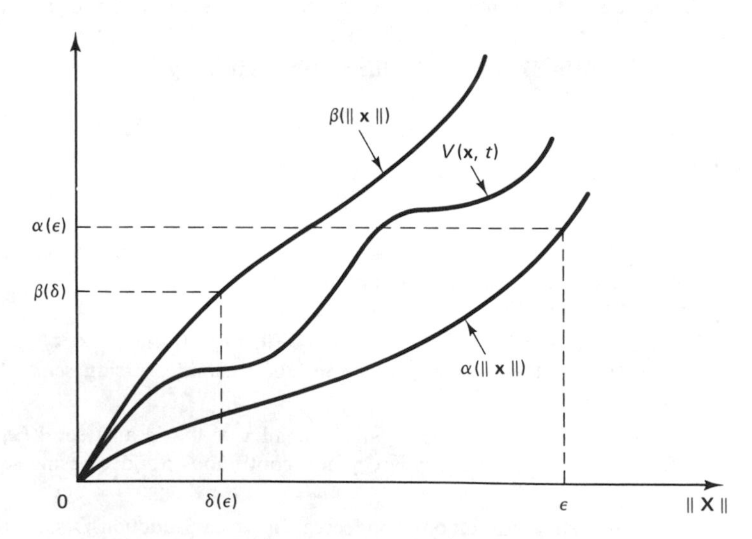

Figure 5–21 Curves $\alpha(\|\mathbf{x}\|)$, $\beta(\|\mathbf{x}\|)$, and $V(\mathbf{x}, t)$.

if $\|\mathbf{x}_0\| \leq \delta$, t_0 being arbitrary, we have

$$\alpha(\epsilon) > \beta(\delta) \geq V(\mathbf{x}_0, t_0) \geq V(\boldsymbol{\phi}(t; \mathbf{x}_0, t_0), t) \geq \alpha(\|\boldsymbol{\phi}(t; \mathbf{x}_0, t_0)\|)$$

for all $t \geq t_0$. Since α is nondecreasing and positive, this implies that

$$\|\boldsymbol{\phi}(t; \mathbf{x}_0, t_0)\| < \epsilon \qquad \text{for} \qquad t \geq t_0, \|\mathbf{x}_0\| \leq \delta$$

Hence, we have shown that for each real number $\epsilon > 0$, there is a real number $\delta > 0$ such that $\|\mathbf{x}_0\| \leq \delta$ implies $\|\boldsymbol{\phi}(t; \mathbf{x}_0, t_0)\| \leq \epsilon$ for all $t \geq t_0$. Thus we have proved uniform stability.

Next, we shall prove that $\|\boldsymbol{\phi}(t; \mathbf{x}_0, t_0)\| \to 0$ when $t \to \infty$ uniformly in t_0 and $\|\mathbf{x}_0\| \leq \delta$. Let us take any $0 < \mu < \|\mathbf{x}_0\|$ and find a $\nu(\mu) > 0$ such that $\beta(\nu) < \alpha(\mu)$. Let us denote by $\epsilon'(\mu, \delta) > 0$ the minimum of the continuous nondecreasing function $\gamma(\|\mathbf{x}\|)$ on the compact set $\nu(\mu) \leq \|\mathbf{x}\| \leq \epsilon(\delta)$. Let us define

$$T(\mu, \delta) = \frac{\beta(\delta)}{\epsilon'(\mu, \delta)} > 0$$

Suppose that $\|\boldsymbol{\phi}(t; \mathbf{x}_0, t_0)\| > \nu$ over the time interval $t_0 \leq t \leq t_1 = t_0 + T$. Then, we have

$$0 < \alpha(\nu) \leq V(\boldsymbol{\phi}(t_1; \mathbf{x}_0, t_0), t_1) \leq V(\mathbf{x}_0, t_0) - (t_1 - t_0)\epsilon' \leq \beta(\delta) - T\epsilon' = 0$$

which is a contradiction. Hence for some t in the interval $t_0 \leq t \leq t_1$, such as an arbitrary t_2, we have

$$\|\mathbf{x}_2\| = \|\boldsymbol{\phi}(t_2; \mathbf{x}_0, t_0)\| = \nu$$

Therefore,

$$\alpha(\|\boldsymbol{\phi}(t; \mathbf{x}_2, t_2)\|) < V(\boldsymbol{\phi}(t; \mathbf{x}_2, t_2), t) \leq V(\mathbf{x}_2, t_2) \leq \beta(\nu) < \alpha(\mu)$$

for all $t \geq t_2$. Hence,

$$\|\boldsymbol{\phi}(t; \mathbf{x}_0, t_0)\| < \mu$$

for all $t \geq t_0 + T(\mu, \delta) \geq t_2$, which proves uniform asymptotic stability. Since $\alpha(\|\mathbf{x}\|) \to \infty$ as $\|\mathbf{x}\| \to \infty$ there exists for arbitrarily large δ a constant $\epsilon(\delta)$ such that $\beta(\delta) < \alpha(\epsilon)$. Moreover, since $\epsilon(\delta)$ does not depend on t_0, the solution $\boldsymbol{\phi}(t; \mathbf{x}_0, t_0)$ is uniformly bounded. We thus have proved uniform asymptotic stability in the large.

Problem A-5–32. In z plane analysis, an $n \times n$ matrix $\mathbf{G}$ whose n eigenvalues are less than unity in magnitude is called a stable matrix. Consider an $n \times n$ Hermitian (or real symmetric) matrix $\mathbf{P}$ that satisfies the following matrix equation:

$$\mathbf{G}^*\mathbf{P}\mathbf{G} - \mathbf{P} = -\mathbf{Q} \tag{5–187}$$

where $\mathbf{Q}$ is a positive definite $n \times n$ Hermitian (or real symmetric) matrix. Prove that if matrix $\mathbf{G}$ is a stable matrix, then a matrix $\mathbf{P}$ that satisfies Eq. (5–187) is unique and is positive definite. Prove that matrix $\mathbf{P}$ can be given by

$$\mathbf{P} = \sum_{k=0}^{\infty} (\mathbf{G}^*)^k \mathbf{Q} \mathbf{G}^k$$

Prove also that although the right-hand side of this last equation is an infinite series, the matrix is finite. Finally, prove that if Eq. (5–187) is satisfied by positive definite matrices $\mathbf{P}$

and **Q**, then matrix **G** is a stable matrix. Assume that all eigenvalues of **G** are distinct and all eigenvectors of **G** are linearly independent.

Solution. Let us assume that there exist two matrices $\mathbf{P}_1$ and $\mathbf{P}_2$ that satisfy Eq. (5–187). Then

$$\mathbf{G}^*\mathbf{P}_1\mathbf{G} - \mathbf{P}_1 = -\mathbf{Q} \tag{5–188}$$

and

$$\mathbf{G}^*\mathbf{P}_2\mathbf{G} - \mathbf{P}_2 = -\mathbf{Q} \tag{5–189}$$

By subtracting Eq. (5–189) from Eq. (5–188), we obtain

$$\mathbf{G}^*\hat{\mathbf{P}}\mathbf{G} - \hat{\mathbf{P}} = \mathbf{0} \tag{5–190}$$

where

$$\hat{\mathbf{P}} = \mathbf{P}_1 - \mathbf{P}_2$$

Notice that if $\hat{\mathbf{P}} \neq \mathbf{0}$, then there exists an eigenvector $\mathbf{x}_i$ of matrix **G** such that

$$\hat{\mathbf{P}}\mathbf{x}_i \neq \mathbf{0}$$

Let us define the eigenvalue that is associated with the eigenvector $\mathbf{x}_i$ to be λ_i. Then

$$\mathbf{G}\mathbf{x}_i = \lambda_i \mathbf{x}_i$$

Hence, from Eq. (5–190), we obtain

$$\mathbf{G}^*\hat{\mathbf{P}}\mathbf{G}\mathbf{x}_i - \hat{\mathbf{P}}\mathbf{x}_i = \mathbf{G}^*\hat{\mathbf{P}}\lambda_i \mathbf{x}_i - \hat{\mathbf{P}}\mathbf{x}_i = (\lambda_i \mathbf{G}^* - \mathbf{I})\hat{\mathbf{P}}\mathbf{x}_i = \mathbf{0} \tag{5–191}$$

Equation (5–191) implies that λ_i^{-1} is an eigenvalue of $\mathbf{G}^*$. Since $|\lambda_i| < 1$, we have $|\lambda_i^{-1}| > 1$. This contradicts the assumption that **G** is a stable matrix. Hence $\hat{\mathbf{P}}$ must be a zero matrix, or it is necessary that

$$\mathbf{P}_1 = \mathbf{P}_2$$

Thus we have proved the uniqueness of the matrix **P**, the solution to Eq. (5–187).

To prove that a matrix **P** that satisfies Eq. (5–187) can be given by

$$\mathbf{P} = \sum_{k=0}^{\infty} (\mathbf{G}^*)^k \mathbf{Q}\mathbf{G}^k \tag{5–192}$$

we may rewrite Eq. (5–192) as follows:

$$\mathbf{P} = (\mathbf{G}^*)^0\mathbf{Q}\mathbf{G}^0 + \sum_{k=1}^{\infty}(\mathbf{G}^*)^k\mathbf{Q}\mathbf{G}^k = \mathbf{Q} + \mathbf{G}^*\left[\sum_{k=0}^{\infty}(\mathbf{G}^*)^k\mathbf{Q}\mathbf{G}^k\right]\mathbf{G}$$

$$= \mathbf{Q} + \mathbf{G}^*\mathbf{P}\mathbf{G}$$

Thus, Eq. (5–187) is satisfied. Since matrix **Q** is a positive definite matrix, from Eq. (5–192) matrix **P** is also positive definite.

We shall now prove that although matrix **P** given by Eq. (5–192) is the sum of an infinite series, it is a finite matrix. Because of the assumptions made in the problem statement, the eigenvalues λ_i are distinct and the eigenvectors of **G** are linearly independent. For the eigenvalue λ_i that is associated with the eigenvector $\mathbf{x}_i$, we have

$$\mathbf{G}\mathbf{x}_i = \lambda_i \mathbf{x}_i$$

By using this relationship, we may simplify $\mathbf{x}_i^* \left[\sum_{k=0}^{\infty} (\mathbf{G}^*)^k \mathbf{Q}\mathbf{G}^k \right] \mathbf{x}_i$. First note that

$$\begin{aligned} \mathbf{x}_i^* (\mathbf{G}^*)^2 \mathbf{Q}\mathbf{G}^2 \mathbf{x}_i &= (\mathbf{x}_i^* \mathbf{G}^*)(\mathbf{G}^*\mathbf{Q}\mathbf{G})(\mathbf{G}\mathbf{x}_i) = \bar{\lambda}_i \mathbf{x}_i^* \mathbf{G}^*\mathbf{Q}\mathbf{G}\lambda_i \mathbf{x}_i \\ &= |\lambda_i|^2 (\mathbf{x}_i^* \mathbf{G}^*)\mathbf{Q}(\mathbf{G}\mathbf{x}_i) = |\lambda_i|^2 (\bar{\lambda}_i \mathbf{x}_i^*)\mathbf{Q}(\lambda_i \mathbf{x}_i) \\ &= |\lambda_i|^2 |\lambda_i|^2 \mathbf{x}_i^* \mathbf{Q}\mathbf{x}_i \end{aligned}$$

Then, by using this type of simplification, we have

$$\begin{aligned} \mathbf{x}_i^* \left[\sum_{k=0}^{\infty} (\mathbf{G}^*)^k \mathbf{Q}\mathbf{G}^k \right] \mathbf{x}_i &= \mathbf{x}_i^* \mathbf{Q}\mathbf{x}_i + \mathbf{x}_i^* \mathbf{G}^*\mathbf{Q}\mathbf{G}\mathbf{x}_i + \mathbf{x}_i^* (\mathbf{G}^*)^2 \mathbf{Q}\mathbf{G}^2 \mathbf{x}_i + \mathbf{x}_i^* (\mathbf{G}^*)^3 \mathbf{Q}\mathbf{G}^3 \mathbf{x}_i + \cdots \\ &= \mathbf{x}_i^* \mathbf{Q}\mathbf{x}_i + \bar{\lambda}_i \lambda_i \mathbf{x}_i^* \mathbf{Q}\mathbf{x}_i + |\lambda_i|^2 |\lambda_i|^2 \mathbf{x}_i^* \mathbf{Q}\mathbf{x}_i + |\lambda_i|^4 |\lambda_i|^2 \mathbf{x}_i^* \mathbf{Q}\mathbf{x}_i + \cdots \\ &= \mathbf{x}_i^* \mathbf{Q}\mathbf{x}_i (1 + |\lambda_i|^2 + |\lambda_i|^4 + |\lambda_i|^6 + \cdots) \\ &= \mathbf{x}_i^* \mathbf{Q}\mathbf{x}_i \frac{1}{1 - |\lambda_i|^2} \end{aligned}$$

This proves that

$$\sum_{k=0}^{\infty} (\mathbf{G}^*)^k \mathbf{Q}\mathbf{G}^k$$

is a finite matrix.

Finally, we shall prove that if Eq. (5–187) is satisfied by positive definite matrices **P** and **Q**, then matrix **G** is a stable matrix. Let us define the eigenvector associated with an eigenvalue λ_i of **G** as $\mathbf{x}_i$. Then

$$\mathbf{G}\mathbf{x}_i = \lambda_i \mathbf{x}_i$$

By premultiplying both sides of Eq. (5–187) by $\mathbf{x}_i^*$ and postmultiplying both sides by $\mathbf{x}_i$, we obtain

$$\mathbf{x}_i^* \mathbf{G}^*\mathbf{P}\mathbf{G}\mathbf{x}_i - \mathbf{x}_i^* \mathbf{P}\mathbf{x}_i = -\mathbf{x}_i^* \mathbf{Q}\mathbf{x}_i$$

Hence

$$\bar{\lambda}_i \mathbf{x}_i^* \mathbf{P}\lambda_i \mathbf{x}_i - \mathbf{x}_i^* \mathbf{P}\mathbf{x}_i = -\mathbf{x}_i^* \mathbf{Q}\mathbf{x}_i$$

or

$$(|\lambda_i|^2 - 1)\mathbf{x}_i^* \mathbf{P}\mathbf{x}_i = -\mathbf{x}_i^* \mathbf{Q}\mathbf{x}_i$$

Since both $\mathbf{x}_i^* \mathbf{P}\mathbf{x}_i$ and $\mathbf{x}_i^* \mathbf{Q}\mathbf{x}_i$ are positive-definite, we have

$$|\lambda_i|^2 - 1 < 0$$

or

$$|\lambda_i| < 1$$

Hence we have proved that matrix **G** is a stable matrix.

It is noted that the proofs and derivations presented here can be extended to the case where matrix **G** involves multiple eigenvalues and multiple eigenvectors.

Problem A-5–33. Consider the following system:

$$\begin{bmatrix} x_1(k+1) \\ x_2(k+1) \end{bmatrix} = \begin{bmatrix} 0 & 0.5 \\ -0.5 & -1 \end{bmatrix} \begin{bmatrix} x_1(k) \\ x_2(k) \end{bmatrix}$$

Determine the stability of the origin of the system.

Solution. Let us choose $\mathbf{Q}$ to be $\mathbf{I}$. Then the Liapunov stability equation given by Eq. (5–155) becomes

$$\begin{bmatrix} 0 & -0.5 \\ 0.5 & -1 \end{bmatrix} \begin{bmatrix} p_{11} & p_{12} \\ p_{12} & p_{22} \end{bmatrix} \begin{bmatrix} 0 & 0.5 \\ -0.5 & -1 \end{bmatrix} - \begin{bmatrix} p_{11} & p_{12} \\ p_{12} & p_{22} \end{bmatrix} = -\begin{bmatrix} 1 & 0 \\ 0 & 1 \end{bmatrix}$$

which yields

$$0.25p_{22} - p_{11} = -1$$

$$-1.25p_{12} + 0.5p_{22} = 0$$

$$0.25p_{11} - p_{12} = -1$$

Solving these three equations for p_{11}, p_{12}, and p_{22}, we obtain

$$p_{11} = \tfrac{52}{27}, \qquad p_{12} = \tfrac{40}{27}, \qquad p_{22} = \tfrac{100}{27}$$

Hence matrix $\mathbf{P}$ is given by the equation

$$\mathbf{P} = \begin{bmatrix} \frac{52}{27} & \frac{40}{27} \\ \frac{40}{27} & \frac{100}{27} \end{bmatrix}$$

Clearly, matrix $\mathbf{P}$ is positive definite. Thus, the origin, $\mathbf{x} = \mathbf{0}$, is asymptotically stable in the large.

Problem A-5–34. Consider the system

$$\mathbf{x}(k+1) = \mathbf{H}(\mathbf{x}(k))\mathbf{x}(k)$$

Assume that there exist positive constants $c_1, c_2, \ldots, c_n$ such that either

$$\text{(1)} \qquad \max_i \left\{ \sum_{j=1}^{n} \frac{c_i}{c_j} |h_{ij}(\mathbf{x})| \right\} < 1 \qquad \text{for all } \mathbf{x}$$

or

$$\text{(2)} \qquad \max_j \left\{ \sum_{i=1}^{n} \frac{c_i}{c_j} |h_{ij}(\mathbf{x})| \right\} < 1 \qquad \text{for all } \mathbf{x}$$

Show that in either case $\mathbf{H}(\mathbf{x})\mathbf{x}$ is a contraction for all $\mathbf{x}$ and therefore the equilibrium state of the system is asymptotically stable in the large.

Solution. In case 1, define the norm by

$$\|\mathbf{x}\| = \max_i \{c_i |x_i|\}$$

then

$$\|\mathbf{H}(\mathbf{x})\mathbf{x}\| = \max_i \left\{ c_i \left| \sum_{j=1}^{n} h_{ij}(\mathbf{x}) x_j \right| \right\} \leq \max_i \left[\sum_{j=1}^{n} \frac{c_i}{c_j} |h_{ij}(\mathbf{x})| \cdot c_j |x_j| \right\}$$

$$\leq \max_i \left\{ \sum_{j=1}^{n} \frac{c_i}{c_j} |h_{ij}(\mathbf{x})| \right\} \max_j \{c_j |x_j|\} < \max_j \{c_j |x_j|\}$$

$$= \|\mathbf{x}\|$$

which verifies that $\mathbf{H}(\mathbf{x})\mathbf{x}$ is a contraction.

In case 2, define the norm by

$$\|\mathbf{x}\| = \sum_{i=1}^{n} c_i |x_i|$$

Then

$$\|\mathbf{H}(\mathbf{x})\mathbf{x}\| = \sum_{i=1}^{n} c_i \left| \sum_{j=1}^{n} h_{ij}(\mathbf{x}) x_j \right|$$

$$\leq \sum_{i=1}^{n} \frac{c_i}{c_j} \sum_{j=1}^{n} |h_{ij}(\mathbf{x})| \cdot c_j |x_j|$$

$$\leq \max_j \left\{ \sum_{i=1}^{n} \frac{c_i}{c_j} |h_{ij}(\mathbf{x})| \right\} \cdot \sum_{j=1}^{n} c_j |x_j| < \sum_{j=1}^{n} c_j |x_j|$$

$$= \|\mathbf{x}\|$$

which shows that $\mathbf{H}(\mathbf{x})\mathbf{x}$ is a contraction.

Now consider a scalar function $V(\mathbf{x}) = \|\mathbf{x}\|$. Clearly, $V(\mathbf{x}) = \|\mathbf{x}\|$ is positive definite and

$$\Delta V(\mathbf{x}(k)) = V(\mathbf{x}(k+1)) - V(\mathbf{x}(k))$$
$$= \|\mathbf{H}(\mathbf{x}(k))\mathbf{x}(k)\| - \|\mathbf{x}(k)\| < 0$$

and

$$\Delta V(\mathbf{0}) = 0$$

Thus, $\Delta V(\mathbf{x})$ is negative definite. Hence $V(\mathbf{x}) = \|\mathbf{x}\|$ is a Liapunov function for the system considered, and by Theorem 5–5 the origin of the system is asymptotically stable in the large.

Problem A-5–35. Prove that if all solutions of

$$\mathbf{x}(k+1) = \mathbf{G}\mathbf{x}(k) \tag{5–193}$$

where $\mathbf{x}$ is an n vector and $\mathbf{G}$ is an $n \times n$ constant matrix, tend to zero as k approaches infinity, then all solutions of the system

$$\mathbf{x}(k+1) = \mathbf{G}\mathbf{x}(k) + \mathbf{H}\mathbf{u}(k) \tag{5–194}$$

where $\mathbf{H}$ is an $n \times r$ constant matrix, are bounded, provided that the input vector $\mathbf{u}(k)$, an r vector, is bounded.

Solution. Since $\mathbf{u}(k)$ is bounded, there exists a positive constant c such that

$$\|\mathbf{u}(k)\| < c \qquad k = 0, 1, 2, \ldots$$

The solution of Eq. (5–194) is given by

$$\mathbf{x}(k) = \mathbf{G}^k\mathbf{x}(0) + \sum_{j=1}^{k} \mathbf{G}^{k-j}\mathbf{H}\mathbf{u}(j-1)$$

Hence,

$$\|\mathbf{x}(k)\| < \|\mathbf{G}\|^k\|\mathbf{x}(0)\| + c\sum_{j=1}^{k} \|\mathbf{G}\|^{k-j}\|\mathbf{H}\| \leq \|\mathbf{G}\|^k\|\mathbf{x}(0)\| + \lim_{k\to\infty} c\sum_{j=1}^{k} \|\mathbf{G}\|^{k-j}\|\mathbf{H}\|$$

Since the origin of the homogeneous system given by Eq. (5–193) is asymptotically stable, there exist positive constants a and b $(0 < b < 1)$ such that

$$0 < \|\mathbf{G}\|^k < ab^k$$

Then

$$\lim_{k\to\infty} \sum_{j=1}^{k} \|\mathbf{G}\|^{k-j} < \lim_{k\to\infty} \sum_{j=1}^{k} ab^{k-j} = a\frac{1}{1-b}$$

Therefore,

$$\|\mathbf{x}(k)\| < a\|\mathbf{x}(0)\| + c\|\mathbf{H}\|a\frac{1}{1-b}$$

We have thus proved that $\|\mathbf{x}(k)\|$ is bounded.

Problem A-5–36. Consider the system defined by the equations

$$x_1(k+1) = 2x_1(k) + 0.5x_2(k) - 5$$

$$x_2(k+1) = 0.8x_2(k) + 2$$

Determine the stability of the equilibrium state.

Solution. Define the equilibrium state as

$$x_1(k) = x_{1e}, \qquad x_2(k) = x_{2e}$$

Then such an equilibrium state can be determined from the following two simultaneous equations:

$$x_{1e} = 2x_{1e} + 0.5x_{2e} - 5$$

$$x_{2e} = 0.8x_{2e} + 2$$

or

$$x_{1e} = 0, \qquad x_{2e} = 10$$

The equilibrium state is thus (0, 10).

Now, let us consider a new coordinate system with the origin at the equilibrium state. Define

$$\hat{x}_1(k) = x_1(k)$$

$$\hat{x}_2(k) = x_2(k) - 10$$

Then the system equations become

$$\hat{x}_1(k+1) = 2\hat{x}_1(k) + 0.5[\hat{x}_2(k) + 10] - 5$$

$$\hat{x}_2(k+1) + 10 = 0.8[\hat{x}_2(k) + 10] + 2$$

or

$$\begin{bmatrix} \hat{x}_1(k+1) \\ \hat{x}_2(k+1) \end{bmatrix} = \begin{bmatrix} 2 & 0.5 \\ 0 & 0.8 \end{bmatrix} \begin{bmatrix} \hat{x}_1(k) \\ \hat{x}_2(k) \end{bmatrix}$$

To determine the stability of the origin of the system in the new coordinate system, let us apply the Liapunov stability equation given by Eq. (5–155):

$$\begin{bmatrix} 2 & 0 \\ 0.5 & 0.8 \end{bmatrix} \begin{bmatrix} p_{11} & p_{12} \\ p_{12} & p_{22} \end{bmatrix} \begin{bmatrix} 2 & 0.5 \\ 0 & 0.8 \end{bmatrix} - \begin{bmatrix} p_{11} & p_{12} \\ p_{12} & p_{22} \end{bmatrix} = -\begin{bmatrix} 9 & 0 \\ 0 & 0.35 \end{bmatrix}$$

where we choose $\mathbf{Q}$ to be a positive definite matrix having elements that simplify the computation involved. Solving this last equation for matrix $\mathbf{P}$, we obtain

$$\mathbf{P} = \begin{bmatrix} p_{11} & p_{12} \\ p_{12} & p_{22} \end{bmatrix} = \begin{bmatrix} -3 & 5 \\ 5 & 10 \end{bmatrix}$$

By applying the Sylvester criterion for positive definiteness, we find that matrix $\mathbf{P}$ is not positive definite. Therefore, the origin (equilibrium state) is not stable.

The instability of the equilibrium state can, of course, be determined by the z transform approach. Let us first eliminate $\hat{x}_2$ from the state equation. Then we have

$$\hat{x}_1(k+2) - 2.8\hat{x}_1(k+1) + 1.6\hat{x}_1(k) = 0$$

The characteristic equation for the system in the z domain is

$$z^2 - 2.8z + 1.6 = 0$$

or

$$(z-2)(z-0.8) = 0$$

Hence

$$z = 2, \qquad z = 0.8$$

Since pole $z = 2$ is located outside the unit circle in the z plane, the origin (equilibrium state) is unstable.

PROBLEMS

Problem B-5–1. Using the direct programming method, obtain a state space representation of the following pulse-transfer-function system:

$$\frac{Y(z)}{U(z)} = \frac{z^{-1} + 2z^{-2}}{1 + 4z^{-1} + 3z^{-2}}$$

Problem B-5–2. Using the nested programming method, obtain a state space representation of the following pulse-transfer-function system:

$$\frac{Y(z)}{U(z)} = \frac{z^{-2} + 4z^{-3}}{1 + 6z^{-1} + 11z^{-2} + 6z^{-3}}$$

Problem B-5-3. Using the direct programming method, obtain a state space representation of the following system:

$$\frac{Y(z)}{U(z)} = \frac{1 + 2z^{-1} + z^{-2}}{1 + 5z^{-1} + 6z^{-2}}$$

Draw a block diagram for the system showing the state variables, output variable, and input variable.

Problem B-5-4. Using the partial-fraction-expansion programming method, obtain a state space representation of the following pulse-transfer-function system:

$$\frac{Y(z)}{U(z)} = \frac{1 + 6z^{-1} + 8z^{-2}}{1 + 4z^{-1} + 3z^{-2}}$$

Problem B-5-5. Obtain a state space representation of the system described by the equation

$$y(k + 2) + y(k + 1) + 0.16y(k) = u(k + 1) + 2u(k)$$

Problem B-5-6. Obtain the state equation and output equation for the system shown in Fig. 5-22.

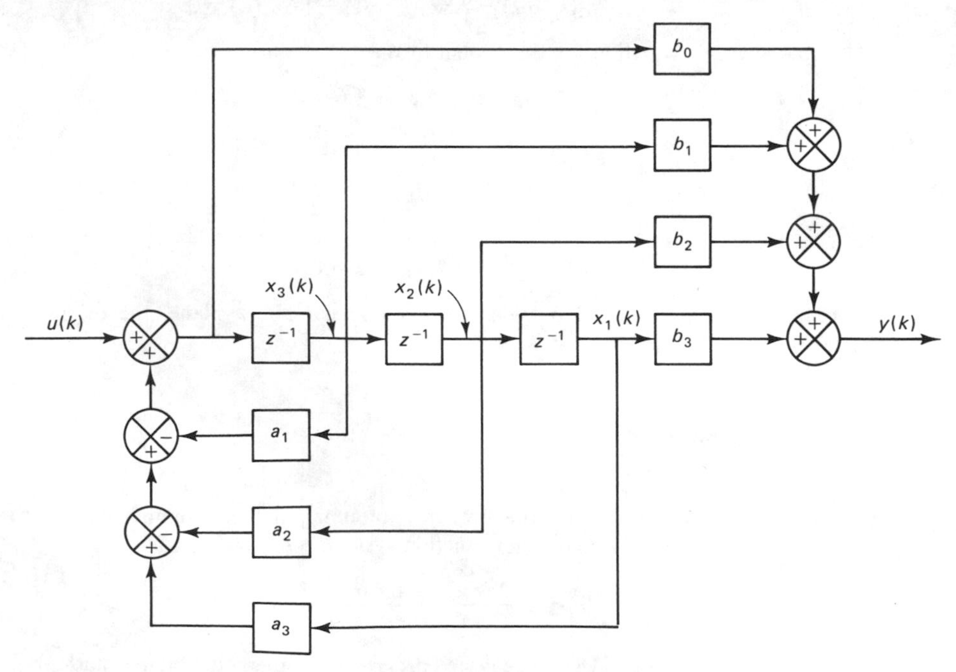

Figure 5-22 Block diagram of a control system.

Problem B-5–7. Obtain the state equation and output equation for the system shown in Fig. 5–23.

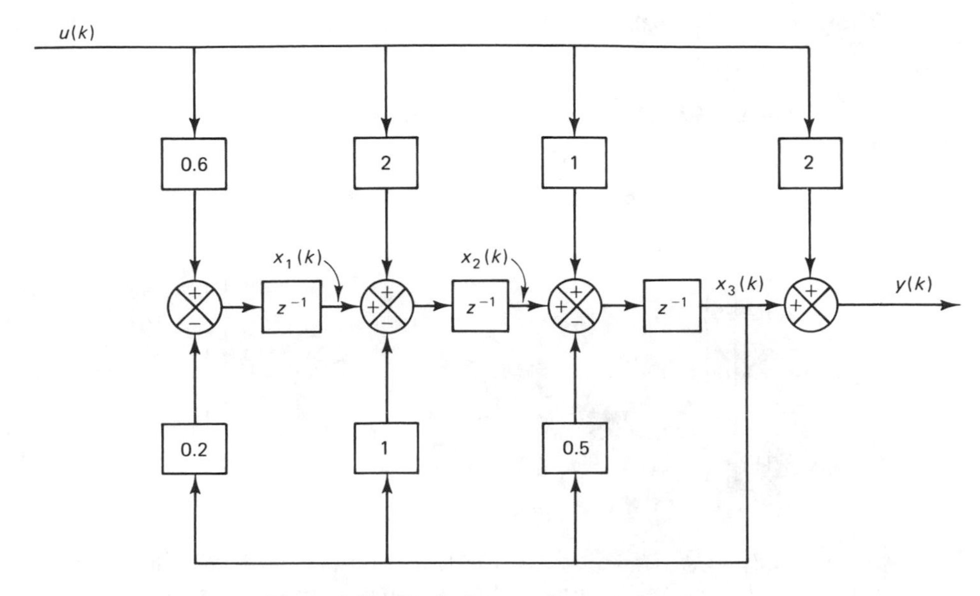

Figure 5–23 Block diagram of a control system.

Problem B-5–8. Obtain a state space representation of the discrete-time control system shown in Fig. 5–24.

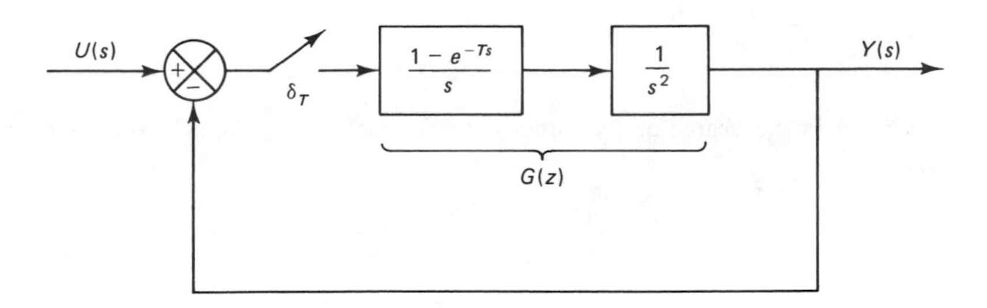

Figure 5–24 Discrete-time control system.

Problem B-5–9. Obtain a state space representation of the following system:

$$\frac{Y(z)}{U(z)} = \frac{z^{-1} + 2z^{-2}}{1 + 0.7z^{-1} + 0.12z^{-2}}$$

Choose state variables such that the state matrix is a diagonal matrix.

Problem B-5–10. Obtain a state space representation of the following pulse-transfer-function system such that the state matrix is a diagonal matrix:

$$\frac{Y(z)}{U(z)} = \frac{1}{(z+1)(z+2)(z+3)}$$

Then, obtain the initial state variables $x_1(0)$, $x_2(0)$, and $x_3(0)$ in terms of $y(0)$, $y(1)$, and $y(2)$.

Problem B-5-11. A state space representation of the scalar difference equation system

$$\begin{aligned} y(k+n) + a_1(k)y(k+n-1) + \cdots + a_n(k)y(k) \\ = b_0(k)u(k+n) + b_1(k)u(k+n-1) + \cdots + b_n(k)u(k) \end{aligned}$$

where $k = 0, 1, 2, \ldots,$ may be given by

$$\begin{bmatrix} x_1(k+1) \\ x_2(k+1) \\ \cdot \\ \cdot \\ \cdot \\ x_{n-1}(k+1) \\ x_n(k+1) \end{bmatrix} = \begin{bmatrix} 0 & 1 & \cdots & 0 & 0 \\ 0 & 0 & \cdots & 0 & 0 \\ \cdot & \cdot & & \cdot & \cdot \\ \cdot & \cdot & & \cdot & \cdot \\ \cdot & \cdot & & \cdot & \cdot \\ 0 & 0 & \cdots & 0 & 1 \\ -a_n(k) & -a_{n-1}(k) & \cdots & -a_2(k) & -a_1(k) \end{bmatrix} \begin{bmatrix} x_1(k) \\ x_2(k) \\ \cdot \\ \cdot \\ \cdot \\ x_{n-1}(k) \\ x_n(k) \end{bmatrix} + \begin{bmatrix} h_1(k) \\ h_2(k) \\ \cdot \\ \cdot \\ \cdot \\ h_{n-1}(k) \\ h_n(k) \end{bmatrix} u(k)$$

$$y(k) = x_1(k) + b_0(k-n)u(k)$$

Determine $h_1(k), h_2(k), \ldots, h_n(k)$ in terms of $a_i(k)$ and $b_j(k)$, where $i = 1, 2, \ldots, n$ and $j = 0, 1, \ldots, n$. Determine also the initial values of the state variables $x_1(0), x_2(0), \ldots, x_n(0)$ in terms of the input sequence $u(0), u(1), \ldots, u(n-1)$ and the output sequence $y(0), y(1), \ldots, y(n-1)$.

Problem B-5-12. Assume that the minimal polynomial of an $n \times n$ matrix $\mathbf{G}$ has only distinct eigenvalues. Show that a function of matrix $\mathbf{G}$, $f(\mathbf{G})$, may be given by

$$f(\mathbf{G}) = \sum_{k=1}^{m} \mathbf{X}_k f(z_k) \tag{5-195}$$

where m is the degree of the minimal polynomial for $\mathbf{G}$, the z_k's are its roots, and the $\mathbf{X}_k$'s are given by

$$\mathbf{X}_k = \frac{(\mathbf{G} - z_1\mathbf{I}) \cdots (\mathbf{G} - z_{k-1}\mathbf{I})(\mathbf{G} - z_{k+1}\mathbf{I}) \cdots (\mathbf{G} - z_m\mathbf{I})}{(z_k - z_1) \cdots (z_k - z_{k-1})(z_k - z_{k+1}) \cdots (z_k - z_m)}$$

Then, using Eq. (5-195), obtain $\mathbf{G}^k$, where

$$\mathbf{G} = \begin{bmatrix} 0 & 1 \\ -0.16 & -1 \end{bmatrix}$$

Problem B-5-13. If the minimal polynomial of an $n \times n$ matrix $\mathbf{G}$ involves only distinct roots, then the inverse of $z\mathbf{I} - \mathbf{G}$ can be given by the following expression:

$$(z\mathbf{I} - \mathbf{G})^{-1} = \sum_{k=1}^{m} \frac{\mathbf{X}_k}{z - z_k} \tag{5-196}$$

where m is the degree of the minimal polynomial of $\mathbf{G}$ and the $\mathbf{X}_k$'s are $n \times n$ matrices determined from

$$g_j(\mathbf{G}) = g_j(z_1)\mathbf{X}_1 + g_j(z_2)\mathbf{X}_2 + \cdots + g_j(z_m)\mathbf{X}_m$$

where

$$g_j(\mathbf{G}) = (\mathbf{G} - z_k\mathbf{I})^{j-1}, \qquad g_j(z) = (z - z_k)^{j-1}$$

where $j = 1, 2, \ldots, m$ and z_k is any one of the roots of the minimal polynomial of **G**.
Using Eq. (5–196), obtain $(z\mathbf{I} - \mathbf{G})^{-1}$ for the following 2×2 matrix **G**:

$$\mathbf{G} = \begin{bmatrix} 0 & 1 \\ 0 & -2 \end{bmatrix}$$

Problem B-5–14. Referring to Prob. B-5–13, modify Eq. (5–196) for the case where the minimal polynomial involves a multiple root of order 2 and other roots that are distinct. Then, using the modified equation obtain $(z\mathbf{I} - \mathbf{G})^{-1}$ for the following 3×3 matrix **G**:

$$\mathbf{G} = \begin{bmatrix} 5 & 4 & 0 \\ 0 & 1 & 0 \\ -4 & 4 & 1 \end{bmatrix}$$

Problem B-5–15. By using Sylvester's interpolation formula given by Eq. (5–117), obtain $e^{\mathbf{A}t}$, where matrix **A** is given by

$$\mathbf{A} = \begin{bmatrix} 0 & 1 & 0 \\ 0 & 0 & 1 \\ 1 & -3 & 3 \end{bmatrix}$$

Problem B-5–16. Show that

$$\exp\left\{\begin{bmatrix} \lambda_1 & 0 & 0 \\ 0 & \lambda_2 & 0 \\ 0 & 0 & \lambda_3 \end{bmatrix} t\right\} = \begin{bmatrix} e^{\lambda_1 t} & 0 & 0 \\ 0 & e^{\lambda_2 t} & 0 \\ 0 & 0 & e^{\lambda_3 t} \end{bmatrix}$$

and

$$\exp\left\{\begin{bmatrix} 0 & 1 & 0 & 0 \\ 0 & 0 & 1 & 0 \\ 0 & 0 & 0 & 1 \\ 0 & 0 & 0 & 0 \end{bmatrix} t\right\} = \begin{bmatrix} 1 & t & \frac{t^2}{2!} & \frac{t^3}{3!} \\ 0 & 1 & t & \frac{t^2}{2!} \\ 0 & 0 & 1 & t \\ 0 & 0 & 0 & 1 \end{bmatrix}$$

Problem B-5–17. Show that

$$\exp\left\{\begin{bmatrix} \lambda & 1 & 0 & 0 \\ 0 & \lambda & 1 & 0 \\ 0 & 0 & \lambda & 1 \\ 0 & 0 & 0 & \lambda \end{bmatrix} t\right\} = e^{\lambda t}\begin{bmatrix} 1 & t & \frac{t^2}{2!} & \frac{t^3}{3!} \\ 0 & 1 & t & \frac{t^2}{2!} \\ 0 & 0 & 1 & t \\ 0 & 0 & 0 & 1 \end{bmatrix}$$

and

$$\exp\left\{\begin{bmatrix} \sigma & \omega \\ -\omega & \sigma \end{bmatrix} t\right\} = \begin{bmatrix} e^{\sigma t}\cos\omega t & e^{\sigma t}\sin\omega t \\ -e^{\sigma t}\sin\omega t & e^{\sigma t}\cos\omega t \end{bmatrix}$$

Problem B-5–18. Obtain the pulse transfer function of the system defined by the equations

$$\mathbf{x}(k+1) = \mathbf{G}\mathbf{x}(k) + \mathbf{H}u(k)$$

$$y(k) = \mathbf{C}\mathbf{x}(k) + Du(k)$$

where

$$\mathbf{G} = \begin{bmatrix} -a_1 & -a_2 & -a_3 \\ 1 & 0 & 0 \\ 0 & 1 & 0 \end{bmatrix}, \qquad \mathbf{H} = \begin{bmatrix} 1 \\ 0 \\ 0 \end{bmatrix}$$

$$\mathbf{C} = [b_1 - a_1 b_0 \quad b_2 - a_2 b_0 \quad b_3 - a_3 b_0], \qquad D = b_0$$

Problem B-5–19. Obtain a state space representation for the system defined by the following pulse-transfer-function matrix:

$$\begin{bmatrix} Y_1(z) \\ Y_2(z) \end{bmatrix} = \begin{bmatrix} \dfrac{1}{1-z^{-1}} & \dfrac{1+z^{-1}}{1-z^{-1}} \\ \dfrac{1}{1+0.6z^{-1}} & \dfrac{1+z^{-1}}{1+0.6z^{-1}} \end{bmatrix} \begin{bmatrix} U_1(z) \\ U_2(z) \end{bmatrix}$$

Problem B-5–20. Obtain the weighting sequence function for the discrete-time system described by the equations

$$\begin{bmatrix} x_1(k+1) \\ x_2(k+1) \end{bmatrix} = \begin{bmatrix} 0 & 1 \\ -0.16 & -1 \end{bmatrix} \begin{bmatrix} x_1(k) \\ x_2(k) \end{bmatrix} + \begin{bmatrix} 0 \\ 1 \end{bmatrix} u(k)$$

$$y(k) = [1 \quad 0] \begin{bmatrix} x_1(k) \\ x_2(k) \end{bmatrix}$$

Problem B-5–21. Show that similar matrices have the same minimal polynomial, that is, that for a nonsingular matrix $\mathbf{T}$, the minimal polynomial of $\mathbf{T}^{-1}\mathbf{A}\mathbf{T}$ and that of $\mathbf{A}$ are the same.

Problem B-5–22. Show that the solution of the matrix differential equation

$$\frac{d}{dt}\mathbf{X}(t) = \mathbf{A}\mathbf{X}(t) + \mathbf{X}(t)\mathbf{B}, \qquad \mathbf{X}(0) = \mathbf{C}$$

where matrices $\mathbf{A}$, $\mathbf{B}$, and $\mathbf{C}$ are constant matrices, is given by

$$\mathbf{X} = e^{\mathbf{A}t}\mathbf{C}e^{\mathbf{B}t}$$

Then obtain the solution of

$$\frac{d}{dt}\mathbf{X}(t) = \mathbf{A}(t)\mathbf{X}(t) + \mathbf{X}(t)\mathbf{A}^*(t), \qquad \mathbf{X}(t_0) = \mathbf{C}$$

where the elements of $\mathbf{A}(t)$ are continuous functions of time.

Problem B-5–23. Obtain the state transition matrix $\mathbf{\Phi}(t)$ for the system

$$\dddot{x} = ax$$

where a is a constant.

Problem B-5–24. Consider the discrete-time state equation

$$\begin{bmatrix} x_1(k+1) \\ x_2(k+1) \end{bmatrix} = \begin{bmatrix} 0 & 1 \\ -0.24 & -1 \end{bmatrix} \begin{bmatrix} x_1(k) \\ x_2(k) \end{bmatrix}$$

Obtain the state transition matrix $\mathbf{\Psi}(k)$.

Problem B-5–25. Consider the continuous-time state equation

$$\begin{bmatrix} \dot{x}_1 \\ \dot{x}_2 \end{bmatrix} = \begin{bmatrix} 0 & 1 \\ -2 & -3 \end{bmatrix} \begin{bmatrix} x_1 \\ x_2 \end{bmatrix}$$

Obtain the state transition matrix $\mathbf{\Phi}(t)$.

Problem B-5–26. Consider the following time-varying discrete-time state equation:

$$\mathbf{x}(k+1) = \mathbf{G}(k)\mathbf{x}(k) + \mathbf{H}(k)\mathbf{u}(k), \qquad \mathbf{x}(0) = \mathbf{x}_0$$

Show that the solution of this equation may be given by

$$\begin{aligned} \mathbf{x}(k) = \mathbf{G}(k-1)\mathbf{G}(k-2) \cdots \mathbf{G}(0)\mathbf{x}_0 + \mathbf{G}(k-1)\mathbf{G}(k-2) \cdots \mathbf{G}(1)\mathbf{H}(0)\mathbf{u}(0) + \cdots \\ + \mathbf{G}(k-1)\mathbf{H}(k-2)\mathbf{u}(k-2) + \mathbf{H}(k-1)\mathbf{u}(k-1) \end{aligned}$$

In particular, show that if $\mathbf{G}(k)$ and $\mathbf{H}(k)$ are constant matrices, then

$$\mathbf{x}(k) = \mathbf{G}^k\mathbf{x}_0 + \mathbf{G}^{k-1}\mathbf{H}\mathbf{u}(0) + \mathbf{G}^{k-2}\mathbf{H}\mathbf{u}(1) + \cdots + \mathbf{G}^2\mathbf{H}\mathbf{u}(k-3) + \mathbf{G}\mathbf{H}\mathbf{u}(k-2) + \mathbf{H}\mathbf{u}(k-1)$$

Problem B-5–27. Consider matrix $\mathbf{A}$ given as follows:

$$\mathbf{A} = \begin{bmatrix} -3 & -1 \\ 2 & 0 \end{bmatrix}$$

Obtain $e^{\mathbf{A}t}$ by three different methods.

Problem B-5–28. By using Sylvester's interpolation formula, compute $e^{\mathbf{A}t}$ where

$$\mathbf{A} = \begin{bmatrix} 0 & 1 & 0 \\ 0 & 0 & 1 \\ -6 & -11 & -6 \end{bmatrix}$$

Problem B-5–29. In the system shown in Fig. 5–25, the input signal is sampled and the sampled signal is fed to the zero-order hold. First obtain a continuous-time state space representation of the plant. Then, using the discretization technique, obtain the discrete-time state space representation of the plant.

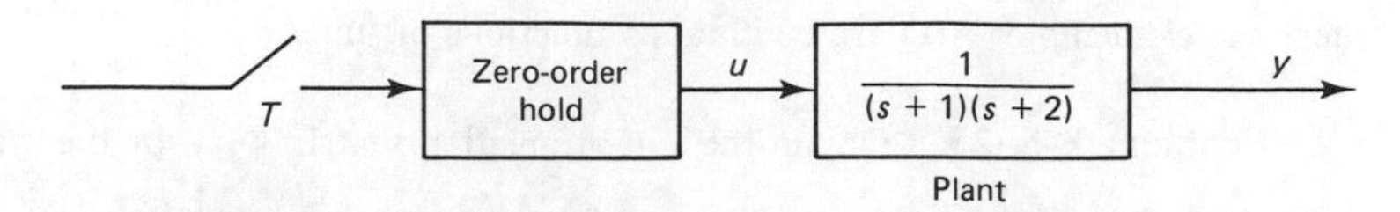

Figure 5–25 Block diagram of the discrete-time system for Prob. B-5–29.

Problem B-5–30. Consider the system governed by the difference equation

$$\mathbf{x}(k+1) = \mathbf{f}(\mathbf{x}(k), \mathbf{u}(k))$$

Assuming that $\mathbf{u}$ is a constant vector, or $\mathbf{u} = \mathbf{c}$ (where $\mathbf{c}$ is a constant vector), find the equilibrium state $\mathbf{x}_e$.

Problem B-5–31. Consider the system defined by

$$\mathbf{x}(k+1) = \mathbf{G}\mathbf{x}(k) + \mathbf{H}\mathbf{u}(k)$$

$$\mathbf{y}(k) = \mathbf{C}\mathbf{x}(k) + \mathbf{D}\mathbf{u}(k)$$

where matrix $\mathbf{G}$ is a stable matrix.

Obtain the steady-state values of $\mathbf{x}(k)$ and $\mathbf{y}(k)$ when $\mathbf{u}(k)$ is a constant vector.

Problem B-5–32. Consider the free discrete-time system

$$\mathbf{x}(k+1) = \mathbf{G}\mathbf{x}(k)$$

Let us assume that the eigenvalues $\lambda_1, \lambda_2, \ldots, \lambda_n$ of $\mathbf{G}$ are distinct. Define the eigenvector corresponding to the eigenvalue λ_i as $\mathbf{v}_i$. Then

$$(\lambda_i \mathbf{I} - \mathbf{G})\mathbf{v}_i = \mathbf{0}$$

The eigenvectors $\mathbf{v}_1, \mathbf{v}_2, \ldots, \mathbf{v}_n$ are linearly independent. Now define the transformation matrix $\mathbf{T}$ as follows:

$$\mathbf{T} = [\mathbf{v}_1 \vdots \mathbf{v}_2 \vdots \cdots \vdots \mathbf{v}_n]$$

Then $\mathbf{T}$ is nonsingular and

$$\mathbf{T}^{-1}\mathbf{G}\mathbf{T} = \begin{bmatrix} \lambda_1 & & & & & 0 \\ & \lambda_2 & & & & \\ & & \cdot & & & \\ & & & \cdot & & \\ & & & & \cdot & \\ 0 & & & & & \lambda_n \end{bmatrix}$$

Define

$$\mathbf{x}(k) = \mathbf{T}\mathbf{z}(k) = [\mathbf{v}_1 \vdots \mathbf{v}_2 \vdots \cdots \vdots \mathbf{v}_n] \begin{bmatrix} z_1(k) \\ z_2(k) \\ \cdot \\ \cdot \\ \cdot \\ z_n(k) \end{bmatrix}$$

$$= \mathbf{v}_1 z_1(k) + \mathbf{v}_2 z_2(k) + \cdots + \mathbf{v}_n z_n(k) \qquad (5\text{–}197)$$

Then the given free discrete-time system becomes

$$\mathbf{T}\mathbf{z}(k+1) = \mathbf{G}\mathbf{T}\mathbf{z}(k)$$

Thus

$$\mathbf{z}(k+1) = \mathbf{T}^{-1}\mathbf{G}\mathbf{T}\mathbf{z}(k) = \begin{bmatrix} \lambda_1 & & & & 0 \\ & \lambda_2 & & & \\ & & \cdot & & \\ & & & \cdot & \\ & & & & \cdot \\ 0 & & & & \lambda_n \end{bmatrix} \mathbf{z}(k)$$

or

$$z_i(k+1) = \lambda_i z_i(k) \qquad i = 1, 2, \ldots, n$$

where $z_i(k)$ is called the ith mode of the system. The solution of $z_i(k)$ is given by the equation

$$z_i(k) = \lambda_i^k z_i(0) \qquad k = 0, 1, 2, \ldots \tag{5–198}$$

The behavior of each mode is determined by λ_i.

By substituting Eq. (5–198) into Eq. (5–197) we obtain

$$\mathbf{x}(k) = \mathbf{v}_1\lambda_1^k z_1(0) + \mathbf{v}_2\lambda_2^k z_2(0) + \cdots + \mathbf{v}_n\lambda_n^k z_n(0)$$

This last expansion of $\mathbf{x}(k)$ is called the *modal expansion* of $\mathbf{x}(k)$. Note that the $z_i(0)$'s are obtained from the relation

$$\mathbf{z}(0) = \mathbf{T}^{-1}\mathbf{x}(0)$$

Examine the behavior of $z_i(k)$ when $|\lambda_i| > 1$, when $\lambda_i = 1$, when $|\lambda_i| < 1$, and when $\lambda_i = -1$. Suppose matrix $\mathbf{G}$ is given by the equation

$$\mathbf{G} = \begin{bmatrix} 0 & 1 \\ 1 & 0 \end{bmatrix}$$

and the initial state

$$\mathbf{x}(0) = \begin{bmatrix} 1 \\ 1 \end{bmatrix}$$

Obtain the modal expansion of $\mathbf{x}(k)$. Examine the behavior of $\mathbf{x}(k)$.

Problem B-5–33. Consider the system defined by

$$\mathbf{x}(k+1) = \mathbf{G}\mathbf{x}(k)$$

where $\mathbf{G}$ is a stable matrix.

Show that for a positive definite (or positive semidefinite) matrix $\mathbf{Q}$

$$J = \sum_{k=0}^{\infty} \mathbf{x}^*(k)\mathbf{Q}\mathbf{x}(k)$$

can be given by

$$J = \mathbf{x}^*(0)\mathbf{P}\mathbf{x}(0)$$

where $\mathbf{P} = \mathbf{Q} + \mathbf{G}^*\mathbf{P}\mathbf{G}$.

Problem B-5-34. Determine the stability of the origin of the following discrete-time system:

$$\begin{bmatrix} x_1(k+1) \\ x_2(k+1) \\ x_3(k+1) \end{bmatrix} = \begin{bmatrix} 1 & 3 & 0 \\ -3 & -2 & -3 \\ 1 & 0 & 0 \end{bmatrix} \begin{bmatrix} x_1(k) \\ x_2(k) \\ x_3(k) \end{bmatrix}$$

Problem B-5-35. Determine the stability of the origin of the following discrete-time system:

$$\begin{bmatrix} x_1((k+1)T) \\ x_2((k+1)T) \end{bmatrix} = \begin{bmatrix} \cos T & \sin T \\ -\sin T & \cos T \end{bmatrix} \begin{bmatrix} x_1(kT) \\ x_2(kT) \end{bmatrix}$$

Problem B-5-36. Consider the system defined by the equations

$$x_1(k+1) = x_1(k) + 0.2x_2(k) + 0.4$$

$$x_2(k+1) = 0.5x_1(k) - 0.5$$

Determine the stability of the equilibrium state.

6

Analysis and Design in State Space

6–1 INTRODUCTION

In the first part of this chapter we present two fundamental concepts of control systems: controllability and observability. Controllability is concerned with the problem of whether it is possible to steer a system from a given initial state to an arbitrary state: a system is said to be controllable if it is possible by means of an unbounded control vector to transfer the system from any initial state to any other state in a finite number of sampling periods. (Thus, the concept of controllability is concerned with the existence of a control vector that can cause the system's state to reach some arbitrary state.)

Observability is concerned with the problem of determining the state of a dynamic system from observations of the output and control vectors in a finite number of sampling periods. A system is said to be observable if, with the system in state $\mathbf{x}(0)$, it is possible to determine this state from the observation of the output and control vectors over a finite number of sampling periods.

The concepts of controllability and observability were introduced by R. E. Kalman. They play an important role in the optimal control of multivariable systems. In fact, the conditions of controllability and observability may govern the existence of a complete solution to an optimal control problem. Specifically, the concept of controllability is the basis for the solutions of the pole placement problem (see Sec. 6–5) and the concept of observability plays an important role for the design of state observers (see Sec. 6–6).

In the second part of this chapter we discuss the pole placement design method and state observers. The design method based on pole placement coupled with state observers is one of the fundamental design methods available to control engineers. If the system is completely state controllable, then the desired closed-loop poles in the z plane (or the roots of the characteristic equation) can be selected and the system that will give such closed-loop poles can be designed. The design approach of placing the closed-loop poles in the desired locations in the z plane is called the *pole placement design technique*; that is, in the pole placement design technique we feed back all state variables so that all poles of the closed-loop system are placed at desired locations. In practical control systems, however, measurement of all state variables may not be possible; in that case, not all state variables will be available for feedback. In order to implement a design based on state feedback, it becomes necessary to estimate the unmeasurable state variables. Such estimation can be done by use of state observers, which will be discussed in detail in this chapter.

The pole placement design process of control systems may be separated into two phases. In the first phase, we design the system assuming all state variables are available for feedback. In the second phase, we design the state observer that estimates all state variables (or only those that are not directly measurable) that are required for feedback to complete the design.

Note that in the preceding design approach the design parameters are the locations of desired closed-loop poles and the sampling period T. (The sampling period T effectively determines the settling time for response.)

In the analysis in this chapter we assume that the disturbances are impulses that take place randomly. The effect of such impulses is to change the system state. Therefore, a disturbance may be represented as an initial state. It is assumed further that the spacing between adjacent disturbances is sufficiently wide that any response to such a disturbance settles down before the next disturbance takes place, so that the system is always ready for the next round.

Thus, our concern in this chapter is primarily with the regulator problem, which applies when the system is subjected to initial conditions. The problem is to reduce the error vector to zero with sufficient speed. In the regulator problem the pole placement formulation of the design boils down to the determination of the desired state feedback gain matrix. The procedure for determining the state feedback gain matrix is first to select suitable locations for all closed-loop poles and then to determine the state feedback gain matrix that yields the specified closed-loop poles so that the effects of the disturbances can be reduced to zero with sufficient speed. In the final state of the design process the state feedback is accomplished by use of the estimated state variables rather than the actual state variables, which are probably not available for direct measurement. If some of the state variables are measurable, then we may use those available state variables and use estimated state variables for those not actually measurable.

In the last part of this chapter we treat a servo design problem that uses integral control coupled with the pole placement technique and the state observer. Note that in the regulator problem we desire to transfer the nonzero error vector (due to disturbance) to the origin. In the servo problem, we require the output to follow the command input. Note that the servo system must follow the command input and at the same time must solve any regulator problem. Consequently, in the design of a servo system we may begin with the design of a regulator system and then modify the regulator system to a servo system.

Outline of the chapter. Section 6–1 has presented an introduction to the materials to be presented in this chapter. Section 6–2 discusses the controllability of linear time-invariant control systems. Section 6–3 treats the observability of such systems. Section 6–4 reviews useful transformations in state space analysis and design that we shall use in the remaining sections of this chapter. The basic state space design method is presented in Secs. 6–5 and 6–6. Section 6–5 presents the pole placement method, the first phase of the design. In the pole placement method we assume that all state variables can be measured and are available for feedback. Section 6–6 discusses the second phase of the design, the design of observers which estimate the state variables that are not actually measurable. Estimation is based on the measurements of the output and control signals. The estimated state variables can be used for state feedback based on the pole placement design. The final section, Sec. 6–7, treats servo systems and discusses the design of such systems; the section concludes with a design example.

6–2 CONTROLLABILITY

A control system is said to be completely state controllable if it is possible to transfer the system from any arbitrary initial state to any desired state (also an arbitrary state) in a finite time period. That is, a control system is controllable if every state variable can be controlled in a finite time period by some unconstrained control signal. If any state variable is independent of the control signal, then it is impossible to control this state variable and therefore the system is uncontrollable.

The solution to an optimal control problem may not exist if the system considered is not controllable. Although most physical systems are controllable, the corresponding mathematical models may not possess the property of controllability. Therefore, it is necessary to know the condition under which a system is controllable. We shall see later in Sec. 6–5 that the concept of controllability plays an important role in arbitrary pole placement of control systems. Now we shall derive this condition in the following.

Complete state controllability for a linear time-invariant discrete-time control system. Consider the discrete-time control system defined by

$$\mathbf{x}((k+1)T) = \mathbf{G}\mathbf{x}(kT) + \mathbf{H}u(kT) \tag{6–1}$$

where

$\mathbf{x}(kT)$ = state vector (n-vector) at kth sampling instant

$u(kT)$ = control signal at kth sampling instant

$\mathbf{G} = n \times n$ matrix

$\mathbf{H} = n \times 1$ matrix

$T =$ sampling period

We assume that $u(kT)$ is constant for $kT \leq t < (k+1)T$.

The discrete-time control system given by Eq. (6–1) is said to be completely state controllable or simply state controllable if there exists a piecewise-constant control signal $u(kT)$ defined over a finite number of sampling periods such that, starting from any initial state, the state $\mathbf{x}(kT)$ can be transferred to the desired state $\mathbf{x}_f$ in at most n sampling periods. (In discussing controllability, the desired state $\mathbf{x}_f$ may be specified as the origin, or $\mathbf{x}_f = \mathbf{0}$. See Prob. A-6–1. Here, however, we assume that $\mathbf{x}_f$ is an arbitrary state in the n-dimensional space, including the origin.)

Using the definition just given, we shall now derive the condition for complete state controllability. Since the solution of Eq. (6–1) is

$$\begin{aligned}\mathbf{x}(nT) &= \mathbf{G}^n\mathbf{x}(0) + \sum_{j=0}^{n-1} \mathbf{G}^{n-j-1}\mathbf{H}u(jT) \\ &= \mathbf{G}^n\mathbf{x}(0) + \mathbf{G}^{n-1}\mathbf{H}u(0) + \mathbf{G}^{n-2}\mathbf{H}u(T) + \cdots + \mathbf{H}u((n-1)T)\end{aligned}$$

we obtain

$$\mathbf{x}(nT) - \mathbf{G}^n\mathbf{x}(0) = [\mathbf{H} \vdots \mathbf{GH} \vdots \cdots \vdots \mathbf{G}^{n-1}\mathbf{H}] \begin{bmatrix} u((n-1)T) \\ u((n-2)T) \\ \cdot \\ \cdot \\ \cdot \\ u(0) \end{bmatrix} \tag{6–2}$$

Since $\mathbf{H}$ is an $n \times 1$ matrix, we find that each of the matrices $\mathbf{H}, \mathbf{GH}, \ldots, \mathbf{G}^{n-1}\mathbf{H}$ is an $n \times 1$ matrix or column vector. If the rank of the following matrix is n, or

$$\text{rank}\,[\mathbf{H} \vdots \mathbf{GH} \vdots \cdots \vdots \mathbf{G}^{n-1}\mathbf{H}] = n \tag{6–3}$$

then n vectors $\mathbf{H}, \mathbf{GH}, \ldots, \mathbf{G}^{n-1}\mathbf{H}$ can span the n-dimensional space. The matrix of Eq. (6–3) is commonly called the *controllability matrix*. (Note that all states that can be reached from the origin are spanned by the columns of the controllability matrix.) Thus, if the rank of the controllability matrix is n, then for an arbitrary state $\mathbf{x}(nT) = \mathbf{x}_f$, there exists a sequence of unbounded control signals $u(0)$, $u(T)$, $\ldots$, $u((n-1)T)$ that satisfies Eq. (6–2). Hence, the condition that the rank of the controllability matrix be n gives a sufficient condition for complete state controllability.

To prove that Eq. (6–3) is also a necessary condition for complete state controllability, let us assume that

$$\text{rank}\,[\mathbf{H} \vdots \mathbf{GH} \vdots \cdots \vdots \mathbf{G}^{n-1}\mathbf{H}] < n$$

Then by use of the Cayley-Hamilton theorem it can be shown that for an arbitrary i, $\mathbf{G}^i\mathbf{H}$ can be expressed as a linear combination of $\mathbf{H}, \mathbf{GH}, \ldots, \mathbf{G}^{n-1}\mathbf{H}$. Consequently, we have for any i

$$\text{rank}\,[\mathbf{H} \vdots \mathbf{GH} \vdots \cdots \vdots \mathbf{G}^{i-1}\mathbf{H}] < n$$

and so the vectors $\mathbf{H}, \mathbf{GH}, \ldots, \mathbf{G}^{i-1}\mathbf{H}$ cannot span the n-dimensional space; and therefore, for some $\mathbf{x}_f$, it is not possible to have $\mathbf{x}(iT) = \mathbf{x}_f$ for all i. Thus, the condition given by Eq. (6–3) is necessary. Consequently, we find the rank condition given by Eq. (6–3) to be a necessary and sufficient condition for complete state controllability.

If the system defined by Eq. (6–1) is completely state controllable, then it is possible to transfer any initial state to any arbitrary state in at most n sampling periods. Note, however, that this is true if and only if the magnitude of $u(kT)$ is unbounded. If the magnitude of $u(kT)$ is bounded, it may take more than n sampling periods. (See Prob. A-6–2.)

Complete state controllability in the case where u(kT) is a vector. If the system is defined by

$$\mathbf{x}((k+1)T) = \mathbf{Gx}(kT) + \mathbf{Hu}(kT)$$

where $\mathbf{x}(kT)$ is an n-vector, $\mathbf{u}(kT)$ is an r-vector, $\mathbf{G}$ is an $n \times n$ matrix, and $\mathbf{H}$ is an $n \times r$ matrix, then it can be proved that the condition for complete state controllability is that the $n \times nr$ matrix

$$[\mathbf{H} \vdots \mathbf{GH} \vdots \cdots \vdots \mathbf{G}^{n-1}\mathbf{H}]$$

be of rank n, or that

$$\text{rank}\,[\mathbf{H} \vdots \mathbf{GH} \vdots \cdots \vdots \mathbf{G}^{n-1}\mathbf{H}] = n$$

Determination of control sequence to bring the initial state to a desired state. If the matrix

$$[\mathbf{H} \vdots \mathbf{GH} \vdots \cdots \vdots \mathbf{G}^{n-1}\mathbf{H}]$$

is of rank n and $u(kT)$ is a scalar, then it is possible to find n linearly independent scalar equations from which a sequence of unbounded control signals $u(kT)$ ($k = 0, 1, 2, \ldots, n-1$) can be uniquely determined such that any initial state $\mathbf{x}(0)$ is transferred to the desired state in n sampling periods. [See Eq. (6–2).]

Note also that if the control signal is not a scalar but a vector, then the sequence of $\mathbf{u}(kT)$ is not unique. Then there exists more than one sequence of control vector $\mathbf{u}(kT)$ to bring the initial state $\mathbf{x}(0)$ to a desired state in not more than n sampling periods.

Alternate form of the condition for complete state controllability. Consider the system defined by

$$\mathbf{x}((k+1)T) = \mathbf{G}\mathbf{x}(kT) + \mathbf{H}\mathbf{u}(kT) \tag{6–4}$$

where

$\mathbf{x}(kT)$ = state vector (n-vector) at kth sampling instant

$\mathbf{u}(kT)$ = control vector (r-vector) at kth sampling instant

$\mathbf{G} = n \times n$ matrix

$\mathbf{H} = n \times r$ matrix

T = sampling period

If the eigenvectors of $\mathbf{G}$ are distinct, then it is possible to find a transformation matrix $\mathbf{P}$ such that

$$\mathbf{P}^{-1}\mathbf{G}\mathbf{P} = \begin{bmatrix} \lambda_1 & & & & 0 \\ & \lambda_2 & & & \\ & & \cdot & & \\ & & & \cdot & \\ & & & & \cdot \\ 0 & & & & \lambda_n \end{bmatrix}$$

Note that if the eigenvalues of $\mathbf{G}$ are distinct, then the eigenvectors of $\mathbf{G}$ are distinct. However, the converse is not true. (For example, an $n \times n$ real symmetric matrix having multiple eigenvalues has n distinct eigenvectors.) Note also that the ith column of the $\mathbf{P}$ matrix is an eigenvector of $\mathbf{G}$ associated with ith eigenvalue λ_i ($i = 1, 2, \ldots, n$). Let us define

$$\mathbf{x}(kT) = \mathbf{P}\hat{\mathbf{x}}(kT) \tag{6–5}$$

Substituting Eq. (6–5) into Eq. (6–4), we obtain

$$\hat{\mathbf{x}}((k+1)T) = \mathbf{P}^{-1}\mathbf{G}\mathbf{P}\hat{\mathbf{x}}(kT) + \mathbf{P}^{-1}\mathbf{H}\mathbf{u}(kT) \tag{6–6}$$

Let us define

$$\mathbf{P}^{-1}\mathbf{H} = \mathbf{F} = (f_{ij})$$

Then, Eq. (6–6) may be written as follows:

$$\hat{x}_1((k+1)T) = \lambda_1\hat{x}_1(kT) + f_{11}u_1(kT) + f_{12}u_2(kT) + \cdots + f_{1r}u_r(kT)$$

$$\hat{x}_2((k+1)T) = \lambda_2\hat{x}_2(kT) + f_{21}u_1(kT) + f_{22}u_2(kT) + \cdots + f_{2r}u_r(kT)$$

$$\vdots$$

$$\hat{x}_n((k+1)T) = \lambda_n\hat{x}_n(kT) + f_{n1}u_1(kT) + f_{n2}u_2(kT) + \cdots + f_{nr}u_r(kT)$$

If the elements of any one row of the $n \times r$ matrix $\mathbf{F}$ are all zero, then the corresponding state variable cannot be controlled by any of the $u_i(kT)$. Hence the condition for complete state controllability is that if the eigenvectors of $\mathbf{G}$ are distinct, then the system is completely state controllable if and only if no row of $\mathbf{P}^{-1}\mathbf{H}$ has all zero elements. It is important to note that to apply this condition for complete state controllability, we must put the matrix $\mathbf{P}^{-1}\mathbf{G}\mathbf{P}$ in Eq. (6–6) into diagonal form.

If the $\mathbf{G}$ matrix in Eq. (6–4) does not possess distinct eigenvectors, then diagonalization is impossible. In such a case, we may transform $\mathbf{G}$ into a Jordan canonical form. If, for example, $\mathbf{G}$ has eigenvalues $\lambda_1, \lambda_1, \lambda_1, \lambda_4, \lambda_4, \lambda_6, \ldots, \lambda_n$ and has $n - 3$ distinct eigenvectors, then the Jordan canonical form of $\mathbf{G}$ is

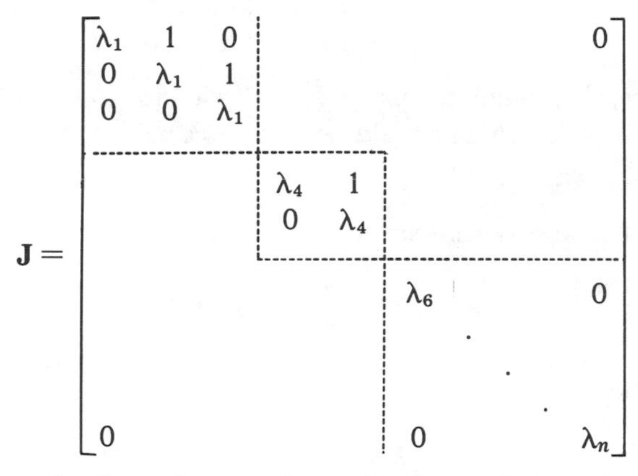

The 3×3 and 2×2 submatrices on the main diagonal are called *Jordan blocks.*

Suppose it is possible to find a transformation matrix $\mathbf{S}$ such that

$$\mathbf{S}^{-1}\mathbf{G}\mathbf{S} = \mathbf{J}$$

If we define a new state vector $\hat{\mathbf{x}}$ by

$$\mathbf{x}(kT) = \mathbf{S}\hat{\mathbf{x}}(kT) \tag{6–7}$$

then substituting Eq. (6–7) into Eq. (6–4) gives

$$\begin{aligned}\hat{\mathbf{x}}((k+1)T) &= \mathbf{S}^{-1}\mathbf{G}\mathbf{S}\hat{\mathbf{x}}(kT) + \mathbf{S}^{-1}\mathbf{H}\mathbf{u}(kT) \\ &= \mathbf{J}\hat{\mathbf{x}}(kT) + \mathbf{S}^{-1}\mathbf{H}\mathbf{u}(kT) \qquad (6\text{–}8)\end{aligned}$$

The conditions for complete state controllability of the system of Eq. (6–8) may then be stated as follows: The system is completely state controllable if and only if (1) no two Jordan blocks in $\mathbf{J}$ of Eq. (6–8) are associated with the same eigenvalues, (2) the elements of any row of $\mathbf{S}^{-1}\mathbf{H}$ that corresponds to the last row of each Jordan block are not all zero, and (3) the elements of each row of $\mathbf{S}^{-1}\mathbf{H}$ that correspond to distinct eigenvalues are not all zero.

Comments. In the preceding conditions for state controllability, it is stated that no two Jordan blocks in $\mathbf{J}$ of Eq. (6–8) should be associated with the same eigenvalues. This point is elaborated as follows.

Consider the following system where the two Jordan blocks are associated with the same eigenvalues λ_1:

$$\begin{bmatrix} x_1(k+1) \\ x_2(k+1) \\ x_3(k+1) \end{bmatrix} = \begin{bmatrix} \lambda_1 & 0 & 0 \\ 0 & \lambda_1 & 1 \\ 0 & 0 & \lambda_1 \end{bmatrix} \begin{bmatrix} x_1(k) \\ x_2(k) \\ x_3(k) \end{bmatrix} + \begin{bmatrix} 1 \\ 1 \\ 1 \end{bmatrix} u(k)$$

Although every state variable is affected by $u(k)$, this system is uncontrollable, since the rank of the controllability matrix

$$[\mathbf{H} \vdots \mathbf{GH} \vdots \mathbf{G}^2\mathbf{H}] = \begin{bmatrix} 1 & \lambda_1 & \lambda_1^2 \\ 1 & \lambda_1 + 1 & \lambda_1^2 + 2\lambda_1 \\ 1 & \lambda_1 & \lambda_1^2 \end{bmatrix}$$

is 2. Hence, in applying the preceding criterion for state controllability, no two Jordan blocks in **J** of Eq. (6–8) should be associated with the same eigenvalues.

Example 6–1.

The following systems are completely state controllable:

1. $$\begin{bmatrix} x_1(k+1) \\ x_2(k+1) \end{bmatrix} = \begin{bmatrix} -1 & 0 \\ 0 & -2 \end{bmatrix} \begin{bmatrix} x_1(k) \\ x_2(k) \end{bmatrix} + \begin{bmatrix} 2 \\ 3 \end{bmatrix} [u(k)]$$

2. $$\begin{bmatrix} x_1(k+1) \\ x_2(k+1) \\ x_3(k+1) \\ x_4(k+1) \\ x_5(k+1) \end{bmatrix} = \begin{bmatrix} -2 & 1 & 0 & & 0 \\ 0 & -2 & 1 & & \\ 0 & 0 & -2 & & \\ & & & -5 & 1 \\ 0 & & & 0 & -5 \end{bmatrix} \begin{bmatrix} x_1(k) \\ x_2(k) \\ x_3(k) \\ x_4(k) \\ x_5(k) \end{bmatrix} + \begin{bmatrix} 0 & 1 \\ 0 & 0 \\ 3 & 0 \\ 0 & 0 \\ 2 & 1 \end{bmatrix} \begin{bmatrix} u_1(k) \\ u_2(k) \end{bmatrix}$$

The following systems are not completely state controllable:

1. $$\begin{bmatrix} x_1(k+1) \\ x_2(k+1) \end{bmatrix} = \begin{bmatrix} -1 & 0 \\ 0 & -2 \end{bmatrix} \begin{bmatrix} x_1(k) \\ x_2(k) \end{bmatrix} + \begin{bmatrix} 2 \\ 0 \end{bmatrix} [u(k)]$$

2. $$\begin{bmatrix} x_1(k+1) \\ x_2(k+1) \\ x_3(k+1) \\ x_4(k+1) \\ x_5(k+1) \end{bmatrix} = \begin{bmatrix} -2 & 1 & 0 & & 0 \\ 0 & -2 & 1 & & \\ 0 & 0 & -2 & & \\ & & & -5 & 1 \\ 0 & & & 0 & -5 \end{bmatrix} \begin{bmatrix} x_1(k) \\ x_2(k) \\ x_3(k) \\ x_4(k) \\ x_5(k) \end{bmatrix} + \begin{bmatrix} 0 & 1 \\ 3 & 0 \\ 0 & 0 \\ 2 & 1 \\ 0 & 0 \end{bmatrix} \begin{bmatrix} u_1(k) \\ u_2(k) \end{bmatrix}$$

Condition for complete state controllability in the *z* plane. The condition for complete state controllability can be stated in terms of pulse transfer functions.

A necessary and sufficient condition for complete state controllability is that no cancellation occur in the pulse transfer function. If cancellation occurs, the system cannot be controlled in the direction of the canceled mode. (See Prob. A-6–5.)

Example 6–2.

Consider the following pulse transfer function:

$$\frac{Y(z)}{U(z)} = \frac{z + 0.2}{(z + 0.8)(z + 0.2)}$$

Clearly, cancellation of factors $(z + 0.2)$ in the numerator and denominator occurs. Thus, one degree of freedom is lost. Because of this cancellation, this system is not completely state controllable.

The same conclusion can be obtained, of course, by writing this pulse transfer function in the form of state equations. A possible state space representation for this system is

$$\begin{bmatrix} x_1(k+1) \\ x_2(k+1) \end{bmatrix} = \begin{bmatrix} 0 & 1 \\ -0.16 & -1 \end{bmatrix} \begin{bmatrix} x_1(k) \\ x_2(k) \end{bmatrix} + \begin{bmatrix} 1 \\ -0.8 \end{bmatrix} u(k)$$

$$y(k) = [1 \quad 0] \begin{bmatrix} x_1(k) \\ x_2(k) \end{bmatrix}$$

Since

$$[\mathbf{H} \vdots \mathbf{GH}] = \begin{bmatrix} 1 & -0.8 \\ -0.8 & 0.64 \end{bmatrix}$$

the rank of $[\mathbf{H} \vdots \mathbf{GH}]$ is 1. Therefore, we arrive at the same conclusion: that the system is not completely state controllable.

Complete output controllability. In the practical design of a control system, we may want to control the output rather than the state of the system. Complete state controllability is neither necessary nor sufficient for controlling the output of the system. For this reason, it is necessary to define separately complete output controllability.

Consider the system defined by the equations

$$\mathbf{x}((k+1)T) = \mathbf{G}\mathbf{x}(kT) + \mathbf{H}u(kT) \tag{6–9}$$

$$\mathbf{y}(kT) = \mathbf{C}\mathbf{x}(kT) \tag{6–10}$$

where

$\mathbf{x}(kT)$ = state vector (n-vector) at kth sampling instant

$u(kT)$ = control signal (scalar) at kth sampling instant

$\mathbf{y}(kT)$ = output vector (m-vector) at kth sampling instant

$\mathbf{G} = n \times n$ matrix

$\mathbf{H} = n \times 1$ matrix

$\mathbf{C} = m \times n$ matrix

The system defined by Eqs. (6–9) and (6–10) is said to be completely output controllable or simply output controllable if it is possible to construct an unconstrained control signal $u(kT)$ defined over a finite number of sampling periods $0 \leq kT < nT$ such that starting from any initial output $\mathbf{y}(0)$, the output $\mathbf{y}(kT)$ can be transferred to the desired point (an arbitrary point) $\mathbf{y}_f$ in the output space in at most n sampling periods.

In what follows we derive the condition for complete output controllability, noting that if a system is completely output controllable, then a piecewise constant control signal exists which will transfer any initial output to any desired point $\mathbf{y}_f$ in the output space in at most n sampling periods. Since the solution of Eq. (6–9) is

$$\mathbf{x}(nT) = \mathbf{G}^n\mathbf{x}(0) + \sum_{j=0}^{n-1} \mathbf{G}^{n-j-1}\mathbf{H}u(jT)$$

we have

$$\begin{aligned}\mathbf{y}(nT) &= \mathbf{C}\mathbf{x}(nT) \\ &= \mathbf{C}\mathbf{G}^n\mathbf{x}(0) + \sum_{j=0}^{n-1} \mathbf{C}\mathbf{G}^{n-j-1}\mathbf{H}u(jT)\end{aligned}$$

or

$$\begin{aligned}\mathbf{y}(nT) - \mathbf{C}\mathbf{G}^n\mathbf{x}(0) &= \sum_{j=0}^{n-1} \mathbf{C}\mathbf{G}^{n-j-1}\mathbf{H}u(jT) \\ &= \mathbf{C}\mathbf{G}^{n-1}\mathbf{H}u(0) + \mathbf{C}\mathbf{G}^{n-2}\mathbf{H}u(T) + \cdots + \mathbf{C}\mathbf{H}u((n-1)T) \\ &= [\mathbf{CH} \vdots \mathbf{CGH} \vdots \cdots \vdots \mathbf{C}\mathbf{G}^{n-1}\mathbf{H}] \begin{bmatrix} u((n-1)T) \\ u((n-2)T) \\ \cdot \\ \cdot \\ \cdot \\ u(0) \end{bmatrix}\end{aligned}$$

Note that $\mathbf{y}(nT) - \mathbf{C}\mathbf{G}^n\mathbf{x}(0) = \mathbf{y}_f - \mathbf{C}\mathbf{G}^n\mathbf{x}(0)$ represents an arbitrary point in the m-dimensional output space. Thus, as in the case of complete state controllability, a necessary and sufficient condition for the system to be completely output-controllable is that vectors $\mathbf{CH}, \mathbf{CGH}, \ldots, \mathbf{C}\mathbf{G}^{n-1}\mathbf{H}$ span the m-dimensional output space, or that

$$\operatorname{rank}[\mathbf{CH} \vdots \mathbf{CGH} \vdots \cdots \vdots \mathbf{C}\mathbf{G}^{n-1}\mathbf{H}] = m \tag{6–11}$$

From this analysis it can be seen that in the system defined by Eqs. (6–9) and (6–10), complete state controllability implies complete output controllability if and only if the m rows of $\mathbf{C}$ are linearly independent.

Next, consider the system defined by the equations

$$\mathbf{x}((k+1)T) = \mathbf{G}\mathbf{x}(kT) + \mathbf{H}\mathbf{u}(kT) \tag{6–12}$$

$$\mathbf{y}(kT) = \mathbf{C}\mathbf{x}(kT) + \mathbf{D}\mathbf{u}(kT) \tag{6–13}$$

where

$\mathbf{x}(kT)$ = state vector (n-vector) at kth sampling intant

$\mathbf{u}(kT)$ = control vector (r-vector) at kth sampling instant

$\mathbf{y}(kT)$ = output vector (m-vector) at kth sampling instant

$\mathbf{G} = n \times n$ matrix

$\mathbf{H} = n \times r$ matrix

$\mathbf{C} = m \times n$ matrix

$\mathbf{D} = m \times r$ matrix

The condition for complete output controllability for this system can be derived as follows. Since the output $\mathbf{y}(nT)$ can be given by the equation

$$\begin{aligned}\mathbf{y}(nT) &= \mathbf{C}\mathbf{x}(nT) + \mathbf{D}\mathbf{u}(nT) \\ &= \mathbf{C}\mathbf{G}^{n}\mathbf{x}(0) + \sum_{j=0}^{n-1} \mathbf{C}\mathbf{G}^{n-j-1}\mathbf{H}\mathbf{u}(jT) + \mathbf{D}\mathbf{u}(nT)\end{aligned}$$

we obtain

$$\begin{aligned}\mathbf{y}(nT) - \mathbf{C}\mathbf{G}^{n}\mathbf{x}(0) &= \sum_{j=0}^{n-1} \mathbf{C}\mathbf{G}^{n-j-1}\mathbf{H}\mathbf{u}(jT) + \mathbf{D}\mathbf{u}(nT) \\ &= \mathbf{C}\mathbf{G}^{n-1}\mathbf{H}\mathbf{u}(0) + \mathbf{C}\mathbf{G}^{n-2}\mathbf{H}\mathbf{u}(T) + \cdots + \mathbf{C}\mathbf{H}\mathbf{u}((n-1)T) \\ &\quad + \mathbf{D}\mathbf{u}(nT) \\ &= [\mathbf{D} \vdots \mathbf{C}\mathbf{H} \vdots \mathbf{C}\mathbf{G}\mathbf{H} \vdots \cdots \vdots \mathbf{C}\mathbf{G}^{n-1}\mathbf{H}] \begin{bmatrix} \mathbf{u}(nT) \\ \mathbf{u}((n-1)T) \\ \cdot \\ \cdot \\ \cdot \\ \mathbf{u}(0) \end{bmatrix}\end{aligned}$$

A necessary and sufficient condition for the system defined by Eqs. (6–12) and (6–13) to be completely output controllable is that the $m \times (n+1)r$ matrix

$$[\mathbf{D} \vdots \mathbf{C}\mathbf{H} \vdots \mathbf{C}\mathbf{G}\mathbf{H} \vdots \cdots \vdots \mathbf{C}\mathbf{G}^{n-1}\mathbf{H}]$$

be of rank m:

$$\text{rank}\, [\mathbf{D} \vdots \mathbf{C}\mathbf{H} \vdots \mathbf{C}\mathbf{G}\mathbf{H} \vdots \cdots \vdots \mathbf{C}\mathbf{G}^{n-1}\mathbf{H}] = m \tag{6–14}$$

It is noted that the presence of matrix $\mathbf{D}$ in the system output equation always helps to establish complete output controllability.

Controllability of a linear time-invariant continuous-time control system. In what follows, we shall briefly state the conditions for complete state controllability and output controllability of linear time-invariant continuous-time control systems. Consider the system defined by

$$\dot{\mathbf{x}} = \mathbf{A}\mathbf{x} + \mathbf{B}\mathbf{u}$$

$$\mathbf{y} = \mathbf{C}\mathbf{x} + \mathbf{D}\mathbf{u}$$

where

$\mathbf{x}$ = state vector (n-vector)

$\mathbf{u}$ = control vector (r-vector)

$\mathbf{y}$ = output vector (m-vector)

$\mathbf{A} = n \times n$ matrix

$\mathbf{B} = n \times r$ matrix

$\mathbf{C} = m \times n$ matrix

$\mathbf{D} = m \times r$ matrix

Complete State Controllability. A necessary and sufficient condition for complete state controllability for this system can be derived in a way similar to what was used in the case of the discrete-time system. Here, we shall present only the result.

The condition for complete state controllability is that the $n \times nr$ matrix

$$[\mathbf{B} \vdots \mathbf{AB} \vdots \cdots \vdots \mathbf{A}^{n-1}\mathbf{B}]$$

be of rank n, or that it contains n linearly independent column vectors. (This matrix is commonly called the controllability matrix for the continuous-time system.) For the proof of this rank condition, see Prob. A-6–6.

The condition for complete state controllability can also be stated in terms of transfer functions or transfer matrices. A necessary and sufficient condition for complete state controllability is that no cancellation occur in the transfer function or transfer matrix. If cancellation occurs, the system cannot be controlled in the direction of the canceled mode.

Output Controllability. As in the case of the discrete-time control system, complete state controllability is neither necessary nor sufficient for controlling the output of a linear time-invariant continuous-time control system. It can be proved that the condition for complete output controllability is that the rank of the $m \times (n + 1)r$ matrix

$$[\mathbf{D} \vdots \mathbf{CB} \vdots \mathbf{CAB} \vdots \mathbf{CA}^2\mathbf{B} \vdots \cdots \vdots \mathbf{CA}^{n-1}\mathbf{B}]$$

be m.

6-3 OBSERVABILITY

In this section, we shall discuss the observability of linear time-invariant control systems. Consider the unforced discrete-time control system defined by

$$\mathbf{x}((k+1)T) = \mathbf{G}\mathbf{x}(kT) \tag{6-15}$$

$$\mathbf{y}(kT) = \mathbf{C}\mathbf{x}(kT) \tag{6-16}$$

where

$\mathbf{x}(kT)$ = state vector (n-vector) at kth sampling instant

$\mathbf{y}(kT)$ = output vector (m-vector) at kth sampling instant

$\mathbf{G} = n \times n$ matrix

$\mathbf{C} = m \times n$ matrix

The system is said to be completely observable if every initial state $\mathbf{x}(0)$ can be determined from the observation of $\mathbf{y}(kT)$ over a finite number of sampling periods. The system, therefore, is completely observable if every transition of the state eventually affects every element of the output vector.

The concept of observability is useful in solving the problem of reconstructing unmeasurable state variables from measurable ones in the minimum number of sampling periods. It will be seen later that state feedback control systems designed by the pole placement method will require feedback of weighted state variables. In practice, however, the difficulty encountered with state feedback control systems is that some of the state variables are not accessible for direct measurement. Then, it becomes necessary to estimate the unmeasurable state variables in order to construct the feedback control signals. In Sec. 6-6 we shall see that the concept of observability plays a dominant role in the design of state observers.

The reason we are considering the unforced system is as follows. If the system is described by the equations

$$\mathbf{x}((k+1)T) = \mathbf{G}\mathbf{x}(kT) + \mathbf{H}\mathbf{u}(kT)$$

$$\mathbf{y}(kT) = \mathbf{C}\mathbf{x}(kT) + \mathbf{D}\mathbf{u}(kT)$$

then

$$\mathbf{x}(kT) = \mathbf{G}^k\mathbf{x}(0) + \sum_{j=0}^{k-1} \mathbf{G}^{k-j-1}\mathbf{H}\mathbf{u}(jT)$$

and $\mathbf{y}(kT)$ is

$$\mathbf{y}(kT) = \mathbf{C}\mathbf{G}^k\mathbf{x}(0) + \sum_{j=0}^{k-1} \mathbf{C}\mathbf{G}^{k-j-1}\mathbf{H}\mathbf{u}(jT) + \mathbf{D}\mathbf{u}(kT)$$

Since the matrices **G, H, C,** and **D** are known and $\mathbf{u}(kT)$ is also known, the second and third terms on the right-hand side of this last equation are known quantities. Therefore, they may be subtracted from the observed value of $\mathbf{y}(kT)$. Hence, for investigating a necessary and sufficient condition for complete observability, it suffices to consider the system described by Eqs. (6–15) and (6–16).

Once $\mathbf{x}(0)$ can be determined from the observation of the output, $\mathbf{x}(k)$ can also be determined, since $u(0), u(T), \ldots, u((k-1)T)$ are known.

Complete observability of discrete-time systems. Consider the system defined by Eqs. (6–15) and (6–16). The system is completely observable if, given the output $\mathbf{y}(kT)$ over a finite number of sampling periods, it is possible to determine the initial state vector $\mathbf{x}(0)$.

In what follows we shall derive the condition for the complete observability of the discrete-time system described by Eqs. (6–15) and (6–16). Since the solution $\mathbf{x}(kT)$ of Eq. (6–15) is

$$\mathbf{x}(kT) = \mathbf{G}^k\mathbf{x}(0)$$

we obtain

$$\mathbf{y}(kT) = \mathbf{C}\mathbf{G}^k\mathbf{x}(0)$$

Complete observability means that, given $\mathbf{y}(0), \mathbf{y}(T), \mathbf{y}(2T), \ldots$, it is possible to determine $x_1(0), x_2(0), \ldots, x_n(0)$. In order to determine n unknowns, we need only n values of $\mathbf{y}(kT)$. Hence, we may use the first n values of $\mathbf{y}(kT)$, or $\mathbf{y}(0)$, $\mathbf{y}(T), \ldots, \mathbf{y}((n-1)T)$ for the determination of $x_1(0), x_2(0), \ldots, x_n(0)$.

For a completely observable system, given

$$\begin{aligned}
\mathbf{y}(0) &= \mathbf{C}\mathbf{x}(0)\\
\mathbf{y}(T) &= \mathbf{C}\mathbf{G}\mathbf{x}(0)\\
&\;\;\vdots\\
\mathbf{y}((n-1)T) &= \mathbf{C}\mathbf{G}^{n-1}\mathbf{x}(0)
\end{aligned}$$

we must be able to determine $x_1(0), x_2(0), \ldots, x_n(0)$. Noting that $\mathbf{y}(kT)$ is an m vector, the preceding n simultaneous equations yield nm equations, all involving $x_1(0), x_2(0), \ldots, x_n(0)$. To obtain a unique set of solutions $x_1(0), x_2(0), \ldots,$ $x_n(0)$ from these nm equations, we must be able to write exactly n linearly independent equations among them. This requires that the $nm \times n$ matrix

$$\begin{bmatrix} \mathbf{C} \\ \hdashline \mathbf{C}\mathbf{G} \\ \hdashline \vdots \\ \hdashline \mathbf{C}\mathbf{G}^{n-1} \end{bmatrix}$$

be of rank n.

Noting that the rank of a matrix and that of the conjugate transpose of the matrix are the same, it is possible to state the condition for complete observability as follows. A necessary and sufficient condition for the system defined by Eqs. (6–15) and (6–16) to be completely observable is that the rank of the $n \times nm$ matrix

$$[\mathbf{C}^* \vdots \mathbf{G}^*\mathbf{C}^* \vdots \cdots \vdots (\mathbf{G}^*)^{n-1}\mathbf{C}^*] \tag{6–17}$$

be n. The matrix given by (6–17) is commonly called the *observability matrix*. [Note that in (6–17), asterisks indicate conjugate transposes. If matrices $\mathbf{C}$ and $\mathbf{G}$ are real, then the conjugate transpose notation such as $\mathbf{G}^*\mathbf{C}^*$ may be changed to the transpose notation such as $\mathbf{G}^T\mathbf{C}^T$.]

Alternate form of the condition for complete observability. Consider the system defined by Eqs. (6–15) and (6–16), repeated here:

$$\mathbf{x}((k+1)T) = \mathbf{G}\mathbf{x}(kT) \tag{6–18}$$

$$\mathbf{y}(kT) = \mathbf{C}\mathbf{x}(kT) \tag{6–19}$$

Suppose the eigenvalues of $\mathbf{G}$ are distinct and a transformation matrix $\mathbf{P}$ transforms $\mathbf{G}$ into a diagonal matrix, so that $\mathbf{P}^{-1}\mathbf{G}\mathbf{P}$ is a diagonal matrix. Let us define

$$\mathbf{x}(kT) = \mathbf{P}\hat{\mathbf{x}}(kT)$$

Then, Eqs. (6–18) and (6–19) can be written as follows:

$$\hat{\mathbf{x}}((k+1)T) = \mathbf{P}^{-1}\mathbf{G}\mathbf{P}\hat{\mathbf{x}}(kT)$$

$$\mathbf{y}(kT) = \mathbf{C}\mathbf{P}\hat{\mathbf{x}}(kT)$$

Hence,

$$\mathbf{y}(nT) = \mathbf{C}\mathbf{P}(\mathbf{P}^{-1}\mathbf{G}\mathbf{P})^n\hat{\mathbf{x}}(0)$$

or

$$\mathbf{y}(nT) = \mathbf{C}\mathbf{P}\begin{bmatrix} \lambda_1^n & & & & \\ & \lambda_2^n & & & \\ & & \cdot & & \\ & & & \cdot & \\ & & & & \lambda_n^n \end{bmatrix}\hat{\mathbf{x}}(0) = \mathbf{C}\mathbf{P}\begin{bmatrix} \lambda_1^n\hat{x}_1(0) \\ \lambda_2^n\hat{x}_2(0) \\ \cdot \\ \cdot \\ \cdot \\ \lambda_n^n\hat{x}_n(0) \end{bmatrix}$$

where $\lambda_1, \lambda_2, \ldots, \lambda_n$ are n distinct eigenvalues of $\mathbf{G}$. The system is completely observable if and only if none of the columns of the $m \times n$ matrix $\mathbf{CP}$ consists of all zero elements. This is because if the ith column of $\mathbf{CP}$ consists of all zero elements, then the state variable $\hat{x}_i(0)$ will not appear in the output equation and therefore cannot be determined from observation of $\mathbf{y}(kT)$. Thus, $\mathbf{x}(0)$, which is related to $\hat{\mathbf{x}}(0)$ by the nonsingular matrix $\mathbf{P}$, cannot be determined.

If the matrix $\mathbf{G}$ involves multiple eigenvalues and cannot be transformed into a diagonal matrix, then by using a suitable transformation matrix **S**, we may transform **G** into the Jordan canonical form:

$$\mathbf{S}^{-1}\mathbf{G}\mathbf{S} = \mathbf{J}$$

where **J** is in the Jordan canonical form. Let us define

$$\mathbf{x}(kT) = \mathbf{S}\hat{\mathbf{x}}(kT)$$

Then, Eqs. (6–18) and (6–19) can be written as follows:

$$\hat{\mathbf{x}}((k+1)T) = \mathbf{S}^{-1}\mathbf{G}\mathbf{S}\hat{\mathbf{x}}(kT) = \mathbf{J}\hat{\mathbf{x}}(kT)$$

$$\mathbf{y}(kT) = \mathbf{C}\mathbf{S}\hat{\mathbf{x}}(kT)$$

Hence

$$\mathbf{y}(nT) = \mathbf{C}\mathbf{S}(\mathbf{S}^{-1}\mathbf{G}\mathbf{S})^n\hat{\mathbf{x}}(0)$$

The system is completely observable if and only if (1) no two Jordan blocks in **J** are associated with the same eigenvalue, (2) none of the columns of **CS** that correspond to the first row of each Jordan block consists of all zero elements, and (3) no columns of **CS** that correspond to distinct eigenvalues consist of all zero elements.

To clarify condition 2, in Example 6–3, which follows, we have enclosed in dashed lines the columns of **CS** that correspond to the first row of each Jordan block.

Example 6–3.

The following systems are completely observable.

1. $$\begin{bmatrix} x_1((k+1)T) \\ x_2((k+1)T) \end{bmatrix} = \begin{bmatrix} -1 & 0 \\ 0 & -2 \end{bmatrix} \begin{bmatrix} x_1(kT) \\ x_2(kT) \end{bmatrix}, \qquad y(kT) = [1 \quad 5] \begin{bmatrix} x_1(kT) \\ x_2(kT) \end{bmatrix}$$

2. $$\begin{bmatrix} x_1((k+1)T) \\ x_2((k+1)T) \\ x_3((k+1)T) \\ x_4((k+1)T) \\ x_5((k+1)T) \end{bmatrix} = \begin{bmatrix} 2 & 1 & 0 & & 0 \\ 0 & 2 & 1 & & \\ 0 & 0 & 2 & & \\ & & & -3 & 1 \\ 0 & & & 0 & -3 \end{bmatrix} \begin{bmatrix} x_1(kT) \\ x_2(kT) \\ x_3(kT) \\ x_4(kT) \\ x_5(kT) \end{bmatrix}$$

$$\begin{bmatrix} y_1(kT) \\ y_2(kT) \end{bmatrix} = \begin{bmatrix} \boxed{1} & 1 & 1 & \boxed{0} & 1 \\ \boxed{0} & 1 & 1 & \boxed{1} & 0 \end{bmatrix} \begin{bmatrix} x_1(kT) \\ x_2(kT) \\ x_3(kT) \\ x_4(kT) \\ x_5(kT) \end{bmatrix}$$

The following systems are not completely observable:

1. $$\begin{bmatrix} x_1((k+1)T) \\ x_2((k+1)T) \end{bmatrix} = \begin{bmatrix} -1 & 0 \\ 0 & -2 \end{bmatrix} \begin{bmatrix} x_1(kT) \\ x_2(kT) \end{bmatrix}, \qquad y(kT) = [0 \quad 1] \begin{bmatrix} x_1(kT) \\ x_2(kT) \end{bmatrix}$$

2. $$\begin{bmatrix} x_1((k+1)T) \\ x_2((k+1)T) \\ x_3((k+1)T) \\ x_4((k+1)T) \\ x_5((k+1)T) \end{bmatrix} = \begin{bmatrix} 2 & 1 & 0 & & 0 \\ 0 & 2 & 1 & & \\ 0 & 0 & 2 & & \\ & & & -3 & 1 \\ 0 & & & 0 & -3 \end{bmatrix} \begin{bmatrix} x_1(kT) \\ x_2(kT) \\ x_3(kT) \\ x_4(kT) \\ x_5(kT) \end{bmatrix}$$

$$\begin{bmatrix} y_1(kT) \\ y_2(kT) \end{bmatrix} = \begin{bmatrix} 1 & 1 & 1 & 0 & 1 \\ 0 & 1 & 1 & 0 & 0 \end{bmatrix} \begin{bmatrix} x_1(kT) \\ x_2(kT) \\ x_3(kT) \\ x_4(kT) \\ x_5(kT) \end{bmatrix}$$

Condition for complete observability in the *z* plane. The condition for complete observability can also be stated in terms of pulse transfer functions. A necessary and sufficient condition for complete observability is that no pole-zero cancellation occur in the pulse transfer function. If cancellation occurs, the canceled mode cannot be observed in the output.

Example 6–4.

Show that the following system is not completely observable:

$$\mathbf{x}((k+1)T) = \mathbf{G}\mathbf{x}(kT) + \mathbf{H}u(kT)$$

$$y(kT) = \mathbf{C}\mathbf{x}(kT)$$

where

$$\mathbf{G} = \begin{bmatrix} 0 & 1 & 0 \\ 0 & 0 & 1 \\ -6 & -11 & -6 \end{bmatrix}, \qquad \mathbf{H} = \begin{bmatrix} 0 \\ 0 \\ 1 \end{bmatrix}, \qquad \mathbf{C} = [4 \quad 5 \quad 1]$$

Note that the control signal $u(kT)$ does not affect the complete observability of the system. In order to examine complete observability, we may simply set $u(kT) = 0$. For this system, we have

$$[\mathbf{C}^* \vdots \mathbf{G}^*\mathbf{C}^* \vdots (\mathbf{G}^*)^2\mathbf{C}^*] = \begin{bmatrix} 4 & -6 & 6 \\ 5 & -7 & 5 \\ 1 & -1 & -1 \end{bmatrix}$$

Notice that

$$\begin{vmatrix} 4 & -6 & 6 \\ 5 & -7 & 5 \\ 1 & -1 & -1 \end{vmatrix} = 0$$

Hence, the rank of the matrix $[\mathbf{C}^* \vdots\ \mathbf{G}^*\mathbf{C}^* \vdots (\mathbf{G}^*)^2\mathbf{C}^*]$ is less than 3. Therefore, the system is not completely observable.

In fact, in this system a pole-zero cancellation occurs in the pulse transfer function of the system. The pulse transfer function between $X_1(z)$ and $U(z)$ is

$$\frac{X_1(z)}{U(z)} = \frac{1}{(z+1)(z+2)(z+3)}$$

and the pulse transfer function between $Y(z)$ and $X_1(z)$ is

$$\frac{Y(z)}{X_1(z)} = (z+1)(z+4)$$

Therefore, the pulse transfer function between output $Y(z)$ and input $U(z)$ is

$$\frac{Y(z)}{U(z)} = \frac{(z+1)(z+4)}{(z+1)(z+2)(z+3)}$$

Clearly, the $(z+1)$ factors in the numerator and denominator cancel each other. This means that there are nonzero initial states $\mathbf{x}(0)$ which cannot be determined from the measurement of $y(kT)$.

Comments. The pulse transfer function has no cancellation if and only if the system is completely state controllable and completely observable. (See Prob. A-6–5.) This means that a canceled transfer function does not carry along all the information characterizing the dynamic system.

Principle of duality. In what follows, we shall examine the relationship between controllability and observability. Consider the system S_1 defined by the equations

$$\mathbf{x}((k+1)T) = \mathbf{G}\mathbf{x}(kT) + \mathbf{H}\mathbf{u}(kT) \tag{6–20}$$

$$\mathbf{y}(kT) = \mathbf{C}\mathbf{x}(kT) \tag{6–21}$$

where

$\mathbf{x}(kT)$ = state vector (n-vector) at kth sampling instant

$\mathbf{u}(kT)$ = control vector (r-vector) at kth sampling instant

$\mathbf{y}(kT)$ = output vector (m-vector) at kth sampling instant

$\mathbf{G} = n \times n$ matrix

$\mathbf{H} = n \times r$ matrix

$\mathbf{C} = m \times n$ matrix

and its dual counterpart, which we call system S_2, defined by the equations

$$\hat{\mathbf{x}}((k+1)T) = (\mathbf{G}^*)^{-1}\hat{\mathbf{x}}(kT) + \mathbf{C}^*\hat{\mathbf{u}}(kT) \tag{6–22}$$

$$\hat{\mathbf{y}}(kT) = \mathbf{H}^*\hat{\mathbf{x}}(kT) \tag{6–23}$$

where

$\hat{\mathbf{x}}(kT)$ = state vector (n-vector) at kth sampling instant

$\hat{\mathbf{u}}(kT)$ = control vector (m-vector) at kth sampling instant

$\hat{\mathbf{y}}(kT)$ = output vector (r-vector) at kth sampling instant

$\mathbf{G}^*$ = conjugate transpose of $\mathbf{G}$

$\mathbf{H}^*$ = conjugate transpose of $\mathbf{H}$

$\mathbf{C}^*$ = conjugate transpose of $\mathbf{C}$

Note that the dual counterpart of a given system is obtained by (1) replacing the free dynamic system by its adjoint system and (2) interchanging input and output constraints. Note also that in discussing the system and its dual counterpart it is assumed that **G** is nonsingular. (If **G** is singular, then duality cannot be defined.) We shall now examine an apparent analogy between controllability and observability. This analogy is referred to as the *principle of duality*, due to Kalman.

The principle of duality states that system S_1 defined by Eqs. (6–20) and (6–21) is completely state controllable (observable) if and only if system S_2 defined by Eqs. (6–22) and (6–23) is completely observable (state controllable). To verify this principle, let us write down the necessary and sufficient conditions for complete state controllability and complete observability for systems S_1 and S_2, respectively.

FOR SYSTEM S_1:

1. A necessary and sufficient condition for complete state controllability is that

$$\text{rank}\,[\mathbf{H} \vdots \mathbf{GH} \vdots \cdots \vdots \mathbf{G}^{n-1}\mathbf{H}] = n$$

2. A necessary and sufficient condition for complete observability is that

$$\text{rank}\,[\mathbf{C}^* \vdots \mathbf{G}^*\mathbf{C}^* \vdots \cdots \vdots (\mathbf{G}^*)^{n-1}\mathbf{C}^*] = n$$

FOR SYSTEM S_2:

1. A necessary and sufficient condition for complete state controllability is that

$$\begin{aligned}&\text{rank}\,[\mathbf{C}^* \vdots (\mathbf{G}^*)^{-1}\mathbf{C}^* \vdots \cdots \vdots (\mathbf{G}^*)^{-n+1}\mathbf{C}^*]\\ &= \text{rank}\,[(\mathbf{G}^*)^{n-1}\mathbf{C}^* \vdots (\mathbf{G}^*)^{n-2}\mathbf{C}^* \vdots \cdots \vdots \mathbf{C}^*]\\ &= \text{rank}\,[\mathbf{C}^* \vdots \mathbf{G}^*\mathbf{C}^* \vdots \cdots \vdots (\mathbf{G}^*)^{n-1}\mathbf{C}^*] = n\end{aligned}$$

2. A necessary and sufficient condition for complete observability is that

$$\begin{aligned}&\text{rank}\,[\mathbf{H} \vdots \mathbf{G}^{-1}\mathbf{H} \vdots \cdots \vdots \mathbf{G}^{-n+1}\mathbf{H}]\\ &= \text{rank}\,[\mathbf{G}^{n-1}\mathbf{H} \vdots \mathbf{G}^{n-2}\mathbf{H} \vdots \cdots \vdots \mathbf{H}]\\ &= \text{rank}\,[\mathbf{H} \vdots \mathbf{GH} \vdots \cdots \vdots \mathbf{G}^{n-1}\mathbf{H}] = n\end{aligned}$$

By comparing these conditions, the truth of the principle of duality is apparent. By use of this principle, the observability of a given system can be checked by testing the state controllability of its dual.

For the linear time-invariant case, instead of system S_2 defined by Eqs. (6–22) and (6–23), it is convenient to consider the following system, defined here as system S_3:

$$\hat{\mathbf{x}}((k+1)T) = \mathbf{G}^*\hat{\mathbf{x}}(kT) + \mathbf{C}^*\hat{\mathbf{u}}(kT) \tag{6–24}$$

$$\hat{\mathbf{y}}(kT) = \mathbf{H}^*\hat{\mathbf{x}}(kT) \tag{6–25}$$

FOR SYSTEM S_3:

1. A necessary and sufficient condition for complete state controllability is that

$$\text{rank}\,[\mathbf{C}^* \vdots \mathbf{G}^*\mathbf{C}^* \vdots \cdots \vdots (\mathbf{G}^*)^{n-1}\mathbf{C}^*] = n$$

2. A necessary and sufficient condition for complete observability is that

$$\text{rank}\,[\mathbf{H} \vdots \mathbf{G}\mathbf{H} \vdots \cdots \vdots \mathbf{G}^{n-1}\mathbf{H}] = n$$

We see that system S_1 being completely state controllable is equivalent to system S_3 being completely observable. And system S_1 being completely observable is equivalent to system S_3 being completely state controllable. We shall later use the dual relationship existing between system S_1 and system S_3 in discussing the dual relationship existing between the pole placement design problem and the observer design problem.

Complete observability of linear time-invariant continuous-time control systems. Finally, we shall briefly state the complete observability condition for the linear time-invariant continuous-time control system. The system is said to be completely observable if every initial state $\mathbf{x}(0)$ can be determined from the observation of $\mathbf{y}(t)$ over a finite time interval. Similar to the case of the discrete-time control system, we need to consider only an unforced system. Consider the system defined by the equation

$$\dot{\mathbf{x}} = \mathbf{A}\mathbf{x}$$

$$\mathbf{y} = \mathbf{C}\mathbf{x}$$

where

$\mathbf{x}$ = state vector (n-vector)

$\mathbf{y}$ = output vector (m-vector)

$\mathbf{A}$ = $n \times n$ matrix

$\mathbf{C}$ = $m \times n$ matrix

As in the case of the discrete-time system, it can be stated that the condition for complete observability is that the rank of the $n \times nm$ matrix

$$[\mathbf{C}^* \vdots \mathbf{A}^*\mathbf{C}^* \vdots \cdots \vdots (\mathbf{A}^*)^{n-1}\mathbf{C}^*]$$

be n. (This $n \times nm$ matrix is commonly called the observability matrix for the continuous-time system.) The proof of this rank condition is given in Prob. A-6–7.

Effects of the discretization of a continuous-time control system on controllability and observability. When a continuous-time control system with complex poles is discretized, the introduction of sampling may impair the controllability and observability of the resulting discretized system. That is, pole-zero cancellation may take place in passing from the continuous-time case to the discrete-time case. Thus the discretized system may lose controllability and observability.

It can be shown that a system which is completely state controllable and observable in the absence of sampling remains completely state controllable and observable after the introduction of sampling if and only if, for every eigenvalue of the characteristic equation, the relation

$$\text{Re}\,\lambda_i = \text{Re}\,\lambda_j$$

implies

$$\text{Im}\,(\lambda_i - \lambda_j) \neq \frac{2n\pi}{T}$$

where T is the sampling period and $n = \pm 1, \pm 2, \ldots$. It is noted that unless the system contains complex poles, pole-zero cancellation will not occur in passing from the continuous-time to the discrete-time case.

Example 6–5.

Consider the following continuous-time control system:

$$\begin{bmatrix} \dot{x}_1 \\ \dot{x}_2 \end{bmatrix} = \begin{bmatrix} 0 & 1 \\ -1 & 0 \end{bmatrix} \begin{bmatrix} x_1 \\ x_2 \end{bmatrix} + \begin{bmatrix} 0 \\ 1 \end{bmatrix} u \tag{6–26}$$

$$y = [1 \quad 0] \begin{bmatrix} x_1 \\ x_2 \end{bmatrix} \tag{6–27}$$

This system is completely state controllable and completely observable, since the rank of the controllability matrix

$$[\mathbf{B} \,\vdots\, \mathbf{AB}] = \begin{bmatrix} 0 & 1 \\ 1 & 0 \end{bmatrix}$$

is 2 and the rank of the observability matrix

$$[\mathbf{C}^* \,\vdots\, \mathbf{A}^*\mathbf{C}^*] = \begin{bmatrix} 1 & 0 \\ 0 & 1 \end{bmatrix}$$

is also 2. Notice that the eigenvalues of the state matrix are

$$\lambda_1 = j, \qquad \lambda_2 = -j$$

The discrete-time control system which is obtained by discretizing the continuous-time control system defined by Eqs. (6–26) and (6–27) may be given as follows:

$$\begin{bmatrix} x_1((k+1)T) \\ x_2((k+1)T) \end{bmatrix} = \begin{bmatrix} \cos T & \sin T \\ -\sin T & \cos T \end{bmatrix} \begin{bmatrix} x_1(kT) \\ x_2(kT) \end{bmatrix} + \begin{bmatrix} 1 - \cos T \\ \sin T \end{bmatrix} u(kT) \tag{6–28}$$

$$y(kT) = [1 \quad 0] \begin{bmatrix} x_1(kT) \\ x_2(kT) \end{bmatrix} \tag{6–29}$$

where T is the sampling period.

Show that the discretized system given by Eqs. (6–28) and (6–29) is completely state controllable and observable if and only if

$$\text{Im}\,(\lambda_1 - \lambda_2) = 1 + 1 \neq \frac{2n\pi}{T}$$

or

$$T \neq n\pi \qquad n = 1, 2, 3, \ldots$$

For the discrete-time control system obtained by discretizing the continuous-time control system we have the following controllability matrix:

$$[\mathbf{H} \vdots \mathbf{GH}] = \begin{bmatrix} 1 - \cos T & \cos T + 1 - 2\cos^2 T \\ \sin T & -\sin T + 2\cos T \sin T \end{bmatrix}$$

Notice that the rank of $[\mathbf{H} \vdots \mathbf{GH}]$ is 2 if and only if $T \neq n\pi$ (where $n = 1, 2, 3, \ldots$). Also, the rank of the observability matrix

$$[\mathbf{C}^* \vdots \mathbf{G}^*\mathbf{C}^*] = \begin{bmatrix} 1 & \cos T \\ 0 & \sin T \end{bmatrix}$$

is 2 if and only if $T \neq n\pi$ (where $n = 1, 2, 3, \ldots$).

From the foregoing analysis, we conclude that the discretized system is completely state controllable and completely observable if and only if $T \neq n\pi$, where $n = 1, 2, 3, \ldots$.

Note that it is always possible to avoid the loss of controllability and observability by choosing a sampling period T sufficiently small compared with the smallest time constant of the system and, of course, not equal to $n\pi$.

6–4 USEFUL TRANSFORMATIONS IN STATE SPACE ANALYSIS AND DESIGN

In this section we shall first review techniques for transforming state space equations into canonical forms. Then we shall review the invariance property of the rank conditions for the controllability matrix and observability matrix.

Transforming state space equations into canonical forms. Consider the discrete-time state equation and output equation

$$\mathbf{x}(k+1) = \mathbf{G}\mathbf{x}(k) + \mathbf{H}u(k) \tag{6–30}$$

$$y(k) = \mathbf{C}\mathbf{x}(k) + Du(k) \tag{6–31}$$

We shall review techniques for transforming the state space equations defined by Eqs. (6–30) and (6–31) into the following three canonical forms:

1. Controllable canonical form
2. Observable canonical form
3. Diagonal or Jordan canonical form

(Note that the diagonal canonical form is a special case of the Jordan canonical form.) It is assumed that the system defined by Eqs. (6–30) and (6–31) is completely state controllable and completely observable.

Controllable canonical form. The system defined by Eqs. (6–30) and (6–31) can be transformed into a controllable canonical form by means of the transformation matrix

$$\mathbf{T} = \mathbf{M}\mathbf{W} \tag{6–32}$$

where

$$\mathbf{M} = [\mathbf{H} \vdots \mathbf{G}\mathbf{H} \vdots \cdots \vdots \mathbf{G}^{n-1}\mathbf{H}] \tag{6–33}$$

and

$$\mathbf{W} = \begin{bmatrix} a_{n-1} & a_{n-2} & \cdots & a_1 & 1 \\ a_{n-2} & a_{n-3} & \cdots & 1 & 0 \\ \cdot & \cdot & & \cdot & \cdot \\ \cdot & \cdot & & \cdot & \cdot \\ \cdot & \cdot & & \cdot & \cdot \\ a_1 & 1 & \cdots & 0 & 0 \\ 1 & 0 & \cdots & 0 & 0 \end{bmatrix} \tag{6–34}$$

The elements a_i shown in matrix $\mathbf{W}$ are coefficients of the characteristic equation

$$|z\mathbf{I} - \mathbf{G}| = z^n + a_1 z^{n-1} + \cdots + a_{n-1}z + a_n = 0$$

It can be shown that

$$\mathbf{T}^{-1}\mathbf{G}\mathbf{T} = (\mathbf{M}\mathbf{W})^{-1}\mathbf{G}(\mathbf{M}\mathbf{W}) = \mathbf{W}^{-1}\mathbf{M}^{-1}\mathbf{G}\mathbf{M}\mathbf{W}$$

$$= \begin{bmatrix} 0 & 1 & 0 & \cdots & 0 \\ 0 & 0 & 1 & \cdots & 0 \\ \cdot & \cdot & \cdot & & \cdot \\ \cdot & \cdot & \cdot & & \cdot \\ \cdot & \cdot & \cdot & & \cdot \\ 0 & 0 & 0 & \cdots & 1 \\ -a_n & -a_{n-1} & -a_{n-2} & \cdots & -a_1 \end{bmatrix} \tag{6–35}$$

and

$$\mathbf{T}^{-1}\mathbf{H} = \begin{bmatrix} 0 \\ 0 \\ \cdot \\ \cdot \\ \cdot \\ 0 \\ 1 \end{bmatrix} \tag{6–36}$$

(For details of the derivations of the preceding two equations, see Probs. A-6–8 and A-6–9.)

Now let us define

$$\mathbf{x}(k) = \mathbf{T}\hat{\mathbf{x}}(k)$$

where the transformation matrix $\mathbf{T}$ is given by Eq. (6–32). Then, Eqs. (6–30) and (6–31) become

$$\hat{\mathbf{x}}(k+1) = \mathbf{T}^{-1}\mathbf{G}\mathbf{T}\hat{\mathbf{x}}(k) + \mathbf{T}^{-1}\mathbf{H}u(k) = \hat{\mathbf{G}}\hat{\mathbf{x}}(k) + \hat{\mathbf{H}}u(k)$$

$$y(k) = \mathbf{C}\mathbf{T}\hat{\mathbf{x}}(k) + Du(k) = \hat{\mathbf{C}}\hat{\mathbf{x}}(k) + \hat{D}u(k)$$

where $\hat{\mathbf{G}} = \mathbf{T}^{-1}\mathbf{G}\mathbf{T}$, $\hat{\mathbf{H}} = \mathbf{T}^{-1}\mathbf{H}$, $\hat{\mathbf{C}} = \mathbf{C}\mathbf{T}$, and $\hat{D} = D$, or

$$\begin{bmatrix} \hat{x}_1(k+1) \\ \hat{x}_2(k+1) \\ \cdot \\ \cdot \\ \cdot \\ \hat{x}_{n-1}(k+1) \\ \hat{x}_n(k+1) \end{bmatrix} = \begin{bmatrix} 0 & 1 & 0 & \cdots & 0 \\ 0 & 0 & 1 & \cdots & 0 \\ \cdot & \cdot & \cdot & & \cdot \\ \cdot & \cdot & \cdot & & \cdot \\ \cdot & \cdot & \cdot & & \cdot \\ 0 & 0 & 0 & \cdots & 1 \\ -a_n & -a_{n-1} & -a_{n-2} & \cdots & -a_1 \end{bmatrix} \begin{bmatrix} \hat{x}_1(k) \\ \hat{x}_2(k) \\ \cdot \\ \cdot \\ \cdot \\ \hat{x}_{n-1}(k) \\ \hat{x}_n(k) \end{bmatrix} + \begin{bmatrix} 0 \\ 0 \\ \cdot \\ \cdot \\ \cdot \\ 0 \\ 1 \end{bmatrix} u(k) \tag{6–37}$$

$$y(k) = [b_n - a_n b_0 \vdots b_{n-1} - a_{n-1} b_0 \vdots \cdots \vdots b_1 - a_1 b_0] \begin{bmatrix} \hat{x}_1(k) \\ \hat{x}_2(k) \\ \cdot \\ \cdot \\ \cdot \\ \hat{x}_n(k) \end{bmatrix} + \hat{D}u(k) \tag{6–38}$$

where the b_k's are those coefficients appearing in the numerator of the following pulse transfer function:

$$\mathbf{C}(z\mathbf{I} - \mathbf{G})^{-1}\mathbf{H} + D = \hat{\mathbf{C}}(z\mathbf{I} - \hat{\mathbf{G}})^{-1}\hat{\mathbf{H}} + \hat{D}$$

$$= \frac{b_0 z^n + b_1 z^{n-1} + \cdots + b_{n-1} z + b_n}{z^n + a_1 z^{n-1} + \cdots + a_{n-1} z + a_n} \tag{6–39}$$

Note that $D = \hat{D} = b_0$. The system given by Eqs. (6–37) and (6–38) is in a controllable canonical form.

Observable canonical form. The system defined by Eqs. (6–30) and (6–31) can be transformed into an observable canonical form by means of the transformation matrix

$$\mathbf{Q} = (\mathbf{W}\mathbf{N}^*)^{-1}$$

where

$$\mathbf{N} = [\mathbf{C}^* \vdots \mathbf{G}^*\mathbf{C}^* \vdots \cdots \vdots (\mathbf{G}^*)^{n-1}\mathbf{C}^*] \tag{6–40}$$

and $\mathbf{W}$ is given by Eq. (6–34). It can be shown that

$$\mathbf{Q}^{-1}\mathbf{G}\mathbf{Q} = \hat{\mathbf{G}} = \begin{bmatrix} 0 & 0 & \cdots & 0 & -a_n \\ 1 & 0 & \cdots & 0 & -a_{n-1} \\ 0 & 1 & \cdots & 0 & -a_{n-2} \\ \cdot & \cdot & & \cdot & \cdot \\ \cdot & \cdot & & \cdot & \cdot \\ \cdot & \cdot & & \cdot & \cdot \\ 0 & 0 & \cdots & 1 & -a_1 \end{bmatrix}$$

$$\mathbf{Q}^{-1}\mathbf{H} = \hat{\mathbf{H}} = \begin{bmatrix} b_n - a_n b_0 \\ b_{n-1} - a_{n-1} b_0 \\ \cdot \\ \cdot \\ \cdot \\ b_1 - a_1 b_0 \end{bmatrix}$$

and

$$\mathbf{C}\mathbf{Q} = \hat{\mathbf{C}} = [0 \quad 0 \quad \cdots \quad 0 \quad 1]$$

where the b_k's are those coefficients appearing in the numerator of the pulse transfer function given by Eq. (6–39). (For details of the derivations of the preceding equations, see Probs. A-6–11 and A-6–12.) Hence, by defining

$$\mathbf{x}(k) = \mathbf{Q}\hat{\mathbf{x}}(k)$$

Eqs. (6–30) and (6–31) become as follows:

$$\hat{\mathbf{x}}(k+1) = \hat{\mathbf{G}}\hat{\mathbf{x}}(k) + \hat{\mathbf{H}}u(k)$$
$$y(k) = \hat{\mathbf{C}}\hat{\mathbf{x}}(k) + \hat{D}u(k)$$

or

$$\begin{bmatrix} \hat{x}_1(k+1) \\ \hat{x}_2(k+1) \\ \cdot \\ \cdot \\ \cdot \\ \hat{x}_{n-1}(k+1) \\ \hat{x}_n(k+1) \end{bmatrix} = \begin{bmatrix} 0 & 0 & \cdots & 0 & -a_n \\ 1 & 0 & \cdots & 0 & -a_{n-1} \\ \cdot & \cdot & & \cdot & \cdot \\ \cdot & \cdot & & \cdot & \cdot \\ \cdot & \cdot & & \cdot & \cdot \\ 0 & 0 & \cdots & 0 & -a_2 \\ 0 & 0 & \cdots & 1 & -a_1 \end{bmatrix} \begin{bmatrix} \hat{x}_1(k) \\ \hat{x}_2(k) \\ \cdot \\ \cdot \\ \cdot \\ \hat{x}_{n-1}(k) \\ \hat{x}_n(k) \end{bmatrix} + \begin{bmatrix} b_n - a_n b_0 \\ b_{n-1} - a_{n-1} b_0 \\ \cdot \\ \cdot \\ \cdot \\ b_2 - a_2 b_0 \\ b_1 - a_1 b_0 \end{bmatrix} u(k) \tag{6–41}$$

and

$$y(k)=[0 \quad 0 \quad \cdots \quad 0 \quad 1]\begin{bmatrix} \hat{x}_1(k) \\ \hat{x}_2(k) \\ \cdot \\ \cdot \\ \cdot \\ \hat{x}_{n-1}(k) \\ \hat{x}_n(k) \end{bmatrix} + Du(k) \tag{6–42}$$

The system defined by Eqs. (6–41) and (6–42) is in an observable canonical form.

Diagonal or Jordan canonical form. If the eigenvalues p_i of matrix $\mathbf{G}$ are distinct, then the corresponding eigenvectors $\boldsymbol{\xi}_1, \boldsymbol{\xi}_2, \ldots, \boldsymbol{\xi}_n$ are distinct. Define the transformation matrix $\mathbf{P}$ as follows:

$$\mathbf{P} = [\boldsymbol{\xi}_1 \vdots \boldsymbol{\xi}_2 \vdots \cdots \vdots \boldsymbol{\xi}_n]$$

Then

$$\mathbf{P}^{-1}\mathbf{G}\mathbf{P} = \begin{bmatrix} p_1 & 0 & \cdots & 0 \\ 0 & p_2 & \cdots & 0 \\ \cdot & \cdot & & \cdot \\ \cdot & \cdot & & \cdot \\ \cdot & \cdot & & \cdot \\ 0 & 0 & & p_n \end{bmatrix}$$

Thus, if we define

$$\mathbf{x}(k) = \mathbf{P}\hat{\mathbf{x}}(k)$$

then Eqs. (6–30) and (6–31) can be given by the equations

$$\hat{\mathbf{x}}(k+1) = \hat{\mathbf{G}}\hat{\mathbf{x}}(k) + \hat{\mathbf{H}}u(k) \tag{6–43}$$

$$y(k) = \hat{\mathbf{C}}\hat{\mathbf{x}}(k) + \hat{D}u(k) \tag{6–44}$$

where $\hat{\mathbf{G}} = \mathbf{P}^{-1}\mathbf{G}\mathbf{P}$, $\hat{\mathbf{H}} = \mathbf{P}^{-1}\mathbf{H}$, $\hat{\mathbf{C}} = \mathbf{C}\mathbf{P}$, and $\hat{D} = D$. Thus, Eqs. (6–43) and (6–44) can be written in the form

$$\begin{bmatrix} \hat{x}_1(k+1) \\ \hat{x}_2(k+1) \\ \cdot \\ \cdot \\ \cdot \\ \hat{x}_n(k+1) \end{bmatrix} = \begin{bmatrix} p_1 & 0 & \cdots & 0 \\ 0 & p_2 & \cdots & 0 \\ \cdot & \cdot & & \cdot \\ \cdot & \cdot & & \cdot \\ \cdot & \cdot & & \cdot \\ 0 & 0 & \cdots & p_n \end{bmatrix} \begin{bmatrix} \hat{x}_1(k) \\ \hat{x}_2(k) \\ \cdot \\ \cdot \\ \cdot \\ \hat{x}_n(k) \end{bmatrix} + \begin{bmatrix} \alpha_1 \\ \alpha_2 \\ \cdot \\ \cdot \\ \cdot \\ \alpha_n \end{bmatrix} u(k) \tag{6–45}$$

$$y(k) = [\beta_1 \quad \beta_2 \quad \cdots \quad \beta_n] \begin{bmatrix} \hat{x}_1(k) \\ \hat{x}_2(k) \\ \cdot \\ \cdot \\ \cdot \\ \hat{x}_n(k) \end{bmatrix} + Du(k) \tag{6–46}$$

where the α_i's and the β_i's are constants such that $\alpha_i \beta_i$ is the residue at the pole $z = p_i$; that is, such that $\alpha_i \beta_i$ will appear in the numerator of the term $1/(z - p_i)$ when the pulse transfer function is expanded into partial fractions as follows:

$$\begin{aligned} \mathbf{C}(z\mathbf{I} - \mathbf{G})^{-1}\mathbf{H} + D &= \hat{\mathbf{C}}(z\mathbf{I} - \hat{\mathbf{G}})\hat{\mathbf{H}} + \hat{D} \\ &= \frac{\alpha_1\beta_1}{z - p_1} + \frac{\alpha_2\beta_2}{z - p_2} + \cdots + \frac{\alpha_n \beta_n}{z - p_n} + D \end{aligned} \tag{6–47}$$

In many cases we choose $\alpha_1 = \alpha_2 = \cdots = \alpha_n = 1$. [Note that the necessary and sufficient condition for the system to be completely state controllable is that $\alpha_i \neq 0$ $(i = 1, 2, \ldots, n)$ and that the condition for it to be completely observable is that $\beta_i \neq 0$ $(i = 1, 2, \ldots, n)$.]

If there are multiple eigenvalues p_i of matrix $\mathbf{G}$, then we choose the transformation matrix $\mathbf{S}$ defined as follows:

$$\mathbf{S} = [\boldsymbol{\eta}_1 \vdots \boldsymbol{\eta}_2 \vdots \cdots \vdots \boldsymbol{\eta}_n]$$

where the $\boldsymbol{\eta}_i$'s are eigenvectors (which correspond to distinct eigenvalues) or generalized eigenvectors (which correspond to multiple eigenvalues). (For details of generalized eigenvectors, see the Appendix.) Then

$$\mathbf{S}^{-1}\mathbf{GS} = \text{matrix in Jordan canonical form}$$

Now if we define

$$\mathbf{x}(k) = \mathbf{S}\hat{\mathbf{x}}(k)$$

then Eqs. (6–30) and (6–31) can be given as follows:

$$\hat{\mathbf{x}}(k+1) = \hat{\mathbf{G}}\hat{\mathbf{x}}(k) + \hat{\mathbf{H}}u(k)$$

$$y(k) = \hat{\mathbf{C}}\hat{\mathbf{x}}(k) + \hat{D}\hat{u}(k)$$

where $\hat{\mathbf{G}} = \mathbf{S}^{-1}\mathbf{GS}$, $\hat{\mathbf{H}} = \mathbf{S}^{-1}\mathbf{H}$, $\hat{\mathbf{C}} = \mathbf{CS}$, and $\hat{D} = D$. If, for example, matrix $\mathbf{G}$ involves an m-multiple eigenvalue p_1 and other eigenvalues $p_{m+1}, p_{m+2}, \ldots, p_n$ that are all distinct and different from p_1 and, in addition, if the rank of $p_1\mathbf{I} - \mathbf{G}$ is $n - 1$ (which implies that the minimal polynomial is identical to the characteristic polynomial), then the state space equations in the Jordan canonical form are given as follows:

$$\begin{bmatrix} \hat{x}_1(k+1) \\ \hat{x}_2(k+1) \\ \cdot \\ \cdot \\ \cdot \\ \hat{x}_m(k+1) \\ \hline \hat{x}_{m+1}(k+1) \\ \cdot \\ \cdot \\ \cdot \\ \hat{x}_n(k+1) \end{bmatrix} = \left[\begin{array}{ccccc|ccc} p_1 & 1 & & & 0 & & & 0 \\ & p_1 & 1 & \cdot & & & & \\ & & \cdot & \cdot & & & & \\ & & & \cdot & \cdot & & & \\ & & & \cdot & 1 & & & \\ 0 & & & & p_1 & & & \\ \hline & & & & & p_{m+1} & & 0 \\ & & & & & & \cdot & \\ & & & & & & \cdot & \\ & & & & & & \cdot & \\ 0 & & & & & 0 & & p_n \end{array}\right] \begin{bmatrix} \hat{x}_1(k) \\ \hat{x}_2(k) \\ \cdot \\ \cdot \\ \cdot \\ \hat{x}_m(k) \\ \hline \hat{x}_{m+1}(k) \\ \cdot \\ \cdot \\ \cdot \\ \hat{x}_n(k) \end{bmatrix} + \begin{bmatrix} 0 \\ 0 \\ \cdot \\ \cdot \\ \cdot \\ \alpha_m \\ \hline \alpha_{m+1} \\ \cdot \\ \cdot \\ \cdot \\ \alpha_n \end{bmatrix} u(k) \tag{6-48}$$

$$y(k) = [\beta_1 \quad \beta_2 \quad \cdots \quad \beta_n] \begin{bmatrix} \hat{x}_1(k) \\ \hat{x}_2(k) \\ \cdot \\ \cdot \\ \cdot \\ \hat{x}_n(k) \end{bmatrix} + Du(k) \tag{6-49}$$

where the α_i's and the β_i's are constants appearing in the pulse transfer function for this system:

$$\begin{aligned} \mathbf{C}(z\mathbf{I} - \mathbf{G})^{-1}\mathbf{H} + D &= \hat{\mathbf{C}}(z\mathbf{I} - \hat{\mathbf{G}})^{-1}\hat{\mathbf{H}} + \hat{D} \\ &= \frac{\alpha_m \beta_1}{(z - p_1)^m} + \frac{\alpha_m \beta_2}{(z - p_1)^{m-1}} + \cdots + \frac{\alpha_m \beta_m}{z - p_1} \\ &\quad + \frac{\alpha_{m+1}\beta_{m+1}}{z - p_{m+1}} + \cdots + \frac{\alpha_n \beta_n}{z - p_n} + D \end{aligned}$$

In many cases we choose $\alpha_m = \alpha_{m+1} = \cdots = \alpha_n = 1$. [Note that the necessary and sufficient condition for the system to be completely state controllable is that $\alpha_i \neq 0$ $(i = m, m+1, \ldots, n)$ and that to be completely observable is that $\beta_i \neq 0$ $(i = 1, m+1, m+2, \ldots, n)$.]

Note that if the rank of $p_1\mathbf{I} - \mathbf{G}$ is $n - s$ (where $2 \leq s \leq n$), that is, if the minimal polynomial is $s - 1$ degree lower than the characteristic polynomial, then $\mathbf{S}^{-1}\mathbf{G}\mathbf{S}$ will have a different Jordan canonical form. (For details, see the Appendix.)

Invariance property of the rank conditions for the controllability matrix and observability matrix. Consider systems related by similarity transformations. Let us define the controllability matrix as $\mathbf{M}$:

$$\mathbf{M} = [\mathbf{H} \vdots \mathbf{G}\mathbf{H} \vdots \cdots \vdots \mathbf{G}^{n-1}\mathbf{H}]$$

Let **P** (an arbitrary $n \times n$ nonsingular matrix) be a similarity transformation matrix and write

$$\mathbf{P}^{-1}\mathbf{G}\mathbf{P} = \tilde{\mathbf{G}}, \qquad \mathbf{P}^{-1}\mathbf{H} = \tilde{\mathbf{H}}$$

Then

$$\mathbf{P}^{-1}\mathbf{G}^2\mathbf{P} = \mathbf{P}^{-1}\mathbf{G}\mathbf{P}\mathbf{P}^{-1}\mathbf{G}\mathbf{P} = \tilde{\mathbf{G}}\tilde{\mathbf{G}} = \tilde{\mathbf{G}}^2$$

$$\mathbf{P}^{-1}\mathbf{G}^3\mathbf{P} = \mathbf{P}^{-1}\mathbf{G}\mathbf{P}\mathbf{P}^{-1}\mathbf{G}\mathbf{P}\mathbf{P}^{-1}\mathbf{G}\mathbf{P} = \tilde{\mathbf{G}}^3$$

$$\vdots$$

$$\mathbf{P}^{-1}\mathbf{G}^{n-1}\mathbf{P} = \tilde{\mathbf{G}}^{n-1}$$

Hence

$$\begin{aligned}\mathbf{P}^{-1}\mathbf{M} &= \mathbf{P}^{-1}[\mathbf{H} \vdots \mathbf{G}\mathbf{H} \vdots \cdots \vdots \mathbf{G}^{n-1}\mathbf{H}] \\ &= [\mathbf{P}^{-1}\mathbf{H} \vdots \mathbf{P}^{-1}\mathbf{G}\mathbf{H} \vdots \cdots \vdots \mathbf{P}^{-1}\mathbf{G}^{n-1}\mathbf{H}] \\ &= [\mathbf{P}^{-1}\mathbf{H} \vdots \mathbf{P}^{-1}\mathbf{G}\mathbf{P}\mathbf{P}^{-1}\mathbf{H} \vdots \cdots \vdots \mathbf{P}^{-1}\mathbf{G}^{n-1}\mathbf{P}\mathbf{P}^{-1}\mathbf{H}] \\ &= [\tilde{\mathbf{H}} \vdots \tilde{\mathbf{G}}\tilde{\mathbf{H}} \vdots \cdots \vdots \tilde{\mathbf{G}}^{n-1}\tilde{\mathbf{H}}] = \tilde{\mathbf{M}}\end{aligned}$$

Since matrix **P** is nonsingular, we have

$$\text{rank } \mathbf{M} = \text{rank } \tilde{\mathbf{M}}$$

Similarly, for the observability matrix define

$$\mathbf{N} = [\mathbf{C}^* \vdots \mathbf{G}^*\mathbf{C}^* \vdots \cdots \vdots (\mathbf{G}^*)^{n-1}\mathbf{C}^*]$$

Let **P** be an arbitrary $n \times n$ nonsingular matrix and write

$$\mathbf{P}^{-1}\mathbf{G}\mathbf{P} = \tilde{\mathbf{G}}, \qquad \mathbf{C}\mathbf{P} = \tilde{\mathbf{C}}$$

Then

$$\begin{aligned}\mathbf{P}^*\mathbf{N} &= \mathbf{P}^*[\mathbf{C}^* \vdots \mathbf{G}^*\mathbf{C}^* \vdots \cdots \vdots (\mathbf{G}^*)^{n-1}\mathbf{C}^*] \\ &= [\mathbf{P}^*\mathbf{C}^* \vdots \mathbf{P}^*\mathbf{G}^*\mathbf{C}^* \vdots \cdots \vdots \mathbf{P}^*(\mathbf{G}^*)^{n-1}\mathbf{C}^*] \\ &= [\tilde{\mathbf{C}}^* \vdots (\mathbf{P}^{-1}\mathbf{G}\mathbf{P})^*\tilde{\mathbf{C}}^* \vdots \cdots \vdots (\mathbf{P}^{-1}\mathbf{G}^{n-1}\mathbf{P})^*\hat{\mathbf{C}}^*] \\ &= [\tilde{\mathbf{C}}^* \vdots \tilde{\mathbf{G}}^*\tilde{\mathbf{C}}^* \vdots \cdots \vdots (\tilde{\mathbf{G}}^*)^{n-1}\tilde{\mathbf{C}}^*] = \tilde{\mathbf{N}}\end{aligned}$$

Hence

$$\text{rank } \mathbf{N} = \text{rank } \tilde{\mathbf{N}}$$

6-5 DESIGN VIA POLE PLACEMENT

In this section we shall present a design method commonly called the *pole placement* or *pole assignment technique*. We assume that all state variables are measurable and are available for feedback. It will be shown that if the system considered is completely state controllable, then poles of the closed-loop system may be placed at any desired locations by means of state feedback through an appropriate state feedback gain matrix.

The present design technique begins with a determination of the desired closed-loop poles based on transient response and/or frequency response requirements such as speed, damping ratio, or bandwidth. Given such considerations, let us assume that we decide that the desired closed-loop poles are to be at $z = \mu_1$, $z = \mu_2$, $\ldots$, $z = \mu_n$. (In choosing the sampling period, care must be exercised so that the desired system will not require unusually large control signals. Otherwise, saturation phenomena will occur in the system. If saturation takes place in the system, the system will become nonlinear and the design method presented here will no longer apply, since the method is applicable only to linear time-invariant systems.) Then, by choosing an appropriate gain matrix for state feedback, it is possible to force the system to have closed-loop poles at the desired locations, provided that the original system is completely state controllable.

In what follows, we shall first treat the case where the control signal is a scalar and prove that a necessary and sufficient condition that the closed-loop poles can be placed at any arbitrary locations in the z plane is that the system be completely state controllable. Then we shall discuss a few methods for determining the required state feedback gain matrix. Finally, we shall deal with the pole placement problem when the control signal is a vector quantity. For such a case it is possible to choose freely more than n parameters; that is, in addition to being able to place n closed-loop poles properly, we have the freedom to satisfy other requirements, if any, of the closed-loop system.

Necessary and sufficient condition for arbitrary pole placement. Consider the open-loop control system shown in Fig. 6–1(a). The state equation is

$$\mathbf{x}(k+1) = \mathbf{G}\mathbf{x}(k) + \mathbf{H}u(k) \tag{6-50}$$

where

$$\mathbf{x}(k) = \text{state vector } (n\text{-vector}) \text{ at } k\text{th sampling instant}$$

$$u(k) = \text{control signal (scalar) at } k\text{th sampling instant}$$

$$\mathbf{G} = n \times n \text{ matrix}$$

$$\mathbf{H} = n \times 1 \text{ matrix}$$

We assume that the magnitude of the control signal $u(k)$ is unbounded. If the control signal $u(k)$ is chosen as

$$u(k) = -\mathbf{K}\mathbf{x}(k)$$

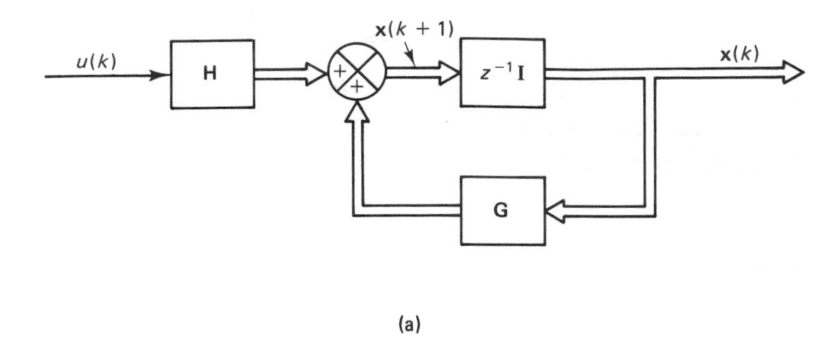

(a)

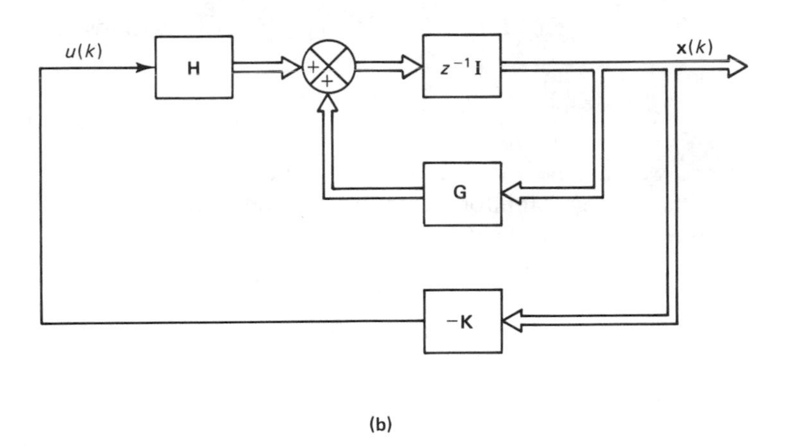

(b)

Figure 6–1 (a) Open-loop control system; (b) closed-loop control system with $u(k) = -\mathbf{K}\mathbf{x}(k)$.

where $\mathbf{K}$ is the state feedback gain matrix (a $1 \times n$ matrix), then the system becomes a closed-loop control system as shown in Fig. 6–1(b) and its state equation becomes

$$\mathbf{x}(k+1) = (\mathbf{G} - \mathbf{H}\mathbf{K})\mathbf{x}(k) \tag{6–51}$$

Note that the eigenvalues of $\mathbf{G} - \mathbf{H}\mathbf{K}$ are the desired closed-loop poles, μ_1, μ_2, $\ldots$, μ_n.

We shall now prove that a necessary and sufficient condition for arbitrary pole placement is that the system be completely state controllable. We shall first derive the necessary condition. We begin by proving that if the system is not completely state controllable, then there are eigenvalues of $\mathbf{G} - \mathbf{H}\mathbf{K}$ that cannot be controlled by state feedback.

Suppose the system of Eq. (6–50) is not completely state controllable. Then, the rank of the controllability matrix is less than n, or

$$\text{rank}\,[\mathbf{H} \vdots \mathbf{GH} \vdots \cdots \vdots \mathbf{G}^{n-1}\mathbf{H}] = q < n \tag{6–52}$$

This means that there are q linearly independent column vectors in the controllability matrix. Let us define such q linearly independent column vectors as $\mathbf{f}_1, \mathbf{f}_2, \ldots, \mathbf{f}_q$. Also, let us choose $n - q$ additional n-vectors $\mathbf{v}_{q+1}, \mathbf{v}_{q+2}, \cdots, \mathbf{v}_n$ such that

$$\mathbf{P} = [\mathbf{f}_1 \vdots \mathbf{f}_2 \vdots \cdots \vdots \mathbf{f}_q \vdots \mathbf{v}_{q+1} \vdots \mathbf{v}_{q+2} \vdots \cdots \vdots \mathbf{v}_n]$$

is of rank n. By using matrix $\mathbf{P}$ as the transformation matrix, let us define

$$\mathbf{P}^{-1}\mathbf{GP} = \hat{\mathbf{G}}, \qquad \mathbf{P}^{-1}\mathbf{H} = \hat{\mathbf{H}}$$

Then, we have

$$\mathbf{GP} = \mathbf{P}\hat{\mathbf{G}}$$

or

$$[\mathbf{Gf}_1 \vdots \cdots \vdots \mathbf{Gf}_q \vdots \mathbf{Gv}_{q+1} \vdots \cdots \vdots \mathbf{Gv}_n] = [\mathbf{f}_1 \vdots \mathbf{f}_2 \vdots \cdots \vdots \mathbf{f}_q \vdots \mathbf{v}_{q+1} \vdots \cdots \vdots \mathbf{v}_n]\,\hat{\mathbf{G}} \tag{6–53}$$

Also

$$\mathbf{H} = \mathbf{P}\hat{\mathbf{H}} = [\mathbf{f}_1 \vdots \mathbf{f}_2 \vdots \cdots \vdots \mathbf{f}_q \vdots \mathbf{v}_{q+1} \vdots \cdots \vdots \mathbf{v}_n]\,\hat{\mathbf{H}} \tag{6–54}$$

Since we have here q linearly independent column vectors $\mathbf{f}_1, \mathbf{f}_2, \ldots, \mathbf{f}_q$, we can use the Cayley-Hamilton theorem to express matrices $\mathbf{Gf}_1, \mathbf{Gf}_2, \ldots, \mathbf{Gf}_q$ in terms of these q vectors. That is,

$$\mathbf{Gf}_1 = g_{11}\mathbf{f}_1 + g_{21}\mathbf{f}_2 + \cdots + g_{q1}\mathbf{f}_q$$

$$\mathbf{Gf}_2 = g_{12}\mathbf{f}_1 + g_{22}\mathbf{f}_2 + \cdots + g_{q2}\mathbf{f}_q$$

$$\vdots$$

$$\mathbf{Gf}_q = g_{1q}\mathbf{f}_1 + g_{2q}\mathbf{f}_2 + \cdots + g_{qq}\mathbf{f}_q$$

Hence, Eq. (6–53) may be written as follows:

$$[\mathbf{Gf}_1 \vdots \cdots \vdots \mathbf{Gf}_q \vdots \mathbf{Gv}_{q+1} \vdots \cdots \vdots \mathbf{Gv}_n]$$

$$= [\mathbf{f}_1 \vdots \cdots \vdots \mathbf{f}_q \vdots \mathbf{v}_{q+1} \vdots \cdots \vdots \mathbf{v}_n]\left[\begin{array}{cccc|cccc} g_{11} & g_{12} & \cdots & g_{1q} & g_{1\,q+1} & g_{1\,q+2} & \cdots & g_{1n} \\ g_{21} & g_{22} & \cdots & g_{2q} & g_{2\,q+1} & g_{2\,q+2} & \cdots & g_{2n} \\ \vdots & \vdots & & \vdots & \vdots & \vdots & & \vdots \\ g_{q1} & g_{q2} & \cdots & g_{qq} & g_{q\,q+1} & g_{q\,q+2} & \cdots & g_{qn} \\ \hline 0 & 0 & \cdots & 0 & g_{q+1\,q+1} & g_{q+1\,q+2} & \cdots & g_{q+1\,n} \\ \vdots & \vdots & & \vdots & \vdots & \vdots & & \vdots \\ 0 & 0 & \cdots & 0 & g_{n\,q+1} & g_{n\,q+2} & \cdots & g_{nn} \end{array}\right]$$

To simplify the notation, let us define

$$\begin{bmatrix} g_{11} & g_{12} & \cdots & g_{1q} \\ g_{21} & g_{22} & \cdots & g_{2q} \\ \vdots & \vdots & & \vdots \\ g_{q1} & g_{q2} & \cdots & g_{qq} \end{bmatrix} = \mathbf{G}_{11}$$

$$\begin{bmatrix} g_{1\,q+1} & g_{1\,q+2} & \cdots & g_{1n} \\ g_{2\,q+1} & g_{2\,q+2} & \cdots & g_{2n} \\ \vdots & \vdots & & \vdots \\ g_{q\,q+1} & g_{q\,q+2} & \cdots & g_{qn} \end{bmatrix} = \mathbf{G}_{12}$$

$$\begin{bmatrix} 0 & 0 & \cdots & 0 \\ \vdots & \vdots & & \vdots \\ 0 & 0 & \cdots & 0 \end{bmatrix} = \mathbf{G}_{21} = (n-q) \times q \text{ zero matrix}$$

$$\begin{bmatrix} g_{q+1\,q+1} & g_{q+1\,q+2} & \cdots & g_{q+1\,n} \\ g_{q+2\,q+1} & g_{q+2\,q+2} & \cdots & g_{q+2\,n} \\ \vdots & \vdots & & \vdots \\ g_{n\,q+1} & g_{n\,q+2} & \cdots & g_{nn} \end{bmatrix} = \mathbf{G}_{22}$$

Then, Eq. (6–53) can be written as follows:

$$[\mathbf{Gf}_1 \vdots \cdots \vdots \mathbf{Gf}_q \vdots \mathbf{Gv}_{q+1} \vdots \cdots \vdots \mathbf{Gv}_n] = [\mathbf{f}_1 \vdots \cdots \vdots \mathbf{f}_q \vdots \mathbf{v}_{q+1} \vdots \cdots \vdots \mathbf{v}_n] \left[\begin{array}{c|c} \mathbf{G}_{11} & \mathbf{G}_{12} \\ \hline \mathbf{0} & \mathbf{G}_{22} \end{array}\right]$$

Thus

$$\hat{\mathbf{G}} = \left[\begin{array}{c|c} \mathbf{G}_{11} & \mathbf{G}_{12} \\ \hline \mathbf{0} & \mathbf{G}_{22} \end{array}\right] \tag{6–55}$$

Next, referring to Eq. (6–54) we have

$$\mathbf{H} = [\mathbf{f}_1 \vdots \mathbf{f}_2 \vdots \cdots \vdots \mathbf{f}_q \vdots \mathbf{v}_{q+1} \vdots \cdots \vdots \mathbf{v}_n]\,\hat{\mathbf{H}} \tag{6–56}$$

Referring to Eq. (6–52), notice that vector $\mathbf{H}$ can be written in terms of q linearly independent column vectors $\mathbf{f}_1, \mathbf{f}_2, \ldots, \mathbf{f}_q$. Thus, we have

$$\mathbf{H} = h_{11}\mathbf{f}_1 + h_{21}\mathbf{f}_2 + \cdots + h_{q1}\mathbf{f}_q$$

Consequently, Eq. (6–56) may be written as follows:

$$h_{11}\mathbf{f}_1 + h_{21}\mathbf{f}_2 + \cdots + h_{q1}\mathbf{f}_q = [\mathbf{f}_1 \vdots \mathbf{f}_2 \vdots \cdots \vdots \mathbf{f}_q \vdots \mathbf{v}_{q+1} \vdots \cdots \vdots \mathbf{v}_n] \begin{bmatrix} h_{11} \\ h_{21} \\ \cdot \\ \cdot \\ \cdot \\ h_{q1} \\ \hdashline 0 \\ \cdot \\ \cdot \\ \cdot \\ 0 \end{bmatrix}$$

Thus

$$\hat{\mathbf{H}} = \begin{bmatrix} \mathbf{H}_{11} \\ \hdashline \mathbf{0} \end{bmatrix} \tag{6–57}$$

where

$$\mathbf{H}_{11} = \begin{bmatrix} h_{11} \\ h_{21} \\ \cdot \\ \cdot \\ \cdot \\ h_{q1} \end{bmatrix}$$

Now consider the closed-loop system equation given by Eq. (6–51). The characteristic equation is

$$|z\mathbf{I} - \mathbf{G} + \mathbf{HK}| = 0$$

Let us define

$$\tilde{\mathbf{K}} = \mathbf{KP}$$

and partition the matrix $\tilde{\mathbf{K}}$ to give

$$\tilde{\mathbf{K}} = [\mathbf{K}_{11} \vdots \mathbf{K}_{12}] \tag{6–58}$$

where $\mathbf{K}_{11}$ is a $1 \times q$ matrix and $\mathbf{K}_{12}$ is a $1 \times (n - q)$ matrix. Now, $1 \times n$ matrix $\mathbf{K}$ can be written as follows:

$$\mathbf{K} = \tilde{\mathbf{K}}\mathbf{P}^{-1} = [\mathbf{K}_{11} \vdots \mathbf{K}_{12}]\, \mathbf{P}^{-1}$$

Then the characteristic equation for the closed-loop system can be written as follows:

$$|z\mathbf{I} - \mathbf{G} + \mathbf{HK}| = |\mathbf{P}^{-1}|\, |z\mathbf{I} - \mathbf{G} + \mathbf{HK}|\, |\mathbf{P}|$$

$$= |z\mathbf{I} - \mathbf{P}^{-1}\mathbf{G}\mathbf{P} + \mathbf{P}^{-1}\mathbf{H}\mathbf{K}\mathbf{P}|$$
$$= |z\mathbf{I} - \hat{\mathbf{G}} + \hat{\mathbf{H}}\tilde{\mathbf{K}}|$$

Substituting Eqs. (6–55), (6–57), and (6–58) into this last equation, we obtain

$$|z\mathbf{I} - \hat{\mathbf{G}} + \hat{\mathbf{H}}\tilde{\mathbf{K}}| = \left| z\begin{bmatrix} \mathbf{I}_q & \mathbf{0} \\ \mathbf{0} & \mathbf{I}_{n-q} \end{bmatrix} - \begin{bmatrix} \mathbf{G}_{11} & \mathbf{G}_{12} \\ \mathbf{0} & \mathbf{G}_{22} \end{bmatrix} + \begin{bmatrix} \mathbf{H}_{11} \\ \mathbf{0} \end{bmatrix} [\mathbf{K}_{11} \vdots \mathbf{K}_{12}] \right|$$
$$= \begin{vmatrix} z\mathbf{I}_q - \mathbf{G}_{11} + \mathbf{H}_{11}\mathbf{K}_{11} & -\mathbf{G}_{12} + \mathbf{H}_{11}\mathbf{K}_{12} \\ \mathbf{0} & z\mathbf{I}_{n-q} - \mathbf{G}_{22} \end{vmatrix}$$
$$= |z\mathbf{I}_q - \mathbf{G}_{11} + \mathbf{H}_{11}\mathbf{K}_{11}|\,|z\mathbf{I}_{n-q} - \mathbf{G}_{22}| \qquad (6\text{–}59)$$

Equation (6–59) shows that matrix $\mathbf{K} = \tilde{\mathbf{K}}\mathbf{P}^{-1}$ has control over the q eigenvalues of $\mathbf{G}_{11} - \mathbf{H}_{11}\mathbf{K}_{11}$, but not over the $n - q$ eigenvalues of $\mathbf{G}_{22}$. That is, there are $n - q$ eigenvalues of $\mathbf{G} - \mathbf{HK}$ that do not depend on matrix $\mathbf{K}$. Hence we have proved that complete state controllability is a necessary condition for controlling the eigenvalues (closed-loop pole locations) of matrix $\mathbf{G} - \mathbf{HK}$.

We shall next derive a sufficient condition. We shall prove that if the system is completely state controllable, then there exists a matrix $\mathbf{K}$ that will make the eigenvalues of $\mathbf{G} - \mathbf{HK}$ as desired, or place the closed-loop poles at the desired locations.

The desired eigenvalues of $\mathbf{G} - \mathbf{HK}$ are $\mu_1, \mu_2, \ldots, \mu_n$; any complex eigenvalues are to occur as conjugate pairs. Noting that the characteristic equation of the original system given by Eq. (6–50) is

$$|z\mathbf{I} - \mathbf{G}| = z^n + a_1 z^{n-1} + a_2 z^{n-2} + \cdots + a_{n-1}z + a_n = 0$$

we define a transformation matrix $\mathbf{T}$ as follows:

$$\mathbf{T} = \mathbf{MW}$$

where

$$\mathbf{M} = [\mathbf{H} \vdots \mathbf{GH} \vdots \cdots \vdots \mathbf{G}^{n-1}\mathbf{H}] \qquad (6\text{–}60)$$

which is of rank n, and where

$$\mathbf{W} = \begin{bmatrix} a_{n-1} & a_{n-2} & \cdots & a_1 & 1 \\ a_{n-2} & a_{n-3} & \cdots & 1 & 0 \\ \cdot & \cdot & & \cdot & \cdot \\ \cdot & \cdot & & \cdot & \cdot \\ \cdot & \cdot & & \cdot & \cdot \\ a_1 & 1 & \cdots & 0 & 0 \\ 1 & 0 & \cdots & 0 & 0 \end{bmatrix} \qquad (6\text{–}61)$$

Then, referring to Eqs. (6–35) and (6–36) we have

$$\mathbf{T}^{-1}\mathbf{G}\mathbf{T} = \hat{\mathbf{G}} = \begin{bmatrix} 0 & 1 & 0 & \cdots & 0 \\ 0 & 0 & 1 & \cdots & 0 \\ \cdot & \cdot & \cdot & & \cdot \\ \cdot & \cdot & \cdot & & \cdot \\ \cdot & \cdot & \cdot & & \cdot \\ 0 & 0 & 0 & \cdots & 1 \\ -a_n & -a_{n-1} & -a_{n-2} & \cdots & -a_1 \end{bmatrix}$$

and

$$\mathbf{T}^{-1}\mathbf{H} = \hat{\mathbf{H}} = \begin{bmatrix} 0 \\ 0 \\ \cdot \\ \cdot \\ \cdot \\ 0 \\ 1 \end{bmatrix}$$

Next, we define

$$\hat{\mathbf{K}} = \mathbf{K}\mathbf{T} = [\delta_n \quad \delta_{n-1} \quad \cdots \quad \delta_1] \tag{6–62}$$

Then

$$\hat{\mathbf{H}}\hat{\mathbf{K}} = \begin{bmatrix} 0 \\ 0 \\ \cdot \\ \cdot \\ \cdot \\ 0 \\ 1 \end{bmatrix} [\delta_n \quad \delta_{n-1} \quad \cdots \quad \delta_1] = \begin{bmatrix} 0 & 0 & \cdots & 0 \\ 0 & 0 & \cdots & 0 \\ \cdot & \cdot & & \cdot \\ \cdot & \cdot & & \cdot \\ \cdot & \cdot & & \cdot \\ 0 & 0 & \cdots & 0 \\ \delta_n & \delta_{n-1} & \cdots & \delta_1 \end{bmatrix}$$

The characteristic equation $|z\mathbf{I} - \mathbf{G} + \mathbf{H}\mathbf{K}|$ becomes as follows:

$$|z\mathbf{I} - \mathbf{G} + \mathbf{H}\mathbf{K}| = |z\mathbf{I} - \hat{\mathbf{G}} + \hat{\mathbf{H}}\hat{\mathbf{K}}|$$

$$= \left| z\begin{bmatrix} 1 & 0 & \cdots & 0 \\ 0 & 1 & \cdots & 0 \\ \cdot & \cdot & & \cdot \\ \cdot & \cdot & & \cdot \\ \cdot & \cdot & & \cdot \\ 0 & 0 & \cdots & 0 \\ 0 & 0 & \cdots & 1 \end{bmatrix} - \begin{bmatrix} 0 & 1 & \cdots & 0 \\ 0 & 0 & \cdots & 0 \\ \cdot & \cdot & & \cdot \\ \cdot & \cdot & & \cdot \\ \cdot & \cdot & & \cdot \\ 0 & 0 & \cdots & 1 \\ -a_n & -a_{n-1} & \cdots & -a_1 \end{bmatrix} + \begin{bmatrix} 0 & 0 & \cdots & 0 \\ 0 & 0 & \cdots & 0 \\ \cdot & \cdot & & \cdot \\ \cdot & \cdot & & \cdot \\ \cdot & \cdot & & \cdot \\ 0 & 0 & \cdots & 0 \\ \delta_n & \delta_{n-1} & \cdots & \delta_1 \end{bmatrix} \right|$$

$$
= \begin{vmatrix} z & -1 & \cdots & 0 \\ 0 & z & \cdots & 0 \\ \cdot & \cdot & & \cdot \\ \cdot & \cdot & & \cdot \\ \cdot & \cdot & & \cdot \\ 0 & 0 & \cdots & -1 \\ a_n + \delta_n & a_{n-1} + \delta_{n-1} & \cdots & z + a_1 + \delta_1 \end{vmatrix}
$$

$$
= z^n + (a_1 + \delta_1)z^{n-1} + \cdots + (a_{n-1} + \delta_{n-1})z + a_n + \delta_n = 0 \tag{6-63}
$$

The characteristic equation with the desired eigenvalues is given by

$$
(z - \mu_1)(z - \mu_2) \cdots (z - \mu_n)
$$
$$
= z^n + \alpha_1 z^{n-1} + \alpha_2 z^{n-2} + \cdots + \alpha_{n-1} z + \alpha_n = 0 \tag{6-64}
$$

Equating the coefficients of equal powers of z of Eqs. (6–63) and (6–64) we obtain

$$
\alpha_1 = a_1 + \delta_1
$$
$$
\alpha_2 = a_2 + \delta_2
$$
$$
\cdot
$$
$$
\cdot
$$
$$
\cdot
$$
$$
\alpha_n = a_n + \delta_n
$$

Hence, from Eq. (6–62) we have

$$
\begin{aligned}
\mathbf{K} &= \hat{\mathbf{K}}\mathbf{T}^{-1} \\
&= [\delta_n \quad \delta_{n-1} \quad \cdots \quad \delta_1]\,\mathbf{T}^{-1} \\
&= [\alpha_n - a_n \vdots \alpha_{n-1} - a_{n-1} \vdots \cdots \vdots \alpha_1 - a_1]\,\mathbf{T}^{-1}
\end{aligned} \tag{6-65}
$$

where the a_i's and the α_i's are known coefficients and $\mathbf{T}$ is a known matrix. Hence, we have determined the required feedback gain matrix $\mathbf{K}$ in terms of known coefficients and a known matrix of the system. This proves the sufficient condition, that is, that if the system defined by Eq. (6–50) is completely state controllable, then it is always possible to determine the required state feedback gain matrix $\mathbf{K}$ for arbitrary pole placement. Hence we have proved that a necessary and sufficient condition for arbitrary pole placement is that the system be completely state controllable.

Ackermann's formula. The expression given by Eq. (6–65) is not the only one used for the determination of the state feedback gain matrix **K.** There are other expressions available. In the following, we shall present one such expression, commonly called *Ackermann's formula*.

Consider the system defined by Eq. (6–50). It is assumed that the system is completely state controllable. By using the state feedback $u(k) = -\mathbf{K}\mathbf{x}(k)$, we wish to place closed-loop poles at $z = \mu_1, z = \mu_2, \ldots, z = \mu_n$. That is, we desire the characteristic equation to be

$$|z\mathbf{I} - \mathbf{G} + \mathbf{HK}| = (z - \mu_1)(z - \mu_2) \cdots (z - \mu_n)$$
$$= z^n + \alpha_1 z^{n-1} + \alpha_2 z^{n-2} + \cdots + \alpha_{n-1} z + \alpha_n = 0$$

Let us define

$$\tilde{\mathbf{G}} = \mathbf{G} - \mathbf{HK}$$

Since the Cayley-Hamilton theorem states that $\tilde{\mathbf{G}}$ satisfies its own characteristic equation, we have

$$\tilde{\mathbf{G}}^n + \alpha_1 \tilde{\mathbf{G}}^{n-1} + \alpha_2 \tilde{\mathbf{G}}^{n-2} + \cdots + \alpha_{n-1}\tilde{\mathbf{G}} + \alpha_n \mathbf{I} = \phi(\tilde{\mathbf{G}}) = \mathbf{0}$$

We shall utilize this last equation to derive Ackermann's formula.

Consider now the following identities:

$$\mathbf{I} = \mathbf{I}$$
$$\tilde{\mathbf{G}} = \mathbf{G} - \mathbf{HK}$$
$$\tilde{\mathbf{G}}^2 = (\mathbf{G} - \mathbf{HK})^2 = \mathbf{G}^2 - \mathbf{GHK} - \mathbf{HK}\tilde{\mathbf{G}}$$
$$\tilde{\mathbf{G}}^3 = (\mathbf{G} - \mathbf{HK})^3 = \mathbf{G}^3 - \mathbf{G}^2\mathbf{HK} - \mathbf{GHK}\tilde{\mathbf{G}} - \mathbf{HK}\tilde{\mathbf{G}}^2$$
$$\vdots$$
$$\tilde{\mathbf{G}}^n = (\mathbf{G} - \mathbf{HK})^n = \mathbf{G}^n - \mathbf{G}^{n-1}\mathbf{HK} - \cdots - \mathbf{HK}\tilde{\mathbf{G}}^{n-1}$$

Multiplying the preceding equations in order by $\alpha_n, \alpha_{n-1}, \ldots, \alpha_0$ (where $\alpha_0 = 1$), respectively, and adding the results, we obtain

$$\alpha_n \mathbf{I} + \alpha_{n-1}\tilde{\mathbf{G}} + \alpha_{n-2}\tilde{\mathbf{G}}^2 + \cdots + \tilde{\mathbf{G}}^n = \alpha_n \mathbf{I} + \alpha_{n-1}\mathbf{G} + \alpha_{n-2}\mathbf{G}^2 + \cdots + \mathbf{G}^n - \alpha_{n-1}\mathbf{HK} - \alpha_{n-2}\mathbf{GHK} - \alpha_{n-2}\mathbf{HK}\tilde{\mathbf{G}} - \cdots - \mathbf{G}^{n-1}\mathbf{HK} - \cdots - \mathbf{HK}\tilde{\mathbf{G}}^{n-1}$$

which can be written as follows:

$$\phi(\tilde{\mathbf{G}}) = \phi(\mathbf{G}) - \alpha_{n-1}\mathbf{HK} - \alpha_{n-2}\mathbf{GHK} - \alpha_{n-2}\mathbf{HK}\tilde{\mathbf{G}} - \cdots - \mathbf{HK}\tilde{\mathbf{G}}^{n-1} - \mathbf{G}^{n-1}\mathbf{HK}$$

$$= \phi(\mathbf{G}) - [\mathbf{H} \vdots \mathbf{GH} \vdots \cdots \vdots \mathbf{G}^{n-1}\mathbf{H}] \begin{bmatrix} \alpha_{n-1}\mathbf{K} + \alpha_{n-2}\mathbf{K}\tilde{\mathbf{G}} + \cdots + \mathbf{K}\tilde{\mathbf{G}}^{n-1} \\ \alpha_{n-2}\mathbf{K} + \alpha_{n-3}\mathbf{K}\tilde{\mathbf{G}} + \cdots + \mathbf{K}\tilde{\mathbf{G}}^{n-2} \\ \vdots \\ \mathbf{K} \end{bmatrix} \tag{6-66}$$

Notice that

$$\phi(\tilde{\mathbf{G}}) = \mathbf{0}$$

Hence Eq. (6–66) may be modified to read

$$\phi(\mathbf{G}) = [\mathbf{H} \vdots \mathbf{GH} \vdots \cdots \vdots \mathbf{G}^{n-1}\mathbf{H}] \begin{bmatrix} \alpha_{n-1}\mathbf{K} + \alpha_{n-2}\mathbf{K}\tilde{\mathbf{G}} + \cdots + \mathbf{K}\tilde{\mathbf{G}}^{n-1} \\ \alpha_{n-2}\mathbf{K} + \alpha_{n-3}\mathbf{K}\tilde{\mathbf{G}} + \cdots + \mathbf{K}\tilde{\mathbf{G}}^{n-2} \\ \cdot \\ \cdot \\ \cdot \\ \mathbf{K} \end{bmatrix} \tag{6–67}$$

Since the system is completely state controllable, the controllability matrix

$$[\mathbf{H} \vdots \mathbf{GH} \vdots \cdots \vdots \mathbf{G}^{n-1}\mathbf{H}]$$

is of rank n and its inverse exists. Then, Eq. (6–67) can be modified to the form

$$\begin{bmatrix} \alpha_{n-1}\mathbf{K} + \alpha_{n-2}\mathbf{K}\tilde{\mathbf{G}} + \cdots + \mathbf{K}\tilde{\mathbf{G}}^{n-1} \\ \alpha_{n-2}\mathbf{K} + \alpha_{n-3}\mathbf{K}\tilde{\mathbf{G}} + \cdots + \mathbf{K}\tilde{\mathbf{G}}^{n-2} \\ \cdot \\ \cdot \\ \cdot \\ \mathbf{K} \end{bmatrix} = [\mathbf{H} \vdots \mathbf{GH} \vdots \cdots \vdots \mathbf{G}^{n-1}\mathbf{H}]^{-1}\phi(\mathbf{G})$$

Premultiplying both sides of this last equation by $[0 \ 0 \ \cdots \ 0 \ 1]$, we obtain

$$[0 \quad 0 \quad \cdots \quad 0 \quad 1] \begin{bmatrix} \alpha_{n-1}\mathbf{K} + \alpha_{n-2}\mathbf{K}\tilde{\mathbf{G}} + \cdots + \mathbf{K}\tilde{\mathbf{G}}^{n-1} \\ \alpha_{n-2}\mathbf{K} + \alpha_{n-3}\mathbf{K}\tilde{\mathbf{G}} + \cdots + \mathbf{K}\tilde{\mathbf{G}}^{n-2} \\ \cdot \\ \cdot \\ \cdot \\ \mathbf{K} \end{bmatrix}$$

$$= [0 \quad 0 \quad \cdots \quad 0 \quad 1][\mathbf{H} \vdots \mathbf{GH} \vdots \cdots \vdots \mathbf{G}^{n-1}\mathbf{H}]^{-1}\phi(\mathbf{G})$$

which can be simplified to

$$\mathbf{K} = [0 \quad 0 \quad \cdots \quad 0 \quad 1][\mathbf{H} \vdots \mathbf{GH} \vdots \cdots \vdots \mathbf{G}^{n-1}\mathbf{H}]^{-1}\phi(\mathbf{G}) \tag{6–68}$$

Equation (6–68) gives the required state feedback gain matrix **K**. It is this particular expression for matrix **K** that is commonly called Ackermann's formula.

Comments. The state feedback gain matrix **K** is determined in such a way that the error (caused by disturbances) will reduce to zero with sufficient speed. Note that the matrix **K** is not unique for a given system, but depends on the desired closed-loop pole locations (which determine the speed of response) selected. The selection of the desired closed-loop poles or the desired characteristic equation is a compromise between the rapidity of the response of the error vector and the sensitivity to disturbances and measurement noises. That is, if we increase the speed of error response, then the adverse effects of disturbances and measurement noises generally increase. In determining the state feedback gain matrix **K** for a given system, it is desirable to examine several matrices **K** based on several different desired characteristic equations and to choose the one that gives the best overall system performance.

Once the desired characteristic equation is selected, there are several different ways to determine the corresponding state feedback gain matrix **K** for the system defined by Eq. (6–50). Four of them are listed as follows:

1. As shown in the preceding discussion, matrix **K** can be given by Eq. (6–65):

$$\mathbf{K} = [\alpha_n - a_n \vdots \alpha_{n-1} - a_{n-1} \vdots \cdots \vdots \alpha_1 - a_1]\mathbf{T}^{-1}$$
$$= [\alpha_n - a_n \vdots \alpha_{n-1} - a_{n-1} \vdots \cdots \vdots \alpha_1 - a_1](\mathbf{MW})^{-1} \qquad (6\text{–}69)$$

where the a_i's are the coefficients of the original system characteristic equation

$$|z\mathbf{I} - \mathbf{G}| = z^n + a_1 z^{n-1} + \cdots + a_{n-1}z + a_n = 0$$

and the α_i's are the coefficients of the desired characteristic equation for the state feedback control system, that is,

$$|z\mathbf{I} - \mathbf{G} + \mathbf{HK}| = z^n + \alpha_1 z^{n-1} + \cdots + \alpha_{n-1}z + \alpha_n = 0$$

Matrix **T** is given by

$$\mathbf{T} = \mathbf{MW}$$

where **M** and **W** are given by Eqs. (6–60) and (6–61), respectively.

If the system state equation is already in the controllable canonical form, the determination of the state feedback gain matrix **K** can be made simple, because the transformation matrix **T** becomes the identity matrix. In this case the desired matrix **K** is simply given as follows:

$$\mathbf{K} = [\alpha_n - a_n \vdots \alpha_{n-1} - a_{n-1} \vdots \cdots \vdots \alpha_1 - a_1] \qquad (6\text{–}70)$$

2. The desired state feedback gain matrix **K** can be given by Ackermann's formula:

$$\mathbf{K} = [0 \quad 0 \quad \cdots \quad 0 \quad 1][\mathbf{H} \vdots \mathbf{GH} \vdots \cdots \vdots \mathbf{G}^{n-1}\mathbf{H}]^{-1}\phi(\mathbf{G}) \qquad (6\text{–}71)$$

where

$$\phi(\mathbf{G}) = \mathbf{G}^n + \alpha_1 \mathbf{G}^{n-1} + \cdots + \alpha_{n-1}\mathbf{G} + \alpha_n \mathbf{I}$$

3. If the desired eigenvalues $\mu_1, \mu_2, \ldots, \mu_n$ are distinct, then the desired state feedback gain matrix **K** can be given as follows:

$$\mathbf{K} = [1 \quad 1 \quad \cdots \quad 1][\boldsymbol{\xi}_1 \vdots \boldsymbol{\xi}_2 \vdots \cdots \vdots \boldsymbol{\xi}_n]^{-1} \qquad (6\text{–}72)$$

where vectors $\boldsymbol{\xi}_1, \boldsymbol{\xi}_2, \cdots, \boldsymbol{\xi}_n$ satisfy the equation

$$\boldsymbol{\xi}_i = (\mathbf{G} - \mu_i \mathbf{I})^{-1}\mathbf{H} \qquad i = 1, 2, \cdots, n$$

Note that the $\boldsymbol{\xi}_i$'s are eigenvectors of matrix $\mathbf{G} - \mathbf{HK}$; that is, $\boldsymbol{\xi}_i$ satisfies the equation

$$(\mathbf{G} - \mathbf{HK})\boldsymbol{\xi}_i = \mu_i \boldsymbol{\xi}_i \qquad i = 1, 2, \cdots, n$$

For the deadbeat response, $\mu_1 = \mu_2 = \cdots = \mu_n = 0$. Equation (6–72) for this case can be simplified as follows:

$$\mathbf{K} = [1 \quad 0 \quad \cdots \quad 0][\boldsymbol{\xi}_1 \vdots \boldsymbol{\xi}_2 \vdots \cdots \vdots \boldsymbol{\xi}_n]^{-1} \tag{6–73}$$

where

$$\boldsymbol{\xi}_1 = \mathbf{G}^{-1}\mathbf{H}, \qquad \boldsymbol{\xi}_2 = \mathbf{G}^{-2}\mathbf{H}, \ldots, \boldsymbol{\xi}_n = \mathbf{G}^{-n}\mathbf{H}$$

[For detailed derivations of Eqs. (6–72) and (6–73), see Probs. A-6–15 and A-6–16.]

4. If the order n of the system is low, substitute $\mathbf{K} = [k_1 \vdots k_2 \vdots \cdots \vdots k_n]$ into the characteristic equation

$$|z\mathbf{I} - \mathbf{G} + \mathbf{HK}| = 0$$

and then match the coefficients of powers in z of this characteristic equation with equal powers in z of the desired characteristic equation

$$z^n + \alpha_1 z^{n-1} + \cdots + \alpha_{n-1} z + \alpha_n = 0$$

Such a direct calculation of matrix $\mathbf{K}$ may be simpler for low-order systems.

Example 6–6.

Consider the system

$$\mathbf{x}(k+1) = \mathbf{Gx}(k) + \mathbf{H}u(k)$$

where

$$\mathbf{G} = \begin{bmatrix} 0 & 1 \\ -0.16 & -1 \end{bmatrix}, \qquad \mathbf{H} = \begin{bmatrix} 0 \\ 1 \end{bmatrix}$$

Note that

$$|z\mathbf{I} - \mathbf{G}| = \begin{vmatrix} z & -1 \\ 0.16 & z+1 \end{vmatrix} = z^2 + z + 0.16$$

Hence

$$a_1 = 1, \qquad a_2 = 0.16$$

Determine a suitable state feedback gain matrix $\mathbf{K}$ such that the system will have the closed-loop poles at

$$z = 0.5 + j0.5, \qquad z = 0.5 - j0.5$$

Let us first examine the rank of the controllability matrix. The rank of

$$[\mathbf{H} \vdots \mathbf{GH}] = \begin{bmatrix} 0 & 1 \\ 1 & -1 \end{bmatrix}$$

is 2. Thus the system is completely state controllable, and therefore arbitrary pole placement is possible. The characteristic equation for the desired system is

$$|z\mathbf{I} - \mathbf{G} + \mathbf{HK}| = (z - 0.5 - j0.5)(z - 0.5 + j0.5) = z^2 - z + 0.5 = 0$$

Hence

$$\alpha_1 = -1, \qquad \alpha_2 = 0.5$$

We shall demonstrate four different ways to determine matrix **K**.

Method 1. From Eq. (6–69), the state feedback gain matrix **K** is given as follows:

$$\mathbf{K} = [\alpha_2 - a_2 \vdots \alpha_1 - a_1]\mathbf{T}^{-1}$$

Notice that the original system is already in a controllable canonical form and therefore the transformation matrix **T** becomes **I**:

$$\mathbf{T} = \mathbf{MW} = [\mathbf{H} \vdots \mathbf{GH}]\begin{bmatrix} a_1 & 1 \\ 1 & 0 \end{bmatrix} = \begin{bmatrix} 0 & 1 \\ 1 & -1 \end{bmatrix}\begin{bmatrix} 1 & 1 \\ 1 & 0 \end{bmatrix} = \begin{bmatrix} 1 & 0 \\ 0 & 1 \end{bmatrix}$$

Hence,

$$\begin{aligned} \mathbf{K} &= [\alpha_2 - a_2 \vdots \alpha_1 - a_1] = [0.5 - 0.16 \vdots -1 - 1] \\ &= [0.34 \quad -2] \end{aligned}$$

Method 2. Referring to Ackermann's formula given by Eq. (6–71), we have

$$\mathbf{K} = [0 \quad 1]\,[\mathbf{H} \vdots \mathbf{GH}]^{-1}\phi(\mathbf{G})$$

where

$$\begin{aligned} \phi(\mathbf{G}) = \mathbf{G}^2 - \mathbf{G} + 0.5\mathbf{I} &= \begin{bmatrix} -0.16 & -1 \\ 0.16 & 0.84 \end{bmatrix} - \begin{bmatrix} 0 & 1 \\ -0.16 & -1 \end{bmatrix} + \begin{bmatrix} 0.5 & 0 \\ 0 & 0.5 \end{bmatrix} \\ &= \begin{bmatrix} 0.34 & -2 \\ 0.32 & 2.34 \end{bmatrix} \end{aligned}$$

Thus

$$\begin{aligned} \mathbf{K} &= [0 \quad 1]\begin{bmatrix} 0 & 1 \\ 1 & -1 \end{bmatrix}^{-1}\begin{bmatrix} 0.34 & -2 \\ 0.32 & 2.34 \end{bmatrix} \\ &= [0.34 \quad -2] \end{aligned}$$

Method 3. From Eq. (6–72), the desired state feedback gain matrix **K** is determined as follows:

$$\mathbf{K} = [1 \quad 1]\,[\boldsymbol{\xi}_1 \quad \boldsymbol{\xi}_2]^{-1}$$

where

$$\boldsymbol{\xi}_i = (\mathbf{G} - \mu_i \mathbf{I})^{-1}\mathbf{H} \qquad i = 1, 2$$

Since $\mu_1 = 0.5 + j0.5$, we have

$$\begin{aligned} \boldsymbol{\xi}_1 &= [\mathbf{G} - (0.5 + j0.5)\mathbf{I}]^{-1}\mathbf{H} \\ &= \begin{bmatrix} -0.5 - j0.5 & 1 \\ -0.16 & -1.5 - j0.5 \end{bmatrix}^{-1}\begin{bmatrix} 0 \\ 1 \end{bmatrix} = \begin{bmatrix} \dfrac{-1}{0.66 + j} \\ \dfrac{-0.5 - j0.5}{0.66 + j} \end{bmatrix} \end{aligned}$$

Similarly, for $\mu_2 = 0.5 - j0.5$, we have

$$\boldsymbol{\xi}_2 = [\mathbf{G} - (0.5 - j0.5)\mathbf{I}]^{-1}\mathbf{H}$$

$$= \begin{bmatrix} -0.5 + j0.5 & 1 \\ -0.16 & -1.5 + j0.5 \end{bmatrix}^{-1} \begin{bmatrix} 0 \\ 1 \end{bmatrix} = \begin{bmatrix} \dfrac{-1}{0.66 - j} \\ \dfrac{-0.5 + j0.5}{0.66 - j} \end{bmatrix}$$

Consequently, we have

$$[\boldsymbol{\xi}_1 \quad \boldsymbol{\xi}_2]^{-1} = \begin{bmatrix} \dfrac{-1}{0.66 + j} & \dfrac{-1}{0.66 - j} \\ \dfrac{-0.5 - j0.5}{0.66 + j} & \dfrac{-0.5 + j0.5}{0.66 - j} \end{bmatrix}^{-1}$$

$$= \begin{bmatrix} \dfrac{0.7178(1 - j)}{1 + j0.66} & \dfrac{-1.4356}{1 + j0.66} \\ \dfrac{-0.7178(1 + j)}{-1 + j0.66} & \dfrac{1.4356}{-1 + j0.66} \end{bmatrix}$$

Hence, the desired state feedback gain matrix $\mathbf{K}$ is determined to be

$$\mathbf{K} = [1 \quad 1]\,[\boldsymbol{\xi}_1 \quad \boldsymbol{\xi}_2]^{-1}$$

$$= [1 \quad 1] \begin{bmatrix} \dfrac{0.7178(1 - j)}{1 + j0.66} & \dfrac{-1.4356}{1 + j0.66} \\ \dfrac{-0.7178(1 + j)}{-1 + j0.66} & \dfrac{1.4356}{-1 + j0.66} \end{bmatrix}$$

$$= [0.34 \quad -2]$$

Method 4. It is noted that for lower-order systems such as this one it may be simpler to substitute

$$\mathbf{K} = [k_1 \quad k_2]$$

into the characteristic equation and to write the equation in terms of undetermined k's. Then this characteristic equation is equated with the desired characteristic equation. The procedure is as follows:

$$|z\mathbf{I} - \mathbf{G} + \mathbf{HK}| = \left| \begin{bmatrix} z & 0 \\ 0 & z \end{bmatrix} - \begin{bmatrix} 0 & 1 \\ -0.16 & -1 \end{bmatrix} + \begin{bmatrix} 0 \\ 1 \end{bmatrix} [k_1 \quad k_2] \right|$$

$$= \begin{vmatrix} z & -1 \\ 0.16 + k_1 & z + 1 + k_2 \end{vmatrix}$$

$$= z^2 + (1 + k_2)z + 0.16 + k_1 = 0$$

Now we equate this characteristic equation with the desired characteristic equation

$$(z - 0.5 - j0.5)(z - 0.5 + j0.5) = 0$$

so that

$$z^2 + (1 + k_2)z + 0.16 + k_1 = z^2 - z + 0.5$$

By comparing the coefficients of equal powers of z, we obtain

$$1 + k_2 = -1, \qquad 0.16 + k_1 = 0.5$$

from which we get

$$k_1 = 0.34, \qquad k_2 = -2$$

Thus the desired state feedback gain matrix $\mathbf{K}$ is given by

$$\mathbf{K} = [k_1 \quad k_2] = [0.34 \quad -2]$$

It is noted that for higher-order systems, the calculations involved by this approach may become laborious. For such a case other methods may be preferred.

Deadbeat response. The pole placement technique is a very powerful approach to designing control systems with desired closed-loop poles. However, in order for arbitrary pole placement to be possible, the system must be completely state controllable and the magnitude of the control signal must be unbounded.

Consider the system defined by

$$\mathbf{x}(k+1) = \mathbf{G}\mathbf{x}(k) + \mathbf{H}u(k)$$

With state feedback $u(k) = -\mathbf{K}\mathbf{x}(k)$, the state equation becomes

$$\mathbf{x}(k+1) = (\mathbf{G} - \mathbf{H}\mathbf{K})\mathbf{x}(k)$$

Note that the solution of this last equation is given by

$$\mathbf{x}(k) = (\mathbf{G} - \mathbf{H}\mathbf{K})^k\mathbf{x}(0) \tag{6–74}$$

If the eigenvalues μ_i of matrix $\mathbf{G} - \mathbf{H}\mathbf{K}$ lie inside the unit circle, then the system is asymptotically stable.

In what follows, we shall show that by choosing all eigenvalues of $\mathbf{G} - \mathbf{H}\mathbf{K}$ to be zero, it is possible to get the deadbeat response, or

$$\mathbf{x}(k) = \mathbf{0} \qquad \text{for } k \geq q \qquad (q \leq n)$$

In discussing deadbeat response, the nilpotent matrix

$$\mathbf{N} = \begin{bmatrix} 0 & 1 & 0 & \cdots & 0 \\ 0 & 0 & 1 & \cdots & 0 \\ \cdot & \cdot & \cdot & & \cdot \\ \cdot & \cdot & \cdot & & \cdot \\ \cdot & \cdot & \cdot & & \cdot \\ 0 & 0 & 0 & \cdots & 1 \\ 0 & 0 & 0 & \cdots & 0 \end{bmatrix}$$

plays an important role. Consider, for example, a 4×4 nilpotent matrix:

$$\mathbf{N} = \begin{bmatrix} 0 & 1 & 0 & 0 \\ 0 & 0 & 1 & 0 \\ 0 & 0 & 0 & 1 \\ 0 & 0 & 0 & 0 \end{bmatrix}$$

Notice that

$$\mathbf{N}^2 = \begin{bmatrix} 0 & 0 & 1 & 0 \\ 0 & 0 & 0 & 1 \\ 0 & 0 & 0 & 0 \\ 0 & 0 & 0 & 0 \end{bmatrix}, \qquad \mathbf{N}^3 = \begin{bmatrix} 0 & 0 & 0 & 1 \\ 0 & 0 & 0 & 0 \\ 0 & 0 & 0 & 0 \\ 0 & 0 & 0 & 0 \end{bmatrix}, \qquad \mathbf{N}^4 = \begin{bmatrix} 0 & 0 & 0 & 0 \\ 0 & 0 & 0 & 0 \\ 0 & 0 & 0 & 0 \\ 0 & 0 & 0 & 0 \end{bmatrix}$$

Similarly, for an $n \times n$ nilpotent matrix $\mathbf{N}$ we have

$$\mathbf{N}^n = \mathbf{0}$$

Now consider the completely state controllable system given by

$$\mathbf{x}(k+1) = \mathbf{G}\mathbf{x}(k) + \mathbf{H}u(k) \tag{6–75}$$

Let us choose the desired pole locations to be at the origin, or choose the desired eigenvalues to be zero: $\mu_1 = \mu_2 = \cdots = \mu_n = 0$. Then we shall show that the response to any initial state $\mathbf{x}(0)$ is deadbeat. Since the characteristic equation with the desired eigenvalues can be given by

$$(z - \mu_1)(z - \mu_2) \cdots (z - \mu_n) = z^n + \alpha_1 z^{n-1} + \cdots + \alpha_{n-1} z + \alpha_n = z^n$$

we obtain

$$\alpha_1 = \alpha_2 = \cdots = \alpha_n = 0$$

and matrix $\mathbf{K}$ given by Eq. (6–65) can be simplified to the following:

$$\begin{aligned} \mathbf{K} &= [\alpha_n - a_n \,\vdots\, \alpha_{n-1} - a_{n-1} \,\vdots\, \cdots \,\vdots\, \alpha_1 - a_1]\mathbf{T}^{-1} \\ &= [-a_n \quad -a_{n-1} \quad \cdots \quad -a_1]\mathbf{T}^{-1} \end{aligned} \tag{6–76}$$

By using the transformation matrix $\mathbf{T}$ given by Eq. (6–32), define

$$\mathbf{x}(k) = \mathbf{T}\hat{\mathbf{x}}(k)$$

Define also

$$\mathbf{T}^{-1}\mathbf{G}\mathbf{T} = \hat{\mathbf{G}}, \qquad \mathbf{T}^{-1}\mathbf{H} = \hat{\mathbf{H}}$$

Then, Eq. (6–75) can be written as

$$\hat{\mathbf{x}}(k+1) = \mathbf{T}^{-1}\mathbf{G}\mathbf{T}\hat{\mathbf{x}}(k) + \mathbf{T}^{-1}\mathbf{H}u(k) = \hat{\mathbf{G}}\hat{\mathbf{x}}(k) + \hat{\mathbf{H}}u(k)$$

If we use the state feedback $u(k) = -\mathbf{K}\mathbf{x}(k) = -\mathbf{K}\mathbf{T}\hat{\mathbf{x}}(k)$, then this last equation becomes

$$\hat{\mathbf{x}}(k+1) = (\hat{\mathbf{G}} - \hat{\mathbf{H}}\mathbf{K}\mathbf{T})\hat{\mathbf{x}}(k)$$

Referring to Eq. (6–76) we have

$$\hat{\mathbf{G}} - \hat{\mathbf{H}}\mathbf{K}\mathbf{T} = \hat{\mathbf{G}} - \hat{\mathbf{H}}[-a_n \quad -a_{n-1} \quad \cdots \quad -a_1]$$

$$= \begin{bmatrix} 0 & 1 & 0 & \cdots & 0 \\ 0 & 0 & 1 & \cdots & 0 \\ \cdot & \cdot & \cdot & & \cdot \\ \cdot & \cdot & \cdot & & \cdot \\ \cdot & \cdot & \cdot & & \cdot \\ 0 & 0 & 0 & \cdots & 1 \\ -a_n & -a_{n-1} & -a_{n-2} & \cdots & -a_1 \end{bmatrix} - \begin{bmatrix} 0 \\ 0 \\ \cdot \\ \cdot \\ \cdot \\ 0 \\ 1 \end{bmatrix} [-a_n \quad -a_{n-1} \quad \cdots \quad -a_1]$$

$$= \begin{bmatrix} 0 & 1 & 0 & \cdots & 0 \\ 0 & 0 & 1 & \cdots & 0 \\ \cdot & \cdot & \cdot & & \cdot \\ \cdot & \cdot & \cdot & & \cdot \\ \cdot & \cdot & \cdot & & \cdot \\ 0 & 0 & 0 & \cdots & 1 \\ -a_n & -a_{n-1} & -a_{n-2} & \cdots & -a_1 \end{bmatrix} - \begin{bmatrix} 0 & 0 & 0 & \cdots & 0 \\ 0 & 0 & 0 & \cdots & 0 \\ \cdot & \cdot & \cdot & & \cdot \\ \cdot & \cdot & \cdot & & \cdot \\ \cdot & \cdot & \cdot & & \cdot \\ 0 & 0 & 0 & \cdots & 0 \\ -a_n & -a_{n-1} & -a_{n-2} & \cdots & -a_1 \end{bmatrix}$$

$$= \begin{bmatrix} 0 & 1 & 0 & \cdots & 0 \\ 0 & 0 & 1 & \cdots & 0 \\ \cdot & \cdot & \cdot & & \cdot \\ \cdot & \cdot & \cdot & & \cdot \\ \cdot & \cdot & \cdot & & \cdot \\ 0 & 0 & 0 & \cdots & 1 \\ 0 & 0 & 0 & \cdots & 0 \end{bmatrix}$$

Thus, $\hat{\mathbf{G}} - \hat{\mathbf{H}}\mathbf{K}\mathbf{T}$ is a nilpotent matrix. Therefore, we have

$$(\hat{\mathbf{G}} - \hat{\mathbf{H}}\mathbf{K}\mathbf{T})^n = \mathbf{0}$$

In terms of the original state $\mathbf{x}(k)$, we have

$$\begin{aligned} \mathbf{x}(n) &= (\mathbf{G} - \mathbf{H}\mathbf{K})^n\mathbf{x}(0) = (\mathbf{T}\hat{\mathbf{G}}\mathbf{T}^{-1} - \mathbf{T}\hat{\mathbf{H}}\mathbf{K})^n\mathbf{x}(0) = [\mathbf{T}(\hat{\mathbf{G}} - \hat{\mathbf{H}}\mathbf{K}\mathbf{T})\mathbf{T}^{-1}]^n\mathbf{x}(0) \\ &= \mathbf{T}(\hat{\mathbf{G}} - \hat{\mathbf{H}}\mathbf{K}\mathbf{T})^n\mathbf{T}^{-1}\mathbf{x}(0) = \mathbf{0} \end{aligned}$$

Thus we have shown that if the desired eigenvalues are all zeros, then any initial state $\mathbf{x}(0)$ can be brought to the origin in at most n sampling periods and the response is deadbeat, provided the control signal $u(k)$ is unbounded.

Comments on deadbeat control. The concept of deadbeat response is unique to discrete-time systems. There is no such thing as deadbeat response in continuous-time systems. In deadbeat control, any nonzero error vector will be driven to zero in at most n sampling periods if the magnitude of the scalar control $u(k)$ is unbounded. The settling time depends on the sampling period, since the response settles down in at most n sampling periods. If the sampling period T is chosen very small, the settling time will also be very small, which implies that the control signal must have an extremely large magnitude. Otherwise, it will not be possible to bring the error response to zero in a short time period.

In deadbeat control, the sampling period is the only design parameter. Thus, if the deadbeat response is desired, the designer must choose the sampling period carefully so that an extremely large control magnitude is not required in normal operation of the system. Note that it is not physically possible to increase the magnitude of the control signal without bound. If the magnitude is increased sufficiently, the saturation phenomenon always takes place. If saturation occurs in the magnitude of the control signal, then the response can no longer be deadbeat. The settling time will be more than n sampling periods. In the actual design of deadbeat control systems, the designer must be aware of the trade-off that must be made between the magnitude of the control signal and the response speed.

Example 6–7.

Consider the system given by

$$\begin{bmatrix} x_1(k+1) \\ x_2(k+1) \end{bmatrix} = \begin{bmatrix} 0 & 1 \\ -0.16 & -1 \end{bmatrix} \begin{bmatrix} x_1(k) \\ x_2(k) \end{bmatrix} + \begin{bmatrix} 0 \\ 1 \end{bmatrix} u(k) \tag{6–77}$$

Determine the state feedback gain matrix $\mathbf{K}$ such that when the control signal is given by

$$u(k) = -\mathbf{K}\mathbf{x}(k)$$

the closed-loop system (regulator system) exhibits the deadbeat response to an initial state $\mathbf{x}(0)$. Assume that the control signal $u(k)$ is unbounded.

Referring to Eq. (6–76), for the deadbeat response we have

$$\mathbf{K} = [-a_2 \quad -a_1]\mathbf{T}^{-1} \tag{6–78}$$

The system given by Eq. (6–77) is already in the controllable canonical form. Therefore, in Eq. (6–78), $\mathbf{T} = \mathbf{I}$. The characteristic equation for the system given by Eq. (6–77) is

$$|z\mathbf{I} - \mathbf{G}| = \begin{vmatrix} z & -1 \\ 0.16 & z+1 \end{vmatrix} = z^2 + z + 0.16 = z^2 + a_1 z + a_2$$

Thus

$$a_1 = 1, \qquad a_2 = 0.16$$

Consequently, Eq. (6–78) becomes

$$\mathbf{K} = [-a_2 \quad -a_1] = [-0.16 \quad -1]$$

This gives the desired state feedback gain matrix.

Let us verify that the response of this system to an arbitrary initial state $\mathbf{x}(0)$ is indeed the deadbeat response. Since the closed-loop state equation becomes

$$\begin{aligned} \begin{bmatrix} x_1(k+1) \\ x_2(k+1) \end{bmatrix} &= \begin{bmatrix} 0 & 1 \\ -0.16 & -1 \end{bmatrix} \begin{bmatrix} x_1(k) \\ x_2(k) \end{bmatrix} + \begin{bmatrix} 0 \\ 1 \end{bmatrix} [0.16 \quad 1] \begin{bmatrix} x_1(k) \\ x_2(k) \end{bmatrix} \\ &= \begin{bmatrix} 0 & 1 \\ 0 & 0 \end{bmatrix} \begin{bmatrix} x_1(k) \\ x_2(k) \end{bmatrix} \end{aligned}$$

if the initial state is given by

$$\begin{bmatrix} x_1(0) \\ x_2(0) \end{bmatrix} = \begin{bmatrix} a \\ b \end{bmatrix}$$

where a and b are arbitrary constants, then we have

$$\begin{bmatrix} x_1(1) \\ x_2(1) \end{bmatrix} = \begin{bmatrix} 0 & 1 \\ 0 & 0 \end{bmatrix} \begin{bmatrix} x_1(0) \\ x_2(0) \end{bmatrix} = \begin{bmatrix} 0 & 1 \\ 0 & 0 \end{bmatrix} \begin{bmatrix} a \\ b \end{bmatrix} = \begin{bmatrix} b \\ 0 \end{bmatrix}$$

$$\begin{bmatrix} x_1(2) \\ x_2(2) \end{bmatrix} = \begin{bmatrix} 0 & 1 \\ 0 & 0 \end{bmatrix} \begin{bmatrix} x_1(1) \\ x_2(1) \end{bmatrix} = \begin{bmatrix} 0 & 1 \\ 0 & 0 \end{bmatrix} \begin{bmatrix} b \\ 0 \end{bmatrix} = \begin{bmatrix} 0 \\ 0 \end{bmatrix}$$

Thus, the state $\mathbf{x}(k)$ for $k = 2, 3, 4, \ldots$ becomes zero and the response is indeed deadbeat.

Pole placement design problem when u(k) is an r-vector. Thus far, we have considered the pole placement design problem when the control signal is a scalar. We shall next consider the general case where the control signal is an r-vector.

Consider the system

$$\mathbf{x}(k+1) = \mathbf{G}\mathbf{x}(k) + \mathbf{H}\mathbf{u}(k) \tag{6–79}$$

where

$\mathbf{x}(k)$ = state vector (n-vector) at kth sampling instant

$\mathbf{u}(k)$ = control vector (r-vector) at kth sampling instant

$\mathbf{G} = n \times n$ matrix

$\mathbf{H} = n \times r$ matrix

We assume that the magnitudes of the r components of $\mathbf{u}(k)$ are unconstrained. As in the case of the system with a scalar control signal, it can be proved that a necessary and sufficient condition for arbitrary pole placement for the system defined by Eq. (6–79) is that the system be completely state controllable.

Let us assume that the system defined by Eq. (6–79) is completely state controllable. In the state feedback control scheme, the control vector $\mathbf{u}(k)$ is chosen as

$$\mathbf{u}(k) = -\mathbf{K}\mathbf{x}(k) \tag{6–80}$$

where $\mathbf{K}$ is the state feedback gain matrix. It is an $r \times n$ matrix. With state feedback the system becomes a closed-loop system and its state equation becomes

$$\mathbf{x}(k+1) = (\mathbf{G} - \mathbf{H}\mathbf{K})\mathbf{x}(k)$$

where we choose matrix $\mathbf{K}$ so that the eigenvalues of $\mathbf{G} - \mathbf{H}\mathbf{K}$ are the desired closed-loop poles $\mu_1, \mu_2, \ldots, \mu_n$.

Note that if the control signal is a vector quantity, the response can be speeded up, because we have more freedom to choose control signals $u_1(k)$, $u_2(k)$, . . . , $u_r(k)$ to speed up the response. For example, in the case of the scalar control $u(k)$, the deadbeat response can be achieved in at most n sampling periods if the state vector is an n-vector. In the case of the vector control $\mathbf{u}(k)$, the deadbeat response can be achieved in less than n sampling periods, as we shall see later.

Preliminary Discussions. Consider the system

$$\mathbf{x}(k+1) = \mathbf{G}\mathbf{x}(k) + \mathbf{H}_1 u(k) \tag{6–81}$$

where

$$\mathbf{x}(k) = \text{state vector } (n\text{-vector})$$

$$u(k) = \text{control signal (scalar)}$$

$$\mathbf{G} = n \times n \text{ matrix}$$

$$\mathbf{H}_1 = n \times 1 \text{ matrix}$$

Assume that the system is completely state controllable. Then the controllability matrix has its inverse. Define

$$[\mathbf{H}_1 \vdots \mathbf{G}\mathbf{H}_1 \vdots \cdots \vdots \mathbf{G}^{n-1}\mathbf{H}_1]^{-1} = \begin{bmatrix} \mathbf{f}_1 \\ \mathbf{f}_2 \\ \cdot \\ \cdot \\ \cdot \\ \mathbf{f}_n \end{bmatrix}$$

where the $\mathbf{f}_i$'s are the row vectors. Then construct a transformation matrix $\mathbf{T}_1$ as follows:

$$\mathbf{T}_1 = \begin{bmatrix} \mathbf{f}_n \\ \mathbf{f}_n\mathbf{G} \\ \cdot \\ \cdot \\ \cdot \\ \mathbf{f}_n\mathbf{G}^{n-1} \end{bmatrix}^{-1} \tag{6–82}$$

where the $\mathbf{f}_n\mathbf{G}^k$ are row vectors ($k = 0, 1, 2, \ldots, n-1$). Then, it can be shown that

$$\mathbf{T}_1^{-1}\mathbf{G}\mathbf{T}_1 = \begin{bmatrix} \mathbf{f}_n \\ \mathbf{f}_n\mathbf{G} \\ \cdot \\ \cdot \\ \cdot \\ \mathbf{f}_n\mathbf{G}^{n-1} \end{bmatrix} \mathbf{G} \begin{bmatrix} \mathbf{f}_n \\ \mathbf{f}_n\mathbf{G} \\ \cdot \\ \cdot \\ \cdot \\ \mathbf{f}_n\mathbf{G}^{n-1} \end{bmatrix}^{-1}$$

$$= \begin{bmatrix} 0 & 1 & 0 & \cdots & 0 \\ 0 & 0 & 1 & \cdots & 0 \\ \cdot & \cdot & \cdot & & \cdot \\ \cdot & \cdot & \cdot & & \cdot \\ \cdot & \cdot & \cdot & & \cdot \\ 0 & 0 & 0 & \cdots & 1 \\ -a_n & -a_{n-1} & -a_{n-2} & \cdots & -a_1 \end{bmatrix} \tag{6–83}$$

and

$$\mathbf{T}_1^{-1}\mathbf{H}_1 = \begin{bmatrix} 0 \\ 0 \\ \cdot \\ \cdot \\ \cdot \\ 0 \\ 1 \end{bmatrix} \tag{6–84}$$

[See Prob. A-6–17 for the derivation of Eqs. (6–83) and (6–84).]

Now if we define

$$\mathbf{x}(k) = \mathbf{T}_1\hat{\mathbf{x}}(k)$$

then Eq. (6–81) becomes

$$\hat{\mathbf{x}}(k+1) = \mathbf{T}_1^{-1}\mathbf{G}\mathbf{T}_1\hat{\mathbf{x}}(k) + \mathbf{T}_1^{-1}\mathbf{H}_1 u(k)$$

or

$$\begin{bmatrix} \hat{x}_1(k+1) \\ \hat{x}_2(k+1) \\ \cdot \\ \cdot \\ \cdot \\ \hat{x}_{n-1}(k+1) \\ \hat{x}_n(k+1) \end{bmatrix} = \begin{bmatrix} 0 & 1 & 0 & \cdots & 0 \\ 0 & 0 & 1 & \cdots & 0 \\ \cdot & \cdot & \cdot & & \cdot \\ \cdot & \cdot & \cdot & & \cdot \\ \cdot & \cdot & \cdot & & \cdot \\ 0 & 0 & 0 & \cdots & 1 \\ -a_n & -a_{n-1} & -a_{n-2} & \cdots & -a_1 \end{bmatrix} \begin{bmatrix} \hat{x}_1(k) \\ \hat{x}_2(k) \\ \cdot \\ \cdot \\ \cdot \\ \hat{x}_{n-1}(k) \\ \hat{x}_n(k) \end{bmatrix} + \begin{bmatrix} 0 \\ 0 \\ \cdot \\ \cdot \\ \cdot \\ 0 \\ 1 \end{bmatrix} u(k) \tag{6–85}$$

We have thus shown that the state equation of Eq. (6–81) can be transformed into the controllable canonical form by use of the transformation matrix $\mathbf{T}_1$ defined by Eq. (6–82).

Design Steps. In what follows we shall discuss the procedure for determining a state feedback gain matrix $\mathbf{K}$ such that the eigenvalues of $\mathbf{G} - \mathbf{HK}$ are the desired values $\mu_1, \mu_2, \ldots, \mu_n$.

The state equation to be considered in the following was given by Eq. (6–79):

$$\mathbf{x}(k+1) = \mathbf{G}\mathbf{x}(k) + \mathbf{H}\mathbf{u}(k)$$

We assume that the rank of the $n \times r$ matrix $\mathbf{H}$ is r. This last equation is equivalent to

$$\mathbf{x}(k+1) = \mathbf{G}\mathbf{x}(k) + [\mathbf{H}_1 \vdots \mathbf{H}_2 \vdots \cdots \vdots \mathbf{H}_r]\,\mathbf{u}(k)$$

where

$$[\mathbf{H}_1 \vdots \mathbf{H}_2 \vdots \cdots \vdots \mathbf{H}_r] = \mathbf{H}, \qquad \mathbf{H}_i = \begin{bmatrix} h_{1i} \\ h_{2i} \\ \cdot \\ \cdot \\ \cdot \\ h_{ni} \end{bmatrix} \qquad i = 1, 2, \ldots, r$$

The procedure for designing the state feedback gain matrix $\mathbf{K}$ involves the following two steps:

Step 1. Extend the transformation process [the process that transforms the state equation given by Eq. (6–81) into the state equation in the controllable canonical form given by Eq. (6–85)] to the case where matrix $\mathbf{H}$ is an $n \times r$ matrix. That is, we transform the given state equation into a controllable canonical form by use of a transformation matrix $\mathbf{T}$, the exact form of which will be given later. By defining

$$\mathbf{x}(k) = \mathbf{T}\hat{\mathbf{x}}(k)$$

the original state equation, Eq. (6–79), can be transformed into

$$\hat{\mathbf{x}}(k+1) = \mathbf{T}^{-1}\mathbf{G}\mathbf{T}\hat{\mathbf{x}}(k) + \mathbf{T}^{-1}\mathbf{H}\mathbf{u}(k) = \hat{\mathbf{G}}\hat{\mathbf{x}}(k) + \hat{\mathbf{H}}\mathbf{u}(k) \qquad (6\text{–}86)$$

where $\hat{\mathbf{G}}$ is in a controllable canonical form. (This controllable canonical form is slightly different from the usual form, as we shall see later.)

Step 2. By use of a state feedback gain matrix $\mathbf{K}$, the control vector can be given by

$$\mathbf{u}(k) = -\mathbf{K}\mathbf{x}(k) = -\mathbf{K}\mathbf{T}\hat{\mathbf{x}}(k)$$

and the system state equation becomes

$$\hat{\mathbf{x}}(k+1) = (\hat{\mathbf{G}} - \hat{\mathbf{H}}\mathbf{K}\mathbf{T})\hat{\mathbf{x}}(k)$$

We choose matrix $\mathbf{K}$ so that matrix $\hat{\mathbf{G}} - \hat{\mathbf{H}}\mathbf{K}\mathbf{T}$ will have the desired eigenvalues $\mu_1, \mu_2, \ldots, \mu_n$.

Pole placement design procedure when u(k) is an r-vector. We shall first discuss the determination of a necessary transformation matrix $\mathbf{T}$ and then determine the state feedback gain matrix $\mathbf{K}$.

Since we have assumed that the rank of matrix $\mathbf{H}$ is r, the component vectors $\mathbf{H}_1, \mathbf{H}_2, \ldots, \mathbf{H}_r$ of matrix $\mathbf{H}$ are linearly independent of each other. Let us write matrix $\mathbf{H}$ as follows:

$$\mathbf{H} = [\mathbf{H}_1 \vdots \mathbf{H}_2 \vdots \cdots \vdots \mathbf{H}_r]$$

Since the system is assumed to be completely state controllable, the rank of the $n \times nr$ controllability matrix

$$[\mathbf{H} \vdots \mathbf{GH} \vdots \cdots \vdots \mathbf{G}^{n-1}\mathbf{H}]$$

is n. The controllability matrix can be written in an expanded form as follows:

$$[\mathbf{H}_1 \vdots \mathbf{H}_2 \vdots \cdots \vdots \mathbf{H}_r \vdots \mathbf{GH}_1 \vdots \mathbf{GH}_2 \vdots \cdots \vdots \mathbf{GH}_r \vdots \cdots \vdots \mathbf{G}^{n-1}\mathbf{H}_1 \vdots \mathbf{G}^{n-1}\mathbf{H}_2 \vdots \cdots \vdots \mathbf{G}^{n-1}\mathbf{H}_r]$$

Let us choose n linearly independent vectors from this $n \times nr$ matrix. Let us begin from the left side of this matrix. Since the first r vectors $\mathbf{H}_1, \mathbf{H}_2, \ldots, \mathbf{H}_r$ are linearly independent of each other, we choose these r vectors first. Then we examine $\mathbf{GH}_1$ if it is linearly independent of the r vectors already chosen. If it is, we have chosen $r + 1$ linearly independent vectors. Next, we examine $\mathbf{GH}_2, \mathbf{GH}_3, \ldots, \mathbf{GH}_r, \ldots$ in the order shown in the expanded controllability matrix until we find altogether n linearly independent vectors. (Since the rank of the controllability matrix is n, there always exist n linearly independent vectors.)

Once we have chosen n linearly independent vectors we rearrange these vectors in the following way:

$$\mathbf{F} = [\mathbf{H}_1 \vdots \mathbf{GH}_1 \vdots \cdots \vdots \mathbf{G}^{n_1-1}\mathbf{H}_1 \vdots \mathbf{H}_2 \vdots \mathbf{GH}_2 \vdots \cdots \vdots \mathbf{G}^{n_2-1}\mathbf{H}_2 \vdots \cdots \vdots \mathbf{H}_r \vdots \mathbf{GH}_r \vdots \cdots \vdots \mathbf{G}^{n_r-1}\mathbf{H}_r] \tag{6-87}$$

The numbers n_i are said to be *Kronecker-invariant* and satisfy the equation

$$n_1 + n_2 + \cdots + n_r = n$$

We shall define the maximum of $n_1, n_2, \ldots, n_r$ as $n_{\min}$:

$$n_{min} = \max(n_1, n_2, \ldots, n_r) \tag{6-88}$$

We shall refer to this equation later in the discussion of deadbeat response. Next, we compute $\mathbf{F}^{-1}$ and define the η_ith row vector as $\mathbf{f}_i$, where

$$\eta_i = n_1 + n_2 + \cdots + n_i \qquad i = 1, 2, \ldots, r$$

Then the required transformation matrix $\mathbf{T}$ can be given by

$$\mathbf{T} = \begin{bmatrix} \mathbf{S}_1 \\ \mathbf{S}_2 \\ \cdot \\ \cdot \\ \cdot \\ \mathbf{S}_r \end{bmatrix}^{-1} \tag{6-89}$$

where

$$\mathbf{S}_i = \begin{bmatrix} \mathbf{f}_i \\ \mathbf{f}_i\mathbf{G} \\ \cdot \\ \cdot \\ \cdot \\ \mathbf{f}_i\mathbf{G}^{n_i-1} \end{bmatrix}$$

Notice that the transformation matrix $\mathbf{T}$ given by Eq. (6–89) is an extension of the transformation matrix given by Eq. (6–82).

In order to simplify the presentation, in what follows we shall consider a simple case where $n = 4$ and $r = 2$. (In this case, only n_1 and n_2 are involved.) (Extension to more general cases is straightforward.) Then the transformation matrix $\mathbf{T}$ becomes a 4×4 matrix. The transformation matrix $\mathbf{T}$ given by Eq. (6–89) becomes

$$\mathbf{T} = \begin{bmatrix} \mathbf{S}_1 \\ \hline \mathbf{S}_2 \end{bmatrix}^{-1}$$

where

$$\mathbf{S}_1 = \begin{bmatrix} \mathbf{f}_1 \\ \cdot \\ \cdot \\ \cdot \\ \mathbf{f}_1\mathbf{G}^{n_1-1} \end{bmatrix}, \qquad \mathbf{S}_2 = \begin{bmatrix} \mathbf{f}_2 \\ \cdot \\ \cdot \\ \cdot \\ \mathbf{f}_2\mathbf{G}^{n_2-1} \end{bmatrix}$$

(Note that in the case of $n = 4$, there are three possibilities for the combinations of n_1 and n_2: $n_1 = 1$, $n_2 = 3$; $n_1 = 2$, $n_2 = 2$; and $n_1 = 3$, $n_2 = 1$.) For example, if $n_1 = 2$ and $n_2 = 2$, then matrices $\hat{\mathbf{G}}$ and $\hat{\mathbf{H}}$ become, respectively, as

$$\hat{\mathbf{G}} = \mathbf{T}^{-1}\mathbf{G}\mathbf{T} = \left[\begin{array}{cc|cc} 0 & 1 & 0 & 0 \\ -a_{11} & -a_{12} & -a_{13} & -a_{14} \\ \hline 0 & 0 & 0 & 1 \\ -a_{21} & -a_{22} & -a_{23} & -a_{24} \end{array}\right] \qquad \text{if } n_1 = 2,\ n_2 = 2 \tag{6–90}$$

and

$$\hat{\mathbf{H}} = \mathbf{T}^{-1}\mathbf{H} = \left[\begin{array}{cc} 0 & 0 \\ 1 & b_{12} \\ \hline 0 & 0 \\ 0 & 1 \end{array}\right] \qquad \begin{array}{l} \text{if } n_1 = 2,\ n_2 = 2 \\ \text{(Note: } b_{12} = \mathbf{f}_1\mathbf{G}\mathbf{H}_2 = 0 \text{ in this case)} \end{array} \tag{6–91}$$

(see Prob. A-6–18). As another example, if $n_1 = 3$ and $n_2 = 1$, then

$$\hat{\mathbf{G}} = \mathbf{T}^{-1}\mathbf{G}\mathbf{T} = \left[\begin{array}{ccc|c} 0 & 1 & 0 & 0 \\ 0 & 0 & 1 & 0 \\ -a_{11} & -a_{12} & -a_{13} & -a_{14} \\ \hline -a_{21} & -a_{22} & -a_{23} & -a_{24} \end{array}\right] \qquad \text{if } n_1 = 3,\ n_2 = 1 \tag{6–92}$$

and

$$\hat{\mathbf{H}} = \mathbf{T}^{-1}\mathbf{H} = \left[\begin{array}{cc} 0 & 0 \\ 0 & 0 \\ 1 & b_{12} \\ \hline 0 & 1 \end{array}\right] \qquad \begin{array}{l} \text{if } n_1 = 3,\ n_2 = 1 \\ \text{(Note: } b_{12} = \mathbf{f}_1\mathbf{G}^2\mathbf{H}_2 \text{ may or} \\ \text{may not be zero)} \end{array} \tag{6–93}$$

(see Prob. A-6–20). In what follows, we shall focus on the case where $n_1 = 2$ and $n_2 = 2$. (Other cases can be handled similarly. For example, for the case where $n_1 = 3$ and $n_2 = 1$, see Probs. A-6–19, A-6–20, and A-6–21.) For the case where $n_1 = 2$ and $n_2 = 2$, matrix $\hat{\mathbf{G}} = \mathbf{T}^{-1}\mathbf{G}\mathbf{T}$ can be given by Eq. (6–90), and the characteristic equation is

$$|z\mathbf{I} - \hat{\mathbf{G}}| = \begin{vmatrix} z & -1 & 0 & 0 \\ a_{11} & z + a_{12} & a_{13} & a_{14} \\ 0 & 0 & z & -1 \\ a_{21} & a_{22} & a_{23} & z + a_{24} \end{vmatrix}$$

$$= \begin{vmatrix} z & -1 \\ a_{11} & z + a_{12} \end{vmatrix} \begin{vmatrix} z & -1 \\ a_{23} & z + a_{24} \end{vmatrix} + \begin{vmatrix} z & -1 \\ a_{21} & a_{22} \end{vmatrix} \begin{vmatrix} a_{13} & a_{14} \\ z & -1 \end{vmatrix}$$

$$= (z^2 + a_{12}z + a_{11})(z^2 + a_{24}z + a_{23}) - (a_{22}z + a_{21})(a_{14}z + a_{13})$$

$$= 0 \tag{6–94}$$

where we have used Laplace's expansion by the minors. (See the Appendix for the details.) From Eq. (6–94) the characteristic equation $|z\mathbf{I} - \hat{\mathbf{G}}| = 0$ becomes

$$|z\mathbf{I} - \hat{\mathbf{G}}| = \left|\begin{array}{c:c} z^2 + a_{12}z + a_{11} & a_{14}z + a_{13} \\ \hdashline a_{22}z + a_{21} & z^2 + a_{24}z + a_{23} \end{array}\right| = 0 \tag{6–95}$$

The eigenvalues of $\hat{\mathbf{G}}$ can be determined by solving this characteristic equation.

Next, we shall determine the state feedback gain matrix $\mathbf{K}$ so that the eigenvalues of $\hat{\mathbf{G}} - \hat{\mathbf{H}}\mathbf{K}$ are $\mu_1, \mu_2, \ldots, \mu_n$, the desired values. Let us define a 2×2 matrix $\mathbf{B}$ such that

$$\mathbf{B} = \begin{bmatrix} 1 & b_{12} \\ 0 & 1 \end{bmatrix}^{-1}$$

(Note that b_{12} is a constant appearing in $\hat{\mathbf{H}}$ matrix.) In the particular case where $n_1 = 2$ and $n_2 = 2$, the value of b_{12} is equal to 0. Thus, $\mathbf{B} = \mathbf{I}$. For more general cases, matrix $\mathbf{B}$ may not be the identity matrix.

Also, define a 2×4 matrix $\boldsymbol{\Delta}$ such that

$$\boldsymbol{\Delta} = \begin{bmatrix} \delta_{11} & \delta_{12} & \delta_{13} & \delta_{14} \\ \delta_{21} & \delta_{22} & \delta_{23} & \delta_{24} \end{bmatrix} \tag{6–96}$$

Then, it will be seen that matrix $\mathbf{K}$ can be given by

$$\mathbf{K} = \mathbf{B}\boldsymbol{\Delta}\mathbf{T}^{-1}$$

and the control vector $\mathbf{u}(k)$ can be given by

$$\mathbf{u}(k) = -\mathbf{B}\boldsymbol{\Delta}\mathbf{T}^{-1}\mathbf{x}(k) = -\mathbf{B}\boldsymbol{\Delta}\hat{\mathbf{x}}(k)$$

Thus, the system state equation given by Eq. (6–86) becomes

$$\hat{\mathbf{x}}(k+1) = \hat{\mathbf{G}}\hat{\mathbf{x}}(k) - \hat{\mathbf{H}}\mathbf{B}\boldsymbol{\Delta}\hat{\mathbf{x}}(k) = (\hat{\mathbf{G}} - \hat{\mathbf{H}}\mathbf{B}\boldsymbol{\Delta})\hat{\mathbf{x}}(k)$$

For the present case, matrix $\hat{\mathbf{H}}\mathbf{B}\boldsymbol{\Delta}$ becomes as follows:

$$\hat{\mathbf{H}}\mathbf{B}\boldsymbol{\Delta} = \begin{bmatrix} 0 & 0 \\ 1 & 0 \\ 0 & 0 \\ 0 & 1 \end{bmatrix} \begin{bmatrix} 1 & 0 \\ 0 & 1 \end{bmatrix}^{-1} \begin{bmatrix} \delta_{11} & \delta_{12} & \delta_{13} & \delta_{14} \\ \delta_{21} & \delta_{22} & \delta_{23} & \delta_{24} \end{bmatrix}$$

$$= \begin{bmatrix} 0 & 0 & 0 & 0 \\ \delta_{11} & \delta_{12} & \delta_{13} & \delta_{14} \\ 0 & 0 & 0 & 0 \\ \delta_{21} & \delta_{22} & \delta_{23} & \delta_{24} \end{bmatrix}$$

Hence

$$\hat{\mathbf{G}} - \hat{\mathbf{H}}\mathbf{B}\boldsymbol{\Delta} = \begin{bmatrix} 0 & 1 & 0 & 0 \\ -a_{11} - \delta_{11} & -a_{12} - \delta_{12} & -a_{13} - \delta_{13} & -a_{14} - \delta_{14} \\ 0 & 0 & 0 & 1 \\ -a_{21} - \delta_{21} & -a_{22} - \delta_{22} & -a_{23} - \delta_{23} & -a_{24} - \delta_{24} \end{bmatrix}$$

Then, referring to Eq. (6–95), the characteristic equation $|z\mathbf{I} - \hat{\mathbf{G}} + \hat{\mathbf{H}}\mathbf{B}\boldsymbol{\Delta}|$ becomes

$$\begin{aligned} |z\mathbf{I} - \hat{\mathbf{G}} + \hat{\mathbf{H}}\mathbf{B}\boldsymbol{\Delta}| &= \left| \begin{array}{c:c} z^2 + (a_{12} + \delta_{12})z + a_{11} + \delta_{11} & (a_{14} + \delta_{14})z + a_{13} + \delta_{13} \\ \hdashline (a_{22} + \delta_{22})z + a_{21} + \delta_{21} & z^2 + (a_{24} + \delta_{24})z + a_{23} + \delta_{23} \end{array} \right| \\ &= [z^2 + (a_{12} + \delta_{12})z + a_{11} + \delta_{11}][z^2 + (a_{24} + \delta_{24})z + a_{23} + \delta_{23}] \\ &\quad - [(a_{14} + \delta_{14})z + a_{13} + \delta_{13}][(a_{22} + \delta_{22})z + a_{21} + \delta_{21}] \\ &= 0 \end{aligned} \tag{6–97}$$

We desire the eigenvalues of $\hat{\mathbf{G}} - \hat{\mathbf{H}}\mathbf{B}\boldsymbol{\Delta}$ to be μ_1, μ_2, μ_3, and μ_4, or the desired characteristic equation to be

$$\begin{aligned} &(z - \mu_1)(z - \mu_2)(z - \mu_3)(z - \mu_4) \\ &= z^4 + \alpha_1 z^3 + \alpha_2 z^2 + \alpha_3 z + \alpha_4 = 0 \end{aligned} \tag{6–98}$$

If we equate the coefficients of equal powers of z of the two characteristic equations, Eqs. (6–97) and (6–98), we obtain the following equations:

$$a_{12} + \delta_{12} + a_{24} + \delta_{24} = \alpha_1$$

$$a_{11} + \delta_{11} + (a_{12} + \delta_{12})(a_{24} + \delta_{24}) + a_{23} + \delta_{23} - (a_{14} + \delta_{14})(a_{22} + \delta_{22}) = \alpha_2$$

$$\begin{aligned} &(a_{11} + \delta_{11})(a_{24} + \delta_{24}) + (a_{12} + \delta_{12})(a_{23} + \delta_{23}) \\ &\qquad - (a_{13} + \delta_{13})(a_{22} + \delta_{22}) - (a_{21} + \delta_{21})(a_{14} + \delta_{14}) = \alpha_3 \end{aligned}$$

$$(a_{11} + \delta_{11})(a_{23} + \delta_{23}) - (a_{13} + \delta_{13})(a_{21} + \delta_{21}) = \alpha_4$$

Notice that we have eight δ variables and four equations. Hence the values of δ_{11}, δ_{12}, δ_{13}, δ_{14}, δ_{21}, δ_{22}, δ_{23}, and δ_{24} cannot be determined uniquely. There are many possible sets of values $\delta_{11}, \delta_{12}, \ldots, \delta_{24}$ and thus matrix $\mathbf{\Delta}$ is not unique. Any matrix $\mathbf{\Delta}$ whose elements satisfy the foregoing four equations is acceptable.

Once matrix $\mathbf{\Delta}$ is chosen, the required state feedback gain matrix $\mathbf{K}$ is given by

$$\mathbf{K} = \mathbf{B\Delta T}^{-1}$$

and the state feedback control vector is

$$\mathbf{u}(k) = -\mathbf{B\Delta T}^{-1}\mathbf{x}(k)$$

and the state equation given by Eq. (6–79) becomes

$$\mathbf{x}(k+1) = \mathbf{Gx}(k) - \mathbf{HB\Delta T}^{-1}\mathbf{x}(k) = (\mathbf{G} - \mathbf{HB\Delta T}^{-1})\mathbf{x}(k)$$

As a matter of course, note that

$$|\mathbf{G} - \mathbf{HB\Delta T}^{-1}| = |\mathbf{T}^{-1}|\,|\mathbf{G} - \mathbf{HB\Delta T}^{-1}|\,|\mathbf{T}| = |\mathbf{T}^{-1}\mathbf{GT} - \mathbf{T}^{-1}\mathbf{HB\Delta}| = |\hat{\mathbf{G}} - \hat{\mathbf{H}}\mathbf{B\Delta}|$$

For a given set of desired eigenvalues $\mu_1, \mu_2, \ldots, \mu_n$ we have the corresponding coefficients $\alpha_1, \alpha_2, \ldots, \alpha_n$ in the characteristic equation $|z\mathbf{I} - \hat{\mathbf{G}} + \hat{\mathbf{H}}\mathbf{B\Delta}| = 0$. For the given $\alpha_1, \alpha_2, \ldots, \alpha_n$, it is possible to choose a matrix $\mathbf{\Delta}$ which is not unique. (This means that we have some freedom to satisfy other requirements, if any.)

If the deadbeat response is desired, we require $\mu_1 = \mu_2 = \mu_3 = \mu_4 = 0$. The desired characteristic equation given by Eq. (6–98) becomes

$$z^4 = 0$$

Notice that if we choose, for example,

$$\mathbf{\Delta} = \begin{bmatrix} -a_{11} & -a_{12} & -a_{13} & -a_{14} \\ * & * & -a_{23} & -a_{24} \end{bmatrix} \tag{6–99}$$

where the elements indicated by asterisks are arbitrary constants, then $\hat{\mathbf{G}} - \hat{\mathbf{H}}\mathbf{B\Delta}$ becomes

$$\hat{\mathbf{G}} - \hat{\mathbf{H}}\mathbf{B\Delta} = \begin{bmatrix} 0 & 1 & 0 & 0 \\ 0 & 0 & 0 & 0 \\ 0 & 0 & 0 & 1 \\ ** & ** & 0 & 0 \end{bmatrix}$$

where the elements indicated by the double asterisks are arbitrary constants.

$$(\hat{\mathbf{G}} - \hat{\mathbf{H}}\mathbf{B\Delta})^2 = \begin{bmatrix} 0 & 0 & 0 & 0 \\ 0 & 0 & 0 & 0 \\ ** & 0 & 0 & 0 \\ 0 & ** & 0 & 0 \end{bmatrix}$$

$$(\hat{\mathbf{G}} - \hat{\mathbf{H}}\mathbf{B\Delta})^3 = \begin{bmatrix} 0 & 0 & 0 & 0 \\ 0 & 0 & 0 & 0 \\ 0 & ** & 0 & 0 \\ 0 & 0 & 0 & 0 \end{bmatrix}$$

and

$$(\hat{\mathbf{G}} - \hat{\mathbf{H}}\mathbf{B}\boldsymbol{\Delta})^4 = \begin{bmatrix} 0 & 0 & 0 & 0 \\ 0 & 0 & 0 & 0 \\ 0 & 0 & 0 & 0 \\ 0 & 0 & 0 & 0 \end{bmatrix}$$

Thus the deadbeat response is obtained. Matrix $\boldsymbol{\Delta}$ given by Eq. (6–99) is not unique because different choices of elements can yield the deadbeat response. Hence, more than one state feedback gain matrix $\mathbf{K}$ exists that will yield the deadbeat response. This is expected, since we have two control signals $u_1(k)$ and $u_2(k)$ available, instead of just one control signal.

It is important to note that if we choose

$$\boldsymbol{\Delta} = \begin{bmatrix} -a_{11} & -a_{12} & -a_{13} & -a_{14} \\ -a_{21} & -a_{22} & -a_{23} & -a_{24} \end{bmatrix} \tag{6–100}$$

then

$$\hat{\mathbf{G}} - \hat{\mathbf{H}}\mathbf{B}\boldsymbol{\Delta} = \begin{bmatrix} 0 & 1 & 0 & 0 \\ 0 & 0 & 0 & 0 \\ 0 & 0 & 0 & 1 \\ 0 & 0 & 0 & 0 \end{bmatrix}$$

and

$$(\hat{\mathbf{G}} - \hat{\mathbf{H}}\mathbf{B}\boldsymbol{\Delta})^2 = \mathbf{0}$$

Thus, $(\hat{\mathbf{G}} - \hat{\mathbf{H}}\mathbf{B}\boldsymbol{\Delta})^k$ becomes zero for $k = 2, 3, 4, \ldots$. The deadbeat response is achieved in two sampling periods. In fact, in general, by choosing the elements of $\boldsymbol{\Delta}$ in the manner given by Eq. (6–100), the deadbeat response can be achieved in $n_{\min}$ steps rather than n steps, where

$$n_{\min} = \max(n_1, n_2, \ldots, n_r)$$

Since $n_1 + n_2 + \cdots + n_r = n$, we note that $n_{\min}$ is always less than n.

Extension to the more general case. Thus far, we have given detailed discussions for the case where $n = 4$ ($n_1 = n_2 = 2$) and $r = 2$. Extension of the preceding discussions to the more general case is straightforward. For example, consider the case where $n = 6$ and $r = 3$. For this case,

$$n_1 + n_2 + n_3 = 6$$

and we have several possible combinations of n_1, n_2, and n_3.

Now consider the case where $n_1 = 3$, $n_2 = 2$, and $n_3 = 1$. The modified 6×6 controllability matrix $\mathbf{F}$ for this case is

$$\mathbf{F} = [\mathbf{H}_1 \vdots \mathbf{G}\mathbf{H}_1 \vdots \mathbf{G}^2\mathbf{H}_1 \vdots \mathbf{H}_2 \vdots \mathbf{G}\mathbf{H}_2 \vdots \mathbf{H}_3]$$

Define

$$\mathbf{F}^{-1} = \begin{bmatrix} *** \\ \hline *** \\ \hline \mathbf{f}_1 \\ \hline *** \\ \hline \mathbf{f}_2 \\ \hline \mathbf{f}_3 \end{bmatrix} \begin{matrix} \left.\vphantom{\begin{matrix} a \\ a \\ a \end{matrix}}\right\} n_1 = 3 \\ \left.\vphantom{\begin{matrix} a \\ a \end{matrix}}\right\} n_2 = 2 \\ \} \; n_3 = 1 \end{matrix}$$

where a row of asterisks denotes a row vector. Then the transformation matrix $\mathbf{T}$ can be formed as follows:

$$\mathbf{T} = \begin{bmatrix} \mathbf{S}_1 \\ \hline \mathbf{S}_2 \\ \hline \mathbf{S}_3 \end{bmatrix}^{-1}$$

where

$$\mathbf{S}_1 = \begin{bmatrix} \mathbf{f}_1 \\ \hline \mathbf{f}_1\mathbf{G} \\ \hline \mathbf{f}_1\mathbf{G}^2 \end{bmatrix}, \qquad \mathbf{S}_2 = \begin{bmatrix} \mathbf{f}_2 \\ \hline \mathbf{f}_2\mathbf{G} \end{bmatrix}, \qquad \mathbf{S}_3 = \mathbf{f}_3$$

Then the matrices $\hat{\mathbf{G}}$ and $\hat{\mathbf{H}}$ will have the following forms:

$$\hat{\mathbf{G}} = \left[\begin{array}{ccc|cc|c} 0 & 1 & 0 & 0 & 0 & 0 \\ 0 & 0 & 1 & 0 & 0 & 0 \\ -a_{11} & -a_{12} & -a_{13} & -a_{14} & -a_{15} & -a_{16} \\ \hline 0 & 0 & 0 & 0 & 1 & 0 \\ -a_{21} & -a_{22} & -a_{23} & -a_{24} & -a_{25} & -a_{26} \\ \hline -a_{31} & -a_{32} & -a_{33} & -a_{34} & -a_{35} & -a_{36} \end{array}\right]$$

$$\hat{\mathbf{H}} = \left[\begin{array}{ccc} 0 & 0 & 0 \\ 0 & 0 & 0 \\ 1 & b_{12} & b_{13} \\ \hline 0 & 0 & 0 \\ 0 & 1 & b_{23} \\ \hline 0 & 0 & 1 \end{array}\right]$$

where $b_{12} = \mathbf{f}_1\mathbf{G}^2\mathbf{H}_2$, $b_{13} = \mathbf{f}_1\mathbf{G}^2\mathbf{H}_3$, and $b_{23} = \mathbf{f}_2\mathbf{G}\mathbf{H}_3$. These values may or may not be zero. (Notice that in matrix $\hat{\mathbf{G}}$ the principal minors are in the controllable canonical form.) The state feedback gain matrix $\mathbf{K}$ is given as follows:

$$\mathbf{K} = \mathbf{B}\boldsymbol{\Delta}\mathbf{T}^{-1}$$

where

$$\mathbf{B} = \begin{bmatrix} 1 & b_{12} & b_{13} \\ 0 & 1 & b_{23} \\ 0 & 0 & 1 \end{bmatrix}^{-1}$$

and

$$\boldsymbol{\Delta} = \begin{bmatrix} \delta_{11} & \delta_{12} & \delta_{13} & \delta_{14} & \delta_{15} & \delta_{16} \\ \delta_{21} & \delta_{22} & \delta_{23} & \delta_{24} & \delta_{25} & \delta_{26} \\ \delta_{31} & \delta_{32} & \delta_{33} & \delta_{34} & \delta_{35} & \delta_{36} \end{bmatrix}$$

Notice that

$$\hat{\mathbf{H}}\mathbf{B} = \begin{bmatrix} 0 & 0 & 0 \\ 0 & 0 & 0 \\ 1 & b_{12} & b_{13} \\ 0 & 0 & 0 \\ 0 & 1 & b_{23} \\ 0 & 0 & 1 \end{bmatrix} \begin{bmatrix} 1 & b_{12} & b_{13} \\ 0 & 1 & b_{23} \\ 0 & 0 & 1 \end{bmatrix}^{-1} = \begin{bmatrix} 0 & 0 & 0 \\ 0 & 0 & 0 \\ 1 & 0 & 0 \\ 0 & 0 & 0 \\ 0 & 1 & 0 \\ 0 & 0 & 1 \end{bmatrix}$$

The effect of postmultiplying matrix $\mathbf{B}$ to matrix $\hat{\mathbf{H}}$ is to eliminate the b_{ij} from the product matrix $\hat{\mathbf{H}}\mathbf{B}$.

Note that if $\mathbf{u}(k)$ is an r-vector, the general form of $\mathbf{B}$ matrix is

$$\mathbf{B} = \begin{bmatrix} 1 & b_{12} & \cdots & b_{1r} \\ 0 & 1 & \cdots & b_{2r} \\ 0 & 0 & \cdots & b_{3r} \\ \cdot & \cdot & & \cdot \\ \cdot & \cdot & & \cdot \\ \cdot & \cdot & & \cdot \\ 0 & 0 & \cdots & 1 \end{bmatrix}^{-1} \tag{6–101}$$

where the constants b_{ij}'s are those which will appear in the $n \times r$ matrix $\hat{\mathbf{H}}$. (The elements of $\hat{\mathbf{H}}\mathbf{B}$ are either 0 or 1.)

Example 6–8.

Consider the system

$$\mathbf{x}(k+1) = \mathbf{G}\mathbf{x}(k) + \mathbf{H}\mathbf{u}(k)$$

where

$$\mathbf{x}(k) = \text{state vector (3-vector)}$$

$$\mathbf{u}(k) = \text{control vector (2-vector)}$$

and

$$\mathbf{G} = \begin{bmatrix} 0 & 1 & 0 \\ 0 & 0 & 1 \\ -0.25 & 0 & 0.5 \end{bmatrix}, \qquad \mathbf{H} = \begin{bmatrix} 0 & 1 \\ 0 & 0 \\ 1 & 0 \end{bmatrix}$$

It is desired to determine the state feedback gain matrix $\mathbf{K}$ so that the response to the initial state $\mathbf{x}(0)$ is deadbeat. Note that with state feedback $\mathbf{u}(k) = -\mathbf{K}\mathbf{x}(k)$ the system equation becomes

$$\mathbf{x}(k+1) = (\mathbf{G} - \mathbf{HK})\mathbf{x}(k) \tag{6–102}$$

We shall first examine the controllability matrix:

$$[\mathbf{H} \vdots \mathbf{GH} \vdots \mathbf{G}^2\mathbf{H}] = [\mathbf{H}_1 \vdots \mathbf{H}_2 \vdots \mathbf{GH}_1 \vdots \mathbf{GH}_2 \vdots \mathbf{G}^2\mathbf{H}_1 \vdots \mathbf{G}^2\mathbf{H}_2]$$

$$= \begin{bmatrix} 0 & 1 & 0 & 0 & 1 & 0 \\ 0 & 0 & 1 & 0 & 0.5 & -0.25 \\ 1 & 0 & 0.5 & -0.25 & 0.25 & -0.125 \end{bmatrix}$$

Clearly, the rank of this controllability matrix is 3. Therefore, arbitrary pole placement is possible. We now choose three linearly independent vectors starting from the left end. These vectors are shown enclosed by dashed lines. (The three linearly independent vectors chosen are $\mathbf{H}_1$, $\mathbf{H}_2$, and $\mathbf{GH}_1$.) Now we rearrange these three vectors according to Eq. (6–87) and define matrix $\mathbf{F}$ as follows:

$$\mathbf{F} = [\mathbf{H}_1 \vdots \mathbf{GH}_1 \vdots \mathbf{H}_2]$$

We note that $n_1 = 2$ and $n_2 = 1$.

Rewriting matrix $\mathbf{F}$, we have

$$\mathbf{F} = \begin{bmatrix} 0 & 0 & 1 \\ 0 & 1 & 0 \\ 1 & 0.5 & 0 \end{bmatrix}$$

The inverse of matrix $\mathbf{F}$ becomes

$$\mathbf{F}^{-1} = \begin{bmatrix} 0 & -0.5 & 1 \\ 0 & 1 & 0 \\ 1 & 0 & 0 \end{bmatrix}$$

We now define the η_ith row vector of $\mathbf{F}^{-1}$ as $\mathbf{f}_i$, where $\eta_1 = n_1$ and $\eta_2 = n_1 + n_2$. Since $n_1 = 2$ and $n_2 = 1$, the vectors $\mathbf{f}_1$ and $\mathbf{f}_2$ are the second and third row vectors, respectively. That is,

$$\mathbf{f}_1 = [0 \quad 1 \quad 0]$$

$$\mathbf{f}_2 = [1 \quad 0 \quad 0]$$

Next, define the transformation matrix $\mathbf{T}$ by

$$\mathbf{T} = \begin{bmatrix} \mathbf{S}_1 \\ \hdashline \mathbf{S}_2 \end{bmatrix}^{-1}$$

where

$$\mathbf{S}_1 = \begin{bmatrix} \mathbf{f}_1 \\ \hdashline \mathbf{f}_1\mathbf{G} \end{bmatrix}, \qquad \mathbf{S}_2 = \mathbf{f}_2$$

Hence

$$\mathbf{T} = \begin{bmatrix} 0 & 1 & 0 \\ 0 & 0 & 1 \\ 1 & 0 & 0 \end{bmatrix}^{-1} = \begin{bmatrix} 0 & 0 & 1 \\ 1 & 0 & 0 \\ 0 & 1 & 0 \end{bmatrix}$$

and

$$\mathbf{T}^{-1} = \begin{bmatrix} 0 & 1 & 0 \\ 0 & 0 & 1 \\ 1 & 0 & 0 \end{bmatrix}$$

With this transformation matrix **T**, we define

$$\mathbf{x}(k) = \mathbf{T}\hat{\mathbf{x}}(k)$$

Then

$$\begin{aligned} \mathbf{T}^{-1}\mathbf{G}\mathbf{T} &= \hat{\mathbf{G}} \\ &= \begin{bmatrix} 0 & 1 & 0 \\ 0 & 0 & 1 \\ 1 & 0 & 0 \end{bmatrix} \begin{bmatrix} 0 & 1 & 0 \\ 0 & 0 & 1 \\ -0.25 & 0 & 0.5 \end{bmatrix} \begin{bmatrix} 0 & 0 & 1 \\ 1 & 0 & 0 \\ 0 & 1 & 0 \end{bmatrix} \\ &= \left[\begin{array}{cc:c} 0 & 1 & 0 \\ 0 & 0.5 & -0.25 \\ \hdashline 1 & 0 & 0 \end{array}\right] \end{aligned}$$

Also,

$$\begin{aligned} \mathbf{T}^{-1}\mathbf{H} &= \hat{\mathbf{H}} \\ &= \begin{bmatrix} 0 & 1 & 0 \\ 0 & 0 & 1 \\ 1 & 0 & 0 \end{bmatrix} \begin{bmatrix} 0 & 1 \\ 0 & 0 \\ 1 & 0 \end{bmatrix} = \left[\begin{array}{cc} 0 & 0 \\ 1 & 0 \\ \hdashline 0 & 1 \end{array}\right] \end{aligned}$$

Next, we determine the state feedback gain matrix **K**, where

$$\mathbf{K} = \mathbf{B}\boldsymbol{\Delta}\mathbf{T}^{-1}$$

From Eq. (6–101), matrix **B** for the present case is a 2×2 matrix. Noting that $b_{12} = 0$, we have

$$\mathbf{B} = \begin{bmatrix} 1 & b_{12} \\ 0 & 1 \end{bmatrix}^{-1} = \begin{bmatrix} 1 & 0 \\ 0 & 1 \end{bmatrix}$$

For the present case, $\boldsymbol{\Delta}$ is a 2×3 matrix:

$$\boldsymbol{\Delta} = \begin{bmatrix} \delta_{11} & \delta_{12} & \delta_{13} \\ \delta_{21} & \delta_{22} & \delta_{23} \end{bmatrix}$$

Now we determine matrix $\hat{\mathbf{G}} - \hat{\mathbf{H}}\mathbf{B}\mathbf{\Delta}$:

$$\hat{\mathbf{G}} - \hat{\mathbf{H}}\mathbf{B}\mathbf{\Delta} = \begin{bmatrix} 0 & 1 & 0 \\ 0 & 0.5 & -0.25 \\ 1 & 0 & 0 \end{bmatrix} - \begin{bmatrix} 0 & 0 \\ 1 & 0 \\ 0 & 1 \end{bmatrix} \begin{bmatrix} 1 & 0 \\ 0 & 1 \end{bmatrix} \begin{bmatrix} \delta_{11} & \delta_{12} & \delta_{13} \\ \delta_{21} & \delta_{22} & \delta_{23} \end{bmatrix}$$

$$= \begin{bmatrix} 0 & 1 & 0 \\ 0 & 0.5 & -0.25 \\ 1 & 0 & 0 \end{bmatrix} - \begin{bmatrix} 0 & 0 & 0 \\ \delta_{11} & \delta_{12} & \delta_{13} \\ \delta_{21} & \delta_{22} & \delta_{23} \end{bmatrix}$$

$$= \begin{bmatrix} 0 & 1 & 0 \\ -\delta_{11} & 0.5 - \delta_{12} & -0.25 - \delta_{13} \\ 1 - \delta_{21} & -\delta_{22} & -\delta_{23} \end{bmatrix}$$

The characteristic equation $|z\mathbf{I} - \hat{\mathbf{G}} + \hat{\mathbf{H}}\mathbf{B}\mathbf{\Delta}| = 0$ is given as follows:

$$|z\mathbf{I} - \hat{\mathbf{G}} + \hat{\mathbf{H}}\mathbf{B}\mathbf{\Delta}| = \begin{vmatrix} z & -1 & 0 \\ \delta_{11} & z - 0.5 + \delta_{12} & 0.25 + \delta_{13} \\ -1 + \delta_{21} & \delta_{22} & z + \delta_{23} \end{vmatrix} = 0$$

Since the deadbeat response is desired, the desired characteristic equation is

$$z^3 = 0$$

Note that the choice of the δ's is not unique and matrix $\mathbf{\Delta}$ is not unique. Suppose we choose the δ's so that

$$\delta_{11} = 0, \qquad \delta_{12} = 0.5, \qquad \delta_{13} = -0.25$$

$$\delta_{21} = 1, \qquad \delta_{22} = 0, \qquad \delta_{23} = 0$$

Then

$$|z\mathbf{I} - \hat{\mathbf{G}} + \hat{\mathbf{H}}\mathbf{B}\mathbf{\Delta}| = \begin{vmatrix} z & -1 & 0 \\ 0 & z & 0 \\ 0 & 0 & z \end{vmatrix} = z^3 = 0$$

and thus

$$\mathbf{\Delta} = \begin{bmatrix} 0 & 0.5 & -0.25 \\ 1 & 0 & 0 \end{bmatrix}$$

is acceptable. Then, matrix $\mathbf{K}$ is obtained as follows:

$$\mathbf{K} = \mathbf{B}\mathbf{\Delta}\mathbf{T}^{-1} = \begin{bmatrix} 1 & 0 \\ 0 & 1 \end{bmatrix} \begin{bmatrix} 0 & 0.5 & -0.25 \\ 1 & 0 & 0 \end{bmatrix} \begin{bmatrix} 0 & 1 & 0 \\ 0 & 0 & 1 \\ 1 & 0 & 0 \end{bmatrix}$$

$$= \begin{bmatrix} -0.25 & 0 & 0.5 \\ 0 & 1 & 0 \end{bmatrix}$$

With this choice of matrix $\mathbf{K}$, $(\hat{\mathbf{G}} - \hat{\mathbf{H}}\mathbf{B}\boldsymbol{\Delta})^k = \mathbf{0}$ for $k \geq n_{\min}$, where

$$n_{\min} = \max(n_1, n_2) = \max(2, 1) = 2$$

In fact,

$$\hat{\mathbf{G}} - \hat{\mathbf{H}}\mathbf{B}\boldsymbol{\Delta} = \begin{bmatrix} 0 & 1 & 0 \\ 0 & 0 & 0 \\ 0 & 0 & 0 \end{bmatrix}$$

$$(\hat{\mathbf{G}} - \hat{\mathbf{H}}\mathbf{B}\boldsymbol{\Delta})^2 = \begin{bmatrix} 0 & 0 & 0 \\ 0 & 0 & 0 \\ 0 & 0 & 0 \end{bmatrix}$$

Thus

$$(\hat{\mathbf{G}} - \hat{\mathbf{H}}\mathbf{B}\boldsymbol{\Delta})^k = \mathbf{0} \qquad k = 2, 3, 4, \ldots$$

Note that

$$\hat{\mathbf{G}} - \hat{\mathbf{H}}\mathbf{B}\boldsymbol{\Delta} = \mathbf{T}^{-1}\mathbf{G}\mathbf{T} - \mathbf{T}^{-1}\mathbf{H}\mathbf{B}\boldsymbol{\Delta} = \mathbf{T}^{-1}\mathbf{G}\mathbf{T} - \mathbf{T}^{-1}\mathbf{H}\mathbf{K}\mathbf{T} = \mathbf{T}^{-1}(\mathbf{G} - \mathbf{H}\mathbf{K})\mathbf{T}$$

Referring to Eq. (6–102) and its solution $\mathbf{x}(k) = (\mathbf{G} - \mathbf{H}\mathbf{K})^k\mathbf{x}(0)$, we have $\mathbf{x}(k) = \mathbf{0}$ for $k = 2, 3, 4, \ldots$, since

$$\mathbf{G} - \mathbf{H}\mathbf{K} = \mathbf{T}(\hat{\mathbf{G}} - \hat{\mathbf{H}}\mathbf{B}\boldsymbol{\Delta})\mathbf{T}^{-1}$$

$$(\mathbf{G} - \mathbf{H}\mathbf{K})^2 = \mathbf{T}(\hat{\mathbf{G}} - \hat{\mathbf{H}}\mathbf{B}\boldsymbol{\Delta})\mathbf{T}^{-1}\mathbf{T}(\hat{\mathbf{G}} - \hat{\mathbf{H}}\mathbf{B}\boldsymbol{\Delta})\mathbf{T}^{-1} = \mathbf{T}(\hat{\mathbf{G}} - \hat{\mathbf{H}}\mathbf{B}\boldsymbol{\Delta})^2\mathbf{T}^{-1} = \mathbf{0}$$

and

$$\mathbf{x}(k) = (\mathbf{G} - \mathbf{H}\mathbf{K})^k\mathbf{x}(0) = \mathbf{0} \qquad k = 2, 3, 4, \ldots$$

We have thus designed the state feedback gain matrix $\mathbf{K}$ so that the system's response to any initial state $\mathbf{x}(0)$ is deadbeat. The state $\mathbf{x}(k)$ can be transferred to the origin in at most 2 sampling periods. [Note that if the control signal $u(k)$ were a scalar, then it would take at most 3 sampling periods, rather than at most 2 sampling periods, for deadbeat response.]

6-6 STATE OBSERVERS

In Sec. 6–5 we discussed a pole placement design method that utilizes the feedback of all state variables to form the desired control vector. In practice, however, not all state variables are available for direct measurement. In many practical cases, only a few state variables of a given system are measurable and the rest are not measurable. For instance, it may be that only the output variables are measurable. Hence, it is necessary to estimate the state variables that are not directly measurable. Such estimation is commonly called *observation*. In a practical system it is necessary to observe or estimate the unmeasurable state variables from the output and control variables.

A state observer, also called a state estimator, is a subsystem in the control system that performs an estimation of the state variables based on the measurements

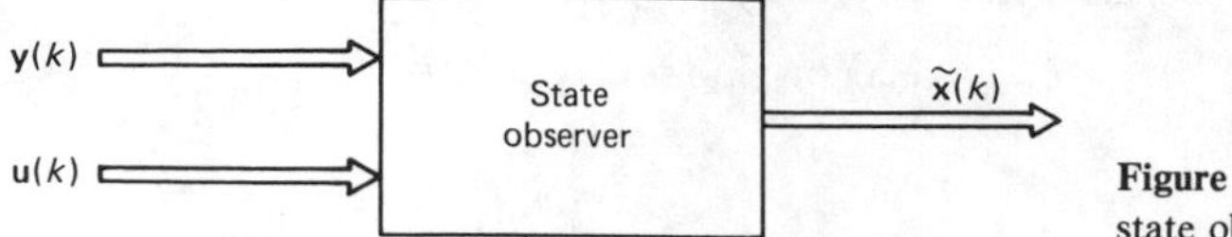

Figure 6–2 Schematic diagram of the state observer.

of the output and control variables. Here, the concept of observability discussed in Sec. 6–3 plays an important role. As we shall see later, state observers can be designed if and only if the observability condition is satisfied.

In the following discussions of state observers, we shall use the notation $\tilde{\mathbf{x}}(k)$ to designate the observed state vector. In many cases the observed state vector $\tilde{\mathbf{x}}(k)$ is used in the state feedback to generate the optimal control vector. Figure 6–2 shows a schematic diagram of a state observer. The state observer will have $\mathbf{y}(k)$ and $\mathbf{u}(k)$ as inputs and $\tilde{\mathbf{x}}(k)$ as output.

In what follows, we shall first discuss the necessary and sufficient condition for state observation and then treat the *full-order state observer*. Full-order state observation means that we observe (estimate) all n state variables regardless of whether some state variables are available for direct measurement. There are times when this will be unnecessary, when we will need observation of only the unmeasurable state variables but not of those that are directly measurable as well. Observation of only the unmeasurable state variables is referred to as *minimum-order state observation*, and we shall discuss it later in this section. Observation of all unmeasurable state variables plus some (but not all) of the measurable state variables is referred to as *reduced-order state observation*.

Necessary and sufficient condition for state observation. Figure 6–3 shows a regulator system with a state observer. We shall discuss a necessary and sufficient condition under which the state vector can be observed (estimated). From Fig. 6–3 we obtain the state and output equations as

$$\mathbf{x}(k+1) = \mathbf{G}\mathbf{x}(k) + \mathbf{H}\mathbf{u}(k) \qquad (6\text{–}103)$$

$$\mathbf{y}(k) = \mathbf{C}\mathbf{x}(k) \qquad (6\text{–}104)$$

where

$\mathbf{x}(k)$ = state vector (n-vector)

$\mathbf{u}(k)$ = control vector (r-vector)

$\mathbf{y}(k)$ = output vector (m-vector)

$\mathbf{G}$ = $n \times n$ nonsingular matrix

$\mathbf{H}$ = $n \times r$ matrix

$\mathbf{C}$ = $m \times n$ matrix

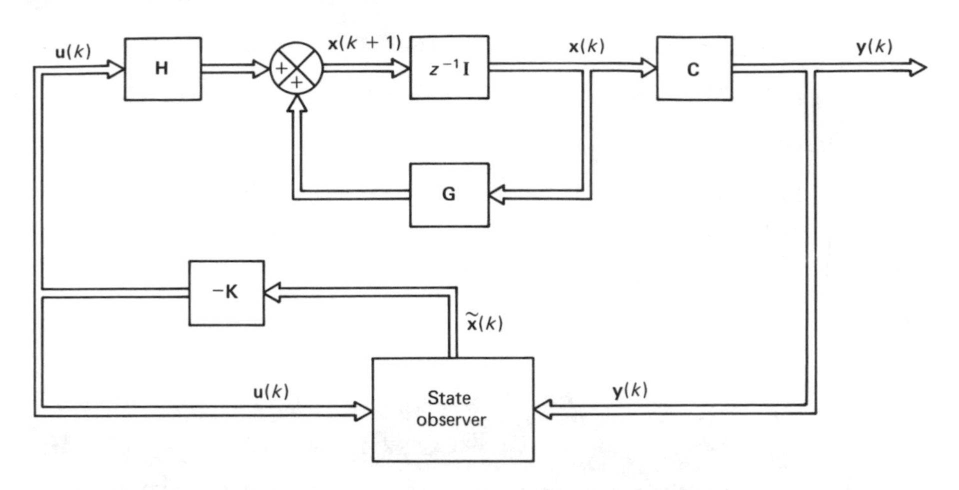

Figure 6–3 Regulator system with a state observer.

To be able to observe (estimate) state variables, we must be able to obtain $\mathbf{x}(k+1)$ in terms of $\mathbf{y}(k)$, $\mathbf{y}(k-1), \ldots, \mathbf{y}(k-n+1)$ and $\mathbf{u}(k)$, $\mathbf{u}(k-1), \ldots, \mathbf{u}(k-n+1)$. From Eq. (6–103) we have

$$\mathbf{G}^{-1}\mathbf{x}(k+1) = \mathbf{x}(k) + \mathbf{G}^{-1}\mathbf{H}\mathbf{u}(k)$$

or

$$\mathbf{x}(k) = \mathbf{G}^{-1}\mathbf{x}(k+1) - \mathbf{G}^{-1}\mathbf{H}\mathbf{u}(k) \tag{6–105}$$

By shifting k by 1, we get

$$\mathbf{x}(k-1) = \mathbf{G}^{-1}\mathbf{x}(k) - \mathbf{G}^{-1}\mathbf{H}\mathbf{u}(k-1) \tag{6–106}$$

By substituting Eq. (6–105) into Eq. (6–106) we obtain

$$\begin{aligned}\mathbf{x}(k-1) &= \mathbf{G}^{-1}[\mathbf{G}^{-1}\mathbf{x}(k+1) - \mathbf{G}^{-1}\mathbf{H}\mathbf{u}(k)] - \mathbf{G}^{-1}\mathbf{H}\mathbf{u}(k-1) \\ &= \mathbf{G}^{-2}\mathbf{x}(k+1) - \mathbf{G}^{-2}\mathbf{H}\mathbf{u}(k) - \mathbf{G}^{-1}\mathbf{H}\mathbf{u}(k-1)\end{aligned}$$

Similarly,

$$\begin{aligned}\mathbf{x}(k-2) &= \mathbf{G}^{-2}\mathbf{x}(k) - \mathbf{G}^{-2}\mathbf{H}\mathbf{u}(k-1) - \mathbf{G}^{-1}\mathbf{H}\mathbf{u}(k-2) \\ &= \mathbf{G}^{-3}\mathbf{x}(k+1) - \mathbf{G}^{-3}\mathbf{H}\mathbf{u}(k) - \mathbf{G}^{-2}\mathbf{H}\mathbf{u}(k-1) - \mathbf{G}^{-1}\mathbf{H}\mathbf{u}(k-2)\end{aligned}$$

$$\vdots$$

$$\begin{aligned}\mathbf{x}(k-n+1) = \mathbf{G}^{-n}\mathbf{x}(k+1) - \mathbf{G}^{-n}\mathbf{H}\mathbf{u}(k) - \mathbf{G}^{-n+1}\mathbf{H}\mathbf{u}(k-1) \\ - \cdots - \mathbf{G}^{-1}\mathbf{H}\mathbf{u}(k-n+1)\end{aligned}$$

By substituting Eq. (6–105) into Eq. (6–104) we obtain

$$\mathbf{y}(k) = \mathbf{C}\mathbf{x}(k) = \mathbf{C}\mathbf{G}^{-1}\mathbf{x}(k+1) - \mathbf{C}\mathbf{G}^{-1}\mathbf{H}\mathbf{u}(k)$$

Similarly,

$$\mathbf{y}(k-1) = \mathbf{C}\mathbf{x}(k-1) = \mathbf{C}\mathbf{G}^{-2}\mathbf{x}(k+1) - \mathbf{C}\mathbf{G}^{-2}\mathbf{H}\mathbf{u}(k) - \mathbf{C}\mathbf{G}^{-1}\mathbf{H}\mathbf{u}(k-1)$$

$$\mathbf{y}(k-2) = \mathbf{C}\mathbf{x}(k-2) = \mathbf{C}\mathbf{G}^{-3}\mathbf{x}(k+1) - \mathbf{C}\mathbf{G}^{-3}\mathbf{H}\mathbf{u}(k) - \mathbf{C}\mathbf{G}^{-2}\mathbf{H}\mathbf{u}(k-1) - \mathbf{C}\mathbf{G}^{-1}\mathbf{H}\mathbf{u}(k-2)$$

$$\vdots$$

$$\mathbf{y}(k-n+1) = \mathbf{C}\mathbf{x}(k-n+1) = \mathbf{C}\mathbf{G}^{-n}\mathbf{x}(k+1) - \mathbf{C}\mathbf{G}^{-n}\mathbf{H}\mathbf{u}(k) - \mathbf{C}\mathbf{G}^{-n+1}\mathbf{H}\mathbf{u}(k-1) - \cdots - \mathbf{C}\mathbf{G}^{-1}\mathbf{H}\mathbf{u}(k-n+1)$$

By combining the preceding n equations into one matrix equation, we get

$$\begin{bmatrix} \mathbf{y}(k) \\ \mathbf{y}(k-1) \\ \cdot \\ \cdot \\ \cdot \\ \mathbf{y}(k-n+1) \end{bmatrix} = \begin{bmatrix} \mathbf{C}\mathbf{G}^{-1} \\ \mathbf{C}\mathbf{G}^{-2} \\ \cdot \\ \cdot \\ \cdot \\ \mathbf{C}\mathbf{G}^{-n} \end{bmatrix} \mathbf{x}(k+1) - \begin{bmatrix} \mathbf{C}\mathbf{G}^{-1}\mathbf{H} & \mathbf{0} & \cdots & \mathbf{0} \\ \mathbf{C}\mathbf{G}^{-2}\mathbf{H} & \mathbf{C}\mathbf{G}^{-1}\mathbf{H} & \cdots & \mathbf{0} \\ \cdot & \cdot & & \cdot \\ \cdot & \cdot & & \cdot \\ \cdot & \cdot & & \cdot \\ \mathbf{C}\mathbf{G}^{-n}\mathbf{H} & \mathbf{C}\mathbf{G}^{-n+1}\mathbf{H} & \cdots & \mathbf{C}\mathbf{G}^{-1}\mathbf{H} \end{bmatrix} \begin{bmatrix} \mathbf{u}(k) \\ \mathbf{u}(k-1) \\ \cdot \\ \cdot \\ \cdot \\ \mathbf{u}(k-n+1) \end{bmatrix}$$

or

$$\begin{bmatrix} \mathbf{C}\mathbf{G}^{-1} \\ \mathbf{C}\mathbf{G}^{-2} \\ \cdot \\ \cdot \\ \cdot \\ \mathbf{C}\mathbf{G}^{-n} \end{bmatrix} \mathbf{x}(k+1) = \begin{bmatrix} \mathbf{y}(k) \\ \mathbf{y}(k-1) \\ \cdot \\ \cdot \\ \mathbf{y}(k-n+1) \end{bmatrix} + \begin{bmatrix} \mathbf{C}\mathbf{G}^{-1}\mathbf{H} & \mathbf{0} & \cdots & \mathbf{0} \\ \mathbf{C}\mathbf{G}^{-2}\mathbf{H} & \mathbf{C}\mathbf{G}^{-1}\mathbf{H} & \cdots & \mathbf{0} \\ \cdot & \cdot & & \cdot \\ \cdot & \cdot & & \cdot \\ \cdot & \cdot & & \cdot \\ \mathbf{C}\mathbf{G}^{-n}\mathbf{H} & \mathbf{C}\mathbf{G}^{-n+1}\mathbf{H} & \cdots & \mathbf{C}\mathbf{G}^{-1}\mathbf{H} \end{bmatrix} \begin{bmatrix} \mathbf{u}(k) \\ \mathbf{u}(k-1) \\ \cdot \\ \cdot \\ \cdot \\ \mathbf{u}(k-n+1) \end{bmatrix} \tag{6–107}$$

Notice that the right-hand side of Eq. (6–107) is entirely known. Hence, $\mathbf{x}(k+1)$ can be determined if and only if

$$\text{rank}\begin{bmatrix} \mathbf{CG}^{-1} \\ \mathbf{CG}^{-2} \\ \cdot \\ \cdot \\ \cdot \\ \mathbf{CG}^{-n} \end{bmatrix} = n \tag{6–108}$$

Since matrix $\mathbf{G}$ is nonsingular, multiplication of each row of the left-hand side of Eq. (6–108) by $\mathbf{G}^n$ does not change the rank condition. Hence, Eq. (6–108) is equivalent to

$$\text{rank}\begin{bmatrix} \mathbf{CG}^{n-1} \\ \mathbf{CG}^{n-2} \\ \cdot \\ \cdot \\ \cdot \\ \mathbf{C} \end{bmatrix} = n$$

which is also equivalent to

$$\text{rank}\,[\mathbf{C}^* \,\vdots\, \mathbf{G}^*\mathbf{C}^* \,\vdots\, \cdots \,\vdots\, (\mathbf{G}^*)^{n-1}\mathbf{C}^*] = n \tag{6–109}$$

Clearly, this is the complete observability condition of the system defined by Eqs. (6–103) and (6–104). [Refer to Eq. (6–17).] This means that if Eq. (6–109) is satisfied (that is, if the system is completely observable), then $\mathbf{x}(k+1)$ can be determined from $\mathbf{y}(k)$, $\mathbf{y}(k-1)$, . . . , $\mathbf{y}(k-n+1)$ and $\mathbf{u}(k)$, $\mathbf{u}(k-1)$, . . . , $\mathbf{u}(k-n+1)$.

As a special case, if $y(k)$ is a scalar and matrix $\mathbf{C}$ is a $1 \times n$ matrix, then $\mathbf{x}(k+1)$ can be obtained by premultiplying both sides of Eq. (6–107) by the inverse of the matrix given in Eq. (6–108), as follows:

$$\mathbf{x}(k+1) = \begin{bmatrix} \mathbf{CG}^{-1} \\ \mathbf{CG}^{-2} \\ \cdot \\ \cdot \\ \cdot \\ \mathbf{CG}^{-n} \end{bmatrix}^{-1} \begin{bmatrix} y(k) \\ y(k-1) \\ \cdot \\ \cdot \\ \cdot \\ y(k-n+1) \end{bmatrix}$$
$$+ \begin{bmatrix} \mathbf{CG}^{-1} \\ \mathbf{CG}^{-2} \\ \cdot \\ \cdot \\ \cdot \\ \mathbf{CG}^{-n} \end{bmatrix}^{-1} \begin{bmatrix} \mathbf{CG}^{-1}\mathbf{H} & \mathbf{0} & \cdots & \mathbf{0} \\ \mathbf{CG}^{-2}\mathbf{H} & \mathbf{CG}^{-1}\mathbf{H} & \cdots & \mathbf{0} \\ \cdot & \cdot & & \cdot \\ \cdot & \cdot & & \cdot \\ \cdot & \cdot & & \cdot \\ \mathbf{CG}^{-n}\mathbf{H} & \mathbf{CG}^{-n+1}\mathbf{H} & \cdots & \mathbf{CG}^{-1}\mathbf{H} \end{bmatrix} \begin{bmatrix} \mathbf{u}(k) \\ \mathbf{u}(k-1) \\ \cdot \\ \cdot \\ \cdot \\ \mathbf{u}(k-n+1) \end{bmatrix} \tag{6–110}$$

Equation (6–110) gives $\mathbf{x}(k+1)$ when $y(k)$ is a scalar.

To summarize, we have shown that the necessary and sufficient condition under which the state can be determined is that the system be completely observable, and we have also shown that if $y(k)$ is a scalar, then $\mathbf{x}(k+1)$ will be given by Eq. (6–110).

As shown in the foregoing analysis, the state $\mathbf{x}(k+1)$ can be determined from Eq. (6–107) provided the system is completely observable. Thus, for a completely observable system, the state vector can be determined in at most n sampling periods. In the presence of external disturbances and measurement noises, however, this approach may not give an accurate determination of the state vector. Hence, to determine the state vector in the presence of disturbances and measurement noises, a different approach is necessary. Also, if matrix $\mathbf{C}$ is not a $1 \times n$ matrix but is an $m \times n$ matrix ($m > 1$), then the inverse of the matrix of Eq. (6–108) cannot be defined and Eq. (6–110) does not apply. To cope with such cases, one very powerful approach for estimating the state vector is to use a dynamic model of the original system, as follows.

Consider the control system defined by Eqs. (6–103) and (6–104). Let us assume that the state $\mathbf{x}(k)$ is to be approximated by the state $\tilde{\mathbf{x}}(k)$ of the dynamic model:

$$\tilde{\mathbf{x}}(k+1) = \mathbf{G}\tilde{\mathbf{x}}(k) + \mathbf{H}\mathbf{u}(k) \tag{6–111}$$

$$\tilde{\mathbf{y}}(k) = \mathbf{C}\tilde{\mathbf{x}}(k) \tag{6–112}$$

where matrices **G, H,** and **C** are the same as those of the original system. Also, let us assume that the dynamic model is subjected to the same control signal $\mathbf{u}(k)$ as the original system. If the initial conditions for the actual system defined by Eqs. (6–103) and (6–104) and the dynamic model defined by Eqs. (6–111) and (6–112) are the same, then the state $\tilde{\mathbf{x}}(k)$ and the state $\mathbf{x}(k)$ will be the same. If the initial conditions are different, then the state $\tilde{\mathbf{x}}(k)$ and the state $\mathbf{x}(k)$ will be different.

If the matrix $\mathbf{G}$ is a stable one, however, $\tilde{\mathbf{x}}(k)$ will approach $\mathbf{x}(k)$ even for different initial conditions, as we shall see. If we denote the difference between $\mathbf{x}(k)$ and $\tilde{\mathbf{x}}(k)$ as $\mathbf{e}(k)$, or define

$$\mathbf{e}(k) = \mathbf{x}(k) - \tilde{\mathbf{x}}(k)$$

then by subtracting Eq. (6–111) from Eq. (6–103), we obtain

$$\mathbf{x}(k+1) - \tilde{\mathbf{x}}(k+1) = \mathbf{G}[\mathbf{x}(k) - \tilde{\mathbf{x}}(k)]$$

or

$$\mathbf{e}(k+1) = \mathbf{G}\mathbf{e}(k)$$

If matrix $\mathbf{G}$ is a stable matrix, then $\mathbf{e}(k)$ will approach zero and $\tilde{\mathbf{x}}(k)$ will approach $\mathbf{x}(k)$. However, the behavior of the error vector, which depends solely on matrix **G,** may not be acceptable. Also, if matrix $\mathbf{G}$ is not a stable matrix, then the error $\mathbf{e}(k)$ will not approach zero. It is therefore necessary to modify the dynamic model defined by Eqs. (6–111) and (6–112).

It is noted that although the state $\mathbf{x}(k)$ may not be measurable, the output

$\mathbf{y}(k)$ is measurable. The dynamic model defined by Eqs. (6–111) and (6–112) does not make use of the measured output $\mathbf{y}(k)$. The performance of the dynamic model can be improved if the difference between the measured output $\mathbf{y}(k)$ and the estimated output $\mathbf{C}\tilde{\mathbf{x}}(k)$ is used to monitor the state $\tilde{\mathbf{x}}(k)$, that is, if the dynamic model of Eq. (6–111) is modified into the following form:

$$\tilde{\mathbf{x}}(k+1) = \mathbf{G}\tilde{\mathbf{x}}(k) + \mathbf{H}\mathbf{u}(k) + \mathbf{K}_e[\mathbf{y}(k) - \mathbf{C}\tilde{\mathbf{x}}(k)]$$

where matrix $\mathbf{K}_e$ serves as a weighting matrix. (This means that the dynamics of the state observer shown in Fig. 6–3 must be given by this last equation.) In the presence of discrepancies between the $\mathbf{G}$ and $\mathbf{H}$ matrices used in the model and those of the actual system, the addition of the difference between the measured output and the estimated output will help reduce the differences between the dynamic model and the actual system.

In what follows we shall discuss details of the observer whose dynamics are characterized by $\mathbf{G}$ and $\mathbf{H}$ matrices and by the additional correction term which consists of the difference between the measured output and the estimated output.

Full-order state observer. The order of the state observer that will be discussed here is the same as that of the system. As stated earlier, such a state observer is called a full-order state observer.

In the following analysis we assume that the actual state $\mathbf{x}(k)$ cannot be measured directly. If the state $\mathbf{x}(k)$ is to be estimated, it is desirable that the observed state or estimated state $\tilde{\mathbf{x}}(k)$ be as close to the actual state $\mathbf{x}(k)$ as possible. Although it is not necessary, it is convenient if the state observer has the same $\mathbf{G}$ and $\mathbf{H}$ matrices as the original system.

It is important to note that in the present analysis, state $\mathbf{x}(k)$ is not available for direct measurement and consequently the observed state $\tilde{\mathbf{x}}(k)$ cannot be compared with the actual state $\mathbf{x}(k)$. Since the output $\mathbf{y}(k) = \mathbf{C}\mathbf{x}(k)$ can be measured, however, it is possible to compare $\tilde{\mathbf{y}}(k) = \mathbf{C}\tilde{\mathbf{x}}(k)$ with $\mathbf{y}(k)$.

Consider the state feedback control system shown in Fig. 6–4. The system equations are

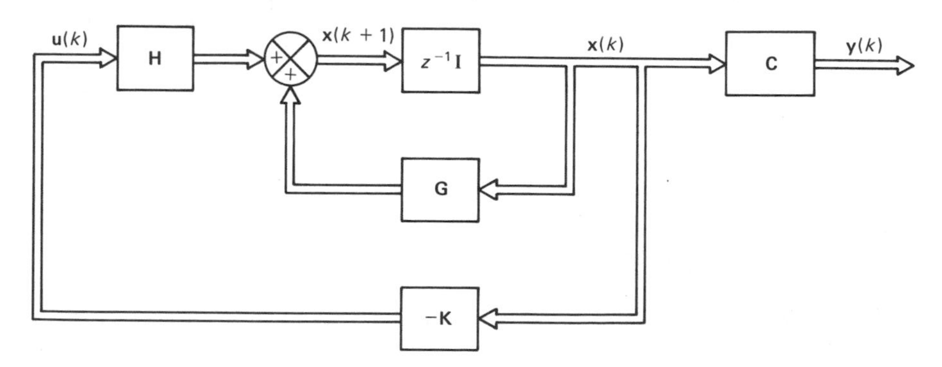

Figure 6–4 State feedback control system.

$$\mathbf{x}(k+1) = \mathbf{G}\mathbf{x}(k) + \mathbf{H}\mathbf{u}(k) \tag{6–113}$$

$$\mathbf{y}(k) = \mathbf{C}\mathbf{x}(k) \tag{6–114}$$

$$\mathbf{u}(k) = -\mathbf{K}\mathbf{x}(k)$$

where

$\mathbf{x}(k) =$ state vector (n-vector)

$\mathbf{u}(k) =$ control vector (r-vector)

$\mathbf{y}(k) =$ output vector (m-vector)

$\mathbf{G} = n \times n$ nonsingular matrix

$\mathbf{H} = n \times r$ matrix

$\mathbf{C} = m \times n$ matrix

$\mathbf{K} =$ state feedback gain matrix ($n \times r$ matrix)

We assume that the system is completely observable and, as stated earlier, $\mathbf{x}(k)$ is not available for direct measurement. Figure 6–5 shows a state observer incorporated into the system of Fig. 6–4. The observed state $\tilde{\mathbf{x}}(k)$ is used to form the control vector $\mathbf{u}(k)$, or

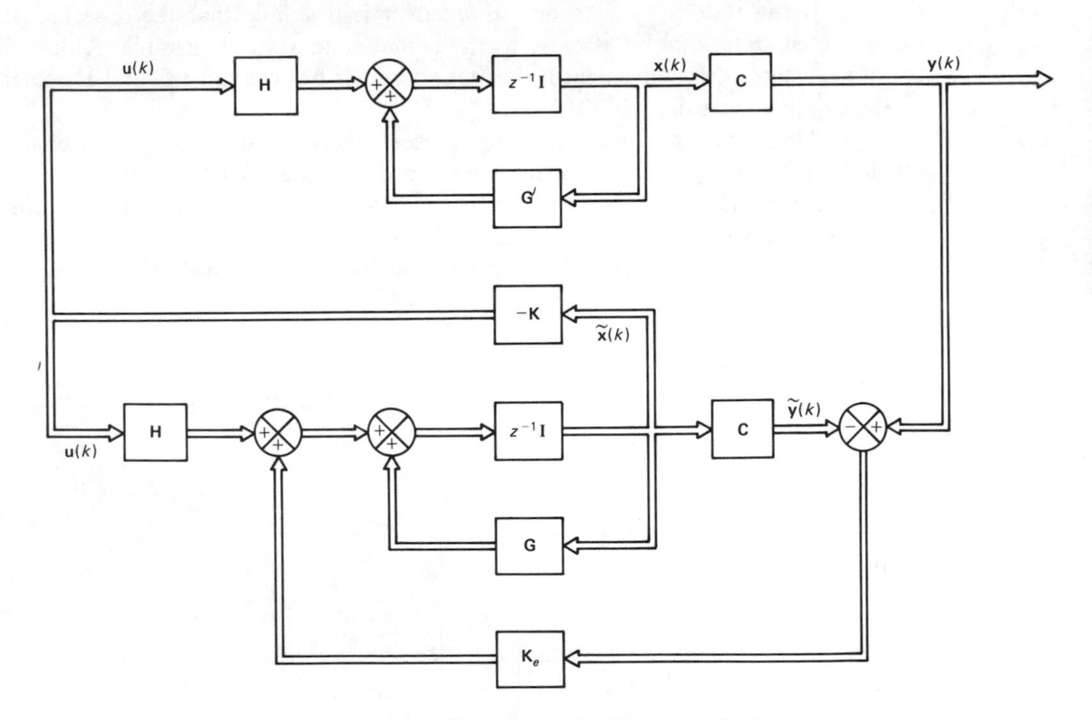

Figure 6–5 Observed-state feedback control system.

$$\mathbf{u}(k) = -\mathbf{K}\tilde{\mathbf{x}}(k) \tag{6–115}$$

From Fig. 6–5 we have

$$\tilde{\mathbf{x}}(k+1) = \mathbf{G}\tilde{\mathbf{x}}(k) + \mathbf{H}\mathbf{u}(k) + \mathbf{K}_e[\mathbf{y}(k) - \tilde{\mathbf{y}}(k)] \tag{6–116}$$

where $\mathbf{K}_e$ is the observer feedback gain matrix (an $n \times m$ matrix). This last equation can be modified to read

$$\tilde{\mathbf{x}}(k+1) = (\mathbf{G} - \mathbf{K}_e\mathbf{C})\tilde{\mathbf{x}}(k) + \mathbf{H}\mathbf{u}(k) + \mathbf{K}_e\mathbf{y}(k) \tag{6–117}$$

The state observer given by Eq. (6–117) is called a *prediction observer*, since the estimate $\tilde{\mathbf{x}}(k+1)$ is one sampling period ahead of the measurement $\mathbf{y}(k)$. The eigenvalues of $\mathbf{G} - \mathbf{K}_e\mathbf{C}$ are commonly called the *observer poles*.

Pulse-transfer-function matrix for the controller. Equations (6–115) and (6–117) define the control vector $\mathbf{u}(k)$ and thus they define the controller for the system. The z transform of Eq. (6–115) gives

$$\mathbf{U}(z) = -\mathbf{K}\tilde{\mathbf{X}}(z) \tag{6–118}$$

The z transform of the observer equation given by Eq. (6–117) is

$$z\tilde{\mathbf{X}}(z) = (\mathbf{G} - \mathbf{K}_e\mathbf{C})\tilde{\mathbf{X}}(z) + \mathbf{H}\mathbf{U}(z) + \mathbf{K}_e\mathbf{Y}(z)$$

From these two z-transformed equations, we obtain

$$z\tilde{\mathbf{X}}(z) = (\mathbf{G} - \mathbf{K}_e\mathbf{C} - \mathbf{H}\mathbf{K})\tilde{\mathbf{X}}(z) + \mathbf{K}_e\mathbf{Y}(z)$$

or

$$\tilde{\mathbf{X}}(z) = (z\mathbf{I} - \mathbf{G} + \mathbf{K}_e\mathbf{C} + \mathbf{H}\mathbf{K})^{-1}\mathbf{K}_e\mathbf{Y}(z)$$

By substituting this last equation into Eq. (6–118), we get

$$\mathbf{U}(z) = [-\mathbf{K}(z\mathbf{I} - \mathbf{G} + \mathbf{K}_e\mathbf{C} + \mathbf{H}\mathbf{K})^{-1}\mathbf{K}_e]\mathbf{Y}(z)$$

This equation gives the pulse-transfer-function matrix for the controller. The characteristic equation for the controller is

$$|z\mathbf{I} - \mathbf{G} + \mathbf{K}_e\mathbf{C} + \mathbf{H}\mathbf{K}| = 0$$

It should be noted that among the roots of this characteristic equation there may be unstable ones.

If both $u(k)$ and $y(k)$ are scalar, then the pulse transfer function is given by

$$\frac{U(z)}{Y(z)} = -\mathbf{K}(z\mathbf{I} - \mathbf{G} + \mathbf{K}_e\mathbf{C} + \mathbf{H}\mathbf{K})^{-1}\mathbf{K}_e$$

Error dynamics of the full-order state observer. Notice that if $\tilde{\mathbf{x}}(k) = \mathbf{x}(k)$, then Eq. (6–117) becomes

$$\tilde{\mathbf{x}}(k+1) = \mathbf{G}\tilde{\mathbf{x}}(k) + \mathbf{H}\mathbf{u}(k)$$

which is identical to the state equation of the system. Thus, if $\tilde{\mathbf{x}}(k) = \mathbf{x}(k)$, then the response of the state observer system is identical to the response of the original system.

In order to obtain the observer error equation, let us subtract Eq. (6–117) from Eq. (6–113):

$$\mathbf{x}(k+1) - \tilde{\mathbf{x}}(k+1) = (\mathbf{G} - \mathbf{K}_e\mathbf{C})\,[\mathbf{x}(k) - \tilde{\mathbf{x}}(k)] \tag{6–119}$$

Now let us define the difference between $\mathbf{x}(k)$ and $\tilde{\mathbf{x}}(k)$ as the error $\mathbf{e}(k)$:

$$\mathbf{e}(k) = \mathbf{x}(k) - \tilde{\mathbf{x}}(k)$$

Then, Eq. (6–119) becomes

$$\mathbf{e}(k+1) = (\mathbf{G} - \mathbf{K}_e\mathbf{C})\mathbf{e}(k) \tag{6–120}$$

From Eq. (6–120) we see that the dynamic behavior of the error signal is determined by the eigenvalues of $\mathbf{G} - \mathbf{K}_e\mathbf{C}$. If matrix $\mathbf{G} - \mathbf{K}_e\mathbf{C}$ is a stable matrix, the error vector will converge to zero for any initial error $\mathbf{e}(0)$. That is, $\tilde{\mathbf{x}}(k)$ will converge to $\mathbf{x}(k)$ regardless of the values of $\mathbf{x}(0)$ and $\tilde{\mathbf{x}}(0)$. If the eigenvalues of $\mathbf{G} - \mathbf{K}_e\mathbf{C}$ are located in such a way that the dynamic behavior of the error vector is adequately fast, then any error will tend to zero with adequate speed. One way to obtain fast response is to use deadbeat response. This can be achieved if all eigenvalues of $\mathbf{G} - \mathbf{K}_e\mathbf{C}$ are chosen to be zero.

Comments. Since the system defined by Eqs. (6–113) and (6–114) is assumed to be completely observable, an arbitrary placement of the eigenvalues of $\mathbf{G} - \mathbf{K}_e\mathbf{C}$ is possible. To explain this further, notice that the eigenvalues of $\mathbf{G} - \mathbf{K}_e\mathbf{C}$ and those of $\mathbf{G}^* - \mathbf{C}^*\mathbf{K}_e^*$ are the same. By use of the principle of duality presented in Sec. 6–3, the condition for complete observability for the system defined by Eqs. (6–113) and (6–114) is the same as the complete state controllability condition for the system

$$\mathbf{x}(k+1) = \mathbf{G}^*\mathbf{x}(k) + \mathbf{C}^*\mathbf{u}(k) \tag{6–121}$$

In Sec. 6–5 we saw that arbitrary pole placement is possible for the system of Eq. (6–121) provided it is completely state controllable or provided the rank of the matrix

$$[\mathbf{C}^* \vdots \mathbf{G}^*\mathbf{C}^* \vdots \cdots \vdots (\mathbf{G}^*)^{n-1}\mathbf{C}^*]$$

is n. [This is the condition for complete observability of the system defined by Eqs. (6–113) and (6–114).] For the system defined by Eq. (6–121), by selecting a set of n desired eigenvalues of $\mathbf{G}^* - \mathbf{C}^*\mathbf{K}$, the state feedback gain matrix $\mathbf{K}$ may be determined. The desired matrix $\mathbf{K}_e$ such that the eigenvalues of $\mathbf{G} - \mathbf{K}_e\mathbf{C}$ are the same as those of $\mathbf{G}^* - \mathbf{C}^*\mathbf{K}$ is related to matrix $\mathbf{K}$ by the equation $\mathbf{K}_e = \mathbf{K}^*$.

Example 6–9.

Consider the system

$$\mathbf{x}(k+1) = \mathbf{G}\mathbf{x}(k) + \mathbf{H}u(k)$$

$$y(k) = \mathbf{C}\mathbf{x}(k)$$

where

$$\mathbf{G} = \begin{bmatrix} 0 & -0.16 \\ 1 & -1 \end{bmatrix}, \qquad \mathbf{H} = \begin{bmatrix} 0 \\ 1 \end{bmatrix}, \qquad \mathbf{C} = [0 \quad 1]$$

Design a full-order state observer, assuming that the system configuration is identical to that shown in Fig. 6–5. The desired eigenvalues of the observer matrix are

$$z = 0.5 + j0.5, \qquad z = 0.5 - j0.5$$

and so the desired characteristic equation is

$$(z - 0.5 - j0.5)(z - 0.5 + j0.5) = z^2 - z + 0.5 = 0$$

Since the configuration of the state observer is specified as shown in Fig. 6–5, the design of the state observer reduces to the determination of an appropriate observer feedback gain matrix $\mathbf{K}_e$. Before we proceed further, let us examine the observability matrix. The rank of

$$[\mathbf{C}^* \vdots \mathbf{G}^*\mathbf{C}^*] = \begin{bmatrix} 0 & 1 \\ 1 & -1 \end{bmatrix}$$

is 2. Hence, the system is completely observable and determination of the desired observer feedback gain matrix is possible.

Referring to Eq. (6–120),

$$\mathbf{e}(k+1) = (\mathbf{G} - \mathbf{K}_e \mathbf{C})\mathbf{e}(k)$$

where

$$\mathbf{e}(k) = \mathbf{x}(k) - \tilde{\mathbf{x}}(k)$$

the characteristic equation of the observer becomes

$$|z\mathbf{I} - \mathbf{G} + \mathbf{K}_e \mathbf{C}| = 0$$

Let us denote the observer feedback gain matrix $\mathbf{K}_e$ as follows:

$$\mathbf{K}_e = \begin{bmatrix} k_1 \\ k_2 \end{bmatrix}$$

Then the characteristic equation becomes

$$\left| z\begin{bmatrix} 1 & 0 \\ 0 & 1 \end{bmatrix} - \begin{bmatrix} 0 & -0.16 \\ 1 & -1 \end{bmatrix} + \begin{bmatrix} k_1 \\ k_2 \end{bmatrix} [0 \quad 1] \right| = \begin{vmatrix} z & 0.16 + k_1 \\ -1 & z + 1 + k_2 \end{vmatrix} = 0$$

which reduces to

$$z^2 + (1 + k_2)z + k_1 + 0.16 = 0 \tag{6–122}$$

Since the desired characteristic equation is

$$z^2 - z + 0.5 = 0$$

by comparing Eq. (6–122) with this last equation, we obtain

$$k_1 = 0.34, \qquad k_2 = -2$$

or

$$\mathbf{K}_e = \begin{bmatrix} 0.34 \\ -2 \end{bmatrix}$$

Note that the dual relationship exists between the system considered in Example 6–6 and the present system. The state feedback gain matrix $\mathbf{K}$ obtained in Example 6–6 was $\mathbf{K} = [0.34 \quad -2]$. The observer feedback gain matrix $\mathbf{K}_e$ obtained here is related to matrix $\mathbf{K}$ by the relationship $\mathbf{K}_e = \mathbf{K}^*$. (Since matrices $\mathbf{K}$ and $\mathbf{K}_e$ are real, we may write the relationship as $\mathbf{K}_e = \mathbf{K}^T$.)

Design of prediction observers. We have thus far discussed full-order prediction observers. They are prediction observers because the estimate $\tilde{\mathbf{x}}(k+1)$ is one sampling period ahead of the measurement $\mathbf{y}(k)$. We solved a simple example problem by assuming the matrix $\mathbf{K}_e$ to exist and determined the characteristic equation $|z\mathbf{I} - \mathbf{G} + \mathbf{K}_e\mathbf{C}| = 0$ to have prescribed eigenvalues. In what follows we shall discuss a more general approach to determine the observer feedback gain matrix $\mathbf{K}_e$.

Consider the system defined by

$$\mathbf{x}(k+1) = \mathbf{G}\mathbf{x}(k) + \mathbf{H}\mathbf{u}(k) \tag{6–123}$$

$$y(k) = \mathbf{C}\mathbf{x}(k) \tag{6–124}$$

where

$\mathbf{x}(k) =$ state vector (n-vector)

$\mathbf{u}(k) =$ control vector (r-vector)

$y(k) =$ output signal (scalar)

$\mathbf{G} = n \times n$ nonsingular matrix

$\mathbf{H} = n \times r$ matrix

$\mathbf{C} = 1 \times n$ matrix

The system is assumed to be completely observable. Thus the inverse of

$$[\mathbf{C}^* \vdots \mathbf{G}^*\mathbf{C}^* \vdots \cdots \vdots (\mathbf{G}^*)^{n-1}\mathbf{C}^*]$$

exists. We also assume that the control law to be used is

$$\mathbf{u}(k) = -\mathbf{K}\tilde{\mathbf{x}}(k)$$

where $\tilde{\mathbf{x}}(k)$ is the observed state and $\mathbf{K}$ is an $r \times n$ matrix. Assume further that the system configuration is the same as that shown in Fig. 6–5.

The state observer dynamics are given by the equation

$$\begin{aligned} \tilde{\mathbf{x}}(k+1) &= \mathbf{G}\tilde{\mathbf{x}}(k) + \mathbf{H}\mathbf{u}(k) + \mathbf{K}_e[y(k) - \tilde{y}(k)] \\ &= (\mathbf{G} - \mathbf{K}_e\mathbf{C})\tilde{\mathbf{x}}(k) + \mathbf{H}\mathbf{u}(k) + \mathbf{K}_e\mathbf{C}\mathbf{x}(k) \end{aligned} \tag{6–125}$$

First define

$$\mathbf{Q} = (\mathbf{WN}^*)^{-1} \tag{6–126}$$

where

$$\mathbf{N} = [\mathbf{C}^* \vdots \mathbf{G}^*\mathbf{C}^* \vdots \cdots \vdots (\mathbf{G}^*)^{n-1}\mathbf{C}^*] \tag{6–127}$$

and

$$\mathbf{W} = \begin{bmatrix} a_{n-1} & a_{n-2} & \cdots & a_1 & 1 \\ a_{n-2} & a_{n-3} & \cdots & 1 & 0 \\ \cdot & \cdot & & \cdot & \cdot \\ \cdot & \cdot & & \cdot & \cdot \\ \cdot & \cdot & & \cdot & \cdot \\ a_1 & 1 & \cdots & 0 & 0 \\ 1 & 0 & \cdots & 0 & 0 \end{bmatrix} \tag{6–128}$$

where $a_1, a_2, \ldots, a_{n-1}$ are coefficients in the characteristic equation of the original state equation given by Eq. (6–123),

$$|z\mathbf{I} - \mathbf{G}| = z^n + a_1 z^{n-1} + \cdots + a_{n-1}z + a_n = 0$$

Next, define

$$\mathbf{x}(k) = \mathbf{Q}\boldsymbol{\xi}(k) \tag{6–129}$$

where $\boldsymbol{\xi}(k)$ is an n-vector. By use of Eq. (6–129), Eqs. (6–123) and (6–124) can be modified to read

$$\boldsymbol{\xi}(k+1) = \mathbf{Q}^{-1}\mathbf{GQ}\boldsymbol{\xi}(k) + \mathbf{Q}^{-1}\mathbf{Hu}(k) \tag{6–130}$$

$$y(k) = \mathbf{CQ}\boldsymbol{\xi}(k) \tag{6–131}$$

where

$$\mathbf{Q}^{-1}\mathbf{GQ} = \begin{bmatrix} 0 & 0 & \cdots & 0 & -a_n \\ 1 & 0 & \cdots & 0 & -a_{n-1} \\ \cdot & \cdot & & \cdot & \cdot \\ \cdot & \cdot & & \cdot & \cdot \\ \cdot & \cdot & & \cdot & \cdot \\ 0 & 0 & \cdots & 1 & -a_1 \end{bmatrix} \tag{6–132}$$

$$\mathbf{CQ} = [0 \quad 0 \quad \cdots \quad 0 \quad 1] \tag{6–133}$$

[Refer to Prob. A-6–12 for the derivations of Eqs. (6–132) and (6–133).]

Now define

$$\tilde{\mathbf{x}}(k) = \mathbf{Q}\tilde{\boldsymbol{\xi}}(k) \tag{6–134}$$

By substituting Eq. (6–134) into Eq. (6–125), we have

$$\tilde{\boldsymbol{\xi}}(k+1) = \mathbf{Q}^{-1}(\mathbf{G} - \mathbf{K}_e\mathbf{C})\mathbf{Q}\tilde{\boldsymbol{\xi}}(k) + \mathbf{Q}^{-1}\mathbf{H}\mathbf{u}(k) + \mathbf{Q}^{-1}\mathbf{K}_e\mathbf{C}\mathbf{Q}\boldsymbol{\xi}(k) \tag{6–135}$$

Subtracting Eq. (6–135) from Eq. (6–130), we obtain

$$\boldsymbol{\xi}(k+1) - \tilde{\boldsymbol{\xi}}(k+1) = (\mathbf{Q}^{-1}\mathbf{G}\mathbf{Q} - \mathbf{Q}^{-1}\mathbf{K}_e\mathbf{C}\mathbf{Q})\,[\boldsymbol{\xi}(k) - \tilde{\boldsymbol{\xi}}(k)] \tag{6–136}$$

Define

$$\mathbf{e}(k) = \boldsymbol{\xi}(k) - \tilde{\boldsymbol{\xi}}(k)$$

Then Eq. (6–136) becomes

$$\mathbf{e}(k+1) = \mathbf{Q}^{-1}(\mathbf{G} - \mathbf{K}_e\mathbf{C})\mathbf{Q}\mathbf{e}(k) \tag{6–137}$$

We require the error dynamics to be stable and $\mathbf{e}(k)$ to reach zero with sufficient speed. The procedure for determining matrix $\mathbf{K}_e$ is first to select the desired observer poles (the eigenvalues of $\mathbf{G} - \mathbf{K}_e\mathbf{C}$) and then to determine matrix $\mathbf{K}_e$ so that it will give the desired poles. If we require $\mathbf{e}(k)$ to reach zero as fast as possible, then we require the error response to be deadbeat, so that we must select all eigenvalues of $\mathbf{G} - \mathbf{K}_e\mathbf{C}$ to be zero.

Notice that

$$\mathbf{Q}^{-1}\mathbf{K}_e = \begin{bmatrix} a_{n-1} & a_{n-2} & \cdots & a_1 & 1 \\ a_{n-2} & a_{n-3} & \cdots & 1 & 0 \\ \cdot & \cdot & & \cdot & \cdot \\ \cdot & \cdot & & \cdot & \cdot \\ \cdot & \cdot & & \cdot & \cdot \\ a_1 & 1 & \cdots & 0 & 0 \\ 1 & 0 & \cdots & 0 & 0 \end{bmatrix} \begin{bmatrix} \mathbf{C} \\ \mathbf{CG} \\ \cdot \\ \cdot \\ \cdot \\ \mathbf{CG}^{n-2} \\ \mathbf{CG}^{n-1} \end{bmatrix} \begin{bmatrix} k_1 \\ k_2 \\ \cdot \\ \cdot \\ \cdot \\ k_{n-1} \\ k_n \end{bmatrix}$$

where

$$\mathbf{K}_e = \begin{bmatrix} k_1 \\ k_2 \\ \cdot \\ \cdot \\ \cdot \\ k_n \end{bmatrix}$$

Since $\mathbf{Q}^{-1}\mathbf{K}_e$ is an n-vector, let us write

$$\mathbf{Q}^{-1}\mathbf{K}_e = \begin{bmatrix} \delta_n \\ \delta_{n-1} \\ \cdot \\ \cdot \\ \cdot \\ \delta_1 \end{bmatrix} \tag{6–138}$$

Then, referring to Eq. (6–133), we have

$$\mathbf{Q}^{-1}\mathbf{K}_e\mathbf{C}\mathbf{Q} = \begin{bmatrix} \delta_n \\ \delta_{n-1} \\ \cdot \\ \cdot \\ \cdot \\ \delta_1 \end{bmatrix} [0 \quad 0 \quad \cdots \quad 1] = \begin{bmatrix} 0 & 0 & \cdots & 0 & \delta_n \\ 0 & 0 & \cdots & 0 & \delta_{n-1} \\ \cdot & \cdot & & \cdot & \cdot \\ \cdot & \cdot & & \cdot & \cdot \\ \cdot & \cdot & & \cdot & \cdot \\ 0 & 0 & \cdots & 0 & \delta_1 \end{bmatrix}$$

and

$$\mathbf{Q}^{-1}(\mathbf{G} - \mathbf{K}_e\mathbf{C})\mathbf{Q} = \mathbf{Q}^{-1}\mathbf{G}\mathbf{Q} - \mathbf{Q}^{-1}\mathbf{K}_e\mathbf{C}\mathbf{Q} = \begin{bmatrix} 0 & 0 & \cdots & 0 & -a_n - \delta_n \\ 1 & 0 & \cdots & 0 & -a_{n-1} - \delta_{n-1} \\ 0 & 1 & \cdots & 0 & -a_{n-2} - \delta_{n-2} \\ \cdot & \cdot & & \cdot & \cdot \\ \cdot & \cdot & & \cdot & \cdot \\ \cdot & \cdot & & \cdot & \cdot \\ 0 & 0 & \cdots & 1 & -a_1 - \delta_1 \end{bmatrix}$$

The characteristic equation

$$|z\mathbf{I} - \mathbf{Q}^{-1}(\mathbf{G} - \mathbf{K}_e\mathbf{C})\mathbf{Q}| = 0$$

becomes

$$\begin{vmatrix} z & 0 & 0 & \cdots & 0 & a_n + \delta_n \\ -1 & z & 0 & \cdots & 0 & a_{n-1} + \delta_{n-1} \\ 0 & -1 & z & \cdots & 0 & a_{n-2} + \delta_{n-2} \\ \cdot & \cdot & \cdot & & \cdot & \cdot \\ \cdot & \cdot & \cdot & & \cdot & \cdot \\ \cdot & \cdot & \cdot & & \cdot & \cdot \\ 0 & 0 & 0 & \cdots & -1 & z + a_1 + \delta_1 \end{vmatrix} = 0$$

or

$$z^n + (a_1 + \delta_1)z^{n-1} + (a_2 + \delta_2)z^{n-2} + \cdots + (a_n + \delta_n) = 0 \qquad (6\text{–}139)$$

It can be seen that each of $\delta_n, \delta_{n-1}, \ldots, \delta_1$ is associated with only one of the coefficients of the characteristic equation.

Suppose the desired characteristic equation for the error dynamics is

$$\begin{aligned} &(z - \mu_1)(z - \mu_2) \cdots (z - \mu_n) \\ &\quad = z^n + \alpha_1 z^{n-1} + \alpha_2 z^{n-2} + \cdots + \alpha_{n-1}z + \alpha_n = 0 \end{aligned} \qquad (6\text{–}140)$$

Note that the desired eigenvalues μ_i (or the locations of the desired closed-loop poles) determine how fast the observed state converges to the actual state of the plant. Comparing the coefficients of equal powers of z in Eqs. (6–139) and (6–140), we obtain

$$a_1 + \delta_1 = \alpha_1$$
$$a_2 + \delta_2 = \alpha_2$$
$$\vdots$$
$$a_n + \delta_n = \alpha_n$$

from which we get

$$\delta_1 = \alpha_1 - a_1$$
$$\delta_2 = \alpha_2 - a_2$$
$$\vdots$$
$$\delta_n = \alpha_n - a_n$$

Then from Eq. (6–138) we have

$$\mathbf{Q}^{-1}\mathbf{K}_e = \begin{bmatrix} \delta_n \\ \delta_{n-1} \\ \cdot \\ \cdot \\ \cdot \\ \delta_1 \end{bmatrix} = \begin{bmatrix} \alpha_n - a_n \\ \alpha_{n-1} - a_{n-1} \\ \cdot \\ \cdot \\ \cdot \\ \alpha_1 - a_1 \end{bmatrix} \tag{6–141}$$

Hence

$$\mathbf{K}_e = \mathbf{Q}\begin{bmatrix} \alpha_n - a_n \\ \alpha_{n-1} - a_{n-1} \\ \cdot \\ \cdot \\ \cdot \\ \alpha_1 - a_1 \end{bmatrix} = (\mathbf{WN^*})^{-1}\begin{bmatrix} \alpha_n - a_n \\ \alpha_{n-1} - a_{n-1} \\ \cdot \\ \cdot \\ \cdot \\ \alpha_1 - a_1 \end{bmatrix} \tag{6–142}$$

Equation (6–142) specifies the necessary observer feedback gain matrix $\mathbf{K}_e$. Figure 6–6 shows an alternate representation of the observed-state feedback control system.

Once we select the desired eigenvalues (or desired characteristic equation) the observer can be designed in a way similar to the method used in the case of the pole placement problem. The desired characteristic equation may be chosen so that the observer responds at least 4 or 5 times faster than the closed-loop system; or in some applications, deadbeat response may be desired.

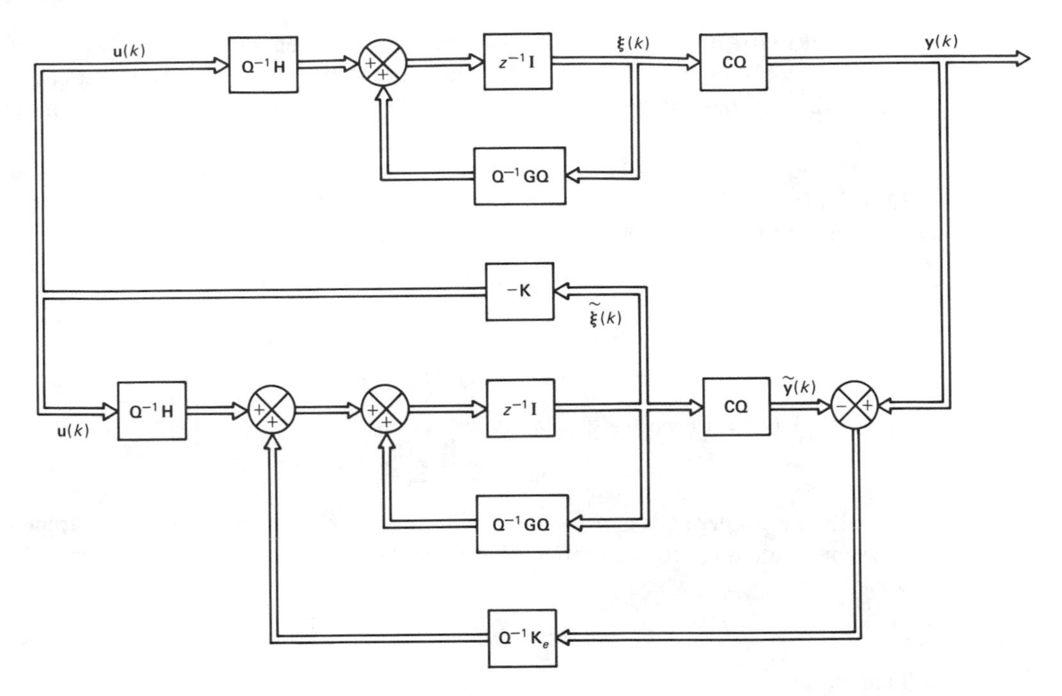

Figure 6–6 Alternate representation of the observed-state feedback control system.

If we wish to have deadbeat response, the desired characteristic equation becomes

$$z^n = 0 \tag{6–143}$$

Comparing Eq. (6–139) with Eq. (6–143), we require that

$$\begin{aligned} a_1 + \delta_1 &= 0 \\ a_2 + \delta_2 &= 0 \\ &\;\,\vdots \\ a_n + \delta_n &= 0 \end{aligned}$$

Hence, for the deadbeat response,

$$\mathbf{K}_e = \begin{bmatrix} k_1 \\ k_2 \\ \cdot \\ \cdot \\ \cdot \\ k_n \end{bmatrix} = \mathbf{Q} \begin{bmatrix} \delta_n \\ \delta_{n-1} \\ \cdot \\ \cdot \\ \cdot \\ \delta_1 \end{bmatrix} = (\mathbf{WN}^*)^{-1} \begin{bmatrix} -a_n \\ -a_{n-1} \\ \cdot \\ \cdot \\ \cdot \\ -a_1 \end{bmatrix} \tag{6–144}$$

Ackermann's formula. The expression given by Eq. (6–142) is not the only one commonly available for the determination of the observer feedback gain matrix $\mathbf{K}_e$. In what follows, we shall derive Ackermann's formula for the determination of $\mathbf{K}_e$.

Consider the completely observable system defined by Eqs. (6–123) and (6–124). Note that in this system the output $y(k)$ is a scalar. Referring to Eq. (6–137), the characteristic equation for the error dynamics is

$$|z\mathbf{I} - \mathbf{Q}^{-1}\mathbf{G}\mathbf{Q} + \mathbf{Q}^{-1}\mathbf{K}_e\mathbf{C}\mathbf{Q}| = 0 \tag{6–145}$$

where $\mathbf{K}_e$ is an $n \times 1$ matrix. Define

$$\mathbf{Q}^{-1}\mathbf{G}\mathbf{Q} = \hat{\mathbf{G}}, \qquad \mathbf{Q}^{-1}\mathbf{K}_e = \hat{\mathbf{K}}_e, \qquad \mathbf{C}\mathbf{Q} = \hat{\mathbf{C}}$$

Then, Eq. (6–145) becomes

$$|z\mathbf{I} - \hat{\mathbf{G}} + \hat{\mathbf{K}}_e\hat{\mathbf{C}}| = 0$$

In the observer design we determine matrix $\hat{\mathbf{K}}_e$ so that this last characteristic equation is identical to the desired characteristic equation for the error vector, which is

$$z^n + \alpha_1 z^{n-1} + \cdots + \alpha_{n-1} z + \alpha_n = 0 \tag{6–146}$$

That is,

$$\begin{aligned} |z\mathbf{I} - \hat{\mathbf{G}} + \hat{\mathbf{K}}_e\hat{\mathbf{C}}| &= (z - \mu_1)(z - \mu_2) \cdots (z - \mu_n) \\ &= z^n + \alpha_1 z^{n-1} + \cdots + \alpha_{n-1} z + \alpha_n = 0 \end{aligned}$$

where $\mu_1, \mu_2, \ldots, \mu_n$ are the eigenvalues of matrix $(\hat{\mathbf{G}} - \hat{\mathbf{K}}_e\hat{\mathbf{C}})$. For physical systems, complex eigenvalues always occur as conjugate complex pairs. In the present analysis we assume that all complex eigenvalues occur as conjugate complex pairs so that the coefficients $\alpha_1, \alpha_2, \ldots, \alpha_n$ of the characteristic equation are real. Then, the characteristic equation for the matrix $(\hat{\mathbf{G}}^* - \hat{\mathbf{C}}^*\hat{\mathbf{K}}_e^*)$ can be given by

$$\begin{aligned} |z\mathbf{I} - \hat{\mathbf{G}}^* + \hat{\mathbf{C}}^*\hat{\mathbf{K}}_e^*| &= (z - \bar{\mu}_1)(z - \bar{\mu}_2) \cdots (z - \bar{\mu}_n) \\ &= z^n + \alpha_1 z^{n-1} + \cdots + \alpha_{n-1} z + \alpha_n = 0 \end{aligned}$$

where $\bar{\mu}_i$ is the complex conjugate of μ_i.

In Sec. 6–5 we derived Ackermann's formula for the determination of the state feedback gain matrix $\mathbf{K}$ for the pole placement design problem. There, we determined matrix $\mathbf{K}$ so that the characteristic equation

$$|z\mathbf{I} - \mathbf{G} + \mathbf{H}\mathbf{K}| = 0$$

would be the same as the desired characteristic equation, Eq. (6–146). Here, in the observer design problem, we wish to determine matrix $\hat{\mathbf{K}}_e^*$ so that the characteristic equation

$$|z\mathbf{I} - \hat{\mathbf{G}}^* + \hat{\mathbf{C}}^*\hat{\mathbf{K}}_e^*| = 0$$

will be the same as the desired characteristic equation given by Eq. (6–146). Clearly, we can see that these two problems are a dual problem. (That is, mathematically, the determination of matrix $\mathbf{K}_e^*$ is the same as the determination of the feedback gain matrix $\mathbf{K}$ in the pole placement problem.) Therefore, it is possible to utilize the results obtained in Sec. 6–5 toward the determination of matrix $\mathbf{K}_e$ for the present problem, as will be shown. (For the direct derivation of $\mathbf{K}_e$, see Prob. A-6–22.)

In the pole placement design problem discussed in Sec. 6–5, for the system equation

$$\mathbf{x}(k+1) = \mathbf{G}\mathbf{x}(k) + \mathbf{H}u(k)$$

with state feedback

$$u(k) = -\mathbf{K}\mathbf{x}(k)$$

the desired matrix $\mathbf{K}$ was obtained as given by Eq. (6–68), repeated here:

$$\mathbf{K} = [0 \quad 0 \quad \cdots \quad 0 \quad 1][\mathbf{H} \vdots \mathbf{G}\mathbf{H} \vdots \cdots \vdots \mathbf{G}^{n-1}\mathbf{H}]^{-1}\phi(\mathbf{G}) \tag{6–147}$$

Here, in the observer design problem, for the state equation

$$\mathbf{x}(k+1) = \hat{\mathbf{G}}^*\mathbf{x}(k) + \hat{\mathbf{C}}^*u(k)$$

with state feedback

$$u(k) = -\hat{\mathbf{K}}_e^*\mathbf{x}(k)$$

the desired matrix $\hat{\mathbf{K}}_e^*$ can therefore be obtained in a form similar to Eq. (6–147), as follows:

$$\hat{\mathbf{K}}_e^* = [0 \quad 0 \quad \cdots \quad 0 \quad 1][\hat{\mathbf{C}}^* \vdots \hat{\mathbf{G}}^*\hat{\mathbf{C}}^* \vdots \cdots \vdots (\hat{\mathbf{G}}^*)^{n-1}\hat{\mathbf{C}}^*]^{-1}\phi(\hat{\mathbf{G}}^*) \tag{6–148}$$

which may be modified to

$$\hat{\mathbf{K}}_e^* = [0 \quad 0 \quad \cdots \quad 0 \quad 1][\mathbf{C}^* \vdots \mathbf{G}^*\mathbf{C}^* \vdots \cdots \vdots (\mathbf{G}^*)^{n-1}\mathbf{C}^*]^{-1}(\mathbf{Q}^*)^{-1}\phi(\hat{\mathbf{G}}^*)$$

By taking the conjugate transpose of both sides of this last equation, we have

$$\hat{\mathbf{K}}_e = [\phi(\hat{\mathbf{G}}^*)]^*\mathbf{Q}^{-1}\begin{bmatrix} \mathbf{C} \\ \mathbf{C}\mathbf{G} \\ \cdot \\ \cdot \\ \cdot \\ \mathbf{C}\mathbf{G}^{n-1} \end{bmatrix}^{-1}\begin{bmatrix} 0 \\ 0 \\ \cdot \\ \cdot \\ \cdot \\ 1 \end{bmatrix}$$

Noting that

$$\mathbf{Q}^{-1}\mathbf{K}_e = \hat{\mathbf{K}}_e$$

we obtain

$$\mathbf{K}_e = \mathbf{Q}\phi(\hat{\mathbf{G}})\mathbf{Q}^{-1}\begin{bmatrix}\mathbf{C}\\ \mathbf{CG}\\ \cdot\\ \cdot\\ \cdot\\ \mathbf{CG}^{n-1}\end{bmatrix}^{-1}\begin{bmatrix}0\\ 0\\ \cdot\\ \cdot\\ \cdot\\ 1\end{bmatrix} \tag{6–149}$$

Notice that since $\hat{\mathbf{G}} = \mathbf{Q}^{-1}\mathbf{GQ}$, we have

$$\mathbf{Q}\hat{\mathbf{G}}^k\mathbf{Q}^{-1} = \mathbf{G}^k \qquad k = 0, 1, 2, \ldots, n$$

Consequently,

$$\begin{aligned}\mathbf{Q}\phi(\hat{\mathbf{G}})\mathbf{Q}^{-1} &= \mathbf{Q}[\hat{\mathbf{G}}^n + \alpha_1\hat{\mathbf{G}}^{n-1} + \cdots + \alpha_{n-1}\hat{\mathbf{G}} + \alpha_n\mathbf{I}]\mathbf{Q}^{-1}\\ &= \mathbf{Q}\hat{\mathbf{G}}^n\mathbf{Q}^{-1} + \alpha_1\mathbf{Q}\hat{\mathbf{G}}^{n-1}\mathbf{Q}^{-1} + \cdots + \alpha_{n-1}\mathbf{Q}\hat{\mathbf{G}}\mathbf{Q}^{-1} + \alpha_n\mathbf{I}\\ &= \mathbf{G}^n + \alpha_1\mathbf{G}^{n-1} + \cdots + \alpha_{n-1}\mathbf{G} + \alpha_n\mathbf{I} = \phi(\mathbf{G})\end{aligned} \tag{6–150}$$

By use of Eq. (6–150), the desired observer feedback gain matrix $\mathbf{K}_e$, given by Eq. (6–149), can be rewritten as follows:

$$\mathbf{K}_e = \phi(\mathbf{G})\begin{bmatrix}\mathbf{C}\\ \mathbf{CG}\\ \cdot\\ \cdot\\ \cdot\\ \mathbf{CG}^{n-1}\end{bmatrix}^{-1}\begin{bmatrix}0\\ 0\\ \cdot\\ \cdot\\ \cdot\\ 1\end{bmatrix} \tag{6–151}$$

where $\phi(\mathbf{G})$ is the desired characteristic polynomial of the error dynamics. The expression for $\mathbf{K}_e$ given by Eq. (6–151) is commonly called Ackermann's formula for the determination of the observer feedback gain matrix $\mathbf{K}_e$.

Summary. The full-order prediction observer is given by Eq. (6–117):

$$\tilde{\mathbf{x}}(k+1) = (\mathbf{G} - \mathbf{K}_e\mathbf{C})\tilde{\mathbf{x}}(k) + \mathbf{Hu}(k) + \mathbf{K}_e\mathbf{y}(k)$$

The observed-state feedback is given by

$$\mathbf{u}(k) = -\mathbf{K}\tilde{\mathbf{x}}(k)$$

If this last equation is substituted into the observer equation, we obtain

$$\tilde{\mathbf{x}}(k+1) = (\mathbf{G} - \mathbf{K}_e\mathbf{C} - \mathbf{HK})\tilde{\mathbf{x}}(k) + \mathbf{K}_e\mathbf{y}(k)$$

This equation defines the full-order prediction observer when the observed-state feedback control is incorporated.

As in the case of the pole placement design, four methods are commonly available for the determination of the observer feedback gain matrix $\mathbf{K}_e$, summarized as follows:

1. Referring to Eq. (6–142), the observer feedback gain matrix $\mathbf{K}_e$ can be given by

$$\mathbf{K}_e = \mathbf{Q}\begin{bmatrix} \alpha_n - a_n \\ \alpha_{n-1} - a_{n-1} \\ \cdot \\ \cdot \\ \cdot \\ \alpha_1 - a_1 \end{bmatrix} = (\mathbf{WN}^*)^{-1}\begin{bmatrix} \alpha_n - a_n \\ \alpha_{n-1} - a_{n-1} \\ \cdot \\ \cdot \\ \cdot \\ \alpha_1 - a_1 \end{bmatrix} \qquad (6\text{–}152)$$

where matrices $\mathbf{N}$ and $\mathbf{W}$ are defined by Eqs. (6–127) and (6–128), respectively. The α_i's are the coefficients of the desired characteristic equation

$$z^n + \alpha_1 z^{n-1} + \cdot \cdot \cdot + \alpha_{n-1} z + \alpha_n = 0$$

and the a_i's are coefficients of the characteristic equation of the original state equation

$$|z\mathbf{I} - \mathbf{G}| = z^n + a_1 z^{n-1} + \cdot \cdot \cdot + a_{n-1} z + a_n = 0$$

Note that if the system is already in an observable canonical form, then the matrix $\mathbf{K}_e$ can be determined easily, because matrix $\mathbf{WN}^*$ becomes an identity matrix and thus $(\mathbf{WN}^*)^{-1} = \mathbf{I}$.

2. The observer feedback gain matrix $\mathbf{K}_e$ may be given by Ackermann's formula, given by Eq. (6–151):

$$\mathbf{K}_e = \phi(\mathbf{G})\begin{bmatrix} \mathbf{C} \\ \mathbf{CG} \\ \cdot \\ \cdot \\ \cdot \\ \mathbf{CG}^{n-1} \end{bmatrix}^{-1}\begin{bmatrix} 0 \\ 0 \\ \cdot \\ \cdot \\ \cdot \\ 1 \end{bmatrix} \qquad (6\text{–}153)$$

where

$$\phi(\mathbf{G}) = \mathbf{G}^n + \alpha_1 \mathbf{G}^{n-1} + \cdot \cdot \cdot + \alpha_{n-1}\mathbf{G} + \alpha_n \mathbf{I}$$

3. If the desired eigenvalues $\mu_1, \mu_2, \ldots, \mu_n$ of matrix $\mathbf{G} - \mathbf{K}_e\mathbf{C}$ are distinct, then the observer feedback gain matrix $\mathbf{K}_e$ may be given by the equation

$$\mathbf{K}_e = \begin{bmatrix} \boldsymbol{\eta}_1 \\ \boldsymbol{\eta}_2 \\ \cdot \\ \cdot \\ \cdot \\ \boldsymbol{\eta}_n \end{bmatrix}^{-1}\begin{bmatrix} 1 \\ 1 \\ \cdot \\ \cdot \\ \cdot \\ 1 \end{bmatrix} \qquad (6\text{–}154)$$

where the $\boldsymbol{\eta}_i$'s are defined as follows:

$$\boldsymbol{\eta}_i = \mathbf{C}(\mathbf{G} - \mu_i \mathbf{I})^{-1}$$

Note that the $\boldsymbol{\eta}_i^*$ are the eigenvectors of matrix $(\mathbf{G} - \mathbf{K}_e\mathbf{C})^*$.

In the special case where we desire the error vector to exhibit deadbeat response, so that $\mu_1 = \mu_2 = \cdots = \mu_n = 0$, Eq. (6–154) can be simplified. The following equation will give the matrix $\mathbf{K}_e$ for the deadbeat response:

$$\mathbf{K}_e = \begin{bmatrix} \boldsymbol{\eta}_1 \\ \boldsymbol{\eta}_2 \\ \cdot \\ \cdot \\ \cdot \\ \boldsymbol{\eta}_n \end{bmatrix}^{-1} \begin{bmatrix} 1 \\ 0 \\ \cdot \\ \cdot \\ \cdot \\ 0 \end{bmatrix} \tag{6–155}$$

where the $\boldsymbol{\eta}_i$'s are given by the equation

$$\boldsymbol{\eta}_i = \mathbf{C}\mathbf{G}^{-i} \qquad i = 1, 2, 3, \ldots, n$$

[For details of the derivations of Eqs. (6–154) and (6–155), see Probs. A-6–15 and A-6–16, which are dual problems.]

4. If the order of the system is low, assume an observer feedback gain matrix $\mathbf{K}_e$ with unknown elements. Then the elements of matrix $\mathbf{K}_e$ may be determined by equating the coefficients of like powers of z of

$$|z\mathbf{I} - \mathbf{G} + \mathbf{K}_e\mathbf{C}|$$

and of the desired characteristic polynomial, which is given by

$$(z - \mu_1)(z - \mu_2)\cdots(z - \mu_n) = z^n + \alpha_1 z^{n-1} + \cdots + \alpha_{n-1}z + \alpha_n$$

where the μ_i's are the desired eigenvalues of $\mathbf{G} - \mathbf{K}_e\mathbf{C}$. If the system is completely observable, the elements of $\mathbf{K}_e$ can be determined in this way.

Example 6–10.

Consider the double integrator system given by the equations

$$\mathbf{x}(k+1) = \mathbf{G}\mathbf{x}(k) + \mathbf{H}u(k)$$

$$y(k) = \mathbf{C}\mathbf{x}(k)$$

where

$$\mathbf{G} = \begin{bmatrix} 1 & T \\ 0 & 1 \end{bmatrix}, \qquad \mathbf{H} = \begin{bmatrix} T^2/2 \\ T \end{bmatrix}, \qquad \mathbf{C} = [1 \quad 0]$$

and T is the sampling period. (See Prob. A-5–29 for the derivation of the discrete-time state space equations for the double integrator system.) Assuming that the observer configuration is the same as that shown in Fig. 6–5, design a state observer for this system. It is desired that the error vector exhibit deadbeat response. Use the four different methods listed in the foregoing discussion.

First we check the observability condition. Notice that the rank of

$$[\mathbf{C}^* \vdots \mathbf{G}^*\mathbf{C}^*] = \begin{bmatrix} 1 & 1 \\ 0 & T \end{bmatrix}$$

is 2. Hence, the system is completely observable. Next, we examine the characteristic equation for the system:

$$|z\mathbf{I}-\mathbf{G}| = \left|\begin{bmatrix} z & 0 \\ 0 & z \end{bmatrix} - \begin{bmatrix} 1 & T \\ 0 & 1 \end{bmatrix}\right| = \begin{vmatrix} z-1 & -T \\ 0 & z-1 \end{vmatrix} = z^2 - 2z + 1 = 0$$

Comparing this characteristic equation with

$$z^2 + a_1 z + a_2 = 0$$

we obtain

$$a_1 = -2, \qquad a_2 = 1$$

Since the deadbeat response is desired, the desired characteristic equation for the error dynamics is

$$z^2 + \alpha_1 z + \alpha_2 = z^2 = 0$$

Thus,

$$\alpha_1 = 0, \qquad \alpha_2 = 0$$

Method 1. Referring to Eq. (6–152), we have

$$\mathbf{K}_e = (\mathbf{WN}^*)^{-1}\begin{bmatrix} \alpha_2 - a_2 \\ \alpha_1 - a_1 \end{bmatrix} = (\mathbf{WN}^*)^{-1}\begin{bmatrix} -1 \\ 2 \end{bmatrix}$$

where **N** and **W**, defined by Eqs. (6–127) and (6–128), respectively, are

$$\mathbf{N} = [\mathbf{C}^* \vdots \mathbf{G}^*\mathbf{C}^*] = \begin{bmatrix} 1 & 1 \\ 0 & T \end{bmatrix}$$

and

$$\mathbf{W} = \begin{bmatrix} a_1 & 1 \\ 1 & 0 \end{bmatrix} = \begin{bmatrix} -2 & 1 \\ 1 & 0 \end{bmatrix}$$

Hence, the observer feedback gain matrix $\mathbf{K}_e$ is obtained as follows:

$$\mathbf{K}_e = \left\{\begin{bmatrix} -2 & 1 \\ 1 & 0 \end{bmatrix}\begin{bmatrix} 1 & 0 \\ 1 & T \end{bmatrix}\right\}^{-1}\begin{bmatrix} -1 \\ 2 \end{bmatrix} = \begin{bmatrix} 0 & 1 \\ \frac{1}{T} & \frac{1}{T} \end{bmatrix}\begin{bmatrix} -1 \\ 2 \end{bmatrix} = \begin{bmatrix} 2 \\ \frac{1}{T} \end{bmatrix}$$

Method 2. From Eq. (6–153), Ackermann's formula, $\mathbf{K}_e$ is given by

$$\mathbf{K}_e = \phi(\mathbf{G})\begin{bmatrix} \mathbf{C} \\ \mathbf{CG} \end{bmatrix}^{-1}\begin{bmatrix} 0 \\ 1 \end{bmatrix}$$

where

$$\phi(\mathbf{G}) = \mathbf{G}^2 + \alpha_1\mathbf{G} + \alpha_2\mathbf{I} = \mathbf{G}^2$$

Hence the observer feedback gain matrix $\mathbf{K}_e$ is obtained as follows:

$$\mathbf{K}_e = \begin{bmatrix} 1 & T \\ 0 & 1 \end{bmatrix}^2\begin{bmatrix} 1 & 0 \\ 1 & T \end{bmatrix}^{-1}\begin{bmatrix} 0 \\ 1 \end{bmatrix}$$

$$= \begin{bmatrix} 1 & 2T \\ 0 & 1 \end{bmatrix}\begin{bmatrix} 1 & 0 \\ -\frac{1}{T} & \frac{1}{T} \end{bmatrix}\begin{bmatrix} 0 \\ 1 \end{bmatrix} = \begin{bmatrix} 2 \\ \frac{1}{T} \end{bmatrix}$$

Method 3. Since deadbeat response is desired, from Eq. (6–155) we have

$$\mathbf{K}_e = \begin{bmatrix} \boldsymbol{\eta}_1 \\ \boldsymbol{\eta}_2 \end{bmatrix}^{-1} \begin{bmatrix} 1 \\ 0 \end{bmatrix}$$

where

$$\boldsymbol{\eta}_1 = \mathbf{CG}^{-1}, \qquad \boldsymbol{\eta}_2 = \mathbf{CG}^{-2}$$

Notice that

$$\mathbf{G}^{-1} = \begin{bmatrix} 1 & -T \\ 0 & 1 \end{bmatrix}$$

The vectors $\boldsymbol{\eta}_1$ and $\boldsymbol{\eta}_2$ are obtained as follows:

$$\boldsymbol{\eta}_1 = [1 \quad 0] \begin{bmatrix} 1 & -T \\ 0 & 1 \end{bmatrix} = [1 \quad -T]$$

$$\boldsymbol{\eta}_2 = [1 \quad 0] \begin{bmatrix} 1 & -T \\ 0 & 1 \end{bmatrix}^2 = [1 \quad 0] \begin{bmatrix} 1 & -2T \\ 0 & 1 \end{bmatrix} = [1 \quad -2T]$$

Thus, the observer feedback gain matrix $\mathbf{K}_e$ is obtained as follows:

$$\mathbf{K}_e = \begin{bmatrix} 1 & -T \\ 1 & -2T \end{bmatrix}^{-1} \begin{bmatrix} 1 \\ 0 \end{bmatrix} = \begin{bmatrix} 2 & -1 \\ \frac{1}{T} & -\frac{1}{T} \end{bmatrix} \begin{bmatrix} 1 \\ 0 \end{bmatrix} = \begin{bmatrix} 2 \\ \frac{1}{T} \end{bmatrix}$$

Method 4. We first assume

$$\mathbf{K}_e = \begin{bmatrix} k_1 \\ k_2 \end{bmatrix}$$

and expand the characteristic equation as follows:

$$|z\mathbf{I} - \mathbf{G} + \mathbf{K}_e\mathbf{C}| = \left| z\begin{bmatrix} 1 & 0 \\ 0 & 1 \end{bmatrix} - \begin{bmatrix} 1 & T \\ 0 & 1 \end{bmatrix} + \begin{bmatrix} k_1 \\ k_2 \end{bmatrix} [1 \quad 0] \right|$$

$$= \begin{vmatrix} z - 1 + k_1 & -T \\ k_2 & z - 1 \end{vmatrix} = z^2 + (k_1 - 2)z + 1 - k_1 + k_2 T = 0$$

Since we desire the deadbeat response, this characteristic equation must be equal to

$$z^2 = 0$$

Thus

$$k_1 = 2, \qquad k_2 = \frac{1}{T}$$

or

$$\mathbf{K}_e = \begin{bmatrix} k_1 \\ k_2 \end{bmatrix} = \begin{bmatrix} 2 \\ 1/T \end{bmatrix}$$

Let us verify that the error vector reduces to zero in at most 2 sampling periods. Note that the coefficient matrix for the error equation becomes

$$\mathbf{G}-\mathbf{K}_e\mathbf{C}=\begin{bmatrix}1 & T\\ 0 & 1\end{bmatrix}-\begin{bmatrix}2\\ \frac{1}{T}\end{bmatrix}[1 \quad 0]=\begin{bmatrix}-1 & T\\ -\frac{1}{T} & 1\end{bmatrix}$$

If the initial state $\mathbf{x}(0)$ is given as

$$\mathbf{x}(0)=\begin{bmatrix}a_1\\ b_1\end{bmatrix}$$

where a_1 and b_1 are arbitrary and $\tilde{\mathbf{x}}(0)$ is assumed as

$$\tilde{\mathbf{x}}(0)=\begin{bmatrix}a_2\\ b_2\end{bmatrix}$$

where a_2 and b_2 are arbitrary, then

$$\mathbf{e}(0)=\mathbf{x}(0)-\tilde{\mathbf{x}}(0)=\begin{bmatrix}a_1-a_2\\ b_1-b_2\end{bmatrix}=\begin{bmatrix}a\\ b\end{bmatrix}$$

where a and b are arbitrary constants. Now Eq. (6–120) becomes

$$\begin{bmatrix}e_1(k+1)\\ e_2(k+1)\end{bmatrix}=\begin{bmatrix}-1 & T\\ -\frac{1}{T} & 1\end{bmatrix}\begin{bmatrix}e_1(k)\\ e_2(k)\end{bmatrix}, \qquad \begin{bmatrix}e_1(0)\\ e_2(0)\end{bmatrix}=\begin{bmatrix}a\\ b\end{bmatrix}$$

The vectors $\mathbf{e}(1)$ and $\mathbf{e}(2)$ are found as follows:

$$\begin{bmatrix}e_1(1)\\ e_2(1)\end{bmatrix}=\begin{bmatrix}-1 & T\\ -\frac{1}{T} & 1\end{bmatrix}\begin{bmatrix}a\\ b\end{bmatrix}=\begin{bmatrix}-a+bT\\ -\frac{1}{T}a+b\end{bmatrix}$$

and

$$\begin{bmatrix}e_1(2)\\ e_2(2)\end{bmatrix}=\begin{bmatrix}-1 & T\\ -\frac{1}{T} & 1\end{bmatrix}\begin{bmatrix}-a+bT\\ -\frac{1}{T}a+b\end{bmatrix}=\begin{bmatrix}a-bT-a+bT\\ \frac{1}{T}a-b-\frac{1}{T}a+b\end{bmatrix}=\begin{bmatrix}0\\ 0\end{bmatrix}$$

Clearly, the error vector $\mathbf{e}(k)$ becomes zero in at most 2 sampling periods. Thus the response is deadbeat. Note that for any initial state $\mathbf{x}(0)$ the observed state vector becomes identical to the actual state vector in at most 2 sampling periods.

Finally, the observer equation is

$$\begin{bmatrix}\tilde{x}_1(k+1)\\ \tilde{x}_2(k+1)\end{bmatrix}=\begin{bmatrix}-1 & T\\ -\frac{1}{T} & 1\end{bmatrix}\begin{bmatrix}\tilde{x}_1(k)\\ \tilde{x}_2(k)\end{bmatrix}+\begin{bmatrix}\frac{1}{2}T^2\\ T\end{bmatrix}u(k)+\begin{bmatrix}2\\ \frac{1}{T}\end{bmatrix}y(k)$$

[Note that this is the equation given by Eq. (6–117).]

Comments on selecting the best $\mathbf{K}_e$. Referring to Fig. 6–5, notice that the feedback signal through the observer feedback gain matrix $\mathbf{K}_e$ serves as a correction signal to the plant model to account for the unknowns in the plant. If significant unknowns are involved, the feedback signal through the matrix $\mathbf{K}_e$ should be relatively large. However, if the output signal is contaminated significantly by disturbances and measurement noises, then the output $\mathbf{y}(k)$ is not reliable and the feedback signal through the matrix $\mathbf{K}_e$ should be relatively small. In determining the matrix $\mathbf{K}_e$ (which depends on the desired eigenvalues $\mu_1, \mu_2, \ldots, \mu_n$), we should carefully examine the effects of disturbances and noises involved in the output $\mathbf{y}(k)$.

Remember that the observer feedback gain matrix $\mathbf{K}_e$ depends on the desired characteristic equation

$$\phi(z) = (z - \mu_1)(z - \mu_2)\cdots(z - \mu_n) = 0$$

The choice of a set of $\mu_1, \mu_2, \ldots, \mu_n$ is, in many instances, not unique. Hence, many different characteristic equations might be chosen as desired characteristic equations. For each desired characteristic equation, we have a different matrix $\mathbf{K}_e$.

In the design of the observer, it is desirable to determine several observer feedback gain matrices $\mathbf{K}_e$ based on several different desired characteristic equations. For each of the several different matrices $\mathbf{K}_e$, simulation tests must be run to evaluate the resulting system performance. Then we select the best $\mathbf{K}_e$ from the viewpoint of overall system performance. In many practical cases the selection of the best matrix $\mathbf{K}_e$ boils down to a compromise between speedy response and sensitivity to disturbances and noises.

Design of a full-order state observer when y(k) is an m vector. Consider the completely observable system given by the equations

$$\mathbf{x}(k+1) = \mathbf{G}\mathbf{x}(k) + \mathbf{H}\mathbf{u}(k)$$

$$\mathbf{y}(k) = \mathbf{C}\mathbf{x}(k)$$

where

$$\mathbf{x}(k) = \text{state vector } (n\text{-vector})$$

$$\mathbf{u}(k) = \text{control vector } (r\text{-vector})$$

$$\mathbf{y}(k) = \text{output vector } (m\text{-vector})$$

$$\mathbf{G} = n \times n \text{ matrix}$$

$$\mathbf{H} = n \times r \text{ matrix}$$

$$\mathbf{C} = m \times n \text{ matrix}$$

The observer state equation can be given by Eq. (6–117):

$$\tilde{\mathbf{x}}(k+1) = (\mathbf{G} - \mathbf{K}_e\mathbf{C})\tilde{\mathbf{x}}(k) + \mathbf{H}\mathbf{u}(k) + \mathbf{K}_e\mathbf{y}(k)$$

The error dynamics are described by Eq. (6–120):

$$\mathbf{e}(k+1) = (\mathbf{G} - \mathbf{K}_e\mathbf{C})\mathbf{e}(k)$$

In the observer design problem we determine the $n \times m$ matrix $\mathbf{K}_e$ so that the eigenvalues of $\mathbf{G} - \mathbf{K}_e\mathbf{C}$ take the desired values $\mu_1, \mu_2, \ldots, \mu_n$ and the characteristic equation is

$$|z\mathbf{I} - \mathbf{G} + \mathbf{K}_e\mathbf{C}| = (z - \mu_1)(z - \mu_2) \cdots (z - \mu_n)$$
$$= z^n + \alpha_1 z^{n-1} + \cdots + \alpha_{n-1} z + \alpha_n = 0 \tag{6–156}$$

As stated earlier, in connection with Ackermann's formula for determining matrix $\mathbf{K}_e$, in physical systems all complex eigenvalues of matrix $(\mathbf{G} - \mathbf{K}_e\mathbf{C})$ occur as conjugate complex pairs. Consequently, Eq. (6–156) is equivalent to

$$|z\mathbf{I} - \mathbf{G}^* + \mathbf{C}^*\mathbf{K}_e^*| = z^n + \alpha_1 z^{n-1} + \cdots + \alpha_{n-1} z + \alpha_n = 0$$

In Sec. 6–5 we discussed the pole placement design problem where the control $\mathbf{u}(k)$ was an r-vector and we determined a matrix $\mathbf{K}$ such that the characteristic equation

$$|z\mathbf{I} - \mathbf{G} + \mathbf{HK}| = 0$$

was identical to the desired characteristic equation

$$z^n + \alpha_1 z^{n-1} + \cdots + \alpha_{n-1} z + \alpha_n = 0 \tag{6–157}$$

[Remember that for the case where $\mathbf{u}(k)$ was an r-vector, matrix $\mathbf{K}$ was found not to be unique.] In the observer design problem we determine matrix $\mathbf{K}_e^*$ so that the characteristic equation

$$|z\mathbf{I} - \mathbf{G}^* + \mathbf{C}^*\mathbf{K}_e^*| = 0$$

is identical to the desired characteristic equation given by Eq. (6–157). (Here, matrix $\mathbf{K}_e^*$ is not unique.) Mathematically, the determination of matrix $\mathbf{K}_e^*$ is the same as the determination of the feedback gain matrix $\mathbf{K}$ in the pole placement process. Thus, these two problems are dual. Since we discussed in detail the pole placement problem when $\mathbf{u}(k)$ was an r-vector, we shall not give detailed discussions of this dual problem. Instead, we refer the reader to Prob. A-6–25, where an example problem for the case where $\mathbf{y}(k)$ is an m-vector is treated.

Effects of the addition of the observer on a closed-loop system. In the pole placement design process we assumed that the true state $\mathbf{x}(k)$ was available for feedback. But in practice the true state $\mathbf{x}(k)$ may not be measurable, so that we will need to use the observed state $\tilde{\mathbf{x}}(k)$. Let us now investigate the effects of the use of the observed state $\tilde{\mathbf{x}}(k)$ rather than the true state $\mathbf{x}(k)$ upon the characteristic equation of a closed-loop control system.

Consider the completely state controllable and completely observable system defined by the equations

$$\mathbf{x}(k+1) = \mathbf{Gx}(k) + \mathbf{Hu}(k)$$
$$\mathbf{y}(k) = \mathbf{Cx}(k)$$

For the state feedback control based on the observed state $\tilde{\mathbf{x}}(k)$ we have

$$\mathbf{u}(k) = -\mathbf{K}\tilde{\mathbf{x}}(k)$$

With this control the state equation becomes

$$\mathbf{x}(k+1) = \mathbf{G}\mathbf{x}(k) - \mathbf{H}\mathbf{K}\tilde{\mathbf{x}}(k) = (\mathbf{G} - \mathbf{H}\mathbf{K})\mathbf{x}(k) + \mathbf{H}\mathbf{K}[\mathbf{x}(k) - \tilde{\mathbf{x}}(k)] \qquad (6\text{–}158)$$

The difference between the actual state $\mathbf{x}(k)$ and the observed state $\tilde{\mathbf{x}}(k)$ has been defined as the error $\mathbf{e}(k)$:

$$\mathbf{e}(k) = \mathbf{x}(k) - \tilde{\mathbf{x}}(k)$$

By substitution of the error vector $\mathbf{e}(k)$, Eq. (6–158) becomes

$$\mathbf{x}(k+1) = (\mathbf{G} - \mathbf{H}\mathbf{K})\mathbf{x}(k) + \mathbf{H}\mathbf{K}\mathbf{e}(k) \qquad (6\text{–}159)$$

Note that the observer error equation was given by Eq. (6–120), repeated here:

$$\mathbf{e}(k+1) = (\mathbf{G} - \mathbf{K}_e\mathbf{C})\mathbf{e}(k) \qquad (6\text{–}160)$$

Combining Eqs. (6–159) and (6–160), we obtain

$$\begin{bmatrix} \mathbf{x}(k+1) \\ \mathbf{e}(k+1) \end{bmatrix} = \begin{bmatrix} \mathbf{G} - \mathbf{H}\mathbf{K} & \mathbf{H}\mathbf{K} \\ \mathbf{0} & \mathbf{G} - \mathbf{K}_e\mathbf{C} \end{bmatrix} \begin{bmatrix} \mathbf{x}(k) \\ \mathbf{e}(k) \end{bmatrix} \qquad (6\text{–}161)$$

Equation (6–161) describes the dynamics of the observed-state feedback control system. The characteristic equation for the system is

$$\begin{vmatrix} z\mathbf{I} - \mathbf{G} + \mathbf{H}\mathbf{K} & -\mathbf{H}\mathbf{K} \\ \mathbf{0} & z\mathbf{I} - \mathbf{G} + \mathbf{K}_e\mathbf{C} \end{vmatrix} = 0$$

or

$$|z\mathbf{I} - \mathbf{G} + \mathbf{H}\mathbf{K}||z\mathbf{I} - \mathbf{G} + \mathbf{K}_e\mathbf{C}| = 0$$

Notice that the closed-loop poles of the observed-state feedback control system consist of the poles due to the pole placement design alone plus the poles due to the observer design alone. This means that the pole placement design and the observer design are independent of each other. They can be designed separately and combined together to form the observed-state feedback control system.

The desired closed-loop poles to be generated by state feedback (pole placement) are chosen in such a way that the system satisfies the performance requirements. The poles of the observer are usually chosen so that the observer response is much faster than the system response. A rule of thumb is to choose an observer response at least 4 to 5 times faster than the system response, or in some cases to choose all observer poles at the origin (for deadbeat response). Since the observer is, in general, not a hardware structure but is programmed on the computer, it is possible to increase the response speed or achieve deadbeat response so that the observed state quickly converges to the true state. The maximum response speed of the observer is generally limited only by noise and sensitivity problems involved in the control system.

Current observer. In the prediction observer the observed state $\tilde{\mathbf{x}}(k)$ is obtained from measurements of the output vector up to $\mathbf{y}(k-1)$ and of the control vector up to $\mathbf{u}(k-1)$. Hence the control vector $\mathbf{u}(k) = -\mathbf{K}\tilde{\mathbf{x}}(k)$ does not utilize the information on the current output $\mathbf{y}(k)$. A different formulation of the state observer is to use $\mathbf{y}(k)$ for the estimation of $\tilde{\mathbf{x}}(k)$. This can be done by separating the observation process into two steps. In the first step we determine $\mathbf{z}(k+1)$, an approximation of $\mathbf{x}(k+1)$ based on $\tilde{\mathbf{x}}(k)$ and $\mathbf{u}(k)$. In the second step, we use $\mathbf{y}(k+1)$ to improve $\mathbf{z}(k+1)$. The improved $\mathbf{z}(k+1)$ is $\tilde{\mathbf{x}}(k+1)$. The state observer based on this formulation is called the *current observer*.

Consider the system defined by the equations

$$\mathbf{x}(k+1) = \mathbf{G}\mathbf{x}(k) + \mathbf{H}\mathbf{u}(k)$$

$$\mathbf{y}(k) = \mathbf{C}\mathbf{x}(k)$$

where

$$\mathbf{x}(k) = \text{state vector } (n\text{-vector})$$

$$\mathbf{u}(k) = \text{control vector } (r\text{-vector})$$

$$\mathbf{y}(k) = \text{output vector } (m\text{-vector})$$

$$\mathbf{G} = n \times n \text{ matrix}$$

$$\mathbf{H} = n \times r \text{ matrix}$$

$$\mathbf{C} = m \times n \text{ matrix}$$

The current observer equations are given by

$$\tilde{\mathbf{x}}(k+1) = \mathbf{z}(k+1) + \mathbf{K}_e[\mathbf{y}(k+1) - \mathbf{C}\mathbf{z}(k+1)] \tag{6–162}$$

$$\mathbf{z}(k+1) = \mathbf{G}\tilde{\mathbf{x}}(k) + \mathbf{H}\mathbf{u}(k) \tag{6–163}$$

Equation (6–163) gives the prediction $\mathbf{z}(k+1)$ based on $\tilde{\mathbf{x}}(k)$ and $\mathbf{u}(k)$ at stage k. Equation (6–162) states that by measuring $\mathbf{y}(k+1)$ we can improve $\mathbf{z}(k+1)$ to obtain $\tilde{\mathbf{x}}(k+1)$.

Define the observer error $\mathbf{e}(k)$ as follows:

$$\mathbf{e}(k) = \mathbf{x}(k) - \tilde{\mathbf{x}}(k)$$

Then

$$\begin{aligned}
\mathbf{e}(k+1) &= \mathbf{x}(k+1) - \tilde{\mathbf{x}}(k+1) \\
&= \mathbf{G}\mathbf{x}(k) + \mathbf{H}\mathbf{u}(k) - \Big(\mathbf{G}\tilde{\mathbf{x}}(k) + \mathbf{H}\mathbf{u}(k) + \mathbf{K}_e\{\mathbf{C}[\mathbf{G}\mathbf{x}(k) + \mathbf{H}\mathbf{u}(k)] \\
&\quad - \mathbf{C}[\mathbf{G}\tilde{\mathbf{x}}(k) + \mathbf{H}\mathbf{u}(k)]\}\Big) \\
&= (\mathbf{G} - \mathbf{K}_e\mathbf{C}\mathbf{G})[\mathbf{x}(k) - \tilde{\mathbf{x}}(k)] \\
&= (\mathbf{G} - \mathbf{K}_e\mathbf{C}\mathbf{G})\mathbf{e}(k)
\end{aligned}$$

Thus the error vector equation for the current observer is similar to that for the prediction observer given by Eq. (6–120). However, a difference appears in the error dynamics. The matrix $\mathbf{K}_e$ can be obtained exactly as in the case of the prediction observer except that matrix $\mathbf{C}$ is replaced by matrix $\mathbf{CG}$. To make it possible for the eigenvalues of $(\mathbf{G} - \mathbf{K}_e\mathbf{CG})$ to be arbitrarily placed, the rank of the matrix

$$\begin{bmatrix} \mathbf{CG} \\ \mathbf{CG}^2 \\ \cdot \\ \cdot \\ \cdot \\ \mathbf{CG}^n \end{bmatrix} = \begin{bmatrix} \mathbf{C} \\ \mathbf{CG} \\ \cdot \\ \cdot \\ \cdot \\ \mathbf{CG}^{n-1} \end{bmatrix} \mathbf{G}$$

must be n. Notice that if matrix $\mathbf{G}$ is nonsingular, then this condition is equivalent to the observability condition, or

$$\text{rank}\,[\mathbf{C}^* \vdots \mathbf{G}^*\mathbf{C}^* \vdots \cdots \vdots (\mathbf{G}^*)^{n-1}\mathbf{C}^*] = n$$

If the rank of the observability matrix is n, then the eigenvalues of $\mathbf{G} - \mathbf{K}_e\mathbf{CG}$ can be arbitrarily located by a proper choice of $\mathbf{K}_e$ and matrix $\mathbf{K}_e$ can be determined in a way similar to what was done in the case of the prediction observer. In determining matrix $\mathbf{K}_e$, we replace matrix $\mathbf{C}$ by $\mathbf{CG}$ in the computations involved. For example, if the output $\mathbf{y}(k)$ is a scalar, then Ackermann's formula as given by Eq. (6–151) is altered to the corresponding form

$$\mathbf{K}_e = \phi(\mathbf{G}) \begin{bmatrix} \mathbf{CG} \\ \mathbf{CG}^2 \\ \cdot \\ \cdot \\ \cdot \\ \mathbf{CG}^{n-1} \\ \mathbf{CG}^n \end{bmatrix}^{-1} \begin{bmatrix} 0 \\ 0 \\ \cdot \\ \cdot \\ \cdot \\ 0 \\ 1 \end{bmatrix} \tag{6–164}$$

However, if matrix $\mathbf{G}$ is singular, then the rank of

$$[\mathbf{G}^*\mathbf{C}^* \vdots (\mathbf{G}^*)^2\mathbf{C}^* \vdots \cdots \vdots (\mathbf{G}^*)^n\mathbf{C}^*]$$

is q, which is less than n. In this case, let us write

$$(\mathbf{G} - \mathbf{K}_e\mathbf{CG})^* = \mathbf{G}^* - \mathbf{G}^*\mathbf{C}^*\mathbf{K}_e^* = \mathbf{G}^* - \mathbf{BK}_e^*$$

where $\mathbf{B} = \mathbf{G}^*\mathbf{C}^*$. Note that matrix $(\mathbf{G}^* - \mathbf{BK}_e^*)$ is of the same form as matrix $(\mathbf{G} - \mathbf{HK})$, which played an important role in the pole placement design.

With an analysis similar to that given in Sec. 6–5 [refer to Eqs. (6–55) and (6–57)] it is possible by use of a suitable transformation matrix $\mathbf{T}$ to transform matrices $\mathbf{G}$ and $\mathbf{B}$ into $\hat{\mathbf{G}}^*$ and $\hat{\mathbf{B}}$, where

$$\hat{\mathbf{G}}^* = \mathbf{T}^{-1}\mathbf{G}^*\mathbf{T} = \left[\begin{array}{c|c} \mathbf{G}_{11}^* & \mathbf{G}_{12}^* \\ \hline \mathbf{0} & \mathbf{G}_{22}^* \end{array}\right], \qquad \hat{\mathbf{B}} = \mathbf{T}^{-1}\mathbf{B} = \left[\begin{array}{c} \mathbf{B}_{11} \\ \hline \mathbf{0} \end{array}\right]$$

and where all eigenvalues of the uncontrollable $(n - q) \times (n - q)$ matrix $\mathbf{G}_{22}^*$ can be made zero. (Hence, the system can be stabilized.) Next, define

$$\mathbf{K}_e^*\mathbf{T} = \hat{\mathbf{K}}_e^* = [\hat{\mathbf{K}}_{e\,11}^* \vdots \hat{\mathbf{K}}_{e\,12}^*]$$

then

$$\hat{\mathbf{G}}^* - \hat{\mathbf{B}}\hat{\mathbf{K}}_e^* = \left[\begin{array}{c|c} \mathbf{G}_{11}^* & \mathbf{G}_{12}^* \\ \hline \mathbf{0} & \mathbf{G}_{22}^* \end{array}\right] - \left[\begin{array}{c} \mathbf{B}_{11} \\ \hline \mathbf{0} \end{array}\right][\hat{\mathbf{K}}_{e\,11}^* \vdots \hat{\mathbf{K}}_{e\,12}^*]$$

$$= \left[\begin{array}{c|c} \mathbf{G}_{11}^* - \mathbf{B}_{11}\hat{\mathbf{K}}_{e\,11}^* & \mathbf{G}_{12}^* - \mathbf{B}_{11}\hat{\mathbf{K}}_{e\,12}^* \\ \hline \mathbf{0} & \mathbf{G}_{22}^* \end{array}\right]$$

Hence, if matrix $\mathbf{G}$ is singular and the rank of the observability matrix is q, then we need to specify only q eigenvalues of the $q \times q$ matrix $\mathbf{G}_{11}^* - \mathbf{B}_{11}\hat{\mathbf{K}}_{e\,11}^*$.

Minimum-order observer. The observers discussed thus far are designed to reconstruct all the state variables. In practice, some of the state variables may be accurately measured. Such accurately measurable state variables need not be estimated. An observer that estimates fewer than n state variables, where n is the dimension of the state vector, is called a *reduced-order observer*. If the order of the reduced-order observer is the minimum possible, the observer is called a *minimum-order observer*.

Suppose the state vector $\mathbf{x}(k)$ is an n-vector and the output vector $\mathbf{y}(k)$ is an m-vector which can be measured. Since m output variables are linear combinations of the state variables, m state variables need not be estimated. We need to estimate only $n - m$ state variables. Then the reduced-order observer becomes an $(n - m)$th-order observer. Such an $(n - m)$th-order observer is the minimum-order observer. Figure 6–7 shows the block diagram of a system with a minimum-order observer.

It is important to note, however, that if the measurement of output variables involves significant noises and is relatively inaccurate, then the use of the full-order observer may result in a better system performance.

The minimum-order observer can be designed by first partitioning the state vector $\mathbf{x}(k)$ into two parts, as follows:

$$\mathbf{x}(k) = \left[\begin{array}{c} \mathbf{x}_a(k) \\ \hline \mathbf{x}_b(k) \end{array}\right]$$

where $\mathbf{x}_a(k)$ is that portion of the state vector that can be directly measured [thus $\mathbf{x}_a(k)$ is an m-vector] and $\mathbf{x}_b(k)$ is the unmeasurable portion of the state vector [thus $\mathbf{x}_b(k)$ is an $(n - m)$-vector]. Then the partitioned state equations become as follows:

$$\left[\begin{array}{c} \mathbf{x}_a(k+1) \\ \hline \mathbf{x}_b(k+1) \end{array}\right] = \left[\begin{array}{c|c} \mathbf{G}_{aa} & \mathbf{G}_{ab} \\ \hline \mathbf{G}_{ba} & \mathbf{G}_{bb} \end{array}\right]\left[\begin{array}{c} \mathbf{x}_a(k) \\ \hline \mathbf{x}_b(k) \end{array}\right] + \left[\begin{array}{c} \mathbf{H}_a \\ \hline \mathbf{H}_b \end{array}\right]\mathbf{u}(k) \tag{6–165}$$

$$\mathbf{y}(k) = [\mathbf{I} \vdots \mathbf{0}]\left[\begin{array}{c} \mathbf{x}_a(k) \\ \hline \mathbf{x}_b(k) \end{array}\right] \tag{6–166}$$

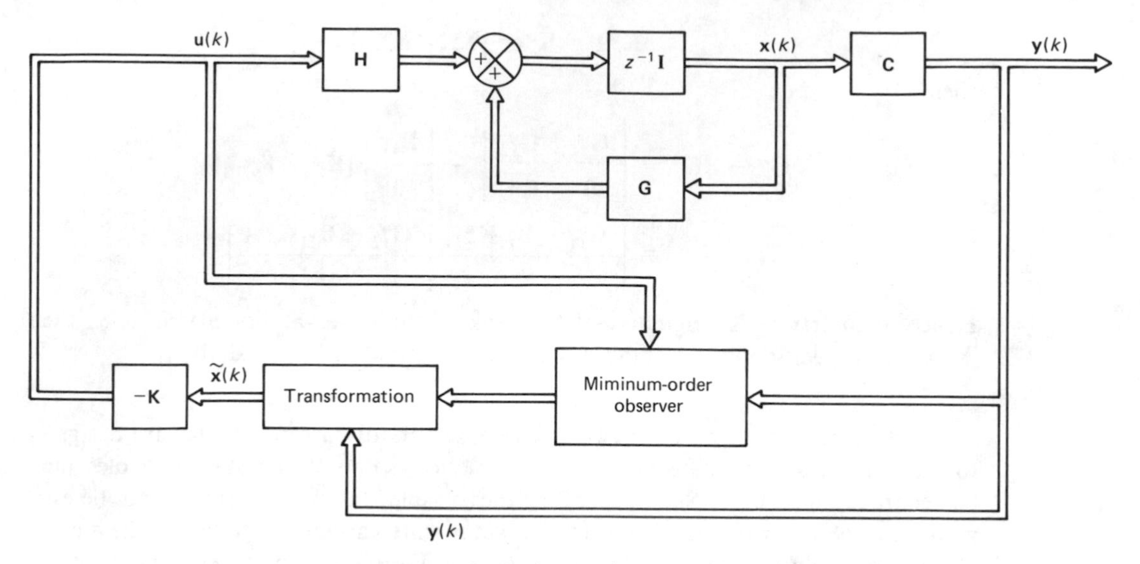

Figure 6–7 Observed-state feedback control system with a minimum-order observer.

where

$$\mathbf{G}_{aa} = m \times m \text{ matrix}$$

$$\mathbf{G}_{ab} = m \times (n - m) \text{ matrix}$$

$$\mathbf{G}_{ba} = (n - m) \times m \text{ matrix}$$

$$\mathbf{G}_{bb} = (n - m) \times (n - m) \text{ matrix}$$

$$\mathbf{H}_{a} = m \times r \text{ matrix}$$

$$\mathbf{H}_{b} = (n - m) \times r \text{ matrix}$$

By rewriting Eq. (6–165), the equation for the measured portion of the state becomes

$$\mathbf{x}_a(k+1) = \mathbf{G}_{aa}\mathbf{x}_a(k) + \mathbf{G}_{ab}\mathbf{x}_b(k) + \mathbf{H}_a\mathbf{u}(k)$$

or

$$\mathbf{x}_a(k+1) - \mathbf{G}_{aa}\mathbf{x}_a(k) - \mathbf{H}_a\mathbf{u}(k) = \mathbf{G}_{ab}\mathbf{x}_b(k) \tag{6–167}$$

where the terms on the left-hand side of the equation can be measured. Equation (6–167) acts as the output equation. In designing the minimum-order observer, we consider the left-hand side of Eq. (6–167) to be known quantities. In fact, Eq. (6–167) relates the measurable quantities and the unmeasurable quantities of the state.

From Eq. (6–165) the equation for the unmeasured portion of the state becomes

$$\mathbf{x}_b(k+1) = \mathbf{G}_{ba}\mathbf{x}_a(k) + \mathbf{G}_{bb}\mathbf{x}_b(k) + \mathbf{H}_b\mathbf{u}(k) \tag{6–168}$$

Equation (6–168) describes the dynamics of the unmeasured portion of the state. Notice that the terms $\mathbf{G}_{ba}\mathbf{x}_a(k)$ and $\mathbf{H}_b\mathbf{u}(k)$ are known quantities.

The design of the minimum-order observer can be facilitated if we utilize the design technique developed for the full-order observer. Let us now compare the state equation for the full-order observer with that for the minimum-order observer. The state equation for the full-order observer is

$$\mathbf{x}(k+1) = \mathbf{G}\mathbf{x}(k) + \mathbf{H}\mathbf{u}(k)$$

and the "state equation" for the minimum-order observer is

$$\mathbf{x}_b(k+1) = \mathbf{G}_{bb}\mathbf{x}_b(k) + [\mathbf{G}_{ba}\mathbf{x}_a(k) + \mathbf{H}_b\mathbf{u}(k)]$$

The output equation for the full-order observer is

$$\mathbf{y}(k) = \mathbf{C}\mathbf{x}(k)$$

and the "output equation" for the minimum-order observer is

$$\mathbf{x}_a(k+1) - \mathbf{G}_{aa}\mathbf{x}_a(k) - \mathbf{H}_a\mathbf{u}(k) = \mathbf{G}_{ab}\mathbf{x}_b(k)$$

The design of the minimum-order observer can be carried out by making the substitutions given in Table 6–1 in the observer equation for the full-order observer given by Eq. (6–117), which we repeat here:

$$\tilde{\mathbf{x}}(k+1) = (\mathbf{G} - \mathbf{K}_e\mathbf{C})\tilde{\mathbf{x}}(k) + \mathbf{H}\mathbf{u}(k) + \mathbf{K}_e\mathbf{y}(k) \tag{6–169}$$

Making the substitution of Table 6–1 in Eq. (6–169), we obtain

$$\begin{aligned}\tilde{\mathbf{x}}_b(k+1) = {} & (\mathbf{G}_{bb} - \mathbf{K}_e\mathbf{G}_{ab})\tilde{\mathbf{x}}_b(k) + \mathbf{G}_{ba}\mathbf{x}_a(k) + \mathbf{H}_b\mathbf{u}(k) \\ & + \mathbf{K}_e[\mathbf{x}_a(k+1) - \mathbf{G}_{aa}\mathbf{x}_a(k) - \mathbf{H}_a\mathbf{u}(k)]\end{aligned} \tag{6–170}$$

where the observer feedback gain matrix $\mathbf{K}_e$ is an $(n-m)\times m$ matrix. Equation (6–170) defines the minimum-order observer.

TABLE 6–1 LIST OF NECESSARY SUBSTITUTIONS FOR WRITING THE OBSERVER EQUATION FOR THE MINIMUM-ORDER STATE OBSERVER

Full-order state observer	Minimum-order state observer
$\tilde{\mathbf{x}}(k)$	$\tilde{\mathbf{x}}_b(k)$
$\mathbf{G}$	$\mathbf{G}_{bb}$
$\mathbf{H}\mathbf{u}(k)$	$\mathbf{G}_{ba}\mathbf{x}_a(k) + \mathbf{H}_b\mathbf{u}(k)$
$\mathbf{y}(k)$	$\mathbf{x}_a(k+1) - \mathbf{G}_{aa}\mathbf{x}_a(k) - \mathbf{H}_a\mathbf{u}(k)$
$\mathbf{C}$	$\mathbf{G}_{ab}$
$\mathbf{K}_e$ ($n \times m$ matrix)	$\mathbf{K}_e$ ($(n-m)\times m$ matrix)

Referring to Eq. (6–166), we have

$$\mathbf{y}(k) = \mathbf{x}_a(k) \tag{6–171}$$

Substituting Eq. (6–171) into Eq. (6–170), we obtain

$$\begin{aligned}\tilde{\mathbf{x}}_b(k+1) &= (\mathbf{G}_{bb} - \mathbf{K}_e\mathbf{G}_{ab})\tilde{\mathbf{x}}_b(k) + \mathbf{K}_e\mathbf{y}(k+1) \\ &\quad + (\mathbf{G}_{ba} - \mathbf{K}_e\mathbf{G}_{aa})\mathbf{y}(k) + (\mathbf{H}_b - \mathbf{K}_e\mathbf{H}_a)\mathbf{u}(k)\end{aligned} \tag{6–172}$$

Notice that in order to estimate $\tilde{\mathbf{x}}_b(k+1)$ we need the measured value of $\mathbf{y}(k+1)$. This is inconvenient and so we may desire some modifications. [In the case of the full-order observer, $\tilde{\mathbf{x}}(k+1)$ can be estimated by use of measurement $\mathbf{y}(k)$ and does not require measurement of $\mathbf{y}(k+1)$. See Eq. (6–117).] Let us rewrite Eq. (6–172) as follows:

$$\begin{aligned}\tilde{\mathbf{x}}_b(k+1) - \mathbf{K}_e\mathbf{y}(k+1) &= (\mathbf{G}_{bb} - \mathbf{K}_e\mathbf{G}_{ab})\tilde{\mathbf{x}}_b(k) + (\mathbf{G}_{ba} - \mathbf{K}_e\mathbf{G}_{aa})\mathbf{y}(k) \\ &\quad + (\mathbf{H}_b - \mathbf{K}_e\mathbf{H}_a)\mathbf{u}(k) \\ &= (\mathbf{G}_{bb} - \mathbf{K}_e\mathbf{G}_{ab})[\tilde{\mathbf{x}}_b(k) - \mathbf{K}_e\mathbf{y}(k)] + (\mathbf{G}_{bb} - \mathbf{K}_e\mathbf{G}_{ab})\mathbf{K}_e\mathbf{y}(k) \\ &\quad + (\mathbf{G}_{ba} - \mathbf{K}_e\mathbf{G}_{aa})\mathbf{y}(k) + (\mathbf{H}_b - \mathbf{K}_e\mathbf{H}_a)\mathbf{u}(k) \\ &= (\mathbf{G}_{bb} - \mathbf{K}_e\mathbf{G}_{ab})[\tilde{\mathbf{x}}_b(k) - \mathbf{K}_e\mathbf{y}(k)] + [(\mathbf{G}_{bb} - \mathbf{K}_e\mathbf{G}_{ab})\mathbf{K}_e \\ &\quad + \mathbf{G}_{ba} - \mathbf{K}_e\mathbf{G}_{aa}]\mathbf{y}(k) + (\mathbf{H}_b - \mathbf{K}_e\mathbf{H}_a)\mathbf{u}(k)\end{aligned} \tag{6–173}$$

Define

$$\mathbf{x}_b(k) - \mathbf{K}_e\mathbf{y}(k) = \mathbf{x}_b(k) - \mathbf{K}_e\mathbf{x}_a(k) = \boldsymbol{\eta}(k) \tag{6–174}$$

and

$$\tilde{\mathbf{x}}_b(k) - \mathbf{K}_e\mathbf{y}(k) = \tilde{\mathbf{x}}_b(k) - \mathbf{K}_e\mathbf{x}_a(k) = \tilde{\boldsymbol{\eta}}(k) \tag{6–175}$$

Then Eq. (6–173) can be written as follows:

$$\begin{aligned}\tilde{\boldsymbol{\eta}}(k+1) &= (\mathbf{G}_{bb} - \mathbf{K}_e\mathbf{G}_{ab})\tilde{\boldsymbol{\eta}}(k) + [(\mathbf{G}_{bb} - \mathbf{K}_e\mathbf{G}_{ab})\mathbf{K}_e + \mathbf{G}_{ba} \\ &\quad - \mathbf{K}_e\mathbf{G}_{aa}]\mathbf{y}(k) + (\mathbf{H}_b - \mathbf{K}_e\mathbf{H}_a)\mathbf{u}(k)\end{aligned} \tag{6–176}$$

Equations (6–175) and (6–176) define the dynamics of the minimum-order observer. Notice that in order to obtain $\tilde{\boldsymbol{\eta}}(k+1)$ we do not need the measured value of $\mathbf{y}(k+1)$.

Let us next obtain the observer error equation. Define

$$\mathbf{e}(k) = \boldsymbol{\eta}(k) - \tilde{\boldsymbol{\eta}}(k) = \mathbf{x}_b(k) - \tilde{\mathbf{x}}_b(k) \tag{6–177}$$

Subtracting Eq. (6–170) from Eq. (6–168), we obtain

$$\begin{aligned}\mathbf{x}_b(k+1) - \tilde{\mathbf{x}}_b(k+1) &= \mathbf{G}_{bb}[\mathbf{x}_b(k) - \tilde{\mathbf{x}}_b(k)] + \mathbf{K}_e\mathbf{G}_{ab}\tilde{\mathbf{x}}_b(k) \\ &\quad - \mathbf{K}_e[\mathbf{x}_a(k+1) - \mathbf{G}_{aa}\mathbf{x}_a(k) - \mathbf{H}_a\mathbf{u}(k)]\end{aligned}$$

By substituting Eq. (6–167) into this last equation, we obtain

$$\begin{aligned}\mathbf{x}_b(k+1) - \tilde{\mathbf{x}}_b(k+1) &= \mathbf{G}_{bb}[\mathbf{x}_b(k) - \tilde{\mathbf{x}}_b(k)] + \mathbf{K}_e\mathbf{G}_{ab}\tilde{\mathbf{x}}_b(k) - \mathbf{K}_e\mathbf{G}_{ab}\mathbf{x}_b(k) \\ &= (\mathbf{G}_{bb} - \mathbf{K}_e\mathbf{G}_{ab})[\mathbf{x}_b(k) - \tilde{\mathbf{x}}_b(k)]\end{aligned}$$

This last equation can be written in the form

$$\mathbf{e}(k+1) = (\mathbf{G}_{bb} - \mathbf{K}_e \mathbf{G}_{ab})\mathbf{e}(k) \tag{6–178}$$

This is the observer error equation. Note that $\mathbf{e}(k)$ is an $(n - m)$-vector. The error dynamics can be determined as desired by following the technique developed for the full-order observer, provided that the rank of matrix

$$\begin{bmatrix} \mathbf{G}_{ab} \\ \mathbf{G}_{ab}\mathbf{G}_{bb} \\ \cdot \\ \cdot \\ \cdot \\ \mathbf{G}_{ab}\mathbf{G}_{bb}^{n-m-1} \end{bmatrix}$$

is $n - m$. (This is the complete observability condition applicable to the minimum-order observer.)

The characteristic equation for the minimum-order observer is obtained from Eq. (6–178) as follows:

$$|z\mathbf{I} - \mathbf{G}_{bb} + \mathbf{K}_e \mathbf{G}_{ab}| = 0 \tag{6–179}$$

The observer feedback gain matrix $\mathbf{K}_e$ can be determined from Eq. (6–179) by first choosing the desired closed-loop pole locations for the minimum-order observer [that is, by placing the roots of the characteristic equation, Eq. (6–179), at the desired locations] and then using the procedure developed for the full-order prediction observer.

If, for example, the output $y(k)$ is a scalar, then $x_a(k)$ is a scalar, $\mathbf{G}_{ab}$ is a $1 \times (n-1)$ matrix, and $\mathbf{G}_{bb}$ is an $(n-1) \times (n-1)$ matrix. For this case, Ackermann's formula as given by Eq. (6–151) may be modified to read

$$\mathbf{K}_e = \phi(\mathbf{G}_{bb}) \begin{bmatrix} \mathbf{G}_{ab} \\ \mathbf{G}_{ab}\mathbf{G}_{bb} \\ \cdot \\ \cdot \\ \cdot \\ \mathbf{G}_{ab}\mathbf{G}_{bb}^{n-3} \\ \mathbf{G}_{ab}\mathbf{G}_{bb}^{n-2} \end{bmatrix}^{-1} \begin{bmatrix} 0 \\ 0 \\ \cdot \\ \cdot \\ \cdot \\ 0 \\ 1 \end{bmatrix} \tag{6–180}$$

where

$$\phi(\mathbf{G}_{bb}) = \mathbf{G}_{bb}^{n-1} + \alpha_1 \mathbf{G}_{bb}^{n-2} + \cdots + \alpha_{n-2}\mathbf{G}_{bb} + \alpha_{n-1}\mathbf{I} \tag{6–181}$$

Summary. Once the observer feedback gain matrix $\mathbf{K}_e$, which is an $(n - m) \times m$ matrix, is determined, then the minimum-order observer can be defined by Eqs. (6–175) and (6–176):

$$\tilde{\mathbf{x}}_b(k) = \tilde{\boldsymbol{\eta}}(k) + \mathbf{K}_e \mathbf{x}_a(k)$$

$$\tilde{\boldsymbol{\eta}}(k+1) = (\mathbf{G}_{bb} - \mathbf{K}_e\mathbf{G}_{ab})\tilde{\boldsymbol{\eta}}(k) + [(\mathbf{G}_{bb} - \mathbf{K}_e\mathbf{G}_{ab})\mathbf{K}_e + \mathbf{G}_{ba} - \mathbf{K}_e\mathbf{G}_{aa}]\mathbf{y}(k) + (\mathbf{H}_b - \mathbf{K}_e\mathbf{H}_a)\mathbf{u}(k)$$

Equivalently, in terms of $\mathbf{e}(k)$ rather than $\tilde{\boldsymbol{\eta}}(k)$, the minimum-order observer can be defined by Eqs. (6–177) and (6–178):

$$\tilde{\mathbf{x}}_b(k) = \mathbf{x}_b(k) - \mathbf{e}(k) \tag{6–182}$$

$$\mathbf{e}(k+1) = (\mathbf{G}_{bb} - \mathbf{K}_e\mathbf{G}_{ab})\mathbf{e}(k) \tag{6–183}$$

Observed-state feedback control system with minimum-order observer. The system considered in the preceding discussions is given by the equations

$$\mathbf{x}(k+1) = \mathbf{G}\mathbf{x}(k) + \mathbf{H}\mathbf{u}(k) \tag{6–184}$$

$$\mathbf{y}(k) = \mathbf{C}\mathbf{x}(k) \tag{6–185}$$

where $\mathbf{x}(k)$ is an n-vector, $\mathbf{u}(k)$ is an r-vector, and $\mathbf{y}(k)$ is an m-vector. Matrices **G**, **H**, and **C** are given by

$$\mathbf{G} = \begin{bmatrix} \mathbf{G}_{aa} & \mathbf{G}_{ab} \\ \mathbf{G}_{ba} & \mathbf{G}_{bb} \end{bmatrix}, \qquad \mathbf{H} = \begin{bmatrix} \mathbf{H}_a \\ \mathbf{H}_b \end{bmatrix}, \qquad \mathbf{C} = [\mathbf{I}_m \,\vdots\, \mathbf{0}]$$

Consider the state feedback control scheme where the fed-back state consists of the measured portion of the state and the observed (estimated) portion of the state obtained by use of the minimum-order observer. Figure 6–8 shows the block diagram for the system. In this system the control vector $\mathbf{u}(k)$ is given by

$$\mathbf{u}(k) = -\mathbf{K}\tilde{\mathbf{x}}(k) \tag{6–186}$$

where $\tilde{\mathbf{x}}(k)$ consists of the measurable state $\mathbf{x}_a(k)$ and unmeasurable (observed) state $\tilde{\mathbf{x}}_b(k)$:

$$\tilde{\mathbf{x}}(k) = \begin{bmatrix} \mathbf{x}_a(k) \\ \tilde{\mathbf{x}}_b(k) \end{bmatrix} = \begin{bmatrix} \mathbf{x}_a(k) \\ \tilde{\boldsymbol{\eta}}(k) + \mathbf{K}_e\mathbf{x}_a(k) \end{bmatrix} \tag{6–187}$$

By substituting Eq. (6–186) into Eq. (6–184), we obtain

$$\mathbf{x}(k+1) = \mathbf{G}\mathbf{x}(k) - \mathbf{H}\mathbf{K}\tilde{\mathbf{x}}(k) = (\mathbf{G} - \mathbf{H}\mathbf{K})\mathbf{x}(k) + \mathbf{H}\mathbf{K}[\mathbf{x}(k) - \tilde{\mathbf{x}}(k)] \tag{6–188}$$

Notice that

$$\mathbf{x}(k) - \tilde{\mathbf{x}}(k) = \begin{bmatrix} \mathbf{x}_a(k) \\ \mathbf{x}_b(k) \end{bmatrix} - \begin{bmatrix} \mathbf{x}_a(k) \\ \tilde{\mathbf{x}}_b(k) \end{bmatrix} = \begin{bmatrix} \mathbf{0} \\ \mathbf{x}_b(k) - \tilde{\mathbf{x}}_b(k) \end{bmatrix} = \begin{bmatrix} \mathbf{0} \\ \mathbf{e}(k) \end{bmatrix}$$

where $\mathbf{e}(k) = \mathbf{x}_b(k) - \tilde{\mathbf{x}}_b(k)$. Define

$$\boldsymbol{\Gamma} = \begin{bmatrix} \mathbf{0} \\ \mathbf{I}_{n-m} \end{bmatrix}$$

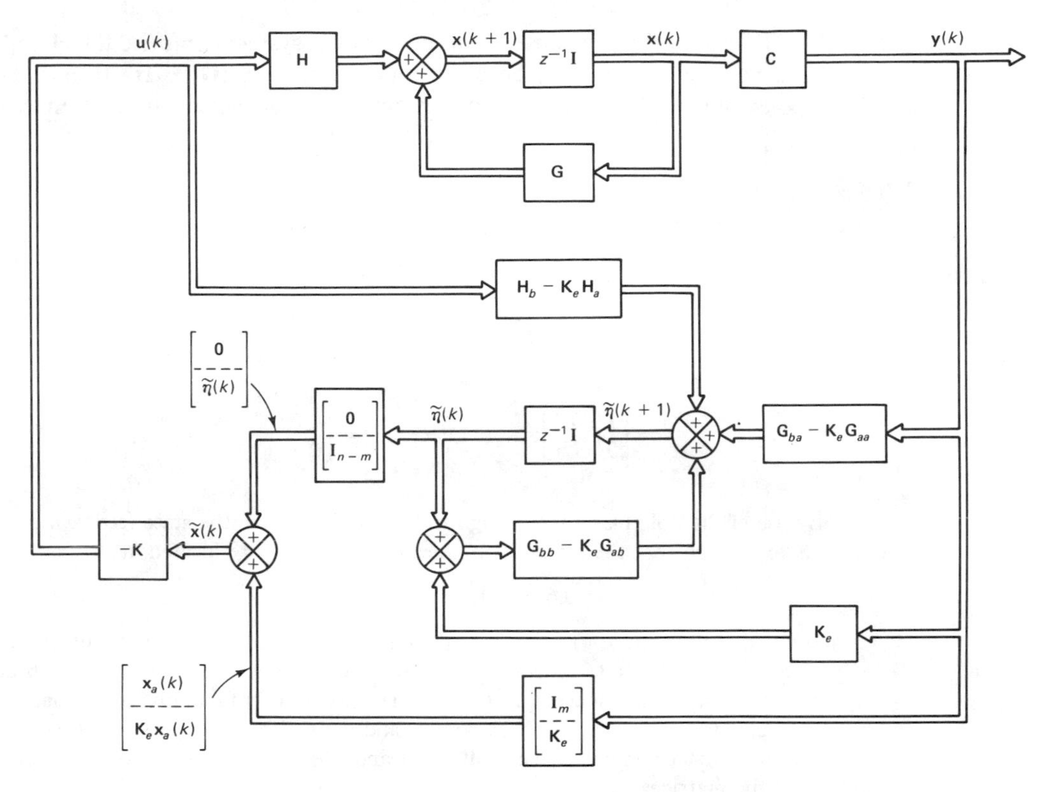

Figure 6–8 State feedback control scheme where the fed-back state consists of the measured portion of the state and the observed portion of the state obtained by use of the minimum-order observer.

Then by use of this matrix $\mathbf{\Gamma}$, Eq. (6–188) can be rewritten as follows:

$$\mathbf{x}(k+1) = (\mathbf{G} - \mathbf{HK})\mathbf{x}(k) + \mathbf{HK\Gamma e}(k) \tag{6–189}$$

Equations (6–189) and (6–183) characterize the state feedback control system where the fed-back state consists of the measured portion of the state, $\mathbf{x}_a(k)$, and the observed portion of the state, $\tilde{\mathbf{x}}_b(k)$, obtained by use of the minimum-order observer. Combining Eqs. (6–189) and (6–183), we have

$$\begin{bmatrix} \mathbf{x}(k+1) \\ \hline \mathbf{e}(k+1) \end{bmatrix} = \left[\begin{array}{c|c} \mathbf{G} - \mathbf{HK} & \mathbf{HK\Gamma} \\ \hline \mathbf{0} & \mathbf{G}_{bb} - \mathbf{K}_e\mathbf{G}_{ab} \end{array}\right] \begin{bmatrix} \mathbf{x}(k) \\ \hline \mathbf{e}(k) \end{bmatrix} \tag{6–190}$$

Equation (6–190) characterizes the dynamics of the system with observed-state feedback and a minimum-order observer. The characteristic equation for the system is

$$\left|\begin{array}{c|c} z\mathbf{I} - \mathbf{G} + \mathbf{HK} & -\mathbf{HK\Gamma} \\ \hline \mathbf{0} & z\mathbf{I} - \mathbf{G}_{bb} + \mathbf{K}_e\mathbf{G}_{ab} \end{array}\right| = |z\mathbf{I} - \mathbf{G} + \mathbf{HK}||z\mathbf{I} - \mathbf{G}_{bb} + \mathbf{K}_e\mathbf{G}_{ab}| = 0 \tag{6–191}$$

Equation (6–191) implies that the closed-loop poles of the system comprise the closed-loop poles due to pole placement [the eigenvalues of matrix $(\mathbf{G} - \mathbf{HK})$] and the closed-loop poles due to the minimum-order observer [the eigenvalues of matrix $(\mathbf{G}_{bb} - \mathbf{K}_e \mathbf{G}_{ab})$].

Example 6–11.

Consider the discrete-time double integrator system defined by the equations

$$\mathbf{x}(k+1) = \mathbf{G}\mathbf{x}(k) + \mathbf{H}u(k) \tag{6–192}$$

$$y(k) = \mathbf{C}\mathbf{x}(k) \tag{6–193}$$

where the sampling period T is assumed to be 0.2 sec, or $T = 0.2$, and

$$\mathbf{G} = \begin{bmatrix} 1 & T \\ 0 & 1 \end{bmatrix} = \begin{bmatrix} 1 & 0.2 \\ 0 & 1 \end{bmatrix}, \qquad \mathbf{H} = \begin{bmatrix} \dfrac{T^2}{2} \\ T \end{bmatrix} = \begin{bmatrix} 0.02 \\ 0.2 \end{bmatrix}, \qquad \mathbf{C} = [1 \quad 0]$$

By use of the pole placement design technique, determine the state feedback gain matrix $\mathbf{K}$ to be such that the closed-loop poles of the system are located at

$$z_1 = 0.6 + j0.4, \qquad z_2 = 0.6 - j0.4$$

Assuming that the output $y(k) = x_1(k)$ is the only state variable that can be measured, design a minimum-order state observer such that the error signal will exhibit a deadbeat response to an arbitrary initial error. Determine the pulse transfer function for the controller (which consists of the state feedback control and the minimum-order state observer).

We shall first examine the controllability and observability of the system. Since the rank of the matrices

$$[\mathbf{H} \vdots \mathbf{GH}] = \begin{bmatrix} 0.02 & 0.06 \\ 0.2 & 0.2 \end{bmatrix}, \qquad [\mathbf{C}^* \vdots \mathbf{G}^*\mathbf{C}^*] = \begin{bmatrix} 1 & 1 \\ 0 & 0.2 \end{bmatrix}$$

is 2 in both cases, the system is completely state controllable and observable.

We shall now solve the pole placement portion of the problem. Since

$$|z\mathbf{I} - \mathbf{G}| = \begin{vmatrix} z-1 & -0.2 \\ 0 & z-1 \end{vmatrix} = z^2 - 2z + 1 = z^2 + a_1 z + a_2 = 0$$

we have

$$a_1 = -2, \qquad a_2 = 1$$

The desired characteristic equation is given by

$$|z\mathbf{I} - \mathbf{G} + \mathbf{HK}| = (z - 0.6 - j0.4)(z - 0.6 + j0.4) = z^2 - 1.2z + 0.52$$
$$= z^2 + \alpha_1 z + \alpha_2 = 0$$

Hence

$$\alpha_1 = -1.2, \qquad \alpha_2 = 0.52$$

From Eq. (6–65), the state feedback gain matrix $\mathbf{K}$ is obtained as follows:

$$\mathbf{K} = [\alpha_2 - a_2 \vdots \alpha_1 - a_1]\mathbf{T}^{-1} = [-0.48 \quad 0.8]\mathbf{T}^{-1} \tag{6–194}$$

where

$$\mathbf{T} = [\mathbf{H} \vdots \mathbf{GH}]\begin{bmatrix} a_1 & 1 \\ 1 & 0 \end{bmatrix} = \begin{bmatrix} 0.02 & 0.06 \\ 0.2 & 0.2 \end{bmatrix}\begin{bmatrix} -2 & 1 \\ 1 & 0 \end{bmatrix}$$

$$= \begin{bmatrix} 0.02 & 0.02 \\ -0.2 & 0.2 \end{bmatrix}$$

and

$$\mathbf{T}^{-1} = \begin{bmatrix} 25 & -2.5 \\ 25 & 2.5 \end{bmatrix}$$

Thus the state feedback gain matrix **K** given by Eq. (6–194) becomes

$$\mathbf{K} = [-0.48 \quad 0.8]\begin{bmatrix} 25 & -2.5 \\ 25 & 2.5 \end{bmatrix} = [8 \quad 3.2]$$

The feedback control signal can then be given by

$$u(k) = -\mathbf{K}\tilde{\mathbf{x}}(k)$$

$$= -[8 \quad 3.2]\begin{bmatrix} x_1(k) \\ \tilde{x}_2(k) \end{bmatrix} = -[8 \quad 3.2]\begin{bmatrix} y(k) \\ \tilde{x}_2(k) \end{bmatrix} \tag{6–195}$$

Next, we shall solve the observer portion of the problem. Since the state $\mathbf{x}(k)$ is a 2-vector and the output $y(k)$ is a scalar, the minimum-order observer is of the first order. Notice that

$$\left[\begin{array}{c|c} G_{aa} & G_{ab} \\ \hline G_{ba} & G_{bb} \end{array}\right] = \left[\begin{array}{c|c} 1 & 0.2 \\ \hline 0 & 1 \end{array}\right], \qquad \left[\begin{array}{c} H_a \\ \hline H_b \end{array}\right] = \left[\begin{array}{c} 0.02 \\ \hline 0.2 \end{array}\right]$$

Since we desire deadbeat response, the desired characteristic equation for the observer is

$$\phi(z) = z = 0$$

Referring to Ackermann's formula as given by Eq. (6–180), we obtain

$$K_e = \phi(G_{bb})[G_{ab}]^{-1}[1] = (1)(0.2)^{-1}(1) = 5$$

Referring to the minimum-order observer equation given by Eq. (6–176), we have

$$\begin{aligned}\tilde{\eta}(k+1) &= (G_{bb} - K_e G_{ab})\tilde{\eta}(k) + [(G_{bb} - K_e G_{ab})K_e + G_{ba} - K_e G_{aa}]y(k) \\ &\quad + (H_b - K_e H_a)u(k) \\ &= (1 - 5 \times 0.2)\tilde{\eta}(k) + [(1 - 5 \times 0.2) \times 5 + 0 - 5 \times 1]y(k) \\ &\quad + (0.2 - 5 \times 0.02)u(k)\end{aligned}$$

which can be simplified to read

$$\tilde{\eta}(k+1) = -5y(k) + 0.1u(k) \tag{6–196}$$

Equation (6–196) defines the minimum-order state observer.

The observed-state feedback control $u(k)$ is now given by

$$u(k) = -\mathbf{K}\tilde{\mathbf{x}}(k) = -8x_1(k) - 3.2\tilde{x}_2(k) = -8y(k) - 3.2\tilde{x}_2(k) \tag{6–197}$$

where, referring to Eq. (6–175),

$$\tilde{x}_2(k) = K_e y(k) + \tilde{\eta}(k) = 5y(k) + \tilde{\eta}(k) \tag{6–198}$$

The block diagram for the system is shown in Fig. 6–9. From Eqs. (6–196), (6–197), and (6–198) we obtain

$$\begin{aligned} u(k+1) &= -8y(k+1) - 3.2[5y(k+1) + \tilde{\eta}(k+1)] \\ &= -24y(k+1) + 16y(k) - 0.32u(k) \end{aligned}$$

or

$$u(k+1) + 0.32u(k) = -24y(k+1) + 16y(k)$$

By taking the z transform of this last equation assuming zero initial conditions, we obtain

$$zU(z) + 0.32U(z) = -24zY(z) + 16Y(z)$$

The pulse transfer function of the controller is

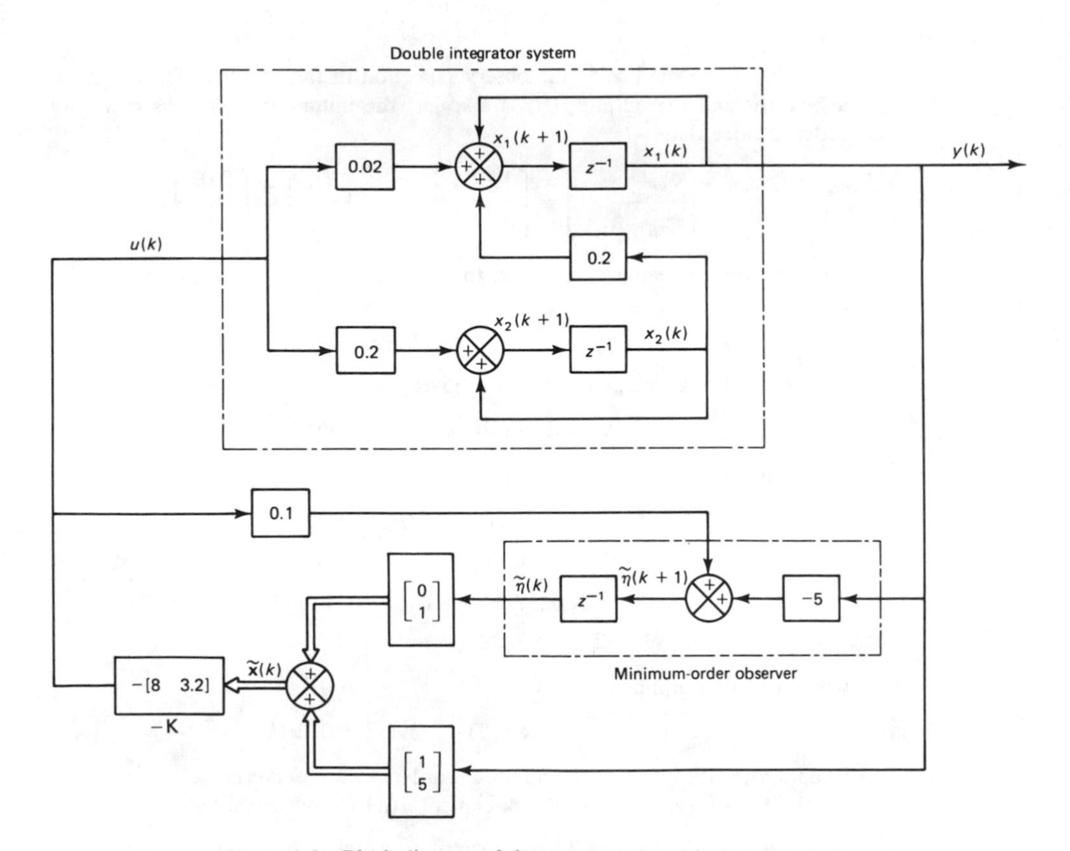

Figure 6–9 Block diagram of the system designed in Example 6–11.

$$\frac{U(z)}{Y(z)} = -D(z) = -24\left(\frac{z-0.6667}{z+0.32}\right) = -24\left(\frac{1-0.6667z^{-1}}{1+0.32z^{-1}}\right) \tag{6–199}$$

By referring to Eq. (5–88), the pulse transfer function of the system defined by Eqs. (6–192) and (6–193) can be obtained as follows:

$$\frac{Y(z)}{U(z)} = G_p(z) = \mathbf{C}(z\mathbf{I}-\mathbf{G})^{-1}\mathbf{H}$$

$$= [1 \quad 0]\begin{bmatrix} z-1 & -0.2 \\ 0 & z-1 \end{bmatrix}^{-1}\begin{bmatrix} 0.02 \\ 0.2 \end{bmatrix}$$

$$= \frac{0.02(z+1)}{(z-1)^2} = \frac{0.02(1+z^{-1})z^{-1}}{(1-z^{-1})^2} \tag{6–200}$$

By using the pulse transfer functions of Eqs. (6–199) and (6–200), the block diagram of Fig. 6–9 may be modified to the form shown in Fig. 6–10.

Using the form given by Eq. (6–191),

$$|z\mathbf{I}-\mathbf{G}+\mathbf{HK}||z-G_{bb}+K_eG_{ab}| = 0$$

we have obtained the following characteristic equation for the system:

$$(z^2-1.2z+0.52)(z-1+5\times 0.2) = (z^2-1.2z+0.52)z = 0$$

The characteristic equation for the closed-loop system shown in Fig. 6–10 is

$$1 + G_p(z)D(z) = 0$$

or

$$1 + \left[\frac{0.02(1+z^{-1})z^{-1}}{(1-z^{-1})^2}\right]\left[24\left(\frac{1-0.6667z^{-1}}{1+0.32z^{-1}}\right)\right] = 0$$

which can be written as follows:

$$1 + \left[\frac{0.02(z+1)}{(z-1)^2}\right]\left[24\left(\frac{z-0.6667}{z+0.32}\right)\right] = 0$$

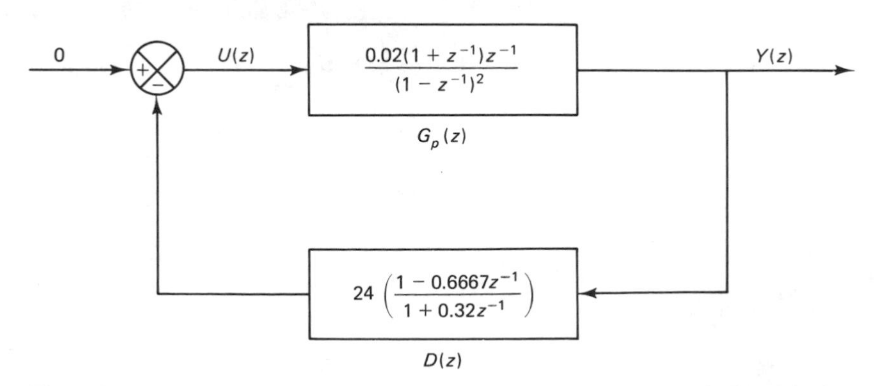

Figure 6–10 Modified form of the block diagram of the system designed in Example 6–11.

And, as a matter of course, this characteristic equation can be simplified to

$$(z^2 - 1.2z + 0.52)z = 0$$

which is the same as that obtained by use of Eq. (6–191).

6–7 SERVO SYSTEMS

In Secs. 6–5 and 6–6 we discussed regulator problems where there were no command signals. The purpose of control was to bring the error vector to zero. Design of such regulator systems can be carried out by use of the pole placement method coupled with state observers.

A control system in which the output must follow command signals is called a *servo system*. In this section we shall briefly discuss servo systems. (We shall treat servo systems again in Chap. 7.)

In the servo system it is generally required that the system have one or more integrators within the closed loop. (Unless the plant to be controlled has an integrating property, it is usually necessary to add one or more integrators within the loop to eliminate steady-state error.)

One way to introduce an integrator in the mathematical model of a closed-loop system is to introduce a new state vector that integrates the difference between the command vector $\mathbf{r}$ and the output vector $\mathbf{y}$. Figure 6–11 shows a possible block diagram configuration for a servo system with state feedback and integral control. The integral controller consists of m integrating elements, one for each command input component. (The command input is an m-vector and has m components.) The integrator can be included as part of the pole placement formulation that was presented in Sec. 6–5.

Servo system with integrator. Consider the servo system shown in Fig. 6–11. The plant is assumed to be completely state controllable and observable. The plant state equation and output equation are

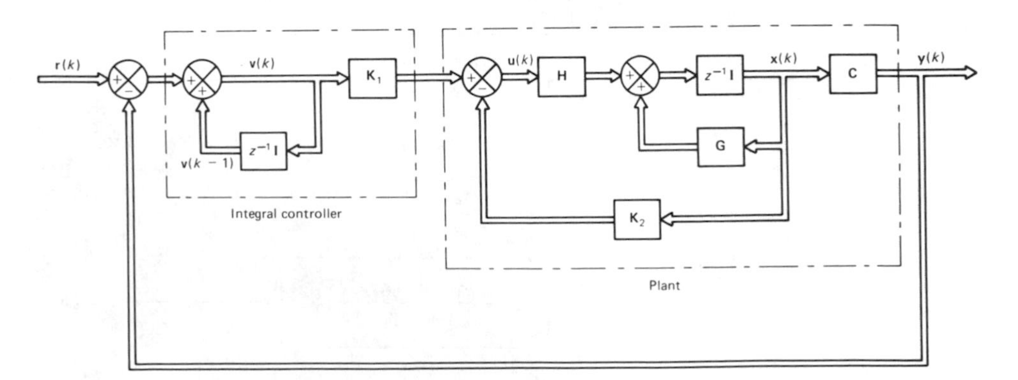

Figure 6–11 Servo system with state feedback and integral control.

$$\mathbf{x}(k+1) = \mathbf{G}\mathbf{x}(k) + \mathbf{H}\mathbf{u}(k) \tag{6–201}$$

$$\mathbf{y}(k) = \mathbf{C}\mathbf{x}(k) \tag{6–202}$$

where

$\mathbf{x}(k)$ = plant state vector (n-vector)

$\mathbf{u}(k)$ = control vector (m-vector)

$\mathbf{y}(k)$ = output vector (m-vector)

$\mathbf{G} = n \times n$ matrix

$\mathbf{H} = n \times m$ matrix

$\mathbf{C} = m \times n$ matrix

(Note that in the present analysis we assume that the dimensions of the output vector and the control vector are the same—that they are both m-vectors.) The integrator state equation is

$$\mathbf{v}(k) = \mathbf{v}(k-1) + \mathbf{r}(k) - \mathbf{y}(k) \tag{6–203}$$

where

$\mathbf{v}(k)$ = actuating error vector (m-vector)

$\mathbf{r}(k)$ = command input vector (m-vector)

Equation (6–203) can be rewritten as follows:

$$\begin{aligned}\mathbf{v}(k+1) &= \mathbf{v}(k) + \mathbf{r}(k+1) - \mathbf{y}(k+1)\\ &= \mathbf{v}(k) + \mathbf{r}(k+1) - \mathbf{C}[\mathbf{G}\mathbf{x}(k) + \mathbf{H}\mathbf{u}(k)]\\ &= -\mathbf{C}\mathbf{G}\mathbf{x}(k) + \mathbf{v}(k) - \mathbf{C}\mathbf{H}\mathbf{u}(k) + \mathbf{r}(k+1)\end{aligned} \tag{6–204}$$

The control vector $\mathbf{u}(k)$ is given by

$$\mathbf{u}(k) = -\mathbf{K}_2\mathbf{x}(k) + \mathbf{K}_1\mathbf{v}(k) \tag{6–205}$$

In our servo system the system configuration is specified in Fig. 6–11. Our design parameters are matrices $\mathbf{K}_1$ and $\mathbf{K}_2$.

In what follows we shall discuss the procedure for determining matrices $\mathbf{K}_1$ and $\mathbf{K}_2$ such that the system has the desired closed-loop poles. From Eqs. (6–201), (6–204), and (6–205) we obtain

$$\begin{aligned}\mathbf{u}(k+1) &= -\mathbf{K}_2\mathbf{x}(k+1) + \mathbf{K}_1\mathbf{v}(k+1)\\ &= (\mathbf{K}_2 - \mathbf{K}_2\mathbf{G} - \mathbf{K}_1\mathbf{C}\mathbf{G})\mathbf{x}(k) + (\mathbf{I}_m - \mathbf{K}_2\mathbf{H} - \mathbf{K}_1\mathbf{C}\mathbf{H})\mathbf{u}(k) + \mathbf{K}_1\mathbf{r}(k+1)\end{aligned} \tag{6–206}$$

Noting that $\mathbf{u}(k)$ is a linear combination of state vectors $\mathbf{x}(k)$ and $\mathbf{v}(k)$, define a new state vector consisting of $\mathbf{x}(k)$ and $\mathbf{u}(k)$ [rather than $\mathbf{x}(k)$ and $\mathbf{v}(k)$]. Then we obtain from Eqs. (6–201) and (6–206) the following state equation:

$$\begin{bmatrix} \mathbf{x}(k+1) \\ \mathbf{u}(k+1) \end{bmatrix} = \begin{bmatrix} \mathbf{G} & \mathbf{H} \\ \mathbf{K}_2 - \mathbf{K}_2\mathbf{G} - \mathbf{K}_1\mathbf{C}\mathbf{G} & \mathbf{I}_m - \mathbf{K}_2\mathbf{H} - \mathbf{K}_1\mathbf{C}\mathbf{H} \end{bmatrix} \begin{bmatrix} \mathbf{x}(k) \\ \mathbf{u}(k) \end{bmatrix} + \begin{bmatrix} \mathbf{0} \\ \mathbf{K}_1 \end{bmatrix} \mathbf{r}(k+1) \tag{6–207}$$

The output equation, Eq. (6–202), can be written as follows:

$$\mathbf{y}(k) = [\mathbf{C} \quad \mathbf{0}] \begin{bmatrix} \mathbf{x}(k) \\ \mathbf{u}(k) \end{bmatrix} \tag{6–208}$$

Note that the closed-loop poles of the system are determined by the system itself and do not depend on the command input $\mathbf{r}(k)$. The eigenvalues of the state matrix in Eq. (6–207) determine the closed-loop poles of the system.

In order to apply the pole placement technique of Sec. 6–5 directly to the design of the present servo system, consider the case where the command vector $\mathbf{r}(k)$ is a constant vector (step input), so that

$$\mathbf{r}(k) = \mathbf{r}$$

Then Eq. (6–207) becomes

$$\begin{bmatrix} \mathbf{x}(k+1) \\ \mathbf{u}(k+1) \end{bmatrix} = \begin{bmatrix} \mathbf{G} & \mathbf{H} \\ \mathbf{K}_2 - \mathbf{K}_2\mathbf{G} - \mathbf{K}_1\mathbf{C}\mathbf{G} & \mathbf{I}_m - \mathbf{K}_2\mathbf{H} - \mathbf{K}_1\mathbf{C}\mathbf{H} \end{bmatrix} \begin{bmatrix} \mathbf{x}(k) \\ \mathbf{u}(k) \end{bmatrix} + \begin{bmatrix} \mathbf{0} \\ \mathbf{K}_1\mathbf{r} \end{bmatrix} \tag{6–209}$$

Notice that for the step input, $\mathbf{x}(k)$, $\mathbf{u}(k)$, and $\mathbf{v}(k)$ approach the constant vector values $\mathbf{x}(\infty)$, $\mathbf{u}(\infty)$, and $\mathbf{v}(\infty)$, respectively. Thus, from Eq. (6–203), we obtain the following equation at steady state:

$$\mathbf{v}(\infty) = \mathbf{v}(\infty) + \mathbf{r} - \mathbf{y}(\infty)$$

or

$$\mathbf{y}(\infty) = \mathbf{r}$$

There is no steady-state error in the output when the command input is a step vector. Also, at steady state, Eq. (6–209) becomes

$$\begin{bmatrix} \mathbf{x}(\infty) \\ \mathbf{u}(\infty) \end{bmatrix} = \begin{bmatrix} \mathbf{G} & \mathbf{H} \\ \mathbf{K}_2 - \mathbf{K}_2\mathbf{G} - \mathbf{K}_1\mathbf{C}\mathbf{G} & \mathbf{I}_m - \mathbf{K}_2\mathbf{H} - \mathbf{K}_1\mathbf{C}\mathbf{H} \end{bmatrix} \begin{bmatrix} \mathbf{x}(\infty) \\ \mathbf{u}(\infty) \end{bmatrix} + \begin{bmatrix} \mathbf{0} \\ \mathbf{K}_1\mathbf{r} \end{bmatrix} \tag{6–210}$$

Let us define the error vectors by

$$\mathbf{x}_e(k) = \mathbf{x}(k) - \mathbf{x}(\infty)$$

$$\mathbf{u}_e(k) = \mathbf{u}(k) - \mathbf{u}(\infty)$$

Then, subtracting Eq. (6–210) from Eq. (6–209), we obtain

$$\begin{bmatrix} \mathbf{x}_e(k+1) \\ \mathbf{u}_e(k+1) \end{bmatrix} = \begin{bmatrix} \mathbf{G} & \mathbf{H} \\ \mathbf{K}_2 - \mathbf{K}_2\mathbf{G} - \mathbf{K}_1\mathbf{C}\mathbf{G} & \mathbf{I}_m - \mathbf{K}_2\mathbf{H} - \mathbf{K}_1\mathbf{C}\mathbf{H} \end{bmatrix} \begin{bmatrix} \mathbf{x}_e(k) \\ \mathbf{u}_e(k) \end{bmatrix} \tag{6–211}$$

The dynamics of the system are determined by the eigenvalues of the state matrix appearing in Eq. (6–211). Equation (6–211) can be modified to read

$$\begin{bmatrix} \mathbf{x}_e(k+1) \\ \mathbf{u}_e(k+1) \end{bmatrix} = \begin{bmatrix} \mathbf{G} & \mathbf{H} \\ \mathbf{0} & \mathbf{0} \end{bmatrix} \begin{bmatrix} \mathbf{x}_e(k) \\ \mathbf{u}_e(k) \end{bmatrix} + \begin{bmatrix} \mathbf{0} \\ \mathbf{I}_m \end{bmatrix} \mathbf{w}(k) \tag{6–212}$$

where

$$\mathbf{w}(k) = [\mathbf{K}_2 - \mathbf{K}_2\mathbf{G} - \mathbf{K}_1\mathbf{C}\mathbf{G} \vdots \mathbf{I}_m - \mathbf{K}_2\mathbf{H} - \mathbf{K}_1\mathbf{C}\mathbf{H}] \begin{bmatrix} \mathbf{x}_e(k) \\ \mathbf{u}_e(k) \end{bmatrix} \tag{6–213}$$

If we define

$$\boldsymbol{\xi}(k) = \begin{bmatrix} \mathbf{x}_e(k) \\ \mathbf{u}_e(k) \end{bmatrix} = (n+m)\text{-vector}$$

$$\hat{\mathbf{G}} = \begin{bmatrix} \mathbf{G} & \mathbf{H} \\ \mathbf{0} & \mathbf{0} \end{bmatrix} = (n+m) \times (n+m) \text{ matrix}$$

$$\hat{\mathbf{H}} = \begin{bmatrix} \mathbf{0} \\ \mathbf{I}_m \end{bmatrix} = (n+m) \times m \text{ matrix}$$

$$\hat{\mathbf{K}} = -[\mathbf{K}_2 - \mathbf{K}_2\mathbf{G} - \mathbf{K}_1\mathbf{C}\mathbf{G} \vdots \mathbf{I}_m - \mathbf{K}_2\mathbf{H} - \mathbf{K}_1\mathbf{C}\mathbf{H}] = m \times (n+m) \text{ matrix} \tag{6–214}$$

then Eqs. (6–212) and (6–213) become, respectively, as

$$\boldsymbol{\xi}(k+1) = \hat{\mathbf{G}}\boldsymbol{\xi}(k) + \hat{\mathbf{H}}\mathbf{w}(k) \tag{6–215}$$

and

$$\mathbf{w}(k) = -\hat{\mathbf{K}}\boldsymbol{\xi}(k) \tag{6–216}$$

Notice that the controllability matrix for the system defined by Eq. (6–215) is

$$[\hat{\mathbf{H}} \vdots \hat{\mathbf{G}}\hat{\mathbf{H}} \vdots \cdots \vdots \hat{\mathbf{G}}^{n+m-1}\hat{\mathbf{H}}] = (n+m) \times m(n+m) \text{ matrix}$$

In terms of **G** and **H**, this controllability matrix can be written as follows:

$$[\hat{\mathbf{H}} \vdots \hat{\mathbf{G}}\hat{\mathbf{H}} \vdots \cdots \vdots \hat{\mathbf{G}}^{n+m-1}\hat{\mathbf{H}}] = \left[\begin{array}{c|c|c|c|c|c|c} \mathbf{0} & \mathbf{H} & \mathbf{GH} & \cdots & \mathbf{G}^{n-1}\mathbf{H} & \cdots & \mathbf{G}^{n+m-2}\mathbf{H} \\ \hline \mathbf{I}_m & \mathbf{0} & \mathbf{0} & \cdots & \mathbf{0} & \cdots & \mathbf{0} \end{array}\right] \tag{6–217}$$

Since the plant state equation given by Eq. (6–201) is assumed to be completely state controllable, the rank of the matrix

$$[\mathbf{H} \vdots \mathbf{G}\mathbf{H} \vdots \cdots \vdots \mathbf{G}^{n-1}\mathbf{H}]$$

is n. Hence the rank of the matrix given by Eq. (6–217) is $n + m$. Consequently, if the plant is completely state controllable, then the system defined by Eq. (6–215) is completely state controllable and therefore the pole placement technique discussed in Sec. 6–5 applies to this case. [Notice that the system equations as given by Eqs. (6–215) and (6–216) are exactly of the same form as Eqs. (6–79) and (6–80).]

Once the desired closed-loop poles are specified, matrix $\hat{\mathbf{K}}$ can be determined by the pole placement technique. Using matrix $\hat{\mathbf{K}}$ thus determined, we can obtain matrices $\mathbf{K}_1$ and $\mathbf{K}_2$ as follows. First, note that

$$[\mathbf{K}_2 \vdots \mathbf{K}_1]\left[\begin{array}{c|c} \mathbf{G}-\mathbf{I}_n & \mathbf{H} \\ \hline \mathbf{CG} & \mathbf{CH} \end{array}\right] = [\mathbf{K}_2\mathbf{G} - \mathbf{K}_2 + \mathbf{K}_1\mathbf{CG} \vdots \mathbf{K}_2\mathbf{H} + \mathbf{K}_1\mathbf{CH}] \tag{6–218}$$

Then, from Eq. (6–214), we have

$$\begin{aligned}\hat{\mathbf{K}} &= [\mathbf{K}_2\mathbf{G} - \mathbf{K}_2 + \mathbf{K}_1\mathbf{CG} \vdots -\mathbf{I}_m + \mathbf{K}_2\mathbf{H} + \mathbf{K}_1\mathbf{CH}] \\ &= [\mathbf{K}_2 \vdots \mathbf{K}_1]\left[\begin{array}{c|c} \mathbf{G}-\mathbf{I}_n & \mathbf{H} \\ \hline \mathbf{CG} & \mathbf{CH} \end{array}\right] + [\mathbf{0} \vdots -\mathbf{I}_m]\end{aligned}$$

Hence, we obtain

$$[\mathbf{K}_2 \vdots \mathbf{K}_1]\left[\begin{array}{c|c} \mathbf{G}-\mathbf{I}_n & \mathbf{H} \\ \hline \mathbf{CG} & \mathbf{CH} \end{array}\right] = \hat{\mathbf{K}} - [\mathbf{0} \vdots -\mathbf{I}_m]$$

or

$$[\mathbf{K}_2 \vdots \mathbf{K}_1] = \left[\hat{\mathbf{K}} + [\mathbf{0} \vdots \mathbf{I}_m]\right]\left[\begin{array}{c|c} \mathbf{G}-\mathbf{I}_n & \mathbf{H} \\ \hline \mathbf{CG} & \mathbf{CH} \end{array}\right]^{-1} \tag{6–219}$$

Equation (6–219) gives the desired matrices $\mathbf{K}_1$ and $\mathbf{K}_2$.

It is noted that when $\mathbf{u}(k)$ is an m-vector and $m > 1$, matrix $\hat{\mathbf{K}}$ is not unique. Consequently, more than one set of matrices $\mathbf{K}_1$ and $\mathbf{K}_2$ can be determined. (Each possible $\hat{\mathbf{K}}$ yields a set of matrices $\mathbf{K}_1$ and $\mathbf{K}_2$.) In general, the set of $\mathbf{K}_1$ and $\mathbf{K}_2$ that gives the best overall system performance must be chosen.

An alternate approach to determining the matrices $\mathbf{K}_1$ and $\mathbf{K}_2$ is to obtain the characteristic equation for the closed-loop system and equate it with the desired characteristic equation. The characteristic equation for the closed-loop system can be obtained from Eq. (6–211), as follows:

$$\left| z\mathbf{I}_{n+m} - \begin{bmatrix} \mathbf{G} & \mathbf{H} \\ \mathbf{K}_2 - \mathbf{K}_2\mathbf{G} - \mathbf{K}_1\mathbf{CG} & \mathbf{I}_m - \mathbf{K}_2\mathbf{H} - \mathbf{K}_1\mathbf{CH} \end{bmatrix} \right| = 0 \tag{6–220}$$

Rewriting Eq. (6–220), we obtain

$$\begin{aligned} &\left|\begin{array}{c|c} z\mathbf{I}_n - \mathbf{G} & -\mathbf{H} \\ \hline -\mathbf{K}_2 + \mathbf{K}_2\mathbf{G} + \mathbf{K}_1\mathbf{CG} & z\mathbf{I}_m - \mathbf{I}_m + \mathbf{K}_2\mathbf{H} + \mathbf{K}_1\mathbf{CH} \end{array}\right| \\ &= \left| \left[\begin{array}{c|c} \mathbf{I}_n & \mathbf{0} \\ \hline -(\mathbf{K}_2 + \mathbf{K}_1\mathbf{C}) & \mathbf{I}_m \end{array}\right] \left[\begin{array}{c|c} z\mathbf{I}_n - \mathbf{G} + \mathbf{H}(\mathbf{K}_2 + \mathbf{K}_1\mathbf{C}) & -\mathbf{H} \\ \hline \mathbf{K}_1\mathbf{C} & z\mathbf{I}_m - \mathbf{I}_m \end{array}\right] \left[\begin{array}{c|c} \mathbf{I}_n & \mathbf{0} \\ \hline \mathbf{K}_2 + \mathbf{K}_1\mathbf{C} & \mathbf{I}_m \end{array}\right] \right| \\ &= \left| \left[\begin{array}{c|c} z\mathbf{I}_n - \mathbf{G} + \mathbf{H}(\mathbf{K}_2 + \mathbf{K}_1\mathbf{C}) & -\mathbf{H} \\ \hline \mathbf{K}_1\mathbf{C} & z\mathbf{I}_m - \mathbf{I}_m \end{array}\right] \right| \end{aligned}$$

$$= \left| \begin{bmatrix} z\mathbf{I}_n - \mathbf{G} + \mathbf{H}\mathbf{K}_2 + \mathbf{H}\mathbf{K}_1\mathbf{C} + \mathbf{H}\mathbf{K}_1\mathbf{C}(z\mathbf{I}_m - \mathbf{I}_m)^{-1} & -\mathbf{H} \\ \mathbf{0} & z\mathbf{I}_m - \mathbf{I}_m \end{bmatrix} \right.$$

$$\left. \cdot \begin{bmatrix} \mathbf{I}_n & \mathbf{0} \\ \mathbf{K}_1\mathbf{C}(z\mathbf{I}_m - \mathbf{I}_m)^{-1} & \mathbf{I}_m \end{bmatrix} \right|$$

$$= \begin{vmatrix} z\mathbf{I}_n - \mathbf{G} + \mathbf{H}\mathbf{K}_2 + \mathbf{H}\mathbf{K}_1\mathbf{C} + \mathbf{H}\mathbf{K}_1\mathbf{C}(z\mathbf{I}_m - \mathbf{I}_m)^{-1} & -\mathbf{H} \\ \mathbf{0} & z\mathbf{I}_m - \mathbf{I}_m \end{vmatrix}$$

$$= |z\mathbf{I}_n - \mathbf{G} + \mathbf{H}\mathbf{K}_2 + \mathbf{H}\mathbf{K}_1\mathbf{C} + \mathbf{H}\mathbf{K}_1\mathbf{C}(z\mathbf{I}_m - \mathbf{I}_m)^{-1}||z\mathbf{I}_m - \mathbf{I}_m| = 0 \quad (6\text{–}221)$$

Equation (6–221) gives the characteristic equation for the system. We determine matrices $\mathbf{K}_1$ and $\mathbf{K}_2$ so that the roots of this characteristic equation assume the desired values. For example, if deadbeat response to a step input is desired, then we determine $\mathbf{K}_1$ and $\mathbf{K}_2$ so that all roots of the characteristic equation are at the origin. As stated earlier, when the control $\mathbf{u}(k)$ is an m-vector (where $m > 1$), matrices $\mathbf{K}_1$ and $\mathbf{K}_2$ are not unique. That is, more than one set of $\mathbf{K}_1$ and $\mathbf{K}_2$ can be obtained.

Finally, it is noted that if not all state variables are measurable, then we need to substitute the observed state variables for the unmeasurable state variables for state feedback purposes. (Also, if the measured state variables are contaminated by noises and therefore are not accurate, then we prefer to use the observed state variables, rather than the actual state variables, for state feedback purposes.) Figure 6–12 shows a block diagram for the servo system with state feedback where the observed state is used in place of the actual state. The effects of the use of the observed state (instead of the actual state) on the performance of a closed-loop control system will be discussed in Chap. 7.

Example 6–12.

Consider the digital control of a plant by use of state feedback and integral control. Assume that the system configuration is the same as that shown in Fig. 6–11. Assume also that the pulse transfer function of the plant is

$$\frac{Y(z)}{U(z)} = \frac{z^{-2} + 0.5z^{-3}}{1 - z^{-1} + 0.01z^{-2} + 0.12z^{-3}} \quad (6\text{–}222)$$

where $Y(z)$ and $U(z)$ are the z transforms of the plant output $y(k)$ and plant input (control signal) $u(k)$, respectively.

Determine an integral gain constant K_1 and a state feedback gain matrix $\mathbf{K}_2$ such that the response to a unit step command input is deadbeat. Assuming that not all state variables are available for direct measurement, and using the system configuration shown in Fig. 6–12 as an example for a block diagram of a system with a state observer, design a state observer such that the observed state approaches the true state as fast as possible.

We shall first obtain a state space representation for the plant pulse transfer function. By comparing the given pulse transfer function with the standard form

$$\frac{Y(z)}{U(z)} = \frac{b_0 + b_1z^{-1} + b_2z^{-2} + b_3z^{-3}}{1 + a_1z^{-1} + a_2z^{-2} + a_3z^{-3}}$$

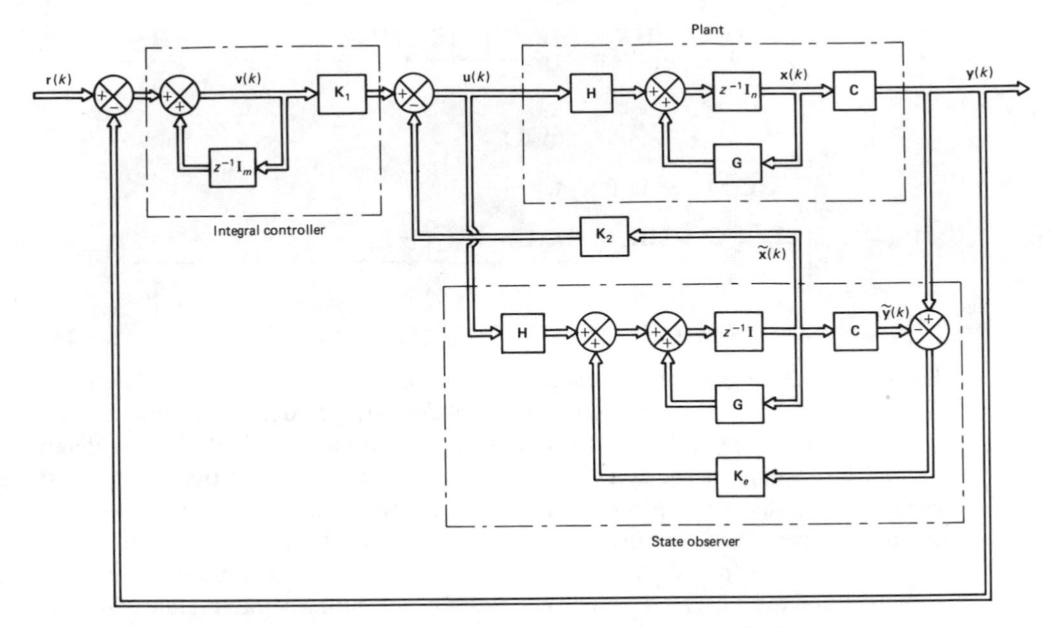

Figure 6–12 Servo system with observed-state feedback.

we find

$$b_0 = 0, \qquad b_1 = 0, \qquad b_2 = 1, \qquad b_3 = 0.5$$

$$a_1 = -1, \qquad a_2 = 0.01, \qquad a_3 = 0.12$$

Then by referring to Eqs. (5–17) and (5–18), we can obtain the following state space equations for the plant:

$$\begin{bmatrix} x_1(k+1) \\ x_2(k+1) \\ x_3(k+1) \end{bmatrix} = \begin{bmatrix} 0 & 1 & 0 \\ 0 & 0 & 1 \\ -0.12 & -0.01 & 1 \end{bmatrix} \begin{bmatrix} x_1(k) \\ x_2(k) \\ x_3(k) \end{bmatrix} + \begin{bmatrix} 0 \\ 0 \\ 1 \end{bmatrix} u(k) \tag{6–223}$$

$$y(k) = [0.5 \quad 1 \quad 0] \begin{bmatrix} x_1(k) \\ x_2(k) \\ x_3(k) \end{bmatrix} \tag{6–224}$$

Determination of Integral Gain Constant K_1 and State Feedback Gain Matrix $\mathbf{K}_2$ for Deadbeat Response. We shall now determine the integral gain constant K_1 and the state feedback gain matrix $\mathbf{K}_2$. In the present system we require the response to the step command input to be deadbeat. (Thus, we must place the closed-loop poles of the system at the origin.)

Referring to Eqs. (6–215) and (6–216), we have

$$\boldsymbol{\xi}(k+1) = \hat{\mathbf{G}}\boldsymbol{\xi}(k) + \hat{\mathbf{H}}w(k)$$

$$w(k) = -\hat{\mathbf{K}}\boldsymbol{\xi}(k)$$

where

$$\hat{\mathbf{G}} = \begin{bmatrix} \mathbf{G} & \vdots & \mathbf{H} \\ \cdots & & \cdots \\ \mathbf{0} & \vdots & 0 \end{bmatrix} = \begin{bmatrix} 0 & 1 & 0 & \vdots & 0 \\ 0 & 0 & 1 & \vdots & 0 \\ -0.12 & -0.01 & 1 & \vdots & 1 \\ \cdots & \cdots & \cdots & & \cdots \\ 0 & 0 & 0 & \vdots & 0 \end{bmatrix}$$

$$\hat{\mathbf{H}} = \begin{bmatrix} 0 \\ 0 \\ 0 \\ \cdots \\ 1 \end{bmatrix}$$

Our problem here is to determine matrix $\hat{\mathbf{K}}$ so that the closed-loop poles of the system are at the origin, or the desired characteristic equation is

$$z^4 = 0$$

By using the pole placement technique discussed in Sec. 6–5, matrix $\hat{\mathbf{K}}$ can be determined easily. Referring to Ackermann's formula as given by Eq. (6–71), we obtain

$$\hat{\mathbf{K}} = [0 \quad 0 \quad 0 \quad 1][\hat{\mathbf{H}} \vdots \hat{\mathbf{G}}\hat{\mathbf{H}} \vdots \hat{\mathbf{G}}^2\hat{\mathbf{H}} \vdots \hat{\mathbf{G}}^3\hat{\mathbf{H}}]^{-1}\phi(\hat{\mathbf{G}})$$

where

$$\phi(\hat{\mathbf{G}}) = \hat{\mathbf{G}}^4$$

Thus

$$\hat{\mathbf{K}} = [0 \quad 0 \quad 0 \quad 1]\begin{bmatrix} 0 & 0 & 0 & 1 \\ 0 & 0 & 1 & 1 \\ 0 & 1 & 1 & 0.99 \\ 1 & 0 & 0 & 0 \end{bmatrix}^{-1}\begin{bmatrix} 0 & 1 & 0 & 0 \\ 0 & 0 & 1 & 0 \\ -0.12 & -0.01 & 1 & 1 \\ 0 & 0 & 0 & 0 \end{bmatrix}^4$$

$$= [0 \quad 0 \quad 0 \quad 1]\begin{bmatrix} 0 & 0 & 0 & 1 \\ 0.01 & -1 & 1 & 0 \\ -1 & 1 & 0 & 0 \\ 1 & 0 & 0 & 0 \end{bmatrix}\begin{bmatrix} -0.12 & -0.13 & 0.99 & 1 \\ -0.1188 & -0.1299 & 0.86 & 0.99 \\ -0.1032 & -0.1274 & 0.7301 & 0.86 \\ 0 & 0 & 0 & 0 \end{bmatrix}$$

$$= [-0.12 \quad -0.13 \quad 0.99 \quad 1] \tag{6–225}$$

Equation (6–225) gives the matrix $\hat{\mathbf{K}}$.

The desired integral gain constant K_1 and the state feedback gain matrix $\mathbf{K}_2$ are obtained from Eq. (6–219):

$$[\mathbf{K}_2 \vdots K_1] = \left[\hat{\mathbf{K}} + [\mathbf{0} \vdots 1]\right]\begin{bmatrix} \mathbf{G} - \mathbf{I}_3 & \vdots & \mathbf{H} \\ \cdots & & \cdots \\ \mathbf{CG} & \vdots & \mathbf{CH} \end{bmatrix}^{-1}$$

$$= [-0.12 \quad -0.13 \quad 0.99 \ \vdots \ 2]\begin{bmatrix} -1 & 1 & 0 & \vdots & 0 \\ 0 & -1 & 1 & \vdots & 0 \\ -0.12 & -0.01 & 0 & \vdots & 1 \\ \cdots & \cdots & \cdots & & \cdots \\ 0 & 0.5 & 1 & \vdots & 0 \end{bmatrix}^{-1}$$

$$= [-0.12 \quad -0.13 \quad 0.99 \ \vdots \ 2] \left[\begin{array}{ccc|c} -1 & -\frac{2}{3} & 0 & \frac{1}{1.5} \\ 0 & -\frac{2}{3} & 0 & \frac{1}{1.5} \\ 0 & \frac{1}{3} & 0 & \frac{1}{1.5} \\ \hline -0.12 & -\frac{0.26}{3} & 1 & \frac{0.13}{1.5} \end{array}\right]$$

$$= [-0.12 \quad 0.3233 \quad 2 \ \vdots \ 0.6667] \tag{6–226}$$

From Eq. (6–226) we obtain the integral gain constant K_1:

$$K_1 = 0.6667 = \tfrac{2}{3} \tag{6–227}$$

The state feedback gain matrix $\mathbf{K}_2$ is given by

$$\mathbf{K}_2 = [-0.12 \quad 0.3233 \quad 2] \tag{6–228}$$

Determining Output $y(k)$. Next, let us determine the output $y(k)$. From Eq. (6–224) we have

$$y(k) = \mathbf{C}\mathbf{x}(k) = [0.5 \quad 1 \quad 0] \begin{bmatrix} x_1(k) \\ x_2(k) \\ x_3(k) \end{bmatrix}$$

In order to obtain output $y(k)$, we shall first determine the state vector $\mathbf{x}(k)$ and signal $v(k)$. From Fig. 6–11, we have

$$\mathbf{x}(k+1) = \mathbf{G}\mathbf{x}(k) + \mathbf{H}u(k) \tag{6–229}$$

$$y(k) = \mathbf{C}\mathbf{x}(k) \tag{6–230}$$

$$v(k) = v(k-1) + r(k) - y(k) \tag{6–231}$$

$$u(k) = -\mathbf{K}_2\mathbf{x}(k) + K_1 v(k) \tag{6–232}$$

Hence from Eqs. (6–229) and (6–232) we obtain

$$\begin{aligned} \mathbf{x}(k+1) &= \mathbf{G}\mathbf{x}(k) + \mathbf{H}u(k) \\ &= (\mathbf{G} - \mathbf{H}\mathbf{K}_2)\mathbf{x}(k) + \mathbf{H}K_1 v(k) \end{aligned} \tag{6–233}$$

Also, from Eqs. (6–230), (6–231), and (6–232) we get

$$\begin{aligned} v(k+1) &= v(k) + r(k+1) - y(k+1) \\ &= v(k) + r(k+1) - \mathbf{C}\mathbf{x}(k+1) \\ &= v(k) + r(k+1) - \mathbf{C}[(\mathbf{G} - \mathbf{H}\mathbf{K}_2)\mathbf{x}(k) + \mathbf{H}K_1 v(k)] \\ &= -(\mathbf{C}\mathbf{G} - \mathbf{C}\mathbf{H}\mathbf{K}_2)\mathbf{x}(k) + (1 - \mathbf{C}\mathbf{H}K_1)v(k) + r(k+1) \end{aligned} \tag{6–234}$$

Combining Eqs. (6–233) and (6–234), we get

$$\begin{bmatrix} \mathbf{x}(k+1) \\ \hline v(k+1) \end{bmatrix} = \left[\begin{array}{c|c} \mathbf{G} - \mathbf{H}\mathbf{K}_2 & \mathbf{H}K_1 \\ \hline -\mathbf{C}\mathbf{G} + \mathbf{C}\mathbf{H}\mathbf{K}_2 & 1 - \mathbf{C}\mathbf{H}K_1 \end{array}\right] \begin{bmatrix} \mathbf{x}(k) \\ \hline v(k) \end{bmatrix} + \begin{bmatrix} \mathbf{0} \\ \hline 1 \end{bmatrix} r(k+1) \tag{6–235}$$

which can be rewritten as

$$\begin{bmatrix} x_1(k+1) \\ x_2(k+1) \\ x_3(k+1) \\ v(k+1) \end{bmatrix} = \begin{bmatrix} 0 & 1 & 0 & 0 \\ 0 & 0 & 1 & 0 \\ 0 & -\frac{1}{3} & -1 & \frac{2}{3} \\ 0 & -\frac{1}{2} & -1 & 1 \end{bmatrix} \begin{bmatrix} x_1(k) \\ x_2(k) \\ x_3(k) \\ v(k) \end{bmatrix} + \begin{bmatrix} 0 \\ 0 \\ 0 \\ 1 \end{bmatrix} r(k+1) \tag{6–236}$$

Since the command input $r(k)$ is a unit step input, we have

$$r(k) = 1 \qquad k = 0, 1, 2, \ldots$$

Let us assume that the initial state is

$$\begin{bmatrix} x_1(0) \\ x_2(0) \\ x_3(0) \\ v(0) \end{bmatrix} = \begin{bmatrix} a \\ b \\ c \\ d \end{bmatrix}$$

where a, b, c, and d are arbitrary. Then, from Eq. (6–236) we have

$$\begin{bmatrix} x_1(1) \\ x_2(1) \\ x_3(1) \\ v(1) \end{bmatrix} = \begin{bmatrix} 0 & 1 & 0 & 0 \\ 0 & 0 & 1 & 0 \\ 0 & -\frac{1}{3} & -1 & \frac{2}{3} \\ 0 & -\frac{1}{2} & -1 & 1 \end{bmatrix} \begin{bmatrix} a \\ b \\ c \\ d \end{bmatrix} + \begin{bmatrix} 0 \\ 0 \\ 0 \\ 1 \end{bmatrix} [1] = \begin{bmatrix} b \\ c \\ -\frac{1}{3}b - c + \frac{2}{3}d \\ -\frac{1}{2}b - c + d + 1 \end{bmatrix}$$

Similarly,

$$\begin{bmatrix} x_1(2) \\ x_2(2) \\ x_3(2) \\ v(2) \end{bmatrix} = \begin{bmatrix} c \\ -\frac{1}{3}b - c + \frac{2}{3}d \\ \frac{2}{3} \\ -\frac{1}{6}b - \frac{1}{2}c + \frac{1}{3}d + 2 \end{bmatrix}$$

$$\begin{bmatrix} x_1(3) \\ x_2(3) \\ x_3(3) \\ v(3) \end{bmatrix} = \begin{bmatrix} -\frac{1}{3}b - c + \frac{2}{3}d \\ \frac{2}{3} \\ \frac{2}{3} \\ \frac{7}{3} \end{bmatrix}$$

and

$$\begin{bmatrix} x_1(k) \\ x_2(k) \\ x_3(k) \\ v(k) \end{bmatrix} = \begin{bmatrix} \frac{2}{3} \\ \frac{2}{3} \\ \frac{2}{3} \\ \frac{7}{3} \end{bmatrix} \qquad k = 4, 5, 6, \ldots$$

The output $y(k)$ is obtained as follows:

$$y(0) = [0.5 \quad 1 \quad 0] \begin{bmatrix} x_1(0) \\ x_2(0) \\ x_3(0) \end{bmatrix} = [0.5 \quad 1 \quad 0] \begin{bmatrix} a \\ b \\ c \end{bmatrix} = \frac{1}{2}a + b$$

Similarly,

$$y(1) = \frac{1}{2}b + c$$

$$y(2) = -\frac{1}{3}b - \frac{1}{2}c + \frac{2}{3}d$$

$$y(3) = -\frac{1}{6}b - \frac{1}{2}c + \frac{1}{3}d + \frac{2}{3}$$

$$y(k) = 1 \qquad k = 4, 5, 6, \ldots$$

Notice that

$$u(k) = -\mathbf{K}_2\mathbf{x}(k) + K_1v(k)$$

$$= -[-0.12 \quad 0.3233 \quad 2]\begin{bmatrix} \frac{2}{3} \\ \frac{2}{3} \\ \frac{2}{3} \end{bmatrix} + (\tfrac{2}{3})(\tfrac{7}{3}) = 0.08670$$

where $k = 4, 5, 6, \ldots$ Since $u(t)$ for $t \geq 4T$ is constant, there is no intersampling oscillation in the output. Thus, the response of the system is deadbeat.

Note that the output $y(k)$ reaches unity in at most 4 sampling periods and will stay there in the absence of disturbances or new command inputs. [See, for example, a sample unit step response sequence shown in Fig. 6–13(*a*).] Under special initial conditions, for example, $a = b = c = 0$ and $d = 1$, the output reaches unity in 3 sampling periods and stays there, or $y(k) = 1$ for $k = 3, 4, 5, \ldots$ [See Fig. 6–13(*b*).]

Design of the State Observer. Next, we shall design a state observer for the system. Since the plant output $y(k)$ is measurable, let us design a minimum-order state observer.

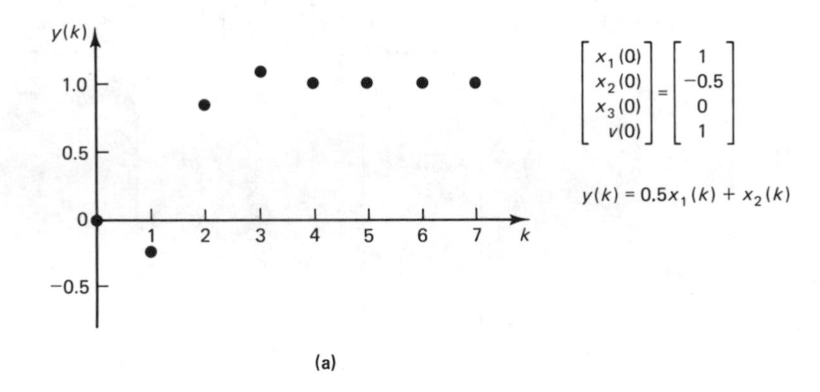

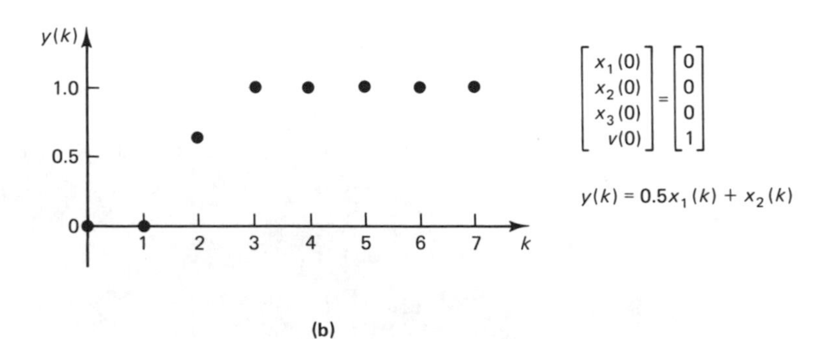

Figure 6–13 Sample unit step response sequences for the servo system with actual- (measured-) state feedback and integral control designed in Example 6–12.

Let us assume that we desire the deadbeat response. In the present system, output matrix $\mathbf{C}$ is given by

$$\mathbf{C} = [0.5 \quad 1 \quad 0]$$

In order to change the output matrix $\mathbf{C}$ from $[0.5 \quad 1 \quad 0]$ to $[1 \quad 0 \quad 0]$, let us make the following transformation:

$$\mathbf{x}(k) = \mathbf{T}\boldsymbol{\xi}(k) \tag{6–237}$$

where

$$\mathbf{T} = \begin{bmatrix} 0 & 0 & 1 \\ 1 & 0 & -0.5 \\ 0 & 1 & 0 \end{bmatrix} \tag{6–238}$$

Note that

$$\mathbf{T}^{-1} = \begin{bmatrix} 0.5 & 1 & 0 \\ 0 & 0 & 1 \\ 1 & 0 & 0 \end{bmatrix} \tag{6–239}$$

Then the plant state equations become

$$\boldsymbol{\xi}(k+1) = \mathbf{T}^{-1}\mathbf{G}\mathbf{T}\boldsymbol{\xi}(k) + \mathbf{T}^{-1}\mathbf{H}u(k) \tag{6–240}$$

$$y(k) = \mathbf{C}\mathbf{T}\boldsymbol{\xi}(k) \tag{6–241}$$

where

$$\mathbf{T}^{-1}\mathbf{G}\mathbf{T} = \begin{bmatrix} 0.5 & 1 & 0 \\ 0 & 0 & 1 \\ 1 & 0 & 0 \end{bmatrix} \begin{bmatrix} 0 & 1 & 0 \\ 0 & 0 & 1 \\ -0.12 & -0.01 & 1 \end{bmatrix} \begin{bmatrix} 0 & 0 & 1 \\ 1 & 0 & -0.5 \\ 0 & 1 & 0 \end{bmatrix}$$

$$= \begin{bmatrix} 0.5 & 1 & -0.25 \\ -0.01 & 1 & -0.115 \\ 1 & 0 & -0.5 \end{bmatrix}$$

$$\mathbf{T}^{-1}\mathbf{H} = \begin{bmatrix} 0.5 & 1 & 0 \\ 0 & 0 & 1 \\ 1 & 0 & 0 \end{bmatrix} \begin{bmatrix} 0 \\ 0 \\ 1 \end{bmatrix} = \begin{bmatrix} 0 \\ 1 \\ 0 \end{bmatrix}$$

$$\mathbf{C}\mathbf{T} = [0.5 \quad 1 \quad 0] \begin{bmatrix} 0 & 0 & 1 \\ 1 & 0 & -0.5 \\ 0 & 1 & 0 \end{bmatrix} = [1 \quad 0 \quad 0]$$

The transformed system equations are as follows:

$$\begin{bmatrix} \xi_1(k+1) \\ \hline \xi_2(k+1) \\ \xi_3(k+1) \end{bmatrix} = \left[\begin{array}{c|cc} 0.5 & 1 & -0.25 \\ \hline -0.01 & 1 & -0.115 \\ 1 & 0 & -0.5 \end{array}\right] \begin{bmatrix} \xi_1(k) \\ \hline \xi_2(k) \\ \xi_3(k) \end{bmatrix} + \begin{bmatrix} 0 \\ \hline 1 \\ 0 \end{bmatrix} u(k) \tag{6–242}$$

$$y(k) = [1 \,\vdots\, 0 \quad 0] \begin{bmatrix} \xi_1(k) \\ \hline \xi_2(k) \\ \xi_3(k) \end{bmatrix} \tag{6–243}$$

Since only one state variable can be measured, we need to observe two state variables. Hence, the order of the minimum-order state observer is 2. From Eqs. (6–165) and (6–242), we have

$$\mathbf{G}_{bb} - \mathbf{K}_e \mathbf{G}_{ab} = \begin{bmatrix} 1 & -0.115 \\ 0 & -0.5 \end{bmatrix} - \begin{bmatrix} k_{e_1} \\ k_{e_2} \end{bmatrix} [1 \quad -0.25]$$

$$= \begin{bmatrix} 1 - k_{e_1} & -0.115 + 0.25k_{e_1} \\ -k_{e_2} & -0.5 + 0.25k_{e_2} \end{bmatrix}$$

The observer characteristic equation is

$$|z\mathbf{I} - \mathbf{G}_{bb} + \mathbf{K}_e \mathbf{G}_{ab}| = \begin{vmatrix} z - 1 + k_{e_1} & 0.115 - 0.25k_{e_1} \\ k_{e_2} & z + 0.5 - 0.25k_{e_2} \end{vmatrix}$$

$$= (z - 1 + k_{e_1})(z + 0.5 - 0.25k_{e_2}) - k_{e_2}(0.115 - 0.25k_{e_1})$$

$$= z^2 + (k_{e_1} - 0.25k_{e_2} - 0.5)z + (0.5k_{e_1} + 0.135k_{e_2} - 0.5)$$

$$= 0 \tag{6–244}$$

Since we desire the deadbeat response, the desired characteristic equation is

$$z^2 = 0$$

Hence we require

$$k_{e_1} - 0.25k_{e_2} - 0.5 = 0$$

$$0.5k_{e_1} + 0.135k_{e_2} - 0.5 = 0$$

Solving these two simultaneous equations for k_{e_1} and k_{e_2}, we obtain

$$\mathbf{K}_e = \begin{bmatrix} k_{e_1} \\ k_{e_2} \end{bmatrix} = \begin{bmatrix} 0.7404 \\ 0.9615 \end{bmatrix} \tag{6–245}$$

Integral Control with State Observer. We have thus considered a design problem in which the observed state variables are fed back in a minor loop and an integral controller is used in the main loop.

In the pole placement part of the design, we used the actual state rather than the observed state. In what follows we shall obtain the system equations for the case where the integral controller and the state observer are used.

The use of the observed state $\tilde{\mathbf{x}}(k)$, where $\tilde{\mathbf{x}}(k) = \mathbf{T}\tilde{\boldsymbol{\xi}}(k)$, in the state feedback control modifies the control signal $u(k)$ as follows. From Eq. (6–232) we have

$$u(k) = -\mathbf{K}_2\tilde{\mathbf{x}}(k) + K_1 v(k) = -\mathbf{K}_2\mathbf{T}\tilde{\boldsymbol{\xi}}(k) + K_1 v(k) \tag{6–246}$$

Define

$$\boldsymbol{\xi}(k) - \tilde{\boldsymbol{\xi}}(k) = \boldsymbol{\epsilon}(k)$$

Then Eq. (6–246) can be written as follows:

$$u(k) = -\mathbf{K}_2\mathbf{T}\boldsymbol{\xi}(k) + K_1 v(k) + \mathbf{K}_2\mathbf{T}\boldsymbol{\epsilon}(k) \tag{6–247}$$

which can be rewritten as

$$u(k)=[-0.3233 \quad -2 \quad 0.2817]\begin{bmatrix}\xi_1(k)\\ \xi_2(k)\\ \xi_3(k)\end{bmatrix}+\tfrac{2}{3}v(k)+[0.3233 \quad 2 \quad -0.2817]\begin{bmatrix}\epsilon_1(k)\\ \epsilon_2(k)\\ \epsilon_3(k)\end{bmatrix} \tag{6–248}$$

By substituting Eq. (6–248) into Eq. (6–242), we obtain

$$\begin{bmatrix}\xi_1(k+1)\\ \xi_2(k+1)\\ \xi_3(k+1)\end{bmatrix}=\begin{bmatrix}0.5 & 1 & -0.25\\ -\frac{1}{3} & -1 & \frac{1}{6}\\ 1 & 0 & -0.5\end{bmatrix}\begin{bmatrix}\xi_1(k)\\ \xi_2(k)\\ \xi_3(k)\end{bmatrix}+\begin{bmatrix}0\\ \frac{2}{3}\\ 0\end{bmatrix}v(k)+\begin{bmatrix}0 & 0 & 0\\ 0.3233 & 2 & -0.2817\\ 0 & 0 & 0\end{bmatrix}\begin{bmatrix}\epsilon_1(k)\\ \epsilon_2(k)\\ \epsilon_3(k)\end{bmatrix} \tag{6–249}$$

Also, Eq. (6–234) can be modified to read

$$v(k+1)=-(\mathbf{CGT}-\mathbf{CHK}_2\mathbf{T})\boldsymbol{\xi}(k)+(1-\mathbf{CHK}_1)v(k)+r(k+1)$$

or

$$v(k+1)=-[0.5 \quad 1 \quad -0.25]\begin{bmatrix}\xi_1(k)\\ \xi_2(k)\\ \xi_3(k)\end{bmatrix}+v(k)+r(k+1) \tag{6–250}$$

Referring to Eq. (6–183) and noting that $\epsilon_1(k)=0$, we can give the observer error dynamics by

$$\begin{bmatrix}\epsilon_2(k+1)\\ \epsilon_3(k+1)\end{bmatrix}=[\mathbf{G}_{bb}-\mathbf{K}_e\mathbf{G}_{ab}]\begin{bmatrix}\epsilon_2(k)\\ \epsilon_3(k)\end{bmatrix}$$

Therefore,

$$\left[\begin{array}{c}\epsilon_1(k+1)\\ \hline \epsilon_2(k+1)\\ \epsilon_3(k+1)\end{array}\right]=\left[\begin{array}{c|cc}1 & 0 & 0\\ \hline 0 & 0.2596 & 0.0701\\ 0 & -0.9615 & -0.2596\end{array}\right]\left[\begin{array}{c}\epsilon_1(k)\\ \hline \epsilon_2(k)\\ \epsilon_3(k)\end{array}\right] \tag{6–251}$$

Combining Eqs. (6–249), (6–250), and (6–251) into one state equation, we obtain

$$\begin{bmatrix}\xi_1(k+1)\\ \xi_2(k+1)\\ \xi_3(k+1)\\ v(k+1)\\ \epsilon_1(k+1)\\ \epsilon_2(k+1)\\ \epsilon_3(k+1)\end{bmatrix}=\begin{bmatrix}0.5 & 1 & -0.25 & 0 & 0 & 0 & 0\\ -\frac{1}{3} & -1 & \frac{1}{6} & \frac{2}{3} & 0.3233 & 2 & -0.2817\\ 1 & 0 & -0.5 & 0 & 0 & 0 & 0\\ -0.5 & -1 & 0.25 & 1 & 0 & 0 & 0\\ 0 & 0 & 0 & 0 & 1 & 0 & 0\\ 0 & 0 & 0 & 0 & 0 & 0.2596 & 0.0701\\ 0 & 0 & 0 & 0 & 0 & -0.9615 & -0.2596\end{bmatrix}\begin{bmatrix}\xi_1(k)\\ \xi_2(k)\\ \xi_3(k)\\ v(k)\\ \epsilon_1(k)\\ \epsilon_2(k)\\ \epsilon_3(k)\end{bmatrix}+\begin{bmatrix}0\\ 0\\ 0\\ 1\\ 0\\ 0\\ 0\end{bmatrix}r(k+1) \tag{6–252}$$

The plant output $y(k)$ can be given by

$$y(k) = [1 \quad 0 \quad 0]\begin{bmatrix} \xi_1(k) \\ \xi_2(k) \\ \xi_3(k) \end{bmatrix} = \xi_1(k) \tag{6-253}$$

It can be shown that for the unit step command input, the response of the system under an arbitrary initial condition,

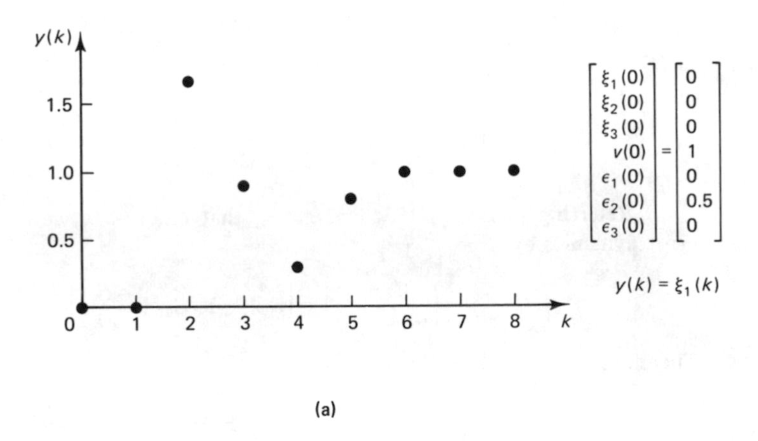

(a)

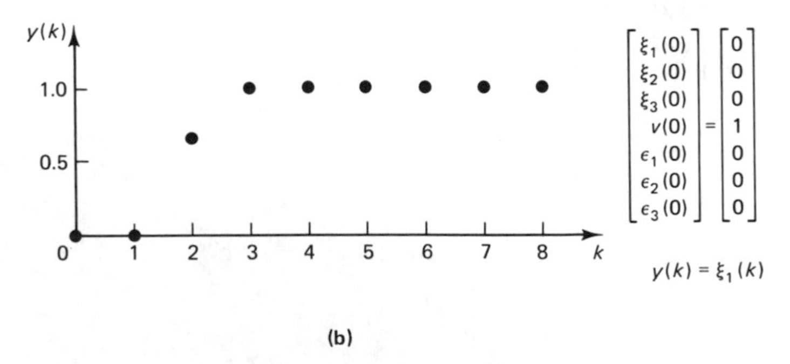

(b)

Figure 6–14 Sample unit step response sequences for the servo system with observed-state feedback and integral control designed in Example 6–12.

$$\begin{bmatrix} \xi_1(0) \\ \xi_2(0) \\ \xi_3(0) \\ v(0) \\ \epsilon_1(0) \\ \epsilon_2(0) \\ \epsilon_3(0) \end{bmatrix} = \begin{bmatrix} a \\ b \\ c \\ d \\ 0 \\ \alpha \\ \beta \end{bmatrix}$$

requires at most 6 sampling periods to complete. (That is, the output reaches unity in at most 6 sampling periods and thereafter it will stay there in the absence of disturbances and new command inputs.)

Notice that, as we have seen earlier, if the actual state can be fed back, the system requires at most 4 sampling periods to complete the unit step response. [The system requires only 3 sampling periods to complete the unit step response if $\xi_1(0) = 0$, $\xi_2(0) = 0$, $\xi_3(0) = 0$, and $v(0) = 1$.] However, if the state observer is used, the response time increases. If the minimum-order state observer (second-order state observer, in the present case) is used, then the system requires at most 6 sampling periods to complete the unit step response. [See, for example, a sample unit step response sequence shown in Fig. 6–14(*a*).] This means that in special cases the response time is much shorter. For example, if the initial conditions are $\xi_1(0) = 0$, $\xi_2(0) = 0$, $\xi_3(0) = 0$, $v(0) = 1$, $\epsilon_2(0) = 0$, and $\epsilon_3(0) = 0$, then the system requires only 3 sampling periods to complete the unit step response. [See Fig. 6–14(*b*).]

REFERENCES

6–1. Åström, K. J., and B. Wittenmark, *Computer Controlled Systems: Theory and Design*. Englewood Cliffs, N.J.: Prentice-Hall, Inc., 1984.

6–2. Butman, S., and R. Sivan (Sussman), "On Cancellations, Controllability and Observability," *IEEE Trans. Automatic Control*, **AC-9** (1964), pp. 317–18.

6–3. Falb, P. L., and M. Athans, "A Direct Constructive Proof of the Criterion for Complete Controllability of Time-Invariant Linear Systems," *IEEE Trans. Automatic Control*, **AC-9** (1964), pp. 189–90.

6–4. Franklin, G. F., and J. D. Powell, *Digital Control of Dynamic Systems*. Reading, Mass: Addison-Wesley Publishing Co., Inc., 1980.

6–5. Gopinath, B., "On the Control of Linear Multiple Input-Output Systems," Bell Syst. Tech. J., **50** (1971), pp. 1063–81.

6–6. Kailath, T., *Linear Systems*. Englewood Cliffs, N.J.: Prentice-hall, Inc., 1980.

6–7. Kalman, R. E., "On the General Theory of Control Systems," *Proc. First Intern. Cong. IFAC*, Moscow, 1960. *Automatic and Remote Control*. London: Butterworth & Co., Ltd., 1961, pp. 481–92.

6–8. Kalman, R. E., Y. C. Ho, and K. S. Narendra, "Controllability of Linear Dynamical Systems," *Contributions to Differential Equations*, **1** (1963), pp. 189–213.

6–9. Katz, P., *Digital Control Using Microprocessors*. London: Prentice-Hall International, Inc., 1981.

6–10. Kreindler, E., and P. E. Sarachik, "On the Concepts of Controllability and Observability of Linear Systems," *IEEE Trans. Automatic Control*, **AC-9** (1964), pp. 129–36.
6–11. Kuo, B. C., *Digital Control Systems*. New York: Holt, Rinehart and Winston, Inc., 1980.
6–12. Leondes, C. T., and M. Novak, "Reduced-Order Observers for Linear Discrete-Time Systems," *IEEE Trans. Automatic Control*, **AC-19** (1974), pp. 42–46.
6–13. Luenberger, D. G., "An Introduction to Observers," *IEEE Trans. Automatic Control*, **AC-16** (1971), pp. 596–602.
6–14. Luenberger, D. G., "Observing the State of a Linear System," *IEEE Trans. Military Electronics*, **MIL-8** (1964), pp. 74–80.
6–15. Ogata, K., *Modern Control Engineering*. Englewood Cliffs, N.J.: Prentice-Hall, Inc., 1970.
6–16. Ogata, K., *State Space Analysis of Control Systems*. Engelwood Cliffs, N.J.: Prentice-Hall, Inc., 1967.
6–17. Phillips, C. L., and H. T. Nagle, Jr., *Digital Control System Analysis and Design*. Englewood Cliffs, N.J.: Prentice-Hall, Inc., 1984.
6–18. Willems, J. C., and S. K. Mitter, "Controllability, Observability, Pole Allocation, and State Reconstruction," *IEEE Trans. Automatic Control*, **AC-16** (1971), pp. 582–595.
6–19. Wonham, W. M., "On Pole Assignment in Multi-input Controllable Linear Systems," *IEEE Trans. Automatic Control*, **AC-12** (1967), pp. 660–65.
6–20. Zadeh, L. A., and C. A. Desoer, *Linear System Theory: The State Space Approach*. New York: McGraw-Hill Book Company, 1963.

EXAMPLE PROBLEMS AND SOLUTIONS

Problem A-6–1. The definition of controllability given in Sec. 6–2 is not the only one used in the literature. Sometimes the following definition is used: A control system is defined to be state controllable if, given an arbitrary initial state $\mathbf{x}(0)$, it is possible to bring the state to the origin of the state space in a finite time interval, provided the control vector is unconstrained (unbounded).

The concept of reachability, similar to the concept of controllability, is available in the literature and is used in the following way: A control system is defined to be reachable if, starting from the origin of the state space, the state can be brought to an arbitrary point in the state space in a finite time period, provided the control vector is unconstrained.

Show that the system

$$\begin{bmatrix} x_1(k+1) \\ x_2(k+1) \end{bmatrix} = \begin{bmatrix} 0 & 0 \\ -1 & 1 \end{bmatrix} \begin{bmatrix} x_1(k) \\ x_2(k) \end{bmatrix} + \begin{bmatrix} 0 \\ 1 \end{bmatrix} u(k)$$

is controllable (in the sense defined in this problem) but is not reachable.

Solution. Rewriting the system state equation, we obtain

$$x_1(k+1) = 0$$

$$x_2(k+1) = -x_1(k) + x_2(k) + u(k)$$

Starting from an arbitrary initial state, we have

$$x_1(1) = 0$$

$$x_2(1) = -x_1(0) + x_2(0) + u(0)$$

Hence by choosing

$$u(0) = x_1(0) - x_2(0)$$

the state can be brought to the origin in one step. Thus, the system is controllable in the sense defined in this problem.

If the state starts from the origin, we have

$$x_1(1) = 0$$

$$x_2(1) = -x_1(0) + x_2(0) + u(0) = -0 + 0 + u(0) = u(0)$$

Although $x_2(1)$ can be brought to an arbitrary point in one step, $x_1(1)$ cannot be controlled. Consequently, the system is not reachable.

Notice that the present system is not controllable in the sense defined in Sec. 6–2 (where the final state is an arbitrary point in the state space including the origin), because the required rank condition is not satisfied, as the following shows:

$$\text{rank}\,[\mathbf{H} \vdots \mathbf{GH}] = \text{rank}\begin{bmatrix} 0 & 0 \\ 1 & 1 \end{bmatrix} = 1 < 2$$

As seen in this problem, controllability and reachability (both defined in this problem) are different. However, if the state matrix $\mathbf{G}$ is nonsingular, then complete state controllability in the sense defined in this problem and complete state reachability mean the same thing. That is, for the system with a nonsingular matrix $\mathbf{G}$, complete state controllability means complete reachability, and vice versa.

Problem A-6–2. Consider the completely state controllable system defined by

$$\mathbf{x}(k+1) = \mathbf{G}\mathbf{x}(k) + \mathbf{H}u(k)$$

where

$$\mathbf{G} = \begin{bmatrix} 1 & 0.6321 \\ 0 & 0.3679 \end{bmatrix}, \qquad \mathbf{H} = \begin{bmatrix} 0.3679 \\ 0.6321 \end{bmatrix}$$

The sampling period is 1 sec. If the control signal $u(k)$ is unbounded, or

$$-\infty \le u(k) \le \infty$$

then an arbitrary initial state $\mathbf{x}(0)$ can be brought to the origin in at most 2 sampling periods by using a piecewise-constant control signal.

Derive the control law to transfer an arbitrary initial state $\mathbf{x}(0)$ to the origin. Determine the region in the state space in which the initial state can be brought to the origin in one sampling period.

If the magnitude of $u(k)$ is bounded, then some initial state cannot be transferred to the origin in two sampling periods. (Three, four, or more sampling periods may be required.) Suppose that

$$|u(k)| \le 1$$

Determine the region of the initial states in the x_1x_2 plane which can be transferred to the origin in one sampling period and two sampling periods, respectively, by using the bounded control signal $|u(k)| \leq 1$.

Solution.

For the case where $u(k)$ is unbounded. Since the system is of the second order, we need at most two sampling periods to transfer any initial state $\mathbf{x}(0)$ to the origin. Noting that

$$\mathbf{x}(1) = \mathbf{G}\mathbf{x}(0) + \mathbf{H}u(0) \tag{6–254}$$

$$\mathbf{x}(2) = \mathbf{0} = \mathbf{G}\mathbf{x}(1) + \mathbf{H}u(1) = \mathbf{G}^2\mathbf{x}(0) + \mathbf{G}\mathbf{H}u(0) + \mathbf{H}u(1)$$

we obtain

$$\mathbf{x}(0) = -\mathbf{G}^{-1}\mathbf{H}u(0) - \mathbf{G}^{-2}\mathbf{H}u(1) \tag{6–255}$$

Substituting Eq. (6–255) into Eq. (6–254), we obtain

$$\mathbf{x}(1) = -\mathbf{G}^{-1}\mathbf{H}u(1) \tag{6–256}$$

Noting that

$$\mathbf{G}^{-1}\mathbf{H} = \begin{bmatrix} 1 & -1.7181 \\ 0 & 2.7181 \end{bmatrix} \begin{bmatrix} 0.3679 \\ 0.6321 \end{bmatrix} = \begin{bmatrix} -0.7181 \\ 1.7181 \end{bmatrix}$$

$$\mathbf{G}^{-2}\mathbf{H} = \begin{bmatrix} 1 & -6.3881 \\ 0 & 7.3881 \end{bmatrix} \begin{bmatrix} 0.3679 \\ 0.6321 \end{bmatrix} = \begin{bmatrix} -3.6700 \\ 4.6700 \end{bmatrix}$$

we obtain from Eqs. (6–255) and (6–256) the following two equations:

$$\begin{bmatrix} x_1(0) \\ x_2(0) \end{bmatrix} = -\begin{bmatrix} -0.7181 \\ 1.7181 \end{bmatrix} u(0) - \begin{bmatrix} -3.6700 \\ 4.6700 \end{bmatrix} u(1) \tag{6–257}$$

$$\begin{bmatrix} x_1(1) \\ x_2(1) \end{bmatrix} = -\begin{bmatrix} -0.7181 \\ 1.7181 \end{bmatrix} u(1) \tag{6–258}$$

Combining Eqs. (6–257) and (6–258), we obtain

$$\begin{bmatrix} x_1(0) & x_1(1) \\ x_2(0) & x_2(1) \end{bmatrix} = \begin{bmatrix} 0.7181 & 3.6700 \\ -1.7181 & -4.6700 \end{bmatrix} \begin{bmatrix} u(0) & u(1) \\ u(1) & 0 \end{bmatrix}$$

which can be modified to

$$\begin{bmatrix} u(0) & u(1) \\ u(1) & 0 \end{bmatrix} = \begin{bmatrix} -1.5820 & -1.2433 \\ 0.5820 & 0.2433 \end{bmatrix} \begin{bmatrix} x_1(0) & x_1(1) \\ x_2(0) & x_2(1) \end{bmatrix} \tag{6–259}$$

from which we obtain

$$u(k) = -1.5820x_1(k) - 1.2433x_2(k) \qquad k = 0, 1$$

This equation gives the required control law. With this control law, any initial state $\mathbf{x}(0)$ can be transferred to the origin in at most 2 sampling periods. (This is the time-optimal control law.)

Let us next find the initial states from which the system state can be tranferred to the origin in one sampling period. By equating $\mathbf{x}(1)$ with $\mathbf{0}$ in Eq. (6–254), we obtain

$$\mathbf{x}(1) = \mathbf{0} = \mathbf{G}\mathbf{x}(0) + \mathbf{H}u(0)$$

from which we get

$$\mathbf{x}(0) = -\mathbf{G}^{-1}\mathbf{H}u(0)$$

or

$$\begin{bmatrix} x_1(0) \\ x_2(0) \end{bmatrix} = \begin{bmatrix} 0.7181 \\ -1.7181 \end{bmatrix} u(0) \tag{6–260}$$

From Eq. (6–260), we find that if the initial state lies on the line

$$1.7181x_1(0) + 0.7181x_2(0) = 0$$

then it can be transferred to the origin in one sampling period. (Otherwise, we require two sampling periods to bring the initial state to the origin.)

For the case where $u(k)$ is bounded, or $|u(k)| \leq 1$. If we require $\mathbf{x}(1) = \mathbf{0}$, then from Eq. (6–260) we have

$$x_1(0) = 0.7181u(0), \qquad x_2(0) = -1.7181u(0)$$

Since $|u(0)| \leq 1$, we obtain

$$|x_1(0)| \leq 0.7181, \qquad |x_2(0)| \leq 1.7181$$

Hence, if the initial state lies on the line segment

$$1.7181x_1(0) + 0.7181x_2(0) = 0, \qquad -0.7181 \leq x_1(0) \leq 0.7181$$

it can be brought to the origin in one sampling period. This line segment is shown in Fig. 6–15 as line AOB.

If we require $\mathbf{x}(2) = \mathbf{0}$, then from Eq. (6–259) we obtain

$$u(0) = -1.5820x_1(0) - 1.2433x_2(0)$$
$$u(1) = 0.5820x_1(0) + 0.2433x_2(0)$$

Since $|u(0)| \leq 1$ and $|u(1)| \leq 1$, we obtain the following four relationships:

$$1.5820x_1(0) + 1.2433x_2(0) \leq 1$$
$$1.5820x_1(0) + 1.2433x_2(0) \geq -1$$
$$0.5820x_1(0) + 0.2433x_2(0) \leq 1$$
$$0.5820x_1(0) + 0.2433x_2(0) \geq -1$$

The region bounded by these four inequalities is shown in Fig. 6–15. If the initial state lies in this region, except on line AOB, then it can be transferred to the origin in two sampling periods. If the initial state lies outside this region, then it will take more than two sampling periods to bring the state to the origin.

Problem A-6–3. Consider the following pulse-transfer-function system:

$$\frac{Y(z)}{U(z)} = \frac{z^{-1}(1 + 0.8z^{-1})}{1 + 1.3z^{-1} + 0.4z^{-2}}$$

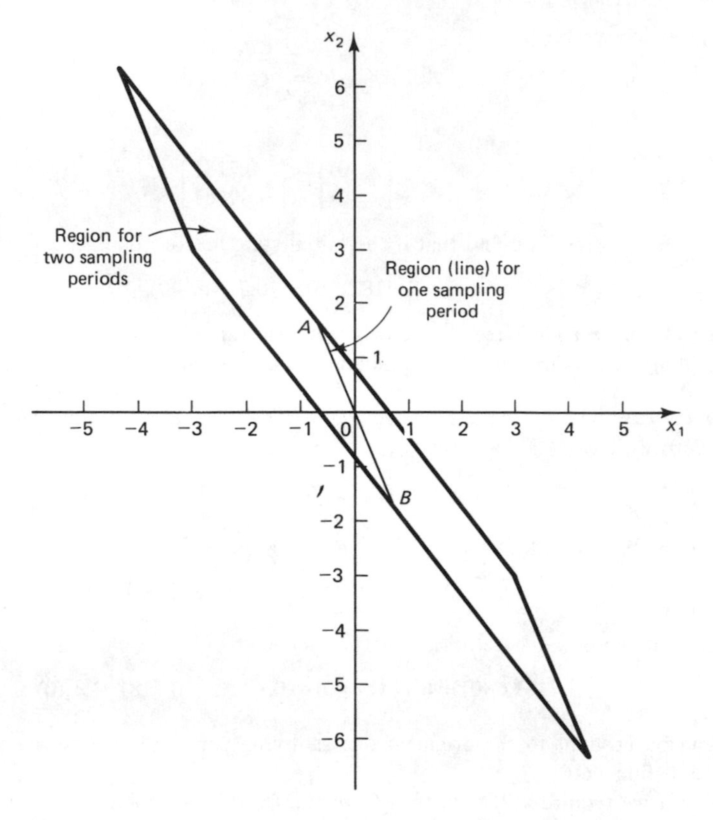

Figure 6–15 Regions from which initial states can be brought to the origin in one or two sampling periods when $u(k)$ is bounded so that $|u(k)| \leq 1$.

A state space representation for this system may be given by

$$\begin{bmatrix} x_1(k+1) \\ x_2(k+1) \end{bmatrix} = \begin{bmatrix} 0 & 1 \\ -0.4 & -1.3 \end{bmatrix} \begin{bmatrix} x_1(k) \\ x_2(k) \end{bmatrix} + \begin{bmatrix} 0 \\ 1 \end{bmatrix} u(k) \tag{6–261}$$

$$y(k) = [0.8 \quad 1] \begin{bmatrix} x_1(k) \\ x_2(k) \end{bmatrix} \tag{6–262}$$

A different state space representation for the same system can be given by

$$\begin{bmatrix} x_1(k+1) \\ x_2(k+1) \end{bmatrix} = \begin{bmatrix} 0 & -0.4 \\ 1 & -1.3 \end{bmatrix} \begin{bmatrix} x_1(k) \\ x_2(k) \end{bmatrix} + \begin{bmatrix} 0.8 \\ 1 \end{bmatrix} u(k) \tag{6–263}$$

$$y(k) = [0 \quad 1] \begin{bmatrix} x_1(k) \\ x_2(k) \end{bmatrix} \tag{6–264}$$

Show that the state space representation defined by Eqs. (6–261) and (6–262) gives a system which is state controllable but not observable. Show, on the other hand, that the state space representation defined by Eqs. (6–263) and (6–264) gives a system which is not completely state controllable but is observable. Explain what causes the apparent difference in the controllability and observability of the same system.

Solution. Consider the discrete-time control system defined by Eqs. (6–261) and (6–262). The rank of the controllability matrix

$$[\mathbf{H} \vdots \mathbf{GH}] = \begin{bmatrix} 0 & 1 \\ 1 & -1.3 \end{bmatrix}$$

is 2. Hence the system is completely state controllable. The rank of the observability matrix

$$[\mathbf{C}^* \vdots \mathbf{G}^*\mathbf{C}^*] = \begin{bmatrix} 0.8 & -0.4 \\ 1 & -0.5 \end{bmatrix}$$

is 1. Hence the system is not observable.

Next, consider the system defined by Eqs. (6–263) and (6–264). The rank of the controllability matrix

$$[\mathbf{H} \vdots \mathbf{GH}] = \begin{bmatrix} 0.8 & -0.4 \\ 1 & -0.5 \end{bmatrix}$$

is 1. Hence the system is not completely state controllable. The rank of the observability matrix

$$[\mathbf{C}^* \vdots \mathbf{G}^*\mathbf{C}^*] = \begin{bmatrix} 0 & 1 \\ 1 & -1.3 \end{bmatrix}$$

is 2. Hence the system is observable.

The apparent difference in the controllability and observability of the same system is caused by the fact that the original system has a pole-zero cancellation in the pulse transfer function:

$$\frac{Y(z)}{U(z)} = \frac{z + 0.8}{z^2 + 1.3z + 0.4} = \frac{z + 0.8}{(z + 0.8)(z + 0.5)}$$

If a pole-zero cancellation occurs in the pulse transfer function, then the controllability and observability vary, depending on how the state variables are chosen.

Note that to be completely state controllable and observable the pulse-transfer-function system must not have any pole-zero cancellation.

Problem A-6–4. Consider the system defined by

$$\mathbf{x}(k+1) = \mathbf{Gx}(k) + \mathbf{H}u(k) \tag{6–265}$$

$$y(k) = \mathbf{Cx}(k) \tag{6–266}$$

where

$\mathbf{x}(k)$ = state vector (n-vector)

$u(k)$ = control signal (scalar)

$y(k)$ = output signal (scalar)

$\mathbf{G} = n \times n$ nonsingular matrix

$\mathbf{H} = n \times 1$ matrix

$\mathbf{C} = 1 \times n$ matrix

Since $y(k)$ is a scalar quantity, it is not possible to determine the vector value $\mathbf{x}(k)$ on the basis of knowledge of $y(k)$ only. However, it may be possible to determine the current vector $\mathbf{x}(k)$ from knowledge of $y(k)$ and the past output values $y(k-1), y(k-2), \ldots$. Determine the condition under which such determination of the current state $\mathbf{x}(k)$ is possible.

Solution. Notice that for $h = 1, 2, 3, \ldots,$

$$\mathbf{x}(k) = \mathbf{G}^h\mathbf{x}(k-h) + \sum_{j=0}^{h-1} \mathbf{G}^j\mathbf{H}u(k-j-1) \tag{6–267}$$

This equation implies that starting from $\mathbf{x}(k-h)$ and using controls $u(k-h), u(k-h+1), \ldots, u(k-1)$, the state will reach $\mathbf{x}(k)$ in h steps. By multiplying both sides of Eq. (6–267) by $\mathbf{G}^{-h}$, we obtain

$$\mathbf{G}^{-h}\mathbf{x}(k) = \mathbf{x}(k-h) + \sum_{j=0}^{h-1} \mathbf{G}^{j-h}\mathbf{H}u(k-j-1)$$

or

$$\mathbf{x}(k-h) = \mathbf{G}^{-h}\mathbf{x}(k) - \sum_{j=0}^{h-1} \mathbf{G}^{j-h}\mathbf{H}u(k-j-1) \tag{6–268}$$

By substituting Eq. (6–268) into Eq. (6–266) we obtain

$$\begin{aligned} y(k-h) &= \mathbf{C}\mathbf{x}(k-h) \\ &= \mathbf{C}\mathbf{G}^{-h}\mathbf{x}(k) - \mathbf{C}\sum_{j=0}^{h-1} \mathbf{G}^{j-h}\mathbf{H}u(k-j-1) \end{aligned}$$

By taking $h = 1, 2, 3, \ldots, n-1$ and substituting in this last equation and writing the result together with the equation $y(k) = \mathbf{C}\mathbf{x}(k)$ in the form of a vector-matrix equation, we obtain

$$\begin{bmatrix} y(k) \\ y(k-1) \\ \cdot \\ \cdot \\ \cdot \\ y(k-n+1) \end{bmatrix} = \begin{bmatrix} \mathbf{C} \\ \mathbf{CG}^{-1} \\ \cdot \\ \cdot \\ \cdot \\ \mathbf{CG}^{-n+1} \end{bmatrix} \begin{bmatrix} x_1(k) \\ x_2(k) \\ \cdot \\ \cdot \\ \cdot \\ x_n(k) \end{bmatrix} + \begin{bmatrix} 0 \\ \mathbf{CG}^{-1}\mathbf{H}u(k-1) \\ \cdot \\ \cdot \\ \cdot \\ \mathbf{C}\sum_{j=0}^{n-2} \mathbf{G}^{j-n+1}\mathbf{H}u(k-j-1) \end{bmatrix}$$

Rearranging, we get

$$\begin{bmatrix} \mathbf{C} \\ \mathbf{CG}^{-1} \\ \cdot \\ \cdot \\ \cdot \\ \mathbf{CG}^{-n+1} \end{bmatrix} \begin{bmatrix} x_1(k) \\ x_2(k) \\ \cdot \\ \cdot \\ \cdot \\ x_n(k) \end{bmatrix} = \begin{bmatrix} y(k) \\ y(k-1) \\ \cdot \\ \cdot \\ \cdot \\ y(k-n+1) \end{bmatrix} - \begin{bmatrix} 0 \\ \mathbf{CG}^{-1}\mathbf{H}u(k-1) \\ \cdot \\ \cdot \\ \cdot \\ \mathbf{C}\sum_{j=0}^{n-2} \mathbf{G}^{j-n+1}\mathbf{H}u(k-j-1) \end{bmatrix}$$

Notice that the right-hand side of this equation represents an arbitrary n-vector. Hence the condition under which the vector $\mathbf{x}(k)$ can be uniquely determined is that the rank of the matrix

$$\begin{bmatrix} \mathbf{C} \\ \mathbf{CG}^{-1} \\ \cdot \\ \cdot \\ \cdot \\ \mathbf{CG}^{-n+1} \end{bmatrix}$$

be n. This condition can be written that the rank of

$$\begin{bmatrix} \mathbf{CG}^{n-1} \\ \mathbf{CG}^{n-2} \\ \cdot \\ \cdot \\ \cdot \\ \mathbf{CG} \\ \mathbf{C} \end{bmatrix}$$

be n, which is equivalent to saying that the rank of the matrix

$$[\mathbf{C}^* \vdots \mathbf{G}^*\mathbf{C}^* \vdots \cdots \vdots (\mathbf{G}^*)^{n-1}\mathbf{C}^*]$$

be n. Notice that this is the condition for complete observability derived in Sec. 6–3. If this rank condition (observability condition) is satisfied, then the current state $\mathbf{x}(k)$ can be determined from the observation of $y(k-n+1), y(k-n+2), \ldots, y(k)$.

Problem A-6–5. Consider the control system defined by

$$\mathbf{x}(k+1) = \mathbf{Gx}(k) + \mathbf{H}u(k) \tag{6–269}$$

$$y(k) = \mathbf{Cx}(k) + D \tag{6–270}$$

where

$\mathbf{x}(k)$ = state vector (n-vector)

$u(k)$ = control signal (scalar)

$y(k)$ = output signal (scalar)

$\mathbf{G} = n \times n$ matrix

$\mathbf{H} = n \times 1$ matrix

$\mathbf{C} = 1 \times n$ matrix

D = scalar (constant)

As stated by Eq. (5–88), the pulse transfer function $F(z)$ can be given as follows:

$$F(z) = \mathbf{C}(z\mathbf{I} - \mathbf{G})^{-1}\mathbf{H} + D$$

Prove that if the system is completely state controllable and completely observable, then there is no pole-zero cancellation in the pulse transfer function $F(z)$.

Solution. Suppose that there is a pole-zero cancellation in the pulse transfer function, even though the system is completely state controllable and completely observable. Consider the following identity equation:

$$\begin{bmatrix} \mathbf{I} & \mathbf{0} \\ \mathbf{C}(z\mathbf{I}-\mathbf{G})^{-1} & 1 \end{bmatrix} \begin{bmatrix} z\mathbf{I}-\mathbf{G} & \mathbf{H} \\ -\mathbf{C} & D \end{bmatrix} = \begin{bmatrix} z\mathbf{I}-\mathbf{G} & \mathbf{H} \\ \mathbf{0} & F(z) \end{bmatrix}$$

Taking the determinant of the left-hand side of the equation and equating it with the determinant of the right-hand side, we obtain

$$\begin{vmatrix} z\mathbf{I}-\mathbf{G} & \mathbf{H} \\ -\mathbf{C} & D \end{vmatrix} = |z\mathbf{I}-\mathbf{G}|F(z)$$

or

$$F(z) = \frac{\begin{vmatrix} z\mathbf{I}-\mathbf{G} & \mathbf{H} \\ -\mathbf{C} & D \end{vmatrix}}{|z\mathbf{I}-\mathbf{G}|}$$

The poles of $F(z)$ are the roots of $|z\mathbf{I} - \mathbf{G}| = 0$, and the zeros of $F(z)$ are the roots of

$$\begin{vmatrix} z\mathbf{I}-\mathbf{G} & \mathbf{H} \\ -\mathbf{C} & D \end{vmatrix} = 0 \tag{6–271}$$

Now suppose that a pole-zero cancellation occurs. Let us assume that $z = z_1$ is a pole of $F(z)$ and is also a zero of $F(z)$, so that cancellation occurs. Then, $z = z_1$ is a root of $|z\mathbf{I} - \mathbf{G}| = 0$. Also, it is a root of the determinant equation given by Eq. (6–271). This means that there exists a vector

$$\begin{bmatrix} \mathbf{v} \\ \hline w \end{bmatrix}$$

where $\mathbf{v}$ is an n-vector and w is a scalar, such that

$$\left[\begin{array}{c|c} z_1\mathbf{I}-\mathbf{G} & \mathbf{H} \\ \hline -\mathbf{C} & D \end{array}\right] \begin{bmatrix} \mathbf{v} \\ \hline w \end{bmatrix} = \begin{bmatrix} \mathbf{0} \\ \hline 0 \end{bmatrix} \tag{6–272}$$

If $w \neq 0$, then from Eq. (6–272) we have

$$(z_1\mathbf{I} - \mathbf{G})\mathbf{v} + \mathbf{H}w = \mathbf{0}$$

or

$$(\mathbf{G} - z_1\mathbf{I})\mathbf{v} = \mathbf{H}w \tag{6–273}$$

Since $z = z_1$ is a root of the characteristic equation, the characteristic polynomial $\phi(z)$ of $\mathbf{G}$ can be written as follows:

$$\phi(z) = (z - z_1)\hat{\phi}(z)$$

or

$$\phi(\mathbf{G}) = (\mathbf{G} - z_1\mathbf{I})\hat{\phi}(\mathbf{G}) = \hat{\phi}(\mathbf{G})(\mathbf{G} - z_1\mathbf{I}) = \mathbf{0}$$

From Eq. (6–273) we have

$$\phi(\mathbf{G})\mathbf{v} = \hat{\phi}(\mathbf{G})(\mathbf{G} - z_1\mathbf{I})\mathbf{v} = \hat{\phi}(\mathbf{G})\mathbf{H}w = \mathbf{0}$$

Hence

$$\hat{\phi}(\mathbf{G})\mathbf{H} = \mathbf{0}$$

Since $\hat{\phi}(z)$ is a polynomial of degree $n - 1$, the fact that $\hat{\phi}(\mathbf{G})\mathbf{H} = \mathbf{0}$ means that vector $\mathbf{G}^{n-1}\mathbf{H}$ can be written in terms of $\mathbf{H}, \mathbf{GH}, \dots, \mathbf{G}^{n-2}\mathbf{H}$. Hence

$$\text{rank}\,[\mathbf{H} \vdots \mathbf{GH} \vdots \cdots \vdots \mathbf{G}^{n-1}\mathbf{H}] < n$$

This contradicts the assumption that the system is completely state controllable. Thus, if the system is completely state controllable, then there is no pole-zero cancellation in the pulse transfer function.

Next, referring to Eq. (6–272), if $w = 0$ and $\mathbf{v} \neq \mathbf{0}$, then we have

$$(z_1\mathbf{I} - \mathbf{G})\mathbf{v} = \mathbf{0} \tag{6–274}$$

$$\mathbf{Cv} = 0 \tag{6–275}$$

From Eq. (6–275) we have

$$\mathbf{v}^*\mathbf{C}^* = 0 \tag{6–276}$$

From Eq. (6–274) we obtain

$$\mathbf{v}^*\mathbf{G}^* = z_1\mathbf{v}^*$$

Hence

$$\mathbf{v}^*\mathbf{G}^*\mathbf{C}^* = z_1\mathbf{v}^*\mathbf{C}^* = 0$$

where we have used Eq. (6–276). Similarly,

$$\mathbf{v}^*(\mathbf{G}^*)^2\mathbf{C}^* = \mathbf{v}^*\mathbf{G}^*\mathbf{G}^*\mathbf{C}^* = z_1\mathbf{v}^*\mathbf{G}^*\mathbf{C}^* = z_1^2\mathbf{v}^*\mathbf{C}^* = 0$$

and

$$\mathbf{v}^*(\mathbf{G}^*)^{k-1}\mathbf{C}^* = z_1^{k-1}\mathbf{v}^*\mathbf{C}^* = 0 \qquad k = 1, 2, 3, \dots, n$$

Hence

$$\mathbf{v}^*\,[\mathbf{C}^* \vdots \mathbf{G}^*\mathbf{C}^* \vdots \cdots \vdots (\mathbf{G}^*)^{n-1}\mathbf{C}^*] = \mathbf{0}$$

or

$$\text{rank}\,[\mathbf{C}^* \vdots \mathbf{G}^*\mathbf{C}^* \vdots \cdots \vdots (\mathbf{G}^*)^{n-1}\mathbf{C}^*] < n$$

This contradicts the assumption that the system is completely observable. Thus, there is no pole-zero cancellation if the system is completely observable.

This completes the proof that if the system is completely state controllable and completely observable, then there is no pole-zero cancellation in the pulse transfer function $F(z)$.

Problem A-6–6. Consider the linear time-invariant continuous-time control system

$$\dot{\mathbf{x}} = \mathbf{Ax} + \mathbf{Bu} \tag{6–277}$$

where

$\mathbf{x}$ = state vector (n-vector)

$\mathbf{u}=$ control vector (r-vector)

$\mathbf{A}=n\times n$ matrix

$\mathbf{B}=n\times r$ matrix

Prove that the system is completely state controllable if and only if the rank of the $n \times nr$ matrix

$$[\mathbf{B}\,\vdots\,\mathbf{AB}\,\vdots\,\cdots\,\vdots\,\mathbf{A}^{n-1}\mathbf{B}]$$

is n.

Solution. The solution of Eq. (6–277) is

$$\mathbf{x}(t_1)=e^{\mathbf{A}t_1}\mathbf{x}(0)+\int_0^{t_1} e^{\mathbf{A}(t_1-\tau)}\mathbf{B}\mathbf{u}(\tau)\,d\tau$$

from which we obtain

$$\mathbf{x}(t_1)-e^{\mathbf{A}t_1}\mathbf{x}(0)=\int_0^{t_1} e^{\mathbf{A}(t_1-\tau)}\mathbf{B}\mathbf{u}(\tau)\,d\tau$$

Since $e^{\mathbf{A}t_1}$ is nonsingular, premultiplying both sides of this last equation by $e^{-\mathbf{A}t_1}$, we obtain

$$e^{-\mathbf{A}t_1}[\mathbf{x}(t_1)-e^{\mathbf{A}t_1}\mathbf{x}(0)]=e^{-\mathbf{A}t_1}\int_0^{t_1} e^{\mathbf{A}(t_1-\tau)}\mathbf{B}\mathbf{u}(\tau)\,d\tau$$

or

$$e^{-\mathbf{A}t_1}\mathbf{x}(t_1)-\mathbf{x}(0)=\int_0^{t_1} e^{-\mathbf{A}\tau}\mathbf{B}\mathbf{u}(\tau)\,d\tau \tag{6–278}$$

Note that $e^{-\mathbf{A}\tau}$ can be written as follows:

$$e^{-\mathbf{A}\tau}=\sum_{k=0}^{p-1}\alpha_k(\tau)\mathbf{A}^k \tag{6–279}$$

where p is the degree of the minimal polynomial of $\mathbf{A}$. Thus, $p \le n$.

Substituting Eq. (6–279) into Eq. (6–278) gives

$$e^{-\mathbf{A}t_1}\mathbf{x}(t_1)-\mathbf{x}(0)=\sum_{k=0}^{p-1}\mathbf{A}^k\mathbf{B}\int_0^{t_1}\alpha_k(\tau)\mathbf{u}(\tau)\,d\tau \tag{6–280}$$

Let us put

$$\int_0^{t_1}\alpha_k(\tau)\mathbf{u}(\tau)\,d\tau=\boldsymbol{\beta}_k=\begin{bmatrix}\beta_{k1}\\ \beta_{k2}\\ \cdot\\ \cdot\\ \cdot\\ \beta_{kr}\end{bmatrix}$$

where the $\boldsymbol{\beta}_k$'s are the r-vectors. Then Eq. (6–280) becomes

$$e^{-\mathbf{A}t_1}\mathbf{x}(t_1) - \mathbf{x}(0) = \sum_{k=0}^{p-1} \mathbf{A}^k \mathbf{B} \boldsymbol{\beta}_k$$

$$= \sum_{k=0}^{p-1} \mathbf{A}^k [\mathbf{B}_1 \vdots \mathbf{B}_2 \vdots \cdots \vdots \mathbf{B}_r] \boldsymbol{\beta}_k$$

$$= \sum_{k=0}^{p-1} \sum_{j=1}^{r} \beta_{kj} \mathbf{A}^k \mathbf{B}_j \qquad (6\text{–}281)$$

From Eq. (6–281) we see that any initial state $\mathbf{x}(0)$ which can be transferred to an arbitrary state $\mathbf{x}(x_1) = \mathbf{x}_f$ must be a linear combination of $\mathbf{B}_j$, $\mathbf{AB}_j$, . . . , $\mathbf{A}^{p-1}\mathbf{B}_j$, where $j = 1, 2, 3, \ldots, r$. Hence, if the rank of the matrix

$$[\mathbf{B}_1 \vdots \mathbf{B}_2 \vdots \cdots \vdots \mathbf{A}^{p-1}\mathbf{B}_{r-1} \vdots \mathbf{A}^{p-1}\mathbf{B}_r] = [\mathbf{B} \vdots \mathbf{AB} \vdots \cdots \vdots \mathbf{A}^{p-1}\mathbf{B}]$$

is n, then there exist n linearly independent vectors in this matrix. Since n linearly independent vectors span the n-dimensional state space, the system is completely state controllable.

Noting that $p \leq n$, let us consider the two cases where $p = n$ and $p < n$ separately. If $p = n$, then the condition becomes that the rank of

$$\mathbf{P} = [\mathbf{B} \vdots \mathbf{AB} \vdots \cdots \vdots \mathbf{A}^{n-1}\mathbf{B}]$$

be n. If $p < n$, the $\mathbf{A}^h\mathbf{B}_j$, where $h = p, p+1, \ldots, n-1$, are linearly dependent on $\mathbf{B}_j$, . . . , $\mathbf{A}^{p-1}\mathbf{B}_j$. Hence, in this case ($p < n$) we have the same conclusion: that if the rank of matrix $\mathbf{P}$ is n, then the system is completely state controllable.

Conversely, suppose that the system is completely state controllable but the rank of matrix $\mathbf{P}$ is $q < n$. Then q linearly independent vectors of matrix $\mathbf{P}$ span the q-dimensional state space. Then there exist, in the n-dimensional state space, initial states that cannot be transferred to an arbitrary state $\mathbf{x}_f$. This contradicts the assumption that the system is completely state controllable. Hence, for complete state controllability, the rank of

$$[\mathbf{B} \vdots \mathbf{AB} \vdots \cdots \vdots \mathbf{A}^{n-1}\mathbf{B}]$$

must be n.

Problem A-6–7. Consider the following linear time-invariant continuous-time control system:

$$\dot{\mathbf{x}} = \mathbf{Ax}$$

$$\mathbf{y} = \mathbf{Cx}$$

where

$\mathbf{x}$ = state vector (n-vector)

$\mathbf{y}$ = output vector (m-vector)

$\mathbf{A}$ = $n \times n$ matrix

$\mathbf{C}$ = $m \times n$ matrix

Prove that a necessary and sufficient conditon for the system to be completely observable is that the rank of the $n \times nm$ matrix

$$[\mathbf{C}^* \vdots \mathbf{A}^*\mathbf{C}^* \vdots \cdots \vdots (\mathbf{A}^*)^{n-1}\mathbf{C}^*]$$

be n.

Solution. Note that

$$\mathbf{x}(t) = e^{\mathbf{A}t}\mathbf{x}(0)$$

Since

$$e^{\mathbf{A}t} = \sum_{k=0}^{p-1} \alpha_k(t)\mathbf{A}^k$$

where p is the degree of the minimal polynomial of $\mathbf{A}$ (that is, $p \le n$), we obtain

$$\mathbf{x}(t) = \sum_{k=0}^{p-1} \alpha_k(t)\mathbf{A}^k\mathbf{x}(0)$$

Hence

$$\mathbf{y}(t) = \mathbf{C}\mathbf{x}(t) = \sum_{k=0}^{p-1} \alpha_k(t)\mathbf{C}\mathbf{A}^k\mathbf{x}(0) = \sum_{k=0}^{p-1} \alpha_k(t)\langle(\mathbf{A}^*)^k\mathbf{C}^*, \mathbf{x}(0)\rangle$$

Notice that matrix $\mathbf{C}^*$ has m columns:

$$\mathbf{C}^* = [\mathbf{C}_1^* \vdots \mathbf{C}_2^* \vdots \cdots \vdots \mathbf{C}_m^*]$$

Thus, the output vector $\mathbf{y}(t)$ has m components, its ith component being

$$y_i(t) = \sum_{k=0}^{p-1} \alpha_k(t)\langle(\mathbf{A}^*)^k\mathbf{C}_i^*, \mathbf{x}(0)\rangle$$

where $i = 1, 2, \ldots, m$.

By taking the inner product of $\alpha_j(t)$ and $y_i(t)$, we obtain pm equations in pm variables $\langle\mathbf{C}_i^*, \mathbf{x}(0)\rangle$, $\langle\mathbf{A}^*\mathbf{C}_i^*, \mathbf{x}(0)\rangle$, $\ldots$, $\langle(\mathbf{A}^*)^{p-1}\mathbf{C}_i^*, \mathbf{x}(0)\rangle$, where $i = 1, 2, \ldots, m$. Noting that the $\alpha_j(t)$'s are linearly independent, it is possible to solve the set of pm equations in pm unknown variables, giving

$$\langle(\mathbf{A}^*)^k\mathbf{C}_i^*, \mathbf{x}(0)\rangle = \beta_{ki} \tag{6–282}$$

Equation (6–282) can be expanded as follows:

$$\varphi_{011}x_1(0) + \varphi_{012}x_2(0) + \cdots + \varphi_{01n}x_n(0) = \beta_{01}$$
$$\varphi_{021}x_1(0) + \varphi_{022}x_2(0) + \cdots + \varphi_{02n}x_n(0) = \beta_{02}$$
$$\vdots$$
$$\varphi_{p-1\,m\,1}x_1(0) + \varphi_{p-1\,m\,2}x_2(0) + \cdots + \varphi_{p-1\,mn}x_n(0) = \beta_{p-1\,m}$$

where $\varphi_{ki\,1}, \varphi_{ki\,2}, \ldots, \varphi_{kin}$ are n components of the row vector φ_{ki} and

$$\boldsymbol{\varphi}_{ki}^* = (\mathbf{A}^*)^k\mathbf{C}_i^* \tag{6–283}$$

These pm equations are equations in n unknown variables $x_1(0), x_2(0), \ldots, x_n(0)$. These equations are necessarily consistent. They have a unique solution if and only if the rank of the $pm \times n$ coefficient matrix

$$\boldsymbol{\Phi} = \begin{bmatrix} \varphi_{011} & \varphi_{012} & \cdots & \varphi_{01n} \\ \varphi_{021} & \varphi_{022} & \cdots & \varphi_{02n} \\ \cdot & \cdot & & \cdot \\ \cdot & \cdot & & \cdot \\ \cdot & \cdot & & \cdot \\ \varphi_{p-1\,m\,1} & \varphi_{p-1\,m\,2} & \cdots & \varphi_{p-1\,mn} \end{bmatrix}$$

is n. The conjugate transpose of $\boldsymbol{\Phi}$ can be written as follows:

$$\boldsymbol{\Phi}^* = \begin{bmatrix} \bar{\varphi}_{011} & \bar{\varphi}_{021} & \cdots & \bar{\varphi}_{p-1\,m\,1} \\ \bar{\varphi}_{012} & \bar{\varphi}_{022} & \cdots & \bar{\varphi}_{p-1\,m\,2} \\ \cdot & \cdot & & \cdot \\ \cdot & \cdot & & \cdot \\ \cdot & \cdot & & \cdot \\ \bar{\varphi}_{01n} & \bar{\varphi}_{02n} & \cdots & \bar{\varphi}_{p-1\,mn} \end{bmatrix}$$

$$= [\boldsymbol{\varphi}_{01}^* \vdots \boldsymbol{\varphi}_{02}^* \vdots \cdots \vdots \boldsymbol{\varphi}_{p-1\,m}^*]$$

$$= [\mathbf{C}_1^* \vdots \mathbf{C}_2^* \vdots \cdots \vdots (\mathbf{A}^*)^{p-1}\mathbf{C}_m^*]$$

$$= [\mathbf{C}^* \vdots \mathbf{A}^*\mathbf{C}^* \vdots \cdots \vdots (\mathbf{A}^*)^{p-1}\mathbf{C}^*]$$

where we use Eq. (6–283).

Hence, the initial state variables $x_1(0), x_2(0), \cdots, x_n(0)$ can be determined uniquely if and only if the rank of the matrix

$$[\mathbf{C}^* \vdots \mathbf{A}^*\mathbf{C}^* \vdots \cdots \vdots (\mathbf{A}^*)^{p-1}\mathbf{C}^*]$$

is n. Since $p \leq n$, let us consider separately the cases where $p = n$ and $p < n$. If $p = n$, then it can be said that the condition is that the rank of

$$\mathbf{Q} = [\mathbf{C}^* \vdots \mathbf{A}^*\mathbf{C}^* \vdots \cdots \vdots (\mathbf{A}^*)^{n-1}\mathbf{C}^*]$$

be n. If $p < n$, then the columns of

$$(\mathbf{A}^*)^h\mathbf{C}^*$$

where $h = p, p + 1, \ldots, n - 1$, are linearly dependent on $\mathbf{C}_1^*, \mathbf{C}_2^*, \ldots, (\mathbf{A}^*)^{p-1}\mathbf{C}_m^*$. In the case of $p < n$, therefore, it can still be stated that the condition of complete observability is that the rank of matrix $\mathbf{Q}$ be n. This completes the proof.

We have thus proved that if the rank of matrix $\mathbf{Q}$ is n, then any initial state $\mathbf{x}(0)$ can be determined by measurement of $\mathbf{y}(t)$ for a finite time interval.

Problem A-6–8. Consider the completely state controllable system

$$\mathbf{x}(k+1) = \mathbf{G}\mathbf{x}(k) + \mathbf{H}u(k)$$

Define the controllability matrix as $\mathbf{M}$:

$$\mathbf{M} = [\mathbf{H} \vdots \mathbf{G}\mathbf{H} \vdots \cdots \vdots \mathbf{G}^{n-1}\mathbf{H}]$$

Show that

$$\mathbf{M}^{-1}\mathbf{G}\mathbf{M} = \begin{bmatrix} 0 & 0 & \cdots & 0 & -a_n \\ 1 & 0 & \cdots & 0 & -a_{n-1} \\ 0 & 1 & \cdots & 0 & -a_{n-2} \\ \cdot & \cdot & & \cdot & \cdot \\ \cdot & \cdot & & \cdot & \cdot \\ \cdot & \cdot & & \cdot & \cdot \\ 0 & 0 & \cdots & 1 & -a_1 \end{bmatrix} \tag{6–284}$$

where $a_1, a_2, \ldots, a_n$ are the coefficients of the characteristic polynomial

$$|z\mathbf{I} - \mathbf{G}| = z^n + a_1 z^{n-1} + \cdots + a_{n-1}z + a_n$$

Solution. Let us consider the case where $n = 3$. We shall show that

$$\mathbf{G}\mathbf{M} = \mathbf{M}\begin{bmatrix} 0 & 0 & -a_3 \\ 1 & 0 & -a_2 \\ 0 & 1 & -a_1 \end{bmatrix} \tag{6–285}$$

The left-hand side of Eq. (6–285) is

$$\mathbf{G}\mathbf{M} = \mathbf{G}[\mathbf{H} \vdots \mathbf{G}\mathbf{H} \vdots \mathbf{G}^2\mathbf{H}] = [\mathbf{G}\mathbf{H} \vdots \mathbf{G}^2\mathbf{H} \vdots \mathbf{G}^3\mathbf{H}]$$

The right-hand side of Eq. (6–285) is

$$[\mathbf{H} \vdots \mathbf{G}\mathbf{H} \vdots \mathbf{G}^2\mathbf{H}]\begin{bmatrix} 0 & 0 & -a_3 \\ 1 & 0 & -a_2 \\ 0 & 1 & -a_1 \end{bmatrix} = [\mathbf{G}\mathbf{H} \vdots \mathbf{G}^2\mathbf{H} \vdots -a_3\mathbf{H} - a_2\mathbf{G}\mathbf{H} - a_1\mathbf{G}^2\mathbf{H}] \tag{6–286}$$

The Cayley-Hamilton theorem states that matrix $\mathbf{G}$ satisfies its own characteristic equation, or

$$\mathbf{G}^n + a_1\mathbf{G}^{n-1} + \cdots + a_{n-1}\mathbf{G} + a_n\mathbf{I} = \mathbf{0}$$

For $n = 3$, we have

$$\mathbf{G}^3 + a_1\mathbf{G}^2 + a_2\mathbf{G} + a_3\mathbf{I} = \mathbf{0} \tag{6–287}$$

Using Eq. (6–287), the third column of the right-hand side of Eq. (6–286) becomes

$$-a_3\mathbf{H} - a_2\mathbf{G}\mathbf{H} - a_1\mathbf{G}^2\mathbf{H} = \mathbf{G}^3\mathbf{H}$$

Thus, Eq. (6–286), the right-hand side of Eq. (6–285), becomes

$$[\mathbf{H} \vdots \mathbf{G}\mathbf{H} \vdots \mathbf{G}^2\mathbf{H}]\begin{bmatrix} 0 & 0 & -a_3 \\ 1 & 0 & -a_2 \\ 0 & 1 & -a_1 \end{bmatrix} = [\mathbf{G}\mathbf{H} \vdots \mathbf{G}^2\mathbf{H} \vdots \mathbf{G}^3\mathbf{H}]$$

Hence, we have shown that Eq. (6–285) is true. Thus,

$$\mathbf{M}^{-1}\mathbf{G}\mathbf{M} = \begin{bmatrix} 0 & 0 & -a_3 \\ 1 & 0 & -a_2 \\ 0 & 1 & -a_1 \end{bmatrix}$$

The preceding derivation can easily be extended to the general case of any positive integer n.

Problem A-6–9. Consider the completely state controllable system

$$\mathbf{x}(k+1) = \mathbf{G}\mathbf{x}(k) + \mathbf{H}u(k)$$

Define

$$\mathbf{M} = [\mathbf{H} \vdots \mathbf{G}\mathbf{H} \vdots \cdots \vdots \mathbf{G}^{n-1}\mathbf{H}]$$

and

$$\mathbf{W} = \begin{bmatrix} a_{n-1} & a_{n-2} & \cdots & a_1 & 1 \\ a_{n-2} & a_{n-3} & \cdots & 1 & 0 \\ \cdot & \cdot & & \cdot & \cdot \\ \cdot & \cdot & & \cdot & \cdot \\ \cdot & \cdot & & \cdot & \cdot \\ a_1 & 1 & \cdots & 0 & 0 \\ 1 & 0 & \cdots & 0 & 0 \end{bmatrix}$$

where the a_i's are coefficients of the characteristic polynomial

$$|z\mathbf{I} - \mathbf{G}| = z^n + a_1 z^{n-1} + \cdots + a_{n-1}z + a_n$$

Define also

$$\mathbf{T} = \mathbf{M}\mathbf{W}$$

Show that

$$\mathbf{T}^{-1}\mathbf{G}\mathbf{T} = \begin{bmatrix} 0 & 1 & 0 & \cdots & 0 \\ 0 & 0 & 1 & \cdots & 0 \\ \cdot & \cdot & \cdot & & \cdot \\ \cdot & \cdot & \cdot & & \cdot \\ \cdot & \cdot & \cdot & & \cdot \\ 0 & 0 & 0 & \cdots & 1 \\ -a_n & -a_{n-1} & -a_{n-2} & \cdots & -a_1 \end{bmatrix}, \qquad \mathbf{T}^{-1}\mathbf{H} = \begin{bmatrix} 0 \\ 0 \\ \cdot \\ \cdot \\ \cdot \\ 0 \\ 1 \end{bmatrix}$$

Solution. Let us consider the case where $n = 3$. We shall show that

$$\mathbf{T}^{-1}\mathbf{G}\mathbf{T} = (\mathbf{M}\mathbf{W})^{-1}\mathbf{G}(\mathbf{M}\mathbf{W}) = \begin{bmatrix} 0 & 1 & 0 \\ 0 & 0 & 1 \\ -a_3 & -a_2 & -a_1 \end{bmatrix} \tag{6–288}$$

Referring to Prob. A-6–8, we have

$$(\mathbf{M}\mathbf{W})^{-1}\mathbf{G}(\mathbf{M}\mathbf{W}) = \mathbf{W}^{-1}(\mathbf{M}^{-1}\mathbf{G}\mathbf{M})\mathbf{W} = \mathbf{W}^{-1}\begin{bmatrix} 0 & 0 & -a_3 \\ 1 & 0 & -a_2 \\ 0 & 1 & -a_1 \end{bmatrix}\mathbf{W}$$

Hence, Eq. (6–288) can be rewritten as follows:

$$\mathbf{W}^{-1}\begin{bmatrix} 0 & 0 & -a_3 \\ 1 & 0 & -a_2 \\ 0 & 1 & -a_1 \end{bmatrix}\mathbf{W} = \begin{bmatrix} 0 & 1 & 0 \\ 0 & 0 & 1 \\ -a_3 & -a_2 & -a_1 \end{bmatrix}$$

Consequently, we need to show that

$$\begin{bmatrix} 0 & 0 & -a_3 \\ 1 & 0 & -a_2 \\ 0 & 1 & -a_1 \end{bmatrix} \mathbf{W} = \mathbf{W} \begin{bmatrix} 0 & 1 & 0 \\ 0 & 0 & 1 \\ -a_3 & -a_2 & -a_1 \end{bmatrix} \tag{6–289}$$

The left-hand side of Eq. (6–289) is

$$\begin{bmatrix} 0 & 0 & -a_3 \\ 1 & 0 & -a_2 \\ 0 & 1 & -a_1 \end{bmatrix} \begin{bmatrix} a_2 & a_1 & 1 \\ a_1 & 1 & 0 \\ 1 & 0 & 0 \end{bmatrix} = \begin{bmatrix} -a_3 & 0 & 0 \\ 0 & a_1 & 1 \\ 0 & 1 & 0 \end{bmatrix}$$

The right-hand side of Eq. (6–289) is

$$\begin{bmatrix} a_2 & a_1 & 1 \\ a_1 & 1 & 0 \\ 1 & 0 & 0 \end{bmatrix} \begin{bmatrix} 0 & 1 & 0 \\ 0 & 0 & 1 \\ -a_3 & -a_2 & -a_1 \end{bmatrix} = \begin{bmatrix} -a_3 & 0 & 0 \\ 0 & a_1 & 1 \\ 0 & 1 & 0 \end{bmatrix}$$

Clearly, Eq. (6–289) holds true. Thus, we have shown that

$$\mathbf{T}^{-1}\mathbf{G}\mathbf{T} = \begin{bmatrix} 0 & 1 & 0 \\ 0 & 0 & 1 \\ -a_3 & -a_2 & -a_1 \end{bmatrix}$$

Next, we shall show that

$$\mathbf{T}^{-1}\mathbf{H} = \begin{bmatrix} 0 \\ 0 \\ 1 \end{bmatrix} \tag{6–290}$$

Note that Eq. (6–290) can be written as follows:

$$\mathbf{H} = \mathbf{T} \begin{bmatrix} 0 \\ 0 \\ 1 \end{bmatrix} = \mathbf{M}\mathbf{W} \begin{bmatrix} 0 \\ 0 \\ 1 \end{bmatrix}$$

This last equation can easily be verified, since

$$\mathbf{T} \begin{bmatrix} 0 \\ 0 \\ 1 \end{bmatrix} = \mathbf{M}\mathbf{W} \begin{bmatrix} 0 \\ 0 \\ 1 \end{bmatrix} = [\mathbf{H} \vdots \mathbf{G}\mathbf{H} \vdots \mathbf{G}^2\mathbf{H}] \begin{bmatrix} a_2 & a_1 & 1 \\ a_1 & 1 & 0 \\ 1 & 0 & 0 \end{bmatrix} \begin{bmatrix} 0 \\ 0 \\ 1 \end{bmatrix}$$

$$= [\mathbf{H} \vdots \mathbf{G}\mathbf{H} \vdots \mathbf{G}^2\mathbf{H}] \begin{bmatrix} 1 \\ 0 \\ 0 \end{bmatrix} = \mathbf{H}$$

Hence

$$\mathbf{T}^{-1}\mathbf{H} = \begin{bmatrix} 0 \\ 0 \\ 1 \end{bmatrix}$$

The derivation shown here can easily be extended to the general case of any positive integer n.

Problem A-6–10. Consider the following system:

$$\mathbf{x}(k+1) = \mathbf{G}\mathbf{x}(k) + \mathbf{H}u(k)$$

where

$$\mathbf{G} = \begin{bmatrix} 0 & 1 & 0 \\ 0 & 0 & 1 \\ -a_3 & -a_2 & -a_1 \end{bmatrix}, \qquad \mathbf{H} = \begin{bmatrix} 0 \\ 0 \\ 1 \end{bmatrix}$$

Notice that the system is in the controllable canonical form.

Define the transformation matrix $\mathbf{T}$ as follows:

$$\mathbf{T} = \mathbf{M}\mathbf{W}$$

where

$$\mathbf{M} = [\mathbf{H} \vdots \mathbf{G}\mathbf{H} \vdots \mathbf{G}^2\mathbf{H}]$$

and

$$\mathbf{W} = \begin{bmatrix} a_2 & a_1 & 1 \\ a_1 & 1 & 0 \\ 1 & 0 & 0 \end{bmatrix}$$

Show that if the system is in the controllable canonical form, then $\mathbf{T} = \mathbf{I}$. Consequently, if the system is in the controllable canonical form, then

$$\mathbf{M}^{-1} = \mathbf{W} = \begin{bmatrix} a_2 & a_1 & 1 \\ a_1 & 1 & 0 \\ 1 & 0 & 0 \end{bmatrix}$$

Solution. Since

$$\mathbf{M} = \begin{bmatrix} 0 & 0 & 1 \\ 0 & 1 & -a_1 \\ 1 & -a_1 & -a_2 + a_1^2 \end{bmatrix}$$

we have

$$\mathbf{T} = \mathbf{M}\mathbf{W} = \begin{bmatrix} 0 & 0 & 1 \\ 0 & 1 & -a_1 \\ 1 & -a_1 & -a_2 + a_1^2 \end{bmatrix} \begin{bmatrix} a_2 & a_1 & 1 \\ a_1 & 1 & 0 \\ 1 & 0 & 0 \end{bmatrix} = \begin{bmatrix} 1 & 0 & 0 \\ 0 & 1 & 0 \\ 0 & 0 & 1 \end{bmatrix} = \mathbf{I}$$

Hence

$$\mathbf{M}^{-1} = \mathbf{W} = \begin{bmatrix} a_2 & a_1 & 1 \\ a_1 & 1 & 0 \\ 1 & 0 & 0 \end{bmatrix}$$

Problem A-6–11. Consider the completely observable system

$$\mathbf{x}(k+1) = \mathbf{G}\mathbf{x}(k)$$

$$y(k) = \mathbf{C}\mathbf{x}(k)$$

Define the observability matrix as $\mathbf{N}$:

$$\mathbf{N} = [\mathbf{C}^* \vdots \mathbf{G}^*\mathbf{C}^* \vdots \cdots \vdots (\mathbf{G}^*)^{n-1}\mathbf{C}^*]$$

Show that

$$\mathbf{N}^*\mathbf{G}(\mathbf{N}^*)^{-1} = \begin{bmatrix} 0 & 1 & 0 & \cdots & 0 \\ 0 & 0 & 1 & \cdots & 0 \\ \cdot & \cdot & \cdot & & \cdot \\ \cdot & \cdot & \cdot & & \cdot \\ \cdot & \cdot & \cdot & & \cdot \\ 0 & 0 & 0 & \cdots & 1 \\ -a_n & -a_{n-1} & -a_{n-2} & \cdots & -a_1 \end{bmatrix} \tag{6–291}$$

where $a_1, a_2, \ldots, a_n$ are the coefficients of the characteristic polynomial

$$|z\mathbf{I} - \mathbf{G}| = z^n + a_1 z^{n-1} + \cdots + a_{n-1}z + a_n$$

Solution. Let us consider the case where $n = 3$. Then Eq. (6–291) can be written as

$$\mathbf{N}^*\mathbf{G}(\mathbf{N}^*)^{-1} = \begin{bmatrix} 0 & 1 & 0 \\ 0 & 0 & 1 \\ -a_3 & -a_2 & -a_1 \end{bmatrix} \tag{6–292}$$

Equation (6–292) may be rewritten as

$$\mathbf{N}^*\mathbf{G} = \begin{bmatrix} 0 & 1 & 0 \\ 0 & 0 & 1 \\ -a_3 & -a_2 & -a_1 \end{bmatrix} \mathbf{N}^* \tag{6–293}$$

We shall show that Eq. (6–293) holds true. The left-hand side of Eq. (6–293) is

$$\mathbf{N}^*\mathbf{G} = \begin{bmatrix} \mathbf{C} \\ \mathbf{CG} \\ \mathbf{CG}^2 \end{bmatrix} \mathbf{G} = \begin{bmatrix} \mathbf{CG} \\ \mathbf{CG}^2 \\ \mathbf{CG}^3 \end{bmatrix}$$

The right-hand side of Eq. (6–293) is

$$\begin{bmatrix} 0 & 1 & 0 \\ 0 & 0 & 1 \\ -a_3 & -a_2 & -a_1 \end{bmatrix} \mathbf{N}^* = \begin{bmatrix} 0 & 1 & 0 \\ 0 & 0 & 1 \\ -a_3 & -a_2 & -a_1 \end{bmatrix} \begin{bmatrix} \mathbf{C} \\ \mathbf{CG} \\ \mathbf{CG}^2 \end{bmatrix} = \begin{bmatrix} \mathbf{CG} \\ \mathbf{CG}^2 \\ -a_3\mathbf{C} - a_2\mathbf{CG} - a_1\mathbf{CG}^2 \end{bmatrix}$$

The Cayley-Hamilton theorem states that matrix $\mathbf{G}$ satisfies its own characteristic equation, or, for the case of $n = 3$,

$$\mathbf{G}^3 + a_1\mathbf{G}^2 + a_2\mathbf{G} + a_3\mathbf{I} = \mathbf{0}$$

Hence

$$-a_1\mathbf{CG}^2 - a_2\mathbf{CG} - a_3\mathbf{C} = \mathbf{CG}^3$$

Consequently,

$$\begin{bmatrix} 0 & 1 & 0 \\ 0 & 0 & 1 \\ -a_3 & -a_2 & -a_1 \end{bmatrix} \mathbf{N}^* = \begin{bmatrix} \mathbf{CG} \\ \mathbf{CG}^2 \\ \mathbf{CG}^3 \end{bmatrix}$$

Thus, we have shown that Eq. (6–293) holds true. Hence

$$\mathbf{N}^*\mathbf{G}(\mathbf{N}^*)^{-1} = \begin{bmatrix} 0 & 1 & 0 \\ 0 & 0 & 1 \\ -a_3 & -a_2 & -a_1 \end{bmatrix}$$

The derivation presented here can be extended to the general case of any positive integer n.

Problem A-6–12. Consider the completely observable system

$$\mathbf{x}(k+1) = \mathbf{G}\mathbf{x}(k) + \mathbf{H}u(k) \tag{6–294}$$

$$y(k) = \mathbf{C}\mathbf{x}(k) + Du(k) \tag{6–295}$$

Define

$$\mathbf{N} = [\mathbf{C}^* \vdots \mathbf{G}^*\mathbf{C}^* \vdots \cdots \vdots (\mathbf{G}^*)^{n-1}\mathbf{C}^*]$$

and

$$\mathbf{W} = \begin{bmatrix} a_{n-1} & a_{n-2} & \cdots & a_1 & 1 \\ a_{n-2} & a_{n-3} & \cdots & 1 & 0 \\ \cdot & \cdot & & \cdot & \cdot \\ \cdot & \cdot & & \cdot & \cdot \\ \cdot & \cdot & & \cdot & \cdot \\ a_1 & 1 & \cdots & 0 & 0 \\ 1 & 0 & \cdots & 0 & 0 \end{bmatrix}$$

where the a_i's are coefficients of the characteristic polynomial

$$|z\mathbf{I} - \mathbf{G}| = z^n + a_1 z^{n-1} + \cdots + a_{n-1}z + a_n$$

Define also

$$\mathbf{Q} = (\mathbf{W}\mathbf{N}^*)^{-1}$$

Show that

$$\mathbf{Q}^{-1}\mathbf{G}\mathbf{Q} = \begin{bmatrix} 0 & 0 & \cdots & 0 & -a_n \\ 1 & 0 & \cdots & 0 & -a_{n-1} \\ 0 & 1 & \cdots & 0 & -a_{n-2} \\ \cdot & \cdot & & \cdot & \cdot \\ \cdot & \cdot & & \cdot & \cdot \\ \cdot & \cdot & & \cdot & \cdot \\ 0 & 0 & \cdots & 1 & -a_1 \end{bmatrix}$$

$$\mathbf{C}\mathbf{Q} = [0 \quad 0 \quad \cdots \quad 0 \quad 1]$$

$$\mathbf{Q}^{-1}\mathbf{H} = \begin{bmatrix} b_n - a_n b_0 \\ b_{n-1} - a_{n-1} b_0 \\ \cdot \\ \cdot \\ \cdot \\ b_1 - a_1 b_0 \end{bmatrix}$$

where the b_k's ($k = 0, 1, 2, \ldots, n$) are those coefficients appearing in the numerator of the pulse transfer function when $\mathbf{C}(z\mathbf{I} - \mathbf{G})^{-1}\mathbf{H} + D$ is written as follows:

$$\mathbf{C}(z\mathbf{I}-\mathbf{G})^{-1}\mathbf{H}+D=\frac{b_0z^n+b_1z^{n-1}+\cdots+b_{n-1}z+b_n}{z^n+a_1z^{n-1}+\cdots+a_{n-1}z+a_n}$$

where $D = b_0$.

Solution. Let us consider the case where $n = 3$. We shall show that

$$\mathbf{Q}^{-1}\mathbf{G}\mathbf{Q}=(\mathbf{W}\mathbf{N}^*)\mathbf{G}(\mathbf{W}\mathbf{N}^*)^{-1}=\begin{bmatrix}0&0&-a_3\\1&0&-a_2\\0&1&-a_1\end{bmatrix} \tag{6–296}$$

Note that by referring to Prob. A-6–11, we have

$$(\mathbf{W}\mathbf{N}^*)\mathbf{G}(\mathbf{W}\mathbf{N}^*)^{-1}=\mathbf{W}[\mathbf{N}^*\mathbf{G}(\mathbf{N}^*)^{-1}]\mathbf{W}^{-1}=\mathbf{W}\begin{bmatrix}0&1&0\\0&0&1\\-a_3&-a_2&-a_1\end{bmatrix}\mathbf{W}^{-1}$$

Hence we need to show that

$$\mathbf{W}\begin{bmatrix}0&1&0\\0&0&1\\-a_3&-a_2&-a_1\end{bmatrix}\mathbf{W}^{-1}=\begin{bmatrix}0&0&-a_3\\1&0&-a_2\\0&1&-a_1\end{bmatrix}$$

or

$$\mathbf{W}\begin{bmatrix}0&1&0\\0&0&1\\-a_3&-a_2&-a_1\end{bmatrix}=\begin{bmatrix}0&0&-a_3\\1&0&-a_2\\0&1&-a_1\end{bmatrix}\mathbf{W} \tag{6–297}$$

The left-hand side of Eq. (6–297) is

$$\mathbf{W}\begin{bmatrix}0&1&0\\0&0&1\\-a_3&-a_2&-a_1\end{bmatrix}=\begin{bmatrix}a_2&a_1&1\\a_1&1&0\\1&0&0\end{bmatrix}\begin{bmatrix}0&1&0\\0&0&1\\-a_3&-a_2&-a_1\end{bmatrix}=\begin{bmatrix}-a_3&0&0\\0&a_1&1\\0&1&0\end{bmatrix}$$

The right-hand side of Eq. (6–297) is

$$\begin{bmatrix}0&0&-a_3\\1&0&-a_2\\0&1&-a_1\end{bmatrix}\mathbf{W}=\begin{bmatrix}0&0&-a_3\\1&0&-a_2\\0&1&-a_1\end{bmatrix}\begin{bmatrix}a_2&a_1&1\\a_1&1&0\\1&0&0\end{bmatrix}=\begin{bmatrix}-a_3&0&0\\0&a_1&1\\0&1&0\end{bmatrix}$$

Thus, Eq. (6–297) holds true. Hence, we have shown that Eq. (6–296) holds true.

Next, we shall show that

$$\mathbf{C}\mathbf{Q}=[0\quad 0\quad 1]$$

or

$$\mathbf{C}(\mathbf{W}\mathbf{N}^*)^{-1}=[0\quad 0\quad 1] \tag{6–298}$$

Notice that

$$[0\quad 0\quad 1](\mathbf{W}\mathbf{N}^*)=[0\quad 0\quad 1]\begin{bmatrix}a_2&a_1&1\\a_1&1&0\\1&0&0\end{bmatrix}\begin{bmatrix}\mathbf{C}\\\mathbf{C}\mathbf{G}\\\mathbf{C}\mathbf{G}^2\end{bmatrix}=[1\quad 0\quad 0]\begin{bmatrix}\mathbf{C}\\\mathbf{C}\mathbf{G}\\\mathbf{C}\mathbf{G}^2\end{bmatrix}=\mathbf{C}$$

Hence, we have shown that

$$[0\quad 0\quad 1]=\mathbf{C}(\mathbf{W}\mathbf{N}^*)^{-1}$$

which is Eq. (6–298).

Next, define

$$\mathbf{x} = \mathbf{Q}\hat{\mathbf{x}}$$

Then Eq. (6–294) becomes

$$\hat{\mathbf{x}}(k+1) = \mathbf{Q}^{-1}\mathbf{G}\mathbf{Q}\hat{\mathbf{x}}(k) + \mathbf{Q}^{-1}\mathbf{H}u(k) \tag{6–299}$$

and Eq. (6–295) becomes

$$y(k) = \mathbf{C}\mathbf{Q}\hat{\mathbf{x}}(k) + Du(k) \tag{6–300}$$

For the case of $n = 3$, Eq. (6–299) becomes as follows:

$$\hat{\mathbf{x}}(k+1) = \begin{bmatrix} 0 & 0 & -a_3 \\ 1 & 0 & -a_2 \\ 0 & 1 & -a_1 \end{bmatrix} \hat{\mathbf{x}}(k) + \begin{bmatrix} \gamma_3 \\ \gamma_2 \\ \gamma_1 \end{bmatrix} u(k)$$

where

$$\begin{bmatrix} \gamma_3 \\ \gamma_2 \\ \gamma_1 \end{bmatrix} = \mathbf{Q}^{-1}\mathbf{H}$$

The pulse transfer function $F(z)$ for the system defined by Eqs. (6–299) and (6–300) is

$$F(z) = (\mathbf{C}\mathbf{Q})(z\mathbf{I} - \mathbf{Q}^{-1}\mathbf{G}\mathbf{Q})^{-1}\mathbf{Q}^{-1}\mathbf{H} + D$$

Noting that

$$\mathbf{C}\mathbf{Q} = [0 \quad 0 \quad 1]$$

we have

$$F(z) = [0 \quad 0 \quad 1] \begin{bmatrix} z & 0 & a_3 \\ -1 & z & a_2 \\ 0 & -1 & z + a_1 \end{bmatrix}^{-1} \begin{bmatrix} \gamma_3 \\ \gamma_2 \\ \gamma_1 \end{bmatrix} + D$$

Note that $D = b_0$. Since

$$\begin{bmatrix} z & 0 & a_3 \\ -1 & z & a_2 \\ 0 & -1 & z + a_1 \end{bmatrix}^{-1} = \frac{1}{z^3 + a_1 z^2 + a_2 z + a_3} \begin{bmatrix} z^2 + a_1 z + a_2 & -a_3 & -a_3 z \\ z + a_1 & z^2 + a_1 z & -a_2 z - a_3 \\ 1 & z & z^2 \end{bmatrix}$$

we have

$$\begin{aligned} F(z) &= \frac{1}{z^3 + a_1 z^2 + a_2 z + a_3} [1 \quad z \quad z^2] \begin{bmatrix} \gamma_3 \\ \gamma_2 \\ \gamma_1 \end{bmatrix} + D \\ &= \frac{\gamma_1 z^2 + \gamma_2 z + \gamma_3}{z^3 + a_1 z^2 + a_2 z + a_3} + b_0 \\ &= \frac{b_0 z^3 + (\gamma_1 + a_1 b_0) z^2 + (\gamma_2 + a_2 b_0) z + (\gamma_3 + a_3 b_0)}{z^3 + a_1 z^2 + a_2 z + a_3} \\ &= \frac{b_0 z^3 + b_1 z^2 + b_2 z + b_3}{z^3 + a_1 z^2 + a_2 z + a_3} \end{aligned}$$

Hence, $\gamma_1 = b_1 - a_1 b_0$, $\gamma_2 = b_2 - a_2 b_0$, and $\gamma_3 = b_3 - a_3 b_0$. Thus, we have shown that

$$\mathbf{Q}^{-1}\mathbf{H} = \begin{bmatrix} \gamma_3 \\ \gamma_2 \\ \gamma_1 \end{bmatrix} = \begin{bmatrix} b_3 - a_3 b_0 \\ b_2 - a_2 b_0 \\ b_1 - a_1 b_0 \end{bmatrix}$$

Note that what we have derived here can easily be extended to the case where n is any positive integer.

Problem A-6–13. Consider the system defined by

$$G(z) = \frac{z+1}{z^2 + z + 0.16} \tag{6–301}$$

Referring to Sec. 6–4, obtain state space representations for this system in the following three different forms:

1. Controllable canonical form
2. Observable canonical form
3. Diagonal canonical form

Solution.

1. *Controllable canonical form.* By comparing Eq. (6–301) with Eq. (6–39), we obtain

$$a_1 = 1, \qquad a_2 = 0.16, \qquad b_0 = 0, \qquad b_1 = 1, \qquad b_2 = 1$$

Hence, referring to Eqs. (6–37) and (6–38), we obtain

$$\begin{bmatrix} x_1(k+1) \\ x_2(k+1) \end{bmatrix} = \begin{bmatrix} 0 & 1 \\ -0.16 & -1 \end{bmatrix} \begin{bmatrix} x_1(k) \\ x_2(k) \end{bmatrix} + \begin{bmatrix} 0 \\ 1 \end{bmatrix} u(k)$$

$$y(k) = \begin{bmatrix} 1 & 1 \end{bmatrix} \begin{bmatrix} x_1(k) \\ x_2(k) \end{bmatrix}$$

2. *Observable canonical form.* Since $a_1 = 1$, $a_2 = 0.16$, $b_0 = 0$, $b_1 = 1$, and $b_2 = 1$, referring to Eqs. (6–41) and (6–42) we obtain

$$\begin{bmatrix} x_1(k+1) \\ x_2(k+1) \end{bmatrix} = \begin{bmatrix} 0 & -0.16 \\ 1 & -1 \end{bmatrix} \begin{bmatrix} x_1(k) \\ x_2(k) \end{bmatrix} + \begin{bmatrix} 1 \\ 1 \end{bmatrix} u(k)$$

$$y(k) = \begin{bmatrix} 0 & 1 \end{bmatrix} \begin{bmatrix} x_1(k) \\ x_2(k) \end{bmatrix}$$

3. *Diagonal canonical form.* Notice that

$$G(z) = \frac{\frac{4}{3}}{z + 0.2} + \frac{-\frac{1}{3}}{z + 0.8}$$

By comparing this last equation with Eq. (6–47), we obtain

$$\alpha_1\beta_1 = \tfrac{4}{3}, \qquad \alpha_2\beta_2 = -\tfrac{1}{3}, \qquad p_1 = -0.2, \qquad p_2 = -0.8, \qquad D = 0$$

Hence, by arbitrarily choosing $\alpha_1 = \alpha_2 = 1$ and referring to Eqs. (6–45) and (6–46), we obtain

$$\begin{bmatrix} x_1(k+1) \\ x_2(k+1) \end{bmatrix} = \begin{bmatrix} -0.2 & 0 \\ 0 & -0.8 \end{bmatrix} \begin{bmatrix} x_1(k) \\ x_2(k) \end{bmatrix} + \begin{bmatrix} 1 \\ 1 \end{bmatrix} u(k)$$

$$y(k) = [\tfrac{4}{3} \quad -\tfrac{1}{3}] \begin{bmatrix} x_1(k) \\ x_2(k) \end{bmatrix}$$

Problem A-6–14. Consider the double integrator system

$$\mathbf{x}((k+1)T) = \mathbf{G}\mathbf{x}(kT) + \mathbf{H}u(kT)$$

where

$$\mathbf{G} = \begin{bmatrix} 1 & T \\ 0 & 1 \end{bmatrix}, \qquad \mathbf{H} = \begin{bmatrix} T^2/2 \\ T \end{bmatrix}$$

and where T is the sampling period. (See Prob. A-5–29 for the derivation of this discrete-time state equation for the double integrator system.) Determine a state feedback gain matrix $\mathbf{K}$ such that the response to an arbitrary initial condition is deadbeat. For the initial state

$$\mathbf{x}(0) = \begin{bmatrix} 1 \\ 1 \end{bmatrix}$$

determine $u(0)$ and $u(T)$ for $T = 0.1$ sec, $T = 1$ sec, and $T = 10$ sec.

Solution. Let us define

$$\mathbf{K} = [k_1 \quad k_2]$$

Then

$$|z\mathbf{I} - \mathbf{G} + \mathbf{H}\mathbf{K}| = \begin{vmatrix} z - 1 + \dfrac{T^2}{2}k_1 & -T + \dfrac{T^2}{2}k_2 \\ Tk_1 & z - 1 + Tk_2 \end{vmatrix}$$

$$= z^2 - \left(2 - \frac{T^2}{2}k_1 - Tk_2\right)z + 1 + \frac{T^2}{2}k_1 - Tk_2 = 0 \tag{6–302}$$

The desired characteristic equation is

$$z^2 = 0 \tag{6–303}$$

Hence by comparing Eqs. (6–302) and (6–303), we obtain

$$2 - \frac{T^2}{2}k_1 - Tk_2 = 0$$

$$1 + \frac{T^2}{2}k_1 - Tk_2 = 0$$

from which we get

$$k_1 = \frac{1}{T^2}, \qquad k_2 = \frac{3}{2T}$$

Hence

$$\mathbf{K} = \begin{bmatrix} \frac{1}{T^2} & \frac{3}{2T} \end{bmatrix}$$

In what follows, we shall show that the response to initial conditions is deadbeat. Assume that the initial state is

$$\begin{bmatrix} x_1(0) \\ x_2(0) \end{bmatrix} = \begin{bmatrix} a \\ b \end{bmatrix}$$

The state feedback equation is

$$\mathbf{x}((k+1)T) = (\mathbf{G} - \mathbf{HK})\mathbf{x}(kT)$$

or

$$\begin{bmatrix} x_1((k+1)T) \\ x_2((k+1)T) \end{bmatrix} = \begin{bmatrix} \frac{1}{2} & \frac{T}{4} \\ -\frac{1}{T} & -\frac{1}{2} \end{bmatrix} \begin{bmatrix} x_1(kT) \\ x_2(kT) \end{bmatrix}$$

Notice that

$$\begin{bmatrix} x_1(T) \\ x_2(T) \end{bmatrix} = \begin{bmatrix} \frac{1}{2} & \frac{T}{4} \\ -\frac{1}{T} & -\frac{1}{2} \end{bmatrix} \begin{bmatrix} x_1(0) \\ x_2(0) \end{bmatrix} = \begin{bmatrix} \frac{1}{2} & \frac{T}{4} \\ -\frac{1}{T} & -\frac{1}{2} \end{bmatrix} \begin{bmatrix} a \\ b \end{bmatrix} = \begin{bmatrix} \frac{1}{2}a + \frac{T}{4}b \\ -\frac{1}{T}a - \frac{1}{2}b \end{bmatrix}$$

$$\begin{bmatrix} x_1(2T) \\ x_2(2T) \end{bmatrix} = \begin{bmatrix} \frac{1}{2} & \frac{T}{4} \\ -\frac{1}{T} & -\frac{1}{2} \end{bmatrix} \begin{bmatrix} \frac{1}{2}a + \frac{T}{4}b \\ -\frac{1}{T}a - \frac{1}{2}b \end{bmatrix} = \begin{bmatrix} 0 \\ 0 \end{bmatrix}$$

Thus, clearly the response is deadbeat.

Now let us determine $u(0)$ and $u(T)$. Notice that

$$u(kT) = -\mathbf{Kx}(kT) = -\begin{bmatrix} \frac{1}{T^2} & \frac{3}{2T} \end{bmatrix} \mathbf{x}(kT)$$

Hence

$$u(0) = -\begin{bmatrix} \frac{1}{T^2} & \frac{3}{2T} \end{bmatrix} \begin{bmatrix} a \\ b \end{bmatrix} = -\frac{1}{T^2}a - \frac{3}{2T}b$$

$$u(T) = -\begin{bmatrix} \frac{1}{T^2} & \frac{3}{2T} \end{bmatrix} \begin{bmatrix} \frac{1}{2}a + \frac{T}{4}b \\ -\frac{1}{T}a - \frac{1}{2}b \end{bmatrix} = \frac{1}{T^2}a + \frac{b}{2T}$$

For $a = 1$ and $b = 1$, we have

$$u(0) = -\frac{1}{T^2} - \frac{3}{2T}, \qquad u(T) = \frac{1}{T^2} + \frac{1}{2T}$$

In particular, for $T = 0.1$ sec,

$$u(0) = -115, \qquad u(T) = u(0.1) = 105$$

For $T = 1$ sec,

$$u(0) = -2.5, \qquad u(T) = u(1) = 1.5$$

For $T = 10$ sec,

$$u(0) = -0.16, \qquad u(T) = u(10) = 0.06$$

Notice that for a small value of the sampling period T, $u(0)$ and $u(T)$ become large. Increasing the value of T reduces the magnitudes of $u(0)$ and $u(T)$ significantly.

Problem A-6–15. Consider the system defined by

$$\mathbf{x}(k+1) = \mathbf{G}\mathbf{x}(k) + \mathbf{H}u(k)$$

where $\mathbf{x}(k)$ is a 3-vector. It is assumed that the system is completely state controllable. By use of the pole placement technique, we wish to design the system to have closed-loop poles at $z = \mu_1$, $z = \mu_2$, and $z = \mu_3$, where the μ_i's are distinct. That is, using the state feedback control

$$u(k) = -\mathbf{K}\mathbf{x}(k)$$

we wish to have

$$|z\mathbf{I} - \mathbf{G} + \mathbf{H}\mathbf{K}| = (z - \mu_1)(z - \mu_2)(z - \mu_3) = z^3 + \alpha_1 z^2 + \alpha_2 z + \alpha_3$$

Show that the desired state feedback gain matrix $\mathbf{K}$ can be given by

$$\mathbf{K} = [1 \quad 1 \quad 1][\boldsymbol{\xi}_1 \quad \boldsymbol{\xi}_2 \quad \boldsymbol{\xi}_3]^{-1} \tag{6–304}$$

where

$$\boldsymbol{\xi}_i = (\mathbf{G} - \mu_i \mathbf{I})^{-1}\mathbf{H} \qquad i = 1, 2, 3 \tag{6–305}$$

Show also that the vectors $\boldsymbol{\xi}_i$ are eigenvectors of matrix $\mathbf{G} - \mathbf{H}\mathbf{K}$, that is, that $\boldsymbol{\xi}_i$ satisfies the equation

$$(\mathbf{G} - \mathbf{H}\mathbf{K})\boldsymbol{\xi}_i = \mu_i \boldsymbol{\xi}_i \qquad i = 1, 2, 3$$

Solution. Let us define

$$\tilde{\mathbf{G}} = \mathbf{G} - \mathbf{H}\mathbf{K}$$

By use of the Cayley-Hamilton theorem, $\tilde{\mathbf{G}}$ satisfies its own characteristic equation:

$$\tilde{\mathbf{G}}^3 + \alpha_1\tilde{\mathbf{G}}^2 + \alpha_2\tilde{\mathbf{G}} + \alpha_3\mathbf{I} = \phi(\tilde{\mathbf{G}}) = \mathbf{0} \tag{6–306}$$

Consider the following identities:

$$\mathbf{I} = \mathbf{I}$$

$$\tilde{\mathbf{G}} = \mathbf{G} - \mathbf{H}\mathbf{K}$$

$$\tilde{\mathbf{G}}^2 = (\mathbf{G} - \mathbf{H}\mathbf{K})^2 = \mathbf{G}^2 - \mathbf{G}\mathbf{H}\mathbf{K} - \mathbf{H}\mathbf{K}\tilde{\mathbf{G}}$$

$$\tilde{\mathbf{G}}^3 = (\mathbf{G} - \mathbf{H}\mathbf{K})^3 = \mathbf{G}^3 - \mathbf{G}^2\mathbf{H}\mathbf{K} - \mathbf{G}\mathbf{H}\mathbf{K}\tilde{\mathbf{G}} - \mathbf{H}\mathbf{K}\tilde{\mathbf{G}}^2$$

Multiplying each of the preceding equations by α_3, α_2, α_1, and α_0 (where $\alpha_0 = 1$) in this order and adding the results, we obtain

$$\begin{aligned}\alpha_3\mathbf{I} + \alpha_2\tilde{\mathbf{G}} + \alpha_1\tilde{\mathbf{G}}^2 + \tilde{\mathbf{G}}^3 = \alpha_3\mathbf{I} + \alpha_2\mathbf{G} + \alpha_1\mathbf{G}^2 + \mathbf{G}^3 - \alpha_2\mathbf{HK} \\ - \alpha_1\mathbf{GHK} - \alpha_1\mathbf{HK}\tilde{\mathbf{G}} - \mathbf{G}^2\mathbf{HK} - \mathbf{GHK}\tilde{\mathbf{G}} - \mathbf{HK}\tilde{\mathbf{G}}^2\end{aligned}$$

By referring to the derivation of Ackermann's formula in Sec. 6–5, this last equation can be reduced to

$$\begin{bmatrix} \alpha_2\mathbf{K} + \alpha_1\mathbf{K}\tilde{\mathbf{G}} + \mathbf{K}\tilde{\mathbf{G}}^2 \\ \alpha_1\mathbf{K} + \mathbf{K}\tilde{\mathbf{G}} \\ \mathbf{K} \end{bmatrix} = [\mathbf{H} \vdots \mathbf{GH} \vdots \mathbf{G}^2\mathbf{H}]^{-1}\phi(\mathbf{G})$$

Premultiplying both sides of this last equation by $[0 \quad 0 \quad 1]$, we obtain

$$[0 \quad 0 \quad 1]\begin{bmatrix} \alpha_2\mathbf{K} + \alpha_1\mathbf{K}\tilde{\mathbf{G}} + \mathbf{K}\tilde{\mathbf{G}}^2 \\ \alpha_1\mathbf{K} + \mathbf{K}\tilde{\mathbf{G}} \\ \mathbf{K} \end{bmatrix} = [0 \quad 0 \quad 1][\mathbf{H} \vdots \mathbf{GH} \vdots \mathbf{G}^2\mathbf{H}]^{-1}\phi(\mathbf{G})$$

or

$$\mathbf{K} = [0 \quad 0 \quad 1][\mathbf{H} \vdots \mathbf{GH} \vdots \mathbf{G}^2\mathbf{H}]^{-1}\phi(\mathbf{G}) \tag{6–307}$$

Noting that

$$\begin{aligned}\phi(\mathbf{G}) &= \mathbf{G}^3 + \alpha_1\mathbf{G}^2 + \alpha_2\mathbf{G} + \alpha_3\mathbf{I} \\ &= (\mathbf{G} - \mu_1\mathbf{I})(\mathbf{G} - \mu_2\mathbf{I})(\mathbf{G} - \mu_3\mathbf{I})\end{aligned}$$

we have

$$\mathbf{K} = [0 \quad 0 \quad 1][\mathbf{H} \vdots \mathbf{GH} \vdots \mathbf{G}^2\mathbf{H}]^{-1}(\mathbf{G} - \mu_1\mathbf{I})(\mathbf{G} - \mu_2\mathbf{I})(\mathbf{G} - \mu_3\mathbf{I})$$

By postmultiplying both sides of this last equation by $\boldsymbol{\xi}_1 = (\mathbf{G} - \mu_1\mathbf{I})^{-1}\mathbf{H}$, we obtain

$$\begin{aligned}\mathbf{K}\boldsymbol{\xi}_1 &= [0 \quad 0 \quad 1][\mathbf{H} \vdots \mathbf{GH} \vdots \mathbf{G}^2\mathbf{H}]^{-1}(\mathbf{G} - \mu_1\mathbf{I})(\mathbf{G} - \mu_2\mathbf{I})(\mathbf{G} - \mu_3\mathbf{I})(\mathbf{G} - \mu_1\mathbf{I})^{-1}\mathbf{H} \\ &= [0 \quad 0 \quad 1][\mathbf{H} \vdots \mathbf{GH} \vdots \mathbf{G}^2\mathbf{H}]^{-1}(\mathbf{G} - \mu_2\mathbf{I})(\mathbf{G} - \mu_3\mathbf{I})\mathbf{H} \end{aligned} \tag{6–308}$$

Let us define

$$(\mathbf{G} - \mu_1\mathbf{I})(\mathbf{G} - \mu_2\mathbf{I}) = \mathbf{G}^2 + \beta_{12}\mathbf{G} + \beta_{13}\mathbf{I}$$

$$(\mathbf{G} - \mu_2\mathbf{I})(\mathbf{G} - \mu_3\mathbf{I}) = \mathbf{G}^2 + \beta_{22}\mathbf{G} + \beta_{23}\mathbf{I}$$

$$(\mathbf{G} - \mu_3\mathbf{I})(\mathbf{G} - \mu_1\mathbf{I}) = \mathbf{G}^2 + \beta_{32}\mathbf{G} + \beta_{33}\mathbf{I}$$

Then, Eq. (6–308) can be written as follows:

$$\begin{aligned}\mathbf{K}\boldsymbol{\xi}_1 &= [0 \quad 0 \quad 1][\mathbf{H} \vdots \mathbf{GH} \vdots \mathbf{G}^2\mathbf{H}]^{-1}(\mathbf{G}^2 + \beta_{22}\mathbf{G} + \beta_{23}\mathbf{I})\mathbf{H} \\ &= [0 \quad 0 \quad 1][\mathbf{H} \vdots \mathbf{GH} \vdots \mathbf{G}^2\mathbf{H}]^{-1}[\mathbf{H} \vdots \mathbf{GH} \vdots \mathbf{G}^2\mathbf{H}]\begin{bmatrix} \beta_{23} \\ \beta_{22} \\ 1 \end{bmatrix} \\ &= [0 \quad 0 \quad 1]\begin{bmatrix} \beta_{23} \\ \beta_{22} \\ 1 \end{bmatrix} = 1\end{aligned}$$

Hence

$$\mathbf{K}\boldsymbol{\xi}_1 = 1$$

Similarly, we obtain

$$\mathbf{K}\boldsymbol{\xi}_2 = 1, \qquad \mathbf{K}\boldsymbol{\xi}_3 = 1$$

Hence

$$\mathbf{K}[\boldsymbol{\xi}_1 \quad \boldsymbol{\xi}_2 \quad \boldsymbol{\xi}_3] = [1 \quad 1 \quad 1]$$

or

$$\mathbf{K} = [1 \quad 1 \quad 1][\boldsymbol{\xi}_1 \quad \boldsymbol{\xi}_2 \quad \boldsymbol{\xi}_3]^{-1} \tag{6–309}$$

Equation (6–309) gives the desired state feedback gain matrix $\mathbf{K}$ in terms of $\boldsymbol{\xi}_1$, $\boldsymbol{\xi}_2$, and $\boldsymbol{\xi}_3$.

To show that the $\boldsymbol{\xi}_i$'s are eigenvectors of matrix $\mathbf{G} - \mathbf{HK}$, notice that

$$\begin{aligned}(\mathbf{G} - \mathbf{HK})\boldsymbol{\xi}_i &= (\mathbf{G} - \mathbf{HK})(\mathbf{G} - \mu_i \mathbf{I})^{-1}\mathbf{H} \\ &= (\mathbf{G} - \mu_i \mathbf{I} + \mu_i \mathbf{I} - \mathbf{HK})(\mathbf{G} - \mu_i \mathbf{I})^{-1}\mathbf{H} \\ &= (\mathbf{G} - \mu_i \mathbf{I})(\mathbf{G} - \mu_i \mathbf{I})^{-1}\mathbf{H} + (\mu_i \mathbf{I} - \mathbf{HK})(\mathbf{G} - \mu_i \mathbf{I})^{-1}\mathbf{H} \\ &= \mathbf{H} + (\mu_i \mathbf{I} - \mathbf{HK})\boldsymbol{\xi}_i \\ &= \mathbf{H} - \mathbf{HK}\boldsymbol{\xi}_i + \mu_i \boldsymbol{\xi}_i \end{aligned} \tag{6–310}$$

As we have shown earlier,

$$\mathbf{K}\boldsymbol{\xi}_i = 1 \qquad i = 1, 2, 3$$

Hence Eq. (6–310) can be simplified to

$$(\mathbf{G} - \mathbf{HK})\boldsymbol{\xi}_i = \mu_i \boldsymbol{\xi}_i \qquad i = 1, 2, 3$$

Thus, vectors $\boldsymbol{\xi}_1$, $\boldsymbol{\xi}_2$, and $\boldsymbol{\xi}_3$ are eigenvectors of matrix $\mathbf{G} - \mathbf{HK}$ corresponding to eigenvalues μ_1, μ_2, and μ_3, respectively.

Problem A-6–16. Consider the system defined by

$$\mathbf{x}(k + 1) = \mathbf{G}\mathbf{x}(k) + \mathbf{H}u(k)$$

where $\mathbf{x}(k)$ is a 3-vector. It is assumed that the system is completely state controllable and that the deadbeat response to the initial state $\mathbf{x}(0)$ is desired. (That is, the desired closed-loop poles must be at the origin, so that $\mu_1 = \mu_2 = \mu_3 = 0$.)

Show that the desired state feedback gain matrix $\mathbf{K}$ can be given by

$$\mathbf{K} = [1 \quad 0 \quad 0][\boldsymbol{\xi}_1 \quad \boldsymbol{\xi}_2 \quad \boldsymbol{\xi}_3]^{-1} \tag{6–311}$$

where

$$\boldsymbol{\xi}_1 = \mathbf{G}^{-1}\mathbf{H}$$

$$\boldsymbol{\xi}_2 = \mathbf{G}^{-2}\mathbf{H}$$

$$\boldsymbol{\xi}_3 = \mathbf{G}^{-3}\mathbf{H}$$

Show also that the vectors $\boldsymbol{\xi}_i$ are generalized eigenvectors of matrix $\mathbf{G} - \mathbf{HK}$, that is, that $\boldsymbol{\xi}_i$ satisfies the equations

$$(\mathbf{G} - \mathbf{HK})\boldsymbol{\xi}_1 = \mathbf{0}$$

$$(\mathbf{G} - \mathbf{HK})\boldsymbol{\xi}_2 = \boldsymbol{\xi}_1$$

$$(\mathbf{G} - \mathbf{HK})\boldsymbol{\xi}_3 = \boldsymbol{\xi}_2$$

Solution. Referring to Eq. (6–307), we have

$$\mathbf{K} = [0 \quad 0 \quad 1][\mathbf{H} \vdots \mathbf{GH} \vdots \mathbf{G}^2\mathbf{H}]^{-1}\phi(\mathbf{G})$$

where

$$\phi(\mathbf{G}) = \mathbf{G}^3$$

Hence

$$\mathbf{K} = [0 \quad 0 \quad 1][\mathbf{H} \vdots \mathbf{GH} \vdots \mathbf{G}^2\mathbf{H}]^{-1}\mathbf{G}^3 \tag{6–312}$$

By postmultiplying both sides of Eq. (6–312) by $\boldsymbol{\xi}_1 = \mathbf{G}^{-1}\mathbf{H}$, we obtain

$$\begin{aligned}
\mathbf{K}\boldsymbol{\xi}_1 &= [0 \quad 0 \quad 1][\mathbf{H} \vdots \mathbf{GH} \vdots \mathbf{G}^2\mathbf{H}]^{-1}\mathbf{G}^3\mathbf{G}^{-1}\mathbf{H} \\
&= [0 \quad 0 \quad 1][\mathbf{H} \vdots \mathbf{GH} \vdots \mathbf{G}^2\mathbf{H}]^{-1}\mathbf{G}^2\mathbf{H} \\
&= [0 \quad 0 \quad 1][\mathbf{H} \vdots \mathbf{GH} \vdots \mathbf{G}^2\mathbf{H}]^{-1}[\mathbf{H} \vdots \mathbf{GH} \vdots \mathbf{G}^2\mathbf{H}]\begin{bmatrix} 0 \\ 0 \\ 1 \end{bmatrix} \\
&= [0 \quad 0 \quad 1]\begin{bmatrix} 0 \\ 0 \\ 1 \end{bmatrix} = 1
\end{aligned}$$

Hence

$$\mathbf{K}\boldsymbol{\xi}_1 = 1$$

By postmultiplying both sides of Eq. (6–312) by $\boldsymbol{\xi}_2 = \mathbf{G}^{-2}\mathbf{H}$, we obtain

$$\begin{aligned}
\mathbf{K}\boldsymbol{\xi}_2 &= [0 \quad 0 \quad 1][\mathbf{H} \vdots \mathbf{GH} \vdots \mathbf{G}^2\mathbf{H}]^{-1}\mathbf{G}^3\mathbf{G}^{-2}\mathbf{H} \\
&= [0 \quad 0 \quad 1][\mathbf{H} \vdots \mathbf{GH} \vdots \mathbf{G}^2\mathbf{H}]^{-1}\mathbf{GH} \\
&= [0 \quad 0 \quad 1][\mathbf{H} \vdots \mathbf{GH} \vdots \mathbf{G}^2\mathbf{H}]^{-1}[\mathbf{H} \vdots \mathbf{GH} \vdots \mathbf{G}^2\mathbf{H}]\begin{bmatrix} 0 \\ 1 \\ 0 \end{bmatrix} = 0
\end{aligned}$$

Hence

$$\mathbf{K}\boldsymbol{\xi}_2 = 0$$

Similarly, by postmultiplying both sides of Eq. (6–312) by $\boldsymbol{\xi}_3 = \mathbf{G}^{-3}\mathbf{H}$, we obtain

$$\begin{aligned}
\mathbf{K}\boldsymbol{\xi}_3 &= [0 \quad 0 \quad 1][\mathbf{H} \vdots \mathbf{GH} \vdots \mathbf{G}^2\mathbf{H}]^{-1}\mathbf{G}^3\mathbf{G}^{-3}\mathbf{H} \\
&= [0 \quad 0 \quad 1][\mathbf{H} \vdots \mathbf{GH} \vdots \mathbf{G}^2\mathbf{H}]^{-1}\mathbf{H} \\
&= [0 \quad 0 \quad 1][\mathbf{H} \vdots \mathbf{GH} \vdots \mathbf{G}^2\mathbf{H}]^{-1}[\mathbf{H} \vdots \mathbf{GH} \vdots \mathbf{G}^2\mathbf{H}]\begin{bmatrix} 1 \\ 0 \\ 0 \end{bmatrix} = 0
\end{aligned}$$

Hence

$$\mathbf{K}\boldsymbol{\xi}_3 = 0$$

Consequently, we have

$$\mathbf{K}[\boldsymbol{\xi}_1 \quad \boldsymbol{\xi}_2 \quad \boldsymbol{\xi}_3] = [1 \quad 0 \quad 0]$$

Hence

$$\mathbf{K} = [1 \quad 0 \quad 0][\boldsymbol{\xi}_1 \quad \boldsymbol{\xi}_2 \quad \boldsymbol{\xi}_3]^{-1}$$

which is Eq. (6–311).

To show that $\boldsymbol{\xi}_1$ is an eigenvector of matrix $\mathbf{G} - \mathbf{HK}$, notice that

$$(\mathbf{G} - \mathbf{HK})\boldsymbol{\xi}_1 = (\mathbf{G} - \mathbf{HK})\mathbf{G}^{-1}\mathbf{H} = \mathbf{H} - \mathbf{HKG}^{-1}\mathbf{H} = \mathbf{H} - \mathbf{HK}\boldsymbol{\xi}_1$$

Since $\mathbf{K}\boldsymbol{\xi}_1 = 1$, we obtain

$$(\mathbf{G} - \mathbf{HK})\boldsymbol{\xi}_1 = \mathbf{0}$$

To show that $\boldsymbol{\xi}_2 = \mathbf{G}^{-2}\mathbf{H}$ is a generalized eigenvector of matrix $\mathbf{G} - \mathbf{HK}$, notice that

$$(\mathbf{G} - \mathbf{HK})\boldsymbol{\xi}_2 = (\mathbf{G} - \mathbf{HK})\mathbf{G}^{-2}\mathbf{H} = \mathbf{G}^{-1}\mathbf{H} - \mathbf{HKG}^{-2}\mathbf{H} = \boldsymbol{\xi}_1 - \mathbf{HK}\boldsymbol{\xi}_2$$

Since $\mathbf{K}\boldsymbol{\xi}_2 = 0$, we obtain

$$(\mathbf{G} - \mathbf{HK})\boldsymbol{\xi}_2 = \boldsymbol{\xi}_1$$

Similarly, to show that $\boldsymbol{\xi}_3 = \mathbf{G}^{-3}\mathbf{H}$ is a generalized eigenvector of matrix $\mathbf{G} - \mathbf{HK}$, notice that

$$(\mathbf{G} - \mathbf{HK})\boldsymbol{\xi}_3 = (\mathbf{G} - \mathbf{HK})\mathbf{G}^{-3}\mathbf{H} = \mathbf{G}^{-2}\mathbf{H} - \mathbf{HKG}^{-3}\mathbf{H} = \boldsymbol{\xi}_2 - \mathbf{HK}\boldsymbol{\xi}_3$$

Since $\mathbf{K}\boldsymbol{\xi}_3 = 0$, we obtain

$$(\mathbf{G} - \mathbf{HK})\boldsymbol{\xi}_3 = \boldsymbol{\xi}_2$$

Problem A-6–17. Consider the system given by

$$\mathbf{x}(k+1) = \mathbf{G}\mathbf{x}(k) + \mathbf{H}_1 u(k)$$

where

$$\mathbf{x}(k) = \text{state vector } (n\text{-vector})$$

$$u(k) = \text{control signal (scalar)}$$

$$\mathbf{G} = n \times n \text{ matrix}$$

$$\mathbf{H}_1 = n \times 1 \text{ matrix}$$

Assume that the system is completely state controllable.

Define

$$[\mathbf{H}_1 \vdots \mathbf{G}\mathbf{H}_1 \vdots \cdots \vdots \mathbf{G}^{n-1}\mathbf{H}_1]^{-1} = \begin{bmatrix} \mathbf{f}_1 \\ \mathbf{f}_2 \\ \cdot \\ \cdot \\ \cdot \\ \mathbf{f}_n \end{bmatrix}$$

where the $\mathbf{f}_i$'s ($i = 1, 2, \ldots, n$) are row vectors and

$$\mathbf{T}_1 = \begin{bmatrix} \mathbf{f}_n \\ \mathbf{f}_n \mathbf{G} \\ \cdot \\ \cdot \\ \cdot \\ \mathbf{f}_n \mathbf{G}^{n-1} \end{bmatrix}^{-1}$$

Show that

$$\mathbf{T}_1^{-1}\mathbf{G}\mathbf{T}_1 = \begin{bmatrix} 0 & 1 & 0 & \cdots & 0 \\ 0 & 0 & 1 & \cdots & 0 \\ \cdot & \cdot & \cdot & & \cdot \\ \cdot & \cdot & \cdot & & \cdot \\ \cdot & \cdot & \cdot & & \cdot \\ 0 & 0 & 0 & \cdots & 1 \\ -a_n & -a_{n-1} & -a_{n-2} & \cdots & -a_1 \end{bmatrix} \tag{6–313}$$

and

$$\mathbf{T}_1^{-1}\mathbf{H}_1 = \begin{bmatrix} 0 \\ 0 \\ \cdot \\ \cdot \\ \cdot \\ 0 \\ 1 \end{bmatrix} \tag{6–314}$$

where the a_i's are the coefficients appearing in the characteristic polynomial of $\mathbf{G}$, or

$$|z\mathbf{I} - \mathbf{G}| = z^n + a_1 z^{n-1} + \cdots + a_{n-1} z + a_n$$

Solution. We shall prove Eqs. (6–313) and (6–314) for the case where $n = 3$. (Extension of the derivation to an arbitrary positive integer n is straightforward.) Thus, we shall derive that

$$\mathbf{T}_1^{-1}\mathbf{G}\mathbf{T}_1 = \begin{bmatrix} 0 & 1 & 0 \\ 0 & 0 & 1 \\ -a_3 & -a_2 & -a_1 \end{bmatrix} \tag{6–315}$$

Since

$$\mathbf{T}_1^{-1} = \begin{bmatrix} \mathbf{f}_3 \\ \mathbf{f}_3 \mathbf{G} \\ \mathbf{f}_3 \mathbf{G}^2 \end{bmatrix}$$

it is possible to rewrite Eq. (6–315) as follows:

$$\mathbf{T}_1^{-1}\mathbf{G} = \begin{bmatrix} 0 & 1 & 0 \\ 0 & 0 & 1 \\ -a_3 & -a_2 & -a_1 \end{bmatrix} \begin{bmatrix} \mathbf{f}_3 \\ \mathbf{f}_3 \mathbf{G} \\ \mathbf{f}_3 \mathbf{G}^2 \end{bmatrix} \tag{6–316}$$

Now consider the conjugate transpose of the right-hand side of Eq. (6–316). Noting that for physical systems, the coefficients $a_1, a_2, \ldots, a_n$ of the characteristic polynomial are real, we have

$$[\mathbf{f}_3^* \vdots \mathbf{G}^*\mathbf{f}_3^* \vdots (\mathbf{G}^*)^2\mathbf{f}_3^*]\begin{bmatrix} 0 & 0 & -a_3 \\ 1 & 0 & -a_2 \\ 0 & 1 & -a_1 \end{bmatrix} = [\mathbf{G}^*\mathbf{f}_3^* \vdots (\mathbf{G}^*)^2\mathbf{f}_3^* \vdots -a_3\mathbf{f}_3^* - a_2\mathbf{G}^*\mathbf{f}_3^* - a_1(\mathbf{G}^*)^2\mathbf{f}_3^*]$$

Note that $\mathbf{G}^*$ satisfies its own characteristic equation:

$$\phi(\mathbf{G}^*) = (\mathbf{G}^*)^3 + a_1(\mathbf{G}^*)^2 + a_2\mathbf{G}^* + a_3\mathbf{I} = \mathbf{0}$$

Hence

$$-[a_3\mathbf{I} + a_2\mathbf{G}^* + a_1(\mathbf{G}^*)^2]\mathbf{f}_3^* = (\mathbf{G}^*)^3\mathbf{f}_3^*$$

Consequently,

$$[\mathbf{f}_3^* \vdots \mathbf{G}^*\mathbf{f}_3^* \vdots (\mathbf{G}^*)^2\mathbf{f}_3^*]\begin{bmatrix} 0 & 0 & -a_3 \\ 1 & 0 & -a_2 \\ 0 & 1 & -a_1 \end{bmatrix} = [\mathbf{G}^*\mathbf{f}_3^* \vdots (\mathbf{G}^*)^2\mathbf{f}_3^* \vdots (\mathbf{G}^*)^3\mathbf{f}_3^*] = \mathbf{G}^*[\mathbf{f}_3^* \vdots \mathbf{G}^*\mathbf{f}_3^* \vdots (\mathbf{G}^*)^2\mathbf{f}_3^*]$$

Taking the conjugate transpose of both sides of this last equation, we obtain

$$\begin{bmatrix} 0 & 1 & 0 \\ 0 & 0 & 1 \\ -a_3 & -a_2 & -a_1 \end{bmatrix}\begin{bmatrix} \mathbf{f}_3 \\ \mathbf{f}_3\mathbf{G} \\ \mathbf{f}_3\mathbf{G}^2 \end{bmatrix} = \begin{bmatrix} \mathbf{f}_3 \\ \mathbf{f}_3\mathbf{G} \\ \mathbf{f}_3\mathbf{G}^2 \end{bmatrix}\mathbf{G} = \mathbf{T}_1^{-1}\mathbf{G}$$

which is Eq. (6–316). Thus, we have shown that Eq. (6–315) is true, or

$$\mathbf{T}_1^{-1}\mathbf{G}\mathbf{T}_1 = \begin{bmatrix} 0 & 1 & 0 \\ 0 & 0 & 1 \\ -a_3 & -a_2 & -a_1 \end{bmatrix}$$

Next, we shall show that

$$\mathbf{T}_1^{-1}\mathbf{H}_1 = \begin{bmatrix} 0 \\ 0 \\ 1 \end{bmatrix}$$

Since

$$[\mathbf{H}_1 \vdots \mathbf{G}\mathbf{H}_1 \vdots \mathbf{G}^2\mathbf{H}_1]^{-1} = \begin{bmatrix} \mathbf{f}_1 \\ \mathbf{f}_2 \\ \mathbf{f}_3 \end{bmatrix}$$

we obtain

$$\mathbf{I} = \begin{bmatrix} \mathbf{f}_1 \\ \mathbf{f}_2 \\ \mathbf{f}_3 \end{bmatrix}[\mathbf{H}_1 \vdots \mathbf{G}\mathbf{H}_1 \vdots \mathbf{G}^2\mathbf{H}_1]$$

or

$$\begin{bmatrix} 1 & 0 & 0 \\ 0 & 1 & 0 \\ 0 & 0 & 1 \end{bmatrix} = \begin{bmatrix} \mathbf{f}_1\mathbf{H}_1 & \mathbf{f}_1\mathbf{G}\mathbf{H}_1 & \mathbf{f}_1\mathbf{G}^2\mathbf{H}_1 \\ \mathbf{f}_2\mathbf{H}_1 & \mathbf{f}_2\mathbf{G}\mathbf{H}_1 & \mathbf{f}_2\mathbf{G}^2\mathbf{H}_1 \\ \mathbf{f}_3\mathbf{H}_1 & \mathbf{f}_3\mathbf{G}\mathbf{H}_1 & \mathbf{f}_3\mathbf{G}^2\mathbf{H}_1 \end{bmatrix}$$

Hence

$$\mathbf{f}_3\mathbf{H}_1 = 0, \qquad \mathbf{f}_3\mathbf{G}\mathbf{H}_1 = 0, \qquad \mathbf{f}_3\mathbf{G}^2\mathbf{H}_1 = 1$$

By using these equations, we obtain

$$\mathbf{T}^{-1}\mathbf{H}_1 = \begin{bmatrix} \mathbf{f}_3 \\ \mathbf{f}_3\mathbf{G} \\ \mathbf{f}_3\mathbf{G}^2 \end{bmatrix} \mathbf{H}_1 = \begin{bmatrix} \mathbf{f}_3\mathbf{H}_1 \\ \mathbf{f}_3\mathbf{G}\mathbf{H}_1 \\ \mathbf{f}_3\mathbf{G}^2\mathbf{H}_1 \end{bmatrix} = \begin{bmatrix} 0 \\ 0 \\ 1 \end{bmatrix}$$

Note that the extension of the derivations presented here to the case of an arbitrary positive integer n can be made easily.

Problem A-6–18. Consider the system

$$\mathbf{x}(k+1) = \mathbf{G}\mathbf{x}(k) + \mathbf{H}\mathbf{u}(k)$$

where

$$\mathbf{x}(k) = \text{state vector (4-vector)}$$

$$\mathbf{u}(k) = \text{control vector (2-vector)}$$

and

$$\mathbf{G} = \begin{bmatrix} -1 & 1 & 0 & 0 \\ 1 & -2 & 1 & 0 \\ 0 & 1 & -1 & 2 \\ 1 & 0 & 0 & 1 \end{bmatrix}, \qquad \mathbf{H} = [\mathbf{H}_1 \vdots \mathbf{H}_2] = \begin{bmatrix} 1 & 0 \\ 0 & 0 \\ 0 & 0 \\ 0 & 1 \end{bmatrix}$$

Referring to Eq. (6–87), obtain matrix **F**. Then by use of the transformation matrix **T** defined by Eq. (6–89), determine matrices $\hat{\mathbf{G}} = \mathbf{T}^{-1}\mathbf{G}\mathbf{T}$ and $\hat{\mathbf{H}} = \mathbf{T}^{-1}\mathbf{H}$. Finally, derive Eq. (6–91).

Solution. We shall first write the controllability matrix as follows:

$$[\mathbf{H}_1 \vdots \mathbf{H}_2 \vdots \mathbf{G}\mathbf{H}_1 \vdots \mathbf{G}\mathbf{H}_2 \vdots \mathbf{G}^2\mathbf{H}_1 \vdots \mathbf{G}^2\mathbf{H}_2 \vdots \mathbf{G}^3\mathbf{H}_1 \vdots \mathbf{G}^3\mathbf{H}_2]$$

$$= \begin{bmatrix} 1 & 0 & -1 & 0 & 2 & 0 & -5 & 2 \\ 0 & 0 & 1 & 0 & -3 & 2 & 11 & -4 \\ 0 & 0 & 0 & 2 & 3 & 0 & -6 & 4 \\ 0 & 1 & 1 & 1 & 0 & 1 & 2 & 1 \end{bmatrix}$$

We now choose four linearly independent vectors from this 4 × 8 matrix, starting from the left end. (These vectors are shown enclosed by dashed lines.) The four linearly independent vectors chosen are $\mathbf{H}_1$, $\mathbf{H}_2$, $\mathbf{G}\mathbf{H}_1$, and $\mathbf{G}\mathbf{H}_2$. Next, we rearrange these four vectors according to Eq. (6–87) and define matrix **F** as follows:

$$\mathbf{F} = [\mathbf{H}_1 \vdots \mathbf{G}\mathbf{H}_1 \vdots \mathbf{H}_2 \vdots \mathbf{G}\mathbf{H}_2]$$

(Note that in this case $n_1 = 2$ and $n_2 = 2$.) Thus

$$\mathbf{F} = \begin{bmatrix} 1 & -1 & 0 & 0 \\ 0 & 1 & 0 & 0 \\ 0 & 0 & 0 & 2 \\ 0 & 1 & 1 & 1 \end{bmatrix}$$

The inverse of this matrix is given by

$$\mathbf{F}^{-1} = \begin{bmatrix} 1 & 1 & 0 & 0 \\ 0 & 1 & 0 & 0 \\ 0 & -1 & -0.5 & 1 \\ 0 & 0 & 0.5 & 0 \end{bmatrix}$$

Since in this case $n_1 = 2$ and $n_2 = 2$ we define the second row vector of $\mathbf{F}^{-1}$ as $\mathbf{f}_1$ and the fourth row vector as $\mathbf{f}_2$. Then

$$\mathbf{f}_1 = [0 \quad 1 \quad 0 \quad 0]$$

$$\mathbf{f}_2 = [0 \quad 0 \quad 0.5 \quad 0]$$

The transformation matrix $\mathbf{T}$ is given by

$$\mathbf{T} = \begin{bmatrix} \mathbf{S}_1 \\ \hdashline \mathbf{S}_2 \end{bmatrix}^{-1}$$

where

$$\mathbf{S}_1 = \begin{bmatrix} \mathbf{f}_1 \\ \hdashline \mathbf{f}_1\mathbf{G} \end{bmatrix}, \qquad \mathbf{S}_2 = \begin{bmatrix} \mathbf{f}_2 \\ \hdashline \mathbf{f}_2\mathbf{G} \end{bmatrix}$$

Hence

$$\mathbf{T} = \begin{bmatrix} \mathbf{f}_1 \\ \mathbf{f}_1\mathbf{G} \\ \mathbf{f}_2 \\ \mathbf{f}_2\mathbf{G} \end{bmatrix}^{-1} = \begin{bmatrix} 0 & 1 & 0 & 0 \\ 1 & -2 & 1 & 0 \\ 0 & 0 & 0.5 & 0 \\ 0 & 0.5 & -0.5 & 1 \end{bmatrix}^{-1} = \begin{bmatrix} 2 & 1 & -2 & 0 \\ 1 & 0 & 0 & 0 \\ 0 & 0 & 2 & 0 \\ -0.5 & 0 & 1 & 1 \end{bmatrix}$$

With this transformation matrix $\mathbf{T}$ we obtain

$$\hat{\mathbf{G}} = \mathbf{T}^{-1}\mathbf{G}\mathbf{T} = \begin{bmatrix} 0 & 1 & 0 & 0 \\ 1 & -2 & 1 & 0 \\ 0 & 0 & 0.5 & 0 \\ 0 & 0.5 & -0.5 & 1 \end{bmatrix} \begin{bmatrix} -1 & 1 & 0 & 0 \\ 1 & -2 & 1 & 0 \\ 0 & 1 & -1 & 2 \\ 1 & 0 & 0 & 1 \end{bmatrix} \begin{bmatrix} 2 & 1 & -2 & 0 \\ 1 & 0 & 0 & 0 \\ 0 & 0 & 2 & 0 \\ -0.5 & 0 & 1 & 1 \end{bmatrix}$$

$$= \begin{bmatrix} 0 & 1 & 0 & 0 \\ -1 & -3 & 2 & 2 \\ 0 & 0 & 0 & 1 \\ 1.5 & 1.5 & -1 & 0 \end{bmatrix}$$

and

$$\hat{\mathbf{H}} = \mathbf{T}^{-1}\mathbf{H} = \begin{bmatrix} 0 & 1 & 0 & 0 \\ 1 & -2 & 1 & 0 \\ 0 & 0 & 0.5 & 0 \\ 0 & 0.5 & -0.5 & 1 \end{bmatrix} \begin{bmatrix} 1 & 0 \\ 0 & 0 \\ 0 & 0 \\ 0 & 1 \end{bmatrix} = \begin{bmatrix} 0 & 0 \\ 1 & 0 \\ 0 & 0 \\ 0 & 1 \end{bmatrix}$$

Notice that when $n_1 = n_2 = 2$, matrix $\hat{\mathbf{G}}$ has the form given by Eq. (6–90) and matrix $\hat{\mathbf{H}}$ has the form given by Eq. (6–91), or

$$\hat{\mathbf{G}} = \left[\begin{array}{cc:cc} 0 & 1 & 0 & 0 \\ -a_{11} & -a_{12} & -a_{13} & -a_{14} \\ \hdashline 0 & 0 & 0 & 1 \\ -a_{21} & -a_{22} & -a_{23} & -a_{24} \end{array}\right], \qquad \hat{\mathbf{H}} = \left[\begin{array}{cc} 0 & 0 \\ 1 & b_{12} \\ \hdashline 0 & 0 \\ 0 & 1 \end{array}\right]$$

(Note that b_{12} is zero in this case.)

Finally, we shall derive Eq. (6–91). Notice that

$$\mathbf{F}^{-1}\mathbf{F} = \begin{bmatrix} \mathbf{m}_1 \\ \mathbf{f}_1 \\ \mathbf{m}_2 \\ \mathbf{f}_2 \end{bmatrix} [\mathbf{H}_1 \quad \mathbf{GH}_1 \quad \mathbf{H}_2 \quad \mathbf{GH}_2]$$

$$= \begin{bmatrix} \mathbf{m}_1\mathbf{H}_1 & \mathbf{m}_1\mathbf{GH}_1 & \mathbf{m}_1\mathbf{H}_2 & \mathbf{m}_1\mathbf{GH}_2 \\ \mathbf{f}_1\mathbf{H}_1 & \mathbf{f}_1\mathbf{GH}_1 & \mathbf{f}_1\mathbf{H}_2 & \mathbf{f}_1\mathbf{GH}_2 \\ \mathbf{m}_2\mathbf{H}_1 & \mathbf{m}_2\mathbf{GH}_1 & \mathbf{m}_2\mathbf{H}_2 & \mathbf{m}_2\mathbf{GH}_2 \\ \mathbf{f}_2\mathbf{H}_1 & \mathbf{f}_2\mathbf{GH}_1 & \mathbf{f}_2\mathbf{H}_2 & \mathbf{f}_2\mathbf{GH}_2 \end{bmatrix}$$

$$= \begin{bmatrix} 1 & 0 & 0 & 0 \\ 0 & 1 & 0 & 0 \\ 0 & 0 & 1 & 0 \\ 0 & 0 & 0 & 1 \end{bmatrix}$$

where $\mathbf{m}_1$ and $\mathbf{m}_2$ are the first row vector and the third row vector of $\mathbf{F}^{-1}$, respectively. Since $\mathbf{F}^{-1}\mathbf{F}$ is an identity matrix, we have $\mathbf{f}_1\mathbf{H}_1 = 0$, $\mathbf{f}_1\mathbf{H}_2 = 0$, $\mathbf{f}_1\mathbf{GH}_1 = 1$, $\mathbf{f}_1\mathbf{GH}_2 = 0$, $\mathbf{f}_2\mathbf{H}_1 = 0$, $\mathbf{f}_2\mathbf{H}_2 = 0$, $\mathbf{f}_2\mathbf{GH}_1 = 0$, and $\mathbf{f}_2\mathbf{GH}_2 = 1$. Thus, we have

$$\hat{\mathbf{H}} = \mathbf{T}^{-1}\mathbf{H} = \begin{bmatrix} \mathbf{f}_1 \\ \mathbf{f}_1\mathbf{G} \\ \mathbf{f}_2 \\ \mathbf{f}_2\mathbf{G} \end{bmatrix} [\mathbf{H}_1 \quad \mathbf{H}_2] = \begin{bmatrix} \mathbf{f}_1\mathbf{H}_1 & \mathbf{f}_1\mathbf{H}_2 \\ \mathbf{f}_1\mathbf{GH}_1 & \mathbf{f}_1\mathbf{GH}_2 \\ \mathbf{f}_2\mathbf{H}_1 & \mathbf{f}_2\mathbf{H}_2 \\ \mathbf{f}_2\mathbf{GH}_1 & \mathbf{f}_2\mathbf{GH}_2 \end{bmatrix} = \begin{bmatrix} 0 & 0 \\ 1 & 0 \\ 0 & 0 \\ 0 & 1 \end{bmatrix}$$

which is Eq. (6–91).

Problem A-6–19. Consider the system defined by Eq. (6–86):

$$\hat{\mathbf{x}}(k+1) = \mathbf{T}^{-1}\mathbf{GT}\hat{\mathbf{x}}(k) + \mathbf{T}^{-1}\mathbf{Hu}(k) = \hat{\mathbf{G}}\hat{\mathbf{x}}(k) + \hat{\mathbf{H}}\mathbf{u}(k)$$

where the transformation matrix $\mathbf{T}$ is defined by Eq. (6–89). Assume that the matrix $\hat{\mathbf{G}}$ is given by Eq. (6–92) and the matrix $\hat{\mathbf{H}}$ is given by Eq. (6–93). That is,

$$\hat{\mathbf{G}} = \left[\begin{array}{ccc:c} 0 & 1 & 0 & 0 \\ 0 & 0 & 1 & 0 \\ -a_{11} & -a_{12} & -a_{13} & -a_{14} \\ \hdashline -a_{21} & -a_{22} & -a_{23} & -a_{24} \end{array}\right], \qquad \hat{\mathbf{H}} = \left[\begin{array}{cc} 0 & 0 \\ 0 & 0 \\ 1 & b_{12} \\ \hdashline 0 & 1 \end{array}\right]$$

Show that

$$|z\mathbf{I} - \hat{\mathbf{G}}| = \begin{vmatrix} z^3 + a_{13}z^2 + a_{12}z + a_{11} & a_{14} \\ a_{23}z^2 + a_{22}z + a_{21} & z + a_{24} \end{vmatrix}$$

and

$$\hat{\mathbf{G}} - \hat{\mathbf{H}}\mathbf{B}\boldsymbol{\Delta} = \begin{bmatrix} 0 & 1 & 0 & 0 \\ 0 & 0 & 1 & 0 \\ -a_{11} - \delta_{11} & -a_{12} - \delta_{12} & -a_{13} - \delta_{13} & -a_{14} - \delta_{14} \\ -a_{21} - \delta_{21} & -a_{22} - \delta_{22} & -a_{23} - \delta_{23} & -a_{24} - \delta_{24} \end{bmatrix}$$

where

$$\mathbf{B} = \begin{bmatrix} 1 & b_{12} \\ 0 & 1 \end{bmatrix}^{-1}, \qquad \boldsymbol{\Delta} = \begin{bmatrix} \delta_{11} & \delta_{12} & \delta_{13} & \delta_{14} \\ \delta_{21} & \delta_{22} & \delta_{23} & \delta_{24} \end{bmatrix}$$

Show also that if we choose, for example,

$$\boldsymbol{\Delta} = \begin{bmatrix} -a_{11} & -a_{12} & -a_{13} & -a_{14} \\ * & * & * & -a_{24} \end{bmatrix} \tag{6–317}$$

where the elements shown by asterisks are arbitrary constants, the system will exhibit the deadbeat response to any initial state $\mathbf{x}(0)$, that is, that

$$(\hat{\mathbf{G}} - \hat{\mathbf{H}}\mathbf{B}\boldsymbol{\Delta})^k = \mathbf{0} \qquad k = 4, 5, 6, \ldots$$

Show also that if we choose

$$\boldsymbol{\Delta} = \begin{bmatrix} -a_{11} & -a_{12} & -a_{13} & -a_{14} \\ -a_{21} & -a_{22} & -a_{23} & -a_{24} \end{bmatrix} \tag{6–318}$$

then

$$(\hat{\mathbf{G}} - \hat{\mathbf{H}}\mathbf{B}\boldsymbol{\Delta})^k = \mathbf{0}$$

for $k \geq n_{\min}$, where

$$n_{\min} = \max(n_1, n_2) = \max(3, 1) = 3$$

Solution. For the case where $\hat{\mathbf{G}}$ is as given by Eq. (6–92), we have

$$|z\mathbf{I} - \hat{\mathbf{G}}| = \begin{vmatrix} z & -1 & 0 & 0 \\ 0 & z & -1 & 0 \\ a_{11} & a_{12} & z + a_{13} & a_{14} \\ a_{21} & a_{22} & a_{23} & z + a_{24} \end{vmatrix}$$

Expanding this determinant using the Laplace's expansion formula, we obtain

$$\begin{aligned} |z\mathbf{I} - \hat{\mathbf{G}}| &= \begin{vmatrix} z & -1 \\ 0 & z \end{vmatrix} \begin{vmatrix} z + a_{13} & a_{14} \\ a_{23} & z + a_{24} \end{vmatrix} - \begin{vmatrix} z & -1 \\ a_{11} & a_{12} \end{vmatrix} \begin{vmatrix} -1 & 0 \\ a_{23} & z + a_{24} \end{vmatrix} \\ &\quad + \begin{vmatrix} z & -1 \\ a_{21} & a_{22} \end{vmatrix} \begin{vmatrix} -1 & 0 \\ z + a_{13} & a_{14} \end{vmatrix} \\ &= z^2[(z + a_{13})(z + a_{24}) - a_{23}a_{14}] + (a_{12}z + a_{11})(z + a_{24}) \\ &\quad + (a_{22}z + a_{21})(-a_{14}) \\ &= (z + a_{24})(z^3 + a_{13}z^2 + a_{12}z + a_{11}) - a_{14}(a_{23}z^2 + a_{22}z + a_{21}) \end{aligned}$$

Hence, the determinant $|z\mathbf{I} - \hat{\mathbf{G}}|$ may be written as follows:

$$|z\mathbf{I} - \hat{\mathbf{G}}| = \begin{vmatrix} z^3 + a_{13}z^2 + a_{12}z + a_{11} & a_{14} \\ a_{23}z^2 + a_{22}z + a_{21} & z + a_{24} \end{vmatrix} \tag{6–319}$$

Next, compute

$$\hat{\mathbf{H}}\mathbf{B}\boldsymbol{\Delta} = \begin{bmatrix} 0 & 0 \\ 0 & 0 \\ 1 & b_{12} \\ 0 & 1 \end{bmatrix} \begin{bmatrix} 1 & b_{12} \\ 0 & 1 \end{bmatrix}^{-1} \begin{bmatrix} \delta_{11} & \delta_{12} & \delta_{13} & \delta_{14} \\ \delta_{21} & \delta_{22} & \delta_{23} & \delta_{24} \end{bmatrix}$$

$$= \begin{bmatrix} 0 & 0 \\ 0 & 0 \\ 1 & 0 \\ 0 & 1 \end{bmatrix} \begin{bmatrix} \delta_{11} & \delta_{12} & \delta_{13} & \delta_{14} \\ \delta_{21} & \delta_{22} & \delta_{23} & \delta_{24} \end{bmatrix} = \begin{bmatrix} 0 & 0 & 0 & 0 \\ 0 & 0 & 0 & 0 \\ \delta_{11} & \delta_{12} & \delta_{13} & \delta_{14} \\ \delta_{21} & \delta_{22} & \delta_{23} & \delta_{24} \end{bmatrix}$$

(Notice that the effect of post-multiplying matrix $\hat{\mathbf{H}}$ by matrix $\mathbf{B}$ is to eliminate b_{12} from the product matrix $\hat{\mathbf{H}}\mathbf{B}$.) Thus,

$$\hat{\mathbf{G}} - \hat{\mathbf{H}}\mathbf{B}\boldsymbol{\Delta} = \begin{bmatrix} 0 & 1 & 0 & 0 \\ 0 & 0 & 1 & 0 \\ -a_{11} - \delta_{11} & -a_{12} - \delta_{12} & -a_{13} - \delta_{13} & -a_{14} - \delta_{14} \\ -a_{21} - \delta_{21} & -a_{22} - \delta_{22} & -a_{23} - \delta_{23} & -a_{24} - \delta_{24} \end{bmatrix}$$

If we choose $\boldsymbol{\Delta}$ as given by Eq. (6–317), then

$$\hat{\mathbf{G}} - \hat{\mathbf{H}}\mathbf{B}\boldsymbol{\Delta} = \begin{bmatrix} 0 & 1 & 0 & 0 \\ 0 & 0 & 1 & 0 \\ 0 & 0 & 0 & 0 \\ * & * & * & 0 \end{bmatrix}$$

where the elements shown by asterisks are arbitrary constants. Notice that

$$(\hat{\mathbf{G}} - \hat{\mathbf{H}}\mathbf{B}\boldsymbol{\Delta})^2 = \begin{bmatrix} 0 & 0 & 1 & 0 \\ 0 & 0 & 0 & 0 \\ 0 & 0 & 0 & 0 \\ 0 & * & * & 0 \end{bmatrix}$$

$$(\hat{\mathbf{G}} - \hat{\mathbf{H}}\mathbf{B}\boldsymbol{\Delta})^3 = \begin{bmatrix} 0 & 0 & 0 & 0 \\ 0 & 0 & 0 & 0 \\ 0 & 0 & 0 & 0 \\ 0 & 0 & * & 0 \end{bmatrix}$$

$$(\hat{\mathbf{G}} - \hat{\mathbf{H}}\mathbf{B}\boldsymbol{\Delta})^4 = \begin{bmatrix} 0 & 0 & 0 & 0 \\ 0 & 0 & 0 & 0 \\ 0 & 0 & 0 & 0 \\ 0 & 0 & 0 & 0 \end{bmatrix}$$

Hence

$$\mathbf{x}(k) = (\mathbf{G} - \mathbf{HK})^k \mathbf{x}(0) = \mathbf{T}(\hat{\mathbf{G}} - \hat{\mathbf{H}}\mathbf{B}\boldsymbol{\Delta})^k \mathbf{T}^{-1}\mathbf{x}(0) = \mathbf{0} \qquad k \geq 4$$

We have thus seen that the deadbeat response is achieved by choosing $\boldsymbol{\Delta}$ as given by Eq. (6–317).

However, if we choose $\boldsymbol{\Delta}$ as given by Eq. (6–318), then the deadbeat response can be achieved in at most 3 sampling periods, because the asterisk appearing in $(\hat{\mathbf{G}} - \hat{\mathbf{H}}\mathbf{B}\boldsymbol{\Delta})^3$ becomes zero and

$$\mathbf{x}(k) = \mathbf{T}(\hat{\mathbf{G}} - \hat{\mathbf{H}}\mathbf{B}\boldsymbol{\Delta})^k \mathbf{T}^{-1}\mathbf{x}(0) = \mathbf{0} \qquad k \geq n_{\min} = 3$$

Problem A-6-20. Consider the following system:

$$\mathbf{x}(k+1) = \mathbf{G}\mathbf{x}(k) + \mathbf{H}\mathbf{u}(k)$$

where

$$\mathbf{x}(k) = \text{state vector (4-vector)}$$

$$\mathbf{u}(k) = \text{control vector (2-vector)}$$

and

$$\mathbf{G} = \begin{bmatrix} -1 & 1 & 0 & 0 \\ 1 & -2 & 1 & 0 \\ 0 & 1 & -1 & 2 \\ 1 & 0 & 0 & 1 \end{bmatrix}, \qquad \mathbf{H} = [\mathbf{H}_1 \vdots \mathbf{H}_2] = \begin{bmatrix} 0 & 1 \\ 1 & 0 \\ 0 & 0 \\ 1 & 0 \end{bmatrix}$$

By use of the state feedback control $\mathbf{u}(k) = -\mathbf{K}\mathbf{x}(k)$, we wish to place the closed-loop poles at the following locations:

$$z_1 = 0.5 + j0.5, \qquad z_2 = 0.5 - j0.5$$

$$z_3 = -0.2, \qquad z_4 = -0.8$$

Determine the required state feedback gain matrix **K.** Then, using the given **G** and **H** matrices, derive Eq. (6–93).

Solution. We shall first examine the controllability matrix:

$$[\mathbf{H} \vdots \mathbf{GH} \vdots \mathbf{G}^2\mathbf{H} \vdots \mathbf{G}^3\mathbf{H}] = [\mathbf{H}_1 \vdots \mathbf{H}_2 \vdots \mathbf{GH}_1 \vdots \mathbf{GH}_2 \vdots \mathbf{G}^2\mathbf{H}_1 \vdots \mathbf{G}^2\mathbf{H}_2 \vdots \mathbf{G}^3\mathbf{H}_1 \vdots \mathbf{G}^3\mathbf{H}_2]$$

$$= \begin{bmatrix} 0 & 1 & 1 & -1 & -3 & 2 & 11 & -5 \\ 1 & 0 & -2 & 1 & 8 & -3 & -22 & 11 \\ 0 & 0 & 3 & 0 & -3 & 3 & 15 & -6 \\ 1 & 0 & 1 & 1 & 2 & 0 & -1 & 2 \end{bmatrix} \tag{6–320}$$

The rank of this controllability matrix is 4. Thus, arbitrary pole placement is possible. Four linearly independent vectors are chosen starting from the left end. (These vectors are shown enclosed by dashed lines.) The four linearly independent vectors chosen are $\mathbf{H}_1$, $\mathbf{H}_2$, $\mathbf{GH}_1$, and $\mathbf{G}^2\mathbf{H}_1$. Now we rearrange these four vectors according to Eq. (6–87) and define matrix **F** as follows:

$$\mathbf{F} = [\mathbf{H}_1 \vdots \mathbf{GH}_1 \vdots \mathbf{G}^2\mathbf{H}_1 \vdots \mathbf{H}_2]$$

We note that $n_1 = 3$ and $n_2 = 1$ in this case. Rewriting matrix **F**, we have

$$\mathbf{F} = \left[\begin{array}{cc|cc} 0 & 1 & -3 & 1 \\ 1 & -2 & 8 & 0 \\ \hline 0 & 3 & -3 & 0 \\ 1 & 1 & 2 & 0 \end{array}\right] = \left[\begin{array}{c|c} \mathbf{A} & \mathbf{B} \\ \hline \mathbf{C} & \mathbf{D} \end{array}\right]$$

Next, we compute $\mathbf{F}^{-1}$. Referring to the Appendix, we have

$$\mathbf{F}^{-1} = \begin{bmatrix} \mathbf{A}^{-1} + \mathbf{A}^{-1}\mathbf{B}(\mathbf{D} - \mathbf{C}\mathbf{A}^{-1}\mathbf{B})^{-1}\mathbf{C}\mathbf{A}^{-1} & -\mathbf{A}^{-1}\mathbf{B}(\mathbf{D} - \mathbf{C}\mathbf{A}^{-1}\mathbf{B})^{-1} \\ -(\mathbf{D} - \mathbf{C}\mathbf{A}^{-1}\mathbf{B})^{-1}\mathbf{C}\mathbf{A}^{-1} & (\mathbf{D} - \mathbf{C}\mathbf{A}^{-1}\mathbf{B})^{-1} \end{bmatrix}$$

$$= \begin{bmatrix} 0 & -1 & -\frac{4}{3} & 2 \\ 0 & \frac{1}{3} & \frac{2}{3} & -\frac{1}{3} \\ 0 & \frac{1}{3} & \frac{1}{3} & -\frac{1}{3} \\ 1 & \frac{2}{3} & \frac{1}{3} & -\frac{2}{3} \end{bmatrix}$$

Since $n_1 = 3$ and $n_2 = 1$, we choose the third row vector as $\mathbf{f}_1$ and the fourth row vector as $\mathbf{f}_2$. (Note that we define the η_ith row vector, where $\eta_i = n_1 + n_2 + \cdots + n_i$, as $\mathbf{f}_i$.) That is,

$$\mathbf{f}_1 = [0 \quad \tfrac{1}{3} \quad \tfrac{1}{3} \quad -\tfrac{1}{3}]$$

$$\mathbf{f}_2 = [1 \quad \tfrac{2}{3} \quad \tfrac{1}{3} \quad -\tfrac{2}{3}]$$

Next, we define the transformation matrix $\mathbf{T}$ by

$$\mathbf{T} = \begin{bmatrix} \mathbf{S}_1 \\ \hline \mathbf{S}_2 \end{bmatrix}^{-1}$$

where

$$\mathbf{S}_1 = \begin{bmatrix} \mathbf{f}_1 \\ \mathbf{f}_1\mathbf{G} \\ \mathbf{f}_1\mathbf{G}^2 \end{bmatrix}, \qquad \mathbf{S}_2 = [\mathbf{f}_2]$$

Hence

$$\mathbf{T} = \begin{bmatrix} 0 & \frac{1}{3} & \frac{1}{3} & -\frac{1}{3} \\ 0 & -\frac{1}{3} & 0 & \frac{1}{3} \\ 0 & \frac{2}{3} & -\frac{1}{3} & \frac{1}{3} \\ 1 & \frac{2}{3} & \frac{1}{3} & -\frac{2}{3} \end{bmatrix}^{-1} = \begin{bmatrix} -1 & 1 & 0 & 1 \\ 1 & 0 & 1 & 0 \\ 3 & 3 & 0 & 0 \\ 1 & 3 & 1 & 0 \end{bmatrix}$$

With this transformation matrix $\mathbf{T}$, if we define

$$\mathbf{x}(k) = \mathbf{T}\hat{\mathbf{x}}(k)$$

then

$$\hat{\mathbf{G}} = \mathbf{T}^{-1}\mathbf{G}\mathbf{T} = \left[\begin{array}{ccc|c} 0 & 1 & 0 & 0 \\ 0 & 0 & 1 & 0 \\ 0 & 3 & -2 & 1 \\ \hline 2 & 0 & 0 & -1 \end{array}\right] \tag{6–321}$$

Also,

$$\hat{\mathbf{H}} = \mathbf{T}^{-1}\mathbf{H} = \begin{bmatrix} 0 & \frac{1}{3} & \frac{1}{3} & -\frac{1}{3} \\ 0 & -\frac{1}{3} & 0 & \frac{1}{3} \\ 0 & \frac{2}{3} & -\frac{1}{3} & \frac{1}{3} \\ 1 & \frac{2}{3} & \frac{1}{3} & -\frac{2}{3} \end{bmatrix} \begin{bmatrix} 0 & 1 \\ 1 & 0 \\ 0 & 0 \\ 1 & 0 \end{bmatrix} = \left[\begin{array}{cc} 0 & 0 \\ 0 & 0 \\ 1 & 0 \\ \hline 0 & 1 \end{array}\right] \tag{6–322}$$

Now we shall determine the state feedback gain matrix $\mathbf{K}$, where

$$\mathbf{K} = \mathbf{B}\boldsymbol{\Delta}\mathbf{T}^{-1}$$

Referring to Eq. (6–101) and noting that $b_{12} = 0$ in this case, matrix $\mathbf{B}$ is a 2×2 matrix given by

$$\mathbf{B} = \begin{bmatrix} 1 & b_{12} \\ 0 & 1 \end{bmatrix}^{-1} = \begin{bmatrix} 1 & 0 \\ 0 & 1 \end{bmatrix} \tag{6–323}$$

For the present case $\mathbf{\Delta}$ is a 2×4 matrix:

$$\mathbf{\Delta} = \begin{bmatrix} \delta_{11} & \delta_{12} & \delta_{13} & \delta_{14} \\ \delta_{21} & \delta_{22} & \delta_{23} & \delta_{24} \end{bmatrix}$$

Hence

$$\hat{\mathbf{G}} - \hat{\mathbf{H}}\mathbf{B}\mathbf{\Delta} = \left[\begin{array}{ccc:c} 0 & 1 & 0 & 0 \\ 0 & 0 & 1 & 0 \\ -\delta_{11} & 3-\delta_{12} & -2-\delta_{13} & 1-\delta_{14} \\ \hdashline 2-\delta_{21} & -\delta_{22} & -\delta_{23} & -1-\delta_{24} \end{array}\right]$$

Referring to Eq. (6–319), we have

$$|z\mathbf{I} - \hat{\mathbf{G}} + \hat{\mathbf{H}}\mathbf{B}\mathbf{\Delta}| = \left|\begin{array}{c:c} z^3 + (2+\delta_{13})z^2 + (-3+\delta_{12})z + \delta_{11} & -1+\delta_{14} \\ \hdashline \delta_{23}z^2 + \delta_{22}z + (-2+\delta_{21}) & z+1+\delta_{24} \end{array}\right| = 0$$

This characteristic equation must be equal to the desired characteristic equation, which is

$$(z-0.5-j0.5)(z-0.5+j0.5)(z+0.2)(z+0.8) = z^4 - 0.34z^2 + 0.34z + 0.08 = 0$$

If we equate the coefficients of the equal powers of z of the two characteristic equations, we will have four equations for the determination of eight δ's. Hence matrix $\mathbf{\Delta}$ is not unique. Suppose we arbitrarily choose

$$\delta_{14} = 0, \qquad \delta_{22} = 0, \qquad \delta_{23} = 0, \qquad \delta_{24} = -1$$

Then

$$|z\mathbf{I} - \hat{\mathbf{G}} + \hat{\mathbf{H}}\mathbf{B}\mathbf{\Delta}| = z^4 + (2+\delta_{13})z^3 + (-3+\delta_{12})z^2 + \delta_{11}z - 2 + \delta_{21} = 0$$

By equating this characteristic equation with the desired characteristic equation, we have

$$\delta_{11} = 0.34$$

$$\delta_{12} = 2.66$$

$$\delta_{13} = -2$$

$$\delta_{21} = 2.08$$

Thus

$$\mathbf{\Delta} = \begin{bmatrix} 0.34 & 2.66 & -2 & 0 \\ 2.08 & 0 & 0 & -1 \end{bmatrix}$$

Then matrix $\mathbf{K}$ is obtained as follows:

$$\mathbf{K} = \mathbf{B}\mathbf{\Delta}\mathbf{T}^{-1} = \begin{bmatrix} 0 & -2.1067 & 0.7800 & 0.1067 \\ -1 & 0.02667 & 0.3600 & -0.02667 \end{bmatrix}$$

With the matrix $\mathbf{K}$ thus determined, state feedback control

$$u(k) = -\mathbf{K}\mathbf{x}(k)$$

will place the closed-loop poles at $z_1 = 0.5 + j0.5$, $z_2 = 0.5 - j0.5$, $z_3 = -0.2$, and $z_4 = -0.8$. It is noted that matrix $\mathbf{K}$ is not unique; there are many other possible matrices for $\mathbf{K}$.

Finally, we shall derive Eq. (6–93). Notice first that

$$\mathbf{F}^{-1}\mathbf{F} = \begin{bmatrix} \mathbf{m}_1 \\ \mathbf{m}_2 \\ \mathbf{f}_1 \\ \mathbf{f}_2 \end{bmatrix} [\mathbf{H}_1 \quad \mathbf{G}\mathbf{H}_1 \quad \mathbf{G}^2\mathbf{H}_1 \quad \mathbf{H}_2]$$

$$= \begin{bmatrix} \mathbf{m}_1\mathbf{H}_1 & \mathbf{m}_1\mathbf{G}\mathbf{H}_1 & \mathbf{m}_1\mathbf{G}^2\mathbf{H}_1 & \mathbf{m}_1\mathbf{H}_2 \\ \mathbf{m}_2\mathbf{H}_1 & \mathbf{m}_2\mathbf{G}\mathbf{H}_1 & \mathbf{m}_2\mathbf{G}^2\mathbf{H}_1 & \mathbf{m}_2\mathbf{H}_2 \\ \mathbf{f}_1\mathbf{H}_1 & \mathbf{f}_1\mathbf{G}\mathbf{H}_1 & \mathbf{f}_1\mathbf{G}^2\mathbf{H}_1 & \mathbf{f}_1\mathbf{H}_2 \\ \mathbf{f}_2\mathbf{H}_1 & \mathbf{f}_2\mathbf{G}\mathbf{H}_1 & \mathbf{f}_2\mathbf{G}^2\mathbf{H}_1 & \mathbf{f}_2\mathbf{H}_2 \end{bmatrix} = \begin{bmatrix} 1 & 0 & 0 & 0 \\ 0 & 1 & 0 & 0 \\ 0 & 0 & 1 & 0 \\ 0 & 0 & 0 & 1 \end{bmatrix}$$

where $\mathbf{m}_1$ and $\mathbf{m}_2$ are the first row vector and second row vector of $\mathbf{F}^{-1}$, respectively. Since $\mathbf{F}^{-1}\mathbf{F}$ is an identity matrix, $\mathbf{f}_1\mathbf{H}_1 = 0$, $\mathbf{f}_1\mathbf{H}_2 = 0$, $\mathbf{f}_1\mathbf{G}\mathbf{H}_1 = 0$, $\mathbf{f}_1\mathbf{G}^2\mathbf{H}_1 = 1$, $\mathbf{f}_2\mathbf{H}_1 = 0$, and $\mathbf{f}_2\mathbf{H}_2 = 1$. From Eq. (6–320) we see that $\mathbf{G}\mathbf{H}_2$ is linearly dependent on $\mathbf{H}_1$, $\mathbf{H}_2$, and $\mathbf{G}\mathbf{H}_1$. Hence $\mathbf{f}_1\mathbf{G}\mathbf{H}_2 = \alpha\mathbf{f}_1\mathbf{H}_1 + \beta\mathbf{f}_1\mathbf{H}_2 + \gamma\mathbf{f}_1\mathbf{G}\mathbf{H}_1 = 0$, where α, β, and γ are constants. Note that $\mathbf{f}_1\mathbf{G}^2\mathbf{H}_2$ may or may not be zero. Consequently,

$$\hat{\mathbf{H}} = \mathbf{T}^{-1}\mathbf{H} = \begin{bmatrix} \mathbf{f}_1 \\ \mathbf{f}_1\mathbf{G} \\ \mathbf{f}_1\mathbf{G}^2 \\ \mathbf{f}_2 \end{bmatrix} [\mathbf{H}_1 \quad \mathbf{H}_2] = \begin{bmatrix} \mathbf{f}_1\mathbf{H}_1 & \mathbf{f}_1\mathbf{H}_2 \\ \mathbf{f}_1\mathbf{G}\mathbf{H}_1 & \mathbf{f}_1\mathbf{G}\mathbf{H}_2 \\ \mathbf{f}_1\mathbf{G}^2\mathbf{H}_1 & \mathbf{f}_1\mathbf{G}^2\mathbf{H}_2 \\ \mathbf{f}_2\mathbf{H}_1 & \mathbf{f}_2\mathbf{H}_2 \end{bmatrix} = \begin{bmatrix} 0 & 0 \\ 0 & 0 \\ 1 & \mathrm{b}_{12} \\ 0 & 1 \end{bmatrix}$$

where $\mathrm{b}_{12} = \mathbf{f}_1\mathbf{G}^2\mathbf{H}_2$. This last equation is Eq. (6–93).

Problem A-6–21. Referring to Prob. A-6–20, consider the same system. Suppose that we desire the deadbeat response to an arbitrary initial state $\mathbf{x}(0)$. Determine the state feedback gain matrix $\mathbf{K}$.

Solution. Referring to Eqs. (6–321), (6–322), and (6–323) in Prob. A-6–20, we have

$$\hat{\mathbf{G}} = \left[\begin{array}{ccc|c} 0 & 1 & 0 & 0 \\ 0 & 0 & 1 & 0 \\ 0 & 3 & -2 & 1 \\ \hline 2 & 0 & 0 & -1 \end{array}\right], \qquad \hat{\mathbf{H}} = \left[\begin{array}{cc} 0 & 0 \\ 0 & 0 \\ 1 & 0 \\ \hline 0 & 1 \end{array}\right]$$

$$\mathbf{B} = \begin{bmatrix} 1 & b_{12} \\ 0 & 1 \end{bmatrix}^{-1} = \begin{bmatrix} 1 & 0 \\ 0 & 1 \end{bmatrix}$$

where b_{12} is zero. For the deadbeat response we choose $\mathbf{\Delta}$ as follows:

$$\mathbf{\Delta} = \begin{bmatrix} -a_{11} & -a_{12} & -a_{13} & -a_{14} \\ -a_{21} & -a_{22} & -a_{23} & -a_{24} \end{bmatrix} = \begin{bmatrix} 0 & 3 & -2 & 1 \\ 2 & 0 & 0 & -1 \end{bmatrix}$$

where the a_{ij}'s are as defined in Eq. (6–92). Then

$$\hat{\mathbf{G}} - \hat{\mathbf{H}}\mathbf{B}\mathbf{\Delta} = \begin{bmatrix} 0 & 1 & 0 & 0 \\ 0 & 0 & 1 & 0 \\ 0 & 0 & 0 & 0 \\ 0 & 0 & 0 & 0 \end{bmatrix}$$

and we find

$$(\hat{\mathbf{G}} - \hat{\mathbf{H}}\mathbf{B}\mathbf{\Delta})^k = \mathbf{0} \qquad k = 3, 4, 5, \ldots$$

The deadbeat response is reached in at most 3 sampling periods. [Note that in this problem, $n_1 = 3$ and $n_2 = 1$. Hence $n_{\min} = \max(n_1, n_2) = 3$.] The desired state feedback gain matrix $\mathbf{K}$ is obtained as follows:

$$\mathbf{K} = \mathbf{B}\mathbf{\Delta}\mathbf{T}^{-1}$$

$$= \begin{bmatrix} 1 & 0 \\ 0 & 1 \end{bmatrix} \begin{bmatrix} 0 & 3 & -2 & 1 \\ 2 & 0 & 0 & -1 \end{bmatrix} \begin{bmatrix} 0 & \frac{1}{3} & \frac{1}{3} & -\frac{1}{3} \\ 0 & -\frac{1}{3} & 0 & \frac{1}{3} \\ 0 & \frac{2}{3} & -\frac{1}{3} & \frac{1}{3} \\ 1 & \frac{2}{3} & \frac{1}{3} & -\frac{2}{3} \end{bmatrix}$$

$$= \begin{bmatrix} 1 & -\frac{5}{3} & 1 & -\frac{1}{3} \\ -1 & 0 & \frac{1}{3} & 0 \end{bmatrix}$$

With this matrix $\mathbf{K}$, the state feedback control

$$\mathbf{u}(k) = -\mathbf{K}\mathbf{x}(k)$$

will place the four closed-loop poles at the origin and thus will produce the deadbeat response to any initial state $\mathbf{x}(0)$.

Problem A-6–22. Consider the system

$$\mathbf{x}(k+1) = \mathbf{G}\mathbf{x}(k)$$

$$y(k) = \mathbf{C}\mathbf{x}(k)$$

where

$\mathbf{x}(k)$ = state vector (n-vector)

$y(k)$ = output signal (scalar)

$\mathbf{G} = n \times n$ nonsingular matrix

$\mathbf{C} = 1 \times n$ matrix

Assume that the system is completely observable.

Derive the following Ackermann's formula for the observer feedback gain matrix $\mathbf{K}_e$:

$$\mathbf{K}_e = \phi(\mathbf{G}) \begin{bmatrix} \mathbf{C} \\ \mathbf{C}\mathbf{G} \\ \cdot \\ \cdot \\ \cdot \\ \mathbf{C}\mathbf{G}^{n-2} \\ \mathbf{C}\mathbf{G}^{n-1} \end{bmatrix}^{-1} \begin{bmatrix} 0 \\ 0 \\ \cdot \\ \cdot \\ \cdot \\ 0 \\ 1 \end{bmatrix} \tag{6–324}$$

Assume that the desired characteristic equation for the error vector is

$$\phi(z) = z^n + \alpha_1 z^{n-1} + \cdots + \alpha_{n-1} z + \alpha_n = 0$$

Solution. From Eq. (6–145), the characteristic equation for the error dynamics is

$$|z\mathbf{I} - \mathbf{Q}^{-1}\mathbf{G}\mathbf{Q} + \mathbf{Q}^{-1}\mathbf{K}_e\mathbf{C}\mathbf{Q}| = 0 \tag{6–325}$$

Define

$$\mathbf{Q}^{-1}\mathbf{G}\mathbf{Q} = \hat{\mathbf{G}}, \qquad \mathbf{Q}^{-1}\mathbf{K}_e = \hat{\mathbf{K}}_e, \qquad \mathbf{C}\mathbf{Q} = \hat{\mathbf{C}}$$

Then Eq. (6–325) becomes

$$|z\mathbf{I} - \hat{\mathbf{G}} + \hat{\mathbf{K}}_e\hat{\mathbf{C}}| = 0$$

which may be modified to

$$|z\mathbf{I} - \hat{\mathbf{G}}^* + \hat{\mathbf{C}}^*\hat{\mathbf{K}}_e^*| = 0$$

where we have taken the conjugate transpose of the matrix $z\mathbf{I} - \hat{\mathbf{G}} + \hat{\mathbf{K}}_e\hat{\mathbf{C}}$ involved in the determinant equation.

Let us define

$$\mathbf{G}_0^* = \hat{\mathbf{G}}^* - \hat{\mathbf{C}}^*\hat{\mathbf{K}}_e^*$$

Since the Cayley-Hamilton theorem states that $\mathbf{G}_0^*$ satisfies its own characteristic equation, we have

$$(\mathbf{G}_0^*)^n + \alpha_1(\mathbf{G}_0^*)^{n-1} + \cdots + \alpha_{n-1}\mathbf{G}_0^* + \alpha_n\mathbf{I} = \phi(\mathbf{G}_0^*) = \mathbf{0} \tag{6–326}$$

Next, consider the following identities:

$$\begin{aligned}
\mathbf{I} &= \mathbf{I} \\
\mathbf{G}_0^* &= \hat{\mathbf{G}}^* - \hat{\mathbf{C}}^*\hat{\mathbf{K}}_e^* \\
(\mathbf{G}_0^*)^2 &= (\hat{\mathbf{G}}^* - \hat{\mathbf{C}}^*\hat{\mathbf{K}}_e^*)^2 = (\hat{\mathbf{G}}^*)^2 - \hat{\mathbf{G}}^*\hat{\mathbf{C}}^*\hat{\mathbf{K}}_e^* - \hat{\mathbf{C}}^*\hat{\mathbf{K}}_e^*\mathbf{G}_0^* \\
(\mathbf{G}_0^*)^3 &= (\hat{\mathbf{G}}^* - \hat{\mathbf{C}}^*\hat{\mathbf{K}}_e^*)^3 = (\hat{\mathbf{G}}^*)^3 - (\hat{\mathbf{G}}^*)^2\hat{\mathbf{C}}^*\hat{\mathbf{K}}_e^* - \hat{\mathbf{G}}^*\hat{\mathbf{C}}^*\hat{\mathbf{K}}_e^*\mathbf{G}_0^* - \hat{\mathbf{C}}^*\hat{\mathbf{K}}_e^*(\mathbf{G}_0^*)^2 \\
&\ \ \vdots \\
(\mathbf{G}_0^*)^n &= (\hat{\mathbf{G}}^* - \hat{\mathbf{C}}^*\hat{\mathbf{K}}_e^*)^n = (\hat{\mathbf{G}}^*)^n - (\hat{\mathbf{G}}^*)^{n-1}\hat{\mathbf{C}}^*\hat{\mathbf{K}}_e^* - \cdots - \hat{\mathbf{C}}^*\hat{\mathbf{K}}_e^*(\mathbf{G}_0^*)^{n-1}
\end{aligned}$$

Multiplying the preceding equations by $\alpha_n, \alpha_{n-1}, \ldots, \alpha_0$ (where $\alpha_0 = 1$) in this order and adding the results, we obtain

$$\begin{aligned}
\alpha_n\mathbf{I} &+ \alpha_{n-1}\mathbf{G}_0^* + \alpha_{n-2}(\mathbf{G}_0^*)^2 + \cdots + (\mathbf{G}_0^*)^n \\
&= \alpha_n\mathbf{I} + \alpha_{n-1}\hat{\mathbf{G}}^* + \alpha_{n-2}(\hat{\mathbf{G}}^*)^2 + \cdots + (\hat{\mathbf{G}}^*)^n - \alpha_{n-1}\hat{\mathbf{C}}^*\hat{\mathbf{K}}_e^* \\
&\quad - \alpha_{n-2}\hat{\mathbf{G}}^*\hat{\mathbf{C}}^*\hat{\mathbf{K}}_e^* - \alpha_{n-2}\hat{\mathbf{C}}^*\hat{\mathbf{K}}_e^*\mathbf{G}_0^* - \cdots - (\hat{\mathbf{G}}^*)^{n-1}\hat{\mathbf{C}}^*\hat{\mathbf{K}}_e^* \\
&\quad - \cdots - \hat{\mathbf{C}}^*\hat{\mathbf{K}}_e^*(\mathbf{G}_0^*)^{n-1}
\end{aligned}$$

which can be written as follows:

$$\phi(\mathbf{G}_0^*) = \phi(\hat{\mathbf{G}}^*) - [\hat{\mathbf{C}}^* \vdots \hat{\mathbf{G}}^*\hat{\mathbf{C}}^* \vdots \cdots \vdots (\hat{\mathbf{G}}^*)^{n-1}\hat{\mathbf{C}}^*] \cdot \begin{bmatrix} \alpha_{n-1}\hat{\mathbf{K}}_e^* + \alpha_{n-2}\hat{\mathbf{K}}_e^*\mathbf{G}_0 + \cdots + \hat{\mathbf{K}}_e^*(\mathbf{G}_0)^{n-1} \\ \alpha_{n-2}\hat{\mathbf{K}}_e^* + \alpha_{n-3}\hat{\mathbf{K}}_e^*\mathbf{G}_0 + \cdots + \hat{\mathbf{K}}_e^*(\mathbf{G}_0)^{n-2} \\ \vdots \\ \hat{\mathbf{K}}_e^* \end{bmatrix} \tag{6-327}$$

By referring to Eq. (6–326), Eq. (6–327) may be simplified to

$$\phi(\hat{\mathbf{G}}^*) = [\hat{\mathbf{C}}^* \vdots \hat{\mathbf{G}}^*\hat{\mathbf{C}}^* \vdots \cdots \vdots (\hat{\mathbf{G}}^*)^{n-1}\hat{\mathbf{C}}^*] \cdot \begin{bmatrix} \alpha_{n-1}\hat{\mathbf{K}}_e^* + \alpha_{n-2}\hat{\mathbf{K}}_e^*\mathbf{G}_0 + \cdots + \hat{\mathbf{K}}_e^*(\mathbf{G}_0)^{n-1} \\ \alpha_{n-2}\hat{\mathbf{K}}_e^* + \alpha_{n-3}\hat{\mathbf{K}}_e^*\mathbf{G}_0 + \cdots + \hat{\mathbf{K}}_e^*(\mathbf{G}_0)^{n-2} \\ \vdots \\ \hat{\mathbf{K}}_e^* \end{bmatrix} \tag{6-328}$$

Notice that

$$\begin{aligned} [\hat{\mathbf{C}}^* \vdots \hat{\mathbf{G}}^*\hat{\mathbf{C}}^* \vdots \cdots \vdots (\hat{\mathbf{G}}^*)^{n-1}\hat{\mathbf{C}}^*] &= [\mathbf{Q}^*\mathbf{C}^* \vdots \mathbf{Q}^*\mathbf{G}^*\mathbf{C}^* \vdots \cdots \vdots \mathbf{Q}^*(\mathbf{G}^*)^{n-1}\mathbf{C}^*] \\ &= \mathbf{Q}^*[\mathbf{C}^* \vdots \mathbf{G}^*\mathbf{C}^* \vdots \cdots \vdots (\mathbf{G}^*)^{n-1}\mathbf{C}^*] \end{aligned} \tag{6-329}$$

Since the system is completely observable, the rank of

$$[\mathbf{C}^* \vdots \mathbf{G}^*\mathbf{C}^* \vdots \cdots \vdots (\mathbf{G}^*)^{n-1}\mathbf{C}^*]$$

is n and its inverse exists. Hence, the matrix at the left-hand side of Eq. (6–329) has the inverse. By premultiplying both sides of Eq. (6–328) by

$$[0 \quad 0 \quad \cdots \quad 0 \quad 1][\hat{\mathbf{C}}^* \vdots \hat{\mathbf{G}}^*\hat{\mathbf{C}}^* \vdots \cdots \vdots (\hat{\mathbf{G}}^*)^{n-1}\hat{\mathbf{C}}^*]^{-1}$$

we obtain

$$[0 \quad 0 \quad \cdots \quad 0 \quad 1][\hat{\mathbf{C}}^* \vdots \hat{\mathbf{G}}^*\hat{\mathbf{C}}^* \vdots \cdots \vdots (\hat{\mathbf{G}}^*)^{n-1}\hat{\mathbf{C}}^*]^{-1}\phi(\hat{\mathbf{G}}^*) = \hat{\mathbf{K}}_e^*$$

By use of Eq. (6–329) this last equation can be written as follows:

$$[0 \quad 0 \quad \cdots \quad 0 \quad 1][\mathbf{C}^* \vdots \mathbf{G}^*\mathbf{C}^* \vdots \cdots \vdots (\mathbf{G}^*)^{n-1}\mathbf{C}^*]^{-1}(\mathbf{Q}^*)^{-1}\phi(\hat{\mathbf{G}}^*) = \hat{\mathbf{K}}_e^* \tag{6-330}$$

By referring to the relationship

$$\mathbf{Q}\phi(\hat{\mathbf{G}})\mathbf{Q}^{-1} = \phi(\mathbf{G})$$

or

$$(\mathbf{Q}^*)^{-1}\phi(\hat{\mathbf{G}}^*)\mathbf{Q}^* = \phi(\mathbf{G}^*)$$

Eq. (6–330) can be written as follows:

$$[0 \quad 0 \quad \cdots \quad 0 \quad 1][\mathbf{C}^* \vdots \mathbf{G}^*\mathbf{C}^* \vdots \cdots \vdots (\mathbf{G}^*)^{n-1}\mathbf{C}^*]^{-1}\phi(\mathbf{G}^*)(\mathbf{Q}^*)^{-1} = \hat{\mathbf{K}}_e^* \tag{6-331}$$

Since

$$\hat{\mathbf{K}}_e = \mathbf{Q}^{-1}\mathbf{K}_e$$

we have

$$\hat{\mathbf{K}}_e^* = \mathbf{K}_e^*(\mathbf{Q}^*)^{-1}$$

Thus Eq. (6–331) becomes

$$[0 \quad 0 \quad \cdots \quad 0 \quad 1][\mathbf{C}^* \vdots \mathbf{G}^*\mathbf{C}^* \vdots \cdots \vdots (\mathbf{G}^*)^{n-1}\mathbf{C}^*]^{-1}\phi(\mathbf{G}^*) = \mathbf{K}_e^*$$

By taking the conjugate transpose of this last equation, we obtain

$$\mathbf{K}_e = \phi(\mathbf{G}) \begin{bmatrix} \mathbf{C} \\ \mathbf{CG} \\ \cdot \\ \cdot \\ \cdot \\ \mathbf{CG}^{n-2} \\ \mathbf{CG}^{n-1} \end{bmatrix}^{-1} \begin{bmatrix} 0 \\ 0 \\ \cdot \\ \cdot \\ \cdot \\ 0 \\ 1 \end{bmatrix}$$

which is Eq. (6–324).

Problem A-6–23. Consider the system defined by

$$\mathbf{x}(k+1) = \mathbf{G}\mathbf{x}(k)$$

$$y(k) = \mathbf{C}\mathbf{x}(k)$$

where $\mathbf{x}(k)$ is a 3-vector and $y(k)$ is a scalar. It is assumed that the system is completely observable. It is desired to determine the observer feedback gain matrix $\mathbf{K}_e$ for a full-order prediction observer such that the error dynamics have characteristic roots at $z = \mu_1$, $z = \mu_2$, and $z = \mu_3$, or

$$|z\mathbf{I} - \mathbf{G} + \mathbf{K}_e\mathbf{C}| = (z - \mu_1)(z - \mu_2)(z - \mu_3)$$
$$= z^3 + \alpha_1 z^2 + \alpha_2 z + \alpha_3$$

Note that the eigenvalues of $\mathbf{G}$ are λ_1, λ_2, and λ_3 and they are different from μ_1, μ_2, and μ_3. We assume that μ_1, μ_2, and μ_3 are distinct.

Show that the matrix $\mathbf{K}_e$ can be given by

$$\mathbf{K}_e = \begin{bmatrix} \mathbf{f}_1 \\ \mathbf{f}_2 \\ \mathbf{f}_3 \end{bmatrix}^{-1} \begin{bmatrix} 1 \\ 1 \\ 1 \end{bmatrix}$$

where

$$\mathbf{f}_i = \mathbf{C}(\mathbf{G} - \mu_i\mathbf{I})^{-1} \qquad i = 1, 2, 3$$

Show also that the $\mathbf{f}_i^*$'s are eigenvectors of matrix $(\mathbf{G} - \mathbf{K}_e\mathbf{C})^*$, that is, that $\mathbf{f}_i^*$ satisfies the equation

$$(\mathbf{G} - \mathbf{K}_e\mathbf{C})^*\mathbf{f}_i^* = \bar{\mu}_i\mathbf{f}_i^* \qquad i = 1, 2, 3$$

where $\bar{\mu}_i$ is the complex conjugate of μ_i. (Note that any complex eigenvalues occur as conjugate pairs.)

Solution. Referring to Ackermann's formula as given by Eq. (6–151) for the observer feedback gain matrix $\mathbf{K}_e$, we have

$$\mathbf{K}_e = \phi(\mathbf{G}) \begin{bmatrix} \mathbf{C} \\ \mathbf{CG} \\ \mathbf{CG}^2 \end{bmatrix}^{-1} \begin{bmatrix} 0 \\ 0 \\ 1 \end{bmatrix}$$

Noting that

$$\phi(\mathbf{G}) = \mathbf{G}^3 + \alpha_1\mathbf{G}^2 + \alpha_2\mathbf{G} + \alpha_3\mathbf{I} = (\mathbf{G} - \mu_1\mathbf{I})(\mathbf{G} - \mu_2\mathbf{I})(\mathbf{G} - \mu_3\mathbf{I})$$

we have

$$\mathbf{K}_e = (\mathbf{G} - \mu_1\mathbf{I})(\mathbf{G} - \mu_2\mathbf{I})(\mathbf{G} - \mu_3\mathbf{I}) \begin{bmatrix} \mathbf{C} \\ \mathbf{CG} \\ \mathbf{CG}^2 \end{bmatrix}^{-1} \begin{bmatrix} 0 \\ 0 \\ 1 \end{bmatrix} \tag{6–332}$$

By premultiplying both sides of this last equation by $\mathbf{f}_1 = \mathbf{C}(\mathbf{G} - \mu_1\mathbf{I})^{-1}$, we obtain

$$\mathbf{f}_1\mathbf{K}_e = \mathbf{C}(\mathbf{G} - \mu_2\mathbf{I})(\mathbf{G} - \mu_3\mathbf{I}) \begin{bmatrix} \mathbf{C} \\ \mathbf{CG} \\ \mathbf{CG}^2 \end{bmatrix}^{-1} \begin{bmatrix} 0 \\ 0 \\ 1 \end{bmatrix} \tag{6–333}$$

Let us define

$$(\mathbf{G} - \mu_1\mathbf{I})(\mathbf{G} - \mu_2\mathbf{I}) = \mathbf{G}^2 + \beta_{12}\mathbf{G} + \beta_{13}\mathbf{I}$$

$$(\mathbf{G} - \mu_2\mathbf{I})(\mathbf{G} - \mu_3\mathbf{I}) = \mathbf{G}^2 + \beta_{22}\mathbf{G} + \beta_{23}\mathbf{I}$$

$$(\mathbf{G} - \mu_3\mathbf{I})(\mathbf{G} - \mu_1\mathbf{I}) = \mathbf{G}^2 + \beta_{32}\mathbf{G} + \beta_{33}\mathbf{I}$$

Then Eq. (6–333) becomes

$$\mathbf{f}_1\mathbf{K}_e = \mathbf{C}(\mathbf{G}^2 + \beta_{22}\mathbf{G} + \beta_{23}\mathbf{I}) \begin{bmatrix} \mathbf{C} \\ \mathbf{CG} \\ \mathbf{CG}^2 \end{bmatrix}^{-1} \begin{bmatrix} 0 \\ 0 \\ 1 \end{bmatrix}$$

$$= [\beta_{23} \quad \beta_{22} \quad 1] \begin{bmatrix} \mathbf{C} \\ \mathbf{CG} \\ \mathbf{CG}^2 \end{bmatrix} \begin{bmatrix} \mathbf{C} \\ \mathbf{CG} \\ \mathbf{CG}^2 \end{bmatrix}^{-1} \begin{bmatrix} 0 \\ 0 \\ 1 \end{bmatrix} = 1$$

Hence

$$\mathbf{f}_1\mathbf{K}_e = 1$$

Similarly, we obtain

$$\mathbf{f}_2\mathbf{K}_e = 1, \qquad \mathbf{f}_3\mathbf{K}_e = 1$$

Thus

$$\begin{bmatrix} \mathbf{f}_1 \\ \mathbf{f}_2 \\ \mathbf{f}_3 \end{bmatrix} \mathbf{K}_e = \begin{bmatrix} 1 \\ 1 \\ 1 \end{bmatrix}$$

or

$$\mathbf{K}_e = \begin{bmatrix} \mathbf{f}_1 \\ \mathbf{f}_2 \\ \mathbf{f}_3 \end{bmatrix}^{-1} \begin{bmatrix} 1 \\ 1 \\ 1 \end{bmatrix} \tag{6–334}$$

Equation (6–334) gives the desired matrix $\mathbf{K}_e$ in terms of $\mathbf{f}_1$, $\mathbf{f}_2$, and $\mathbf{f}_3$, where

$$\mathbf{f}_i = \mathbf{C}(\mathbf{G} - \mu_i \mathbf{I})^{-1} \qquad i = 1, 2, 3$$

To show that the $\mathbf{f}_i^*$'s are eigenvectors of matrix $(\mathbf{G} - \mathbf{K}_e \mathbf{C})^*$, notice that

$$\begin{aligned}
(\mathbf{G} - \mathbf{K}_e \mathbf{C})^* \mathbf{f}_i^* &= (\mathbf{G} - \mathbf{K}_e \mathbf{C})^* [\mathbf{C}(\mathbf{G} - \mu_i \mathbf{I})^{-1}]^* \\
&= (\mathbf{G}^* - \mathbf{C}^* \mathbf{K}_e^*)(\mathbf{G}^* - \bar{\mu}_i \mathbf{I})^{-1} \mathbf{C}^* \\
&= (\mathbf{G}^* - \bar{\mu}_i \mathbf{I} + \bar{\mu}_i \mathbf{I} - \mathbf{C}^* \mathbf{K}_e^*)(\mathbf{G}^* - \bar{\mu}_i \mathbf{I})^{-1} \mathbf{C}^* \\
&= \mathbf{C}^* + (\bar{\mu}_i \mathbf{I} - \mathbf{C}^* \mathbf{K}_e^*) \mathbf{f}_i^* \\
&= \mathbf{C}^* - \mathbf{C}^* \mathbf{K}_e^* \mathbf{f}_i^* + \bar{\mu}_i \mathbf{f}_i^*
\end{aligned} \tag{6–335}$$

As we have shown earlier,

$$\mathbf{f}_i \mathbf{K}_e = 1$$

Hence

$$\mathbf{K}_e^* \mathbf{f}_i^* = 1$$

Thus Eq. (6–335) becomes

$$(\mathbf{G} - \mathbf{K}_e \mathbf{C})^* \mathbf{f}_i^* = \mathbf{C}^* - \mathbf{C}^* + \bar{\mu}_i \mathbf{f}_i^* = \bar{\mu}_i \mathbf{f}_i^*$$

We have thus shown that vectors $\mathbf{f}_1^*$, $\mathbf{f}_2^*$, and $\mathbf{f}_3^*$ are eigenvectors of matrix $(\mathbf{G} - \mathbf{K}_e \mathbf{C})^*$ corresponding to eigenvalues μ_1, μ_2, and μ_3, respectively.

Problem A-6–24. In Prob. A-6–23 we obtained the observer feedback gain matrix $\mathbf{K}_e$ for the case where the eigenvalues μ_1, μ_2, and μ_3 of $\mathbf{G} - \mathbf{K}_e \mathbf{C}$ were distinct. Suppose that we desire the deadbeat response for the error vector. Then we require that $\mu_1 = \mu_2 = \mu_3 = 0$. Show that for this case, matrix $\mathbf{K}_e$ can be given as follows:

$$\mathbf{K}_e = \begin{bmatrix} \mathbf{f}_3 \\ \mathbf{f}_2 \\ \mathbf{f}_1 \end{bmatrix}^{-1} \begin{bmatrix} 0 \\ 0 \\ 1 \end{bmatrix} \quad \text{or} \quad \mathbf{K}_e = \begin{bmatrix} \mathbf{f}_1 \\ \mathbf{f}_2 \\ \mathbf{f}_3 \end{bmatrix}^{-1} \begin{bmatrix} 1 \\ 0 \\ 0 \end{bmatrix}$$

where

$$\mathbf{f}_1 = \mathbf{C}\mathbf{G}^{-1}, \qquad \mathbf{f}_2 = \mathbf{C}\mathbf{G}^{-2}, \qquad \mathbf{f}_3 = \mathbf{C}\mathbf{G}^{-3}$$

[Note that vectors $\mathbf{f}_1$, $\mathbf{f}_2$, and $\mathbf{f}_3$ given here are the eigenvectors or generalized eigenvectors of matrix $(\mathbf{G} - \mathbf{K}_e \mathbf{C})^*$.] The system is assumed to be completely observable.

Solution. Referring to Eq. (6–332), we have

$$\mathbf{K}_e = (\mathbf{G} - \mu_1 \mathbf{I})(\mathbf{G} - \mu_2 \mathbf{I})(\mathbf{G} - \mu_3 \mathbf{I}) \begin{bmatrix} \mathbf{C} \\ \mathbf{C}\mathbf{G} \\ \mathbf{C}\mathbf{G}^2 \end{bmatrix}^{-1} \begin{bmatrix} 0 \\ 0 \\ 1 \end{bmatrix}$$

By taking $\mu_1 = \mu_2 = \mu_3 = 0$ and substituting accordingly in this last equation, we obtain

$$\mathbf{K}_e = \mathbf{G}^3 \begin{bmatrix} \mathbf{C} \\ \mathbf{CG} \\ \mathbf{CG}^2 \end{bmatrix}^{-1} \begin{bmatrix} 0 \\ 0 \\ 1 \end{bmatrix}$$

which can be rewritten as follows:

$$\mathbf{K}_e = \begin{bmatrix} \mathbf{CG}^{-3} \\ \mathbf{CG}^{-2} \\ \mathbf{CG}^{-1} \end{bmatrix}^{-1} \begin{bmatrix} 0 \\ 0 \\ 1 \end{bmatrix} = \begin{bmatrix} \mathbf{f}_3 \\ \mathbf{f}_2 \\ \mathbf{f}_1 \end{bmatrix}^{-1} \begin{bmatrix} 0 \\ 0 \\ 1 \end{bmatrix} \tag{6–336}$$

Equation (6–336) gives the desired observer feedback gain matrix $\mathbf{K}_e$ where $\mu_1 = \mu_2 = \mu_3 = 0$.

Notice that Eq. (6–336) can be modified to read

$$\begin{bmatrix} \mathbf{CG}^{-3} \\ \mathbf{CG}^{-2} \\ \mathbf{CG}^{-1} \end{bmatrix} \mathbf{K}_e = \begin{bmatrix} 0 \\ 0 \\ 1 \end{bmatrix}$$

which is equivalent to the following three equations:

$$\mathbf{CG}^{-3}\mathbf{K}_e = \mathbf{f}_3\mathbf{K}_e = 0$$

$$\mathbf{CG}^{-2}\mathbf{K}_e = \mathbf{f}_2\mathbf{K}_e = 0$$

$$\mathbf{CG}^{-1}\mathbf{K}_e = \mathbf{f}_1\mathbf{K}_e = 1$$

Hence, we obtain

$$\begin{bmatrix} \mathbf{f}_1\mathbf{K}_e \\ \mathbf{f}_2\mathbf{K}_e \\ \mathbf{f}_3\mathbf{K}_e \end{bmatrix} = \begin{bmatrix} \mathbf{f}_1 \\ \mathbf{f}_2 \\ \mathbf{f}_3 \end{bmatrix} \mathbf{K}_e = \begin{bmatrix} 1 \\ 0 \\ 0 \end{bmatrix}$$

or

$$\mathbf{K}_e = \begin{bmatrix} \mathbf{f}_1 \\ \mathbf{f}_2 \\ \mathbf{f}_3 \end{bmatrix}^{-1} \begin{bmatrix} 1 \\ 0 \\ 0 \end{bmatrix}$$

which also gives the desired observer feedback gain matrix when $\mu_1 = \mu_2 = \mu_3 = 0$.

Problem A-6–25. Consider the system

$$\mathbf{x}(k+1) = \mathbf{G}\mathbf{x}(k)$$

$$\mathbf{y}(k) = \mathbf{C}\mathbf{x}(k)$$

where

$\mathbf{x}(k)$ = state vector (3-vector)

$\mathbf{y}(k)$ = output vector (2-vector)

and

$$\mathbf{G} = \begin{bmatrix} 0 & 0 & -0.25 \\ 1 & 0 & 0 \\ 0 & 1 & 0.5 \end{bmatrix}, \qquad \mathbf{C} = \begin{bmatrix} 0 & 0 & 1 \\ 1 & 0 & 0 \end{bmatrix}$$

It is desired to design a full-order prediction observer such that the error response is deadbeat.

Solution. We shall first examine the observability condition. The rank of the observability matrix

$$[\mathbf{C}^* \vdots \mathbf{G}^*\mathbf{C}^* \vdots (\mathbf{G}^*)^2\mathbf{C}^*] = \begin{bmatrix} 0 & 1 & 0 & 0 & 1 & 0 \\ 0 & 0 & 1 & 0 & 0.5 & -0.25 \\ 1 & 0 & 0.5 & -0.25 & 0.25 & -0.125 \end{bmatrix}$$

is 3. Hence, arbitrary observer pole placement is possible.

The equation for the observer error dynamics is given by Eq. (6–120):

$$\mathbf{e}(k+1) = (\mathbf{G} - \mathbf{K}_e\mathbf{C})\mathbf{e}(k)$$

In the present problem we need to determine a 3×2 matrix $\mathbf{K}_e$ such that the eigenvalues of $\mathbf{G} - \mathbf{K}_e\mathbf{C}$ are all zeros, or the characteristic equation is

$$|z\mathbf{I} - \mathbf{G} + \mathbf{K}_e\mathbf{C}| = z^3 = 0$$

This equation is equivalent to

$$|z\mathbf{I} - \mathbf{G}^* + \mathbf{C}^*\mathbf{K}_e^*| = z^3 = 0$$

In solving the present problem let us consider the dual counterpart of this problem, that is, to convert the observer design problem into a pole placement problem. In Sec. 6–5 we discussed the pole placement design problem where the control signal $\mathbf{u}(k)$ was an r-vector. There, we derived details of the design procedure. We can utilize such a pole placement design procedure for the design of state observers. Consider the system equation

$$\boldsymbol{\xi}(k+1) = \mathbf{G}^*\boldsymbol{\xi}(k) + \mathbf{C}^*\mathbf{v}(k) \tag{6–337}$$

and the state feedback control vector $\mathbf{v}(k)$, where

$$\mathbf{v}(k) = -\mathbf{K}_e^*\boldsymbol{\xi}(k) \tag{6–338}$$

Our original problem (an observer design problem) can be converted into the determination of a matrix $\mathbf{K}_e^*$ such that the eigenvalues of matrix $\mathbf{G}^* - \mathbf{C}^*\mathbf{K}_e^*$ are all zeros. Note that

$$\mathbf{G}^* = \begin{bmatrix} 0 & 1 & 0 \\ 0 & 0 & 1 \\ -0.25 & 0 & 0.5 \end{bmatrix}, \qquad \mathbf{C}^* = \begin{bmatrix} 0 & 1 \\ 0 & 0 \\ 1 & 0 \end{bmatrix}$$

Rewriting Eqs. (6–337) and (6–338), we have

$$\begin{bmatrix} \xi_1(k+1) \\ \xi_2(k+1) \\ \xi_3(k+1) \end{bmatrix} = \begin{bmatrix} 0 & 1 & 0 \\ 0 & 0 & 1 \\ -0.25 & 0 & 0.5 \end{bmatrix} \begin{bmatrix} \xi_1(k) \\ \xi_2(k) \\ \xi_3(k) \end{bmatrix} + \begin{bmatrix} 0 & 1 \\ 0 & 0 \\ 1 & 0 \end{bmatrix} \begin{bmatrix} v_1(k) \\ v_2(k) \end{bmatrix} \tag{6–339}$$

$$\mathbf{v}(k) = -\mathbf{K}_e^*\boldsymbol{\xi}(k) = -\mathbf{K}\boldsymbol{\xi}(k) \tag{6–340}$$

where $\mathbf{K} = \mathbf{K}_e^*$.

In Example 6–8 we solved the pole placement problem for the system given by Eqs. (6–339) and (6–340), where we required the response to the initial state to be deadbeat. So the dual counterpart of this problem was solved in Example 6–8. Using the same procedure as in that example, we can determine the matrix $\mathbf{K}$. Rather than repeating the design details, we may directly utilize the result obtained there. The state feedback gain matrix $\mathbf{K}$ was determined to be

$$\mathbf{K}=\begin{bmatrix}-0.25 & 0 & 0.5\\ 0 & 1 & 0\end{bmatrix}$$

Therefore,

$$\mathbf{K}_e^*=\begin{bmatrix}-0.25 & 0 & 0.5\\ 0 & 1 & 0\end{bmatrix}$$

Consequently, the desired observer feedback gain matrix $\mathbf{K}_e$ can be given by

$$\mathbf{K}_e=\begin{bmatrix}-0.25 & 0\\ 0 & 1\\ 0.5 & 0\end{bmatrix}$$

With this observer feedback gain matrix $\mathbf{K}_e$, the error vector will exhibit deadbeat response. To verify this statement, let us evaluate $(\mathbf{G}-\mathbf{K}_e\mathbf{C})^k$, where $k=1, 2, 3, \ldots$. Since

$$\mathbf{G}-\mathbf{K}_e\mathbf{C}=\begin{bmatrix}0 & 0 & -0.25\\ 1 & 0 & 0\\ 0 & 1 & 0.5\end{bmatrix}-\begin{bmatrix}-0.25 & 0\\ 0 & 1\\ 0.5 & 0\end{bmatrix}\begin{bmatrix}0 & 0 & 1\\ 1 & 0 & 0\end{bmatrix}$$
$$=\begin{bmatrix}0 & 0 & 0\\ 0 & 0 & 0\\ 0 & 1 & 0\end{bmatrix}$$

we have

$$(\mathbf{G}-\mathbf{K}_e\mathbf{C})^2=\begin{bmatrix}0 & 0 & 0\\ 0 & 0 & 0\\ 0 & 0 & 0\end{bmatrix}$$

Hence

$$(\mathbf{G}-\mathbf{K}_e\mathbf{C})^k=\mathbf{0} \qquad k=2, 3, 4, \ldots$$

Thus, any error vector reduces to zero in at most two sampling periods and the response is deadbeat. Therefore, the observed state becomes identical to the actual state in at most two sampling periods.

Problem A-6–26. Consider the system

$$\mathbf{x}(k+1)=\mathbf{G}\mathbf{x}(k)+\mathbf{H}u(k)$$
$$y(k)=\mathbf{C}\mathbf{x}(k)$$

where

$$\mathbf{G}=\begin{bmatrix}0.16 & 2.16\\ -0.16 & -1.16\end{bmatrix}, \qquad \mathbf{H}=\begin{bmatrix}-1\\ 1\end{bmatrix}, \qquad \mathbf{C}=[1 \quad 1]$$

The output $y(k)$ can be measured. Determine the state feedback gain matrix $\mathbf{K}$ such that the closed-loop poles of the system are located at

$$z_1=0.6+j0.4, \qquad z_2=0.6-j0.4$$

Then design a minimum-order observer for the system. [Since $y(k)$ is a linear combination of $x_1(k)$ and $x_2(k)$, we need observe only one state variable.] For the error dynamics it is desired that the response to an initial error be deadbeat. (Assume that the system configuration is the same as that shown in Fig. 6–8.) Then, obtain the pulse transfer function $U(z)/Y(z)$. Also, obtain the pulse transfer function of the plant and draw a block diagram for the system.

Solution. First, we shall examine the controllability and observability conditions for the system. The rank of each of the matrices

$$[\mathbf{H} \vdots \mathbf{GH}] = \begin{bmatrix} -1 & 2 \\ 1 & -1 \end{bmatrix}, \qquad [\mathbf{C}^* \vdots \mathbf{G}^*\mathbf{C}^*] = \begin{bmatrix} 1 & 0 \\ 1 & 1 \end{bmatrix}$$

is 2. Hence an arbitrary pole placement and observer design are possible.

In order to apply the method for designing the minimum-order observer presented in Sec. 6–6, let us transform matrix $\mathbf{C}$ from $[1 \quad 1]$ to $[1 \quad 0]$. To do so, define

$$\mathbf{x}(k) = \mathbf{T}\boldsymbol{\xi}(k) = \begin{bmatrix} 1 & -1 \\ 0 & 1 \end{bmatrix} \boldsymbol{\xi}(k)$$

Then the system equations become

$$\boldsymbol{\xi}(k+1) = \mathbf{T}^{-1}\mathbf{G}\mathbf{T}\boldsymbol{\xi}(k) + \mathbf{T}^{-1}\mathbf{H}u(k) = \hat{\mathbf{G}}\boldsymbol{\xi}(k) + \hat{\mathbf{H}}u(k)$$
$$y(k) = \mathbf{C}\mathbf{T}\boldsymbol{\xi}(k) = \hat{\mathbf{C}}\boldsymbol{\xi}(k)$$

where

$$\hat{\mathbf{G}} = \mathbf{T}^{-1}\mathbf{G}\mathbf{T} = \begin{bmatrix} 1 & 1 \\ 0 & 1 \end{bmatrix} \begin{bmatrix} 0.16 & 2.16 \\ -0.16 & -1.16 \end{bmatrix} \begin{bmatrix} 1 & -1 \\ 0 & 1 \end{bmatrix} = \begin{bmatrix} 0 & 1 \\ -0.16 & -1 \end{bmatrix}$$

$$\hat{\mathbf{H}} = \mathbf{T}^{-1}\mathbf{H} = \begin{bmatrix} 1 & 1 \\ 0 & 1 \end{bmatrix} \begin{bmatrix} -1 \\ 1 \end{bmatrix} = \begin{bmatrix} 0 \\ 1 \end{bmatrix}$$

$$\hat{\mathbf{C}} = \mathbf{C}\mathbf{T} = [1 \quad 1] \begin{bmatrix} 1 & -1 \\ 0 & 1 \end{bmatrix} = [1 \quad 0]$$

Thus, the system equations in the new state variables are

$$\begin{bmatrix} \xi_1(k+1) \\ \xi_2(k+1) \end{bmatrix} = \begin{bmatrix} 0 & 1 \\ -0.16 & -1 \end{bmatrix} \begin{bmatrix} \xi_1(k) \\ \xi_2(k) \end{bmatrix} + \begin{bmatrix} 0 \\ 1 \end{bmatrix} u(k) \tag{6–341}$$

$$y(k) = [1 \quad 0] \begin{bmatrix} \xi_1(k) \\ \xi_2(k) \end{bmatrix} \tag{6–342}$$

where $\xi_1(k)$ is the measurable state variable and $\xi_2(k)$ is the unmeasurable state variable which must be estimated. Notice that Eqs. (6–341) and (6–342) are in the forms of Eqs. (6–165) and (6–166), respectively.

In the following design, we shall use the state equation and the output equation given by Eqs. (6–341) and (6–342). The state feedback control is given by

$$u(k) = -\mathbf{K}\boldsymbol{\xi}(k)$$

We shall now design the state feedback gain matrix $\mathbf{K}$. Since the characteristic equation for the system of Eq. (6–341) is

$$|z\mathbf{I} - \hat{\mathbf{G}}| = \begin{vmatrix} z & -1 \\ 0.16 & z+1 \end{vmatrix} = z^2 + z + 0.16 = z^2 + a_1 z + a_2 = 0$$

we have

$$a_1 = 1, \qquad a_2 = 0.16$$

The desired characteristic equation for the system is

$$|z\mathbf{I} - \hat{\mathbf{G}} + \hat{\mathbf{H}}\mathbf{K}| = (z - 0.6 - j0.4)(z - 0.6 + j0.4) = z^2 - 1.2z + 0.52$$
$$= z^2 + \alpha_1 z + \alpha_2 = 0$$

Hence

$$\alpha_1 = -1.2, \qquad \alpha_2 = 0.52$$

Since the system defined by Eqs. (6–341) and (6–342) is in the controllable canonical form, the required matrix **K** can be given by Eq. (6–70):

$$\mathbf{K} = [\alpha_2 - a_2 \ \vdots \ \alpha_1 - a_1] = [0.52 - 0.16 \ \vdots \ -1.2 - 1] = [0.36 \quad -2.2]$$

Next, we shall design the minimum-order observer. Since the minimum-order observer is of the first order, we know the desired deadbeat characteristic polynomial to be

$$\phi(z) = z$$

From the system equations given by Eqs. (6–341) and (6–342), we have

$$G_{aa} = 0, \quad G_{ab} = 1, \quad G_{ba} = -0.16, \quad G_{bb} = -1, \quad H_a = 0, \quad H_b = 1$$

Referring to Ackermann's formula as given by Eq. (6–180), we obtain

$$K_e = \phi(G_{bb})[G_{ab}]^{-1}[1] = (-1)(1)^{-1}(1) = -1$$

Then the observer equations defined by Eqs. (6–182) and (6–183) become as follows:

$$\tilde{\xi}_2(k) = \xi_2(k) - e(k)$$
$$e(k+1) = (G_{bb} - K_e G_{ab})e(k) = (-1+1)e(k) = 0e(k)$$

Thus, the error becomes zero in at most one sampling period.

Next, let us obtain the equation for the minimum-order observer. In terms of $\tilde{\eta}(k)$, we have from Eq. (6–176)

$$\tilde{\eta}(k+1) = (G_{bb} - K_e G_{ab})\tilde{\eta}(k) + [(G_{bb} - K_e G_{ab})K_e + G_{ba} - K_e G_{aa}]y(k)$$
$$+ (H_b - K_e H_a)u(k)$$
$$= (-1 + 1 \times 1)\tilde{\eta}(k) + [(-1 + 1 \times 1)(-1) - 0.16 + 1 \times 0]y(k)$$
$$+ (1 + 1 \times 0)u(k)$$
$$= -0.16y(k) + u(k)$$

Hence, the minimum-order observer equation is

$$\tilde{\eta}(k) = -0.16y(k-1) + u(k-1)$$

The observed-state feedback control $u(k)$ is given by

$$u(k) = -\mathbf{K}\boldsymbol{\xi}(k) = -[0.36 \quad -2.2]\begin{bmatrix}\xi_1(k)\\ \xi_2(k)\end{bmatrix}$$
$$= -0.36\xi_1(k) + 2.2\xi_2(k)$$
$$= -0.36y(k) + 2.2\xi_2(k)$$

Referring to Eq. (6–175) we have

$$\xi_2(k) = K_e y(k) + \tilde{\eta}(k) = (-1)y(k) - 0.16y(k-1) + u(k-1)$$

Hence

$$u(k+1) = -0.36y(k+1) + 2.2[-y(k+1) - 0.16y(k) + u(k)]$$

or

$$u(k+1) - 2.2u(k) = -2.56y(k+1) - 0.352y(k)$$

By taking the z transform of this last equation with $u(0) = 0$ and $y(0) = 0$, we get

$$zU(z) - 2.2U(z) = -2.56zY(z) - 0.352Y(z)$$

The pulse transfer function relating $u(k)$ and $y(k)$ is

$$\frac{U(z)}{Y(z)} = -D(z) = \frac{-2.56z - 0.352}{z - 2.2} = -\frac{2.56(1 + 0.1375z^{-1})}{1 - 2.2z^{-1}}$$

The pulse transfer function $G_p(z)$ of the plant is

$$G_p(z) = \hat{\mathbf{C}}(z\mathbf{I} - \hat{\mathbf{G}})^{-1}\hat{\mathbf{H}} = [1 \quad 0]\begin{bmatrix} z & -1 \\ 0.16 & z+1\end{bmatrix}^{-1}\begin{bmatrix}0\\1\end{bmatrix}$$
$$= \frac{z^{-2}}{1 + z^{-1} + 0.16z^{-2}}$$

Hence, the block diagram for the system may be drawn as shown in Fig. 6–16.

From the block diagram, the characteristic equation for the system is

$$1 + \frac{z^{-2}}{1 + z^{-1} + 0.16z^{-2}} \frac{2.56(1 + 0.1375z^{-1})}{1 - 2.2z^{-1}} = 0$$

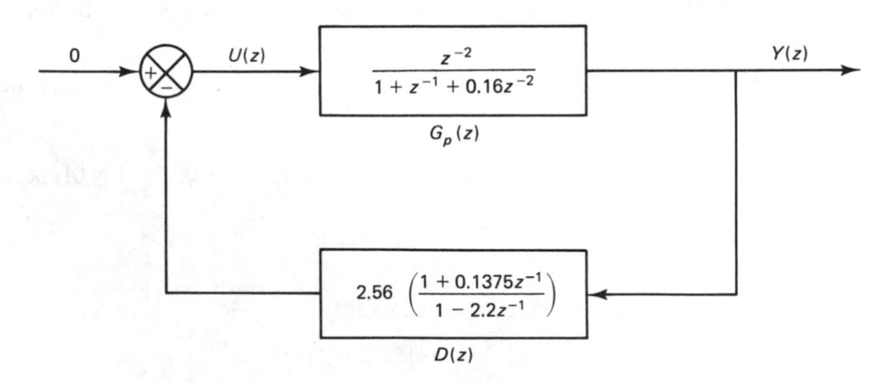

Figure 6–16 Block diagram for the system designed in Prob. A-6–26.

which can be reduced to

$$(z^2 - 1.2z + 0.52)z = 0$$

This characteristic equation is, of course, the same as

$$|z\mathbf{I} - \hat{\mathbf{G}} + \hat{\mathbf{H}}\mathbf{K}|\ |z - G_{bb} + K_e G_{ab}| = 0$$

where

$$|z\mathbf{I} - \hat{\mathbf{G}} + \hat{\mathbf{H}}\mathbf{K}| = z^2 - 1.2z + 0.52$$

$$|z - G_{bb} + K_e G_{ab}| = z$$

Problem A-6–27. Consider the system

$$\mathbf{x}(k+1) = \mathbf{G}\mathbf{x}(k) + \mathbf{H}u(k)$$

$$y(k) = \mathbf{C}\mathbf{x}(k)$$

where

$$\mathbf{x}(k) = \text{state vector (3-vector)}$$

$$u(k) = \text{control signal (scalar)}$$

$$y(k) = \text{output signal (scalar)}$$

and

$$\mathbf{G} = \begin{bmatrix} 0 & 1 & 0 \\ 0 & 0 & 1 \\ -0.5 & -0.2 & 1.1 \end{bmatrix}, \qquad \mathbf{H} = \begin{bmatrix} 0 \\ 0 \\ 1 \end{bmatrix}, \qquad \mathbf{C} = [0 \quad 1 \quad 0]$$

(1) Determine the state feedback gain matrix **K** such that the system will exhibit a deadbeat response to any initial state. Assuming that the state is completely measurable so that the actual state $\mathbf{x}(k)$ can be fed back for control, or that

$$u(k) = -\mathbf{K}\mathbf{x}(k)$$

determine the response of the system to the initial state

$$x(0) = \begin{bmatrix} a \\ b \\ c \end{bmatrix}$$

where a, b, and c are arbitrary constants.

(2) Assuming that only a portion of the state vector is measurable, that is, that only the output $y(k)$ is measurable, design a minimum-order observer such that the response to the observer error is deadbeat. Assume that the system configuration is the same as that shown in Fig. 6–8.

(3) Assuming that the observed state is used for feedback, obtain the response of the system to

$$\mathbf{x}(0) = \begin{bmatrix} a \\ b \\ c \end{bmatrix}, \qquad \mathbf{e}(0) = \begin{bmatrix} \alpha \\ \beta \end{bmatrix}$$

where $\mathbf{e}(0)$ is the initial observer error for the minimum-order observer and a, b, c, α, and β are arbitrary constants.

Solution. Notice that the system is completely state controllable and observable.

(1) The required state feedback gain matrix **K** for deadbeat response can be obtained easily, as follows. Let us define

$$\mathbf{K} = [k_1 \quad k_2 \quad k_3]$$

Then

$$|z\mathbf{I} - \mathbf{G} + \mathbf{HK}| = \begin{vmatrix} z & -1 & 0 \\ 0 & z & -1 \\ k_1 + 0.5 & k_2 + 0.2 & z + k_3 - 1.1 \end{vmatrix}$$

$$= z^3 + (k_3 - 1.1)z^2 + (k_2 + 0.2)z + k_1 + 0.5 = 0$$

By equating this characteristic equation with the desired characteristic equation (for deadbeat response)

$$z^3 = 0$$

we obtain

$$\mathbf{K} = [k_1 \quad k_2 \quad k_3] = [-0.5 \quad -0.2 \quad 1.1]$$

With this matrix **K**, the system equation becomes as follows:

$$\mathbf{x}(k+1) = \mathbf{Gx}(k) + \mathbf{H}u(k) = (\mathbf{G} - \mathbf{HK})\mathbf{x}(k)$$

or

$$\begin{bmatrix} x_1(k+1) \\ x_2(k+1) \\ x_3(k+1) \end{bmatrix} = \begin{bmatrix} 0 & 1 & 0 \\ 0 & 0 & 1 \\ 0 & 0 & 0 \end{bmatrix} \begin{bmatrix} x_1(k) \\ x_2(k) \\ x_3(k) \end{bmatrix}$$

The response of this system to an arbitrary initial state becomes as follows:

$$\begin{bmatrix} x_1(1) \\ x_2(1) \\ x_3(1) \end{bmatrix} = \begin{bmatrix} 0 & 1 & 0 \\ 0 & 0 & 1 \\ 0 & 0 & 0 \end{bmatrix} \begin{bmatrix} a \\ b \\ c \end{bmatrix} = \begin{bmatrix} b \\ c \\ 0 \end{bmatrix}$$

$$\begin{bmatrix} x_1(2) \\ x_2(2) \\ x_3(2) \end{bmatrix} = \begin{bmatrix} 0 & 1 & 0 \\ 0 & 0 & 1 \\ 0 & 0 & 0 \end{bmatrix} \begin{bmatrix} b \\ c \\ 0 \end{bmatrix} = \begin{bmatrix} c \\ 0 \\ 0 \end{bmatrix}$$

$$\begin{bmatrix} x_1(3) \\ x_2(3) \\ x_3(3) \end{bmatrix} = \begin{bmatrix} 0 & 1 & 0 \\ 0 & 0 & 1 \\ 0 & 0 & 0 \end{bmatrix} \begin{bmatrix} c \\ 0 \\ 0 \end{bmatrix} = \begin{bmatrix} 0 \\ 0 \\ 0 \end{bmatrix}$$

or

$$\mathbf{x}(k) = \mathbf{0} \qquad k = 3, 4, 5, \ldots$$

Clearly, the response is deadbeat.

(2) We shall now design a minimum-order observer assuming that only the output $y(k)$ is measurable. We shall first transform the state vector $\mathbf{x}(k)$ into a new state vector $\boldsymbol{\xi}(k)$

such that the output matrix $\mathbf{C}$ is transformed from $[0 \quad 1 \quad 0]$ to $[1 \quad 0 \quad 0]$. The following matrix $\mathbf{T}$ will accomplish the required transformation:

$$\mathbf{T} = \begin{bmatrix} 0 & 1 & 0 \\ 1 & 0 & 0 \\ 0 & 0 & 1 \end{bmatrix}$$

Thus, we define

$$\mathbf{x}(k) = \mathbf{T}\boldsymbol{\xi}(k)$$

Then the system equations become

$$\boldsymbol{\xi}(k+1) = \mathbf{T}^{-1}\mathbf{G}\mathbf{T}\boldsymbol{\xi}(k) + \mathbf{T}^{-1}\mathbf{H}u(k) = \hat{\mathbf{G}}\boldsymbol{\xi}(k) + \hat{\mathbf{H}}u(k)$$

$$y(k) = \mathbf{C}\mathbf{T}\boldsymbol{\xi}(k) = \hat{\mathbf{C}}\boldsymbol{\xi}(k)$$

where

$$\hat{\mathbf{G}} = \mathbf{T}^{-1}\mathbf{G}\mathbf{T} = \begin{bmatrix} 0 & 1 & 0 \\ 1 & 0 & 0 \\ 0 & 0 & 1 \end{bmatrix} \begin{bmatrix} 0 & 1 & 0 \\ 0 & 0 & 1 \\ -0.5 & -0.2 & 1.1 \end{bmatrix} \begin{bmatrix} 0 & 1 & 0 \\ 1 & 0 & 0 \\ 0 & 0 & 1 \end{bmatrix}$$

$$= \begin{bmatrix} 0 & 0 & 1 \\ 1 & 0 & 0 \\ -0.2 & -0.5 & 1.1 \end{bmatrix} = \begin{bmatrix} \hat{G}_{aa} & \hat{\mathbf{G}}_{ab} \\ \hat{\mathbf{G}}_{ba} & \hat{\mathbf{G}}_{bb} \end{bmatrix}$$

$$\hat{\mathbf{H}} = \mathbf{T}^{-1}\mathbf{H} = \begin{bmatrix} 0 & 1 & 0 \\ 1 & 0 & 0 \\ 0 & 0 & 1 \end{bmatrix} \begin{bmatrix} 0 \\ 0 \\ 1 \end{bmatrix} = \begin{bmatrix} 0 \\ 0 \\ 1 \end{bmatrix}$$

$$\hat{\mathbf{C}} = \mathbf{C}\mathbf{T} = [0 \quad 1 \quad 0] \begin{bmatrix} 0 & 1 & 0 \\ 1 & 0 & 0 \\ 0 & 0 & 1 \end{bmatrix} = [1 \quad 0 \quad 0]$$

The transformed system is thus given by

$$\begin{bmatrix} \xi_1(k+1) \\ \hdashline \xi_2(k+1) \\ \xi_3(k+1) \end{bmatrix} = \left[\begin{array}{c:cc} 0 & 0 & 1 \\ \hdashline 1 & 0 & 0 \\ -0.2 & -0.5 & 1.1 \end{array}\right] \begin{bmatrix} \xi_1(k) \\ \hdashline \xi_2(k) \\ \xi_3(k) \end{bmatrix} + \begin{bmatrix} 0 \\ \hdashline 0 \\ 1 \end{bmatrix} u(k)$$

$$y(k) = [1 \,\vdots\, 0 \quad 0] \begin{bmatrix} \xi_1(k) \\ \hdashline \xi_2(k) \\ \xi_3(k) \end{bmatrix}$$

Since only one state variable, $\xi_1(k)$, can be measured, we need to observe two state variables. Hence, the order of the minimum-order observer is 2. Since

$$\hat{\mathbf{G}}_{bb} - \mathbf{K}_e \hat{\mathbf{G}}_{ab} = \begin{bmatrix} 0 & 0 \\ -0.5 & 1.1 \end{bmatrix} - \begin{bmatrix} k_{e_1} \\ k_{e_2} \end{bmatrix} [0 \quad 1] = \begin{bmatrix} 0 & -k_{e_1} \\ -0.5 & 1.1 - k_{e_2} \end{bmatrix}$$

the observer characteristic equation becomes

$$|z\mathbf{I} - \hat{\mathbf{G}}_{bb} + \mathbf{K}_e \hat{\mathbf{G}}_{ab}| = \begin{vmatrix} z & k_{e_1} \\ 0.5 & z - 1.1 + k_{e_2} \end{vmatrix} = z^2 + (k_{e_2} - 1.1)z - 0.5k_{e_1}$$

The desired characteristic equation (for deadbeat response) is

$$z^2 = 0$$

Hence we obtain

$$k_{e_1} = 0, \qquad k_{e_2} = 1.1$$

or

$$\mathbf{K}_e = \begin{bmatrix} 0 \\ 1.1 \end{bmatrix}$$

(3) The equation for the state feedback control system with a minimum-order observer is given by Eq. (6–190):

$$\begin{bmatrix} \mathbf{x}(k+1) \\ \hline \mathbf{e}(k+1) \end{bmatrix} = \left[\begin{array}{c|c} \mathbf{G} - \mathbf{HK} & \mathbf{HK\Gamma} \\ \hline \mathbf{0} & \mathbf{G}_{bb} - \mathbf{K}_e\mathbf{G}_{ab} \end{array}\right] \begin{bmatrix} \mathbf{x}(k) \\ \hline \mathbf{e}(k) \end{bmatrix} \tag{6–343}$$

Let us rewrite Eq. (6–343) in terms of the new state vector $\boldsymbol{\xi}(k)$ and error vector $\mathbf{e}(k)$. Noting that the observed state is used for feedback, that is

$$u(k) = -\hat{\mathbf{K}}\tilde{\boldsymbol{\xi}}(k)$$

we have

$$\begin{aligned} \boldsymbol{\xi}(k+1) &= \hat{\mathbf{G}}\boldsymbol{\xi}(k) + \hat{\mathbf{H}}u(k) \\ &= \hat{\mathbf{G}}\boldsymbol{\xi}(k) - \hat{\mathbf{H}}\hat{\mathbf{K}}\tilde{\boldsymbol{\xi}}(k) \\ &= (\hat{\mathbf{G}} - \hat{\mathbf{H}}\hat{\mathbf{K}})\boldsymbol{\xi}(k) + \hat{\mathbf{H}}\hat{\mathbf{K}}[\boldsymbol{\xi}(k) - \tilde{\boldsymbol{\xi}}(k)] \\ &= (\hat{\mathbf{G}} - \hat{\mathbf{H}}\hat{\mathbf{K}})\boldsymbol{\xi}(k) + \hat{\mathbf{H}}\hat{\mathbf{K}}\boldsymbol{\Gamma}\mathbf{e}(k) \end{aligned}$$

where

$$\boldsymbol{\Gamma} = \begin{bmatrix} 0 & 0 \\ \hline 1 & 0 \\ 0 & 1 \end{bmatrix}$$

and

$$\hat{\mathbf{K}} = \mathbf{KT} = [-0.5 \quad -0.2 \quad 1.1] \begin{bmatrix} 0 & 1 & 0 \\ 1 & 0 & 0 \\ 0 & 0 & 1 \end{bmatrix} = [-0.2 \quad -0.5 \quad 1.1]$$

Hence Eq. (6–343) can be modified to read

$$\begin{bmatrix} \boldsymbol{\xi}(k+1) \\ \hline \mathbf{e}(k+1) \end{bmatrix} = \left[\begin{array}{c|c} \hat{\mathbf{G}} - \hat{\mathbf{H}}\hat{\mathbf{K}} & \hat{\mathbf{H}}\hat{\mathbf{K}}\boldsymbol{\Gamma} \\ \hline \mathbf{0} & \hat{\mathbf{G}}_{bb} - \mathbf{K}_e\hat{\mathbf{G}}_{ab} \end{array}\right] \begin{bmatrix} \boldsymbol{\xi}(k) \\ \hline \mathbf{e}(k) \end{bmatrix}$$

or

$$\begin{bmatrix} \xi_1(k+1) \\ \xi_2(k+1) \\ \xi_3(k+1) \\ e_1(k+1) \\ e_2(k+1) \end{bmatrix} = \begin{bmatrix} 0 & 0 & 1 & 0 & 0 \\ 1 & 0 & 0 & 0 & 0 \\ 0 & 0 & 0 & -0.5 & 1.1 \\ 0 & 0 & 0 & 0 & 0 \\ 0 & 0 & 0 & -0.5 & 0 \end{bmatrix} \begin{bmatrix} \xi_1(k) \\ \xi_2(k) \\ \xi_3(k) \\ e_1(k) \\ e_2(k) \end{bmatrix}$$

The response of this system to the given initial condition can be obtained as follows. First note that the assumed initial condition is

$$\begin{bmatrix} x_1(0) \\ x_2(0) \\ x_3(0) \\ e_1(0) \\ e_2(0) \end{bmatrix} = \begin{bmatrix} a \\ b \\ c \\ \alpha \\ \beta \end{bmatrix}$$

Hence

$$\begin{bmatrix} \xi_1(0) \\ \xi_2(0) \\ \xi_3(0) \\ e_1(0) \\ e_2(0) \end{bmatrix} = \begin{bmatrix} b \\ a \\ c \\ \alpha \\ \beta \end{bmatrix}, \qquad \begin{bmatrix} \xi_1(1) \\ \xi_2(1) \\ \xi_3(1) \\ e_1(1) \\ e_2(1) \end{bmatrix} = \begin{bmatrix} c \\ b \\ -0.5\alpha + 1.1\beta \\ 0 \\ -0.5\alpha \end{bmatrix}$$

$$\begin{bmatrix} \xi_1(2) \\ \xi_2(2) \\ \xi_3(2) \\ e_1(2) \\ e_2(2) \end{bmatrix} = \begin{bmatrix} -0.5\alpha + 1.1\beta \\ c \\ -0.55\alpha \\ 0 \\ 0 \end{bmatrix}, \qquad \begin{bmatrix} \xi_1(3) \\ \xi_2(3) \\ \xi_3(3) \\ e_1(3) \\ e_2(3) \end{bmatrix} = \begin{bmatrix} -0.55\alpha \\ -0.5\alpha + 1.1\beta \\ 0 \\ 0 \\ 0 \end{bmatrix}$$

$$\begin{bmatrix} \xi_1(4) \\ \xi_2(4) \\ \xi_3(4) \\ e_1(4) \\ e_2(4) \end{bmatrix} = \begin{bmatrix} 0 \\ -0.55\alpha \\ 0 \\ 0 \\ 0 \end{bmatrix}, \qquad \begin{bmatrix} \xi_1(5) \\ \xi_2(5) \\ \xi_3(5) \\ e_1(5) \\ e_2(5) \end{bmatrix} = \begin{bmatrix} 0 \\ 0 \\ 0 \\ 0 \\ 0 \end{bmatrix}$$

The response is clearly deadbeat. For any initial condition, the settling time is at most 5 sampling periods. (This means that at most 2 sampling periods are needed for the error vector to become zero and, additionally, at most 3 sampling periods are needed for the state vector to become zero.)

Problem A-6–28. Referring to the servo system design problem discussed in Example 6–12, consider first the problem of determining an integral gain constant K_1 and a state feedback gain matrix $\mathbf{K}_2$ by use of the characteristic equation given by Eq. (6–220) or Eq. (6–221) such that the unit step response is deadbeat. Then, consider a design for a full-order prediction observer such that the response to the observer error is deadbeat. Defining the observer feedback gain matrix as $\mathbf{K}_e$, determine this matrix by equating the coefficients of the powers of z of

$$|z\mathbf{I} - \mathbf{G} + \mathbf{K}_e\mathbf{C}| = 0$$

and those of like powers of z in the desired characteristic equation, which is

$$z^4 = 0$$

Solution. Let us define

$$\mathbf{K}_2 = [k_1 \quad k_2 \quad k_3]$$

Noting that

$$\mathbf{G} = \begin{bmatrix} 0 & 1 & 0 \\ 0 & 0 & 1 \\ -0.12 & -0.01 & 1 \end{bmatrix}, \qquad \mathbf{H} = \begin{bmatrix} 0 \\ 0 \\ 1 \end{bmatrix}, \qquad \mathbf{C} = [0.5 \quad 1 \quad 0]$$

we can give Eq. (6–221) as follows:

$$|z\mathbf{I}_3 - \mathbf{G} + \mathbf{H}\mathbf{K}_2 + \mathbf{H}K_1\mathbf{C} + \mathbf{H}K_1\mathbf{C}(zI_1 - I_1)^{-1}|\,|zI_1 - I_1|$$
$$= |z\mathbf{I}_3 - \mathbf{G} + \mathbf{H}\mathbf{K}_2 + \mathbf{H}K_1\mathbf{C}[1 + (z-1)^{-1}]|\,|z-1|$$

$$= \left| \begin{bmatrix} z & 0 & 0 \\ 0 & z & 0 \\ 0 & 0 & z \end{bmatrix} - \begin{bmatrix} 0 & 1 & 0 \\ 0 & 0 & 1 \\ -0.12 & -0.01 & 1 \end{bmatrix} + \begin{bmatrix} 0 \\ 0 \\ 1 \end{bmatrix} [k_1 \quad k_2 \quad k_3] \right.$$
$$\left. + \begin{bmatrix} 0 \\ 0 \\ 1 \end{bmatrix} [K_1]\,[0.5 \quad 1 \quad 0]\left(1 + \frac{1}{z-1}\right) \right| (z-1)$$

$$= \begin{vmatrix} z & -1 & 0 \\ 0 & z & -1 \\ 0.12 + k_1 + \dfrac{0.5K_1 z}{z-1} & 0.01 + k_2 + \dfrac{K_1 z}{z-1} & z - 1 + k_3 \end{vmatrix} (z-1)$$

$$= \begin{vmatrix} z & -1 & 0 \\ 0 & z & -1 \\ (0.12 + k_1)(z-1) + 0.5K_1 z & (0.01 + k_2)(z-1) + K_1 z & (z-1)^2 + k_3(z-1) \end{vmatrix}$$

$$= z^4 + (-2 + k_3)z^3 + (1.01 + k_2 - k_3 + K_1)z^2 + (0.11 + k_1 - k_2 + 0.5K_1)z - 0.12 - k_1 = 0$$

This characteristic equation must be equal to

$$z^4 = 0$$

Hence, we require

$$-2 + k_3 = 0$$
$$1.01 + k_2 - k_3 + K_1 = 0$$
$$0.11 + k_1 - k_2 + 0.5K_1 = 0$$
$$-0.12 - k_1 = 0$$

from which we get

$$K_1 = \tfrac{2}{3}, \qquad k_1 = -0.12, \qquad k_2 = \tfrac{0.97}{3}, \qquad k_3 = 2$$

or

$$K_1 = \tfrac{2}{3}, \qquad \mathbf{K} = [-0.12 \quad 0.3233 \quad 2]$$

[As a matter of course, these values agree with those given by Eqs. (6–227) and (6–228).]

Next, we shall design a full-order prediction observer. Define

$$\mathbf{K}_e = \begin{bmatrix} k_{e\,1} \\ k_{e\,2} \\ k_{e\,3} \end{bmatrix}$$

Then

$$\mathbf{G} - \mathbf{K}_e\mathbf{C} = \begin{bmatrix} 0 & 1 & 0 \\ 0 & 0 & 1 \\ -0.12 & -0.01 & 1 \end{bmatrix} - \begin{bmatrix} k_{e\,1} \\ k_{e\,2} \\ k_{e\,3} \end{bmatrix} [0.5 \quad 1 \quad 0]$$

$$= \begin{bmatrix} -0.5k_{e\,1} & 1 - k_{e\,1} & 0 \\ -0.5k_{e\,2} & -k_{e\,2} & 1 \\ -0.12 - 0.5k_{e\,3} & -0.01 - k_{e\,3} & 1 \end{bmatrix}$$

and we have

$$|z\mathbf{I} - \mathbf{G} + \mathbf{K}_e\mathbf{C}| = \begin{vmatrix} z + 0.5k_{e\,1} & -1 + k_{e\,1} & 0 \\ 0.5k_{e\,2} & z + k_{e\,2} & -1 \\ 0.12 + 0.5k_{e\,3} & 0.01 + k_{e\,3} & z - 1 \end{vmatrix}$$

$$= z^3 + (-1 + 0.5k_{e\,1} + k_{e\,2})z^2 + (0.01 - 0.5k_{e\,1} - 0.5k_{e\,2} + k_{e\,3})z$$
$$+ 0.12 - 0.115k_{e\,1} - 0.5k_{e\,2} + 0.5k_{e\,3} = 0$$

This characteristic equation must be equal to the desired characteristic equation

$$z^3 = 0$$

Hence, we require

$$-1 + 0.5k_{e\,1} + k_{e\,2} = 0$$

$$0.01 - 0.5k_{e\,1} - 0.5k_{e\,2} + k_{e\,3} = 0$$

$$0.12 - 0.115k_{e\,1} - 0.5k_{e\,2} + 0.5k_{e\,3} = 0$$

Solving these three simultaneous equations for $k_{e\,1}$, $k_{e\,2}$, and $k_{e\,3}$, we obtain

$$\mathbf{K}_e = \begin{bmatrix} k_{e\,1} \\ k_{e\,2} \\ k_{e\,3} \end{bmatrix} = \begin{bmatrix} 0.5192 \\ 0.7404 \\ 0.6198 \end{bmatrix}$$

This matrix gives the desired observer feedback gain matrix $\mathbf{K}_e$.

Remember that the design of the integral gain constant K_1 and the state feedback gain matrix $\mathbf{K}_2$ (a pole placement problem) and the design of the observer feedback gain matrix $\mathbf{K}_e$ (an observer problem) are independent problems. That is, matrix $\mathbf{K}_e$ does not depend on K_1 and $\mathbf{K}_2$, and vice versa.

PROBLEMS

Problem B-6-1. Consider the system defined by

$$\begin{bmatrix} x_1(k+1) \\ x_2(k+1) \end{bmatrix} = \begin{bmatrix} a & b \\ c & d \end{bmatrix} \begin{bmatrix} x_1(k) \\ x_2(k) \end{bmatrix} + \begin{bmatrix} 1 \\ 1 \end{bmatrix} u(k)$$

$$y(k) = [1 \quad 0] \begin{bmatrix} x_1(k) \\ x_2(k) \end{bmatrix}$$

Determine the conditions on a, b, c, and d for complete state controllability and complete observability.

Problem B-6–2. The control system defined by

$$\begin{bmatrix} x_1(k+1) \\ x_2(k+1) \end{bmatrix} = \begin{bmatrix} 0 & 1 \\ -0.16 & -1 \end{bmatrix} \begin{bmatrix} x_1(k) \\ x_2(k) \end{bmatrix} + \begin{bmatrix} 1 \\ 0.5 \end{bmatrix} u(k)$$

$$\begin{bmatrix} x_1(0) \\ x_2(0) \end{bmatrix} = \begin{bmatrix} 1 \\ -1 \end{bmatrix}$$

is completely state controllable. Determine a sequence of control signals $u(0)$ and $u(1)$ such that the state $\mathbf{x}(2)$ becomes

$$\begin{bmatrix} x_1(2) \\ x_2(2) \end{bmatrix} = \begin{bmatrix} -1 \\ 2 \end{bmatrix}$$

Problem B-6–3. Consider the system

$$\begin{bmatrix} x_1(k+1) \\ x_2(k+1) \end{bmatrix} = \begin{bmatrix} 0 & 1 \\ -0.16 & -1 \end{bmatrix} \begin{bmatrix} x_1(k) \\ x_2(k) \end{bmatrix} + \begin{bmatrix} 1 \\ -0.8 \end{bmatrix} u(k)$$

$$\begin{bmatrix} x_1(0) \\ x_2(0) \end{bmatrix} = \begin{bmatrix} 1 \\ -1 \end{bmatrix}$$

Determine whether it is possible to bring the state to

1. $$\begin{bmatrix} x_1(2) \\ x_2(2) \end{bmatrix} = \begin{bmatrix} 0 \\ -0.008 \end{bmatrix}$$

2. $$\begin{bmatrix} x_1(2) \\ x_2(2) \end{bmatrix} = \begin{bmatrix} -1 \\ 2 \end{bmatrix}$$

Problem B-6–4. Consider the system

$$\begin{bmatrix} x_1(k+1) \\ x_2(k+1) \\ x_3(k+1) \end{bmatrix} = \begin{bmatrix} 0 & 1 & 0 \\ 0 & 0 & 1 \\ a & b & -\frac{a}{b} \end{bmatrix} \begin{bmatrix} x_1(k) \\ x_2(k) \\ x_3(k) \end{bmatrix} + \begin{bmatrix} 0 \\ 1 \\ 0 \end{bmatrix} u(k)$$

Starting from the initial state

$$\mathbf{x}(0) = \begin{bmatrix} 1 \\ 1 \\ 1 \end{bmatrix}$$

determine whether or not the state $\mathbf{x}(3)$ can be brought to the origin. Also, determine whether or not the state can be brought to

$$\mathbf{x}(3) = \begin{bmatrix} 1 \\ 1 \\ 1 \end{bmatrix}$$

if the initial state is $\mathbf{x}(0) = \mathbf{0}$.

Problem B-6–5. For the system defined by

$$\begin{bmatrix} x_1(k+1) \\ x_2(k+1) \end{bmatrix} = \begin{bmatrix} 0 & 1 \\ -0.16 & -1 \end{bmatrix} \begin{bmatrix} x_1(k) \\ x_2(k) \end{bmatrix} + \begin{bmatrix} 0 \\ 1 \end{bmatrix} u(k)$$

$$y(k) = [1 \quad 0] \begin{bmatrix} x_1(k) \\ x_2(k) \end{bmatrix}$$

assume that the following outputs are observed:

$$y(0) = 1, \qquad y(1) = 2$$

The control signals given are

$$u(0) = 2, \qquad u(1) = -1$$

Determine the initial state $\mathbf{x}(0)$. Also, determine states $\mathbf{x}(1)$ and $\mathbf{x}(2)$.

Problem B-6–6. Show that the system

$$\mathbf{x}(k+1) = \mathbf{G}[\mathbf{x}(k) + \mathbf{C}^* u(k)]$$

$$y(k) = \mathbf{C}\mathbf{x}(k)$$

where

$$\mathbf{x}(k) = \text{state vector (4-vector)}$$

$$u(k) = \text{control signal (scalar)}$$

$$y(k) = \text{output signal (scalar)}$$

and

$$\mathbf{G} = \begin{bmatrix} 0 & 1 & 0 & 0 \\ 0 & 0 & 1 & 0 \\ 0 & 0 & 0 & 1 \\ 1 & 0 & 0 & 0 \end{bmatrix}, \qquad \mathbf{C} = [1 \quad 0 \quad 0 \quad 0]$$

is completely state controllable and observable.

Show also that given any initial state $\mathbf{x}(0)$, every state vector can be brought to the origin in at most 4 sampling periods if and only if the control signal is given by

$$u(k) = -\mathbf{C}\mathbf{x}(k)$$

Problem B-6–7. Show that the following system

$$\begin{bmatrix} \dot{x}_1 \\ \dot{x}_2 \\ \dot{x}_3 \end{bmatrix} = \begin{bmatrix} -1 & -2 & 0 \\ 0 & -3 & 1 \\ 1 & 0 & -1 \end{bmatrix} \begin{bmatrix} x_1 \\ x_2 \\ x_3 \end{bmatrix} + \begin{bmatrix} 1 \\ 0 \\ 1 \end{bmatrix} u$$

$$y = [1 \quad 1 \quad 0] \begin{bmatrix} x_1 \\ x_2 \\ x_3 \end{bmatrix}$$

is completely state controllable and observable.

Problem B-6–8. Consider the continuous-time control system

$$\begin{bmatrix} \dot{x}_1 \\ \dot{x}_2 \end{bmatrix} = \begin{bmatrix} 0 & 1 \\ -25 & -6 \end{bmatrix} \begin{bmatrix} x_1 \\ x_2 \end{bmatrix} + \begin{bmatrix} 0 \\ 1 \end{bmatrix} u$$

$$y = [3 \quad 1] \begin{bmatrix} x_1 \\ x_2 \end{bmatrix}$$

This system is completely state controllable and observable. Note that the eigenvalues of the state matrix are

$$\lambda_1 = -3 + j4, \qquad \lambda_2 = -3 - j4$$

Thus this system involves complex poles.

As stated in Sec. 6–3, a system which is completely state controllable and observable in the absence of sampling remains completely state controllable and observable after the introduction of sampling if and only if, for every eigenvalue of the state matrix (root of the characteristic equation),

$$\text{Re}\,\lambda_i = \text{Re}\,\lambda_j$$

implies

$$\text{Im}\,(\lambda_i - \lambda_j) \neq \frac{2\pi n}{T}$$

where T is the sampling period and $n = \pm 1, \pm 2, \ldots$.

Consider the discretized version of this system. Show that for this system if the sampling period T is equal to $\pi n/4$ (where $n = 1, 2, 3, \ldots$), then the discretized system is uncontrollable and unobservable.

Problem B-6–9. Consider the pulse-transfer-function system

$$G(z) = \frac{z^{-1}(1 + z^{-1})}{(1 + 0.5z^{-1})(1 - 0.5z^{-1})}$$

Referring to Sec. 6–4, obtain the state space representation of the system in the following forms:

1. Controllable canonical form
2. Observable canonical form
3. Diagonal canonical form

Problem B-6–10. Consider the pulse-transfer-function system

$$G(z) = \frac{1 + 0.8z^{-1}}{1 - z^{-1} + 0.5z^{-2}}$$

Obtain the state space representation of the system in the following forms:

1. Controllable canonical form
2. Observable canonical form
3. Diagonal canonical form

Problem B-6–11. Consider the following system given in the controllable canonical form:

$$\begin{bmatrix} x_1(k+1) \\ x_2(k+1) \\ x_3(k+1) \end{bmatrix} = \begin{bmatrix} 0 & 1 & 0 \\ 0 & 0 & 1 \\ -a_3 & -a_2 & -a_1 \end{bmatrix} \begin{bmatrix} x_1(k) \\ x_2(k) \\ x_3(k) \end{bmatrix} + \begin{bmatrix} 0 \\ 0 \\ 1 \end{bmatrix} u(k)$$

$$y(k) = [b_3 - a_3 b_0 \;\vdots\; b_2 - a_2 b_0 \;\vdots\; b_1 - a_1 b_0] \begin{bmatrix} x_1(k) \\ x_2(k) \\ x_3(k) \end{bmatrix} + b_0 u(k)$$

It is desired to transform the system equations into the observable canonical form by means of the transformation of the state vector:

$$\mathbf{x} = \mathbf{Q}\hat{\mathbf{x}}$$

Determine a transformation matrix $\mathbf{Q}$ that will give the desired observable canonical form.

Problem B-6–12. Consider the double integrator system

$$\mathbf{x}((k+1)T) = \mathbf{G}\mathbf{x}(kT) + \mathbf{H}u(kT)$$

where

$$\mathbf{G} = \begin{bmatrix} 1 & T \\ 0 & 1 \end{bmatrix}, \qquad \mathbf{H} = \begin{bmatrix} T^2/2 \\ T \end{bmatrix}$$

and T is the sampling period. (See Prob. A-5–29 for the derivation of this discrete-time state equation for the double integrator system.)

It is desired that the closed-loop poles be located at $z = \mu_1$ and $z = \mu_2$. Assuming that the state feedback control

$$u(kT) = -\mathbf{K}\mathbf{x}(kT)$$

is used, determine the state feedback gain matrix $\mathbf{K}$.

Problem B-6–13. Consider the system defined by

$$\begin{bmatrix} x_1(k+1) \\ x_2(k+1) \\ x_3(k+1) \end{bmatrix} = \begin{bmatrix} 0 & 1 & 0 \\ 0 & 0 & 1 \\ -0.16 & 0.84 & 0 \end{bmatrix} \begin{bmatrix} x_1(k) \\ x_2(k) \\ x_3(k) \end{bmatrix} + \begin{bmatrix} 1 \\ 1 \\ 1 \end{bmatrix} u(k)$$

Determine the state feedback gain matrix $\mathbf{K}$ such that when the control signal is given by

$$u(k) = -\mathbf{K}\mathbf{x}(k)$$

the closed-loop system will exhibit the deadbeat response to any initial state $\mathbf{x}(0)$.

Problem B-6–14. Consider the system defined by

$$\begin{bmatrix} x_1(k+1) \\ x_2(k+1) \end{bmatrix} = \begin{bmatrix} 0 & 1 \\ -0.16 & -1 \end{bmatrix} \begin{bmatrix} x_1(k) \\ x_2(k) \end{bmatrix} + \begin{bmatrix} 0 & 1 \\ 1 & 1 \end{bmatrix} \begin{bmatrix} u_1(k) \\ u_2(k) \end{bmatrix}$$

Determine the state feedback gain matrix **K** such that when the control vector is given by

$$\mathbf{u}(k) = -\mathbf{K}\mathbf{x}(k)$$

the closed-loop system will exhibit the deadbeat response to any initial state $\mathbf{x}(0)$.

Problem B-6–15. For the system defined by

$$\begin{bmatrix} x_1(k+1) \\ x_2(k+1) \\ x_3(k+1) \end{bmatrix} = \begin{bmatrix} 0 & 1 & 0 \\ 0 & 0 & 1 \\ -0.16 & 0.84 & 0 \end{bmatrix} \begin{bmatrix} x_1(k) \\ x_2(k) \\ x_3(k) \end{bmatrix} + \begin{bmatrix} 0 & 1 \\ 0 & 1 \\ 1 & 0 \end{bmatrix} \begin{bmatrix} u_1(k) \\ u_2(k) \end{bmatrix}$$

determine a state feedback gain matrix **K** such that when the control vector is given by

$$\mathbf{u}(k) = -\mathbf{K}\mathbf{x}(k)$$

the closed-loop system will exhibit the deadbeat response to any initial state $\mathbf{x}(0)$.

Problem B-6–16. Consider the following system:

$$\mathbf{x}(k+1) = \mathbf{G}\mathbf{x}(k) + \mathbf{H}\mathbf{u}(k)$$

where

$$\mathbf{x}(k) = \text{state vector (4-vector)}$$

$$\mathbf{u}(k) = \text{control vector (2-vector)}$$

and

$$\mathbf{G} = \begin{bmatrix} -1 & 1 & 0 & 0 \\ 1 & -2 & 1 & 0 \\ 0 & 1 & -1 & 2 \\ 1 & 0 & 0 & 1 \end{bmatrix}, \qquad \mathbf{H} = [\mathbf{H}_1 \vdots \mathbf{H}_2] = \begin{bmatrix} 0 & 0 \\ 1 & 0 \\ 0 & 0 \\ 0 & 1 \end{bmatrix}$$

By use of the state feedback control $\mathbf{u}(k) = -\mathbf{K}\mathbf{x}(k)$, it is desired to place the closed-loop poles at the following locations:

$$z_1 = -0.4 + j0.3, \qquad z_2 = -0.4 - j0.3, \qquad z_3 = -0.5, \qquad z_4 = -0.8$$

Determine the required state feedback gain matrix **K**.

Problem B-6–17. Consider the same system as that given in Prob. B-6–16. By use of the state feedback control $\mathbf{u}(k) = -\mathbf{K}\mathbf{x}(k)$, it is desired to have the deadbeat response to any initial state $\mathbf{x}(0)$. Determine the state feedback gain matrix **K**.

Problem B-6–18. Consider the system

$$\mathbf{x}(k+1) = \mathbf{G}\mathbf{x}(k) + \mathbf{H}\mathbf{u}(k)$$

where

$$\mathbf{x}(k) = \text{state vector (4-vector).}$$

$$\mathbf{u}(k) = \text{control vector (2-vector).}$$

and

$$\mathbf{G} = \begin{bmatrix} 1 & 1 & 0 & 0 \\ -2 & -1 & 1 & 0 \\ -2 & 0 & 3 & -2 \\ 0 & -1 & 1 & -2 \end{bmatrix}, \qquad \mathbf{H} = \begin{bmatrix} 0 & 0 \\ 0 & 0 \\ 1 & 0 \\ 1 & 1 \end{bmatrix}$$

Assuming that the state feedback control

$$\mathbf{u}(k) = -\mathbf{K}\mathbf{x}(k)$$

is to be used, determine the state feedback gain matrix $\mathbf{K}$ such that the response to the initial state $\mathbf{x}(0)$ is deadbeat.

Problem B-6–19. Consider the system

$$\mathbf{x}(k+1) = \mathbf{G}\mathbf{x}(k) + \mathbf{H}u(k)$$

$$y(k) = \mathbf{C}\mathbf{x}(k)$$

where

$\mathbf{x}(k)$ = state vector (2-vector)

$u(k)$ = control signal (scalar)

$y(k)$ = output signal (scalar)

and

$$\mathbf{G} = \begin{bmatrix} 0 & 1 \\ -0.16 & -1 \end{bmatrix}, \qquad \mathbf{H} = \begin{bmatrix} 0 \\ 1 \end{bmatrix}, \qquad \mathbf{C} = [1 \quad 1]$$

Design a current observer for the system. It is desired that the response to the initial observer error be deadbeat.

Problem B-6–20. Consider the system given in Prob. B-6–19. Assuming that the output is measurable, design a minimum-order observer for the system. It is desired that the response to the initial observer error be deadbeat.

Problem B-6–21. For the system

$$\mathbf{x}(k+1) = \mathbf{G}\mathbf{x}(k)$$

$$y(k) = \mathbf{C}\mathbf{x}(k)$$

where

$\mathbf{x}(k)$ = state vector (2-vector)

$y(k)$ = output signal (scalar)

and

$$\mathbf{G} = \begin{bmatrix} 0 & 1 \\ 0 & -0.5 \end{bmatrix}, \qquad \mathbf{C} = [1 \quad 1]$$

design a current observer. It is desired that the response to the initial observer error be sufficiently fast.

Problem B-6–22. Consider the system

$$\mathbf{x}(k+1) = \mathbf{G}\mathbf{x}(k) + \mathbf{H}u(k)$$
$$y(k) = \mathbf{C}\mathbf{x}(k)$$

where

$$\mathbf{x}(k) = \text{state vector (3-vector)}$$
$$u(k) = \text{control signal (scalar)}$$
$$y(k) = \text{output signal (scalar)}$$

and

$$\mathbf{G} = \begin{bmatrix} 0 & 0 & -0.25 \\ 1 & 0 & 0 \\ 0 & 1 & 0.5 \end{bmatrix}, \qquad \mathbf{H} = \begin{bmatrix} 1 \\ 0 \\ 1 \end{bmatrix}, \qquad \mathbf{C} = [1 \quad 0 \quad 0]$$

Assuming that the output $y(k)$ is measurable, design a minimum-order observer such that the response to the initial observer error is deadbeat.

Problem B-6–23. Consider the system

$$\mathbf{x}(k+1) = \mathbf{G}\mathbf{x}(k) + \mathbf{H}u(k)$$
$$\mathbf{y}(k) = \mathbf{C}\mathbf{x}(k)$$

where

$$\mathbf{x}(k) = \text{state vector (3-vector)}$$
$$u(k) = \text{control signal (scalar)}$$
$$\mathbf{y}(k) = \text{output vector (2-vector)}$$

and

$$\mathbf{G} = \begin{bmatrix} 0 & 0 & -0.25 \\ 1 & 0 & 0 \\ 0 & 1 & 0.5 \end{bmatrix}, \qquad \mathbf{H} = \begin{bmatrix} 1 \\ 0 \\ 1 \end{bmatrix}, \qquad \mathbf{C} = \begin{bmatrix} 0 & 0 & 1 \\ 1 & 0 & 0 \end{bmatrix}$$

Assuming that the output $\mathbf{y}(k)$ is measurable, design a minimum-order observer such that the response to the initial observer error is deadbeat.

Problem B-6–24. Consider the system

$$\mathbf{x}(k+1) = \mathbf{G}\mathbf{x}(k) + \mathbf{H}\mathbf{u}(k)$$
$$\mathbf{y}(k) = \mathbf{C}\mathbf{x}(k)$$

It is assumed that a full-order prediction observer is used to estimate the actual state $\mathbf{x}(k)$. If the observed state $\tilde{\mathbf{x}}(k)$ is used for the state feedback, or if

$$\mathbf{u}(k) = -\mathbf{K}\tilde{\mathbf{x}}(k)$$

then the system state equation becomes

$$\mathbf{x}(k+1) = \mathbf{G}\mathbf{x}(k) - \mathbf{H}\mathbf{K}\tilde{\mathbf{x}}(k)$$

The full-order prediction observer equation is given by Eq. (6–117), which we repeat here:

$$\begin{aligned}\tilde{\mathbf{x}}(k+1) &= (\mathbf{G} - \mathbf{K}_e\mathbf{C})\tilde{\mathbf{x}}(k) + \mathbf{H}\mathbf{u}(k) + \mathbf{K}_e\mathbf{y}(k) \\ &= (\mathbf{G} - \mathbf{K}_e\mathbf{C} - \mathbf{H}\mathbf{K})\tilde{\mathbf{x}}(k) + \mathbf{K}_e\mathbf{C}\mathbf{x}(k)\end{aligned}$$

Define

$$\mathbf{e}(k) = \mathbf{x}(k) - \tilde{\mathbf{x}}(k)$$

Show that

$$\begin{bmatrix}\mathbf{x}(k+1) \\ \mathbf{e}(k+1)\end{bmatrix} = \begin{bmatrix}\mathbf{G} - \mathbf{H}\mathbf{K} & \mathbf{H}\mathbf{K} \\ \mathbf{0} & \mathbf{G} - \mathbf{K}_e\mathbf{C}\end{bmatrix}\begin{bmatrix}\mathbf{x}(k) \\ \mathbf{e}(k)\end{bmatrix}$$

Problem B-6–25. Consider the system

$$\mathbf{x}(k+1) = \mathbf{G}\mathbf{x}(k) + \mathbf{H}\mathbf{u}(k)$$

$$\mathbf{y}(k) = \mathbf{C}\mathbf{x}(k)$$

where

$$\mathbf{G} = \begin{bmatrix} -1 & 1 & 0 & 0 \\ 1 & -2 & 1 & 0 \\ 0 & 1 & -1 & 2 \\ 1 & 0 & 0 & 1 \end{bmatrix}, \quad \mathbf{H} = \begin{bmatrix} 0 & 1 \\ 1 & 0 \\ 0 & 0 \\ 1 & 0 \end{bmatrix}, \quad \mathbf{C} = \begin{bmatrix} 1 & 0 & 0 & 0 \\ 0 & 1 & 0 & 0 \end{bmatrix}$$

Assuming that only $x_1(k)$ and $x_2(k)$ are directly measurable, design a minimum-order observer such that the observer error vector exhibits deadbeat response.

In Prob. A-6–21 we solved a pole placement problem for this system. There we determined the state feedback gain matrix **K** for the case where the response to an arbitrary initial state $\mathbf{x}(0)$ was deadbeat. Using the state feedback gain matrix **K** obtained in Prob. A-6–21, draw a block diagram for the entire system when the control law is given by

$$\mathbf{u}(k) = -\mathbf{K}\tilde{\mathbf{x}}(k)$$

where $\tilde{\mathbf{x}}(k)$ is the observed state vector obtained by use of the minimum-order observer, or

$$\tilde{\mathbf{x}}(k) = \begin{bmatrix} x_1(k) \\ x_2(k) \\ \tilde{x}_3(k) \\ \tilde{x}_4(k) \end{bmatrix}$$

Problem B-6–26. In Example 6–12 we designed a minimum-order observer for a servo system. Considering the same servo system, assume that the output $y(k)$ is not reliable on account of measurement noises. Using Ackermann's formula as given by Eq. (6–153), design a full-order prediction observer such that the observer error response is deadbeat. (Assume that the system configuration is the same as that shown in Fig. 6–12.)

Problem B-6–27. Figure 6–17 shows a servo system where the integral controller has a time delay of one sampling period. (Compare this system with the servo system shown in Fig. 6–11.)

Determine the feedforward gain K_1 and the feedback gain K_2 such that the response to the unit step input $r(k) = 1$ (where $k = 0, 1, 2, \ldots$) is deadbeat. Plot the response $y(k)$ to this input.

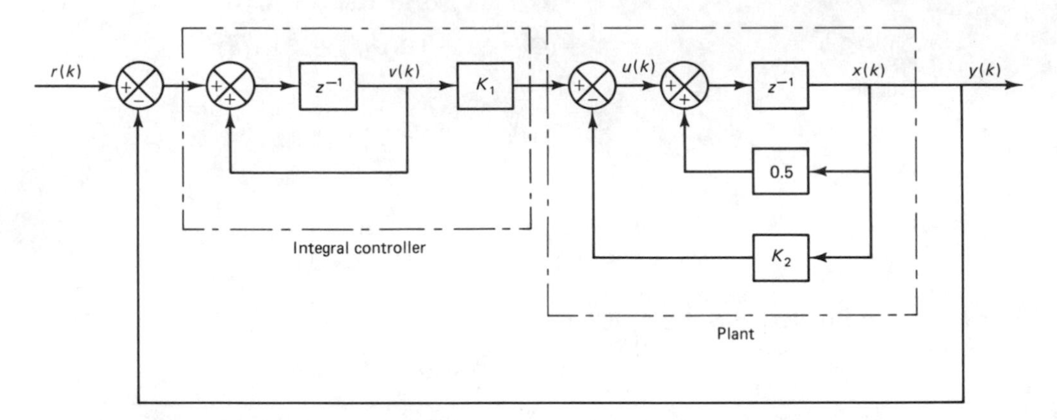

Figure 6–17 Servo system with state feedback and integral control involving a unit delay in the feedforward path.

Problem B-6–28. Consider the servo system shown in Fig. 6–18. (This system is similar to that shown in Fig. 6–17, except that the integral controller has a unit delay element in the minor loop.) Determine the feedforward gain K_1 and the feedback gain K_2 such that the response to the unit step input $r(k) = 1$ (where $k = 0, 1, 2, \ldots$) is deadbeat. Plot the response $y(k)$ to this input.

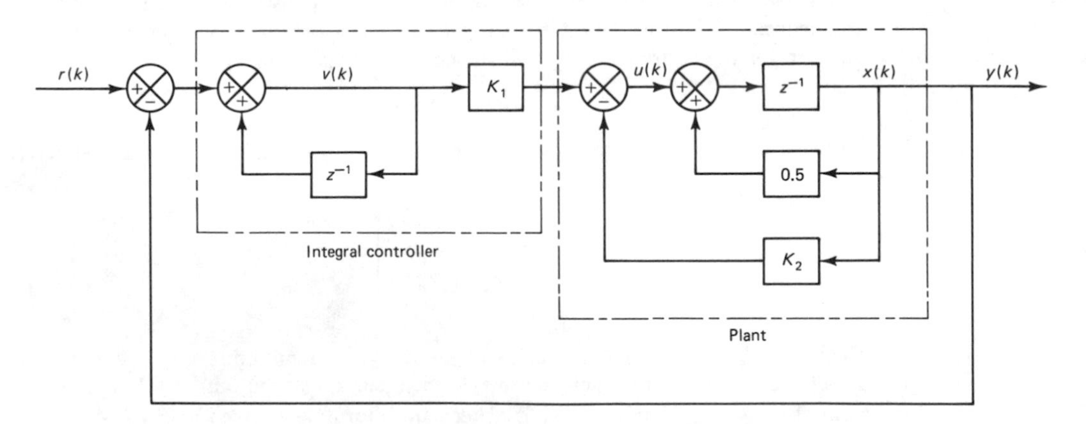

Figure 6–18 Servo system with state feedback and integral control involving a unit delay in the minor loop.

7

Optimal Control Systems

7-1 INTRODUCTION

Problems of optimal control have received a great deal of attention from control engineers. An optimal control system—a system whose design "optimizes" (minimizes or maximizes, as the case may be) the value of a function chosen as the *performance index*—differs from an ideal one in that the former is the best attainable in the presence of physical constraints whereas the latter may well be an unattainable goal.

Performance indexes. In designing an optimal control system, we need to find a rule for determining the present control decision, subject to certain constraints, so as to minimize some measure of the deviation from ideal behavior. That measure is usually provided by the chosen performance index, which is a function whose value we consider to be an indication of how well the actual performance of the system matches the desired performance. In most cases, the behavior of a system is optimized by choosing the control vector $\mathbf{u}(k)$ in such a way that the performance index is minimized (or maximized, depending on the nature of the performance index chosen). The selection of an appropriate performance index is important, because, to a large degree, it determines the nature of the resulting optimal control system. That is, whether the resulting control system will be linear, nonlinear, stationary, or time-varying will depend on the form of the performance index. The control engineer thus formulates this index on the basis of the requirements the system must meet and takes it into account in determining the nature of the resulting system. The requirements of the design usually include not only performance specifications, but also, to ensure physical realizability, restrictions on the form of control to be used.

The optimization process not only should provide optimal control laws and parameter configurations but should also predict the degradation in performance due to any departure of the performance index function from its minimum (or maximum) value that results when nonoptimal control laws are applied.

Choosing the most appropriate performance index for a given problem is very difficult, especially in complex systems. To a considerable degree, the use of optimization theory in system design has been hampered by the conflict between analytical feasibility and practical utility in the selection of the performance index. It is desirable that the criteria for optimal control originate not from a mathematical but from an application point of view. In general, however, the choice of a performance index involves a compromise between a meaningful evaluation of system performance and a tractable mathematical problem.

Formulation of optimization problems. The problem of optimization of a control system may be formulated if the following information is given:

1. System equations
2. Class of allowable control vectors
3. Constraints on the problem
4. Performance index
5. System parameters

The solution of an optimal control problem is to determine the optimal control vector $\mathbf{u}(k)$ within the class of allowable control vectors. This vector $\mathbf{u}(k)$ depends on

1. The nature of the performance index
2. The nature of the constraints
3. The initial state or initial output
4. The desired state or desired output

Except for special cases, the optimal control problem may be so complicated for an analytical solution that a computational solution has to be obtained.

Questions concerning the existence of solutions to optimal control problems. It has been stated that the optimal control problem, given any initial state $\mathbf{x}(0)$, consists of finding an allowable control vector $\mathbf{u}(k)$ that transfers the state to the desired region of the state space and for which the performance index is minimized.

It is important to mention that in some cases a particular combination of plant, desired state, performance index, and constraints makes optimal control impossible. This is a matter of requiring performance beyond the physical capabilities of the system.

Questions regarding the existence of an optimal control vector are important, since they serve to inform the designer whether or not optimal control is possible for a given system and given set of constraints. Two of the most important among these questions are those of controllability and observability, which were presented in Chap. 6.

Comments on optimal control systems. The system whose design minimizes (or maximizes) the selected performance index is, by definition, optimal. It is evident that the performance index, in reality, determines the configuration of the system. It is very important to point out that a control system that is optimal under one performance index is, in general, not optimal under other performance indexes. In addition, hardware realization of a particular optimal control law may be quite difficult and expensive. Hence, it may be pointless to devote too much expense to implementing an optimal controller which is the best only in some narrow, individualistic sense. A control system is seldom designed to perform a single task that is completely specified beforehand. Instead, it is designed to perform a task selected at random from a complete repertoire of possible tasks. In practical systems, then, it may be more sensible to seek approximate optimal control laws which are not rigidly tied to a single performance index.

Strictly speaking, we should realize that a mathematically obtained optimal control system gives, in most practical situations, the highest possible performance under the given performance index and is more a measuring stick than a practical goal. Therefore, before we decide whether to build an optimal control system or something inferior but simpler, we should carefully evaluate a measure of the degree to which the performance of the complex, optimal control system exceeds that of a simpler, suboptimal one. Unless it can be justified, we should not build an extremely complicated and expensive optimal control system.

Once the ultimate degree of performance is found by use of optimal control theory, we should make efforts to design a simple system that is close to optimal. Keeping this in mind, we build a prototype physical system, test it, and modify it until a satisfactory system is obtained which has performance characteristics close to the optimal control system we have worked out in theory.

Analytically solvable optimal control problems provide good insight into optimal structures and algorithms that may be applied to practical cases. An example of analytically solvable optimal control problems is the problem of optimal control of linear systems based on quadratic performance indexes. Quadratic performance indexes have been used very frequently in practical control systems as measures of system performance.

Quadratic optimal control. Let us consider the control system defined by

$$\mathbf{x}(k+1) = \mathbf{G}\mathbf{x}(k) + \mathbf{H}\mathbf{u}(k)$$

where

$\mathbf{x}(k)$ = state vector (n-vector)

$\mathbf{u}(k)$ = control vector (r-vector)

$\mathbf{G} = n \times n$ matrix

$\mathbf{H} = n \times r$ matrix

In the quadratic optimal control problem we desire to determine a law for the control vector $\mathbf{u}(k)$ such that a given quadratic performance index is minimized.

An example of a quadratic performance index is

$$J = \frac{1}{2}\mathbf{x}^*(N)\mathbf{S}\mathbf{x}(N) + \frac{1}{2}\sum_{k=0}^{N-1}[\mathbf{x}^*(k)\mathbf{Q}\mathbf{x}(k) + \mathbf{u}^*(k)\mathbf{R}\mathbf{u}(k)]$$

where matrices $\mathbf{S}$ and $\mathbf{Q}$ are positive definite or positive semidefinite Hermitian matrices and $\mathbf{R}$ is a positive definite Hermitian matrix. The first term on the right-hand side of this last equation accounts for the importance of the final state. The first term in the summation brackets accounts for the relative importance of the error during the control process, and the second term accounts for the expenditure of the energy of the control signals. We assume that the control vector $\mathbf{u}(k)$ is unconstrained.

It will be shown in Sec. 7–2 that the optimal control law is given by

$$\mathbf{u}(k) = -\mathbf{K}(k)\mathbf{x}(k)$$

where $\mathbf{K}(k)$ is a time-varying $r \times n$ matrix. If $N = \infty$, then $\mathbf{K}(k)$ becomes a constant $r \times n$ matrix. The design of optimal control systems based on such quadratic performance indexes boils down to the determination of matrix $\mathbf{K}(k)$.

The major characteristic of the optimal control law based on a quadratic performance index is that it is a linear function of the state vector $\mathbf{x}(k)$. Such a state feedback requires that all state variables be available for feedback. It is advantageous, therefore, to represent the system in terms of measurable state variables. If not all state variables can be measured, we need to estimate or observe the unmeasurable state variables. We then use the measured and observed state variables to generate optimal control signals.

There are many different approaches to the solution of quadratic optimal control problems. In this chapter, we present a commonly used approach based on the minimization technique using Lagrange multipliers. For the steady-state quadratic optimal control problem, we also present the Liapunov approach. It will be shown in Sec. 7–3 that there is a direct relationship between Liapunov functions and quadratic performance indexes.

Note that when an optimal control system is designed in state space, it is important to check the frequency response characteristics. Sometimes a specific compensation for noise effects may be needed. Then, it may become necessary to modify the optimal configuration and accept a suboptimal configuration, or it may become necessary to modify the performance index.

Outline of the chapter. Section 7–1 has presented introductory material. Section 7–2 presents a basic quadratic optimal control problem and its solution. Section

7–3 treats the steady-state quadratic optimal control problem. Here we include the Liapunov approach to the solution of the quadratic optimal control problem. Section 7–4 discusses the quadratic optimal control of servo systems. Section 7–5 is concerned with the system identification technique based on the least-squares method. Finally, Sec. 7–6 gives a brief treatment of Kalman filter problems.

7–2 QUADRATIC OPTIMAL CONTROL

Quadratic optimal control problems can be solved by many different approaches. In this section we shall solve the basic quadratic optimal control problem by the conventional minimization method using Lagrange multipliers.

Quadratic optimal control problem. The quadratic optimal control problem may be stated as follows. Given a linear discrete-time control system

$$\mathbf{x}(k+1) = \mathbf{G}\mathbf{x}(k) + \mathbf{H}\mathbf{u}(k), \qquad \mathbf{x}(0) = \mathbf{c} \tag{7–1}$$

where it is assumed to be completely state controllable and where

$$\mathbf{x}(k) = \text{state vector } (n\text{-vector})$$

$$\mathbf{u}(k) = \text{control vector } (r\text{-vector})$$

$$\mathbf{G} = n \times n \text{ nonsingular matrix}$$

$$\mathbf{H} = n \times r \text{ matrix}$$

find the optimal control sequence $\mathbf{u}(0), \mathbf{u}(1), \mathbf{u}(2), \ldots, \mathbf{u}(N-1)$ that minimizes a quadratic performance index. An example of the quadratic performance indexes for a finite time process ($0 \le k \le N$) is

$$J = \frac{1}{2}\mathbf{x}^*(N)\mathbf{S}\mathbf{x}(N) + \frac{1}{2}\sum_{k=0}^{N-1}[\mathbf{x}^*(k)\mathbf{Q}\mathbf{x}(k) + \mathbf{u}^*(k)\mathbf{R}\mathbf{u}(k)] \tag{7–2}$$

where

$\mathbf{Q} = n \times n$ positive definite or positive semidefinite Hermitian matrix (or real symmetric matrix)

$\mathbf{R} = r \times r$ positive definite Hermitian matrix (or real symmetric matrix)

$\mathbf{S} = n \times n$ positive definite or positive semidefinite Hermitian matrix (or real symmetric matrix)

Matrices **Q, R,** and **S** are selected to weigh the relative importance of the performance measures caused by the state vector $\mathbf{x}(k)$ $(k = 0, 1, 2, \ldots, N-1)$, the control vector $\mathbf{u}(k)$ $(k = 0, 1, 2, \ldots, N-1)$, and the final state $\mathbf{x}(N)$, respectively.

The initial state of the system is at some arbitrary state $\mathbf{x}(0) = \mathbf{c}$. The final state $\mathbf{x}(N)$ may be fixed, in which case the term $\frac{1}{2}\mathbf{x}^*(N)\mathbf{S}\mathbf{x}(N)$ is removed from the performance index of Eq. (7–2) and instead the terminal condition $\mathbf{x}(N) = \mathbf{x}_f$ is imposed, where $\mathbf{x}_f$ is the fixed terminal state. If the final state $\mathbf{x}(N)$ is not fixed, then the first term in Eq. (7–2) represents the weight of the performance measure due to the final state. Note that in the minimization problem the inclusion of the term $\frac{1}{2}\mathbf{x}^*(N)\mathbf{S}\mathbf{x}(N)$ in the performance index J implies that we desire the final state $\mathbf{x}(N)$ to be as close to the origin as possible.

Solution by the conventional minimization method using Lagrange multipliers. The quadratic optimal control problem is a minimization problem involving a function of several variables. Thus, it can be solved by the conventional minimization method. The minimization problem subjected to equality constraints may be solved by adjoining the constraints to the function to be minimized by use of Lagrange multipliers.

In the present optimization problem, we minimize J as given by Eq. (7–2), repeated here,

$$J = \frac{1}{2}\mathbf{x}^*(N)\mathbf{S}\mathbf{x}(N) + \frac{1}{2}\sum_{k=0}^{N-1}[\mathbf{x}^*(k)\mathbf{Q}\mathbf{x}(k) + \mathbf{u}^*(k)\mathbf{R}\mathbf{u}(k)] \tag{7–3}$$

when it is subjected to the constraint equation specified by Eq. (7–1),

$$\mathbf{x}(k+1) = \mathbf{G}\mathbf{x}(k) + \mathbf{H}\mathbf{u}(k) \tag{7–4}$$

where $k = 0, 1, 2, \ldots, N-1$, and where the initial condition on the state vector is specified as

$$\mathbf{x}(0) = \mathbf{c} \tag{7–5}$$

Now, by using a set of Lagrange multipliers $\boldsymbol{\lambda}(1), \boldsymbol{\lambda}(2), \ldots, \boldsymbol{\lambda}(N)$, we define a new performance index L as follows:

$$\begin{aligned} L = {} & \frac{1}{2}\mathbf{x}^*(N)\mathbf{S}\mathbf{x}(N) + \frac{1}{2}\sum_{k=0}^{N-1}\{[\mathbf{x}^*(k)\mathbf{Q}\mathbf{x}(k) + \mathbf{u}^*(k)\mathbf{R}\mathbf{u}(k)] \\ & + \boldsymbol{\lambda}^*(k+1)[\mathbf{G}\mathbf{x}(k) + \mathbf{H}\mathbf{u}(k) - \mathbf{x}(k+1)] \\ & + [\mathbf{G}\mathbf{x}(k) + \mathbf{H}\mathbf{u}(k) - \mathbf{x}(k+1)]^*\boldsymbol{\lambda}(k+1)\} \end{aligned} \tag{7–6}$$

The reason for writing the terms involving the Lagrange multiplier in the form shown in Eq. (7–6) is to ensure that $L = L^*$. (L is a real scalar quantity.) Note that

$$\boldsymbol{\lambda}^*(0)[\mathbf{c} - \mathbf{x}(0)] + [\mathbf{c} - \mathbf{x}(0)]^*\boldsymbol{\lambda}(0)$$

may be added to the performance index L. However, we shall not do so, to simplify the presentation. It is a well-known fact that minimization of the function L defined by Eq. (7–6) is equivalent to minimization of J as defined by Eq. (7–3) when it is subjected to the equality constraint defined by Eq. (7–4).

In order to minimize the function L, we need to differentiate L with respect to each component of vectors $\mathbf{x}(k)$, $\mathbf{u}(k)$, and $\boldsymbol{\lambda}(k)$ and set the results equal to zero.

From the computational viewpoint, however, it is convenient to differentiate L with respect to $\bar{x}_i(k)$, $\bar{u}_i(k)$, and $\bar{\lambda}_i(k)$, where $\bar{x}_i(k)$, $\bar{u}_i(k)$, and $\bar{\lambda}_i(k)$ are, respectively, the complex conjugates of $x_i(k)$, $u_i(k)$, and $\lambda_i(k)$. (Note that the signal and its complex conjugate contain the same mathematical information.) Thus we set

$$\frac{\partial L}{\partial \bar{x}_i(k)} = 0 \qquad i = 1, 2, \ldots, n;\ k = 1, 2, \ldots, N$$

$$\frac{\partial L}{\partial \bar{u}_i(k)} = 0 \qquad i = 1, 2, \ldots, r;\ k = 0, 1, \ldots, N-1$$

$$\frac{\partial L}{\partial \bar{\lambda}_i(k)} = 0 \qquad i = 1, 2, \ldots, n;\ k = 1, 2, \ldots, N$$

These equations are necessary conditions for L to have a minimum. Note that the simplified expressions for the preceding partial derivative equations are

$$\frac{\partial L}{\partial \bar{\mathbf{x}}(k)} = \mathbf{0} \qquad k = 1, 2, \ldots, N \tag{7–7}$$

$$\frac{\partial L}{\partial \bar{\mathbf{u}}(k)} = \mathbf{0} \qquad k = 0, 1, \ldots, N-1 \tag{7–8}$$

$$\frac{\partial L}{\partial \bar{\boldsymbol{\lambda}}(k)} = \mathbf{0} \qquad k = 1, 2, \ldots, N \tag{7–9}$$

Referring to the Appendix (See Probs. A-7 and A-8) for partial differentiation of complex quadratic and bilinear forms with respect to vector variables, we have

$$\frac{\partial}{\partial \bar{\mathbf{x}}} \mathbf{x}^*\mathbf{A}\mathbf{x} = \mathbf{A}\mathbf{x} \qquad \text{and} \qquad \frac{\partial}{\partial \bar{\mathbf{x}}} \mathbf{x}^*\mathbf{A}\mathbf{y} = \mathbf{A}\mathbf{y}$$

Then, Eqs. (7–7), (7–8), and (7–9) may be obtained as follows:

$$\frac{\partial L}{\partial \bar{\mathbf{x}}(k)} = \mathbf{0}: \quad \mathbf{Q}\mathbf{x}(k) + \mathbf{G}^*\boldsymbol{\lambda}(k+1) - \boldsymbol{\lambda}(k) = \mathbf{0} \qquad k = 1, 2, \ldots, N-1 \tag{7–10}$$

$$\frac{\partial L}{\partial \bar{\mathbf{x}}(N)} = \mathbf{0}: \quad \mathbf{S}\mathbf{x}(N) - \boldsymbol{\lambda}(N) = \mathbf{0} \tag{7–11}$$

$$\frac{\partial L}{\partial \bar{\mathbf{u}}(k)} = \mathbf{0}: \quad \mathbf{R}\mathbf{u}(k) + \mathbf{H}^*\boldsymbol{\lambda}(k+1) = \mathbf{0} \qquad k = 0, 1, \ldots, N-1 \tag{7–12}$$

$$\frac{\partial L}{\partial \bar{\boldsymbol{\lambda}}(k)} = \mathbf{0}: \quad \mathbf{G}\mathbf{x}(k-1) + \mathbf{H}\mathbf{u}(k-1) - \mathbf{x}(k) = \mathbf{0} \qquad k = 1, 2, \ldots, N \tag{7–13}$$

Equation (7–13) is simply the system state equation. Equation (7–11) specifies the final value of the Lagrange multiplier. Note that the Lagrange multiplier $\boldsymbol{\lambda}(k)$ is often called a *covector* or *adjoint vector.*

Now we shall simplify the equations just obtained. From Eq. (7–10) we have

$$\boldsymbol{\lambda}(k) = \mathbf{Qx}(k) + \mathbf{G}^*\boldsymbol{\lambda}(k+1) \qquad k = 1, 2, 3, \ldots, N-1 \tag{7–14}$$

with the final condition $\boldsymbol{\lambda}(N) = \mathbf{Sx}(N)$. By solving Eq. (7–12) for $\mathbf{u}(k)$ and noting that $\mathbf{R}^{-1}$ exists, we obtain

$$\mathbf{u}(k) = -\mathbf{R}^{-1}\mathbf{H}^*\boldsymbol{\lambda}(k+1) \qquad k = 0, 1, 2, \ldots, N-1 \tag{7–15}$$

Equation (7–13) can be rewritten as

$$\mathbf{x}(k+1) = \mathbf{Gx}(k) + \mathbf{Hu}(k) \qquad k = 0, 1, 2, \ldots, N-1 \tag{7–16}$$

which is simply the state equation. Substitution of Eq. (7–15) into Eq. (7–16) results in

$$\mathbf{x}(k+1) = \mathbf{Gx}(k) - \mathbf{HR}^{-1}\mathbf{H}^*\boldsymbol{\lambda}(k+1) \tag{7–17}$$

with the initial condition $\mathbf{x}(0) = \mathbf{c}$.

In order to obtain the solution to the minimization problem we need to solve Eqs. (7–14) and (7–17) simultaneously. Notice that for the system equation, Eq. (7–16), the initial condition $\mathbf{x}(0)$ is specified, while for the Lagrange multiplier equation, Eq. (7–14), the final condition $\boldsymbol{\lambda}(N)$ is specified. Thus, the problem here becomes a two-point boundary-value problem.

If the two-point boundary-value problem is solved, then the optimal values for the state vector and Lagrange multiplier vector may be determined and the optimal control vector $\mathbf{u}(k)$ may be obtained in the open-loop form. However, if we employ the Riccati transformation, the optimal control vector $\mathbf{u}(k)$ can be obtained in the following closed-loop, or feedback, form:

$$\mathbf{u}(k) = -\mathbf{K}(k)\mathbf{x}(k)$$

where $\mathbf{K}(k)$ is the $r \times n$ feedback matrix.

In what follows, we shall obtain the optimal control vector $\mathbf{u}(k)$ in the closed-loop form by first obtaining the Riccati equation. Assume that $\boldsymbol{\lambda}(k)$ can be written in the following form:

$$\boldsymbol{\lambda}(k) = \mathbf{P}(k)\mathbf{x}(k) \tag{7–18}$$

where $\mathbf{P}(k)$ is an $n \times n$ Hermitian matrix (or an $n \times n$ real symmetric matrix). Substitution of Eq. (7–18) into Eq. (7–14) results in

$$\mathbf{P}(k)\mathbf{x}(k) = \mathbf{Qx}(k) + \mathbf{G}^*\mathbf{P}(k+1)\mathbf{x}(k+1) \tag{7–19}$$

and substitution of Eq. (7–18) into Eq. (7–17) gives

$$\mathbf{x}(k+1) = \mathbf{Gx}(k) - \mathbf{HR}^{-1}\mathbf{H}^*\mathbf{P}(k+1)\mathbf{x}(k+1) \tag{7–20}$$

Notice that Eqs. (7–19) and (7–20) do not involve $\boldsymbol{\lambda}(k)$ and thus we have eliminated $\boldsymbol{\lambda}(k)$. The transformation process employed here is called the Riccati transformation. It is of extreme importance in solving such a two-point boundary-value problem.

From Eq. (7–20) we have

$$[\mathbf{I} + \mathbf{H}\mathbf{R}^{-1}\mathbf{H}^*\mathbf{P}(k+1)]\mathbf{x}(k+1) = \mathbf{G}\mathbf{x}(k) \tag{7–21}$$

For completely state controllable systems, it can be shown that $\mathbf{P}(k+1)$ is positive definite or positive semidefinite. For at least a positive semidefinite matrix $\mathbf{P}(k+1)$, we have

$$|\mathbf{I}_n + \mathbf{H}\mathbf{R}^{-1}\mathbf{H}^*\mathbf{P}(k+1)| = |\mathbf{I}_r + \mathbf{H}^*\mathbf{P}(k+1)\mathbf{H}\mathbf{R}^{-1}| = |\mathbf{I}_r + \mathbf{R}^{-1}\mathbf{H}^*\mathbf{P}(k+1)\mathbf{H}|$$
$$= |\mathbf{R}^{-1}|\,|\mathbf{R} + \mathbf{H}^*\mathbf{P}(k+1)\mathbf{H}| \neq 0$$

where we have used the relationship

$$|\mathbf{I}_n + \mathbf{A}\mathbf{B}| = |\mathbf{I}_r + \mathbf{B}\mathbf{A}| \qquad \mathbf{A} = n \times r \text{ matrix, } \mathbf{B} = r \times n \text{ matrix}$$

(See the Appendix.) Hence, the inverse of $\mathbf{I} + \mathbf{H}\mathbf{R}^{-1}\mathbf{H}^*\mathbf{P}(k+1)$ exists. Consequently, Eq. (7–21) can be written as follows:

$$\mathbf{x}(k+1) = [\mathbf{I} + \mathbf{H}\mathbf{R}^{-1}\mathbf{H}^*\mathbf{P}(k+1)]^{-1}\mathbf{G}\mathbf{x}(k) \tag{7–22}$$

By substituting Eq. (7–22) into Eq. (7–19), we obtain

$$\mathbf{P}(k)\mathbf{x}(k) = \mathbf{Q}\mathbf{x}(k) + \mathbf{G}^*\mathbf{P}(k+1)[\mathbf{I} + \mathbf{H}\mathbf{R}^{-1}\mathbf{H}^*\mathbf{P}(k+1)]^{-1}\mathbf{G}\mathbf{x}(k)$$

or

$$\{\mathbf{P}(k) - \mathbf{Q} - \mathbf{G}^*\mathbf{P}(k+1)[\mathbf{I} + \mathbf{H}\mathbf{R}^{-1}\mathbf{H}^*\mathbf{P}(k+1)]^{-1}\mathbf{G}\}\mathbf{x}(k) = \mathbf{0}$$

This last equation must hold for all $\mathbf{x}(k)$. Hence, we must have

$$\mathbf{P}(k) = \mathbf{Q} + \mathbf{G}^*\mathbf{P}(k+1)[\mathbf{I} + \mathbf{H}\mathbf{R}^{-1}\mathbf{H}^*\mathbf{P}(k+1)]^{-1}\mathbf{G} \tag{7–23}$$

Equation (7–23) may be modified. By using the matrix inversion lemma

$$(\mathbf{A} + \mathbf{B}\mathbf{D})^{-1} = \mathbf{A}^{-1} - \mathbf{A}^{-1}\mathbf{B}(\mathbf{I} + \mathbf{D}\mathbf{A}^{-1}\mathbf{B})^{-1}\mathbf{D}\mathbf{A}^{-1}$$

and making the substitutions

$$\mathbf{A} = \mathbf{I}, \qquad \mathbf{B} = \mathbf{H}\mathbf{R}^{-1}, \qquad \mathbf{D} = \mathbf{H}^*\mathbf{P}(k+1)$$

we obtain

$$[\mathbf{I} + \mathbf{H}\mathbf{R}^{-1}\mathbf{H}^*\mathbf{P}(k+1)]^{-1} = \mathbf{I} - \mathbf{H}\mathbf{R}^{-1}[\mathbf{I} + \mathbf{H}^*\mathbf{P}(k+1)\mathbf{H}\mathbf{R}^{-1}]^{-1}\mathbf{H}^*\mathbf{P}(k+1)$$
$$= \mathbf{I} - \mathbf{H}[\mathbf{R} + \mathbf{H}^*\mathbf{P}(k+1)\mathbf{H}]^{-1}\mathbf{H}^*\mathbf{P}(k+1)$$

Hence, Eq. (7–23) can be modified to

$$\begin{aligned}\mathbf{P}(k) = {} & \mathbf{Q} + \mathbf{G}^*\mathbf{P}(k+1)\mathbf{G} \\ & - \mathbf{G}^*\mathbf{P}(k+1)\mathbf{H}[\mathbf{R} + \mathbf{H}^*\mathbf{P}(k+1)\mathbf{H}]^{-1}\mathbf{H}^*\mathbf{P}(k+1)\mathbf{G}\end{aligned} \tag{7–24}$$

Equation (7–24) is called the Riccati equation. Referring to Eqs. (7–11) and (7–18), notice that at $k = N$ we have

$$\mathbf{P}(N)\mathbf{x}(N) = \boldsymbol{\lambda}(N) = \mathbf{S}\mathbf{x}(N)$$

or

$$\mathbf{P}(N) = \mathbf{S} \tag{7–25}$$

Hence, Eq. (7–23) or (7–24) can be solved uniquely backward from $k = N$ to $k = 0$. That is, we can obtain $\mathbf{P}(N)$, $\mathbf{P}(N-1)$, . . . , $\mathbf{P}(0)$ starting from $\mathbf{P}(N)$, which is known.

By referring to Eqs. (7–14) and (7–18), the optimal control vector $\mathbf{u}(k)$, given by Eq. (7–15), now becomes

$$\begin{aligned}\mathbf{u}(k) &= -\mathbf{R}^{-1}\mathbf{H}^*\boldsymbol{\lambda}(k+1) = -\mathbf{R}^{-1}\mathbf{H}^*(\mathbf{G}^*)^{-1}[\boldsymbol{\lambda}(k) - \mathbf{Q}\mathbf{x}(k)] \\ &= -\mathbf{R}^{-1}\mathbf{H}^*(\mathbf{G}^*)^{-1}[\mathbf{P}(k) - \mathbf{Q}]\mathbf{x}(k) = -\mathbf{K}(k)\mathbf{x}(k) \end{aligned} \tag{7–26}$$

where

$$\mathbf{K}(k) = \mathbf{R}^{-1}\mathbf{H}^*(\mathbf{G}^*)^{-1}[\mathbf{P}(k) - \mathbf{Q}] \tag{7–27}$$

Equation (7–26) gives the closed-loop form, or feedback form, for the optimal control vector $\mathbf{u}(k)$. Notice that the optimal control vector is proportional to the state vector.

Note that the optimal control vector $\mathbf{u}(k)$ can be given in a few different forms. Referring to Eqs. (7–18) and (7–22), $\mathbf{u}(k)$ may be given by

$$\begin{aligned}\mathbf{u}(k) &= -\mathbf{R}^{-1}\mathbf{H}^*\boldsymbol{\lambda}(k+1) = -\mathbf{R}^{-1}\mathbf{H}^*\mathbf{P}(k+1)\mathbf{x}(k+1) \\ &= -\mathbf{R}^{-1}\mathbf{H}^*\mathbf{P}(k+1)[\mathbf{I} + \mathbf{H}\mathbf{R}^{-1}\mathbf{H}^*\mathbf{P}(k+1)]^{-1}\mathbf{G}\mathbf{x}(k) \\ &= -\mathbf{R}^{-1}\mathbf{H}^*[\mathbf{P}^{-1}(k+1) + \mathbf{H}\mathbf{R}^{-1}\mathbf{H}^*]^{-1}\mathbf{G}\mathbf{x}(k) \\ &= -\mathbf{K}(k)\mathbf{x}(k)\end{aligned} \tag{7–28}$$

where

$$\mathbf{K}(k) = \mathbf{R}^{-1}\mathbf{H}^*[\mathbf{P}^{-1}(k+1) + \mathbf{H}\mathbf{R}^{-1}\mathbf{H}^*]^{-1}\mathbf{G} \tag{7–29}$$

A slightly different form of the optimal control vector $\mathbf{u}(k)$ can be given by

$$\begin{aligned}\mathbf{u}(k) &= -[\mathbf{R} + \mathbf{H}^*\mathbf{P}(k+1)\mathbf{H}]^{-1}\mathbf{H}^*\mathbf{P}(k+1)\mathbf{G}\mathbf{x}(k) \\ &= -\mathbf{K}(k)\mathbf{x}(k)\end{aligned} \tag{7–30}$$

where

$$\mathbf{K}(k) = [\mathbf{R} + \mathbf{H}^*\mathbf{P}(k+1)\mathbf{H}]^{-1}\mathbf{H}^*\mathbf{P}(k+1)\mathbf{G} \tag{7–31}$$

The equivalence of the expressions for the optimal control vector $\mathbf{u}(k)$ given by Eqs. (7–26), (7–28), and (7–30) can be shown easily; see Prob. A-7–1.

Equation (7–26), (7–28), or (7–30) clearly indicates that the optimal control law requires feedback of the state vector with time-varying gain $\mathbf{K}(k)$. Figure 7–1 shows the optimal control scheme of the system based on the quadratic performance index. It is important to point out that a time-varying gain $\mathbf{K}(k)$ can be computed before the process begins, once the system state matrix $\mathbf{G}$, control matrix $\mathbf{H}$, and weighting matrices $\mathbf{Q}$, $\mathbf{R}$, and $\mathbf{S}$ are given. Consequently, $\mathbf{K}(k)$ can be precomputed off-line and stored for future use. Note that the initial state $\mathbf{x}(0)$ does not enter the computation for $\mathbf{K}(k)$. The optimal control vector $\mathbf{u}(k)$ at each stage can be determined immediately by premultiplying the state vector $\mathbf{x}(k)$ by $-\mathbf{K}(k)$.

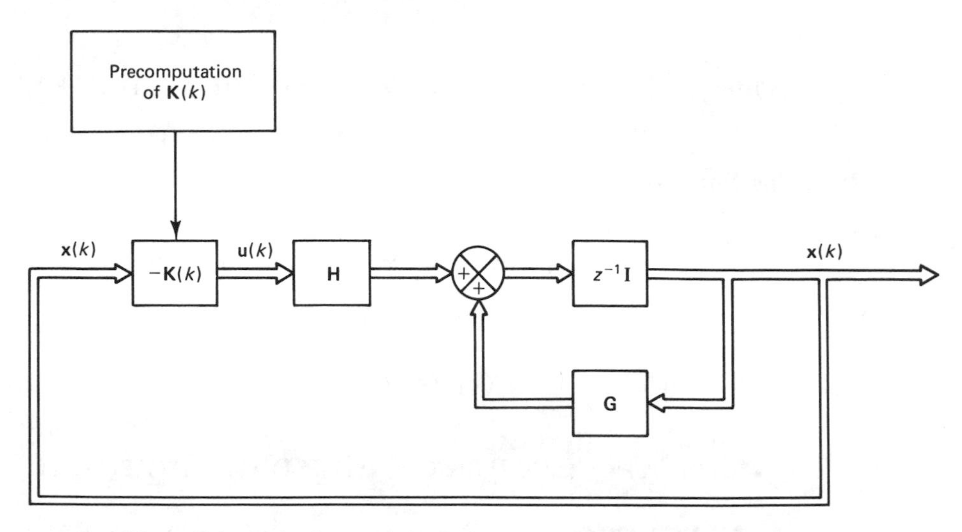

Figure 7–1 Optimal control system based on a quadratic performance index.

Note that a property of the feedback gain matrix $\mathbf{K}(k)$ is that it is almost constant, except near the end of the process at $k = N$. (See Example 7–1, which follows shortly, and Prob. A-7–3.)

Evaluation of the minimum performance index. We shall next evaluate the minimum value of the performance index:

$$\min J = \min \left\{ \frac{1}{2} \mathbf{x}^*(N)\mathbf{S}\mathbf{x}(N) + \frac{1}{2} \sum_{k=0}^{N-1} [\mathbf{x}^*(k)\mathbf{Q}\mathbf{x}(k) + \mathbf{u}^*(k)\mathbf{R}\mathbf{u}(k)] \right\}$$

Premultiplying both sides of Eq. (7–19) by $\mathbf{x}^*(k)$, we have

$$\mathbf{x}^*(k)\mathbf{P}(k)\mathbf{x}(k) = \mathbf{x}^*(k)\mathbf{Q}\mathbf{x}(k) + \mathbf{x}^*(k)\mathbf{G}^*\mathbf{P}(k+1)\mathbf{x}(k+1)$$

Substituting Eq. (7–21) into this last equation, we obtain

$$\begin{aligned} \mathbf{x}^*(k)\mathbf{P}(k)\mathbf{x}(k) &= \mathbf{x}^*(k)\mathbf{Q}\mathbf{x}(k) + \mathbf{x}^*(k+1)[\mathbf{I} + \mathbf{H}\mathbf{R}^{-1}\mathbf{H}^*\mathbf{P}(k+1)]^*\mathbf{P}(k+1)\mathbf{x}(k+1) \\ &= \mathbf{x}^*(k)\mathbf{Q}\mathbf{x}(k) + \mathbf{x}^*(k+1)[\mathbf{I} + \mathbf{P}(k+1)\mathbf{H}\mathbf{R}^{-1}\mathbf{H}^*]\mathbf{P}(k+1)\mathbf{x}(k+1) \end{aligned}$$

Hence

$$\begin{aligned} \mathbf{x}^*(k)\mathbf{Q}\mathbf{x}(k) = {} & \mathbf{x}^*(k)\mathbf{P}(k)\mathbf{x}(k) - \mathbf{x}^*(k+1)\mathbf{P}(k+1)\mathbf{x}(k+1) \\ & -\mathbf{x}^*(k+1)\mathbf{P}(k+1)\mathbf{H}\mathbf{R}^{-1}\mathbf{H}^*\mathbf{P}(k+1)\mathbf{x}(k+1) \end{aligned} \qquad (7\text{–}32)$$

Also, from Eqs. (7–15) and (7–18) we have

$$\mathbf{u}(k) = -\mathbf{R}^{-1}\mathbf{H}^*\mathbf{P}(k+1)\mathbf{x}(k+1)$$

Hence

$$\mathbf{u}^*(k)\mathbf{R}\mathbf{u}(k) = [-\mathbf{x}^*(k+1)\mathbf{P}(k+1)\mathbf{H}\mathbf{R}^{-1}]\mathbf{R}[-\mathbf{R}^{-1}\mathbf{H}^*\mathbf{P}(k+1)\mathbf{x}(k+1)]$$
$$= \mathbf{x}^*(k+1)\mathbf{P}(k+1)\mathbf{H}\mathbf{R}^{-1}\mathbf{H}^*\mathbf{P}(k+1)\mathbf{x}(k+1) \tag{7–33}$$

By adding Eqs. (7–32) and (7–33), we have

$$\mathbf{x}^*(k)\mathbf{Q}\mathbf{x}(k) + \mathbf{u}^*(k)\mathbf{R}\mathbf{u}(k)$$
$$= \mathbf{x}^*(k)\mathbf{P}(k)\mathbf{x}(k) - \mathbf{x}^*(k+1)\mathbf{P}(k+1)\mathbf{x}(k+1) \tag{7–34}$$

By substituting Eq. (7–34) into Eq. (7–3), we obtain

$$J_{\min} = \frac{1}{2}\mathbf{x}^*(N)\mathbf{S}\mathbf{x}(N) + \frac{1}{2}\sum_{k=0}^{N-1}[\mathbf{x}^*(k)\mathbf{P}(k)\mathbf{x}(k) - \mathbf{x}^*(k+1)\mathbf{P}(k+1)\mathbf{x}(k+1)]$$
$$= \frac{1}{2}\mathbf{x}^*(N)\mathbf{S}\mathbf{x}(N) + \frac{1}{2}[\mathbf{x}^*(0)\mathbf{P}(0)\mathbf{x}(0) - \mathbf{x}^*(1)\mathbf{P}(1)\mathbf{x}(1) + \mathbf{x}^*(1)\mathbf{P}(1)\mathbf{x}(1)$$
$$- \mathbf{x}^*(2)\mathbf{P}(2)\mathbf{x}(2) + \cdots + \mathbf{x}^*(N-1)\mathbf{P}(N-1)\mathbf{x}(N-1) - \mathbf{x}^*(N)\mathbf{P}(N)\mathbf{x}(N)]$$
$$= \frac{1}{2}\mathbf{x}^*(N)\mathbf{S}\mathbf{x}(N) + \frac{1}{2}\mathbf{x}^*(0)\mathbf{P}(0)\mathbf{x}(0) - \frac{1}{2}\mathbf{x}^*(N)\mathbf{P}(N)\mathbf{x}(N) \tag{7–35}$$

Notice that from Eq. (7–25) we have $\mathbf{P}(N) = \mathbf{S}$. Hence, Eq. (7–35) becomes

$$J_{\min} = \frac{1}{2}\mathbf{x}^*(0)\mathbf{P}(0)\mathbf{x}(0) \tag{7–36}$$

Thus, the minimum value of the performance index J is given by Eq. (7–36). It is a function of $\mathbf{P}(0)$ and the initial state $\mathbf{x}(0)$.

Example 7–1.

Consider the discrete-time control system defined by

$$x(k+1) = 0.3679x(k) + 0.6321u(k), \qquad x(0) = 1$$

Determine the optimal control law to minimize the following performance index:

$$J = \frac{1}{2}[x(10)]^2 + \frac{1}{2}\sum_{k=0}^{9}[x^2(k) + u^2(k)]$$

Note that in this example, $S = 1$, $Q = 1$, and $R = 1$. Also, determine the minimum value of the performance index J.

Referring to Eq. (7–23), we obtain $P(k)$ as follows:

$$P(k) = 1 + (0.3679)P(k+1)[1 + (0.6321)(1)(0.6321)P(k+1)]^{-1}(0.3679)$$

which can be simplified to

$$P(k) = 1 + 0.1354P(k+1)[1 + 0.3996P(k+1)]^{-1}$$

The boundary condition for $P(k)$ is specified by Eq. (7–25), and in this example,

$$P(N) = P(10) = S = 1$$

We now compute $P(k)$ backward from $k = 9$ to $k = 0$:

$$P(9) = 1 + 0.1354 \times 1(1 + 0.3996 \times 1)^{-1} = 1.0967$$

$$P(8) = 1 + 0.1354 \times 1.0967(1 + 0.3996 \times 1.0967)^{-1} = 1.1032$$

$$P(7) = 1 + 0.1354 \times 1.1032(1 + 0.3996 \times 1.1032)^{-1} = 1.1037$$

$$P(6) = 1 + 0.1354 \times 1.1037(1 + 0.3996 \times 1.1037)^{-1} = 1.1037$$

$$P(k) = 1.1037 \qquad k = 5, 4, 3, 2, 1, 0$$

Notice that the values of $P(k)$ rapidly approach the steady-state value. The steady-state value P_{ss} can be obtained from

$$P_{ss} = 1 + 0.1354P_{ss}(1 + 0.3996P_{ss})^{-1}$$

or

$$0.3996P_{ss}^2 + 0.4650P_{ss} - 1 = 0$$

Solving this last equation for P_{ss}, we have

$$P_{ss} = 1.1037 \qquad \text{or} \qquad -2.2674$$

Since $P(k)$ must be positive, we find the steady-state value for $P(k)$ to be 1.1037.

The feedback gain $K(k)$ can be computed from Eq. (7–27):

$$K(k) = (1)(0.6321)(0.3679)^{-1}[P(k) - 1] = 1.7181[P(k) - 1]$$

By substituting the values of $P(k)$ we have obtained, we get

$$K(10) = 1.7181(1 - 1) = 0$$

$$K(9) = 1.7181(1.0967 - 1) = 0.1661$$

$$K(8) = 1.7181(1.1032 - 1) = 0.1773$$

$$K(7) = 1.7181(1.1037 - 1) = 0.1782$$

$$K(6) = K(5) = \cdots = K(0) = 0.1782$$

The optimal control law is given by

$$u(k) = -K(k)x(k)$$

Since

$$x(k + 1) = 0.3679x(k) + 0.6321u(k) = [0.3679 - 0.6321K(k)]x(k)$$

we obtain

$$\begin{aligned}x(1) &= [0.3679 - 0.6321K(0)]x(0)\\ &= (0.3679 - 0.6321 \times 0.1782) \times 1 = 0.2553\\ x(2) &= (0.3679 - 0.6321 \times 0.1782) \times 0.2553 = 0.0652\\ x(3) &= (0.3679 - 0.6321 \times 0.1782) \times 0.0652 = 0.0166\\ x(4) &= (0.3679 - 0.6321 \times 0.1782) \times 0.0166 = 0.00424\end{aligned}$$

The values of $x(k)$ for $k = 5, 6, \ldots, 10$ approach zero rapidly.

The optimal control sequence $u(k)$ is now obtained as follows:

$$u(0) = -K(0)x(0) = -0.1782 \times 1 = -0.1782$$

$$u(1) = -K(1)x(1) = -0.1782 \times 0.2553 = -0.0455$$

$$u(2) = -K(2)x(2) = -0.1782 \times 0.0652 = -0.0116$$

$$u(3) = -K(3)x(3) = -0.1782 \times 0.0166 = -0.00296$$

$$u(4) = -K(4)x(4) = -0.1782 \times 0.00424 = -0.000756$$

$$u(k) \cong 0 \qquad k = 5, 6, \ldots, 10$$

The values of $P(k)$, $K(k)$, $x(k)$, and $u(k)$ are plotted in Fig. 7–2. Notice that the values of $P(k)$ and $K(k)$ are constant except for the final few stages.

Finally, the minimum value of the performance index J can be obtained from Eq. (7–36):

$$J_{\min} = \frac{1}{2} x^*(0)P(0)x(0) = \frac{1}{2}(1 \times 1.1037 \times 1) = 0.55185$$

Discretized quadratic optimal control problem. We shall next consider the quadratic optimal control of a discretized control system. Consider the continuous-time control system

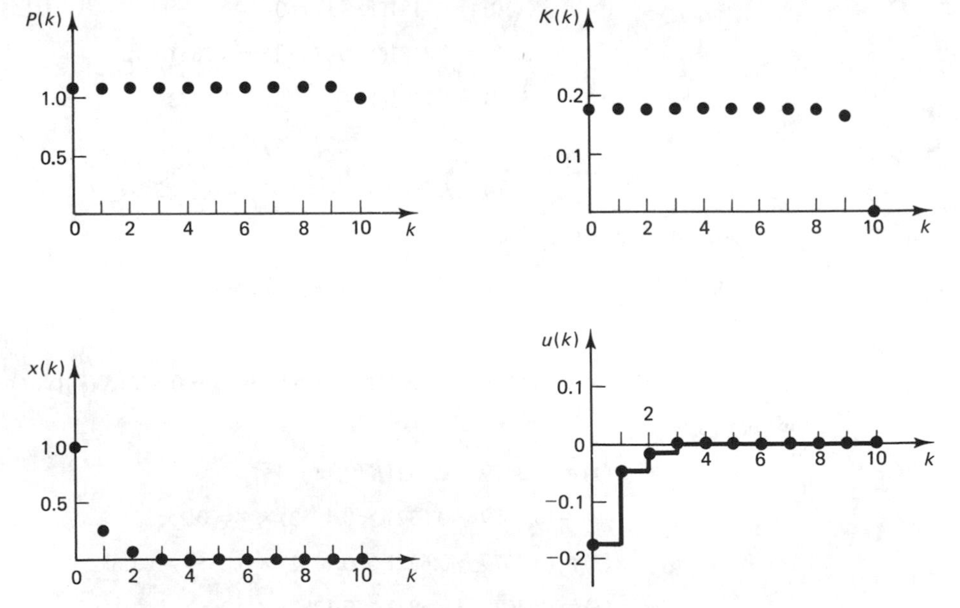

Figure 7–2 Plots of $P(k)$ vs. k, $x(k)$ vs. k, $K(k)$ vs. k, and $u(k)$ vs. k for the system considered in Example 7–1.

$$\dot{\mathbf{x}} = \mathbf{A}\mathbf{x} + \mathbf{B}\mathbf{u} \tag{7–37}$$

where

$$\mathbf{u}(t) = \mathbf{u}(kT) \qquad kT \leq t < (k+1)T$$

and the performance index to be minimized is

$$J = \frac{1}{2}\mathbf{x}^*(t_f)\mathbf{S}\mathbf{x}(t_f) + \frac{1}{2}\int_0^{t_f}[\mathbf{x}^*(t)\mathbf{Q}\mathbf{x}(t) + \mathbf{u}^*(t)\mathbf{R}\mathbf{u}(t)]\, dt \tag{7–38}$$

Suppose the continuous-time control system is approximated by its discrete equivalent. The discretized system equation is

$$\mathbf{x}((k+1)T) = \mathbf{G}(T)\mathbf{x}(kT) + \mathbf{H}(T)\mathbf{u}(kT)$$

and the discretized performance index when $t_f = NT$ will become as follows:

$$J = \frac{1}{2}\mathbf{x}^*(NT)\mathbf{S}\mathbf{x}^*(NT) + \frac{1}{2}\sum_{k=0}^{N-1}[\mathbf{x}^*(kT)\mathbf{Q}_1\mathbf{x}(kT) + 2\mathbf{x}^*(kT)\mathbf{M}_1\mathbf{u}(kT) + \mathbf{u}^*(kT)\mathbf{R}_1\mathbf{u}(kT)] \tag{7–39}$$

It is noted that the integral term in Eq. (7–38) is not replaced by

$$\frac{1}{2}\sum_{k=0}^{N-1}[\mathbf{x}^*(kT)\mathbf{Q}\mathbf{x}(kT) + \mathbf{u}^*(kT)\mathbf{R}\mathbf{u}(kT)]$$

but is modified to include a cross term involving $\mathbf{x}(kT)$ and $\mathbf{u}(kT)$. Also, matrices $\mathbf{Q}$ and $\mathbf{R}$ are modified. In what follows, we shall consider the discretized quadratic optimal control problem by use of a simple example.

Consider the continuous-time system defined by

$$\dot{x}(t) = ax(t) + bu(t) \tag{7–40}$$

where a and b are constants and

$$u(t) = u(kT) \qquad kT \leq t < (k+1)T$$

The performance index to be minimized is

$$J = \frac{1}{2}x^2(NT) + \frac{1}{2}\int_0^{NT}[Qx^2(t) + Ru^2(t)]\, dt \tag{7–41}$$

Let us discretize the system equation and the performance index and formulate the discretized quadratic optimal control problem.

Equation (7–40) may be discretized as follows:

$$x((k+1)T) = G(T)x(kT) + H(T)u(kT)$$

where

$$G(T) = e^{aT}$$

$$H(T) = \int_0^T e^{a(T-\tau)} b\, d\tau = \frac{b}{a}(e^{aT} - 1)$$

or

$$x((k+1)T) = e^{aT}x(kT) + \frac{b}{a}(e^{aT} - 1)u(kT) \tag{7–42}$$

The performance index J given by Eq. (7–41) may be discretized. First, rewrite J as

$$J_1 = \frac{1}{2}x^2(NT) + \frac{1}{2}\sum_{k=0}^{N-1}\int_{kT}^{(k+1)T}[Qx^2(t) + Ru^2(t)]\, dt$$

Noting that the solution $x(t)$ for $kT \leq t < (k+1)T$ can be written as

$$x(t) = e^{a(t-kT)}x(kT) + \int_{kT}^{t} e^{a(t-\tau)}bu(\tau)\, d\tau$$
$$= \xi(t-kT)x(kT) + \eta(t-kT)u(kT)$$

where

$$\xi(t-kT) = e^{a(t-kT)}$$

$$\eta(t-kT) = \int_{kT}^{t}\xi(t-\tau)b\, d\tau = \frac{b}{a}[e^{a(t-kT)} - 1]$$

the performance index J_1 can be written as follows:

$$J_1 = \frac{1}{2}x^2(NT) + \frac{1}{2}\sum_{k=0}^{N-1}\int_{kT}^{(k+1)T}\{Q[\xi(t-kT)x(kT) + \eta(t-kT)u(kT)]^2 + Ru^2(kT)\}\, dt$$
$$= \frac{1}{2}x^2(NT) + \frac{1}{2}\sum_{k=0}^{N-1}\int_{kT}^{(k+1)T}[Q\xi^2(t-kT)x^2(kT)$$
$$+ 2Q\xi(t-kT)\eta(t-kT)x(kT)u(kT)$$
$$+ Q\eta^2(t-kT)u^2(kT) + Ru^2(kT)]\, dt$$
$$= \frac{1}{2}x^2(NT) + \frac{1}{2}\sum_{k=0}^{N-1}[Q_1x^2(kT) + 2M_1x(kT)u(kT) + R_1u^2(kT)] \tag{7–43}$$

where

$$Q_1 = \int_{kT}^{(k+1)T} Q\xi^2(t-kT)\, dt$$

$$M_1 = \int_{kT}^{(k+1)T} Q\xi(t-kT)\eta(t-kT)\,dt$$

$$R_1 = \int_{kT}^{(k+1)T} [Q\eta^2(t-kT)+R]\,dt$$

Notice that Q_1, M_1, and R_1 may be simplified as follows:

$$Q_1 = \int_{kT}^{(k+1)T} Qe^{2a(t-kT)}\,dt = \frac{Q}{2a}(e^{2aT}-1)$$

$$M_1 = \int_{kT}^{(k+1)T} Qe^{a(t-kT)}\frac{b}{a}[e^{a(t-kT)}-1]\,dt = \frac{bQ}{2a^2}(e^{aT}-1)^2$$

$$R_1 = \int_{kT}^{(k+1)T}\left[Q\left\{\frac{b}{a}[e^{a(t-kT)}-1]\right\}^2 + R\right]dt$$

$$= \frac{b^2Q}{2a^3}[(e^{aT}-3)(e^{aT}-1)+2aT]+RT$$

Summarizing, the present discretized quadratic optimal control problem may be stated as follows. Given the discretized system equation

$$x((k+1)T) = G(T)x(kT) + H(T)u(kT)$$

where

$$G(T) = e^{aT} \qquad \text{and} \qquad H(T) = \frac{b}{a}(e^{aT}-1)$$

find the optimal control sequence $u(0)$, $u(T)$, . . . , $u((N-1)T)$ such that the following performance index is minimized:

$$J_1 = \frac{1}{2}x^2(NT) + \frac{1}{2}\sum_{k=0}^{N-1}[Q_1x^2(kT) + 2M_1x(kT)u(kT) + R_1u^2(kT)]$$

Such a performance index including a cross term involving $x(kT)$ and $u(kT)$ can be modified to a form that does not include a cross term, and the solution to the discretized quadratic optimal control problem can then be obtained in a manner similar to that for the quadratic optimal control problem presented earlier in this section. This subject is presented in the following.

Performance index including a cross term involving x(*k*) and u(*k*). Next, we shall consider the quadratic optimal control problem where the system is as given by Eq. (7–1), which was

$$\mathbf{x}(k+1) = \mathbf{G}\mathbf{x}(k) + \mathbf{H}\mathbf{u}(k), \qquad \mathbf{x}(0) = \mathbf{c}$$

and the performance index is given by

$$J = \frac{1}{2}\mathbf{x}^*(N)\mathbf{S}\mathbf{x}(N) + \frac{1}{2}\sum_{k=0}^{N-1}[\mathbf{x}^*(k)\mathbf{Q}\mathbf{x}(k) + 2\mathbf{x}^*(k)\mathbf{M}\mathbf{u}(k) + \mathbf{u}^*(k)\mathbf{R}\mathbf{u}(k)] \tag{7–44}$$

where $\mathbf{Q}$ and $\mathbf{S}$ are $n \times n$ positive definite or positive semidefinite Hermitian matrices, $\mathbf{R}$ is an $r \times r$ positive definite Hermitian matrix, and $\mathbf{M}$ is an $n \times r$ matrix such that matrix

$$\begin{bmatrix} \mathbf{Q} & \mathbf{M} \\ \mathbf{M}^* & \mathbf{R} \end{bmatrix}$$

is positive definite. This means that

$$\begin{aligned}[\mathbf{x}^*(k) \quad \mathbf{u}^*(k)]\begin{bmatrix} \mathbf{Q} & \mathbf{M} \\ \mathbf{M}^* & \mathbf{R} \end{bmatrix}\begin{bmatrix} \mathbf{x}(k) \\ \mathbf{u}(k) \end{bmatrix} &= \mathbf{x}^*(k)\mathbf{Q}\mathbf{x}(k) + \mathbf{x}^*(k)\mathbf{M}\mathbf{u}(k) + \mathbf{u}^*(k)\mathbf{M}^*\mathbf{x}(k) + \mathbf{u}^*(k)\mathbf{R}\mathbf{u}(k) \\ &= \mathbf{x}^*(k)\mathbf{Q}\mathbf{x}(k) + 2\mathbf{x}^*(k)\mathbf{M}\mathbf{u}(k) + \mathbf{u}^*(k)\mathbf{R}\mathbf{u}(k)\end{aligned}$$

is positive definite. Note that the performance index J given by Eq. (7–44) includes a cross term involving $\mathbf{x}(k)$ and $\mathbf{u}(k)$.

In order to obtain the optimal control vector $\mathbf{u}(k)$, let us define

$$\hat{\mathbf{Q}} = \mathbf{Q} - \mathbf{M}\mathbf{R}^{-1}\mathbf{M}^* \tag{7–45}$$

and eliminate $\mathbf{Q}$ from the performance index J. Then Eq. (7–44) becomes

$$\begin{aligned} J &= \frac{1}{2}\mathbf{x}^*(N)\mathbf{S}\mathbf{x}(N) + \frac{1}{2}\sum_{k=0}^{N-1}\{\mathbf{x}^*(k)[\hat{\mathbf{Q}} + \mathbf{M}\mathbf{R}^{-1}\mathbf{M}^*]\mathbf{x}(k) \\ &\quad + 2\mathbf{x}^*(k)\mathbf{M}\mathbf{u}(k) + \mathbf{u}^*(k)\mathbf{R}\mathbf{u}(k)\} \\ &= \frac{1}{2}\mathbf{x}^*(N)\mathbf{S}\mathbf{x}(N) + \frac{1}{2}\sum_{k=0}^{N-1}[\mathbf{x}^*(k)\hat{\mathbf{Q}}\mathbf{x}(k) + \mathbf{x}^*(k)\mathbf{M}\mathbf{R}^{-1}\mathbf{M}^*\mathbf{x}(k) \\ &\quad + 2\mathbf{x}^*(k)\mathbf{M}\mathbf{u}(k) + \mathbf{u}^*(k)\mathbf{R}\mathbf{u}(k)] \\ &= \frac{1}{2}\mathbf{x}^*(N)\mathbf{S}\mathbf{x}(N) + \frac{1}{2}\sum_{k=0}^{N-1}\{\mathbf{x}^*(k)\hat{\mathbf{Q}}\mathbf{x}(k) \\ &\quad + [\mathbf{x}^*(k)\mathbf{M}\mathbf{R}^{-1} + \mathbf{u}^*(k)]\mathbf{R}[\mathbf{R}^{-1}\mathbf{M}^*\mathbf{x}(k) + \mathbf{u}(k)]\} \end{aligned} \tag{7–46}$$

Define

$$\mathbf{v}(k) = \mathbf{R}^{-1}\mathbf{M}^*\mathbf{x}(k) + \mathbf{u}(k) \tag{7–47}$$

Then Eq. (7–46) can be written as follows:

$$J = \frac{1}{2}\mathbf{x}^*(N)\mathbf{S}\mathbf{x}(N) + \frac{1}{2}\sum_{k=0}^{N-1}[\mathbf{x}^*(k)\hat{\mathbf{Q}}\mathbf{x}(k) + \mathbf{v}^*(k)\mathbf{R}\mathbf{v}(k)] \tag{7–48}$$

Notice that Eq. (7–48) no longer involves the cross term. We have effectively eliminated the cross term involving $\mathbf{x}(k)$ and $\mathbf{u}(k)$.

By substituting Eq. (7–47) into the system equation, Eq. (7–1), we obtain

$$\begin{aligned}\mathbf{x}(k+1) &= \mathbf{G}\mathbf{x}(k) + \mathbf{H}[\mathbf{v}(k) - \mathbf{R}^{-1}\mathbf{M}^*\mathbf{x}(k)]\\ &= (\mathbf{G} - \mathbf{H}\mathbf{R}^{-1}\mathbf{M}^*)\mathbf{x}(k) + \mathbf{H}\mathbf{v}(k)\\ &= \hat{\mathbf{G}}\mathbf{x}(k) + \mathbf{H}\mathbf{v}(k) \end{aligned} \tag{7–49}$$

where

$$\hat{\mathbf{G}} = \mathbf{G} - \mathbf{H}\mathbf{R}^{-1}\mathbf{M}^* \tag{7–50}$$

Note that the quadratic optimal control of the system given by Eq. (7–1) with the performance index given by Eq. (7–44) is equivalent to the quadratic optimal control of the system given by Eq. (7–49) with the performance index given by Eq. (7–48). Hence, the optimal control vector $\mathbf{v}(k)$ that minimizes the performance index given by Eq. (7–48) can be given as follows. Referring to Eq. (7–26), (7–28), or (7–30), we have

$$\mathbf{v}(k) = -\mathbf{R}^{-1}\mathbf{H}^*(\hat{\mathbf{G}}^*)^{-1}[\hat{\mathbf{P}}(k) - \hat{\mathbf{Q}}]\mathbf{x}(k) \tag{7–51}$$

or

$$\mathbf{v}(k) = -\mathbf{R}^{-1}\mathbf{H}^*[\hat{\mathbf{P}}^{-1}(k+1) + \mathbf{H}\mathbf{R}^{-1}\mathbf{H}^*]^{-1}\hat{\mathbf{G}}\mathbf{x}(k) \tag{7–52}$$

or

$$\mathbf{v}(k) = -[\mathbf{R} + \mathbf{H}^*\hat{\mathbf{P}}(k+1)\mathbf{H}]^{-1}\mathbf{H}^*\hat{\mathbf{P}}(k+1)\hat{\mathbf{G}}\mathbf{x}(k) \tag{7–53}$$

where $\hat{\mathbf{P}}(k)$ is a modified version of Eq. (7–23), or

$$\hat{\mathbf{P}}(k) = \hat{\mathbf{Q}} + \hat{\mathbf{G}}^*\hat{\mathbf{P}}(k+1)[\mathbf{I} + \mathbf{H}\mathbf{R}^{-1}\mathbf{H}^*\hat{\mathbf{P}}(k+1)]^{-1}\hat{\mathbf{G}}, \qquad \hat{\mathbf{P}}(N) = \mathbf{S} \tag{7–54}$$

The optimal control vector $\mathbf{u}(k)$ can then be given by

$$\mathbf{u}(k) = \mathbf{v}(k) - \mathbf{R}^{-1}\mathbf{M}^*\mathbf{x}(k) \tag{7–55}$$

where $\mathbf{v}(k)$ is given by Eq. (7–51), (7–52), or (7–53). Whichever expression for $\mathbf{v}(k)$ is used, Eq. (7–55) may be reduced to the following form:

$$\mathbf{u}(k) = -[\mathbf{R} + \mathbf{H}^*\hat{\mathbf{P}}(k+1)\mathbf{H}]^{-1}[\mathbf{H}^*\hat{\mathbf{P}}(k+1)\mathbf{G} + \mathbf{M}^*]\mathbf{x}(k) \tag{7–56}$$

(See Prob. A-7–2 for details.)

Example 7–2.

Consider the continuous-time control system

$$\dot{x}(t) = -x(t) + u(t), \qquad x(0) = 1 \tag{7–57}$$

where

$$u(t) = u(kT) \qquad kT \le t < (k+1)T$$

and the performance index

$$J = \frac{1}{2}x^2(NT) + \frac{1}{2}\int_0^{NT} [x^2(t) + u^2(t)]\, dt \tag{7-58}$$

where $T = 1$ sec and $N = 10$. Discretize the system equation and the performance index. Then, determine the optimal control sequence $u(kT)$ for $k = 0, 1, 2, \ldots, 9$; this will be the control sequence for which the performance index is minimum. Also, obtain the minimum value of J.

Referring to Eqs. (7–40) and (7–42), the discretized system equation is

$$x((k+1)T) = e^{aT}x(kT) + \frac{b}{a}(e^{aT} - 1)u(kT)$$

where $a = -1$, $b = 1$, and $T = 1$. Thus, the system equation becomes

$$x(k+1) = 0.3679x(k) + 0.6321u(k), \qquad x(0) = 1 \tag{7-59}$$

The discretized performance index becomes

$$J_1 = \frac{1}{2}x^2(N) + \frac{1}{2}\sum_{k=0}^{N-1} [Q_1 x^2(k) + 2M_1 x(k)u(k) + R_1 u^2(k)] \tag{7-60}$$

where Q_1, M_1, and R_1 are given by

$$Q_1 = \frac{Q}{2a}(e^{2aT} - 1) = \frac{1}{-2}(e^{-2} - 1) = 0.4323$$

$$M_1 = \frac{bQ}{2a^2}(e^{aT} - 1)^2 = \frac{1}{2}(e^{-1} - 1)^2 = 0.1998$$

$$R_1 = \frac{b^2Q}{2a^3}[(e^{aT} - 3)(e^{aT} - 1) + 2aT] + RT$$

$$= \frac{1}{2(-1)^3}[(e^{-1} - 3)(e^{-1} - 1) - 2] + 1 = 1.1681$$

Thus, the performance index given by Eq. (7–60) can be written as follows:

$$J_1 = \frac{1}{2}x^2(10) + \frac{1}{2}\sum_{k=0}^{9} [0.4323x^2(k) + 0.3996x(k)u(k) + 1.1681u^2(k)] \tag{7-61}$$

Therefore, our problem becomes as follows. Given the system equation, Eq. (7–59), find the optimal control sequence $u(k)$ where $k = 0, 1, 2, \ldots, 9$ such that the performance index given by Eq. (7–61) is minimum.

Now, comparing Eqs. (7–44) and (7–61), we have

$$S = 1, \qquad Q = 0.4323, \qquad M = 0.1998, \qquad R = 1.1681$$

Notice that

$$\begin{bmatrix} Q & M \\ M^* & R \end{bmatrix} = \begin{bmatrix} 0.4323 & 0.1998 \\ 0.1998 & 1.1681 \end{bmatrix}$$

is positive definite. The next step is to modify J_1 as given by Eq. (7–61) into the form given by Eq. (7–48). Since $\hat{Q} = Q - MR^{-1}M^* = 0.3981$,

$$J_1 = \frac{1}{2}x^2(10) + \frac{1}{2}\sum_{k=0}^{9}[0.3981x^2(k) + 1.1681v^2(k)] \tag{7–62}$$

The optimal control signal $u(k)$ can be found from Eq. (7–55):

$$u(k) = v(k) - R^{-1}M^*x(k)$$

which can be written in the form given by Eq. (7–56):

$$u(k) = -[R + H^*\hat{P}(k+1)H]^{-1}[H^*\hat{P}(k+1)G + M^*]x(k) \tag{7–63}$$

where

$$G = 0.3679 \qquad \text{and} \qquad H = 0.6321$$

Equation (7–63) can be rewritten as follows:

$$\begin{aligned} u(k) &= -[1.1681 + 0.3996\hat{P}(k+1)]^{-1}[0.2325\hat{P}(k+1) + 0.1998]x(k) \\ &= -\frac{0.2325\hat{P}(k+1) + 0.1998}{1.1681 + 0.3996\hat{P}(k+1)}x(k) = -K(k)x(k) \end{aligned} \tag{7–64}$$

where

$$K(k) = \frac{0.2325\hat{P}(k+1) + 0.1998}{1.1681 + 0.3996\hat{P}(k+1)} \tag{7–65}$$

Note that $\hat{P}(k)$ is as given by Eq. (7–54), or

$$\hat{P}(k) = \hat{Q} + \hat{G}^*\hat{P}(k+1)[1 + HR^{-1}H^*\hat{P}(k+1)]^{-1}\hat{G} \tag{7–66}$$

where $\hat{P}(N) = \hat{P}(10) = 1$ and

$$\hat{Q} = Q - MR^{-1}M^* = 0.3981$$

$$\hat{G} = G - HR^{-1}M^* = 0.3679 - 0.1081 = 0.2598$$

Equation (7–66) can be simplified into the following form:

$$\hat{P}(k) = 0.3981 + \frac{0.06750\hat{P}(k+1)}{1 + 0.3421\hat{P}(k+1)} \tag{7–67}$$

We shall now compute $\hat{P}(k)$ with the boundary condition $\hat{P}(10) = 1$. Using Eq. (7–67), we find $\hat{P}(k)$ backward from $k = 9$ to $k = 0$. The results are tabulated in Table 7–1. Using the values of $\hat{P}(k)$ just obtained, we compute $K(k)$ from Eq. (7–65). The results are also shown in Table 7–1. Next, we compute $x(k)$. By substituting Eq. (7–64) into Eq. (7–59) and eliminating $u(k)$ from these two equations, we obtain

$$x(k+1) = \frac{0.3035}{1.1681 + 0.3996\hat{P}(k+1)}x(k), \qquad x(0) = 1 \tag{7–68}$$

Starting with $x(0) = 1$, the values of $x(k)$ can be computed from Eq. (7–68) by using the values of $\hat{P}(k)$ already obtained. The computed results are shown in Table 7–1. Once we get the values of $K(k)$ and $x(k)$, the optimal control signal $u(k)$ can be obtained from Eq. (7–64), or

$$u(k) = -K(k)x(k)$$

TABLE 7–1 VALUES OF $\hat{P}(k)$, $K(k)$, $x(k)$, AND $u(k)$ FOR THE SYSTEM CONSIDERED IN EXAMPLE 7–2

k	$\hat{P}(k)$	$K(k)$	$x(k)$	$u(k)$
0	0.4230	0.2230	1.0000	−0.2230
1	0.4230	0.2230	0.2270	−0.05062
2	0.4230	0.2230	0.05152	−0.01149
3	0.4230	0.2230	0.01169	−0.002607
4	0.4230	0.2230	0.002653	−0.0005916
5	0.4230	0.2230	0.000602	−0.0001342
6	0.4230	0.2230	0.0001366	−0.0000304
7	0.4231	0.2231	$\cong 0$	$\cong 0$
8	0.4243	0.2257	$\cong 0$	$\cong 0$
9	0.4484	0.2758	$\cong 0$	$\cong 0$

The results are also shown in Table 7–1.

Finally, referring to Eq. (7–36), the minimum value of J_1 can be obtained as follows:

$$J_{1,\min} = \frac{1}{2}\hat{P}(0)x^2(0) = \frac{1}{2} \times 0.4230 \times 1^2 = 0.2115$$

7–3 STEADY-STATE QUADRATIC OPTIMAL CONTROL

Consider the control system defined by Eq. (7–1):

$$\mathbf{x}(k+1) = \mathbf{G}\mathbf{x}(k) + \mathbf{H}\mathbf{u}(k) \tag{7–69}$$

We have seen that when the control process is finite (when N is finite), the feedback gain matrix $\mathbf{K}(k)$ becomes a time-varying matrix.

Let us now consider the quadratic optimal control problem where the process continues without bound, or where $N = \infty$ (that is, where the process is an infinite-stage process). As N approaches infinity, the optimal control solution becomes a steady-state solution and the time-varying gain matrix $\mathbf{K}(k)$ becomes a constant gain matrix. Such a constant gain matrix $\mathbf{K}(k)$ is called a steady-state gain matrix and is written as $\mathbf{K}$.

For $N = \infty$, the performance index may be modified to

$$J = \frac{1}{2}\sum_{k=0}^{\infty}[\mathbf{x}^*(k)\mathbf{Q}\mathbf{x}(k) + \mathbf{u}^*(k)\mathbf{R}\mathbf{u}(k)] \tag{7–70}$$

The term $\frac{1}{2}\mathbf{x}^*(N)\mathbf{S}\mathbf{x}(N)$, which appeared in Eq. (7–2), is not included in this representation of J. This is because if the optimal control system is stable so that the value of J converges to a constant, $\mathbf{x}(\infty)$ becomes zero and $\frac{1}{2}\mathbf{x}^*(\infty)\mathbf{S}\mathbf{x}(\infty) = 0$. To test whether the optimal control system based on a quadratic performance index is stable, we substitute

$$\mathbf{u}(k) = -\mathbf{K}\mathbf{x}(k)$$

into Eq. (7–69). Then the system equation becomes

$$\mathbf{x}(k+1) = (\mathbf{G} - \mathbf{H}\mathbf{K})\mathbf{x}(k) \tag{7–71}$$

Kalman has shown that the requirement that $\mathbf{G} - \mathbf{HK}$ be a stable matrix (that is, that the system be asymptotically stable) is equivalent to the requirement that

$$\text{rank}[(\mathbf{Q}^{1/2})^* \vdots \mathbf{G}^*(\mathbf{Q}^{1/2})^* \vdots \cdots \vdots (\mathbf{G}^*)^{n-1}(\mathbf{Q}^{1/2})^*] = n \tag{7–72}$$

where $\mathbf{Q}^{1/2}$ is defined by

$$(\mathbf{Q}^{1/2})^*\mathbf{Q}^{1/2} = \mathbf{Q} \tag{7–73}$$

This rank condition may conveniently be applied to check whether the matrix $\mathbf{G} - \mathbf{HK}$ is a stable one. It is important to note that although practical optimal control systems that minimize quadratic performance indexes are almost always asymptotically stable, in an academic situation a system that is not asymptotically stable may minimize a given performance index and could therefore be called "optimal." This occurs when matrix $\mathbf{Q}$ is positive semidefinite and state matrix $\mathbf{G}$ is such that the rank of the matrix in Eq. (7–72) is less than n. (See Prob. A-7–5.)

Let us now define the steady-state matrix $\mathbf{P}(k)$ as $\mathbf{P}$. Referring to Eq. (7–23), matrix $\mathbf{P}$ can be determined as follows:

$$\begin{aligned}\mathbf{P} &= \mathbf{Q} + \mathbf{G}^*\mathbf{P}(\mathbf{I} + \mathbf{H}\mathbf{R}^{-1}\mathbf{H}^*\mathbf{P})^{-1}\mathbf{G} \\ &= \mathbf{Q} + \mathbf{G}^*(\mathbf{P}^{-1} + \mathbf{H}\mathbf{R}^{-1}\mathbf{H}^*)^{-1}\mathbf{G}\end{aligned} \tag{7–74}$$

Clearly, matrix $\mathbf{P}$ is determined by matrices $\mathbf{G}$, $\mathbf{H}$, $\mathbf{Q}$, and $\mathbf{R}$. A slightly different expression for $\mathbf{P}$ can be derived from Eq. (7–24):

$$\mathbf{P} = \mathbf{Q} + \mathbf{G}^*\mathbf{P}\mathbf{G} - \mathbf{G}^*\mathbf{P}\mathbf{H}(\mathbf{R} + \mathbf{H}^*\mathbf{P}\mathbf{H})^{-1}\mathbf{H}^*\mathbf{P}\mathbf{G} \tag{7–75}$$

The steady-state gain matrix $\mathbf{K}$ can be obtained in terms of $\mathbf{P}$ as follows. From Eq. (7–27),

$$\mathbf{K} = \mathbf{R}^{-1}\mathbf{H}^*(\mathbf{G}^*)^{-1}(\mathbf{P} - \mathbf{Q}) \tag{7–76}$$

From Eq. (7–29),

$$\mathbf{K} = \mathbf{R}^{-1}\mathbf{H}^*(\mathbf{P}^{-1} + \mathbf{H}\mathbf{R}^{-1}\mathbf{H}^*)^{-1}\mathbf{G} \tag{7–77}$$

Still another expression for $\mathbf{K}$ is possible. From Eq. (7–31),

$$\mathbf{K} = (\mathbf{R} + \mathbf{H}^*\mathbf{P}\mathbf{H})^{-1}\mathbf{H}^*\mathbf{P}\mathbf{G} \tag{7–78}$$

The optimal control law for steady-state operation is given by

$$\mathbf{u}(k) = -\mathbf{K}\mathbf{x}(k)$$

If, for example, Eq. (7–78) is substituted into this last equation, we obtain

$$\mathbf{u}(k) = -(\mathbf{R} + \mathbf{H}^*\mathbf{P}\mathbf{H})^{-1}\mathbf{H}^*\mathbf{P}\mathbf{G}\mathbf{x}(k) \tag{7–79}$$

and the control system becomes an optimal regulator system:

$$\begin{aligned}\mathbf{x}(k+1) &= [\mathbf{G} - \mathbf{H}(\mathbf{R} + \mathbf{H}^*\mathbf{P}\mathbf{H})^{-1}\mathbf{H}^*\mathbf{P}\mathbf{G}]\mathbf{x}(k) \\ &= (\mathbf{I} + \mathbf{H}\mathbf{R}^{-1}\mathbf{H}^*\mathbf{P})^{-1}\mathbf{G}\mathbf{x}(k)\end{aligned} \tag{7–80}$$

where we have used the matrix inversion lemma,

$$(\mathbf{A} + \mathbf{B}\mathbf{C})^{-1} = \mathbf{A}^{-1} - \mathbf{A}^{-1}\mathbf{B}(\mathbf{I} + \mathbf{C}\mathbf{A}^{-1}\mathbf{B})^{-1}\mathbf{C}\mathbf{A}$$

with $\mathbf{A} = \mathbf{I}$, $\mathbf{B} = \mathbf{H}$, and $\mathbf{C} = \mathbf{R}^{-1}\mathbf{H}^*\mathbf{P}$. (Refer to the Appendix.)

Note that if the rank condition given by Eq. (7–72) is satisfied, then the regulator system given by Eq. (7–80) is asymptotically stable. The performance index J associated with the steady-state optimal control law can be obtained from Eq. (7–36) by substituting $\mathbf{P}$ for $\mathbf{P}(0)$:

$$J_{\min} = \frac{1}{2}\mathbf{x}^*(0)\mathbf{P}\mathbf{x}(0) \tag{7–81}$$

In many practical systems, instead of using a time-varying gain matrix $\mathbf{K}(k)$ we approximate such a gain matrix by the constant gain matrix $\mathbf{K}$. Deviations from the optimal performance due to the approximation will appear only near the end of the control process.

Steady-state Riccati equation. In implementing the steady-state (or time-invariant) optimal controller we require the steady-state solution of the Riccati equation. There are several ways to obtain the steady-state solution.

One way to solve the steady-state Riccati equation given by Eq. (7–75),

$$\mathbf{P} = \mathbf{Q} + \mathbf{G}^*\mathbf{P}\mathbf{G} - \mathbf{G}^*\mathbf{P}\mathbf{H}(\mathbf{R} + \mathbf{H}^*\mathbf{P}\mathbf{H})^{-1}\mathbf{H}^*\mathbf{P}\mathbf{G}$$

is to start with the following non-steady-state Riccati equation, which was given by Eq. (7–24):

$$\begin{aligned}\mathbf{P}(k) = \mathbf{Q} &+ \mathbf{G}^*\mathbf{P}(k+1)\mathbf{G} \\ &- \mathbf{G}^*\mathbf{P}(k+1)\mathbf{H}[\mathbf{R} + \mathbf{H}^*\mathbf{P}(k+1)\mathbf{H}]^{-1}\mathbf{H}^*\mathbf{P}(k+1)\mathbf{G}\end{aligned} \tag{7–82}$$

By reversing the direction of time we may modify Eq. (7–82) to read

$$\mathbf{P}(k+1) = \mathbf{Q} + \mathbf{G}^*\mathbf{P}(k)\mathbf{G} - \mathbf{G}^*\mathbf{P}(k)\mathbf{H}[\mathbf{R} + \mathbf{H}^*\mathbf{P}(k)\mathbf{H}]^{-1}\mathbf{H}^*\mathbf{P}(k)\mathbf{G} \tag{7–83}$$

and begin the solution with $\mathbf{P}(0) = \mathbf{0}$ and iterate the equation until a stationary solution is obtained. In computing the numerical solution, it is important to note that matrix $\mathbf{P}$ is either a Hermitian or a real symmetric matrix and is positive definite.

In what follows we present two additional approaches to the solution of steady-state quadratic optimal control problems. One approach is based on the generalized eigenvalue, and the other is based on the Liapunov method.

Preliminary materials in the generalized eigenvalue approach. Referring to Eq. (7–17) we have

$$\mathbf{G}\mathbf{x}(k) = \mathbf{x}(k+1) + \mathbf{H}\mathbf{R}^{-1}\mathbf{H}^*\boldsymbol{\lambda}(k+1) \tag{7–84}$$

Also, referring to Eq. (7–14), we get

$$-\mathbf{Q}\mathbf{x}(k) + \boldsymbol{\lambda}(k) = \mathbf{G}^*\boldsymbol{\lambda}(k+1) \tag{7–85}$$

Equations (7–84) and (7–85) may be combined into a vector-matrix equation:

$$\begin{bmatrix} \mathbf{G} & \mathbf{0} \\ -\mathbf{Q} & \mathbf{I}_n \end{bmatrix} \begin{bmatrix} \mathbf{x}(k) \\ \boldsymbol{\lambda}(k) \end{bmatrix} = \begin{bmatrix} \mathbf{I}_n & \mathbf{H}\mathbf{R}^{-1}\mathbf{H}^* \\ \mathbf{0} & \mathbf{G}^* \end{bmatrix} \begin{bmatrix} \mathbf{x}(k+1) \\ \boldsymbol{\lambda}(k+1) \end{bmatrix} \tag{7–86}$$

The z transform of Eq. (7–86) gives

$$\begin{bmatrix} \mathbf{G} & \mathbf{0} \\ -\mathbf{Q} & \mathbf{I}_n \end{bmatrix} \begin{bmatrix} \mathbf{X}(z) \\ \boldsymbol{\Lambda}(z) \end{bmatrix} = \begin{bmatrix} \mathbf{I}_n & \mathbf{H}\mathbf{R}^{-1}\mathbf{H}^* \\ \mathbf{0} & \mathbf{G}^* \end{bmatrix} \begin{bmatrix} z\mathbf{X}(z) \\ z\boldsymbol{\Lambda}(z) \end{bmatrix}$$

which can be written as

$$\begin{bmatrix} \mathbf{G} & \mathbf{0} \\ -\mathbf{Q} & \mathbf{I}_n \end{bmatrix} \boldsymbol{\xi} = z \begin{bmatrix} \mathbf{I}_n & \mathbf{H}\mathbf{R}^{-1}\mathbf{H}^* \\ \mathbf{0} & \mathbf{G}^* \end{bmatrix} \boldsymbol{\xi} \tag{7–87}$$

where

$$\boldsymbol{\xi} = \begin{bmatrix} \mathbf{X}(z) \\ \boldsymbol{\Lambda}(z) \end{bmatrix}$$

which is a $2n$ vector. Define

$$\mathbf{X} = \begin{bmatrix} \mathbf{G} & \mathbf{0} \\ -\mathbf{Q} & \mathbf{I}_n \end{bmatrix}, \qquad \mathbf{Y} = \begin{bmatrix} \mathbf{I}_n & \mathbf{H}\mathbf{R}^{-1}\mathbf{H}^* \\ \mathbf{0} & \mathbf{G}^* \end{bmatrix}$$

Then Eq. (7–87) can be written as follows:

$$\mathbf{X}\boldsymbol{\xi} = z\mathbf{Y}\boldsymbol{\xi} \tag{7–88}$$

Let us replace z by μ_i and $\boldsymbol{\xi}$ by $\boldsymbol{\xi}_i$. Then Eq. (7–88) may be written as

$$\mathbf{X}\boldsymbol{\xi}_i = \mu_i \mathbf{Y}\boldsymbol{\xi}_i \tag{7–89}$$

The μ_i's and the $\boldsymbol{\xi}_i$'s that satisfy Eq. (7–89) are called *generalized eigenvalues* and *generalized eigenvectors*, respectively. Note that if the $2n \times 2n$ matrix $\mathbf{Y}$ is nonsingular, then Eq. (7–89) may be written as

$$\mathbf{Y}^{-1}\mathbf{X}\boldsymbol{\xi}_i = \mu_i \boldsymbol{\xi}_i$$

or

$$\mathbf{M}\boldsymbol{\xi}_i = \mu_i \boldsymbol{\xi}_i$$

where $\mathbf{M} = \mathbf{Y}^{-1}\mathbf{X}$. Thus, if matrix $\mathbf{Y}$ is nonsingular, then the generalized eigenvalues of matrices $\mathbf{X}$ and $\mathbf{Y}$ are identical to the eigenvalues of matrix $\mathbf{M}$.

Define

$$\boldsymbol{\xi}_i = \begin{bmatrix} \mathbf{v}_i \\ \mathbf{w}_i \end{bmatrix}$$

Then Eq. (7–89) becomes

$$\begin{bmatrix} \mathbf{G} & \mathbf{0} \\ -\mathbf{Q} & \mathbf{I}_n \end{bmatrix} \begin{bmatrix} \mathbf{v}_i \\ \mathbf{w}_i \end{bmatrix} = \mu_i \begin{bmatrix} \mathbf{I}_n & \mathbf{H}\mathbf{R}^{-1}\mathbf{H}^* \\ \mathbf{0} & \mathbf{G}^* \end{bmatrix} \begin{bmatrix} \mathbf{v}_i \\ \mathbf{w}_i \end{bmatrix} \tag{7–90}$$

which can be written as

$$\begin{bmatrix} \mathbf{G} - \mu_i \mathbf{I}_n & -\mu_i \mathbf{H}\mathbf{R}^{-1}\mathbf{H}^* \\ -\mathbf{Q} & \mathbf{I}_n - \mu_i \mathbf{G}^* \end{bmatrix} \begin{bmatrix} \mathbf{v}_i \\ \mathbf{w}_i \end{bmatrix} = \begin{bmatrix} \mathbf{0} \\ \mathbf{0} \end{bmatrix}$$

from which we obtain

$$(\mathbf{G} - \mu_i \mathbf{I}_n)\mathbf{v}_i - \mu_i \mathbf{H}\mathbf{R}^{-1}\mathbf{H}^*\mathbf{w}_i = \mathbf{0} \tag{7–91}$$

$$-\mathbf{Q}\mathbf{v}_i + (\mathbf{I}_n - \mu_i \mathbf{G}^*)\mathbf{w}_i = \mathbf{0} \tag{7–92}$$

Assume that Eq. (7–90) holds true. Let us see whether there exists a generalized eigenvector

$$\begin{bmatrix} \mathbf{g}_i \\ \mathbf{h}_i \end{bmatrix}$$

which is a $2n$ vector, such that

$$\begin{bmatrix} \mathbf{G} & \mathbf{0} \\ -\mathbf{Q} & \mathbf{I}_n \end{bmatrix}^* \begin{bmatrix} \mathbf{g}_i \\ \mathbf{h}_i \end{bmatrix} = \frac{1}{\mu_i} \begin{bmatrix} \mathbf{I}_n & \mathbf{H}\mathbf{R}^{-1}\mathbf{H}^* \\ \mathbf{0} & \mathbf{G}^* \end{bmatrix}^* \begin{bmatrix} \mathbf{g}_i \\ \mathbf{h}_i \end{bmatrix} \tag{7–93}$$

where we assume $\mu_i \neq 0$, or

$$\begin{bmatrix} \mathbf{G}^* & -\mathbf{Q} \\ \mathbf{0} & \mathbf{I}_n \end{bmatrix} \begin{bmatrix} \mathbf{g}_i \\ \mathbf{h}_i \end{bmatrix} = \frac{1}{\mu_i} \begin{bmatrix} \mathbf{I}_n & \mathbf{0} \\ \mathbf{H}\mathbf{R}^{-1}\mathbf{H}^* & \mathbf{G} \end{bmatrix} \begin{bmatrix} \mathbf{g}_i \\ \mathbf{h}_i \end{bmatrix} \tag{7–94}$$

Now Eq. (7–94) can be expanded as follows:

$$\mathbf{G}^*\mathbf{g}_i - \mathbf{Q}\mathbf{h}_i = \frac{1}{\mu_i}\mathbf{g}_i$$

$$\mathbf{I}_n \mathbf{h}_i = \frac{1}{\mu_i}(\mathbf{H}\mathbf{R}^{-1}\mathbf{H}^*\mathbf{g}_i + \mathbf{G}\mathbf{h}_i)$$

which may be written as

$$-\mu_i \mathbf{Q}\mathbf{h}_i - (\mathbf{I}_n - \mu_i \mathbf{G}^*)\mathbf{g}_i = \mathbf{0} \tag{7–95}$$

$$(\mu_i \mathbf{I}_n - \mathbf{G})\mathbf{h}_i - \mathbf{H}\mathbf{R}^{-1}\mathbf{H}^*\mathbf{g}_i = \mathbf{0} \tag{7–96}$$

Comparing Eqs. (7–92) and (7–91) with Eqs. (7–95) and (7–96), respectively, we notice that if we choose

$$\mathbf{g}_i = -\mathbf{w}_i, \qquad \mathbf{h}_i = \frac{1}{\mu_i}\mathbf{v}_i$$

then these two sets of equations become identical to each other. Hence, there exist vectors $\mathbf{g}_i$ and $\mathbf{h}_i$ that satisfy Eq. (7–93):

$$\begin{bmatrix} \mathbf{G} & \mathbf{0} \\ -\mathbf{Q} & \mathbf{I}_n \end{bmatrix}^* \begin{bmatrix} -\mathbf{w}_i \\ \dfrac{1}{\mu_i}\mathbf{v}_i \end{bmatrix} = \frac{1}{\mu_i}\begin{bmatrix} \mathbf{I}_n & \mathbf{H}\mathbf{R}^{-1}\mathbf{H}^* \\ \mathbf{0} & \mathbf{G}^* \end{bmatrix}^* \begin{bmatrix} -\mathbf{w}_i \\ \dfrac{1}{\mu_i}\mathbf{v}_i \end{bmatrix} \tag{7–97}$$

Let us summarize what we have stated. We have shown that if

$$\mathbf{X}\boldsymbol{\xi}_i = \mu_i \mathbf{Y}\boldsymbol{\xi}_i$$

where

$$\boldsymbol{\xi}_i = \begin{bmatrix} \mathbf{v}_i \\ \mathbf{w}_i \end{bmatrix}$$

then we have

$$\mathbf{X}^*\boldsymbol{\eta}_i = \frac{1}{\mu_i}\mathbf{Y}^*\boldsymbol{\eta}_i$$

where

$$\boldsymbol{\eta}_i = \begin{bmatrix} -\mathbf{w}_i \\ \dfrac{1}{\mu_i}\mathbf{v}_i \end{bmatrix}$$

Note that the set of matrices $\mathbf{X}^*$ and $\mathbf{Y}^*$ has the same generalized eigenvalues as the set of matrices $\mathbf{X}$ and $\mathbf{Y}$. If μ_i^{-1} is a generalized eigenvalue of the set of matrices $\mathbf{X}^*$ and $\mathbf{Y}^*$, then μ_i^{-1} is also a generalized eigenvalue of the set of matrices $\mathbf{X}$ and $\mathbf{Y}$.

From the preceding analysis, a set of matrices $\mathbf{X}$ and $\mathbf{Y}$ possesses generalized eigenvalues μ_i and μ_i^{-1} (or the generalized eigenvalues of a set of matrices $\mathbf{X}$ and $\mathbf{Y}$ are $\mu_1, \mu_2, \ldots, \mu_n$ and $\mu_1^{-1}, \mu_2^{-1}, \ldots, \mu_n^{-1}$) provided none of the μ_i's is zero. (If matrix $\mathbf{G}$ is nonsingular, then none of the μ_i's is zero.) In the case where matrix $\mathbf{G}$ is nonsingular, the eigenvalues of $\mathbf{Z} = \mathbf{Y}^{-1}\mathbf{X}$ can also be shown to be $\mu_1, \mu_2, \ldots, \mu_n$ and $\mu_1^{-1}, \mu_2^{-1}, \ldots, \mu_n^{-1}$. (See Prob. A-7–4.) If matrix $\mathbf{G}$ is singular, then the generalized eigenvalues involve $\mu_i = 0$; that is, the generalized eigenvalues of a set of matrices $\mathbf{X}$ and $\mathbf{Y}$ are $\mu_1, \mu_2, \ldots, \mu_k, 0, 0, \ldots, 0, \mu_1^{-1}, \mu_2^{-1}, \ldots, \mu_k^{-1}$.

Solution of the steady-state Riccati equation by the generalized eigenvalue approach. Referring to Eq. (7–90), we have

$$\begin{bmatrix} \mathbf{G} & \mathbf{0} \\ -\mathbf{Q} & \mathbf{I}_n \end{bmatrix} \begin{bmatrix} \mathbf{v}_i \\ \mathbf{w}_i \end{bmatrix} = \mu_i \begin{bmatrix} \mathbf{I}_n & \mathbf{HR}^{-1}\mathbf{H}^* \\ \mathbf{0} & \mathbf{G}^* \end{bmatrix} \begin{bmatrix} \mathbf{v}_i \\ \mathbf{w}_i \end{bmatrix} \tag{7–98}$$

It has been shown that if matrix $\mathbf{G}$ is nonsingular, then the generalized eigenvalues are

$$\mu_1, \mu_2, \ldots, \mu_n, \mu_1^{-1}, \mu_2^{-1}, \ldots, \mu_n^{-1}$$

Let us assume that the μ_i's are distinct. Let us also assume that $|\mu_i| < 1$. Then $\mu_1, \mu_2, \ldots, \mu_n$ are inside the unit circle and $\mu_1^{-1}, \mu_2^{-1}, \ldots, \mu_n^{-1}$ are outside the unit circle.

It is a well-known fact that if $\mu_1, \mu_2, \ldots, \mu_n$ are distinct, then $\mathbf{v}_1, \mathbf{v}_2, \ldots, \mathbf{v}_n$ are linearly independent. If the μ_i's involve multiple eigenvalues, then $\mathbf{v}_1, \mathbf{v}_2, \ldots, \mathbf{v}_n$ may or may not be linearly independent. If matrix $\mathbf{G}$ is singular, then the generalized eigenvalues are

$$\mu_1, \mu_2, \ldots, \mu_k, 0, 0, \ldots, 0, \mu_1^{-1}, \mu_2^{-1}, \ldots, \mu_k^{-1}$$

In this case, for $\mu_1, \mu_2, \ldots, \mu_k, \mu_{k+1} = \mu_{k+2} = \cdots \mu_n = 0$, the vectors $\mathbf{v}_1, \mathbf{v}_2, \ldots, \mathbf{v}_n$ may or may not be linearly independent.

In the following analysis, we shall limit our discussion to the case where $\mathbf{v}_1, \mathbf{v}_2, \ldots, \mathbf{v}_n$ are linearly independent, regardless of whether matrix $\mathbf{G}$ is nonsingular or singular. Let us assume that n-vector $\mathbf{w}_i$ is related to n-vector $\mathbf{v}_i$ by some $n \times n$ Hermitian matrix $\mathbf{P}$ as follows:

$$\mathbf{w}_i = \mathbf{P}\mathbf{v}_i$$

Then, Eq. (7–98) becomes

$$\begin{bmatrix} \mathbf{G} & \mathbf{0} \\ -\mathbf{Q} & \mathbf{I}_n \end{bmatrix} \begin{bmatrix} \mathbf{v}_i \\ \mathbf{P}\mathbf{v}_i \end{bmatrix} = \mu_i \begin{bmatrix} \mathbf{I}_n & \mathbf{HR}^{-1}\mathbf{H}^* \\ \mathbf{0} & \mathbf{G}^* \end{bmatrix} \begin{bmatrix} \mathbf{v}_i \\ \mathbf{P}\mathbf{v}_i \end{bmatrix} \tag{7–99}$$

which may be expanded as follows:

$$\mathbf{G}\mathbf{v}_i = \mu_i (\mathbf{I}_n + \mathbf{HR}^{-1}\mathbf{H}^*\mathbf{P})\mathbf{v}_i \tag{7–100}$$

$$-\mathbf{Q}\mathbf{v}_i + \mathbf{P}\mathbf{v}_i = \mu_i \mathbf{G}^*\mathbf{P}\mathbf{v}_i \tag{7–101}$$

From Eq. (7–100) we have

$$\mu_i \mathbf{v}_i = (\mathbf{I}_n + \mathbf{HR}^{-1}\mathbf{H}^*\mathbf{P})^{-1}\mathbf{G}\mathbf{v}_i$$

By premultiplying this last equation by $\mathbf{G}^*\mathbf{P}$, we obtain

$$\mu_i \mathbf{G}^*\mathbf{P}\mathbf{v}_i = \mathbf{G}^*\mathbf{P}(\mathbf{I}_n + \mathbf{HR}^{-1}\mathbf{H}^*\mathbf{P})^{-1}\mathbf{G}\mathbf{v}_i \tag{7–102}$$

From Eqs. (7–101) and (7–102), we obtain

$$\mathbf{G}^*\mathbf{P}(\mathbf{I}_n + \mathbf{HR}^{-1}\mathbf{H}^*\mathbf{P})^{-1}\mathbf{G}\mathbf{v}_i = -\mathbf{Q}\mathbf{v}_i + \mathbf{P}\mathbf{v}_i$$

or

$$[\mathbf{P} - \mathbf{Q} - \mathbf{G}^*\mathbf{P}(\mathbf{I}_n + \mathbf{HR}^{-1}\mathbf{H}^*\mathbf{P})^{-1}\mathbf{G}]\mathbf{v}_i = \mathbf{0}$$

Since we have assumed that $\mathbf{v}_1, \mathbf{v}_2, \ldots, \mathbf{v}_n$ are linearly independent, we must have

$$\mathbf{P} - \mathbf{Q} - \mathbf{G}^*\mathbf{P}(\mathbf{I}_n + \mathbf{H}\mathbf{R}^{-1}\mathbf{H}^*\mathbf{P})^{-1}\mathbf{G} = \mathbf{0}$$

or

$$\mathbf{P} = \mathbf{Q} + \mathbf{G}^*\mathbf{P}(\mathbf{I}_n + \mathbf{H}\mathbf{R}^{-1}\mathbf{H}^*\mathbf{P})^{-1}\mathbf{G} \tag{7–103}$$

Since

$$(\mathbf{I}_n + \mathbf{H}\mathbf{R}^{-1}\mathbf{H}^*\mathbf{P})^{-1} = \mathbf{I}_n - \mathbf{H}(\mathbf{R} + \mathbf{H}^*\mathbf{P}\mathbf{H})^{-1}\mathbf{H}^*\mathbf{P}$$

Eq. (7–103) can be written as follows:

$$\begin{aligned}\mathbf{P} &= \mathbf{Q} + \mathbf{G}^*\mathbf{P}[\mathbf{I}_n - \mathbf{H}(\mathbf{R} + \mathbf{H}^*\mathbf{P}\mathbf{H})^{-1}\mathbf{H}^*\mathbf{P}]\mathbf{G} \\ &= \mathbf{Q} + \mathbf{G}^*\mathbf{P}\mathbf{G} - \mathbf{G}^*\mathbf{P}\mathbf{H}(\mathbf{R} + \mathbf{H}^*\mathbf{P}\mathbf{H})^{-1}\mathbf{H}^*\mathbf{P}\mathbf{G} \end{aligned} \tag{7–104}$$

Equation (7–104) is the steady-state Riccati equation given by Eq. (7–75). Hence, if

$$\mathbf{w}_i = \mathbf{P}\mathbf{v}_i \tag{7–105}$$

then matrix $\mathbf{P}$ satisfies Eq. (7–75). For $i = 1, 2, \ldots, n$, Eq. (7–105) can be written as follows:

$$[\mathbf{w}_1 \vdots \mathbf{w}_2 \vdots \cdots \vdots \mathbf{w}_n] = [\mathbf{P}\mathbf{v}_1 \vdots \mathbf{P}\mathbf{v}_2 \vdots \cdots \vdots \mathbf{P}\mathbf{v}_n] = \mathbf{P}[\mathbf{v}_1 \vdots \mathbf{v}_2 \vdots \cdots \vdots \mathbf{v}_n]$$

Consequently, if $\mathbf{v}_1, \mathbf{v}_2, \ldots, \mathbf{v}_n$ are linearly independent, then the inverse of $[\mathbf{v}_1 \vdots \mathbf{v}_2 \vdots \cdots \vdots \mathbf{v}_n]$ exists and matrix $\mathbf{P}$ can be given by

$$\mathbf{P} = [\mathbf{w}_1 \vdots \mathbf{w}_2 \vdots \cdots \vdots \mathbf{w}_n][\mathbf{v}_1 \vdots \mathbf{v}_2 \vdots \cdots \vdots \mathbf{v}_n]^{-1} \tag{7–106}$$

If matrix $\mathbf{P}$ as given by Eq. (7–106) is positive definite, then it is the solution of the steady-state Riccati equation, since matrix $\mathbf{P}$ satisfies Eq. (7–75).

It is important to note that we have limited our discussion here to the case where vectors $\mathbf{v}_1, \mathbf{v}_2, \ldots, \mathbf{v}_n$ are linearly independent.

Example 7–3.

Consider the system

$$\begin{bmatrix} x_1(k+1) \\ x_2(k+1) \end{bmatrix} = \begin{bmatrix} 1 & 1 \\ 1 & 0 \end{bmatrix} \begin{bmatrix} x_1(k) \\ x_2(k) \end{bmatrix} + \begin{bmatrix} 1 \\ 0 \end{bmatrix} u(k) \tag{7–107}$$

where

$$\begin{bmatrix} x_1(0) \\ x_2(0) \end{bmatrix} = \begin{bmatrix} 1 \\ 0 \end{bmatrix}$$

and the performance index is given by

$$J = \frac{1}{2} \sum_{k=0}^{\infty} [\mathbf{x}^*(k)\mathbf{Q}\mathbf{x}(k) + u^*(k)Ru(k)] \tag{7–108}$$

where

$$\mathbf{Q} = \begin{bmatrix} 1 & 0 \\ 0 & 1 \end{bmatrix}, \qquad R = 1$$

Determine the optimal control law to minimize the performance index. Also, determine the minimum value of J.

From Eq. (7–107), we have

$$\mathbf{G} = \begin{bmatrix} 1 & 1 \\ 1 & 0 \end{bmatrix}, \qquad \mathbf{H} = \begin{bmatrix} 1 \\ 0 \end{bmatrix}$$

Notice that matrix $\mathbf{G}$ is nonsingular. Referring to Eq. (7–90), we have

$$\begin{bmatrix} \mathbf{G} & \mathbf{0} \\ -\mathbf{Q} & \mathbf{I}_n \end{bmatrix} \begin{bmatrix} \mathbf{v}_i \\ \mathbf{w}_i \end{bmatrix} = \mu_i \begin{bmatrix} \mathbf{I}_n & \mathbf{H}\mathbf{R}^{-1}\mathbf{H}^* \\ \mathbf{0} & \mathbf{G}^* \end{bmatrix} \begin{bmatrix} \mathbf{v}_i \\ \mathbf{w}_i \end{bmatrix}$$

or

$$\begin{bmatrix} 1 & 1 & 0 & 0 \\ 1 & 0 & 0 & 0 \\ -1 & 0 & 1 & 0 \\ 0 & -1 & 0 & 1 \end{bmatrix} \begin{bmatrix} v_{i1} \\ v_{i2} \\ w_{i1} \\ w_{i2} \end{bmatrix} = \mu_i \begin{bmatrix} 1 & 0 & 1 & 0 \\ 0 & 1 & 0 & 0 \\ 0 & 0 & 1 & 1 \\ 0 & 0 & 1 & 0 \end{bmatrix} \begin{bmatrix} v_{i1} \\ v_{i2} \\ w_{i1} \\ w_{i2} \end{bmatrix}$$

which can be rewritten as

$$\begin{bmatrix} 1-\mu_i & 1 & -\mu_i & 0 \\ 1 & -\mu_i & 0 & 0 \\ -1 & 0 & 1-\mu_i & -\mu_i \\ 0 & -1 & -\mu_i & 1 \end{bmatrix} \begin{bmatrix} v_{i1} \\ v_{i2} \\ w_{i1} \\ w_{i2} \end{bmatrix} = \begin{bmatrix} 0 \\ 0 \\ 0 \\ 0 \end{bmatrix} \tag{7–109}$$

The generalized eigenvalues can be found from

$$\begin{vmatrix} 1-\mu_i & 1 & -\mu_i & 0 \\ 1 & -\mu_i & 0 & 0 \\ -1 & 0 & 1-\mu_i & -\mu_i \\ 0 & -1 & -\mu_i & 1 \end{vmatrix} = 0$$

The Laplace's expansion by the minors gives

$$\begin{vmatrix} 1-\mu_i & 1 \\ 1 & -\mu_i \end{vmatrix} \begin{vmatrix} 1-\mu_i & -\mu_i \\ -\mu_i & 1 \end{vmatrix} + \begin{vmatrix} 1 & -\mu_i \\ -1 & 0 \end{vmatrix} \begin{vmatrix} -\mu_i & 0 \\ -\mu_i & 1 \end{vmatrix} - \begin{vmatrix} 1 & -\mu_i \\ 0 & -1 \end{vmatrix} \begin{vmatrix} -\mu_i & 0 \\ 1-\mu_i & -\mu_i \end{vmatrix}$$
$$= -\mu_i^4 + 5\mu_i^2 - 1$$

Hence we have four values of μ_i:

$$\mu_i = 0.45685, \qquad -0.45685, \qquad 2.18890, \qquad -2.18890$$

Next, let us obtain the generalized eigenvectors for $\mu_1 = 0.45685$ and $\mu_2 = -0.45685$. For $\mu_1 = 0.45685$, Eq. (7–109) becomes

$$\begin{bmatrix} 0.54315 & 1 & -0.45685 & 0 \\ 1 & -0.45685 & 0 & 0 \\ -1 & 0 & 0.54315 & -0.45685 \\ 0 & -1 & -0.45685 & 1 \end{bmatrix} \begin{bmatrix} v_{11} \\ v_{12} \\ w_{11} \\ w_{12} \end{bmatrix} = \begin{bmatrix} 0 \\ 0 \\ 0 \\ 0 \end{bmatrix}$$

Solving this last equation, we obtain

$$\begin{bmatrix} v_{11} \\ v_{12} \end{bmatrix} = \begin{bmatrix} 0.45685a \\ a \end{bmatrix}, \qquad \begin{bmatrix} w_{11} \\ w_{12} \end{bmatrix} = \begin{bmatrix} 2.73205a \\ 2.24814a \end{bmatrix}$$

where a is a constant. By arbitrarily choosing $a = 1$, we obtain

$$\begin{bmatrix} v_{11} \\ v_{12} \end{bmatrix} = \begin{bmatrix} 0.45685 \\ 1 \end{bmatrix}, \qquad \begin{bmatrix} w_{11} \\ w_{12} \end{bmatrix} = \begin{bmatrix} 2.73205 \\ 2.24814 \end{bmatrix}$$

For $\mu_2 = -0.45685$, Eq. (7–109) becomes

$$\begin{bmatrix} 1.45685 & 1 & 0.45685 & 0 \\ 1 & 0.45685 & 0 & 0 \\ -1 & 0 & 1.45685 & 0.45685 \\ 0 & -1 & 0.45685 & 1 \end{bmatrix} \begin{bmatrix} v_{21} \\ v_{22} \\ w_{21} \\ w_{22} \end{bmatrix} = \begin{bmatrix} 0 \\ 0 \\ 0 \\ 0 \end{bmatrix}$$

Solving this last equation, we obtain

$$\begin{bmatrix} v_{21} \\ v_{22} \end{bmatrix} = \begin{bmatrix} -0.45685b \\ b \end{bmatrix}, \qquad \begin{bmatrix} w_{21} \\ w_{22} \end{bmatrix} = \begin{bmatrix} -0.73205b \\ 1.33444b \end{bmatrix}$$

where b is a constant. By arbitrarily choosing $b = 1$, we have

$$\begin{bmatrix} v_{21} \\ v_{22} \end{bmatrix} = \begin{bmatrix} -0.45685 \\ 1 \end{bmatrix}, \qquad \begin{bmatrix} w_{21} \\ w_{22} \end{bmatrix} = \begin{bmatrix} -0.73205 \\ 1.33444 \end{bmatrix}$$

Hence, matrix $\mathbf{P}$ is obtained from

$$\begin{aligned} \mathbf{P} &= [\mathbf{w}_1 \quad \mathbf{w}_2][\mathbf{v}_1 \quad \mathbf{v}_2]^{-1} \\ &= \begin{bmatrix} 2.73205 & -0.73205 \\ 2.24814 & 1.33444 \end{bmatrix} \begin{bmatrix} 0.45685 & -0.45685 \\ 1 & 1 \end{bmatrix}^{-1} \\ &= \begin{bmatrix} 2.73205 & -0.73205 \\ 2.24814 & 1.33444 \end{bmatrix} \begin{bmatrix} 1.09445 & 0.5 \\ -1.09445 & 0.5 \end{bmatrix} \\ &= \begin{bmatrix} 3.7913 & 1.0000 \\ 1.0000 & 1.7913 \end{bmatrix} \end{aligned} \tag{7–110}$$

Notice that matrix $\mathbf{P}$ is positive definite. Equation (7–110) gives the required solution of the steady-state Riccati equation, Eq. (7–75).

Referring to Eq. (7–79), we have

$$\begin{aligned} u(k) &= -(R + \mathbf{H}^*\mathbf{P}\mathbf{H})^{-1}\mathbf{H}^*\mathbf{P}\mathbf{G}\,\mathbf{x}(k) \\ &= -\left\{1 + [1 \quad 0] \begin{bmatrix} 3.7913 & 1.0000 \\ 1.0000 & 1.7913 \end{bmatrix} \begin{bmatrix} 1 \\ 0 \end{bmatrix}\right\}^{-1} \\ &\quad \cdot [1 \quad 0] \begin{bmatrix} 3.7913 & 1.0000 \\ 1.0000 & 1.7913 \end{bmatrix} \begin{bmatrix} 1 & 1 \\ 1 & 0 \end{bmatrix} \mathbf{x}(k) \\ &= -(1 + 3.7913)^{-1}[4.7913 \quad 3.7913]\mathbf{x}(k) \\ &= -[1 \quad 0.7913]\mathbf{x}(k) \end{aligned} \tag{7–111}$$

Equation (7–111) gives the optimal control law.

The closed-loop system becomes

$$\mathbf{x}(k+1) = \mathbf{G}\mathbf{x}(k) + \mathbf{H}u(k)$$
$$= \begin{bmatrix} 1 & 1 \\ 1 & 0 \end{bmatrix} \mathbf{x}(k) - \begin{bmatrix} 1 \\ 0 \end{bmatrix} [1 \quad 0.7913]\, \mathbf{x}(k)$$
$$= \begin{bmatrix} 0 & 0.2087 \\ 1 & 0 \end{bmatrix} \mathbf{x}(k) \tag{7–112}$$

Equation (7–112) gives the optimal closed-loop operation for the system. The closed-loop poles are at $\mu_1 = 0.45685$ and $\mu_2 = -0.45685$.

The minimum value of J is obtained from Eq. (7–81), as follows:

$$J_{\min} = \frac{1}{2} \mathbf{x}^*(0)\mathbf{P}\mathbf{x}(0) = \frac{1}{2} [1 \quad 0] \begin{bmatrix} 3.7913 & 1.0000 \\ 1.0000 & 1.7913 \end{bmatrix} \begin{bmatrix} 1 \\ 0 \end{bmatrix}$$
$$= 1.8956$$

Liapunov approach to the solution of the steady-state quadratic optimal regulator problem. In what follows we shall present the Liapunov approach to the parameter optimization problem and the steady-state quadratic optimal regulator problem. As we shall see, there is a direct relationship between Liapunov functions and quadratic performance indexes.

Let us consider the system

$$\mathbf{x}(k+1) = \mathbf{G}\mathbf{x}(k) \tag{7–113}$$

where matrix $\mathbf{G}$ involves one or more adjustable parameters and all eigenvalues of $\mathbf{G}$ lie inside the unit circle, or the origin $\mathbf{x} = \mathbf{0}$ is asymptotically stable. Let us assume that we desire to minimize the following performance index by adjusting the parameter (or parameters):

$$J = \frac{1}{2} \sum_{k=0}^{\infty} \mathbf{x}^*(k)\mathbf{Q}\mathbf{x}(k) \tag{7–114}$$

where $\mathbf{Q}$ is a positive definite or positive semidefinite Hermitian (or real symmetric) matrix. We shall show that a Liapunov function can be utilized for solving this problem.

For the system of Eq. (7–113) a Liapunov function may be given by

$$V(\mathbf{x}(k)) = \mathbf{x}^*(k)\mathbf{P}\mathbf{x}(k)$$

where $\mathbf{P}$ is a positive definite Hermitian (or real symmetric) matrix and

$$\Delta V(\mathbf{x}(k)) = V(\mathbf{x}(k+1)) - V(\mathbf{x}(k))$$
$$= \mathbf{x}^*(k+1)\mathbf{P}\mathbf{x}(k+1) - \mathbf{x}^*(k)\mathbf{P}\mathbf{x}(k)$$

Let us set

$$\mathbf{x}^*(k)\mathbf{Q}\mathbf{x}(k) = -[\mathbf{x}^*(k+1)\mathbf{P}\mathbf{x}(k+1) - \mathbf{x}^*(k)\mathbf{P}\mathbf{x}(k)] \tag{7–115}$$

Notice that Eq. (7–115) can be rewritten as follows:

$$\mathbf{x}^*(k)\mathbf{Q}\mathbf{x}(k) = -\{[\mathbf{G}\mathbf{x}(k)]^*\mathbf{P}[\mathbf{G}\mathbf{x}(k)] - \mathbf{x}^*(k)\mathbf{P}\mathbf{x}(k)\}$$
$$= -\mathbf{x}^*(k)[\mathbf{G}^*\mathbf{P}\mathbf{G} - \mathbf{P}]\mathbf{x}(k)$$

By the second method of Liapunov, we know that for a given matrix **Q** there exists a matrix **P**, since matrix **G** is stable, such that

$$\mathbf{G}^*\mathbf{P}\mathbf{G} - \mathbf{P} = -\mathbf{Q} \tag{7–116}$$

Hence we can determine the elements of **P** from this equation.

The performance index J can be evaluated as follows:

$$J = \frac{1}{2}\sum_{k=0}^{\infty} \mathbf{x}^*(k)\mathbf{Q}\mathbf{x}(k) = \frac{1}{2}\sum_{k=0}^{\infty} [\mathbf{x}^*(k)\mathbf{P}\mathbf{x}(k) - \mathbf{x}^*(k+1)\mathbf{P}\mathbf{x}(k+1)]$$
$$= \frac{1}{2}\mathbf{x}^*(0)\mathbf{P}\mathbf{x}(0) \tag{7–117}$$

where **P** is a function of the adjustable parameter(s). In obtaining Eq. (7–117) we used the condition that $\mathbf{x}(\infty) \to \mathbf{0}$, since all eigenvalues of **G** lie inside the unit circle. Thus, the performance index J can be obtained in terms of the initial condition $\mathbf{x}(0)$ and matrix **P**, which is related to matrices **G** and **Q** by Eq. (7–116). Minimization of the performance index J can be accomplished by minimizing $\mathbf{x}^*(0)\mathbf{P}\mathbf{x}(0)$ with respect to the parameter in question.

It is important to note that the optimal value of the parameter depends, in general, upon the initial condition $\mathbf{x}(0)$. However, if $\mathbf{x}(0)$ involves only one nonzero component, for example, if $x_1(0) \neq 0$ and the other initial conditions are zero, then the optimal value of the parameter does not depend on the numerical value of $x_1(0)$. (See Example 7–4.)

Liapunov approach to the solution of the steady-state quadratic optimal control problem. We shall now consider the optimal control problem where, given the system equation

$$\mathbf{x}(k+1) = \mathbf{G}\mathbf{x}(k) + \mathbf{H}\mathbf{u}(k) \tag{7–118}$$

we wish to determine the matrix **K** of the optimal control law

$$\mathbf{u}(k) = -\mathbf{K}\mathbf{x}(k) \tag{7–119}$$

such that the performance index

$$J = \frac{1}{2}\sum_{k=0}^{\infty} [\mathbf{x}^*(k)\mathbf{Q}\mathbf{x}(k) + \mathbf{u}^*(k)\mathbf{R}\mathbf{u}(k)] \tag{7–120}$$

is minimized, where **Q** is a positive definite or positive semidefinite Hermitian (or real symmetric) matrix and **R** is a positive definite Hermitian (or real symmetric) matrix.

Substituting Eq. (7–119) into Eq. (7–118), we obtain

$$\mathbf{x}(k+1) = \mathbf{G}\mathbf{x}(k) - \mathbf{H}\mathbf{K}\mathbf{x}(k) = (\mathbf{G} - \mathbf{H}\mathbf{K})\mathbf{x}(k) \tag{7–121}$$

Substituting Eq. (7–119) into Eq. (7–120) yields

$$\begin{aligned} J &= \frac{1}{2}\sum_{k=0}^{\infty} [\mathbf{x}^*(k)\mathbf{Q}\mathbf{x}(k) + \mathbf{x}^*(k)\mathbf{K}^*\mathbf{R}\mathbf{K}\mathbf{x}(k)] \\ &= \frac{1}{2}\sum_{k=0}^{\infty} \mathbf{x}^*(k)(\mathbf{Q} + \mathbf{K}^*\mathbf{R}\mathbf{K})\mathbf{x}(k) \end{aligned} \tag{7–122}$$

In the following analysis, we assume that the matrix $\mathbf{G} - \mathbf{HK}$ is stable, or that the eigenvalues of $\mathbf{G} - \mathbf{HK}$ lie inside the unit circle. [For the test of the stability of matrix $\mathbf{G} - \mathbf{HK}$, refer to Eq. (7–72).] Then, a Liapunov function exists that is positive definite and whose derivative is negative definite. Following the discussion given in solving the parameter optimization problem, we set

$$\mathbf{x}^*(k)(\mathbf{Q} + \mathbf{K}^*\mathbf{R}\mathbf{K})\mathbf{x}(k) = -[\mathbf{x}^*(k+1)\mathbf{P}\mathbf{x}(k+1) - \mathbf{x}^*(k)\mathbf{P}\mathbf{x}(k)] \tag{7–123}$$

By referring to Eq. (7–121), Eq. (7–123) can be modified to

$$\begin{aligned} &\mathbf{x}^*(k)(\mathbf{Q} + \mathbf{K}^*\mathbf{R}\mathbf{K})\mathbf{x}(k) \\ &\qquad = -[(\mathbf{G} - \mathbf{H}\mathbf{K})\mathbf{x}(k)]^*\mathbf{P}[(\mathbf{G} - \mathbf{H}\mathbf{K})\mathbf{x}(k)] + \mathbf{x}^*(k)\mathbf{P}\mathbf{x}(k) \\ &\qquad = -\mathbf{x}^*(k)[(\mathbf{G} - \mathbf{H}\mathbf{K})^*\mathbf{P}(\mathbf{G} - \mathbf{H}\mathbf{K}) - \mathbf{P}]\mathbf{x}(k) \end{aligned} \tag{7–124}$$

Comparing the two sides of Eq. (7–124) and noting that this equation must hold true for any $\mathbf{x}(k)$, we require that

$$\mathbf{Q} + \mathbf{K}^*\mathbf{R}\mathbf{K} = -(\mathbf{G} - \mathbf{H}\mathbf{K})^*\mathbf{P}(\mathbf{G} - \mathbf{H}\mathbf{K}) + \mathbf{P} \tag{7–125}$$

Note that by the second method of Liapunov, for a stable matrix $\mathbf{G} - \mathbf{HK}$, there exists a positive definite matrix $\mathbf{P}$ which satisfies Eq. (7–125).

Equation (7–125) can be modified as follows:

$$\mathbf{Q} + \mathbf{K}^*\mathbf{R}\mathbf{K} + (\mathbf{G}^* - \mathbf{K}^*\mathbf{H}^*)\mathbf{P}(\mathbf{G} - \mathbf{H}\mathbf{K}) - \mathbf{P} = \mathbf{0}$$

or

$$\mathbf{Q} + \mathbf{G}^*\mathbf{P}\mathbf{G} - \mathbf{P} + \mathbf{K}^*(\mathbf{R} + \mathbf{H}^*\mathbf{P}\mathbf{H})\mathbf{K} - (\mathbf{K}^*\mathbf{H}^*\mathbf{P}\mathbf{G} + \mathbf{G}^*\mathbf{P}\mathbf{H}\mathbf{K}) = \mathbf{0}$$

This last equation can further be modified as follows:

$$\begin{aligned} \mathbf{Q} + \mathbf{G}^*\mathbf{P}\mathbf{G} - \mathbf{P} &+ [(\mathbf{R} + \mathbf{H}^*\mathbf{P}\mathbf{H})^{1/2}\mathbf{K} - (\mathbf{R} + \mathbf{H}^*\mathbf{P}\mathbf{H})^{-1/2}\mathbf{H}^*\mathbf{P}\mathbf{G}]^* \cdot [(\mathbf{R} + \mathbf{H}^*\mathbf{P}\mathbf{H})^{1/2}\mathbf{K} \\ &- (\mathbf{R} + \mathbf{H}^*\mathbf{P}\mathbf{H})^{-1/2}\mathbf{H}^*\mathbf{P}\mathbf{G}] - \mathbf{G}^*\mathbf{P}\mathbf{H}(\mathbf{R} + \mathbf{H}^*\mathbf{P}\mathbf{H})^{-1}\mathbf{H}^*\mathbf{P}\mathbf{G} = \mathbf{0} \end{aligned} \tag{7–126}$$

Minimization of J with respect to $\mathbf{K}$ requires minimization of the left-hand side of Eq. (7–126) with respect to $\mathbf{K}$. (See Prob. A-7–7.) Since

$$\begin{aligned} &[(\mathbf{R} + \mathbf{H}^*\mathbf{P}\mathbf{H})^{1/2}\mathbf{K} - (\mathbf{R} + \mathbf{H}^*\mathbf{P}\mathbf{H})^{-1/2}\mathbf{H}^*\mathbf{P}\mathbf{G}]^*[(\mathbf{R} + \mathbf{H}^*\mathbf{P}\mathbf{H})^{1/2}\mathbf{K} \\ &\qquad\qquad - (\mathbf{R} + \mathbf{H}^*\mathbf{P}\mathbf{H})^{-1/2}\mathbf{H}^*\mathbf{P}\mathbf{G}] \end{aligned}$$

is nonnegative, the miminum occurs when it is zero, or when

$$(\mathbf{R}+\mathbf{H}^*\mathbf{P}\mathbf{H})^{1/2}\mathbf{K}=(\mathbf{R}+\mathbf{H}^*\mathbf{P}\mathbf{H})^{-1/2}\mathbf{H}^*\mathbf{P}\mathbf{G}$$

Hence, we obtain

$$\mathbf{K}=(\mathbf{R}+\mathbf{H}^*\mathbf{P}\mathbf{H})^{-1}\mathbf{H}^*\mathbf{P}\mathbf{G} \tag{7–127}$$

Substitution of Eq. (7–127) into Eq. (7–126) gives

$$\mathbf{P}=\mathbf{Q}+\mathbf{G}^*\mathbf{P}\mathbf{G}-\mathbf{G}^*\mathbf{P}\mathbf{H}(\mathbf{R}+\mathbf{H}^*\mathbf{P}\mathbf{H})^{-1}\mathbf{H}^*\mathbf{P}\mathbf{G} \tag{7–128}$$

Matrix $\mathbf{P}$ must satisfy Eq. (7–128), which is the same as Eq. (7–75). Equation (7–128) can be modified to read

$$\mathbf{P}=\mathbf{Q}+\mathbf{G}^*\mathbf{P}[\mathbf{I}-\mathbf{H}(\mathbf{I}+\mathbf{R}^{-1}\mathbf{H}^*\mathbf{P}\mathbf{H})^{-1}\mathbf{R}^{-1}\mathbf{H}^*\mathbf{P}]\mathbf{G} \tag{7–129}$$

By use of the matrix inversion lemma

$$(\mathbf{I}+\mathbf{H}\mathbf{R}^{-1}\mathbf{H}^*\mathbf{P})^{-1}=\mathbf{I}-\mathbf{H}(\mathbf{I}+\mathbf{R}^{-1}\mathbf{H}^*\mathbf{P}\mathbf{H})^{-1}\mathbf{R}^{-1}\mathbf{H}^*\mathbf{P}$$

Eq. (7–129) may be modified to

$$\mathbf{P}=\mathbf{Q}+\mathbf{G}^*\mathbf{P}(\mathbf{I}+\mathbf{H}\mathbf{R}^{-1}\mathbf{H}^*\mathbf{P})^{-1}\mathbf{G} \tag{7–130}$$

Matrix $\mathbf{P}$ may be determined from Eq. (7–130).

Finally, the minimum value of J can be obtained as follows. Referring to Eq. (7–122) and noting that $\mathbf{x}(\infty)=\mathbf{0}$, we obtain the minimum value of the performance index J as follows:

$$\begin{aligned} J_{\min} &= \frac{1}{2}\sum_{k=0}^{\infty}\mathbf{x}^*(k)(\mathbf{Q}+\mathbf{K}^*\mathbf{R}\mathbf{K})\mathbf{x}(k) \\ &= \frac{1}{2}\sum_{k=0}^{\infty}[\mathbf{x}^*(k)\mathbf{P}\mathbf{x}(k)-\mathbf{x}^*(k+1)\mathbf{P}\mathbf{x}(k+1)] \\ &= \frac{1}{2}\mathbf{x}^*(0)\mathbf{P}\mathbf{x}(0) \end{aligned}$$

where we have used Eq. (7–123).

Example 7–4.

Consider the system

$$\begin{bmatrix} x_1(k+1) \\ x_2(k+1) \end{bmatrix}=\begin{bmatrix} 1 & 1 \\ a & -1 \end{bmatrix}\begin{bmatrix} x_1(k) \\ x_2(k) \end{bmatrix}, \qquad \begin{bmatrix} x_1(0) \\ x_2(0) \end{bmatrix}=\begin{bmatrix} 1 \\ 0 \end{bmatrix}$$

where $-0.25 \le a < 0$. We desire to determine an optimal value of a that will minimize the following performance index:

$$J=\frac{1}{2}\sum_{k=0}^{\infty}\mathbf{x}^*(k)\mathbf{Q}\mathbf{x}(k)$$

where $\mathbf{Q}=\mathbf{I}$.

From Eq. (7–117) the performance index J in terms of the system parameter a is given by

$$J = \frac{1}{2}\mathbf{x}^*(0)\mathbf{P}\mathbf{x}(0)$$

where $\mathbf{P}$ involves the system parameter a. We now determine $\mathbf{P}$ from Eq. (7–116):

$$\mathbf{G}^*\mathbf{P}\mathbf{G} - \mathbf{P} = -\mathbf{Q}$$

or

$$\begin{bmatrix} 1 & a \\ 1 & -1 \end{bmatrix}\begin{bmatrix} p_{11} & p_{12} \\ p_{12} & p_{22} \end{bmatrix}\begin{bmatrix} 1 & 1 \\ a & -1 \end{bmatrix} - \begin{bmatrix} p_{11} & p_{12} \\ p_{12} & p_{22} \end{bmatrix} = -\begin{bmatrix} 1 & 0 \\ 0 & 1 \end{bmatrix}$$

which can be simplified to

$$\begin{bmatrix} 2ap_{12} + a^2p_{22} & p_{11} + (a-2)p_{12} - ap_{22} \\ p_{11} + (a-2)p_{12} - ap_{22} & p_{11} - 2p_{12} \end{bmatrix} = \begin{bmatrix} -1 & 0 \\ 0 & -1 \end{bmatrix}$$

This last equation results in the following three equations:

$$2ap_{12} + a^2p_{22} = -1$$

$$p_{11} + (a-2)p_{12} - ap_{22} = 0$$

$$p_{11} - 2p_{12} = -1$$

Solving these three equations for the p_{ij}'s, we obtain

$$\mathbf{P} = \begin{bmatrix} -\dfrac{1 + 0.5a^2}{a(1 + 0.5a)} & \dfrac{0.5(a-1)}{a(1+0.5a)} \\ \dfrac{0.5(a-1)}{a(1+0.5a)} & -\dfrac{1.5}{a(1+0.5a)} \end{bmatrix}$$

Since $-0.25 \leq a < 0$, $\mathbf{P}$ is positive definite.

The performance index J becomes

$$J = \frac{1}{2}\mathbf{x}^*(0)\mathbf{P}\mathbf{x}(0) = \frac{1}{2}[1 \quad 0]\begin{bmatrix} p_{11} & p_{12} \\ p_{12} & p_{22} \end{bmatrix}\begin{bmatrix} 1 \\ 0 \end{bmatrix} = \frac{1}{2}p_{11}$$

$$= -\frac{1 + 0.5a^2}{2a(1 + 0.5a)}$$

The minimum of J occurs at endpoint $a = -0.25$. Thus, the minimum value of J is found to be

$$J_{\min} = -\frac{1 + 0.5(-0.25)^2}{2(-0.25)(1 - 0.5 \times 0.25)} = 2.3571$$

Example 7–5.

Consider the system discussed in Example 7–3,

$$\begin{bmatrix} x_1(k+1) \\ x_2(k+1) \end{bmatrix} = \begin{bmatrix} 1 & 1 \\ 1 & 0 \end{bmatrix}\begin{bmatrix} x_1(k) \\ x_2(k) \end{bmatrix} + \begin{bmatrix} 1 \\ 0 \end{bmatrix}[u(k)], \qquad \begin{bmatrix} x_1(0) \\ x_2(0) \end{bmatrix} = \begin{bmatrix} 1 \\ 0 \end{bmatrix}$$

and the performance index

$$J = \frac{1}{2} \sum_{k=0}^{\infty} [\mathbf{x}^*(k)\mathbf{Q}\mathbf{x}(k) + u^*(k)Ru(k)]$$

where $\mathbf{Q} = \mathbf{I}$ and $R = 1$. From Eq. (7–127), the optimal control law that minimizes the performance index J is given by

$$u(k) = -\mathbf{K}\mathbf{x}(k) = -(\mathbf{R} + \mathbf{H}^*\mathbf{P}\mathbf{H})^{-1}\mathbf{H}^*\mathbf{P}\mathbf{G}\mathbf{x}(k)$$

where matrix $\mathbf{P}$ may be obtained from Eq. (7–130):

$$\mathbf{P} = \mathbf{Q} + \mathbf{G}^*\mathbf{P}(\mathbf{I} + \mathbf{H}R^{-1}\mathbf{H}^*\mathbf{P})^{-1}\mathbf{G} \tag{7–131}$$

Let us determine matrix $\mathbf{P}$ for the present system.

Noting that

$$\mathbf{G} = \begin{bmatrix} 1 & 1 \\ 1 & 0 \end{bmatrix}$$

which is nonsingular, we may modify Eq. (7–131) to read

$$(\mathbf{P} - \mathbf{Q})\mathbf{G}^{-1}(\mathbf{I} + \mathbf{H}R^{-1}\mathbf{H}^*\mathbf{P}) = \mathbf{G}^*\mathbf{P} \tag{7–132}$$

Matrix $\mathbf{P}$ in this problem is a real symmetric matrix. Hence, Eq. (7–132) can be written as follows:

$$\left\{\begin{bmatrix} p_{11} & p_{12} \\ p_{12} & p_{22} \end{bmatrix} - \begin{bmatrix} 1 & 0 \\ 0 & 1 \end{bmatrix}\right\}\begin{bmatrix} 0 & 1 \\ 1 & -1 \end{bmatrix}\left\{\begin{bmatrix} 1 & 0 \\ 0 & 1 \end{bmatrix} + \begin{bmatrix} 1 \\ 0 \end{bmatrix}[1][1 \quad 0]\begin{bmatrix} p_{11} & p_{12} \\ p_{12} & p_{22} \end{bmatrix}\right\}$$
$$= \begin{bmatrix} 1 & 1 \\ 1 & 0 \end{bmatrix}\begin{bmatrix} p_{11} & p_{12} \\ p_{12} & p_{22} \end{bmatrix}$$

or

$$\begin{bmatrix} p_{12}(1 + p_{11}) & p_{12}^2 + p_{11} - 1 - p_{12} \\ (p_{22} - 1)(1 + p_{11}) & (p_{22} - 1)p_{12} + p_{12} - p_{22} + 1 \end{bmatrix} = \begin{bmatrix} p_{11} + p_{12} & p_{12} + p_{22} \\ p_{11} & p_{12} \end{bmatrix}$$

This last equation yields four scalar equations. Note, however, that only three of them are linearly independent. The four scalar equations are as follows:

$$p_{12}(1 + p_{11}) = p_{11} + p_{12}$$

$$p_{12}^2 + p_{11} - 1 - p_{12} = p_{12} + p_{22}$$

$$(p_{22} - 1)(1 + p_{11}) = p_{11}$$

$$(p_{22} - 1)p_{12} + p_{12} - p_{22} + 1 = p_{12}$$

From the first of these four equations, we obtain

$$p_{12} = 1 \tag{7–133}$$

From the second of the four equations, we have

$$p_{22} = p_{11} - 2 \tag{7–134}$$

Then the third of the four equations gives

$$p_{11}^2 - 3p_{11} - 3 = 0 \tag{7–135}$$

[By substituting Eqs. (7–133) and (7–134) into the last of the four equations, we find that it is always satisfied.] By solving Eq. (7–135), we find

$$p_{11} = 3.7913 \qquad \text{or} \qquad -0.7913$$

Since matrix $\mathbf{P}$ must be positive definite, we choose $p_{11} = 3.7913$. Then,

$$p_{22} = p_{11} - 2 = 3.7913 - 2 = 1.7913$$

Consequently, matrix $\mathbf{P}$ is found to be as follows:

$$\mathbf{P} = \begin{bmatrix} 3.7913 & 1.0000 \\ 1.0000 & 1.7913 \end{bmatrix}$$

7–4 QUADRATIC OPTIMAL CONTROL OF SERVO SYSTEMS

In Sec. 6–7 we discussed the design of servo systems based on the pole placement technique. In this section, we shall discuss quadratic optimal control of servo systems.

Consider the servo system shown in Fig. 7–3. The input $r(k)$ is assumed to be a unit step sequence. The system equations are

$$\mathbf{x}(k+1) = \mathbf{G}\mathbf{x}(k) + \mathbf{H}u(k) \tag{7–136}$$

$$u(k) = K_1 v(k) - \mathbf{K}_2\mathbf{x}(k) \tag{7–137}$$

$$y(k) = \mathbf{C}\mathbf{x}(k) \tag{7–138}$$

$$v(k) = r(k) - y(k) + v(k-1) \tag{7–139}$$

Let us assume that the initital state is

$$\mathbf{x}(0) = \mathbf{0}$$

First, define

$$\mathbf{x}_e(k) = \mathbf{x}(k) - \mathbf{x}(\infty) \tag{7–140}$$

$$u_e(k) = u(k) - u(\infty) \tag{7–141}$$

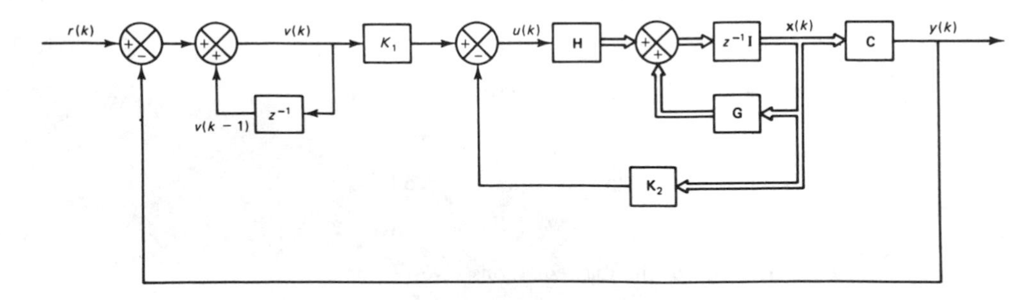

Figure 7–3 Servo system.

Referring to Eqs. (6–212) and (6–213), the system defined by Eqs. (7–136), (7–137), and (7–139) can be described by the equations

$$\begin{bmatrix} \mathbf{x}_e(k+1) \\ u_e(k+1) \end{bmatrix} = \begin{bmatrix} \mathbf{G} & \mathbf{H} \\ \mathbf{0} & 0 \end{bmatrix} \begin{bmatrix} \mathbf{x}_e(k) \\ u_e(k) \end{bmatrix} + \begin{bmatrix} \mathbf{0} \\ 1 \end{bmatrix} w(k) \tag{7–142}$$

where

$$w(k) = [\mathbf{K}_2 - \mathbf{K}_2\mathbf{G} - K_1\mathbf{C}\mathbf{G} \,\vdots\, 1 - \mathbf{K}_2\mathbf{H} - K_1\mathbf{C}\mathbf{H}] \begin{bmatrix} \mathbf{x}_e(k) \\ u_e(k) \end{bmatrix} = -\hat{\mathbf{K}} \begin{bmatrix} \mathbf{x}_e(k) \\ u_e(k) \end{bmatrix} \tag{7–143}$$

and where

$$\hat{\mathbf{K}} = -[\mathbf{K}_2 - \mathbf{K}_2\mathbf{G} - K_1\mathbf{C}\mathbf{G} \,\vdots\, 1 - \mathbf{K}_2\mathbf{H} - K_1\mathbf{C}\mathbf{H}]$$

Let us design the servo system in such a way that the following performance index is minimized:

$$J = \frac{1}{2} \sum_{k=0}^{\infty} [\mathbf{x}_e^*(k)\mathbf{Q}\mathbf{x}_e(k) + w^*(k)Rw(k)] \tag{7–144}$$

Notice that the design problem here becomes that of the determination of the feedforward gain K_1 and the feedback gain matrix $\mathbf{K}_2$ so that the performance index becomes minimum.

As we have seen earlier, the quadratic optimal control law is given by the weighted feedback of the state vector, as given by Eq. (7–143). In a way that follows the Liapunov approach to the steady-state quadratic optimal control problem, we shall determine matrix $\hat{\mathbf{K}}$ by use of the Liapunov equation given by Eq. (7–125).

Equations (7–142) and (7–143) can be written in the form

$$\boldsymbol{\xi}(k+1) = \hat{\mathbf{G}}\boldsymbol{\xi}(k) + \hat{\mathbf{H}}w(k) = (\hat{\mathbf{G}} - \hat{\mathbf{H}}\hat{\mathbf{K}})\boldsymbol{\xi}(k) \tag{7–145}$$

$$w(k) = -\hat{\mathbf{K}}\boldsymbol{\xi}(k) \tag{7–146}$$

where

$$\boldsymbol{\xi}(k) = \begin{bmatrix} \mathbf{x}_e(k) \\ u_e(k) \end{bmatrix}, \qquad \hat{\mathbf{G}} = \begin{bmatrix} \mathbf{G} & \mathbf{H} \\ \mathbf{0} & 0 \end{bmatrix}, \qquad \hat{\mathbf{H}} = \begin{bmatrix} \mathbf{0} \\ 1 \end{bmatrix}$$

and thus Eq. (7–144) becomes

$$J = \frac{1}{2} \sum_{k=0}^{\infty} [\boldsymbol{\xi}^*(k)\hat{\mathbf{Q}}\boldsymbol{\xi}(k) + w^*(k)Rw(k)] \tag{7–147}$$

where

$$\hat{\mathbf{Q}} = \begin{bmatrix} \mathbf{Q} & \mathbf{0} \\ \mathbf{0} & 0 \end{bmatrix} \tag{7–148}$$

We wish to minimize this performance index by the proper choice of matrix $\hat{\mathbf{K}}$. By substituting Eq. (7–146) into Eq. (7–147), we have

$$J = \frac{1}{2}\sum_{k=0}^{\infty} [\boldsymbol{\xi}^*(k)\hat{\mathbf{Q}}\boldsymbol{\xi}(k) + \boldsymbol{\xi}^*(k)\hat{\mathbf{K}}^* R\,\hat{\mathbf{K}}\boldsymbol{\xi}(k)]$$

$$= \frac{1}{2}\sum_{k=0}^{\infty} [\boldsymbol{\xi}^*(k)(\hat{\mathbf{Q}} + \hat{\mathbf{K}}^* R\,\hat{\mathbf{K}})\boldsymbol{\xi}(k)] \tag{7–149}$$

In the following analysis, we assume that the matrix $\hat{\mathbf{G}} - \hat{\mathbf{H}}\hat{\mathbf{K}}$ in Eq. (7–145) is stable. Following the Liapunov approach discussed in Sec. 7–3, we set

$$\boldsymbol{\xi}^*(k)(\hat{\mathbf{Q}} + \hat{\mathbf{K}}^* R\,\hat{\mathbf{K}})\boldsymbol{\xi}(k) = -[\boldsymbol{\xi}^*(k+1)\hat{\mathbf{P}}\boldsymbol{\xi}(k+1) - \boldsymbol{\xi}^*(k)\hat{\mathbf{P}}\boldsymbol{\xi}(k)] \tag{7–150}$$

Note that by the second method of Liapunov, for a stable matrix $\hat{\mathbf{G}} - \hat{\mathbf{H}}\hat{\mathbf{K}}$, there exists a positive definite matrix $\hat{\mathbf{P}}$ which satisfies Eq. (7–150). Equation (7–150) can be rewritten as follows:

$$\boldsymbol{\xi}^*(k)(\hat{\mathbf{Q}} + \hat{\mathbf{K}}^* R\,\hat{\mathbf{K}})\boldsymbol{\xi}(k) = -\boldsymbol{\xi}^*(k)[(\hat{\mathbf{G}} - \hat{\mathbf{H}}\hat{\mathbf{K}})^*\hat{\mathbf{P}}(\hat{\mathbf{G}} - \hat{\mathbf{H}}\hat{\mathbf{K}}) - \hat{\mathbf{P}}]\boldsymbol{\xi}(k) \tag{7–151}$$

Comparing the two sides of Eq. (7–151) and noting that this equation must hold true for any $\boldsymbol{\xi}(k)$, we require that

$$\hat{\mathbf{Q}} + \hat{\mathbf{K}}^* R\,\hat{\mathbf{K}} = -(\hat{\mathbf{G}} - \hat{\mathbf{H}}\hat{\mathbf{K}})^*\hat{\mathbf{P}}(\hat{\mathbf{G}} - \hat{\mathbf{H}}\hat{\mathbf{K}}) + \hat{\mathbf{P}}$$

which can be written as

$$\hat{\mathbf{Q}} + \hat{\mathbf{G}}^*\hat{\mathbf{P}}\hat{\mathbf{G}} - \hat{\mathbf{P}} + \hat{\mathbf{K}}^*(R + \hat{\mathbf{H}}^*\hat{\mathbf{P}}\hat{\mathbf{H}})\hat{\mathbf{K}} - (\hat{\mathbf{K}}^*\hat{\mathbf{H}}^*\hat{\mathbf{P}}\hat{\mathbf{G}} + \hat{\mathbf{G}}^*\hat{\mathbf{P}}\hat{\mathbf{H}}\hat{\mathbf{K}}) = \mathbf{0} \tag{7–152}$$

The optimal feedback gain matrix $\hat{\mathbf{K}}$ must minimize the left-hand side of Eq. (7–152). Since Eq. (7–152) can be written in the form

$$\hat{\mathbf{Q}} + \hat{\mathbf{G}}^*\hat{\mathbf{P}}\hat{\mathbf{G}} - \hat{\mathbf{P}} + [(R + \hat{\mathbf{H}}^*\hat{\mathbf{P}}\hat{\mathbf{H}})^{1/2}\hat{\mathbf{K}} - (R + \hat{\mathbf{H}}^*\hat{\mathbf{P}}\hat{\mathbf{H}})^{-1/2}\hat{\mathbf{H}}^*\hat{\mathbf{P}}\hat{\mathbf{G}}]^* \cdot [(R + \hat{\mathbf{H}}^*\hat{\mathbf{P}}\hat{\mathbf{H}})^{1/2}\hat{\mathbf{K}}$$
$$- (R + \hat{\mathbf{H}}^*\hat{\mathbf{P}}\hat{\mathbf{H}})^{-1/2}\hat{\mathbf{H}}^*\hat{\mathbf{P}}\hat{\mathbf{G}}] - \hat{\mathbf{G}}^*\hat{\mathbf{P}}\hat{\mathbf{H}}(R + \hat{\mathbf{H}}^*\hat{\mathbf{P}}\hat{\mathbf{H}})^{-1}\hat{\mathbf{H}}^*\hat{\mathbf{P}}\hat{\mathbf{G}} = \mathbf{0} \tag{7–153}$$

the minimum of the left-hand side of this equation can be obtained when

$$(R + \hat{\mathbf{H}}^*\hat{\mathbf{P}}\hat{\mathbf{H}})^{1/2}\hat{\mathbf{K}} - (R + \hat{\mathbf{H}}^*\hat{\mathbf{P}}\hat{\mathbf{H}})^{-1/2}\hat{\mathbf{H}}^*\hat{\mathbf{P}}\hat{\mathbf{G}} = \mathbf{0}$$

Thus,

$$\hat{\mathbf{K}} = (R + \hat{\mathbf{H}}^*\hat{\mathbf{P}}\hat{\mathbf{H}})^{-1}\hat{\mathbf{H}}^*\hat{\mathbf{P}}\hat{\mathbf{G}} \tag{7–154}$$

Equation (7–154) gives the optimal feedback gain matrix $\hat{\mathbf{K}}$. By substituting Eq. (7–154) into Eq. (7–153) we obtain the following Riccati equation:

$$\hat{\mathbf{Q}} + \hat{\mathbf{G}}^*\hat{\mathbf{P}}\hat{\mathbf{G}} - \hat{\mathbf{P}} - \hat{\mathbf{G}}^*\hat{\mathbf{P}}\hat{\mathbf{H}}(R + \hat{\mathbf{H}}^*\hat{\mathbf{P}}\hat{\mathbf{H}})^{-1}\hat{\mathbf{H}}^*\hat{\mathbf{P}}\hat{\mathbf{G}} = \mathbf{0} \tag{7–155}$$

which can be rewritten as

$$\begin{bmatrix} \mathbf{Q} & \mathbf{0} \\ \mathbf{0} & 0 \end{bmatrix} + \begin{bmatrix} \mathbf{G}^*\mathbf{P}_{11}\mathbf{G} & \mathbf{G}^*\mathbf{P}_{11}\mathbf{H} \\ \mathbf{H}^*\mathbf{P}_{11}\mathbf{G} & \mathbf{H}^*\mathbf{P}_{11}\mathbf{H} \end{bmatrix} - \begin{bmatrix} \mathbf{P}_{11} & \mathbf{P}_{12} \\ \mathbf{P}_{12}^* & P_{22} \end{bmatrix}$$
$$- \begin{bmatrix} \mathbf{G}^*\mathbf{P}_{12}(R + P_{22})^{-1}\mathbf{P}_{12}^*\mathbf{G} & \mathbf{G}^*\mathbf{P}_{12}(R + P_{22})^{-1}\mathbf{P}_{12}^*\mathbf{H} \\ \mathbf{H}^*\mathbf{P}_{12}(R + P_{22})^{-1}\mathbf{P}_{12}^*\mathbf{G} & \mathbf{H}^*\mathbf{P}_{12}(R + P_{22})^{-1}\mathbf{P}_{12}^*\mathbf{H} \end{bmatrix} = \begin{bmatrix} \mathbf{0} & \mathbf{0} \\ \mathbf{0} & 0 \end{bmatrix}$$

This last equation is equivalent to the following three equations:

$$\mathbf{Q} + \mathbf{G}^*\mathbf{P}_{11}\mathbf{G} - \mathbf{P}_{11} - \mathbf{G}^*\mathbf{P}_{12}(R + P_{22})^{-1}\mathbf{P}_{12}^*\mathbf{G} = \mathbf{0} \tag{7–156}$$

$$\mathbf{G}^*\mathbf{P}_{11}\mathbf{H} - \mathbf{P}_{12} - \mathbf{G}^*\mathbf{P}_{12}(R + P_{22})^{-1}\mathbf{P}_{12}^*\mathbf{H} = \mathbf{0} \tag{7–157}$$

$$\mathbf{H}^*\mathbf{P}_{11}\mathbf{H} - P_{22} - \mathbf{H}^*\mathbf{P}_{12}(R + P_{22})^{-1}\mathbf{P}_{12}^*\mathbf{H} = 0 \tag{7–158}$$

which can be rewritten as follows:

$$\mathbf{P}_{11} = \mathbf{Q} + \mathbf{G}^*[\mathbf{P}_{11} - \mathbf{P}_{12}(R + P_{22})^{-1}\mathbf{P}_{12}^*]\mathbf{G} \tag{7–159}$$

$$\mathbf{P}_{12} = \mathbf{G}^*[\mathbf{P}_{11} - \mathbf{P}_{12}(R + P_{22})^{-1}\mathbf{P}_{12}^*]\mathbf{H} \tag{7–160}$$

$$P_{22} = \mathbf{H}^*[\mathbf{P}_{11} - \mathbf{P}_{12}(R + P_{22})^{-1}\mathbf{P}_{12}^*]\mathbf{H} \tag{7–161}$$

Define

$$\mathbf{P}_{11} - \mathbf{P}_{12}(R + P_{22})^{-1}\mathbf{P}_{12}^* = \mathbf{P} \tag{7–162}$$

Then, Eqs. (7–159), (7–160), and (7–161) become as follows:

$$\mathbf{P}_{11} = \mathbf{Q} + \mathbf{G}^*\mathbf{P}\mathbf{G} \tag{7–163}$$

$$\mathbf{P}_{12} = \mathbf{G}^*\mathbf{P}\mathbf{H} \tag{7–164}$$

$$P_{22} = \mathbf{H}^*\mathbf{P}\mathbf{H} \tag{7–165}$$

Substituting Eqs. (7–163), (7–164), and (7–165) into Eq. (7–162), we obtain

$$\mathbf{Q} + \mathbf{G}^*\mathbf{P}\mathbf{G} - \mathbf{G}^*\mathbf{P}\mathbf{H}(R + \mathbf{H}^*\mathbf{P}\mathbf{H})^{-1}\mathbf{H}^*\mathbf{P}\mathbf{G} = \mathbf{P} \tag{7–166}$$

Notice that matrix $\mathbf{P}$ is the solution of Eq. (7–166), the Riccati equation. Referring to Eq. (7–154), we have

$$w(k) = -\hat{\mathbf{K}}\boldsymbol{\xi}(k) = -(R + \hat{\mathbf{H}}^*\hat{\mathbf{P}}\hat{\mathbf{H}})^{-1}\hat{\mathbf{H}}^*\hat{\mathbf{P}}\hat{\mathbf{G}}\begin{bmatrix}\mathbf{x}_e(k)\\ u_e(k)\end{bmatrix} \tag{7–167}$$

Referring to Eq. (6–219), K_1 and $\mathbf{K}_2$ can be determined from the following equation:

$$[\mathbf{K}_2 \vdots K_1] = \{(R + \hat{\mathbf{H}}^*\hat{\mathbf{P}}\hat{\mathbf{H}})^{-1}\hat{\mathbf{H}}^*\hat{\mathbf{P}}\hat{\mathbf{G}} + [0 \vdots 1]\}\begin{bmatrix}\mathbf{G} - \mathbf{I}_n & \mathbf{H}\\ \mathbf{CG} & \mathbf{CH}\end{bmatrix}^{-1} \tag{7–168}$$

Note that $\mathbf{x}_e(k)$ and $u_e(k)$ are defined as follows [see Eqs. (7–140) and (7–141)]:

$$\mathbf{x}_e(k) = \mathbf{x}(k) - \mathbf{x}(\infty)$$

$$u_e(k) = u(k) - u(\infty)$$

Referring to Eq. (6–210) and noting that we have assumed the initial state to be $\mathbf{x}(0) = \mathbf{0}$ and that from Eq. (7–137) $u(0) = K_1 v(0) = \mathbf{K}_1 r$, we obtain

$$\begin{bmatrix} \mathbf{x}_e(0) \\ u_e(0) - u(0) \end{bmatrix} = -\begin{bmatrix} \mathbf{x}(\infty) \\ u(\infty) \end{bmatrix} = -\begin{bmatrix} \mathbf{I}-\mathbf{G} & -\mathbf{H} \\ -\mathbf{K}_2 + \mathbf{K}_2\mathbf{G} + K_1\mathbf{C}\mathbf{G} & \mathbf{K}_2\mathbf{H} + K_1\mathbf{C}\mathbf{H} \end{bmatrix}^{-1} \begin{bmatrix} \mathbf{0} \\ K_1 r \end{bmatrix}$$

$$= -\left[\begin{array}{c|c} \mathbf{I}-\mathbf{G} & -\mathbf{H} \\ \hline K_1\mathbf{C} & 0 \end{array}\right]^{-1} \left[\begin{array}{c|c} \mathbf{I} & \mathbf{0} \\ \hline -\mathbf{K}_2 - K_1\mathbf{C} & 1 \end{array}\right]^{-1} \left[\begin{array}{c} \mathbf{0} \\ \hline K_1 r \end{array}\right]$$

$$= -\left[\begin{array}{c|c} \mathbf{I}-\mathbf{G} & -\mathbf{H} \\ \hline K_1\mathbf{C} & 0 \end{array}\right]^{-1} \left[\begin{array}{c} \mathbf{0} \\ \hline K_1 r \end{array}\right]$$

or

$$\begin{bmatrix} \mathbf{x}_e(0) \\ u_e(0) \end{bmatrix} = -\begin{bmatrix} (\mathbf{I}-\mathbf{G})^{-1}\mathbf{H}[\mathbf{C}(\mathbf{I}-\mathbf{G})^{-1}\mathbf{H}]^{-1} r \\ -K_1 r + [\mathbf{C}(\mathbf{I}-\mathbf{G})^{-1}\mathbf{H}]^{-1} r \end{bmatrix} \tag{7–169}$$

The minimum value of the performance index can be given as follows:

$$J_{\min} = \frac{1}{2}\boldsymbol{\xi}^*(0)\hat{\mathbf{P}}\boldsymbol{\xi}(0) = \frac{1}{2}[\mathbf{x}_e^*(0) \quad u_e^*(0)] \begin{bmatrix} \mathbf{P}_{11} & \mathbf{P}_{12} \\ \mathbf{P}_{12}^* & P_{22} \end{bmatrix} \begin{bmatrix} \mathbf{x}_e(0) \\ u_e(0) \end{bmatrix} \tag{7–170}$$

where $\mathbf{P}_{11}$, $\mathbf{P}_{12}$, and P_{22} are given by Eqs. (7–163), (7–164), and (7–165), respectively, and $[\mathbf{x}_e^*(0) \quad u_e^*(0)]^*$ is given by Eq. (7–169).

In summary, the optimal control law for the present system is given by Eq. (7–167), and the necessary feedforward gain K_1 and feedback gain matrix $\mathbf{K}_2$ can be determined from Eq. (7–168).

Example 7–6.

Consider the servo system shown in Fig. 7–4. The system equations are

$$x(k+1) = 0.5x(k) + 2u(k) \tag{7–171}$$

$$u(k) = K_1 v(k) - K_2 x(k) \tag{7–172}$$

$$y(k) = x(k) \tag{7–173}$$

$$v(k) = r(k) - y(k) + v(k-1) \tag{7–174}$$

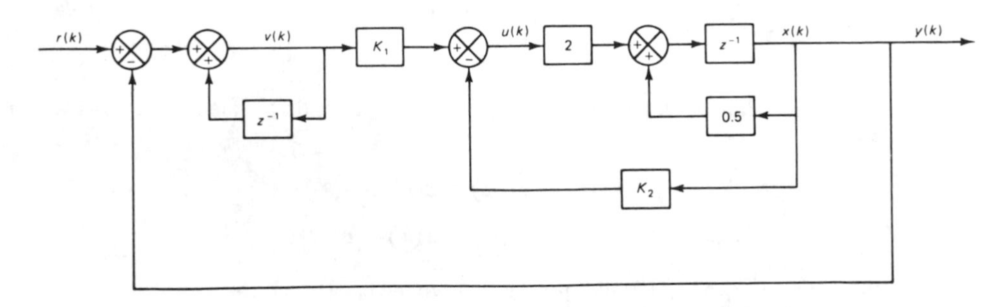

Figure 7–4 Servo system.

The initial state is assumed to be

$$x(0) = 0$$

and the input $r(k)$ is a unit step sequence, or $r(k) = 1$ for $k = 0, 1, 2, \ldots$ Define

$$x_e(k) = x(k) - x(\infty)$$

$$u_e(k) = u(k) - u(\infty)$$

Referring to Eqs. (7–142) and (7–143), we have

$$\begin{bmatrix} x_e(k+1) \\ u_e(k+1) \end{bmatrix} = \begin{bmatrix} 0.5 & 2 \\ 0 & 0 \end{bmatrix} \begin{bmatrix} x_e(k) \\ u_e(k) \end{bmatrix} + \begin{bmatrix} 0 \\ 1 \end{bmatrix} w(k)$$

where

$$w(k) = -\hat{\mathbf{K}} \begin{bmatrix} x_e(k) \\ u_e(k) \end{bmatrix}$$

We wish to determine values of the feedforward gain K_1 and feedback gain K_2 such that the following performance index is minimum:

$$J = \frac{1}{2} \sum_{k=0}^{\infty} [Qx_e^2(k) + Rw^2(k)] = \frac{1}{2} \sum_{k=0}^{\infty} [x_e^2(k) + w^2(k)]$$

where $Q = 1$ and $R = 1$.

We shall first determine P by solving the Riccati equation given by Eq. (7–166):

$$1 + (0.5)P(0.5) - (0.5)P(2)[1 + (2)P(2)]^{-1}(2)P(0.5) = P$$

which can be simplified to

$$1 + 0.25P - P(1 + 4P)^{-1}P = P$$

or

$$4P^2 - 3.25P - 1 = 0$$

which yields

$$P = 1.0505 \qquad \text{or} \qquad -0.2380$$

Since P must be positive, we choose $P = 1.0505$. Then, from Eqs. (7–163), (7–164), and (7–165), we have

$$P_{11} = 1.2626$$

$$P_{12} = 1.0505$$

$$P_{22} = 4.2020$$

Hence,

$$\hat{\mathbf{P}} = \begin{bmatrix} 1.2626 & 1.0505 \\ 1.0505 & 4.2020 \end{bmatrix}$$

Referring to Eq. (7–154), the optimal matrix $\hat{\mathbf{K}}$ is determined as follows:

$$\hat{\mathbf{K}} = (R + \hat{\mathbf{H}}^*\hat{\mathbf{P}}\hat{\mathbf{H}})^{-1}\hat{\mathbf{H}}^*\hat{\mathbf{P}}\hat{\mathbf{G}}$$

$$= \left\{1 + [0 \quad 1]\begin{bmatrix}1.2626 & 1.0505\\1.0505 & 4.2020\end{bmatrix}\begin{bmatrix}0\\1\end{bmatrix}\right\}^{-1}[0 \quad 1]\begin{bmatrix}1.2626 & 1.0505\\1.0505 & 4.2020\end{bmatrix}\begin{bmatrix}0.5 & 2\\0 & 0\end{bmatrix}$$

$$= [0.1010 \,\vdots\, 0.4039]$$

Referring to Eq. (7–168), we have

$$[K_2 \vdots K_1] = \{(R + \hat{\mathbf{H}}^*\hat{\mathbf{P}}\hat{\mathbf{H}})^{-1}\hat{\mathbf{H}}^*\hat{\mathbf{P}}\hat{\mathbf{G}} + [0 \,\vdots\, 1]\}\begin{bmatrix}G - I & H\\CG & CH\end{bmatrix}^{-1}$$

$$= [0.1010 \,\vdots\, 1.4039]\begin{bmatrix}0.5 - 1 & 2\\0.5 & 2\end{bmatrix}^{-1}$$

$$= [0.2500 \,\vdots\, 0.4520]$$

Hence, we have determined the optimal gains K_1 and K_2 to be as follows:

$$K_1 = 0.4520 \qquad \text{and} \qquad K_2 = 0.2500$$

Next, let us evaluate the minimum value of J. From Eq. (7–169) we have

$$\begin{bmatrix}x_e(0)\\u_e(0)\end{bmatrix} = -\begin{bmatrix}(1 - G)^{-1}H[C(1 - G)^{-1}H]^{-1}r\\- K_1 r + [C(1 - G)^{-1}H]^{-1}r\end{bmatrix}$$

$$= -\begin{bmatrix}(1 - 0.5)^{-1}(2)[(1 - 0.5)^{-1}(2)]^{-1}(1)\\- (0.4520)(1) + [(1 - 0.5)^{-1}(2)]^{-1}(1)\end{bmatrix} = -\begin{bmatrix}1\\- 0.202\end{bmatrix}$$

Hence the minimum value of J is given by Eq. (7–170), or

$$J_{\min} = \frac{1}{2}[-1 \quad 0.202]\begin{bmatrix}1.2626 & 1.0505\\1.0505 & 4.2020\end{bmatrix}\begin{bmatrix}-1\\0.202\end{bmatrix} = 0.5048$$

7–5 SYSTEM IDENTIFICATION

In the analysis and design of control systems, it is necessary to have a mathematical model of the given plant. Such a mathematical model, called simply a model, must describe the system dynamics as completely as possible.

There are basically two approaches to obtaining a mathematical model of a plant. One approach is to obtain a model based on physical laws; the other is to use experimentation. In general, however, it may not be possible to obtain an accurate model by applying physical laws only. Some of the plant parameters must be determined by experiments. Thus the two approaches need to be combined to get a reasonably good mathematical model.

Constructing mathematical models and estimating optimal parameter values by experimental means is called *system identification*. A mathematical model involving parameters is referred to as a *parametric description* of the plant.

Note that the determination of a mathematical model of a system by a frequency response approach is not a system identification problem. This is because in that approach the transfer function of a plant is obtained as a function of ω. The frequency response curve obtained is nonparametric, because no "parameters" are involved. We shall not discuss such an approach in this section.

The designer should remember that in determining a mathematical model of a plant, it is desirable to obtain a simple model that satisfies the specific objectives of the analysis and design. That is, within a class of mathematical models we should choose a model which has the smallest number of parameters that will satisfactorily describe the plant dynamics from the viewpoint of the given objectives. Also, a model should be such that the parameters will be uniquely determined by the observed data.

Parametric description of plants. In obtaining a parametric description of a plant, reasonable knowledge of both the plant and any disturbances entering the plant is required. In the discussion to be presented in this section we limit our analysis to the linear discrete-time single-input–single-output system. Also, we assume the model to be time-invariant.

In mathematically formulating the identification problem we must introduce a criterion function or performance index that will measure how well the model in question fits the experimental data. In this section we choose the performance index to be the sum of error squares. The parameter estimation problem is then formulated as an optimization problem. The optimal model is the one that minimizes the performance index. In what follows we shall treat this optimization problem by the least-squares method. This method is applicable to parametric models that are linear in their parameters.

Real plant versus mathematical model. Although the real plant can never be fully known to the designer, we assume that its dynamics can be described by $G(z, \xi)$, where ξ represents those aspects of the real plant not fully knowable. We assume that the plant's mathematical model $G(z)$ is capable of exhibiting input-output behavior that closely resembles that of the real plant under the limited circumstances where it is observed. The difference between the output of the real plant and that of the model is the error. If a set of plant parameters is chosen to minimize the sum of the squared errors, then this particular set of parameters is optimal. An optimal set of parameters constitutes the "best" estimate of the plant parameters.

System identification by the least-squares method. Consider the real plant and its model shown in Fig. 7–5. The real plant involves unknown aspects, and its dynamics may be expressed as $G(z, \xi)$, where ξ represents those aspects unknown to the designer. Let us assume that by an analysis based on physical laws the plant dynamics may be approximated by the following pulse transfer function:

$$G(z) = \frac{b_0 + b_1 z^{-1} + b_2 z^{-2} + \cdots + b_n z^{-n}}{1 + a_1 z^{-1} + a_2 z^{-2} + \cdots + a_n z^{-n}} \tag{7–175}$$

The coefficients $a_1, a_2, \ldots, a_n$ and $b_0, b_1, \ldots, b_n$ of the denominator and numerator polynomials are the parameters of the system. We apply the input sequence to the plant and observe the output sequence. It is assumed that the input sequence $\{u(k)\}$ excites all modes of the plant dynamics.

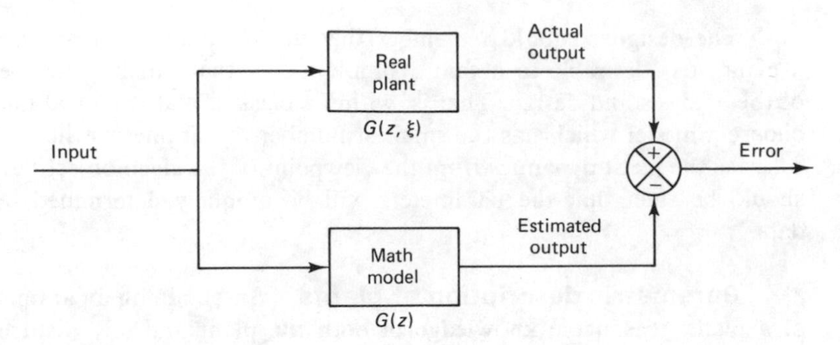

Figure 7–5 Block diagram showing the difference (error) between the output of the real plant and that of the mathematical model.

In the following analysis we assume that the completely known input sequence $u(0), u(1), \ldots, u(N)$ is applied to the plant and the corresponding plant output sequence $y(0), y(1), \ldots, y(N)$ is observed. In the present identification problem we determine from $\{u(k)\}$ and $\{y(k)\}$ the optimal estimates of the system parameters $a_1, a_2, \ldots, a_n, b_0, b_1, \ldots, b_n$ that approximate the true parameter values.

Let us determine the least-squares estimates of the parameters $a_1, a_2, \ldots, a_n, b_0, b_1, \ldots, b_n$. From Eq. (7–175), the output $y(k)$ is estimated on the basis of $y(k-1), y(k-2), \ldots, y(k-n), u(k), u(k-1), \ldots, u(k-n)$ according to the following equation:

$$\tilde{y}(k) = -a_1 y(k-1) - a_2 y(k-2) - \cdots - a_n y(k-n) + b_0 u(k) + b_1 u(k-1) + \cdots + b_n u(k-n) \qquad (7\text{–}176)$$

where $\tilde{y}(k)$ is the estimated value of $y(k)$. Note that the estimate $\tilde{y}(k)$ is made using the actual plant outputs $y(k-n), y(k-n+1), \ldots, y(k-1)$. Define error $\epsilon(k)$ as the difference between the actual output $y(k)$ and the estimated output $\tilde{y}(k)$:

$$\epsilon(k) = y(k) - \tilde{y}(k) = y(k) + a_1 y(k-1) + a_2 y(k-2) + \cdots + a_n y(k-n) - b_0 u(k) - b_1 u(k-1) - \cdots - b_n u(k-n) \qquad (7\text{–}177)$$

The error $\epsilon(k)$ depends on the measured values of $y(k), y(k-1), \ldots, y(k-n)$ and $u(k), u(k-1), \ldots, u(k-n)$. Equation (7–177) can be rewritten as follows:

$$y(k) = -a_1 y(k-1) - a_2 y(k-2) - \cdots - a_n y(k-n) + b_0 u(k) + b_1 u(k-1) + \ldots + b_n u(k-n) + \epsilon(k) \qquad (7\text{–}178)$$

Let us assume that the input and actual output are measured for $0 \le k \le N$. Note that since $y(k)$ depends on past data up to n sampling periods earlier, the error $\epsilon(k)$ is defined only for $k \ge n$. By substituting $k = n, n+1, \ldots, N$ into Eq. (7–178) and combining the resulting $N - n + 1$ equations into the vector-matrix equation, we obtain

$$
\begin{bmatrix} y(n) \\ y(n+1) \\ \cdot \\ \cdot \\ \cdot \\ y(N) \end{bmatrix} =
\left[\begin{array}{cccc|cccc} y(n-1) & y(n-2) & \cdots & y(0) & u(n) & u(n-1) & \cdots & u(0) \\ y(n) & y(n-1) & \cdots & y(1) & u(n+1) & u(n) & \cdots & u(1) \\ \cdot & \cdot & & \cdot & \cdot & \cdot & & \cdot \\ \cdot & \cdot & & \cdot & \cdot & \cdot & & \cdot \\ \cdot & \cdot & & \cdot & \cdot & \cdot & & \cdot \\ y(N-1) & y(N-2) & \cdots & y(N-n) & u(N) & u(N-1) & \cdots & u(N-n) \end{array}\right]
\cdot \begin{bmatrix} -a_1 \\ -a_2 \\ \cdot \\ \cdot \\ \cdot \\ -a_n \\ \hline b_0 \\ b_1 \\ \cdot \\ \cdot \\ \cdot \\ b_n \end{bmatrix} + \begin{bmatrix} \epsilon(n) \\ \epsilon(n+1) \\ \cdot \\ \cdot \\ \cdot \\ \epsilon(N) \end{bmatrix} \tag{7–179}
$$

Let us define

$$
\mathbf{y}(N) = \begin{bmatrix} y(n) \\ y(n+1) \\ \cdot \\ \cdot \\ \cdot \\ y(N) \end{bmatrix}, \qquad
\boldsymbol{\epsilon}(N) = \begin{bmatrix} \epsilon(n) \\ \epsilon(n+1) \\ \cdot \\ \cdot \\ \cdot \\ \epsilon(N) \end{bmatrix}, \qquad
\mathbf{x}(N) = \begin{bmatrix} -a_1 \\ -a_2 \\ \cdot \\ \cdot \\ \cdot \\ -a_n \\ b_0 \\ b_1 \\ \cdot \\ \cdot \\ \cdot \\ b_n \end{bmatrix} = \begin{bmatrix} -a_1(N) \\ -a_2(N) \\ \cdot \\ \cdot \\ \cdot \\ -a_n(N) \\ b_0(N) \\ b_1(N) \\ \cdot \\ \cdot \\ \cdot \\ b_n(N) \end{bmatrix}
$$

where the values $a_i(N)$ and $b_j(N)$ represent, respectively, the estimated values of the a_i's and b_j's based on observations up to and including $k = N$. [Thus, the estimated values of a_i and b_j based on observations up to and including $k = N + 1$ will be denoted by $a_i(N + 1)$ and $b_j(N + 1)$, and the vector consisting of the values of $-a_i(N + 1)$ and $b_j(N + 1)$ will be denoted by $\mathbf{x}(N + 1)$.] Define also

$$\mathbf{C}(N) = \left[\begin{array}{ccc|ccc} y(n-1) & \cdots & y(0) & u(n) & \cdots & u(0) \\ y(n) & \cdots & y(1) & u(n+1) & \cdots & u(1) \\ \cdot & & \cdot & \cdot & & \cdot \\ \cdot & & \cdot & \cdot & & \cdot \\ \cdot & & \cdot & \cdot & & \cdot \\ y(N-1) & \cdots & y(N-n) & u(N) & \cdots & u(N-n) \end{array}\right] \tag{7-180}$$

Then, Eq. (7–179) can be written as follows:

$$\mathbf{y}(N) = \mathbf{C}(N)\mathbf{x}(N) + \boldsymbol{\epsilon}(N) \tag{7-181}$$

Define the performance index as

$$J_N = \frac{1}{2}\sum_{k=n}^{N} \epsilon^2(k) = \frac{1}{2}\boldsymbol{\epsilon}^*(N)\boldsymbol{\epsilon}(N) \tag{7-182}$$

Then, our problem becomes that of determining $\mathbf{x}(N)$ such that the parameter values of $a_1, a_2, \ldots, a_n, b_0, b_1, \ldots, b_n$ will best fit the observed data.

In the following analysis we assume that the input sequence $\{u(k)\}$ is such that for $N > n$, $\mathbf{C}^*(N)\mathbf{C}(N)$ is nonsingular. Then, it will be shown that the optimal $\tilde{\mathbf{x}}(N)$ can be given by the equation

$$\tilde{\mathbf{x}}(N) = [\mathbf{C}^*(N)\mathbf{C}(N)]^{-1}\mathbf{C}^*(N)\mathbf{y}(N) \tag{7-183}$$

provided that the inverse of $\mathbf{C}^*(N)\mathbf{C}(N)$ exists. Note that this inverse exists if the input sequence $\{u(k)\}$ is sufficiently time-varying. [Notice that if $\{u(k)\}$ is a constant function, such as a step function, then the $u(k)$'s for $k > 0$ become constant and $\mathbf{C}^*(N)\mathbf{C}(N)$ becomes singular. For this reason, $\{u(k)\}$ must vary sufficiently with time.]

Let us now derive Eq. (7–183). Since

$$\begin{aligned} J_N &= \tfrac{1}{2}\boldsymbol{\epsilon}^*(N)\boldsymbol{\epsilon}(N) = \tfrac{1}{2}[\mathbf{y}(N) - \mathbf{C}(N)\mathbf{x}(N)]^*[\mathbf{y}(N) - \mathbf{C}(N)\mathbf{x}(N)] \\ &= \tfrac{1}{2}[-\mathbf{x}^*(N)\mathbf{C}^*(N) + \mathbf{y}^*(N)][\mathbf{y}(N) - \mathbf{C}(N)\mathbf{x}(N)] \\ &= \tfrac{1}{2}[-\mathbf{x}^*(N)\mathbf{C}^*(N)\mathbf{y}(N) + \mathbf{y}^*(N)\mathbf{y}(N) + \mathbf{x}^*(N)\mathbf{C}^*(N)\mathbf{C}(N)\mathbf{x}(N) - \mathbf{y}^*(N)\mathbf{C}(N)\mathbf{x}(N)] \end{aligned}$$

to minimize J_N with respect to $\mathbf{x}(N)$, we set

$$\frac{\partial J_N}{\partial \tilde{\mathbf{x}}(N)} = \frac{1}{2}[\mathbf{C}^*(N)\mathbf{C}(N)\mathbf{x}(N) - \mathbf{C}^*(N)\mathbf{y}(N)] = \mathbf{0} \tag{7-184}$$

Let us denote the $\mathbf{x}(N)$ that satisfies Eq. (7–184) as $\tilde{\mathbf{x}}(N)$. Then, we have

$$\mathbf{C}^*(N)\mathbf{C}(N)\tilde{\mathbf{x}}(N) = \mathbf{C}^*(N)\mathbf{y}(N) \tag{7-185}$$

Since we have assumed that $\mathbf{C}^*(N)\mathbf{C}(N)$ is nonsingular, the inverse of $\mathbf{C}^*(N)\mathbf{C}(N)$ exists. Hence, solving Eq. (7–185) for $\tilde{\mathbf{x}}(N)$, we obtain

$$\tilde{\mathbf{x}}(N) = [\mathbf{C}^*(N)\mathbf{C}(N)]^{-1}\mathbf{C}^*(N)\mathbf{y}(N)$$

which is Eq. (7–183). Thus, Eq. (7–183) defines the optimal $\tilde{\mathbf{x}}(N)$, or the optimal set of parameters $a_1, a_2, \ldots, a_n, b_0, b_1, \ldots, b_n$.

If only one set of parameters makes J_N minimum, that is, if only one optimal $\tilde{\mathbf{x}}(N)$ exists, then the system is called an *identifiable* one.

Recursive formula. In many practical cases the observations are obtained sequentially and it is desirable to obtain least-squares estimates sequentially as N increases. The process of obtaining estimates sequentially is called *recursive identification*. Recursive identification may show an improvement in the parameter estimates as N is increased.

In obtaining a recursive formula we use the result obtained for N observations to make the estimate for $N + 1$ observations. Let us assume that we have the estimate for N observations. Now consider the case where we add one more observation. As a new set of data $u(N + 1)$, $y(N + 1)$ is made available, a row is added to the matrix $\mathbf{C}(N)$, an element $y(N + 1)$ is added to vector $\mathbf{y}(N)$, and an element $\epsilon(N + 1)$ is added to vector $\boldsymbol{\epsilon}(N)$. The vector $\mathbf{x}(N)$ is changed to $\mathbf{x}(N + 1)$. Thus, Eq. (7–181) becomes as follows:

$$\begin{bmatrix} \mathbf{y}(N) \\ y(N+1) \end{bmatrix} = \begin{bmatrix} \mathbf{C}(N) \\ \mathbf{c}(N+1) \end{bmatrix} \mathbf{x}(N+1) + \begin{bmatrix} \boldsymbol{\epsilon}(N) \\ \epsilon(N+1) \end{bmatrix} \tag{7–186}$$

where

$$\mathbf{c}(N+1) = [y(N) \vdots y(N-1) \vdots \cdots \vdots y(N-n+1) \vdots u(N+1) \vdots u(N) \vdots \cdots \vdots u(N-n+1)]$$

From Eq. (7–183), the optimal $\tilde{\mathbf{x}}(N + 1)$ can be given by the equation

$$\tilde{\mathbf{x}}(N+1) = \left\{ [\mathbf{C}^*(N) \vdots \mathbf{c}^*(N+1)] \begin{bmatrix} \mathbf{C}(N) \\ \mathbf{c}(N+1) \end{bmatrix} \right\}^{-1} [\mathbf{C}^*(N) \vdots \mathbf{c}^*(N+1)] \begin{bmatrix} \mathbf{y}(N) \\ y(N+1) \end{bmatrix} \tag{7–187}$$

Equation (7–187) can be simplified by first noting that

$$\left\{ [\mathbf{C}^*(N) \vdots \mathbf{c}^*(N+1)] \begin{bmatrix} \mathbf{C}(N) \\ \mathbf{c}(N+1) \end{bmatrix} \right\}^{-1} = [\mathbf{C}^*(N)\mathbf{C}(N) + \mathbf{c}^*(N+1)\mathbf{c}(N+1)]^{-1}$$

By using the matrix inversion lemma

$$(\mathbf{A} + \mathbf{BD})^{-1} = \mathbf{A}^{-1} - \mathbf{A}^{-1}\mathbf{B}(\mathbf{I} + \mathbf{DA}^{-1}\mathbf{B})^{-1}\mathbf{DA}^{-1}$$

and identifying

$$\mathbf{A} = \mathbf{C}^*(N)\mathbf{C}(N), \qquad \mathbf{B} = \mathbf{c}^*(N+1), \qquad \mathbf{D} = \mathbf{c}(N+1)$$

we obtain

$$[\mathbf{C}^*(N)\mathbf{C}(N) + \mathbf{c}^*(N+1)\mathbf{c}(N+1)]^{-1}$$

$$= [\mathbf{C}^*(N)\mathbf{C}(N)]^{-1} - \frac{[\mathbf{C}^*(N)\mathbf{C}(N)]^{-1}\mathbf{c}^*(N+1)\mathbf{c}(N+1)[\mathbf{C}^*(N)\mathbf{C}(N)]^{-1}}{1 + \mathbf{c}(N+1)[\mathbf{C}^*(N)\mathbf{C}(N)]^{-1}\mathbf{c}^*(N+1)} \quad (7\text{–}188)$$

Hence

$$\tilde{\mathbf{x}}(N+1) = \left\{[\mathbf{C}^*(N)\mathbf{C}(N)]^{-1} - \frac{[\mathbf{C}^*(N)\mathbf{C}(N)]^{-1}\mathbf{c}^*(N+1)\mathbf{c}(N+1)[\mathbf{C}^*(N)\mathbf{C}(N)]^{-1}}{1 + \mathbf{c}(N+1)[\mathbf{C}^*(N)\mathbf{C}(N)]^{-1}\mathbf{c}^*(N+1)}\right\}[\mathbf{C}^*(N)\mathbf{y}(N) + \mathbf{c}^*(N+1)y(N+1)] \quad (7\text{–}189)$$

Referring to Eq. (7–183), note that

$$[\mathbf{C}^*(N)\mathbf{C}(N)]^{-1}\mathbf{C}^*(N)\mathbf{y}(N) = \tilde{\mathbf{x}}(N) \quad (7\text{–}190)$$

By substituting Eq. (7–190) into Eq. (7–189) we obtain

$$\begin{aligned}\tilde{\mathbf{x}}(N+1) &= \tilde{\mathbf{x}}(N) - \frac{[\mathbf{C}^*(N)\mathbf{C}(N)]^{-1}\mathbf{c}^*(N+1)\mathbf{c}(N+1)}{1 + \mathbf{c}(N+1)[\mathbf{C}^*(N)\mathbf{C}(N)]^{-1}\mathbf{c}^*(N+1)}\tilde{\mathbf{x}}(N) \\ &\quad + [\mathbf{C}^*(N)\mathbf{C}(N)]^{-1}\mathbf{c}^*(N+1)y(N+1) \\ &\quad - \frac{[\mathbf{C}^*(N)\mathbf{C}(N)]^{-1}\mathbf{c}^*(N+1)\mathbf{c}(N+1)[\mathbf{C}^*(N)\mathbf{C}(N)]^{-1}}{1 + \mathbf{c}(N+1)[\mathbf{C}^*(N)\mathbf{C}(N)]^{-1}\mathbf{c}^*(N+1)}\mathbf{c}^*(N+1)y(N+1) \\ &= \tilde{\mathbf{x}}(N) + \frac{[\mathbf{C}^*(N)\mathbf{C}(N)]^{-1}\mathbf{c}^*(N+1)}{1 + \mathbf{c}(N+1)[\mathbf{C}^*(N)\mathbf{C}(N)]^{-1}\mathbf{c}^*(N+1)}[y(N+1) - \mathbf{c}(N+1)\tilde{\mathbf{x}}(N)]\end{aligned} \quad (7\text{–}191)$$

Now define

$$\mathbf{K}(N+1) = \frac{[\mathbf{C}^*(N)\mathbf{C}(N)]^{-1}\mathbf{c}^*(N+1)}{1 + \mathbf{c}(N+1)[\mathbf{C}^*(N)\mathbf{C}(N)]^{-1}\mathbf{c}^*(N+1)} \quad (7\text{–}192)$$

Then, Eq. (7–191) becomes

$$\tilde{\mathbf{x}}(N+1) = \tilde{\mathbf{x}}(N) + \mathbf{K}(N+1)[y(N+1) - \mathbf{c}(N+1)\tilde{\mathbf{x}}(N)] \quad (7\text{–}193)$$

Equation (7–193) gives the desired recursive formula. It gives the update estimate when one more observation is included. By use of Eq. (7–193) the next estimate $\tilde{\mathbf{x}}(N+1)$ is obtained as the current estimate $\tilde{\mathbf{x}}(N)$ plus a correction term which is proportional to the difference between the observed output $y(N+1)$ and the estimated output $\mathbf{c}(N+1)\tilde{\mathbf{x}}(N)$.

Let us define

$$\mathbf{P}(N) = [\mathbf{C}^*(N)\mathbf{C}(N)]^{-1} \quad (7\text{–}194)$$

Then

$$\mathbf{K}(N+1) = \frac{\mathbf{P}(N)\mathbf{c}^*(N+1)}{1 + \mathbf{c}(N+1)\mathbf{P}(N)\mathbf{c}^*(N+1)} \quad (7\text{–}195)$$

From Eq. (7–188) we have

$$\mathbf{P}(N+1)=\mathbf{P}(N)-\frac{\mathbf{P}(N)\mathbf{c}^*(N+1)\mathbf{c}(N+1)\mathbf{P}(N)}{1+\mathbf{c}(N+1)\mathbf{P}(N)\mathbf{c}^*(N+1)} \tag{7–196}$$

The matrix $\mathbf{K}(N+1)$ given by Eq. (7–195) can be computed by use of Eq. (7–196). Notice that in computing $\mathbf{P}(N+1)$ no matrix inversion is involved, since

$$1+\mathbf{c}(N+1)\mathbf{P}(N)\mathbf{c}^*(N+1)$$

is a scalar. This is because we are considering a single-input–single-output system.

Minimization of weighted error squares. The performance index given by Eq. (7–182) is based on the assumption that all errors have equal weight. If, however, there is a reason to give different weights to different errors, the performance index may be modified to read

$$J_N=\frac{1}{2}\boldsymbol{\epsilon}^*(N)\mathbf{W}(N)\boldsymbol{\epsilon}(N)$$

where $\mathbf{W}(N)$ is a weighting matrix which is positive definite. Such a weighted least-squares problem can be handled in a way similar to the case presented in this section.

Example 7–7.

Consider a plant and its mathematical model as shown in Fig. 7–6. Assume that a mathematical model of the plant may be given by the following pulse transfer function:

$$G(z)=b_0+b_1z^{-1}$$

The input to the plant and model alike is $u(k)$. The output of the plant is $y(k)$, and that of the mathematical model is $\tilde{y}(k)$, where

$$\tilde{y}(k)=b_0u(k)+b_1u(k-1)$$

The difference between $y(k)$ and $\tilde{y}(k)$ is the error $\epsilon(k)$:

$$\epsilon(k)=y(k)-\tilde{y}(k)=y(k)-b_0u(k)-b_1u(k-1)$$

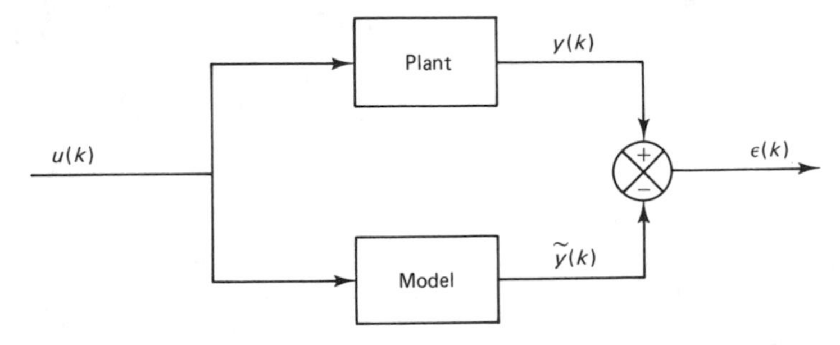

Figure 7–6 Block diagram showing the outputs of a plant and its mathematical model. The difference in the outputs produces the error signal $\epsilon(k)$.

The input data $\{u(k)\}$ and output data $\{y(k)\}$ are shown in Table 7–2.

Using the least-squares method, determine the optimal estimates of the parameters b_0 and b_1 for $N = 7, 8, \ldots, 14$.

Let us begin with the case where $N = 7$. Equation (7–179) becomes as follows:

$$\begin{bmatrix} y(1) \\ y(2) \\ y(3) \\ y(4) \\ y(5) \\ y(6) \\ y(7) \end{bmatrix} = \begin{bmatrix} u(1) & u(0) \\ u(2) & u(1) \\ u(3) & u(2) \\ u(4) & u(3) \\ u(5) & u(4) \\ u(6) & u(5) \\ u(7) & u(6) \end{bmatrix} \begin{bmatrix} b_0 \\ b_1 \end{bmatrix} + \begin{bmatrix} \epsilon(1) \\ \epsilon(2) \\ \epsilon(3) \\ \epsilon(4) \\ \epsilon(5) \\ \epsilon(6) \\ \epsilon(7) \end{bmatrix} \tag{7–197}$$

By substituting the numerical values from Table 7–2 into Eq. (7–197), we obtain

TABLE 7–2 INPUT DATA AND OUTPUT DATA FOR THE SYSTEM CONSIDERED IN EXAMPLE 7–7

k	$u(k)$	$y(k)$
0	1.0	0.9
1	0.8	2.5
2	0.6	2.4
3	0.4	1.3
4	0.2	1.2
5	0	0.8
6	0.2	0
7	0.4	0.9
8	0.6	1.4
9	0.8	1.9
10	1.0	2.3
11	0.8	2.4
12	0.6	2.3
13	0.4	1.3
14	0.2	1.2

$$\begin{bmatrix} 2.5 \\ 2.4 \\ 1.3 \\ 1.2 \\ 0.8 \\ 0 \\ 0.9 \end{bmatrix} = \begin{bmatrix} 0.8 & 1.0 \\ 0.6 & 0.8 \\ 0.4 & 0.6 \\ 0.2 & 0.4 \\ 0 & 0.2 \\ 0.2 & 0 \\ 0.4 & 0.2 \end{bmatrix} \begin{bmatrix} b_0 \\ b_1 \end{bmatrix} + \begin{bmatrix} \epsilon(1) \\ \epsilon(2) \\ \epsilon(3) \\ \epsilon(4) \\ \epsilon(5) \\ \epsilon(6) \\ \epsilon(7) \end{bmatrix} \tag{7–198}$$

In the least-squares approach, we desire to minimize the following performance index:

$$J_N = \frac{1}{2} \sum_{k=n}^{N} \epsilon^2(k)$$

where in the present case $n = 1$ and $N = 7, 8, \ldots, 14$. The optimal estimates for b_0 and b_1 based on observations up to and including $k = N$ can be given by Eq. (7–183),

$$\tilde{\mathbf{x}}(N) = [\mathbf{C}^*(N)\mathbf{C}(N)]^{-1}\mathbf{C}^*(N)\mathbf{y}(N) \tag{7–199}$$

provided $\mathbf{C}^*(N)\mathbf{C}(N)$ is nonsingular.

Let us now compute $\mathbf{C}^*(7)\mathbf{C}(7)$:

$$\mathbf{C}^*(7)\mathbf{C}(7) = \begin{bmatrix} 0.8 & 0.6 & 0.4 & 0.2 & 0 & 0.2 & 0.4 \\ 1.0 & 0.8 & 0.6 & 0.4 & 0.2 & 0 & 0.2 \end{bmatrix} \begin{bmatrix} 0.8 & 1.0 \\ 0.6 & 0.8 \\ 0.4 & 0.6 \\ 0.2 & 0.4 \\ 0 & 0.2 \\ 0.2 & 0 \\ 0.4 & 0.2 \end{bmatrix}$$

$$= \begin{bmatrix} 1.4 & 1.68 \\ 1.68 & 2.24 \end{bmatrix}$$

Since $\mathbf{C}^*(7)\mathbf{C}(7)$ is nonsingular, we get

$$[\mathbf{C}^*(7)\mathbf{C}(7)]^{-1} = \begin{bmatrix} 7.14286 & -5.35714 \\ -5.35714 & 4.46429 \end{bmatrix}$$

Then the optimal estimate $\tilde{\mathbf{x}}(7)$ can be given by Eq. (7–199):

$$\tilde{\mathbf{x}}(7) = [\mathbf{C}^*(7)\mathbf{C}(7)]^{-1}\mathbf{C}^*(7)\mathbf{y}(7)$$

$$= \begin{bmatrix} 7.14286 & -5.35714 \\ -5.35714 & 4.46429 \end{bmatrix} \begin{bmatrix} 0.8 & 0.6 & 0.4 & 0.2 & 0 & 0.2 & 0.4 \\ 1.0 & 0.8 & 0.6 & 0.4 & 0.2 & 0 & 0.2 \end{bmatrix} \begin{bmatrix} 2.5 \\ 2.4 \\ 1.3 \\ 1.2 \\ 0.8 \\ 0 \\ 0.9 \end{bmatrix}$$

$$= \begin{bmatrix} 0.32143 \\ 2.44643 \end{bmatrix}$$

To obtain the optimal estimates $\tilde{\mathbf{x}}(8), \tilde{\mathbf{x}}(9), \ldots, \tilde{\mathbf{x}}(14)$, we use Eq. (7–193),

$$\tilde{\mathbf{x}}(N+1) = \tilde{\mathbf{x}}(N) + \mathbf{K}(N+1)[y(N+1) - \mathbf{c}(N+1)\tilde{\mathbf{x}}(N)]$$

where, from Eqs. (7–195) and (7–194), we have

$$\mathbf{K}(N+1) = \frac{\mathbf{P}(N)\mathbf{c}^*(N+1)}{1 + \mathbf{c}(N+1)\mathbf{P}(N)\mathbf{c}^*(N+1)}$$

$$\mathbf{P}(N) = [\mathbf{C}^*(N)\mathbf{C}(N)]^{-1}$$

and, from Eq. (7–196),

$$\mathbf{P}(N+1) = \mathbf{P}(N) - \frac{\mathbf{P}(N)\mathbf{c}^*(N+1)\mathbf{c}(N+1)\mathbf{P}(N)}{1 + \mathbf{c}(N+1)\mathbf{P}(N)\mathbf{c}^*(N+1)}$$

Let us determine $\tilde{\mathbf{x}}(8)$. First, note that

$$\mathbf{c}(8) = [0.6 \quad 0.4] \qquad \text{and} \qquad \mathbf{P}(7) = \begin{bmatrix} 7.14286 & -5.35714 \\ -5.35714 & 4.46429 \end{bmatrix}$$

Hence

$$\begin{aligned} \mathbf{K}(8) &= \frac{\mathbf{P}(7)\mathbf{c}^*(8)}{1 + \mathbf{c}(8)\mathbf{P}(7)\mathbf{c}^*(8)} \\ &= \frac{\begin{bmatrix} 7.14286 & -5.35714 \\ -5.35714 & 4.46429 \end{bmatrix}\begin{bmatrix} 0.6 \\ 0.4 \end{bmatrix}}{1 + [0.6 \quad 0.4]\begin{bmatrix} 7.14286 & -5.35714 \\ -5.35714 & 4.46429 \end{bmatrix}\begin{bmatrix} 0.6 \\ 0.4 \end{bmatrix}} \\ &= \begin{bmatrix} 1.25000 \\ -0.83333 \end{bmatrix} \end{aligned}$$

Then, the optimal estimate $\tilde{\mathbf{x}}(8)$ can be determined as follows:

$$\begin{aligned} \tilde{\mathbf{x}}(8) &= \tilde{\mathbf{x}}(7) + \mathbf{K}(8)[y(8) - \mathbf{c}(8)\tilde{\mathbf{x}}(7)] \\ &= \begin{bmatrix} 0.32143 \\ 2.44643 \end{bmatrix} + \begin{bmatrix} 1.25000 \\ -0.83333 \end{bmatrix}\left\{1.4 - [0.6 \quad 0.4]\begin{bmatrix} 0.32143 \\ 2.44643 \end{bmatrix}\right\} = \begin{bmatrix} 0.60714 \\ 2.25596 \end{bmatrix} \end{aligned}$$

Next we obtain $\mathbf{P}(8)$:

$$\mathbf{P}(8) = \mathbf{P}(7) - \frac{\mathbf{P}(7)\mathbf{c}^*(8)\mathbf{c}(8)\mathbf{P}(7)}{1 + \mathbf{c}(8)\mathbf{P}(7)\mathbf{c}^*(8)} = \begin{bmatrix} 4.46429 & -3.57143 \\ -3.57143 & 3.27381 \end{bmatrix}$$

Then we compute $\mathbf{K}(9)$. Noting that

$$\mathbf{c}(9) = [0.8 \quad 0.6]$$

we have

$$\mathbf{K}(9) = \frac{\mathbf{P}(8)\mathbf{c}^*(9)}{1 + \mathbf{c}(9)\mathbf{P}(8)\mathbf{c}^*(9)} = \begin{bmatrix} 0.88889 \\ -0.55553 \end{bmatrix}$$

Hence, $\tilde{\mathbf{x}}(9)$ is obtained as follows:

$$\tilde{\mathbf{x}}(9) = \tilde{\mathbf{x}}(8) + \mathbf{K}(9)[y(9) - \mathbf{c}(9)\tilde{\mathbf{x}}(8)]$$

TABLE 7–3 VALUES OF b_0 AND b_1 OBTAINED FOR $N = 7, 8, \ldots, 14$ FOR THE SYSTEM CONSIDERED IN EXAMPLE 7–7

N	b_0	b_1
7	0.32143	2.44643
8	0.60714	2.25596
9	0.66111	2.22223
10	0.57272	2.27273
11	0.66216	2.11622
12	0.60638	2.19909
13	0.68144	2.10437
14	0.61072	2.18139

$$= \begin{bmatrix} 0.60714 \\ 2.25596 \end{bmatrix} + \begin{bmatrix} 0.88889 \\ -0.55553 \end{bmatrix} \left\{ 1.9 - [0.8 \quad 0.6] \begin{bmatrix} 0.60714 \\ 2.25596 \end{bmatrix} \right\} = \begin{bmatrix} 0.66111 \\ 2.22223 \end{bmatrix}$$

Repeating the same process, we obtain the results shown in Table 7–3. A computer solution can easily be carried out for such a problem.

7–6 KALMAN FILTERS

The technique of Kalman filters is a very general filtering technique which can be applied to the solution of such problems as optimal estimation, prediction, noise filtering, and stochastic optimal control. Time-varying optimal solutions to these problems are commonly known as *Kalman filters*. Kalman filters can be easily programmed on a digital computer. Adaptive gain tuning capability is the characteristic of the Kalman filter.

Different from other estimation filters such as Wiener filters (which can be applied to only stationary processes and thus cannot include the effect of initial conditions), Kalman filters can be applied to both stationary and nonstationary processes and can include the initial conditions of processes in estimation, prediction, filtering, or stochastic optimal control algorithms.

In Sect. 6–6 the observer gain matrix $\mathbf{K}_e$ was determined in such a way that the observer would have the desired eigenvalues. In this section we shall present the Kalman filter approach to the determination of matrix $\mathbf{K}_e$. [As we shall see later, the estimator gain matrix for a general case becomes time-varying, or takes the form $\mathbf{K}_e(k)$.] Here, we specifically consider noises (which have certain properties which we shall describe later) present in the system and measurement process. The approach here is basically to minimize the mean square of the estimation error. It is noted that the optimal filters designed by this approach have the same basic structure as that given by Eq. (6–116) or that given by Eqs. (6–162) and (6–163).

Note that there are direct relationships between Kalman filters and the time-varying optimal solution to the identification problem presented in Sec. 7–5. This will be pointed out later in discussing example problems. (See Examples 7–8 and 7–9.)

Smoothing, filtering, and prediction. Consider the following discrete-time state equation and output equation:

$$\mathbf{x}(k+1) = \mathbf{G}(k)\mathbf{x}(k) + \mathbf{H}(k)\mathbf{u}(k) + \mathbf{w}(k) \tag{7–200}$$

$$\mathbf{y}(k) = \mathbf{C}(k)\mathbf{x}(k) + \boldsymbol{\epsilon}(k) \tag{7–201}$$

where $\mathbf{w}(k)$ is the system noise and $\boldsymbol{\epsilon}(k)$ is the measurement noise. Assume that output measurements $\mathbf{y}(0), \mathbf{y}(1), \ldots, \mathbf{y}(k)$ are known. Here, state $\mathbf{x}(k)$ is not known but the generating process and measuring process are known and are as given by Eqs. (7–200) and (7–201), respectively. That is, in Eqs. (7–200) and (7–201) matrices $\mathbf{G}(k)$, $\mathbf{H}(k)$, and $\mathbf{C}(k)$ are deterministic and known and the statistical properties of $\mathbf{w}(k)$ and $\boldsymbol{\epsilon}(k)$ are also known.

The problem of estimating $\mathbf{x}(k)$, given $\mathbf{y}(0), \mathbf{y}(1), \ldots, \mathbf{y}(k)$, can be subdivided into three cases:

Smoothing problem: The problem of determining the optimal estimate $\tilde{\mathbf{x}}(n)$, where $0 \leq n < k$

Filtering problem: The problem of determining the current optimal estimate $\tilde{\mathbf{x}}(k)$

Prediction problem: The problem of determining the optimal estimate $\tilde{\mathbf{x}}(n)$, where $n > k$

In this section we shall be concerned with the prediction problem and the filtering problem. In the prediction problem we determine the optimal estimate $\tilde{\mathbf{x}}(k+1)$ on the basis of the knowledge of the output sequence $\mathbf{y}(0), \mathbf{y}(1), \ldots, \mathbf{y}(k)$. And in the filtering problem here we determine the optimal current estimate $\tilde{\mathbf{x}}(k)$ also on the basis of the knowledge of the output sequence $\mathbf{y}(0), \mathbf{y}(1), \ldots, \mathbf{y}(k)$. The criterion for the optimal estimate is the minimum covariance of the estimation error.

System equations and assumptions. The system we shall consider in this section is defined by Eqs. (7–200) and (7–201), which we repeat here:

$$\mathbf{x}(k+1) = \mathbf{G}(k)\mathbf{x}(k) + \mathbf{H}(k)\mathbf{u}(k) + \mathbf{w}(k) \tag{7–202}$$

$$\mathbf{y}(k) = \mathbf{C}(k)\mathbf{x}(k) + \boldsymbol{\epsilon}(k) \tag{7–203}$$

where $\mathbf{w}(k)$ represents the system noise sequence and $\boldsymbol{\epsilon}(k)$ the measurement noise sequence. We assume that the initial state $\mathbf{x}(0)$ and noises $\mathbf{w}(k)$ and $\boldsymbol{\epsilon}(k)$ satisfy the following conditions:

$$E[\mathbf{x}(0)] = \bar{\mathbf{x}}(0) \qquad \text{(deterministic)} \tag{7–204}$$

$$E[\mathbf{w}(k)] = \mathbf{0} \tag{7–205}$$

$$E[\boldsymbol{\epsilon}(k)] = \mathbf{0} \tag{7–206}$$

$$E\{[\mathbf{x}(0) - \bar{\mathbf{x}}(0)][\mathbf{x}(0) - \bar{\mathbf{x}}(0)]^*\} = \mathbf{P}_0 \tag{7–207}$$

$$E[\mathbf{w}(j)\boldsymbol{\epsilon}^*(k)] = \mathbf{0} \tag{7–208}$$

$$E[\mathbf{w}(j)\mathbf{w}^*(k)] = \mathbf{Q}(k)\delta_{jk} \qquad [\mathbf{Q}(k) \text{ is positive semidefinite}] \tag{7–209}$$

$$E[\boldsymbol{\epsilon}(j)\boldsymbol{\epsilon}^*(k)] = \mathbf{R}(k)\delta_{jk} \qquad [\mathbf{R}(k) \text{ is positive definite}] \tag{7–210}$$

$$E\{[\mathbf{x}(0) - \bar{\mathbf{x}}(0)]\mathbf{w}^*(k)\} = \mathbf{0} \tag{7–211}$$

$$E\{[\mathbf{x}(0) - \bar{\mathbf{x}}(0)]\boldsymbol{\epsilon}^*(k)\} = \mathbf{0} \tag{7–212}$$

where $\delta_{jk} = 1$ if $j = k$ and $\delta_{jk} = 0$ if $j \neq k$.

In the following discussion, given $\mathbf{y}(0), \mathbf{y}(1), \ldots, \mathbf{y}(k)$, we wish to determine the optimal estimate $\tilde{\mathbf{x}}(k+1)$ such that an $n \times n$ positive definite matrix

$$\mathbf{P}(k+1) = E[\mathbf{e}(k+1)\mathbf{e}^*(k+1)]$$

is minimum (for the prediction-type Kalman filter) or to determine the optimal estimate $\tilde{\mathbf{x}}(k)$ such that

$$\mathbf{P}(k) = E[\mathbf{e}(k)\mathbf{e}^*(k)]$$

is minimum (for the current estimation-type Kalman filter). The estimation error $\mathbf{e}(k)$ is defined by the equation

$$\mathbf{e}(k) = \mathbf{x}(k) - \tilde{\mathbf{x}}(k) \tag{7–213}$$

Note that "$\mathbf{P}(k)$ is minimum" means that the quadratic form $\boldsymbol{\alpha}^*\mathbf{P}(k)\boldsymbol{\alpha}$ is minimum, where $\boldsymbol{\alpha}$ is an arbitrary n-vector. Note also that when $\mathbf{P}(k)$ is minimum, then

$$J = E[\mathbf{e}^*(k)\mathbf{e}(k)] = \operatorname{tr} \mathbf{P}(k)$$

becomes minimum. That is, the mean square error $E[\mathbf{e}^*(k)\mathbf{e}(k)]$ becomes minimum when $\mathbf{P}(k)$ is minimum. {It is important to distinguish between $E[\mathbf{e}(k)\mathbf{e}^*(k)]$, which is an $n \times n$ matrix, and $E[\mathbf{e}^*(k)\mathbf{e}(k)]$, which is a scalar.}

Derivation of prediction-type Kalman filter equations. For the system defined by Eqs. (7–202) and (7–203) our problem here is to determine the optimal estimate $\tilde{\mathbf{x}}(k+1)$ on the basis of the known outputs $\mathbf{y}(0), \mathbf{y}(1), \ldots, \mathbf{y}(k)$, so that the covariance matrix of the estimation error at time $k + 1$, $E[\mathbf{e}(k+1)\mathbf{e}^*(k+1)]$, is minimum. That is, we want the quadratic form $\boldsymbol{\alpha}^* E[\mathbf{e}(k+1)\mathbf{e}^*(k+1)]\boldsymbol{\alpha}$ to be minimum, where $\boldsymbol{\alpha}$ is an arbitrary n-vector.

Let the prediction-type Kalman filter have the form

$$\begin{aligned}\tilde{\mathbf{x}}(k+1) &= \mathbf{G}(k)\tilde{\mathbf{x}}(k) + \mathbf{H}(k)\mathbf{u}(k) + \mathbf{K}_e(k)[\mathbf{y}(k) - \tilde{\mathbf{y}}(k)] \\ &= \mathbf{G}(k)\tilde{\mathbf{x}}(k) + \mathbf{H}(k)\mathbf{u}(k) + \mathbf{K}_e(k)[\mathbf{y}(k) - \mathbf{C}(k)\tilde{\mathbf{x}}(k)] \qquad (7\text{–}214)\end{aligned}$$

We shall now determine the matrix $\mathbf{K}_e(k)$ such that $E[\mathbf{e}(k+1)\mathbf{e}^*(k+1)]$ is minimum. Note that the estimation error $\mathbf{e}(k+1)$ can be given as follows:

$$\begin{aligned}\mathbf{e}(k+1) &= \mathbf{x}(k+1) - \tilde{\mathbf{x}}(k+1) \\ &= \mathbf{G}(k)\mathbf{x}(k) + \mathbf{w}(k) - [\mathbf{G}(k) - \mathbf{K}_e(k)\mathbf{C}(k)]\tilde{\mathbf{x}}(k) \\ &\quad - \mathbf{K}_e(k)[\mathbf{C}(k)\mathbf{x}(k) + \boldsymbol{\epsilon}(k)] \\ &= [\mathbf{G}(k) - \mathbf{K}_e(k)\mathbf{C}(k)]\mathbf{e}(k) - \mathbf{K}_e(k)\boldsymbol{\epsilon}(k) + \mathbf{w}(k)\end{aligned}$$

or

$$\mathbf{e}(k+1) = \mathbf{L}(k)\mathbf{e}(k) - \mathbf{v}(k) \qquad (7\text{–}215)$$

where

$$\mathbf{L}(k) = \mathbf{G}(k) - \mathbf{K}_e(k)\mathbf{C}(k) \qquad (7\text{–}216)$$

$$\mathbf{v}(k) = \mathbf{K}_e(k)\boldsymbol{\epsilon}(k) - \mathbf{w}(k) = \text{white noise} \qquad (7\text{–}217)$$

Notice that

$$\begin{aligned}E[\mathbf{v}(k)] &= E[\mathbf{K}_e(k)\boldsymbol{\epsilon}(k) - \mathbf{w}(k)] \\ &= \mathbf{K}_e(k)E[\boldsymbol{\epsilon}(k)] - E[\mathbf{w}(k)] = \mathbf{0} \qquad (7\text{–}218)\end{aligned}$$

Also,

$$\begin{aligned}E[\mathbf{v}(j)\mathbf{v}^*(k)] &= E\{[\mathbf{K}_e(j)\boldsymbol{\epsilon}(j) - \mathbf{w}(j)][\mathbf{K}_e(k)\boldsymbol{\epsilon}(k) - \mathbf{w}(k)]^*\} \\ &= \mathbf{K}_e(j)E[\boldsymbol{\epsilon}(j)\boldsymbol{\epsilon}^*(k)]\mathbf{K}_e^*(k) - E[\mathbf{w}(j)\boldsymbol{\epsilon}^*(k)\mathbf{K}_e^*(k)] \\ &\quad - E[\mathbf{K}_e(j)\boldsymbol{\epsilon}(j)\mathbf{w}^*(k)] + E[\mathbf{w}(j)\mathbf{w}^*(k)] \\ &= \mathbf{K}_e(j)\mathbf{R}(k)\delta_{jk}\mathbf{K}_e^*(k) + \mathbf{Q}(k)\delta_{jk} \\ &= [\mathbf{Q}(k) + \mathbf{K}_e(j)\mathbf{R}(k)\mathbf{K}_e^*(k)]\delta_{jk}\end{aligned}$$

Hence

$$E[\mathbf{v}(k)\mathbf{v}^*(k)] = \mathbf{Q}(k) + \mathbf{K}_e(k)\mathbf{R}(k)\mathbf{K}_e^*(k) \qquad (7\text{–}219)$$

$$E[\mathbf{v}(j)\mathbf{v}^*(k)] = \mathbf{0} \qquad j \neq k \qquad (7\text{–}220)$$

Next, consider Eq. (7–215). By taking the expected values of both sides of Eq. (7–215), we have

$$E[\mathbf{e}(k+1)] = E[\mathbf{L}(k)\mathbf{e}(k)] - E[\mathbf{v}(k)] \tag{7–221}$$

Since $\mathbf{L}(k)$ is deterministic, referring to Eq. (7–218), Eq. (7–221) becomes

$$\bar{\mathbf{e}}(k+1) = \mathbf{L}(k)\bar{\mathbf{e}}(k) \tag{7–222}$$

Equation (7–222) relates the mean value of $\mathbf{e}(k+1)$ and that of $\mathbf{e}(k)$.

Now define

$$\mathbf{M}(k) = E\{[\mathbf{e}(k) - \bar{\mathbf{e}}(k)][\mathbf{e}(k) - \bar{\mathbf{e}}(k)]^*\} \tag{7–223}$$

Then, referring to Eqs. (7–215) and (7–222), we have

$$\begin{aligned}\mathbf{M}(k+1) &= E\{[\mathbf{e}(k+1) - \bar{\mathbf{e}}(k+1)][\mathbf{e}(k+1) - \bar{\mathbf{e}}(k+1)]^*\} \\ &= E\Big[\{\mathbf{L}(k)[\mathbf{e}(k) - \bar{\mathbf{e}}(k)] - \mathbf{v}(k)\}\{\mathbf{L}(k)[\mathbf{e}(k) - \bar{\mathbf{e}}(k)] - \mathbf{v}(k)\}^*\Big] \\ &= \mathbf{L}(k)\mathbf{M}(k)\mathbf{L}^*(k) - \mathbf{L}(k)E\{[\mathbf{e}(k) - \bar{\mathbf{e}}(k)]\mathbf{v}^*(k)\} \\ &\quad - E\{\mathbf{v}(k)[\mathbf{e}(k) - \bar{\mathbf{e}}(k)]^*\}\mathbf{L}^*(k) + \mathbf{Q}(k) + \mathbf{K}_e(k)\mathbf{R}(k)\mathbf{K}_e^*(k)\end{aligned} \tag{7–224}$$

where we have used Eq. (7–219). Note that referring to Eqs. (7–211), (7–212), (7–217), and (7–220), we have

$$\begin{aligned}&E\{[\mathbf{e}(k) - \bar{\mathbf{e}}(k)]\mathbf{v}^*(k)\} \\ &\quad = E[\{\mathbf{L}(k-1)[\mathbf{e}(k-1) - \bar{\mathbf{e}}(k-1)] - \mathbf{v}(k-1)\}\mathbf{v}^*(k)] \\ &\quad = E\{\mathbf{L}(k-1)[\mathbf{e}(k-1) - \bar{\mathbf{e}}(k-1)]\mathbf{v}^*(k)\} - E[\mathbf{v}(k-1)\mathbf{v}^*(k)] \\ &\quad = \mathbf{L}(k-1)E\{[\mathbf{e}(k-1) - \bar{\mathbf{e}}(k-1)]\mathbf{v}^*(k)\} \\ &\quad = \mathbf{L}(k-1)\mathbf{L}(k-2)E\{[\mathbf{e}(k-2) - \bar{\mathbf{e}}(k-2)]\mathbf{v}^*(k)\} \\ &\quad = \cdots \\ &\quad = \mathbf{L}(k-1)\mathbf{L}(k-2)\cdots\mathbf{L}(0)E\{[\mathbf{e}(0) - \bar{\mathbf{e}}(0)]\mathbf{v}^*(k)\} \\ &\quad = \mathbf{L}(k-1)\mathbf{L}(k-2)\cdots\mathbf{L}(0)E\{[\mathbf{x}(0) - \bar{\mathbf{x}}(0)]\mathbf{v}^*(k)\} \\ &\quad = \mathbf{0}\end{aligned} \tag{7–225}$$

Hence Eq. (7–224) becomes

$$\mathbf{M}(k+1) = \mathbf{L}(k)\mathbf{M}(k)\mathbf{L}^*(k) + \mathbf{Q}(k) + \mathbf{K}_e(k)\mathbf{R}(k)\mathbf{K}_e^*(k) \tag{7–226}$$

By substituting Eq. (7–216) into Eq. (7–226), we obtain

$$\begin{aligned}\mathbf{M}(k+1) = {}&[\mathbf{G}(k) - \mathbf{K}_e(k)\mathbf{C}(k)]\mathbf{M}(k)[\mathbf{G}(k) - \mathbf{K}_e(k)\mathbf{C}(k)]^* \\ &+ \mathbf{Q}(k) + \mathbf{K}_e(k)\mathbf{R}(k)\mathbf{K}_e^*(k)\end{aligned} \tag{7–227}$$

Equation (7–223) can also be written as follows:

$$\begin{aligned}\mathbf{M}(k) &= E[\mathbf{e}(k)\mathbf{e}^*(k)] - \bar{\mathbf{e}}(k)E[\mathbf{e}^*(k)] - E[\mathbf{e}(k)]\bar{\mathbf{e}}^*(k) + \bar{\mathbf{e}}(k)\bar{\mathbf{e}}^*(k) \\ &= E[\mathbf{e}(k)\mathbf{e}^*(k)] - \bar{\mathbf{e}}(k)\bar{\mathbf{e}}^*(k)\end{aligned} \tag{7–228}$$

Define

$$\mathbf{P}(k) = E[\mathbf{e}(k)\mathbf{e}^*(k)] \tag{7–229}$$

Then, Eq. (7–228) becomes

$$\mathbf{M}(k) = \mathbf{P}(k) - \bar{\mathbf{e}}(k)\bar{\mathbf{e}}^*(k) \tag{7–230}$$

Since

$$\bar{\mathbf{e}}(0) = \bar{\mathbf{x}}(0) - \tilde{\mathbf{x}}(0)$$

if we choose

$$\tilde{\mathbf{x}}(0) = \bar{\mathbf{x}}(0) \tag{7–231}$$

then $\bar{\mathbf{e}}(0) = \mathbf{0}$. Referring to Eq. (7–222), which was

$$\bar{\mathbf{e}}(k+1) = \mathbf{L}(k)\bar{\mathbf{e}}(k)$$

we have

$$\bar{\mathbf{e}}(1) = \bar{\mathbf{e}}(2) = \cdots = \mathbf{0}$$

Thus,

$$\bar{\mathbf{e}}(k) = \mathbf{0}$$

This means that the mean value of the estimation error is zero for all $k \geq 0$, independent of $\mathbf{K}_e(k)$, if we choose $\tilde{\mathbf{x}}(0) = \bar{\mathbf{x}}(0)$. Then Eq. (7–230) becomes

$$\mathbf{M}(k) = \mathbf{P}(k) \tag{7–232}$$

By substituting Eq. (7–232) into Eq. (7–227), we obtain

$$\begin{aligned}\mathbf{P}(k+1) = {} & [\mathbf{G}(k) - \mathbf{K}_e(k)\mathbf{C}(k)]\mathbf{P}(k)[\mathbf{G}(k) - \mathbf{K}_e(k)\mathbf{C}(k)]^* \\ & + \mathbf{Q}(k) + \mathbf{K}_e(k)\mathbf{R}(k)\mathbf{K}_e^*(k)\end{aligned} \tag{7–233}$$

It is desired to minimize $\mathbf{P}(k+1)$ by properly choosing $\mathbf{K}_e(k)$. That is, we wish to minimize the quadratic form $\boldsymbol{\alpha}^*\mathbf{P}(k+1)\boldsymbol{\alpha}$, where $\boldsymbol{\alpha}$ is an n-vector, by properly choosing $\mathbf{K}_e(k)$. Since Eq. (7–233) can be rewritten as

$$\begin{aligned}\mathbf{P}(k+1) = {} & \mathbf{G}(k)\mathbf{P}(k)\mathbf{G}^*(k) - \mathbf{G}(k)\mathbf{P}(k)\mathbf{C}^*(k)\mathbf{K}_e^*(k) \\ & - \mathbf{K}_e(k)\mathbf{C}(k)\mathbf{P}(k)\mathbf{G}^*(k) + \mathbf{K}_e(k)\mathbf{C}(k)\mathbf{P}(k)\mathbf{C}^*(k)\mathbf{K}_e^*(k) \\ & + \mathbf{Q}(k) + \mathbf{K}_e(k)\mathbf{R}(k)\mathbf{K}_e^*(k) \\ = {} & \mathbf{Q}(k) + \mathbf{G}(k)\mathbf{P}(k)\mathbf{G}^*(k) + \mathbf{K}_e(k)[\mathbf{R}(k) + \mathbf{C}(k)\mathbf{P}(k)\mathbf{C}^*(k)]\mathbf{K}_e^*(k) \\ & - \mathbf{K}_e(k)\mathbf{C}(k)\mathbf{P}(k)\mathbf{G}^*(k) - \mathbf{G}(k)\mathbf{P}(k)\mathbf{C}^*(k)\mathbf{K}_e^*(k) \\ = {} & \mathbf{Q}(k) + \mathbf{G}(k)\mathbf{P}(k)\mathbf{G}^*(k) + \{\mathbf{K}_e(k) - \mathbf{G}(k)\mathbf{P}(k)\mathbf{C}^*(k)[\mathbf{R}(k) \\ & + \mathbf{C}(k)\mathbf{P}(k)\mathbf{C}^*(k)]^{-1}\} \\ & \cdot [\mathbf{R}(k) + \mathbf{C}(k)\mathbf{P}(k)\mathbf{C}^*(k)]\{\mathbf{K}_e(k) - \mathbf{G}(k)\mathbf{P}(k)\mathbf{C}^*(k)[\mathbf{R}(k) \\ & + \mathbf{C}(k)\mathbf{P}(k)\mathbf{C}^*(k)]^{-1}\}^* \\ & - \mathbf{G}(k)\mathbf{P}(k)\mathbf{C}^*(k)[\mathbf{R}(k) + \mathbf{C}(k)\mathbf{P}(k)\mathbf{C}^*(k)]^{-1}\mathbf{C}(k)\mathbf{P}(k)\mathbf{G}^*(k)\end{aligned} \tag{7–234}$$

the minimum of the quadratic form $\boldsymbol{\alpha}^*\mathbf{P}(k+1)\boldsymbol{\alpha}$ occurs when

$$\mathbf{K}_e(k) = \mathbf{G}(k)\mathbf{P}(k)\mathbf{C}^*(k)[\mathbf{R}(k) + \mathbf{C}(k)\mathbf{P}(k)\mathbf{C}^*(k)]^{-1} \tag{7–235}$$

By substituting Eq. (7–235) into $\boldsymbol{\alpha}^*\mathbf{P}(k+1)\boldsymbol{\alpha}$, we obtain the minimum value of the quadratic form $\boldsymbol{\alpha}^*\mathbf{P}(k+1)\boldsymbol{\alpha}$. The corresponding $\mathbf{P}(k+1)$ is given by substituting Eq. (7–235) into Eq. (7–234), as follows:

$$\begin{aligned}\mathbf{P}(k+1) &= \mathbf{Q}(k) + \mathbf{G}(k)\mathbf{P}(k)\mathbf{G}^*(k) \\ &\quad -\mathbf{G}(k)\mathbf{P}(k)\mathbf{C}^*(k)[\mathbf{R}(k) + \mathbf{C}(k)\mathbf{P}(k)\mathbf{C}^*(k)]^{-1}\mathbf{C}(k)\mathbf{P}(k)\mathbf{G}^*(k)\end{aligned} \tag{7–236}$$

Notice that Eq. (7–236) is a Riccati equation. Notice also that since $\bar{\mathbf{e}}(0) = \mathbf{0}$, referring to Eq. (7–207) we have

$$\begin{aligned}\mathbf{P}(0) &= E[\mathbf{e}(0)\mathbf{e}^*(0)] = E\{[\mathbf{e}(0) - \bar{\mathbf{e}}(0)][\mathbf{e}(0) - \bar{\mathbf{e}}(0)]^*\} \\ &= E\{[\mathbf{x}(0) - \bar{\mathbf{x}}(0)][\mathbf{x}(0) - \bar{\mathbf{x}}(0)]^*\} = \mathbf{P}_0\end{aligned}$$

Prediction-type Kalman filter. We shall now summarize what we have derived so far. The optimal estimation scheme based on Eqs. (7–214), (7–235), and (7–236) gives the minimum covariance of the estimation error. [The covariance of the estimation error is given by Eq. (7–236).] Equations (7–214), (7–235), and (7–236) are repeated here:

$$\tilde{\mathbf{x}}(k+1) = \mathbf{G}(k)\tilde{\mathbf{x}}(k) + \mathbf{H}(k)\mathbf{u}(k) + \mathbf{K}_e(k)[\mathbf{y}(k) - \mathbf{C}(k)\tilde{\mathbf{x}}(k)] \tag{7–237}$$

$$\mathbf{K}_e(k) = \mathbf{G}(k)\mathbf{P}(k)\mathbf{C}^*(k)[\mathbf{R}(k) + \mathbf{C}(k)\mathbf{P}(k)\mathbf{C}^*(k)]^{-1} \tag{7–238}$$

$$\begin{aligned}\mathbf{P}(k+1) &= \mathbf{Q}(k) + \mathbf{G}(k)\mathbf{P}(k)\mathbf{G}^*(k) - \mathbf{G}(k)\mathbf{P}(k)\mathbf{C}^*(k)[\mathbf{R}(k) \\ &\quad + \mathbf{C}(k)\mathbf{P}(k)\mathbf{C}^*(k)]^{-1}\mathbf{C}(k)\mathbf{P}(k)\mathbf{G}^*(k) \\ &= \mathbf{Q}(k) + [\mathbf{G}(k) - \mathbf{K}_e(k)\mathbf{C}(k)]\mathbf{P}(k)\mathbf{G}^*(k)\end{aligned} \tag{7–239}$$

where

$$\tilde{\mathbf{x}}(0) = \bar{\mathbf{x}}(0) \qquad \text{and} \qquad \mathbf{P}(0) = \mathbf{P}_0$$

Equations (7–237), (7–238), and (7–239) define the prediction-type Kalman filter. Figure 7–7 is a block diagram for the prediction-type Kalman filter.

Example 7–8.

Consider the system defined by the equations

$$\mathbf{x}(k+1) = \mathbf{x}(k)$$

$$y(k) = \mathbf{C}(k)\mathbf{x}(k) + \epsilon(k)$$

where

$$E[\epsilon^2(k)] = R(k) = \sigma^2$$

We wish to estimate state $\mathbf{x}(k)$ and to obtain the prediction-type Kalman filter equation on the basis of Eqs. (7–237), (7–238), and (7–239).

Notice that in this system

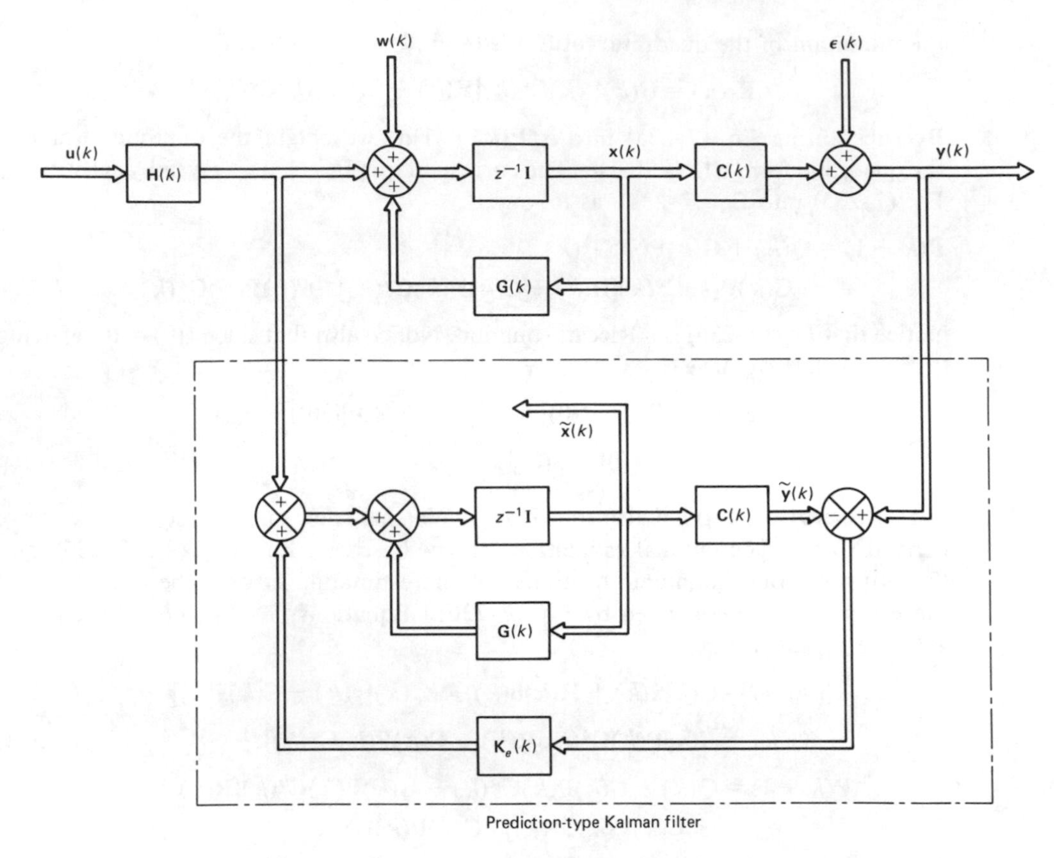

Figure 7–7 Block diagram for the prediction-type Kalman filter.

$$\mathbf{G}(k) = \mathbf{I} \qquad \text{and} \qquad \mathbf{Q}(k) = \mathbf{0}$$

Hence Eq. (7–237) becomes

$$\tilde{\mathbf{x}}(k+1) = \tilde{\mathbf{x}}(k) + \mathbf{K}_e(k)[y(k) - \mathbf{C}(k)\tilde{\mathbf{x}}(k)] \tag{7–240}$$

where $\tilde{\mathbf{x}}(0) = \bar{\mathbf{x}}(0)$. Since $R(k) = \sigma^2$, Eq. (7–238) becomes

$$\begin{aligned}\mathbf{K}_e(k) &= \mathbf{P}(k)\mathbf{C}^*(k)[\sigma^2 + \mathbf{C}(k)\mathbf{P}(k)\mathbf{C}^*(k)]^{-1} \\ &= \frac{\mathbf{P}(k)\mathbf{C}^*(k)}{\sigma^2 + \mathbf{C}(k)\mathbf{P}(k)\mathbf{C}^*(k)}\end{aligned} \tag{7–241}$$

Also, noting that $\mathbf{Q}(k) = \mathbf{0}$ and $R(k) = \sigma^2$, Eq. (7–239) becomes

$$\begin{aligned}\mathbf{P}(k+1) &= \mathbf{P}(k) - \mathbf{P}(k)\mathbf{C}^*(k)[\sigma^2 + \mathbf{C}(k)\mathbf{P}(k)\mathbf{C}^*(k)]^{-1}\mathbf{C}(k)\mathbf{P}(k) \\ &= \mathbf{P}(k) - \frac{\mathbf{P}(k)\mathbf{C}^*(k)\mathbf{C}(k)\mathbf{P}(k)}{\sigma^2 + \mathbf{C}(k)\mathbf{P}(k)\mathbf{C}^*(k)}\end{aligned} \tag{7–242}$$

Equations (7–240), (7–241), and (7–242) define the prediction-type Kalman filter. Comparing Eqs. (7–240), (7–241), and (7–242) with Eqs. (7–193), (7–195), and (7–196), respectively, we notice that they are of the same form with the exception of differences in the arguments of the gain matrix, the output vector, and the **C** matrix. [Note that the current estimation-type Kalman filter equations for the present system become identical with Eqs. (7–193), (7–195), and (7–196) if $\sigma^2 = 1$. See Example 7–9.]

Current estimation-type Kalman filter. Suppose that the state equation and output equation are as given by Eqs. (7–202) and (7–203):

$$\mathbf{x}(k+1) = \mathbf{G}(k)\mathbf{x}(k) + \mathbf{H}(k)\mathbf{u}(k) + \mathbf{w}(k) \tag{7–243}$$

$$\mathbf{y}(k) = \mathbf{C}(k)\mathbf{x}(k) + \boldsymbol{\epsilon}(k) \tag{7–244}$$

where $\mathbf{w}(k)$ and $\boldsymbol{\epsilon}(k)$ are assumed to satisfy all conditions given by Eqs. (7–204) through (7–212).

Our problem here is as follows. Given $\mathbf{y}(0), \mathbf{y}(1), \ldots, \mathbf{y}(k)$, determine the optimal estimate $\tilde{\mathbf{x}}(k)$ such that

$$\mathbf{P}(k) = E\,[\mathbf{e}(k)\mathbf{e}^*(k)] \tag{7–245}$$

is minimum, where $\mathbf{e}(k) = \mathbf{x}(k) - \tilde{\mathbf{x}}(k)$. That is, we wish to determine the optimal estimate $\tilde{\mathbf{x}}(k)$ such that the quadratic form $\boldsymbol{\alpha}^*\mathbf{P}(k)\boldsymbol{\alpha}$, where $\boldsymbol{\alpha}$ is an arbitrary n-vector, is minimum. It can be shown that the current estimation-type Kalman filter that minimizes $\mathbf{P}(k)$ as given by Eq. (7–245) can be given by the equations

$$\tilde{\mathbf{x}}(k+1) = \mathbf{z}(k+1) + \mathbf{K}_e(k+1)[\mathbf{y}(k+1) - \mathbf{C}(k+1)\mathbf{z}(k+1)] \tag{7–246}$$

$$\mathbf{z}(k+1) = \mathbf{G}(k)\tilde{\mathbf{x}}(k) + \mathbf{H}(k)\mathbf{u}(k) \tag{7–247}$$

[Compare Eqs. (7–246) and (7–247) with the equations for the current observer given by Eqs. (6–162) and (6–163).] From Eq. (7–246) we obtain

$$\tilde{\mathbf{x}}(0) = \mathbf{z}(0) + \mathbf{K}_e(0)[\mathbf{y}(0) - \mathbf{C}(0)\mathbf{z}(0)]$$

If we choose

$$\mathbf{z}(0) = \bar{\mathbf{x}}(0)$$

which is deterministic, then $\tilde{\mathbf{x}}(0)$ can be given by the equation

$$\tilde{\mathbf{x}}(0) = \bar{\mathbf{x}}(0) + \mathbf{K}_e(0)[\mathbf{y}(0) - \mathbf{C}(0)\bar{\mathbf{x}}(0)]$$

Referring to Eqs. (7–244) and (7–246), the estimation error vector $\mathbf{e}(k+1)$ can be given by

$$\begin{aligned}
\mathbf{e}(k+1) &= \mathbf{x}(k+1) - \tilde{\mathbf{x}}(k+1) \\
&= \mathbf{x}(k+1) - \mathbf{z}(k+1) - \mathbf{K}_e(k+1)[\mathbf{y}(k+1) - \mathbf{C}(k+1)\mathbf{z}(k+1)] \\
&= [\mathbf{K}_e(k+1)\mathbf{C}(k+1) - \mathbf{I}]\mathbf{z}(k+1) \\
&\quad - \mathbf{K}_e(k+1)[\mathbf{C}(k+1)\mathbf{x}(k+1) + \boldsymbol{\epsilon}(k+1)] + \mathbf{x}(k+1)
\end{aligned}$$

$$= [\mathbf{K}_e(k+1)\mathbf{C}(k+1) - \mathbf{I}][\mathbf{z}(k+1) - \mathbf{x}(k+1)] - \mathbf{K}_e(k+1)\boldsymbol{\epsilon}(k+1) \tag{7–248}$$

Equation (7–248) can be rewritten as follows:

$$\begin{aligned}\mathbf{e}(k+1) &= [\mathbf{K}_e(k+1)\mathbf{C}(k+1) - \mathbf{I}][\mathbf{G}(k)\tilde{\mathbf{x}}(k) + \mathbf{H}(k)\mathbf{u}(k) - \mathbf{G}(k)\mathbf{x}(k) \\ &\quad - \mathbf{H}(k)\mathbf{u}(k) - \mathbf{w}(k)] - \mathbf{K}_e(k+1)\boldsymbol{\epsilon}(k+1) \\ &= [\mathbf{I} - \mathbf{K}_e(k+1)\mathbf{C}(k+1)]\mathbf{G}(k)\mathbf{e}(k) \\ &\quad + [\mathbf{I} - \mathbf{K}_e(k+1)\mathbf{C}(k+1)]\mathbf{w}(k) - \mathbf{K}_e(k+1)\boldsymbol{\epsilon}(k+1)\end{aligned}$$

or

$$\mathbf{e}(k+1) = [\mathbf{I} - \mathbf{K}_e(k+1)\mathbf{C}(k+1)]\mathbf{G}(k)\mathbf{e}(k) + \boldsymbol{\xi}(k) \tag{7–249}$$

where

$$\boldsymbol{\xi}(k) = [\mathbf{I} - \mathbf{K}_e(k+1)\mathbf{C}(k+1)]\mathbf{w}(k) - \mathbf{K}_e(k+1)\boldsymbol{\epsilon}(k+1)$$

which is statistically independent of $\mathbf{e}(k)$. Hence

$$E[\mathbf{e}(k+1)] = [\mathbf{I} - \mathbf{K}_e(k+1)\mathbf{C}(k+1)]\mathbf{G}(k)E[\mathbf{e}(k)] \tag{7–250}$$

Notice that from Eq. (7–248) we have

$$\mathbf{e}(0) = [\mathbf{K}_e(0)\mathbf{C}(0) - \mathbf{I}][\mathbf{z}(0) - \mathbf{x}(0)] - \mathbf{K}_e(0)\boldsymbol{\epsilon}(0)$$

Hence

$$E[\mathbf{e}(0)] = [\mathbf{K}_e(0)\mathbf{C}(0) - \mathbf{I}][\mathbf{z}(0) - \bar{\mathbf{x}}(0)] \tag{7–251}$$

Note that

$$\begin{aligned}E[\boldsymbol{\xi}(k)\boldsymbol{\xi}^*(k)] &= [\mathbf{I} - \mathbf{K}_e(k+1)\mathbf{C}(k+1)]\mathbf{Q}(k)[\mathbf{I} - \mathbf{K}_e(k+1)\mathbf{C}(k+1)]^* \\ &\quad + \mathbf{K}_e(k+1)\mathbf{R}(k+1)\mathbf{K}_e^*(k+1)\end{aligned} \tag{7–252}$$

Since we have chosen

$$\mathbf{z}(0) = \bar{\mathbf{x}}(0)$$

Eq. (7–251) gives us $\bar{\mathbf{e}}(0) = \mathbf{0}$. Then, from Eq. (7–250) we find

$$\bar{\mathbf{e}}(1) = \bar{\mathbf{e}}(2) = \cdots = \mathbf{0}$$

or

$$\bar{\mathbf{e}}(k) = \mathbf{0} \qquad k = 0, 1, 2, \ldots$$

Define

$$\mathbf{P}(k+1) = E[\mathbf{e}(k+1)\mathbf{e}^*(k+1)]$$

Then, referring to Eqs. (7–249) and (7–252), we obtain

$$\begin{aligned}\mathbf{P}(k+1) &= [\mathbf{I}-\mathbf{K}_e(k+1)\mathbf{C}(k+1)]\mathbf{G}(k)\mathbf{P}(k)\mathbf{G}^*(k)[\mathbf{I}-\mathbf{K}_e(k+1)\mathbf{C}(k+1)]^* \\ &\quad + [\mathbf{I}-\mathbf{K}_e(k+1)\mathbf{C}(k+1)]\mathbf{Q}(k)[\mathbf{I}-\mathbf{K}_e(k+1)\mathbf{C}(k+1)]^* \\ &\quad + \mathbf{K}_e(k+1)\mathbf{R}(k+1)\mathbf{K}_e^*(k+1) \\ &= [\mathbf{I}-\mathbf{K}_e(k+1)\mathbf{C}(k+1)][\mathbf{G}(k)\mathbf{P}(k)\mathbf{G}^*(k)+\mathbf{Q}(k)][\mathbf{I}-\mathbf{K}_e(k+1)\mathbf{C}(k+1)]^* \\ &\quad + \mathbf{K}_e(k+1)\mathbf{R}(k+1)\mathbf{K}_e^*(k+1) \qquad (7\text{–}253)\end{aligned}$$

Define

$$\mathbf{N}(k+1) = \mathbf{G}(k)\mathbf{P}(k)\mathbf{G}^*(k)+\mathbf{Q}(k) \qquad (7\text{–}254)$$

Then Eq. (7–253) becomes

$$\begin{aligned}\mathbf{P}(k+1) &= [\mathbf{I}-\mathbf{K}_e(k+1)\mathbf{C}(k+1)]\mathbf{N}(k+1)[\mathbf{I}-\mathbf{K}_e(k+1)\mathbf{C}(k+1)]^* \\ &\quad + \mathbf{K}_e(k+1)\mathbf{R}(k+1)\mathbf{K}_e^*(k+1) \\ &= \mathbf{N}(k+1) + \mathbf{K}_e(k+1)[\mathbf{R}(k+1) \\ &\quad + \mathbf{C}(k+1)\mathbf{N}(k+1)\mathbf{C}^*(k+1)]\mathbf{K}_e^*(k+1) \\ &\quad - \mathbf{N}(k+1)\mathbf{C}^*(k+1)\mathbf{K}_e^*(k+1) - \mathbf{K}_e(k+1)\mathbf{C}(k+1)\mathbf{N}(k+1) \\ &= \mathbf{N}(k+1) + \{\mathbf{K}_e(k+1) - \mathbf{N}(k+1)\mathbf{C}^*(k+1)[\mathbf{R}(k+1) \\ &\quad + \mathbf{C}(k+1)\mathbf{N}(k+1)\mathbf{C}^*(k+1)]^{-1}\}[\mathbf{R}(k+1) \\ &\quad + \mathbf{C}(k+1)\mathbf{N}(k+1)\mathbf{C}^*(k+1)] \cdot \{\mathbf{K}_e(k+1) \\ &\quad - \mathbf{N}(k+1)\mathbf{C}^*(k+1)[\mathbf{R}(k+1)+\mathbf{C}(k+1)\mathbf{N}(k+1)\mathbf{C}^*(k+1)]^{-1}\}^* \\ &\quad - \mathbf{N}(k+1)\mathbf{C}^*(k+1)[\mathbf{R}(k+1) \\ &\quad + \mathbf{C}(k+1)\mathbf{N}(k+1)\mathbf{C}^*(k+1)]^{-1}\mathbf{C}(k+1)\mathbf{N}(k+1)\end{aligned}$$

Note that the quadratic form $\boldsymbol{\alpha}^*\mathbf{P}(k+1)\boldsymbol{\alpha}$ becomes minimum when

$$\mathbf{K}_e(k+1) = \mathbf{N}(k+1)\mathbf{C}^*(k+1)[\mathbf{R}(k+1)+\mathbf{C}(k+1)\mathbf{N}(k+1)\mathbf{C}^*(k+1)]^{-1} \qquad (7\text{–}255)$$

and the corresponding $\mathbf{P}(k+1)$ becomes as follows:

$$\begin{aligned}\mathbf{P}(k+1) &= \mathbf{N}(k+1) - \mathbf{N}(k+1)\mathbf{C}^*(k+1)[\mathbf{R}(k+1) \\ &\quad + \mathbf{C}(k+1)\mathbf{N}(k+1)\mathbf{C}^*(k+1)]^{-1}\mathbf{C}(k+1)\mathbf{N}(k+1) \\ &= [\mathbf{I}-\mathbf{K}_e(k+1)\mathbf{C}(k+1)]\mathbf{N}(k+1) \qquad (7\text{–}256)\end{aligned}$$

Now let us summarize what we have obtained. The current estimation-type Kalman filter can be given by Eqs. (7–246), (7–247), (7–254), (7–255), and (7–256), or by the equations

$$\tilde{\mathbf{x}}(k+1) = \mathbf{z}(k+1) + \mathbf{K}_e(k+1)[\mathbf{y}(k+1) - \mathbf{C}(k+1)\mathbf{z}(k+1)] \qquad (7\text{–}257)$$

$$\mathbf{z}(k+1) = \mathbf{G}(k)\tilde{\mathbf{x}}(k) + \mathbf{H}(k)\mathbf{u}(k) \qquad (7\text{–}258)$$

where

$$\mathbf{K}_e(k+1) = \mathbf{N}(k+1)\mathbf{C}^*(k+1)[\mathbf{R}(k+1) + \mathbf{C}(k+1)\mathbf{N}(k+1)\mathbf{C}^*(k+1)]^{-1} \tag{7–259}$$

$$\mathbf{P}(k+1) = \mathbf{N}(k+1) - \mathbf{N}(k+1)\mathbf{C}^*(k+1)[\mathbf{R}(k+1) + \mathbf{C}(k+1)\mathbf{N}(k+1)\mathbf{C}^*(k+1)]^{-1}\mathbf{C}(k+1)\mathbf{N}(k+1) \tag{7–260}$$

$$\mathbf{N}(k+1) = \mathbf{G}(k)\mathbf{P}(k)\mathbf{G}^*(k) + \mathbf{Q}(k) \tag{7–261}$$

Equations (7–257) through (7–261) define the current estimation-type Kalman filter. Figure 7–8 shows the block diagram for this Kalman filter. The initial conditions involved in this type of Kalman filter are given as follows:

$$\tilde{\mathbf{x}}(0) = \bar{\mathbf{x}}(0) + \mathbf{K}_e(0)[\mathbf{y}(0) - \mathbf{C}(0)\bar{\mathbf{x}}(0)] \tag{7–262}$$

$$\mathbf{z}(0) = \bar{\mathbf{x}}(0) \tag{7–263}$$

$$\mathbf{K}_e(0) = \mathbf{P}_0\mathbf{C}^*(0)[\mathbf{R}(0) + \mathbf{C}(0)\mathbf{N}(0)\mathbf{C}^*(0)]^{-1} \tag{7–264}$$

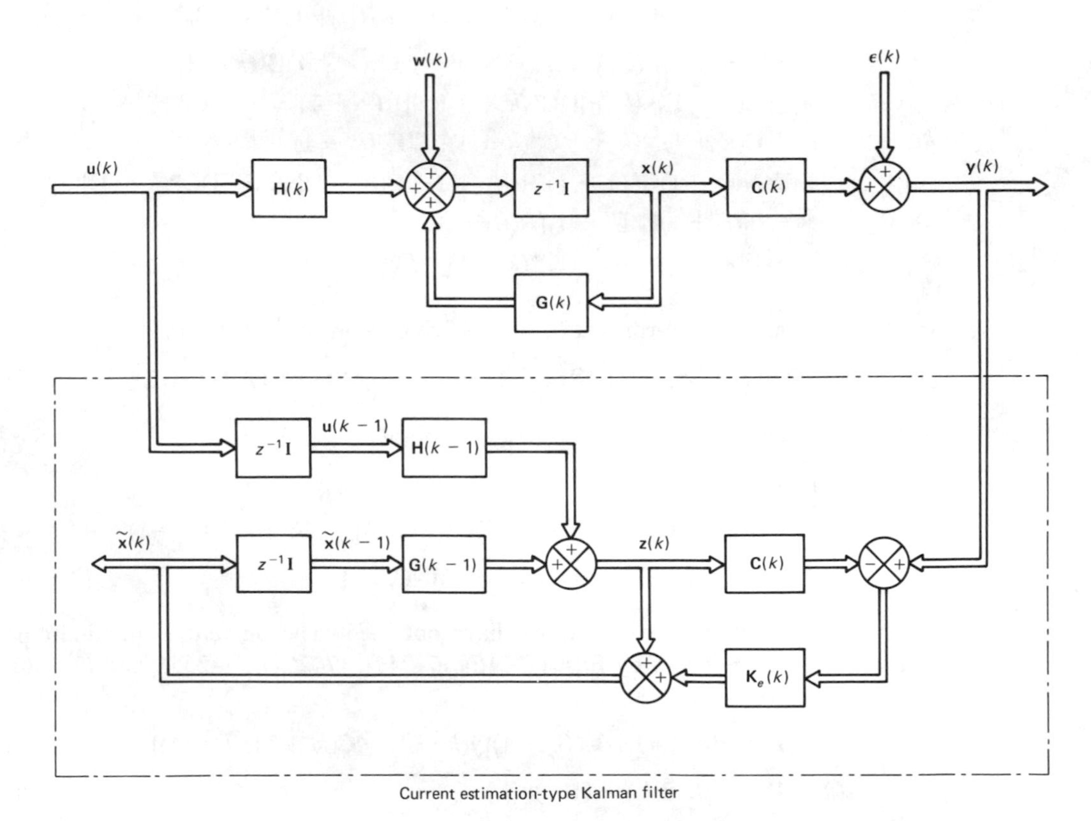

Figure 7–8 Block diagram for the current estimation-type Kalman filter.

$$\mathbf{P}(0) = [\mathbf{I} - \mathbf{K}_e(0)\mathbf{C}(0)]\mathbf{P}_0 \tag{7–265}$$

$$\mathbf{N}(0) = \mathbf{P}_0 \tag{7–266}$$

Note that matrix $\mathbf{N}(k+1)$ satisfies the following equation:

$$\begin{aligned}\mathbf{N}(k+1) = \mathbf{Q}(k) + \mathbf{G}(k)\mathbf{N}(k)\mathbf{G}^*(k) - \mathbf{G}(k)\mathbf{N}(k)\mathbf{C}^*(k)[\mathbf{R}(k) \\ + \mathbf{C}(k)\mathbf{N}(k)\mathbf{C}^*(k)]^{-1}\mathbf{C}(k)\mathbf{N}(k)\mathbf{G}^*(k)\end{aligned} \tag{7–267}$$

which is of the same form as Eq. (7–239).

Example 7–9.

Consider the same system discussed in Example 7–8:

$$\mathbf{x}(k+1) = \mathbf{x}(k)$$

$$y(k) = \mathbf{C}(k)\mathbf{x}(k) + \epsilon(k)$$

where

$$E[\epsilon^2(k)] = R(k) = \sigma^2$$

Obtain the current estimation-type Kalman filter equations.

Since $\mathbf{G}(k) = \mathbf{I}$ and $\mathbf{Q}(k) = \mathbf{0}$, Eqs. (7–257) and (7–258) can be combined into one equation as follows:

$$\tilde{\mathbf{x}}(k+1) = \tilde{\mathbf{x}}(k) + \mathbf{K}_e(k+1)[y(k+1) - \mathbf{C}(k+1)\tilde{\mathbf{x}}(k)] \tag{7–268}$$

Knowing that $\mathbf{Q}(k) = \mathbf{0}$ and $R(k+1) = \sigma^2$, and referring to Eq. (7–261), we can rewrite Eq. (7–259) as follows:

$$\begin{aligned}\mathbf{K}_e(k+1) &= \mathbf{P}(k)\mathbf{C}^*(k+1)[\sigma^2 + \mathbf{C}(k+1)\mathbf{P}(k)\mathbf{C}^*(k+1)]^{-1} \\ &= \frac{\mathbf{P}(k)\mathbf{C}^*(k+1)}{\sigma^2 + \mathbf{C}(k+1)\mathbf{P}(k)\mathbf{C}^*(k+1)}\end{aligned} \tag{7–269}$$

Also, Eq. (7–256) or Eq. (7–260) becomes

$$\begin{aligned}\mathbf{P}(k+1) &= [\mathbf{I} - \mathbf{K}_e(k+1)\mathbf{C}(k+1)]\mathbf{P}(k) \\ &= \mathbf{P}(k) - \frac{\mathbf{P}(k)\mathbf{C}^*(k+1)\mathbf{C}(k+1)\mathbf{P}(k)}{\sigma^2 + \mathbf{C}(k+1)\mathbf{P}(k)\mathbf{C}^*(k+1)}\end{aligned} \tag{7–270}$$

Comparing Eqs. (7–268) and (7–193), notice that they are of identical form. Also, Eqs. (7–269) and (7–270) are of identical form, respectively, with Eqs. (7–195) and (7–196) if $\sigma^2 = 1$. Thus, the time-varying optimal current estimation solution (current estimation-type Kalman filter) and the time-varying optimal solution to the identification problem based on the least-squares error are identical.

Steady-state Kalman filters. Thus far we have discussed non-steady-state Kalman filters. We shall next consider steady-state Kalman filters.

Consider the system described by the equations

$$\mathbf{x}(k+1) = \mathbf{G}\mathbf{x}(k) + \mathbf{H}\mathbf{u}(k) + \mathbf{w}(k) \tag{7–271}$$

$$\mathbf{y}(k) = \mathbf{C}\mathbf{x}(k) + \boldsymbol{\epsilon}(k) \tag{7–272}$$

This system model differs from that given by Eqs. (7–202) and (7–203) in that the state matrix **G**, input matrix **H**, and output matrix **C** are constant matrices and the noises $\mathbf{w}(k)$ and $\boldsymbol{\epsilon}(k)$ are steady-state noises, so that $\mathbf{R}(k) = \mathbf{R}$ and $\mathbf{Q}(k) = \mathbf{Q}$.

We shall first treat the prediction-type Kalman filter. Referring to Eqs. (7–237), (7–238), and (7–239), the steady-state prediction-type Kalman filter can be given by the equations

$$\tilde{\mathbf{x}}(k+1) = \mathbf{G}\tilde{\mathbf{x}}(k) + \mathbf{H}\mathbf{u}(k) + \mathbf{K}_e(k)[\mathbf{y}(k) - \mathbf{C}\tilde{\mathbf{x}}(k)]$$

$$\mathbf{K}_e(k) = \mathbf{G}\mathbf{P}(k)\mathbf{C}^*[\mathbf{R} + \mathbf{C}\mathbf{P}(k)\mathbf{C}^*]^{-1}$$

$$\begin{aligned}\mathbf{P}(k+1) &= \mathbf{Q} + \mathbf{G}\mathbf{P}(k)\mathbf{G}^* - \mathbf{G}\mathbf{P}(k)\mathbf{C}^*[\mathbf{R} + \mathbf{C}\mathbf{P}(k)\mathbf{C}^*]^{-1}\mathbf{C}\mathbf{P}(k)\mathbf{G}^* \\ &= \mathbf{Q} + [\mathbf{G} - \mathbf{K}_e(k)\mathbf{C}]\mathbf{P}(k)\mathbf{G}^*\end{aligned}$$

Notice that matrix $\mathbf{P}(k+1)$ can be rewritten as follows:

$$\begin{aligned}\mathbf{P}(k+1) &= [\mathbf{G} - \mathbf{K}_e(k)\mathbf{C}]\mathbf{P}(k)[\mathbf{G} - \mathbf{K}_e(k)\mathbf{C}]^* \\ &\quad + \mathbf{K}_e(k)[\mathbf{R} + \mathbf{C}\mathbf{P}(k)\mathbf{C}^*]\mathbf{K}_e^*(k) - \mathbf{K}_e(k)\mathbf{C}\mathbf{P}(k)\mathbf{C}^*\mathbf{K}_e^*(k) + \mathbf{Q} \\ &= [\mathbf{G} - \mathbf{K}_e(k)\mathbf{C}]\mathbf{P}(k)[\mathbf{G} - \mathbf{K}_e(k)\mathbf{C}]^* + \mathbf{K}_e(k)\mathbf{R}\mathbf{K}_e^*(k) + \mathbf{Q} \qquad (7\text{–}273)\end{aligned}$$

Let us examine the Kalman filter gain $\mathbf{K}_e(k)$ and the covariance of estimation error $\mathbf{P}(k)$ as k approaches infinity. It can be shown that if the system is observable, then matrix $\mathbf{P}(k+1)$ as given by Eq. (7–273) approaches a positive semidefinite matrix **P**. If the system is controllable, matrix $\mathbf{P}(k)$, with the initial value $\mathbf{P}(0)$, which is positive semidefinite, converges to a unique positive definite matrix **P**. In this case $\mathbf{K}_e(k)$ approaches a constant matrix $\mathbf{K}_e$. Then $\mathbf{K}_e$ and **P** can be given as follows:

$$\mathbf{K}_e = \mathbf{G}\mathbf{P}\mathbf{C}^*(\mathbf{R} + \mathbf{C}\mathbf{P}\mathbf{C}^*)^{-1}$$

$$\mathbf{P} = \mathbf{Q} + \mathbf{G}\mathbf{P}\mathbf{G}^* - \mathbf{G}\mathbf{P}\mathbf{C}^*(\mathbf{R} + \mathbf{C}\mathbf{P}\mathbf{C}^*)^{-1}\mathbf{C}\mathbf{P}\mathbf{G}^*$$

By referring to Eq. (7–273), it is possible to rewrite this last equation as follows:

$$\mathbf{P} = (\mathbf{G} - \mathbf{K}_e\mathbf{C})\mathbf{P}(\mathbf{G} - \mathbf{K}_e\mathbf{C})^* + \mathbf{K}_e\mathbf{R}\mathbf{K}_e^* + \mathbf{Q}$$

Notice that this equation has the same form as the Liapunov equation given by Eq. (7–125). Since $\mathbf{K}_e\mathbf{R}\mathbf{K}_e^* + \mathbf{Q}$ is not necessarily positive definite, it is not possible to say that $\mathbf{G} - \mathbf{K}_e\mathbf{C}$ is a stable matrix. It can be proved, however, that if the system is controllable and observable, then $\mathbf{G} - \mathbf{K}_e\mathbf{C}$ is a stable matrix.

Steady-State Prediction-Type Kalman Filter. From what we have just presented, the steady-state prediction-type Kalman filter can be given by

$$\tilde{\mathbf{x}}(k+1) = \mathbf{G}\tilde{\mathbf{x}}(k) + \mathbf{H}\mathbf{u}(k) + \mathbf{K}_e[\mathbf{y}(k) - \mathbf{C}\tilde{\mathbf{x}}(k)] \qquad (7\text{–}274)$$

where

$$\mathbf{K}_e = \mathbf{G}\mathbf{P}\mathbf{C}^*(\mathbf{R} + \mathbf{C}\mathbf{P}\mathbf{C}^*)^{-1} \qquad (7\text{–}275)$$

$$\mathbf{P} = \mathbf{Q} + \mathbf{G}\mathbf{P}\mathbf{G}^* - \mathbf{G}\mathbf{P}\mathbf{C}^*(\mathbf{R} + \mathbf{C}\mathbf{P}\mathbf{C}^*)^{-1}\mathbf{C}\mathbf{P}\mathbf{G}^* \qquad (7\text{–}276)$$

Equation (7–275) gives the steady-state gain $\mathbf{K}_e$, and Eq. (7–276) gives the covariance matrix of the estimation error. Equation (7–276) is the algebraic Riccati equation.

Steady-State Current Estimation-Type Kalman Filter. The steady-state current estimation-type Kalman filter can be obtained in much the same way as the steady-state prediction-type filter. From Eqs. (7–257), (7–258), (7–259), (7–260), and (7–261), the steady-state current estimation-type Kalman filter can be given by

$$\tilde{\mathbf{x}}(k+1) = \mathbf{z}(k+1) + \mathbf{K}_e[\mathbf{y}(k+1) - \mathbf{C}\mathbf{z}(k+1)] \tag{7–277}$$

$$\mathbf{z}(k+1) = \mathbf{G}\tilde{\mathbf{x}}(k) + \mathbf{H}\mathbf{u}(k) \tag{7–278}$$

where

$$\mathbf{K}_e = \mathbf{N}\mathbf{C}^*(\mathbf{R} + \mathbf{C}\mathbf{N}\mathbf{C}^*)^{-1} \tag{7–279}$$

$$\mathbf{N} = \mathbf{Q} + \mathbf{G}\mathbf{N}\mathbf{G}^* - \mathbf{G}\mathbf{N}\mathbf{C}^*(\mathbf{R} + \mathbf{C}\mathbf{N}\mathbf{C}^*)^{-1}\mathbf{C}\mathbf{N}\mathbf{G}^* \tag{7–280}$$

Notice that the equation for $\mathbf{N}$ [given by Eq. (7–280)] is of the same form as the equation for $\mathbf{P}$ given by Eq. (7–276).

Summary. In this section, we have presented both the prediction-type Kalman filter and the current estimation-type Kalman filter. Each of these optimal estimation schemes, which provide optimal estimates of the state, will become an integral part of the optimal controller.

Unless the state of the control system can be precisely measured, the optimal control strategy can be divided into two stages. In the first stage, we make the best estimate of the state based on the observed or measured sequence of the output. (To obtain an optimal estimate of the state, we may use one of the Kalman filters presented here or we may use the approach presented in Sec. 6–6 to obtain an estimate of the state.) In the second stage, we determine the optimal state feedback law, where we use the estimated state for feedback purposes.

The optimal controller using a Kalman filter that we have presented in this section can be used even if there are no noises or disturbances present in the system.

REFERENCES

7–1. Anderson, B. D. O., and J. B. Moore, *Optimal Filtering*. Englewood Cliffs, N.J.: Prentice-Hall, Inc., 1979.

7–2. Åström, K. J., and P. Eykhoff, "System Identification: A Survey," *Automatica*, **7** (1971), pp. 123–62.

7–3. Åström, K. J., and B. Wittenmark, *Computer Controlled Systems: Theory and Design*. Englewood Cliffs, N.J.: Prentice-Hall, Inc., 1984.

7–4. Athans, M., "The Role and Use of the Stochastic Linear-Quadratic-Gaussian Problem in Control System Design," *IEEE Trans. Automatic Control*, **AC-16** (1971), pp. 529–52.

7–5. Chan, S. W., G. C. Goodwin, and K. S. Sin, "Convergence Properties of the Riccati Difference Equation in Optimal Filtering of Nonstabilizable Systems," *IEEE Trans. Automatic Control*, **AC-29** (1984), pp. 110–18.

7–6. Dorato, P., and A. H. Levis, "Optimal Linear Regulators: The Discrete-Time Case," *IEEE Trans. Automatic Control*, **AC-16** (1971), pp. 613–20.

7–7. Fortmann, T. E., "A Matrix Inversion Identity," *IEEE Trans. Automatic Control*, **AC-15** (1970), p. 599.

7–8. Franklin, G. F., and J. D. Powell, *Digital Control of Dynamic Systems*. Reading, Mass: Addison-Wesley Publishing Co., Inc., 1980.

7–9. Kailath, T., "An Innovations Approach to Least-Squares Estimation, Part I: Linear Filtering in Additive White Noise," *IEEE Trans. Automatic Control*, **AC-13** (1968), pp. 646–55.

7–10. Kailath, T., "A View of Three Decades of Linear Filtering Theory," *IEEE Trans. Information Theory*, **IT-20** (1974), pp. 146–81.

7–11. Kailath, T., and P. Frost, "An Innovations Approach to Least-Squares Estimation, Part II: Linear Smoothing in Additive White Noise," *IEEE Trans. Automatic Control*, **AC-13** (1968), pp. 655–60.

7–12. Kalman, R. E., "A New Approach to Linear Filtering and Prediction Problems," *ASME Trans. Ser. D, J. Basic Engineering*, **82** (1960), pp. 34–45.

7–13. Kalman, R. E., and R. S. Bucy, "New Results in Linear Filtering and Prediction Theory," *ASME J. Basic Engineering, ser*, **D, 83** (1961), pp. 95–108.

7–14. Katz, P., *Digital Control Using Microprocessors*. London: Prentice-Hall International, Inc., 1981.

7–15. Kuo, B. C., *Digital Control Systems*. New York: Holt, Rinehart and Winston, Inc., 1980.

7–16. Lee, E. B., and L. Markus, *Foundations of Optimal Control Theory*. New York: John Wiley & Sons, Inc., 1967.

7–17. Ljung, L., "Analysis of a General Recursive Prediction Error Identification Algorithm," *Automatica*, **17** (1981), pp. 89–99.

7–18. Ogata, K., *Modern Control Engineering*. Englewood Cliffs, N.J.: Prentice-Hall, Inc., 1970.

7–19. Ogata, K., *State Space Analysis of Control Systems*. Englewood Cliffs, N.J.: Prentice-Hall, Inc., 1967.

7–20. Pappas, T., A. J. Laub, and N. R. Sandell, Jr., "On the Numerical Solution of the Discrete-Time Algebraic Riccati Equation," *IEEE Trans. Automatic Control*, **AC-25** (1980), pp. 631–41.

7–21. Payne, H. J., and L. M. Silverman, "On the Discrete Time Algebraic Riccati Equation," *IEEE Trans. Automatic Control*, **AC-18** (1973), pp. 226–34.

7–22. Phillips, C. L., and H. T. Nagle, Jr., *Digital Control System Analysis and Design*. Englewood Cliffs, N.J.: Prentice-Hall, Inc., 1984.

7–23. Price, C. F., "An Analysis of the Divergence Problem in the Kalman Filter," *IEEE Trans. Automatic Control*, **AC-13** (1968), pp. 699–702.

7–24. Sage, A. P., and J. M. Melsa, *Estimation Theory with Applications to Communications and Control*. New York: McGraw-Hill Book Company, 1971.

7–25. Schlee, F. H., C. J. Standish, and N. F. Toda, "Divergence in the Kalman Filter," *AIAA J.*, **5** (1967), pp. 1114–20.

7–26. Sorensen, H. W., "Least-Squares Estimation: From Gauss to Kalman," *IEEE Spectrum*, **7**, no. 7 (July 1970), pp. 63–68.

7–27. Van Dooren, P., "A Generalized Eigenvalue Approach for Solving Riccati Equations," *SIAM J. Scientific and Statistical Computing*, **2** (1981), pp. 121–35.

EXAMPLE PROBLEMS AND SOLUTIONS

Problem A-7–1. Consider the discrete-time control system

$$\mathbf{x}(k+1) = \mathbf{G}\mathbf{x}(k) + \mathbf{H}\mathbf{u}(k)$$

where

$\mathbf{x}(k)$ = state vector (n-vector)

$\mathbf{u}(k)$ = control vector (r-vector)

$\mathbf{G} = n \times n$ nonsingular matrix

$\mathbf{H} = n \times r$ matrix

We wish to find the optimal control vector that will minimize the following performance index:

$$J = \frac{1}{2}\mathbf{x}^*(N)\mathbf{S}\mathbf{x}(N) + \frac{1}{2}\sum_{k=0}^{N-1}[\mathbf{x}^*(k)\mathbf{Q}\mathbf{x}(k) + \mathbf{u}^*(k)\mathbf{R}\mathbf{u}(k)]$$

where $\mathbf{Q}$ and $\mathbf{S}$ are $n \times n$ positive definite or positive semidefinite Hermitian matrices and $\mathbf{R}$ is an $r \times r$ positive definite Hermitian matrix.

In Sec. 7–2 we obtained the optimal control vector $\mathbf{u}(k)$ in the form given by Eq. (7–26):

$$\mathbf{u}(k) = -\mathbf{R}^{-1}\mathbf{H}^*(\mathbf{G}^*)^{-1}[\mathbf{P}(k) - \mathbf{Q}]\mathbf{x}(k) \tag{7–281}$$

where $\mathbf{P}(k)$ is given by Eqs. (7–23) and (7–25):

$$\mathbf{P}(k) = \mathbf{Q} + \mathbf{G}^*\mathbf{P}(k+1)[\mathbf{I} + \mathbf{H}\mathbf{R}^{-1}\mathbf{H}^*\mathbf{P}(k+1)]^{-1}\mathbf{G}, \qquad \mathbf{P}(N) = \mathbf{S} \tag{7–282}$$

1. Show that the optimal control vector $\mathbf{u}(k)$ can be modified to read

$$\mathbf{u}(k) = -\mathbf{R}^{-1}\mathbf{H}^*[\mathbf{P}^{-1}(k+1) + \mathbf{H}\mathbf{R}^{-1}\mathbf{H}^*]^{-1}\mathbf{G}\mathbf{x}(k) \tag{7–283}$$

where

$$\mathbf{P}(k) = \mathbf{Q} + \mathbf{G}^*[\mathbf{P}^{-1}(k+1) + \mathbf{H}\mathbf{R}^{-1}\mathbf{H}^*]^{-1}\mathbf{G}, \qquad \mathbf{P}(N) = \mathbf{S} \tag{7–284}$$

2. Show that the optimal control vector $\mathbf{u}(k)$ can also be given by

$$\mathbf{u}(k) = -[\mathbf{R} + \mathbf{H}^*\mathbf{P}(k+1)\mathbf{H}]^{-1}\mathbf{H}^*\mathbf{P}(k+1)\mathbf{G}\mathbf{x}(k) \tag{7–285}$$

where

$$\mathbf{P}(k) = \mathbf{Q} + \mathbf{G}^*\mathbf{P}(k+1)\mathbf{G} - \mathbf{G}^*\mathbf{P}(k+1)\mathbf{H}[\mathbf{R} + \mathbf{H}^*\mathbf{P}(k+1)\mathbf{H}]^{-1}\mathbf{H}^*\mathbf{P}(k+1)\mathbf{G}, \qquad \mathbf{P}(N) = \mathbf{S} \tag{7–286}$$

3. Show that the three different expressions for $\mathbf{P}(k)$ given by Eqs. (7–282), (7–284), and (7–286) are equivalent.

Solution.

1. We shall first show that Eqs. (7–281) and (7–283) are equivalent. Referring to Eq. (7–23),

$$\begin{aligned}(\mathbf{G}^*)^{-1}[\mathbf{P}(k)-\mathbf{Q}] &= (\mathbf{G}^*)^{-1}\mathbf{G}^*\mathbf{P}(k+1)[\mathbf{I}+\mathbf{H}\mathbf{R}^{-1}\mathbf{H}^*\mathbf{P}(k+1)]^{-1}\mathbf{G}\\ &= \mathbf{P}(k+1)[\mathbf{I}+\mathbf{H}\mathbf{R}^{-1}\mathbf{H}^*\mathbf{P}(k+1)]^{-1}\mathbf{G}\\ &= [\mathbf{P}^{-1}(k+1)+\mathbf{H}\mathbf{R}^{-1}\mathbf{H}^*]^{-1}\mathbf{G}\end{aligned}$$

Hence

$$\begin{aligned}\mathbf{u}(k) &= -\mathbf{R}^{-1}\mathbf{H}^*(\mathbf{G}^*)^{-1}[\mathbf{P}(k)-\mathbf{Q}]\mathbf{x}(k)\\ &= -\mathbf{R}^{-1}\mathbf{H}^*[\mathbf{P}^{-1}(k+1)+\mathbf{H}\mathbf{R}^{-1}\mathbf{H}^*]^{-1}\mathbf{G}\mathbf{x}(k)\end{aligned}$$

and we have shown that Eqs. (7–281) and (7–283) are equivalent.

2. To show that Eqs. (7–283) and (7–285) are equivalent, note that

$$\begin{aligned}&[\mathbf{R}+\mathbf{H}^*\mathbf{P}(k+1)\mathbf{H}]^{-1}\mathbf{H}^*\mathbf{P}(k+1)[\mathbf{P}^{-1}(k+1)+\mathbf{H}\mathbf{R}^{-1}\mathbf{H}^*]\\ &\quad= [\mathbf{R}+\mathbf{H}^*\mathbf{P}(k+1)\mathbf{H}]^{-1}\mathbf{H}^*[\mathbf{I}+\mathbf{P}(k+1)\mathbf{H}\mathbf{R}^{-1}\mathbf{H}^*]\\ &\quad= [\mathbf{R}+\mathbf{H}^*\mathbf{P}(k+1)\mathbf{H}]^{-1}[\mathbf{R}+\mathbf{H}^*\mathbf{P}(k+1)\mathbf{H}]\mathbf{R}^{-1}\mathbf{H}^*\\ &\quad= \mathbf{R}^{-1}\mathbf{H}^*\end{aligned}$$

Hence

$$\mathbf{R}^{-1}\mathbf{H}^*[\mathbf{P}^{-1}(k+1)+\mathbf{H}\mathbf{R}^{-1}\mathbf{H}^*]^{-1} = [\mathbf{R}+\mathbf{H}^*\mathbf{P}(k+1)\mathbf{H}]^{-1}\mathbf{H}^*\mathbf{P}(k+1)$$

and consequently

$$\begin{aligned}\mathbf{u}(k) &= -\mathbf{R}^{-1}\mathbf{H}^*[\mathbf{P}^{-1}(k+1)+\mathbf{H}\mathbf{R}^{-1}\mathbf{H}^*]^{-1}\mathbf{G}\mathbf{x}(k)\\ &= -[\mathbf{R}+\mathbf{H}^*\mathbf{P}(k+1)\mathbf{H}]^{-1}\mathbf{H}^*\mathbf{P}(k+1)\mathbf{G}\mathbf{x}(k)\end{aligned}$$

We have thus shown that Eqs. (7–283) and (7–285) are equivalent.

3. Next, we shall prove that Eqs. (7–282) and (7–284) are equivalent. If we note that

$$\mathbf{P}(k+1)[\mathbf{I}+\mathbf{H}\mathbf{R}^{-1}\mathbf{H}^*\mathbf{P}(k+1)]^{-1} = [\mathbf{P}^{-1}(k+1)+\mathbf{H}\mathbf{R}^{-1}\mathbf{H}^*]^{-1}$$

then the equivalence of Eqs. (7–282) and (7–284) is apparent.

To show that Eqs. (7–284) and (7–286) are equivalent, notice that

$$\begin{aligned}&[\mathbf{P}^{-1}(k+1)+\mathbf{H}\mathbf{R}^{-1}\mathbf{H}^*]\{\mathbf{P}(k+1)-\mathbf{P}(k+1)\mathbf{H}[\mathbf{R}+\mathbf{H}^*\mathbf{P}(k+1)\mathbf{H}]^{-1}\mathbf{H}^*\mathbf{P}(k+1)\}\\ &= \mathbf{I}+\mathbf{H}\mathbf{R}^{-1}\mathbf{H}^*\mathbf{P}(k+1)-\mathbf{H}[\mathbf{R}+\mathbf{H}^*\mathbf{P}(k+1)\mathbf{H}]^{-1}\mathbf{H}^*\mathbf{P}(k+1)\\ &\quad -\mathbf{H}\mathbf{R}^{-1}\mathbf{H}^*\mathbf{P}(k+1)\mathbf{H}[\mathbf{R}+\mathbf{H}^*\mathbf{P}(k+1)\mathbf{H}]^{-1}\mathbf{H}^*\mathbf{P}(k+1)\\ &= \mathbf{I}-\mathbf{H}\{-\mathbf{R}^{-1}+[\mathbf{R}+\mathbf{H}^*\mathbf{P}(k+1)\mathbf{H}]^{-1}+\mathbf{R}^{-1}\mathbf{H}^*\mathbf{P}(k+1)\mathbf{H}[\mathbf{R}+\mathbf{H}^*\mathbf{P}(k+1)\mathbf{H}]^{-1}\}\mathbf{H}^*\mathbf{P}(k+1)\\ &= \mathbf{I}-\mathbf{H}\{[\mathbf{I}+\mathbf{R}^{-1}\mathbf{H}^*\mathbf{P}(k+1)\mathbf{H}][\mathbf{R}+\mathbf{H}^*\mathbf{P}(k+1)\mathbf{H}]^{-1}-\mathbf{R}^{-1}\}\mathbf{H}^*\mathbf{P}(k+1)\\ &= \mathbf{I}-\mathbf{H}\{\mathbf{R}^{-1}[\mathbf{R}+\mathbf{H}^*\mathbf{P}(k+1)\mathbf{H}][\mathbf{R}+\mathbf{H}^*\mathbf{P}(k+1)\mathbf{H}]^{-1}-\mathbf{R}^{-1}\}\mathbf{H}^*\mathbf{P}(k+1)\\ &= \mathbf{I}-\mathbf{H}[\mathbf{R}^{-1}-\mathbf{R}^{-1}]\mathbf{H}^*\mathbf{P}(k+1) = \mathbf{I}\end{aligned}$$

Hence

$$[\mathbf{P}^{-1}(k+1)+\mathbf{H}\mathbf{R}^{-1}\mathbf{H}^*]^{-1} = \mathbf{P}(k+1)-\mathbf{P}(k+1)\mathbf{H}[\mathbf{R}+\mathbf{H}^*\mathbf{P}(k+1)\mathbf{H}]^{-1}\mathbf{H}^*\mathbf{P}(k+1)$$

and we have

$$\mathbf{P}(k) = \mathbf{Q} + \mathbf{G}^*[\mathbf{P}^{-1}(k+1) + \mathbf{H}\mathbf{R}^{-1}\mathbf{H}^*]^{-1}\mathbf{G}$$
$$= \mathbf{Q} + \mathbf{G}^*\mathbf{P}(k+1)\mathbf{G} - \mathbf{G}^*\mathbf{P}(k+1)\mathbf{H}[\mathbf{R} + \mathbf{H}^*\mathbf{P}(k+1)\mathbf{H}]^{-1}\mathbf{H}^*\mathbf{P}(k+1)\mathbf{G}$$

Thus, we have shown that Eqs. (7–284) and (7–286) are equivalent.

Problem A-7–2. For the quadratic optimal control problem where the system is as given by Eq. (7–1) and the performance index is as given by Eq. (7–44), we have found in Sec. 7–2 that the optimal control vector $\mathbf{u}(k)$ can be given by the equation

$$\mathbf{u}(k) = \mathbf{v}(k) - \mathbf{R}^{-1}\mathbf{M}^*\mathbf{x}(k) \tag{7–287}$$

where $\mathbf{v}(k)$ is given by Eq. (7–51), (7–52), or (7–53) as follows:

$$\mathbf{v}(k) = -\mathbf{R}^{-1}\mathbf{H}^*(\hat{\mathbf{G}}^*)^{-1}[\hat{\mathbf{P}}(k) - \hat{\mathbf{Q}}]\mathbf{x}(k)$$

or

$$\mathbf{v}(k) = -\mathbf{R}^{-1}\mathbf{H}^*[\hat{\mathbf{P}}^{-1}(k+1) + \mathbf{H}\mathbf{R}^{-1}\mathbf{H}^*]^{-1}\hat{\mathbf{G}}\mathbf{x}(k)$$

or

$$\mathbf{v}(k) = -[\mathbf{R} + \mathbf{H}^*\hat{\mathbf{P}}(k+1)\mathbf{H}]^{-1}\mathbf{H}^*\hat{\mathbf{P}}(k+1)\hat{\mathbf{G}}\mathbf{x}(k) \tag{7–288}$$

where

$$\hat{\mathbf{P}}(k) = \hat{\mathbf{Q}} + \hat{\mathbf{G}}^*\hat{\mathbf{P}}(k+1)[\mathbf{I} + \mathbf{H}\mathbf{R}^{-1}\mathbf{H}^*\hat{\mathbf{P}}(k+1)]^{-1}\hat{\mathbf{G}}, \qquad \hat{\mathbf{P}}(N) = \mathbf{S}$$

and

$$\hat{\mathbf{G}} = \mathbf{G} - \mathbf{H}\mathbf{R}^{-1}\mathbf{M}^* \qquad \text{and} \qquad \hat{\mathbf{Q}} = \mathbf{Q} - \mathbf{M}\mathbf{R}^{-1}\mathbf{M}^*$$

Show that the optimal control vector $\mathbf{u}(k)$ can be expressed as follows:

$$\mathbf{u}(k) = -[\mathbf{R} + \mathbf{H}^*\hat{\mathbf{P}}(k+1)\mathbf{H}]^{-1}[\mathbf{H}^*\hat{\mathbf{P}}(k+1)\mathbf{G} + \mathbf{M}^*]\mathbf{x}(k) \tag{7–289}$$

Solution. The equivalence of the right-hand sides of the three expressions for $\mathbf{v}(k)$ was shown in Prob. A-7–1. Hence, we may derive Eq. (7–289) using, for example, Eq. (7–288).

From Eqs. (7–287) and (7–288), we have

$$\begin{aligned}
\mathbf{u}(k) &= \mathbf{v}(k) - \mathbf{R}^{-1}\mathbf{M}^*\mathbf{x}(k) \\
&= -[\mathbf{R} + \mathbf{H}^*\hat{\mathbf{P}}(k+1)\mathbf{H}]^{-1}\mathbf{H}^*\hat{\mathbf{P}}(k+1)\hat{\mathbf{G}}\mathbf{x}(k) - \mathbf{R}^{-1}\mathbf{M}^*\mathbf{x}(k) \\
&= -\{[\mathbf{R} + \mathbf{H}^*\hat{\mathbf{P}}(k+1)\mathbf{H}]^{-1}\mathbf{H}^*\hat{\mathbf{P}}(k+1)[\mathbf{G} - \mathbf{H}\mathbf{R}^{-1}\mathbf{M}^*] \\
&\quad + [\mathbf{R} + \mathbf{H}^*\hat{\mathbf{P}}(k+1)\mathbf{H}]^{-1}[\mathbf{R} + \mathbf{H}^*\hat{\mathbf{P}}(k+1)\mathbf{H}]\mathbf{R}^{-1}\mathbf{M}^*\}\mathbf{x}(k) \\
&= -[\mathbf{R} + \mathbf{H}^*\hat{\mathbf{P}}(k+1)\mathbf{H}]^{-1}[\mathbf{H}^*\hat{\mathbf{P}}(k+1)\mathbf{G} - \mathbf{H}^*\hat{\mathbf{P}}(k+1)\mathbf{H}\mathbf{R}^{-1}\mathbf{M}^* \\
&\quad + \mathbf{M}^* + \mathbf{H}^*\hat{\mathbf{P}}(k+1)\mathbf{H}\mathbf{R}^{-1}\mathbf{M}^*]\mathbf{x}(k) \\
&= -[\mathbf{R} + \mathbf{H}^*\hat{\mathbf{P}}(k+1)\mathbf{H}]^{-1}[\mathbf{H}^*\hat{\mathbf{P}}(k+1)\mathbf{G} + \mathbf{M}^*]\mathbf{x}(k)
\end{aligned}$$

which is Eq. (7–289).

Problem A-7–3. Consider the discrete-time control system defined by

$$\mathbf{x}(k+1) = \mathbf{G}\mathbf{x}(k) + \mathbf{H}u(k)$$

where

$$\mathbf{G} = \begin{bmatrix} 1 & 1 \\ 1 & 0 \end{bmatrix}, \qquad \mathbf{H} = \begin{bmatrix} 1 \\ 0 \end{bmatrix}, \qquad \mathbf{x}(0) = \begin{bmatrix} 1 \\ 0 \end{bmatrix}$$

Determine the optimal control sequence $u(k)$ that will minimize the following performance index:

$$J = \frac{1}{2}\mathbf{x}^*(8)\mathbf{S}\mathbf{x}(8) + \frac{1}{2}\sum_{k=0}^{7} [\mathbf{x}^*(k)\mathbf{Q}\mathbf{x}(k) + u^*(k)Ru(k)]$$

where

$$\mathbf{Q} = \begin{bmatrix} 1 & 0 \\ 0 & 1 \end{bmatrix}, \qquad R = 1, \qquad \mathbf{S} = \begin{bmatrix} 1 & 0 \\ 0 & 1 \end{bmatrix}$$

Solution. Referring to Eq. (7–23), we have

$$\begin{aligned}\mathbf{P}(k) &= \mathbf{Q} + \mathbf{G}^*\mathbf{P}(k+1)[\mathbf{I} + \mathbf{H}\mathbf{R}^{-1}\mathbf{H}^*\mathbf{P}(k+1)]^{-1}\mathbf{G} \\ &= \begin{bmatrix} 1 & 0 \\ 0 & 1 \end{bmatrix} + \begin{bmatrix} 1 & 1 \\ 1 & 0 \end{bmatrix}\begin{bmatrix} p_{11}(k+1) & p_{12}(k+1) \\ p_{12}(k+1) & p_{22}(k+1) \end{bmatrix} \\ &\quad \times \left\{\begin{bmatrix} 1 & 0 \\ 0 & 1 \end{bmatrix} + \begin{bmatrix} 1 & 0 \\ 0 & 0 \end{bmatrix}\begin{bmatrix} p_{11}(k+1) & p_{12}(k+1) \\ p_{12}(k+1) & p_{22}(k+1) \end{bmatrix}\right\}^{-1}\begin{bmatrix} 1 & 1 \\ 1 & 0 \end{bmatrix}\end{aligned}$$

The boundary condition for $\mathbf{P}(k)$ is specified by Eq. (7–25) and is given by

$$\mathbf{P}(N) = \mathbf{P}(8) = \mathbf{S} = \begin{bmatrix} 1 & 0 \\ 0 & 1 \end{bmatrix}$$

Now we compute $\mathbf{P}(k)$ backward from $\mathbf{P}(7)$ to $\mathbf{P}(0)$:

$$\begin{aligned}\mathbf{P}(7) &= \begin{bmatrix} 1 & 0 \\ 0 & 1 \end{bmatrix} + \begin{bmatrix} 1 & 1 \\ 1 & 0 \end{bmatrix}\begin{bmatrix} 1 & 0 \\ 0 & 1 \end{bmatrix}\left\{\begin{bmatrix} 1 & 0 \\ 0 & 1 \end{bmatrix}\right. \\ &\quad \left. + \begin{bmatrix} 1 & 0 \\ 0 & 0 \end{bmatrix}\begin{bmatrix} 1 & 0 \\ 0 & 1 \end{bmatrix}\right\}^{-1}\begin{bmatrix} 1 & 1 \\ 1 & 0 \end{bmatrix} \\ &= \begin{bmatrix} \frac{5}{2} & \frac{1}{2} \\ \frac{1}{2} & \frac{3}{2} \end{bmatrix} = \begin{bmatrix} 2.5 & 0.5 \\ 0.5 & 1.5 \end{bmatrix}\end{aligned}$$

$$\begin{aligned}\mathbf{P}(6) &= \begin{bmatrix} 1 & 0 \\ 0 & 1 \end{bmatrix} + \begin{bmatrix} 1 & 1 \\ 1 & 0 \end{bmatrix}\begin{bmatrix} 2.5 & 0.5 \\ 0.5 & 1.5 \end{bmatrix}\left\{\begin{bmatrix} 1 & 0 \\ 0 & 1 \end{bmatrix}\right. \\ &\quad \left. + \begin{bmatrix} 1 & 0 \\ 0 & 0 \end{bmatrix}\begin{bmatrix} 2.5 & 0.5 \\ 0.5 & 1.5 \end{bmatrix}\right\}^{-1}\begin{bmatrix} 1 & 1 \\ 1 & 0 \end{bmatrix} \\ &= \begin{bmatrix} \frac{24}{7} & \frac{6}{7} \\ \frac{6}{7} & \frac{12}{7} \end{bmatrix} = \begin{bmatrix} 3.4286 & 0.8571 \\ 0.8571 & 1.7143 \end{bmatrix}\end{aligned}$$

Similarly, $\mathbf{P}(5)$, $\mathbf{P}(4)$, . . . , $\mathbf{P}(0)$ can be computed as shown in Table 7–4.

Next, we shall determine the feedback gain matrix $\mathbf{K}(k)$. Referring to Eq. (7–27), matrix $\mathbf{K}(k)$ can be given as follows:

$$\begin{aligned}\mathbf{K}(k) &= R^{-1}\mathbf{H}^*(\mathbf{G}^*)^{-1}[\mathbf{P}(k) - \mathbf{Q}] \\ &= [1][1 \quad 0]\begin{bmatrix} 1 & 1 \\ 1 & 0 \end{bmatrix}^{-1}[\mathbf{P}(k) - \mathbf{Q}]\end{aligned}$$

TABLE 7–4 TABLE SHOWING **P**(k), **K**(k), **x**(k), AND $u(k)$ FOR $k = 0, 1, 2, \ldots, 8$, RESPECTIVELY, FOR THE SYSTEM CONSIDERED IN PROB. A-7–3

k	$\mathbf{P}(k)$	$\mathbf{K}(k)$	$\mathbf{x}(k)$	$u(k)$
0	$\begin{bmatrix} 3.7913 & 1.0000 \\ 1.0000 & 1.7913 \end{bmatrix}$	$[1.0000 \quad 0.7913]$	$\begin{bmatrix} 1.0000 \\ 0.0000 \end{bmatrix}$	-1.0000
1	$\begin{bmatrix} 3.7911 & 0.9999 \\ 0.9999 & 1.7913 \end{bmatrix}$	$[0.9999 \quad 0.7913]$	$\begin{bmatrix} 0.0000 \\ 1.0000 \end{bmatrix}$	-0.7913
2	$\begin{bmatrix} 3.7905 & 0.9997 \\ 0.9997 & 1.7911 \end{bmatrix}$	$[0.9997 \quad 0.7911]$	$\begin{bmatrix} 0.2087 \\ 0.0000 \end{bmatrix}$	-0.2087
3	$\begin{bmatrix} 3.7877 & 0.9986 \\ 0.9986 & 1.7905 \end{bmatrix}$	$[0.9986 \quad 0.7905]$	$\begin{bmatrix} 0.0001 \\ 0.2087 \end{bmatrix}$	-0.1651
4	$\begin{bmatrix} 3.7740 & 0.9932 \\ 0.9932 & 1.7877 \end{bmatrix}$	$[0.9932 \quad 0.7877]$	$\begin{bmatrix} 0.0437 \\ 0.0001 \end{bmatrix}$	-0.0435
5	$\begin{bmatrix} 3.7097 & 0.9677 \\ 0.9677 & 1.7742 \end{bmatrix}$	$[0.9677 \quad 0.7742]$	$\begin{bmatrix} 0.0003 \\ 0.0437 \end{bmatrix}$	-0.0342
6	$\begin{bmatrix} 3.4286 & 0.8571 \\ 0.8571 & 1.7143 \end{bmatrix}$	$[0.8571 \quad 0.7143]$	$\begin{bmatrix} 0.0099 \\ 0.0003 \end{bmatrix}$	-0.0087
7	$\begin{bmatrix} 2.5000 & 0.5000 \\ 0.5000 & 1.5000 \end{bmatrix}$	$[0.5000 \quad 0.5000]$	$\begin{bmatrix} 0.0015 \\ 0.0099 \end{bmatrix}$	-0.0057
8	$\begin{bmatrix} 1.0000 & 0.0000 \\ 0.0000 & 1.0000 \end{bmatrix}$	$[0.0000 \quad 0.0000]$	$\begin{bmatrix} 0.0057 \\ 0.0015 \end{bmatrix}$	0.0000

$$
\begin{aligned}
&= [0 \quad 1] \begin{bmatrix} p_{11}(k) - 1 & p_{12}(k) \\ p_{12}(k) & p_{22}(k) - 1 \end{bmatrix} \\
&= [p_{12}(k) \quad p_{22}(k) - 1]
\end{aligned}
$$

Thus,

$$
\mathbf{K}(8) = [p_{12}(8) \quad p_{22}(8) - 1] = [0.0000 \quad 0.0000]
$$

$$
\mathbf{K}(7) = [p_{12}(7) \quad p_{22}(7) - 1] = [0.5000 \quad 0.5000]
$$

Similarly, $\mathbf{K}(6)$, $\mathbf{K}(5)$, . . . , $\mathbf{K}(0)$ can be computed to give the values shown in Table 7–4.

Next, we shall compute $\mathbf{x}(k)$. Let us write

$$
\mathbf{K}(k) = [K_1(k) \quad K_2(k)]
$$

Then

$$
\mathbf{x}(k+1) = \mathbf{G}\mathbf{x}(k) + \mathbf{H}u(k)
$$

$$= [\mathbf{G} - \mathbf{H}\mathbf{K}(k)]\mathbf{x}(k)$$

$$= \begin{bmatrix} 1 - K_1(k) & 1 - K_2(k) \\ 1 & 0 \end{bmatrix} \begin{bmatrix} x_1(k) \\ x_2(k) \end{bmatrix}$$

Since the initial state is

$$\mathbf{x}(0) = \begin{bmatrix} 1 \\ 0 \end{bmatrix}$$

$\mathbf{x}(k)$, where $k = 1, 2, \ldots, 8$, can be obtained as follows:

$$\mathbf{x}(1) = \begin{bmatrix} 1 - 1 & 1 - 0.7913 \\ 1 & 0 \end{bmatrix} \begin{bmatrix} 1 \\ 0 \end{bmatrix} = \begin{bmatrix} 0.0000 \\ 1.0000 \end{bmatrix}$$

$$\mathbf{x}(2) = \begin{bmatrix} 1 - 0.9999 & 1 - 0.7913 \\ 1 & 0 \end{bmatrix} \begin{bmatrix} 0.0000 \\ 1.0000 \end{bmatrix} = \begin{bmatrix} 0.2087 \\ 0.0000 \end{bmatrix}$$

Similarly, $\mathbf{x}(3), \mathbf{x}(4), \ldots, \mathbf{x}(8)$ can be computed. The results are shown in Table 7–4.

Finally, the optimal control sequence $u(k)$ can be obtained from Eq. (7–28):

$$u(k) = -\mathbf{K}(k)\mathbf{x}(k)$$

That is,

$$u(0) = -\mathbf{K}(0)\mathbf{x}(0) = -[1 \quad 0.7913] \begin{bmatrix} 1 \\ 0 \end{bmatrix} = -1.0000$$

$$u(1) = -\mathbf{K}(1)\mathbf{x}(1) = -[0.9999 \quad 0.7913] \begin{bmatrix} 0 \\ 1 \end{bmatrix} = -0.7913$$

Similarly, $u(2), u(3), \ldots, u(8)$ can be computed to give the values shown in Table 7–4.

As mentioned earlier, the feedback gain matrix $\mathbf{K}(k)$ is constant except for the last several values of k. This means that if the number of stages is not 8 but 100, then $\mathbf{K}(0)$, $\mathbf{K}(1), \ldots, \mathbf{K}(93)$ will be constant matrices and $\mathbf{K}(94), \mathbf{K}(95), \ldots, \mathbf{K}(100)$ will vary. This fact is important, because if the number of stages N is sufficiently large, then the feedback gain matrix becomes a constant matrix and so the designer is able to use a constant feedback gain matrix to approximate the time-varying optimal gain matrix.

The minimum value of J is obtained from Eq. (7–36), as follows:

$$J_{\min} = \frac{1}{2}\mathbf{x}^*(0)\mathbf{P}(0)\mathbf{x}(0) = \frac{1}{2}[1 \quad 0] \begin{bmatrix} 3.7913 & 1.0000 \\ 1.0000 & 1.7913 \end{bmatrix} \begin{bmatrix} 1 \\ 0 \end{bmatrix}$$

$$= 1.8956$$

Problem A-7–4. Consider the following equation:

$$\begin{bmatrix} \mathbf{I}_n & \mathbf{H}\mathbf{R}^{-1}\mathbf{H}^* \\ \mathbf{0} & \mathbf{G}^* \end{bmatrix} \begin{bmatrix} \mathbf{x}(k+1) \\ \boldsymbol{\lambda}(k+1) \end{bmatrix} = \begin{bmatrix} \mathbf{G} & \mathbf{0} \\ -\mathbf{Q} & \mathbf{I}_n \end{bmatrix} \begin{bmatrix} \mathbf{x}(k) \\ \boldsymbol{\lambda}(k) \end{bmatrix}$$

If matrix $\mathbf{G}$ is nonsingular, this equation can be written as follows:

$$\begin{bmatrix} \mathbf{x}(k+1) \\ \boldsymbol{\lambda}(k+1) \end{bmatrix} = \begin{bmatrix} \mathbf{I}_n & \mathbf{H}\mathbf{R}^{-1}\mathbf{H}^* \\ \mathbf{0} & \mathbf{G}^* \end{bmatrix}^{-1} \begin{bmatrix} \mathbf{G} & \mathbf{0} \\ -\mathbf{Q} & \mathbf{I}_n \end{bmatrix} \begin{bmatrix} \mathbf{x}(k) \\ \boldsymbol{\lambda}(k) \end{bmatrix}$$

$$= \begin{bmatrix} \mathbf{I}_n & -\mathbf{HR}^{-1}\mathbf{H}^*(\mathbf{G}^*)^{-1} \\ \mathbf{0} & (\mathbf{G}^*)^{-1} \end{bmatrix} \begin{bmatrix} \mathbf{G} & \mathbf{0} \\ -\mathbf{Q} & \mathbf{I}_n \end{bmatrix} \begin{bmatrix} \mathbf{x}(k) \\ \boldsymbol{\lambda}(k) \end{bmatrix}$$

$$= \begin{bmatrix} \mathbf{G} + \mathbf{HR}^{-1}\mathbf{H}^*(\mathbf{G}^*)^{-1}\mathbf{Q} & -\mathbf{HR}^{-1}\mathbf{H}^*(\mathbf{G}^*)^{-1} \\ -(\mathbf{G}^*)^{-1}\mathbf{Q} & (\mathbf{G}^*)^{-1} \end{bmatrix} \begin{bmatrix} \mathbf{x}(k) \\ \boldsymbol{\lambda}(k) \end{bmatrix}$$

Define

$$\mathbf{Z} = \begin{bmatrix} \mathbf{G} + \mathbf{HR}^{-1}\mathbf{H}^*(\mathbf{G}^*)^{-1}\mathbf{Q} & -\mathbf{HR}^{-1}\mathbf{H}^*(\mathbf{G}^*)^{-1} \\ -(\mathbf{G}^*)^{-1}\mathbf{Q} & (\mathbf{G}^*)^{-1} \end{bmatrix}$$

Show that if μ_i is an eigenvalue of $\mathbf{Z}$, then so is μ_i^{-1}. That is, show that the eigenvalues of $\mathbf{Z}$ are $\mu_1, \mu_2, \ldots, \mu_n$ and $\mu_1^{-1}, \mu_2^{-1}, \ldots, \mu_n^{-1}$. ($\mathbf{R}$ is a positive definite Hermitian matrix.)

Solution. Since matrix $\mathbf{G}$ is nonsingular and matrix $\mathbf{Z}$ can be written in the form

$$\mathbf{Z} = \begin{bmatrix} \mathbf{I}_n & -\mathbf{HR}^{-1}\mathbf{H}^*(\mathbf{G}^*)^{-1} \\ \mathbf{0} & (\mathbf{G}^*)^{-1} \end{bmatrix} \begin{bmatrix} \mathbf{G} & \mathbf{0} \\ -\mathbf{Q} & \mathbf{I}_n \end{bmatrix} \tag{7–290}$$

none of the eigenvalues of $\mathbf{Z}$ is zero, and so $\mu_i \neq 0$.

Let us assume that μ_i is an eigenvalue of matrix $\mathbf{Z}$ and that $\boldsymbol{\xi}_i$ is the eigenvector of $\mathbf{Z}$ corresponding to μ_i. Then

$$\mathbf{Z}\boldsymbol{\xi}_i = \mu_i \boldsymbol{\xi}_i \tag{7–291}$$

Since $\mathbf{Z}$ is a $2n \times 2n$ matrix, $\boldsymbol{\xi}_i$ is a $2n$-vector. Define

$$\boldsymbol{\xi}_i = \begin{bmatrix} \mathbf{v}_i \\ \mathbf{w}_i \end{bmatrix}$$

Then

$$\mathbf{Z}\begin{bmatrix} \mathbf{v}_i \\ \mathbf{w}_i \end{bmatrix} = \mu_i \begin{bmatrix} \mathbf{v}_i \\ \mathbf{w}_i \end{bmatrix}$$

or

$$\begin{bmatrix} \mathbf{G} + \mathbf{HR}^{-1}\mathbf{H}^*(\mathbf{G}^*)^{-1}\mathbf{Q} & -\mathbf{HR}^{-1}\mathbf{H}^*(\mathbf{G}^*)^{-1} \\ -(\mathbf{G}^*)^{-1}\mathbf{Q} & (\mathbf{G}^*)^{-1} \end{bmatrix} \begin{bmatrix} \mathbf{v}_i \\ \mathbf{w}_i \end{bmatrix} = \mu_i \begin{bmatrix} \mathbf{v}_i \\ \mathbf{w}_i \end{bmatrix}$$

Hence

$$[\mathbf{G} + \mathbf{HR}^{-1}\mathbf{H}^*(\mathbf{G}^*)^{-1}\mathbf{Q}]\mathbf{v}_i - \mathbf{HR}^{-1}\mathbf{H}^*(\mathbf{G}^*)^{-1}\mathbf{w}_i = \mu_i \mathbf{v}_i \tag{7–292}$$

$$-(\mathbf{G}^*)^{-1}\mathbf{Q}\mathbf{v}_i + (\mathbf{G}^*)^{-1}\mathbf{w}_i = \mu_i \mathbf{w}_i \tag{7–293}$$

Taking conjugate transposes of both sides of Eqs. (7–292) and (7–293), respectively, we obtain

$$\mathbf{v}_i^*[\mathbf{G}^* + \mathbf{QG}^{-1}\mathbf{HR}^{-1}\mathbf{H}^*] - \mathbf{w}_i^*\mathbf{G}^{-1}\mathbf{HR}^{-1}\mathbf{H}^* = \bar{\mu}_i \mathbf{v}_i^* \tag{7–294}$$

$$-\mathbf{v}_i^*\mathbf{QG}^{-1} + \mathbf{w}_i^*\mathbf{G}^{-1} = \bar{\mu}_i \mathbf{w}_i^* \tag{7–295}$$

Now consider the following matrix:

$$[\mathbf{w}_i^* \quad -\mathbf{v}_i^*]\mathbf{Z}^{-1}$$

Note that from Eq. (7–290) we have

$$\mathbf{Z}^{-1} = \begin{bmatrix} \mathbf{G}^{-1} & \mathbf{0} \\ \mathbf{QG}^{-1} & \mathbf{I}_n \end{bmatrix} \begin{bmatrix} \mathbf{I}_n & \mathbf{HR}^{-1}\mathbf{H}^* \\ \mathbf{0} & \mathbf{G}^* \end{bmatrix} = \begin{bmatrix} \mathbf{G}^{-1} & \mathbf{G}^{-1}\mathbf{HR}^{-1}\mathbf{H}^* \\ \mathbf{QG}^{-1} & \mathbf{QG}^{-1}\mathbf{HR}^{-1}\mathbf{H}^* + \mathbf{G}^* \end{bmatrix}$$

Hence

$$[\mathbf{w}_i^* \quad -\mathbf{v}_i^*]\mathbf{Z}^{-1} = [\mathbf{w}_i^* \quad -\mathbf{v}_i^*]\begin{bmatrix} \mathbf{G}^{-1} & \mathbf{G}^{-1}\mathbf{H}\mathbf{R}^{-1}\mathbf{H}^* \\ \mathbf{Q}\mathbf{G}^{-1} & \mathbf{Q}\mathbf{G}^{-1}\mathbf{H}\mathbf{R}^{-1}\mathbf{H}^* + \mathbf{G}^* \end{bmatrix}$$

$$= \left[\mathbf{w}_i^*\mathbf{G}^{-1} - \mathbf{v}_i^*\mathbf{Q}\mathbf{G}^{-1} \vdots \mathbf{w}_i^*\mathbf{G}^{-1}\mathbf{H}\mathbf{R}^{-1}\mathbf{H}^* - \mathbf{v}_i^*(\mathbf{Q}\mathbf{G}^{-1}\mathbf{H}\mathbf{R}^{-1}\mathbf{H}^* + \mathbf{G}^*)\right] \tag{7–296}$$

From Eqs. (7–294) and (7–295), the right-hand side of Eq. (7–296) is

$$[\bar{\mu}_i \mathbf{w}_i^* \quad -\bar{\mu}_i \mathbf{v}_i^*]$$

Hence

$$[\mathbf{w}_i^* \quad -\mathbf{v}_i^*]\mathbf{Z}^{-1} = [\bar{\mu}_i \mathbf{w}_i^* \quad -\bar{\mu}_i v_i^*] \tag{7–297}$$

Taking the conjugate transposes of both sides of Eq. (7–297), we obtain

$$(\mathbf{Z}^*)^{-1}\begin{bmatrix} \mathbf{w}_i \\ -\mathbf{v}_i \end{bmatrix} = \mu_i \begin{bmatrix} \mathbf{w}_i \\ -\mathbf{v}_i \end{bmatrix}$$

or

$$\mathbf{Z}^*\begin{bmatrix} \mathbf{w}_i \\ -\mathbf{v}_i \end{bmatrix} = \frac{1}{\mu_i}\begin{bmatrix} \mathbf{w}_i \\ -\mathbf{v}_i \end{bmatrix} \tag{7–298}$$

Equation (7–298) implies that if μ_i is an eigenvalue of matrix $\mathbf{Z}$, then μ_i^{-1} is an eigenvalue of $\mathbf{Z}^*$. Since $\mathbf{Z}^*$ and $\mathbf{Z}$ have the same eigenvalues, μ_i^{-1} is an eigenvalue of $\mathbf{Z}$. Consequently, if μ_i is an eigenvalue of matrix $\mathbf{Z}$, then so is μ_i^{-1}. Thus, the eigenvalues of matrix $\mathbf{Z}$ are μ_1, $\mu_2, \ldots, \mu_n$ and $\mu_1^{-1}, \mu_2^{-1}, \ldots, \mu_n^{-1}$.

Problem A-7–5. Consider the system

$$\begin{bmatrix} x_1(k+1) \\ x_2(k+1) \end{bmatrix} = \begin{bmatrix} 0 & 0 \\ 1 & 1 \end{bmatrix}\begin{bmatrix} x_1(k) \\ x_2(k) \end{bmatrix} + \begin{bmatrix} 1 \\ 0 \end{bmatrix}u(k), \qquad \begin{bmatrix} x_1(0) \\ x_2(0) \end{bmatrix} = \begin{bmatrix} 1 \\ 1 \end{bmatrix} \tag{7–299}$$

and the performance index

$$J = \frac{1}{2}\sum_{k=0}^{\infty}[\mathbf{x}^*(k)\mathbf{Q}\mathbf{x}(k) + u^*(k)Ru(k)] \tag{7–300}$$

where

$$\mathbf{Q} = \begin{bmatrix} 1 & 0 \\ 0 & 0 \end{bmatrix}, \qquad R = 1$$

Determine the optimal control law to minimize the performance index. Also, determine the minimum value of J.

Solution. From Eq. (7–299) we have

$$\mathbf{G} = \begin{bmatrix} 0 & 0 \\ 1 & 1 \end{bmatrix}, \qquad \mathbf{H} = \begin{bmatrix} 1 \\ 0 \end{bmatrix}$$

Notice that $\mathbf{G}$ is singular. Referring to Eq. (7–90), we have

$$\begin{bmatrix} \mathbf{G} & \mathbf{0} \\ -\mathbf{Q} & \mathbf{I}_n \end{bmatrix}\begin{bmatrix} \mathbf{v}_i \\ \mathbf{w}_i \end{bmatrix} = \mu_i \begin{bmatrix} \mathbf{I}_n & \mathbf{H}\mathbf{R}^{-1}\mathbf{H}^* \\ \mathbf{0} & \mathbf{G}^* \end{bmatrix}\begin{bmatrix} \mathbf{v}_i \\ \mathbf{w}_i \end{bmatrix}$$

or

$$\begin{bmatrix} 0 & 0 & 0 & 0 \\ 1 & 1 & 0 & 0 \\ -1 & 0 & 1 & 0 \\ 0 & 0 & 0 & 1 \end{bmatrix} \begin{bmatrix} v_{i1} \\ v_{i2} \\ w_{i1} \\ w_{i2} \end{bmatrix} = \mu_i \begin{bmatrix} 1 & 0 & 1 & 0 \\ 0 & 1 & 0 & 0 \\ 0 & 0 & 0 & 1 \\ 0 & 0 & 0 & 1 \end{bmatrix} \begin{bmatrix} v_{i1} \\ v_{i2} \\ w_{i1} \\ w_{i2} \end{bmatrix}$$

which can be rewritten as follows:

$$\begin{bmatrix} -\mu_i & 0 & -\mu_i & 0 \\ 1 & 1-\mu_i & 0 & 0 \\ -1 & 0 & 1 & -\mu_i \\ 0 & 0 & 0 & 1-\mu_i \end{bmatrix} \begin{bmatrix} v_{i1} \\ v_{i2} \\ w_{i1} \\ w_{i2} \end{bmatrix} = \begin{bmatrix} 0 \\ 0 \\ 0 \\ 0 \end{bmatrix} \tag{7–301}$$

The generalized eigenvalues can be found from

$$\begin{vmatrix} -\mu_i & 0 & -\mu_i & 0 \\ 1 & 1-\mu_i & 0 & 0 \\ -1 & 0 & 1 & -\mu_i \\ 0 & 0 & 0 & 1-\mu_i \end{vmatrix} = 0$$

The Laplace's expansion by the minors gives

$$\begin{vmatrix} -\mu_i & 0 \\ 1 & 1-\mu_i \end{vmatrix} \begin{vmatrix} 1 & -\mu_i \\ 0 & 1-\mu_i \end{vmatrix} + \begin{vmatrix} 1 & 1-\mu_i \\ -1 & 0 \end{vmatrix} \begin{vmatrix} -\mu_i & 0 \\ 0 & 1-\mu_i \end{vmatrix} = -2\mu_i(1-\mu_i)^2 = 0$$

Hence we have

$$\mu_i = 1, 0, 1$$

Next, let us obtain the generalized eigenvectors for $\mu_1 = 1$ and $\mu_2 = 0$. For $\mu_1 = 1$, Eq. (7–301) becomes

$$\begin{bmatrix} -1 & 0 & -1 & 0 \\ 1 & 0 & 0 & 0 \\ -1 & 0 & 1 & -1 \\ 0 & 0 & 0 & 0 \end{bmatrix} \begin{bmatrix} v_{11} \\ v_{12} \\ w_{11} \\ w_{12} \end{bmatrix} = \begin{bmatrix} 0 \\ 0 \\ 0 \\ 0 \end{bmatrix}$$

Solving this last equation, we obtain

$$\begin{bmatrix} v_{11} \\ v_{12} \end{bmatrix} = \begin{bmatrix} 0 \\ a \end{bmatrix}, \qquad \begin{bmatrix} w_{11} \\ w_{12} \end{bmatrix} = \begin{bmatrix} 0 \\ 0 \end{bmatrix}$$

By choosing $a = 1$, we have

$$\begin{bmatrix} v_{11} \\ v_{12} \end{bmatrix} = \begin{bmatrix} 0 \\ 1 \end{bmatrix}, \qquad \begin{bmatrix} w_{11} \\ w_{12} \end{bmatrix} = \begin{bmatrix} 0 \\ 0 \end{bmatrix}$$

For $\mu_2 = 0$, Eq. (7–301) becomes

$$\begin{bmatrix} 0 & 0 & 0 & 0 \\ 1 & 1 & 0 & 0 \\ -1 & 0 & 1 & 0 \\ 0 & 0 & 0 & 1 \end{bmatrix} \begin{bmatrix} v_{21} \\ v_{22} \\ w_{21} \\ w_{22} \end{bmatrix} = \begin{bmatrix} 0 \\ 0 \\ 0 \\ 0 \end{bmatrix}$$

Solving this equation, we obtain

$$\begin{bmatrix} v_{21} \\ v_{22} \end{bmatrix} = \begin{bmatrix} b \\ -b \end{bmatrix}, \qquad \begin{bmatrix} w_{21} \\ w_{22} \end{bmatrix} = \begin{bmatrix} b \\ 0 \end{bmatrix}$$

By choosing $b = 1$, we have

$$\begin{bmatrix} v_{21} \\ v_{22} \end{bmatrix} = \begin{bmatrix} 1 \\ -1 \end{bmatrix}, \qquad \begin{bmatrix} w_{21} \\ w_{22} \end{bmatrix} = \begin{bmatrix} 1 \\ 0 \end{bmatrix}$$

Hence matrix $\mathbf{P}$ is obtained from Eq. (7–106), as follows:

$$\mathbf{P} = [\mathbf{w}_1 \quad \mathbf{w}_2][\mathbf{v}_1 \quad \mathbf{v}_2]^{-1} = \begin{bmatrix} 0 & 1 \\ 0 & 0 \end{bmatrix} \begin{bmatrix} 0 & 1 \\ 1 & -1 \end{bmatrix}^{-1} = \begin{bmatrix} 1 & 0 \\ 0 & 0 \end{bmatrix} \tag{7–302}$$

Equation (7–302) gives the required solution of the steady-state Riccati equation, Eq. (7–75).

Referring to Eq. (7–79), we have

$$\begin{aligned} u(k) &= -(R + \mathbf{H}^*\mathbf{P}\mathbf{H})^{-1}\mathbf{H}^*\mathbf{P}\mathbf{G}\mathbf{x}(k) \\ &= -(1+1)^{-1}[1 \quad 0] \begin{bmatrix} 1 & 0 \\ 0 & 0 \end{bmatrix} \begin{bmatrix} 0 & 0 \\ 1 & 1 \end{bmatrix} \mathbf{x}(k) \\ &= -2^{-1}[0 \quad 0]\mathbf{x}(k) = 0 \end{aligned} \tag{7–303}$$

Equation (7–303) gives the optimal control law.

The closed-loop system now becomes

$$\mathbf{x}(k+1) = \mathbf{G}\mathbf{x}(k) + \mathbf{H}u(k) = \begin{bmatrix} 0 & 0 \\ 1 & 1 \end{bmatrix} \mathbf{x}(k) \tag{7–304}$$

Equation (7–304) gives the optimal closed-loop operation for the system. The closed-loop poles are at $\mu_1 = 1$ and $\mu_2 = 0$. The closed-loop system is not asymptotically stable.

The minimum value of J is obtained from Eq. (7–81), as follows:

$$J_{\min} = \frac{1}{2}\mathbf{x}^*(0)\mathbf{P}\mathbf{x}(0) = \frac{1}{2}[1 \quad 1] \begin{bmatrix} 1 & 0 \\ 0 & 0 \end{bmatrix} \begin{bmatrix} 1 \\ 1 \end{bmatrix} = \frac{1}{2}$$

Although the system is not asymptotically stable, the performance index becomes finite and is minimum. In fact, since $u(k) = 0$ for $k = 0, 1, 2, \ldots$, the system equation becomes

$$\begin{aligned} x_1(k+1) &= 0 \\ x_2(k+1) &= x_1(k) + x_2(k) \end{aligned}$$

or

$$\begin{aligned} x_1(0) &= 1, \qquad x_1(k) = 0 \qquad k = 1, 2, 3, \ldots \\ x_2(0) &= 1, \qquad x_2(k) = 2 \qquad k = 1, 2, 3, \ldots \end{aligned}$$

Notice that the performance index becomes finite, because it involves $x_1(k)$, but does not include $x_2(k)$.

Remember that the stability of the quadratic optimal control system can be checked by use of Eq. (7–72). In this problem, since

$$\mathbf{Q} = \begin{bmatrix} 1 & 0 \\ 0 & 0 \end{bmatrix}$$

we have

$$\text{rank}[(\mathbf{Q}^{1/2})^* \vdots \mathbf{G}^*(\mathbf{Q}^{1/2})^*] = \text{rank}\left[\begin{array}{cc|cc} 1 & 0 & 0 & 0 \\ 0 & 0 & 0 & 0 \end{array}\right] = 1 < 2$$

Hence the system is not asymptotically stable, although it minimizes the given performance index.

Problem A-7-6. Consider the same system as discussed in Prob. A-7-5. Obtain matrix **P** by use of Eq. (7-130).

Solution. Equation (7-130) which we repeat here:

$$\mathbf{P} = \mathbf{Q} + \mathbf{G}^*\mathbf{P}(\mathbf{I} + \mathbf{H}\mathbf{R}^{-1}\mathbf{H}^*\mathbf{P})^{-1}\mathbf{G}$$

becomes as follows:

$$\begin{bmatrix} p_{11} & p_{12} \\ p_{12} & p_{22} \end{bmatrix} = \begin{bmatrix} 1 & 0 \\ 0 & 0 \end{bmatrix}$$

$$+ \begin{bmatrix} 0 & 1 \\ 0 & 1 \end{bmatrix} \begin{bmatrix} p_{11} & p_{12} \\ p_{12} & p_{22} \end{bmatrix} \left\{ \begin{bmatrix} 1 & 0 \\ 0 & 1 \end{bmatrix} + \begin{bmatrix} 1 \\ 0 \end{bmatrix} [1][1 \quad 0] \begin{bmatrix} p_{11} & p_{12} \\ p_{12} & p_{22} \end{bmatrix} \right\}^{-1} \begin{bmatrix} 0 & 0 \\ 1 & 1 \end{bmatrix}$$

$$= \begin{bmatrix} 1 & 0 \\ 0 & 0 \end{bmatrix} + \begin{bmatrix} p_{12} & p_{22} \\ p_{12} & p_{22} \end{bmatrix} \begin{bmatrix} \dfrac{1}{1+p_{11}} & -\dfrac{p_{12}}{1+p_{11}} \\ 0 & 1 \end{bmatrix} \begin{bmatrix} 0 & 0 \\ 1 & 1 \end{bmatrix}$$

$$= \begin{bmatrix} 1 - \dfrac{p_{12}^2}{1+p_{11}} + p_{22} & -\dfrac{p_{12}^2}{1+p_{11}} + p_{22} \\ -\dfrac{p_{12}^2}{1+p_{11}} + p_{22} & -\dfrac{p_{12}^2}{1+p_{11}} + p_{22} \end{bmatrix} \tag{7-305}$$

Equation (7-305) yields the following three equations:

$$p_{11} = 1 - \frac{p_{12}^2}{1+p_{11}} + p_{22} \tag{7-306}$$

$$p_{12} = -\frac{p_{12}^2}{1+p_{11}} + p_{22} \tag{7-307}$$

$$p_{22} = -\frac{p_{12}^2}{1+p_{11}} + p_{22} \tag{7-308}$$

Equations (7-307) and (7-308) yield

$$p_{12} = p_{22}$$

Equation (7-308) gives

$$\frac{p_{12}^2}{1+p_{11}} = 0$$

Hence

$$p_{12} = 0$$

By substituting 0 for p_{12} and p_{22} in Eq. (7-306), we obtain

$$p_{11} = 1$$

Thus, we have determined matrix **P** to be as follows:

$$\mathbf{P} = \begin{bmatrix} 1 & 0 \\ 0 & 0 \end{bmatrix}$$

Matrix **P** is positive semidefinite.

Problem A-7-7. Consider the scalar control system

$$x(k+1) = gx(k) + hu(k) \tag{7-309}$$

and the performance index

$$J = \frac{1}{2} \sum_{k=0}^{\infty} [qx^2(k) + ru^2(k)] \tag{7-310}$$

where $q > 0$ and $r > 0$. It was shown in Sec. 7-3 that the optimal control law that will minimize the performance index J can be given by

$$u(k) = -Kx(k) \tag{7-311}$$

Substituting Eq. (7-311) into Eq. (7-309), we obtain

$$x(k+1) = (g - hK)x(k) \tag{7-312}$$

By substituting Eq. (7-311) into Eq. (7-310), we have

$$J = \frac{1}{2} \sum_{k=0}^{\infty} (q + rK^2)x^2(k)$$

Using the Liapunov approach and referring to Eq. (7-123), we set

$$(q + rK^2)x^2(k) = -[px^2(k+1) - px^2(k)] \tag{7-313}$$

By substituting Eq. (7-312) into Eq. (7-313), we obtain

$$(q + rK^2)x^2(k) = [-p(g - hK)^2 + p]x^2(k)$$

or

$$[q + rK^2 + p(g - hK)^2 - p]x^2(k) = 0$$

This last equation must hold true for any $x(k)$. Hence we require that

$$q + rK^2 + p(g - hK)^2 - p = 0 \tag{7-314}$$

Show that the optimal control law can be given by

$$u(k) = -Kx(k) = -ghp(r + ph^2)^{-1}x(k)$$

or

$$K = ghp(r + ph^2)^{-1} \tag{7-315}$$

Also show that p can be determined as a positive root of the following equation:

$$q - p + g^2rp(r + ph^2)^{-1} = 0 \tag{7-316}$$

Solution. By referring to Eq. (7-117), the performance index J can be given as follows:

$$J = \tfrac{1}{2}px^2(0)$$

To minimize this value of J for a given $x(0)$ with respect to K, we set

$$\frac{\partial p}{\partial K} = 0 \tag{7–317}$$

where p is as given by Eq. (7–314). Notice that in Eq. (7–314), $q + rK^2 > 0$. Hence, $1 - (g - hK)^2 \neq 0$. Therefore, p can be given as follows:

$$p = \frac{q + rK^2}{1 - (g - hK)^2} \tag{7–318}$$

By differentiating p with respect to K and equating the result to zero, we obtain

$$\frac{\partial p}{\partial K} = \frac{2rK[1 - (g - hK)^2] - (q + rK^2)[2(g - hK)h]}{[1 - (g - hK)^2]^2} = 0$$

which yields

$$rK[1 - (g - hK)^2] - (q + rK^2)(g - hK)h = 0$$

Hence, we obtain

$$\frac{q + rK^2}{1 - (g - hK)^2} = \frac{rK}{h(g - hK)} \tag{7–319}$$

From Eqs. (7–318) and (7–319) we get

$$p = \frac{rK}{h(g - hK)} \tag{7–320}$$

Solving Eq. (7–320) for K and noting that $r + ph^2 > 0$, we have

$$K = \frac{ghp}{r + ph^2} = ghp(r + ph^2)^{-1} \tag{7–321}$$

which is Eq. (7–315).

By substituting Eq. (7–321) into Eq. (7–314),

$$q + \frac{g^2h^2p^2r}{(r + ph^2)^2} + p\left(\frac{gr}{r + ph^2}\right)^2 - p = 0$$

which can be simplified to

$$q - p + g^2rp(r + ph^2)^{-1} = 0$$

which is Eq. (7–316).

The same results can also be obtained in the following way. First note that Eq. (7–314) can be modified as follows:

$$q + (r + ph^2)K^2 - 2ghpK + pg^2 - p = 0$$

or

$$q + pg^2 - p + \left(\sqrt{r + ph^2}\,K - \frac{ghp}{\sqrt{r + ph^2}}\right)^2 - \frac{g^2h^2p^2}{r + ph^2} = 0 \tag{7–322}$$

Then, considering this last equation as a function of K, the minimum of the left-hand side of this last equation with respect to K occurs when

$$\sqrt{r+ph^2}\,K - \frac{ghp}{\sqrt{1+ph^2}} = 0$$

or

$$K = ghp(r+ph^2)^{-1} \tag{7–323}$$

which is Eq. (7–315).

By substituting Eq. (7–323) into Eq. (7–322), we obtain

$$q + pg^2 - p - \frac{g^2h^2p^2}{r+ph^2} = 0$$

which can be simplified as follows:

$$q - p + g^2rp(r+ph^2)^{-1} = 0$$

which is Eq. (7–316).

Problem A-7–8. Consider the following linear continuous-time system:

$$\dot{\mathbf{x}} = \mathbf{A}\mathbf{x}$$

where $\mathbf{A}$ has one or more adjustable parameters. Matrix $\mathbf{A}$ is assumed to be a stable matrix. (That is, the eigenvalues of $\mathbf{A}$ have negative real parts.) We wish to minimize the quadratic performance index given by

$$J = \frac{1}{2}\int_0^\infty \mathbf{x}^*\mathbf{Q}\mathbf{x}\,dt$$

where $\mathbf{Q}$ is a positive definite or positive semidefinite Hermitian (or real symmetric) matrix. The problem thus becomes that of determining the value(s) of the adjustable parameter(s) so as to minimize the performance index.

Show that a Liapunov funciton can be used effectively in the solution of this problem.

Solution. Let us assume that

$$\mathbf{x}^*\mathbf{Q}\mathbf{x} = -\frac{d}{dt}(\mathbf{x}^*\mathbf{P}\mathbf{x})$$

where $\mathbf{P}$ is a positive definite Hermitian (or real symmetric) matrix. Then we obtain

$$\mathbf{x}^*\mathbf{Q}\mathbf{x} = -\dot{\mathbf{x}}^*\mathbf{P}\mathbf{x} - \mathbf{x}^*\mathbf{P}\dot{\mathbf{x}} = -\mathbf{x}^*\mathbf{A}^*\mathbf{P}\mathbf{x} - \mathbf{x}^*\mathbf{P}\mathbf{A}\mathbf{x} = -\mathbf{x}^*(\mathbf{A}^*\mathbf{P} + \mathbf{P}\mathbf{A})\mathbf{x}$$

Since $\mathbf{A}$ is a stable matrix, by the second method of Liapunov we know that for a given $\mathbf{Q}$ there exists a $\mathbf{P}$ such that

$$\mathbf{A}^*\mathbf{P} + \mathbf{P}\mathbf{A} = -\mathbf{Q} \tag{7–324}$$

Hence, it is possible to determine the elements of $\mathbf{P}$ from Eq. (7–324).

The performance index J can be evaluated as follows:

$$J = \frac{1}{2}\int_0^\infty \mathbf{x}^*\mathbf{Q}\mathbf{x}\,dt = -\frac{1}{2}\mathbf{x}^*\mathbf{P}\mathbf{x}\Big|_0^\infty = -\frac{1}{2}\mathbf{x}^*(\infty)\mathbf{P}\mathbf{x}(\infty) + \frac{1}{2}\mathbf{x}^*(0)\mathbf{P}\mathbf{x}(0)$$

Since all eigenvalues of $\mathbf{A}$ have negative real parts, we have $\mathbf{x}(\infty) \rightarrow \mathbf{0}$. Therefore, we obtain

$$J = \frac{1}{2}\mathbf{x}^*(0)\mathbf{P}\mathbf{x}(0)$$

Thus the performance index J can be obtained in terms of the initial condition $\mathbf{x}(0)$ and $\mathbf{P}$, which is related to $\mathbf{A}$ and $\mathbf{Q}$ by Eq. (7–324). If, for example, a system parameter is to be adjusted so as to minimize the performance index J, then it can be accomplished by minimizing $\frac{1}{2}\mathbf{x}^*(0)\mathbf{P}\mathbf{x}(0)$ with respect to the parameter in question. Since $\mathbf{x}(0)$ is the given initial condition and $\mathbf{Q}$ is also given, $\mathbf{P}$ is a function of the elements of $\mathbf{A}$. Hence, this minimization process will result in the optimal value of the adjustable parameter. See Prob. A-7–9.

Problem A-7–9. Consider the control system shown in Fig. 7–9. Determine the value of the damping ratio $\zeta > 0$ so that when the system is subjected to a unit step input $r(t) = 1(t)$, the following performance index is minimized:

$$J = \frac{1}{2}\int_0^\infty \mathbf{x}^*(t)\mathbf{Q}\mathbf{x}(t)\, dt$$

where

$$\mathbf{x} = \begin{bmatrix} x_1 \\ x_2 \end{bmatrix} = \begin{bmatrix} x \\ \dot{x} \end{bmatrix}, \qquad \mathbf{Q} = \begin{bmatrix} 1 & 0 \\ 0 & \mu \end{bmatrix} \qquad \mu > 0$$

The system is assumed to be at rest initially.

Solution. From Fig. 7–9, we obtain the following equation for the system:

$$\ddot{c} + 2\zeta\dot{c} + c = r$$

Noting that $x = r - c$ and the initial conditions are equal to zero, we have

$$\ddot{x} + 2\zeta\dot{x} + x = 0 \qquad t > 0$$

By choosing $x_1 = x$ and $x_2 = \dot{x}$, the state space equation can be given by

$$\begin{bmatrix} \dot{x}_1 \\ \dot{x}_2 \end{bmatrix} = \begin{bmatrix} 0 & 1 \\ -1 & -2\zeta \end{bmatrix}\begin{bmatrix} x_1 \\ x_2 \end{bmatrix}, \qquad \begin{bmatrix} x_1(0) \\ x_2(0) \end{bmatrix} = \begin{bmatrix} 1 \\ 0 \end{bmatrix}$$

or

$$\dot{\mathbf{x}} = \mathbf{A}\mathbf{x}$$

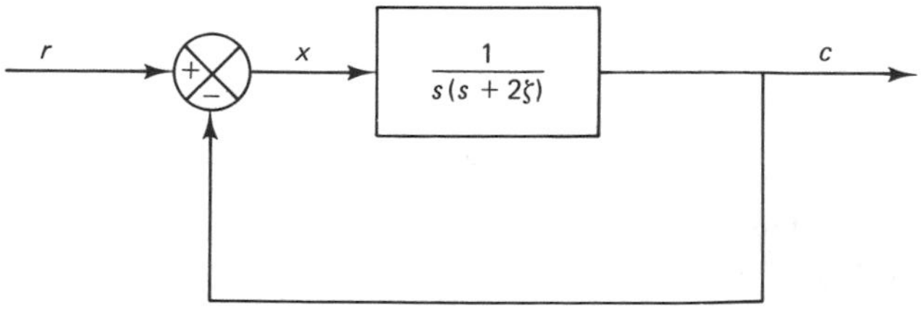

Figure 7–9 Control system of Prob. A-7–9.

where

$$\mathbf{A} = \begin{bmatrix} 0 & 1 \\ -1 & -2\zeta \end{bmatrix}$$

Since **A** is a stable matrix, the value of J is given by

$$J = \frac{1}{2}\mathbf{x}^*(0)\mathbf{P}\mathbf{x}(0)$$

where **P** is determined from

$$\mathbf{A}^*\mathbf{P} + \mathbf{P}\mathbf{A} = -\mathbf{Q}$$

or

$$\begin{bmatrix} 0 & -1 \\ 1 & -2\zeta \end{bmatrix}\begin{bmatrix} p_{11} & p_{12} \\ p_{12} & p_{22} \end{bmatrix} + \begin{bmatrix} p_{11} & p_{12} \\ p_{12} & p_{22} \end{bmatrix}\begin{bmatrix} 0 & 1 \\ -1 & -2\zeta \end{bmatrix} = -\begin{bmatrix} 1 & 0 \\ 0 & \mu \end{bmatrix}$$

This equation results in the following three equations:

$$-2p_{12} = -1$$

$$p_{11} - 2\zeta p_{12} - p_{22} = 0$$

$$2p_{12} - 4\zeta p_{22} = -\mu$$

Solving these three equations for the p_{ij}'s, we obtain

$$\mathbf{P} = \begin{bmatrix} p_{11} & p_{12} \\ p_{12} & p_{22} \end{bmatrix} = \begin{bmatrix} \zeta + \dfrac{1+\mu}{4\zeta} & \dfrac{1}{2} \\ \dfrac{1}{2} & \dfrac{1+\mu}{4\zeta} \end{bmatrix}$$

Thus, the performance index J becomes as follows:

$$J = \frac{1}{2}\mathbf{x}^*(0)\mathbf{P}\mathbf{x}(0) = \frac{1}{2}\left[\left(\zeta + \frac{1+\mu}{4\zeta}\right)x_1^2(0) + x_1(0)x_2(0) + \frac{1+\mu}{4\zeta}x_2^2(0)\right]$$

Substituting the initial conditions $x_1(0) = 1$ and $x_2(0) = 0$ into this last equation, we obtain

$$J = \frac{1}{2}\left(\zeta + \frac{1+\mu}{4\zeta}\right)$$

To minimize the performance index J with respect to ζ, we set

$$\frac{\partial J}{\partial \zeta} = \frac{1}{2}\left(1 - \frac{1+\mu}{4\zeta^2}\right) = 0$$

which yields

$$\zeta = \frac{\sqrt{1+\mu}}{2}$$

Thus, the optimal value of ζ is $\frac{1}{2}\sqrt{1+\mu}$. For example, if $\mu = 1$, then the optimal value of ζ is $\frac{1}{2}\sqrt{2}$, or 0.707.

Problem A-7–10. Consider the following continuous-time system:

$$\dot{x} = ax + bu \tag{7–325}$$

where $a < 0$ and the performance index is given by

$$J = \frac{1}{2}\int_0^\infty (qx^2 + ru^2)\, dt \tag{7–326}$$

where $q > 0$ and $r > 0$. The optimal control law that will minimize the performance index J can be given by

$$u = -Kx \tag{7–327}$$

Substituting Eq. (7–327) into Eq. (7–325) gives

$$\dot{x} = (a - bK)x \tag{7–328}$$

Also, substituting Eq. (7–327) into Eq. (7–326) gives

$$J = \frac{1}{2}\int_0^\infty (q + rK^2)x^2\, dt \tag{7–329}$$

Using the Liapunov approach, we set

$$(q + rK^2)x^2 = -\frac{d}{dt}(px^2)$$

or

$$(q + rK^2)x^2 = -2px\dot{x} = -2p(a - bK)x^2$$

which can be simplified to

$$[q + rK^2 + 2p(a - bK)]x^2 = 0$$

This last equation must hold true for any $x(t)$. Hence, we require

$$q + rK^2 + 2p(a - bK) = 0 \tag{7–330}$$

Note that by the second method of Liapunov we know that for a given $q + rK^2$ there exists a p such that

$$(a - bK)p + p(a - bK) = -q - rK^2$$

which is the same as Eq. (7–330). Hence, there exists a p that satisfies Eq. (7–330).

Show that the optimal control law can be given by

$$u = -Kx = -\frac{pb}{r}x$$

and p can be determined as a positive root of the following equation:

$$q + 2ap - \frac{p^2b^2}{r} = 0 \tag{7–331}$$

Solution. For a stable system we have $x(\infty) = 0$. Hence the performance index can be evaluated as follows:

$$J = \frac{1}{2}\int_0^\infty (q + rK^2)x^2\, dt = \frac{1}{2}\int_0^\infty -\frac{d}{dt}(px^2)\, dt$$

$$= \frac{1}{2}[-px^2(\infty) + px^2(0)] = \frac{1}{2}px^2(0)$$

To minimize the value of J [for a given $x(0)$] with respect to K, we set

$$\frac{\partial p}{\partial K} = 0 \tag{7–332}$$

where, referring to Eq. (7–330),

$$p = -\frac{q + rK^2}{2(a - bK)} \tag{7–333}$$

Thus

$$\frac{\partial p}{\partial K} = -\frac{2rK(a - bK) - (q + rK^2)(-b)}{2(a - bK)^2} = 0$$

which yields

$$2rK(a - bK) + b(q + rK^2) = 0$$

Hence we have

$$\frac{q + rK^2}{2(a - bK)} = -\frac{rK}{b} \tag{7–334}$$

From Eqs. (7–333) and (7–334) we obtain

$$p = \frac{rK}{b}$$

or

$$K = \frac{pb}{r} \tag{7–335}$$

By substituting Eq. (7–335) into Eq. (7–330), we obtain

$$q + 2pa - \frac{p^2b^2}{r} = 0 \tag{7–336}$$

which is Eq. (7–331). The value of p can be determined as a positive root of the quadratic equation given by Eq. (7–336).

The same results can be obtained in a different way. First note that Eq. (7–330) can be modified as follows:

$$q + 2pa + \left(\sqrt{r}\,K - \frac{pb}{\sqrt{r}}\right)^2 - \frac{p^2b^2}{r} = 0 \tag{7–337}$$

Then, considering this last equation as a function of K, the minimum of the left-hand side of this last equation with respect to K occurs when

$$\sqrt{r}\,K - \frac{pb}{\sqrt{r}} = 0$$

or

$$K = \frac{pb}{r} \tag{7-338}$$

which is Eq. (7–335). By substituting Eq. (7–338) into Eq. (7–337) we obtain

$$q + 2pa - \frac{p^2 b^2}{r} = 0$$

which is Eq. (7–331).

Problem A-7–11. Consider the linear continuous-time system

$$\dot{\mathbf{x}} = \mathbf{A}\mathbf{x} + \mathbf{B}\mathbf{u} \tag{7-339}$$

Determine the matrix **K** of the optimal control vector

$$\mathbf{u} = -\mathbf{K}\mathbf{x} \tag{7-340}$$

so as to minimize the performance index

$$J = \frac{1}{2}\int_0^\infty (\mathbf{x}^*\mathbf{Q}\mathbf{x} + \mathbf{u}^*\mathbf{R}\mathbf{u})\, dt \tag{7-341}$$

where **Q** is a positive definite or positive semidefinite Hermitian (or real symmetric) matrix and **R** is a positive definite Hermitian (or real symmetric) matrix.

By substituting Eq. (7–340) into Eq. (7–339), we obtain

$$\dot{\mathbf{x}} = (\mathbf{A} - \mathbf{B}\mathbf{K})\mathbf{x} \tag{7-342}$$

where we assume that matrix $\mathbf{A} - \mathbf{BK}$ is stable. Substituting Eq. (7–340) into Eq. (7–341), we obtain

$$J = \frac{1}{2}\int_0^\infty (\mathbf{x}^*\mathbf{Q}\mathbf{x} + \mathbf{x}^*\mathbf{K}^*\mathbf{R}\mathbf{K}\mathbf{x})\, dt = \frac{1}{2}\int_0^\infty \mathbf{x}^*(\mathbf{Q} + \mathbf{K}^*\mathbf{R}\mathbf{K})\mathbf{x}\, dt$$

For a stable system, a Liapunov function exists. Let us define $\mathbf{x}^*\mathbf{P}\mathbf{x}$ as a Liapunov function. Then, we may set

$$\mathbf{x}^*(\mathbf{Q} + \mathbf{K}^*\mathbf{R}\mathbf{K})\mathbf{x} = -\frac{d}{dt}(\mathbf{x}^*\mathbf{P}\mathbf{x}) \tag{7-343}$$

Derive the optimal control law $\mathbf{u}(t)$. Also, derive an equation from which matrix **P** can be obtained.

Solution. From Eq. (7–343) we obtain

$$\mathbf{x}^*(\mathbf{Q} + \mathbf{K}^*\mathbf{R}\mathbf{K})\mathbf{x} = -\dot{\mathbf{x}}^*\mathbf{P}\mathbf{x} - \mathbf{x}^*\mathbf{P}\dot{\mathbf{x}} = -\mathbf{x}^*[(\mathbf{A} - \mathbf{B}\mathbf{K})^*\mathbf{P} + \mathbf{P}(\mathbf{A} - \mathbf{B}\mathbf{K})]\mathbf{x}$$

Comparing both sides of this last equation and noting that this equation must hold true for any **x**, we require that

$$(\mathbf{A} - \mathbf{B}\mathbf{K})^*\mathbf{P} + \mathbf{P}(\mathbf{A} - \mathbf{B}\mathbf{K}) = -(\mathbf{Q} + \mathbf{K}^*\mathbf{R}\mathbf{K}) \tag{7-344}$$

Since **R** is positive definite, we may write

$$\mathbf{R} = \mathbf{T}^*\mathbf{T}$$

where **T** is a nonsingular matrix. Then, Eq. (7–344) can be written as

$$(\mathbf{A}^* - \mathbf{K}^*\mathbf{B}^*)\mathbf{P} + \mathbf{P}(\mathbf{A} - \mathbf{B}\mathbf{K}) + \mathbf{Q} + \mathbf{K}^*\mathbf{T}^*\mathbf{T}\mathbf{K} = \mathbf{0}$$

which can be rewritten in turn as

$$\mathbf{A}^*\mathbf{P} + \mathbf{P}\mathbf{A} + [\mathbf{T}\mathbf{K} - (\mathbf{T}^*)^{-1}\mathbf{B}^*\mathbf{P}]^*[\mathbf{T}\mathbf{K} - (\mathbf{T}^*)^{-1}\mathbf{B}^*\mathbf{P}] - \mathbf{P}\mathbf{B}\mathbf{R}^{-1}\mathbf{B}^*\mathbf{P} + \mathbf{Q} = \mathbf{0} \tag{7–345}$$

The minimization of J with respect to **K** requires the minimization of the left-hand side of Eq. (7–345) with respect to **K.** (See Prob. A-7–10.)

Since

$$[\mathbf{T}\mathbf{K} - (\mathbf{T}^*)^{-1}\mathbf{B}^*\mathbf{P}]^*[\mathbf{T}\mathbf{K} - (\mathbf{T}^*)^{-1}\mathbf{B}^*\mathbf{P}]$$

is nonnegative, the minimum occurs when it is zero, or when

$$\mathbf{T}\mathbf{K} = (\mathbf{T}^*)^{-1}\mathbf{B}^*\mathbf{P}$$

Hence

$$\mathbf{K} = \mathbf{T}^{-1}(\mathbf{T}^*)^{-1}\mathbf{B}^*\mathbf{P} = \mathbf{R}^{-1}\mathbf{B}^*\mathbf{P} \tag{7–346}$$

The optimal control law is given by

$$\mathbf{u}(t) = -\mathbf{K}\mathbf{x}(t) = -\mathbf{R}^{-1}\mathbf{B}^*\mathbf{P}\mathbf{x}(t) \tag{7–347}$$

Matrix **P** in Eq. (7–346) is determined as the unique positive definite solution of Eq. (7–344), or the following algebraic Riccati equation which is obtained by substituting Eq. (7–346) into Eq. (7–345):

$$\mathbf{A}^*\mathbf{P} + \mathbf{P}\mathbf{A} - \mathbf{P}\mathbf{B}\mathbf{R}^{-1}\mathbf{B}^*\mathbf{P} + \mathbf{Q} = \mathbf{0} \tag{7–348}$$

[Note that more than one **P** may satisfy Eq. (7–348). However, only one of the **P**'s is positive definite. We choose the positive-definite **P** as the desired solution.]

Problem A-7–12. In Sec. 7-3 it was shown that the system defined by

$$\mathbf{x}(k+1) = \mathbf{G}\mathbf{x}(k) + \mathbf{H}\mathbf{u}(k) \tag{7–349}$$

will minimize the steady-state quadratic performance index

$$J = \frac{1}{2}\sum_{k=0}^{\infty} [\mathbf{x}^*(k)\mathbf{Q}\mathbf{x}(k) + \mathbf{u}^*(k)\mathbf{R}\mathbf{u}(k)] \tag{7–350}$$

where **Q** is a positive definite or positive semidefinite Hermitian (or real symmetric) matrix and **R** is a positive definite Hermitian (or real symmetric) matrix, if the following control law is used:

$$\mathbf{u}(k) = -\mathbf{K}\mathbf{x}(k)$$

where

$$\mathbf{K} = (\mathbf{R} + \mathbf{H}^*\mathbf{P}\mathbf{H})^{-1}\mathbf{H}^*\mathbf{P}\mathbf{G} \tag{7–351}$$

and matrix **P** is the solution of the following Riccati equation:

$$\mathbf{P} = \mathbf{Q} + \mathbf{G}^*\mathbf{P}\mathbf{G} - \mathbf{G}^*\mathbf{P}\mathbf{H}(\mathbf{R} + \mathbf{H}^*\mathbf{P}\mathbf{H})^{-1}\mathbf{H}^*\mathbf{P}\mathbf{G} \tag{7–352}$$

The minimum value of the performance index in this case was found to be

$$J_{\min} = \frac{1}{2}\mathbf{x}^*(0)\mathbf{P}\mathbf{x}(0) \tag{7–353}$$

Consider now the regulator system with observed-state feedback shown in Fig. 7–10. The system equations are

$$\mathbf{x}(k+1) = \mathbf{G}\mathbf{x}(k) + \mathbf{H}\mathbf{u}(k) \tag{7–354}$$

$$\mathbf{u}(k) = -\mathbf{K}\tilde{\mathbf{x}}(k) \tag{7–355}$$

$$\tilde{\mathbf{x}}(k+1) = \mathbf{G}\tilde{\mathbf{x}}(k) + \mathbf{H}\mathbf{u}(k) + \mathbf{K}_e[\mathbf{y}(k) - \mathbf{C}\tilde{\mathbf{x}}(k)] \tag{7–356}$$

In Eq. (7–355) we assume that we use the same feedback gain matrix $\mathbf{K}$ as given by Eq. (7–351).

Determine the minimum value of the performance index when the system uses the observed-state feedback $\mathbf{u}(k) = -\mathbf{K}\tilde{\mathbf{x}}(k)$.

Solution. Define the error vector $\mathbf{e}(k)$ as follows:

$$\mathbf{e}(k) = \mathbf{x}(k) - \tilde{\mathbf{x}}(k) \tag{7–357}$$

Then, from Eqs. (7–354) and (7–356) we obtain

$$\begin{aligned}\mathbf{e}(k+1) &= \mathbf{x}(k+1) - \tilde{\mathbf{x}}(k+1)\\ &= \mathbf{G}[\mathbf{x}(k) - \tilde{\mathbf{x}}(k)] - \mathbf{K}_e\mathbf{C}[\mathbf{x}(k) - \tilde{\mathbf{x}}(k)]\\ &= (\mathbf{G} - \mathbf{K}_e\mathbf{C})\mathbf{e}(k)\end{aligned} \tag{7–358}$$

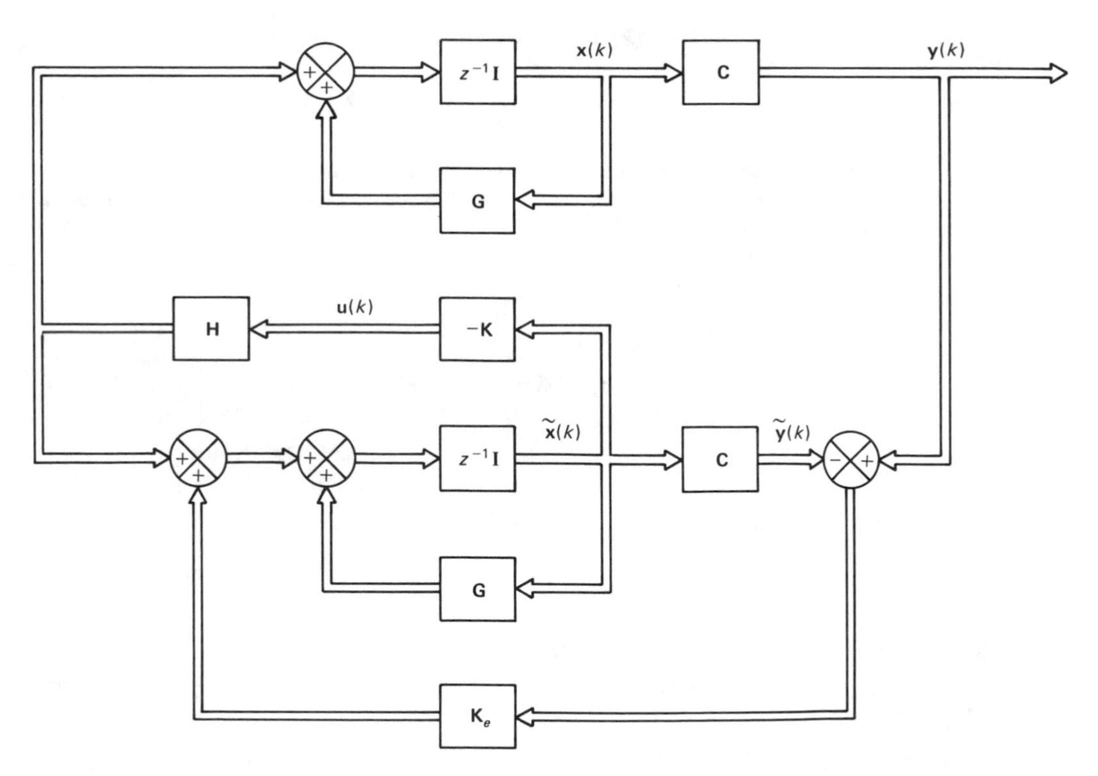

Figure 7–10 Regulator system with observed-state feedback.

Equation (7–354) can be modified as follows:

$$\begin{aligned}\mathbf{x}(k+1) &= \mathbf{G}\mathbf{x}(k) - \mathbf{H}\mathbf{K}\tilde{\mathbf{x}}(k) \\ &= (\mathbf{G} - \mathbf{H}\mathbf{K})\mathbf{x}(k) + \mathbf{H}\mathbf{K}\mathbf{e}(k) \end{aligned} \tag{7–359}$$

Combining Eqs. (7–358) and (7–359) into one equation, we have

$$\begin{bmatrix}\mathbf{x}(k+1) \\ \mathbf{e}(k+1)\end{bmatrix} = \begin{bmatrix}\mathbf{G} - \mathbf{H}\mathbf{K} & \mathbf{H}\mathbf{K} \\ \mathbf{0} & \mathbf{G} - \mathbf{K}_e\mathbf{C}\end{bmatrix}\begin{bmatrix}\mathbf{x}(k) \\ \mathbf{e}(k)\end{bmatrix} \tag{7–360}$$

Equation (7–360) may be considered to be a new state equation for the system, where the state vector is

$$\begin{bmatrix}\mathbf{x}(k) \\ \mathbf{e}(k)\end{bmatrix}$$

which is a $2n$-vector.

The performance index can then be rewritten in terms of the new state vector as follows:

$$\begin{aligned}J &= \frac{1}{2}\sum_{k=0}^{\infty}[\mathbf{x}^*(k)\mathbf{Q}\mathbf{x}(k) + \mathbf{u}^*(k)\mathbf{R}\mathbf{u}(k)] \\ &= \frac{1}{2}\sum_{k=0}^{\infty}[\mathbf{x}^*(k)\mathbf{Q}\mathbf{x}(k) + \tilde{\mathbf{x}}^*(k)\mathbf{K}^*\mathbf{R}\mathbf{K}\tilde{\mathbf{x}}(k)] \\ &= \frac{1}{2}\sum_{k=0}^{\infty}\{\mathbf{x}^*(k)\mathbf{Q}\mathbf{x}(k) + [\mathbf{x}^*(k) - \mathbf{e}^*(k)]\mathbf{K}^*\mathbf{R}\mathbf{K}[\mathbf{x}(k) - \mathbf{e}(k)]\} \\ &= \frac{1}{2}\sum_{k=0}^{\infty}\left\{[\mathbf{x}^*(k) \quad \mathbf{e}^*(k)]\begin{bmatrix}\mathbf{Q} + \mathbf{K}^*\mathbf{R}\mathbf{K} & -\mathbf{K}^*\mathbf{R}\mathbf{K} \\ -\mathbf{K}^*\mathbf{R}\mathbf{K} & \mathbf{K}^*\mathbf{R}\mathbf{K}\end{bmatrix}\begin{bmatrix}\mathbf{x}(k) \\ \mathbf{e}(k)\end{bmatrix}\right\} \end{aligned} \tag{7–361}$$

Referring to the Liapunov approach to the steady-state quadratic optimal control problem discussed in Sec. 7–3, if the state matrix of Eq. (7–360) is a stable matrix, then there exists a positive definite matrix $\hat{\mathbf{P}}$, where

$$\hat{\mathbf{P}} = \begin{bmatrix}\mathbf{P}_{11} & \mathbf{P}_{12} \\ \mathbf{P}_{12}^* & \mathbf{P}_{22}\end{bmatrix} \tag{7–362}$$

such that

$$\begin{bmatrix}\mathbf{G} - \mathbf{H}\mathbf{K} & \mathbf{H}\mathbf{K} \\ \mathbf{0} & \mathbf{G} - \mathbf{K}_e\mathbf{C}\end{bmatrix}^*\begin{bmatrix}\mathbf{P}_{11} & \mathbf{P}_{12} \\ \mathbf{P}_{12}^* & \mathbf{P}_{22}\end{bmatrix}\begin{bmatrix}\mathbf{G} - \mathbf{H}\mathbf{K} & \mathbf{H}\mathbf{K} \\ \mathbf{0} & \mathbf{G} - \mathbf{K}_e\mathbf{C}\end{bmatrix} - \begin{bmatrix}\mathbf{P}_{11} & \mathbf{P}_{12} \\ \mathbf{P}_{12}^* & \mathbf{P}_{22}\end{bmatrix} = -\begin{bmatrix}\mathbf{Q} + \mathbf{K}^*\mathbf{R}\mathbf{K} & -\mathbf{K}^*\mathbf{R}\mathbf{K} \\ -\mathbf{K}^*\mathbf{R}\mathbf{K} & \mathbf{K}^*\mathbf{R}\mathbf{K}\end{bmatrix} \tag{7–363}$$

[refer to Eq. (7–116)] and the minimum value of the performance index can be given by

$$J = \frac{1}{2}[\mathbf{x}^*(0) \quad \mathbf{e}^*(0)]\begin{bmatrix}\mathbf{P}_{11} & \mathbf{P}_{12} \\ \mathbf{P}_{12}^* & \mathbf{P}_{22}\end{bmatrix}\begin{bmatrix}\mathbf{x}(0) \\ \mathbf{e}(0)\end{bmatrix} \tag{7–364}$$

[refer to Eq. (7–117)]. Equation (7–363) can be simplified to the form

$$\left[\begin{matrix}(\mathbf{G}^* - \mathbf{K}^*\mathbf{H}^*)\mathbf{P}_{11}(\mathbf{G} - \mathbf{H}\mathbf{K}) & (\mathbf{G}^* - \mathbf{K}^*\mathbf{H}^*)[\mathbf{P}_{11}\mathbf{H}\mathbf{K} + \mathbf{P}_{12}(\mathbf{G} - \mathbf{K}_e\mathbf{C})] \\ [\mathbf{K}^*\mathbf{H}^*\mathbf{P}_{11} + (\mathbf{G}^* - \mathbf{C}^*\mathbf{K}_e^*)\mathbf{P}_{12}^*](\mathbf{G} - \mathbf{H}\mathbf{K}) & \begin{matrix}\mathbf{K}^*\mathbf{H}^*[\mathbf{P}_{11}\mathbf{H}\mathbf{K} + \mathbf{P}_{12}(\mathbf{G} - \mathbf{K}_e\mathbf{C})] \\ + (\mathbf{G}^* - \mathbf{C}^*\mathbf{K}_e^*)[\mathbf{P}_{12}^*\mathbf{H}\mathbf{K} + \mathbf{P}_{22}(\mathbf{G} - \mathbf{K}_e\mathbf{C})]\end{matrix}\end{matrix}\right]$$

$$-\begin{bmatrix} \mathbf{P}_{11} & \mathbf{P}_{12} \\ \mathbf{P}_{12}^* & \mathbf{P}_{22} \end{bmatrix} = -\begin{bmatrix} \mathbf{Q} + \mathbf{K}^*\mathbf{R}\mathbf{K} & -\mathbf{K}^*\mathbf{R}\mathbf{K} \\ -\mathbf{K}^*\mathbf{R}\mathbf{K} & \mathbf{K}^*\mathbf{R}\mathbf{K} \end{bmatrix}$$

which yields the following three equations:

$$\mathbf{P}_{11} = \mathbf{Q} + \mathbf{K}^*\mathbf{R}\mathbf{K} + (\mathbf{G}^* - \mathbf{K}^*\mathbf{H}^*)\mathbf{P}_{11}(\mathbf{G} - \mathbf{H}\mathbf{K}) \tag{7–365}$$

$$\mathbf{P}_{12} = -\mathbf{K}^*\mathbf{R}\mathbf{K} + (\mathbf{G}^* - \mathbf{K}^*\mathbf{H}^*)[\mathbf{P}_{11}\mathbf{H}\mathbf{K} + \mathbf{P}_{12}(\mathbf{G} - \mathbf{K}_e\mathbf{C})] \tag{7–366}$$

$$\mathbf{P}_{22} = \mathbf{K}^*\mathbf{R}\mathbf{K} + \mathbf{K}^*\mathbf{H}^*[\mathbf{P}_{11}\mathbf{H}\mathbf{K} + \mathbf{P}_{12}(\mathbf{G} - \mathbf{K}_e\mathbf{C})]$$
$$+ (\mathbf{G}^* - \mathbf{C}^*\mathbf{K}_e^*)[\mathbf{P}_{12}^*\mathbf{H}\mathbf{K} + \mathbf{P}_{22}(\mathbf{G} - \mathbf{K}_e\mathbf{C})] \tag{7–367}$$

Notice that if $\mathbf{P}$ is substituted for $\mathbf{P}_{11}$ in Eq. (7–365), the equation is satisfied, because by making that substitution and also substituting Eq. (7–351) into Eq. (7–365), we obtain

$$\mathbf{P} - [\mathbf{Q} + \mathbf{K}^*\mathbf{R}\mathbf{K} + (\mathbf{G}^* - \mathbf{K}^*\mathbf{H}^*)\mathbf{P}(\mathbf{G} - \mathbf{H}\mathbf{K})]$$
$$= \mathbf{P} - [\mathbf{Q} + \mathbf{K}^*(\mathbf{R} + \mathbf{H}^*\mathbf{P}\mathbf{H})\mathbf{K} + \mathbf{G}^*\mathbf{P}\mathbf{G} - \mathbf{K}^*\mathbf{H}^*\mathbf{P}\mathbf{G} - \mathbf{G}^*\mathbf{P}\mathbf{H}\mathbf{K}]$$
$$= \mathbf{P} - (\mathbf{Q} + \mathbf{K}^*\mathbf{H}^*\mathbf{P}\mathbf{G} + \mathbf{G}^*\mathbf{P}\mathbf{G} - \mathbf{K}^*\mathbf{H}^*\mathbf{P}\mathbf{G} - \mathbf{G}^*\mathbf{P}\mathbf{H}\mathbf{K})$$
$$= \mathbf{P} - [\mathbf{Q} + \mathbf{G}^*\mathbf{P}\mathbf{G} - \mathbf{G}^*\mathbf{P}\mathbf{H}(\mathbf{R} + \mathbf{H}^*\mathbf{P}\mathbf{H})^{-1}\mathbf{H}^*\mathbf{P}\mathbf{G}] = \mathbf{0}$$

where we have used Eq. (7–352). Thus, we set $\mathbf{P}_{11} = \mathbf{P}$. Then matrix $\hat{\mathbf{P}}$ can be written as follows:

$$\hat{\mathbf{P}} = \begin{bmatrix} \mathbf{P} & \mathbf{P}_{12} \\ \mathbf{P}_{12}^* & \mathbf{P}_{22} \end{bmatrix}$$

Note that we have determined a positive-definite matrix $\hat{\mathbf{P}}$ with Eqs. (7–365), (7–366), and (7–367).

Next, let us simplify Eq. (7–366). Since

$$-\mathbf{K}^*\mathbf{R}\mathbf{K} + (\mathbf{G}^* - \mathbf{K}^*\mathbf{H}^*)\mathbf{P}\mathbf{H}\mathbf{K} = -\mathbf{K}^*(\mathbf{R} + \mathbf{H}^*\mathbf{P}\mathbf{H})\mathbf{K} + \mathbf{G}^*\mathbf{P}\mathbf{H}\mathbf{K}$$
$$= -\mathbf{G}^*\mathbf{P}\mathbf{H}(\mathbf{R} + \mathbf{H}^*\mathbf{P}\mathbf{H})^{-1}(\mathbf{R} + \mathbf{H}^*\mathbf{P}\mathbf{H})\mathbf{K} + \mathbf{G}^*\mathbf{P}\mathbf{H}\mathbf{K} = \mathbf{0}$$

Eq. (7–366) can be simplified to

$$\mathbf{P}_{12} = (\mathbf{G}^* - \mathbf{K}^*\mathbf{H}^*)\mathbf{P}_{12}(\mathbf{G} - \mathbf{K}_e\mathbf{C})$$

Notice that if $\mathbf{P}_{12} = \mathbf{0}$, then this last equation is clearly satisfied. Hence, we let $\mathbf{P}_{12} = \mathbf{0}$ in matrix $\hat{\mathbf{P}}$:

$$\hat{\mathbf{P}} = \begin{bmatrix} \mathbf{P}_{11} & \mathbf{P}_{12} \\ \mathbf{P}_{12}^* & \mathbf{P}_{22} \end{bmatrix} = \begin{bmatrix} \mathbf{P} & \mathbf{0} \\ \mathbf{0} & \mathbf{P}_{22} \end{bmatrix} \tag{7–368}$$

By taking $\mathbf{P}_{11} = \mathbf{P}$ and $\mathbf{P}_{12} = \mathbf{0}$ and substituting in Eq. (7–367), we have

$$\mathbf{P}_{22} = \mathbf{K}^*(\mathbf{R} + \mathbf{H}^*\mathbf{P}\mathbf{H})\mathbf{K} + (\mathbf{G}^* - \mathbf{C}^*\mathbf{K}_e^*)\mathbf{P}_{22}(\mathbf{G} - \mathbf{K}_e\mathbf{C})$$
$$= \mathbf{G}^*\mathbf{P}\mathbf{H}(\mathbf{R} + \mathbf{H}^*\mathbf{P}\mathbf{H})^{-1}\mathbf{H}^*\mathbf{P}\mathbf{G} + (\mathbf{G}^* - \mathbf{C}^*\mathbf{K}_e^*)\mathbf{P}_{22}(\mathbf{G} - \mathbf{K}_e\mathbf{C}) \tag{7–369}$$

Referring to Eqs. (7–364) and (7–368), the minimum value of the performance index is obtained as follows:

$$\hat{J}_{\min} = \frac{1}{2}[\mathbf{x}^*(0) \quad \mathbf{e}^*(0)]\begin{bmatrix} \mathbf{P} & \mathbf{0} \\ \mathbf{0} & \mathbf{P}_{22} \end{bmatrix}\begin{bmatrix} \mathbf{x}(0) \\ \mathbf{e}(0) \end{bmatrix}$$

$$= \frac{1}{2}[\mathbf{x}^*(0)\mathbf{P}\mathbf{x}(0) + \mathbf{e}^*(0)\mathbf{P}_{22}\mathbf{e}(0)] \tag{7–370}$$

where positive definite matrix $\mathbf{P}_{22}$ can be determined by solving Eq. (7–369).

Comparing the performance index given by Eq. (7–370) with that given by Eq. (7–353), we notice that

$$\hat{J}_{\min} - J_{\min} = \frac{1}{2}[\mathbf{e}^*(0)\mathbf{P}_{22}\mathbf{e}(0)] \tag{7–371}$$

The increase in the minimum value of the performance index J due to the observed-state feedback instead of the actual-state feedback is $\frac{1}{2}[\mathbf{e}^*(0)\mathbf{P}_{22}\mathbf{e}(0)]$. The degradation of the performance index depends on matrix $\mathbf{P}_{22}$, which can be determined from Eq. (7–369), and the initial error $\mathbf{e}(0)$.

Problem A-7–13. If an nth-order linear single-input–single-output discrete-time control system is completely state controllable, we need at most n sampling periods to bring an arbitrary initial state to the desired final state, provided the control vector is not constrained. Hence, if we allow N (where $N > n$) sampling periods, then we have extra freedom to satisfy additional constraints.

The amount of control energy needed depends on the time period (number of sampling periods) allowed for control. If the number of sampling periods allowed is n, the order of the system, then the time-optimal control sequence $u(0)$, $u(1)$, . . . , $u(n-1)$ is unique. However, if N sampling periods ($N > n$) are allowed, then more than one possible control sequence is possible. Each possible control sequence requires a certain amount of control energy. In many industrial applications, if many control sequences are possible, it is desirable to accomplish control tasks using the minimum amount of control energy.

In this problem, we treat the problem of transferring the state from an arbitrary initial state to the desired final state (which we assume to be the origin of the state space) in N sampling periods and at the same time using the minimum control energy.

Consider the discrete-time control system defined by

$$\mathbf{x}(k+1) = \mathbf{G}\mathbf{x}(k) + \mathbf{H}u(k) \tag{7–372}$$

where

$\mathbf{x}(k)$ = state vector (n-vector) at kth sampling instant

$u(k)$ = control signal (scalar) at kth sampling instant

$\mathbf{G} = n \times n$ nonsingular matrix

$\mathbf{H} = n \times 1$ matrix

Determine the control law that will bring the system state from an arbitrary initial state to the origin in N sampling periods (where $N > n$) using a minimum amount of control energy, where the control energy is measured by

$$\frac{1}{2}\sum_{k=0}^{N-1} u^2(k)$$

Assume that the system is completely state controllable.

Solution. Referring to Eq. (5–57), the state $\mathbf{x}(N)$ of Eq. (7–372) can be given by

$$\mathbf{x}(N) = \mathbf{G}^N\mathbf{x}(0) + \mathbf{G}^{N-1}\mathbf{H}u(0) + \mathbf{G}^{N-2}\mathbf{H}u(1) + \cdots + \mathbf{GH}u(N-2) + \mathbf{H}u(N-1)$$

Substituting $\mathbf{0}$ for $\mathbf{x}(N)$ in this last equation yields

$$\mathbf{x}(0) = -\mathbf{G}^{-1}\mathbf{H}u(0) - \mathbf{G}^{-2}\mathbf{H}u(1) - \cdots - \mathbf{G}^{-N+1}\mathbf{H}u(N-2) - \mathbf{G}^{-N}\mathbf{H}u(N-1) \tag{7–373}$$

Define

$$\mathbf{f}_i = \mathbf{G}^{-i}\mathbf{H} \tag{7–374}$$

Then, Eq. (7–373) becomes

$$\mathbf{x}(0) = -\mathbf{f}_1u(0) - \mathbf{f}_2u(1) - \cdots - \mathbf{f}_{N-1}u(N-2) - \mathbf{f}_Nu(N-1) \tag{7–375}$$

Since the system is completely state controllable, the vectors $\mathbf{f}_1, \mathbf{f}_2, \ldots, \mathbf{f}_n$ are linearly independent. (The remaining $N-n$ vectors can be expressed as linear combinations of these n linearly independent vectors.) Equation (7–375) can be rewritten as

$$\mathbf{x}(0) = -\mathbf{FU} \tag{7–376}$$

where

$$\mathbf{F} = [\mathbf{f}_1 \vdots \mathbf{f}_2 \vdots \cdots \vdots \mathbf{f}_N], \qquad \mathbf{U} = \begin{bmatrix} u(0) \\ u(1) \\ \cdot \\ \cdot \\ \cdot \\ u(N-1) \end{bmatrix}$$

We shall now find the control sequence that satisfies Eq. (7–376) and at the same time minimizes the total control energy. Note that matrix $\mathbf{F}$ is an $n \times N$ matrix and has rank n. Since $\mathbf{F}$ is not a square matrix, the inverse of matrix $\mathbf{F}$ is not defined. Notice that since $N > n$, the number of unknown control signals $u(0), u(1), \ldots, u(N-1)$ in Eq. (7–376) is greater than the number n of component scalar equations. A set of scalar equations in such a situation is said to be *underdetermined* and possesses an indefinite number of solutions. However, in the present case we have a constraint that a set of N unknown variables $u(0), u(1), \ldots, u(N-1)$ gives a minimum norm:

$$\frac{1}{2}\sum_{k=0}^{N-1} u^2(k) = \text{minimum}$$

Then, as seen in the Appendix (Sec. 8), there is a unique solution. Such a unique solution gives the control sequence that brings an arbitrary initial state $\mathbf{x}(0)$ to the origin in N sampling periods and in so doing minimizes the total energy of control.

The minimizing solution in such a problem, where the number of unknown variables is greater than the number of equations, can be obtained in terms of the right pseudoinverse (refer to the Appendix). The right pseudoinverse is defined as follows:

$$\mathbf{F}^{RM} = \mathbf{F}^*(\mathbf{FF}^*)^{-1} \tag{7–377}$$

By using the right pseudoinverse, the minimum-energy control sequence $u(0), u(1), \ldots, u(N-1)$ that transfers an arbitrary initial state $\mathbf{x}(0)$ to the origin can be given by

$$\mathbf{U} = -\mathbf{F}^{RM}\mathbf{x}(0) = -\mathbf{F}^*(\mathbf{FF}^*)^{-1}\mathbf{x}(0) \tag{7–378}$$

Note that $\mathbf{F}^*(\mathbf{FF}^*)^{-1}$ is an $N \times n$ matrix. Hence, $\mathbf{F}^*(\mathbf{FF}^*)^{-1}$ postmultiplied by $\mathbf{x}(0)$ is an $N \times 1$ matrix. Equation (7–378) can be rewritten as follows:

$$\begin{bmatrix} u(0) \\ u(1) \\ \cdot \\ \cdot \\ \cdot \\ u(N-1) \end{bmatrix} = -\mathbf{F}^*(\mathbf{FF}^*)^{-1}\mathbf{x}(0) \tag{7–379}$$

The control sequence given by Eq. (7–379) will bring an arbitrary initial state to the origin in N sampling periods and will require the minimum control energy among all possible control sequences requiring N sampling periods.

Problem A-7–14. Consider the system

$$\mathbf{x}(k+1) = \mathbf{Gx}(k) + \mathbf{H}u(k) \tag{7–380}$$

where

$$\mathbf{G} = \begin{bmatrix} 1 & 0.6321 \\ 0 & 0.3679 \end{bmatrix}, \qquad \mathbf{H} = \begin{bmatrix} 0.3679 \\ 0.6321 \end{bmatrix}, \qquad \begin{bmatrix} x_1(0) \\ x_2(0) \end{bmatrix} = \begin{bmatrix} 5 \\ -5 \end{bmatrix}$$

It is desired to bring the initial state to the origin in 3 sampling periods. (The sampling period is assumed to be 1 sec.) Among infinitely many possible choices for the control sequence, determine the optimal control sequence that will minimize the control energy, or will minimize the following performance index:

$$J = \frac{1}{2}\sum_{k=0}^{2} u^2(k) \tag{7–381}$$

Solution. From Eq. (7–375), the initial state $\mathbf{x}(0)$ can be written as follows:

$$\mathbf{x}(0) = -\mathbf{f}_1 u(0) - \mathbf{f}_2 u(1) - \mathbf{f}_3 u(2)$$

where

$$\mathbf{f}_1 = \mathbf{G}^{-1}\mathbf{H} = \begin{bmatrix} -0.7181 \\ 1.7181 \end{bmatrix}, \qquad \mathbf{f}_2 = \mathbf{G}^{-2}\mathbf{H} = \begin{bmatrix} -3.6700 \\ 4.6700 \end{bmatrix}, \qquad \mathbf{f}_3 = \mathbf{G}^{-3}\mathbf{H} = \begin{bmatrix} -11.6935 \\ 12.6935 \end{bmatrix}$$

Hence,

$$\begin{bmatrix} x_1(0) \\ x_2(0) \end{bmatrix} = -\begin{bmatrix} -0.7181 \\ 1.7181 \end{bmatrix} u(0) - \begin{bmatrix} -3.6700 \\ 4.6700 \end{bmatrix} u(1) - \begin{bmatrix} -11.6935 \\ 12.6935 \end{bmatrix} u(2)$$

or

$$\begin{bmatrix} x_1(0) \\ x_2(0) \end{bmatrix} = -\begin{bmatrix} -0.7181 & -3.6700 & -11.6935 \\ 1.7181 & 4.6700 & 12.6935 \end{bmatrix} \begin{bmatrix} u(0) \\ u(1) \\ u(2) \end{bmatrix} \tag{7–382}$$

By use of the right pseudoinverse, we can give the minimum norm solution to Eq. (7–382) as follows:

$$\begin{bmatrix} u(0) \\ u(1) \\ u(2) \end{bmatrix} = -\mathbf{F}^{RM}\mathbf{x}(0) = -\mathbf{F}^*(\mathbf{FF}^*)^{-1} \begin{bmatrix} x_1(0) \\ x_2(0) \end{bmatrix}$$

where

$$\mathbf{F}=\begin{bmatrix} -0.7181 & -3.6700 & -11.6935 \\ 1.7181 & 4.6700 & 12.6935 \end{bmatrix}$$

The right pseudoinverse $\mathbf{F}^{RM}$ is determined as follows:

$$\mathbf{F}^{RM}=\mathbf{F}^*(\mathbf{F}\mathbf{F}^*)^{-1}=\begin{bmatrix} -0.7181 & 1.7181 \\ -3.6700 & 4.6700 \\ -11.6935 & 12.6935 \end{bmatrix}\begin{bmatrix} 150.723 & -166.804 \\ -166.804 & 185.886 \end{bmatrix}^{-1}$$

$$=\begin{bmatrix} 0.7904 & 0.7185 \\ 0.4996 & 0.4734 \\ -0.2908 & -0.1927 \end{bmatrix}$$

Hence

$$\begin{bmatrix} u(0) \\ u(1) \\ u(2) \end{bmatrix}=-\begin{bmatrix} 0.7904 & 0.7185 \\ 0.4996 & 0.4734 \\ -0.2908 & -0.1927 \end{bmatrix}\begin{bmatrix} 5 \\ -5 \end{bmatrix}=\begin{bmatrix} -0.3595 \\ -0.1310 \\ 0.4905 \end{bmatrix} \tag{7–383}$$

The control sequence given by Eq. (7–383) will bring the state to the origin in 3 sampling periods and also will minimize the control energy given by Eq. (7–381).

By using the optimal control sequence given by Eq. (7–383), the state can be transferred as follows:

$$\begin{bmatrix} x_1(1) \\ x_2(1) \end{bmatrix}=\begin{bmatrix} 1 & 0.6321 \\ 0 & 0.3679 \end{bmatrix}\begin{bmatrix} 5 \\ -5 \end{bmatrix}+\begin{bmatrix} 0.3679 \\ 0.6321 \end{bmatrix}[-0.3595]=\begin{bmatrix} 1.7072 \\ -2.0667 \end{bmatrix}$$

$$\begin{bmatrix} x_1(2) \\ x_2(2) \end{bmatrix}=\begin{bmatrix} 1 & 0.6321 \\ 0 & 0.3679 \end{bmatrix}\begin{bmatrix} 1.7072 \\ -2.0667 \end{bmatrix}+\begin{bmatrix} 0.3679 \\ 0.6321 \end{bmatrix}[-0.1310]=\begin{bmatrix} 0.3526 \\ -0.8431 \end{bmatrix}$$

$$\begin{bmatrix} x_1(3) \\ x_2(3) \end{bmatrix}=\begin{bmatrix} 1 & 0.6321 \\ 0 & 0.3679 \end{bmatrix}\begin{bmatrix} 0.3526 \\ -0.8431 \end{bmatrix}+\begin{bmatrix} 0.3679 \\ 0.6321 \end{bmatrix}[0.4905]=\begin{bmatrix} 0 \\ 0 \end{bmatrix}$$

The minimum energy required for this control is

$$J_{\min}=\frac{1}{2}\sum_{k=0}^{2} u^2(k)=\frac{1}{2}[u^2(0)+u^2(1)+u^2(2)]=\frac{1}{2}[(-0.3595)^2+(-0.1310)^2+(0.4905)^2]$$
$$=0.1935$$

It is interesting to compare the minimum energy obtained here with the energy required for time-optimal control of this system. The time-optimal control requires 2 sampling periods. In Prob. A-6-2, the time-optimal control sequence $u(0)$ and $u(1)$, where the sampling period was 1 sec, was found to be

$$u(0)=-1.5820x_1(0)-1.2433x_2(0)$$
$$u(1)=0.5820x_1(0)+0.2433x_2(0)$$

[Refer to Eq. (6–259).] By taking $x_1(0) = 5$ and $x_2(0) = -5$ and substituting these values in these two equations, we obtain $u(0) = -1.6935$ and $u(1) = 1.6935$. Hence, for time-optimal control the total energy required is

$$\frac{1}{2}[u^2(0)+u^2(1)]=\frac{1}{2}[(-1.6935)^2+(1.6935)^2]=2.8679$$

Notice that by allowing the control duration to be 3 sampling periods (3 sec), rather than 2 sampling periods (2 sec) the energy required can be reduced remarkably.

Problem A-7-15. Consider the measurement of a constant value x. The measurement process involves noise effects. Let us assume that the system can be described by the following equations:

$$x(k+1) = x(k)$$
$$y(k) = x(k) + \epsilon(k)$$

where $\epsilon(k)$ is a white noise sequence with zero mean value and covariance $E[\epsilon^2(k)] = R = \sigma^2$. Determine the current estimation-type Kalman filter for this system.

Solution. From Eqs. (7-260) and (7-261) we have

$$\begin{aligned} P(k+1) &= P(k) - P(k)[\sigma^2 + P(k)]^{-1}P(k) \\ &= \frac{P(k)\sigma^2}{P(k)+\sigma^2} \end{aligned} \tag{7-384}$$

The solution of Eq. (7-384) can be given by

$$P(k) = \frac{P(0)\sigma^2}{kP(0)+\sigma^2}$$

Kalman filter gain $K_e(k)$ can be obtained from Eqs. (7-259) and (7-261), as follows:

$$\begin{aligned} K_e(k+1) &= P(k)[\sigma^2 + P(k)]^{-1} = \frac{P(k)}{P(k)+\sigma^2} \\ &= \frac{P(0)}{(k+1)P(0)+\sigma^2} \end{aligned} \tag{7-385}$$

Referring to Eqs. (7-264), (7-265), and (7-266), we have

$$K_e(0) = P_0[\sigma^2 + P_0]^{-1} = \frac{P_0}{P_0+\sigma^2}$$

$$P(0) = [1 - K_e(0)]P_0$$

Hence

$$P(0) = \frac{\sigma^2 P_0}{P_0+\sigma^2}$$

In terms of P_0, Eq. (7-385) can be rewritten as

$$K_e(k+1) = \frac{P_0}{(k+2)P_0+\sigma^2}$$

or

$$K_e(k) = \frac{P_0}{(k+1)P_0+\sigma^2} \tag{7-386}$$

Notice that as k approaches infinity, Kalman filter gain $K_e(k)$ approaches zero.

Referring to Eq. (7–257), the current estimation-type Kalman filter can now be given as follows:

$$\tilde{x}(k+1) = \tilde{x}(k) + K_e(k+1)[y(k+1) - \tilde{x}(k)]$$

or

$$\tilde{x}(k+1) = \tilde{x}(k) + \frac{P_0}{(k+2)P_0 + \sigma^2}[y(k+1) - \tilde{x}(k)] \tag{7–387}$$

Problem A-7–16. Referring to Prob. A-7–15, consider the same process of measuring a constant-value x. The measurement process involves noise effects. Let us assume that the measuring system can be given by the equations

$$x(k+1) = x(k)$$

$$y(k) = x(k) + \epsilon(k)$$

where $\epsilon(k)$ is a white noise sequence with zero mean value and covariance

$$E[\epsilon^2(k)] = R = 1$$

Let us assume that we make N measurements of x. Assuming that we employ the current estimation-type Kalman filter and make the optimal estimate of $x(k)$, show that as N approaches infinity, the optimal estimate of $x(N)$ can be given by

$$\lim_{N\to\infty} \tilde{x}(N) = \lim_{N\to\infty} \left\{\frac{1}{N+1}[y(0) + y(1) + \cdots + y(N)]\right\}$$

Solution. Referring to the solution of Prob. A-7–15 and noting that $R = \sigma^2 = 1$, we have from Eq. (7–386) the following Kalman filter gain $K_e(k)$:

$$K_e(k) = \frac{P_0}{(k+1)P_0 + 1}$$

From Eq. (7–387), the current estimation-type Kalman filter becomes

$$\begin{aligned}\tilde{x}(k+1) &= \tilde{x}(k) + \frac{P_0}{(k+2)P_0 + 1}[y(k+1) - \tilde{x}(k)] \\ &= \frac{(k+1)P_0 + 1}{(k+2)P_0 + 1}\tilde{x}(k) + \frac{P_0}{(k+2)P_0 + 1}y(k+1)\end{aligned} \tag{7–388}$$

Referring to Eq. (7–262), we have

$$\begin{aligned}\tilde{x}(0) &= \bar{x}(0) + K_e(0)[y(0) - C(0)\bar{x}(0)] \\ &= \bar{x}(0) + \frac{P_0}{P_0 + 1}[y(0) - \bar{x}(0)] \\ &= \frac{1}{P_0 + 1}\bar{x}(0) + \frac{P_0}{P_0 + 1}y(0)\end{aligned}$$

From Eq. (7–388), we have for $k = 0$

$$\tilde{x}(1) = \frac{P_0 + 1}{2P_0 + 1}\tilde{x}(0) + \frac{P_0}{2P_0 + 1}y(1)$$

$$= \frac{P_0 + 1}{2P_0 + 1}\left[\frac{1}{P_0 + 1}\tilde{x}(0) + \frac{P_0}{P_0 + 1}y(0)\right] + \frac{P_0}{2P_0 + 1}y(1)$$

$$= \frac{1}{2P_0 + 1}\tilde{x}(0) + \frac{P_0}{2P_0 + 1}[y(0) + y(1)]$$

For $k = 1$,

$$\tilde{x}(2) = \frac{2P_0 + 1}{3P_0 + 1}\tilde{x}(1) + \frac{P_0}{3P_0 + 1}y(2)$$

$$= \frac{2P_0 + 1}{3P_0 + 1}\left\{\frac{1}{2P_0 + 1}\tilde{x}(0) + \frac{P_0}{2P_0 + 1}[y(0) + y(1)]\right\} + \frac{P_0}{3P_0 + 1}y(2)$$

$$= \frac{1}{3P_0 + 1}\tilde{x}(0) + \frac{P_0}{3P_0 + 1}[y(0) + y(1) + y(2)]$$

For $k = 2$,

$$\tilde{x}(3) = \frac{1}{4P_0 + 1}\tilde{x}(0) + \frac{P_0}{4P_0 + 1}[y(0) + y(1) + y(2) + y(3)]$$

Similarly, for $k = N$, we have

$$\tilde{x}(N) = \frac{1}{(N + 1)P_0 + 1}\tilde{x}(0) + \frac{P_0}{(N + 1)P_0 + 1}[y(0) + y(1) + \cdots + y(N)] \qquad (7\text{–}389)$$

As N approaches infinity, Eq. (7–389) becomes

$$\lim_{N\to\infty} \tilde{x}(N) = \lim_{N\to\infty}\left\{\frac{1}{N + 1}[y(0) + y(1) + \cdots + y(N)]\right\}$$

Problem A-7–17. Consider the system

$$\mathbf{x}(k + 1) = \mathbf{G}\mathbf{x}(k) + \mathbf{H}\mathbf{u}(k) + \mathbf{w}(k)$$

where $\mathbf{w}(k)$ is the system noise sequence. We assume that

$$E[\mathbf{w}(k)] = \mathbf{0}$$

$$E[\mathbf{w}(j)\mathbf{w}^*(k)] = \mathbf{Q}_w\,\delta_{jk} \qquad (\mathbf{Q}_w \text{ is positive semidefinite})$$

$$E\{[\mathbf{x}(0) - \bar{\mathbf{x}}(0)]\mathbf{w}^*(k)\} = \mathbf{0}$$

where $\delta_{jk} = 1$ if $j = k$ and $\delta_{jk} = 0$ if $j \neq k$.

Obtain the optimal control law $\mathbf{u}(k)$ that will minimize the following performance index:

$$J = \frac{1}{2}E\left\{\mathbf{x}^*(N)\mathbf{S}\mathbf{x}(N) + \sum_{k=0}^{N-1}[\mathbf{x}^*(k)\mathbf{Q}\mathbf{x}(k) + \mathbf{u}^*(k)\mathbf{R}\mathbf{u}(k)]\right\}$$

Also, obtain the minimum value of this performance index.

Solution. Assume that Eqs. (7–24), (7–25), and (7–31) apply to this case. These equations are repeated as follows:

$$\mathbf{P}(k) = \mathbf{Q} + \mathbf{G}^*\mathbf{P}(k+1)\mathbf{G} - \mathbf{G}^*\mathbf{P}(k+1)\mathbf{H}[\mathbf{R} + \mathbf{H}^*\mathbf{P}(k+1)\mathbf{H}]^{-1}\mathbf{H}^*\mathbf{P}(k+1)\mathbf{G} \tag{7–390}$$

$$\mathbf{P}(N) = \mathbf{S} \tag{7–391}$$

$$\mathbf{K}(k) = [\mathbf{R} + \mathbf{H}^*\mathbf{P}(k+1)\mathbf{H}]^{-1}\mathbf{H}^*\mathbf{P}(k+1)\mathbf{G} \tag{7–392}$$

By using Eq. (7–392), Eq. (7–390) can be written as follows:

$$\mathbf{P}(k) = \mathbf{G}^*\mathbf{P}(k+1)\mathbf{G} + \mathbf{Q} - \mathbf{K}^*(k)[\mathbf{R} + \mathbf{H}^*\mathbf{P}(k+1)\mathbf{H}]\mathbf{K}(k)$$

Now consider the difference between $\mathbf{x}^*(k+1)\mathbf{P}(k+1)\mathbf{x}(k+1)$ and $\mathbf{x}^*(k)\mathbf{P}(k)\mathbf{x}(k)$:

$$\begin{aligned}
&\mathbf{x}^*(k+1)\mathbf{P}(k+1)\mathbf{x}(k+1) - \mathbf{x}^*(k)\mathbf{P}(k)\mathbf{x}(k) \\
&\quad = [\mathbf{Gx}(k) + \mathbf{Hu}(k) + \mathbf{w}(k)]^*\mathbf{P}(k+1)[\mathbf{Gx}(k) + \mathbf{Hu}(k) + \mathbf{w}(k)] \\
&\qquad - \mathbf{x}^*(k)\{[\mathbf{G}^*\mathbf{P}(k+1)\mathbf{G} + \mathbf{Q}] - \mathbf{K}^*(k)[\mathbf{R} + \mathbf{H}^*\mathbf{P}(k+1)\mathbf{H}]\mathbf{K}(k)\}\mathbf{x}(k) \\
&\quad = \mathbf{u}^*(k)\mathbf{H}^*\mathbf{P}(k+1)\mathbf{Gx}(k) + \mathbf{x}^*(k)\mathbf{G}^*\mathbf{P}(k+1)\mathbf{Hu}(k) \\
&\qquad + \mathbf{u}^*(k)[\mathbf{R} + \mathbf{H}^*\mathbf{P}(k+1)\mathbf{H}]\mathbf{u}(k) + \mathbf{x}^*(k)\mathbf{K}^*(k)[\mathbf{R} + \mathbf{H}^*\mathbf{P}(k+1)\mathbf{H}]\mathbf{K}(k)\mathbf{x}(k) \\
&\qquad - \mathbf{x}^*(k)\mathbf{Qx}(k) - \mathbf{u}^*(k)\mathbf{Ru}(k) + \mathbf{w}^*(k)\mathbf{P}(k+1)[\mathbf{Gx}(k) + \mathbf{Hu}(k)] \\
&\qquad + [\mathbf{Gx}(k) + \mathbf{Hu}(k)]^*\mathbf{P}(k+1)\mathbf{w}(k) + \mathbf{w}^*(k)\mathbf{P}(k+1)\mathbf{w}(k) \\
&\quad = [\mathbf{u}(k) + \mathbf{K}(k)\mathbf{x}(k)]^*[\mathbf{R} + \mathbf{H}^*\mathbf{P}(k+1)\mathbf{H}][\mathbf{u}(k) + \mathbf{K}(k)\mathbf{x}(k)] \\
&\qquad - \mathbf{x}^*(k)\mathbf{Qx}(k) - \mathbf{u}^*(k)\mathbf{Ru}(k) + \mathbf{w}^*(k)\mathbf{P}(k+1)[\mathbf{Gx}(k) + \mathbf{Hu}(k)] \\
&\qquad + [\mathbf{Gx}(k) + \mathbf{Hu}(k)]^*\mathbf{P}(k+1)\mathbf{w}(k) + \mathbf{w}^*(k)\mathbf{P}(k+1)\mathbf{w}(k)
\end{aligned}$$

Now the performance index J can be written as follows:

$$\begin{aligned}
J &= \frac{1}{2}E\sum_{k=0}^{N-1}[\mathbf{x}^*(k+1)\mathbf{P}(k+1)\mathbf{x}(k+1) - \mathbf{x}^*(k)\mathbf{P}(k)\mathbf{x}(k) \\
&\qquad + \mathbf{x}^*(k)\mathbf{Qx}(k) + \mathbf{u}^*(k)\mathbf{Ru}(k) + \mathbf{x}^*(0)\mathbf{P}(0)\mathbf{x}(0)] \\
&= \frac{1}{2}E\sum_{k=0}^{N-1}\{[\mathbf{u}(k) + \mathbf{K}(k)\mathbf{x}(k)]^*[\mathbf{R} + \mathbf{H}^*\mathbf{P}(k+1)\mathbf{H}][\mathbf{u}(k) + \mathbf{K}(k)\mathbf{x}(k)] \\
&\qquad + \mathbf{x}^*(0)\mathbf{P}(0)\mathbf{x}(0) + \mathbf{w}^*(k)\mathbf{P}(k+1)[\mathbf{Gx}(k) + \mathbf{Hu}(k)] \\
&\qquad + [\mathbf{Gx}(k) + \mathbf{Hu}(k)]^*\mathbf{P}(k+1)\mathbf{w}(k) + \mathbf{w}^*(k)\mathbf{P}(k+1)\mathbf{w}(k)\}
\end{aligned}$$

Since the control vector $\mathbf{u}(k)$ and the noise vector $\mathbf{w}(k)$ are uncorrelated, we have

$$\begin{aligned}
J = \frac{1}{2}\Big[\mathbf{x}^*(0)\mathbf{P}(0)\mathbf{x}(0) + E\sum_{k=0}^{N-1}\{&[\mathbf{u}(k) + \mathbf{K}(k)\mathbf{x}(k)]^* \\
&\cdot [\mathbf{R} + \mathbf{H}^*\mathbf{P}(k+1)\mathbf{H}][\mathbf{u}(k) + \mathbf{K}(k)\mathbf{x}(k)] + \mathbf{w}^*(k)\mathbf{P}(k+1)\mathbf{w}(k)\}\Big]
\end{aligned}$$

Since $\mathbf{R} + \mathbf{H}^*\mathbf{P}(k+1)\mathbf{H}$ is positive definite, the control vector that minimizes J is given by the equation

$$\mathbf{u}(k) = -\mathbf{K}(k)\mathbf{x}(k)$$

and the minimum value of the performance index can be given as follows:

$$J_{\min} = \frac{1}{2}\mathbf{x}^*(0)\mathbf{P}(0)\mathbf{x}(0) + \frac{1}{2}\sum_{k=0}^{N-1} E[\mathbf{w}^*(k)\mathbf{P}(k+1)\mathbf{w}(k)]$$

$$= \frac{1}{2}\mathbf{x}^*(0)\mathbf{P}(0)\mathbf{x}(0) + \frac{1}{2}\sum_{k=0}^{N-1} \text{tr}\,[\mathbf{P}(k+1)\mathbf{Q}_w] \qquad (7\text{–}393)$$

Notice that the optimal control law is the same regardless of whether or not noise $\mathbf{w}(k)$ is present. The only effect of the noise is to increase the value of the performance index, as shown in Eq. (7–393).

Problem A-7–18. Consider the system defined by the equations

$$x(k+1) = -0.5x(k) + w(k)$$

$$y(k) = x(k) + \epsilon(k)$$

where the system noise $w(k)$ and measurement noise $\epsilon(k)$ are white noise sequences with zero mean values and covariances

$$E[w^2(k)] = Q = 1.5$$

$$E[\epsilon^2(k)] = R = 0.5$$

Obtain the steady-state current estimation-type Kalman filter.

Solution. From Eqs. (7–277) and (7–278), the Kalman filter equation is

$$\tilde{x}(k+1) = -0.5\tilde{x}(k) + K_e[y(k+1) + 0.5\tilde{x}(k)] \qquad (7\text{–}394)$$

where the steady-state Kalman filter gain K_e is given by Eq. (7–279), or

$$K_e = N(R+N)^{-1} = \frac{N}{R+N} = \frac{N}{0.5+N} \qquad (7\text{–}395)$$

Note that N is given by Eq. (7–280) as

$$N = Q + (-0.5)N(-0.5) - (-0.5)N(R+N)^{-1}N(-0.5)$$

or

$$N = 1.5 + 0.25N - \frac{0.25N^2}{0.5+N} \qquad (7\text{–}396)$$

Equation (7–396) can be modified to read

$$N^2 - 1.125N - 0.75 = 0$$

Solving this equation for N, we obtain

$$N = 1.5952 \qquad \text{or} \qquad -0.4702$$

Since N must be positive, we choose

$$N = 1.5952$$

Using this value of N, the Kalman filter gain K_e given by Eq. (7–395) can be determined as follows:

$$K_e = \frac{N}{0.5 + N} = \frac{1.5952}{0.5 + 1.5952} = 0.7614$$

Then the Kalman filter equation given by Eq. (7–394) becomes as follows:

$$\begin{aligned}\tilde{x}(k+1) &= -0.5\tilde{x}(k) + 0.7614[y(k+1) + 0.5\tilde{x}(k)] \\ &= -0.1193\tilde{x}(k) + 0.7614y(k+1)\end{aligned}$$

PROBLEMS

Problems B-7–1. Consider the discrete-time system

$$\mathbf{x}(k+1) = \mathbf{G}\mathbf{x}(k) + \mathbf{H}u(k)$$

where

$$\mathbf{G} = \begin{bmatrix} 0 & 1 \\ -0.5 & 1 \end{bmatrix}, \qquad \mathbf{H} = \begin{bmatrix} 1 \\ 1 \end{bmatrix}, \qquad \mathbf{x}(0) = \begin{bmatrix} 2 \\ 2 \end{bmatrix}$$

Determine the optimal control sequence $u(k)$ that will minimize the following performance index:

$$J = \frac{1}{2}\mathbf{x}^*(8)\mathbf{S}\mathbf{x}(8) + \frac{1}{2}\sum_{k=0}^{7}[\mathbf{x}^*(k)\mathbf{Q}\mathbf{x}(k) + u^*(k)Ru(k)]$$

where

$$\mathbf{Q} = \begin{bmatrix} 1 & 0 \\ 0 & 1 \end{bmatrix}, \qquad R = 1, \qquad \mathbf{S} = \begin{bmatrix} 1 & 0 \\ 0 & 1 \end{bmatrix}$$

Problem B-7–2. Consider the system

$$\mathbf{x}(k+1) = \mathbf{G}\mathbf{x}(k) + \mathbf{H}u(k)$$

where

$$\mathbf{G} = \begin{bmatrix} 0 & 0 \\ -0.5 & 1 \end{bmatrix}, \qquad \mathbf{H} = \begin{bmatrix} 1 \\ 0 \end{bmatrix}, \qquad \mathbf{x}(0) = \begin{bmatrix} 2 \\ 2 \end{bmatrix}$$

and the performance index

$$J = \frac{1}{2}\sum_{k=0}^{\infty}[\mathbf{x}^*(k)\mathbf{Q}\mathbf{x}(k) + u^*(k)Ru(k)]$$

where

$$\mathbf{Q} = \begin{bmatrix} 1 & 0 \\ 0 & 0.5 \end{bmatrix}, \qquad R = 1$$

Determine the optimal control law to minimize the performance index. Also, determine the minimum value of J.

Problem B-7–3. Consider the system defined by

$$\begin{bmatrix} x_1(k+1) \\ x_2(k+1) \end{bmatrix} = \begin{bmatrix} 1 & 1 \\ a & -1 \end{bmatrix} \begin{bmatrix} x_1(k) \\ x_2(k) \end{bmatrix}, \qquad \begin{bmatrix} x_1(0) \\ x_2(0) \end{bmatrix} = \begin{bmatrix} 1 \\ 1 \end{bmatrix}$$

where $-1 \leq a < 0$. Determine the value of a such that the performance index

$$J = \frac{1}{2} \sum_{k=0}^{\infty} \mathbf{x}^*(k)\mathbf{Q}\mathbf{x}(k)$$

where

$$\mathbf{Q} = \begin{bmatrix} 1 & 0 \\ 0 & 0.5 \end{bmatrix}$$

is minimized.

Problem B-7–4. A discrete-time control system is described by the equation

$$x(k+1) = 0.3679x(k) + 0.6321u(k)$$

Determine the optimal control law to minimize the following performance index:

$$J = \frac{1}{2} \sum_{k=0}^{\infty} [x^2(k) + u^2(k)]$$

Also, determine the minimum value of the performance index J.

Problem B-7–5. Consider the continuous-time control system

$$\dot{x}(t) = -x(t) + u(t), \qquad x(0) = 1$$

where

$$u(t) = u(kT) \qquad kT \leq t < (k+1)T$$

and the performance index

$$J = \frac{1}{2} \int_0^{\infty} [x^2(t) + u^2(t)]\, dt$$

Discretize the system equation and the performance index. Then, determine the optimal control law that minimizes the discretized performance index. Also, determine the minimum value of J. Assume that the sampling period T is one second.

Problem B-7–6. A continuous-time control system is described by the equation

$$\begin{bmatrix} \dot{x}_1 \\ \dot{x}_2 \end{bmatrix} = \begin{bmatrix} 0 & 1 \\ 0 & 0 \end{bmatrix} \begin{bmatrix} x_1 \\ x_2 \end{bmatrix} + \begin{bmatrix} 0 \\ 1 \end{bmatrix} u$$

It is desired to determine the optimal control signal $u(t)$ such that the performance index

$$J = \frac{1}{2} \int_0^{\infty} (\mathbf{x}^*\mathbf{Q}\mathbf{x} + u^*u)\, dt$$

is minimized, where

$$\mathbf{Q} = \begin{bmatrix} 1 & 0 \\ 0 & \mu \end{bmatrix} \qquad \mu \geq 0$$

Using the Liapunov approach, determine the optimal control signal $u(t)$.

Problem B-7-7. Consider the system defined by the equations

$$\mathbf{x}(k+1) = \mathbf{G}\mathbf{x}(k) + \mathbf{H}\mathbf{u}(k)$$

$$\mathbf{y}(k) = \mathbf{C}\mathbf{x}(k)$$

where $\mathbf{x}(k)$ is an n-vector, $\mathbf{u}(k)$ is an r-vector, $\mathbf{y}(k)$ is an m-vector, $\mathbf{G}$ is an $n \times n$ matrix, $\mathbf{H}$ is an $n \times r$ matrix, and $\mathbf{C}$ is an $m \times n$ matrix. The performance index is

$$J = \frac{1}{2} \sum_{k=0}^{\infty} [\mathbf{x}^*(k)\mathbf{Q}\mathbf{x}(k) + \mathbf{u}^*(k)\mathbf{R}\mathbf{u}(k)]$$

where $\mathbf{Q}$ is an $n \times n$ positive definite Hermitian matrix and $\mathbf{R}$ is an $r \times r$ positive definite Hermitian matrix. Let us define the optimal control law that minimizes the performance index as $\mathbf{u}(k) = -\mathbf{K}\mathbf{x}(k)$.

Show that if the system is completely state controllable and observable, then the following algebraic Riccati equation:

$$\mathbf{P} = \mathbf{Q} + \mathbf{G}\mathbf{P}\mathbf{G}^* - \mathbf{G}\mathbf{P}\mathbf{C}^*(\mathbf{R} + \mathbf{C}\mathbf{P}\mathbf{C}^*)^{-1}\mathbf{C}\mathbf{P}\mathbf{G}^*$$

has a unique positive definite solution. Show also that the optimal closed-loop system is stable, or $\mathbf{G} - \mathbf{H}\mathbf{K}$ is a stable matrix.

Problem B-7-8. Consider the system

$$\begin{bmatrix} x_1(k+1) \\ x_2(k+1) \end{bmatrix} = \begin{bmatrix} 1 & 1 \\ 1 & 0 \end{bmatrix} \begin{bmatrix} x_1(k) \\ x_2(k) \end{bmatrix} + \begin{bmatrix} 1 \\ 1 \end{bmatrix} u(k), \qquad \begin{bmatrix} x_1(0) \\ x_2(0) \end{bmatrix} = \begin{bmatrix} 1 \\ 1 \end{bmatrix}$$

It is desired to bring the initial state to the origin in 3 sampling periods. Determine the optimal control law to minimize the control energy measured by

$$J = \frac{1}{2} \sum_{k=0}^{2} u^2(k)$$

Problem B-7-9. Consider the problem discussed in Example 7-7. Assume that the mathematical model of the plant is given by

$$G(z) = \frac{b_0 + b_1 z^{-1}}{1 + a_1 z^{-1}}$$

The input to the plant and to the model is $u(k)$. The output of the plant is $y(k)$, and that of the mathematical model is $\tilde{y}(k)$, where

$$\tilde{y}(k) = -a_1 y(k-1) + b_0 u(k) + b_1 u(k-1)$$

The difference between $y(k)$ and $\tilde{y}(k)$ is the error $\epsilon(k)$:

$$\epsilon(k) = y(k) - \tilde{y}(k) = y(k) + a_1 y(k-1) - b_0 u(k) - b_1 u(k-1)$$

Using the input data $\{u(k)\}$ and output data $\{y(k)\}$ given in Table 7-2, determine the optimal estimates of the parameters a_1, b_0, and b_1 for $N = 7, 8, \ldots, 14$. Use the least-squares method.

Problem B-7–10. Consider the system

$$\mathbf{x}(k+1) = \mathbf{G}(k)\mathbf{x}(k) + \mathbf{F}(k)\mathbf{w}(k)$$

where $\mathbf{x}$ is an n-vector, $\mathbf{G}(k)$ is an $n \times n$ matrix, $\mathbf{F}(k)$ is an $n \times r$ matrix, and $\mathbf{w}(k)$ is an r-dimensional random noise vector. Assume that $\mathbf{w}(k)$ is a white noise sequence with the following mean and covariance:

$$E[\mathbf{w}(k)] = \bar{\mathbf{w}}(k)$$

$$E\{[\mathbf{w}(j) - \bar{\mathbf{w}}(j)][\mathbf{w}(k) - \bar{\mathbf{w}}(k)]^*\} = \mathbf{Q}(k)\delta_{jk}$$

where $\mathbf{Q}(k)$ is an $r \times r$ positive semidefinite matrix. We assume that $\mathbf{G}(k)$ and $\mathbf{F}(k)$ are deterministic.

We also assume that state $\mathbf{x}(0)$ satisfies the following:

$$E[\mathbf{x}(0)] = \bar{\mathbf{x}}(0) \qquad \text{(deterministic)}$$

$$E\{[\mathbf{x}(0) - \bar{\mathbf{x}}(0)][\mathbf{x}(0) - \bar{\mathbf{x}}(0)]^*\} = \mathbf{P}(0)$$

The initial state $\mathbf{x}(0)$ is assumed to be uncorrelated with the noise sequence $\mathbf{w}(k)$.

Show that

$$\begin{aligned}\mathbf{P}(k+1) &= E\{[\mathbf{x}(k+1) - \bar{\mathbf{x}}(k+1)][\mathbf{x}(k+1) - \bar{\mathbf{x}}(k+1)]^*\} \\ &= \mathbf{G}(k)\mathbf{P}(k)\mathbf{G}^*(k) + \mathbf{F}(k)\mathbf{Q}(k)\mathbf{F}^*(k)\end{aligned}$$

Problem B-7–11. Consider a measurement process defined by

$$\mathbf{y} = \mathbf{C}\mathbf{x} + \boldsymbol{\epsilon}$$

where $\mathbf{x}$ is an n-vector, $\mathbf{y}$ is an m-vector, $\mathbf{C}$ is an $m \times n$ matrix, and $\boldsymbol{\epsilon}$ is an m-vector. Assume that signal $\mathbf{x}$ is uncorrelated with noise $\boldsymbol{\epsilon}$. Assume also that

$$E[\mathbf{x}] = \bar{\mathbf{x}}$$

$$E[(\mathbf{x} - \bar{\mathbf{x}})(\mathbf{x} - \bar{\mathbf{x}})^*] = \mathbf{N} = n \times n \text{ positive definite matrix}$$

$$E[(\boldsymbol{\epsilon} - \bar{\boldsymbol{\epsilon}})(\boldsymbol{\epsilon} - \bar{\boldsymbol{\epsilon}})^*] = \mathbf{R} = m \times m \text{ positive definite matrix}$$

Then, we obtain

$$E[\mathbf{y}] = E[\mathbf{C}\mathbf{x} + \boldsymbol{\epsilon}] = \mathbf{C}E[\mathbf{x}] + E[\boldsymbol{\epsilon}] = \mathbf{C}\bar{\mathbf{x}} + \bar{\boldsymbol{\epsilon}}$$

$$\begin{aligned}E[(\mathbf{y} - \bar{\mathbf{y}})(\mathbf{y} - \bar{\mathbf{y}})^*] &= E[\mathbf{C}(\mathbf{x} - \bar{\mathbf{x}}) + (\boldsymbol{\epsilon} - \bar{\boldsymbol{\epsilon}})][\mathbf{C}(\mathbf{x} - \bar{\mathbf{x}}) + (\boldsymbol{\epsilon} - \bar{\boldsymbol{\epsilon}})]^* \\ &= \mathbf{C}E[(\mathbf{x} - \bar{\mathbf{x}})(\mathbf{x} - \bar{\mathbf{x}})^*]\mathbf{C}^* + E[(\boldsymbol{\epsilon} - \bar{\boldsymbol{\epsilon}})(\boldsymbol{\epsilon} - \bar{\boldsymbol{\epsilon}})^*] \\ &= \mathbf{C}\mathbf{N}\mathbf{C}^* + \mathbf{R}\end{aligned}$$

Now consider the optimal estimate of $\mathbf{x}$ as a function of $\mathbf{y}$. On the basis of the minimum mean-square error criterion, define the optimal estimate of $\mathbf{x}$ as $\tilde{\mathbf{x}}$. Note that the optimal estimate $\tilde{\mathbf{x}}$ minimizes $E[\mathbf{e}\mathbf{e}^*]$, where $\mathbf{e} = \mathbf{x} - \tilde{\mathbf{x}}$. Minimizing $E[\mathbf{e}\mathbf{e}^*]$ means that we minimize the quadratic form $\boldsymbol{\alpha}^*E[\mathbf{e}\mathbf{e}^*]\boldsymbol{\alpha}$, where $\boldsymbol{\alpha}$ is an arbitrary n-vector.

Show that $\tilde{\mathbf{x}}$ can be given by the equation

$$\tilde{\mathbf{x}} = \bar{\mathbf{x}} + \mathbf{NC}^*(\mathbf{R} + \mathbf{CNC}^*)^{-1}[\mathbf{y} - (\mathbf{C}\bar{\mathbf{x}} + \bar{\boldsymbol{\epsilon}})]$$

and that the covariance $\mathbf{P}$ of the error vector in this case is given by

$$\mathbf{P} = \mathbf{N} - \mathbf{NC}^*(\mathbf{R} + \mathbf{CNC}^*)^{-1}\mathbf{CN}$$

Problem B-7-12. Referring to Prob. B-7-11, the optimal estimate of the vector $\mathbf{x}$ based on the minimum mean-square error criterion is given by

$$\tilde{\mathbf{x}} = \bar{\mathbf{x}} + \mathbf{F}[\mathbf{y} - (\mathbf{C}\bar{\mathbf{x}} + \bar{\boldsymbol{\epsilon}})]$$

where

$$\mathbf{F} = \mathbf{NC}^*(\mathbf{R} + \mathbf{CNC}^*)^{-1}$$

and the covariance $\mathbf{P}$ of the error $\mathbf{e}$, where

$$\mathbf{e} = \mathbf{x} - \tilde{\mathbf{x}}$$

is given by

$$\mathbf{P} = \mathbf{N} - \mathbf{NC}^*(\mathbf{R} + \mathbf{CNC}^*)^{-1}\mathbf{CN}$$

Show that covariance $\mathbf{P}$ can be given by the equation

$$\mathbf{P} = (\mathbf{N}^{-1} + \mathbf{C}^*\mathbf{R}^{-1}\mathbf{C})^{-1}$$

and matrix $\mathbf{F}$ can be given by

$$\mathbf{F} = \mathbf{PC}^*\mathbf{R}^{-1}$$

with the result that $\tilde{\mathbf{x}}$ can also be given by

$$\tilde{\mathbf{x}} = \bar{\mathbf{x}} + \mathbf{PC}^*\mathbf{R}^{-1}[\mathbf{y} - (\mathbf{C}\bar{\mathbf{x}} + \bar{\boldsymbol{\epsilon}})]$$

Problem B-7-13. Consider the system

$$x(k+1) = x(k) + w(k)$$

$$y(k) = Cx(k) + \epsilon(k)$$

Assume that $w(k)$ and $\epsilon(k)$ are independent white noises and

$$E[w(k)] = 0, \qquad E[w^2(k)] = 1$$

$$E[\epsilon(k)] = 0, \qquad E[\epsilon^2(k)] = 1$$

Show that the current estimation-type Kalman filter can be given by the equation

$$\tilde{x}(k+1) = [1 - C^2P(k+1)]\tilde{x}(k) + CP(k+1)y(k+1)$$

where

$$P(k+1) = \frac{1}{C^2 + [1 + P(k)]^{-1}}$$

with the initial conditions

$$\tilde{x}(0) = [1 - C^2P(0)]\bar{x}(0) + CP(0)y(0)$$

$$P(0) = \frac{1}{P_0^{-1} + C^2}$$

$$N(0) = P_0$$

Note that

$$N(k+1) = P(k) + 1$$

Show also that

$$P(\infty) = -\frac{1}{2} + \sqrt{\frac{1}{4} + \frac{1}{C^2}}$$

Problem B-7–14. Consider the system

$$\mathbf{x}(k+1) = \mathbf{G}\mathbf{x}(k) + \mathbf{H}u(k) + \mathbf{w}(k)$$

$$y(k) = \mathbf{C}\mathbf{x}(k) + \epsilon(k)$$

where

$$\mathbf{G} = \begin{bmatrix} 0 & 1 \\ -0.5 & -1 \end{bmatrix}, \qquad \mathbf{H} = \begin{bmatrix} 1 \\ 1 \end{bmatrix}, \qquad \mathbf{C} = [1 \quad 0]$$

Assume that both the system noise $\mathbf{w}(k)$ and the measurement noise $\epsilon(k)$ are white noise sequences and all assumptions given by Eqs. (7–204) through (7–212) apply to this system. Assume also that

$$E[\mathbf{w}(k)\mathbf{w}^*(k)] = \mathbf{Q} = \begin{bmatrix} 1 & 0 \\ 0 & 0 \end{bmatrix}$$

$$E[\epsilon^2(k)] = R = 0.2$$

$$E[\mathbf{x}(0)] = \bar{\mathbf{x}}(0)$$

$$E\{[\mathbf{x}(0) - \bar{\mathbf{x}}(0)][\mathbf{x}(0) - \bar{\mathbf{x}}(0)]^*\} = (0.1)\mathbf{I}$$

Design both the prediction-type and the current estimation-type Kalman filters.

Problem B-7–15. Consider the modeling of a measuring process which involves an unintentional bias error. Suppose that the measurement process of a constant-value x is modeled by

$$x(k+1) = x(k) + \alpha \qquad (\alpha \text{ is constant})$$

$$y(k) = x(k) + \epsilon(k)$$

where α is a bias signal and the measurement noise $\epsilon(k)$ is a white noise sequence with the following mean and covariance:

$$E[\epsilon(k)] = 0, \qquad E[\epsilon^2(k)] = R = \sigma^2$$

Suppose also that we design a Kalman filter using an optimal gain (corresponding to the case where $\alpha = 0$).

Show that in this case the error involved in the estimation by the Kalman filter diverges and the estimation process becomes useless. (Such a case can be avoided by carefully modeling the process to eliminate constant bias signals.)

Problem B-7–16. A discrete-time system is defined by

$$x(k+1) = -0.8x(k) + w(k)$$

$$y(k) = x(k) + \epsilon(k)$$

where the system noise $w(k)$ and measurement noise $\epsilon(k)$ are white noise sequences with zero mean values and the following covariances:

$$E[w^2(k)] = Q = 1.2, \qquad E[\epsilon^2(k)] = R = 0.3$$

Design both the prediction-type and current estimation-type Kalman filters.

Problem B-7–17. Consider the system

$$\dot{\mathbf{x}} = \mathbf{A}\mathbf{x} + \mathbf{B}\mathbf{u} + \mathbf{w}$$

$$\mathbf{y} = \mathbf{C}\mathbf{x} + \boldsymbol{\epsilon}$$

where $\mathbf{w}(t)$ is the system noise and $\boldsymbol{\epsilon}(t)$ is the measurement noise. We assume that both $\mathbf{w}(t)$ and $\boldsymbol{\epsilon}(t)$ are white noises. We also assume that the system is completely state controllable and observable. We further assume that

$$E[\mathbf{w}(t)] = \mathbf{0}$$

$$E[\boldsymbol{\epsilon}(t)] = \mathbf{0}$$

$$E[\mathbf{w}(t)\mathbf{w}^*(\tau)] = \mathbf{Q}\delta(t-\tau) \qquad (\mathbf{Q} \text{ is positive semidefinite})$$

$$E[\boldsymbol{\epsilon}(t)\boldsymbol{\epsilon}^*(\tau)] = \mathbf{R}\delta(t-\tau) \qquad (\mathbf{R} \text{ is positive definite})$$

$$E[\mathbf{w}(t)\boldsymbol{\epsilon}^*(\tau)] = \mathbf{0}$$

Define $\tilde{\mathbf{x}}(t)$ as the optimal estimate of state $\mathbf{x}(t)$ in that the mean square error

$$J = E[\mathbf{e}^*(t)\mathbf{e}(t)]$$

where $\mathbf{e}(t) = \mathbf{x}(t) - \tilde{\mathbf{x}}(t)$, is minimum. {Note that the mean square error becomes minimum when the covariance $\mathbf{P}$ of the error signal is minimum, since $J = E[\mathbf{e}^*(t)\mathbf{e}(t)] = \text{tr } \mathbf{P}$.} We assume that $\mathbf{e}(0)$ is uncorrelated with noises $\mathbf{w}(t)$ and $\boldsymbol{\epsilon}(t)$.

Show that the optimal estimate $\tilde{\mathbf{x}}(t)$ can be determined from the following Kalman filter equation:

$$\dot{\tilde{\mathbf{x}}}(t) = \mathbf{A}\tilde{\mathbf{x}}(t) + \mathbf{B}\mathbf{u}(t) + \mathbf{K}_e[\mathbf{y}(t) - \mathbf{C}\tilde{\mathbf{x}}(t)]$$

where

$$\mathbf{K}_e = \mathbf{P}\mathbf{C}^*\mathbf{R}^{-1}$$

and that the $n \times n$ matrix $\mathbf{P}$ is the unique positive definite solution of the following Riccati equation:

$$\mathbf{P}\mathbf{A}^* + \mathbf{A}\mathbf{P} - \mathbf{P}\mathbf{C}^*\mathbf{R}^{-1}\mathbf{C}\mathbf{P} + \mathbf{Q} = \mathbf{0}$$

Problem B-7–18. Consider the system

$$\dot{\mathbf{x}} = \mathbf{A}\mathbf{x} + \mathbf{B}\mathbf{u}$$

In Prob. A-7–11, it was shown that the optimal control vector

$$\mathbf{u} = -\mathbf{K}\mathbf{x}$$

that minimizes the performance index

$$J = \frac{1}{2}\int_0^\infty (\mathbf{x}^*\mathbf{Q}\mathbf{x} + \mathbf{u}^*\mathbf{R}\mathbf{u})\, dt$$

where $\mathbf{Q}$ is a positive definite Hermitian matrix and $\mathbf{R}$ is also a positive definite Hermitian matrix, is given by

$$\mathbf{u}(t) = -\mathbf{K}\mathbf{x}(t) = -\mathbf{R}^{-1}\mathbf{B}^*\mathbf{P}\mathbf{x}(t)$$

where matrix $\mathbf{P}$ is the unique positive definite solution of the following Riccati equation:

$$\mathbf{A}^*\mathbf{P} + \mathbf{P}\mathbf{A} - \mathbf{P}\mathbf{B}\mathbf{R}^{-1}\mathbf{B}^*\mathbf{P} + \mathbf{Q} = \mathbf{0}$$

Show that the solution to the estimation problem given in Prob. B-7–17 can also be obtained by determining the optimal feedback gain matrix $\mathbf{K}$ (that is, $\mathbf{u} = -\mathbf{K}\mathbf{x}$) of the dual counterpart of the system

$$\dot{\mathbf{x}} = \mathbf{A}^*\mathbf{x} + \mathbf{C}^*\mathbf{u}$$

with the performance index

$$J = \frac{1}{2}\int_0^\infty (\mathbf{x}^*\mathbf{Q}\mathbf{x} + \mathbf{u}^*\mathbf{R}\mathbf{u})\, dt$$

where

$$\mathbf{Q} = E[\mathbf{w}(t)\mathbf{w}^*(t)], \qquad \mathbf{R} = E[\boldsymbol{\epsilon}(t)\boldsymbol{\epsilon}^*(t)]$$

and by transposing this $\mathbf{K}$ matrix to obtain the optimal estimation gain matrix $\mathbf{K}_e$, so that

$$\mathbf{K}_e = \mathbf{K}^*$$

APPENDIX

Vector-Matrix Analysis

1 DEFINITIONS

Conjugate matrix. The conjugate of a matrix $\mathbf{A}$ is the matrix in which each element is the complex conjugate of the corresponding element of $\mathbf{A}$. The conjugate of $\mathbf{A}$ is denoted by $\bar{\mathbf{A}} = [\bar{a}_{ij}]$, where $\bar{a}_{ij}$ is the complex conjugate of a_{ij}. For example, if $\mathbf{A}$ is given by

$$\mathbf{A} = \begin{bmatrix} 0 & 1 & 0 \\ -1+j & -3-j3 & -1+j4 \\ -1+j & -1 & -2+j3 \end{bmatrix} \tag{1}$$

then

$$\bar{\mathbf{A}} = \begin{bmatrix} 0 & 1 & 0 \\ -1-j & -3+j3 & -1-j4 \\ -1-j & -1 & -2-j3 \end{bmatrix}$$

Transpose. If the rows and columns of an $n \times m$ matrix $\mathbf{A}$ are interchanged, then the resulting $m \times n$ matrix is called the transpose of $\mathbf{A}$. The transpose of $\mathbf{A}$ is denoted by $\mathbf{A}^T$. That is, if $\mathbf{A}$ is given by

$$\mathbf{A} = \begin{bmatrix} a_{11} & a_{12} & \cdots & a_{1m} \\ a_{21} & a_{22} & \cdots & a_{2m} \\ \cdot & \cdot & & \cdot \\ \cdot & \cdot & & \cdot \\ \cdot & \cdot & & \cdot \\ a_{n1} & a_{n2} & \cdots & a_{nm} \end{bmatrix}$$

then $\mathbf{A}^T$ is given by

$$\mathbf{A}^T = \begin{bmatrix} a_{11} & a_{21} & \cdots & a_{n1} \\ a_{12} & a_{22} & \cdots & a_{n2} \\ \cdot & \cdot & & \cdot \\ \cdot & \cdot & & \cdot \\ \cdot & \cdot & & \cdot \\ a_{1m} & a_{2m} & \cdots & a_{nm} \end{bmatrix}$$

Note that $(\mathbf{A}^T)^T = \mathbf{A}$. It is easy to verify that if $\mathbf{A} + \mathbf{B}$ and $\mathbf{AB}$ can be defined, then

$$(\mathbf{A} + \mathbf{B})^T = \mathbf{A}^T + \mathbf{B}^T, \qquad (\mathbf{AB})^T = \mathbf{B}^T\mathbf{A}^T$$

Conjugate transpose. The conjugate transpose is the conjugate of the transpose of a matrix. Given a matrix $\mathbf{A} = [a_{ij}]$, the conjugate transpose is denoted by $\bar{\mathbf{A}}^T$ or $\mathbf{A}^*$; that is,

$$\bar{\mathbf{A}}^T = \mathbf{A}^* = [\bar{a}_{ji}]$$

For example, if $\mathbf{A}$ is as given by Eq. (1), then

$$\bar{\mathbf{A}}^T = \mathbf{A}^* = \begin{bmatrix} 0 & -1-j & -1-j \\ 1 & -3+j3 & -1 \\ 0 & -1-j4 & -2-j3 \end{bmatrix}$$

Clearly, the conjugate of $\mathbf{A}^T$ is the same as the transpose of $\bar{\mathbf{A}}$. Note that $(\mathbf{A}^*)^* = \mathbf{A}$. It can easily be shown that if $\mathbf{A} + \mathbf{B}$ and $\mathbf{AB}$ can be defined, then

$$(\mathbf{A} + \mathbf{B})^* = \mathbf{A}^* + \mathbf{B}^*, \qquad (\mathbf{AB})^* = \mathbf{B}^*\mathbf{A}^*$$

Note also that if c is a complex number, then

$$(c\mathbf{A})^* = \bar{c}\mathbf{A}^*$$

If $\mathbf{A}$ is a real matrix (a matrix whose elements are all real), the conjugate transpose $\mathbf{A}^*$ is the same as the transpose $\mathbf{A}^T$.

Symmetric matrix and skew-symmetric matrix. A symmetric matrix is a matrix which is equal to its transpose. That is, for a symmetric matrix $\mathbf{A}$,

$$\mathbf{A}^T = \mathbf{A} \qquad \text{or} \qquad a_{ji} = a_{ij}$$

If a matrix $\mathbf{A}$ is equal to the negative of its transpose, or

$$\mathbf{A}^T = -\mathbf{A} \qquad \text{or} \qquad a_{ji} = -a_{ij}$$

then it is called a *skew-symmetric* matrix.

If $\mathbf{A}$ is any square matrix, then $\mathbf{A} + \mathbf{A}^T$ is a symmetric matrix and $\mathbf{A} - \mathbf{A}^T$ is a skew-symmetric matrix. For example, if $\mathbf{A}$ is given by

$$\mathbf{A} = \begin{bmatrix} 1 & 2 & 3 \\ 4 & 5 & 6 \\ 7 & 8 & 9 \end{bmatrix}$$

then

$$\mathbf{A} + \mathbf{A}^T = \begin{bmatrix} 2 & 6 & 10 \\ 6 & 10 & 14 \\ 10 & 14 & 18 \end{bmatrix} = \text{symmetric matrix}$$

and

$$\mathbf{A} - \mathbf{A}^T = \begin{bmatrix} 0 & -2 & -4 \\ 2 & 0 & -2 \\ 4 & 2 & 0 \end{bmatrix} = \text{skew-symmetric matrix}$$

Notice that if $\mathbf{A}$ is a rectangular matrix, then $\mathbf{A}^T\mathbf{A} = \mathbf{B}$ is a symmetric matrix. Notice also that the inverse of a symmetric matrix, if the inverse exists, is symmetric. To establish this fact take the transpose of $\mathbf{B}\mathbf{B}^{-1} = \mathbf{I}$. We have $(\mathbf{B}^{-1})^T\mathbf{B}^T = \mathbf{I}^T = \mathbf{I}$. Noting that $\mathbf{B} = \mathbf{B}^T$, we have $(\mathbf{B}^{-1})^T\mathbf{B}^T = (\mathbf{B}^{-1})^T\mathbf{B} = \mathbf{I} = \mathbf{B}^{-1}\mathbf{B}$. Hence, $\mathbf{B}^{-1} = (\mathbf{B}^{-1})^T$. Thus, the inverse of a symmetric matrix is symmetric.

Orthogonal matrix. A matrix $\mathbf{A}$ is called an orthogonal matrix if it is real and satisfies the relationship $\mathbf{A}^T\mathbf{A} = \mathbf{A}\mathbf{A}^T = \mathbf{I}$. (This implies that $|\mathbf{A}| = \pm 1$ and thus $\mathbf{A}$ is nonsingular.) Examples of orthogonal matrices are

$$\mathbf{A} = \begin{bmatrix} \cos\theta & -\sin\theta \\ \sin\theta & \cos\theta \end{bmatrix}, \qquad \mathbf{B} = \begin{bmatrix} 0.6 & 0.8 & 0 \\ -0.8 & 0.6 & 0 \\ 0 & 0 & 1 \end{bmatrix}$$

In an orthogonal matrix the inverse is exactly equal to the transpose.

$$\mathbf{A}^{-1} = \mathbf{A}^T$$

To demonstrate, let us obtain $\mathbf{A}^{-1}$ and $\mathbf{B}^{-1}$ for the matrices $\mathbf{A}$ and $\mathbf{B}$ that were just given:

$$\mathbf{A}^{-1} = \begin{bmatrix} \cos\theta & \sin\theta \\ -\sin\theta & \cos\theta \end{bmatrix} = \mathbf{A}^T, \qquad \mathbf{B}^{-1} = \begin{bmatrix} 0.6 & -0.8 & 0 \\ 0.8 & 0.6 & 0 \\ 0 & 0 & 1 \end{bmatrix} = \mathbf{B}^T$$

If $\mathbf{A}$ and $\mathbf{B}$ are $n \times n$ orthogonal matrices, then so are $\mathbf{A}^{-1}$, $\mathbf{A}^T$, and $\mathbf{AB}$. This may be seen as follows. Since $\mathbf{A}$ is orthogonal, $\mathbf{A}\mathbf{A}^T = \mathbf{I}$ and $(\mathbf{A}^T)^T\mathbf{A}^T = \mathbf{I}$. Hence, $\mathbf{A}^T$ is orthogonal. Since $\mathbf{A}^{-1} = \mathbf{A}^T$, $\mathbf{A}^{-1}$ is also orthogonal. Since $\mathbf{B}^T = \mathbf{B}^{-1}$, $\mathbf{A}^T = \mathbf{A}^{-1}$, $(\mathbf{AB})^T = \mathbf{B}^T\mathbf{A}^T$, and $(\mathbf{AB})^{-1} = \mathbf{B}^{-1}\mathbf{A}^{-1}$, we have $(\mathbf{AB})^T = (\mathbf{AB})^{-1}$, and so $\mathbf{AB}$ is orthogonal.

Hermitian matrix and skew-Hermitian matrix. A matrix whose elements are complex quantities is called a *complex* matrix. If a complex matrix **A** satisfies the relationship

$$\mathbf{A}^* = \mathbf{A} \qquad \text{or} \qquad a_{ij} = \bar{a}_{ji}$$

where $\bar{a}_{ji}$ is the complex conjugate of a_{ji}, then **A** is called a *Hermitian* matrix. Note that a Hermitian matrix must be square and that the main diagonal elements must be real. An example is

$$\mathbf{A} = \begin{bmatrix} 1 & 4+j3 & j5 \\ 4-j3 & 2 & 2+j \\ -j5 & 2-j & 0 \end{bmatrix}$$

If a Hermitian matrix **A** is written as $\mathbf{A} = \mathbf{B} + j\mathbf{C}$, where **B** and **C** are real matrices, then

$$\mathbf{B} = \mathbf{B}^T, \qquad \mathbf{C} = -\mathbf{C}^T$$

In the foregoing example,

$$\mathbf{A} = \mathbf{B} + j\mathbf{C} = \begin{bmatrix} 1 & 4 & 0 \\ 4 & 2 & 2 \\ 0 & 2 & 0 \end{bmatrix} + j \begin{bmatrix} 0 & 3 & 5 \\ -3 & 0 & 1 \\ -5 & -1 & 0 \end{bmatrix}$$

Notice that the inverse of a Hermitian matrix **A** is Hermitian, or $\mathbf{A}^{-1} = (\mathbf{A}^{-1})^*$. Notice also that every square matrix can be expressed uniquely as $\mathbf{A} = \mathbf{G} + j\mathbf{H}$, where **G** and **H** are Hermitian and are given by

$$\mathbf{G} = \frac{1}{2}(\mathbf{A} + \mathbf{A}^*), \qquad \mathbf{H} = \frac{1}{2j}(\mathbf{A} - \mathbf{A}^*)$$

The fact that **G** and **H** are Hermitian is seen as follows:

$$\mathbf{G}^* = \frac{1}{2}(\mathbf{A}^* + \mathbf{A}) = \mathbf{G}, \qquad \mathbf{H}^* = -\frac{1}{2j}(\mathbf{A}^* - \mathbf{A}) = \mathbf{H}$$

It can be easily verified that for $n \times n$ Hermitian matrices **A** and **B**, the matrices $\mathbf{A} + \mathbf{B}$, $\mathbf{A} - \mathbf{B}$, and $\mathbf{AB} + \mathbf{BA}$ are also Hermitian. The product **AB** is Hermitian, however, if and only if **A** and **B** commute, since $\mathbf{AB} = \mathbf{A}^*\mathbf{B}^* = (\mathbf{BA})^*$. The determinant of a Hermitian matrix is always real, since

$$|\mathbf{A}| = |\mathbf{A}^*| = |\bar{\mathbf{A}}^T| = |\bar{\mathbf{A}}|$$

If a matrix **A** satisfies the relationship $\mathbf{A}^* = -\mathbf{A}$, then **A** is called a *skew-Hermitian* matrix. An example is

$$\mathbf{A} = \begin{bmatrix} j5 & -2+j3 & -4+j6 \\ 2+j3 & j4 & -2+j2 \\ 4+j6 & 2+j2 & j \end{bmatrix}$$

Note that a skew-Hermitian matrix must be square and that the main diagonal elements must be imaginary or zero.

If a skew-Hermitian matrix $\mathbf{A}$ is written as $\mathbf{A} = \mathbf{B} + j\mathbf{C}$, where $\mathbf{B}$ and $\mathbf{C}$ are real matrices, then

$$\mathbf{B} = -\mathbf{B}^T, \qquad \mathbf{C} = \mathbf{C}^T$$

In the present example,

$$\mathbf{A} = \mathbf{B} + j\mathbf{C} = \begin{bmatrix} 0 & -2 & -4 \\ 2 & 0 & -2 \\ 4 & 2 & 0 \end{bmatrix} + j \begin{bmatrix} 5 & 3 & 6 \\ 3 & 4 & 2 \\ 6 & 2 & 1 \end{bmatrix}$$

Unitary matrix. A unitary matrix is a complex matrix in which the inverse is equal to the conjugate of the transpose. That is,

$$\mathbf{A}^{-1} = \mathbf{A}^* \qquad \text{or} \qquad \mathbf{A}\mathbf{A}^* = \mathbf{A}^*\mathbf{A} = \mathbf{I}$$

An example of a unitary matrix is

$$\mathbf{A} = \begin{bmatrix} \frac{1}{\sqrt{15}}(2+j) & \frac{1}{\sqrt{15}}(3+j) \\ \frac{1}{\sqrt{15}}(-3+j) & \frac{1}{\sqrt{15}}(2-j) \end{bmatrix}$$

To demonstrate, let us compute $\mathbf{A}^{-1}$ and $\mathbf{A}^*$. Since for a unitary matrix $\mathbf{A}$ the determinant of $\mathbf{A}$ is equal to unity, or $|\mathbf{A}| = 1$, we obtain

$$\mathbf{A}^{-1} = \begin{bmatrix} \frac{1}{\sqrt{15}}(2-j) & -\frac{1}{\sqrt{15}}(3+j) \\ \frac{1}{\sqrt{15}}(3-j) & \frac{1}{\sqrt{15}}(2+j) \end{bmatrix}$$

The conjugate transpose $\mathbf{A}^*$ is given by

$$\mathbf{A}^* = \begin{bmatrix} \frac{1}{\sqrt{15}}(2-j) & -\frac{1}{\sqrt{15}}(3+j) \\ \frac{1}{\sqrt{15}}(3-j) & \frac{1}{\sqrt{15}}(2+j) \end{bmatrix}$$

Hence we have verified that $\mathbf{A}^{-1} = \mathbf{A}^*$.

Orthogonal matrices satisfy the relationship $\mathbf{A}\mathbf{A}^* = \mathbf{A}^*\mathbf{A} = \mathbf{I}$; hence they are unitary.

Notice that if $\mathbf{A}$ is unitary, then so is the inverse $\mathbf{A}^{-1}$. To see this, notice that since $\mathbf{A}\mathbf{A}^* = \mathbf{A}^*\mathbf{A} = \mathbf{I}$, we obtain

$$(\mathbf{A}^*)^{-1}\mathbf{A}^{-1} = (\mathbf{A}^{-1})^*(\mathbf{A}^{-1}) = \mathbf{I}$$

$$\mathbf{A}^{-1}(\mathbf{A}^*)^{-1} = (\mathbf{A}^{-1})(\mathbf{A}^{-1})^* = \mathbf{I}$$

If $n \times n$ matrices **A** and **B** are unitary, then so is matrix **AB.** To prove this, notice that since $\mathbf{AA}^* = \mathbf{A}^*\mathbf{A} = \mathbf{I}$ and $\mathbf{BB}^* = \mathbf{B}^*\mathbf{B} = \mathbf{I}$, we have

$$(\mathbf{AB})(\mathbf{AB})^* = \mathbf{ABB}^*\mathbf{A}^* = \mathbf{AA}^* = \mathbf{I}$$

$$(\mathbf{AB})^*(\mathbf{AB}) = \mathbf{B}^*\mathbf{A}^*\mathbf{AB} = \mathbf{B}^*\mathbf{B} = \mathbf{I}$$

Hence, it is shown that **AB** is unitary.

Normal matrix. A matrix which commutes with its conjugate transpose is called a *normal* matrix. Specifically, for a normal matrix **A**,

$$\mathbf{AA}^* = \mathbf{A}^*\mathbf{A} \qquad \text{if } \mathbf{A} \text{ is a complex matrix}$$

$$\mathbf{AA}^T = \mathbf{A}^T\mathbf{A} \qquad \text{if } \mathbf{A} \text{ is a real matrix}$$

Notice that if **A** is normal and **U** is unitary, then $\mathbf{U}^{-1}\mathbf{AU}$ is also normal, since

$$\begin{aligned}(\mathbf{U}^{-1}\mathbf{AU})(\mathbf{U}^{-1}\mathbf{AU})^* &= \mathbf{U}^{-1}\mathbf{AUU}^*\mathbf{A}^*(\mathbf{U}^{-1})^* = \mathbf{U}^{-1}\mathbf{AA}^*(\mathbf{U}^{-1})^* = \mathbf{U}^*\mathbf{A}^*\mathbf{AU} \\ &= \mathbf{U}^*\mathbf{A}^*(\mathbf{U}^{-1})^*\mathbf{U}^{-1}\mathbf{AU} = (\mathbf{U}^{-1}\mathbf{AU})^*(\mathbf{U}^{-1}\mathbf{AU})\end{aligned}$$

A matrix is normal if it is a real symmetric, a Hermitian, a real skew-symmetric, a skew-Hermitian, a unitary, or an orthogonal matrix.

Summary. In summarizing the definitions of various matrices, the reader may find the following list useful:

$\mathbf{A}^T = \mathbf{A}$	**A** is symmetric
$\mathbf{A}^T = -\mathbf{A}$	**A** is skew-symmetric
$\mathbf{AA}^T = \mathbf{A}^T\mathbf{A} = \mathbf{I}$	**A** is orthogonal
$\mathbf{A}^* = \mathbf{A}$	**A** is Hermitian
$\mathbf{A}^* = -\mathbf{A}$	**A** is skew-Hermitian
$\mathbf{AA}^* = \mathbf{A}^*\mathbf{A} = \mathbf{I}$	**A** is unitary
$\mathbf{AA}^* = \mathbf{A}^*\mathbf{A}$ or $\mathbf{AA}^T = \mathbf{A}^T\mathbf{A}$	**A** is normal

2 DETERMINANTS

Determinants of a 2 × 2 matrix, a 3 × 3 matrix, and a 4 × 4 matrix. For a 2×2 matrix **A** we have

$$|\mathbf{A}| = \begin{vmatrix} a_1 & a_2 \\ b_1 & b_2 \end{vmatrix} = a_1b_2 - b_1a_2$$

For a 3×3 matrix **A**,

$$|\mathbf{A}| = \begin{vmatrix} a_1 & a_2 & a_3 \\ b_1 & b_2 & b_3 \\ c_1 & c_2 & c_3 \end{vmatrix} = a_1b_2c_3 + b_1c_2a_3 + c_1a_2b_3 - c_1b_2a_3 - b_1a_2c_3 - a_1b_3c_2$$

For a 4×4 matrix $\mathbf{A}$,

$$\begin{aligned} |\mathbf{A}| &= \begin{vmatrix} a_1 & a_2 & a_3 & a_4 \\ b_1 & b_2 & b_3 & b_4 \\ c_1 & c_2 & c_3 & c_4 \\ d_1 & d_2 & d_3 & d_4 \end{vmatrix} \\ &= \begin{vmatrix} a_1 & a_2 \\ b_1 & b_2 \end{vmatrix} \begin{vmatrix} c_3 & c_4 \\ d_3 & d_4 \end{vmatrix} - \begin{vmatrix} a_1 & a_2 \\ c_1 & c_2 \end{vmatrix} \begin{vmatrix} b_3 & b_4 \\ d_3 & d_4 \end{vmatrix} \\ &\quad + \begin{vmatrix} a_1 & a_2 \\ d_1 & d_2 \end{vmatrix} \begin{vmatrix} b_3 & b_4 \\ c_3 & c_4 \end{vmatrix} + \begin{vmatrix} b_1 & b_2 \\ c_1 & c_2 \end{vmatrix} \begin{vmatrix} a_3 & a_4 \\ d_3 & d_4 \end{vmatrix} \\ &\quad - \begin{vmatrix} b_1 & b_2 \\ d_1 & d_2 \end{vmatrix} \begin{vmatrix} a_3 & a_4 \\ c_3 & c_4 \end{vmatrix} + \begin{vmatrix} c_1 & c_2 \\ d_1 & d_2 \end{vmatrix} \begin{vmatrix} a_3 & a_4 \\ b_3 & b_4 \end{vmatrix} \end{aligned}$$

(This expansion is called Laplace's expansion by the minors.)

Properties of the determinant. The determinant of an $n \times n$ matrix has the following properties:

1. If two rows (or two columns) of the determinant are interchanged, only the sign of the determinant is changed.
2. The determinant is invariant under the addition of a scalar multiple of a row (or a column) to another row (or column).
3. If an $n \times n$ matrix has two identical rows (or columns), then the determinant is zero.
4. For an $n \times n$ matrix $\mathbf{A}$,

$$|\mathbf{A}^T| = |\mathbf{A}|, \qquad |\mathbf{A}^*| = |\bar{\mathbf{A}}|$$

5. The determinant of a product of two square matrices $\mathbf{A}$ and $\mathbf{B}$ is the product of their determinants:

$$|\mathbf{AB}| = |\mathbf{A}|\,|\mathbf{B}| = |\mathbf{BA}|$$

6. If a row (or a column) is multipled by a scalar k, then the determinant is multiplied by k.
7. If all elements of an $n \times n$ matrix are multiplied by k, then the determinant is multiplied by k^n; that is,

$$|k\mathbf{A}| = k^n|\mathbf{A}|$$

8. If the eigenvalues of $\mathbf{A}$ are $\lambda_i \quad (i = 1, 2, \ldots, n)$, then

$$|\mathbf{A}| = \lambda_1\lambda_2 \ldots \lambda_n$$

Hence $|\mathbf{A}| \neq 0$ implies $\lambda_i \neq 0$ for $i = 1, 2, \ldots, n$. (For details of the eigenvalue, see Sec. 6.)

9. If matrices **A, B, C,** and **D** are an $n \times n$, an $n \times m$, an $m \times n$, and an $m \times m$ matrix, respectively, then

$$\begin{vmatrix} \mathbf{A} & \mathbf{B} \\ \mathbf{0} & \mathbf{D} \end{vmatrix} = \begin{vmatrix} \mathbf{A} & \mathbf{0} \\ \mathbf{C} & \mathbf{D} \end{vmatrix} = |\mathbf{A}|\,|\mathbf{D}| \qquad \text{if } |\mathbf{A}| \neq 0 \text{ and } |\mathbf{D}| \neq 0 \tag{2}$$

$$\begin{vmatrix} \mathbf{A} & \mathbf{B} \\ \mathbf{0} & \mathbf{D} \end{vmatrix} = \begin{vmatrix} \mathbf{A} & \mathbf{0} \\ \mathbf{C} & \mathbf{D} \end{vmatrix} = 0 \qquad \text{if } |\mathbf{A}| = 0 \text{ or } |\mathbf{D}| = 0 \text{ or } |\mathbf{A}| = |\mathbf{D}| = 0$$

Also,

$$\begin{vmatrix} \mathbf{A} & \mathbf{B} \\ \mathbf{C} & \mathbf{D} \end{vmatrix} = \begin{cases} |\mathbf{A}|\,|\mathbf{D} - \mathbf{C}\mathbf{A}^{-1}\mathbf{B}| & \text{if } |\mathbf{A}| \neq 0 \quad (3) \\ |\mathbf{D}|\,|\mathbf{A} - \mathbf{B}\mathbf{D}^{-1}\mathbf{C}| & \text{if } |\mathbf{D}| \neq 0 \quad (4) \end{cases}$$

[For the derivation of Eq. (2), see Prob. A-1. For derivations of Eqs. (3) and (4), refer to Prob. A-2.]

10. For an $n \times m$ matrix **A** and an $m \times n$ matrix **B,**

$$|\mathbf{I}_n + \mathbf{A}\mathbf{B}| = |\mathbf{I}_m + \mathbf{B}\mathbf{A}| \tag{5}$$

(For the proof, see Prob. A-3.) In particular, for $m = 1$, that is, for an $n \times 1$ matrix **A** and a $1 \times n$ matrix **B,** we have

$$|\mathbf{I}_n + \mathbf{A}\mathbf{B}| = 1 + \mathbf{B}\mathbf{A} \tag{6}$$

Equations (2) through (6) are useful in computing the determinants of matrices of large order.

3 INVERSION OF MATRICES

Nonsingular matrix and singular matrix. A square matrix **A** is called a nonsingular matrix if a matrix **B** exists such that $\mathbf{BA} = \mathbf{AB} = \mathbf{I}$. If such a matrix **B** exists, then it is denoted by $\mathbf{A}^{-1}$. $\mathbf{A}^{-1}$ is called the *inverse* of **A.** The inverse matrix $\mathbf{A}^{-1}$ exists if $|\mathbf{A}|$ is nonzero. If $\mathbf{A}^{-1}$ does not exist, **A** is said to be *singular*.

If **A** and **B** are nonsingular matrices, then the product **AB** is a nonsingular matrix and

$$(\mathbf{AB})^{-1} = \mathbf{B}^{-1}\mathbf{A}^{-1}$$

Also,

$$(\mathbf{A}^T)^{-1} = (\mathbf{A}^{-1})^T$$

and

$$(\mathbf{A}^*)^{-1} = (\mathbf{A}^{-1})^*$$

Properties of the inverse matrix. The inverse of a matrix has the following properties:

1. If k is a nonzero scalar and $\mathbf{A}$ is an $n \times n$ nonsingular matrix, then

$$(k\mathbf{A})^{-1} = \frac{1}{k}\mathbf{A}^{-1}$$

2. The determinant of $\mathbf{A}^{-1}$ is the inverse of the determinant of $\mathbf{A}$, or

$$|\mathbf{A}^{-1}| = \frac{1}{|\mathbf{A}|}$$

This can be verified easily as follows:

$$|\mathbf{A}\mathbf{A}^{-1}| = |\mathbf{A}|\,|\mathbf{A}^{-1}| = 1$$

Useful formulas for finding the inverse of a matrix.

1. For a 2×2 matrix $\mathbf{A}$, where

$$\mathbf{A} = \begin{bmatrix} a & b \\ c & d \end{bmatrix} \qquad ad - bc \neq 0$$

the inverse matrix is given by

$$\mathbf{A}^{-1} = \frac{1}{ad - bc}\begin{bmatrix} d & -b \\ -c & a \end{bmatrix}$$

2. For a 3×3 matrix $\mathbf{A}$, where

$$\mathbf{A} = \begin{bmatrix} a & b & c \\ d & e & f \\ g & h & i \end{bmatrix} \qquad |\mathbf{A}| \neq 0$$

the inverse matrix is given by

$$\mathbf{A}^{-1} = \frac{1}{|\mathbf{A}|}\begin{bmatrix} \begin{vmatrix} e & f \\ h & i \end{vmatrix} & -\begin{vmatrix} b & c \\ h & i \end{vmatrix} & \begin{vmatrix} b & c \\ e & f \end{vmatrix} \\ -\begin{vmatrix} d & f \\ g & i \end{vmatrix} & \begin{vmatrix} a & c \\ g & i \end{vmatrix} & -\begin{vmatrix} a & c \\ d & f \end{vmatrix} \\ \begin{vmatrix} d & e \\ g & h \end{vmatrix} & -\begin{vmatrix} a & b \\ g & h \end{vmatrix} & \begin{vmatrix} a & b \\ d & e \end{vmatrix} \end{bmatrix}$$

3. If $\mathbf{A}$, $\mathbf{B}$, $\mathbf{C}$, and $\mathbf{D}$ are, respectively, an $n \times n$, an $n \times m$, an $m \times n$, and an $m \times m$ matrix, then

$$(\mathbf{A} + \mathbf{BDC})^{-1} = \mathbf{A}^{-1} - \mathbf{A}^{-1}\mathbf{B}(\mathbf{D}^{-1} + \mathbf{C}\mathbf{A}^{-1}\mathbf{B})^{-1}\mathbf{C}\mathbf{A}^{-1} \tag{7}$$

provided the indicated inverses exist. Equation (7) is commonly referred to as the *matrix inversion lemma*. (For the proof, see Prob. A-4.)

If $\mathbf{D} = \mathbf{I}_m$, then Eq. (7) simplifies to

$$(\mathbf{A}+\mathbf{BC})^{-1}=\mathbf{A}^{-1}-\mathbf{A}^{-1}\mathbf{B}(\mathbf{I}_m+\mathbf{CA}^{-1}\mathbf{B})^{-1}\mathbf{CA}^{-1}$$

In this last equation, if $\mathbf{B}$ and $\mathbf{C}$ are an $n \times 1$ matrix and a $1 \times n$ matrix, respectively, then

$$(\mathbf{A}+\mathbf{BC})^{-1}=\mathbf{A}^{-1}-\frac{\mathbf{A}^{-1}\mathbf{BCA}^{-1}}{1+\mathbf{CA}^{-1}\mathbf{B}} \tag{8}$$

Equation (8) is useful in that if an $n \times n$ matrix $\mathbf{X}$ can be written as $\mathbf{A} + \mathbf{BC}$, where $\mathbf{A}$ is an $n \times n$ matrix whose inverse is known and $\mathbf{BC}$ is a product of a column vector and a row vector, then $\mathbf{X}^{-1}$ can be obtained easily in terms of the known $\mathbf{A}^{-1}$, $\mathbf{B}$, and $\mathbf{C}$.

4. If $\mathbf{A}$, $\mathbf{B}$, $\mathbf{C}$, and $\mathbf{D}$ are, respectively, an $n \times n$, an $n \times m$, an $m \times n$, and an $m \times m$ matrix, then

$$\begin{bmatrix}\mathbf{A} & \mathbf{B}\\ \mathbf{C} & \mathbf{D}\end{bmatrix}^{-1}=\begin{bmatrix}\mathbf{A}^{-1}+\mathbf{A}^{-1}\mathbf{B}(\mathbf{D}-\mathbf{CA}^{-1}\mathbf{B})^{-1}\mathbf{CA}^{-1} & -\mathbf{A}^{-1}\mathbf{B}(\mathbf{D}-\mathbf{CA}^{-1}\mathbf{B})^{-1}\\ -(\mathbf{D}-\mathbf{CA}^{-1}\mathbf{B})^{-1}\mathbf{CA}^{-1} & (\mathbf{D}-\mathbf{CA}^{-1}\mathbf{B})^{-1}\end{bmatrix} \tag{9}$$

provided $|\mathbf{A}| \neq 0$ and $|\mathbf{D} - \mathbf{CA}^{-1}\mathbf{B}| \neq 0$, or

$$\begin{bmatrix}\mathbf{A} & \mathbf{B}\\ \mathbf{C} & \mathbf{D}\end{bmatrix}^{-1}=\begin{bmatrix}(\mathbf{A}-\mathbf{BD}^{-1}\mathbf{C})^{-1} & -(\mathbf{A}-\mathbf{BD}^{-1}\mathbf{C})^{-1}\mathbf{BD}^{-1}\\ -\mathbf{D}^{-1}\mathbf{C}(\mathbf{A}-\mathbf{BD}^{-1}\mathbf{C})^{-1} & \mathbf{D}^{-1}\mathbf{C}(\mathbf{A}-\mathbf{BD}^{-1}\mathbf{C})^{-1}\mathbf{BD}^{-1}+\mathbf{D}^{-1}\end{bmatrix} \tag{10}$$

provided $|\mathbf{D}| \neq 0$ and $|\mathbf{A} - \mathbf{BD}^{-1}\mathbf{C}| \neq 0$. In particular, if $\mathbf{C} = \mathbf{0}$ or $\mathbf{B} = \mathbf{0}$, then Eqs. (9) and (10) can be simplified as follows:

$$\begin{bmatrix}\mathbf{A} & \mathbf{B}\\ \mathbf{0} & \mathbf{D}\end{bmatrix}^{-1}=\begin{bmatrix}\mathbf{A}^{-1} & -\mathbf{A}^{-1}\mathbf{BD}^{-1}\\ \mathbf{0} & \mathbf{D}^{-1}\end{bmatrix} \tag{11}$$

or

$$\begin{bmatrix}\mathbf{A} & \mathbf{0}\\ \mathbf{C} & \mathbf{D}\end{bmatrix}^{-1}=\begin{bmatrix}\mathbf{A}^{-1} & \mathbf{0}\\ -\mathbf{D}^{-1}\mathbf{CA}^{-1} & \mathbf{D}^{-1}\end{bmatrix} \tag{12}$$

[For the derivation of Eqs. (8) through (12), refer to Probs. A-5 and A-6.]

4 RULES OF MATRIX OPERATIONS

In this section we shall review some of the rules of algebraic operations with matrices and then give definitions of the derivative and the integral of matrices. Then the rules of differentiation of matrices will be presented.

Note that matrix algebra differs from ordinary number algebra in that matrix multiplication is not commutative and cancellation of matrices is not valid.

Multiplication of a matrix by a scalar. The product of a matrix and a scalar is a matrix in which each element is multiplied by the scalar. That is,

$$k\mathbf{A} = \begin{bmatrix} ka_{11} & ka_{12} & \cdots & ka_{1m} \\ ka_{21} & ka_{22} & \cdots & ka_{2m} \\ \cdot & \cdot & & \cdot \\ \cdot & \cdot & & \cdot \\ \cdot & \cdot & & \cdot \\ ka_{n1} & ka_{n2} & \cdots & ka_{nm} \end{bmatrix}$$

Multiplication of a matrix by a matrix. Multiplication of a matrix by a matrix is possible between matrices in which the number of columns in the first matrix is equal to the number of rows in the second. Otherwise, multiplication is not defined.

Consider the product of an $n \times m$ matrix $\mathbf{A}$ and an $m \times r$ matrix $\mathbf{B}$:

$$\mathbf{AB} = \begin{bmatrix} a_{11} & a_{12} & \cdots & a_{1m} \\ a_{21} & a_{22} & \cdots & a_{2m} \\ \cdot & \cdot & & \cdot \\ \cdot & \cdot & & \cdot \\ \cdot & \cdot & & \cdot \\ a_{n1} & a_{n2} & \cdots & a_{nm} \end{bmatrix} \begin{bmatrix} b_{11} & b_{12} & \cdots & b_{1r} \\ b_{21} & b_{22} & \cdots & b_{2r} \\ \cdot & \cdot & & \cdot \\ \cdot & \cdot & & \cdot \\ \cdot & \cdot & & \cdot \\ b_{m1} & b_{m2} & \cdots & b_{mr} \end{bmatrix}$$

$$= \begin{bmatrix} c_{11} & c_{12} & \cdots & c_{1r} \\ c_{21} & c_{22} & \cdots & c_{2r} \\ \cdot & \cdot & & \cdot \\ \cdot & \cdot & & \cdot \\ \cdot & \cdot & & \cdot \\ c_{n1} & c_{n2} & \cdots & c_{nr} \end{bmatrix}$$

where

$$c_{ik} = \sum_{j=1}^{m} a_{ij} b_{jk}$$

Thus multiplication of an $n \times m$ matrix by an $m \times r$ matrix yields an $n \times r$ matrix. It should be noted that, in general, matrix multiplication is not commutative; that is

$$\mathbf{AB} \neq \mathbf{BA} \qquad \text{in general}$$

For example,

$$\mathbf{AB} = \begin{bmatrix} a_{11} & a_{12} \\ a_{21} & a_{22} \end{bmatrix} \begin{bmatrix} b_{11} & b_{12} \\ b_{21} & b_{22} \end{bmatrix} = \begin{bmatrix} a_{11}b_{11} + a_{12}b_{21} & a_{11}b_{12} + a_{12}b_{22} \\ a_{21}b_{11} + a_{22}b_{21} & a_{21}b_{12} + a_{22}b_{22} \end{bmatrix}$$

and

$$\mathbf{BA} = \begin{bmatrix} b_{11} & b_{12} \\ b_{21} & b_{22} \end{bmatrix} \begin{bmatrix} a_{11} & a_{12} \\ a_{21} & a_{22} \end{bmatrix} = \begin{bmatrix} b_{11}a_{11} + b_{12}a_{21} & b_{11}a_{12} + b_{12}a_{22} \\ b_{21}a_{11} + b_{22}a_{21} & b_{21}a_{12} + b_{22}a_{22} \end{bmatrix}$$

Thus, in general, $\mathbf{AB} \neq \mathbf{BA}$. Hence, the order of multiplication is significant and must be preserved. If **AB** = **BA**, matrices **A** and **B** are said to commute. In the preceding matrices **A** and **B**, if, for example, $a_{12} = a_{21} = b_{12} = b_{21} = 0$, then **A** and **B** commute.

For $n \times n$ diagonal matrices **A** and **B**,

$$\mathbf{AB} = [a_{ij}\delta_{ij}][b_{ij}\delta_{ij}] = \begin{bmatrix} a_{11}b_{11} & & & & 0 \\ & a_{22}b_{22} & & & \\ & & \cdot & & \\ & & & \cdot & \\ & & & & \cdot \\ 0 & & & & a_{nn}b_{nn} \end{bmatrix}$$

If **A**, **B**, and **C** are an $n \times m$ matrix, an $m \times r$ matrix, and an $r \times p$ matrix, respectively, then the following associativity law holds true:

$$(\mathbf{AB})\mathbf{C} = \mathbf{A}(\mathbf{BC})$$

This may be proved as follows:

$$(i, k)\text{th element of } \mathbf{AB} = \sum_{j=1}^{m} a_{ij}b_{jk}$$

$$(j, h)\text{th element of } \mathbf{BC} = \sum_{k=1}^{r} b_{jk}c_{kh}$$

$$\begin{aligned}(i, h)\text{th element of } (\mathbf{AB})\mathbf{C} &= \sum_{k=1}^{r}\left(\sum_{j=1}^{m} a_{ij}b_{jk}\right)c_{kh} = \sum_{j=1}^{m}\sum_{k=1}^{r}(a_{ij}b_{jk})c_{kh} \\ &= \sum_{j=1}^{m}\sum_{k=1}^{r} a_{ij}(b_{jk}c_{kh}) = \sum_{j=1}^{m} a_{ij}\left[\sum_{k=1}^{r} b_{jk}c_{kh}\right] \\ &= (i, h)\text{th element of } \mathbf{A}(\mathbf{BC})\end{aligned}$$

Since the associativity of multiplication of matrices holds true, we have

$$\mathbf{ABCD} = (\mathbf{AB})(\mathbf{CD}) = \mathbf{A}(\mathbf{BCD}) = (\mathbf{ABC})\mathbf{D}$$

$$\mathbf{A}^{m+n} = \mathbf{A}^m\mathbf{A}^n \qquad m, n = 1, 2, 3, \ldots$$

If **A** and **B** are $n \times m$ matrices and **C** and **D** are $m \times r$ matrices, then the following distributivity law holds true:

$$(\mathbf{A} + \mathbf{B})(\mathbf{C} + \mathbf{D}) = \mathbf{AC} + \mathbf{AD} + \mathbf{BC} + \mathbf{BD}$$

This can be proved by comparing the (i, j)th element of $(\mathbf{A} + \mathbf{B})(\mathbf{C} + \mathbf{D})$ and the (i, j)th element of $(\mathbf{AC} + \mathbf{AD} + \mathbf{BC} + \mathbf{BD})$.

Remarks on cancellation of matrices. Cancellation of matrices is not valid in matrix algebra. Consider the product of two singular matrices **A** and **B**. Take, for example,

$$\mathbf{A} = \begin{bmatrix} 2 & 1 \\ 6 & 3 \end{bmatrix} \neq \mathbf{0}, \qquad \mathbf{B} = \begin{bmatrix} 1 & -2 \\ -2 & 4 \end{bmatrix} \neq \mathbf{0}$$

Then

$$\mathbf{AB} = \begin{bmatrix} 2 & 1 \\ 6 & 3 \end{bmatrix} \begin{bmatrix} 1 & -2 \\ -2 & 4 \end{bmatrix} = \begin{bmatrix} 0 & 0 \\ 0 & 0 \end{bmatrix} = \mathbf{0}$$

Clearly, $\mathbf{AB} = \mathbf{0}$ implies neither $\mathbf{A} = \mathbf{0}$ nor $\mathbf{B} = \mathbf{0}$. In fact, $\mathbf{AB} = \mathbf{0}$ implies one of the following three:

1. $\mathbf{A} = \mathbf{0}$.
2. $\mathbf{B} = \mathbf{0}$.
3. Both **A** and **B** are singular.

It can easily be proved that if both **A** and **B** are nonzero matrices and $\mathbf{AB} = \mathbf{0}$, then both **A** and **B** must be singular. Assume that **B** is nonzero and **A** is not singular. Then $|\mathbf{A}| \neq 0$ and $\mathbf{A}^{-1}$ exists. Then we obtain

$$\mathbf{A}^{-1}\mathbf{AB} = \mathbf{B} = \mathbf{0}$$

which contradicts the assumption that **B** is nonzero. In this way we can prove that both **A** and **B** must be singular if $\mathbf{A} \neq \mathbf{0}$ and $\mathbf{B} \neq \mathbf{0}$.

Similarly, notice that if **A** is singular, then neither $\mathbf{AB} = \mathbf{AC}$ nor $\mathbf{BA} = \mathbf{CA}$ implies $\mathbf{B} = \mathbf{C}$. If, however, **A** is a nonsingular matrix, then $\mathbf{AB} = \mathbf{AC}$ implies $\mathbf{B} = \mathbf{C}$ and $\mathbf{BA} = \mathbf{CA}$ also implies $\mathbf{B} = \mathbf{C}$.

Derivative and integral of a matrix. The derivative of an $n \times m$ matrix $\mathbf{A}(t)$ is defined by the matrix whose (i, j)th element is the derivative of the (i, j)th element of the original matrix, provided that all the elements $a_{ij}(t)$ have derivatives with respect to t:

$$\frac{d}{dt}\mathbf{A}(t) = \begin{bmatrix} \frac{d}{dt}a_{11}(t) & \cdots & \frac{d}{dt}a_{1m}(t) \\ \vdots & & \vdots \\ \frac{d}{dt}a_{n1}(t) & \cdots & \frac{d}{dt}a_{nm}(t) \end{bmatrix}$$

In the case of an n-dimensional vector $\mathbf{x}(t)$,

$$\frac{d}{dt}\mathbf{x}(t) = \begin{bmatrix} \frac{d}{dt}x_1(t) \\ \cdot \\ \cdot \\ \cdot \\ \frac{d}{dt}x_n(t) \end{bmatrix}$$

Similarly, the integral of an $n \times m$ matrix $\mathbf{A}(t)$ with respect to t is defined by the matrix whose (i, j)th element is the integral of the (i, j)th element of the original matrix, or

$$\int \mathbf{A}(t)\,dt = \begin{bmatrix} \int a_{11}(t)\,dt & \cdots & \int a_{1m}(t)\,dt \\ \cdot & & \cdot \\ \cdot & & \cdot \\ \cdot & & \cdot \\ \int a_{n1}(t)\,dt & \cdots & \int a_{nm}(t)\,dt \end{bmatrix}$$

provided that the $a_{ij}(t)$'s are integrable as functions of t.

Differentiation of a matrix. If the elements of matrices $\mathbf{A}$ and $\mathbf{B}$ are functions of t, then

$$\frac{d}{dt}(\mathbf{A}+\mathbf{B}) = \frac{d}{dt}\mathbf{A} + \frac{d}{dt}\mathbf{B} \tag{13}$$

$$\frac{d}{dt}(\mathbf{AB}) = \frac{d\mathbf{A}}{dt}\mathbf{B} + \mathbf{A}\frac{d\mathbf{B}}{dt} \tag{14}$$

If $k(t)$ is a scalar and is a function of t, then

$$\frac{d}{dt}[\mathbf{A}k(t)] = \frac{d\mathbf{A}}{dt}k(t) + \mathbf{A}\frac{dk(t)}{dt} \tag{15}$$

Also,

$$\int_a^b \frac{d\mathbf{A}}{dt}\mathbf{B}\,dt = \mathbf{AB}\Big|_a^b - \int_a^b \mathbf{A}\frac{d\mathbf{B}}{dt}\,dt \tag{16}$$

It is important to note that the derivative of $\mathbf{A}^{-1}$ is given by

$$\frac{d}{dt}\mathbf{A}^{-1} = -\mathbf{A}^{-1}\frac{d\mathbf{A}}{dt}\mathbf{A}^{-1} \tag{17}$$

Equation (17) can be derived easily by differentiating $\mathbf{AA}^{-1}$ with respect to t. Since

$$\frac{d}{dt}\mathbf{A}\mathbf{A}^{-1} = \frac{d\mathbf{A}}{dt}\mathbf{A}^{-1} + \mathbf{A}\frac{d\mathbf{A}^{-1}}{dt}$$

and also

$$\frac{d}{dt}\mathbf{A}\mathbf{A}^{-1} = \frac{d}{dt}\mathbf{I} = \mathbf{0}$$

we obtain

$$\mathbf{A}\frac{d\mathbf{A}^{-1}}{dt} = -\frac{d\mathbf{A}}{dt}\mathbf{A}^{-1}$$

or

$$\mathbf{A}^{-1}\mathbf{A}\frac{d\mathbf{A}^{-1}}{dt} = \frac{d\mathbf{A}^{-1}}{dt} = -\mathbf{A}^{-1}\frac{d\mathbf{A}}{dt}\mathbf{A}^{-1}$$

which is the desired result.

Derivatives of a scalar function with respect to a vector. If $J(\mathbf{x})$ is a scalar function of a vector $\mathbf{x}$, then

$$\frac{\partial J}{\partial \mathbf{x}} = \begin{bmatrix} \dfrac{\partial J}{\partial x_1} \\ \cdot \\ \cdot \\ \cdot \\ \dfrac{\partial J}{\partial x_n} \end{bmatrix}, \qquad \frac{\partial^2 J}{\partial \mathbf{x}^2} = \begin{bmatrix} \dfrac{\partial^2 J}{\partial^2 x_1} & \dfrac{\partial^2 J}{\partial x_1\,\partial x_2} & \cdots & \dfrac{\partial^2 J}{\partial x_1\,\partial x_n} \\ \cdot & \cdot & & \cdot \\ \cdot & \cdot & & \cdot \\ \cdot & \cdot & & \cdot \\ \dfrac{\partial^2 J}{\partial x_n\,\partial x_1} & \dfrac{\partial^2 J}{\partial x_n\,\partial x_2} & \cdots & \dfrac{\partial^2 J}{\partial x_n^2} \end{bmatrix}$$

Also, for a scalar function $V(\mathbf{x}(t))$, we have

$$\frac{d}{dt}V(\mathbf{x}(t)) = \left(\frac{\partial V}{\partial \mathbf{x}}\right)^T \frac{d\mathbf{x}}{dt}$$

Jacobian. If an $m \times 1$ matrix $\mathbf{f}(\mathbf{x})$ is a vector function of an n-vector $\mathbf{x}$ (*note*: an n-vector is meant as an n-dimensional vector), then

$$\frac{\partial \mathbf{f}}{\partial \mathbf{x}} = \begin{bmatrix} \dfrac{\partial f_1}{\partial x_1} & \dfrac{\partial f_2}{\partial x_1} & \cdots & \dfrac{\partial f_m}{\partial x_1} \\ \dfrac{\partial f_1}{\partial x_2} & \dfrac{\partial f_2}{\partial x_2} & \cdots & \dfrac{\partial f_m}{\partial x_2} \\ \cdot & \cdot & & \cdot \\ \cdot & \cdot & & \cdot \\ \cdot & \cdot & & \cdot \\ \dfrac{\partial f_1}{\partial x_n} & \dfrac{\partial f_2}{\partial x_n} & \cdots & \dfrac{\partial f_m}{\partial x_n} \end{bmatrix} \tag{18}$$

Such an $n \times m$ matrix is called a *Jacobian*.

Notice that by using this definition of the Jacobian, we have

$$\frac{\partial}{\partial \mathbf{x}} \mathbf{A}\mathbf{x} = \mathbf{A}^T \tag{19}$$

The fact that Eq. (19) holds true can be easily seen from the following example. If **A** and **x** are given by

$$\mathbf{A} = \begin{bmatrix} a_{11} & a_{12} & a_{13} \\ a_{21} & a_{22} & a_{23} \end{bmatrix}, \qquad \mathbf{x} = \begin{bmatrix} x_1 \\ x_2 \\ x_3 \end{bmatrix}$$

then

$$\mathbf{A}\mathbf{x} = \begin{bmatrix} a_{11} & a_{12} & a_{13} \\ a_{21} & a_{22} & a_{23} \end{bmatrix} \begin{bmatrix} x_1 \\ x_2 \\ x_3 \end{bmatrix} = \begin{bmatrix} a_{11}x_1 + a_{12}x_2 + a_{13}x_3 \\ a_{21}x_1 + a_{22}x_2 + a_{23}x_3 \end{bmatrix} = \begin{bmatrix} f_1 \\ f_2 \end{bmatrix}$$

and

$$\frac{\partial}{\partial \mathbf{x}} \mathbf{A}\mathbf{x} = \begin{bmatrix} \dfrac{\partial f_1}{\partial x_1} & \dfrac{\partial f_2}{\partial x_1} \\ \dfrac{\partial f_1}{\partial x_2} & \dfrac{\partial f_2}{\partial x_2} \\ \dfrac{\partial f_1}{\partial x_3} & \dfrac{\partial f_2}{\partial x_3} \end{bmatrix} = \begin{bmatrix} a_{11} & a_{21} \\ a_{12} & a_{22} \\ a_{13} & a_{23} \end{bmatrix} = \mathbf{A}^T$$

Also, we have the following useful formula. For an $n \times n$ real matrix **A** and a real n-vector **x**,

$$\frac{\partial}{\partial \mathbf{x}} \mathbf{x}^T \mathbf{A}\mathbf{x} = \mathbf{A}\mathbf{x} + \mathbf{A}^T \mathbf{x} \tag{20}$$

In addition, if matrix **A** is a real symmetric matrix, then

$$\frac{\partial}{\partial \mathbf{x}} \mathbf{x}^T \mathbf{A}\mathbf{x} = 2\mathbf{A}\mathbf{x}$$

Note that if **A** is an $n \times n$ Hermitian matrix and **x** is a complex n-vector, then

$$\frac{\partial}{\partial \bar{\mathbf{x}}} \mathbf{x}^* \mathbf{A}\mathbf{x} = \mathbf{A}\mathbf{x} \tag{21}$$

[For derivations of Eqs. (20) and (21), see Prob. A-7.]

For an $n \times m$ real matrix **A**, a real n-vector **x**, and a real m-vector **y**, we have

$$\frac{\partial}{\partial \mathbf{x}} \mathbf{x}^T \mathbf{A}\mathbf{y} = \mathbf{A}\mathbf{y} \tag{22}$$

$$\frac{\partial}{\partial \mathbf{y}} \mathbf{x}^T \mathbf{A} \mathbf{y} = \mathbf{A}^T \mathbf{x} \tag{23}$$

Similarly, for an $n \times m$ complex matrix $\mathbf{A}$, a complex n-vector $\mathbf{x}$, and a complex m-vector $\mathbf{y}$, we have

$$\frac{\partial}{\partial \bar{\mathbf{x}}} \mathbf{x}^* \mathbf{A} \mathbf{y} = \mathbf{A} \mathbf{y} \tag{24}$$

$$\frac{\partial}{\partial \mathbf{y}} \mathbf{x}^* \mathbf{A} \mathbf{y} = \mathbf{A}^T \bar{\mathbf{x}} \tag{25}$$

[For derivations of Eqs. (22) through (25), refer to Prob. A-8.] Note that Eq. (25) is equivalent to the following equation:

$$\overline{\frac{\partial}{\partial \mathbf{y}} \mathbf{x}^* \mathbf{A} \mathbf{y}} = \mathbf{A}^* \mathbf{x}$$

5 VECTORS AND VECTOR ANALYSIS

Linear dependence and independence of vectors. Vectors $\mathbf{x}_1$, $\mathbf{x}_2$, . . . , $\mathbf{x}_n$ are said to be *linearly independent* if the equation

$$c_1 \mathbf{x}_1 + c_2 \mathbf{x}_2 + \cdots + c_n \mathbf{x}_n = \mathbf{0}$$

where $c_1, c_2, \ldots, c_n$ are constants, implies that $c_1 = c_2 = \cdots = c_n = 0$. Conversely, vectors $\mathbf{x}_1$, $\mathbf{x}_2$, . . . , $\mathbf{x}_n$ are said to be *linearly dependent* if and only if $\mathbf{x}_i$ can be expressed as a linear combination of $\mathbf{x}_j$ ($j = 1, 2, \ldots, n; j \neq i$).

It is important to note that if vectors $\mathbf{x}_1$, $\mathbf{x}_2$, . . . , $\mathbf{x}_n$ are linearly independent and vectors $\mathbf{x}_1$, $\mathbf{x}_2$, . . . , $\mathbf{x}_n$, $\mathbf{x}_{n+1}$ are linearly dependent, then $\mathbf{x}_{n+1}$ can be expressed as a unique linear combination of $\mathbf{x}_1$, $\mathbf{x}_2$, . . . , $\mathbf{x}_n$.

Necessary and sufficient conditions for linear independence of vectors. It can be proved that the necessary and sufficient conditions for n-vectors $\mathbf{x}_i$ ($i = 1, 2, \ldots, m$) to be linearly independent are that

1. $m \leq n$.
2. There exists at least one nonzero m-column determinant of the $n \times m$ matrix whose columns consist of $\mathbf{x}_1$, $\mathbf{x}_2$, . . . , $\mathbf{x}_m$.

Hence, for n vectors $\mathbf{x}_1$, $\mathbf{x}_2$, . . . , $\mathbf{x}_n$ the necessary and sufficient condition for linear independence is

$$|\mathbf{A}| \neq 0$$

where $\mathbf{A}$ is the $n \times n$ matrix whose ith column is made up of the components of $\mathbf{x}_i$ ($i = 1, 2, \ldots, n$).

Inner product. Any rule which assigns to each pair of vectors **x** and **y** in a vector space a scalar quantity is called an *inner product* or *scalar product* and is given the symbol $\langle \mathbf{x}, \mathbf{y} \rangle$, provided that the following four axioms are satisfied:

1. $$\langle \mathbf{y}, \mathbf{x} \rangle = \overline{\langle \mathbf{x}, \mathbf{y} \rangle}$$

   where the bar denotes the conjugate of a complex number

2. $$\langle c\mathbf{x}, \mathbf{y} \rangle = \bar{c}\langle \mathbf{x}, \mathbf{y} \rangle = \langle \mathbf{x}, \bar{c}\mathbf{y} \rangle$$

   where c is a complex number

3. $$\langle \mathbf{x} + \mathbf{y}, \mathbf{z} + \mathbf{w} \rangle = \langle \mathbf{x}, \mathbf{z} \rangle + \langle \mathbf{y}, \mathbf{z} \rangle + \langle \mathbf{x}, \mathbf{w} \rangle + \langle \mathbf{y}, \mathbf{w} \rangle$$

4. $$\langle \mathbf{x}, \mathbf{x} \rangle > 0 \qquad \text{for } \mathbf{x} \neq \mathbf{0}$$

In any finite dimensional vector space, there are many different definitions of the inner product, all having the properties required by the definition.

In this book, unless the contrary is stated, we shall adopt the following definition of the inner product: The inner product of a pair of n-vectors **x** and **y** in a vector space V is given by

$$\langle \mathbf{x}, \mathbf{y} \rangle = \bar{x}_1 y_1 + \bar{x}_2 y_2 + \cdot \cdot \cdot + \bar{x}_n y_n = \sum_{i=1}^{n} \bar{x}_i y_i \tag{26}$$

where the summation is a complex number and where the $\bar{x}_i$'s are the complex conjugates of the x_i's. This definition clearly satisfies the four axioms. The inner product can then be expressed as follows:

$$\langle \mathbf{x}, \mathbf{y} \rangle = \mathbf{x}^* \mathbf{y}$$

where $\mathbf{x}^*$ denotes the conjugate transpose of **x**. Also,

$$\langle \mathbf{x}, \mathbf{y} \rangle = \overline{\langle \mathbf{y}, \mathbf{x} \rangle} = \overline{\mathbf{y}^* \mathbf{x}} = \mathbf{y}^T \bar{\mathbf{x}} = \mathbf{x}^* \mathbf{y} \tag{27}$$

The inner product of two n-vectors **x** and **y** with real components is therefore given by

$$\langle \mathbf{x}, \mathbf{y} \rangle = x_1 y_1 + x_2 y_2 + \cdot \cdot \cdot + x_n y_n = \sum_{i=1}^{n} x_i y_i \tag{28}$$

In this case, clearly we have

$$\langle \mathbf{x}, \mathbf{y} \rangle = \mathbf{x}^T \mathbf{y} = \mathbf{y}^T \mathbf{x} \qquad \text{for real vectors } \mathbf{x} \text{ and } \mathbf{y}$$

It is noted that the real or complex vector **x** is said to be *normalized* if $\langle \mathbf{x}, \mathbf{x} \rangle = 1$.

It is also noted that for an n-vector **x**, $\mathbf{x}^*\mathbf{x}$ is a nonnegative scalar but $\mathbf{x}\mathbf{x}^*$ is an $n \times n$ matrix. That is,

$$\begin{aligned} \mathbf{x}^* \mathbf{x} = \langle \mathbf{x}, \mathbf{x} \rangle &= \bar{x}_1 x_1 + \bar{x}_2 x_2 + \cdot \cdot \cdot + \bar{x}_n x_n \\ &= |x_1|^2 + |x_2|^2 + \cdot \cdot \cdot + |x_n|^2 \end{aligned}$$

and

$$\mathbf{xx}^* = \begin{bmatrix} x_1\bar{x}_1 & x_1\bar{x}_2 & \cdots & x_1\bar{x}_n \\ x_2\bar{x}_1 & x_2\bar{x}_2 & \cdots & x_2\bar{x}_n \\ \cdot & \cdot & & \cdot \\ \cdot & \cdot & & \cdot \\ \cdot & \cdot & & \cdot \\ x_n\bar{x}_1 & x_n\bar{x}_2 & \cdots & x_n\bar{x}_n \end{bmatrix}$$

Notice that for an $n \times n$ complex matrix $\mathbf{A}$ and complex n-vectors $\mathbf{x}$ and $\mathbf{y}$, the inner product of $\mathbf{x}$ and $\mathbf{Ay}$ and that of $\mathbf{A}^*\mathbf{x}$ and $\mathbf{y}$ are the same, or

$$\langle \mathbf{x}, \mathbf{Ay} \rangle = \mathbf{x}^*\mathbf{Ay}, \qquad \langle \mathbf{A}^*\mathbf{x}, \mathbf{y} \rangle = \mathbf{x}^*\mathbf{Ay}$$

Similarly, for an $n \times n$ real matrix $\mathbf{A}$ and real n-vectors $\mathbf{x}$ and $\mathbf{y}$, the inner product of $\mathbf{x}$ and $\mathbf{Ay}$ and that of $\mathbf{A}^T\mathbf{x}$ and $\mathbf{y}$ are the same, or

$$\langle \mathbf{x}, \mathbf{Ay} \rangle = \mathbf{x}^T\mathbf{Ay}, \qquad \langle \mathbf{A}^T\mathbf{x}, \mathbf{y} \rangle = \mathbf{x}^T\mathbf{Ay}$$

Unitary transformation. If $\mathbf{A}$ is a unitary matrix (that is, if $\mathbf{A}^{-1} = \mathbf{A}^*$), then the inner product $\langle \mathbf{x}, \mathbf{x} \rangle$ is invariant under the linear transformation $\mathbf{x} = \mathbf{Ay}$, because

$$\langle \mathbf{x}, \mathbf{x} \rangle = \langle \mathbf{Ay}, \mathbf{Ay} \rangle = \langle \mathbf{y}, \mathbf{A}^*\mathbf{Ay} \rangle = \langle \mathbf{y}, \mathbf{A}^{-1}\mathbf{Ay} \rangle = \langle \mathbf{y}, \mathbf{y} \rangle$$

Such a transformation $\mathbf{x} = \mathbf{Ay}$, where $\mathbf{A}$ is a unitary matrix, which transforms $\Sigma_{i=1}^n \bar{x}_i x_i$ into $\Sigma_{i=1}^n \bar{y}_i y_i$, is called a *unitary transformation.*

Orthogonal transformation. If $\mathbf{A}$ is an orthogonal matrix (that is, if $\mathbf{A}^{-1} = \mathbf{A}^T$), then the inner product $\langle \mathbf{x}, \mathbf{x} \rangle$ is invariant under the linear transformation $\mathbf{x} = \mathbf{Ay}$, because

$$\langle \mathbf{x}, \mathbf{x} \rangle = \langle \mathbf{Ay}, \mathbf{Ay} \rangle = \langle \mathbf{y}, \mathbf{A}^T\mathbf{Ay} \rangle = \langle \mathbf{y}, \mathbf{A}^{-1}\mathbf{Ay} \rangle = \langle \mathbf{y}, \mathbf{y} \rangle$$

Such a transformation $\mathbf{x} = \mathbf{Ay}$, which transforms $\Sigma_{i=1}^n x_i^2$ into $\Sigma_{i=1}^n y_i^2$, is called an *orthogonal transformation*.

Norms of a vector. Once we define the inner product, we can use this inner product to define norms of a vector $\mathbf{x}$. The concept of a norm is somewhat similar to that of the absolute value. A norm is a function which assigns to every vector $\mathbf{x}$ in a given vector space a real number denoted by $\|\mathbf{x}\|$ such that

1. $\|\mathbf{x}\| > 0 \qquad \text{for } \mathbf{x} \neq \mathbf{0}$
2. $\|\mathbf{x}\| = 0 \qquad \text{if and only if } \mathbf{x} = \mathbf{0}$
3. $\|k\mathbf{x}\| = |k| \, \|\mathbf{x}\|$

   where k is a scalar and $|k|$ is the absolute value of k
4. $\|\mathbf{x} + \mathbf{y}\| \le \|\mathbf{x}\| + \|\mathbf{y}\| \qquad \text{for all } \mathbf{x} \text{ and } \mathbf{y}$
5. $|\langle \mathbf{x}, \mathbf{y} \rangle| \le \|\mathbf{x}\| \, \|\mathbf{y}\| \qquad \text{(Schwarz inequality)}$

Several different definitions of norms are commonly used in the literature. However, the following definition is widely used. A norm of a vector is defined as the nonnegative square root of $\langle \mathbf{x}, \mathbf{x} \rangle$:

$$\|\mathbf{x}\| = \langle \mathbf{x}, \mathbf{x} \rangle^{1/2} = (\mathbf{x}^*\mathbf{x})^{1/2} = \sqrt{|x_1|^2 + |x_2|^2 + \cdots + |x_n|^2} \tag{29}$$

If $\mathbf{x}$ is a real vector, the quantity $\|\mathbf{x}\|^2$ can be interpreted geometrically as the square of the distance from the origin to the point represented by the vector $\mathbf{x}$. Note that

$$\|\mathbf{x} - \mathbf{y}\| = \langle \mathbf{x} - \mathbf{y}, \mathbf{x} - \mathbf{y} \rangle^{1/2} = \sqrt{(x_1 - y_1)^2 + (x_2 - y_2)^2 + \cdots + (x_n - y_n)^2}$$

The five properties of norms listed earlier may be obvious, except perhaps the last two inequalities. These two inequalities may be proved as follows. From the definitions of the inner product and the norm, we have

$$\begin{aligned}\|\lambda\mathbf{x} + \mathbf{y}\|^2 &= \langle \lambda\mathbf{x} + \mathbf{y}, \lambda\mathbf{x} + \mathbf{y} \rangle = \langle \lambda\mathbf{x}, \lambda\mathbf{x} \rangle + \langle \mathbf{y}, \lambda\mathbf{x} \rangle + \langle \lambda\mathbf{x}, \mathbf{y} \rangle \\ &\quad + \langle \mathbf{y}, \mathbf{y} \rangle \\ &= \bar{\lambda}\lambda\|\mathbf{x}\|^2 + \lambda\langle \mathbf{y}, \mathbf{x} \rangle + \bar{\lambda}\langle \mathbf{x}, \mathbf{y} \rangle + \|\mathbf{y}\|^2 \\ &= \bar{\lambda}(\lambda\|\mathbf{x}\|^2 + \langle \mathbf{x}, \mathbf{y} \rangle) + \lambda\overline{\langle \mathbf{x}, \mathbf{y} \rangle} + \|\mathbf{y}\|^2 \geq 0\end{aligned}$$

If we choose

$$\lambda = -\frac{\langle \mathbf{x}, \mathbf{y} \rangle}{\|\mathbf{x}\|^2} \qquad \text{for } \mathbf{x} \neq \mathbf{0}$$

then

$$\lambda\overline{\langle \mathbf{x}, \mathbf{y} \rangle} + \|\mathbf{y}\|^2 = -\frac{\langle \mathbf{x}, \mathbf{y} \rangle\overline{\langle \mathbf{x}, \mathbf{y} \rangle}}{\|\mathbf{x}\|^2} + \|\mathbf{y}\|^2 \geq 0$$

and

$$\|\mathbf{x}\|^2\|\mathbf{y}\|^2 \geq \langle \mathbf{x}, \mathbf{y} \rangle\overline{\langle \mathbf{x}, \mathbf{y} \rangle} = |\langle \mathbf{x}, \mathbf{y} \rangle|^2 \qquad \text{for } \mathbf{x} \neq \mathbf{0}$$

For $\mathbf{x} = \mathbf{0}$, clearly,

$$\|\mathbf{x}\|^2\|\mathbf{y}\|^2 = |\langle \mathbf{x}, \mathbf{y} \rangle|^2$$

Therefore, we obtain the Schwarz inequality,

$$|\langle \mathbf{x}, \mathbf{y} \rangle| \leq \|\mathbf{x}\|\,\|\mathbf{y}\| \tag{30}$$

By use of the Schwarz inequality we obtain the following inequality:

$$\|\mathbf{x} + \mathbf{y}\| \leq \|\mathbf{x}\| + \|\mathbf{y}\| \tag{31}$$

This can be proved easily, since

$$\begin{aligned}\|\mathbf{x} + \mathbf{y}\|^2 &= \langle \mathbf{x} + \mathbf{y}, \mathbf{x} + \mathbf{y} \rangle \\ &= \langle \mathbf{x}, \mathbf{x} \rangle + \langle \mathbf{x}, \mathbf{y} \rangle + \langle \mathbf{y}, \mathbf{x} \rangle + \langle \mathbf{y}, \mathbf{y} \rangle \\ &= \|\mathbf{x}\|^2 + \langle \mathbf{x}, \mathbf{y} \rangle + \overline{\langle \mathbf{x}, \mathbf{y} \rangle} + \|\mathbf{y}\|^2\end{aligned}$$

$$= \|\mathbf{x}\|^2 + \|\mathbf{y}\|^2 + 2 \operatorname{Re} \langle \mathbf{x}, \mathbf{y} \rangle$$
$$\leq \|\mathbf{x}\|^2 + \|\mathbf{y}\|^2 + 2 |\langle \mathbf{x}, \mathbf{y} \rangle|$$
$$\leq \|\mathbf{x}\|^2 + \|\mathbf{y}\|^2 + 2 \|\mathbf{x}\| \|\mathbf{y}\|$$
$$= (\|\mathbf{x}\| + \|\mathbf{y}\|)^2$$

Equations (26) through (31) are useful in modern control theory.

As stated earlier, there are different definitions of norms used in the literature. Three such definitions of norms follow.

1. A norm $\|\mathbf{x}\|$ may be defined as follows:

$$\|\mathbf{x}\| = [(\mathbf{Tx})^*(\mathbf{Tx})]^{1/2} = (\mathbf{x}^*\mathbf{T}^*\mathbf{Tx})^{1/2} = (\mathbf{x}^*\mathbf{Qx})^{1/2}$$
$$= \left[\sum_{i=1}^{n} \sum_{j=1}^{n} q_{ij}\bar{x}_i x_j \right]^{1/2} \geq 0$$

The matrix $\mathbf{Q} = \mathbf{T}^*\mathbf{T}$ is Hermitian, since $\mathbf{Q}^* = \mathbf{T}^*\mathbf{T} = \mathbf{Q}$. The norm $\|\mathbf{x}\| = (\mathbf{x}^*\mathbf{Qx})^{1/2}$ is a generalized form of $(\mathbf{x}^*\mathbf{x})^{1/2}$ which can be written as $(\mathbf{x}^*\mathbf{Ix})^{1/2}$.

2. A norm may be defined as the sum of the magnitudes of all the components x_i:

$$\|\mathbf{x}\| = \sum_{i=1}^{n} |x_i|$$

3. A norm may be defined as the maximum of the magnitudes of all the components x_i:

$$\|\mathbf{x}\| = \max_i \{|x_i|\}$$

It can be shown that the various norms just defined are equivalent. Among these definitions of norms, norm $(\mathbf{x}^*\mathbf{x})^{1/2}$ is most commonly used in explicit calculations.

Norms of a matrix. The concept of norms of a vector can be extended to matrices. There are several different definitions of norms of a matrix. Some of them follow.

1. A norm $\|\mathbf{A}\|$ of an $n \times n$ matrix $\mathbf{A}$ may be defined by

$$\|\mathbf{A}\| = \min k$$

such that

$$\|\mathbf{Ax}\| \leq k\|\mathbf{x}\|$$

For the norm $(\mathbf{x}^*\mathbf{x})^{1/2}$, this definition is equivalent to

$$\|\mathbf{A}\|^2 = \max_{\mathbf{x}} \{\mathbf{x}^*\mathbf{A}^*\mathbf{Ax};\ \mathbf{x}^*\mathbf{x} = 1\}$$

which means that $\|\mathbf{A}\|^2$ is the maximum of the "absolute value" of the vector $\mathbf{Ax}$ when $\mathbf{x}^*\mathbf{x} = 1$.

2. A norm of an $n \times n$ matrix $\mathbf{A}$ may be defined by

$$\|\mathbf{A}\| = \sum_{i=1}^{n} \sum_{j=1}^{n} |a_{ij}|$$

where $|a_{ij}|$ is the absolute value of a_{ij}.

3. A norm may be defined by

$$\|\mathbf{A}\| = \left(\sum_{i=1}^{n} \sum_{j=1}^{n} |a_{ij}|^2 \right)^{1/2}$$

4. Another definition of a norm is given by

$$\|\mathbf{A}\| = \max_i \left(\sum_{j=1}^{n} |a_{ij}| \right)$$

Note that all definitions of norms of an $n \times n$ matrix $\mathbf{A}$ have the following properties:

1. $$\|\mathbf{A}\| = \|\mathbf{A}^*\| \qquad \text{or} \qquad \|\mathbf{A}\| = \|\mathbf{A}^T\|$$
2. $$\|\mathbf{A} + \mathbf{B}\| \le \|\mathbf{A}\| + \|\mathbf{B}\|$$
3. $$\|\mathbf{AB}\| \le \|\mathbf{A}\| \, \|\mathbf{B}\|$$
4. $$\|\mathbf{Ax}\| \le \|\mathbf{A}\| \, \|\mathbf{x}\|$$

Orthogonality of vectors. If the inner product of two vectors $\mathbf{x}$ and $\mathbf{y}$ is zero, or $\langle \mathbf{x}, \mathbf{y} \rangle = 0$, then vectors $\mathbf{x}$ and $\mathbf{y}$ are said to be *orthogonal to each other*. For example, vectors

$$\mathbf{x}_1 = \begin{bmatrix} 1 \\ 1 \\ 0 \end{bmatrix}, \qquad \mathbf{x}_2 = \begin{bmatrix} 0 \\ 0 \\ 1 \end{bmatrix}, \qquad \mathbf{x}_3 = \begin{bmatrix} 1 \\ -1 \\ 0 \end{bmatrix}$$

are othogonal in pairs and thus form an orthogonal set.

In an n-dimensional vector space, vectors $\mathbf{x}_1, \mathbf{x}_2, \ldots, \mathbf{x}_n$ defined by

$$\mathbf{x}_1 = \begin{bmatrix} 1 \\ 0 \\ \cdot \\ \cdot \\ \cdot \\ 0 \end{bmatrix}, \qquad \mathbf{x}_2 = \begin{bmatrix} 0 \\ 1 \\ \cdot \\ \cdot \\ \cdot \\ 0 \end{bmatrix}, \quad \ldots, \qquad \mathbf{x}_n = \begin{bmatrix} 0 \\ 0 \\ \cdot \\ \cdot \\ \cdot \\ 1 \end{bmatrix}$$

satisfy the conditions $\langle \mathbf{x}_i, \mathbf{x}_j \rangle = \delta_{ij}$, or

$$\langle \mathbf{x}_i, \mathbf{x}_i \rangle = 1$$

$$\langle \mathbf{x}_i, \mathbf{x}_j \rangle = 0 \qquad i \neq j$$

where $i, j = 1, 2, \ldots, n$. Such a set of vectors is said to be *orthonormal*, since the vectors are orthogonal to each other and each vector is normalized.

A nonzero vector $\mathbf{x}$ can be normalized by dividing $\mathbf{x}$ by $\|\mathbf{x}\|$. The normalized vector $\mathbf{x}/\|\mathbf{x}\|$ is a unit vector. Unit vectors $\mathbf{x}_1, \mathbf{x}_2, \ldots, \mathbf{x}_n$ form an orthonormal set if they are orthogonal in pairs.

Consider a unitary matrix $\mathbf{A}$. By partitioning $\mathbf{A}$ into column vectors $\mathbf{A}_1, \mathbf{A}_2, \ldots, \mathbf{A}_n$, we have

$$\mathbf{A}^*\mathbf{A} = \begin{bmatrix} \mathbf{A}_1^* \\ \hdashline \mathbf{A}_2^* \\ \hdashline \cdot \\ \cdot \\ \cdot \\ \hdashline \mathbf{A}_n^* \end{bmatrix} [\mathbf{A}_1 \vdots \mathbf{A}_2 \vdots \cdots \vdots \mathbf{A}_n]$$

$$= \begin{bmatrix} \mathbf{A}_1^*\mathbf{A}_1 & \mathbf{A}_1^*\mathbf{A}_2 & \cdots & \mathbf{A}_1^*\mathbf{A}_n \\ \mathbf{A}_2^*\mathbf{A}_1 & \mathbf{A}_2^*\mathbf{A}_2 & \cdots & \mathbf{A}_2^*\mathbf{A}_n \\ \cdot & \cdot & & \cdot \\ \cdot & \cdot & & \cdot \\ \cdot & \cdot & & \cdot \\ \mathbf{A}_n^*\mathbf{A}_1 & \mathbf{A}_n^*\mathbf{A}_2 & \cdots & \mathbf{A}_n^*\mathbf{A}_n \end{bmatrix}$$

$$= \begin{bmatrix} 1 & 0 & \cdots & 0 \\ 0 & 1 & \cdots & 0 \\ \cdot & \cdot & & \cdot \\ \cdot & \cdot & & \cdot \\ \cdot & \cdot & & \cdot \\ 0 & 0 & \cdots & 1 \end{bmatrix}$$

it follows that

$$\mathbf{A}_i^*\mathbf{A}_i = \langle \mathbf{A}_i, \mathbf{A}_i \rangle = 1$$
$$\mathbf{A}_i^*\mathbf{A}_j = \langle \mathbf{A}_i, \mathbf{A}_j \rangle = 0 \qquad i \neq j$$

Thus we see that the column vectors (or row vectors) of a unitary matrix $\mathbf{A}$ are orthonormal. The same is true for orthogonal matrices, since they are unitary.

6 EIGENVALUES, EIGENVECTORS, AND SIMILARITY TRANSFORMATION

In this section we shall first review important properties of the rank of a matrix and then give definitions of eigenvalues and eigenvectors. Finally, we shall discuss Jordan canonical forms, similarity transformation, and similar matrices.

Rank of a matrix. We shall list important properties of the rank of a matrix.

1. The rank of a matrix is invariant under the interchange of two rows (or columns), or the addition of a scalar multiple of a row (or column) to another row (or column), or the multiplication of any row (or column) by a nonzero scalar.
2. For an $n \times m$ matrix $\mathbf{A}$
$$\text{rank } \mathbf{A} \leq \min(n, m)$$
3. For an $n \times n$ matrix $\mathbf{A}$, a necessary and sufficient condition for rank $\mathbf{A} = n$ is that $|\mathbf{A}| \neq 0$.
4. For an $n \times m$ matrix $\mathbf{A}$,
$$\text{rank } \mathbf{A}^* = \text{rank } \mathbf{A} \qquad \text{or} \qquad \text{rank } \mathbf{A}^T = \text{rank } \mathbf{A}$$
5. The rank of a product of two matrices $\mathbf{AB}$ cannot exceed the rank of $\mathbf{A}$ or the rank of $\mathbf{B}$; that is,
$$\text{rank } \mathbf{AB} \leq \min(\text{rank } \mathbf{A}, \text{rank } \mathbf{B})$$
Hence, if $\mathbf{A}$ is an $n \times 1$ matrix and $\mathbf{B}$ is a $1 \times m$ matrix, then rank $\mathbf{AB} = 1$ unless $\mathbf{AB} = \mathbf{0}$. If a matrix has rank 1, then this matrix can be expressed as a product of a column vector and a row vector.
6. For an $n \times n$ matrix $\mathbf{A}$ (where $|\mathbf{A}| \neq 0$) and an $n \times m$ matrix $\mathbf{B}$,
$$\text{rank } \mathbf{AB} = \text{rank } \mathbf{B}$$
Similarly, for an $m \times m$ matrix $\mathbf{A}$ (where $|\mathbf{A}| \neq 0$) and an $n \times m$ matrix $\mathbf{B}$,
$$\text{rank } \mathbf{BA} = \text{rank } \mathbf{B}$$

Eigenvalues of a square matrix. For an $n \times n$ matrix $\mathbf{A}$, the determinant

$$|\lambda \mathbf{I} - \mathbf{A}|$$

is called the *characteristic polynomial* of $\mathbf{A}$. It is an nth-degree polynomial in λ. The characteristic equation is given by

$$|\lambda \mathbf{I} - \mathbf{A}| = 0$$

If the determinant $|\lambda \mathbf{I} - \mathbf{A}|$ is expanded, the characteristic equation becomes

$$|\lambda \mathbf{I} - \mathbf{A}| = \begin{vmatrix} \lambda - a_{11} & -a_{12} & \cdots & -a_{1n} \\ -a_{21} & \lambda - a_{22} & \cdots & -a_{2n} \\ \cdot & \cdot & & \cdot \\ \cdot & \cdot & & \cdot \\ \cdot & \cdot & & \cdot \\ -a_{n1} & -a_{n2} & \cdots & \lambda - a_{nn} \end{vmatrix}$$

$$= \lambda^n + a_1 \lambda^{n-1} + \cdots + a_{n-1}\lambda + a_n = 0$$

The n roots of the characteristic equation are called the *eigenvalues* of **A.** They are also called the *characteristic roots*.

It is noted that an $n \times n$ real matrix **A** does not necessarily possess real eigenvalues. However, for an $n \times n$ real matrix **A,** the characteristic equation $|\lambda \mathbf{I} - \mathbf{A}| = 0$ is a polynomial with real coefficients, and therefore any complex eigenvalues must occur in conjugate pairs; that is, if $\alpha + j\beta$ is an eigenvalue of **A,** then $\alpha - j\beta$ is also an eigenvalue of **A.**

There is an important relationship between the eigenvalues of an $n \times n$ matrix **A** and those of $\mathbf{A}^{-1}$. If we assume the eigenvalues of **A** to be λ_i and those of $\mathbf{A}^{-1}$ to be μ_i, then

$$\mu_i = \lambda_i^{-1} \qquad i = 1, 2, \ldots, n$$

That is, if λ_i is an eigenvalue of **A,** then λ_i^{-1} is an eigenvalue of $\mathbf{A}^{-1}$. To prove this, notice that the characteristic equation for matrix **A** can be written as

$$|\lambda \mathbf{I} - \mathbf{A}| = |\lambda \mathbf{A}^{-1} - \mathbf{I}|\,|\mathbf{A}| = |\lambda|\,|\mathbf{A}^{-1} - \lambda^{-1}\mathbf{I}|\,|\mathbf{A}| = 0$$

or

$$|\lambda^{-1}\mathbf{I} - \mathbf{A}^{-1}| = 0$$

By assumption, the characteristic equation for the inverse matrix $\mathbf{A}^{-1}$ is

$$|\mu \mathbf{I} - \mathbf{A}^{-1}| = 0$$

By comparing the last two equations, we see that

$$\mu = \lambda^{-1}$$

Hence, if λ is an eigenvalue of **A,** then $\mu = \lambda^{-1}$ is an eigenvalue of $\mathbf{A}^{-1}$.

Finally, note that it is possible to prove that, for two square matrices **A** and **B,**

$$|\lambda \mathbf{I} - \mathbf{AB}| = |\lambda \mathbf{I} - \mathbf{BA}|$$

(For the proof, see Prob. A-9.)

Eigenvectors of an $n \times n$ matrix. Any nonzero vector $\mathbf{x}_i$ such that

$$\mathbf{A}\mathbf{x}_i = \lambda_i \mathbf{x}_i$$

is said to be an *eigenvector* associated with an eigenvalue λ_i of **A,** where **A** is an $n \times n$ matrix. Since the components of $\mathbf{x}_i$ are determined from n linear homogeneous algebraic equations within a constant factor, if $\mathbf{x}_i$ is an eigenvector, then for any scalar $\alpha \neq 0$, $\alpha\mathbf{x}_i$ is also an eigenvector. The eigenvector is said to be a *normalized* eigenvector if its length or absolute value is unity.

Similar matrices. The $n \times n$ matrices **A** and **B** are said to be *similar* if a nonsingular matrix **P** exists such that

$$\mathbf{P}^{-1}\mathbf{A}\mathbf{P} = \mathbf{B}$$

The matrix $\mathbf{B}$ is said to be obtained from $\mathbf{A}$ by a *similarity transformation*, in which $\mathbf{P}$ is the transformation matrix. Notice that $\mathbf{A}$ can be obtained from $\mathbf{B}$ by a similarity transformation with a transformation matrix $\mathbf{P}^{-1}$, since

$$\mathbf{A} = \mathbf{PBP}^{-1} = (\mathbf{P}^{-1})^{-1}\mathbf{B}(\mathbf{P}^{-1})$$

Diagonalization of matrices. If an $n \times n$ matrix $\mathbf{A}$ has n distinct eigenvalues, then there are n linearly independent eigenvectors. If matrix $\mathbf{A}$ has a multiple eigenvalue of multiplicity k, then there are at least one and not more than k linearly independent eigenvectors associated with this eigenvalue.

If an $n \times n$ matrix has n linearly independent eigenvectors, it can be diagonalized by a similarity transformation. However, a matrix that does not have a complete set of n linearly independent eigenvectors cannot be diagonalized. Such a matrix can be transformed into a Jordan canonical form.

Jordan canonical form. A $k \times k$ matrix $\mathbf{J}$ is said to be in the Jordan canonical form if

$$\mathbf{J} = \begin{bmatrix} \mathbf{J}_{p_1} & & & & \mathbf{0} \\ & \mathbf{J}_{p_2} & & & \\ & & \cdot & & \\ & & & \cdot & \\ & & & & \cdot \\ \mathbf{0} & & & & \mathbf{J}_{p_s} \end{bmatrix}$$

where the $\mathbf{J}_{p_i}$'s are $p_i \times p_i$ matrices of the form

$$\mathbf{J}_{p_i} = \begin{bmatrix} \lambda & 1 & 0 & \cdots & 0 & 0 \\ 0 & \lambda & 1 & \cdots & 0 & 0 \\ \cdot & \cdot & \cdot & & \cdot & \cdot \\ \cdot & \cdot & \cdot & & \cdot & \cdot \\ \cdot & \cdot & \cdot & & \cdot & \cdot \\ 0 & 0 & 0 & \cdots & \lambda & 1 \\ 0 & 0 & 0 & \cdots & 0 & \lambda \end{bmatrix}$$

The matrices $\mathbf{J}_{p_i}$ are called p_ith order Jordan blocks. Note that the λ in $\mathbf{J}_{p_i}$ and that in $\mathbf{J}_{p_j}$ may or may not be the same, and that

$$p_1 + p_2 + \cdots + p_s = k$$

For example, in a 7×7 matrix $\mathbf{J}$, if $p_1 = 3$, $p_2 = 2$, $p_3 = 1$, $p_4 = 1$, and the eigenvalues of $\mathbf{J}$ are $\lambda_1, \lambda_1, \lambda_1, \lambda_1, \lambda_1, \lambda_6, \lambda_7$, then the Jordan canonical form may be given by

$$\mathbf{J} = \begin{bmatrix} \mathbf{J}_3(\lambda_1) & & & \mathbf{0} \\ & \mathbf{J}_2(\lambda_1) & & \\ & & \mathbf{J}_1(\lambda_6) & \\ \mathbf{0} & & & \mathbf{J}_1(\lambda_7) \end{bmatrix} = \left[\begin{array}{ccc:cc:c:c} \lambda_1 & 1 & 0 & & & & 0 \\ 0 & \lambda_1 & 1 & & & & \\ 0 & 0 & \lambda_1 & & & & \\ \hdashline & & & \lambda_1 & 1 & & \\ & & & 0 & \lambda_1 & & \\ \hdashline & & & & & \lambda_6 & \\ \hdashline 0 & & & & & & \lambda_7 \end{array}\right]$$

Notice that a diagonal matrix is a special case of the Jordan canonical form.

Jordan canonical forms have the properties that the elements on the main diagonal of the matrix are the eigenvalues of **A** and that the elements immediately above (or below) the main diagonal are either 1 or 0 and all other elements are zeros.

The determination of the exact form of the Jordan block may not be simple. To illustrate some possible structures, consider a 3×3 matrix having a triple eigenvalue of λ_1. Then any one of the following Jordan canonical forms is possible:

$$\begin{bmatrix} \lambda_1 & 1 & 0 \\ 0 & \lambda_1 & 1 \\ 0 & 0 & \lambda_1 \end{bmatrix}, \quad \left[\begin{array}{cc:c} \lambda_1 & 1 & 0 \\ 0 & \lambda_1 & 0 \\ \hdashline 0 & 0 & \lambda_1 \end{array}\right], \quad \left[\begin{array}{c:c:c} \lambda_1 & 0 & 0 \\ \hdashline 0 & \lambda_1 & 0 \\ \hdashline 0 & 0 & \lambda_1 \end{array}\right]$$

Each of the three preceding matrices has the same characteristic equation $(\lambda - \lambda_1)^3 = 0$. The first one corresponds to the case where there exists only one linearly independent eigenvector, since by denoting the first matrix by **A** and solving the following equation for **x**:

$$(\mathbf{A} - \lambda_1 \mathbf{I})\mathbf{x} = \mathbf{0}$$

we obtain only one eigenvector:

$$\mathbf{x} = \begin{bmatrix} a \\ 0 \\ 0 \end{bmatrix} \qquad a = \text{nonzero constant}$$

The second and third of these matrices have, respectively, two and three linearly independent eigenvectors. (Notice that only the diagonal matrix has three linearly independent eigenvectors.)

As we have seen, if a $k \times k$ matrix **A** has a k-multiple eigenvalue, then the following can be shown:

1. If the rank of $\lambda \mathbf{I} - \mathbf{A}$ is $k - s$ (where $1 \le s \le k$), then there exist s linearly independent eigenvectors associated with λ.

2. There are s Jordan blocks corresponding to the s eigenvectors.
3. The sum of the orders p_i of the Jordan blocks equals the multiplicity k.

Therefore, as demonstrated in the preceding three 3×3 matrices, even if the multiplicity of eigenvalue is the same, the number of Jordan blocks and their orders may be different depending on the structure of matrix **A**.

Similarity transformation when an $n \times n$ matrix has distinct eigenvalues. If n eigenvalues of **A** are distinct, there exists one eigenvector associated with each eigenvalue λ_i. It can be proved that such n eigenvectors $\mathbf{x}_1, \mathbf{x}_2, \ldots, \mathbf{x}_n$ are linearly independent.

Let us define an $n \times n$ matrix **P** such that

$$\mathbf{P} = [\mathbf{P}_1 \vdots \mathbf{P}_2 \vdots \cdots \vdots \mathbf{P}_n] = [\mathbf{x}_1 \vdots \mathbf{x}_2 \vdots \cdots \vdots \mathbf{x}_n]$$

where column vector $\mathbf{P}_i$ is equal to column vector $\mathbf{x}_i$, or

$$\mathbf{P}_i = \mathbf{x}_i \qquad i = 1, 2, \ldots, n$$

Matrix **P** defined in this way is nonsingular, and $\mathbf{P}^{-1}$ exists. Noting that eigenvectors $\mathbf{x}_1, \mathbf{x}_2, \ldots, \mathbf{x}_n$ satisfy the equations

$$\begin{aligned} \mathbf{A}\mathbf{x}_1 &= \lambda_1 \mathbf{x}_1 \\ \mathbf{A}\mathbf{x}_2 &= \lambda_2 \mathbf{x}_2 \\ &\vdots \\ \mathbf{A}\mathbf{x}_n &= \lambda_n \mathbf{x}_n \end{aligned}$$

we may combine these n equations into one, as follows:

$$\mathbf{A}[\mathbf{x}_1 \vdots \mathbf{x}_2 \vdots \cdots \vdots \mathbf{x}_n] = [\mathbf{x}_1 \vdots \mathbf{x}_2 \vdots \cdots \vdots \mathbf{x}_n] \begin{bmatrix} \lambda_1 & & & 0 \\ & \lambda_2 & & \\ & & \ddots & \\ 0 & & & \lambda_n \end{bmatrix}$$

or, in terms of matrix **P**,

$$\mathbf{A}\mathbf{P} = \mathbf{P} \begin{bmatrix} \lambda_1 & & & 0 \\ & \lambda_2 & & \\ & & \ddots & \\ 0 & & & \lambda_n \end{bmatrix}$$

By premultiplying this last equation by $\mathbf{P}^{-1}$, we obtain

$$\mathbf{P}^{-1}\mathbf{A}\mathbf{P} = \begin{bmatrix} \lambda_1 & & & & 0 \\ & \lambda_2 & & & \\ & & \cdot & & \\ & & & \cdot & \\ & & & & \cdot \\ 0 & & & & \lambda_n \end{bmatrix} = \text{diag}\,(\lambda_1, \lambda_2, \ldots, \lambda_n)$$

Thus matrix $\mathbf{A}$ is transformed into a diagonal matrix by a similarity transformation.

The process that transforms matrix $\mathbf{A}$ into a diagonal matrix is called the *diagonalization* of matrix $\mathbf{A}$.

As noted earlier, a scalar multiple of eigenvector $\mathbf{x}_i$ is also an eigenvector, since $\alpha\mathbf{x}_i$ satisfies the following equation:

$$\mathbf{A}(\alpha\mathbf{x}_i) = \lambda_i(\alpha\mathbf{x}_i)$$

Consequently, we may choose an α such that the transformation matrix $\mathbf{P}$ becomes as simple as possible.

To summarize, if the eigenvalues of an $n \times n$ matrix $\mathbf{A}$ are distinct, then there are exactly n eigenvectors and they are linearly independent. A transformation matrix $\mathbf{P}$ that transforms $\mathbf{A}$ into a diagonal matrix can be constructed from such n linearly independent eigenvectors.

Similarity transformation when an $n \times n$ matrix has multiple eigenvalues. Let us assume that an $n \times n$ matrix $\mathbf{A}$ involves a k-multiple eigenvalue λ_1 and other eigenvalues $\lambda_{k+1}, \lambda_{k+2}, \ldots, \lambda_n$ that are all distinct and different from λ_1. That is, the eigenvalues of $\mathbf{A}$ are

$$\lambda_1, \lambda_1, \ldots, \lambda_1, \lambda_{k+1}, \lambda_{k+2}, \ldots, \lambda_n$$

We shall first consider the case where the rank of $\lambda_1\mathbf{I} - \mathbf{A}$ is $n - 1$. For such a case there exists only one Jordan block for the multiple eigenvalue λ_1 and there is only one eigenvector associated with this multiple eigenvalue. The order of the Jordan block is k, which is the same as the order of multiplicity of the eigenvalue λ_1.

Note that when an $n \times n$ matrix $\mathbf{A}$ does not possess n linearly independent eigenvectors, it cannot be diagonalized, but can be reduced to a Jordan canonical form.

In the present case, only one linearly independent eigenvector exists for λ_1. We shall now investigate whether it is possible to find $k - 1$ vectors which are somehow associated with this eigenvalue and which are linearly independent of the eigenvectors. Without proof, we shall show that this is possible. First, note that the eigenvector $\mathbf{x}_1$ is a vector that satisfies the equation

$$(\mathbf{A} - \lambda_1\mathbf{I})\mathbf{x}_1 = \mathbf{0}$$

so that $\mathbf{x}_1$ is annihilated by $\mathbf{A} - \lambda_1\mathbf{I}$. Since we do not have enough vectors that are annihilated by $\mathbf{A} - \lambda_1\mathbf{I}$, we seek vectors that are annihilated by $(\mathbf{A} - \lambda_1\mathbf{I})^2$, $(\mathbf{A} -$

$\lambda_1\mathbf{I})^3$, and so on, until we obtain $k-1$ vectors. The $k-1$ vectors determined in this way are called *generalized eigenvectors*.

Let us define the desired $k-1$ generalized eigenvectors as $\mathbf{x}_2, \mathbf{x}_3, \ldots, \mathbf{x}_k$. Then these $k-1$ generalized eigenvectors can be determined from the equations

$$\begin{aligned}
(\mathbf{A}-\lambda_1\mathbf{I})\mathbf{x}_1 &= \mathbf{0} \\
(\mathbf{A}-\lambda_1\mathbf{I})^2\mathbf{x}_2 &= \mathbf{0} \\
&\vdots \\
(\mathbf{A}-\lambda_1\mathbf{I})^k\mathbf{x}_k &= \mathbf{0}
\end{aligned} \tag{32}$$

which can be rewritten as

$$\begin{aligned}
(\mathbf{A}-\lambda_1\mathbf{I})\mathbf{x}_1 &= \mathbf{0} \\
(\mathbf{A}-\lambda_1\mathbf{I})\mathbf{x}_2 &= \mathbf{x}_1 \\
&\vdots \\
(\mathbf{A}-\lambda_1\mathbf{I})\mathbf{x}_k &= \mathbf{x}_{k-1}
\end{aligned}$$

Notice that

$$(\mathbf{A}-\lambda_1\mathbf{I})^{k-1}\mathbf{x}_k = (\mathbf{A}-\lambda_1\mathbf{I})^{k-2}\mathbf{x}_{k-1} = \cdots = (\mathbf{A}-\lambda_1\mathbf{I})\mathbf{x}_2 = \mathbf{x}_1$$

or

$$(\mathbf{A}-\lambda_1\mathbf{I})^{k-1}\mathbf{x}_k = \mathbf{x}_1 \tag{33}$$

The eigenvector $\mathbf{x}_1$ and the $k-1$ generalized eigenvectors $\mathbf{x}_2, \mathbf{x}_3, \ldots, \mathbf{x}_k$ determined in this way form a set of k linearly independent vectors.

A proper way to determine the generalized eigenvectors is to start with $\mathbf{x}_k$. That is, we first determine the $\mathbf{x}_k$ that will satisfy Eq. (32) and at the same time will yield a nonzero vector $(\mathbf{A}-\lambda_1\mathbf{I})^{k-1}\mathbf{x}_k$. Any such nonzero vector can be considered as a possible eigenvector $\mathbf{x}_1$. Therefore, in order to find eigenvector $\mathbf{x}_1$ we apply a row reduction process to $(\mathbf{A}-\lambda_1\mathbf{I})^k$ and find k linearly independent vectors satisfying Eq. (32). Then these vectors are tested to find one that yields a nonzero vector on the right-hand side of Eq. (33). Note that if we start with $\mathbf{x}_1$, then we must make arbitrary choices at each step along the way to determine $\mathbf{x}_2, \mathbf{x}_3, \ldots, \mathbf{x}_k$. This is time-consuming and inconvenient. For this reason, this approach is not recommended.

To summarize what we have discussed so far, the eigenvector $\mathbf{x}_1$ and the generalized eigenvectors $\mathbf{x}_2, \mathbf{x}_3, \ldots, \mathbf{x}_k$ satisfy the following equations:

$$\mathbf{A}\mathbf{x}_1 = \lambda_1 \mathbf{x}_1$$
$$\mathbf{A}\mathbf{x}_2 = \mathbf{x}_1 + \lambda_1 \mathbf{x}_2$$
$$\vdots$$
$$\mathbf{A}\mathbf{x}_k = \mathbf{x}_{k-1} + \lambda_1 \mathbf{x}_k$$

The eigenvectors $\mathbf{x}_{k+1}, \mathbf{x}_{k+2}, \ldots, \mathbf{x}_n$ associated with distinct eigenvalues λ_{k+1}, $\lambda_{k+2}, \ldots, \lambda_n$, respectively, can be determined from

$$\mathbf{A}\mathbf{x}_{k+1} = \lambda_{k+1} \mathbf{x}_{k+1}$$
$$\mathbf{A}\mathbf{x}_{k+2} = \lambda_{k+2} \mathbf{x}_{k+2}$$
$$\vdots$$
$$\mathbf{A}\mathbf{x}_n = \lambda_n \mathbf{x}_n$$

Now define

$$\mathbf{S} = [\mathbf{S}_1 \vdots \mathbf{S}_2 \vdots \cdots \vdots \mathbf{S}_n] = [\mathbf{x}_1 \vdots \mathbf{x}_2 \vdots \cdots \vdots \mathbf{x}_n]$$

where the n column vectors of $\mathbf{S}$ are linearly independent. Thus, matrix $\mathbf{S}$ is nonsingular. Then, combining the preceding eigenvector equations and generalized eigenvector equations into one, we obtain

$$\mathbf{A}[\mathbf{x}_1 \vdots \mathbf{x}_2 \vdots \cdots \vdots \mathbf{x}_k \vdots \mathbf{x}_{k+1} \vdots \cdots \vdots \mathbf{x}_n]$$

$$= [\mathbf{x}_1 \vdots \mathbf{x}_2 \vdots \cdots \vdots \mathbf{x}_k \vdots \mathbf{x}_{k+1} \vdots \cdots \vdots \mathbf{x}_n] \left[\begin{array}{ccccc|ccc} \lambda_1 & 1 & & & 0 & & & 0 \\ & \lambda_1 & 1 & & & & & \\ & & \ddots & \ddots & & & & \\ & & & \ddots & 1 & & & \\ 0 & & & & \lambda_1 & 0 & & \\ \hline & & & & 0 & \lambda_{k+1} & & 0 \\ & & & & & & \ddots & \\ 0 & & & & & 0 & & \lambda_n \end{array}\right]$$

Hence

$$\mathbf{AS} = \mathbf{S}\left[\begin{array}{c|cccc} \mathbf{J}_k(\lambda_1) & & & & 0 \\ \hline & \lambda_{k+1} & & & \\ & & \cdot & & \\ & & & \cdot & \\ & & & & \cdot \\ 0 & & & & \lambda_n \end{array}\right]$$

By premultiplying this last equation by $\mathbf{S}^{-1}$, we obtain

$$\mathbf{S}^{-1}\mathbf{AS} = \left[\begin{array}{c|cccc} \mathbf{J}_k(\lambda_1) & & & & 0 \\ \hline & \lambda_{k+1} & & & \\ & & \cdot & & \\ & & & \cdot & \\ & & & & \cdot \\ 0 & & & & \lambda_n \end{array}\right]$$

In the preceding discussion we considered the case where the rank of $\lambda_1\mathbf{I} - \mathbf{A}$ was $n - 1$. Next, we shall consider the case where the rank of $\lambda_1\mathbf{I} - \mathbf{A}$ is $n - s$ (where $2 \le s \le n$). Since we assumed that matrix $\mathbf{A}$ involves the k-multiple eigenvalue λ_1 and other eigenvalues $\lambda_{k+1}, \lambda_{k+2}, \ldots, \lambda_n$ that are all distinct and different from λ_1, we have s linearly independent eigenvectors associated with eigenvalue λ_1. Hence, there are s Jordan blocks corresponding to eigenvalue λ_1.

For notational convenience, let us define the s linearly independent eigenvectors associated with eigenvalue λ_1 as $\mathbf{v}_{11}, \mathbf{v}_{21}, \ldots, \mathbf{v}_{s1}$. We shall define the generalized eigenvectors associated with $\mathbf{v}_{i1}$ as $\mathbf{v}_{i2}, \mathbf{v}_{i3}, \ldots, \mathbf{v}_{ip_i}$, where $i = 1, 2, \ldots, s$. Then, there are altogether k such vectors (eigenvectors and generalized eigenvectors), which are

$$\mathbf{v}_{11}, \mathbf{v}_{12}, \ldots, \mathbf{v}_{1p_1}, \mathbf{v}_{21}, \mathbf{v}_{22}, \ldots, \mathbf{v}_{2p_2}, \ldots, \mathbf{v}_{s1}, \mathbf{v}_{s2}, \ldots, \mathbf{v}_{sp_s}$$

The generalized eigenvectors are determined from

$$\begin{array}{lcl}
(\mathbf{A} - \lambda_1\mathbf{I})\mathbf{v}_{11} = \mathbf{0}, & \cdots & (\mathbf{A} - \lambda_1\mathbf{I})\mathbf{v}_{s1} = \mathbf{0} \\
(\mathbf{A} - \lambda_1\mathbf{I})\mathbf{v}_{12} = \mathbf{v}_{11}, & \cdots & (\mathbf{A} - \lambda_1\mathbf{I})\mathbf{v}_{s2} = \mathbf{v}_{s1} \\
\vdots & & \vdots \\
(\mathbf{A} - \lambda_1\mathbf{I})\mathbf{v}_{1p_1} = \mathbf{v}_{1\,p_1-1}, & \cdots & (\mathbf{A} - \lambda_1\mathbf{I})\mathbf{v}_{sp_s} = \mathbf{v}_{s\,p_s-1}
\end{array}$$

where the s eigenvectors $\mathbf{v}_{11}, \mathbf{v}_{21}, \ldots, \mathbf{v}_{s1}$ are linearly independent and

$$p_1 + p_2 + \cdots + p_s = k$$

Note that $p_1, p_2, \ldots, p_s$ represent the order of each of the s Jordan blocks. (For the determination of the generalized eigenvectors, we follow the method discussed earlier. For an example showing the details of such a determination, see Prob. A-11.)

Let us define an $n \times k$ matrix consisting of $\mathbf{v}_{11}, \mathbf{v}_{12}, \ldots, \mathbf{v}_{sp_s}$ as

$$\begin{aligned}\mathbf{S}(\lambda_1) &= [\mathbf{v}_{11} \vdots \mathbf{v}_{12} \vdots \cdots \vdots \mathbf{v}_{1p_1} \vdots \cdots \vdots \mathbf{v}_{s1} \vdots \mathbf{v}_{s2} \vdots \cdots \vdots \mathbf{v}_{sp_s}] \\ &= [\mathbf{x}_1 \vdots \mathbf{x}_2 \vdots \cdots \vdots \mathbf{x}_{p_1} \vdots \cdots \vdots \mathbf{x}_k] \\ &= [\mathbf{S}_1 \vdots \mathbf{S}_2 \vdots \cdots \vdots \mathbf{S}_k]\end{aligned}$$

and define

$$\begin{aligned}\mathbf{S} &= [\mathbf{S}(\lambda_1) \vdots \mathbf{S}_{k+1} \vdots \mathbf{S}_{k+2} \vdots \cdots \vdots \mathbf{S}_n] \\ &= [\mathbf{S}_1 \vdots \mathbf{S}_2 \vdots \cdots \vdots \mathbf{S}_n]\end{aligned}$$

where

$$\mathbf{S}_{k+1} = \mathbf{x}_{k+1}, \qquad \mathbf{S}_{k+2} = \mathbf{x}_{k+2}, \ldots, \mathbf{S}_n = \mathbf{x}_n$$

Note that $\mathbf{x}_{k+1}, \mathbf{x}_{k+2}, \ldots, \mathbf{x}_n$ are eigenvectors associated with eigenvalues $\lambda_{k+1}, \lambda_{k+2}, \ldots, \lambda_n$, respectively. Matrix $\mathbf{S}$ defined in this way is nonsingular. Now we obtain

$$\mathbf{AS} = \mathbf{S}\left[\begin{array}{cccc:cccc} \mathbf{J}_{p_1}(\lambda_1) & & & 0 & & & & 0 \\ & \mathbf{J}_{p_2}(\lambda_1) & & & & & & \\ & & \ddots & & & & & \\ 0 & & & \mathbf{J}_{p_s}(\lambda_1) & 0 & & & \\ \hdashline & & & 0 & \lambda_{k+1} & & & 0 \\ & & & & & \ddots & & \\ 0 & & & & 0 & & & \lambda_n \end{array}\right]$$

where $\mathbf{J}_{p_i}(\lambda_1)$ is in the form

$$\mathbf{J}_{p_i}(\lambda_1) = \begin{bmatrix} \lambda_1 & 1 & & & 0 \\ & \lambda_1 & 1 & & \\ & & \ddots & \ddots & \\ & & & \ddots & 1 \\ 0 & & & & \lambda_1 \end{bmatrix}$$

which is a $p_i \times p_i$ matrix. Hence,

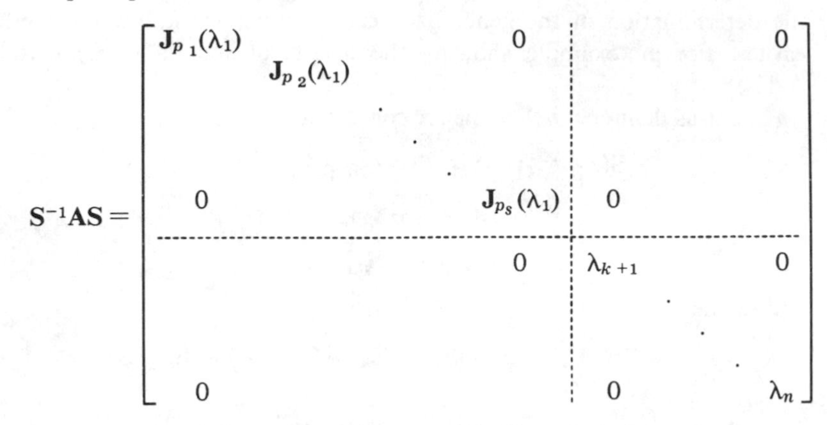

Thus, as we have shown, by using a set of n linearly independent vectors (eigenvectors and generalized eigenvectors) any $n \times n$ matrix can be reduced to a Jordan canonical form by a similarity transformation.

Similarity transformation when an $n \times n$ matrix is normal. First, recall that a matrix is normal if it is a real symmetric, a Hermitian, a real skew-symmetric, a skew-Hermitian, an orthogonal, or a unitary matrix.

Assume that an $n \times n$ normal matrix has a k-multiple eigenvalue λ_1 and that its other $n - k$ eigenvalues are distinct and different from λ_1. Then the rank of $\mathbf{A} - \lambda_1\mathbf{I}$ becomes $n - k$. (Refer to Prob. A-12 for the proof.) If the rank of $\mathbf{A} - \lambda_1\mathbf{I}$ is $n - k$, there are k linearly independent eigenvectors $\mathbf{x}_1, \mathbf{x}_2, \ldots, \mathbf{x}_k$ that satisfy the equation

$$(\mathbf{A} - \lambda_1\mathbf{I})\mathbf{x}_i = \mathbf{0} \qquad i = 1, 2, \ldots, k$$

Therefore, there exist k Jordan blocks for eigenvalue λ_1. Since the number of Jordan blocks is the same as the multiplicity number of eigenvalue λ_1, all k Jordan blocks become first-order. Since the remaining $n - k$ eigenvalues are distinct, the eigenvectors associated with these eigenvalues are linearly independent. Hence the $n \times n$ normal matrix possesses altogether n linearly independent eigenvectors and the Jordan canonical form of the normal matrix becomes a diagonal matrix.

It can be proved that if $\mathbf{A}$ is an $n \times n$ normal matrix, then regardless of whether or not the eigenvalues include multiple eigenvalues, there exists an $n \times n$ unitary matrix $\mathbf{U}$ such that

$$\mathbf{U}^{-1}\mathbf{A}\mathbf{U} = \mathbf{U}^*\mathbf{A}\mathbf{U} = \mathbf{D} = \text{diag}\,(\lambda_1, \lambda_2, \ldots, \lambda_n)$$

where $\mathbf{D}$ is a diagonal matrix with n eigenvalues as diagonal elements.

Trace of an $n \times n$ matrix. The trace of an $n \times n$ matrix $\mathbf{A}$ is defined as follows:

$$\text{trace of } \mathbf{A} = \text{tr } \mathbf{A} = \sum_{i=1}^{n} a_{ii}$$

The trace of an $n \times n$ matrix $\mathbf{A}$ has the following properties:

1. $$\text{tr } \mathbf{A}^T = \text{tr } \mathbf{A}$$
2. For $n \times n$ matrices $\mathbf{A}$ and $\mathbf{B}$,
$$\text{tr } (\mathbf{A} + \mathbf{B}) = \text{tr } \mathbf{A} + \text{tr } \mathbf{B}$$
3. If the eigenvalues of $\mathbf{A}$ are denoted by $\lambda_1, \lambda_2, \ldots, \lambda_n$, then
$$\text{tr } \mathbf{A} = \lambda_1 + \lambda_2 + \cdots + \lambda_n \tag{34}$$
4. For an $n \times m$ matrix $\mathbf{A}$ and and $m \times n$ matrix $\mathbf{B}$, regardless of whether $\mathbf{AB} = \mathbf{BA}$ or $\mathbf{AB} \neq \mathbf{BA}$, we have
$$\text{tr } \mathbf{AB} = \text{tr } \mathbf{BA} = \sum_{i=1}^{n} \sum_{j=1}^{m} a_{ij} b_{ji}$$
If $m = 1$, then by writing $\mathbf{A}$ and $\mathbf{B}$ as $\mathbf{a}$ and $\mathbf{b}$, respectively, we have
$$\text{tr } \mathbf{ab} = \mathbf{ba}$$
Hence, for an $n \times n$ matrix $\mathbf{C}$, we have
$$\mathbf{a}^T\mathbf{C}\mathbf{a} = \text{tr } \mathbf{a}\mathbf{a}^T\mathbf{C}$$

Note that Eq. (34) may be proved as follows. By use of a similarity transformation we have

$$\mathbf{P}^{-1}\mathbf{AP} = \mathbf{D} = \text{diagonal matrix}$$

or

$$\mathbf{S}^{-1}\mathbf{AS} = \mathbf{J} = \text{Jordan canonical form}$$

That is,

$$\mathbf{A} = \mathbf{PDP}^{-1} \qquad \text{or} \qquad \mathbf{A} = \mathbf{SJS}^{-1}$$

Hence by using property 4 listed here we have

$$\text{tr } \mathbf{A} = \text{tr } \mathbf{PDP}^{-1} = \text{tr } \mathbf{P}^{-1}\mathbf{PD} = \text{tr } \mathbf{D} = \lambda_1 + \lambda_2 + \cdots + \lambda_n$$

Similarly,

$$\text{tr } \mathbf{A} = \text{tr } \mathbf{SJS}^{-1} = \text{tr } \mathbf{S}^{-1}\mathbf{SJ} = \text{tr } \mathbf{J} = \lambda_1 + \lambda_2 + \cdots + \lambda_n$$

Invariant properties under similarity transformation. If an $n \times n$ matrix $\mathbf{A}$ can be reduced to a similar matrix which has a simple form, then important properties of $\mathbf{A}$ can be readily observed. A property of a matrix is said to be invariant if it is possessed by all similar matrices. For example, the determinant and the characteristic polynomial are invariant under a similarity transformation, as shown in the following. Suppose that $\mathbf{P}^{-1}\mathbf{AP} = \mathbf{B}$. Then

$$|\mathbf{B}| = |\mathbf{P}^{-1}\mathbf{AP}| = |\mathbf{P}^{-1}|\,|\mathbf{A}|\,|\mathbf{P}| = |\mathbf{A}|\,|\mathbf{P}^{-1}|\,|\mathbf{P}| = |\mathbf{A}|\,|\mathbf{P}^{-1}\mathbf{P}|$$
$$= |\mathbf{A}|\,|\mathbf{I}| = |\mathbf{A}|$$

and

$$|\lambda\mathbf{I} - \mathbf{B}| = |\lambda\mathbf{I} - \mathbf{P}^{-1}\mathbf{AP}| = |\mathbf{P}^{-1}(\lambda\mathbf{I})\mathbf{P} - \mathbf{P}^{-1}\mathbf{AP}|$$
$$= |\mathbf{P}^{-1}(\lambda\mathbf{I} - \mathbf{A})\mathbf{P}| = |\mathbf{P}^{-1}|\,|\lambda\mathbf{I} - \mathbf{A}|\,|\mathbf{P}|$$
$$= |\lambda\mathbf{I} - \mathbf{A}|\,|\mathbf{P}^{-1}|\,|\mathbf{P}| = |\lambda\mathbf{I} - \mathbf{A}|$$

Notice that the trace of a matrix is also invariant under similarity transformation, as was shown earlier:

$$\operatorname{tr} \mathbf{A} = \operatorname{tr} \mathbf{P}^{-1}\mathbf{AP}$$

The property of symmetry of a matrix, however, is not invariant.

Notice that only invariant properties of matrices present intrinsic characteristics of the class of similar matrices. To determine the invariant properties of a matrix **A** we examine the Jordan canonical form of **A**, since the similarity of two matrices can be defined in terms of the Jordan canonical form: The necessary and sufficient condition for $n \times n$ matrices **A** and **B** to be similar is that the Jordan canonical form of **A** and that of **B** be identical.

7 QUADRATIC FORMS

Quadratic forms. For an $n \times n$ real symmetric matrix **A** and a real n-vector **x**, the form

$$\mathbf{x}^T\mathbf{A}\mathbf{x} = \sum_{i=1}^{n}\sum_{j=1}^{n} a_{ij}x_i x_j \qquad a_{ji} = a_{ij}$$

is called a *real quadratic form* in x_i. Note that $\mathbf{x}^T\mathbf{A}\mathbf{x}$ is a real scalar quantity.

Any real quadratic form can always be written as $\mathbf{x}^T\mathbf{A}\mathbf{x}$. For example,

$$x_1^2 - 2x_1x_2 + 4x_1x_3 + x_2^2 + 8x_3^2 = [x_1 \quad x_2 \quad x_3]\begin{bmatrix} 1 & -1 & 2 \\ -1 & 1 & 0 \\ 2 & 0 & 8 \end{bmatrix}\begin{bmatrix} x_1 \\ x_2 \\ x_3 \end{bmatrix}$$

It is worthwhile to mention that for an $n \times n$ real matrix **A**, if we define

$$\mathbf{B} = \tfrac{1}{2}(\mathbf{A} + \mathbf{A}^T) \qquad \text{and} \qquad \mathbf{C} = \tfrac{1}{2}(\mathbf{A} - \mathbf{A}^T)$$

then

$$\mathbf{A} = \mathbf{B} + \mathbf{C}$$

Notice that

$$\mathbf{B}^T = \mathbf{B} \qquad \text{and} \qquad \mathbf{C}^T = -\mathbf{C}$$

Hence an $n \times n$ real matrix $\mathbf{A}$ can be expressed as a sum of a real symmetric and a real skew-symmetric matrix. Noting that since $\mathbf{x}^T\mathbf{C}\mathbf{x}$ is a real scalar quantity, we have

$$\mathbf{x}^T\mathbf{C}\mathbf{x} = (\mathbf{x}^T\mathbf{C}\mathbf{x})^T = \mathbf{x}^T\mathbf{C}^T\mathbf{x} = -\mathbf{x}^T\mathbf{C}\mathbf{x}$$

consequently, we have

$$\mathbf{x}^T\mathbf{C}\mathbf{x} = 0$$

This means that a quadratic form for a real skew-symmetric matrix is zero. Hence

$$\mathbf{x}^T\mathbf{A}\mathbf{x} = \mathbf{x}^T(\mathbf{B} + \mathbf{C})\mathbf{x} = \mathbf{x}^T\mathbf{B}\mathbf{x}$$

and we see that the real quadratic form $\mathbf{x}^T\mathbf{A}\mathbf{x}$ involves only the symmetric component $\mathbf{x}^T\mathbf{B}\mathbf{x}$. This is the reason why the real quadratic form is defined only for a real symmetric matrix.

For a Hermitian matrix $\mathbf{A}$ and a complex n-vector $\mathbf{x}$, the form

$$\mathbf{x}^*\mathbf{A}\mathbf{x} = \sum_{i=1}^{n}\sum_{j=1}^{n} a_{ij}\bar{x}_i x_j \qquad a_{ji} = \bar{a}_{ij}$$

is called a *complex quadratic form*, or Hermitian form. Notice that the scalar quantity $\mathbf{x}^*\mathbf{A}\mathbf{x}$ is real, because

$$\overline{\mathbf{x}^*\mathbf{A}\mathbf{x}} = \mathbf{x}^T\bar{\mathbf{A}}\bar{\mathbf{x}} = (\mathbf{x}^T\bar{\mathbf{A}}\bar{\mathbf{x}})^T = \bar{\mathbf{x}}^T\bar{\mathbf{A}}^T\mathbf{x} = \mathbf{x}^*\mathbf{A}\mathbf{x}$$

Bilinear forms. For an $n \times m$ real matrix $\mathbf{A}$, a real n-vector $\mathbf{x}$, and a real m-vector $\mathbf{y}$, the form

$$\mathbf{x}^T\mathbf{A}\mathbf{y} = \sum_{i=1}^{n}\sum_{j=1}^{m} a_{ij}x_i y_j$$

is called a *real bilinear form* in x_i and y_j. $\mathbf{x}^T\mathbf{A}\mathbf{y}$ is a real scalar quantity.

For an $n \times m$ complex matrix $\mathbf{A}$, a complex n-vector $\mathbf{x}$, and a complex m-vector $\mathbf{y}$, the form

$$\mathbf{x}^*\mathbf{A}\mathbf{y} = \sum_{i=1}^{n}\sum_{j=1}^{m} a_{ij}\bar{x}_i y_j$$

is called a *complex bilinear form*. $\mathbf{x}^*\mathbf{A}\mathbf{y}$ is a complex scalar quantity.

Definiteness and semidefiniteness. A quadratic form $\mathbf{x}^T\mathbf{A}\mathbf{x}$ where $\mathbf{A}$ is a real symmetric matrix (or a Hermitian form $\mathbf{x}^*\mathbf{A}\mathbf{x}$ where $\mathbf{A}$ is a Hermitian matrix) is said to be positive definite if

$$\mathbf{x}^T\mathbf{A}\mathbf{x} > 0 \qquad (\text{or } \mathbf{x}^*\mathbf{A}\mathbf{x} > 0) \qquad \text{for } \mathbf{x} \neq \mathbf{0}$$

$$\mathbf{x}^T\mathbf{A}\mathbf{x} = 0 \qquad (\text{or } \mathbf{x}^*\mathbf{A}\mathbf{x} = 0) \qquad \text{for } \mathbf{x} = \mathbf{0}$$

$\mathbf{x}^T\mathbf{A}\mathbf{x}$ (or $\mathbf{x}^*\mathbf{A}\mathbf{x}$) is said to be positive semidefinite if

$$\mathbf{x}^T\mathbf{A}\mathbf{x} \geq 0 \qquad (\text{or } \mathbf{x}^*\mathbf{A}\mathbf{x} \geq 0) \qquad \text{for } \mathbf{x} \neq \mathbf{0}$$

$$\mathbf{x}^T\mathbf{A}\mathbf{x} = 0 \qquad (\text{or } \mathbf{x}^*\mathbf{A}\mathbf{x} = 0) \qquad \text{for } \mathbf{x} = \mathbf{0}$$

$\mathbf{x}^T\mathbf{A}\mathbf{x}$ (or $\mathbf{x}^*\mathbf{A}\mathbf{x}$) is said to be negative definite if

$$\mathbf{x}^T\mathbf{A}\mathbf{x} < 0 \qquad (\text{or } \mathbf{x}^*\mathbf{A}\mathbf{x} < 0) \qquad \text{for } \mathbf{x} \neq \mathbf{0}$$

$$\mathbf{x}^T\mathbf{A}\mathbf{x} = 0 \qquad (\text{or } \mathbf{x}^*\mathbf{A}\mathbf{x} = 0) \qquad \text{for } \mathbf{x} = \mathbf{0}$$

$\mathbf{x}^T\mathbf{A}\mathbf{x}$ (or $\mathbf{x}^*\mathbf{A}\mathbf{x}$) is said to be negative semidefinite if

$$\mathbf{x}^T\mathbf{A}\mathbf{x} \leq 0 \qquad (\text{or } \mathbf{x}^*\mathbf{A}\mathbf{x} \leq 0) \qquad \text{for } \mathbf{x} \neq \mathbf{0}$$

$$\mathbf{x}^T\mathbf{A}\mathbf{x} = 0 \qquad (\text{or } \mathbf{x}^*\mathbf{A}\mathbf{x} = 0) \qquad \text{for } \mathbf{x} = \mathbf{0}$$

If $\mathbf{x}^T\mathbf{A}\mathbf{x}$ (or $\mathbf{x}^*\mathbf{A}\mathbf{x}$) can be of either sign, then $\mathbf{x}^T\mathbf{A}\mathbf{x}$ (or $\mathbf{x}^*\mathbf{A}\mathbf{x}$) is said to be indefinite.

Note that if $\mathbf{x}^T\mathbf{A}\mathbf{x}$ or $\mathbf{x}^*\mathbf{A}\mathbf{x}$ is positive (or negative) definite, then we say that $\mathbf{A}$ is a positive (or negative) definite matrix. Similarly, matrix $\mathbf{A}$ is called a positive (or negative) semidefinite matrix if $\mathbf{x}^T\mathbf{A}\mathbf{x}$ or $\mathbf{x}^*\mathbf{A}\mathbf{x}$ is positive (or negative) semidefinite; matrix $\mathbf{A}$ is called an indefinite matrix if $\mathbf{x}^T\mathbf{A}\mathbf{x}$ or $\mathbf{x}^*\mathbf{A}\mathbf{x}$ is indefinite.

Note also that the eigenvalues of an $n \times n$ real symmetric or Hermitian matrix are real. (For the proof, see Prob. A-13.) It can be shown that an $n \times n$ real symmetric or Hermitian matrix $\mathbf{A}$ is a positive-definite matrix if all eigenvalues λ_i ($i = 1, 2, \ldots, n$) are positive. Matrix $\mathbf{A}$ is positive semidefinite if all eigenvalues are nonnegative, or $\lambda_i \geq 0$ ($i = 1, 2, \ldots, n$), and at least one of them is zero.

Notice that if $\mathbf{A}$ is a positive definite matrix, then $|\mathbf{A}| \neq 0$, because all eigenvalues are positive. Hence, the inverse matrix always exists for a positive definite matrix.

In the process of determining the stability of an equilibrium state we frequently encounter a scalar function $V(\mathbf{x})$. A scalar function $V(\mathbf{x})$, which is a function of $x_1, x_2, \ldots, x_n$, is said to be positive definite if

$$V(\mathbf{x}) > 0 \qquad \text{for } \mathbf{x} \neq \mathbf{0}$$

$$V(\mathbf{0}) = 0$$

$V(\mathbf{x})$ is said to be positive semidefinite if

$$V(\mathbf{x}) \geq 0 \qquad \text{for } \mathbf{x} \neq \mathbf{0}$$

$$V(\mathbf{0}) = 0$$

If $-V(\mathbf{x})$ is positive definite (or positive semidefinite), then $V(\mathbf{x})$ is said to be negative definite (or negative semidefinite).

Necessary and sufficient conditions for the quadratic form $\mathbf{x}^T\mathbf{A}\mathbf{x}$ (or the Hermitian form $\mathbf{x}^*\mathbf{A}\mathbf{x}$) to be positive definite, negative definite, positive semidefinite, or negative semidefinite have been given by J. J. Sylvester. Sylvester's criteria follow.

Sylvester's criterion for positive definiteness of a quadratic form or Hermitian form. A necessary and sufficient condition for a quadratic form $\mathbf{x}^T\mathbf{A}\mathbf{x}$ (or a Hermitian form $\mathbf{x}^*\mathbf{A}\mathbf{x}$), where $\mathbf{A}$ is an $n \times n$ real symmetric matrix (or Hermitian matrix), to be positive definite is that the determinant of $\mathbf{A}$ be positive and the successive principal minors of the determinant of $\mathbf{A}$ (the determinants of the $k \times k$ matrices in the top left-hand corner of matrix $\mathbf{A}$, where $k = 1, 2, \ldots, n - 1$) be positive; that is, we must have

$$a_{11} > 0, \qquad \begin{vmatrix} a_{11} & a_{12} \\ a_{21} & a_{22} \end{vmatrix} > 0, \qquad \begin{vmatrix} a_{11} & a_{12} & a_{13} \\ a_{21} & a_{22} & a_{23} \\ a_{31} & a_{32} & a_{33} \end{vmatrix} > 0, \ldots, |\mathbf{A}| > 0$$

where

$$a_{ij} = a_{ji} \qquad \text{for real symmetric matrix } \mathbf{A}$$

$$a_{ij} = \bar{a}_{ji} \qquad \text{for Hermitian matrix } \mathbf{A}$$

Sylvester's criterion for negative definiteness of a quadratic form or Hermitian form. A necessary and sufficient condition for a quadratic form $\mathbf{x}^T\mathbf{A}\mathbf{x}$ (or a Hermitian form $\mathbf{x}^*\mathbf{A}\mathbf{x}$), where $\mathbf{A}$ is an $n \times n$ real symmetric matrix (or Hermitian matrix), to be negative definite is that the determinant of $\mathbf{A}$ be positive if n is even and negative if n is odd, and that the successive principal minors of even order be positive and the successive principal minors of odd order be negative; that is, we must have

$$a_{11} < 0, \qquad \begin{vmatrix} a_{11} & a_{12} \\ a_{21} & a_{22} \end{vmatrix} > 0, \qquad \begin{vmatrix} a_{11} & a_{12} & a_{13} \\ a_{21} & a_{22} & a_{23} \\ a_{31} & a_{32} & a_{33} \end{vmatrix} < 0, \ldots$$

$$|\mathbf{A}| > 0 \qquad (n \text{ even})$$

$$|\mathbf{A}| < 0 \qquad (n \text{ odd})$$

where

$$a_{ij} = a_{ji} \qquad \text{for real symmetric matrix } \mathbf{A}$$

$$a_{ij} = \bar{a}_{ji} \qquad \text{for Hermitian matrix } \mathbf{A}$$

[This condition can be derived by requiring that $\mathbf{x}^T(-\mathbf{A})\mathbf{x}$ be positive definite.]

Sylvester's criterion for positive semidefiniteness of a quadratic form or Hermitian form. A necessary and sufficient condition for a quadratic form $\mathbf{x}^T\mathbf{A}\mathbf{x}$ (or a Hermitian form $\mathbf{x}^*\mathbf{A}\mathbf{x}$), where $\mathbf{A}$ is a real symmetric matrix (or a Hermitian matrix), to be positive semidefinite is that $\mathbf{A}$ be singular ($|\mathbf{A}| = 0$) and all the principal minors be nonnegative:

$$a_{ii} \geq 0, \qquad \begin{vmatrix} a_{ii} & a_{ij} \\ a_{ji} & a_{jj} \end{vmatrix} \geq 0, \qquad \begin{vmatrix} a_{ii} & a_{ij} & a_{ik} \\ a_{ji} & a_{jj} & a_{jk} \\ a_{ki} & a_{kj} & a_{kk} \end{vmatrix} \geq 0, \ldots, |\mathbf{A}| = 0$$

where $i < j < k$ and

$$a_{ij} = a_{ji} \qquad \text{for real symmetric matrix } \mathbf{A}$$

$$a_{ij} = \bar{a}_{ji} \qquad \text{for Hermitian matrix } \mathbf{A}$$

(It is important to point out that in the positive semidefiniteness test or negative semidefiniteness test, we must check the signs of all the principal minors, not just successive principal minors. See Prob. A-15.)

Sylvester's criterion for negative semidefiniteness of a quadratic form or a Hermitian form. A necessary and sufficient condition for a quadratic form $\mathbf{x}^T\mathbf{A}\mathbf{x}$ (or a Hermitian form $\mathbf{x}^*\mathbf{A}\mathbf{x}$), where $\mathbf{A}$ is an $n \times n$ real symmetric matrix (or Hermitian matrix), to be negative semidefinite is that $\mathbf{A}$ be singular ($|\mathbf{A}| = 0$) and that all the principal minors of even order be nonnegative and those of odd order be nonpositive:

$$a_{ii} \leq 0, \qquad \begin{vmatrix} a_{ii} & a_{ij} \\ a_{ji} & a_{jj} \end{vmatrix} \geq 0, \qquad \begin{vmatrix} a_{ii} & a_{ij} & a_{ik} \\ a_{ji} & a_{jj} & a_{jk} \\ a_{ki} & a_{kj} & a_{kk} \end{vmatrix} \leq 0, \; . \; . \; . \; , \; |\mathbf{A}| = 0$$

where $i < j < k$ and

$$a_{ij} = a_{ji} \qquad \text{for real symmetric matrix } \mathbf{A}$$

$$a_{ij} = \bar{a}_{ji} \qquad \text{for Hermitian matrix } \mathbf{A}$$

8 PSEUDOINVERSES

The concept of pseudoinverses of a matrix is a generalization of the notion of an inverse. It is useful for finding a "solution" to a set of algebraic equations in which the number of unknown variables and the number of independent linear equations are not equal.

In what follows, we shall consider pseudoinverses which enable us to determine minimum norm solutions.

Minimum norm solution that minimizes $\|\mathbf{x}\|$. Consider a linear algebraic equation

$$x_1 + 5x_2 = 1$$

Since we have two variables and only one equation, no unique solution exists. Instead, there exist an infinite number of solutions. Graphically, any point on line $x_1 + 5x_2 = 1$, as shown in Fig. 1, is a possible solution. However, if we decide to pick the point that is closest to the origin, the solution becomes unique.

Consider the vector-matrix equation

$$\mathbf{A}\mathbf{x} = \mathbf{b} \tag{35}$$

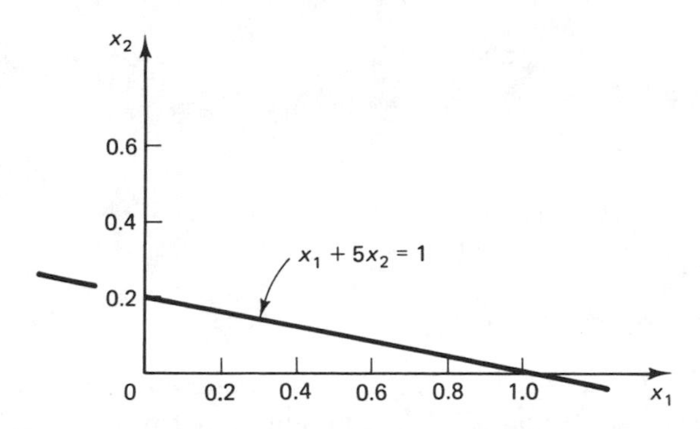

Figure 1 Line $x_1 + 5x_2 = 1$ on the x_1x_2 plane.

where $\mathbf{A}$ is an $n \times m$ matrix, $\mathbf{x}$ is an m-vector, and $\mathbf{b}$ is an n-vector. We assume that $m > n$ (that is, that the number of unknown variables is greater than the number of equations) and that the equation has an infinite number of solutions. Let us find the unique solution $\mathbf{x}$ that is located closest to the origin, or that has the minimum norm $\|\mathbf{x}\|$.

Let us define the minimum norm solution as $\mathbf{x}^o$. That is, $\mathbf{x}^o$ satisfies the condition that $\mathbf{A}\mathbf{x}^o = \mathbf{b}$ and $\|\mathbf{x}^o\| \leq \|\mathbf{x}\|$ for all $\mathbf{x}$ that satisfy $\mathbf{A}\mathbf{x} = \mathbf{b}$. This means that the solution point $\mathbf{x}^o$ is nearest to the origin of the m-dimensional space among all possible solutions of Eq. (35). We shall obtain such a minimum norm solution in the following.

Right pseudoinverse matrix. For a vector-matrix equation

$$\mathbf{A}\mathbf{x} = \mathbf{b}$$

where $\mathbf{A}$ is an $n \times m$ matrix having rank n, $\mathbf{x}$ is an m-vector, and $\mathbf{b}$ is an n-vector, the solution that minimizes the norm $\|\mathbf{x}\|$ is given by

$$\mathbf{x}^o = \mathbf{A}^{RM}\mathbf{b}$$

where $\mathbf{A}^{RM} = \mathbf{A}^T(\mathbf{A}\mathbf{A}^T)^{-1}$.

This can be proved as follows. First, note that norm $\|\mathbf{x}\|$ can be written as follows:

$$\|\mathbf{x}\| = \|\mathbf{x} - \mathbf{x}^o + \mathbf{x}^o\| = \|\mathbf{x}^o\| + \|\mathbf{x} - \mathbf{x}^o\| + 2(\mathbf{x}^o)^T(\mathbf{x} - \mathbf{x}^o)$$

The last term, $2(\mathbf{x}^o)^T(\mathbf{x} - \mathbf{x}^o)$, can be shown to be zero, since

$$\begin{aligned}
(\mathbf{x}^o)^T(\mathbf{x} - \mathbf{x}^o) &= [\mathbf{A}^T(\mathbf{A}\mathbf{A}^T)^{-1}\mathbf{b}]^T\,[\mathbf{x} - \mathbf{A}^T(\mathbf{A}\mathbf{A}^T)^{-1}\mathbf{b}] \\
&= \mathbf{b}^T(\mathbf{A}\mathbf{A}^T)^{-1}\mathbf{A}[\mathbf{x} - \mathbf{A}^T(\mathbf{A}\mathbf{A}^T)^{-1}\mathbf{b}] \\
&= \mathbf{b}^T(\mathbf{A}\mathbf{A}^T)^{-1}\,[\mathbf{A}\mathbf{x} - (\mathbf{A}\mathbf{A}^T)(\mathbf{A}\mathbf{A}^T)^{-1}\mathbf{b}] \\
&= \mathbf{b}^T(\mathbf{A}\mathbf{A}^T)^{-1}(\mathbf{b} - \mathbf{b}) \\
&= 0
\end{aligned}$$

Hence

$$\|\mathbf{x}\| = \|\mathbf{x}^o\| + \|\mathbf{x} - \mathbf{x}^o\|$$

which can be rewritten as

$$\|\mathbf{x}\| - \|\mathbf{x}^o\| = \|\mathbf{x} - \mathbf{x}^o\|$$

Since $\|\mathbf{x} - \mathbf{x}^o\| \geq 0$, we obtain

$$\|\mathbf{x}\| \geq \|\mathbf{x}^o\|$$

Thus, we have shown that $\mathbf{x}^o$ is the solution that gives the minimum norm $\|\mathbf{x}\|$.

The matrix $\mathbf{A}^{RM} = \mathbf{A}^T(\mathbf{A}\mathbf{A}^T)^{-1}$ that yields the minimum norm solution ($\|\mathbf{x}^o\|$ = minimum) is called the *right pseudoinverse* or *minimal right inverse* of $\mathbf{A}$.

Summary on the right pseudoinverse matrix. The right pseudoinverse $\mathbf{A}^{RM}$ gives the solution $\mathbf{x}^o = \mathbf{A}^{RM}\mathbf{b}$ that minimizes the norm, or gives $\|\mathbf{x}^o\|$ = minimum. Note that the right pseudoinverse $\mathbf{A}^{RM}$ is an $m \times n$ matrix, since $\mathbf{A}$ is an $n \times m$ matrix and

$$\begin{aligned}\mathbf{A}^{RM} &= \mathbf{A}^T(\mathbf{A}\mathbf{A}^T)^{-1} \\ &= (m \times n \text{ matrix})(n \times n \text{ matrix})^{-1} \\ &= m \times n \text{ matrix} \qquad m > n\end{aligned}$$

Notice that the dimension of $\mathbf{A}\mathbf{A}^T$ is smaller than the dimension of vector $\mathbf{x}$, which is m. Notice also that the right pseudoinverse $\mathbf{A}^{RM}$ possesses the property that it is indeed an "inverse" matrix if premultiplied by $\mathbf{A}$:

$$\mathbf{A}\mathbf{A}^{RM} = \mathbf{A}[\mathbf{A}^T(\mathbf{A}\mathbf{A}^T)^{-1}] = \mathbf{A}\mathbf{A}^T(\mathbf{A}\mathbf{A}^T)^{-1} = \mathbf{I}_n$$

Solution that minimizes $\|\mathbf{Ax} - \mathbf{b}\|$. Consider a vector-matrix equation

$$\mathbf{Ax} = \mathbf{b} \tag{36}$$

where $\mathbf{A}$ is an $n \times m$ matrix, $\mathbf{x}$ is an m-vector, and $\mathbf{b}$ is an n-vector. Here, we assume that $n > m$. That is, the number of unknown variables is smaller than the number of equations. In the classical sense, there may or may not exist any solution.

If no solution exists, we may wish to find a unique "solution" that minimizes the norm $\|\mathbf{Ax} - \mathbf{b}\|$. Let us define a "solution" to Eq. (36) that will minimize $\|\mathbf{Ax} - \mathbf{b}\|$ as $\mathbf{x}^o$. In other words, $\mathbf{x}^o$ satisfies the condition

$$\|\mathbf{Ax} - \mathbf{b}\| \geq \|\mathbf{Ax}^o - \mathbf{b}\| \qquad \text{for all } \mathbf{x}$$

Note that $\mathbf{x}^o$ is not a solution in the classical sense, since it does not satisfy the original vector-matrix equation $\mathbf{Ax} = \mathbf{b}$. Therefore we may call $\mathbf{x}^o$ an "approximate solution," in that it minimizes norm $\|\mathbf{Ax} - \mathbf{b}\|$. We shall obtain such an approximate solution in the following.

Left pseudoinverse matrix. For a vector-matrix equation

$$\mathbf{A}\mathbf{x} = \mathbf{b}$$

where $\mathbf{A}$ is an $n \times m$ matrix having rank m, $\mathbf{x}$ is an m-vector, and $\mathbf{b}$ is an n-vector, the vector $\mathbf{x}^o$ that minimizes the norm $\|\mathbf{A}\mathbf{x} - \mathbf{b}\|$ is given by

$$\mathbf{x}^o = \mathbf{A}^{LM}\mathbf{b} = (\mathbf{A}^T\mathbf{A})^{-1}\mathbf{A}^T\mathbf{b}$$

where $\mathbf{A}^{LM} = (\mathbf{A}^T\mathbf{A})^{-1}\mathbf{A}^T$.

To verify this, first note that

$$\begin{aligned}\|\mathbf{A}\mathbf{x} - \mathbf{b}\| &= \|\mathbf{A}(\mathbf{x} - \mathbf{x}^o) + \mathbf{A}\mathbf{x}^o - \mathbf{b}\| \\ &= \|\mathbf{A}(\mathbf{x} - \mathbf{x}^o)\| + \|\mathbf{A}\mathbf{x}^o - \mathbf{b}\| + 2[\mathbf{A}(\mathbf{x} - \mathbf{x}^o)]^T(\mathbf{A}\mathbf{x}^o - \mathbf{b})\end{aligned}$$

The last term can be shown to be zero, as follows:

$$\begin{aligned}[\mathbf{A}(\mathbf{x} - \mathbf{x}^o)]^T(\mathbf{A}\mathbf{x}^o - \mathbf{b}) &= (\mathbf{x} - \mathbf{x}^o)^T\mathbf{A}^T[\mathbf{A}(\mathbf{A}^T\mathbf{A})^{-1}\mathbf{A}^T - \mathbf{I}_n]\mathbf{b} \\ &= (\mathbf{x} - \mathbf{x}^o)^T[(\mathbf{A}^T\mathbf{A})(\mathbf{A}^T\mathbf{A})^{-1}\mathbf{A}^T - \mathbf{A}^T]\mathbf{b} \\ &= (\mathbf{x} - \mathbf{x}^o)^T(\mathbf{A}^T - \mathbf{A}^T)\mathbf{b} \\ &= 0\end{aligned}$$

Hence

$$\|\mathbf{A}\mathbf{x} - \mathbf{b}\| = \|\mathbf{A}(\mathbf{x} - \mathbf{x}^o)\| + \|\mathbf{A}\mathbf{x}^o - \mathbf{b}\|$$

Noting that $\|\mathbf{A}(\mathbf{x} - \mathbf{x}^o)\| \geq 0$, we obtain

$$\|\mathbf{A}\mathbf{x} - \mathbf{b}\| - \|\mathbf{A}\mathbf{x}^o - \mathbf{b}\| = \|\mathbf{A}(\mathbf{x} - \mathbf{x}^o)\| \geq 0$$

or

$$\|\mathbf{A}\mathbf{x} - \mathbf{b}\| \geq \|\mathbf{A}\mathbf{x}^o - \mathbf{b}\|$$

Thus

$$\mathbf{x}^o = \mathbf{A}^{LM}\mathbf{b} = (\mathbf{A}^T\mathbf{A})^{-1}\mathbf{A}^T\mathbf{b}$$

minimizes $\|\mathbf{A}\mathbf{x} - \mathbf{b}\|$.

The matrix $\mathbf{A}^{LM} = (\mathbf{A}^T\mathbf{A})^{-1}\mathbf{A}^T$ is called the *left pseudoinverse* or *minimal left inverse* of matrix $\mathbf{A}$. Note that $\mathbf{A}^{LM}$ is indeed the inverse matrix of $\mathbf{A}$, in that if postmultiplied by $\mathbf{A}$, it will give an identity matrix $\mathbf{I}_m$:

$$\mathbf{A}^{LM}\mathbf{A} = (\mathbf{A}^T\mathbf{A})^{-1}\mathbf{A}^T\mathbf{A} = (\mathbf{A}^T\mathbf{A})^{-1}(\mathbf{A}^T\mathbf{A}) = \mathbf{I}_m$$

For illustrative applications of the use of the right and left pseudoinverses for obtaining minimum norm solutions of vector-matrix equations, refer to Probs. A-16 and A-17.

REFERENCES

1. Bellman, R., *Introduction to Matrix Analysis*. New York: McGraw-Hill Book Company, 1960.
2. Gantmacher, F. R., *Theory of Matrices*, Vols. I and II. New York: Chelsea Publishing Co., Inc., 1959.
3. Halmos, P. R., *Finite Dimensional Vector Spaces*. Princeton, N.J.: D. Van Nostrand Company, 1958.
4. Noble, B., and J. Daniel, *Applied Linear Algebra*, 2nd ed. Englewood Cliffs, N.J.: Prentice-Hall, Inc., 1977.
5. Ogata, K., *State Space Analysis of Control Systems*. Englewood Cliffs, N.J.: Prentice-Hall, Inc., 1967.
6. Strang, G., *Linear Algebra and Its Applications*. New York: Academic Press, Inc., 1976.
7. Turnbull, H. W., and A. C. Aitken, *An Introduction to the Theory of Canonical Matrices*. London: Blackie and Son, Ltd., 1932.

EXAMPLE PROBLEMS AND SOLUTIONS

Problem A-1. Show that if matrices **A, B, C,** and **D** are an $n \times n$, an $n \times m$, an $m \times n$, and an $m \times m$ matrix, respectively, and if $|\mathbf{A}| \neq 0$ and $|\mathbf{D}| \neq 0$, then

$$\begin{vmatrix} \mathbf{A} & \mathbf{B} \\ \mathbf{0} & \mathbf{D} \end{vmatrix} = \begin{vmatrix} \mathbf{A} & \mathbf{0} \\ \mathbf{C} & \mathbf{D} \end{vmatrix} = |\mathbf{A}|\,|\mathbf{D}| \neq 0 \qquad \text{if } |\mathbf{A}| \neq 0 \text{ and } |\mathbf{D}| \neq 0$$

Solution. Since matrix **A** is nonsingular, we have

$$\begin{bmatrix} \mathbf{A} & \mathbf{B} \\ \mathbf{0} & \mathbf{D} \end{bmatrix} = \begin{bmatrix} \mathbf{A} & \mathbf{0} \\ \mathbf{0} & \mathbf{I} \end{bmatrix} \begin{bmatrix} \mathbf{I} & \mathbf{0} \\ \mathbf{0} & \mathbf{D} \end{bmatrix} \begin{bmatrix} \mathbf{I} & \mathbf{A}^{-1}\mathbf{B} \\ \mathbf{0} & \mathbf{I} \end{bmatrix}$$

Hence,

$$\begin{vmatrix} \mathbf{A} & \mathbf{B} \\ \mathbf{0} & \mathbf{D} \end{vmatrix} = \begin{vmatrix} \mathbf{A} & \mathbf{0} \\ \mathbf{0} & \mathbf{I} \end{vmatrix} \begin{vmatrix} \mathbf{I} & \mathbf{0} \\ \mathbf{0} & \mathbf{D} \end{vmatrix} \begin{vmatrix} \mathbf{I} & \mathbf{A}^{-1}\mathbf{B} \\ \mathbf{0} & \mathbf{I} \end{vmatrix} = |\mathbf{A}|\,|\mathbf{D}|$$

Similarly, since **D** is nonsingular, we get

$$\begin{vmatrix} \mathbf{A} & \mathbf{0} \\ \mathbf{C} & \mathbf{D} \end{vmatrix} = \begin{vmatrix} \mathbf{A} & \mathbf{0} \\ \mathbf{0} & \mathbf{I} \end{vmatrix} \begin{vmatrix} \mathbf{I} & \mathbf{0} \\ \mathbf{0} & \mathbf{D} \end{vmatrix} \begin{vmatrix} \mathbf{I} & \mathbf{0} \\ \mathbf{D}^{-1}\mathbf{C} & \mathbf{I} \end{vmatrix} = |\mathbf{A}|\,|\mathbf{D}|$$

Problem A-2. Show that if matrices **A, B, C,** and **D** are an $n \times n$, an $n \times m$, an $m \times n$, and an $m \times m$ matrix, respectively, then

$$\begin{vmatrix} \mathbf{A} & \mathbf{B} \\ \mathbf{C} & \mathbf{D} \end{vmatrix} = \begin{cases} |\mathbf{A}|\,|\mathbf{D} - \mathbf{C}\mathbf{A}^{-1}\mathbf{B}| & \text{if } |\mathbf{A}| \neq 0 \\ |\mathbf{D}|\,|\mathbf{A} - \mathbf{B}\mathbf{D}^{-1}\mathbf{C}| & \text{if } |\mathbf{D}| \neq 0 \end{cases}$$

Solution. If $|\mathbf{A}| \neq 0$, the matrix

$$\begin{bmatrix} \mathbf{A} & \mathbf{B} \\ \mathbf{C} & \mathbf{D} \end{bmatrix}$$

can be written as a product of two matrices:

$$\begin{bmatrix} \mathbf{A} & \mathbf{0} \\ \mathbf{C} & \mathbf{I}_m \end{bmatrix} \quad \text{and} \quad \begin{bmatrix} \mathbf{I}_n & \mathbf{A}^{-1}\mathbf{B} \\ \mathbf{0} & \mathbf{D}-\mathbf{C}\mathbf{A}^{-1}\mathbf{B} \end{bmatrix}$$

or

$$\begin{bmatrix} \mathbf{A} & \mathbf{B} \\ \mathbf{C} & \mathbf{D} \end{bmatrix} = \begin{bmatrix} \mathbf{A} & \mathbf{0} \\ \mathbf{C} & \mathbf{I}_m \end{bmatrix} \begin{bmatrix} \mathbf{I}_n & \mathbf{A}^{-1}\mathbf{B} \\ \mathbf{0} & \mathbf{D}-\mathbf{C}\mathbf{A}^{-1}\mathbf{B} \end{bmatrix}$$

Hence

$$\begin{aligned} \begin{vmatrix} \mathbf{A} & \mathbf{B} \\ \mathbf{C} & \mathbf{D} \end{vmatrix} &= \begin{vmatrix} \mathbf{A} & \mathbf{0} \\ \mathbf{C} & \mathbf{I}_m \end{vmatrix} \begin{vmatrix} \mathbf{I}_n & \mathbf{A}^{-1}\mathbf{B} \\ \mathbf{0} & \mathbf{D}-\mathbf{C}\mathbf{A}^{-1}\mathbf{B} \end{vmatrix} \\ &= |\mathbf{A}|\,|\mathbf{I}_m|\,|\mathbf{I}_n|\,|\mathbf{D}-\mathbf{C}\mathbf{A}^{-1}\mathbf{B}| \\ &= |\mathbf{A}|\,|\mathbf{D}-\mathbf{C}\mathbf{A}^{-1}\mathbf{B}| \end{aligned}$$

Similarly, if $|\mathbf{D}| \neq 0$, then

$$\begin{bmatrix} \mathbf{A} & \mathbf{B} \\ \mathbf{C} & \mathbf{D} \end{bmatrix} = \begin{bmatrix} \mathbf{I}_n & \mathbf{B} \\ \mathbf{0} & \mathbf{D} \end{bmatrix} \begin{bmatrix} \mathbf{A}-\mathbf{B}\mathbf{D}^{-1}\mathbf{C} & \mathbf{0} \\ \mathbf{D}^{-1}\mathbf{C} & \mathbf{I}_m \end{bmatrix}$$

and therefore

$$\begin{aligned} \begin{vmatrix} \mathbf{A} & \mathbf{B} \\ \mathbf{C} & \mathbf{D} \end{vmatrix} &= \begin{vmatrix} \mathbf{I}_n & \mathbf{B} \\ \mathbf{0} & \mathbf{D} \end{vmatrix} \begin{vmatrix} \mathbf{A}-\mathbf{B}\mathbf{D}^{-1}\mathbf{C} & \mathbf{0} \\ \mathbf{D}^{-1}\mathbf{C} & \mathbf{I}_m \end{vmatrix} \\ &= |\mathbf{I}_n|\,|\mathbf{D}|\,|\mathbf{A}-\mathbf{B}\mathbf{D}^{-1}\mathbf{C}|\,|\mathbf{I}_m| \\ &= |\mathbf{D}|\,|\mathbf{A}-\mathbf{B}\mathbf{D}^{-1}\mathbf{C}| \end{aligned}$$

Problem A-3. For an $n \times m$ matrix $\mathbf{A}$ and an $m \times n$ matrix $\mathbf{B}$, show that

$$|\mathbf{I}_n + \mathbf{A}\mathbf{B}| = |\mathbf{I}_m + \mathbf{B}\mathbf{A}|$$

Solution. Consider the following matrix:

$$\begin{bmatrix} \mathbf{I}_n & -\mathbf{A} \\ \mathbf{B} & \mathbf{I}_m \end{bmatrix}$$

Referring to Prob. A-2,

$$\begin{vmatrix} \mathbf{A} & \mathbf{B} \\ \mathbf{C} & \mathbf{D} \end{vmatrix} = \begin{cases} |\mathbf{A}|\,|\mathbf{D}-\mathbf{C}\mathbf{A}^{-1}\mathbf{B}| & \text{if } |\mathbf{A}| \neq 0 \\ |\mathbf{D}|\,|\mathbf{A}-\mathbf{B}\mathbf{D}^{-1}\mathbf{C}| & \text{if } |\mathbf{D}| \neq 0 \end{cases}$$

Hence

$$\begin{vmatrix} \mathbf{I}_n & -\mathbf{A} \\ \mathbf{B} & \mathbf{I}_m \end{vmatrix} = \begin{cases} |\mathbf{I}_n|\,|\mathbf{I}_m + \mathbf{B}\mathbf{A}| = |\mathbf{I}_m + \mathbf{B}\mathbf{A}| \\ |\mathbf{I}_m|\,|\mathbf{I}_n + \mathbf{A}\mathbf{B}| = |\mathbf{I}_n + \mathbf{A}\mathbf{B}| \end{cases}$$

and we have

$$|\mathbf{I}_n + \mathbf{A}\mathbf{B}| = |\mathbf{I}_m + \mathbf{B}\mathbf{A}|$$

Problem A-4. If $\mathbf{A}$, $\mathbf{B}$, $\mathbf{C}$, and $\mathbf{D}$ are, respectively, an $n \times n$, an $n \times m$, an $m \times n$, and an $m \times m$ matrix, then we have the following matrix inversion lemma:

$$(\mathbf{A}+\mathbf{BDC})^{-1}=\mathbf{A}^{-1}-\mathbf{A}^{-1}\mathbf{B}(\mathbf{D}^{-1}+\mathbf{CA}^{-1}\mathbf{B})^{-1}\mathbf{CA}^{-1}$$

where we assume the indicated inverses to exist. Prove this matrix inversion lemma.

Solution. Let us premultiply both sides of the equation by $(\mathbf{A}+\mathbf{BDC})$:

$$(\mathbf{A}+\mathbf{BDC})(\mathbf{A}+\mathbf{BDC})^{-1}=(\mathbf{A}+\mathbf{BDC})[\mathbf{A}^{-1}-\mathbf{A}^{-1}\mathbf{B}(\mathbf{D}^{-1}+\mathbf{CA}^{-1}\mathbf{B})^{-1}\mathbf{CA}^{-1}]$$

or

$$\begin{aligned}\mathbf{I}&=\mathbf{I}+\mathbf{BDCA}^{-1}-\mathbf{B}(\mathbf{D}^{-1}+\mathbf{CA}^{-1}\mathbf{B})^{-1}\mathbf{CA}^{-1}-\mathbf{BDCA}^{-1}\mathbf{B}(\mathbf{D}^{-1}+\mathbf{CA}^{-1}\mathbf{B})^{-1}\mathbf{CA}^{-1}\\&=\mathbf{I}+\mathbf{BDCA}^{-1}-(\mathbf{B}+\mathbf{BDCA}^{-1}\mathbf{B})(\mathbf{D}^{-1}+\mathbf{CA}^{-1}\mathbf{B})^{-1}\mathbf{CA}^{-1}\\&=\mathbf{I}+\mathbf{BDCA}^{-1}-\mathbf{BD}(\mathbf{D}^{-1}+\mathbf{CA}^{-1}\mathbf{B})(\mathbf{D}^{-1}+\mathbf{CA}^{-1}\mathbf{B})^{-1}\mathbf{CA}^{-1}\\&=\mathbf{I}+\mathbf{BDCA}^{-1}-\mathbf{BDCA}^{-1}\\&=\mathbf{I}\end{aligned}$$

Hence we have proved the matrix inversion lemma.

Problem A-5. Prove that if **A, B, C,** and **D** are, respectively, an $n\times n$, an $n\times m$, an $m\times n$, and an $m\times m$ matrix, then

$$\begin{bmatrix}\mathbf{A}&\mathbf{B}\\\mathbf{0}&\mathbf{D}\end{bmatrix}^{-1}=\begin{bmatrix}\mathbf{A}^{-1}&-\mathbf{A}^{-1}\mathbf{BD}^{-1}\\\mathbf{0}&\mathbf{D}^{-1}\end{bmatrix}\tag{37}$$

provided $|\mathbf{A}|\neq 0$ and $|\mathbf{D}|\neq 0$.

Prove also that

$$\begin{bmatrix}\mathbf{A}&\mathbf{0}\\\mathbf{C}&\mathbf{D}\end{bmatrix}^{-1}=\begin{bmatrix}\mathbf{A}^{-1}&\mathbf{0}\\-\mathbf{D}^{-1}\mathbf{CA}^{-1}&\mathbf{D}^{-1}\end{bmatrix}\tag{38}$$

provided $|\mathbf{A}|\neq 0$ and $|\mathbf{D}|\neq 0$.

Solution. Note that

$$\begin{bmatrix}\mathbf{A}^{-1}&-\mathbf{A}^{-1}\mathbf{BD}^{-1}\\\mathbf{0}&\mathbf{D}^{-1}\end{bmatrix}\begin{bmatrix}\mathbf{A}&\mathbf{B}\\\mathbf{0}&\mathbf{D}\end{bmatrix}=\begin{bmatrix}\mathbf{I}_n&\mathbf{A}^{-1}\mathbf{B}-\mathbf{A}^{-1}\mathbf{B}\\\mathbf{0}&\mathbf{I}_m\end{bmatrix}=\begin{bmatrix}\mathbf{I}_n&\mathbf{0}\\\mathbf{0}&\mathbf{I}_m\end{bmatrix}$$

Hence, Eq. (37) is proved.

Similarly,

$$\begin{bmatrix}\mathbf{A}^{-1}&\mathbf{0}\\-\mathbf{D}^{-1}\mathbf{CA}^{-1}&\mathbf{D}^{-1}\end{bmatrix}\begin{bmatrix}\mathbf{A}&\mathbf{0}\\\mathbf{C}&\mathbf{D}\end{bmatrix}=\begin{bmatrix}\mathbf{I}_n&\mathbf{0}\\-\mathbf{D}^{-1}\mathbf{C}+\mathbf{D}^{-1}\mathbf{C}&\mathbf{I}_m\end{bmatrix}=\begin{bmatrix}\mathbf{I}_n&\mathbf{0}\\\mathbf{0}&\mathbf{I}_m\end{bmatrix}$$

Hence, we have proved Eq. (38).

Problem A-6. Prove that if **A, B, C,** and **D** are, respectively, an $n\times n$, an $n\times m$, an $m\times n$, and an $m\times m$ matrix, then

$$\begin{bmatrix}\mathbf{A}&\mathbf{B}\\\mathbf{C}&\mathbf{D}\end{bmatrix}^{-1}=\begin{bmatrix}\mathbf{A}^{-1}+\mathbf{A}^{-1}\mathbf{B}(\mathbf{D}-\mathbf{CA}^{-1}\mathbf{B})^{-1}\mathbf{CA}^{-1}&-\mathbf{A}^{-1}\mathbf{B}(\mathbf{D}-\mathbf{CA}^{-1}\mathbf{B})^{-1}\\-(\mathbf{D}-\mathbf{CA}^{-1}\mathbf{B})^{-1}\mathbf{CA}^{-1}&(\mathbf{D}-\mathbf{CA}^{-1}\mathbf{B})^{-1}\end{bmatrix}$$

provided $|\mathbf{A}|\neq 0$ and $|\mathbf{D}-\mathbf{CA}^{-1}\mathbf{B}|\neq 0$.

Prove also that

$$\begin{bmatrix} \mathbf{A} & \mathbf{B} \\ \mathbf{C} & \mathbf{D} \end{bmatrix}^{-1} = \begin{bmatrix} (\mathbf{A}-\mathbf{BD}^{-1}\mathbf{C})^{-1} & -(\mathbf{A}-\mathbf{BD}^{-1}\mathbf{C})^{-1}\mathbf{BD}^{-1} \\ -\mathbf{D}^{-1}\mathbf{C}(\mathbf{A}-\mathbf{BD}^{-1}\mathbf{C})^{-1} & \mathbf{D}^{-1}\mathbf{C}(\mathbf{A}-\mathbf{BD}^{-1}\mathbf{C})^{-1}\mathbf{BD}^{-1}+\mathbf{D}^{-1} \end{bmatrix}$$

provided $|\mathbf{D}| \neq 0$ and $|\mathbf{A} - \mathbf{BD}^{-1}\mathbf{C}| \neq 0$.

Solution. First, note that

$$\begin{bmatrix} \mathbf{A} & \mathbf{B} \\ \mathbf{C} & \mathbf{D} \end{bmatrix} = \begin{bmatrix} \mathbf{A} & \mathbf{0} \\ \mathbf{C} & \mathbf{I}_m \end{bmatrix} \begin{bmatrix} \mathbf{I}_n & \mathbf{A}^{-1}\mathbf{B} \\ \mathbf{0} & \mathbf{D}-\mathbf{CA}^{-1}\mathbf{B} \end{bmatrix} \tag{39}$$

By taking the inverse of both sides of Eq. (39), we obtain

$$\begin{bmatrix} \mathbf{A} & \mathbf{B} \\ \mathbf{C} & \mathbf{D} \end{bmatrix}^{-1} = \begin{bmatrix} \mathbf{I}_n & \mathbf{A}^{-1}\mathbf{B} \\ \mathbf{0} & \mathbf{D}-\mathbf{CA}^{-1}\mathbf{B} \end{bmatrix}^{-1} \begin{bmatrix} \mathbf{A} & \mathbf{0} \\ \mathbf{C} & \mathbf{I}_m \end{bmatrix}^{-1}$$

By referring to Prob. A-5, we find

$$\begin{bmatrix} \mathbf{I}_n & \mathbf{A}^{-1}\mathbf{B} \\ \mathbf{0} & \mathbf{D}-\mathbf{CA}^{-1}\mathbf{B} \end{bmatrix}^{-1} = \begin{bmatrix} \mathbf{I}_n & -\mathbf{A}^{-1}\mathbf{B}(\mathbf{D}-\mathbf{CA}^{-1}\mathbf{B})^{-1} \\ \mathbf{0} & (\mathbf{D}-\mathbf{CA}^{-1}\mathbf{B})^{-1} \end{bmatrix}$$

and

$$\begin{bmatrix} \mathbf{A} & \mathbf{0} \\ \mathbf{C} & \mathbf{I}_m \end{bmatrix}^{-1} = \begin{bmatrix} \mathbf{A}^{-1} & \mathbf{0} \\ -\mathbf{CA}^{-1} & \mathbf{I}_m \end{bmatrix}$$

Hence

$$\begin{aligned} \begin{bmatrix} \mathbf{A} & \mathbf{B} \\ \mathbf{C} & \mathbf{D} \end{bmatrix}^{-1} &= \begin{bmatrix} \mathbf{I}_n & \mathbf{A}^{-1}\mathbf{B} \\ \mathbf{0} & \mathbf{D}-\mathbf{CA}^{-1}\mathbf{B} \end{bmatrix}^{-1} \begin{bmatrix} \mathbf{A} & \mathbf{0} \\ \mathbf{C} & \mathbf{I}_m \end{bmatrix}^{-1} \\ &= \begin{bmatrix} \mathbf{I}_n & -\mathbf{A}^{-1}\mathbf{B}(\mathbf{D}-\mathbf{CA}^{-1}\mathbf{B})^{-1} \\ \mathbf{0} & (\mathbf{D}-\mathbf{CA}^{-1}\mathbf{B})^{-1} \end{bmatrix} \begin{bmatrix} \mathbf{A}^{-1} & \mathbf{0} \\ -\mathbf{CA}^{-1} & \mathbf{I}_m \end{bmatrix} \\ &= \begin{bmatrix} \mathbf{A}^{-1}+\mathbf{A}^{-1}\mathbf{B}(\mathbf{D}-\mathbf{CA}^{-1}\mathbf{B})^{-1}\mathbf{CA}^{-1} & -\mathbf{A}^{-1}\mathbf{B}(\mathbf{D}-\mathbf{CA}^{-1}\mathbf{B})^{-1} \\ -(\mathbf{D}-\mathbf{CA}^{-1}\mathbf{B})^{-1}\mathbf{CA}^{-1} & (\mathbf{D}-\mathbf{CA}^{-1}\mathbf{B})^{-1} \end{bmatrix} \end{aligned}$$

provided $|\mathbf{A}| \neq 0$ and $|\mathbf{D} - \mathbf{CA}^{-1}\mathbf{B}| \neq 0$.

Similarly, notice that

$$\begin{bmatrix} \mathbf{A} & \mathbf{B} \\ \mathbf{C} & \mathbf{D} \end{bmatrix} = \begin{bmatrix} \mathbf{I}_n & \mathbf{B} \\ \mathbf{0} & \mathbf{D} \end{bmatrix} \begin{bmatrix} \mathbf{A}-\mathbf{BD}^{-1}\mathbf{C} & \mathbf{0} \\ \mathbf{D}^{-1}\mathbf{C} & \mathbf{I}_m \end{bmatrix} \tag{40}$$

By taking the inverse of both sides of Eq. (40) and referring to Prob. A-5, we obtain

$$\begin{aligned} \begin{bmatrix} \mathbf{A} & \mathbf{B} \\ \mathbf{C} & \mathbf{D} \end{bmatrix}^{-1} &= \begin{bmatrix} \mathbf{A}-\mathbf{BD}^{-1}\mathbf{C} & \mathbf{0} \\ \mathbf{D}^{-1}\mathbf{C} & \mathbf{I}_m \end{bmatrix}^{-1} \begin{bmatrix} \mathbf{I}_n & \mathbf{B} \\ \mathbf{0} & \mathbf{D} \end{bmatrix}^{-1} \\ &= \begin{bmatrix} (\mathbf{A}-\mathbf{BD}^{-1}\mathbf{C})^{-1} & \mathbf{0} \\ -\mathbf{D}^{-1}\mathbf{C}(\mathbf{A}-\mathbf{BD}^{-1}\mathbf{C})^{-1} & \mathbf{I}_m \end{bmatrix} \begin{bmatrix} \mathbf{I}_n & -\mathbf{BD}^{-1} \\ \mathbf{0} & \mathbf{D}^{-1} \end{bmatrix} \\ &= \begin{bmatrix} (\mathbf{A}-\mathbf{BD}^{-1}\mathbf{C})^{-1} & -(\mathbf{A}-\mathbf{BD}^{-1}\mathbf{C})^{-1}\mathbf{BD}^{-1} \\ -\mathbf{D}^{-1}\mathbf{C}(\mathbf{A}-\mathbf{BD}^{-1}\mathbf{C})^{-1} & \mathbf{D}^{-1}\mathbf{C}(\mathbf{A}-\mathbf{BD}^{-1}\mathbf{C})^{-1}\mathbf{BD}^{-1}+\mathbf{D}^{-1} \end{bmatrix} \end{aligned}$$

provided $|\mathbf{D}| \neq 0$ and $|\mathbf{A} - \mathbf{BD}^{-1}\mathbf{C}| \neq 0$.

Problem A-7. For an $n \times n$ real matrix $\mathbf{A}$ and real n-vectors $\mathbf{x}$ and $\mathbf{y}$, show that

(1)
$$\frac{\partial}{\partial \mathbf{x}} \mathbf{y}^T \mathbf{x} = \mathbf{y}$$

(2)
$$\frac{\partial}{\partial \mathbf{x}} \mathbf{x}^T \mathbf{A} \mathbf{x} = \mathbf{A}\mathbf{x} + \mathbf{A}^T \mathbf{x}$$

For an $n \times n$ Hermitian matrix $\mathbf{A}$ and a complex n-vector $\mathbf{x}$, show that

(3)
$$\frac{\partial}{\partial \bar{\mathbf{x}}} \mathbf{x}^* \mathbf{A} \mathbf{x} = \mathbf{A}\mathbf{x}$$

Solution.

(1) Note that

$$\mathbf{y}^T \mathbf{x} = y_1 x_1 + y_2 x_2 + \cdots + y_n x_n$$

which is a scalar quantity. Hence

$$\frac{\partial}{\partial \mathbf{x}} \mathbf{y}^T \mathbf{x} = \begin{bmatrix} \frac{\partial}{\partial x_1} \mathbf{y}^T \mathbf{x} \\ \cdot \\ \cdot \\ \cdot \\ \frac{\partial}{\partial x_n} \mathbf{y}^T \mathbf{x} \end{bmatrix} = \begin{bmatrix} y_1 \\ \cdot \\ \cdot \\ \cdot \\ y_n \end{bmatrix} = \mathbf{y}$$

(2) Notice that

$$\mathbf{x}^T \mathbf{A} \mathbf{x} = \sum_{i=1}^{n} \sum_{j=1}^{n} a_{ij} x_i x_j$$

which is a scalar quantity. Hence

$$\frac{\partial}{\partial \mathbf{x}} \mathbf{x}^T \mathbf{A} \mathbf{x} = \begin{bmatrix} \frac{\partial}{\partial x_1} \left(\sum_{i=1}^{n} \sum_{j=1}^{n} a_{ij} x_i x_j \right) \\ \cdot \\ \cdot \\ \cdot \\ \frac{\partial}{\partial x_n} \left(\sum_{i=1}^{n} \sum_{j=1}^{n} a_{ij} x_i x_j \right) \end{bmatrix} = \begin{bmatrix} \sum_{j=1}^{n} a_{1j} x_j + \sum_{i=1}^{n} a_{i1} x_i \\ \cdot \\ \cdot \\ \cdot \\ \sum_{j=1}^{n} a_{nj} x_j + \sum_{i=1}^{n} a_{in} x_i \end{bmatrix}$$

$$= \mathbf{A}\mathbf{x} + \mathbf{A}^T \mathbf{x}$$

which is Eq. (20).

If matrix $\mathbf{A}$ is a real symmetric matrix, then

$$\frac{\partial}{\partial \mathbf{x}} \mathbf{x}^T \mathbf{A} \mathbf{x} = 2\mathbf{A}\mathbf{x} \qquad \text{if } \mathbf{A} = \mathbf{A}^T$$

(3) For a Hermitian matrix **A**, we have

$$\mathbf{x}^*\mathbf{A}\mathbf{x} = \sum_{i=1}^{n}\sum_{j=1}^{n} a_{ij}\bar{x}_i x_j$$

and

$$\frac{\partial}{\partial \bar{\mathbf{x}}}\mathbf{x}^*\mathbf{A}\mathbf{x} = \begin{bmatrix} \frac{\partial}{\partial \bar{x}_1}\left(\sum_{i=1}^{n}\sum_{j=1}^{n} a_{ij}\bar{x}_i x_j\right) \\ \cdot \\ \cdot \\ \cdot \\ \frac{\partial}{\partial \bar{x}_n}\left(\sum_{i=1}^{n}\sum_{j=1}^{n} a_{ij}\bar{x}_i x_j\right) \end{bmatrix} = \begin{bmatrix} \sum_{j=1}^{n} a_{1j}x_j \\ \cdot \\ \cdot \\ \cdot \\ \sum_{j=1}^{n} a_{nj}x_j \end{bmatrix} = \mathbf{A}\mathbf{x}$$

which is Eq. (21).

Note that

$$\frac{\partial}{\partial \mathbf{x}}\mathbf{x}^*\mathbf{A}\mathbf{x} = \begin{bmatrix} \frac{\partial}{\partial x_1}\left(\sum_{i=1}^{n}\sum_{j=1}^{n} a_{ij}\bar{x}_i x_j\right) \\ \cdot \\ \cdot \\ \cdot \\ \frac{\partial}{\partial x_n}\left(\sum_{i=1}^{n}\sum_{j=1}^{n} a_{ij}\bar{x}_i x_j\right) \end{bmatrix} = \begin{bmatrix} \sum_{i=1}^{n} a_{i1}\bar{x}_i \\ \cdot \\ \cdot \\ \cdot \\ \sum_{i=1}^{n} a_{in}\bar{x}_i \end{bmatrix} = \mathbf{A}^T\bar{\mathbf{x}}$$

Therefore,

$$\overline{\frac{\partial}{\partial \mathbf{x}}\mathbf{x}^*\mathbf{A}\mathbf{x}} = \mathbf{A}^*\mathbf{x} = \mathbf{A}\mathbf{x}$$

Problem A-8. For an $n \times m$ complex matrix **A**, a complex n-vector **x**, and a complex m-vector **y**, show that

(1)
$$\frac{\partial}{\partial \bar{\mathbf{x}}}\mathbf{x}^*\mathbf{A}\mathbf{y} = \mathbf{A}\mathbf{y}$$

(2)
$$\frac{\partial}{\partial \mathbf{y}}\mathbf{x}^*\mathbf{A}\mathbf{y} = \mathbf{A}^T\bar{\mathbf{x}}$$

Solution.

(1) Notice that

$$\mathbf{x}^*\mathbf{A}\mathbf{y} = \sum_{i=1}^{n}\sum_{j=1}^{m} a_{ij}\bar{x}_i y_j$$

Hence

$$\frac{\partial}{\partial \bar{\mathbf{x}}}\mathbf{x}^*\mathbf{A}\mathbf{y} = \begin{bmatrix} \dfrac{\partial}{\partial \bar{x}_1}\left(\sum_{i=1}^{n}\sum_{j=1}^{m} a_{ij}\bar{x}_i y_j\right) \\ \cdot \\ \cdot \\ \cdot \\ \dfrac{\partial}{\partial \bar{x}_n}\left(\sum_{i=1}^{n}\sum_{j=1}^{m} a_{ij}\bar{x}_i y_j\right) \end{bmatrix} = \begin{bmatrix} \sum_{j=1}^{m} a_{1j}y_j \\ \cdot \\ \cdot \\ \cdot \\ \sum_{j=1}^{m} a_{nj}y_j \end{bmatrix} = \mathbf{A}\mathbf{y}$$

which is Eq. (24).

(2)

$$\frac{\partial}{\partial \mathbf{y}}\mathbf{x}^*\mathbf{A}\mathbf{y} = \begin{bmatrix} \dfrac{\partial}{\partial y_1}\left(\sum_{i=1}^{n}\sum_{j=1}^{m} a_{ij}\bar{x}_i y_j\right) \\ \cdot \\ \cdot \\ \cdot \\ \dfrac{\partial}{\partial y_m}\left(\sum_{i=1}^{n}\sum_{j=1}^{m} a_{ij}\bar{x}_i y_j\right) \end{bmatrix} = \begin{bmatrix} \sum_{i=1}^{n} a_{i1}\bar{x}_i \\ \cdot \\ \cdot \\ \cdot \\ \sum_{i=1}^{n} a_{im}\bar{x}_i \end{bmatrix} = \mathbf{A}^T\bar{\mathbf{x}}$$

which is Eq. (25).

Similarly, for an $n \times m$ real matrix **A**, a real n-vector **x**, and a real m-vector **y**, we have

$$\frac{\partial}{\partial \mathbf{x}}\mathbf{x}^T\mathbf{A}\mathbf{y} = \mathbf{A}\mathbf{y}, \qquad \frac{\partial}{\partial \mathbf{y}}\mathbf{x}^T\mathbf{A}\mathbf{y} = \mathbf{A}^T\mathbf{x}$$

which are Eqs. (22) and (23), respectively.

Problem A-9. Given two $n \times n$ matrices **A** and **B**, prove that the eigenvalues of **AB** and those of **BA** are the same, even if $\mathbf{AB} \neq \mathbf{BA}$.

Solution. First, we shall consider the case where **A** (or **B**) is nonsingular. In this case,

$$|\lambda\mathbf{I} - \mathbf{BA}| = |\lambda\mathbf{I} - \mathbf{A}^{-1}(\mathbf{AB})\mathbf{A}| = |\mathbf{A}^{-1}(\lambda\mathbf{I} - \mathbf{AB})\mathbf{A}| = |\mathbf{A}^{-1}|\,|\lambda\mathbf{I} - \mathbf{AB}|\,|\mathbf{A}| = |\lambda\mathbf{I} - \mathbf{AB}|$$

Next, we shall consider the case where both **A** and **B** are singular. There exist $n \times n$ nonsingular matrices **P** and **Q** such that

$$\mathbf{PAQ} = \begin{bmatrix} \mathbf{I}_r & \mathbf{0} \\ \mathbf{0} & \mathbf{0} \end{bmatrix}$$

where $\mathbf{I}_r$ is the $r \times r$ identity matrix and r is the rank of **A**, $r < n$. We have

$$|\lambda\mathbf{I} - \mathbf{BA}| = |\lambda\mathbf{I} - \mathbf{Q}^{-1}\mathbf{BAQ}| = |\lambda\mathbf{I} - \mathbf{Q}^{-1}\mathbf{BP}^{-1}\mathbf{PAQ}|$$

$$= \left|\lambda\mathbf{I} - \begin{bmatrix} \mathbf{G}_{11} & \mathbf{G}_{12} \\ \mathbf{G}_{21} & \mathbf{G}_{22} \end{bmatrix}\begin{bmatrix} \mathbf{I}_r & \mathbf{0} \\ \mathbf{0} & \mathbf{0} \end{bmatrix}\right|$$

where

$$\mathbf{Q}^{-1}\mathbf{B}\mathbf{P}^{-1} = \begin{bmatrix} \mathbf{G}_{11} & \mathbf{G}_{12} \\ \mathbf{G}_{21} & \mathbf{G}_{22} \end{bmatrix}$$

Then

$$|\lambda\mathbf{I} - \mathbf{BA}| = \left|\lambda\mathbf{I} - \begin{bmatrix} \mathbf{G}_{11} & \mathbf{0} \\ \mathbf{G}_{21} & \mathbf{0} \end{bmatrix}\right| = \begin{vmatrix} \lambda\mathbf{I}_r - \mathbf{G}_{11} & \mathbf{0} \\ -\mathbf{G}_{21} & \lambda\mathbf{I}_{n-r} \end{vmatrix}$$

$$= |\lambda\mathbf{I}_r - \mathbf{G}_{11}|\,|\lambda\mathbf{I}_{n-r}|$$

Also,

$$|\lambda\mathbf{I} - \mathbf{AB}| = |\lambda\mathbf{I} - \mathbf{PABP}^{-1}| = |\lambda\mathbf{I} - \mathbf{PAQQ}^{-1}\mathbf{BP}^{-1}|$$

$$= \left|\lambda\mathbf{I} - \begin{bmatrix} \mathbf{I}_r & \mathbf{0} \\ \mathbf{0} & \mathbf{0} \end{bmatrix}\begin{bmatrix} \mathbf{G}_{11} & \mathbf{G}_{12} \\ \mathbf{G}_{21} & \mathbf{G}_{22} \end{bmatrix}\right|$$

$$= \left|\lambda\mathbf{I} - \begin{bmatrix} \mathbf{G}_{11} & \mathbf{G}_{12} \\ \mathbf{0} & \mathbf{0} \end{bmatrix}\right|$$

$$= \begin{vmatrix} \lambda\mathbf{I}_r - \mathbf{G}_{11} & -\mathbf{G}_{12} \\ \mathbf{0} & \lambda\mathbf{I}_{n-r} \end{vmatrix}$$

$$= |\lambda\mathbf{I}_r - \mathbf{G}_{11}|\,|\lambda\mathbf{I}_{n-r}|$$

Hence, we have proved that

$$|\lambda\mathbf{I} - \mathbf{BA}| = |\lambda\mathbf{I} - \mathbf{AB}|$$

or that the eigenvalues of **AB** and **BA** are the same regardless of whether $\mathbf{AB} = \mathbf{BA}$ or $\mathbf{AB} \neq \mathbf{BA}$.

Problem A-10. Show that the following 2×2 matrix **A** has two distinct eigenvalues and that the eigenvectors are linearly independent of each other:

$$\mathbf{A} = \begin{bmatrix} 1 & 1 \\ 0 & 2 \end{bmatrix}$$

Then normalize the eigenvectors.

Solution. The eigenvalues are obtained from

$$|\lambda\mathbf{I} - \mathbf{A}| = \begin{vmatrix} \lambda - 1 & -1 \\ 0 & \lambda - 2 \end{vmatrix} = (\lambda - 1)(\lambda - 2) = 0$$

as

$$\lambda_1 = 1 \qquad \text{and} \qquad \lambda_2 = 2$$

Thus matrix **A** has two distinct eigenvalues.

There are two eigenvectors $\mathbf{x}_1$ and $\mathbf{x}_2$, associated with λ_1 and λ_2, respectively. If we define

$$\mathbf{x}_1 = \begin{bmatrix} x_{11} \\ x_{21} \end{bmatrix}, \qquad \mathbf{x}_2 = \begin{bmatrix} x_{12} \\ x_{22} \end{bmatrix}$$

then the eigenvector $\mathbf{x}_1$ can be found from

$$\mathbf{A}\mathbf{x}_1 = \lambda_1\mathbf{x}_1$$

or

$$(\lambda_1\mathbf{I} - \mathbf{A})\mathbf{x}_1 = \mathbf{0}$$

Noting that $\lambda_1 = 1$, we have

$$\begin{bmatrix} 1-1 & -1 \\ 0 & 1-2 \end{bmatrix} \begin{bmatrix} x_{11} \\ x_{21} \end{bmatrix} = \begin{bmatrix} 0 \\ 0 \end{bmatrix}$$

which gives

$$x_{11} = \text{arbitrary constant} \qquad \text{and} \qquad x_{21} = 0$$

Hence, eigenvector $\mathbf{x}_1$ may be written as follows:

$$\mathbf{x}_1 = \begin{bmatrix} x_{11} \\ x_{21} \end{bmatrix} = \begin{bmatrix} c_1 \\ 0 \end{bmatrix}$$

where $c_1 \neq 0$ is an arbitrary constant.

Similarly, for the eigenvector $\mathbf{x}_2$, we have

$$\mathbf{A}\mathbf{x}_2 = \lambda_2\mathbf{x}_2$$

or

$$(\lambda_2\mathbf{I} - \mathbf{A})\mathbf{x}_2 = \mathbf{0}$$

Noting that $\lambda_2 = 2$, we obtain

$$\begin{bmatrix} 2-1 & -1 \\ 0 & 2-2 \end{bmatrix} \begin{bmatrix} x_{12} \\ x_{22} \end{bmatrix} = \begin{bmatrix} 0 \\ 0 \end{bmatrix}$$

from which we get

$$x_{12} - x_{22} = 0$$

Hence the eigenvector associated with $\lambda_2 = 2$ may be selected as

$$\mathbf{x}_2 = \begin{bmatrix} x_{12} \\ x_{22} \end{bmatrix} = \begin{bmatrix} c_2 \\ c_2 \end{bmatrix}$$

where $c_2 \neq 0$ is an arbitrary constant.

The two eigenvectors are therefore given by

$$\mathbf{x}_1 = \begin{bmatrix} c_1 \\ 0 \end{bmatrix} \qquad \text{and} \qquad \mathbf{x}_2 = \begin{bmatrix} c_2 \\ c_2 \end{bmatrix}$$

The fact that eigenvectors $\mathbf{x}_1$ and $\mathbf{x}_2$ are linearly independent can be seen from the fact that the determinant of the matrix $[\mathbf{x}_1 \;\; \mathbf{x}_2]$ is nonzero:

$$\begin{vmatrix} c_1 & c_2 \\ 0 & c_2 \end{vmatrix} \neq 0$$

To normalize the eigenvectors, we choose $c_1 = 1$ and $c_2 = 1/\sqrt{2}$, or

$$\mathbf{x}_1 = \begin{bmatrix} 1 \\ \\ 0 \end{bmatrix}, \qquad \mathbf{x}_2 = \begin{bmatrix} \dfrac{1}{\sqrt{2}} \\ \\ \dfrac{1}{\sqrt{2}} \end{bmatrix}$$

Clearly, the absolute value of each eigenvector becomes unity and therefore the eigenvectors are normalized.

Problem A-11. Obtain a transformation matrix $\mathbf{T}$ that transforms the matrix

$$\mathbf{A} = \begin{bmatrix} 0 & 1 & 0 & 3 \\ 0 & -1 & 1 & 1 \\ 0 & 0 & 0 & 1 \\ 0 & 0 & -1 & -2 \end{bmatrix}$$

into a Jordan canonical form.

Solution. The characteristic equation is

$$|\lambda\mathbf{I} - \mathbf{A}| = \begin{vmatrix} \lambda & -1 & 0 & -3 \\ 0 & \lambda+1 & -1 & -1 \\ 0 & 0 & \lambda & -1 \\ 0 & 0 & 1 & \lambda+2 \end{vmatrix} = \begin{vmatrix} \lambda & -1 \\ 0 & \lambda+1 \end{vmatrix} \begin{vmatrix} \lambda & -1 \\ 1 & \lambda+2 \end{vmatrix}$$

$$= (\lambda + 1)^3\lambda = 0$$

Hence matrix $\mathbf{A}$ involves eigenvalues

$$\lambda_1 = -1, \qquad \lambda_2 = -1, \qquad \lambda_3 = -1, \qquad \lambda_4 = 0$$

For the multiple eigenvalue -1, we have

$$\lambda_1\mathbf{I} - \mathbf{A} = \begin{bmatrix} -1 & -1 & 0 & -3 \\ 0 & 0 & -1 & -1 \\ 0 & 0 & -1 & -1 \\ 0 & 0 & 1 & 1 \end{bmatrix}$$

which is of rank 2, or rank $(4 - 2)$. From the rank condition we see that there must be two Jordan blocks for eigenvalue -1, that is, one $p_1 \times p_1$ Jordan block and one $p_2 \times p_2$ Jordan block, where $p_1 + p_2 = 3$. Notice that for $p_1 + p_2 = 3$, there is only one combination (2 and 1) for the orders of p_1 and p_2. Let us choose

$$p_1 = 2 \qquad \text{and} \qquad p_2 = 1$$

Then there are one eigenvector and one generalized eigenvector for Jordan block $\mathbf{J}_{p_1}$ and one eigenvector for Jordan block $\mathbf{J}_{p_2}$.

Let us define an eigenvector and a generalized eigenvector for Jordan block $\mathbf{J}_{p_1}$ as $\mathbf{v}_{11}$ and $\mathbf{v}_{12}$, respectively, and an eigenvector for Jordan block $\mathbf{J}_{p_2}$ as $\mathbf{v}_{21}$. Then, there must be vectors $\mathbf{v}_{11}$, $\mathbf{v}_{12}$, and $\mathbf{v}_{21}$ that satisfy the following equations:

$$(\mathbf{A} - \lambda_1\mathbf{I})\mathbf{v}_{11} = \mathbf{0}, \qquad (\mathbf{A} - \lambda_1\mathbf{I})\mathbf{v}_{21} = \mathbf{0}$$

$$(\mathbf{A} - \lambda_1\mathbf{I})\mathbf{v}_{12} = \mathbf{v}_{11}$$

For $\lambda_1 = -1$, $\mathbf{A} - \lambda_1\mathbf{I}$ can be given as follows:

$$\mathbf{A} - \lambda_1\mathbf{I} = \begin{bmatrix} 1 & 1 & 0 & 3 \\ 0 & 0 & 1 & 1 \\ 0 & 0 & 1 & 1 \\ 0 & 0 & -1 & -1 \end{bmatrix}$$

Noting that

$$(\mathbf{A} - \lambda_1\mathbf{I})^2 = \begin{bmatrix} 1 & 1 & -2 & 1 \\ 0 & 0 & 0 & 0 \\ 0 & 0 & 0 & 0 \\ 0 & 0 & 0 & 0 \end{bmatrix}$$

we determine vector $\mathbf{v}_{12}$ to be such that it will satisfy the equation

$$(\mathbf{A} - \lambda_1\mathbf{I})^2\mathbf{v}_{12} = \mathbf{0}$$

and at the same time will make $(\mathbf{A} - \lambda_1\mathbf{I})\mathbf{v}_{12}$ nonzero. An example of such a generalized eigenvector $\mathbf{v}_{12}$ can be found to be

$$\mathbf{v}_{12} = \begin{bmatrix} -a \\ 0 \\ 0 \\ a \end{bmatrix} \qquad a = \text{arbitrary nonzero constant}$$

The eigenvector $\mathbf{v}_{11}$ is then found to be a nonzero vector $(\mathbf{A} - \lambda_1\mathbf{I})\mathbf{v}_{12}$:

$$\mathbf{v}_{11} = (\mathbf{A} - \lambda_1\mathbf{I})\mathbf{v}_{12} = \begin{bmatrix} 2a \\ a \\ a \\ -a \end{bmatrix}$$

Since a is an arbitrary nonzero constant, let us choose $a = 1$. Then we have

$$\mathbf{v}_{11} = \begin{bmatrix} 2 \\ 1 \\ 1 \\ -1 \end{bmatrix} \qquad \text{and} \qquad \mathbf{v}_{12} = \begin{bmatrix} -1 \\ 0 \\ 0 \\ 1 \end{bmatrix}$$

Next, we determine $\mathbf{v}_{21}$ so that $\mathbf{v}_{21}$ and $\mathbf{v}_{11}$ are linearly independent. For $\mathbf{v}_{21}$ we may choose

$$\mathbf{v}_{21} = \begin{bmatrix} b + 3c \\ -b \\ c \\ -c \end{bmatrix}$$

Since b and c are arbitrary constants, let us choose, for example, $b = 1$ and $c = 0$. Then

$$\mathbf{v}_{21} = \begin{bmatrix} 1 \\ -1 \\ 0 \\ 0 \end{bmatrix}$$

Clearly, $\mathbf{v}_{11}$, $\mathbf{v}_{12}$, and $\mathbf{v}_{21}$ are linearly independent. Let us define

$$\mathbf{v}_{11} = \mathbf{x}_1, \qquad \mathbf{v}_{12} = \mathbf{x}_2, \qquad \mathbf{v}_{21} = \mathbf{x}_3$$

and

$$\mathbf{T}(\lambda_1) = [\mathbf{v}_{11} \vdots \mathbf{v}_{12} \vdots \mathbf{v}_{21}] = [\mathbf{x}_1 \vdots \mathbf{x}_2 \vdots \mathbf{x}_3] = \begin{bmatrix} 2 & -1 & 1 \\ 1 & 0 & -1 \\ 1 & 0 & 0 \\ -1 & 1 & 0 \end{bmatrix}$$

For the distinct eigenvalue $\lambda_4 = 0$, the eigenvector $\mathbf{x}_4$ can be determined from

$$(\mathbf{A} - \lambda_4\mathbf{I})\mathbf{x}_4 = \mathbf{0}$$

Noting that

$$\mathbf{A} - \lambda_4\mathbf{I} = \mathbf{A} = \begin{bmatrix} 0 & 1 & 0 & 3 \\ 0 & -1 & 1 & 1 \\ 0 & 0 & 0 & 1 \\ 0 & 0 & -1 & -2 \end{bmatrix}$$

we find

$$\mathbf{x}_4 = \begin{bmatrix} d \\ 0 \\ 0 \\ 0 \end{bmatrix}$$

where d is an arbitrary constant. By choosing $d = 1$, we have

$$\mathbf{T}(\lambda_4) = \mathbf{x}_4 = \begin{bmatrix} 1 \\ 0 \\ 0 \\ 0 \end{bmatrix}$$

Thus the transformation matrix $\mathbf{T}$ can be written as follows:

$$\mathbf{T} = [\mathbf{T}(\lambda_1) \vdots \mathbf{T}(\lambda_4)] = \begin{bmatrix} 2 & -1 & 1 & 1 \\ 1 & 0 & -1 & 0 \\ 1 & 0 & 0 & 0 \\ -1 & 1 & 0 & 0 \end{bmatrix}$$

Then

$$\mathbf{T}^{-1}\mathbf{A}\mathbf{T} = \begin{bmatrix} 0 & 0 & 1 & 0 \\ 0 & 0 & 1 & 1 \\ 0 & -1 & 1 & 0 \\ 1 & 1 & -2 & 1 \end{bmatrix} \begin{bmatrix} 0 & 1 & 0 & 3 \\ 0 & -1 & 1 & 1 \\ 0 & 0 & 0 & 1 \\ 0 & 0 & -1 & -2 \end{bmatrix} \begin{bmatrix} 2 & -1 & 1 & 1 \\ 1 & 0 & -1 & 0 \\ 1 & 0 & 0 & 0 \\ -1 & 1 & 0 & 0 \end{bmatrix}$$

$$= \left[\begin{array}{cc:c:c} -1 & 1 & 0 & 0 \\ 0 & -1 & 0 & 0 \\ \hdashline 0 & 0 & -1 & 0 \\ \hdashline 0 & 0 & 0 & 0 \end{array}\right] = \text{diag}\,[\mathbf{J}_2(-1), \mathbf{J}_1(-1), \mathbf{J}_1(0)]$$

Problem A-12. Assume that an $n \times n$ normal matrix $\mathbf{A}$ has a k-multiple eigenvalue λ_1. Prove that the rank of $\mathbf{A} - \lambda_1 \mathbf{I}$ is $n - k$.

Solution. Suppose that the rank of $\mathbf{A} - \lambda_1 \mathbf{I}$ is $n - m$. Then the equation

$$(\mathbf{A} - \lambda_1 \mathbf{I})\mathbf{x} = \mathbf{0} \tag{41}$$

will have m linearly independent vector solutions. Let us choose m such vectors so that they are orthogonal to each other and normalized. That is, vectors $\mathbf{x}_1, \mathbf{x}_2, \ldots, \mathbf{x}_m$ will satisfy Eq. (41) and will be orthonormal.

Let us consider $n - m$ vectors $\mathbf{x}_{m+1}, \mathbf{x}_{m+2}, \ldots, \mathbf{x}_n$ such that all n vectors

$$\mathbf{x}_1, \mathbf{x}_2, \ldots, \mathbf{x}_n$$

will be orthonormal to each other. Then, matrix $\mathbf{U}$, defined by

$$\mathbf{U} = [\mathbf{x}_1 \vdots \mathbf{x}_2 \vdots \cdots \vdots \mathbf{x}_n]$$

is a unitary matrix.

Since for $1 \le i \le m$, we have

$$\mathbf{A}\mathbf{x}_i = \lambda_1 \mathbf{x}_i$$

and therefore we can write

$$\mathbf{A}\mathbf{U} = \mathbf{U}\begin{bmatrix} \lambda_1 \mathbf{I}_m & \mathbf{B} \\ \mathbf{0} & \mathbf{C} \end{bmatrix}$$

or

$$\mathbf{U}^*\mathbf{A}\mathbf{U} = \begin{bmatrix} \lambda_1 \mathbf{I}_m & \mathbf{B} \\ \mathbf{0} & \mathbf{C} \end{bmatrix}$$

Noting that

$$\begin{aligned} \|\mathbf{A}\mathbf{x}_i - \lambda \mathbf{x}_i\|^2 &= \langle (\mathbf{A} - \lambda \mathbf{I})\mathbf{x}_i, (\mathbf{A} - \lambda \mathbf{I})\mathbf{x}_i \rangle \\ &= \langle (\mathbf{A}^* - \bar{\lambda}\mathbf{I})(\mathbf{A} - \lambda \mathbf{I})\mathbf{x}_i, \mathbf{x}_i \rangle \\ &= \langle (\mathbf{A} - \lambda \mathbf{I})(\mathbf{A}^* - \bar{\lambda}\mathbf{I})\mathbf{x}_i, \mathbf{x}_i \rangle \\ &= \langle (\mathbf{A}^* - \bar{\lambda}\mathbf{I})\mathbf{x}_i, (\mathbf{A}^* - \bar{\lambda}\mathbf{I})\mathbf{x}_i \rangle \\ &= \|\mathbf{A}^*\mathbf{x}_i - \bar{\lambda}\mathbf{x}_i\|^2 \\ &= 0 \end{aligned}$$

we have

$$\mathbf{A}^*\mathbf{x}_i = \bar{\lambda}\mathbf{x}_i$$

Therefore, we can write

$$\mathbf{A}^*\mathbf{U} = \mathbf{U}\begin{bmatrix} \bar{\lambda}_1 \mathbf{I}_m & \mathbf{B}_1 \\ \mathbf{0} & \mathbf{C}_1 \end{bmatrix}$$

or

$$\mathbf{U}^*\mathbf{A}^*\mathbf{U} = \begin{bmatrix} \bar{\lambda}_1 \mathbf{I}_m & \mathbf{B}_1 \\ \mathbf{0} & \mathbf{C}_1 \end{bmatrix}$$

Hence

$$\begin{bmatrix} \lambda_1 \mathbf{I}_m & \mathbf{B} \\ \mathbf{0} & \mathbf{C} \end{bmatrix} = \mathbf{U}^* \mathbf{A} \mathbf{U} = (\mathbf{U}^* \mathbf{A}^* \mathbf{U})^* = \begin{bmatrix} \bar{\lambda}_1 \mathbf{I}_m & \mathbf{B}_1 \\ \mathbf{0} & \mathbf{C}_1 \end{bmatrix}^* = \begin{bmatrix} \lambda_1 \mathbf{I}_m & \mathbf{0} \\ \mathbf{B}_1^* & \mathbf{C}_1^* \end{bmatrix}$$

Comparing the left- and right-hand sides of this last equation, we obtain

$$\mathbf{B} = \mathbf{0}$$

Hence we get

$$\mathbf{A} = \mathbf{U} \begin{bmatrix} \lambda_1 \mathbf{I}_m & \mathbf{0} \\ \mathbf{0} & \mathbf{C} \end{bmatrix} \mathbf{U}^*$$

Then

$$\mathbf{A} - \lambda \mathbf{I} = \mathbf{U} \begin{bmatrix} (\lambda_1 - \lambda) \mathbf{I}_m & \mathbf{0} \\ \mathbf{0} & \mathbf{C} - \lambda \mathbf{I}_{n-m} \end{bmatrix} \mathbf{U}^*$$

The determinant of this last equation is

$$|\mathbf{A} - \lambda \mathbf{I}| = (\lambda_1 - \lambda)^m |\mathbf{C} - \lambda \mathbf{I}_{n-m}| \tag{42}$$

On the other hand we have

$$\text{rank } (\mathbf{A} - \lambda_1 \mathbf{I}) = n - m = \text{rank} \left\{ \mathbf{U} \begin{bmatrix} \mathbf{0} & \mathbf{0} \\ \mathbf{0} & \mathbf{C} - \lambda_1 \mathbf{I}_{n-m} \end{bmatrix} \mathbf{U}^* \right\}$$

$$= \text{rank} \begin{bmatrix} \mathbf{0} & \mathbf{0} \\ \mathbf{0} & \mathbf{C} - \lambda_1 \mathbf{I}_{n-m} \end{bmatrix} = \text{rank } (\mathbf{C} - \lambda_1 \mathbf{I}_{n-m})$$

Hence, we conclude that the rank of $\mathbf{C} - \lambda_1 \mathbf{I}_{n-m}$ is $n - m$. Consequently,

$$|\mathbf{C} - \lambda_1 \mathbf{I}_{n-m}| \neq 0$$

and from Eq. (42), λ_1 is shown to be the m-multiple eigenvalue of $|\mathbf{A} - \lambda \mathbf{I}| = 0$. Since λ_1 is the k-multiple eigenvalue of $\mathbf{A}$, we must have $m = k$. Therefore, the rank of $\mathbf{A} - \lambda_1 \mathbf{I}$ is $n - k$.

Note that since the rank of $\mathbf{A} - \lambda_1 \mathbf{I}$ is $n - k$, the equation

$$(\mathbf{A} - \lambda_1 \mathbf{I}) \mathbf{x}_i = \mathbf{0}$$

will have k linearly independent eigenvectors $\mathbf{x}_1, \mathbf{x}_2, \ldots, \mathbf{x}_k$.

Problem A-13. Prove that the eigenvalues of an $n \times n$ Hermitian matrix and of an $n \times n$ real symmetric matrix are real. Prove also that the eigenvalues of a skew-Hermitian matrix and of a real skew-symmetric matrix are either zero or purely imaginary.

Solution. Let us define any eigenvalue of an $n \times n$ Hermitian matrix $\mathbf{A}$ by $\lambda = \alpha + j\beta$. There exists a vector $\mathbf{x} \neq \mathbf{0}$ such that

$$\mathbf{A}\mathbf{x} = (\alpha + j\beta)\mathbf{x}$$

Transposing this last equation, we obtain

$$\mathbf{x}^* \mathbf{A}^* = (\alpha - j\beta)\mathbf{x}^*$$

Since $\mathbf{A}$ is Hermitian, $\mathbf{A}^* = \mathbf{A}$. Therefore,

$$\mathbf{x}^*\mathbf{A}\mathbf{x} = (\alpha - j\beta)\mathbf{x}^*\mathbf{x}$$

On the other hand, since $\mathbf{A}\mathbf{x} = (\alpha + j\beta)\mathbf{x}$, we have

$$\mathbf{x}^*\mathbf{A}\mathbf{x} = (\alpha + j\beta)\mathbf{x}^*\mathbf{x}$$

Hence we obtain

$$[(\alpha - j\beta) - (\alpha + j\beta)]\mathbf{x}^*\mathbf{x} = 0$$

or

$$-2j\beta\mathbf{x}^*\mathbf{x} = 0$$

Since $\mathbf{x}^*\mathbf{x} \neq 0$ (for $\mathbf{x} \neq \mathbf{0}$), we conclude that

$$\beta = 0$$

This proves that any eigenvalue of an $n \times n$ Hermitian matrix $\mathbf{A}$ is real. It follows that the eigenvalues of a real symmetric matrix are also real, since it is Hermitian.

To prove the second half of the problem, notice that if $\mathbf{B}$ is skew-Hermitian, then $j\mathbf{B}$ is Hermitian. Hence, the eigenvalues of $j\mathbf{B}$ are real, which implies that the eigenvalues of $\mathbf{B}$ are either zero or purely imaginary.

The eigenvalues of a real skew-symmetric matrix are also either zero or purely imaginary, since a real skew-symmetric matrix is skew-Hermitian.

Note that in the real skew-symmetric matrix, purely imaginary eigenvalues always occur in conjugate pairs, since the coefficients of the characteristic equation are real. Note also that an $n \times n$ real skew-symmetric matrix is singular if n is odd, since such a matrix must include at least one zero eigenvalue.

Problem A-14. Examine whether or not the following 3×3 matrix $\mathbf{A}$ is positive-definite:

$$\mathbf{A} = \begin{bmatrix} 2 & 2 & -1 \\ 2 & 6 & 0 \\ -1 & 0 & 1 \end{bmatrix}$$

Solution. We shall demonstrate three different ways to test the positive definiteness of matrix $\mathbf{A}$.

1. We may first apply Sylvester's criterion for positive definiteness of a quadratic form $\mathbf{x}^T\mathbf{A}\mathbf{x}$. For the given matrix $\mathbf{A}$, we have

$$2 > 0, \qquad \begin{vmatrix} 2 & 2 \\ 2 & 6 \end{vmatrix} > 0, \qquad \begin{vmatrix} 2 & 2 & -1 \\ 2 & 6 & 0 \\ -1 & 0 & 1 \end{vmatrix} > 0$$

Thus, the successive principal minors are all positive. Hence matrix $\mathbf{A}$ is positive definite.

2. We may examine the positive definiteness of $\mathbf{x}^T\mathbf{A}\mathbf{x}$. Since

$$\mathbf{x}^T\mathbf{A}\mathbf{x} = [x_1 \; x_2 \; x_3] \begin{bmatrix} 2 & 2 & -1 \\ 2 & 6 & 0 \\ -1 & 0 & 1 \end{bmatrix} \begin{bmatrix} x_1 \\ x_2 \\ x_3 \end{bmatrix}$$

$$= 2x_1^2 + 4x_1x_2 - 2x_1x_3 + 6x_2^2 + x_3^2$$
$$= (x_1 - x_3)^2 + (x_1 + 2x_2)^2 + 2x_2^2$$

we find that $\mathbf{x}^T\mathbf{A}\mathbf{x}$ is positive except at the origin ($\mathbf{x} = \mathbf{0}$). Hence, we conclude that matrix $\mathbf{A}$ is positive definite.

3. We may examine the eigenvalues of matrix $\mathbf{A}$. Note that

$$|\lambda\mathbf{I} - \mathbf{A}| = \lambda^3 - 9\lambda^2 + 15\lambda - 2$$
$$= (\lambda - 2)(\lambda - 0.1459)(\lambda - 6.8541)$$

Hence

$$\lambda_1 = 2, \qquad \lambda_2 = 0.1459, \qquad \lambda_3 = 6.8541$$

Since all eigenvalues are positive, we conclude that $\mathbf{A}$ is a positive definite matrix.

Problem A-15. Examine whether the following matrix $\mathbf{A}$ is positive semidefinite:

$$\mathbf{A} = \begin{bmatrix} 1 & 2 & 1 \\ 2 & 4 & 2 \\ 1 & 2 & 0 \end{bmatrix}$$

Solution. In the positive semidefiniteness test, we need to examine the signs of all principal minors in addition to the sign of the determinant of the given matrix, which must be zero; that is, $|\mathbf{A}|$ must be equal to 0.

For the 3×3 matrix

$$\begin{bmatrix} a_{11} & a_{12} & a_{13} \\ a_{21} & a_{22} & a_{23} \\ a_{31} & a_{32} & a_{33} \end{bmatrix}$$

there are six principal minors:

$$a_{11}, \qquad a_{22}, \qquad a_{33}, \qquad \begin{vmatrix} a_{11} & a_{12} \\ a_{21} & a_{22} \end{vmatrix}, \qquad \begin{vmatrix} a_{22} & a_{23} \\ a_{32} & a_{33} \end{vmatrix}, \qquad \begin{vmatrix} a_{11} & a_{13} \\ a_{31} & a_{33} \end{vmatrix}$$

We need to examine the signs of all six principal minors and the sign of $|\mathbf{A}|$.

For the given matrix $\mathbf{A}$,

$$a_{11} = 1 > 0$$
$$a_{22} = 4 > 0$$
$$a_{33} = 0$$
$$\begin{vmatrix} a_{11} & a_{12} \\ a_{21} & a_{22} \end{vmatrix} = \begin{vmatrix} 1 & 2 \\ 2 & 4 \end{vmatrix} = 0$$
$$\begin{vmatrix} a_{22} & a_{23} \\ a_{32} & a_{33} \end{vmatrix} = \begin{vmatrix} 4 & 2 \\ 2 & 0 \end{vmatrix} = -4 < 0$$
$$\begin{vmatrix} a_{11} & a_{13} \\ a_{31} & a_{33} \end{vmatrix} = \begin{vmatrix} 1 & 1 \\ 1 & 0 \end{vmatrix} = -1 < 0$$

$$\begin{vmatrix} a_{11} & a_{12} & a_{13} \\ a_{21} & a_{22} & a_{23} \\ a_{31} & a_{32} & a_{33} \end{vmatrix} = \begin{vmatrix} 1 & 2 & 1 \\ 2 & 4 & 2 \\ 1 & 2 & 0 \end{vmatrix} = 0$$

Clearly, two of the principal minors are negative. Hence we conclude that matrix **A** is not positive semidefinite.

It is important to note that had we tested the signs of only the successive principal minors and the determinant of **A**,

$$1 > 0, \qquad \begin{vmatrix} 1 & 2 \\ 2 & 4 \end{vmatrix} = 0, \qquad |\mathbf{A}| = \begin{vmatrix} 1 & 2 & 1 \\ 2 & 4 & 2 \\ 1 & 2 & 0 \end{vmatrix} = 0$$

we would have reached a wrong conclusion that matrix **A** is positive semidefinite.

In fact, for the given matrix **A**,

$$|\lambda\mathbf{I} - \mathbf{A}| = \begin{vmatrix} \lambda - 1 & -2 & -1 \\ -2 & \lambda - 4 & -2 \\ -1 & -2 & \lambda \end{vmatrix} = (\lambda^2 - 5\lambda - 5)\lambda$$

$$= (\lambda - 5.8541)\lambda(\lambda + 0.8541)$$

and so the eigenvalues are

$$\lambda_1 = 5.8541, \qquad \lambda_2 = 0, \qquad \lambda_3 = -0.8541$$

(Clearly, matrix **A** is an indefinite matrix.)

If matrix **A** is diagonalized, we have

$$\mathbf{P}^{-1}\mathbf{A}\mathbf{P} = \mathbf{D} = \begin{bmatrix} 5.8541 & 0 & 0 \\ 0 & 0 & 0 \\ 0 & 0 & -0.8541 \end{bmatrix}$$

In order for matrix **A** to be positive semidefinite, all eigenvalues must be nonnegative and at least one of them must be zero.

Problem A-16. Consider the equation

$$x_1 + 5x_2 = 1 \tag{43}$$

or, in the vector-matrix form,

$$\mathbf{A}\mathbf{x} = b$$

where

$$\mathbf{A} = [1 \quad 5], \qquad \mathbf{x} = \begin{bmatrix} x_1 \\ x_2 \end{bmatrix}, \qquad b = 1$$

Find the solution that will give the minimum norm $\|\mathbf{x}\|$, that is, the solution closest to the origin.

Solution. The minimum norm solution is given by

$$\mathbf{x}^o = \mathbf{A}^{RM} b$$

where the right pseudoinverse $\mathbf{A}^{RM}$ is given by

$$\mathbf{A}^{RM} = \mathbf{A}^T(\mathbf{A}\mathbf{A}^T)^{-1}$$

In the present example, the right pseudoinverse matrix becomes as follows:

$$A^{RM} = \begin{bmatrix} 1 \\ 5 \end{bmatrix} \left\{ [1 \quad 5] \begin{bmatrix} 1 \\ 5 \end{bmatrix} \right\}^{-1} = \begin{bmatrix} 1 \\ 5 \end{bmatrix} (26)^{-1} = \begin{bmatrix} \frac{1}{26} \\ \frac{5}{26} \end{bmatrix}$$

Hence the minimum norm solution becomes

$$\mathbf{x}^{\mathrm{o}} = \begin{bmatrix} x_1^{\mathrm{o}} \\ x_2^{\mathrm{o}} \end{bmatrix} = \begin{bmatrix} \frac{1}{26} \\ \frac{5}{26} \end{bmatrix}$$

It is instructive to examine the minimum norm solution graphically. In Fig. 2 the minimum norm solution is located at point P. This point is the intersection of line $x_1 + 5x_2 = 1$ and the line that is perpendicular to it and passes through the origin. This perpendicular line can be written as

$$x_1 - \tfrac{1}{5}x_2 = 0 \tag{44}$$

The solution of the simultaneous Eqs. (43) and (44) gives point P:

$$x_1 = \tfrac{1}{26}, \qquad x_2 = \tfrac{5}{26}$$

which is the same as the result obtained by use of the right pseudoinverse matrix.

Problem A-17. Consider a vector-matrix equation

$$\mathbf{A}\mathbf{x} = \mathbf{b}$$

where

$$\mathbf{A} = \begin{bmatrix} 1 & 1 \\ 1 & 2 \\ 1 & 4 \end{bmatrix}, \qquad \mathbf{x} = \begin{bmatrix} x_1 \\ x_2 \end{bmatrix}, \qquad \mathbf{b} = \begin{bmatrix} 1 \\ 2 \\ 2 \end{bmatrix}$$

Clearly, no solution exists in the classical sense.

Find the minimum norm solution such that norm $\|\mathbf{A}\mathbf{x} - \mathbf{b}\|$ is minimum.

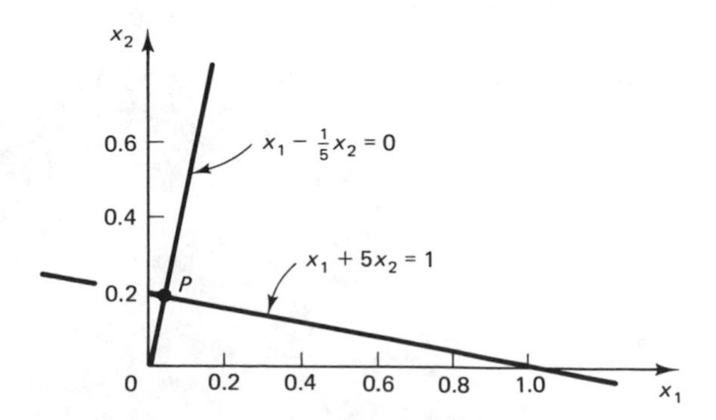

Figure 2 Graphical representation of the minimum norm solution for Eq. (43) in Prob. A-16.

Solution. The desired minimum norm solution is given by

$$\mathbf{x}^o = \mathbf{A}^{LM}\mathbf{b} = (\mathbf{A}^T\mathbf{A})^{-1}\mathbf{A}^T\mathbf{b}$$

Hence

$$\mathbf{x}^o = \left\{\begin{bmatrix} 1 & 1 & 1 \\ 1 & 2 & 4 \end{bmatrix}\begin{bmatrix} 1 & 1 \\ 1 & 2 \\ 1 & 4 \end{bmatrix}\right\}^{-1}\begin{bmatrix} 1 & 1 & 1 \\ 1 & 2 & 4 \end{bmatrix}\begin{bmatrix} 1 \\ 2 \\ 2 \end{bmatrix} = \begin{bmatrix} 1 \\ \frac{2}{7} \end{bmatrix}$$

This problem can, of course, be solved in several different ways. The use of the left pseudoinverse matrix $\mathbf{A}^{LM}$ is one approach, as just demonstrated. An additional method, based on an ordinary minimization method, follows.

Noting that minimizing $\|\mathbf{Ax} - \mathbf{b}\|$ is the same as minimizing $\|\mathbf{Ax} - \mathbf{b}\|^2$, let us minimize $\|\mathbf{Ax} - \mathbf{b}\|^2$. We shall begin the solution by writing $\|\mathbf{Ax} - \mathbf{b}\|^2$ as follows:

$$\|\mathbf{Ax} - \mathbf{b}\|^2 = (x_1 + x_2 - 1)^2 + (x_1 + 2x_2 - 2)^2 + (x_1 + 4x_2 - 2)^2$$

Let

$$L = (x_1 + x_2 - 1)^2 + (x_1 + 2x_2 - 2)^2 + (x_1 + 4x_2 - 2)^2$$

Then, by differentiating L with respect to x_1 and x_2 respectively, and equating each of the resulting equations to zero we obtain

$$\frac{\partial L}{\partial x_1} = 2(x_1 + x_2 - 1) + 2(x_1 + 2x_2 - 2) + 2(x_1 + 4x_2 - 2) = 0$$

$$\frac{\partial L}{\partial x_2} = 2(x_1 + x_2 - 1) + 4(x_1 + 2x_2 - 2) + 8(x_1 + 4x_2 - 2) = 0$$

which can be simplified to read

$$3x_1 + 7x_2 - 5 = 0$$

$$7x_1 + 21x_2 - 13 = 0$$

The solution to these two simultaneous equations is

$$x_1 = 1, \qquad x_2 = \tfrac{2}{7}$$

or

$$\mathbf{x}^o = \begin{bmatrix} 1 \\ \frac{2}{7} \end{bmatrix}$$

which is the same as the solution obtained by use of the left pseudoinverse of $\mathbf{A}$.

Index

C

D

N

O

P

Q

R

S

T

U

V

W

Z